Table of Atomic Masses*

Element	Symbol	Atomic Number	Atomic Mass	Element	Symbol	Atomic Number	Atomic Mass
Actinium	Ac	89	(227)†	Mercury	Hg	80	200.6
Aluminum	Al	13	26.98	Molybdenum	Mo	42	95.94
Americium	Am	95	(243)	Neodymium	Nd	60	144.2
Antimony	Sb	51	121.8	Neon	Ne	10	20.18
Argon	Ar	18	39.95	Neptunium	Np	93	(237)
Arsenic	As	33	74.92	Nickel	Ni	28	58.70
Astatine	At	85	(210)	Niobium	Nb	41	92.91
Barium	Ba	56	137.3	Nitrogen	N	7	14.01
Berkelium	Bk	97	(247)	Nobelium	No	102	(259)
Beryllium	Be	4	9.012	Osmium	Os	76	190.2
Bismuth	Bi	83	209.0	Oxygen	O	8	16.00
Boron	B	5	10.81	Palladium	Pd	46	106.4
Bromine	Br	35	79.90	Phosphorus	P	15	30.97
Cadmium	Cd	48	112.4	Platinum	Pt	78	195.1
Calcium	Ca	20	40.08	Plutonium	Pu	94	(244)
Californium	Cf	98	(251)	Polonium	Po	84	(209)
Carbon	C	6	12.01	Potassium	K	19	39.10
Cerium	Ce	58	140.1	Praseodymium	Pr	59	140.9
Cesium	Cs	55	132.9	Promethium	Pm	61	(145)
Chlorine	Cl	17	35.45	Protactinium	Pa	91	(231)
Chromium	Cr	24	52.00	Radium	Ra	88	226.0
Cobalt	Co	27	58.93	Radon	Rn	86	(222)
Copper	Cu	29	63.55	Rhenium	Re	75	186.2
Curium	Cm	96	(247)	Rhodium	Rh	45	102.9
Dysprosium	Dy	66	162.5	Rubidium	Rb	37	85.47
Einsteinium	Es	99	(252)	Ruthenium	Ru	44	101.1
Erbium	Er	68	167.3	Samarium	Sm	62	150.4
Europium	Eu	63	152.0	Scandium	Sc	21	44.96
Fermium	Fm	100	(257)	Selenium	Se	34	78.96
Fluorine	F	9	19.00	Silicon	Si	14	28.09
Francium	Fr	87	(223)	Silver	Ag	47	107.9
Gadolinium	Gd	64	157.3	Sodium	Na	11	22.99
Gallium	Ga	31	69.72	Strontium	Sr	38	87.62
Germanium	Ge	32	72.59	Sulfur	S	16	32.06
Gold	Au	79	197.0	Tantalum	Ta	73	180.9
Hafnium	Hf	72	178.5	Technetium	Tc	43	(98)
Helium	He	2	4.003	Tellurium	Te	52	127.6
Holmium	Ho	67	164.9	Terbium	Tb	65	158.9
Hydrogen	H	1	1.008	Thallium	Tl	81	204.4
Indium	In	49	114.8	Thorium	Th	90	232.0
Iodine	I	53	126.9	Thulium	Tm	69	168.9
Iridium	Ir	77	192.2	Tin	Sn	50	118.7
Iron	Fe	26	55.85	Titanium	Ti	22	47.90
Krypton	Kr	36	83.80	Tungsten	W	74	183.9
Lanthanum	La	57	138.9	Uranium	U	92	238.0
Lawrencium	Lr	103	(260)	Vanadium	V	23	50.94
Lead	Pb	82	207.2	Xenon	Xe	54	131.3
Lithium	Li	3	6.941	Ytterbium	Yb	70	173.0
Lutetium	Lu	71	175.0	Yttrium	Y	39	88.91
Magnesium	Mg	12	24.31	Zinc	Zn	30	65.38
Manganese	Mn	25	54.94	Zirconium	Zr	40	91.22
Mendelevium	Md	101	(258)				

*The values given here are to four significant figures. A table of more accurate atomic masses is given in Appendix F.

†A value given in parentheses denotes the mass of the longest-lived isotope.

General Chemistry

Inge S. Zelling

General Chemistry

DONALD A. McQUARRIE
University of California, Davis

PETER A. ROCK
University of California, Davis

W. H. FREEMAN AND COMPANY
New York

Printed in the United States of America

2 3 4 5 6 7 8 9 0

Library of Congress Cataloging in Publication Data

McQuarrie, Donald A. (Donald Allan)
General chemistry.

Includes index.
1. Chemistry. I. Rock, Peter A., 1939–
II. Title.
QD31.2.M388 1984 540 83-20832
ISBN 0-7167-1499-X

Preface

At one time, general chemistry texts were essentially a litany of descriptive chemistry with very little theory. Then theoretical considerations were included in ever-increasing amounts until the point was reached where some students knew how to construct hybrid orbitals to describe the bonding in methane but did not know that methane is a colorless, odorless, combustible gas. In recent years it has been widely acknowledged that there is a need to achieve a more suitable balance between descriptive and theoretical chemistry. Although we strongly support this trend, we also find it difficult to present lectures exclusively devoted to the descriptive chemistry of the elements. Probably like many other instructors, we relegated most of the material in the descriptive chemistry chapters to assigned reading. In this text we have attempted to solve the problem of "what to do with descriptive chemistry in the general chemistry course" in several ways. The group properties of the chemical elements and the periodic table are introduced in Chapter 2. Furthermore, Chapter 4, which follows an introduction to stoichiometry in Chapter 3, deals with simple chemical reactions and is mostly descriptive chemistry. This early introduction allows us to integrate descriptive chemistry throughout the text. When a compound is used to illustrate a principle or a type of calculation, we often comment on one or more of its chemical or physical properties. Many of the in-text worked examples and the end-of-chapter problems tell something about the compound being used. A feature that is unique to this text, however, is the use of *Interchapters*. Most of the Interchapters are short, elementary discussions of particular elements or groups of elements. The Interchapters are meant to be inviting and interesting by presenting chemistry in an everyday context. We have found that students are especially interested in industrial applications, and we have presented many such applications in the Interchapters. The Interchapters contain no new principles, and so they may be covered in any order or even omitted entirely; but they are brief enough and focused enough to be easily given as assigned reading.

Another unique feature of the text is the frequent use of color photographs. This is a novel feature in general chemistry texts, but not so in biology or astronomy, where many of the texts use color photographs throughout. Chemistry is at least as colorful as these subjects, and our intent is to present the subject in this light. The reader can see first-hand that bromine is a dark red liquid, that chlorine is a pale yellow gas, and that the pH of a solution can be estimated by means of indicators. We have used color functionally to show elements, compounds, reactions and other chemical events. By doing so we hope to make the study of chemistry more lively and enjoyable. Most of the photographs have been set up and taken specifically for this text by Chip Clark, a scientific photographer for the Smithsonian Institution.

There are other novel features within these pages: (1) We feel that first-year students should become proficient in writing Lewis formulas, and so we have devoted much of Chapter 10 to Lewis formulas. This chapter can be covered in two lectures. (2) Chapter 11, The Shapes of Molecules, develops the valence-shell-electron-pair-repulsion (VSEPR) theory. We feel that VSEPR theory is easy to understand, easy to apply, and amazingly reliable. It reinforces the writing of Lewis formulas; it introduces first-year students to a large number of compounds that ordinarily they never see; and it is our experience that students enjoy VSEPR theory because of its simplicity and predictive power. For the instructor who for one reason or another does not wish to present VSEPR, most of the chapter may be omitted. (3) Our section headings are declarative topic statements rather than brief terms or phrases. This form allows headings to focus on the primary objective or result of each section within a chapter; we feel that just reading the section headings will give a student a good overview of a chapter. We have found such declarative section headings to be a strong pedagogical aid. (4) We have made a great effort to use actual chemical compounds and real data in formulating in-chapter examples and end-of-chapter problems. We have avoided discussions of unspecified substances A and B or reactions such as $A + B \rightarrow C$; rather, we have used actual chemical species in order to make the presentation real and vivid.

There are other features of this text that, though not necessarily novel, we think deserve mention: (1) We have taken what might be called the experimental approach. We introduce and discuss experimental observations and data *before* developing the theory to tie these data together. (2) We have ordered the material with the companion laboratory course in mind. We introduce stoichiometry, solutions, and the more elementary properties of acids and bases early to accommodate a variety of laboratory schedules. (3) There is no consensus concerning the relative placement of the chapters on chemical equilibrium and chemical kinetics in general chemistry texts. To this end, we introduced the basic idea of both rate and equilibrium (particularly the concept of dynamic equilibrium) in our discussion of vapor pressure in Chapter 13 and in the process of dissolution of solids in liquids in Chapter 14, preceding our formal treatment of chemical equilibrium and chemical kinetics. (4) We use SI units almost exclusively. Authors of textbooks today face a dilemma with regard to units. Although SI units are endorsed by numerous organizations and journals, there are many instructors who are reluctant, or even hostile, to change to SI units. Neither of us were strong advocates of SI units before writing this text, but in the process we have found that with a very small effort we became comfortable with them. Thus, we use joules instead of calories and picometers instead of Ångstroms. One SI unit that we could not readily adjust to, however, is the pascal, the SI unit of pressure. We generally have expressed pressure in units of atmospheres or torr (mm Hg), although we have a separate section of gas-law problems involving pascals for instructors who have made a complete transition to SI units. We also have not expressed density in $kg \cdot m^{-3}$. (5) Throughout the text we have used marginal notes both to reinforce certain points and to introduce information that we feel may be interesting, but peripheral to the discussion in the text. (6) Each chapter ends with a Summary, a list of Terms You Should Know (with the page numbers on which the terms are introduced) and a list of Equations You Should Know How To Use. (7) The problems at the ends of the chapters are arranged in matched pairs, both dealing with the same principle or operation; in this way, students who have difficulty with a particular problem can work a similar but new problem to test understanding. The problems are ordered by topic as covered in the chapter and are graded within each topic, with the easier problems coming first. (8) The answers to the odd-numbered problems are at the end of the text; the detailed solutions to the odd-numbered problems and the answers to the even-numbered problems are given in the Study Guide/Solutions Manual that accompanies this text; and the detailed solutions to the even-numbered problems are given in the Instructor's Manual.

We firmly believe that the Study Guide/Solutions Manual that accompanies this text is of real benefit to the student. Students have most difficulty with the numerical problems in general chemistry and so we have designed the Study Guide/Solutions Manual with this in mind. For each chapter in the text, the Study Guide/Solutions Manual has

A. Outline of Chapter (section headings and short descriptive sentences)

B. Self-Test (about 40 fill-in-the-blank questions; no numerical problems)

C. Calculations You Should Know How To Do

D. Solutions to the Odd-numbered Problems (detailed solutions are unquestionably the most valuable feature for the student)

E. Answers to the Even-numbered Problems (answers only)

F. Answers to the Self-Test.

There also is a Glossary at the end of the Study Guide, which is cross-referenced to the text.

Included in the ancillary package for this text is a laboratory manual, and accompanying instructor's manual, by Julian Roberts, Lee Hollenberg, and James Postma. This laboratory manual is derived from the popular Frantz-Malm series and provides 40 experiments that have been thoroughly class-tested for safety and reproducibility.

ACKNOWLEDGMENTS

Numerous teachers of chemistry have made significant contributions to various drafts of our manuscript. We are grateful for their efforts and support and would like to acknowledge their assistance on this project:

David L. Adams, North Shore Community College
Robert C. Atkins, James Madison University
Robert J. Balahura, University of Guelph
Otto T. Benfey, Guilford College
Larry E. Bennett, San Diego State University
David W. Brooks, University of Nebraska-Lincoln
Bruce W. Brown, Portland State University
George Brubaker, Illinois Institute of Technology
Ian S. Butler, McGill University
Harvey F. Carroll, Kingsborough Community College, CUNY
Ronald J. Clark, Florida State University
John M. D'Auria, Simon Fraser University
Derek A. Davenport, Purdue University
Daniel R. Decious, California State University, Sacramento
Robert Desiderato, North Texas State University
Timothy C. Donnelly, University of California, Davis
Frank J. Gomba, United States Naval Academy
Charles G. Haas, Jr., Pennsylvania State University
Edward D. Harris, Texas A&M University
Henry M. Hellman, New York University
Forrest C. Hentz, Jr., North Carolina State University
Earl S. Huyser, University of Kansas
Joseph E. Ledbetter, California State University, San Jose
Edward C. Lingafelter, University of Washington
William M. Litchman, University of New Mexico
Saundra Y. McGuire, University of Tennessee
Arlene M. McPherson, Tulane University
John M. Newey, American River College
Dennis G. Peters, Indiana University
Grace S. Petrie, Nassau Community College
Henry Po, California State University, Long Beach
James M. Postma, California State University, Chico
W. H. Reinmuth, Columbia University
Randall J. Remmel, University of Alabama in Birmingham
Don Roach, Miami-Dade Community College
Charles B. Rose, University of Nevada
Barbara Sawrey, San Diego State University
William M. Scovell, Bowling Green State University
Donald Showalter, University of Wisconsin
R. T. Smedberg, American River College
James C. Thompson, University of Toronto
Russell F. Trimble, Southern Illinois University
Carl Trindle, University of Virginia
Carl A. von Frankenberg, University of Delaware
E. J. Wells, Simon Frazer University
Helmut Wieser, University of Calgary

There are a number of people involved in this project to whom we give special thanks—Bruce Armbruster for convincing us that we should undertake this project and for directing us to W. H. Freeman and Company, Donna Salmon for producing most of the beautiful artwork in the book, Travis Amos for researching most of the photographs and for numerous brilliant suggestions, Chip Clark for setting up and taking many of the photographs of chemicals and chemical reactions, Ruth Allen for always being there in a most gracious and supportive manner, Jane Grosinger for producing a marvelous calendar, Linda Chaput for always being behind the scenes giving us her enthusiastic support, Lee Walters, from whom we could not expect more in commitment or in support, for directing all the aspects of the development and production of this book and for pacifying us in critical moments with patience and understanding, and to Neil Patterson and Howard Boyer for having the courage and foresight to undertake this project. We should also like to thank Carole McQuarrie and Joe Ledbetter for reading all the manuscript, all the galleys, and all the page proofs and for doing all the problems. Their generous assistance was almost equivalent and important as the role of coauthors. We finally thank Elaine Rock for promptly and accurately typing the numerous drafts as are inevitable in a project of this magnitude.

DONALD A. McQUARRIE
PETER A. ROCK

Contents in Brief

Contents

22. Entropy and Gibbs Free Energy 857

23. Transition Metal Complexes 907

24. Nuclear and Radiochemistry 943

General Chemistry

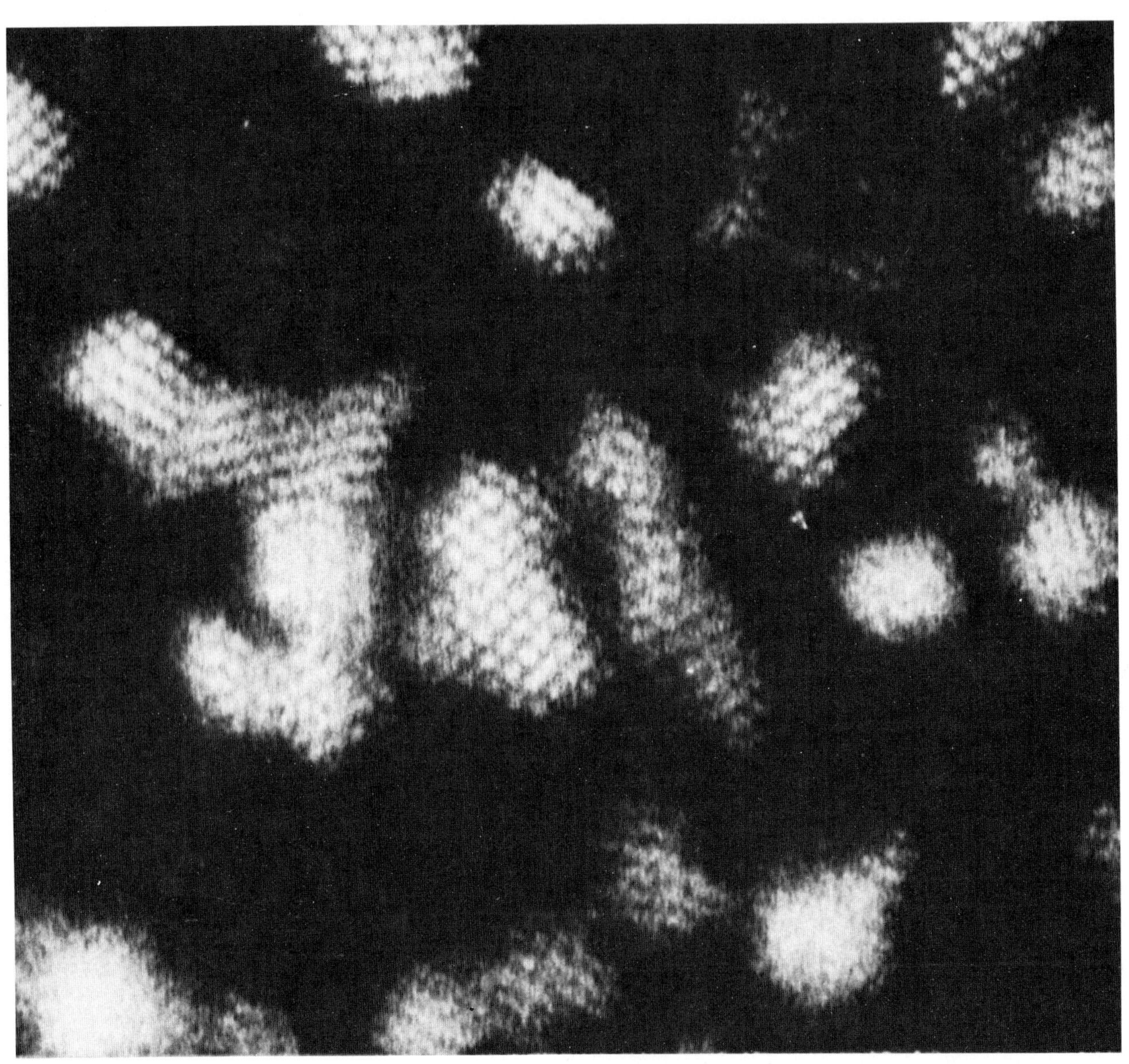

Although atoms are much too small to be seen even with the most powerful optical microscopes, they can be photographed with modern, sophisticated instruments. The above photograph shows small clusters of uranium atoms on a very thin carbon surface. The uranium atoms appear as white dots on the dark background.

1 / Atoms and Molecules

Chemistry is the study of the properties of various substances and of how they react with each other. To understand the nature of these substances better, we must begin by examining the atomic theory, one of the fundamental theories of chemistry. The atomic theory pictures all substances as consisting of atoms or of groups of atoms called molecules. Proposed in the early 1800's by John Dalton, the atomic theory gives a simple picture of chemical reactions and provides explanations for many chemical observations. As we trace the development of the atomic theory, we shall see the role it plays in understanding chemistry. We begin with a picture of the atom as a solid, structureless sphere and conclude the chapter with the nuclear model of the atom as a small, massive nucleus containing protons and neutrons with electrons swirling around the nucleus.

1-1. WHY SHOULD YOU STUDY CHEMISTRY?

You and about 400,000 other students in the United States and Canada are about to begin your first college course in chemistry. Although most of you do not plan to become professional chemists (only about 10,000 students graduate each year in the United States with a bachelor's degree in chemistry), your proposed major field of study probably requires at least one year of college chemistry. A knowledge of elementary chemistry is so necessary in many fields that general chemistry is one of the courses with the largest enrollment at most colleges.

Chemistry plays a pervasive role in all our lives. Hundreds of materials that you and your family use directly and indirectly every day are products of chemical research. The development of fertilizers, which is one of the major areas of the chemical industry, has profoundly affected agricultural production in developed countries. Another major area of the chemical industry is the pharmaceutical, or drug, industry. Who among us has not used an antibiotic to cure an infection or various drugs to alleviate the pain associated with dental work, accidents, or surgery? Modern medicine, which rests firmly upon chemistry, has increased our life expectancy by 15 years since the 1920's. It is hard to believe that little over a century ago many people actually died from simple infections and other diseases that we seldom hear of anymore.

Perhaps the chemical products most familiar to all of us are plastics. The annual production of synthetic fibers in the United States exceeds 10 billion pounds. About 50 percent of industrial chemists are involved with the development or production of plastics. There is hardly an activity in your daily life that does not include some plastic product. Names such as nylon, Formica, Saran, Teflon, Hollofil, Gore-Tex, polyester, and silicone are familiar to most of us. Chemistry also plays a major role in materials science—from the manufacture of computer chips to paper and wood products to structural metals such as steel and lightweight titanium and aluminum alloys for ships and aircraft.

Regardless of your reasons for studying chemistry, it is important to remember that it is a requirement for your major because the people who work in your field consider it necessary and useful. With a reasonable effort on your part, we are confident that you will find chemistry both interesting and enjoyable.

Table 1-1 Elemental composition of the earth's surface, which includes the crust, oceans, and the atmosphere

Element	*Percent by mass*
oxygen	49.1
silicon	26.1
aluminum	7.5
iron	4.7
calcium	3.4
sodium	2.6
potassium	2.4
magnesium	1.9
hydrogen	0.88
titanium	0.58
chlorine	0.19
carbon	0.09
all others	0.56

1-2. ELEMENTS ARE THE SIMPLEST SUBSTANCES

Almost all of the millions of different chemicals known today can be broken down into simpler substances. Any substance that cannot be broken down into simpler substances is called an *element*. This is strictly an operational definition, but it does lend itself to experimental testing. Pure substances that can be broken down into simpler substances are called *compounds*. Before the early 1800's, many substances were incorrectly classified as elements because methods to break them down had not yet been developed, but these errors have gradually been rectified over the years. Although our definition of an element is a satisfactory working definition, we shall learn later that the modern definition is that *an element* is a substance that contains only atoms with the same nuclear charge.

There are 108 known chemical elements. Some of them are very rare; fewer than half of the elements constitute 99.99 percent of all substances. Table 1-1 lists the most common elements found in the earth's crust, the oceans, and the atmosphere. Note that only ten elements make up over 99 percent of the total. Oxygen and silicon are the most common elements because they are the major constituents of sand, soil, and rocks. Oxygen also occurs as a free element in the atmosphere and in combination with hydrogen in water. Table 1-2 lists the most common elements found in the human body. Note that only ten elements constitute over 99.8 percent of the total mass of the human body.

Table 1-2 Elemental composition of the human body

Element	*Percent by mass*
oxygen	64.6
carbon	18.0
hydrogen	10.0
nitrogen	3.1
calcium	1.9
phosphorus	1.1
chlorine	0.40
potassium	0.36
sulfur	0.25
sodium	0.11
magnesium	0.03
iron	0.005
zinc	0.002
copper	0.0004
tin	0.0001
manganese	0.0001
iodine	0.0001

1-3. ABOUT THREE FOURTHS OF THE ELEMENTS ARE METALS

One broad classification of the elements is into *metals* and *nonmetals*. We are all familiar with the properties of metals. They have a characteristic luster; can be rolled or hammered into sheets, drawn into wires, melted, and cast into various shapes; and are usually good conductors of electricity and heat. About three fourths of the elements are metals. All the metals except mercury are solids at room temperature (about 20°C). Mercury is a shiny, silver-colored liquid at room temperature and used to be called quicksilver.

Table 1-3 lists some metals and their *chemical symbols*. Chemical symbols are abbreviations used to designate the elements and are usually the first one or two letters in the name of the element. Some chemical symbols do not seem to correspond at all to the names because the symbols are derived from the Latin names of those ele-

Table 1-3 Some common metals and their chemical symbols

Element	*Symbol*	*Element*	*Symbol*
aluminum	Al	mercury	Hg
barium	Ba	nickel	Ni
cadmium	Cd	platinum	Pt
calcium	Ca	potassium	K
chromium	Cr	silver	Ag
cobalt	Co	sodium	Na
copper	Cu	strontium	Sr
gold	Au	tin	Sn
iron	Fe	titanium	Ti
lead	Pb	tungsten	W
lithium	Li	uranium	U
magnesium	Mg	zinc	Zn
manganese	Mn		

Table 1-4 Elements whose symbol corresponds to the Latin name

Element	*Symbol*	*Latin name*
antimony	Sb	stibium
copper	Cu	cuprum
gold	Au	aurum
iron	Fe	ferrum
lead	Pb	plumbum
mercury	Hg	hydrargyrum
potassium	K	kalium
silver	Ag	argentum
sodium	Na	natrium
tin	Sn	stannum

ments (Table 1-4). It is necessary to memorize the chemical symbols of the more common elements because we shall be using them throughout this book.

Unlike metals, nonmetals vary greatly in their appearance. Over half of the nonmetals are gases at room temperature, and the others are solids except for bromine, which is a red-brown, corrosive liquid. In contrast to metals, nonmetals are poor conductors of electricity and heat, cannot be rolled into sheets or drawn into wires, and do not have a characteristic luster. Table 1-5 lists several common nonmetals, their chemical symbols, and their appearances. Note that several of the symbols of the nonmetallic elements in Table 1-5 have a 2 subscript. This indicates that these elements—hydrogen (H_2), nitrogen (N_2), oxygen (O_2), fluorine (F_2), chlorine (Cl_2), bromine (Br_2), and iodine (I_2)—exist in nature as two atoms joined together.

Table 1-5 Some common nonmetals and their appearance at room temperature

Element	*Symbol*[a]	*Appearance*
	Gases	
hydrogen	H_2	colorless
helium	He	colorless
nitrogen	N_2	colorless
oxygen	O_2	colorless
fluorine	F_2	pale yellow
neon	Ne	colorless
chlorine	Cl_2	green-yellow
argon	Ar	colorless
krypton	Kr	colorless
xenon	Xe	colorless
	Liquids	
bromine	Br_2	red-brown
	Solids	
carbon	C	black (in the form of coal or graphite)
phosphorus	P	pale yellow or red
sulfur	S	lemon yellow
iodine	I_2	violet-black

[a]The subscript 2 tells us that, at room temperature, the element exists as a diatomic molecule, that is, a molecule consisting of two atoms.

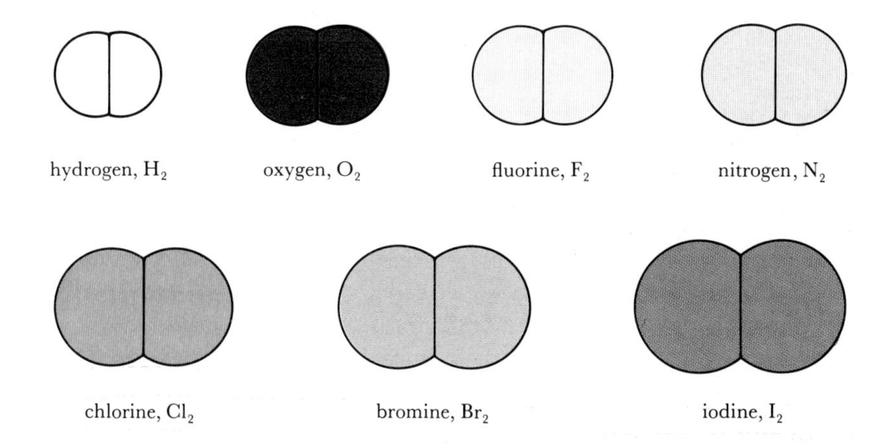

Figure 1-1 Scale models of molecules of hydrogen, oxygen, fluorine, chlorine, bromine, and iodine. These substances exist as diatomic molecules in their natural state but are still classified as elements because the molecules consist of identical atoms. We shall discuss atoms in Section 1-6.

A unit of two or more atoms that are joined together is called a *molecule,* and a molecule consisting of two atoms is called a *diatomic molecule*. Scale models of some diatomic molecules are shown in Figure 1-1.

1-4. ANTOINE LAVOISIER WAS THE FOUNDER OF MODERN CHEMISTRY

Although chemistry was beginning to develop as a science by the eighteenth century, it still lacked one ingredient essential for becoming a modern science. That ingredient was *quantitative measurement.* A quantitative measurement is one in which the result is expressed as a number. For example, the determination that the mass of 1.00 cm^3 (cubic centimeter) of gold is 19.3 g (grams) or that 1.25 g of calcium reacts with 1.00 g of sulfur is a quantitative measurement. Compare these statements to *qualitative observations,* where we note general characteristics, such as color and odor. An example of a qualitative statement is that lead is much denser than aluminum; the quantitative statement is that the mass of 1.00 cm^3 of lead is 11.3 g while that of 1.00 cm^3 of aluminum is 2.70 g.

It was the French scientist Antoine Lavoisier (Figure 1-2) who first appreciated the importance of carrying out quantitative measurements. Lavoisier designed special balances that were more accurate than ever before and discovered the *law of conservation of mass:* in an ordinary chemical reaction, the total mass of the reacting substances is equal to the total mass of the products formed. A *natural law,* such as the law of conservation of mass, is a summary of a large number of experimental observations. A law is not an explanation of the observed facts but is a concise summary statement of many observations.

Figure 1-2 The French chemist Antoine Lavoisier and his wife and colleague, Marie-Anne Pierrette. Marie-Anne assisted Antoine in much of his work and illustrated and helped write his famous book, *Elementary Treatise on Chemistry.* Because of his financial connection with a much-hated tax-collecting firm, Lavoisier was denounced, arrested, and guillotined in 1794 by supporters of the French Revolution.

Lavoisier's influence on the development of chemistry as a modern science cannot be overstated. In 1789, he published his *Elementary Treatise on Chemistry,* in which he presented a unified picture of chemi-

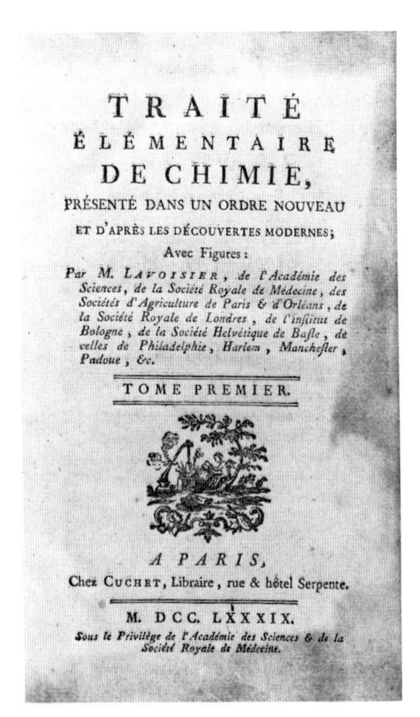
TRAITÉ ÉLÉMENTAIRE DE CHIMIE, PRÉSENTÉ DANS UN ORDRE NOUVEAU ET D'APRÈS LES DÉCOUVERTES MODERNES; Avec Figures: Par M. LAVOISIER, de l'Académie des Sciences, de la Société Royale de Médecine, des Sociétés d'Agriculture de Paris & d'Orléans, de la Société Royale de Londres, de l'Institut de Bologne, de la Société Helvétique de Basle, de celles de Philadelphie, Harlem, Manchester, Padoue, &c.

TOME PREMIER.

A PARIS, Chez CUCHET, Libraire, rue & hôtel Serpente.

M. DCC. LXXXIX.

Sous le Privilège de l'Académie des Sciences & de la Société Royale de Médecine.

The title page to Lavoisier's textbook of chemistry.

cal knowledge. The *Elementary Treatise on Chemistry* was translated into many languages and was the first textbook of chemistry based on quantitative experiments.

1-5. THE RELATIVE AMOUNT OF EACH ELEMENT IN A COMPOUND IS ALWAYS THE SAME

The quantitative approach pioneered by Lavoisier was used in the chemical analysis of compounds. The quantitative chemical analysis of a great many compounds led to the *law of constant composition:* the relative amount of each element in a particular compound is always the same, regardless of the source of the compound or how the compound is prepared. For example, if calcium metal is heated with sulfur, the compound called calcium sulfide is formed. We can specify the relative amounts of calcium and sulfur in calcium sulfide by the mass percentage of each element. The mass percentages of calcium and sulfur in calcium sulfide are defined as

$$\begin{matrix}\text{mass percentage of calcium} \\ \text{in calcium sulfide}\end{matrix} = \frac{\text{mass of calcium}}{\text{mass of calcium sulfide}} \times 100$$

$$\begin{matrix}\text{mass percentage of sulfur} \\ \text{in calcium sulfide}\end{matrix} = \frac{\text{mass of sulfur}}{\text{mass of calcium sulfide}} \times 100$$

Suppose we analyze 1.630 g of calcium sulfide and find that it consists of 0.906 g of calcium and 0.724 g of sulfur. Then the mass percentages of calcium and sulfur are

Although many people use the terms mass and weight interchangeably, these terms are not the same. Mass is the inherent amount of material of an object, whereas weight is the force of attraction of the object to a large body such as the earth or the moon. An object on the moon weighs about one-eighth as much as it does on earth, but its mass is the same in both places. We shall use the term mass throughout the book.

$$\begin{aligned}\begin{matrix}\text{mass percentage of calcium} \\ \text{in calcium sulfide}\end{matrix} &= \frac{\text{mass of calcium}}{\text{mass of calcium sulfide}} \times 100 \\ &= \frac{0.906\text{ g}}{1.630\text{ g}} \times 100 = 55.6\%\end{aligned}$$

$$\begin{aligned}\begin{matrix}\text{mass percentage of sulfur} \\ \text{in calcium sulfide}\end{matrix} &= \frac{\text{mass of sulfur}}{\text{mass of calcium sulfide}} \times 100 \\ &= \frac{0.724\text{ g}}{1.630\text{ g}} \times 100 = 44.4\%\end{aligned}$$

The law of constant composition says that the mass percentage of calcium in calcium sulfide is 55.6 percent whether the calcium sulfide is prepared by heating a large amount of calcium with a small amount of sulfur or by heating a small amount of calcium with a large amount of sulfur. The mass percentage of calcium in calcium sulfide is always 55.6 percent, and the mass percentage of sulfur in calcium sulfide is always 44.4 percent.

Example 1-1: Suppose we analyze 2.83 g of a compound of lead and sulfur and find that it consists of 2.45 g of lead and 0.380 g of sulfur. Calculate the mass percentages of lead and sulfur in the compound, which is called lead sulfide.

Solution: The mass percentage of lead in lead sulfide is

$$\begin{aligned}\text{mass percentage of lead in lead sulfide} &= \frac{\text{mass of lead}}{\text{mass of lead sulfide}} \times 100 \\ &= \frac{2.45\ \text{g}}{2.83\ \text{g}} \times 100 = 86.6\%\end{aligned}$$

The mass percentage of sulfur in lead sulfide is

$$\begin{aligned}\text{mass percentage of sulfur in lead sulfide} &= \frac{\text{mass of sulfur}}{\text{mass of lead sulfide}} \times 100 \\ &= \frac{0.380\ \text{g}}{2.83\ \text{g}} \times 100 = 13.4\%\end{aligned}$$

The law of constant composition assures us that the mass percentage of lead in lead sulfide is independent of the source of the lead sulfide. The principal source of lead sulfide is the ore *galena.*

Problems 1-5 to 1-10 deal with mass percentages in chemical compounds.

1.6 DALTON'S ATOMIC THEORY EXPLAINS THE LAW OF CONSTANT COMPOSITION

By the end of the eighteenth century, many compounds had been analyzed and a large amount of experimental data had been accumulated. A theory was needed to bring all these data into a single framework. A *theory* is one or more hypotheses that can be used to explain a law or a number of experimental observations. In 1803, John Dalton, an English schoolteacher, proposed an *atomic theory* that provided a simple and beautiful explanation of the law of constant composition. We can express the postulates of Dalton's atomic theory in modern terms as follows:

1. Matter is composed of small, indivisible particles called *atoms.*
2. The atoms of a given element all have the same mass and are identical in all respects, including chemical behavior.
3. The atoms of different elements differ in mass and in chemical behavior.
4. Chemical *compounds* are composed of two or more different atoms joined together in simple, fixed ratios. The particle that results when two or more atoms join together is called a *molecule.* The

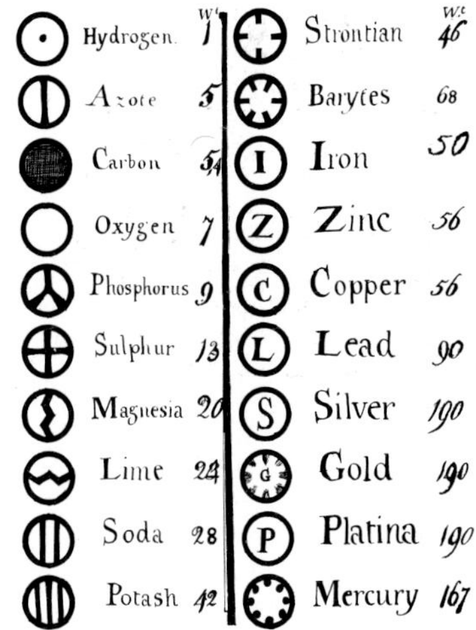

Dalton's symbols for chemical elements. Some of these are now known to be compounds, not elements.

atoms in a molecule do not necessarily have to be different. If the atoms are the same, it is a molecule of an element. If the atoms are different, it is a molecule of a compound.

5. In a chemical reaction, the atoms involved are rearranged to form different molecules; no atoms are created or destroyed.

As we shall see, some of these postulates were later modified, but the main features of Dalton's atomic theory are accepted today.

The law of constant composition follows nicely from Dalton's atomic theory. Consider calcium sulfide, which we know consists of 55.6 percent calcium and 44.4 percent sulfur by mass. Suppose that there is one calcium atom for each sulfur atom in calcium sulfide. Because we know that the relative masses of a calcium atom and a sulfur atom are 55.6 and 44.4, we know that the ratio of the mass of a calcium atom to that of a sulfur atom is

$$\frac{\text{mass of a calcium atom}}{\text{mass of a sulfur atom}} = \frac{55.6}{44.4} = 1.25$$

or

$$\text{mass of a calcium atom} = 1.25 \times \text{mass of a sulfur atom}$$

Thus, even though we cannot determine the mass of any individual atom, we can use the quantitative results of chemical analysis to determine the *relative* masses of atoms. Of course, our result for calcium and sulfur is based on the assumption that there is one atom of calcium for each atom of sulfur in calcium sulfide.

Let's consider another compound, hydrogen chloride. Quantitative chemical analysis shows that the mass percentages of hydrogen and chlorine in hydrogen chloride are 2.76 percent and 97.24 percent, respectively. Once again, assuming that one atom of hydrogen is combined with one atom of chlorine, we find that

$$\frac{\text{mass of a chlorine atom}}{\text{mass of a hydrogen atom}} = \frac{97.24}{2.76} = 35.2$$

or

$$\text{mass of a chlorine atom} = 35.2 \times \text{mass of a hydrogen atom}$$

By continuing in this manner with other compounds, it is possible to build up a table of relative atomic masses. We define a quantity called *atomic mass*, which is the mass of a given atom relative to some particular atom. Being relative quantities, atomic masses have no units. Nevertheless, it is often convenient to assign a unit called an *atomic mass unit* (amu) to atomic masses. Thus, for example, we can say that the atomic mass of carbon is 12.01 or 12.01 amu; both statements are correct.

At one time the mass of hydrogen, the lightest atom, was arbitrarily given the value of exactly 1 and used as the standard by which all other atomic masses were expressed. As discussed later in this chapter, however, a form of carbon is now used as the standard. Thus, today the atomic mass of hydrogen (listed on the inside front cover) is 1.008 instead of exactly 1.

Atomic masses are relative masses.

1-7. MOLECULES ARE GROUPS OF ATOMS JOINED TOGETHER

Dalton's atomic theory postulated that an element is a substance that consists of identical atoms and that a compound is a substance that consists of identical molecules, or groups of atoms. Although Dalton did not realize it at the time, some of the elements consist of molecules containing identical atoms. As noted in Table 1-5, the elements hydrogen, nitrogen, oxygen, fluorine, chlorine, bromine, and iodine exist as diatomic molecules of the same kind of atoms (see Figure 1-1). Consequently, these substances are classified as elements. Compounds, on the other hand, are made up of molecules containing different kinds of atoms. Some examples of molecules of compounds are

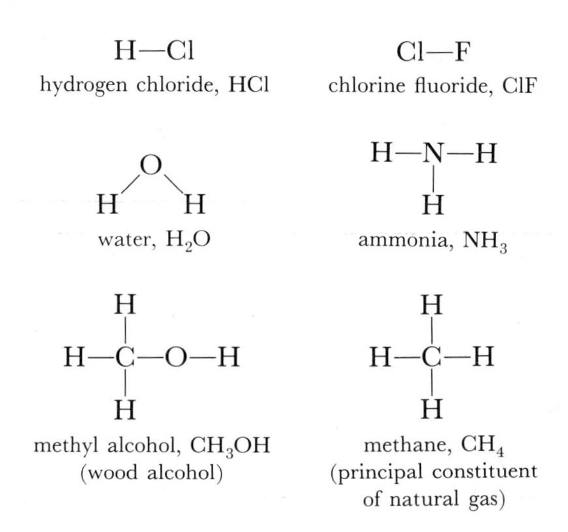

These formulas indicate how the atoms are joined together in the molecules. We shall learn how to write such formulas in Chapter 10.

Molecular models of these molecules are shown in Figure 1-3.

Dalton's atomic theory provides a nice pictorial view of chemical reactions. Recall that Dalton proposed that, in a chemical reaction, the atoms in the reactant molecules are separated and then rearranged into product molecules. According to this view, the chemical

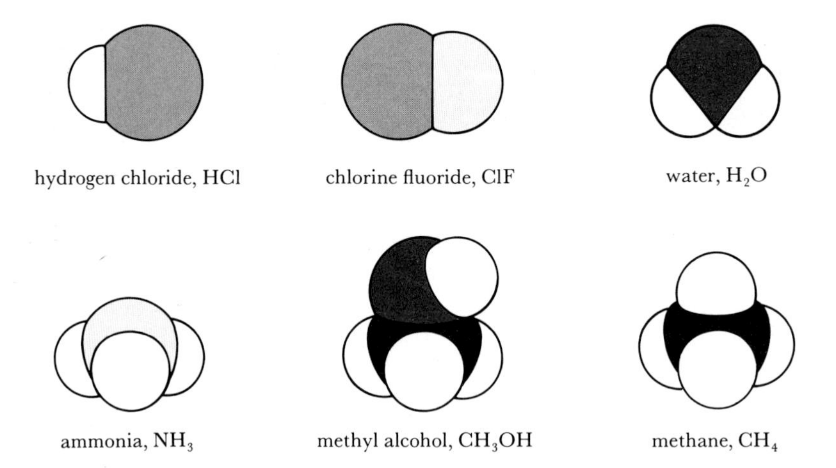

Figure 1-3 Scale molecular models of hydrogen chloride, chlorine fluoride, water, ammonia, methyl alcohol, and methane.

reaction between hydrogen and oxygen to form water involves the following rearrangement:

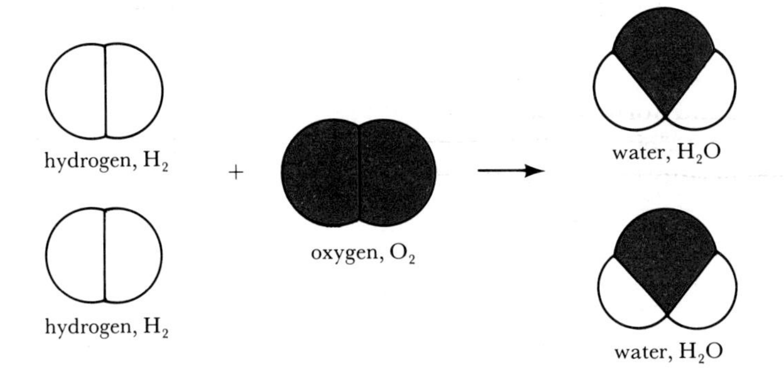

Note that completely different molecules and hence completely different substances are formed in a chemical reaction. Hydrogen and oxygen are gases, whereas water is a liquid.

As another example, consider the burning of carbon in oxygen to form carbon dioxide:

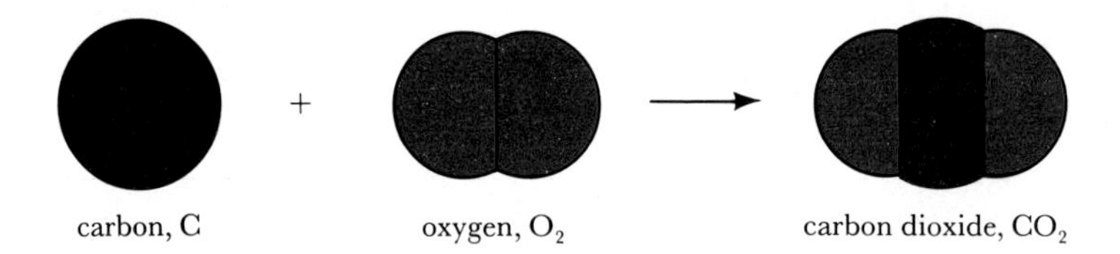

Once again, note that a completely new substance is formed. Carbon is a black solid; the product, carbon dioxide, is a colorless gas.

As a final example, consider the reaction between steam (hot gaseous water) and red-hot carbon to form hydrogen and carbon monoxide:

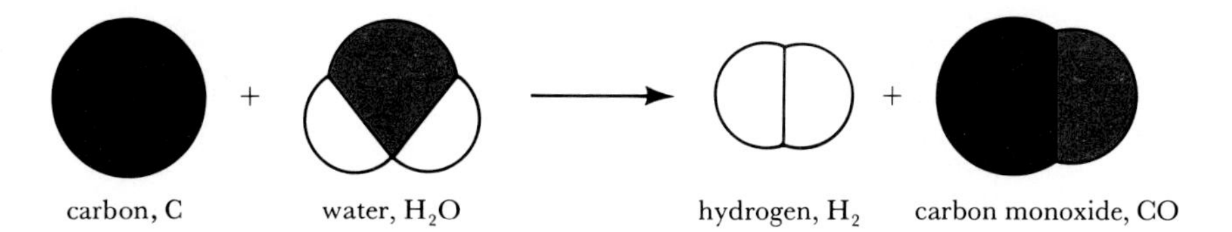

Notice that in each of the three reactions pictured here, the numbers of each kind of atom do not change. Atoms are neither created nor destroyed in chemical reactions; they are simply rearranged into new molecules.

1-8. THE ASSIGNMENT OF NAMES TO COMPOUNDS IS CALLED CHEMICAL NOMENCLATURE

At this point we discuss the system of naming compounds that consist of just two elements. Such compounds are called *binary compounds.* If the two elements that make up a compound are a metal and a nonmetal (Tables 1-3 through 1-5), the compound is named by first naming the metal and then the nonmetal, with the ending of the name of the nonmetal changed to *-ide*. For example, the name of the compound formed between calcium and sulfur is calcium sulf*ide.* Because calcium sulfide consists of one atom of calcium for each atom of sulfur, we write the *chemical formula* of calcium sulfide as CaS; in other words, we simply join the chemical symbols of the two elements. In a different case, calcium combines with *two* atoms of chlorine to form calcium chlor*ide;* thus, the formula of calcium chloride is $CaCl_2$. Note that the number of atoms is indicated by a subscript. The subscript 2 in $CaCl_2$ means that there are two chlorine atoms per calcium atom in calcium chloride. Table 1-6 lists the *-ide* nomenclature for some common nonmetals.

Table 1-6 The *-ide* nomenclature of some common nonmetals

Element	*-ide nomenclature*
oxygen	oxide
nitrogen	nitride
sulfur	sulfide
fluorine	fluoride
chlorine	chloride
bromine	bromide
iodine	iodide
phosphorus	phosphide
hydrogen	hydride
carbon	carbide

Example 1-2: Name the following compounds:
(a) KI (b) Al_2O_3 (c) Na_2S (d) $BaCl_2$ (e) Li_3N

Solution: Use Table 1-6 for the correct *-ide* nomenclature.
(a) KI, potassium iodide
(b) Al_2O_3, aluminum oxide
(c) Na_2S, sodium sulfide
(d) $BaCl_2$, barium chloride
(e) Li_3N, lithium nitride

Many binary compounds are combinations of two nonmetals. For example, let's consider CO and CO_2. We cannot call both of these compounds carbon oxide because the name is ambiguous. When two or more compounds can result from the same two nonmetallic elements, we distinguish among them by means of Greek numerical prefixes:

CO	carbon *mono*xide
CO_2	carbon *di*oxide

Some other examples are

SO_2	sulfur *di*oxide
SO_3	sulfur *tri*oxide
SF_4	sulfur *tetra*fluoride
SF_6	sulfur *hexa*fluoride
PCl_3	phosphorus *tri*chloride
PCl_5	phosphorus *penta*chloride

The Greek prefixes used are summarized in Table 1-7.

Example 1-3: Name the following compounds:
(a) BrF_3 and BrF_5
(b) XeF_2 and XeF_4
(c) N_2O, NO, N_2O_3, NO_2, and N_2O_5

Solution:
(a) Because there is more than one compound formed by bromine and fluorine, we must distinguish between them by using Greek prefixes. Bromine is written first in the formulas, and so we name them

BrF_3	bromine trifluoride
BrF_5	bromine pentafluoride

(b) Xenon is written first in the formulas, and so we have

XeF_2	xenon difluoride
XeF_4	xenon tetrafluoride

(c) This series of compounds represents various oxides of nitrogen. Following the rules we have developed, their names are

N_2O	dinitrogen oxide
NO	nitrogen oxide (the prefix mono- is often dropped)
N_2O_3	dinitrogen trioxide
NO_2	nitrogen dioxide
N_2O_5	dinitrogen pentoxide

Problems 1-11 to 1-16 ask you to name compounds from their chemical formulas.

Dinitrogen oxide is commonly called by its less systematic name, nitrous oxide. Nitrous oxide was the first known general anesthetic (laughing gas) and is used as a foam propellant for whipped cream and shaving cream. Except for N_2O_5, which is a solid, all of the oxides of nitrogen are gases at room temperature and atmospheric pressure.

Table 1-7 Greek prefixes used to indicate the number of atoms of a given type in a molecule

Number	*Prefix*	*Example*
1	*mono-*	carbon monoxide, CO
2	*di-*	carbon dioxide, CO_2
3	*tri-*	sulfur trioxide, SO_3
4	*tetra-*	carbon tetrachloride, CCl_4
5	*penta-*	phosphorus pentachloride, PCl_5
6	*hexa-*	sulfur hexafluoride, SF_6

At this point you should understand how to name binary compounds if you are given the formula. In the next chapter, we shall learn how to write a correct formula from the name.

1-9. MOLECULAR MASS IS THE SUM OF THE MASSES OF THE ATOMS IN A MOLECULE

The sum of the atomic masses of the atoms in a molecule is called the *molecular mass* of the substance. For example, a water molecule, H_2O, consists of two atoms of hydrogen and one atom of oxygen. Therefore, the molecular mass of water is

$$\begin{aligned}\text{molecular mass of } H_2O &= 2(\text{atomic mass of H}) + (\text{atomic mass of O})\\ &= 2(1.008) + (16.00)\\ &= 18.02\end{aligned}$$

The molecular mass of dinitrogen pentoxide, N_2O_5, is

$$\begin{aligned}\text{molecular mass of } N_2O_5 &= 2(\text{atomic mass of N}) + 5(\text{atomic mass of O})\\ &= 2(14.01) + 5(16.00)\\ &= 108.02\end{aligned}$$

The following example shows how to use atomic and molecular masses to calculate the mass percentage composition of compounds.

Example 1-4: Using the atomic masses given in the table on the inside front cover, calculate the mass percentage of lead and sulfur in the compound lead sulfide, PbS.

Solution: As the formula PbS indicates, lead sulfide consists of one atom of lead for each atom of sulfur. The molecular mass of lead sulfide is

$$\begin{aligned}\text{molecular mass of lead sulfide} &= \text{atomic mass of lead} + \text{atomic mass of sulfur}\\ &= 207.2 + 32.06\\ &= 239.3\end{aligned}$$

The two mass percentages in Example 1-4 do not add up to 100.00% because of a slight round-off error.

The mass percentages of lead and sulfur in lead sulfide are

$$\text{mass percentage of lead} = \frac{\text{atomic mass of lead}}{\text{molecular mass of lead sulfide}} \times 100$$

$$= \frac{207.2}{239.3} \times 100 = 86.59\%$$

$$\text{mass percentage of sulfur} = \frac{\text{atomic mass of sulfur}}{\text{molecular mass of lead sulfide}} \times 100$$

$$= \frac{32.06}{239.3} \times 100 = 13.40\%$$

Problems 1-21 to 1-28 are similar to Example 1-4.

Note that this is the same result we got in Example 1-1. The table of atomic masses must be consistent with experimental values of mass percentages.

One of the great attractions of Dalton's atomic theory was that he was able to use it to devise a table of atomic masses that could then be used in chemical calculations like those in Example 1-4.

1-10. THE LAW OF MULTIPLE PROPORTIONS IS EXPLAINED BY THE ATOMIC THEORY

The two compounds carbon monoxide, CO, and carbon dioxide, CO_2, can be used to illustrate another law that was discovered around the same time as the law of constant composition—the *law of multiple proportions:* if two elements combine in more than one way, then the mass of one element that combines with a fixed mass of the other element will always be in the ratio of small, whole numbers. For example, of the two oxides of carbon, one is 42.9 percent carbon and 57.1 percent oxygen and the other is 27.3 percent carbon and 72.7 percent oxygen. Let's now compute the oxygen-to-carbon mass ratios in the two oxides of carbon. For numerical convenience, we take 100.0 g of each compound and write

$$\text{oxide I} \quad \frac{57.1\text{ g oxygen}}{42.9\text{ g carbon}} = \frac{1.33\text{ g oxygen}}{1.00\text{ g carbon}}$$

$$\text{oxide II} \quad \frac{72.7\text{ g oxygen}}{27.3\text{ g carbon}} = \frac{2.66\text{ g oxygen}}{1.00\text{ g carbon}}$$

Thus, in oxide I, 1.33 g of oxygen combines with 1.00 g of carbon, and in oxide II, 2.66 g of oxygen combines with 1.00 g of carbon. In these oxides, the ratio of the mass of oxygen that combines with 1.00 g of carbon is 2.66/1.33, or 2, a small, whole number.

The atomic theory explains the law of multiple proportions in simple terms: compounds differ in the number of atoms of each kind that combine. In the case of the oxides of carbon, the data suggest that one atom of oxygen combines with one atom of carbon in oxide I and that two atoms of oxygen combine with one atom of carbon in oxide II.

Example 1-5: Two oxides of sulfur are known. One of them (I) is a gas that results when sulfur is burned in oxygen. The other (II) must be prepared by special methods and yields sulfuric acid when dissolved in water. The mass percentages of sulfur and oxygen in these two oxides are

Compound	*Mass percentage of sulfur/%*	*Mass percentage of oxygen/%*
oxide I	50.0	50.0
oxide II	40.0	60.0

Show how these data illustrate the law of multiple proportions.

Solution: We take a 100.0-g sample of each oxide and compute the oxygen-to-sulfur mass ratios in the two oxides:

$$\text{oxide I} \quad \frac{50.0\text{ g oxygen}}{50.0\text{ g sulfur}} = \frac{1.00\text{ g oxygen}}{1.00\text{ g sulfur}}$$

$$\text{oxide II} \quad \frac{60.0\text{ g oxygen}}{40.0\text{ g sulfur}} = \frac{1.50\text{ g oxygen}}{1.00\text{ g sulfur}}$$

The ratio of the masses of oxygen that combine with 1.00 g of sulfur is 1.50/1.00. The smallest whole-number ratio that corresponds to 1.50/1.00 is 3/2, which illustrates the law of multiple proportions. Assuming that oxide I has two atoms of oxygen combined with one atom of sulfur, we conclude that oxide II has three atoms of oxygen combined with one atom of sulfur. In fact, the two compounds are sulfur dioxide, SO_2, and sulfur trioxide, SO_3.

Problems 1-29 to 1-34 deal with the law of multiple proportions.

1-11. THE DISCOVERY OF SUBATOMIC PARTICLES CHANGED OUR CONCEPT OF THE ATOM

For most of the nineteenth century, atoms were considered to be indivisible, stable particles. Toward the end of the century, however,

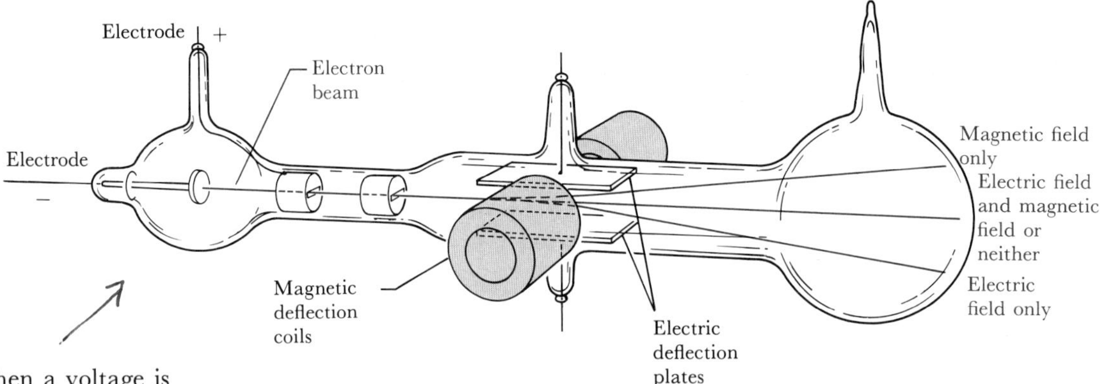

Figure 1-4 When a voltage is applied across electrodes that are sealed in a partially evacuated glass tube, the space between the electrodes glows. Thomson showed that this glow discharge consisted of a stream of identical negatively charged particles, now called electrons. He was able to do this by deflecting the stream of particles in electric and magnetic fields.

new experiments indicated that the atom is composed of even smaller, *subatomic particles.* One of the first of these experiments was done by the English physicist J. J. Thomson in 1897. Some years earlier, it had been discovered that an electric discharge (glowing current) flows between metallic electrodes that are sealed in a partially evacuated glass tube, as shown in Figure 1-4. Using the apparatus pictured in Figure 1-5, Thomson deflected the electric discharge with electric and magnetic fields and showed that it was actually a stream of identical, negatively charged particles and that the mass of each particle was only 1/1837 that of a hydrogen atom. Because the hydrogen atom is the lightest atom, he correctly reasoned that these particles, which are now called *electrons,* are constituents of atoms. The electron was the first subatomic particle to be discovered.

If an atom contains electrons, which are negatively charged particles, then it also must contain positively charged particles because atoms are electrically neutral. The total amount of negative charge in a neutral atom must be balanced by an equal amount of positive charge. In the early 1900's, a New Zealand-born chemist and physi-

Figure 1-5 The apparatus that Thomson used to discover the electron. A schematic diagram of his experiment is shown in Figure 1-4.

cist, Ernest Rutherford, discovered the *proton*, a subatomic particle that has a positive charge equal in magnitude to that of the electron but opposite in sign. Its mass is almost the same as the mass of the hydrogen atom, about 1836 times the mass of the electron.

1-12. MOST OF THE MASS OF AN ATOM IS CONCENTRATED IN THE NUCLEUS

We have just learned that atoms contain electrons and protons. Because atoms are electrically neutral, the number of electrons must equal the number of protons in any given atom. The question is, how are the protons and electrons arranged within an atom? The first person to answer this question was Rutherford, the discoverer of the proton.

To gain a sense of Rutherford's experiment, we must first mention another discovery of the 1890's, radioactivity. About the same time that Thomson discovered the electron, the French scientist Henri Becquerel discovered *radioactivity*, the process by which certain atoms spontaneously break apart. Becquerel showed that uranium atoms are *radioactive*. Shortly after Becquerel's discovery, Marie and Pierre Curie, working in Paris, discovered other radioactive elements, which they named radium (emits rays) and polonium (after Poland, Marie Curie's native country).

Radioactivity is discussed in Chapter 24.

Rutherford studied the radiation emitted by a sample of radium. He placed radium in a cavity drilled into a lead brick in such a way that the emerging radiation would form a fine beam. He then allowed the beam to pass through an electric field, and the beam was split into three separate beams, as shown in Figure 1-6. Rutherford called these beams *α-rays* (alpha rays), *β-rays* (beta rays), and *γ-rays*

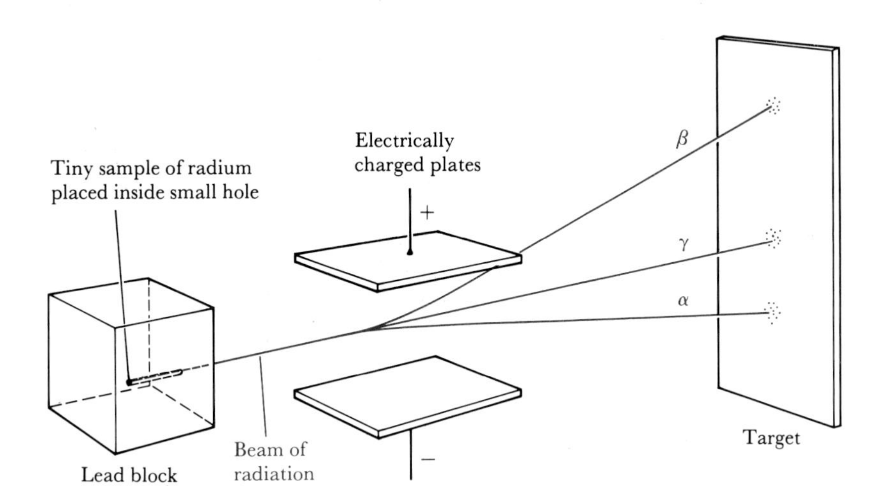

Figure 1-6 A schematic illustration of the experimental setup used by Rutherford to show that the emissions from radium consist of three distinct components, which he called α-rays, β-rays, and γ-rays.

Notice that the third column in Table 1-8 is headed mass/amu. This slash notation means that the dimensionless numbers in the column are masses divided by the unit amu. Symbolically, we write that the mass of an α-particle is

$$\text{mass/amu} = 4.00$$

Multiplying both sides by amu, we have

$$\text{mass} = 4.00 \text{ amu}$$

We use this notation to label columns in tables and axes in figures because of its unambiguous nature and algebraic convenience.

Table 1-8 Properties of the three radioactive emissions discovered by Rutherford

Original name	*Modern name*	*Mass/amu*	*Charge*[a]
α-ray	α-particle	4.00	+2
β-ray	β-particle (electron)	5.49×10^{-4}	−1
γ-ray	γ-ray	0	0

[a]Relative to the charge on a proton.

(gamma rays). He determined the charge on the particles in each separate beam by noting where the beam was deflected. He found the mass of the particles in each beam by measuring the amount of deflection (heavier particles are deflected less than lighter particles). With such experiments Rutherford showed that α-rays consist of particles (now called *α-particles*) having a charge equal to that of two protons and a mass equal to the mass of a helium atom (4.00 amu), that β-rays are just a beam of electrons (called *β-particles* when they result from radioactive disintegrations), and that γ-rays are very similar to X-rays. Table 1-8 summarizes the properties of these three common radioactive emissions.

Rutherford became intrigued with the idea of using α-particles as subatomic projectiles. In a now-famous experiment, one of Rutherford's students, Ernest Marsden, took a piece of gold and rolled it into an extremely thin foil (gold is very *malleable,* meaning that it can easily be rolled into a thin foil). He then directed a beam of α-particles at the gold foil and observed the paths of the particles by watching them strike a fluorescent screen surrounding the foil (Figure 1-7). Contrary to expectations, most of the particles passed right through the foil, but a few were deflected through large angles (path-

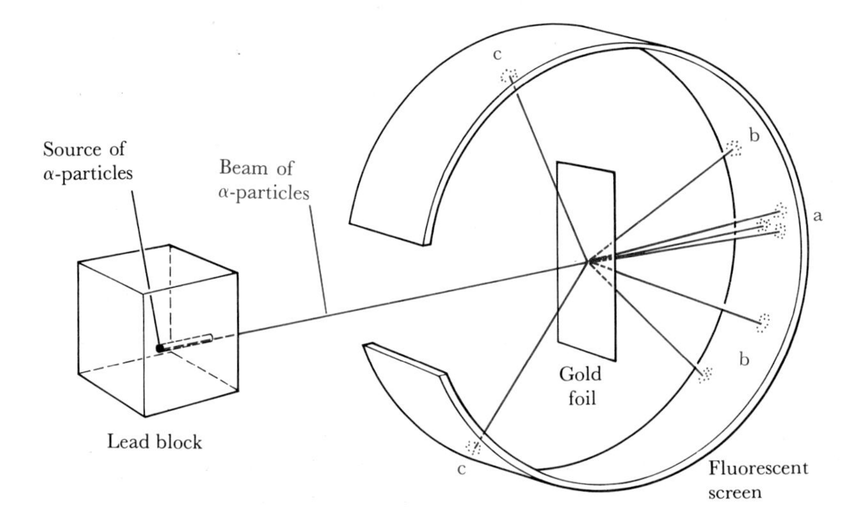

Figure 1-7 In 1911, Rutherford and Marsden set up an experiment in which a thin gold foil was bombarded with α-particles. Most of the particles passed through the foil (pathway a). Some were deflected only slightly (pathway b) when they passed near a gold nucleus in the foil, and a few were deflected backwards (pathway c).

way c in Figure 1-7). Rutherford interpreted this unexpected result by saying that an atom is mostly empty space and that all of the positive charge and essentially all of the mass of an atom are concentrated in a very small volume in the center of the atom, which he called the *nucleus*. Most α-particles passed through the gold foil; the few that were deflected through large angles were the result of collisions of the α-particles with gold nuclei. Those α-particles that were deflected through intermediate angles (pathway b in Figure 1-7) had passed near gold nuclei and were repelled by their positive charge. Because α-particles are positively charged, they would be expected to be repelled by a positively charged nucleus because like charges repel each other. By counting the numbers of α-particles deflected in various directions, Rutherford was able to show that the diameter of a nucleus is about 1/10,000 to 1/100,000 times the diameter of an atom.

The size of the nucleus relative to the size of the whole atom can be grasped from the following analogy. If an atom could be enlarged so that its nucleus were the size of a pea, then the entire atom would be about the size of Yankee Stadium. The electrons in an atom are located throughout the space surrounding the nucleus. Just how the electrons are arranged in an atom is taken up in later chapters.

1-13. ATOMS CONSIST OF PROTONS, NEUTRONS, AND ELECTRONS

Our picture of the atom is not yet complete. Later experiments suggested that the mass of a nucleus could not be attributed to the protons alone. It was postulated in the 1920's, and experimentally verified by James Chadwick in 1932, that there is another particle in the nucleus. This particle has essentially the same mass as a proton and is called a *neutron* because it is electrically neutral.

Rutherford proposed the existence of neutrons in 1920.

The modern picture of an atom, then, consists of three types of particles—electrons, protons, and neutrons. The properties of these three subatomic particles are

Particle	*Charge*[a]	*Mass/amu*	*Where located*
proton	+1	1.0073	in nucleus
neutron	0	1.0087	in nucleus
electron	−1	5.49×10^{-4}	outside nucleus

[a]Relative to the charge on a proton.

In this table, the charges are expressed in terms of the charge on a proton and the masses are in atomic mass units. (See Table 1-8 for an explanation of the mass/amu column heading notation.)

The number of protons in an atom is called the *atomic number*, denoted by Z, of that atom. In a neutral atom, the number of electrons is equal to the number of protons. The total number of protons and neutrons in an atom is called the *mass number*, denoted by A, of that atom. As we shall learn, the differences between atoms are a result of the different atomic numbers. Each element can be characterized by its atomic number. For example, hydrogen has an atomic number of 1 (1 proton in the nucleus), helium has an atomic number of 2 (2 protons in the nucleus), and uranium has an atomic number of 92 (92 protons in the nucleus). The table of the elements given on the inside front cover of this book lists the atomic numbers and atomic masses of all the known elements.

1-14. ISOTOPES CONTAIN THE SAME NUMBER OF PROTONS BUT DIFFERENT NUMBERS OF NEUTRONS

Because nuclei are made up of protons and neutrons, each of which has a mass of approximately 1 amu, you might expect that atomic masses should be approximately equal to whole numbers. Although many atomic masses are approximately whole numbers (for example, C is 12.0 and F is 19.0), many others are not. Chlorine ($Z = 17$) has an atomic mass of 35.45, magnesium ($Z = 12$) has an atomic mass of 24.3, and copper ($Z = 29$) has an atomic mass of 63.55. The explanation lies in the fact that many elements consist of two or more *isotopes*, which are atoms that contain the same number of protons and electrons but different numbers of neutrons. Realize that it is the number of protons that characterizes a particular element, and so nuclei of the same element may have varying numbers of neutrons. For example, the most common isotope of the simplest element, hydrogen, contains one proton and one electron. There is another, less common isotope of hydrogen that contains one proton, one neutron, and one electron. These two isotopes are identical chemically; they both undergo the same chemical reactions. The heavier isotope is called heavy hydrogen or, more commonly, *deuterium*, and is denoted by the symbol D. Water that is made from deuterium is called *heavy water* and is often denoted by D_2O.

Heavy water is used in nuclear reactors.

An isotope is specified by its atomic number and its mass number. The notation used to designate isotopes is the chemical symbol of the element written with its atomic number as a left subscript and its mass number as a left superscript:

$$\begin{matrix}\text{mass number} \rightarrow \\ \text{atomic number} \rightarrow\end{matrix} {}^{A}_{Z}\text{X} \leftarrow \text{chemical symbol}$$

For example, an ordinary hydrogen atom is denoted ${}^{1}_{1}\text{H}$ and a deuterium atom is denoted ${}^{2}_{1}\text{H}$.

Example 1-6: Two isotopes that play a major role in nuclear energy and nuclear weapons are uranium-235 ($^{235}_{92}U$) and plutonium-239 ($^{239}_{94}Pu$). How many protons, neutrons, and electrons are there in each of these isotopes?

Solution: The isotope ($^{235}_{92}U$) contains 92 protons and consequently 92 electrons. The number of neutrons is its mass number minus its atomic number, or $235 - 92 = 143$ neutrons.

Plutonium-239 contains 94 protons, 94 electrons, and $239 - 94 = 145$ neutrons.

Example 1-7: Fill in the blanks:

	Symbol	*Atomic number*	*Number of neutrons*	*Mass number*
(a)	___	17	___	37
(b)	___	___	52	90
(c)	$^{?}_{?}Co$	___	___	60

Solution:

(a) The number of neutrons is the mass number minus the atomic number, or $37 - 17 = 20$ neutrons. The element with atomic number 17 is chlorine (see inside front cover), and so the symbol of this isotope is $^{37}_{17}Cl$, called chlorine-37.

(b) The number of protons is the mass number minus the number of neutrons, or $90 - 52 = 38$ protons. The element with atomic number 38 is strontium, and so the symbol is $^{90}_{38}Sr$. Strontium-90 is a radioactive isotope present in the fallout from nuclear explosions.

(c) According to the symbol, the element is cobalt, whose atomic number is 27. The symbol of the particular isotope is $^{60}_{27}Co$, and the isotope has $60 - 27 = 33$ neutrons. Cobalt-60 is used as a γ-radiation source for the treatment of cancer.

Problems 1-39 to 1-42 are similar to Example 1-7.

Although one of the postulates of Dalton's atomic theory was that all the atoms of a given element have the same mass, we see that this is not so. Isotopes of the same element have different masses. Several common natural isotopes and their corresponding masses are given in Table 1-9. Of particular note is that the isotopic mass of carbon-12 is exactly 12. This is the convention on which the modern atomic mass scale is based. All atomic masses are given relative to the mass of carbon-12, which is defined by international convention to be exactly 12.

Table 1-9 Naturally occurring isotopes of some common elements

Element	*Isotope*	*Natural abundance/%*	*Isotopic mass*	*Protons*	*Neutrons*	*Mass number*
hydrogen	$^{1}_{1}H$	99.985	1.0078	1	0	1
	$^{2}_{1}H$	0.015	2.0141	1	1	2
	$^{3}_{1}H$	trace	3.0160	1	2	3
helium	$^{3}_{2}He$	1.4×10^{-4}	3.0160	2	1	3
	$^{4}_{2}He$	99.99986	4.0026	2	2	4
carbon	$^{12}_{6}C$	98.89	12.0000	6	6	12
	$^{13}_{6}C$	1.11	13.0034	6	7	13
	$^{14}_{6}C$	trace	14.0032	6	8	14
oxygen	$^{16}_{8}O$	99.758	15.9949	8	8	16
	$^{17}_{8}O$	0.038	16.9991	8	9	17
	$^{18}_{8}O$	0.204	17.9992	8	10	18
fluorine	$^{19}_{9}F$	100.0	18.9984	9	10	19
magnesium	$^{24}_{12}Mg$	78.99	23.9850	12	12	24
	$^{25}_{12}Mg$	10.00	24.9858	12	13	25
	$^{26}_{12}Mg$	11.01	25.9826	12	14	26
chlorine	$^{35}_{17}Cl$	75.77	34.9689	17	18	35
	$^{37}_{17}Cl$	24.23	36.9659	17	20	37

Data like these in Table 1-9 are available for all the elements.

Note also that helium has two isotopes, helium-3 and helium-4. The atomic number of helium is 2, which means that a helium nucleus has two protons and a nuclear charge of +2. A helium-4 nucleus has a charge of +2 and an atomic mass of 4, the same as an α-particle (Table 1-8). In fact, an α-particle is simply the nucleus of a helium-4 isotope.

As Table 1-9 implies, many elements occur in nature as a mixture of isotopes with a *natural abundance* for each isotope. Although naturally occurring chlorine consists of two isotopes, they occur as 75.77 percent $^{35}_{17}Cl$ and 24.23 percent $^{37}_{17}Cl$, independent of any natural source of the chlorine. Chlorine obtained from, say, salt deposits in Africa or Australia has essentially the same isotopic composition as that given in Table 1-9. The observed atomic mass of chlorine is the sum of the masses of each isotope, each weighted by its natural abundance. If we use the isotopic masses and natural abundances of chlorine given in Table 1-9, then we obtain

$$\begin{matrix}\text{observed atomic mass}\\ \text{of chlorine}\end{matrix} = (34.97)\left(\frac{75.77}{100}\right) + (36.97)\left(\frac{24.23}{100}\right) = 35.45$$

This is the atomic mass of chlorine given on the inside front cover of the book. The factors 75.77/100 and 24.23/100 must be included in order to take into account the relative natural abundance of each isotope.

Example 1-8: Naturally occurring neon is a mixture of three isotopes with the following isotopic masses and natural abundances:

Mass number	*Isotopic mass*	*Natural abundance/%*
20	19.99	90.51
21	20.99	0.27
22	21.99	9.22

Calculate the observed atomic mass of neon.

Solution: The observed atomic mass is the sum of the masses of the three isotopes weighted by their abundance:

Problems 1-43 to 1-54 deal with percentages of isotopes.

$$\begin{matrix}\text{observed atomic}\\ \text{mass of neon}\end{matrix} = (19.99)\left(\frac{90.51}{100}\right) + (20.99)\left(\frac{0.27}{100}\right) + (21.99)\left(\frac{9.22}{100}\right)$$

$$= 20.18$$

Small variations in natural abundances limit the precision with which atomic masses can be specified. The masses of individual isotopes are known much more accurately than atomic masses given in the periodic table. In the next section, we discuss how isotopic masses are determined.

1-15. ISOTOPES CAN BE SEPARATED BY A MASS SPECTROMETER

The mass and percentage of each isotope in an element can be determined by using a *mass spectrometer.* When a gas is bombarded with electrons from an outside source, electrons are knocked out of the neutral atoms in the gas, producing positively charged, atomic-sized particles called *ions.* Ions are atoms that have either a deficiency

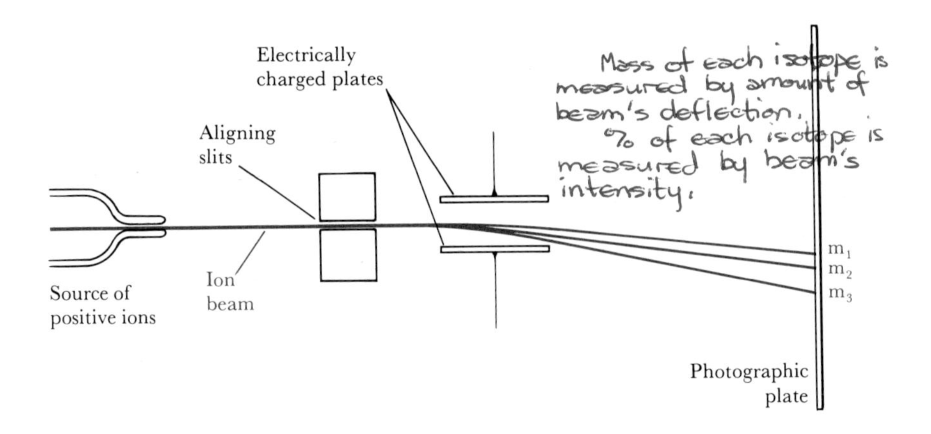

Figure 1-8 Schematic diagram of a mass spectrometer. A gas is bombarded with electrons in a process that knocks electrons out of the gas atoms or molecules and produces positively charged ions. These ions are then accelerated by an applied electric field and form a sharply focused beam as they pass through the two aligning slits. The beam is then allowed to pass through an electric or a magnetic field. The ions of different masses are deflected to different extents, and the beam is split according to the masses of the various ions in the beam. The particles of different masses strike a detector, such as a photographic plate, at different places, and plate exposure at various places is proportional to the number of particles having a particular mass.

of electrons (in which case the ions are positively charged) or extra electrons (in which case the ions are negatively charged). The positively charged gas ions produced by electron bombardment are accelerated by an electric field and passed through slits to form a narrow, well-focused beam (Figure 1-8). The beam of ions is then passed through an electric field, which deflects the ions according to their mass. As in Rutherford's experiment in which he identified the three components of the emissions from radium, heavier ions are deflected less than lighter ions. Thus, the original beam of ions is split into several separate beams, one for each isotope of the gas. The intensities of the separated ion beams, which can be determined experimentally, are a direct measure of the number of ions in each beam. In this manner, it is possible to determine not only the mass of each isotope of any element (by the amount of deflection of each beam) but also the percentage of each isotope (by the intensity of each beam).

We shall encounter ions throughout our study of chemistry, and so we introduce a notation for them here. An atom that has lost one electron has a net charge of $+1$; one that has lost two electrons has a charge of $+2$; one that has gained an electron has a charge of -1; and so on. We denote an ion by the chemical symbol of the element with a right-hand superscript to indicate its charge:

singly charged sodium ion	Na^{+}
singly charged neon ion	Ne^{+}
doubly charged magnesium ion	Mg^{2+}
singly charged chloride ion	Cl^{-}
doubly charged sulfide ion	S^{2-}

Note from these examples that the names of negative ions have the *-ide* ending characteristic of the second element in binary compounds.

Example 1-9: How many electrons are there in Na^+ and F^-?

Solution: From the table on the inside front cover, we see that the atomic number of sodium is 11. The Na^+ ion is a sodium atom that is lacking one electron, and so Na^+ has 10 electrons.

The atomic number of fluorine is 9. The F^- ion has one electron more than a fluorine atom, and so F^- has 10 electrons.

Notice that both Na^+ and F^- have 10 electrons. Species that contain the same number of electrons are said to be *isoelectronic*.

1-16. THERE ARE UNCERTAINTIES ASSOCIATED WITH MEASURED QUANTITIES

Before we leave this chapter, we should discuss a matter of technical importance regarding chemical calculations.

To carry out scientific calculations, you must know how to handle *significant figures, units of measurement, and unit conversions*. The determination of the number of significant figures in a calculated quantity and the manipulation of the units of physical quantities in calculations are straightforward procedures. If you learn a few simple rules on significant figures and always include the units of various quantities, then you will find that your calculations will go smoothly.

The counting of objects is the only type of experiment that can be carried out with complete accuracy, that is, without any inherent error. Let's consider the problem of determining how many pennies there are in a jar. It is possible to determine the exact number of pennies simply by counting them. Suppose that there are 1542 pennies in the jar. The number 1542 is exact; there is no uncertainty associated with it. It is a different matter, however, when we determine the mass of 1542 pennies with a balance (Figure 1-9). Suppose that the balance we use is capable of measuring the mass of an object to the nearest one-tenth of a gram and that the experimentally determined mass of the 1542 pennies is 4776.2 ± 0.1 g. The $\pm$ 0.1 denotes the experimental uncertainty in the measured mass of the pennies. The actual mass lies somewhere between 4776.1 and 4776.3 g. We cannot tell from the result 4776.2 ± 0.1 g if the mass is, say, 4776.15 g, or 4776.24 g, or some other mass in the range 4776.1 to 4776.3. We would need a more accurate balance to determine the mass of the pennies to the nearest hundredth of a gram ($\pm$0.01 g), and a still more accurate balance to determine the mass to the nearest milligram (mg, $\pm$0.001 g).

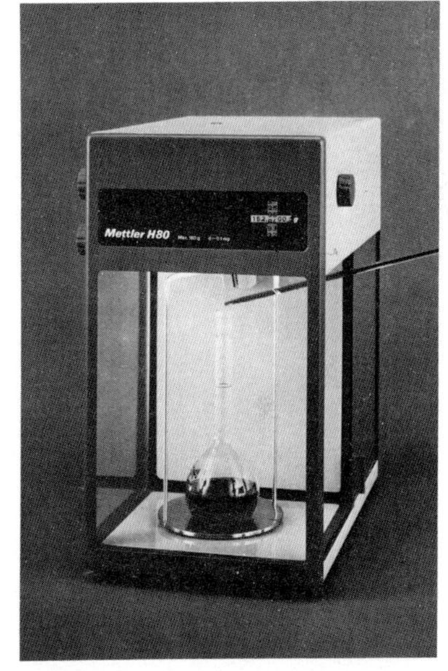

Figure 1-9 A modern analytical balance, used to determine mass. The mass determination is made by turning a dial to find the known mass (using a set of standard masses enclosed within the balance) that balances the unknown mass. When the two masses just balance, they are equal within the accuracy of the balance.

Denoting the uncertainty of a measured quantity by the $\pm$ notation is desirable in precise work, but such a notation is too cumbersome for our purposes. We indicate the accuracy of measured quantities by the number of *significant figures* used to express a result. All digits in a numerical result are significant if only the last digit has some uncertainty. A result expressed as 4776.2 g means that only the 2 has some uncertainty and that there are five significant figures in the result. The result 4776 g means that the 6 has some uncertainty and that there are four significant figures.

Zeros in a measured quantity require special consideration. Zeros are not taken as significant figures if they serve only to position the decimal point, as is illustrated in the following examples:

	Result	*Number of significant figures*	*Comment*
(a)	0.0056	2	The three zeros are used only to position the decimal point and are thus not significant figures.
(b)	5.6×10^{-3}	2	Compare with (a); the 10^{-3} simply positions the decimal point.
(c)	38.70	4	The zero is not necessary to position the decimal point and thus must be regarded as a significant figure.
(d)	100.0	4	All the zeros are significant in this case.

In certain cases, it is not clear just how many significant figures are implied. Consider the number 100. With the number presented as 100, we might mean that the value of the number is *exactly* 100. On the other hand, the two zeros might not be significant and we might mean that the value is *approximately* 100, say 100 ± 10, for example. The number of significant figures in such a case is uncertain. Usually the number of significant figures can be deduced from the statement of a problem. It is the writer's obligation to indicate clearly the number of significant figures.

Example 1-10: State the number of significant figures in each of the following numbers:
(a) 0.0312 (b) 0.03120 (c) 3120

Solution

(a) 0.0312 has three significant figures: 3, 1, and 2.
(b) 0.03120 has four significant figures: 3, 1, 2, and 0.
(c) 3120 has four significant figures if we mean *exactly* 3120 and three significant figures if we mean 3120 ± 10 or so.

1-17. CALCULATED NUMBERS SHOULD SHOW THE CORRECT NUMBER OF SIGNIFICANT FIGURES

In multiplication and division, the calculated result should not be expressed to more significant figures than the factor in the calculation with the least number of significant figures. For example, if we perform the multiplication

$$8.3143 \times 298.2$$

on a hand calculator, the following result comes up on the calculator display:

$$2479.3243$$

Your hand calculator usually will give many more digits than are significant.

Not all these figures are significant. The correct result is 2479 because the factor 298.2 has only four significant figures and thus the result cannot have more than four. The extra figures are not significant and should be discarded.

Example 1-11: Determine the result to the correct number of significant figures:

$$y = \frac{3.00 \times 0.08205 \times 298}{0.93}$$

Solution: Using a hand calculator, we obtain

$$y = 78.873871$$

The factor 0.93 has the least number of significant figures—only two. Thus the calculated result should not be expressed to more than two significant figures, and the correct result is $y = 79$, which is 78.873871 rounded off to two significant figures.

Problems 1-63 to 1-68 deal with significant figures.

The number of figures after the decimal point in a number resulting from addition or subtraction can have no more figures after the decimal point than the least number of figures after the decimal point in any of the numbers that are being added or subtracted. Consider the sum

$$\begin{array}{r} 6.939 \\ +1.00797 \\ \hline 7.94697 \end{array} \quad \text{round off to } 7.947$$

The last two digits in 7.94697 are not significant because we know the value of the first number in the addition, 6.939, only to three digits beyond the decimal place. Thus the result cannot be accurate to more than three digits past the decimal. Therefore our result expressed to the correct number of significant figures is 7.947.

In rounding off insignificant figures, we use the following convention. If the figure following the last figure retained is a 5, 6, 7, 8, or 9, then the preceding figure should be increased by 1; otherwise (that is, for 0, 1, 2, 3, and 4), the preceding figure should be left unchanged. Thus, rounding off the following numbers to three significant figures, we obtain

$$27.35 \rightarrow 27.4$$
$$27.34 \rightarrow 27.3$$

Note that in half of the cases (0, 1, 2, 3, 4) we discard the insignificant digit and in the other half (5, 6, 7, 8, 9) we increase the preceding digit by 1 when we discard the insignificant digits.

1-18. THE VALUE OF A PHYSICAL QUANTITY DEPENDS ON THE UNITS

When a number represents a measurement, the units of that measurement must always be indicated. For example, if we measure the thickness of a wire to be 1.35 millimeters (mm), we express the result as 1.35 mm. To say that the thickness of the wire is 1.35 would be completely meaningless.

The preferred system of units used in scientific work is the *metric system*. There are several sets of units in the metric system, and in recent years there has been a worldwide movement to express all measurements in terms of just one set of metric units called SI units (for *S*ystème *I*nternational). The metric system, and SI units in particular, are described in Appendix B. We discuss the individual units in detail when we encounter them in the text.

The metric system is used in almost every country in the world and by all scientists.

Only numbers that have the same units can be added or subtracted. If we add 2.12 centimeters (cm) and 4.73 cm, we obtain 6.85 cm. If we wish to add 76.4 cm to 1.19 meters (m), we must first convert centimeters to meters or meters to centimeters. We convert from one unit to another by using a *unit conversion factor.* Suppose we want to convert meters to centimeters. From Appendix B we find that

$$1\ \text{m} = 100\ \text{cm} \tag{1-1}$$

Equation (1-1) is a definition and is exact; there is no limit to the number of significant figures on either side of the equation. If we divide both sides of Equation (1-1) by 1 m, we get

$$1 = \frac{100\ \text{cm}}{1\ \text{m}} \tag{1-2}$$

Equation (1-2) is called a *unit conversion factor* because we can use it to convert meters to centimeters. A unit conversion factor, as expressed in Equation (1-2), is equal to unity, and thus we can multiply by a unit conversion factor to obtain any quantity of interest. If we multiply 1.19 m by Equation (1-2), we obtain

$$(1.19\,\cancel{\text{m}})\left(\frac{100\text{ cm}}{1\,\cancel{\text{m}}}\right) = 119\text{ cm}$$

Notice that the units of meters cancel, giving the final result in centimeters. To convert centimeters to meters, we use the reciprocal of Equation (1-2):

$$(76.4\,\cancel{\text{cm}})\left(\frac{1\text{ m}}{100\,\cancel{\text{cm}}}\right) = 0.764\text{ m}$$

Notice in this case that the units of centimeters cancel, giving the final result in meters. Using these results, we see that the sum of 76.4 cm and 1.19 m is

$$76.4\text{ cm} + (1.19\,\cancel{\text{m}})\left(\frac{100\text{ cm}}{1\,\cancel{\text{m}}}\right) = 195\text{ cm}$$

or

$$(76.4\,\cancel{\text{cm}})\left(\frac{1\text{ m}}{100\,\cancel{\text{cm}}}\right) + 1.19\text{ m} = 1.95\text{ m}$$

Notice that both results are given to three significant figures.

As another example of converting from one set of units to another, let's convert 55 miles per hour (mph) to kilometers (km) per hour (h). From the inside back cover, we find that

$$1\text{ mile} = 1.61\text{ km} \tag{1-3}$$

Dividing both sides of Equation (1-3) by 1 mile yields the unit conversion factor:

$$1 = \frac{1.61\text{ km}}{1\text{ mile}} \tag{1-4}$$

Equation (1-4) is the unit conversion factor that can be used to convert a speed given in miles per hour to a speed in kilometers per hour. Thus

$$\left(\frac{55\,\cancel{\text{miles}}}{\text{h}}\right)\left(\frac{1.61\text{ km}}{1\,\cancel{\text{mile}}}\right) = \frac{89\text{ km}}{\text{h}}$$

Note that the use of the proper units for each quantity provides an internal check on the correctness of the calculation. We must *multiply* 55 mph by

$$\frac{1.61 \text{ km}}{1 \text{ mile}}$$

in order to obtain the result in the desired units of km/h. Note that if we had used the conversion

$$1 \text{ km} = 0.62 \text{ mile}$$

from the inside back cover, then the unit conversion would be

$$\left(\frac{55 \cancel{\text{miles}}}{\text{hour}}\right)\left(\frac{1 \text{ km}}{0.62 \cancel{\text{mile}}}\right) = 89 \frac{\text{km}}{\text{h}}$$

so that the units of miles cancel out.

Example 1-12: One quart (qt) of liquid is equivalent to 0.946 liter (L), that is, 1 qt = 0.946 L. Compute the number of liters in 10.0 U.S. gallons (gal) of gasoline.

Solution: There are 4 qt in 1 U.S. gal, and thus there are

$$(10.0 \cancel{\text{gal}})\left(\frac{4 \text{ qt}}{\cancel{\text{gal}}}\right) = 40.0 \text{ qt}$$

Problems 1-69 to 1-74 ask you to convert quantities from one set of units to another.

in 10.0 gal. To convert from quarts to liters, we use the conversion factor 0.946 L/qt and compute:

$$(40.0 \cancel{\text{qt}})\left(\frac{0.946 \text{ L}}{\cancel{\text{qt}}}\right) = 37.8 \text{ L}$$

There are 37.8 L in 10.0 gal of gasoline. Note that we have expressed our result to three significant figures. The 4 in 4 qt/gal is exactly 4, by definition, and so does not place any limit on the number of significant figures in the final result.

Many quantities are expressed in *compound units*. To see what we mean by a compound unit, consider the quantity density. The density of a substance is defined as the ratio of the mass of the substance to its volume:

$$\text{density} = \frac{\text{mass}}{\text{volume}} \tag{1-5}$$

Thus we say that the *dimensions* of density are mass per unit volume, which can be expressed in a variety of units. If we express the mass in grams and the volume in cubic centimeters, then the units of density are grams per cubic centimeter. For example, the density of ice is 0.92 g/cm^3, where the slash denotes "per." From algebra we know that

$$\frac{1}{a^n} = a^{-n}$$

Exponents are reviewed in Appendix A.

where n is an exponent. Thus

$$\frac{1}{\text{cm}^3} = \text{cm}^{-3}$$

Therefore we also can express density as $g \cdot cm^{-3}$ instead of g/cm^3. In the expression $g \cdot cm^{-3}$, the dot serves to separate the g from the cm^{-3}. The use of dots in compound units is an SI convention (Appendix B) and is used to avoid ambiguities. For example, m·s denotes meter-second, whereas ms (without the dot) denotes millisecond.

We now consider an example of converting quantities in compound units. It has been estimated that all the gold that has ever been mined would occupy a cube 17 m on a side. Given that the density of gold is 18.9 $g \cdot cm^{-3}$, let's calculate the mass of all this gold. The volume of a cube 17 m on a side is

$$\text{volume} = (17\text{ m})^3 = 4913\text{ m}^3 \qquad (1\text{-}6)$$

Although the volume calculated in Equation (1-5) is good to only two significant figures (because the value 17 m is good to only two significant figures), we shall carry extra significant figures through the calculation and then round off the final result to two significant figures. The mass of the gold is obtained by multiplying the density by the volume, but before doing this, we must convert cubic meters to cubic centimeters (because the density is given in $g \cdot cm^{-3}$). From Appendix B, we find that

$$1\text{ m} = 100\text{ cm}$$

By cubing both sides of this expression, we obtain

$$1\text{ m}^3 = 10^6\text{ cm}^3$$

and so the unit conversion factor is

$$1 = \frac{10^6\text{ cm}^3}{1\text{ m}^3}$$

Thus the volume in Equation (1-5) is

$$\text{volume} = (4913\ \cancel{\text{m}^3})\frac{10^6\ \text{cm}^3}{1\ \cancel{\text{m}^3}}$$
$$= 4.913 \times 10^9\ \text{cm}^3$$

If we multiply the volume of the gold by its density, we obtain the mass:

$$\text{mass} = \text{volume} \times \text{density}$$
$$= (4.913 \times 10^9\ \text{cm}^3)(18.9\ \text{g}\cdot\text{cm}^{-3})$$
$$= 9.3 \times 10^{10}\ \text{g}$$

The result is rounded off to two significant figures because, as we mentioned before, the side of the cube (17 m) is given to only two significant figures. In obtaining this result, we have used the fact that

$$(\text{cm}^3)(\text{cm}^{-3}) = 1$$

The price of gold varies from day to day.

Let's see what this mass of gold would be worth at $400 per troy ounce (oz). (Gold is sold by the troy ounce, which is about 10 percent heavier than the avoirdupois ounce, which is the unit used for foods.) There are 31.1 g in 1 troy oz and so the unit conversion factor is

$$1 = \frac{1\ \text{troy oz}}{31.1\ \text{g}}$$

The mass of gold in troy ounces is

$$\text{mass} = (9.3 \times 10^{10}\ \cancel{\text{g}})\left(\frac{1\ \text{troy oz}}{31.1\ \cancel{\text{g}}}\right)$$
$$= 3.0 \times 10^9\ \text{troy oz}$$

At $400 per troy ounce, the value of all the gold ever mined is

$$\text{value} = (3.0 \times 10^9\ \cancel{\text{troy oz}})\left(\frac{\$400}{1\ \cancel{\text{troy oz}}}\right)$$
$$= 1.2 \times 10^{12} = 1.2\ \text{trillion dollars}$$

If you carefully set up the necessary conversion factors and make certain that the appropriate units cancel to give the units you need for the answer, then you cannot go wrong in making unit conversions. With a little practice, the manipulation of units will become easy.

SUMMARY

The beginning of modern chemistry occurred in the late eighteenth century when Lavoisier, considered to be the founder of modern chemistry, introduced quantitative measurements into chemical research. Lavoisier's work led directly to the discovery of the law of constant composition and then to Dalton's atomic theory. Dalton was able to use the atomic theory to determine the relative masses of atoms and molecules and to use these values in interpreting the results of chemical analyses. Dalton also was able to present a simple and clear atomic interpretation of the law of multiple proportions. According to the atomic theory, the atoms in reactant molecules are separated and rearranged into product molecules in a chemical reaction. Because atoms are neither created nor destroyed in chemical reactions, chemical reactions obey the law of conservation of mass.

Elements are substances that consist of only one kind of atom. There are 108 known elements, about three quarters of which are metals. Elements combine to form compounds, whose constituent particles are called molecules, which are groups of atoms joined together. Chemists represent elements by chemical symbols and compounds by chemical formulas. The system of naming compounds is called chemical nomenclature.

Protons and neutrons form the nucleus of the atom, the small center containing all the positive charge and essentially all the mass of the atom. The number of protons in an atom is the atomic number (Z) of that atom. The total number of protons and neutrons in an atom is the mass number (A) of that atom.

Each element can be characterized by its atomic number. Nuclei with the same number of protons but different numbers of neutrons are called isotopes. Most elements occur naturally as mixtures of isotopes, and atomic masses are weighted averages of the isotopic masses. Isotopic abundances and masses can be determined with a mass spectrometer. Because isotopes of an element have the same atomic number, they are chemically identical and undergo the same reactions.

In a neutral atom, the number of electrons is equal to the number of protons. When an atom loses or gains electrons, it is called an ion. Positive ions have a deficiency of electrons, and negative ions have extra electrons.

To carry out scientific calculations, it is important to understand significant figures, units of measurement, and unit conversions. Significant figures represent the exactness of a measurement. Units must always be included with numbers that represent measurements or else the numbers are meaningless. When different units are used in a measurement, one unit must be converted to the other before they can be used together in calculations. Unit conversions are carried out using unit conversion factors.

TERMS YOU SHOULD KNOW*

element 2
compound 2
metal 3
nonmetal 3
chemical symbol 3
atom 5 and 7
molecule 5 and 7
diatomic molecule 5
quantitative measurement 5
qualitative observation 5
law of conservation of mass 5
natural law 5
law of constant composition 6
atomic theory 7
atomic mass 8
atomic mass unit (amu) 8
binary compound 11
chemical formula 11
chemical nomenclature 11
molecular mass 13
law of multiple proportions 14
subatomic particle 16
electron 16
proton 17
radioactivity 17
radioactive 17
α-particle 18
β-particle 18
γ-ray 18
malleable 18
nucleus 19
neutron 19
atomic number 20
mass number 20
isotope 20
deuterium 20
heavy water 20
natural abundance 22
mass spectrometer 23
ion 23
isoelectronic 25
significant figure 25
metric system 28
unit conversion factor 29
compound unit 30
density 30

*These terms are listed in the order in which they appear in the text. Page numbers refer to the pages on which the terms are introduced. A complete glossary of these terms can be found in the *Study Guide/Solutions Manual* accompanying this text.

PROBLEMS*

CHEMICAL SYMBOLS

1-1. Give the chemical symbol for the following elements:

(a) cadmium Cd
(b) indium In
(c) lead Pb
(d) tin Sn
(e) mercury Hg
(f) xenon Xe
(g) copper Cu
(h) potassium K
(i) uranium U
(j) phosphorus P

1-2. Give the chemical symbol for the following elements:

(a) vanadium
(b) gold
(c) zinc
(d) magnesium
(e) iron
(f) cesium
(g) bromine
(h) krypton
(i) antimony
(j) arsenic

*Problems are grouped by subject area and arranged in matched pairs (1-1 and 1-2, for example) such that both involve the same principle or operation. Answers to odd-numbered problems are given in Appendix G. Solutions to odd-numbered problems and answers to even-numbered problems can be found in the *Study Guide/Solutions Manual* accompanying this text.

1-3. Give the element for the following chemical symbols:

(a) Se
(b) Au
(c) Ag
(d) Pd
(e) Li
(f) H
(g) Si
(h) V
(i) Pu
(j) He

1-4. Give the element for the following chemical symbols:

(a) Pt
(b) Sr
(c) Pb
(d) W
(e) Ca
(f) Cr
(g) Ni
(h) Sn
(i) S
(j) C

MASS PERCENTAGES IN COMPOUNDS

1-5. A 0.436-g sample of a compound of potassium and chlorine contains 0.229 g of potassium and 0.207 g of chlorine. Calculate the mass percentages of potassium and chlorine in the compound.

1-6. The compound lanthanum oxide is used in the production of optical glass and the fluorescent phosphors used to coat television screens. An 8.29-g sample is found to contain 7.08 g of lanthanum and 1.21 g of oxygen. Calculate the mass percentages of lanthanum and oxygen in lanthanum oxide.

1-7. A 1.28-g sample of calcium is burned in oxygen to produce 1.79 g of calcium oxide (also called lime), a compound widely used in the chemical industry. Calculate the mass percentages of calcium and oxygen in calcium oxide.

1-8. Stannous fluoride, an active ingredient in toothpaste that helps to prevent cavities, contains tin and fluorine. A 1.793-g sample was found to contain 1.358 g of tin. Calculate the mass percentages of tin and fluorine in stannous fluoride.

1-9. Potassium cyanide, KCN, is used in extracting gold and silver from their ores. A 12.63-mg sample is found to contain 7.58 mg of potassium, 2.33 mg of carbon, and 2.72 mg of nitrogen. Calculate the mass percentages of potassium, carbon, and nitrogen in potassium cyanide.

1-10. Ethyl alcohol, the alcohol in alcoholic beverages, is a compound of carbon, hydrogen, and oxygen. A 3.70-g sample of ethyl alcohol contains 1.93 g of carbon and 0.49 g of hydrogen. Calculate the mass percentages of carbon, hydrogen, and oxygen in ethyl alcohol.

NOMENCLATURE

1-11. Name the following binary compounds:

(a) LiCl
(b) MgO
(c) AlF_3
(d) Na_3P
(e) KI

1-12. Name the following binary compounds:

(a) BaF_2
(b) Mg_3N_2
(c) RbBr
(d) CsCl
(e) CaS

1-13. Name the following binary compounds:

(a) CaC_2
(b) GaAs
(c) Be_3N_2
(d) K_2O
(e) SrF_2

1-14. Name the following binary compounds:

(a) Al_2O_3
(b) MgF_2
(c) AlN
(d) MgSe
(e) Li_3P

1-15. Name the following pairs of compounds:

(a) IF_3 and IF_5
(b) ICl and ICl_3
(c) NO_2 and N_2O_4
(d) AsF_3 and AsF_5
(e) ClO and ClO_2

1-16. Name the following pairs of compounds:

(a) $SbCl_3$ and $SbCl_5$
(b) ICl_3 and ICl_5
(c) KrF_2 and KrF_4
(d) SeO_2 and SeO_3
(e) CS and CS_2

MOLECULAR MASSES

1-17. Compute the molecular mass for the following hydrocarbons:

(a) CH_4 (methane, natural gas)
(b) C_3H_8 (propane, fuel)
(c) C_4H_{10} (butane, fuel)
(d) C_8H_{18} (octane, gasoline component)

1-19. Calculate the molecular mass for the following ores:

(a) $CaWO_4$ (scheelite, an ore of tungsten)
(b) Fe_3O_4 (magnetite)
(c) Na_3AlF_6 (cryolite)
(d) $Be_3Al_2Si_6O_{18}$ (beryl)
(e) Zn_2SiO_4 (willemite)

1-18 Compute the molecular mass for the following flavoring agents:

(a) $C_5H_{10}O_2$ (apple)
(b) $C_8H_{16}O_2$ (pineapple)
(c) $C_6H_{12}O_2$ (peach)
(d) $C_9H_{18}O_2$ (pear)

1-20. Calculate the molecular mass for the following vitamins:

(a) $C_{20}H_{30}O$ (vitamin A)
(b) $C_{12}H_{17}ClN_4OS$ (vitamin B_1, thiamine)
(c) $C_{17}H_{20}N_4O_6$ (vitamin B_2, riboflavin)
(d) $C_{56}H_{88}O_2$ (vitamin D_1)
(e) $C_6H_8O_6$ (vitamin C, ascorbic acid)

MASS PERCENTAGES FROM ATOMIC MASSES

1-21. Using the atomic masses given on the inside front cover, calculate the mass percentages of zinc and sulfur in zinc sulfide, which consists of one atom of zinc and one atom of sulfur.

1-23. Using the table of atomic masses given on the inside front cover, calculate the mass percentages of carbon and hydrogen in methane, CH_4, the chief component of natural gas.

1-25. Ordinary table sugar, whose common chemical name is sucrose, has the chemical formula $C_{12}H_{22}O_{11}$. Calculate the mass percentages of carbon, hydrogen, and oxygen in sucrose.

1-27. Which of the following compounds has the highest mass percentage of nitrogen:

(a) N_2O_3
(b) HNO_3
(c) NH_3
(d) NH_4Cl
(e) PbN_6

1-22. Using the atomic masses given on the inside front cover, calculate the mass percentages of nitrogen and oxygen in nitrogen oxide, which consists of one atom of nitrogen and one atom of oxygen.

1-24. Using the table of atomic masses given on the inside front cover, calculate the mass percentages of silicon and oxygen in silicon dioxide, SiO_2, the main compound of sand.

1-26. A key compound in the production of aluminum metal is Na_3AlF_6. Calculate the mass percentages of sodium, aluminum, and fluorine in this compound.

1-28. Lithium—in the form of lithium carbonate (Li_2CO_3), lithium acetate ($LiC_2H_3O_2$), lithium citrate ($Li_3C_6H_5O_7$), or lithium sulfate (Li_2SO_4)—is used in the treatment of manic-depressive disorders. All these compounds are equally effective, but lithium carbonate is used most often because it contains a greater mass percentage of lithium. Calculate and compare the mass percentage of lithium in each of these four compounds.

MULTIPLE PROPORTIONS

1-29. Two fluorides of xenon are known. The mass percentages of xenon and fluorine in these two fluorides are

Compound	*Mass percentage of xenon*	*Mass percentage of fluorine*
I	77.56	22.44
II	63.34	36.66

Show how these data illustrate the law of multiple proportions and suggest an explanation based upon the atomic theory.

$\frac{22.44\ g\ F}{77.56\ g\ Xe} = \frac{0.2893\ g\ F}{1\ g\ Xe}$ $\frac{36.66\ g\ F}{63.34\ g\ Xe} = \frac{0.5788\ g\ F}{1\ g\ Xe}$

ratio $= \frac{0.5788\ g\ F/1\ g\ Xe}{0.2893\ g\ F/1\ g\ Xe} = 2$

Compound I : XeF_2
Compound II : XeF_4

1-30. Bromine forms several compounds with fluorine. One of them (I) has mass percentages of 58.36 percent bromine and 41.64 percent fluorine. Another one (II) has mass percentages of 45.68 percent bromine and 54.32 percent fluorine. Show how these data illustrate the law of multiple proportions and suggest an explanation based upon the atomic theory.

1-31. Two oxides of hydrogen are known. One of them (I) is water. The other (II) is the extensively used bleaching agent hydrogen peroxide. The mass percentages of hydrogen and oxygen in these two oxides are

Compound	*Mass percentage of hydrogen*	*Mass percentage of oxygen*
I	11.2	88.8
II	5.94	94.1

Show how these data illustrate the law of multiple proportions and suggest an explanation based upon the atomic theory.

$\frac{11.2\ g\ H}{88.8\ g\ O} = \frac{0.126\ g\ H}{1\ g\ O}$ $\frac{5.94\ g\ H}{94.1\ g\ O} = \frac{0.0631\ g\ H}{1\ g\ O}$

$\frac{0.126\ g\ H}{0.0631\ g\ H} = 2$

Compound I : H_2O
Compound II : H_2O_2

1-32. Two chlorides of tin are known. The mass percentages of tin and chlorine in the two chlorides are

Compound	*Mass percentage of tin*	*Mass percentage of chlorine*
I	62.60	37.40
II	45.56	54.43

Show how these data illustrate the law of multiple proportions and suggest an explanation based upon the atomic theory.

1-33. Two 6.35-g samples of copper reacted completely with oxygen gas in separate experiments under different conditions. In one experiment the mass of the product was 7.95 g, and in the other the mass of the product was 7.15 g. Are these results consistent with the law of multiple proportions?

Cu = 6.35 g O = 1.60 g
Cu = 6.35 g O = 0.80 g

$\frac{0.80\ g\ O}{6.35\ g\ Cu} = \frac{0.126\ g\ O}{1\ g\ Cu}$

$\frac{1.60\ g\ O}{6.32\ g\ Cu} = \frac{0.252\ g\ O}{1\ g\ Cl}$ $\frac{0.252\ g\ O}{0.126\ g\ O} = 2$

Compound I : CuO
Compound II : Cu_2O

1-34. Two 5.58-g samples of iron reacted completely with chlorine gas in separate experiments under different conditions. In one experiment (I) the mass of the product was 12.68 g, and in the other (II) the mass of the product was 16.23 g. Are these results consistent with the law of multiple proportions?

PROTONS, NEUTRONS, AND ELECTRONS

1-35. Two isotopes of iodine are used in the diagnosis of thyroid malfunction. One is $^{131}_{53}I$, and the other is $^{125}_{53}I$. How many electrons, protons, and neutrons are there in each of these atoms?

$^{131}_{53}I$: 53 electrons, 53 protons, 78 neutrons
$^{125}_{53}I$: 53 electrons, 53 protons, 72 neutrons

1-36. Two isotopes that are important in dating archaeological finds and geological specimens are $^{14}_{6}C$ and $^{206}_{82}Pb$. How many protons, neutrons, and electrons are there in each of these isotopes?

1-37. The following isotopes are used widely in medicine or industry:

(a) plutonium-239 (c) potassium-43
(b) cobalt-60 (d) uranium-235

How many protons, neutrons, and electrons are there in each of these isotopes? Do you know what any of these isotopes are used for?

1-38. The following isotopes do not occur naturally and are produced in nuclear reactors:

(a) phosphorus-30 (c) iron-55
(b) technetium-97 (d) americium-240

How many protons, neutrons, and electrons are there in each of these isotopes?

1-39. Fill in the blanks in the following table:

Symbol	*Atomic number*	*Number of neutrons*	*Mass number*
$^{12}_{6}C$	___	___	___
$^{?}_{?}S$	___	___	32
___	79	___	196
___	___	10	20

1-40. Fill in the blanks in the following table:

Symbol	*Atomic number*	*Number of neutrons*	*Mass number*
$^{?}_{?}Ca$	—	—	48
—	40	—	90
—	—	78	131
$^{?}_{?}Mo$	—	57	—

1-41. Fill in the blanks in the following table:

Symbol	*Atomic number*	*Number of neutrons*	*Mass number*
___	11	12	___
___	___	122	202
___	94	___	239
$^{?}_{?}Cf$	___	___	249

1-42. Fill in the blanks in the following table:

Symbol	*Atomic number*	*Number of neutrons*	*Mass number*
$^{39}_{19}K$	—	—	—
$^{?}_{?}Fe$	—	—	56
—	36	—	84
—	—	70	120

ISOTOPIC COMPOSITION

1-43. Naturally occurring oxygen consists of three isotopes with the atomic masses and abundances given in Table 1-9. Calculate the atomic mass of oxygen.

1-44. Naturally occurring magnesium consists of three isotopes with the atomic masses and abundances given in Table 1-9. Calculate the atomic mass of magnesium.

1-45. Naturally occurring carbon is a mixture of two isotopes with the atomic masses and abundances given in Table 1-9. Calculate the atomic mass of carbon.

1-46. Naturally occurring silicon consists of three isotopes with the following atomic masses and abundances:

Mass number	*Atomic mass*	*Abundance/%*
28	27.977	92.23
29	28.977	4.67
30	29.974	3.10

Calculate the atomic mass of naturally occurring silicon.

1-47. Naturally occurring bromine consists of two isotopes, ^{79}Br and ^{81}Br, whose atomic masses are 78.9183 and 80.9163, respectively. Given that the observed atomic mass of bromine is 79.904, calculate the percentages of ^{79}Br and ^{81}Br in naturally occurring bromine. $79.904 = 78.9183\left(\frac{x}{100}\right) + 80.9163\left(\frac{100-x}{100}\right)$

$x = 50.65\%$ $\quad ^{79}Br = 50.65\%$ $\quad ^{81}Br = 49.35\%$

1-48. Naturally occurring boron consists of two isotopes with the atomic masses 10.013 and 11.009. The observed atomic mass of boron is 10.811. Calculate the abundance of each isotope.

1-49. Nitrogen has two naturally occurring isotopes:

Isotope	*Mass*
$^{14}_{7}N$	14.0031
$^{15}_{7}N$	15.0001

The atomic mass of nitrogen is 14.0067. Use these data to compute the percentage of ^{15}N in naturally occurring nitrogen. $14.0067 = 14.0031\left(\frac{100-x}{100}\right) + 15.0001\left(\frac{x}{100}\right)$

$x = 0.36\%$ $\quad ^{15}_{7}N = 0.36\%$

1-50. Naturally occurring europium consists of two isotopes, ^{151}Eu and ^{153}Eu, whose atomic masses are 150.9199 and 152.9212, respectively. Given that the observed atomic mass of europium is 151.96, calculate its isotopic percentage composition.

1-51. Naturally occurring silicon consists of three isotopes, ^{28}Si, ^{29}Si, and ^{30}Si, whose atomic masses are 27.9769, 28.9765, and 29.9738, respectively. The most abundant isotope is ^{28}Si, which accounts for 92.23 percent of naturally occurring silicon. Given that the observed atomic mass of silicon is 28.0855, calculate the percentages of ^{29}Si and ^{30}Si in nature.

$28.0855 = 27.9769\left(\frac{92.23}{100}\right) + 28.9765\left(\frac{x}{100}\right) + 29.9738\left(\frac{7.77-x}{100}\right)$

$x = 4.67\%$

$^{29}Si = 4.67\%$

$^{30}Si = 3.10\%$

1-52. Naturally occurring lithium consists of two isotopes:

Isotope	*Atomic mass*	*Abundance/%*
$^{6}_{3}Li$	6.0151	7.42
$^{7}_{3}Li$	7.0160	92.58

Lithium-6 is extracted from natural lithium for use in the manufacture of nuclear weapons. (a) Compute the atomic mass of naturally occurring lithium. (b) Compute the percentage of $^{6}_{3}Li$ in a lithium sample with an atomic mass of 7.000.

1-53. Compute the number of grams of $^{2}_{1}H$, $^{17}_{8}O$, and $^{18}_{8}O$ in 1 metric ton (= 1000 kg = 2200 lb) of pure water. (Use data in Table 1-9.)

$2H + O = H_2O$

$2(1.008) + 16.00 = 18.02\,g/mol$

$10^6\,g \times \frac{mol}{18.02\,g} = 5.55 \times 10^4$ mol water

1.11×10^5 mol H; 5.55×10^4 O

1.11×10^5 mol H $\times \left(\frac{0.015}{100}\right) = 16.65$ mol $^{2}_{1}H$

5.55×10^4 mol O $\times \left(\frac{0.038}{100}\right) = 21.1$ mol $^{17}_{8}O$

5.55×10^4 mol O $\times \left(\frac{0.204}{100}\right) = 113.2$ mol $^{18}_{8}O$

$16.65\,m\left(\frac{2.0141\,g}{mole}\right) = 34\,g$ $^{2}_{1}H$ $\qquad 113.2$ mol $\left(\frac{17.9992\,g}{1\,mol}\right) = 2040\,g$

$21.1\,mol\left(\frac{16.9991\,g}{mole}\right) = 360\,g$ $^{17}_{8}O$

1-54. Compute the number of grams of $^{13}_{6}C$ in 1 metric ton of sucrose, $C_{12}H_{22}O_{11}$. (Use data in Table 1-9.)

IONS

1-55. How many electrons are there in the following ions:

(a) K^+ 18
(b) Cl^- 18
(c) Ca^{2+} 18
(d) S^{2-} 18
(e) N^{3-} 10

1-56. How many electrons are there in the following ions:

(a) Br^-
(b) P^{3-}
(c) Zn^{2+}
(d) Ag^+
(e) Pb^{4+}

1-57. Determine the number of electrons in the following ions:

(a) Zr^{2+} 38
(b) Th^+ 89
(c) In^{2+} 47
(d) Y^{3+} 36
(e) Sc^{3+} 18

1-58. Determine the number of electrons in the following ions:

(a) I^-
(b) Te^{2-}
(c) La^{3+}
(d) Au^+
(e) Ir^{3+}

1-59. Give three ions that are isoelectronic with each of the following:

(a) Na^+
(b) Ar
(c) Xe
(d) C
(e) Cl^-

1-60. Give three ions that are isoelectronic with each of the following:

(a) F
(b) Be
(c) Ba
(d) Ne
(e) N

ATOMIC AND NUCLEAR SIZE

1-61. An approximate, theoretical formula for the radius of a nucleus is

$$r = (1.3 \times 10^{-15})A^{1/3}\ \text{m}$$

where A is the mass number of the nucleus. Given the following atomic radii, calculate the ratio of the atomic diameter to the nuclear diameter:

Atom	*Radius/pm*[a]
carbon-12	77
argon-40	94
silver-108	144
radium-226	220

[a]The "pm" denotes picometers (1 pm = 10^{-12} m).

1-62. Assume that the diameter of an atom is about 10^4 times larger than the diameter of the nucleus. If the nucleus were magnified to the size of a ping-pong ball (approximately 3 cm diameter), estimate the diameter of the atom. Construct an analogy in terms of everyday objects.

SIGNIFICANT FIGURES

1-63. Determine the number of significant figures in each of the following:

(a) 7.510
(b) 0.00797
(c) 3.65×10^{-5}
(d) the 1980 U.S. population of about 226,000,000
(e) the number 2

1-64. Determine the number of significant figures in each of the following:

(a) 578
(b) 0.000578
(c) There are 1000 m in 1 km.
(d) The distance from the earth to the sun is 93,000,000 miles.
(e) the two numbers in the first two sentences of Section 1-1

1-65. Use the atomic masses given in Appendix F to compute molecular masses to the greatest number of significant figures justified by the data:

(a) H_2O
(b) PbO_2
(c) $AlCl_3$
(d) $^{98}TcBr_2$

1-66. Use the atomic masses given in Appendix F to compute molecular masses to the greatest number of significant figures justified by the data:

(a) CH_4
(b) CaF_2
(c) $TiCl_4$
(d) $^{243}AmCl_3$

1-67. Calculate the following to the correct number of significant figures:

(a) $37.3654 + 0.0629 + 19.34 =$ 56.77

(b) $219.567 - 0.068 =$ 219.499

(c) $(2.674)^3 =$ 19.12

(d) $\frac{6.837}{19} =$ 0.3598 or 0.36

(e) $\frac{49.767 - 1.008}{92.3} + 7.1 =$ 7.6

1-68. Calculate the following to the correct number of significant figures:

(a) $213.3642 + 17.54 + 32978 =$

(b) $373.26 - 119 =$

(c) $(6.0220 \times 10^{23})(5.6 \times 10^{-2}) =$

(d) $\frac{(6.626196 \times 10^{-34})(2.997925 \times 10^{8})}{(1.38062 \times 10^{-23})} =$

(e) $(9.109558 \times 10^{-31} + 1.67252 \times 10^{-27} - 1.67482 \times 10^{-27})(2.997925 \times 10^{8})^2 =$

UNIT CONVERSIONS

1-69. Use the information from the inside back cover to make the following conversions, expressing your results to the correct number of significant figures:

(a) 1.00 qt to liters $1\,qt \times \frac{1\,L}{1.0567\,qt} = 0.9462$

(b) $3.00 \times 10^8\ m \cdot s^{-1}$ to miles per hour $3\times10^8\ m/s \times \frac{1\,km}{1000\,m} \times \frac{1\,mi}{1.6093\,km} \times \frac{60\,sec}{min} \times \frac{60\,min}{hr} = 6.71\times10^8\ mi/hr$

(c) $1.9872\ cal \cdot K^{-1} \cdot mol^{-1}$ to $J \cdot K^{-1} \cdot mol^{-1}$ $\frac{1.9872\,cal}{K \cdot mol} \times \frac{4.184\,J}{cal} = 8.314\,J/K \cdot mol$

1-70. Use the information from the inside back cover to make the following conversions, expressing your results to the correct number of significant figures:

(a) 325 ft to meters

(b) 1.54 Å to picometers and to nanometers

(c) 175 pounds to kilograms

1-71. A lightyear is the distance light travels in 1 year. The speed of light is $3.00 \times 10^8\ m \cdot s^{-1}$. Compute the distance in meters and in miles that light travels in 1 year. $3\times10^8\ m/s \times \frac{60\,sec}{min} \times \frac{60\,min}{hr} \times \frac{24\,hr}{1\,day} \times \frac{7\,day}{week} \times \frac{52\,weeks}{year} = 9.43\times10^{15}\ m/year$

$9.46\times10^{15}\ m/year \times \frac{1\,km}{1000\,m} \times \frac{1\,mi}{1.6093\,km} = 5.88\times10^{12}\ mi/year$

1-72. In older U.S. cars, total cylinder volume is expressed in cubic inches. Compute the total cylinder volume in liters of a 454-cu.-in. engine.

1-73. You are shopping for groceries and see that one bottle of soda has a volume of 750 mL and costs 67 cents, whereas an eight-pack of 16-oz bottles costs \$3.00. Which is the better buy?

16 oz · 8 = 128 oz

$128\,oz \left(\frac{0.94633\,L}{32\,oz}\right)\left(\frac{10^3\,mL}{L}\right) = 3785\,mL$

① cost/mL $= \frac{3\,dollars}{3785\,mL} \times \frac{100\,cents}{dollar} = 0.0793$ cents/mL

② cost/mL $= \frac{67\,cents}{750\,mL} = 0.089$ cents/mL

eight-pack is better buy.

1-74. Compute the speed in meters per second of a 90-mph fastball. The pitcher's mound on a regulation baseball diamond is 60 ft, 6 in. from home plate. Compute the time in seconds that it takes a 90-mph fastball to travel from the pitcher's mound to home plate.

Starting at the upper left and going in a clock-wise direction, we have the elements sulfur, copper, bromine, mercury and iodine.

2 / The Chemical Elements and the Periodic Table

There are 108 known chemical elements, and each undergoes a variety of characteristic chemical reactions. The study of these chemical reactions is simplified because there are patterns in the chemical properties of the elements. Instead of having to learn the chemical reactions of each individual element, we shall divide the elements into groups and study the reactions of each group.

The arrangement of the elements into groups can be displayed in a form called the *periodic table of the elements*. The periodic table is the most useful guide to the study of chemistry and is the central theme of this chapter. Before introducing the periodic table, let us first discuss the chemical properties of a few elements.

2-1. NEW SUBSTANCES ARE FORMED IN CHEMICAL REACTIONS

There are many different types of chemical reactions. Some of the simplest are those in which a metal and a nonmetal react directly with each other. For example, consider the reaction between sodium and chlorine. Sodium is a very reactive metal. It reacts spontaneously with the oxygen and water vapor in the air and must be stored under kerosene, an unreactive, oily liquid. Chlorine, a very reactive nonmetal, is a greenish yellow, irritating, toxic gas that attacks many metals. When sodium metal is dropped into a container of chlorine gas, there is a vigorous, spontaneous reaction. The product of the reaction is a white, crystalline solid, sodium chloride, which is ordinary table salt. We can represent the reaction of sodium with chlorine as

sodium metal	+ chlorine gas →	sodium chloride	(2-1)
very reactive metal	very reactive nonmetal	ordinary table salt	

Reaction (2-1) illustrates the fact that the chemical properties of a product of a chemical reaction need bear no resemblance to the chemical properties of the reactants. *Entirely new substances are formed in chemical reactions.* Another chemical reaction that illustrates this fact is the reaction between hydrogen and oxygen to form water. Both hydrogen and oxygen are colorless, odorless gases. They form an explosive mixture that is set off easily by a spark or a flame. The reaction between hydrogen and oxygen can be written

hydrogen gas	+ oxygen gas →	water	(2-2)
colorless gas	colorless gas	colorless liquid	

The properties of water are radically different from those of hydrogen or oxygen.

Because hydrogen is the least dense gas, being about 15 times less dense than air, balloons and dirigibles used to be filled with hydrogen. This practice was discontinued after 1937, however, when the hydrogen-filled dirigible *Hindenburg* exploded (Figure 2-1).

Figure 2-1 The explosion of the German dirigible *Hindenburg* during landing in Lakehurst, New Jersey. Today helium, a nonflammable lighter-than-air gas, is used in place of hydrogen in lighter-than-air craft in order to eliminate the possibility of a similar disaster.

2-2. CHEMICAL REACTIONS ARE REPRESENTED BY CHEMICAL EQUATIONS

It is cumbersome to write out the full names of the reactants and products in chemical reactions as we have been doing. Consequently, chemists have devised a shorthand way of describing the chemical changes that occur in chemical reactions. A chemical reaction is expressed in terms of the chemical formulas of the reactants and the products. The *reactants* are the substances that react with each other, and the *products* are the substances formed in the reaction. An arrow is used to indicate that the reactants are converted to products. For example, the reaction between sodium metal and chlorine gas to form sodium chloride (Figure 2-2) is expressed by

$$Na(s) + Cl_2(g) \rightarrow NaCl(s) \qquad \text{(not balanced)} \qquad (2\text{-}3)$$

Figure 2-2 When sodium metal (a very reactive metal) reacts with chlorine gas (a very reactive nonmetal), the product is sodium chloride (ordinary table salt), illustrating the fact that entirely new substances are formed in chemical reactions.

The symbol (s) after a chemical formula tells us that the substance is a *s*olid, and the symbol (g) tells us that the substance is a *g*as.

The representation of a chemical reaction by writing the chemical formulas of the reactants and products separated by an arrow as we have done above is called a *chemical equation*. As it stands, Equation (2-3) is not a *balanced* chemical equation because the number of chlorine atoms is not the same on the left (reactants) and on the right (product) sides of the arrow.

2-3. A CHEMICAL EQUATION MUST BE BALANCED

An essential feature of chemical reactions is the conservation of each type of atom. Although new substances are formed in chemical reactions as a result of new arrangements of the atoms, *the individual atoms of various types are neither created nor destroyed in a chemical reaction.* A chemical equation must always be balanced; that is, it must have the same number of each type of atom on both sides.

Note that Equation (2-3) contains two atoms of chlorine on the left but only one atom of chlorine on the right. The law of conservation of matter requires that all the atoms that enter into a chemical reaction appear in the products. We can *balance* Equation (2-3) with respect to the chlorine atoms by placing a "2" in front of the NaCl(s) on the right-hand side of the reaction:

$$Na(s) + Cl_2(g) \rightarrow 2NaCl(s) \qquad \text{(not balanced)}$$

Now there are two sodium atoms on the right but only one on the left. If we place a 2 in front of the Na(s) on the left, we obtain a *balanced chemical equation* for the reaction of sodium with chlorine:

$$2Na(s) + Cl_2(g) \rightarrow 2NaCl(s) \qquad \text{(balanced)} \qquad (2\text{-}4)$$

Both sides of Equation (2-4) contain the same number of each kind of atom.

We balance chemical equations by placing the appropriate numbers, called *balancing coefficients,* in front of the chemical formulas. The chemical formulas of the reactants and products themselves are fixed and should not be altered by changing the subscripts. Equation (2-3) should not be balanced by changing NaCl to $NaCl_2$. The chemical formula of sodium chloride is NaCl, not $NaCl_2$. In fact, there is no such compound as $NaCl_2$.

Let's consider now the reaction of hydrogen with oxygen. Recall that hydrogen and oxygen exist as diatomic molecules. We first write Reaction (2-2) as an unbalanced chemical equation:

$$H_2(g) + O_2(g) \rightarrow H_2O(l) \qquad \text{(not balanced)}$$

where the (l) after the formula tells us that the substance is a *l*iquid. In this equation, there are two oxygen atoms on the left (in O_2) and one oxygen atom on the right (in H_2O). If we place a 2 in front of the $H_2O(l)$, then the equation is balanced with respect to oxygen atoms:

$$H_2(g) + O_2(g) \rightarrow 2H_2O(l) \qquad \text{(not balanced)}$$

Now there are four hydrogen atoms on the right (2×2) and only two on the left. We balance the hydrogen atoms by placing a 2 in front of the $H_2(g)$:

$$2H_2(g) + O_2(g) \rightarrow 2H_2O(l) \qquad \text{(balanced)}$$

to obtain the balanced equation for the reaction of hydrogen with oxygen. Once again, note that a chemical equation is balanced by placing coefficients *in front of* the formulas of the reactants and products; the formulas themselves are not altered. Figure 2-3 is a schematic representation of the reaction between hydrogen and oxygen.

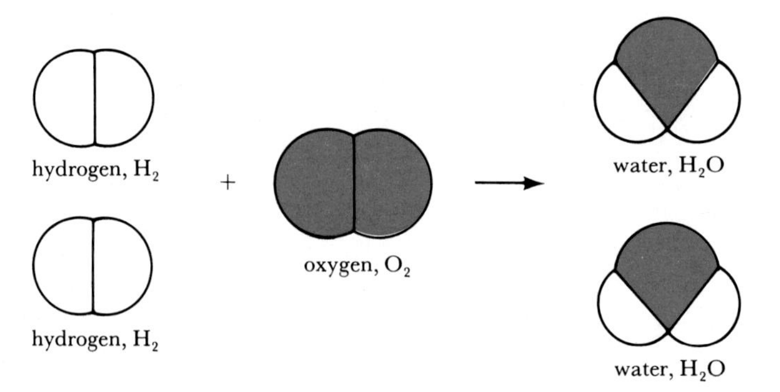

Figure 2-3 The chemical reaction between hydrogen and oxygen. The figure illustrates that a chemical reaction is a rearrangement of the atoms in reactant molecules into product molecules. The numbers of each kind of atom are the same before and after the reaction. The spatial arrangement of atoms in molecules is discussed in Chapter 11.

Example 2-1: Magnesium is a low-density metal that reacts with oxygen, producing a bright flame in the process. The emission of light during the reaction is the basis of using magnesium in photographic flash bulbs, which contain magnesium filaments and oxygen gas. The reactants and products of the reaction are

$Mg(s)$	$+\ O_2(g)\ \rightarrow$	$MgO(s)$	(not balanced)
magnesium metal	oxygen gas	magnesium oxide, a white powder	

Balance this chemical equation.

Solution: The two oxygen atoms on the left require that there be two oxygen atoms on the right, and so we place a 2 in front of $MgO(s)$. Now there are two magnesium atoms on the right but only one on the left. By placing a 2 in front of $Mg(s)$, we have a balanced chemical equation:

$$2Mg(s) + O_2(g) \rightarrow 2MgO(s)$$

Magnesium oxide (magnesia) is used in the manufacture of fire-resistant bricks and the antacid, Milk of Magnesia.

Magnesium metal burns in oxygen.

Example 2-2: Sodium metal reacts vigorously with water. The reactants and products of the reaction are

$Na(s)$	$+\ H_2O(l)\ \rightarrow$	$NaOH(s)$	$+$	$H_2(g)$
sodium metal	water	sodium hydroxide		hydrogen gas

Balance this chemical equation.

Solution: Let's balance the equation with respect to hydrogen atoms first. If we place a 2 in front of $NaOH(s)$ and a 2 in front of $H_2O(l)$, there will be four hydrogen atoms on each side of the equation:

$$Na(s) + 2H_2O(l) \rightarrow 2NaOH(s) + H_2(g) \qquad \text{(not balanced)}$$

Now we need to balance the sodium atoms. If we place a 2 in front of $Na(s)$, we have a balanced equation:

$$2Na(s) + 2H_2O(l) \rightarrow 2NaOH(s) + H_2(g) \qquad \text{(balanced)}$$

Sodium hydroxide is a white, translucent solid used in the manufacture of paper and soaps and in petroleum refining. It is extremely corrosive to the skin and other tissues and is sometimes called *caustic soda* or *lye.*

This method of balancing chemical equations is called *balancing by inspection.* With a little practice, you can become proficient in balancing certain types of chemical equations this way. (Problems 2-1 through 2-10 should be worked for practice in balancing equations.)

Figure 2-4 Lithium floating on oil, which in turn is floating on water. Lithium has the lowest density of any metal.

2-4. SOME ELEMENTS HAVE SIMILAR CHEMICAL PROPERTIES

By the 1860's, more than 60 elements had been discovered and many chemists had begun to notice some pattern in the chemical properties of certain elements. For example, consider the three metals lithium (Li), sodium (Na), and potassium (K). All three of these metals are less dense than water (Figure 2-4), are soft enough to be cut with a knife, have fairly low melting points (below 200°C), and are very reactive. They all react spontaneously with oxygen and water. Just as sodium reacts vigorously with chlorine, so do lithium and potassium:

(a)

(b)

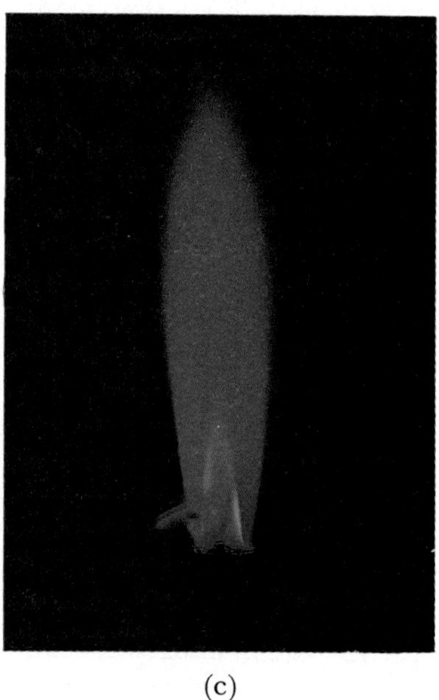

(c)

Figure 2-5 Compounds of the elements lithium (a), sodium (b), and potassium (c) impart a characteristic color to a flame.

$$2Li(s) + Cl_2(g) \rightarrow 2LiCl(s)$$
$$2Na(s) + Cl_2(g) \rightarrow 2NaCl(s)$$
$$2K(s) + Cl_2(g) \rightarrow 2KCl(s)$$

The product in all three cases is a white, unreactive, crystalline solid that dissolves readily in water.

Example 2-2 shows that sodium reacts with water to produce sodium hydroxide (NaOH) and hydrogen. Lithium and potassium undergo a similar reaction (Figure 2-6):

$$2Li(s) + 2H_2O(l) \rightarrow 2LiOH(s) + H_2(g)$$
$$2K(s) + 2H_2O(l) \rightarrow 2KOH(s) + H_2(g)$$

Sylvite is a potassium chloride mineral that is found in extensive deposits in ancient lake and sea beds. It is used as a major source of potassium and its compounds.

Just like sodium hydroxide, lithium hydroxide and potassium hydroxide are corrosive, white, translucent solids. Table 2-1 lists some of the reactions of lithium, sodium, and potassium. Since these three metals have similar chemical properties, they can be considered as a group. Lithium, sodium, and potassium are called *alkali metals*. The hydroxides of these metals are *alkaline*, a property discussed in Chapter 4.

There are other groups of elements that share similar chemical properties. For example, magnesium (Mg), calcium (Ca), strontium (Sr), and barium (Ba) have many chemical properties in common. As a group, these metals are called *alkaline earth metals*. They all burn brightly when heated in oxygen to form white oxides:

$$2Mg(s) + O_2(g) \rightarrow 2MgO(s)$$
$$2Ca(s) + O_2(g) \rightarrow 2CaO(s)$$
$$2Sr(s) + O_2(g) \rightarrow 2SrO(s)$$
$$2Ba(s) + O_2(g) \rightarrow 2BaO(s)$$

Figure 2-6 Potassium (Group 1) reacts violently with water.

Table 2-1 Some reactions of the alkali metals lithium, sodium, and potassium

Reactants[a]	*Products*	*Type of compound*
$2M(s) + F_2(g) \rightarrow$	$2MF(s)$	metal fluoride
$2M(s) + Cl_2(g) \rightarrow$	$2MCl(s)$	metal chloride
$2M(s) + Br_2(l) \rightarrow$	$2MBr(s)$	metal bromide
$2M(s) + I_2(s) \rightarrow$	$2MI(s)$	metal iodide
$2M(s) + S(s) \rightarrow$	$M_2S(s)$	metal sulfide
$2M(s) + H_2(g) \rightarrow$	$2MH(s)$	metal hydride
$2M(s) + 2H_2O(l) \rightarrow$	$2MOH(s) + H_2(g)$	metal hydroxide + hydrogen

[a]In each case, M stands for Li, Na, or K.

Table 2-2 Some reactions of the alkaline earth metals magnesium, calcium, strontium, and barium

Reactants[a]	*Products*	*Type of compound*
$M(s) + Cl_2(g) \longrightarrow$	$MCl_2(s)$	metal chloride
$M(s) + Br_2(l) \longrightarrow$	$MBr_2(s)$	metal bromide
$M(s) + I_2(s) \longrightarrow$	$MI_2(s)$	metal iodide
$2M(s) + O_2(g) \longrightarrow$	$2MO(s)$	metal oxide
$M(s) + S(s) \longrightarrow$	$MS(s)$	metal sulfide
$3M(s) + N_2(g) \longrightarrow$	$M_3N_2(s)$	metal nitride
$M(s) + H_2(g) \longrightarrow$	$MH_2(s)$	metal hydride
$M(s) + H_2O(g) \longrightarrow$	$MO(s) + H_2(g)$	metal oxide + hydrogen
$M(s) + 2HCl(g) \longrightarrow$	$MCl_2(s) + H_2(g)$	metal chloride + hydrogen

[a]In each case, M stands for Mg, Ca, Sr, or Ba.

Figure 2-7 Calcium (Group 2) reacts slowly with cold water to produce calcium hydroxide and $H_2(g)$.

Although magnesium, calcium, strontium, and barium react slowly with cold water (Figure 2-7), they react rapidly with steam at high temperatures to yield a metal oxide and hydrogen gas:

$$Mg(s) + H_2O(g) \rightarrow MgO(s) + H_2(g)$$
$$Ca(s) + H_2O(g) \rightarrow CaO(s) + H_2(g)$$
$$Sr(s) + H_2O(g) \rightarrow SrO(s) + H_2(g)$$
$$Ba(s) + H_2O(g) \rightarrow BaO(s) + H_2(g)$$

Table 2-2 lists some of the reactions of magnesium, calcium, strontium, and barium. We see, then, that these four metals have similar chemical properties and can be placed into a group, just as lithium, sodium, and potassium can.

Example 2-3: By referring to Tables 2-1 and 2-2, write the chemical equations for the reactions of sodium with sulfur and calcium with sulfur.

Solution: From Table 2-1, we find that with $M(s) = Na(s)$

$$2Na(s) + S(s) \rightarrow \underset{\text{sodium sulfide}}{Na_2S(s)}$$

and from Table 2-2, with $M(s) = Ca(s)$

$$Ca(s) + S(s) \rightarrow \underset{\text{calcium sulfide}}{CaS(s)}$$

Sodium sulfide is used in the manufacture of rubber and in metal refining. Calcium sulfide is used to make luminous paints and as a food preservative.

Another group of elements having similar chemical properties consists of the nonmetals fluorine (F_2), chlorine (Cl_2), bromine (Br_2), and iodine (I_2) (Figure 2-8). As a group, these elements are called *halogens*, meaning "salt formers." The halogens react with the alkali metals to give white, crystalline solids called *halides*. The specific names of the halides are fluoride, chloride, bromide, and iodide. For example, in the case of sodium we have

$$2Na(s) + F_2(g) \rightarrow 2NaF(s) \quad \text{sodium fluoride}$$
$$2Na(s) + Cl_2(g) \rightarrow 2NaCl(s) \quad \text{sodium chloride}$$
$$2Na(s) + Br_2(l) \rightarrow 2NaBr(s) \quad \text{sodium bromide}$$
$$2Na(s) + I_2(s) \rightarrow 2NaI(s) \quad \text{sodium iodide}$$

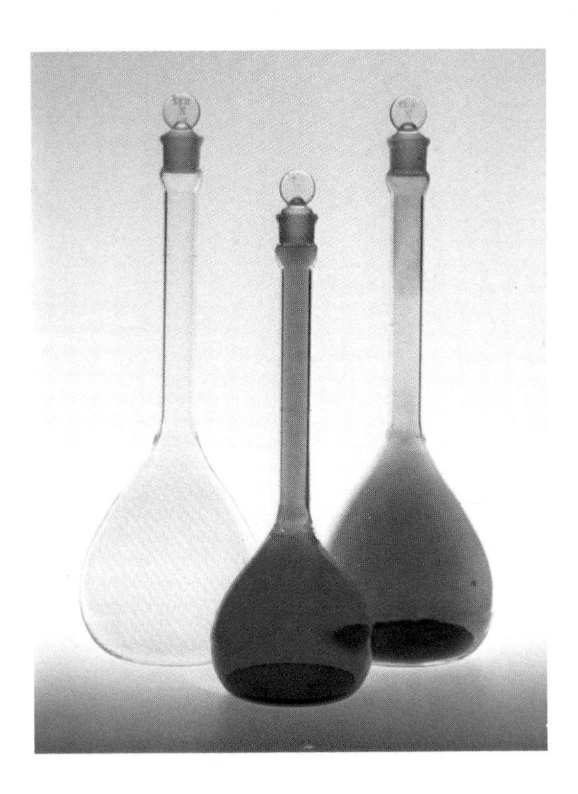

Figure 2-8 Chlorine, bromine, and iodine.

When antimony powder is placed in an atmosphere of chlorine, a vigorous reaction takes place.

Table 2-3 Some reactions of the halogens fluorine, chlorine, bromine, and iodine

Reactants[a]	*Products*	*Type of compound*
$2Li(s) + X_2 \rightarrow$	$2LiX(s)$	lithium halide
$2Na(s) + X_2 \rightarrow$	$2NaX(s)$	sodium halide
$2K(s) + X_2 \rightarrow$	$2KX(s)$	potassium halide
$Mg(s) + X_2 \rightarrow$	$MgX_2(s)$	magnesium halide
$Ca(s) + X_2 \rightarrow$	$CaX_2(s)$	calcium halide
$Sr(s) + X_2 \rightarrow$	$SrX_2(s)$	strontium halide
$Ba(s) + X_2 \rightarrow$	$BaX_2(s)$	barium halide
$2P(s) + 3X_2 \rightarrow$	$2PX_3$	phosphorus trihalide
$2As(s) + 3X_2 \rightarrow$	$2AsX_3$	arsenic trihalide
$2Sb(s) + 3X_2 \rightarrow$	$2SbX_3$	antimony trihalide
$H_2(g) + X_2 \rightarrow$	$2HX(g)$	hydrogen halide

[a]In each case, X stands for F, Cl, Br, or I.

Table 2-2 shows that the halogens react with the alkaline earth metals to give $MCl_2(s)$, $MBr_2(s)$, and $MI_2(s)$. Table 2-3 summarizes some of the chemical reactions of fluorine, chlorine, bromine, and iodine.

2-5. THE ELEMENTS SHOW A PERIODIC PATTERN WHEN LISTED IN ORDER OF INCREASING ATOMIC NUMBER

After Dalton proposed the atomic theory, the concept of atomic mass and the experimental determination of atomic masses took on increasing importance. In 1869, the Russian chemist Dmitri Mendeleev arranged the elements in order of increasing atomic mass and was able to show that the chemical properties of the elements exhibit periodic behavior. To illustrate the idea of Mendeleev's observation, let's start with the element lithium and arrange the succeeding elements in order of increasing atomic mass, as shown in Table 2-4. If we look at Table 2-4 carefully, we see that the chemical properties of the elements show a remarkably repetitive, or periodic, pattern. The variations in properties as we go from lithium to neon are repeated as we go from sodium to argon. The repeating pattern (or periodicity) is seen more clearly if we arrange the elements horizontally:

Table 2-4 The chemical properties of some elements, listed in order of increasing atomic mass

Atomic mass	*Element*	*Symbol*	*Properties*	*Formula of halogen compound*[a]
6.9	lithium	Li	very reactive metal	LiX
9.0	beryllium	Be	reactive metal	BeX_2
10.8	boron	B	semimetal[b]	BX_3
12.0	carbon	C	nonmetallic solid	CX_4
14.0	nitrogen	N	nonmetallic diatomic gas	NX_3
16.0	oxygen	O	nonmetallic, moderately reactive diatomic gas	OX_2
19.0	fluorine	F	very reactive diatomic gas	FX
20.2	neon	Ne	very unreactive monatomic gas	none
23.0	sodium	Na	very reactive metal	NaX
24.3	magnesium	Mg	reactive metal	MgX_2
27.0	aluminum	Al	metal	AlX_3
28.1	silicon	Si	semimetal[b]	SiX_4
31.0	phosphorus	P	nonmetallic solid	PX_3
32.1	sulfur	S	nonmetallic solid	SX_2
35.5	chlorine	Cl	very reactive diatomic gas	ClX
39.9	argon	Ar	very unreactive monatomic gas	none

[a]X stands for F, Cl, Br, or I.
[b]Boron and silicon are called semimetals because they have properties that are intermediate between those of the metals and those of the nonmetals.

Li	Be	B	C	N	O	F	Ne
lithium	beryllium	boron	carbon	nitrogen	oxygen	fluorine	neon
Na	Mg	Al	Si	P	S	Cl	Ar
sodium	magnesium	aluminum	silicon	phosphorus	sulfur	chlorine	argon

Notice that elements with similar chemical properties, for example, lithium and sodium, are placed in the same column.

Figure 2-9 presents a modern version of Mendeleev's *periodic table of the elements*. The modern version is more complicated than the abbreviated version shown above, primarily because it contains all 108 known elements instead of just the 16 listed in Table 2-4. In the

Row	1	2																									3	4	5	6	7	8	Row
1	1 H																															2 He	1
2	3 Li	4 Be																									5 B	6 C	7 N	8 O	9 F	10 Ne	2
3	11 Na	12 Mg																									13 Al	14 Si	15 P	16 S	17 Cl	18 Ar	3
4	19 K	20 Ca															21 Sc	22 Ti	23 V	24 Cr	25 Mn	26 Fe	27 Co	28 Ni	29 Cu	30 Zn	31 Ga	32 Ge	33 As	34 Se	35 Br	36 Kr	4
5	37 Rb	38 Sr															39 Y	40 Zr	41 Nb	42 Mo	43 Tc	44 Ru	45 Rh	46 Pd	47 Ag	48 Cd	49 In	50 Sn	51 Sb	52 Te	53 I	54 Xe	5
6	55 Cs	56 Ba	57 La	58 Ce	59 Pr	60 Nd	61 Pm	62 Sm	63 Eu	64 Gd	65 Tb	66 Dy	67 Ho	68 Er	69 Tm	70 Yb	71 Lu	72 Hf	73 Ta	74 W	75 Re	76 Os	77 Ir	78 Pt	79 Au	80 Hg	81 Ti	82 Pb	83 Bi	84 Po	85 At	86 Rn	6
7	87 Fr	88 Ra	89 Ac	90 Th	91 Pa	92 U	93 Np	94 Pu	95 Am	96 Cm	97 Bk	98 Cl	99 Es	100 Fm	101 Md	102 No	103 Lr	104 Unq	105 Unp	106 Unh	107 Uns		109 Une										7

Figure 2-9 A modern version of the periodic table of the elements. The elements are ordered according to increasing atomic number, and their chemical properties show a periodic pattern. Elements that appear in the same column have similar chemical properties. The symbols for elements 104 and beyond are explained in Appendix E.

modern periodic table in Figure 2-9, the elements are arranged in order of increasing atomic number instead of increasing atomic mass. With a few exceptions, the order is the same in both cases. The idea of atomic number was not developed until the early 1900's, about 40 years after Mendeleev's first periodic table.

2-6. ELEMENTS IN THE SAME COLUMN IN THE PERIODIC TABLE HAVE SIMILAR CHEMICAL PROPERTIES

Notice in Figure 2-9 that lithium, sodium, and potassium occur in the far left column of the periodic table. All the elements in that column have similar chemical properties. Although we have not discussed rubidium (Rb), cesium (Cs), or francium (Fr), the fact that these elements occur in the same column as lithium, sodium, and potassium suggests that they undergo similar chemical reactions. Francium is a radioactive element not found in nature, but rubidium and cesium are light, soft, very reactive metals. Rubidium and cesium react vigorously with the halogens, water, hydrogen, oxygen, and many other substances. By analogy with the reactions of sodium that we discussed earlier, we can predict that, for example,

$$2Rb(s) + Cl_2(g) \rightarrow 2RbCl(s)$$
$$2Cs(s) + 2H_2O(l) \rightarrow 2CsOH(s) + H_2(g)$$

and that other reactions of rubidium and cesium are similar to those given in Table 2-1; these predictions are correct.

Example 2-4: Predict the product of the reaction between rubidium and hydrogen.

Solution: According to the periodic table and Table 2-1, rubidium reacts with hydrogen to form rubidium hydride, RbH. The balanced equation for the reaction is

$$2Rb(s) + H_2(g) \rightarrow 2RbH(s)$$

The far left column in the periodic table is labeled 1, and so the elements in that column are referred to as the *Group 1 metals*. As noted previously, the Group 1 metals are also called the alkali metals.

The metals in the column labeled 2 are called the *Group 2 metals* or the alkaline earth metals. All the elements in Group 2 are reactive metals and undergo similar chemical reactions.

Example 2-5: Predict whether radium reacts with oxygen and, if so, write the balanced chemical equation.

Solution: Radium is in Group 2 in the periodic table, and according to Table 2-2, the Group 2 metals react with oxygen to produce an oxide whose formula is MO. Thus we predict correctly that

$$2Ra(s) + O_2(g) \rightarrow \underset{\text{radium oxide}}{2RaO(s)}$$

Radium is a radioactive element, and so radium oxide is a radioactive compound. It is used in the radiation treatment of tumors.

Elements in the same column in the rest of the periodic table also share similar chemical properties. This similarity is particularly strong in the columns headed by numbers. The halogens, which we have seen behave similarly, occur in Group 7. The extreme right column of the periodic table (Group 8) contains the *noble gases*, which are characterized primarily by their lack of chemical reactivity. Prior to 1962, they were called the *inert gases* because no compounds of these gases were known. In 1962, however, xenon was shown to form compounds with fluorine and oxygen, the most reactive nonmetals.

Figure 2-10 Each noble gas produces a particular color in fluorescent lights. Helium gives a yellowish flesh tone, neon is orange-red, argon is lavender (mercury + argon = blue), krypton is silver white, and xenon is blue. Other colors are produced from fluorescent powders and coated tubes.

Krypton fluorides are also known, but no stable compounds of helium, neon, or argon have been made yet, and the noble gas elements are generally very unreactive.

Example 2-6: Phosphorus is a nonmetallic solid that occurs in white, red, and black forms. White phosphorus spontaneously bursts into flame in the presence of oxygen to produce the oxide P_4O_6. Use the periodic table to predict the reaction between arsenic (atomic number 33) and oxygen.

Solution: Arsenic occurs in the same group (5) as phosphorus, and so we predict, by analogy with phosphorus, that arsenic reacts with oxygen to produce As_4O_6. The balanced chemical equation for the reaction is

$$4As(s) + 3O_2(g) \rightarrow As_4O_6(s)$$

Arsenic compounds are well known to be poisonous, the lethal dose of As_4O_6 being about 0.1 g for an average adult male. Small amounts of arsenic compounds, however, promote the growth of red blood cells in bone marrow, and the human body normally contains about 5 mg of arsenic.

Problems 2-11 to 2-16 ask you to predict the products of chemical reactions.

2-7. MENDELEEV PREDICTED THE EXISTENCE AND PROPERTIES OF ELEMENTS NOT YET DISCOVERED

The periodic table shown in Figure 2-9 differs from the original table proposed by Mendeleev. A number of elements had not yet been discovered in 1869, and Mendeleev had the genius to leave gaps in his periodic table to accommodate undiscovered elements. For example, in 1869 the element following zinc in atomic mass was arse-

Table 2-5 Comparison of Mendeleev's predictions and actual experimental values for the properties of gallium

Property	*Predicted*	*Observed*
atomic mass	69	69.7
density/$g \cdot cm^{-3}$	6.0	5.9
melting point	low	30°C
boiling point	high	2000°C
formula of oxide	M_2O_3	Ga_2O_3

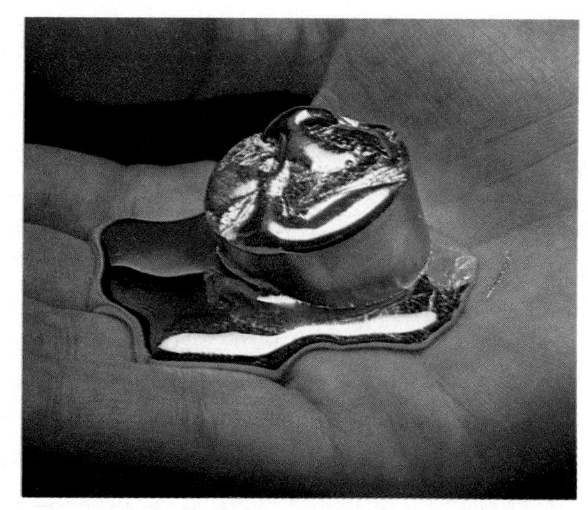

Figure 2-11 Gallium metal has a melting point of 30°C, and so a piece of gallium melts when held in the hand (human body temperature is 37°C). Only two elements are liquids at room temperature (20°C). One is the metal mercury, and the other is the nonmetal bromine.

nic. Yet Mendeleev felt that arsenic belonged in Group 5 rather than in Group 3. He boldly proposed that there are two as yet undiscovered elements between zinc and arsenic and predicted many of the properties of these two elements prior to their discovery. Table 2-5 compares Mendeleev's 1869 predictions with the actual properties of the element gallium (atomic number 31), which was discovered in 1875. Gallium has an unusually low melting point for a metal (30°C) and melts when held in the hand (Figure 2-11). Two metals, gallium and cesium, melt between 20 and 30°C. Cesium belongs to Group 1 and so is much too reactive to be held in the hand.

2-8. THE PERIODIC TABLE ORGANIZES OUR STUDY OF THE ELEMENTS

The periodic table is the most important and useful concept in chemistry. Almost every general chemistry classroom and laboratory has a periodic table hanging on the wall. You may have noticed that the periodic table hanging in your classroom or laboratory is different from the one given in Figure 2-9. A more common version of the periodic table is shown in Figure 2-12. The only difference between Figures 2-9 and 2-12 is that elements 57 through 70 (*the lanthanide series*) and elements 89 through 102 (*the actinide series*) have been cut out and placed at the bottom of the table. Some versions of the periodic table place lanthanum (La) and actinium (Ac) in the table where we have lutetium (Lu) and lawrencium (Lr) and place elements 58 through 71 (cerium through lutetium) and elements 90 through 103 (thorium through lawrencium) below. In addition, many versions have headings for every column. For example, the periodic table in your lecture room or laboratory probably has the heading 3A or 3B for the third column. The differences between the table shown in Figure 2-12 and any others are minor and of no real consequence.

The elements beyond uranium in the periodic table are called transuranium elements. With the exception of recently discovered traces of plutonium, these elements do not occur in nature but have been created in the laboratory. Shown here is the first visible sample of americium produced, having a mass of only a few micrograms.

1	2											3	4	5	6	7	8
1 **H** 1.008																	2 **He** 4.003
3 **Li** 6.941	4 **Be** 9.012											5 **B** 10.81	6 **C** 12.01	7 **N** 14.01	8 **O** 16.00	9 **F** 19.00	10 **Ne** 20.18
11 **Na** 22.99	12 **Mg** 24.31											13 **Al** 26.98	14 **Si** 28.09	15 **P** 30.97	16 **S** 32.06	17 **Cl** 35.45	18 **Ar** 39.95
19 **K** 39.10	20 **Ca** 40.08	21 **Sc** 44.96	22 **Ti** 47.90	23 **V** 50.94	24 **Cr** 52.00	25 **Mn** 54.94	26 **Fe** 55.85	27 **Co** 58.93	28 **Ni** 58.70	29 **Cu** 63.55	30 **Zn** 65.38	31 **Ga** 69.72	32 **Ge** 72.59	33 **As** 74.92	34 **Se** 78.96	35 **Br** 79.90	36 **Kr** 83.80
37 **Rb** 85.47	38 **Sr** 87.62	39 **Y** 88.91	40 **Zr** 91.22	41 **Nb** 92.91	42 **Mo** 95.94	43 **Tc** (98)	44 **Ru** 101.1	45 **Rh** 102.9	46 **Pd** 106.4	47 **Ag** 107.9	48 **Cd** 112.4	49 **In** 114.8	50 **Sn** 118.7	51 **Sb** 121.8	52 **Te** 127.6	53 **I** 126.9	54 **Xe** 131.3
55 **Cs** 132.9	56 **Ba** 137.3	71 **Lu** 175.0	72 **Hf** 178.5	73 **Ta** 180.9	74 **W** 183.9	75 **Re** 186.2	76 **Os** 190.2	77 **Ir** 192.2	78 **Pt** 195.1	79 **Au** 197.0	80 **Hg** 200.6	81 **Tl** 204.4	82 **Pb** 207.2	83 **Bi** 209.0	84 **Po** (209)	85 **At** (210)	86 **Rn** (222)
87 **Fr** (223)	88 **Ra** (226.0)	103 **Lr** (260)	104 **Unq**	105 **Unp**	106 **Unh**	107 **Uns**	108	109 **Une**									

Lanthanide series

57 **La** 138.9	58 **Ce** 140.1	59 **Pr** 140.9	60 **Nd** 144.2	61 **Pm** (145)	62 **Sm** 150.4	63 **Eu** 152.0	64 **Gd** 157.3	65 **Tb** 158.9	66 **Dy** 162.5	67 **Ho** 164.9	68 **Er** 167.3	69 **Tm** 168.9	70 **Yb** 173.0

Actinide series

89 **Ac** (227)	90 **Th** 232.0	91 **Pa** (231)	92 **U** 238.0	93 **Np** (237)	94 **Pu** (244)	95 **Am** (243)	96 **Cm** (247)	97 **Bk** (247)	98 **Cf** (251)	99 **Es** (252)	100 **Fm** (257)	101 **Md** (258)	102 **No** (259)

Figure 2-12 A common version of the periodic table. In this version the lanthanide series (elements 57 through 70) and the actinide series (elements 89 through 102) have been placed at the bottom of the table, for two reasons. First, the separation leads to a more compact table. Second, and more importantly, all the elements in each series have exceptionally similar chemical properties, and, in effect, all can be assigned to just one position in the periodic table. The number under the symbol for each element is the atomic mass.

Most periodic tables have the elements cerium (58) through lutetium (71) and thorium (90) through lawrencium (103) placed below. See, however, the paper by William B. Jensen in the *Journal of Chemical Education* [*59*, 634 (1982)].

The periodic table contains all the known chemical elements and shows the periodic relationships among them. Elements in the same column are said to belong to the same *group* or *family*. The horizontal rows in the periodic table often are called *periods*. Your progress in learning basic chemistry will be greatly aided by an understanding and appreciation of the periodic table. In the remainder of this chapter we shall study some of the general features and uses of the periodic table.

2-9. ELEMENTS ARE ARRANGED AS MAIN-GROUP ELEMENTS, TRANSITION METALS, AND INNER TRANSITION METALS

Metals and nonmetals occur in separate regions of the periodic table, as shown in Figure 2-13. The nonmetals are on the right-hand side and are separated from the metals by a zigzag line. As might be expected, the elements on the border between metals and nonmetals (shaded elements in Figure 2-13) have properties that are intermediate between those of metals and those of nonmetals. Such elements are called *semimetals* and are brittle, dull solids (Figure 2-14). Because semimetals conduct electricity and heat less well than metals but

Figure 2-13 The position of metals and nonmetals in the periodic table. The nonmetals appear only at the far right of the table, and the metals appear at the left. The shaded elements along the steplike border between metals and nonmetals are the semimetals.

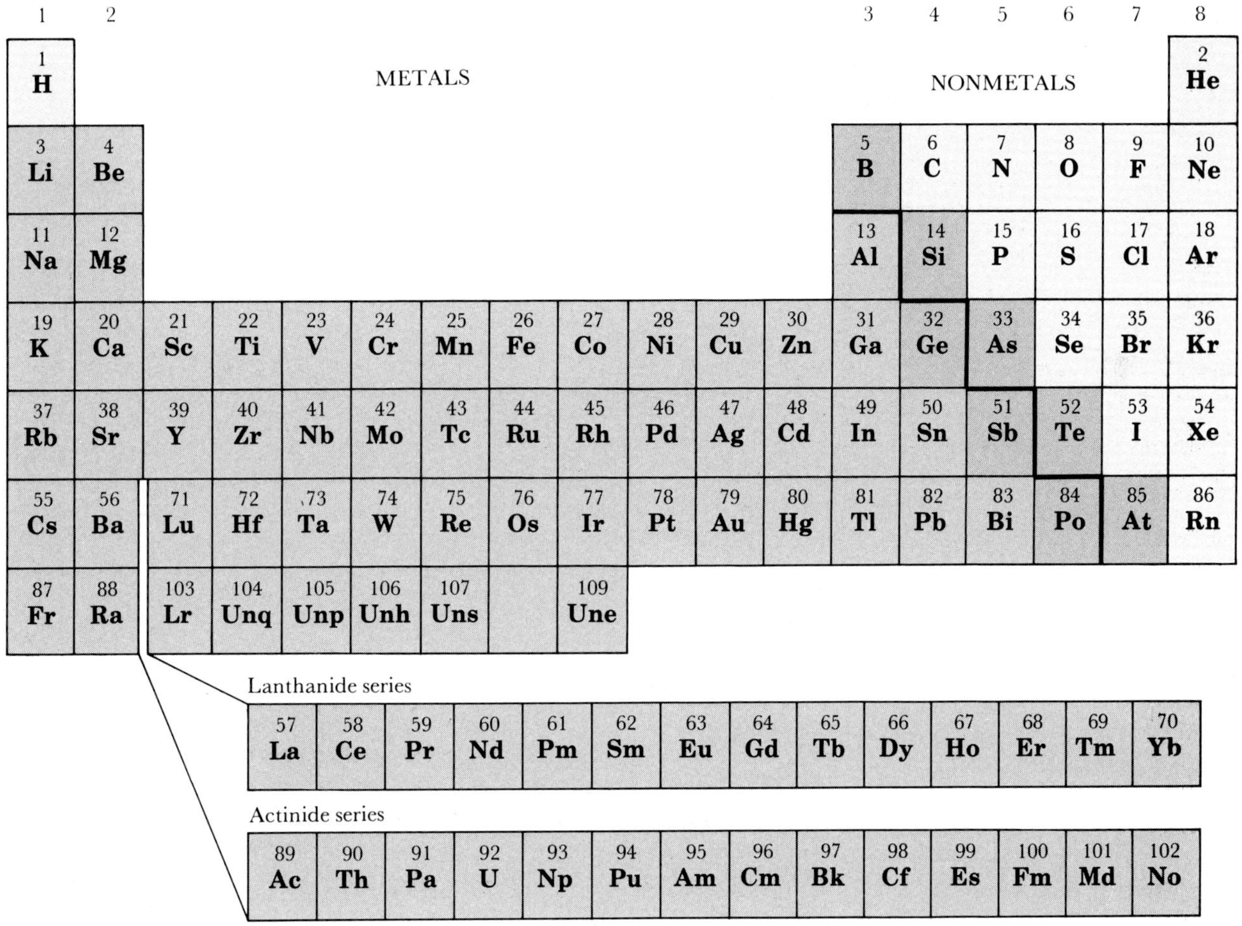

METALS NONMETALS

1	2											3	4	5	6	7	8
1 H																	2 He
3 Li	4 Be											5 B	6 C	7 N	8 O	9 F	10 Ne
11 Na	12 Mg											13 Al	14 Si	15 P	16 S	17 Cl	18 Ar
19 K	20 Ca	21 Sc	22 Ti	23 V	24 Cr	25 Mn	26 Fe	27 Co	28 Ni	29 Cu	30 Zn	31 Ga	32 Ge	33 As	34 Se	35 Br	36 Kr
37 Rb	38 Sr	39 Y	40 Zr	41 Nb	42 Mo	43 Tc	44 Ru	45 Rh	46 Pd	47 Ag	48 Cd	49 In	50 Sn	51 Sb	52 Te	53 I	54 Xe
55 Cs	56 Ba	71 Lu	72 Hf	73 Ta	74 W	75 Re	76 Os	77 Ir	78 Pt	79 Au	80 Hg	81 Tl	82 Pb	83 Bi	84 Po	85 At	86 Rn
87 Fr	88 Ra	103 Lr	104 Unq	105 Unp	106 Unh	107 Uns		109 Une									

Lanthanide series

57 La	58 Ce	59 Pr	60 Nd	61 Pm	62 Sm	63 Eu	64 Gd	65 Tb	66 Dy	67 Ho	68 Er	69 Tm	70 Yb

Actinide series

89 Ac	90 Th	91 Pa	92 U	93 Np	94 Pu	95 Am	96 Cm	97 Bk	98 Cf	99 Es	100 Fm	101 Md	102 No

Figure 2-14 The semimetals are brittle solids. Boron (*top*) and silicon (*bottom*) are shown here.

better than nonmetals, they are called *semiconductors*. The semimetals silicon and germanium are widely used in the manufacture of semiconducting devices and transistors (see Interchapter H).

For the elements in groups headed by a number, there is a continuous progression from metallic to nonmetallic properties as we move from left to right across the periodic table. The most reactive metals are the Group 1 metals, and the most reactive nonmetals are the halogens (Group 7). The noble gases, which are unusually nonreactive, occupy the far right column (8). The elements in groups headed by numbers in Figure 2-12 are called the *main-group elements*.

The elements in the groups not headed by numbers are called *transition metals*. Within any row of transition metals, the chemical properties are more similar across the row than is the case for the main-group elements. Many of the transition metals are probably already familiar to you. Iron, nickel, chromium, tungsten, and titanium are widely used in alloys for structural materials and play a key role in the world's technology. The precious metals—gold, platinum, and silver—are used as hard currency, in jewelry, and in high-quality electronic circuits. The transition metals vary greatly in abundance. Iron and titanium are plentiful, whereas rhenium (Re) and hafnium (Hf) are rare.

Figure 2-15 Crystals of the chlorides of the transition metals scandium (Sc) through zinc (Zn). Most compounds of the transition metals are colored. The origin of the color is discussed in Chapter 23.

The characteristics of the transition metals vary from group to group; yet they all are characterized by high densities and high melting points. In addition, unlike the compounds of the Group 1 and 2 metals, many compounds of the transition metals are colored. Figure 2-15 shows the chlorides of the first transition-metal series, scandium (Sc) through zinc (Zn). The metals with the greatest densities, iridium (Ir), 22.65 $g \cdot cm^{-3}$, and osmium (Os), 22.61 $g \cdot cm^{-3}$, and the highest melting points, tungsten (W), 3410°C, are transition metals.

The two series that begin with lanthanum ($Z = 57$) and actinium ($Z = 89$) in Figure 2-12 are called the *inner transition metals*. The elements in each of these two series have remarkably similar chemical properties. The *lanthanide series* is also called the *rare-earth series* because its members were once thought to occur only in very small quantities. The actinides are radioactive elements, most of which do not occur in nature but are produced in nuclear reactions. Figure 2-16 indicates the position in the periodic table of the three main classes of elements.

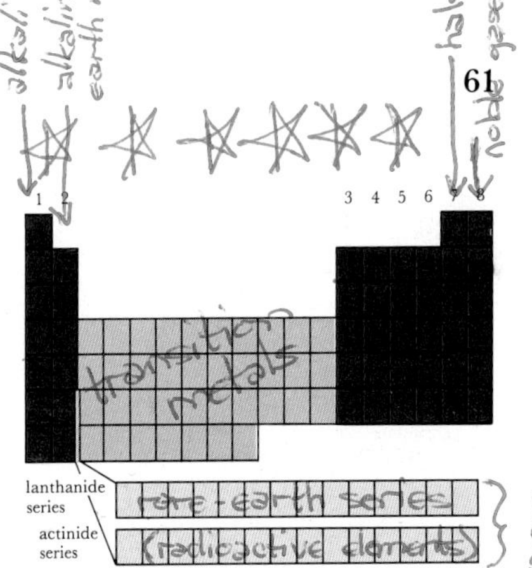

Figure 2-16 The usual form of the periodic table, showing the main-group elements (red), the transition metals (pink), and the inner transition metals (gray). The main-group elements occur in columns headed by numbers, the transition metals occur in the other columns, and the inner transition metals are located at the bottom of the table.

Example 2-7: By referring to the periodic table, classify each of the following as a main-group element, a transition metal, or an inner transition metal. If it is a main-group element, indicate its group and whether it is a metal or a nonmetal.

Zr Eu Se Mo In U Xe

Solution:

Symbol	*Name*	*Atomic number*	*Classification*
Zr	zirconium	40	transition metal
Eu	europium	63	inner transition metal (rare earth)
Se	selenium	34	main-group element (Group 6 nonmetal)
Mo	molybdenum	42	transition metal
In	indium	49	main group element (Group 3 metal)
U	uranium	92	inner transition metal (actinide series)
Xe	xenon	54	main-group element (Group 8 noble gas)

2-10. THE PERIODIC TABLE CONTAINS SOME IRREGULARITIES

Even though the periodic table is our most important guide to chemistry, it would be overly optimistic to expect that the great diversity of the chemical reactions of 108 elements could be summarized or condensed into a single diagram. To begin with, hydrogen is unusual because it does not fit nicely into any group. It is placed sometimes in Group 1 with the alkali metals and sometimes in Group 7 with the halogens. Some versions of the periodic table place hydrogen in both groups. Hydrogen is not a metal like the Group 1 metals; yet it forms many compounds whose formulas are similar to those of the Group 1 metal compounds. For example, we have

HCl	hydrogen chloride
$NaCl$	sodium chloride
H_2S	hydrogen sulfide
Na_2S	sodium sulfide

Sodium chloride and sodium sulfide are white, crystalline solids, and hydrogen chloride and hydrogen sulfide are suffocating, toxic gases.

Hydrogen is a diatomic gas like the halogens and forms many compounds similar to the halogen compounds:

NaH	sodium hydride
$NaCl$	sodium chloride
NH_3	ammonia
NCl_3	nitrogen trichloride
CH_4	methane
CCl_4	carbon tetrachloride

This analogy is superficial, however, because hydrogen and the halogens undergo many different chemical reactions.

Another irregularity in the periodic table concerns the first element of each group. Although the members of a given group usually undergo similar chemical reactions, the first member is somewhat nontypical. For example, lithium, the first member of the alkali metals, reacts directly with nitrogen at room temperature to form lithium nitride:

$$6Li(s) + N_2(g) \rightarrow 2Li_3N(s)$$

whereas the rest of the alkali metals do not react readily with nitrogen. The Group 5 elements phosphorus, arsenic, antimony, and bismuth react directly with chlorine:

$$2P(s) + 3Cl_2(g) \rightarrow 2PCl_3(l)$$
$$2As(s) + 3Cl_2(g) \rightarrow 2AsCl_3(l)$$
$$2Sb(s) + 3Cl_2(g) \rightarrow 2SbCl_3(s)$$
$$2Bi(s) + 3Cl_2(g) \rightarrow 2BiCl_3(s)$$

but nitrogen does not react directly with Cl_2. Five of the Group 2 metals react vigorously with dilute acids, but the first one, beryllium, reacts very slowly.

2-11. MANY ATOMS FORM IONS THAT HAVE A NOBLE-GAS ELECTRON ARRANGEMENT

When Mendeleev formulated his version of the periodic table in 1869, some scientists pictured atoms as little solid spheres. However, the very existence of atoms was not accepted by all scientists at that time. It wasn't until the discovery of the electron by Thomson in 1897 and the proposal of the nuclear model of the atom by Rutherford in 1911 that the idea of the atom was generally accepted. As the structure of the atom became better understood, a great deal of research was directed toward an explanation of the periodic properties of the elements in terms of atomic structure.

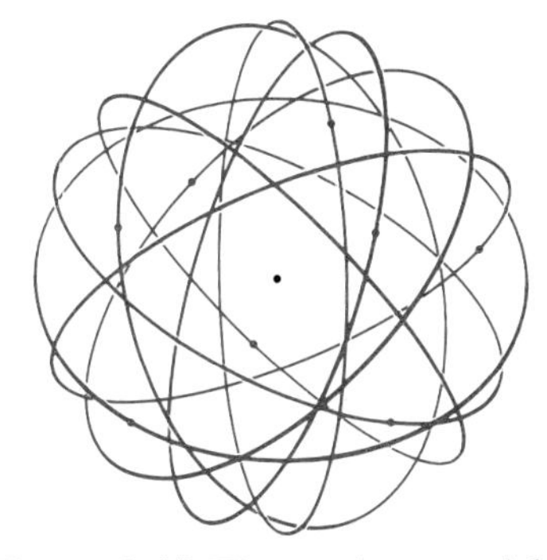

Figure 2-17 The nuclear model of the atom. The nucleus is very small and located at the center. The electrons are located in the space around the nucleus. We shall see in Chapter 7 that electrons do not travel around a nucleus in well-defined orbits, as shown in this commonly seen but rather fanciful picture of an atom.

The nuclear model of the atom pictures it as a very small, central nucleus surrounded by electrons (Figure 2-17). When two atoms collide and undergo a chemical reaction, the outermost electrons of each atom interact and are primarily responsible for the reaction that occurs. In later chapters, we learn that the electrons are arranged in an atom in an orderly fashion that is characteristic of that atom. The electron arrangement in an atom is important because it determines the chemical properties of that element. We study the connection between the chemical properties of an element and electron arrangement in Chapters 7 through 12, but even now we can deduce some important features of electron arrangement by using the periodic table as a guide.

The noble gases are unusual in that they are very nonreactive. Only a few noble-gas compounds exist. This lack of chemical reactivity suggests that the noble-gas atoms have an exceptionally stable arrangement of electrons about their nuclei. If we accept the stability of the noble gases as a working postulate, then we can account for a number of chemical properties of the elements we have discussed so far. Let's consider sodium, which we know is a very reactive metal. The atomic number of sodium is 11, and the atomic number of neon is 10. Thus, sodium follows the noble gas neon in the periodic table and has one more electron than a neon atom. If a sodium atom loses one electron, then it becomes a sodium ion, Na^+. A sodium ion has the same number of electrons around its nucleus as does a neon atom, and so we assume that a sodium ion, like a neon atom, is exceptionally stable. Thus there is a tendency, or a driving force, for a sodium atom to lose an electron. We can depict the loss of an electron by a sodium atom as

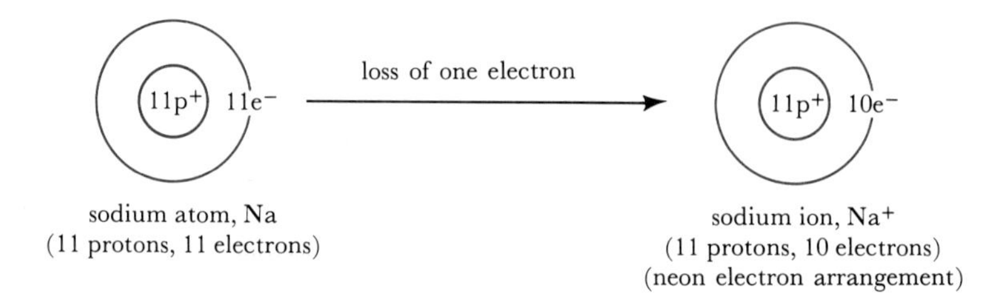

where p^+ denotes a proton and e^- denotes an electron.

In a similar manner, the other elements in Group 1 can lose one electron to become +1 ions and achieve the electron arrangement of the preceding noble gas. For example,

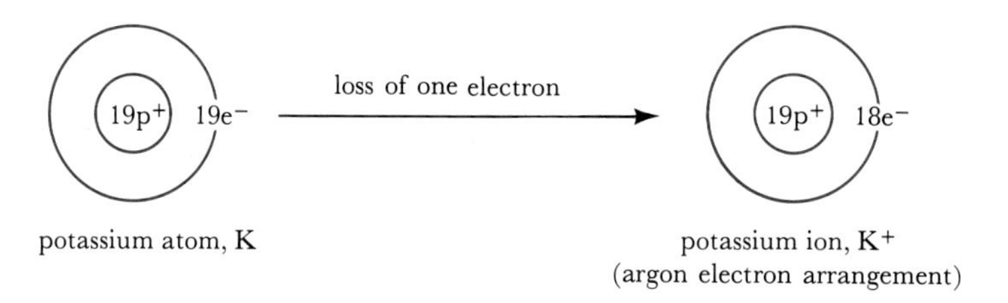

Example 2-8: Predict the charge on a calcium ion.

Solution: Calcium belongs to Group 2. A calcium atom has two more electrons than an argon atom. If it loses two electrons, then the Ca^{2+} ion achieves the relatively stable argon electron arrangement. When a calcium atom loses two electrons, the charge of the resulting calcium ion is +2:

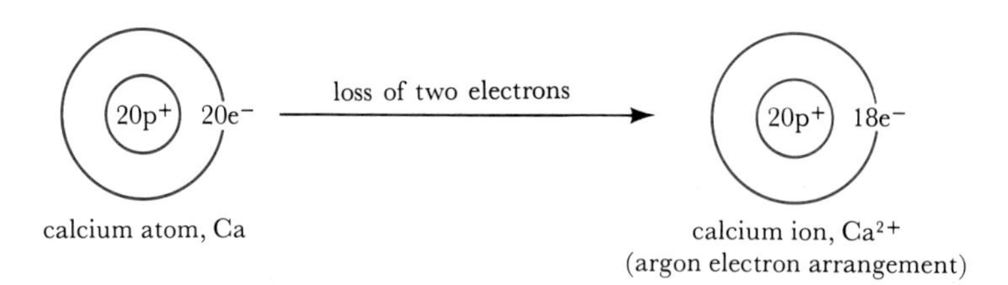

We can also conclude that the other Group 2 elements form relatively stable +2 ions.

The atoms of an element that directly precedes a noble gas in the periodic table can gain electrons to achieve the noble-gas electron arrangement. For example, fluorine directly precedes neon and so we expect

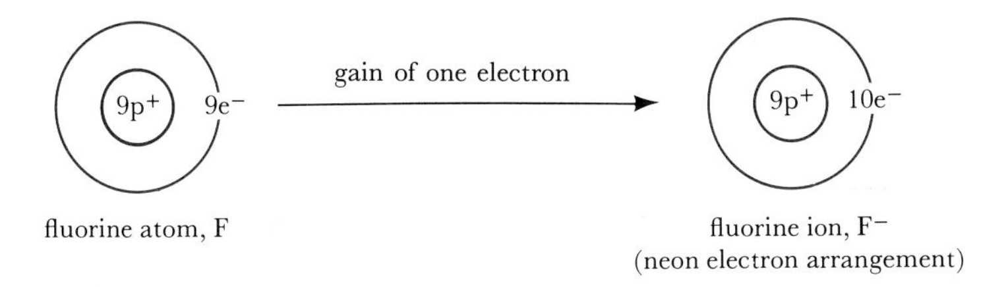

Note that each of the halogens comes right before a noble gas in the periodic table, and so we predict that all the halogen atoms gain one electron to form halide ions with a charge of −1.

Example 2-9: Predict the charge on an oxide ion.

Solution: Oxygen occurs two positions before neon in the periodic table, and so an oxygen atom has two electrons fewer than a neon atom. An oxygen atom can achieve a neon electron arrangement by gaining two electrons to become an oxide ion, written O^{2-}:

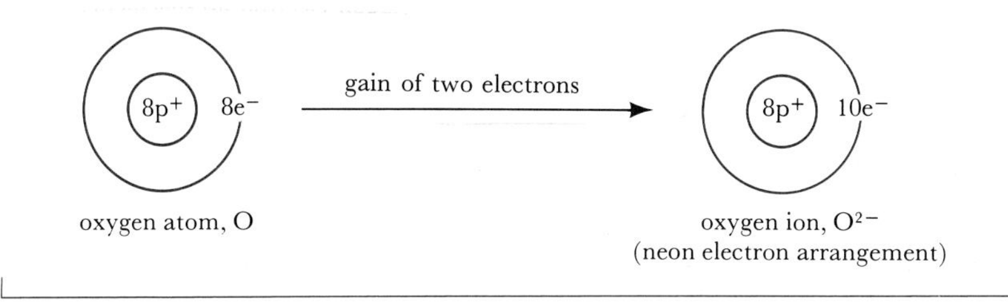

The other elements in Group 6 form −2 ions, such as a sulfide ion, S^{2-}, and a selenide ion, Se^{2-}.

2-12. IONIC CHARGES CAN BE USED TO WRITE CHEMICAL FORMULAS

You may have noticed that metal atoms lose electrons to become positive ions and nonmetal atoms gain electrons to become negative ions. This observation accounts for the chemical formulas of many of the compounds that form between the reactive metals and reactive nonmetals.

The chemical formula of potassium bromide is KBr; it is *not* KBr_2, K_2Br, or anything other than KBr. If we realize that a potassium atom readily gives up one electron and that a bromine atom readily takes on one electron, then we see that the chemical combination KBr is a natural result. We may depict the reaction by

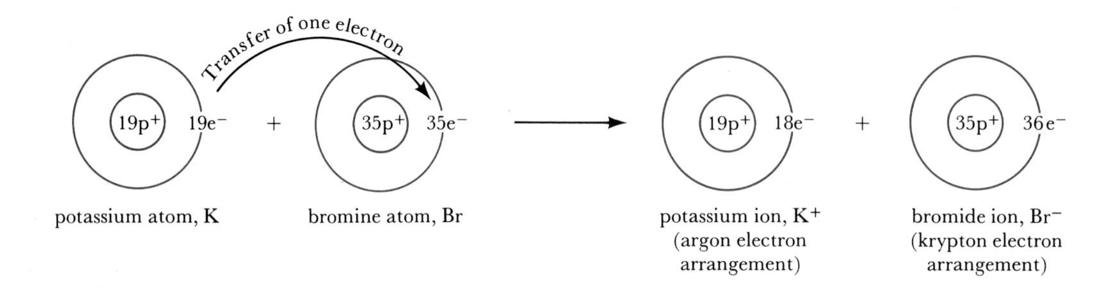

The compound KBr is called an *ionic compound* because it consists of positively charged potassium ions and negatively charged bromide ions, K^+Br^-. Note that the *formula unit,* K^+Br^-, has no net charge. It is conventional to write the formula of potassium bromide simply KBr instead of K^+Br^-.

Example 2-10: Predict the chemical formula of rubidium iodide.

Solution: Rubidium is in Group 1, and a rubidium atom loses one electron to become a rubidium ion, Rb^+, which has a krypton electron arrangement. Iodine is in Group 7, and so an iodine atom gains one electron to become an iodide ion, I^-, which has a xenon electron arrangement. The formation of rubidium iodide may be pictured as the transfer of one electron from a rubidium atom to an iodine atom, resulting in the ionic compound Rb^+I^-. Thus the chemical formula of rubidium iodide is RbI.

Example 2-11: Predict the chemical formula of calcium chloride.

Solution: Calcium belongs to Group 2, and so a calcium atom loses two electrons to form a calcium ion, Ca^{2+}, which has an argon electron arrangement. Chlorine belongs to Group 7, and so a chlorine atom gains one electron to form a chloride ion, Cl^-. A chloride ion also has an argon electron arrangement and so has no tendency to gain more electrons. Thus *two* chlorine atoms are required to take on the two electrons that a calcium atom loses. The formula unit of calcium chloride consists of one calcium ion and two chloride ions, and the chemical formula is $CaCl_2$. Note that the total positive and negative charges in the formula unit are equal and $CaCl_2$ has no net charge.

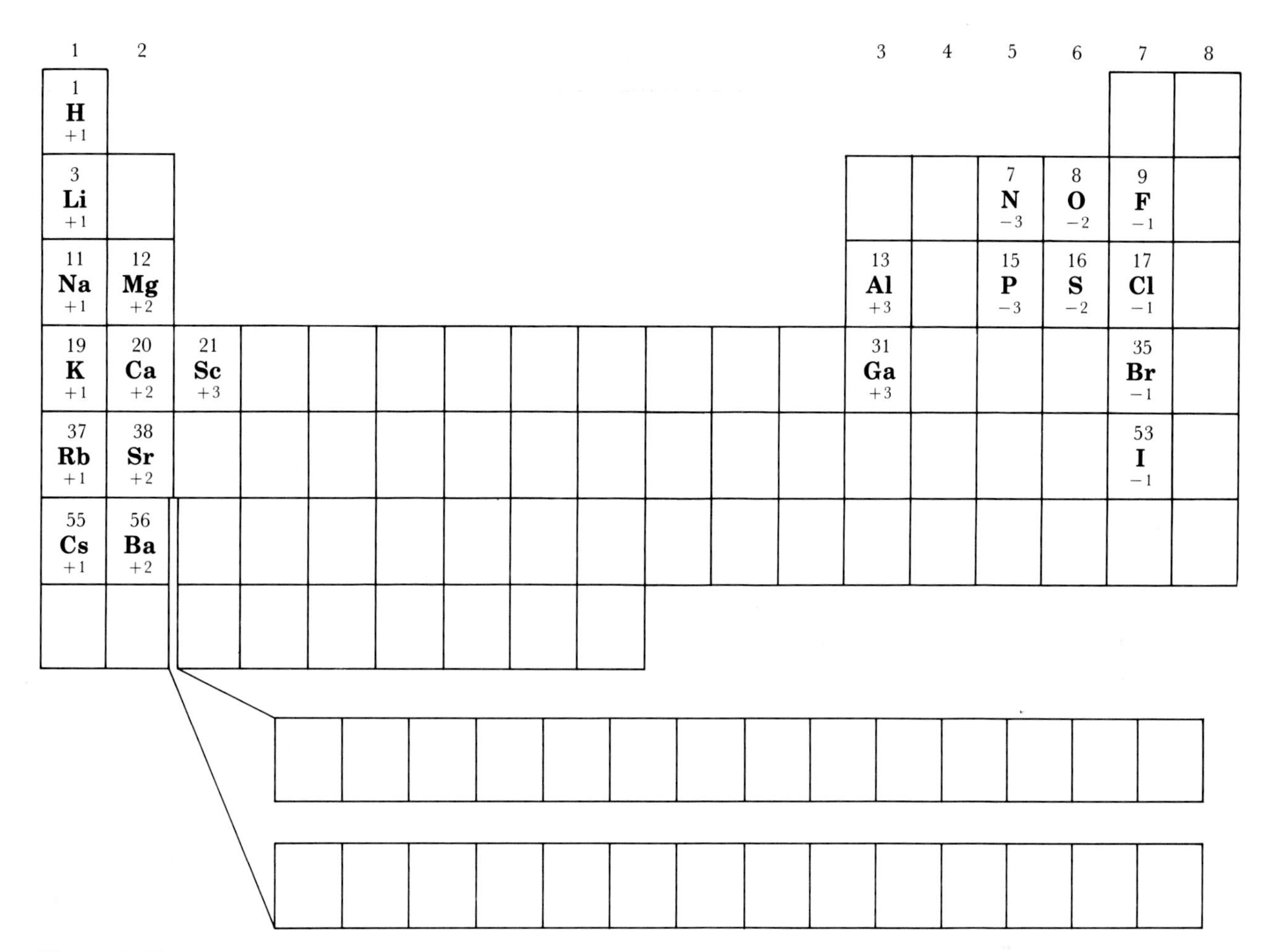

Figure 2-18 The ionic charges of some metals and nonmetals. The ionic charges of Group 1 elements are all +1, Group 2 elements are all +2, and Group 3 elements are all +3. Group 6 and 7 elements gain electrons to become negative ions, with Group 6 elements gaining two electrons to become −2 ions and Group 7 elements gaining one electron to become −1 ions. In every case depicted here, the ions have a noble-gas electron arrangement.

Using this method, we could deduce the formulas of many compounds, but it is simpler to give a set of rules for writing correct chemical formulas directly. We assign a positive or negative number, called the *ionic charge,* to an element. The ionic charges of some common elements are shown in Figure 2-18. Note that the ionic charges in Figure 2-18 are simply the charges on the ions that have a noble-gas electron arrangement. Also note the correspondence between the ionic charge of an element and its position in the periodic table.

A correct chemical formula is obtained by combining elements in Figure 2-18 such that the total positive and negative charges are equal. For example, the correct formula of calcium fluoride is CaF_2 because a calcium ion has an ionic charge of $+2$ and a fluoride ion has an ionic charge of -1. It requires two fluoride ions to balance the $+2$ ionic charge of one calcium ion. Some other chemical formulas of ionic compounds are

LiI	lithium iodide
AlF_3	aluminum fluoride
BaS	barium sulfide
Ga_2S_3	gallium sulfide
Na_2O	sodium oxide

Example 2-12: Use Figure 2-18 to write the chemical formula of (a) sodium sulfide, (b) calcium iodide, and (c) aluminum oxide.

Solution:

(a) According to Figure 2-18, the ionic charge of sodium is $+1$ and the ionic charge of sulfur is -2. We must combine sodium and sulfur into sodium sulfide in such a way that the total positive charge is equal to the total negative charge in the formula unit. Thus we combine two sodium ions and one sulfur ion, and the formula of sodium sulfide is Na_2S.

(b) The ionic charge of calcium is $+2$, and the ionic charge of iodine is -1. Consequently, we must combine one calcium ion with two iodine ions, and CaI_2 is the chemical formula of calcium iodide.

(c) The ionic charge of aluminum is $+3$, and that of oxygen is -2. We must combine two aluminum ions (total charge on two Al^{3+} ions is $2 \times (+3) = +6$) with three oxygen ions (total charge on three O^{2-} ions is $3 \times (-2) = -6$) in order to obtain a neutral formula unit for aluminum oxide. Thus Al_2O_3 is the formula of aluminum oxide. Notice that the subscript "2" on the Al in Al_2O_3 is equal to the ionic charge (without the minus sign) on oxygen and the subscript "3" on the O is equal to the ionic charge on aluminum. This can be illustrated schematically as

$$\begin{array}{c} Al_2O_3 \\ \nwarrow\!\!\nearrow \\ Al^{③+} \quad O^{②-} \end{array}$$

Problems 2-35 to 2-40 ask you to write the chemical formulas of compounds.

The use of ionic charges to write chemical formulas is limited, and we shall learn a more general method for writing chemical formulas later. Nevertheless, the ionic charges listed in Figure 2-18 can be used to write correct chemical formulas for many binary compounds. In Chapter 8, where we study the origin of these ionic charges, we shall see that a great deal of the chemistry of the elements can be understood on the basis of electron arrangement.

2-13. OXIDATION-REDUCTION REACTIONS INVOLVE A TRANSFER OF ELECTRONS

Many chemical reactions involve changes in ionic charge. Consider the reaction between sodium and sulfur to give sodium sulfide:

$$2Na(s) + S(s) \rightarrow Na_2S(s)$$

A neutral sodium atom has no charge (a charge of zero). The sodium ion in Na_2S has a charge of $+1$, and so the charge on sodium changes from 0 to $+1$ in the reaction. Similarly, the charge on sulfur changes from 0 to -2.

When the charge on an atom increases, the atom is said to be *oxidized.* When the charge on an atom decreases, the atom is said to be *reduced.* In the above reaction, sodium is oxidized and sulfur is reduced. The reaction itself is an example of an *oxidation-reduction reaction.* Notice that each sodium atom loses one electron and each sulfur atom gains two electrons; thus it requires two sodium atoms to react with each sulfur atom. The total number of electrons lost by the element that is oxidized must equal the total number of electrons gained by the element that is reduced.

Example 2-13: In the reaction

$$2Mg(s) + O_2(g) \rightarrow 2MgO(s)$$

which atom is oxidized and which is reduced?

Problems 2-45 to 2-48 deal with oxidation-reduction reactions.

Solution: Magnesium atoms are neutral, and so have no charge. An oxygen molecule is neutral, and each oxygen atom in O_2 has no charge. In MgO, the ionic charges on magnesium and oxygen are $+2$ and -2, respectively. The charge on magnesium changes from 0 to $+2$, and the charge on each oxygen changes from 0 to -2. Thus, magnesium is oxidized and oxygen is reduced in this reaction.

The concept of oxidation and reduction plays an important role in chemistry, and we shall study oxidation-reduction reactions in more depth in Chapter 20.

SUMMARY

Chemical reactions are written in terms of chemical formulas. The numbers of each kind of atom are the same on both sides of a balanced chemical equation (law of conservation of mass), although atoms are rearranged so that new substances are formed in chemical reactions.

When the elements are ordered according to atomic number, there is a repetitive, or periodic, pattern of chemical properties. These periodic patterns of chemical behavior are displayed in the periodic table of the elements, which contains all the known chemical elements. Elements are denoted as main-group elements, transition metals, or inner transition metals. Most elements are metals and appear on the left side of the periodic table. The nonmetals appear on the right side of the periodic table, and semimetals appear on the border between metals and nonmetals.

It is possible to infer from the periodic table that all Group 1 metal ions have a charge of +1 and that all halogen ions have a charge of −1; the ions of both Group 1 and Group 7 elements have a noble-gas electron arrangement. The correct formulas of many binary compounds are based on the fact that the compounds have a zero net charge; thus, the relative numbers of positive and negative ions must be adjusted such that the net charge on the formula unit is zero.

Many chemical reactions involve changes in ionic charge. When its charge increases, an atom is said to be oxidized. When its charge decreases, the atom is reduced. Such reactions are known as oxidation-reduction reactions.

TERMS YOU SHOULD KNOW

chemical reaction 44
reactant 45
product 45
balanced chemical equation 45
balancing coefficient 46
alkali metal 49
alkaline earth metal 49
halogen 51
halide 51
periodic table of the elements 54
Group 1 metal 55
Group 2 metal 55
noble gas (inert gas) 55
group (family) 58
period 58
semimetal 59
semiconductor 60
main-group element 60
transition metal 60
inner transition metal 61
lanthanide series 61
rare-earth element 61
actinide series 61
ionic compound 66
formula unit 66
ionic charge 67
oxidation 69
reduction 69
oxidation-reduction reaction 69

PROBLEMS

BALANCING EQUATIONS

2-1. Balance the following equations. (The symbol (*aq*) means that the compound is dissolved in water. The *aq* stands for *aq*ueous.)

(a) $AgClO_3(aq) + CaBr_2(aq) \rightarrow AgBr(s) + Ca(ClO_3)_2(aq)$
(b) $Ba(s) + HNO_3(aq) \rightarrow Ba(NO_3)_2(aq) + H_2(g)$
(c) $H_2SO_4(aq) + KOH(aq) \rightarrow K_2SO_4(aq) + H_2O(l)$
(d) $C_3H_8(g) + O_2(g) \rightarrow CO(g) + H_2O(g)$
(e) $CaH_2(s) + H_2O(l) \rightarrow Ca(OH)_2(aq) + H_2(g)$

2-3. A reaction in which two substances combine to produce a single product is called a *combination reaction*. Balance the equations for the following combination reactions:

(a) $NO(g) + Br_2(g) \rightarrow NOBr(g)$
(b) $Na(s) + O_2(g) \rightarrow Na_2O_2(s)$
(c) $P(s) + Br_2(l) \rightarrow PBr_3(l)$
(d) $H_2(g) + N_2(g) \rightarrow NH_3(g)$
(e) $MgO(s) + SiO_2(s) \rightarrow MgSiO_3(s)$

2-5. A reaction in which a substance burns in oxygen to form an oxide is called a *combustion reaction*. Balance the equations for the following combustion reactions:

(a) $CH_4(g) + O_2(g) \rightarrow CO_2(g) + H_2O(l)$
(b) $CO(g) + O_2(g) \rightarrow CO_2(g)$
(c) $C_3H_8(g) + O_2(g) \rightarrow CO_2(g) + H_2O(l)$
(d) $C_6H_{12}O_6(s) + O_2(g) \rightarrow CO_2(g) + H_2O(l)$
(e) $Sr(s) + O_2(g) \rightarrow SrO(s)$

2-7. Hot lithium metal is one of the most reactive substances known; it reacts vigorously with nitrogen, oxygen, water, and hydrogen and with carbon dioxide in the presence of water. Balance the following chemical equations:

(a) $Li(s) + N_2(g) \rightarrow Li_3N(s)$
(b) $Li(s) + O_2(g) \rightarrow Li_2O(s)$
(c) $Li(s) + H_2O(g) \rightarrow LiOH(s) + H_2(g)$
(d) $Li(s) + H_2(g) \rightarrow LiH(s)$
(e) $Li(s) + CO_2(g) + H_2O(g) \rightarrow LiHCO_3(s) + H_2(g)$

2-2. Balance the following equations. (The symbol (*aq*) means that the compound is dissolved in water. The *aq* stands for *aq*ueous.)

(a) $Li_3N(s) + H_2O(l) \rightarrow LiOH(s) + NH_3(g)$
(b) $Al_4C_3(s) + HCl(aq) \rightarrow AlCl_3(aq) + CH_4(g)$
(c) $H_2S(g) + NaOH(aq) \rightarrow Na_2S(aq) + H_2O(l)$
(d) $HCl(aq) + CaCO_3(s) \rightarrow CaCl_2(aq) + CO_2(g) + H_2O(l)$
(e) $CoO(s) + O_2(g) \rightarrow Co_2O_3(s)$

2-4. A reaction in which one substance breaks down into two or more substances is called a *decomposition reaction*. Balance the equations for the following decomposition reactions:

(a) $N_2H_4(g) \rightarrow NH_3(g) + N_2(g)$
(b) $GeO_2(s) \rightarrow GeO(g) + O_2(g)$
(c) $KHF_2(s) \rightarrow KF(s) + H_2(g) + F_2(g)$
(d) $H_2O_2(l) \rightarrow H_2O(l) + O_2(g)$
(e) $N_2O(g) \rightarrow N_2(g) + O_2(g)$

2-6. A reaction in which atoms in a compound are replaced by other atoms is called a *replacement reaction*. Balance the equations for the following replacement reactions:

(a) $AgNO_3(aq) + Cu(s) \rightarrow Cu(NO_3)_2(aq) + Ag(s)$
(b) $Zn(s) + HCl(aq) \rightarrow ZnCl_2(aq) + H_2(g)$
(c) $KI(aq) + Br_2(l) \rightarrow KBr(aq) + I_2(s)$
(d) $ZnS(s) + O_2(g) \rightarrow ZnO(s) + SO_2(g)$
(e) $GaBr_3(aq) + Cl_2(g) \rightarrow GaCl_3(aq) + Br_2(l)$

2-8. Balance the following chemical equations:

(a) $Ca(s) + H_2(g) \rightarrow CaH_2(s)$
(b) $S(s) + O_2(g) \rightarrow SO_3(g)$
(c) $PCl_5(s) + H_2O(l) \rightarrow H_3PO_4(l) + HCl(g)$
(d) $P_4O_{10}(s) + H_2O(l) \rightarrow H_3PO_4(l)$
(e) $Sb(s) + Cl_2(g) \rightarrow SbCl_3(s)$

2-9. Balance the following chemical equations and name the reactants and products in each case:

(a) $Al(s) + Cl_2(g) \rightarrow AlCl_3(s)$
(b) $Al(s) + O_2(g) \rightarrow Al_2O_3(s)$
(c) $COCl_2(g) + Na(s) \rightarrow CO(g) + NaCl(s)$
  phosgene
(d) $Be(s) + O_2(g) \rightarrow BeO(s)$
(e) $K(s) + S(l) \rightarrow K_2S(s)$

2-10. Balance the following equations and name the reactants and products in each case:

(a) $NaH(s) + H_2O(l) \rightarrow NaOH(s) + H_2(g)$
(b) $Li_3N(s) + D_2O(l) \rightarrow LiOD(s) + ND_3(g)$
(c) $NaN_3(s) \rightarrow Na(s) + N_2(g)$
(d) $LiD(s) + D_2O(l) \rightarrow LiOD(s) + D_2(g)$
(e) $LiOH(s) + HCl(g) \rightarrow LiCl(s) + H_2O(g)$

PREDICTION OF PRODUCTS

2-11. Using Tables 2-1 and 2-2, complete and balance the following equations:

(a) $K(s) + H_2O(l) \rightarrow$
(b) $Mg(s) + N_2(g) \rightarrow$
(c) $Ca(s) + O_2(g) \rightarrow$
(d) $Na(s) + I_2(s) \rightarrow$
(e) $Sr(s) + HCl(g) \rightarrow$

2-12. Using Tables 2-1 and 2-2, complete and balance the following equations:

(a) $K(s) + Cl_2(g) \rightarrow$
(b) $Sr(s) + S(s) \rightarrow$
(c) $Ba(s) + H_2O(g) \rightarrow$
(d) $Li(s) + H_2(g) \rightarrow$
(e) $Na(s) + Br_2(l) \rightarrow$

2-13. Using Tables 2-1 and 2-2, complete and balance the following equations:

(a) $Mg(s) + Br_2(l) \rightarrow$
(b) $Ba(s) + O_2(g) \rightarrow$
(c) $Ba(s) + S(s) \rightarrow$
(d) $Mg(s) + HCl(g) \rightarrow$
(e) $Sr(s) + H_2(g) \rightarrow$

2-14. Using Tables 2-1 and 2-2, complete and balance the following equations:

(a) $Li(s) + I_2(s) \rightarrow$
(b) $Na(s) + S(s) \rightarrow$
(c) $K(s) + H_2(g) \rightarrow$
(d) $Li(s) + F_2(g) \rightarrow$
(e) $K(s) + Br_2(l) \rightarrow$

2-15. Complete and balance the following equations:

(a) $Na(s) + H_2(g) \rightarrow$
(b) $Ba(s) + H_2(g) \rightarrow$
(c) $K(s) + F_2(g) \rightarrow$
(d) $Ba(s) + Br_2(l) \rightarrow$
(e) $Ca(s) + N_2(g) \rightarrow$

2-16. Complete and balance the following equations:

(a) $P(s) + Cl_2(g) \rightarrow$
(b) $H_2(g) + F_2(g) \rightarrow$
(c) $Sb(s) + Cl_2(g) \rightarrow$
(d) $As(s) + Br_2(l) \rightarrow$
(e) $P(s) + I_2(s) \rightarrow$

PERIODIC TABLE

2-17. When tellurium and iodine were assigned positions in the periodic table, tellurium was placed first even though its atomic mass is greater than that of iodine. Since atomic numbers were not known at this time, on what basis do you think assignment was made?

2-18. Argon had not been discovered when Mendeleev arranged the elements in order of atomic mass, but if it had been known, he would have placed it before potassium even though the atomic mass of argon is greater than that of potassium. Suggest a plausible explanation for positioning argon before potassium.

2-19. Astatine is a radioactive halogen that concentrates in the thyroid gland. Predict the following properties of astatine:

(a) physical state at 25 °C (solid, liquid, or gas) solid
(b) formula of sodium salt NaAt
(c) color of sodium salt white
(d) formula of gaseous astatine molecules At_2
(e) color of solid astatine black

2-20. Radon is a radioactive noble gas that has been used as a tracer in detecting gas leaks. Predict the following properties of radon:

(a) color
(b) odor
(c) formula of gaseous molecules
(d) reaction with water

2-21. By referring to the periodic table, classify each of the following as a main-group element, a transition metal, or an inner transition metal. If a main-group element, indicate which group and whether the element is a metal or a nonmetal:

In	Er	Ar	Y	Rh	Cf	Be
main-group metal	inner transition metal	main-group nonmetal	transition metal	transition metal	inner transition metal	main-group metal

2-22. By referring to the periodic table, classify each of the following as a main-group element, a transition metal, or an inner transition metal. If a main-group element, indicate which group and whether the element is a metal or a nonmetal:

Te P Mn Kr W Pb Ga

2-23. Without using any references, list as many elements as you can from memory and classify each as a metal or a nonmetal. Check your results and score yourself as follows:

more than 95	hall-of-famer
80 to 95	major leaguer
60 to 79	triple A player
40 to 59	semipro player
fewer than 40	little leaguer

2-24. Classify the following elements as a main-group element, a transition metal, or an inner transition metal. If a main-group element, indicate which group and whether the element is a metal or a nonmetal:

sodium carbon helium iron copper zinc

2-25. Antimony (Sb) is an extremely brittle, bluish white semimetal. It is used for making infrared devices and in alloys of lead. Predict the reaction that occurs between antimony and oxygen.

$4Sb + 3O_2 \rightarrow Sb_4O_6$

2-26. Krypton is a colorless, odorless gas present in the atmosphere. It does not react with acids, bases, halogens, alkali metals, alkaline metals, oxygen, or water. Suggest to which family of elements it belongs.

2-27. Radium is a brilliant white radioactive metal that was discovered by Pierre and Marie Curie. It was isolated from the mineral pitchblende from North Bohemia. It took 7 tons of pitchblende to recover only one g of radium. Predict the reaction of radium with

(a) oxygen $2Ra + O_2 \rightarrow 2RaO$
(b) chlorine $Ra + Cl_2 \rightarrow RaCl_2$
(c) hydrogen chloride $Ra + 2HCl \rightarrow RaCl_2 + H_2$
(d) hydrogen $Ra + H_2 \rightarrow RaH_2$
(e) sulfur $Ra + S \rightarrow RaS$

2-28. A radioactive isotope of strontium, $^{90}_{38}Sr$, is produced in nuclear explosions and is one of the dangerous components of nuclear fallout. Using the periodic table, suggest a reason for its impact on living organisms.

IONIC CHARGES

2-29. Determine the number of electrons in each of the following:

(a) Fe^{2+} 24
(b) Se^{2-} 36
(c) N^{3-} 10
(d) Ca^{2+} 18
(e) O^{2-} 10

2-30. Determine the number of electrons in each of the following:

(a) Mn^{2+}
(b) P^{3-}
(c) Cs^{+}
(d) La^{3+}
(e) I^{-}

2-31. Indicate which of the following ions have a noble-gas electron arrangement and, for such ions, identify the corresponding noble gas:

(a) K^{+} yes, argon
(b) Sr^{+} no
(c) O^{-} no
(d) N^{3-} yes, neon
(e) Al^{3+} yes, neon
(f) I^{-} yes, xenon

2-32. Indicate which of the following ions have a noble-gas electron arrangement and, for such ions, identify the corresponding noble gas:

(a) Ba^{2+}
(b) Ca^{+}
(c) Br^{-}
(d) P^{3-}
(e) Se^{2-}
(f) Fe^{2+}

2-33. Determine the charge on each atom in the following compounds:

(a) K_2S K^{+} S^{2-}
(b) AlN Al^{3+} N^{3-}
(c) AlF_3 Al^{3+} F^{-}
(d) CsI Cs^{+} I^{-}
(e) MgSe Mg^{2+} Se^{2-}

2-34. Determine the charge on each atom in the following compounds:

(a) Li_2O
(b) CaS
(c) Mg_3N_2
(d) Al_2S_3
(e) KI

CHEMICAL FORMULAS

2-35. Write the chemical formula for

(a) gallium oxide Ga_2O_3
(b) aluminum chloride $AlCl_3$
(c) lithium oxide Li_2O
(d) hydrogen bromide HBr
(e) strontium iodide SrI_2

2-36. Write the chemical formula for

(a) aluminum sulfide
(b) sodium oxide
(c) gallium bromide
(d) barium fluoride
(e) potassium iodide

2-37. Write the chemical formulas for the following compounds:

(a) lithium nitride Li_3N
(b) rubidium iodide RbI
(c) gallium sulfide Ga_2S_3
(d) barium oxide BaO
(e) magnesium iodide MgI_2

2-38. Write the chemical formulas for the following compounds:

(a) cesium oxide
(b) sodium selenide
(c) strontium sulfide
(d) lithium sulfide
(e) calcium iodide

2-39. Write the formula of the binary compound formed from each of the following pairs of ions:

(a) Ga^{3+} and O^{2-} Ga_2O_3
(b) Zn^{2+} and I^{-} ZnI_2
(c) Fe^{2+} and S^{2-} FeS
(d) Ru^{3+} and Cl^{-} $RuCl_3$
(e) Ag^{+} and S^{2-} Ag_2S

2-40. Write the formula of the binary compound formed from each of the following pairs of ions:

(a) Pt^{4+} and F^{-}
(b) $Au^{3+}$and O^{2-}
(c) Fe^{3+} and Se^{2-}
(d) Ba^{2+} and At^{-}
(e) Zn^{2+} and N^{3-}

WRITING FORMULAS AND BALANCING EQUATIONS

2-41. For each of the following reactions, write the chemical formulas and balance the equation:

(a) potassium + water → potassium hydroxide + hydrogen
(b) potassium hydride + water → potassium hydroxide + hydrogen
(c) silicon dioxide + carbon → silicon carbide + carbon monoxide
(d) silicon dioxide + hydrogen fluoride → silicon tetrafluoride + water
(e) phosphorus + chlorine → phosphorus trichloride

2-42. For each of the following combination reactions, write the chemical formulas and balance the equation:

(a) sodium + sulfur → sodium sulfide
(b) calcium + bromine → calcium bromide
(c) barium + oxygen → barium oxide
(d) sulfur dioxide + oxygen → sulfur trioxide
(e) magnesium + nitrogen → magnesium nitride

2-43. For each of the following reactions, write the chemical formulas and balance the equations:

(a) sodium + hydrogen → sodium hydride
(b) aluminum + sulfur → aluminum sulfide
(c) steam + carbon → carbon monoxide + hydrogen
(d) carbon + hydrogen → methane
(e) phosphorus trichloride + chlorine → phosphorus pentachloride

2-44. For each of the following reactions, write the chemical formulas and balance the equations:

(a) carbon monoxide + oxygen → carbon dioxide
(b) cesium + bromine → cesium bromide
(c) nitrogen monoxide + oxygen → nitrogen dioxide
(d) ammonia + oxygen → nitrogen monoxide + water
(e) gallium + arsenic → gallium arsenide

OXIDATION AND REDUCTION

2-45. Indicate which element is oxidized and which is reduced in the following reactions:

(a) $Ca(s) + Cl_2(g) \rightarrow CaCl_2(s)$
(b) $4Al(s) + 3O_2(g) \rightarrow 2Al_2O_3(s)$
(c) $2Rb(s) + Br_2(l) \rightarrow 2RbBr(s)$
(d) $2Na(s) + S(s) \rightarrow Na_2S(s)$

2-46. Indicate which element is oxidized and which is reduced in the following reactions:

(a) $2Li(s) + Se(s) \rightarrow Li_2Se(s)$
(b) $2Sc(s) + 3I_2(s) \rightarrow 2ScI_3(s)$
(c) $Ga(s) + P(s) \rightarrow GaP(s)$
(d) $2K(s) + F_2(g) \rightarrow 2KF(s)$

2-47. In the oxidation-reduction reactions in Problem 2-45, indicate how many electrons are transferred in the formation of one formula unit of product.

2-48. In the oxidation-reduction reactions in Problem 2-46, indicate how many electrons are transferred in the formation of one formula unit of product.

INTERCHAPTER A

Oxygen

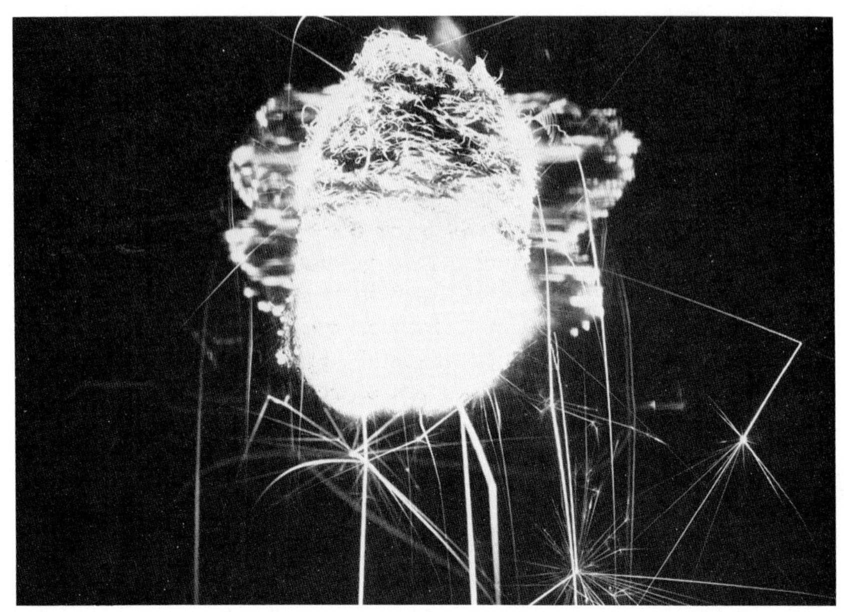

Although steel wool does not burn in air, it burns vigorously in pure oxygen.

Oxygen (atomic number 8, atomic mass 15.9994) is the most abundant element on earth and the third most abundant element in the universe, ranking behind only hydrogen and helium. Most rocks contain a large amount of oxygen. For example, sand is predominantly silicon dioxide (SiO_2) and consists of more than 50 percent oxygen by mass. Almost 90 percent of the mass of the oceans and two-thirds of the mass of the human body are oxygen. Air is 21 percent oxygen by volume. We can live weeks without food, days without water, but only minutes without oxygen.

A-1. THIRTY-FIVE BILLION POUNDS OF OXYGEN ARE SOLD ANNUALLY IN THE UNITED STATES

Oxygen in air exists as the diatomic molecule O_2. It is a colorless, odorless, tasteless gas with a boiling point of $-183°C$ and a freezing point of $-218^\circ C$. Although oxygen is colorless in the gas state, both liquid and solid oxygen are pale blue.

Industrially, oxygen is produced by the fractional distillation of liquid air (Interchapter B), a method which exploits the difference in the boiling points of nitrogen and oxygen, the principal components of air. The nitrogen can be separated from the oxygen because nitrogen boils at $-196^\circ C$, whereas oxygen boils at $-183^\circ C$. The pure oxygen thus obtained is compressed in steel cylinders to a pressure of about 150 atmospheres. Approximately 35 billion pounds of oxygen are sold annually in the United States, making it the fifth most important industrial chemical. The major commercial use of oxygen is in the blast furnaces used to manufacture steel. (To save transportation costs, the oxygen is produced on site.) Oxygen is also used in hospitals, in oxyhydrogen and oxyacetylene torches for welding metals, at athletic events, and to facilitate breathing at high altitudes and under water. Tremendous quantities of oxygen are used directly from air as a reactant in the combustion of hydrocarbon fuels, which supply 93 percent of the energy consumed in the United States. In terms of total usage (pure oxygen and oxygen used directly from air), oxygen is the number two chemical, ranking behind only water.

A-2. OXYGEN IN THE EARTH'S ATMOSPHERE IS PRODUCED BY PHOTOSYNTHESIS

Most of the oxygen in the atmosphere is the result of *photosynthesis,* the process by which green plants combine $CO_2(g)$ and $H_2O(g)$ into carbohydrates and $O_2(g)$ under the influence of visible light. The carbohydrates appear in the plants as starch, cellulose, and sugars. The overall reaction is

$$CO_2(g) + H_2O(g) \xrightarrow{\text{visible light}} \text{carbohydrate} + O_2(g)$$

Although many details of photosynthesis have yet to be worked out, it is known that the light is first absorbed by the chlorophyll molecules of the plants. These green chlorophyll pigments are located in the plant cells in special compartments called *chloroplasts.* In one year, more than 10^{10} metric tons of carbon is incorporated into carbohydrates by photosynthesis. It is well established that in the hundreds of millions of years that plant life has existed on earth, photosynthesis has produced much more oxygen than the amount now present in the atmosphere. Whether enough has been produced to account for all the oxygen in the earth's crust has not been established.

Photosynthesis is an active area of chemical research.

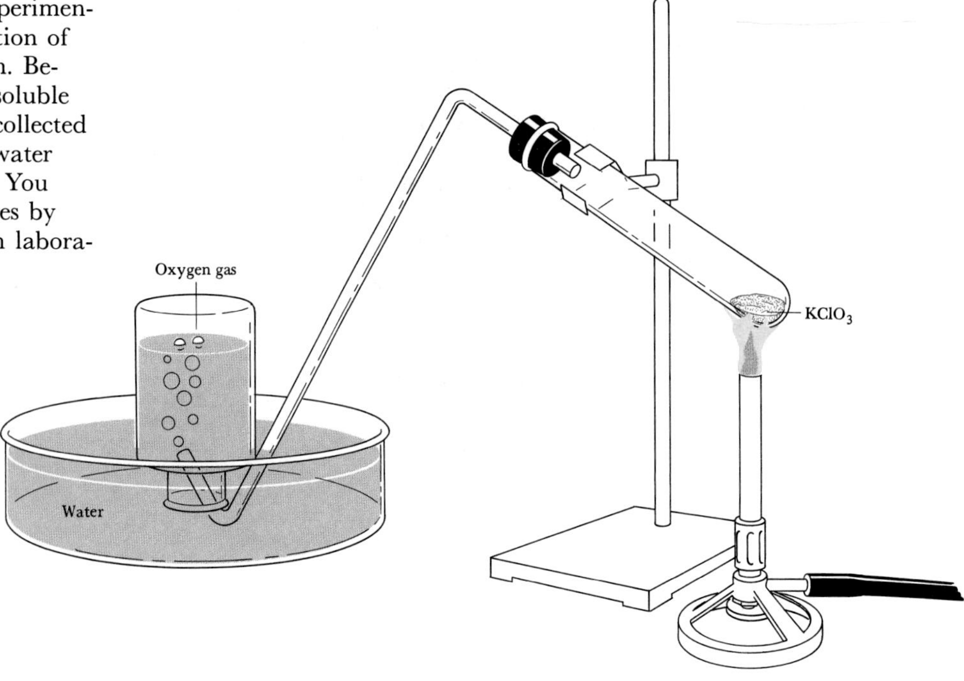

Figure A-1 A typical experimental setup for the production of small amounts of oxygen. Because it is only slightly soluble in water, the oxygen is collected by the displacement of water from an inverted bottle. You probably will collect gases by this method in your own laboratory work.

A-3. OXYGEN CAN BE PREPARED IN THE LABORATORY BY VARIOUS METHODS

The most frequently used method for preparing oxygen in the laboratory involves the thermal decomposition of potassium chlorate, $KClO_3$ (Figure A-1). The chemical equation for the reaction is

$$2KClO_3(s) \rightarrow 2KCl(s) + 3O_2(g)$$

This reaction requires a fairly high temperature (400°C), but if a small amount of manganese dioxide, MnO_2, is added, the reaction occurs rapidly at a lower temperature (250°C). The manganese dioxide speeds up the reaction, yet is not a reactant itself. We shall meet examples of this behavior often. A substance that facilitates a reaction and yet is not consumed in the reaction is called a *catalyst*.

Catalysts are discussed in Chapter 16.

The production of oxygen by the thermal decomposition of $KClO_3$ can result in an explosion, particularly if the reaction mixture contains any organic impurities. An alternate method for the laboratory preparation of oxygen is to add sodium peroxide, Na_2O_2, to water:

In this equation *aq* means *aq*ueous.

$$2Na_2O_2(s) + 2H_2O(l) \rightarrow 4NaOH(aq) + O_2(g)$$

This rapid and convenient reaction does not require heat. However,

Figure A-2 Apparatus used for the electrolysis of water. A voltage is applied across a pair of metal electrodes placed in water containing a small amount of sulfuric acid. Hydrogen gas is evolved at one electrode and oxygen gas at the other. Notice that the volume of hydrogen produced is twice the volume of oxygen produced.

sodium peroxide in contact with organic matter (such as a laboratory towel) presents a significant fire hazard.

Oxygen can be prepared by passing an electric current through water in an apparatus like the one shown in Figure A-2. The decomposition of a substance by an electric current is called *electrolysis*. The chemical equation for the electrolytic decomposition of water is

$$2H_2O(l) \xrightarrow{\text{electrolysis}} 2H_2(g) + O_2(g)$$

The energy necessary to decompose water into hydrogen and oxygen by electrolysis is derived from the applied voltage. Electrolysis is described in more detail in Chapter 21.

A-4. OXYGEN REACTS DIRECTLY WITH MOST OTHER ELEMENTS

Oxygen is a very reactive element. It reacts directly with all the other elements except the halogens, the noble gases, and some of the less reactive metals to form a wide variety of compounds. Only fluorine reacts with more elements than oxygen. Compounds containing oxygen constitute 31 of the top 50 industrial chemicals. The ranking of the first few oxygen-containing compounds, in terms of quantities sold, is shown in Table A-1.

Table A-1. Major oxygen-containing compounds sold in the United States

Rank	*Compound*	*Name*	*Annual U.S. production/ billions of pounds*
1	H_2SO_4	sulfuric acid	85
3	CaO	calcium oxide (lime)	36
5	O_2	oxygen	35
7	$NaOH$	sodium hydroxide	22
9	H_3PO_4	phosphoric acid	20
10	HNO_3	nitric acid	18
11	NH_4NO_3	ammonium nitrate	18
12	Na_2CO_3	sodium carbonate	17
13	H_2NCONH_2	urea	15
18	CH_3OH	methanol	9
19	CO_2	carbon dioxide	8

Source: Chemical and Engineering News, June 14, 1982, p. 33.

Oxygen forms oxides with many elements. Most metals react rather slowly with oxygen at ordinary temperatures but react more rapidly as the temperature is increased. For example, iron, in the form of steel wool, burns vigorously in pure oxygen but does not burn in air (Frontispiece).

Fuels are compounds that burn in oxygen and release large quantities of heat. (Such a reaction is called a *combustion reaction.*) Many fuels are *hydrocarbons,* which are compounds that contain only carbon and hydrogen. Methane, the main constituent of natural gas, burns in oxygen according to the equation

$$\underset{\text{methane}}{CH_4(g)} + 2O_2(g) \rightarrow CO_2(g) + 2H_2O(g)$$

All hydrocarbons burn in oxygen to give carbon dioxide and water. Gasoline is a mixture of hydrocarbons. Using octane, C_8H_{18}, as a typical hydrocarbon in gasoline, we write the combustion of gasoline as

$$\underset{\text{octane}}{2C_8H_{18}(l)} + 25O_2(g) \rightarrow 16CO_2(g) + 18H_2O(g)$$

The energy released in this reaction is used to power machinery and to produce electricity (Chapter 6).

A mixture of acetylene and oxygen is burned in the oxyacetylene torch. The chemical equation for the combustion of acetylene is

$$\underset{\text{acetylene}}{2C_2H_2(g)} + 5O_2(g) \rightarrow 4CO_2(g) + 2H_2O(g)$$

The flame temperature of an oxyacetylene welding torch is about 2400°C, which is sufficient to melt iron and steel. A combustion reaction with which we are all familiar is the burning of a candle. The wax in a candle is composed of long-chain hydrocarbons, such as $C_{20}H_{42}$. The molten wax rises up the wick to the combustion zone the way ink rises in a piece of blotting paper.

Different flames have different temperatures. The temperature of a hot region of a candle flame is about 1200°C, that of a Bunsen burner flame is about 1800°C, and that of the flame of an oxyacetylene torch is about 2400°C.

A-5. SOME METALS REACT WITH OXYGEN TO YIELD PEROXIDES

Although most metals yield oxides when they react with oxygen, some of the more reactive metals, such as sodium, potassium, and rubidium, yield *peroxides* and *superoxides*. Peroxides are compounds in which the negative ion is the peroxide ion, O_2^{2-}, and superoxides are compounds in which the negative ion is the superoxide ion, O_2^-. For example,

$$2Na(s) + O_2(g) \rightarrow \underset{\text{sodium peroxide}}{Na_2O_2(s)}$$

$$K(s) + O_2(g) \rightarrow \underset{\text{potassium superoxide}}{KO_2(s)}$$

One of the most important peroxides is hydrogen peroxide, H_2O_2, a colorless, syrupy liquid that explodes violently when heated. Dilute aqueous solutions of hydrogen peroxide are fairly safe to use. A 3% aqueous solution is sold in drugstores and used as a mild antiseptic and as a bleach (Figure A-3). It is sold in brown bottles because hydrogen peroxide decomposes in light according to the reaction

$$2H_2O_2(aq) \xrightarrow{\text{light}} 2H_2O(l) + O_2(g)$$

More concentrated solutions (30%) of hydrogen peroxide are used industrially. Some of the industrial applications of hydrogen peroxide are as a bleaching agent for feathers, hair, flour, bone, and textile fibers; in renovating old paintings and engravings; in the artificial aging of wines and liquor; in refining oils and fats; and in photography as a fixative eliminant. Concentrated solutions of hydrogen peroxide are extremely corrosive and explosive and must be handled with great care.

BAKER SAF-T-DATA™ System

HEALTH 2 MODERATE | FLAMMABILITY 0 NONE | REACTIVITY 3 SEVERE | CONTACT 4 EXTREME

LABORATORY PROTECTIVE EQUIPMENT

GOGGLES & SHIELD | LAB COAT & APRON | VENT HOOD | PROPER GLOVES

STORAGE COLOR: YELLOW

DANGER!

STRONG OXIDIZER. CAUSES BURNS. MAY CAUSE EYE INJURY.
EFFECTS MAY BE DELAYED.

Do not get in eyes, on skin, on clothing. Avoid contact with combustible materials. Drying of product on clothing or combustible materials may cause fire. Avoid contamination from any source, metals, dust, and organic materials that may cause rapid decomposition, generation of large quantities of oxygen gas and high pressure. In case of fire involving product, use water only. Flush spills or leaks away by flooding with water applied quickly to entire spill or leak.
FIRST AID: Call a Physician. In case of contact, immediately flush eyes or skin with plenty of water for at least 15 minutes. Remove and wash contaminated clothing and shoes promptly and thoroughly.

DOT Description: Hydrogen Peroxide, Solution UN 2014 401-2300
IMO Description: Hydrogen Peroxide
CAS NO: 7722-84-1 Material Safety Data Sheet Available EPA-HW: Ignitables

NFPA 0 2 1 OXY

© J.T. Baker Chemical Co.
Phillipsburg, NJ 08865

J.T.Baker®

4 L 2186-3

Hydrogen Peroxide, 30%
(Superoxol®)

H_2O_2 (Contains ~1 ppm $Na_2SnO_3 \cdot 3H_2O$ as a Preservative) FW 34.0

'BAKER ANALYZED'® Reagent

ACTUAL ANALYSIS, LOT 316845 MEETS A.C.S. SPECIFICATIONS

Assay (H_2O_2)(by $KMnO_4$ titrn)		31.3	%
Color (APHA)	<	5	
Residue after Evaporation		0.0008	%
Titrable Acid		0.0001	meq/g
Trace Impurities (in ppm):			
Chloride (Cl)	<	1	
Nitrate (NO_3)		2	
Phosphate (PO_4)	<	1	
Sulfate (SO_4)	<	5	
Ammonium (NH_4)	<	3	
Heavy Metals (as Pb)	<	0.3	
Iron (Fe)		0.002	

USE BEFORE: 10/84 DATE OF MANUFACTURE: 4/83

Figure A-3 The labels on reagent bottles carry information concerning the handling of the reagents. Here we see that 30% hydrogen peroxide is very reactive and corrosive.

A-6. OZONE IS A TRIATOMIC OXYGEN MOLECULE

When a spark is passed through oxygen, some of the oxygen is converted to ozone, O_3:

$$3O_2(g) \rightarrow 2O_3(g)$$

Ozone is a light blue gas at room temperature. It has a sharp, characteristic odor that occurs after electrical storms or near high-voltage generators. Liquid ozone (boiling point $-112°C$) is a deep blue, explosive liquid. Ozone is so reactive that it cannot be transported, but must be generated as needed. Relatively unreactive metals, such as silver and mercury, which do not react with oxygen, react with ozone to form oxides. Ozone is used as a bleaching agent and is being considered as a replacement for chlorine in water treatment because of the environmental problem involving chlorinated hydrocarbons.

Oxygen and ozone are called *allotropes*. Allotropes are two different forms of an element that have a different number or arrangement of the atoms in the molecules. Many other elements have allotropic forms. For example, graphite and diamond are allotropes. They both consist of carbon atoms, but the atoms are arranged differently in the two substances (Section 13-12).

Graphite and diamond are two forms of elemental carbon.

Ozone plays a vital role in the earth's atmosphere. The action of sunlight on oxygen in the upper atmosphere leads to the production of ozone:

$$O_2(g) \xrightarrow{\text{sunlight}} 2O(g)$$

$$O_2(g) + O(g) \rightarrow O_3(g)$$

The ozone produced in the upper atmosphere absorbs the ultraviolet radiation from sunlight that would otherwise destroy most life on earth. Without ozone in the upper atmosphere, there could be no life as we know it on earth. Interchapter E deals with the earth's atmosphere and describes the role that ozone plays in screening us from the sun's ultraviolet radiation.

QUESTIONS

A-1. Sand is predominantly what oxide?

A-2. How is oxygen produced commercially?

A-3. What is the source of most of the oxygen in the earth's atmosphere?

A-4. What is the overall reaction of photosynthesis?

A-5. Give two methods used to produce small quantities of oxygen in the laboratory.

A-6. What is a catalyst?

A-7. What is combustion?

A-8. What is the reaction for the combustion of the principal component of natural gas?

A-9. What is the heat-producing reaction of an oxyacetylene torch?

A-10. Describe the chemical process involved in the burning of a candle.

A-11. What is an allotrope?

A-12. What are the two allotropes of oxygen?

Starting at the upper left and going in a clockwise direction, we have molar quantities of acetylsalicylic acid (aspirin) (180 g), water (18.0 g), sucrose (table sugar) (342 g), mercury (201 g), iron (55.9 g), sodium chloride (58.5 g) and iodine (254 g). One mole of a substance is that quantity containing the number of grams numerically equal to the formula mass.

3 / Chemical Calculations

It is important to be able to calculate how much of a reaction product can be obtained from a given amount of reactant or how much reactant should be used to obtain a desired amount of product. In this chapter, we develop a systematic method for carrying out a variety of chemical calculations. For example, we shall determine simplest chemical formulas using mass percentages obtained from chemical analysis. We'll combine chemical analysis data with molecular mass information to determine molecular formulas. After discussing the determination of chemical formulas, we'll treat calculations involving chemical reactions. For example, we'll calculate how much of a compound can be prepared starting with a given amount of an element contained in that compound.

3-1. THE QUANTITY OF A SUBSTANCE THAT IS EQUAL TO ITS FORMULA MASS IN GRAMS IS CALLED A MOLE

In Chapter 1 we learned that the atomic mass of an element is a relative quantity; it is the mass of one atom of the element relative to the mass of one atom of carbon-12, which has a mass of exactly 12 amu. Consider these four elements:

Element	*Relative atomic mass*
helium, He	4
carbon, C	12
titanium, Ti	48
molybdenum, Mo	96

This table shows that one carbon atom has a mass three times that of one helium atom, one titanium atom has a mass four times that of one carbon atom, and one molybdenum atom has a mass twice that of one titanium atom, eight times that of one carbon atom, and 24 times that of one helium atom. It is important to realize that we have not deduced the mass of any one atom; at this point we can determine only relative masses.

Consider 12 g of carbon and 48 g of titanium. One titanium atom has a mass four times that of one carbon atom, and therefore 48 g of titanium atoms must contain the same number of atoms as 12 g of carbon. Similarly, one molybdenum atom has twice the mass of one titanium atom, and so 96 g of molybdenum must contain the same number of atoms as 48 g of titanium. We conclude that 12 g of carbon, 48 g of titanium, and 96 g of molybdenum all contain the same number of atoms. If we continue this line of reasoning, then we find that the quantity of any element that is numerically equal to its atomic mass in grams contains the same number of atoms as the corresponding quantity of any other element. Thus, 10.8 g of boron, 23.0 g of sodium, 63.6 g of copper, and 200.6 g of mercury all contain the same number of atoms.

All the substances we have considered so far are *atomic substances,* that is, substances whose constituent particles are atoms. Now consider the following *molecular substances:*

Substance	*Relative molecular mass*
methane, CH_4	$12.0 + (4 \times 1.0) = 16.0$
oxygen, O_2	$2 \times 16.0 = 32.0$
ozone, O_3	$3 \times 16.0 = 48.0$

Recall from Chapter 1 that the molecular mass of a substance is the sum of the atomic masses in the chemical formula. Like atomic masses, molecular masses are relative. A molecule of oxygen, O_2, has a mass (32.0 amu) twice that of a molecule of methane (16.0 amu). A molecule of ozone has a mass (48.0 amu) three times that of a molecule of methane. Using the same reasoning we used for atomic substances, we conclude that 16.0 g of methane, 32.0 g of oxygen, and 48.0 g of ozone must all contain the same number of *molecules.* In addition, because the atomic mass of titanium is equal to the molecular mass of ozone, the number of *atoms* in 48.0 g of titanium must equal the number of *molecules* in 48.0 g of ozone.

By using the term *formula mass* instead of atomic mass or molecular mass, we can combine all the above statements and say that *one formula mass in grams of any substance contains the same number of formula units as the corresponding quantity of any other substance.* Thus, 4.0 g of helium, 12.0 g of carbon, 16.0 g of methane, and 32.0 g of oxygen all

A formula unit is the collection of atoms or ions indicated by the chemical formula.

contain the same number of formula units. The formula units are atoms in the cases of helium and carbon and molecules in the cases of methane and oxygen. We can now state our preliminary definition of a mole: *the quantity of a substance that is equal to its formula mass in grams is called a mole* (abbreviated *mol*). The frontispiece to this chapter shows one mole of a number of common substances.

The mole is one of the most important concepts in chemistry.

As you perform chemical calculations, you will often need to determine the number of moles in a given mass of a substance. For example, we might need to calculate the number of moles of methane in 50.0 g of methane. The formula mass of CH_4 is 16.0, and so there are 16.0 g per mole of methane. To determine how many moles there are in 50.0 g, we must convert grams of methane to moles of methane. We carry out this conversion using the unit conversion factor

$$1 = \frac{1 \text{ mol } CH_4}{16.0 \text{ g } CH_4}$$

Thus

$$\text{mol } CH_4 = (50.0 \text{ g } CH_4)\left(\frac{1 \text{ mol } CH_4}{16.0 \text{ g } CH_4}\right) = 3.13 \text{ mol } CH_4$$

Notice that we express our result to three significant figures and that the answer is expressed in moles of CH_4, as required.

You should also be able to calculate the mass of a certain number of moles of a substance. For example, let's calculate the mass of 2.16 mol of sodium chloride. The formula mass of NaCl is 58.44. The mass of NaCl in 2.16 mol is

$$\text{g of NaCl} = (2.16 \text{ mol NaCl})\left(\frac{58.44 \text{ g NaCl}}{1 \text{ mol NaCl}}\right) = 126 \text{ g NaCl}$$

Notice that we express the final result to three significant figures and that the units yield grams of NaCl.

Example 3-1: Calculate the number of moles in (a) 15.0 g of argon, (b) 1.60 mg of nitrogen, and (c) 3.00 kg of ammonia.

Solution: (a) Argon exists as a monatomic gas whose formula mass is 40.0. Consequently, the number of moles of argon in 15.0 g is

$$\text{mol Ar} = (15.0 \text{ g Ar})\left(\frac{1 \text{ mol Ar}}{40.0 \text{ g Ar}}\right) = 0.375 \text{ mol Ar}$$

(b) Nitrogen exists as a diatomic molecule, and so the formula mass is $2 \times 14.0 = 28.0$. The number of moles of N_2 in 1.60 mg is

$$\text{mol } N_2 = (1.60 \text{ mg } N_2)\left(\frac{10^{-3}\text{ g}}{1\text{ mg}}\right)\left(\frac{1\text{ mol } N_2}{28.0\text{ g } N_2}\right) = 5.71 \times 10^{-5}\text{ mol } N_2$$

Notice that we have to convert milligrams to grams before dividing by 28.0 g.

(c) The chemical formula of ammonia is NH_3, and so its formula mass is $14.0 + (3 \times 1.0) = 17.0$. The number of moles of ammonia contained in 3.00 kg is

$$\text{mol } NH_3 = (3.00\text{ kg } NH_3)\left(\frac{10^3\text{ g}}{1\text{ kg}}\right)\left(\frac{1\text{ mol } NH_3}{17.0\text{ g } NH_3}\right) = 176\text{ mol } NH_3$$

Notice that we have to convert kilograms to grams before dividing by 17.0 g.

Problems 3-3 to 3-6 are similar to Example 3-1.

Example 3-1 illustrates an important point. *In order to calculate the number of moles in a given mass of a chemical compound, it is necessary to know the chemical formula of the compound. A mole of any compound is defined only in terms of its chemical formula.*

3-2. ONE MOLE OF ANY SUBSTANCE CONTAINS AVOGADRO'S NUMBER OF FORMULA UNITS

It has been determined experimentally that 1 mol of any substance contains 6.022×10^{23} formula units. This number is called *Avogadro's number* after Amedeo Avogadro, an Italian scientist who was one of the earliest proponents of the atomic theory. Not only do we say that 1 mol of any substance contains Avogadro's number of formula units, but another widely used definition of a mole is that 1 mol is that mass of a substance that contains Avogadro's number of formula units or "elementary entities." For example, the molecular mass of water is 18.02, and thus 18.02 g of water contains 6.022×10^{23} molecules.

Avogadro's number is an enormous number, much larger than any number you may have ever encountered. In order to appreciate the magnitude of Avogadro's number, let's compute how many years it would take to spend Avogadro's number of dollars at the rate of one million dollars per second. There are 3.15×10^7 s in 1 year; thus the number of years required is

Avogadro's number is 6.022×10^{23} formula units/mole.

$$(6.022 \times 10^{23}\text{ dollars})\left(\frac{1\text{ s}}{10^6\text{ dollars}}\right)\left(\frac{1\text{ year}}{3.15 \times 10^7\text{ s}}\right) = 1.91 \times 10^{10}\text{ years}$$

or 19.1 billion years (1 billion = 10^9). This is over four times longer than the estimated age of the earth (4.6 billion years) and roughly equal to the conjectured age of the universe (13 to 20 billion years). This illustrates how large Avogadro's number is and, consequently, how small atoms and molecules are. Look again at the Frontispiece to this chapter. Each of those samples contains 6.022×10^{23} formula units of the indicated substance.

We can use Avogadro's number to calculate the mass of one atom or molecule.

Example 3-2: Using Avogadro's number, calculate the mass of (a) one nitrogen atom and (b) one nitrogen molecule.

Solution: (a) The formula mass, or atomic mass, of nitrogen is 14.01 g, and so 1 mol of atomic nitrogen has a mass of 14.01 g. There are 6.022×10^{23} nitrogen atoms in 1 mol (14.01 g) of atomic nitrogen. The mass of one nitrogen atom is given by

$$\begin{aligned}\text{mass of one nitrogen atom} &= \frac{\text{mass of Avogadro's number of nitrogen atoms}}{\text{Avogadro's number}} \\ &= \frac{\text{mass corresponding to 1 mol of N}}{\text{Avogadro's number}} \\ &= \left(\frac{14.01\text{ g N}}{1\text{ mol N}}\right)\left(\frac{1\text{ mol N}}{6.022 \times 10^{23}\text{ N atoms}}\right) \\ &= 2.327 \times 10^{-23}\text{ g}\cdot\text{atom}^{-1}\end{aligned}$$

This is an extremely small mass. The mass of one of the smallest bacteria is about 10^{-13} g, which is 10 billion times larger than the mass of a nitrogen atom.

(b) The formula of molecular nitrogen is N_2, and so its formula mass, or molecular mass, is 28.02. Thus there are 6.022×10^{23} molecules of nitrogen in 28.02 g of nitrogen. The mass of one nitrogen molecule is

$$\begin{aligned}\text{mass of one nitrogen molecule} &= \frac{\text{mass of Avogadro's number of nitrogen molecules}}{\text{Avogadro's number}} \\ &= \frac{\text{mass corresponding to 1 mol of N}_2}{\text{Avogadro's number}} \\ &= \frac{28.02\text{ g N}_2}{6.022 \times 10^{23}\text{ N}_2\text{ molecules}} \\ &= 4.653 \times 10^{-23}\text{ g}\cdot\text{molecule}^{-1}\end{aligned}$$

Note that the mass of one nitrogen molecule is twice the mass of one nitrogen atom, as you should expect.

Avogadro's number can also be used to calculate the number of atoms or molecules in a given mass of a substance. The next two examples illustrate this type of calculation.

Example 3-3: How many lead atoms are there in 50.0 g of lead?

Solution: We know that there are 6.022×10^{23} atoms in 1 mol of lead. The atomic mass of lead is 207.2, and so 50.0 g of lead corresponds to

$$\text{mol Pb} = (50.0\text{ g Pb})\left(\frac{1\text{ mol Pb}}{207.2\text{ g Pb}}\right) = 0.241\text{ mol Pb}$$

The number of lead atoms in 0.241 mol is

$$\text{number of Pb atoms} = (0.241\text{ mol Pb})\left(\frac{6.022 \times 10^{23}\text{ Pb atoms}}{1\text{ mol Pb}}\right)$$

$$= 1.45 \times 10^{23}\text{ Pb atoms}$$

Example 3-4: Calculate how many methane molecules and how many hydrogen and carbon atoms there are in 25.0 g of methane.

Solution: The formula mass of CH_4 is $12.0 + (4 \times 1.0) = 16.0$, and so 25.0 g of CH_4 corresponds to

$$\text{moles of } CH_4 = (25.0\text{ g } CH_4)\left(\frac{1\text{ mol } CH_4}{16.0\text{ g } CH_4}\right) = 1.56\text{ mol } CH_4$$

Since 1 mol of CH_4 contains 6.022×10^{23} molecules, 1.56 mol must contain

$$\text{molecules of } CH_4 = (1.56\text{ mol } CH_4)\left(\frac{6.022 \times 10^{23}\text{ molecules}}{1\text{ mol}}\right)$$

$$= 9.39 \times 10^{23}\ CH_4\text{ molecules}$$

Each molecule of methane contains one carbon atom and four hydrogen atoms, and so

Problems 3-7 to 3-18 involve Avogadro's number.

$$\text{number of C atoms} = (9.39 \times 10^{23}\ CH_4\text{ molecules})\left(\frac{1\text{ C atom}}{1\ CH_4\text{ molecule}}\right)$$

$$= 9.39 \times 10^{23}\text{ C atoms}$$

$$\text{number of H atoms} = (9.39 \times 10^{23}\ CH_4\text{ molecules})\left(\frac{4\text{ H atoms}}{1\ CH_4\text{ molecule}}\right)$$

$$= 3.76 \times 10^{24}\text{ H atoms}$$

Table 3-1 Some relationships between molar quantities

Substance	*Formula*	*Formula mass*	*Mass of 1 mol/g*	*Number of particles in 1 mol*
atomic chlorine	Cl	35.45	35.45	6.022×10^{23} chlorine atoms
chlorine gas	Cl_2	70.90	70.90	6.022×10^{23} chlorine molecules
				12.044×10^{23} chlorine atoms
water	H_2O	18.02	18.02	6.022×10^{23} water molecules
				12.044×10^{23} hydrogen atoms
				6.022×10^{23} oxygen atoms
sodium chloride	NaCl	58.44	58.44	6.022×10^{23} NaCl formula units
				6.022×10^{23} sodium ions
				6.022×10^{23} chloride ions
barium fluoride	BaF_2	175.3	175.3	6.022×10^{23} BaF_2 formula units
				6.022×10^{23} barium ions
				12.044×10^{23} fluoride ions
nitrate ion	NO_3^-	62.01	62.01	6.022×10^{23} nitrate ions
				6.022×10^{23} nitrogen atoms
				18.066×10^{23} oxygen atoms

Table 3-1 summarizes some relationships between molar quantities. In reading the table, recall that sodium chloride is an ionic compound, with one formula unit consisting of one sodium ion, Na^+, and one chloride ion, Cl^-. Similarly, the formula unit of barium fluoride consists of one barium ion, Ba^{2+}, and *two* fluoride ions, F^-.

We conclude this section with the official SI definition of mole. "The mole is the amount of substance of a system which contains as many elementary entities as there are atoms in exactly 0.012 kilogram of carbon-12. When the mole is used, the elementary entities must be specified; they may be atoms, molecules, ions, electrons, other particles, or specified groups of such particles." Note that because the atomic mass of ^{12}C is exactly 12 by definition, a mole of ^{12}C contains exactly 12 g (= 0.012 kg) of carbon. The SI definition of mole is equivalent to the other definitions given in this section.

3-3. SIMPLEST FORMULAS CAN BE DETERMINED BY CHEMICAL ANALYSIS

The word *stoichiometry* is derived from the Greek words *stoicheia,* meaning simplest components or parts, and *metrein,* meaning to measure.

Stoichiometry (stoi'kē om'i trē) is the calculation of the quantities of elements or compounds involved in chemical reactions. The concept of a mole is central to carrying out stoichiometric calculations. For example, we can use the concept of a mole to determine the simplest chemical formula of a substance. Zinc oxide is found by chemical analysis to be 80.3 percent (by mass) zinc and 19.7 percent (by mass) oxygen. When working with mass percentages in chemical calculations, it is convenient to consider a 100-g sample so that the mass percentages can be easily converted to grams. For example, a 100-g sample of zinc oxide contains 80.3 g of zinc and 19.7 g of oxygen. We can write this schematically as

$$80.3 \text{ g Zn} \Bumpeq 19.7 \text{ g O}$$

where the symbol $\Bumpeq$ means "stoichiometrically equivalent to" or, in this case, "combines with." If we divide 80.3 g by the atomic mass of zinc (65.38) and 19.7 g by the atomic mass of oxygen (16.00), then we find that

$$\text{mol Zn} = (80.3 \text{ g Zn})\left(\frac{1 \text{ mol Zn}}{65.38 \text{ g Zn}}\right) = 1.23 \text{ mol Zn}$$

and

$$\text{mol O} = (19.7 \text{ g O})\left(\frac{1 \text{ mol O}}{16.00 \text{ g O}}\right) = 1.23 \text{ mol O}$$

Thus, we have

$$1.23 \text{ mol Zn} \Bumpeq 1.23 \text{ mol O}$$

or, correspondingly,

$$1.00 \text{ mol Zn} \Bumpeq 1.00 \text{ mol O}$$

By dividing both sides by Avogadro's number, we get

$$1.00 \text{ atom of Zn} \Bumpeq 1.00 \text{ atom of O}$$

This says that one atom of zinc combines with one atom of oxygen and that the *simplest chemical formula* of zinc oxide is therefore ZnO. We call ZnO the simplest chemical formula of zinc oxide because chemical analysis provides us with only the *ratios* of atoms in a compound. Mass percentages alone cannot be used to distinguish among ZnO, Zn_2O_2, Zn_3O_3, or any other multiple of ZnO. Simplest formulas are often called *empirical formulas*. The following example illustrates another calculation of a simplest, or empirical, formula.

Example 3-5: The chemical name for rubbing alcohol is isopropyl alcohol. Chemical analysis shows that pure isopropyl alcohol is 60.0 percent carbon, 13.4 percent hydrogen, and 26.6 percent oxygen by mass. Determine the empirical formula of isopropyl alcohol.

Solution: As usual, we take a 100-g sample and write

$$60.0\text{ g C} \Leftrightarrow 13.4\text{ g H} \Leftrightarrow 26.6\text{ g O}$$

We divide each value by the corresponding atomic mass and get

$$\begin{aligned}(60.0\text{ g C})\left(\frac{1\text{ mol C}}{12.01\text{ g C}}\right) &= 5.00\text{ mol C} \Leftrightarrow (13.4\text{ g H})\left(\frac{1\text{ mol H}}{1.008\text{ g H}}\right)\\ &= 13.3\text{ mol H} \Leftrightarrow (26.6\text{ g O})\left(\frac{1\text{ mol O}}{16.00\text{ g O}}\right)\\ &= 1.66\text{ mol O}\end{aligned}$$

or

$$5.00\text{ mol C} \Leftrightarrow 13.3\text{ mol H} \Leftrightarrow 1.66\text{ mol O}$$

To find a simple, whole-number relationship for these values, we divide through by the smallest value (1.66) and get

$$3.01\text{ mol C} \Leftrightarrow 8.01\text{ mol H} \Leftrightarrow 1.00\text{ mol O}$$

Realizing that 3.01 and 8.01 can be rounded off to 3 and 8 within probable experimental error, we find that an isopropyl alcohol molecule consists of three carbon atoms, eight hydrogen atoms, and one oxygen atom and thus the empirical formula is C_3H_8O.

The next example illustrates an experimental procedure for determining empirical formulas.

Example 3-6: A 0.450-g sample of magnesium metal is reacted completely in a nitrogen atmosphere to produce 0.623 g of magnesium nitride. Use these data to determine the empirical formula of magnesium nitride.

Solution: The 0.623 g of magnesium nitride contains 0.450 g of magnesium, and so the mass of nitrogen in the product is

$$\text{mass of N in product} = 0.623\text{ g} - 0.450\text{ g} = 0.173\text{ g N}$$

We can convert 0.450 g of Mg to moles of Mg by dividing by its atomic mass (24.31):

$$\text{mol Mg} = (0.450 \text{ g Mg})\left(\frac{1 \text{ mol Mg}}{24.31 \text{ g Mg}}\right) = 0.0185 \text{ mol Mg}$$

Similarly, by dividing the mass of nitrogen by its atomic mass (14.01), we obtain

$$\text{mol N} = (0.173 \text{ g N})\left(\frac{1 \text{ mol N}}{14.01 \text{ g N}}\right) = 0.0123 \text{ mol N}$$

Thus we have the relation

$$0.0185 \text{ mol Mg} \approx 0.0123 \text{ mol N}$$

Problems 3-19 to 3-26 deal with empirical formulas.

If we divide both quantities by 0.0123, then we obtain

$$1.50 \text{ mol Mg} \approx 1.00 \text{ mol N}$$

Multiplying both sides by 2 to get whole numbers, we get

$$3.00 \text{ mol Mg} \approx 2.00 \text{ mol N}$$

Thus, we see that 3.00 mol of magnesium combines with 2.00 mol of nitrogen. Thus the empirical formula for magnesium nitride is Mg_3N_2 .

3-4. EMPIRICAL FORMULAS CAN BE USED TO DETERMINE AN UNKNOWN ATOMIC MASS

If we know the empirical formula of a compound, then we can determine the atomic mass of one of the elements in the compound if the atomic masses of the others are known. This is a standard experiment in many general chemistry laboratory courses.

Example 3-7: The empirical formula of magnesium oxide is MgO. A weighed quantity of magnesium, 0.490 g, is burned in oxygen, and the MgO produced is found to have a mass of 0.813 g. Given that the atomic mass of oxygen is 16.00, determine the atomic mass of magnesium.

Solution: The mass of oxygen in the magnesium oxide is

$$\text{mass of O in sample} = 0.813 \text{ g MgO} - 0.490 \text{ g Mg} = 0.323 \text{ g O}$$

The number of moles of O is

$$\text{mol O} = (0.323 \text{ g O})\left(\frac{1 \text{ mol O}}{16.00 \text{ g O}}\right) = 0.0202 \text{ mol O}$$

The empirical formula MgO tells us that 0.0202 mol of Mg is combined with 0.0202 mol of O, and so we have

$$0.490 \text{ g Mg} \approx 0.0202 \text{ mol Mg}$$

The atomic mass of magnesium can be obtained if we determine how many grams of magnesium correspond to 1.00 mol. To determine this, we divide the above stoichiometric correspondence by 0.0202 to get

$$24.3 \text{ g Mg} \approx 1.00 \text{ mol Mg}$$

or that the atomic mass of magnesium is 24.3, in excellent agreement with the accepted value.

You can see from these examples that it is possible to determine the empirical formula of a compound if the atomic masses are known and that it is possible to determine the atomic mass of an element if the empirical formula of one of its compounds and the atomic masses of the other elements that make up the compound are known. We are faced with a dilemma here. Atomic masses are determined if empirical formulas are known (Example 3-7), but atomic masses must be known to determine empirical formulas (Examples 3-5 and 3-6). This was a serious problem in the early 1800's, shortly after Dalton formulated his atomic theory, because it was necessary to guess the empirical formulas of compounds—an incorrect guess led to an incorrect atomic mass. We shall see in Chapter 5 that it was the quantitative study of gases and of reactions between gases that was the key to resolving the difficulty in determining reliable values of atomic masses.

3-5. AN EMPIRICAL FORMULA ALONG WITH THE MOLECULAR MASS DETERMINES THE MOLECULAR FORMULA

Suppose that the chemical analysis of a compound gives 85.7 percent carbon and 14.3 percent hydrogen by mass. We then have

$$85.7 \text{ g C} \approx 14.3 \text{ g H}$$

$$7.14 \text{ mol C} \approx 14.2 \text{ mol H}$$

$$1 \text{ mol C} \approx 2 \text{ mol H}$$

and conclude that the empirical formula is CH_2. However, the actual formula might be C_2H_4, C_3H_6, or, generally, C_nH_{2n}. The chemical analysis gives us only ratios of numbers of atoms. If we know the molecular mass from another experiment, however, then we can determine the *molecular formula* unambiguously. For example, suppose we know that the molecular mass of our compound is 42. By listing the various possible formulas,

Formula	*Formula mass*
CH_2	14
C_2H_4	28
C_3H_6	42
C_4H_8	56

we see that the molecular formula of the compound is C_3H_6. This is why the formula deduced from chemical analysis is called the empirical formula or the simplest formula. It must be supplemented by molecular mass data to determine the molecular formula.

Example 3-8: There are many compounds, called *hydrocarbons,* that consist of only carbon and hydrogen. Gasoline is a mixture of over 100 different hydrocarbons. Chemical analysis of one of the constituents of gasoline yields 92.30 percent carbon and 7.70 percent hydrogen by mass. (a) Determine the simplest formula of this compound. (b) Given that the molecular mass is 78, determine the molecular formula.

Problems 3-31 to 3-34 ask you to determine molecular formulas.

Solution: (a) The determination of the simplest formula can be summarized by

$$92.30 \text{ g C} \approx 7.70 \text{ g H}$$

$$7.69 \text{ mol C} \approx 7.64 \text{ mol H}$$

$$1 \text{ mol C} \approx 1 \text{ mol H}$$

The simplest formula is CH, which has a formula mass of 13. (b) The molecular mass must be a multiple of 13, and indeed, 78 is 6 times 13. Thus, the molecular formula is C_6H_6.

3-6. THE COEFFICIENTS IN CHEMICAL EQUATIONS CAN BE INTERPRETED AS NUMBERS OF MOLES

A subject of great practical importance in chemistry is the determination of what quantity of product can be obtained from a given quantity of reactants. For example, the reaction between hydrogen and nitrogen to produce ammonia is described by the equation

$$3H_2(g) + N_2(g) \rightarrow 2NH_3(g)$$

The production of ammonia from hydrogen and nitrogen requires high pressure and temperature and a catalyst to speed up the reaction (Interchapter C).

where the coefficients in the equation are called balancing coefficients or *stoichiometric coefficients.* We might wish to know how much NH_3 is produced when 100 g of H_2 reacts with N_2. To answer questions like this, we interpret the above equation in terms of moles rather than molecules. The molecular interpretation of the hydrogen-nitrogen reaction is

3 molecules of hydrogen + 1 molecule of nitrogen $\rightarrow$
2 molecules of ammonia

If we multiply both sides of this equation by Avogadro's number, then we obtain

$3(6.022 \times 10^{23})$ molecules of hydrogen
$+ 6.022 \times 10^{23}$ molecules of nitrogen $\rightarrow$
$2(6.022 \times 10^{23})$ molecules of ammonia

If we use the fact that 6.022×10^{23} molecules corresponds to 1 mol, then we also have

$$3 \text{ mol of } H_2 + 1 \text{ mol of } N_2 \rightarrow 2 \text{ mol of } NH_3$$

This is an important result. It tells us that the balancing coefficients are the relative numbers of moles of each substance in a balanced chemical equation.

We can also interpret the hydrogen-nitrogen reaction in terms of masses. If we convert moles to masses by multiplying by the appropriate molecular masses, then we get

$$6.05 \text{ g } H_2 + 28.02 \text{ g } N_2 \rightarrow 34.07 \text{ g } NH_3$$

Note that the total mass is the same on the two sides of the equation, in accord with the law of conservation of mass. Table 3-2 summarizes

Table 3-2 The various interpretations of two chemical reactions

Interpretation	$3H_2$	$+ N_2$	$\rightarrow 2NH_3$
molecular	3 molecules	+ 1 molecule	$\rightarrow$ 2 molecules
molar	3 mol	+ 1 mol	$\rightarrow$ 2 mol
mass	6.05 g	+ 28.02 g	$\rightarrow$ 34.07 g
Interpretation	$2Na$	$+ Cl_2$	$\rightarrow 2NaCl$
molecular	2 atoms	+ 1 molecule	$\rightarrow$ 2 ion pairs or 2 formula units
molar	2 mol	+ 1 mol	$\rightarrow$ 2 mol
mass	45.98 g	+ 70.90 g	$\rightarrow$ 116.88 g

the various interpretations of the hydrogen-nitrogen reaction as well as those of the sodium-chlorine reaction.

We are now ready to calculate how much ammonia is produced when a given quantity of nitrogen or hydrogen is used.

Example 3-9: (a) How many moles of NH_3 are produced from 10.0 mol of N_2? (b) How many grams of NH_3 are produced from 280 g of N_2?

Solution: (a) According to the balanced chemical equation,

$$3H_2(g) + N_2(g) \rightarrow 2NH_3(g)$$

Calculations like the ones in Example 3-9 require that you understand the mole concept.

2 mol of NH_3 are produced from each mole of N_2; thus, the stoichiometric unit conversion factor is

$$1 = \frac{2 \text{ mol } NH_3}{1 \text{ mol } N_2}$$

Therefore 10.0 mol of N_2 yields

$$(10.0 \text{ mol } N_2)\left(\frac{2 \text{ mol } NH_3}{1 \text{ mol } N_2}\right) = 20.0 \text{ mol } NH_3$$

(b) The molecular mass of N_2 is 28.0, and so 280 g of N_2 corresponds to

$$(280 \text{ g } N_2)\left(\frac{1 \text{ mol } N_2}{28.0 \text{ g } N_2}\right) = 10.0 \text{ mol } N_2$$

We have just seen that 10.0 mol of N_2 yields 20.0 mol of NH_3. The molecular mass of NH_3 is 17.0, and so 20.0 mol corresponds to

$$\text{g } NH_3 = (20.0 \text{ mol } NH_3)\left(\frac{17.0 \text{ g } NH_3}{1 \text{ mol } NH_3}\right) = 340 \text{ g } NH_3$$

Thus, 280 g of N_2 yields 340 g of NH_3.

Example 3-10: Propane, C_3H_8, a common fuel, burns in oxygen according to the equation

$$C_3H_8(g) + 5O_2(g) \rightarrow 3CO_2(g) + 4H_2O(l)$$

(a) How many grams of O_2 are required to burn 75.0 g of C_3H_8? (b) How many grams of H_2O and CO_2 are produced?

Solution: (a) The chemical equation states that 5 mol of O_2 is required to burn 1 mol of C_3H_8. The molecular mass of C_3H_8 is 44.09, and so 75.0 g of C_3H_8 corresponds to

$$\text{mol } C_3H_8 = (75.0 \text{ g } C_3H_8)\left(\frac{1 \text{ mol } C_3H_8}{44.09 \text{ g } C_3H_8}\right) = 1.70 \text{ mol}$$

The number of moles of O_2 required is

$$\text{mol } O_2 = (1.70 \text{ mol } C_3H_8)\left(\frac{5 \text{ mol } O_2}{1 \text{ mol } C_3H_8}\right) = 8.50 \text{ mol } O_2$$

To find out how many grams of O_2 this is, we multiply the number of moles by the molecular mass:

$$\text{g } O_2 = (8.50 \text{ mol } O_2)\left(\frac{32.00 \text{ g } O_2}{1 \text{ mol } O_2}\right) = 272 \text{ g } O_2$$

Thus, we see that 272 g of O_2 is required to burn 75.0 g of C_3H_8.
(b) According to the reaction, 3 mol of CO_2 and 4 mol of H_2O are produced for each mole of C_3H_8 burned. Therefore the number of moles of CO_2 produced is

$$\text{mol } CO_2 = (1.70 \text{ mol } C_3H_8)\left(\frac{3 \text{ mol } CO_2}{1 \text{ mol } C_3H_8}\right) = 5.10 \text{ mol } CO_2$$

The number of moles of H_2O produced is

$$\text{mol } H_2O = (1.70 \text{ mol } C_3H_8)\left(\frac{4 \text{ mol } H_2O}{1 \text{ mol } C_3H_8}\right) = 6.80 \text{ mol } H_2O$$

The moles of CO_2 and H_2O are converted to grams by multiplying by the respective molecular masses:

Problems 3-35 to 3-46 ask you to do calculations involving chemical equations. You will be doing many calculations like these throughout your study of chemistry.

$$\text{g } CO_2 = (5.10 \text{ mol } CO_2)\left(\frac{44.01 \text{ g } CO_2}{1 \text{ mol } CO_2}\right) = 224 \text{ g } CO_2$$

$$\text{g } H_2O = (6.80 \text{ mol } H_2O)\left(\frac{18.02 \text{ g } H_2O}{1 \text{ mol } H_2O}\right) = 123 \text{ g } H_2O$$

These are the quantities of CO_2 and H_2O produced when 75.0 g of propane are burned.

We can summarize the results of this example by

$$\begin{array}{llll} C_3H_8(g) & + \; 5O_2(g) & \rightarrow \; 3CO_2(g) & + \; 4H_2O(l) \\ 75 \text{ g} & + \; 272 \text{ g} & \rightarrow \; 224 \text{ g} & + \; 123 \text{ g} \end{array}$$

Notice that the total masses on the two sides of the chemical reaction are the same, as they must be according to the principle of conservation of mass.

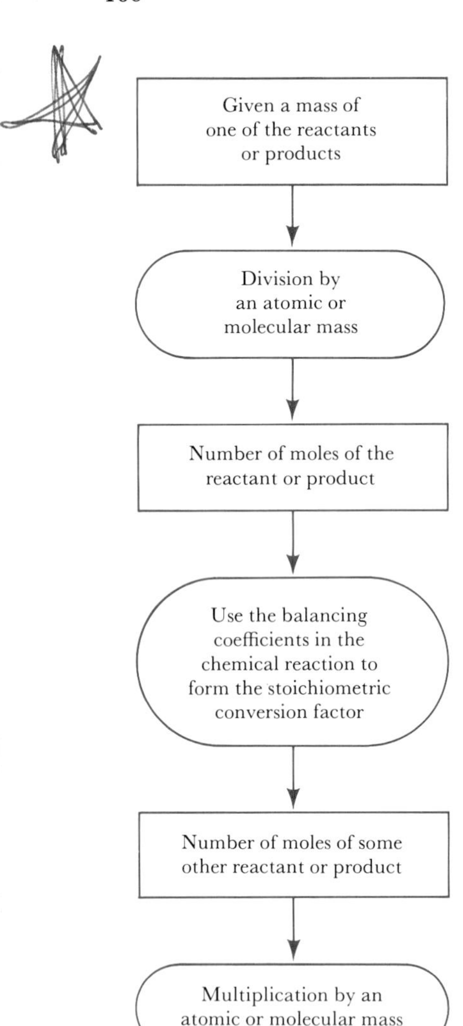

Figure 3-1 This flow diagram describes the procedure for calculations involving chemical equations. The essence of the method is to realize that we convert from moles of one substance to moles of another substance in a chemical equation by using the balancing coefficients.

3-7. CALCULATIONS INVOLVING CHEMICAL REACTIONS ARE CARRIED OUT IN TERMS OF MOLES

For calculations involving chemical reactions, the procedure is to first convert mass to moles, then convert moles of one substance to moles of another by using the balancing coefficients in the chemical equation, and then convert moles into mass. An understanding of the flow chart in Figure 3-1 will allow you to do any calculations involving chemical equations. The following calculation illustrates the use of Figure 3-1.

Sulfuric acid, H_2SO_4, usually in the form of an aqueous solution, is the most widely used and important industrial chemical. More than 40 million tons of sulfuric acid are produced annually in the United States, and most sulfuric acid is made by the *contact process*. First sulfur is burned in oxygen to produce sulfur dioxide:

$$S(s) + O_2(g) \rightarrow SO_2(g) \tag{3-1}$$

Then the $SO_2(g)$ is mixed with more $O_2(g)$ and passed over vanadium pentoxide, which increases the rate of the reaction:

$$2SO_2(g) + O_2(g) \xrightarrow[500°C]{V_2O_5} 2SO_3(g) \tag{3-2}$$

Vanadium pentoxide is a *catalyst* for this reaction. We shall learn more about catalysts in later chapters. At this point it is sufficient to realize that a catalyst is a substance that increases the rate of a reaction but is not consumed in the reaction. We denote a catalyst by placing the formula for it over the arrow in the chemical equation. The $SO_3(g)$ produced in Equation (3-2) is dissolved in sulfuric acid, which is then reacted with water. The overall reaction is

$$SO_3(g) + H_2O(l) \rightarrow H_2SO_4(l) \tag{3-3}$$

Let's calculate how much $H_2SO_4(l)$ can be produced from 1 metric ton of sulfur. (A *metric ton* is equal to 1000 kg, or 2200 pounds.) From Equation (3-1), we see that 1 mol of $S(s)$ yields 1 mol of $SO_2(g)$; Equation (3-2) tells us that each mole of $SO_2(g)$ that reacts yields 1 mol of $SO_3(g)$; Equation (3-3) shows that 1 mol of $SO_3(g)$ produces 1 mol of $H_2SO_4(l)$. We can summarize these statements as follows:

$$1 \text{ mol } S(s) \approx 1 \text{ mol } SO_2(g) \approx 1 \text{ mol } SO_3(g) \approx 1 \text{ mol } H_2SO_4(l)$$

Thus we see that 1 mol of $H_2SO_4(l)$ is produced from 1 mol of $S(s)$:

$$1 \text{ mol S}(s) \approx 1 \text{ mol H}_2\text{SO}_4(l)$$

This result should not be surprising because one molecule of H_2SO_4 contains one atom of sulfur. One metric ton of $S(s)$ corresponds to

$$\text{mol S} = (1.00 \times 10^6 \text{ g S})\left(\frac{1 \text{ mol S}}{32.06 \text{ g S}}\right) = 3.12 \times 10^4 \text{ mol S}$$

The chemical equations show that

$$3.12 \times 10^4 \text{ mol S}(s) \approx 3.12 \times 10^4 \text{ mol H}_2\text{SO}_4(l)$$

The molecular mass of H_2SO_4 is 98.1, and so the quantity of H_2SO_4 produced is

$$\text{g H}_2\text{SO}_4 = (3.12 \times 10^4 \text{ mol H}_2\text{SO}_4)\left(\frac{98.08 \text{ g H}_2\text{SO}_4}{1 \text{ mol H}_2\text{SO}_4}\right)$$

$$= 3.06 \times 10^6 \text{ g H}_2\text{SO}_4$$

$$\text{kg H}_2\text{SO}_4 = (3.06 \times 10^6 \text{ g H}_2\text{SO}_4)\left(\frac{1 \text{ kg}}{10^3 \text{ g}}\right)$$

$$= 3.06 \times 10^3 \text{ kg H}_2\text{SO}_4$$

$$\text{tons H}_2\text{SO}_4 = (3.06 \times 10^3 \text{ kg H}_2\text{SO}_4)\left(\frac{1 \text{ metric ton}}{10^3 \text{ kg}}\right)$$

$$= 3.06 \text{ metric tons H}_2\text{SO}_4$$

Example 3-11: The density of pure sulfuric acid is $1.83 \text{ kg}\cdot\text{L}^{-1}$. What volume of sulfuric acid is produced from 1 metric ton of sulfur?

Note that a density of $1.83 \text{ kg}\cdot\text{L}^{-1}$ is equal to a density of $1.83 \text{ g}\cdot\text{mL}^{-1}$.

Solution: From the previous calculation, we know that

$$1 \text{ metric ton S}(s) = 10^3 \text{ kg S}(s) \approx 3.06 \times 10^3 \text{ kg H}_2\text{SO}_4(l)$$

Solving the density equation (Equation 1-6)

$$d = \frac{m}{V}$$

for V, we have

$$V = \frac{m}{d} = \frac{3.06 \times 10^3 \text{ kg}}{1.83 \text{ kg}\cdot\text{L}^{-1}} = 1.67 \times 10^3 \text{ L}$$

Thus we see that 3.06 metric tons of $H_2SO_4(l)$ occupies a volume of 1670 L.

3-8. WHEN TWO OR MORE SUBSTANCES REACT, THE QUANTITY OF PRODUCT IS DETERMINED BY THE LIMITING REACTANT

If you look back over the examples in this chapter, you will notice that in no case did we start out with stated quantities of two reactants. Let's consider an example in which we do. Cadmium sulfide, which is used in light meters, solar cells, and other light-sensitive devices, can be made by the direct combination of the two elements:

$$\mathrm{Cd}(s) + \mathrm{S}(s) \rightarrow \mathrm{CdS}(s) \tag{3-4}$$

How much CdS is produced if we start out with 2.00 g of cadmium and 2.00 g of sulfur? The number of moles of each element is

$$\text{mol Cd} = (2.00 \text{ g Cd})\left(\frac{1 \text{ mol Cd}}{112.4 \text{ g Cd}}\right) = 0.0178 \text{ mol Cd}$$

$$\text{mol S} = (2.00 \text{ g S})\left(\frac{1 \text{ mol S}}{32.06 \text{ g S}}\right) = 0.0624 \text{ mol S}$$

According to the chemical equation, 1 mol of cadmium requires 1 mol of sulfur, and so the 0.0178 mol of cadmium requires 0.0178 mol of sulfur. There is excess sulfur; only 0.0178 mol of sulfur reacts, and (0.0624 − 0.0178) mol = 0.0446 mol of sulfur remains. The cadmium reacts completely, and the moles of cadmium consumed determines how much CdS is produced. The reactant that is consumed completely and thereby limits the amount of product formed is called the *limiting reactant*, and any other reactants are called *excess reactants*. The initial quantity of limiting reactant is used to calculate how much product is formed.

In Equation (3-4), 0.0178 mol of cadmium reacts with 0.0178 mol of sulfur to produce 0.0178 mol of CdS. The mass of CdS produced is

$$\text{g CdS} = (0.0178 \text{ mol CdS})\left(\frac{144.5 \text{ g CdS}}{1 \text{ mol CdS}}\right) = 2.57 \text{ g CdS}$$

The unused sulfur (0.0446 mol) has a mass of

$$\text{g unused S} = (0.0446 \text{ mol S})\left(\frac{32.06 \text{ g S}}{1 \text{ mol S}}\right) = 1.43 \text{ g S}$$

Note that before the reaction there is 2.00 g of cadmium and 2.00 g of sulfur, or 4.00 g of reactants. After the reaction, there is 2.57 g of cadmium sulfide and 1.43 g of sulfur, or 4.00 g of products, as required by the law of conservation of mass.

When the masses of two or more reactants are given in a problem, the limiting reactant must be determined and is the only one to be used in the calculation of the quantity of product obtained.

Example 3-12: A mixture is prepared from 25.0 g of aluminum and 85.0 g of Fe_2O_3. The reaction that occurs is described by the equation

$$Fe_2O_3(s) + 2Al(s) \rightarrow Al_2O_3(s) + 2Fe(l)$$

How much iron is produced in the reaction?

The reaction in Example 3-12 is called a *thermite reaction,* and evolves enough heat to melt the iron that is produced (see Figure 4-9).

Solution: Because the masses of both reactants are given, we must check to see which, if either, is a limiting reactant. The number of moles of Al and Fe_2O_3 available is

$$\text{mol Al} = (25.0 \text{ g Al})\left(\frac{1 \text{ mol Al}}{26.98 \text{ g Al}}\right) = 0.927 \text{ mol Al}$$

$$\text{mol } Fe_2O_3 = (85.0 \text{ g } Fe_2O_3)\left(\frac{1 \text{ mol } Fe_2O_3}{159.7 \text{ g } Fe_2O_3}\right) = 0.532 \text{ mol } Fe_2O_3$$

From the balanced chemical equation, we see that 0.927 mol of aluminum consumes only

$$(0.927 \text{ mol Al})\left(\frac{1 \text{ mol } Fe_2O_3}{2 \text{ mol Al}}\right) = 0.464 \text{ mol } Fe_2O_3$$

Thus, we see that the Fe_2O_3 is in excess. The amount of Fe_2O_3 in excess is

$$\text{mol excess } Fe_2O_3 = (0.532 - 0.464) \text{ mol } Fe_2O_3 = 0.068 \text{ mol } Fe_2O_3$$

The aluminum is the limiting reactant and the one to use in calculating how much iron is produced. From the balanced equation,

$$2 \text{ mol Al} \approx 2 \text{ mol Fe}$$

or

$$0.927 \text{ mol Al} \approx 0.927 \text{ mol Fe}$$

The mass of iron corresponding to 0.927 mol is

$$\text{g Fe} = (0.927 \text{ mol Fe})\left(\frac{55.85 \text{ g Fe}}{1 \text{ mol Fe}}\right) = 51.8 \text{ g Fe}$$

There are many instances in which it is important to add reactants in stoichiometric proportions so as not to have any reactants left over. The propulsion of rockets and space vehicles serves as a good example. The Lunar Lander rocket engines were powered by a reaction similar to

$$\underset{\text{dinitrogen tetroxide}}{N_2O_4(l)} + \underset{\text{hydrazine}}{2N_2H_4(l)} \rightarrow 3N_2(g) + 4H_2O(g)$$

Dinitrogen tetroxide and hydrazine react explosively when brought into contact. The two reactants are kept in separate tanks and pumped through pipes into the rocket engines, where they react. The gases produced (H_2O is a gas at the exhaust temperatures of the rocket engines) exit through the exhaust chamber of the engine and propel the rocket forward. The cost of carrying materials into space is enormous, and the two fuels must be combined in the correct proportions. It would be wasteful to carry any excess reactant.

Example 3-13: Suppose that a rocket is powered by the reaction between liquid dinitrogen tetroxide and hydrazine. If the tanks are designed to hold 50.0 metric tons of dinitrogen tetroxide, how much hydrazine should be carried?

Solution: Recall that a metric ton is 1000 kg. The molecular mass of N_2O_4 is 92.0, and the number of moles of N_2O_4 carried is

$$\text{mol } N_2O_4 = \left(\frac{50.0 \text{ metric tons}}{N_2O_4}\right)\left(\frac{10^3 \text{ kg}}{1 \text{ metric ton}}\right)\left(\frac{10^3 \text{ g}}{1 \text{ kg}}\right)\left(\frac{1 \text{ mol } N_2O_4}{92.0 \text{ g } N_2O_4}\right)$$
$$= 5.43 \times 10^5 \text{ mol } N_2O_4$$

According to the balanced chemical equation, it requires 2 mol of hydrazine for every mole of dinitrogen tetroxide. Therefore

$$\text{mol } N_2H_4 \text{ required} = \left(\frac{2 \text{ mol } N_2H_4}{1 \text{ mol } N_2O_4}\right)(5.43 \times 10^5 \text{ mol } N_2O_4)$$
$$= 1.09 \times 10^6 \text{ mol } N_2H_4$$

Problems 3-47 to 3-52 deal with limiting reactants.

The mass of hydrazine required is

$$\text{mass } N_2H_4 \text{ required} = (1.09 \times 10^6 \text{ mol } N_2H_4)\left(\frac{32.05 \text{ g } N_2H_4}{1 \text{ mol } N_2H_4}\right)$$
$$= 3.49 \times 10^7 \text{ g } N_2H_4$$
$$= 34.9 \text{ metric tons } N_2H_4$$

3-9. MANY REACTIONS TAKE PLACE IN SOLUTION

Many important chemical and biological processes take place in solution, particularly in aqueous solution. A *solution* is a mixture of two or more substances that is uniform at the molecular level. A solution must be *homogeneous*, meaning that it has the same properties from one region to another. The most common examples of solutions involve a solid, such as NaCl, dissolved in water. The resulting solution is clear and homogeneous. From a molecular point of view, the sodium ions (Na^+) and chloride ions (Cl^-) are uniformly dispersed throughout the water.

The solid that is dissolved is called the *solute*, and the liquid in which it is dissolved is called the *solvent*. In the case of NaCl dissolved in water, NaCl is the solute and water is the solvent. We designate a species in aqueous solution by writing (*aq*) after the species. For example, $Na^+(aq)$ and $Cl^-(aq)$ represent sodium ions and chloride ions in water. The process of dissolving NaCl(*s*) in water is represented by the equation

$$\mathrm{NaCl}(s) \xrightarrow[\mathrm{H_2O}(l)]{} \mathrm{Na^+}(aq) + \mathrm{Cl^-}(aq)$$

where $H_2O(l)$ *under* the arrow tells us that water is the solvent.

3-10. THE CONCENTRATION OF A SOLUTION CAN BE EXPRESSED IN TERMS OF MOLARITY

The *concentration* of solute in a solution describes the quantity of solute dissolved in a given quantity of solvent. The most common method of expressing the concentration of a solute is *molarity*, which is represented by the symbol M. Molarity is defined as the number of moles of solute per liter of solution:

$$\text{molarity} = \frac{\text{moles of solute}}{\text{liters of solution}} \tag{3-5}$$

In terms of symbols, Equation (3-5) is

$$M = \frac{n}{V} \tag{3-6}$$

where M is the molarity of the solution, n is the number of moles of solute dissolved in the solution, and V is the volume of the solution in liters. To see how to use Equation (3-6), let's calculate the molarity of a solution prepared by dissolving 62.3 g of sucrose $C_{12}H_{22}O_{11}$ in enough water to form 0.500 L of solution. The formula mass of sucrose is 342, and so 62.3 g corresponds to

$$(62.3\text{ g sucrose})\left(\frac{1\text{ mol sucrose}}{342\text{ g sucrose}}\right) = 0.182\text{ mol sucrose}$$

This is the number of moles of sucrose dissolved in 0.500 L of solution. The molarity of the solution is

$$M = \frac{n}{V} = \frac{0.182\text{ mol}}{0.500\text{ L}} = 0.364\text{ mol}\cdot\text{L}^{-1} = 0.364\text{ M}$$

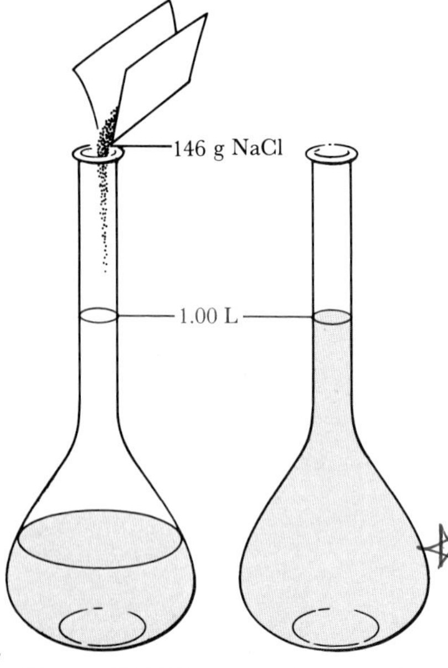

Figure 3-2 The procedure used to prepare one L of a solution of a certain molarity, such as 2.50 M NaCl. The 2.50 mol of NaCl (146 g) is weighed out and placed in a one-liter volumetric flask that is only partially filled with water. The NaCl is dissolved, and then more water is added to bring the final volume up to the mark on the flask.

The definition of molarity involves the volume of the *solution, not* the volume of the solvent. Suppose we wish to prepare 1 L of a 2.50 M aqueous solution of NaCl. We would prepare the solution by weighing out 2.50 mol (146 g) of NaCl, dissolving it in *less* than 1 L of water, say, about 500 mL, and then adding water until the final volume of the *solution* is exactly 1 L (Figure 3-2). Realize that it would be incorrect to add 2.50 mol of NaCl to 1 L of water; the final volume of such a solution is not exactly 1 L because of the added NaCl (its volume is 1.04 L).

Solutions of known molarity must always be made by the procedure described in Figure 3-2. For such measurements we use a *volumetric flask,* which is a precision piece of glassware used to prepare precise volumes. The following example illustrates the procedure for making up a solution of a specified molarity.

Example 3-14: Potassium bromide, KBr, is used as a sedative and as an anticonvulsive agent. Explain how you would prepare 250 mL of a 0.600 M aqueous KBr solution.

Solution: From Equation (3-6) and the specified volume and concentration, we can calculate the number of moles of KBr required. Equation (3-6) can be written as

$$n = MV$$

and so

$$\begin{aligned} \text{mol KBr} &= (0.600\ \text{M})(0.250\ \text{L}) \\ &= (0.600\ \text{mol}\cdot\text{L}^{-1})(0.250\ \text{L}) \\ &= 0.150\ \text{mol} \end{aligned}$$

We can convert moles to grams by multiplying by the formula mass of KBr (119.0):

$$\begin{aligned} \text{g KBr} &= (0.150\ \text{mol KBr})\left(\frac{119.0\ \text{g KBr}}{1\ \text{mol KBr}}\right) \\ &= 17.9\ \text{g KBr} \end{aligned}$$

Problems 3-53 to 3-58 involve preparation of solutions.

To prepare the solution, we dissolve 17.9 g of KBr in a 250-mL volumetric flask that is partially filled with distilled water, shake the flask until the salt is dissolved, and then dilute the solution to the 250-mL mark on the flask and shake again to assure uniformity. *Do not* add the KBr to 250 mL of water because the volume of the resulting solution will not necessarily be 250 mL.

The next example illustrates a calculation involving a reaction that takes place in solution.

Example 3-15: Zinc reacts with hydrochloric acid, HCl(*aq*), according to the equation

$$Zn(s) + 2HCl(aq) \rightarrow ZnCl_2(aq) + H_2(g)$$

Calculate how many grams of zinc react with 50.0 mL of 6.00 M HCl(*aq*).

Solution: According to the equation for the reaction, 1 mol of Zn reacts with 2 mol of HCl(*aq*). We can use Equation (3-6) to calculate how many moles of HCl there are in 50.0 mL of a 6.00 M HCl solution:

$$\text{mol HCl} = MV = (6.00\text{ M})(50.0\text{ mL})\left(\frac{1\text{ L}}{1000\text{ mL}}\right)$$
$$= (6.00\text{ mol}\cdot\text{L}^{-1})(0.0500\text{ L})$$
$$= 0.300\text{ mol}$$

By referring to the equation, we see that 0.150 mol of Zn reacts with 0.300 mol of HCl. The mass that corresponds to 0.150 mol of Zn (atomic mass 65.38) is

Problems 3-59 to 3-68 ask you to do calculations involving solutions.

$$\text{g Zn} = (0.150\text{ mol Zn})\left(\frac{65.38\text{ g Zn}}{1\text{ mol Zn}}\right)$$
$$= 9.81\text{ g Zn}$$

Although we have done a number of different types of calculations in this chapter, they are all unified by the concept of a mole, the chapter's central theme. The mole is one of the most important and useful concepts in chemistry.

SUMMARY

Stoichiometric calculations are based on the concept of a mole. The quantity of a substance that is numerically equal to its formula mass in grams is called a mole of that substance. Thus in order to calculate the number of moles in a given mass of a substance, it is necessary to know its chemical formula. Another definition of a mole is that mass of a substance that contains Avogadro's number (6.022×10^{23}) of formula units or elementary entities. Using the concept of a mole and Avogadro's number, it is possible to calculate the mass of individual atoms and molecules, to calculate how many atoms and molecules there are in a given mass, to determine chemical formulas from chemical analysis, and to calculate quantities of substances involved in chemical reactions.

TERMS YOU SHOULD KNOW

atomic substance 86
molecular substance 86
formula mass 86
formula unit 86
mole 87
mol 87
Avogadro's number (6.022×10^{23}) 88
stoichiometry 92
$\approx$ (stoichiometrically equivalent to) 92
simplest formula 92
empirical formula 93
molecular formula 95
hydrocarbon 96
stoichiometric coefficient 96
catalyst 100
metric ton 100
limiting reactant 102
excess reactant 102
solution 104
homogeneous 104
solute 105
solvent 105
concentration 105
molarity 105
volumetric flask 106

EQUATIONS YOU SHOULD KNOW HOW TO USE

$$M = \frac{n}{V} \qquad (3\text{-}6)$$

PROBLEMS

FORMULA MASS

3-1. Calculate the formula mass of

(a) H_2O
(b) $FeSO_4$
(c) $BaCl_2$
(d) $C_6H_{12}O_6$ (glucose)

3-2. Compute the formula mass of

(a) $C_9H_8O_4$ (aspirin)
(b) $C_{56}H_{88}O_2$ (vitamin D_1)
(c) C_3H_5BrO (bromoacetone; chemical war gas)
(d) $C_2H_3Cl_3O_2$ (chloral hydrate; "knockout drops")

NUMBER OF MOLES

3-3. Calculate the number of moles in 250 g of each of the following oils:

(a) clove	$C_{10}H_{12}O_2$
(b) cinnamon	$C_{18}H_{14}O_3$
(c) peppermint	$C_{10}H_{20}O$

3-4. The chemical formulas of some common insecticides are

(a) malathion	$C_{10}H_{19}O_6PS_2$
(b) parathion	$C_{10}H_{14}NO_5PS$
(c) methoxychlor	$C_{16}H_{15}Cl_3O_2$

Calculate the number of moles in 1.00 kg of each.

3-5. Calculate the number of moles in

(a) 28.0 g of H_2O (1 ounce)
(b) 200 mg of diamond (C) (1 carat)
(c) 454 g of NaCl (1 pound)

3-6. Compute the number of moles in the recommended daily allowance of the following substances:

(a) 60 mg of vitamin C, $C_6H_8O_6$
(b) 1.5 mg of vitamin A, $C_{20}H_{30}O$
(c) 6.0 μg of vitamin B_{12}, $C_{63}H_{88}CoN_{14}O_{14}P$

AVOGADRO'S NUMBER

3-7. A baseball has a mass of 142 g. Calculate the mass of Avogadro's number of baseballs (1 mol) and compare your result with the mass of the earth, 5.975×10^{24} kg.

3-8. The U.S. population is about 230 million. If Avogadro's number of dollars was distributed equally among the population, how many dollars would each person receive?

3-9. One method of determining Avogadro's number involves measuring how much electricity is required to plate out a certain mass of metal. The units of the amount of charge are *coulombs*. It requires 894 coulombs of electricity to plate out 1.00 g of silver from silver ions, Ag^+. Using the fact that the charge on an electron is 1.602×10^{-19} coulomb, calculate Avogadro's number.

3-10. Make up an analogy to explain to a friend just how large Avogadro's number is.

3-11. Compute the mass of

(a) one NH_3 molecule
(b) one $C_6H_{12}O_6$ (glucose) molecule
(c) one Fe atom

3-12. Compute the mass of

(a) one O_2 molecule
(b) one $FeSO_4$ formula unit
(c) one $C_{12}H_{22}O_{11}$ (sucrose) molecule

3-13. Calculate the mass of

(a) 200 copper atoms
(b) 10^{16} ammonia molecules
(c) 10^6 fluorine atoms
(d) 10^6 fluorine molecules (F_2)

3-14. Calculate the mass of

(a) 100 molecules of nitroglycerin, $C_3H_5N_3O_9$
(b) 5000 molecules of TNT, $C_7H_5N_3O_6$
(c) 10^{10} molecules of octane, C_8H_{18}
(d) 10 molecules of ozone, O_3

3-15. A 10.0-g sample of H_2O contains ____ moles of H_2O, ____ molecules of H_2O, and a total of ____ atoms.

3-16. Calculate how many methanol (CH_3OH) molecules and how many hydrogen, carbon, and oxygen atoms there are in 20.0 g of methanol (commonly called wood alcohol).

3-17. Male silkworm moths are attracted to an organic sex attractant with a molecular mass of 238. If 4.10×10^{-6} g of this attractant is released into $4.10 \times 10^3\ m^3$ of air, how many molecules of attractant are there in $1.00\ cm^3$ of air?

3-18. A trace amount of nerve gas escapes into the air. A local newspaper reports that we breathe 5.24×10^8 nerve gas molecules per breath of air (500 mL). Is there any reason for panic if the molecular mass of the gas is 238 and the substance is harmless in concentrations of less than $1.0\ pg \cdot L^{-1}$?

EMPIRICAL FORMULA

3-19. Calcium carbide produces acetylene when water is added to it. The acetylene evolved is burned to provide the light source on spelunkers' helmets. Chemical analysis shows that calcium carbide is 62.5 percent (by mass) calcium and 37.5 percent (by mass) carbon. What is its empirical formula?

3-20. Rust occurs when iron metal reacts with the oxygen in the air. Chemical analysis shows that dry rust is 69.9 percent iron and 30.1 percent oxygen by mass. Determine its empirical formula.

3-21. A 1.23-g sample of copper metal is reacted completely with chlorine gas to produce 2.61 g of copper chloride. Determine the empirical formula for this chloride.

3-22. A 3.78-g sample of iron metal is reacted with sulfur to produce 5.95 g of iron sulfide. Determine the empirical formula of this compound.

3-23. A 2.00-g sample of bromine was reacted completely with excess fluorine. The mass of the compound formed was 4.37 g. Determine its empirical formula.

3-24. A 5.00-g sample of aluminum metal is burned in an oxygen atmosphere to provide 9.45 g of aluminum oxide. Use these data to determine the empirical formula of aluminum oxide.

3-25. Freons are gases used as refrigerants. Chemical analysis shows that a certain Freon is 9.9 percent carbon, 58.7 percent chlorine, and 31.4 percent fluorine by mass. Determine its empirical formula.

3-26. Lead sulfate is one of the components in lead storage batteries. Chemical analysis shows that it is 68.3 percent lead, 10.6 percent sulfur, and 21.1 percent oxygen by mass. What is its empirical formula?

DETERMINATION OF ATOMIC MASS

3-27. A 2.885-g sample of metal is reacted with oxygen to yield 3.365 g of the oxide M_2O_3. Compute the atomic mass of the element M.

3-28. An element forms a chloride whose formula is XCl_4, which is known to consist of 75.0 percent chlorine by mass. Find the atomic mass of X and identify it.

3-29. A sample of a compound with the formula $MCl_2 \cdot 2H_2O$ has a mass of 0.642 g. When the compound is heated to remove the water of hydration (represented by $\cdot 2H_2O$ in the formula), 0.0949 g of water is collected. What element is M?

3-30. The formula of an acid is only partially known as HXO_3. The mass of 0.0133 mol of this acid is 1.123 g. Find the atomic mass of X and identify the element represented by X.

MOLECULAR FORMULAS

3-31. Acetone is an important chemical solvent; a familiar home use is as a nail polish remover. Chemical analysis shows that acetone is 62.0 percent carbon, 10.4 percent hydrogen, and 27.5 percent oxygen by mass. Determine the empirical formula of acetone. In a separate experiment, the molecular mass is found to be 58.1. What is the molecular formula of acetone?

3-32. Glucose, one of the main sources of energy used by living organisms, has a molecular mass of 180.2. Chemical analysis shows that glucose is 40.0 percent carbon, 6.71 percent hydrogen, and 53.3 percent oxygen by mass. Determine its molecular formula.

3-33. A class of compounds called sodium metaphosphates was used as additives to detergents to improve their cleaning ability. One of them has a molecular mass of 612. Chemical analysis shows that this sodium metaphosphate consists of 22.5 percent sodium, 30.4 percent phosphorus, and 47.1 percent oxygen by mass. Determine the molecular formula of this compound.

3-34. A hemoglobin sample was found to be 0.373 percent (by mass) iron. Given that there are four iron atoms per hemoglobin molecule, estimate the molecular mass of hemoglobin.

CALCULATIONS INVOLVING CHEMICAL REACTIONS

3-35. Elemental hydrogen (H_2), which may be used as a synthetic fuel in the future, does not exist in nature. Hydrogen can be produced from coal by a process called coal gasification, in which coal is heated to a temperature of 325 to 625°C to drive off volatile substances and produce a carbon residue called *char*. The char is then reacted with steam at 725°C according to the reaction

$$\underset{\text{char}}{C(s)} + \underset{\text{steam}}{H_2O(g)} \rightarrow \underset{\text{synthesis gas}}{CO(g) + H_2(g)}$$

How many metric tons of H_2 can be obtained from 1.00 metric ton of char?

3-36. Benzene reacts with chlorine in the presence of the catalyst aluminum chloride to produce chlorobenzene according to the equation

$$C_6H_6(l) + Cl_2(g) \xrightarrow{AlCl_3} C_6H_5Cl(l) + HCl(g)$$

Chlorobenzene is used as a solvent and in the synthesis of pesticides. How many grams of chlorobenzene can be obtained from 150 g of benzene?

3-37. Chlorine is produced industrially by the electrolysis of brine, which is a solution of naturally occurring salt and consists mainly of sodium chloride:

$$2NaCl(aq) + 2H_2O(l) \xrightarrow{\text{elec}} 2NaOH(aq) + Cl_2(g) + H_2(g)$$

The other products, sodium hydroxide and hydrogen, are also valuable commercial compounds. How many kilograms of each product can be obtained from the electrolysis of 1.00 kg of salt that is 95 percent sodium chloride by mass?

3-38. Calcium phosphate reacts with silicon dioxide, which is found in sand, according to the reaction

$$2Ca_3(PO_4)_2(s) + 6SiO_2(s) \rightarrow 6CaSiO_3(s) + P_4O_{10}(g)$$

How many grams of P_4O_{10} can be produced from 1.00 kg of sand that is 60 percent by mass SiO_2? This reaction is used to extract phosphorous oxide from phosphate-containing rocks.

3-39. The most common ore of arsenic is mispickel, FeSAs. Upon heating this ore, free arsenic is obtained:

$$FeSAs(s) \rightarrow FeS(s) + As(s)$$

How many grams of FeSAs are required to produce 10.0 g of arsenic?

3-40. Glucose is used as an energy source by the human body. The overall reaction in the body is

$$C_6H_{12}O_6(aq) + 6O_2(g) \rightarrow 6CO_2(g) + 6H_2O(l)$$

Calculate the number of grams of oxygen required to convert 28 g of glucose to CO_2 and H_2O. Also compute the number of grams of CO_2 produced.

3-41. An ore is analyzed for its lead content as follows. A sample is dissolved in water; then sodium sulfate is added to precipitate the lead as lead sulfate ($PbSO_4$). The reaction can be written

$$Pb^{2+}(aq) + SO_4^{2-}(aq) \rightarrow PbSO_4(s)$$

It was found that 13.73 g of lead sulfate was precipitated from a sample of ore having a mass of 53.92 g. How many grams of lead are there in the sample? What is the mass percentage of lead in the ore?

3-42. An ore is assayed for its silver content as follows. A sample is dissolved in water; then sodium chloride is added to precipitate the silver according to the reaction

$$Ag^+(aq) + Cl^-(aq) \rightarrow AgCl(s)$$

It was found that 3.74 g of silver chloride was precipitated from a sample of ore having a mass of 24.31 g. What is the mass percentage of silver in the ore?

3-43. Zinc is produced from its principal ore, sphalerite (ZnS), by the two-step process

$$2ZnS(s) + 3O_2(g) \rightarrow 2ZnO(s) + 2SO_2(g)$$
$$ZnO(s) + C(s) \rightarrow Zn(s) + CO(g)$$

How many kilograms of zinc can be produced from 2.00×10^5 kg of ZnS?

3-44. Titanium is produced from its principal ore, rutile (TiO_2), by the two-step process

$$TiO_2(s) + 2Cl_2(g) + 2C(s) \rightarrow TiCl_4(g) + 2CO(g)$$
$$TiCl_4(g) + 2Mg(s) \rightarrow Ti(s) + 2MgCl_2(s)$$

How many kilograms of titanium can be produced from 4.10×10^3 kg of TiO_2?

3-45. Nitric acid, HNO_3, is made commercially from ammonia by the Ostwald process, which was developed by the German chemist Wilhelm Ostwald. The process consists of three steps:

$$4NH_3(g) + 5O_2(g) \rightarrow 4NO(g) + 6H_2O(g)$$
$$2NO(g) + O_2(g) \rightarrow 2NO_2(g)$$
$$3NO_2(g) + H_2O(l) \rightarrow 2HNO_3(aq) + NO(g)$$

How many kilograms of nitric acid can be produced from 6.40×10^4 kg of ammonia?

3-46. Antimony is usually found in nature as the mineral stibnite, Sb_2S_3. Pure antimony can be obtained by first converting the sulfide to an oxide and then heating the oxide with coke (carbon). The reactions are

$$2Sb_2S_3(s) + 9O_2(g) \rightarrow Sb_4O_6(s) + 6SO_2(g)$$
$$Sb_4O_6(s) + 6C(s) \rightarrow 4Sb(s) + 6CO(g)$$

How many grams of antimony are formed from 500 g of stibnite?

LIMITING REACTANT

3-47. Potassium nitrate is widely used as a fertilizer because it provides two essential elements, potassium and nitrogen. It is made by mixing potassium chloride and nitric acid in the presence of oxygen according to the equation

$$4KCl(aq) + 4HNO_3(aq) + O_2(g) \rightarrow 4KNO_3(aq) + 2Cl_2(g) + 2H_2O(l)$$

How many kilograms of potassium nitrate will be produced from 81.6 kg of potassium chloride and 40.8 kg of nitric acid? An important by-product is chlorine. How many kilograms of chlorine will be produced?

3-48. Phosphorus is directly under nitrogen in the periodic table and forms a compound similar to ammonia. The compound has the chemical formula PH_3 and is called phosphine. It can be prepared by the reaction

$$P_4(s) + 3NaOH(aq) + 3H_2O(l) \rightarrow PH_3(g) + 3NaH_2PO_2(aq)$$

If 0.610 g of phosphorus and 0.250 g of NaOH are reacted with $H_2O(l)$ in excess, how many grams of phosphine will be obtained?

3-49. Sodium hydroxide reacts with phosphoric acid according to the equation

$$3NaOH(aq) + H_3PO_4(aq) \rightarrow Na_3PO_4(aq) + 3H_2O(l)$$

Suppose that 60.0 g of sodium hydroxide is added to 20.0 g of phosphoric acid. How much Na_3PO_4 will be produced?

60g NaOH / 40.00 g/mol = 1.50 mol NaOH   20 g H_3PO_4 / 97.99 g/mol = 0.204 mol H_3PO_4 (limiting reactant)

3-51. A 2.00-g sample of Fe_3O_4 is reacted with 7.50 g of O_2 to produce Fe_2O_3:

$$4Fe_3O_4(s) + O_2(g) \rightarrow 6Fe_2O_3(s)$$

Compute the number of grams of Fe_2O_3 produced.

2 g Fe_3O_4 / 231.55 g/mol = 8.64×10^{-3} mol Fe_3O_4 (limiting reactant)

7.5 g O_2 / 32 g/mol = 0.234 mol O_2

8.64×10^{-3} mol Fe_3O_4 × 6 mol Fe_2O_3 / 4 mol Fe_3O_4 = 1.296×10^{-2} mol Fe_2O_3

1.296×10^{-2} mol Fe_2O_3 × 159.70 g/mol = 2.07 g Fe_2O_3

PREPARATION OF SOLUTIONS

3-53. A saturated solution of calcium hydroxide, $Ca(OH)_2$, contains 0.185 g per 100 mL of solution. Calculate the molarity of a saturated calcium hydroxide solution.

M = mol/L

0.185 g $Ca(OH)_2$ / 74.10 g/mol = 2.50×10^{-3} mol   M = 2.50×10^{-3} mol / 0.1 L   M = 0.0250

3-55. Sodium hydroxide is extremely soluble in water. A saturated solution contains 572 g of NaOH per liter of solution. Calculate the molarity of a saturated NaOH solution.

572 g / 40.00 g/mol = 14.3 mol   M = 14.3 mol / 1 L   M = 14.3

3-57. Explain how you would prepare 1.00 L of a 0.250 M aqueous solution of sucrose, $C_{12}H_{22}O_{11}$. This solution is frequently used in biological experiments.

M = mol/L   0.250 M = mol / 1 L   mol = 0.250   0.250 mol × 342.3 g/mol = 85.6 g $C_{12}H_{22}O_{11}$

Add 85.6 g $C_{12}H_{22}O_{11}$ to ½ L of water, then fill to 1 liter.

3-50. Bromine can be prepared by adding $Cl_2(g)$ to an aqueous solution of sodium bromide. The reaction is

$$2NaBr(aq) + Cl_2(g) \rightarrow Br_2(l) + 2NaCl(aq)$$

How many grams of bromine are formed if 25.0 g of NaBr and 25.0 g of Cl_2 are reacted?

0.204 mol Na_3PO_4 × 163.94 g/mol = 33.4 g Na_3PO_4

3-52. The following is one of the side reactions in the manufacture of rayon from wood pulp:

$$3CS_2(g) + 6NaOH(aq) \rightarrow 2Na_2CS_3(aq) + Na_2CO_3(aq) + 3H_2O(l)$$

How many grams of each product are formed when 1.00 kg of each reactant is used?

3-54. Calculate the molarity of a saturated solution of sodium hydrogen carbonate, $NaHCO_3$ (baking soda), which contains 69.0 g in 1.00 L of solution.

3-56. A cup of coffee may contain as much as 300 mg of caffeine, $C_8H_{10}N_4O_2$. Compute the molarity of caffeine in one cup of coffee (4 cups = 0.946 L).

3-58. How would you prepare 50.0 mL of 0.200 M $CuSO_4$ solution, starting with solid $CuSO_4 \cdot 5H_2O$?

CALCULATIONS INVOLVING SOLUTIONS

3-59. Zinc reacts with hydrochloric acid according to

$$Zn(s) + 2HCl(aq) \rightarrow ZnCl_2(aq) + H_2(g)$$

How many milliliters of 3.00 M HCl are required to react with 4.33 g of Zn?

4.33 g Zn / 65.38 g/mol = 6.623×10^{-2} mol Zn 1.325×10^{-1} mol HCl

M = mol/L   3 M = 1.325×10^{-1} mol HCl / L   L = 0.0442   44.2 mL

3-61. Sodium hypochlorite, NaClO, is used as a bleaching agent in many commercial bleaches. Sodium hypochlorite can be prepared by the reaction

$$Cl_2(g) + 2NaOH(aq) \rightarrow NaClO(aq) + NaCl(aq) + H_2O(l)$$

How many grams of Cl_2 are required to react with 10.0 L of 5.00 M NaOH?

5 M NaOH = mol / 10 L   mol = 50   50 mol NaOH

25 mol Cl_2

25 mol Cl_2 × 70.90 g/mol = 1770 g Cl_2

3-60. Bromine is obtained commercially from natural brines from wells in Michigan and Arkansas by the reaction

$$Cl_2(g) + 2NaBr(aq) \rightarrow 2NaCl(aq) + Br_2(l)$$

If the concentration of NaBr is 4.00×10^{-3} M, how many grams of bromine can be obtained per cubic meter of brine? How many grams of chlorine are required?

3-62. Silver chloride can be dissolved in an aqueous solution of ammonia according to

$$AgCl(s) + 2NH_3(aq) \rightarrow Ag(NH_3)_2^+(aq) + Cl^-(aq)$$

How many liters of 0.100 M NH_3 solution would be required to dissolve 0.231 g of AgCl?

3-63. Suppose that we wish to react 15.0 g of solid potassium hydroxide, KOH, with 1.00 M aqueous nitric acid. How many milliliters of HNO_3 should be added? The reaction is

$$KOH(s) + HNO_3(aq) \rightarrow KNO_3(aq) + H_2O(l).$$

$\frac{15 g\ KOH}{56.11 g/mol} = 0.267$ mol KOH = 0.267 mol HNO_3

$M = \frac{mol}{L}$ L = 0.267

$1M = \frac{0.267 \text{ mol } HNO_3}{L}$ 267 mL HNO_3

3-64. Suppose we wish to know the concentration of an aqueous acetic acid solution, such as vinegar. A volume of 75.2 mL of the solution reacts with exactly 4.00 g of solid sodium hydroxide. What is the molarity of the solution? The reaction is

$$HC_2H_3O_2(aq) + NaOH(s) \rightarrow NaC_2H_3O_2(aq) + H_2O(l)$$

acetic acid; sodium acetate

3-65. Milk of Magnesia is a suspension of $Mg(OH)_2$ in water. Stomach acid is about 0.10 M HCl. Compute the number of grams of $Mg(OH)_2$ required to react with 47 mL of stomach acid. The reaction is

$$Mg(OH)_2(aq) + 2HCl(aq) \rightarrow MgCl_2(aq) + 2H_2O(l).$$

$M = \frac{mol}{L}$ mol = 0.0047 HCl = 4.7×10^{-3} mol

$0.10 M = \frac{mol}{0.047 L}$ 2.4×10^{-3} mol $Mg(OH)_2 \times \frac{58.33 g}{mole}$ = 0.14 g $Mg(OH)_2$

3-66. Suppose we wish to react 10.0 g of oxalic acid, $H_2C_2O_4$, with 0.50 M aqueous sodium hydroxide. How many milliliters of the NaOH solution must be added? The reaction is

$$H_2C_2O_4(s) + 2NaOH(aq) \rightarrow Na_2C_2O_4(aq) + 2H_2O(l)$$

3-67. Tetrodotoxin, $C_{11}H_{17}N_3O_8$, the puffer fish poison, will kill a human being when its concentration in the blood is as low as 1.0×10^{-8} M. Taking the total blood volume in a 200-lb human as 8.0 pints, estimate how many micrograms of tetrodotoxin will produce a concentration of 1.0×10^{-8} M in the blood (1 L = 1.06 qt and 1 qt = 2 pt).

$8 pt \times \frac{1 qt}{2 pt} \times \frac{1 L}{1.06 qt} = 3.8 L$

$M = \frac{n}{L}$

$1.0 \times 10^{-8} M = \frac{n}{3.8 L}$ $n = 3.8 \times 10^{-8}$ mol

3.8×10^{-8} mol $\times$ 319 g/mol = 1.2×10^{-5} g $C_{11}H_{17}N_3O_8$

= 12 μg $C_{11}H_{17}N_3O_8$

3-68. The LD_{50} value of a toxic substance is the amount (usually expressed as a ratio of milligrams of substance to kilograms of body weight) of the substance that results in the death of 50 percent of the test animals. The LD_{50} for caffeine, $C_8H_{10}N_4O_2$, administered orally to rats is 200 mg/kg. How many milliliters of a 0.11 M aqueous solution of caffeine constitutes an LD_{50} dose for a 450-g rat?

INTERCHAPTER B

Separation of Mixtures

Potassium compounds are obtained commercially by evaporation of brines. The above photo is a solar evaporation pond. The blue color is from a dye that is mixed with the solution to absorb heat faster and hence speed up evaporation.

When determining the physical and chemical characteristics of an element or a compound, chemists must be certain that the substance is pure. Most substances in nature, however, are mixtures, and so it is necessary for chemists to be able to separate a mixture into its various pure components. In this interchapter, we discuss some of the techniques that chemists use to separate mixtures and to purify substances.

B-1. SOME MIXTURES OF SOLIDS CAN BE SEPARATED BY DISSOLVING ONE OR MORE COMPONENTS

Let's consider the problem of separating a mixture composed of sugar, sand, iron filings, and gold dust into its four pure components (Figure B-1). The first thing to recognize about the mixture is that it is *heterogeneous;* that is, it is not uniform from point to point. The heterogeneity of the mixture can be clearly seen with the aid of a microscope (Figure B-2). The mixture could be separated into its four components by using a tweezers, a microscope, and a lot of time

Figure B-1 (a) A mixture of sugar, sand, iron filings, and gold dust.

(b) The pure, separated components of the mixture.

Figure B-2 A microscopic view of a mixture of sugar, sand, iron filings, and gold dust. Note that the mixture is heterogeneous, that is, not uniform from point to point and that each of the four components is clearly distinguishable.

Figure B-3 A magnet can be used to separate iron filings from a mixture of sugar, sand, iron filings, and gold dust. The iron filings are attracted by the magnet, but the other three components are not.

and patience; however, a much more rapid separation can be achieved with other methods. The iron filings can be separated from the mixture by using a magnet (Figure B-3), which attracts the magnetic iron particles but has no effect on the other three components. The same technique is used on a much larger scale in waste recycling to separate ferrous metals (iron, steel, nickel) from nonferrous refuse, such as aluminum, glass, paper, and plastics.

After the iron has been removed from our mixture, the sugar can be separated by adding water. Only the sugar dissolves in the water, leaving the sand and gold particles at the bottom of the container. The solution-plus-solid mixture can then be separated by *filtration* (Figure B-4). The sugar-water solution passes readily through the small pores in the filter paper, but the solid particles are too large to pass through and are trapped on the paper. The solid sugar can be

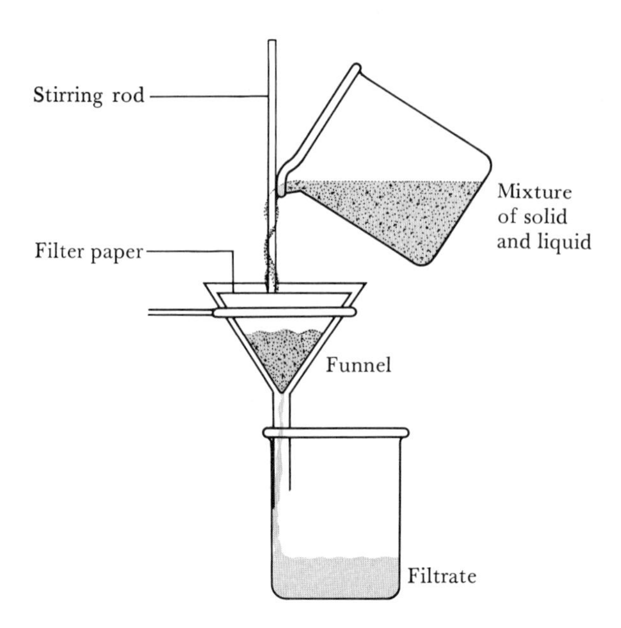

Figure B-4 Filtration can be used to separate a liquid from a solid. The liquid passes through filter paper, but the solid particles are too large to do so. Filter paper is available in a wide range of pore sizes, down to pores small enough (2.5×10^{-8} m) to remove bacteria (the smallest bacteria are about 1×10^{-7} m in diameter).

recovered from the sugar-water solution by evaporating the water, a process that leaves the recrystallized solid sugar in the container.

The sand and gold dust can be separated by panning or by sluice-box techniques, which rely on the differences in density of the two solids to achieve a separation. In simple panning, water is added to the mixture of sand and gold and the slurry is swirled in a shallow, saucer-shaped metal pan. The dense ($18.9\ g\cdot cm^{-3}$) gold particles collect near the center of the pan, whereas the less dense sand particles (2 to $3\ g\cdot cm^{-3}$) are swirled out of the pan. In the sluice-box technique, running water is passed over an agitated sand-gold mixture and the less dense sand particles rise higher in the water than the gold and are swept away in the stream of water (Figure B-5).

Figure B-5 Gold miners used sluice boxes to separate particles of gold from sand and crushed rock. Water channeled through the box carries the less dense sand and rock along and leaves denser gold particles behind.

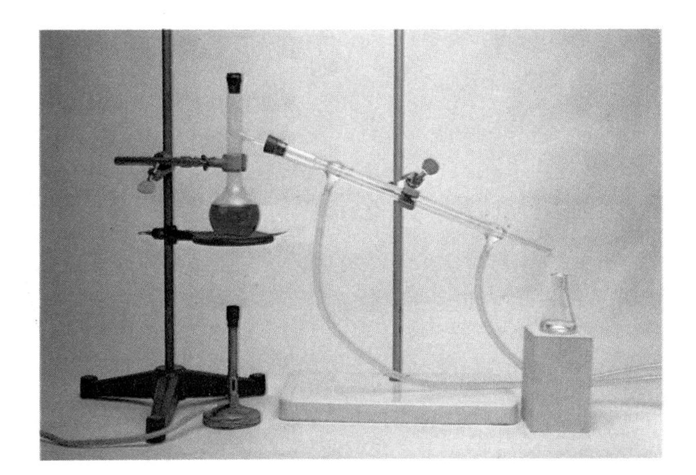

Figure B-6 A simple distillation apparatus can be used to separate a solid from a liquid in which it is dissolved. The solution in the distillation flask is heated, and the liquid is vaporized. The vapors rise in the distillation flask and pass into the condenser (the long, horizontal tube with the two hoses connected to it). The condenser is surrounded by a water jacket through which cooling water circulates. The vapor is cooled and condenses as it flows down the condenser tube and is collected in the flask at the right. The solid component of the solution remains behind in the distillation flask.

A similar mechanism occurs in stream beds that carry gold-bearing ores eroded from the banks of the stream. The larger gold particles tend to collect in depressions in the stream bed. Gold miners sometimes lay several logs in the stream bed at right angles to the flow. The dense gold particles tend to move along near the bottom of the stream over the logs and are trapped when they drop between the logs.

When fine gold particles are firmly attached to sand particles, the gold can be separated by shaking the mixture with liquid mercury, in which the gold dissolves. The sand, which floats on the mercury, is removed. The solution of gold in mercury is then separated by *distillation,* a process in which the mercury is boiled away and the solid gold is left behind. The mercury vapors are condensed (cooled and thereby converted back to liquid) in a *condenser.* A simple but typical distillation apparatus is shown in Figure B-6. Mercury distillation is usually carried out in an iron flask, and the mercury is collected and reused to extract more gold. Another example of distillation is the extraction of fresh water from seawater; the dissolved salts remain behind in the distillation flask after the water is boiled away.

B-2. VOLATILE LIQUIDS CAN BE SEPARATED BY FRACTIONAL DISTILLATION

The simple distillation apparatus shown in Figure B-6 is suitable for the separation of a liquid from a solution when a solid is dissolved in the liquid. The liquid is the only component that vaporizes or, in

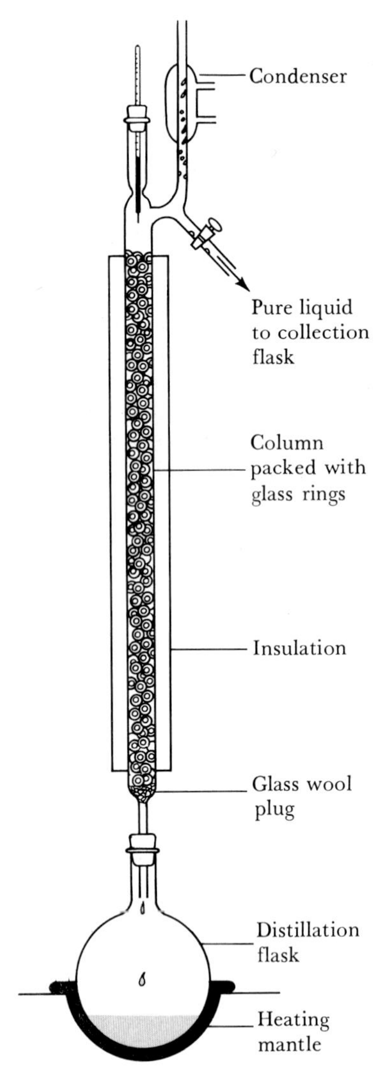

Figure B-7 A simple fractional distillation column. Repeated condensation and re-evaporation occurs along the entire column. The vapor becomes progressively richer in the more volatile component as it moves up the column.

other words, is the only *volatile* component. The idea is that the liquid is boiled away, leaving the solid behind. If a solution contains two or more volatile components, however, such as alcohol and water, then the components can be separated by taking advantage of differences in boiling point. The separation of a solution with two or more volatile components is achieved by *fractional distillation.*

The vapor phase over a solution of two volatile components contains both components, but is richer than is the liquid solution in the more volatile component, that is, the one with the lower boiling point. If this vapor is condensed and then re-evaporated, the resulting vapor will be even richer in the more volatile component. If this condensation-evaporation process is repeated many times, a separation of the two volatile components is achieved. Such a process is carried out automatically in a single fractional distillation column (Figure B-7). A fractional distillation column differs from an ordinary distillation column in that the former is packed with glass beads, glass rings, or glass wool. The packing material provides a large surface area for the repeated condensation-evaporation process.

Remarkable separations can be achieved with elaborate fractional distillation units. Fractional distillation techniques are used to separate heavy water (D_2O) from regular water. The heavy water is used on a large scale in nuclear reactors and as a coolant in heavy-water nuclear power plants. Regular water has a normal boiling point of 100.00°C, whereas heavy water has a normal boiling point of 101.42°C. Only 0.015 percent of the hydrogen atoms in regular water are the deuterium isotope. Nonetheless, a modern heavy-water distillation plant produces almost pure D_2O from regular water at a total cost of around $400 per kilogram of D_2O. These distillation plants have over 300 successive distillation stages and require an input of over 1 metric ton of water per gram of D_2O produced. Canada, which uses D_2O extensively in its nuclear reactors, has two heavy-water plants with a combined D_2O output capability of 1600 tons per year.

B-3. LIQUID OR GAS MIXTURES CAN BE SEPARATED BY CHROMATOGRAPHY

One of the most versatile and powerful separation techniques used by chemists is *chromatography,* of which there are several types. Gas-liquid chromatography (GLC) is used to separate mixtures of gases or volatile liquids. The mixture to be separated is vaporized and passed through a long, narrow column that is packed with fine solid particles thinly coated with some nonvolatile liquid. The gaseous mixture is swept through the column by an unreactive gas, such as helium (called the *carrier gas*). The different vapors move along the column at different rates because of their different degrees of interaction with the liquid coating on the packed solid particles. Those

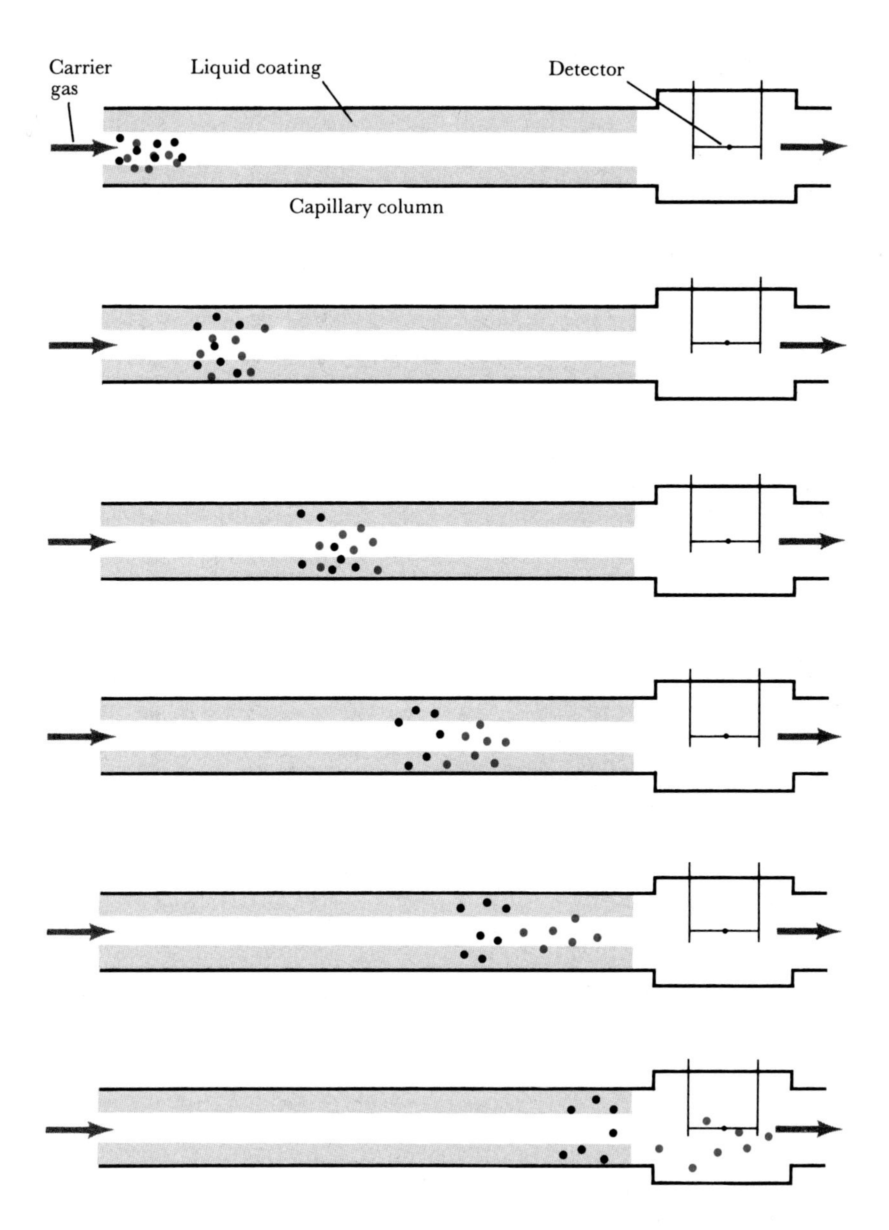

Figure B-8 Chromatographic separation takes place when a sample mixture (black and red circles) is driven by an unreactive carrier gas through a narrow column packed with a solid that is coated with a nonvolatile liquid. The sample mixture is separated because the liquid coating dissolves (and absorbs) various components of the sample to different degrees. After separation, each of the components passes through a detector and its presence is recorded on a chart, as in Figure B-9.

vapors that interact least with the liquid coating travel through the column most rapidly and exit first. The vapors that interact most strongly with the liquid coating lag behind and exit last. This process is shown schematically in Figure B-8. As each component emerges, it passes through a detector and its presence is recorded on a moving piece of chart paper. The result is called a *chromatogram.* Gas chromatography is a powerful tool for separating complex mixtures in which the components are only very slightly different from each other (Figure B-9). Gasoline mixtures containing over 100 different compounds can be separated and characterized by gas chromatography.

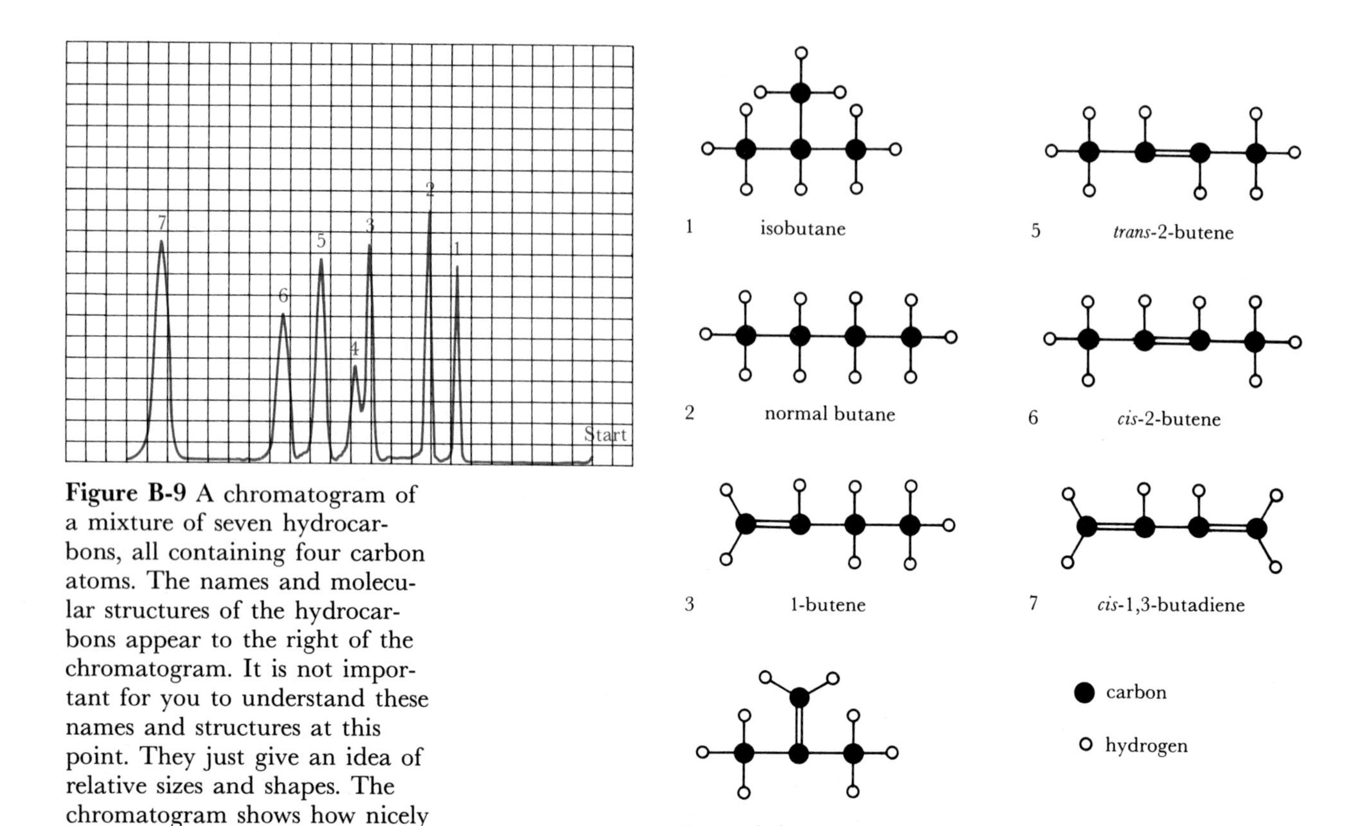

Figure B-9 A chromatogram of a mixture of seven hydrocarbons, all containing four carbon atoms. The names and molecular structures of the hydrocarbons appear to the right of the chromatogram. It is not important for you to understand these names and structures at this point. They just give an idea of relative sizes and shapes. The chromatogram shows how nicely these compounds are separated by a chromatography column. The separation of hydrocarbons is a common problem in the oil and synthetic rubber industries.

Other types of chromatography are *paper chromatography* and *thin-layer chromatography* (TLC). Both methods are used to separate compounds in solution. They are based upon the fact that if a piece of porous paper is hung with one end dipping in a liquid, then the liquid will wick up along the paper. A drop of solution containing the substances to be separated is placed on one edge of the paper, and the liquid is allowed to pass over the drop. Those components that interact least strongly with the paper and most strongly with the liquid are drawn farthest up the paper with the moving liquid. This technique is used often in biochemistry to separate the products of a reaction. Figure B-10 shows a mixture of amino acids separated by thin-layer chromatography.

The name chromatography is derived from the original chromatographic separation method used in 1906 by Mikhail Tswett, a Russian botanist. Tswett separated a chlorophyll solution into various colored bands (chroma means "color") by passing the solution through a column packed with pulverized calcium carbonate. Because this method involves a liquid moving past a stationary solid, it is called *liquid-solid chromatography*. The different chlorophyll components move through the column at different rates because of their different degrees of interaction with the calcium carbonate. As additional solvent is passed through the column, the chlorophyll compo-

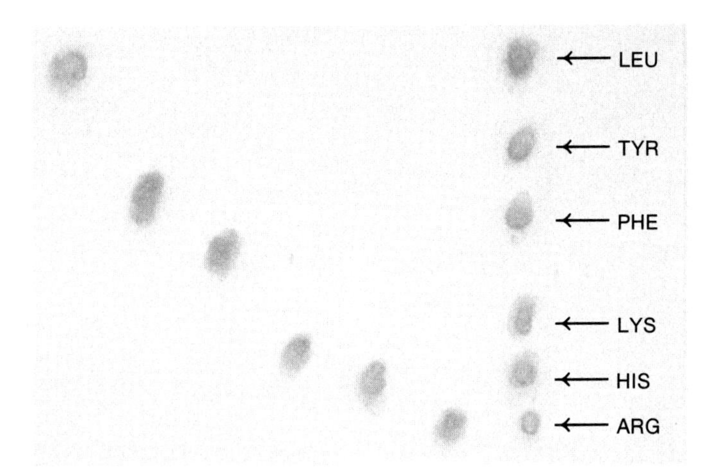

Figure B-10 Thin-layer chromatographic separation of an amino acid mixture. The identities of the particular amino acids are established by using known samples of the various amino acids in a comparison experiment.

Figure B-11 This liquid-solid chromatographic separation of a chlorophyll mixture shows the column with chlorophyll just added, the column with separated colored bands, and the first colored band coming off the column.

nents that are more strongly adsorbed on the surface of the calcium carbonate crystals move more slowly down the column and the components are separated (Figure B-11).

These are just a few of the methods that chemists use to separate mixtures and purify substances. It is possible to achieve extremely high degrees of purity. The purity required in the production of semiconductors is 99.99999 percent, and such purities are achieved routinely by semiconductor manufacturers.

QUESTIONS

B-1. Describe filtration.

B-2. What are the contrasting physical properties of gold and sand on which panning for gold depends?

B-3. Describe distillation.

B-4. What is the role of the condenser in a distillation apparatus like that shown in Figure B-6?

B-5. What is meant when a liquid is said to be volatile?

B-6. What is the procedure for separating a mixture of two volatile liquids?

B-7. Describe how fractional distillation works.

B-8. Describe how two gases or vapors can be separated by gas-liquid chromatography.

When colorless aqueous solutions of cadmium nitrate, $Cd(NO_3)_2$, and sodium sulfide, Na_2S, are mixed, a yellow-orange precipitate of cadmium sulfide, CdS, forms spontaneously.

4 / Chemical Reactions

Throughout your study of chemistry you will encounter many chemical reactions. One classification scheme for chemical reactions consists of four categories: (1) combination reactions, (2) decomposition reactions, (3) single-replacement reactions, and (4) double-replacement reactions.

Although not all chemical reactions fall into these four categories, a large number do, and it is helpful in first learning about chemical reactions to try to classify them into types rather than to view each one separately. In this chapter, we describe each of these four classes of reactions by way of examples. Our discussion of reaction types leads naturally to chemical nomenclature, to relative reactivities of metals, and to an introduction to the chemistry of acids and bases.

4-1. THE REACTION OF TWO SUBSTANCES TO FORM A SINGLE PRODUCT IS A COMBINATION REACTION

The simplest example of a *combination reaction* is that of a metal with a nonmetal. We saw several such reactions in Chapter 2. For example,

Metal		*Nonmetal*		*Ionic compound*
$2Na(s)$	+	$Cl_2(g)$	→	$2NaCl(s)$
$6Li(s)$	+	$N_2(g)$	→	$2Li_3N(s)$
$4Al(s)$	+	$3O_2(g)$	→	$2Al_2O_3(s)$

Figure 4-1 Sulfur burns in oxygen with a blue flame. The production of SO_2 by burning sulfur is a key reaction in the manufacture of sulfuric acid. Sulfur dioxide is also produced when sulfur-containing coal or petroleum is burned.

In each of these examples, the combination reaction occurs between a main-group metal and a main-group nonmetal. Electrons are transferred from the metal atoms to the nonmetal atoms, resulting in ionic compounds in which the ions have a noble-gas electron arrangement. As we saw in Chapter 2, the chemical formulas of the products can be written by using the ionic charges given in Figure 2-18. Note that these combination reactions are also oxidation-reduction reactions.

Many combination reactions occur between two nonmetals. For example, carbon burns in oxygen to give carbon dioxide:

$$C(s) + O_2(g) \rightarrow CO_2(g)$$

Sulfur burns in oxygen with a blue flame (Figure 4-1) to form sulfur dioxide, a colorless, toxic, foul-smelling gas:

$$S(s) + O_2(g) \rightarrow SO_2(g)$$

Reactions in which a substance is burned in oxygen are called *combustion reactions*, and these combination reactions of carbon and sulfur with oxygen are two examples of combustion reactions.

Some examples of combination reactions between nonmetals other than oxygen are

$$S(s) + 3F_2(g) \rightarrow \underset{\text{sulfur hexafluoride}}{SF_6(g)}$$

$$H_2(g) + Cl_2(g) \rightarrow \underset{\text{hydrogen chloride}}{2HCl(g)}$$

Carbon dioxide, sulfur dioxide, sulfur hexafluoride, and hydrogen chloride are *molecular compounds* (as opposed to ionic compounds). Molecular compounds are composed of neutral molecules, rather than ions, and are generally much more volatile than ionic compounds; that is, they are more readily vaporized than ionic solids. Thus the molecular compounds CO_2 and SO_2 are gases at 25°C, whereas the ionic compounds Na_2O and CaO are solids even at high temperatures.

4-2. THERE ARE MANY STABLE POLYATOMIC IONS

The combination reactions presented so far are combinations of two elements. There are also combination reactions between compounds. For example, the reaction between sodium hydroxide and carbon dioxide,

$$NaOH(s) + CO_2(g) \rightarrow \underset{\text{sodium hydrogen carbonate}}{NaHCO_3(s)}$$

Sodium hydrogen carbonate is sometimes called sodium bicarbonate.

is a combination reaction. This reaction can be used to remove CO_2 from air.

Sulfur dioxide can be removed from fuel combustion gases by a combination reaction with calcium oxide:

$$CaO(s) + SO_2(g) \rightarrow \underset{\text{calcium sulfite}}{CaSO_3(s)}$$

Sulfur trioxide reacts with magnesium oxide to form solid magnesium sulfate:

$$MgO(s) + SO_3(g) \rightarrow \underset{\text{magnesium sulfate}}{MgSO_4(s)}$$

Solid ammonium chloride forms when the gases ammonia and hydrogen chloride are mixed (Figure 4-2):

$$NH_3(g) + HCl(g) \rightarrow \underset{\text{ammonium chloride}}{NH_4Cl(s)}$$

None of the products in these four reactions are *binary compounds*, which are compounds composed of only two elements. Nevertheless, when, for example, $MgSO_4$ is dissolved in water, only two ions per $MgSO_4$ formula unit result:

Figure 4-2 Ammonia and hydrogen chloride are colorless gases. They react to produce the solid white compound ammonium chloride. The white cloud in the picture consists of small ammonium chloride particles formed where the gaseous NH_3 and HCl come into contact with each other. The $NH_3(g)$ comes from one bottle and the $HCl(g)$ comes from the other.

Table 4-1 Some common polyatomic ions

Positive ions	
ammonium	NH_4^+
mercury(I)[a]	Hg_2^{2+}
Negative ions	
acetate	$C_2H_3O_2^-$
carbonate	CO_3^{2-}
chlorate	ClO_3^-
chromate	CrO_4^{2-}
cyanide	CN^-
hydrogen carbonate ("bicarbonate")	HCO_3^-
hydrogen sulfate ("bisulfate")	HSO_4^-
hydroxide	OH^-
hypochlorite	ClO^-
nitrate	NO_3^-
nitrite	NO_2^-
perchlorate	ClO_4^-
permanganate	MnO_4^-
phosphate	PO_4^{3-}
sulfate	SO_4^{2-}
sulfite	SO_3^{2-}
thiosulfate	$S_2O_3^{2-}$

[a]Note that Hg_2^{2+} contains two Hg^+ ions joined together.

$$\underset{\text{magnesium sulfate}}{MgSO_4(s)} \xrightarrow[H_2O(l)]{} \underset{\text{magnesium ion}}{Mg^{2+}(aq)} + \underset{\text{sulfate ion}}{SO_4^{2-}(aq)}$$

(Recall that $H_2O(l)$ placed *below* the reaction arrow indicates that water is the solvent.) The SO_4^{2-} ion remains intact as a unit in the solution. Magnesium sulfate, $MgSO_4$, consists of Mg^{2+} ions and SO_4^{2-} ions. A *polyatomic ion* is an ion that contains more than one atom. There are a number of groups of atoms, such as SO_4^{2-}, that form stable polyatomic ions. The SO_4^{2-} ion is a *sulfate ion,* and thus $MgSO_4$ is magnesium sulfate. Table 4-1 lists a number of important polyatomic ions and their charges. Compounds containing the ions in Table 4-1 are named according to the rules for binary compounds, with the polyatomic ions treated as simple monatomic ions. For example, NaOH is called sodium hydroxide, KCN is called potassium cyanide, and NH_4Cl is called ammonium chloride.

Example 4-1: Name the following compounds:

$KMnO_4$ $CaCrO_4$ $AlPO_4$ NH_4NO_3

Solution: From Table 4-1, we see that MnO_4^- is called the permanganate ion, and so $KMnO_4$ is called potassium permanganate. The CrO_4^{2-} is called the chromate ion, and so $CaCrO_4$ is called calcium chromate. Similarly, PO_4^{3-} is called the phosphate ion, and so $AlPO_4$ is called aluminum phosphate. The fourth compound is made up of NH_4^+ and NO_3^- and so is called ammonium nitrate.

Example 4-2: Write the formula for each of the following compounds:

sodium sulfite calcium chlorate
ammonium sulfate barium hydroxide

Solution: The formula for sodium sulfite is written by combining Na^+ and SO_3^{2-}. Because SO_3^{2-} has an ionic charge of -2, it requires two Na^+ for each SO_3^{2-}, and so Na_2SO_3 is the formula for sodium sulfite.

Calcium chlorate requires two ClO_3^- for each Ca^{2+} and so the formula is $Ca(ClO_3)_2$. Note that the entire chlorate ion is enclosed in parentheses and the subscript 2 lies outside the parentheses. One formula unit of $Ca(ClO_3)_2$ contains one calcium atom, two chlorine atoms, and six oxygen atoms.

Ammonium sulfate contains two ammonium ions, NH_4^+, and one sulfate ion, SO_4^{2-}, and so its formula is $(NH_4)_2SO_4$. Once again, note the use of parentheses. One formula unit of $(NH_4)_2SO_4$ contains two nitrogen atoms, eight hydrogen atoms, one sulfur atom, and four oxygen atoms.

Barium hydroxide involves Ba^{2+} and OH^- ions and thus has the formula $Ba(OH)_2$.

Problems 4-1 to 4-12 deal with nomenclature and writing chemical formulas.

4-3. TRANSITION-METAL IONS HAVE MORE THAN ONE POSSIBLE IONIC CHARGE

Most transition metals can form ions having more than one possible charge, and so it is necessary to indicate the charge of the metal when naming transition-metal compounds. The ionic charges of some transition metals are given in Table 4-2. For those metals with more than one possible charge, the charge is indicated by a roman numeral after the name of the metal. For example, according to Table 4-2, iron has two possible charges, +2 and +3. The two chlorides of iron are

$FeCl_2$	iron(II) chloride
$FeCl_3$	iron(III) chloride

An older nomenclature that is still in use is shown in Table 4-3. Using these older names, we have

Compound	*Modern name*	*Old name*
$FeCl_2$	iron(II) chloride	ferrous chloride
$FeCl_3$	iron(III) chloride	ferric chloride

Note that the *-ous* ending indicates the lower ionic charge and the *-ic* ending indicates the higher ionic charge. A compound that became famous as an ingredient in fluoride-containing toothpastes is stan-

Table 4-2 Some common metals whose ions have more than one possible ionic charge

Metal	*Ionic charges*	*Symbols*	*Names*
chromium	+2 and +3	Cr^{2+} and Cr^{3+}	chromium(II) and chromium(III)
cobalt	+2 and +3	Co^{2+} and Co^{3+}	cobalt(II) and cobalt(III)
copper	+1 and +2	Cu^{+} and Cu^{2+}	copper(I) and copper(II)
gold	+1 and +3	Au^{+} and Au^{3+}	gold(I) and gold(III)
iron	+2 and +3	Fe^{2+} and Fe^{3+}	iron(II) and iron(III)
mercury	+1 and +2	Hg_2^{2+} and Hg^{2+}	mercury(I) and mercury(II)
tin	+2 and +4	Sn^{2+} and Sn^{4+}	tin(II) and tin(IV)

Compounds of transition-metal ions of different ionic charges often have different colors. Note here the pale blue of CuCl and the darker blue of $CuCl_2$.

Table 4-3 Older nomenclature for metal ions having more than one possible ionic charge

Ion	*Old name*	*Modern (systematic) name*
Cr^{2+}	chromous	chromium(II)
Cr^{3+}	chromic	chromium(III)
Co^{2+}	cobaltous	cobalt(II)
Co^{3+}	cobaltic	cobalt(III)
Cu^{+}	cuprous	copper(I)
Cu^{2+}	cupric	copper(II)
Au^{+}	aurous	gold(I)
Au^{3+}	auric	gold(III)
Fe^{2+}	ferrous	iron(II)
Fe^{3+}	ferric	iron(III)
Hg_2^{2+}	mercurous	mercury(I)
Hg^{2+}	mercuric	mercury(II)
Sn^{2+}	stannous	tin(II)
Sn^{4+}	stannic	tin(IV)

nous fluoride, SnF_2. The modern name for SnF_2 is tin(II) fluoride. In this book, we shall in most cases use the modern (systematic) names of compounds or ions. The older names are given in Table 4-3 only for reference.

4-4. MANY OXIDES COMBINE WITH WATER TO PRODUCE ACIDS OR BASES

Because of the central role that water plays in chemistry, the combination reactions of metal oxides and nonmetal oxides with water are particularly important. If we dissolve solid sodium oxide in water, the water acts not only as the solvent but also as a reactant:

$$Na_2O(s) + H_2O(l) \xrightarrow[H_2O(l)]{} 2NaOH(aq)$$

Similarly

$$BaO(s) + H_2O(l) \xrightarrow[H_2O(l)]{} Ba(OH)_2(aq)$$

In aqueous solution, NaOH(*aq*) exists as $Na^+(aq)$ and $OH^-(aq)$ and $Ba(OH)_2(aq)$ exists as $Ba^{2+}(aq)$ and $2OH^-(aq)$. Thus we can write

$$Na_2O(s) + H_2O(l) \xrightarrow[H_2O(l)]{} 2Na^+(aq) + 2OH^-(aq)$$

and

$$BaO(s) + H_2O(l) \xrightarrow[H_2O(l)]{} Ba^{2+}(aq) + 2OH^-(aq)$$

basic anhydride base

Compounds (such as sodium hydroxide and barium hydroxide) that yield hydroxide ions when dissolved in water are called *bases*. Oxides (such as sodium oxide and barium oxide) that yield bases when dissolved in water are called *basic anhydrides* (the word anhydride means without water). Metallic oxides are always basic anhydrides.

Many metal oxides are not soluble in water, and so there is no reaction when the oxide is placed in contact with water. For example, aluminum oxide, Al_2O_3, which is what makes the surface of aluminum doors and window frames dull, does not dissolve in water, and so we write

Scanning electron micrograph of aluminum oxide, Al_2O_3. (×4500)

$$Al_2O_3(s) + H_2O(l) \rightarrow \text{no reaction}$$

The only metal oxides that are soluble in water are the Group 1 metal oxides and some of the Group 2 metal oxides.

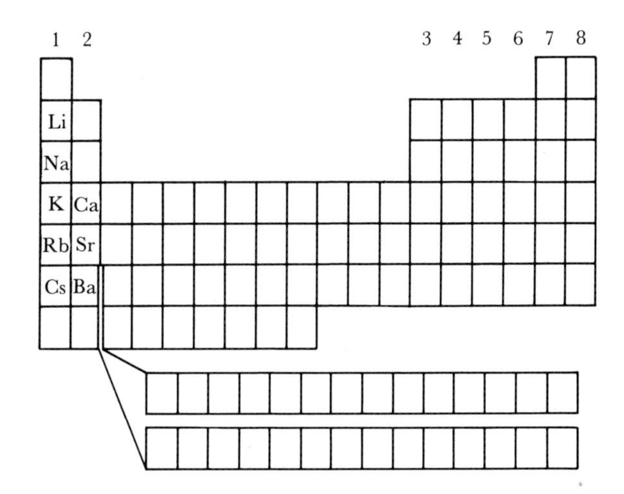

The metals whose oxides are water soluble

Example 4-3: Write the chemical equation for the reaction for the dissolution of potassium oxide, K_2O, in water.

Solution: Potassium is a Group 1 metal, and so its oxide reacts with water much as sodium oxide does. Using the above reaction of Na_2O with water as a guide, we write

$$K_2O(s) + H_2O(l) \xrightarrow[H_2O(l)]{} 2K^+(aq) + 2OH^-(aq)$$

Because hydroxide ions are produced when K_2O is dissolved in water, K_2O is a basic anhydride.

Water-soluble metal oxides yield bases when dissolved in water; water-soluble nonmetal oxides yield acids when dissolved in water. An *acid* is a compound that yields *hydrogen ions* when dissolved in water. Experimental evidence suggests that H^+ in aqueous solution exists in several forms, such as H_3O^+ and $H_9O_4^+$. The species H_3O^+, called the *hydronium ion*, appears to be the dominant species, however. Note that a hydronium ion is a combination of a hydrogen ion, H^+, and a water molecule. For simplicity, we denote a hydrogen ion in water by $H^+(aq)$ instead of by $H_3O^+(aq)$. The (aq) notation emphasizes that the hydrogen ion is in contact with one or more water molecules.

Oxides that yield acids when dissolved in water are called *acidic anhydrides*. The acidic anhydride of nitric acid is dinitrogen pentoxide:

$$\underset{\text{dinitrogen pentoxide}}{N_2O_5(s)} + H_2O(l) \rightarrow \underset{\text{nitric acid}}{2HNO_3(l)}$$

Nitric acid is an acid because it yields $H^+(aq)$ when dissolved in water:

$$HNO_3(l) \xrightarrow[H_2O(l)]{} H^+(aq) + NO_3^-(aq)$$

It is not possible, in most cases, to tell if a compound is an acid or not simply from its chemical formula. The fact that a compound contains hydrogen atoms does not guarantee that it will yield $H^+(aq)$ ions in water. For example, H_2 is not an acid. The hydrogen atoms, or actually the protons, that result in $H^+(aq)$ ions when an acid is dissolved in water are called *acidic hydrogen atoms,* or *acidic protons.* Thus, nitric acid, HNO_3, is said to have one acidic proton, whereas sulfuric acid, H_2SO_4, is said to have two acidic protons:

$$H_2SO_4(l) \xrightarrow[H_2O(l)]{} 2H^+(aq) + SO_4^{2-}(aq)$$

The names of some common polyatomic acids and their corresponding anions are given in Table 4-4.

Table 4-4 Some common polyatomic acids and their anions

Acid	*Formula*	*Anion*	*Formula*
acetic	$HC_2H_3O_2$	acetate	$C_2H_3O_2^-$
carbonic	H_2CO_3	carbonate	CO_3^{2-}
nitric	HNO_3	nitrate	NO_3^-
perchloric	$HClO_4$	perchlorate	ClO_4^-
phosphoric	H_3PO_4	phosphate	PO_4^{3-}
sulfuric	H_2SO_4	sulfate	SO_4^{2-}

Example 4-4: Write a chemical equation showing the dissolution of perchloric acid in water.

Solution: According to Table 4-4, perchloric acid is $HClO_4$ and the corresponding anion is ClO_4^-. Consequently, we write

$$HClO_4(l) \xrightarrow[H_2O(l)]{} H^+(aq) + ClO_4^-(aq)$$

The acids listed in Table 4-4 are called *oxyacids* because they contain oxygen atoms. There is another group of acids called *binary acids.* Binary acids consist of two elements, one of which must be hydrogen. Binary acids are produced by dissolving certain binary hydrogen compounds (those containing only hydrogen and one other element) in water. The most important binary acid is hydrochloric acid, which is made by dissolving hydrogen chloride gas in water:

$$HCl(g) \xrightarrow[H_2O(l)]{} \underbrace{H^+(aq) + Cl^-(aq)}_{\text{hydrochloric acid}}$$

Some binary acids are given in Table 4-5.

Table 4-5 Some binary acids

Acid	*Ions in aqueous solution*	*Corresponding gas*
hydrazoic	$H^+(aq)$ $N_3^-(aq)$	hydrogen azide, HN_3
hydrobromic	$H^+(aq)$ $Br^-(aq)$	hydrogen bromide, HBr
hydrochloric	$H^+(aq)$ $Cl^-(aq)$	hydrogen chloride, HCl
hydroiodic	$H^+(aq)$ $I^-(aq)$	hydrogen iodide, HI
hydrosulfuric	$H^+(aq)$ $S^{2-}(aq)$	hydrogen sulfide, H_2S

Figure 4-3 When mercury(II) oxide is heated, it decomposes into elemental mercury and oxygen gas. The red compound shown here is mercury(II) oxide. The elemental liquid mercury has condensed on the walls of the test tube. This reaction was one of the earliest methods used to produce oxygen in the laboratory. It is no longer used, however, because of the toxicity of the mercury vapor produced.

4-5. IN A DECOMPOSITION REACTION, A SUBSTANCE IS BROKEN DOWN INTO TWO OR MORE SIMPLER SUBSTANCES

Decomposition reactions are the opposite of combination reactions because they involve the breaking down of a substance into simpler substances. They are easy to recognize because there is usually only one reactant and more than one product.

When heated, many metal oxides decompose by giving off oxygen gas. Some examples are

$$2HgO(s) \xrightarrow{\text{high T}} 2Hg(l) + O_2(g)$$

$$2Ag_2O(s) \xrightarrow{\text{high T}} 4Ag(s) + O_2(g)$$

The first of these reactions was once used to produce oxygen in the laboratory (Figure 4-3).

Many metal carbonates decompose to the metal oxide and CO_2 gas upon heating. For example,

$$CaCO_3(s) \rightarrow CaO(s) + CO_2(g)$$

$$NiCO_3(s) \rightarrow NiO(s) + CO_2(g)$$

Calcium carbonate occurs in nature as the principal constituent of limestone and seashells (Figures 4-4 and 4-5).

Compounds that decompose rapidly and produce large volumes of gases in the process can be used as explosives. For example, ammonium nitrate decomposes by the reaction

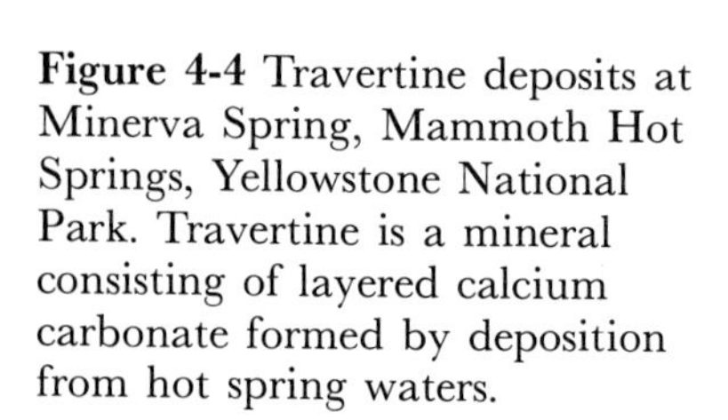

Figure 4-4 Travertine deposits at Minerva Spring, Mammoth Hot Springs, Yellowstone National Park. Travertine is a mineral consisting of layered calcium carbonate formed by deposition from hot spring waters.

$$2NH_4NO_3(s) \xrightarrow{200°C} 2N_2(g) + 4H_2O(g) + O_2(g)$$

The H_2O produced is shown as a gas because the energy that is released in this reaction vaporizes the water. In 1947, a large barge loaded with NH_4NO_3 exploded in the harbor of Texas City, Texas, killing 576 people and causing extensive damage. With proper precautions, however, NH_4NO_3 can be handled safely and is used extensively as a fertilizer.

Figure 4-5 Calcium carbonate is the principal constituent of eggshells. (a) A magnification of an egg shell showing the shell and the underlying membrane. (300×) (b) Further enlargement of the surface of the shell itself showing the crystalline structure. (12,000×)

Example 4-5: When potassium chlorate is heated, oxygen gas is evolved and solid potassium chloride formed. This reaction is used to generate small quantities of oxygen in the laboratory. Write a balanced chemical equation for the process.

Solution: The chemical formulas of the reactant and products are

$$KClO_3(s) \rightarrow KCl(s) + O_2(g)$$

Note that the equation as written is not balanced with respect to oxygen atoms; there are three oxygen atoms on the left and two on the right. We can balance the equation with respect to oxygen by placing a 2 in front of $KClO_3(s)$ ($2 \times 3 = 6$ oxygen atoms) and a 3 in front of $O_2(g)$ ($3 \times 2 = 6$ oxygen atoms):

$$2KClO_3(s) \rightarrow KCl(s) + 3O_2(g)$$

We now note that there are two K atoms and two Cl atoms on the left and only one K atom and one Cl atom on the right; thus we place a 2 in front of $KCl(s)$ to obtain the balanced equation:

$$2KClO_3(s) \rightarrow 2KCl(s) + 3O_2(g)$$

Note that the reaction is a decomposition reaction. It is called a *thermal decomposition* reaction because the decomposition is brought about by heating.

4-6. IN A SINGLE-REPLACEMENT REACTION, ONE ELEMENT IN A COMPOUND IS REPLACED BY ANOTHER

Titanium is used to make lightweight, high-strength alloys for airplanes and missiles. Titanium metal is prepared by reacting titanium tetrachloride with molten magnesium:

$$2Mg(l) + TiCl_4(g) \rightarrow 2MgCl_2(s) + Ti(s)$$

Figure 4-6 The reaction between iron metal and a dilute aqueous solution of sulfuric acid. The bubbles generated at the surface of the iron nail are hydrogen gas. The iron replaces the hydrogen in H_2SO_4 and thereby enters the solution as $FeSO_4(aq)$.

Note that the magnesium takes the place of the titanium in the chloride. A reaction in which an element in a compound is replaced by another element is called a *single-replacement reaction* or a *substitution reaction.*

An important type of single-replacement reaction involves the reaction between a reactive metal, such as iron, and a dilute solution of an acid, such as sulfuric acid. As the reaction takes place, bubbles appear at the surface of the iron (Figure 4-6) and the iron slowly dissolves. The equation for the reaction that occurs is

$$\mathrm{Fe}(s) + 2\mathrm{H}^+(aq) + \mathrm{SO}_4^{2-}(aq) \rightarrow \mathrm{Fe}^{2+}(aq) + \mathrm{SO}_4^{2-}(aq) + \mathrm{H}_2(g)$$

Although the sulfuric acid and iron(II) sulfate exist as ions in aqueous solution, the equation often is written in the following simplified manner:

$$\mathrm{Fe}(s) + \mathrm{H_2SO_4}(aq) \rightarrow \mathrm{FeSO_4}(aq) + \mathrm{H_2}(g)$$

Hydrogen gas is only slightly soluble in water and so appears as bubbles that escape from the solution. The iron metal is consumed because it reacts with the H_2SO_4 to produce the soluble compound $FeSO_4$. Note that the iron replaces the hydrogen atoms in H_2SO_4 to yield $FeSO_4$. One of the properties of acids is that dilute solutions of acids attack reactive metals to produce hydrogen gas.

Example 4-6: Magnesium is a reactive Group 2 metal. Write the chemical equation that describes the reaction of magnesium with hydrobromic acid, HBr(aq).

Solution: Magnesium reacts with hydrobromic acid to produce hydrogen. The chemical equation for the reaction is

$$\mathrm{Mg}(s) + 2\mathrm{HBr}(aq) \rightarrow \mathrm{MgBr_2}(aq) + \mathrm{H_2}(g)$$

All the reactions between reactive metals and dilute acids may be pictured as the replacement of the hydrogen atoms of the acid by the metal.

4-7. METALS CAN BE ORDERED IN TERMS OF RELATIVE REACTIVITY

Silver nitrate dissolves in water to form a colorless, clear solution. If we place a copper wire in the $AgNO_3$ solution, then the solution becomes blue (Figure 4-7). In addition, a deposit of silver metal

(a) (b)

Figure 4-7 A solution of silver nitrate, $AgNO_3$, in which a copper wire has been placed turns blue. (a) The original silver nitrate solution is clear and colorless. (b) When a copper wire is added, the solution slowly turns blue as a result of the formation of $Cu(NO_3)_2(aq)$.

"whiskers" forms on the copper wire (Figure 4-8). The equation for the reaction is

$$\underset{\text{colorless}}{Cu(s) + 2AgNO_3(aq)} \rightarrow \underset{\text{blue}}{Cu(NO_3)_2(aq)} + 2Ag(s)$$

Copper(II) nitrate dissolved in water forms a blue solution. We can see from the chemical equation that the copper replaces the silver in the nitrate compound. On the basis of this reaction, we conclude that copper metal is more reactive than silver metal because copper metal displaces silver from its nitrate salt.

If we carry out a similar reaction with zinc metal and $Cu(NO_3)_2(aq)$ solution, then we find that

$$Zn(s) + Cu(NO_3)_2(aq) \rightarrow Cu(s) + Zn(NO_3)_2(aq)$$

We can conclude from this reaction that zinc is more reactive than copper. Thus, on the basis of these reactions we can arrange zinc, copper, and silver in order of their relative reactivities:

zinc is more reactive than copper
copper is more reactive than silver

Figure 4-8 When a copper wire is placed in a silver nitrate solution, the copper replaces the silver and "whiskers" of elemental silver deposit onto the wire in the form of needle-like growths.

Table 4-6 Reactivity series for some common metals

increasing reactivity ↑	Metal	
	K	react directly with cold water and vigorously with dilute acids
	Ca	
	Al	do not react with cold water, but do react with dilute acids
	Zn	
	Fe	
	Sn	
	Pb	
	Cu	do not react with water or dilute acids
	Hg	
	Ag	
	Au	

Thus we have for the order of reactivities of Zn, Cu, and Ag

	↑ increasing reactivity
Zn	
Cu	
Ag	

By performing additional experiments similar to those just described, we can determine the position of other metals in the *reactivity series* of the metals (Table 4-6). In general, a metal will displace from a compound any metal that lies below it in the activity series.

Example 4-7: Predict whether or not a reaction occurs in the following cases and complete and balance the chemical equation if a reaction does occur:

$$Zn(s) + HgCl_2(aq) \rightarrow$$

$$Zn(s) + Ca(ClO_4)_2(aq) \rightarrow$$

Solution: Zinc lies above mercury in the reactivity series given in Table 4-6, and so zinc replaces mercury from its chloride compound:

$$Zn(s) + HgCl_2(aq) \rightarrow ZnCl_2(aq) + Hg(l)$$

Zinc lies below calcium in the reactivity series and so will not replace calcium from the perchlorate compound:

$$Zn(s) + Ca(ClO_4)_2(aq) \rightarrow \text{no reaction}$$

When a zinc rod is placed in a copper nitrate solution, the zinc replaces the copper and elemental copper forms.

4-8. SINGLE-REPLACEMENT REACTIONS DO NOT HAVE TO TAKE PLACE IN SOLUTION

A spectacular example of a single-replacement reaction that does not take place in solution is the reaction between aluminum metal and iron(III) oxide:

$$2Al(s) + Fe_2O_3(s) \rightarrow 2Fe(l) + Al_2O_3(s)$$

Once this reaction is initiated, it proceeds vigorously, producing so much heat that the iron is formed as a liquid. The reaction of aluminum metal with a metal oxide is called a *thermite reaction* and has numerous applications. Thermite reactions (Figure 4-9) were once used to weld railroad rails and are used in thermite grenades, which are employed by the military to destroy heavy equipment.

Figure 4-9 The thermite reaction between aluminum powder and iron(III) oxide is initiated by a heat source, such as the combustion of a magnesium wire ribbon.

In a thermite reaction the reaction temperature can exceed 3500°C.

Although the single-replacement reactions we have discussed so far involve the replacement of one metal by another, there are many other types of single-replacement reactions. One particularly important type is the reaction between a metal oxide and carbon:

$$3C(s) + 2Fe_2O_3(s) \rightarrow 4Fe(s) + 3CO_2(g)$$

$$C(s) + 2ZnO(s) \rightarrow 2Zn(s) + CO_2(g)$$

Reactions such as these are used in the large-scale production of metals from their ores, which are often either metal oxides or compounds readily convertible to oxides.

4-9. THE REACTIVITY ORDER OF THE HALOGENS IS $F_2 > Cl_2 > Br_2 > I_2$

The reactions of the nonmetals are too diversified to allow for a single table similar to Table 4-6. However, the relative reactivities of the halogens can be established by means of single-replacement reactions. For example, if chlorine gas is bubbled into an aqueous solution of NaBr, then free bromine is produced:

$$Cl_2(g) + 2NaBr(aq) \rightarrow 2NaCl(aq) + Br_2(aq)$$

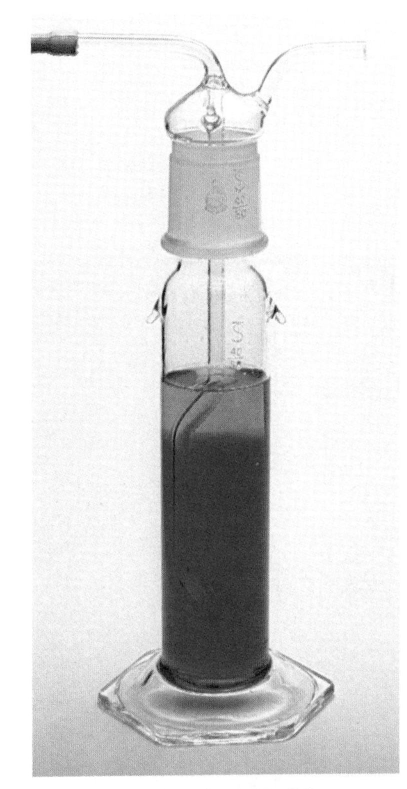

When $Cl_2(g)$ is bubbled into an aqueous NaBr solution, the chlorine replaces the bromine in NaBr, forming NaCl(aq) and reddish $Br_2(aq)$.

Figure 4-10 Bromine processing plant. The red color is produced by the bromine gas.

Bromine can be swept out of the resulting solution with a stream of air and collected. Bromine is a dense, brownish-red, corrosive liquid which, in the form of bromide salts, constitutes 0.015 percent of the total dissolved solids in seawater. About half a million tons of bromine are produced annually from seawater using the above reaction.

In another single-replacement reaction, bromine replaces iodide in solution:

$$Br_2(l) + 2NaI(aq) \rightarrow 2NaBr(aq) + I_2(s)$$

Solid iodine is purple gray and gaseous iodine is purple.

Iodine is a purple-gray, volatile solid that is only slightly soluble in water. The concentration of iodide salts in seawater is too low to make this a good source for the commercial production of iodine, but iodine does occur in sufficient concentrations for commercial recovery in certain oil-well brines and in some seaweed and marine sponges.

Fluorine is not only the most reactive halogen, but is also the most reactive element. Fluorine readily displaces chlorine from chlorides; for example,

$$2KCl(s) + F_2(g) \rightarrow 2KF(s) + Cl_2(g)$$

Thus the reactivity order of the halogens is

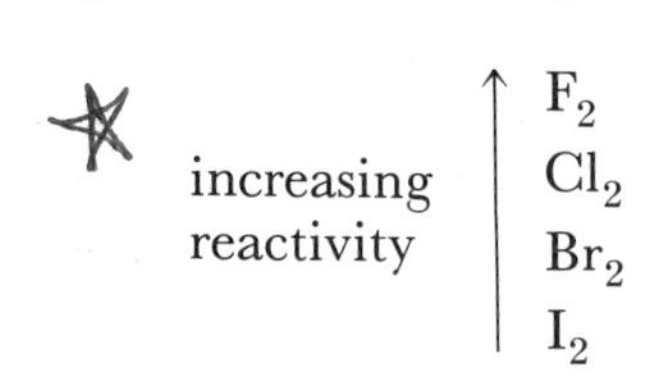

F_2
Cl_2
Br_2
I_2

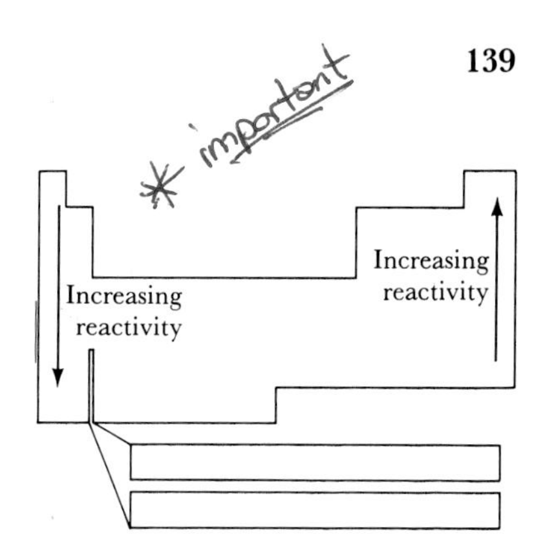

Generally, the reactivity of nonmetals increases as we go *up* a group in the periodic table. This is in sharp contrast to the reactivity of metals, which increases as we go *down* a group.

Example 4-8: Predict the product of the reaction when bromine liquid is added to an aqueous solution of calcium iodide.

Solution: Bromine is more reactive than I_2 and so will replace the iodide in CaI_2:

$$Br_2(l) + CaI_2(aq) \rightarrow CaBr_2(aq) + I_2(s)$$

Example 4-9: Will bromine replace fluorine in an aqueous potassium fluoride solution?

Solution: Fluorine is the most reactive halogen, and so none of the other halogens will replace free fluorine from fluoride salts. Thus we predict that

$$Br_2(l) + KF(aq) \rightarrow \text{no reaction}$$

which is correct.

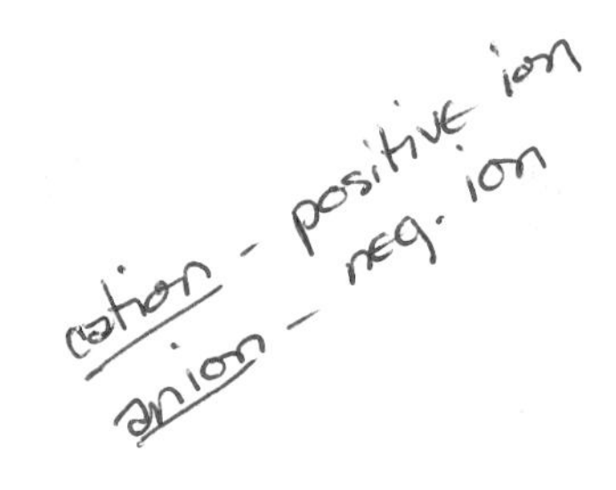

4-10. A DOUBLE-REPLACEMENT REACTION IS ONE OF THE TYPE AB + CD → AD + BC

A simple and impressive *double-replacement reaction* is the reaction between an aqueous solution of NaCl and an aqueous solution of $AgNO_3$. Both solutions are clear; yet when they are mixed, a white *precipitate* forms immediately. A precipitate is an insoluble product of a reaction that occurs in solution. In the chemical reaction between sodium chloride and silver nitrate,

$$NaCl(aq) + AgNO_3(aq) \rightarrow NaNO_3(aq) + AgCl(s)$$

the white precipitate that forms is silver chloride. The reaction is a double-replacement reaction because the two cations, $Na^+(aq)$ and $Ag^+(aq)$, can be visualized as exchanging anions:

$$NaCl(aq) + AgNO_3(aq) \rightarrow NaNO_3(aq) + AgCl(s)$$

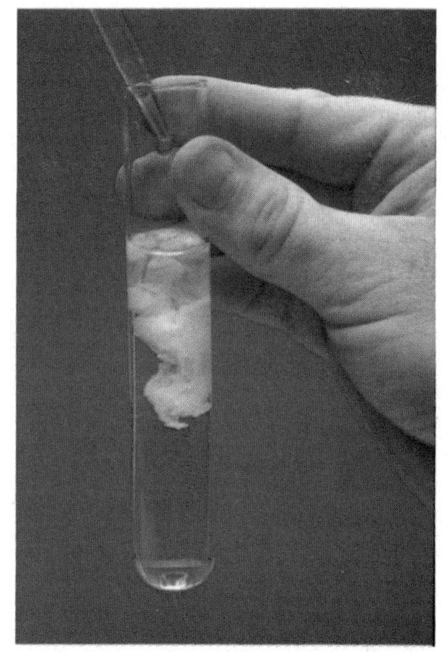

The reaction of NaCl(*aq*) with $AgNO_3(aq)$ yields the white precipitate AgCl(*s*).

It is instructive to analyze this reaction in terms of the ions involved. Sodium chloride and silver nitrate are soluble ionic compounds, and so, in aqueous solution, they consist of $Na^+(aq)$ and $Cl^-(aq)$ ions and $Ag^+(aq)$ and $NO_3^-(aq)$ ions, respectively. At the very instant the NaCl and $AgNO_3$ solutions are mixed, these four kinds of ions exist in one solution. As we shall learn in later chapters, the ions in a solution constantly move around and collide with water molecules and with each other. If a $Na^+(aq)$ ion collides with a $Cl^-(aq)$ ion, the ions simply drift apart because NaCl is soluble in water. Similarly, a collision between $Ag^+(aq)$ and $NO_3^-(aq)$ is of no consequence because $AgNO_3$ is also soluble in water. When a $Ag^+(aq)$ ion collides with a $Cl^-(aq)$ ion, however, AgCl is formed, and it is not soluble in water. Consequently, the AgCl precipitates out of the solution. The $Ag^+(aq)$ and $Cl^-(aq)$ are depleted from the solution and appear as a white precipitate. The $Na^+(aq)$ and $NO_3^-(aq)$ remain in solution because $NaNO_3$ is soluble in water. We say that the driving force of the chemical reaction between NaCl and $AgNO_3$ is the formation of the AgCl precipitate.

The formation of a precipitate drives a double replacement reaction involving salts.

It is convenient to use *net ionic equations* to describe double-replacement reactions that occur in solution. In the reaction between NaCl and $AgNO_3$, the $Na^+(aq)$ and $NO_3^-(aq)$ ions do not participate in the reaction, which is the formation of solid AgCl from $Ag^+(aq)$ and $Cl^-(aq)$ ions. We say that the $Na^+(aq)$ and $NO_3^-(aq)$ ions are *spectator ions;* that is, they are not involved directly in the formation of the AgCl precipitate. Thus, if we write the chemical equation in terms of ions (an *ionic equation*),

$$Na^+(aq) + Cl^-(aq) + Ag^+(aq) + NO_3^-(aq) \rightarrow Na^+(aq) + NO_3^-(aq) + AgCl(s)$$

then we see that the spectator ions, $Na^+(aq)$ and $NO_3^-(aq)$, appear on both sides of the equation. Because they appear on both sides of the ionic equation, the spectator ions can be canceled:

Spectator ions do not appear in a net ionic equation.

$$\cancel{Na^+(aq)} + Cl^-(aq) + Ag^+(aq) + \cancel{NO_3^-(aq)} \rightarrow \cancel{Na^+(aq)} + \cancel{NO_3^-(aq)} + AgCl(s)$$

to give the net ionic equation:

$$Ag^+(aq) + Cl^-(aq) \rightarrow AgCl(s)$$

The net ionic equation describes the essence of the reaction, namely, the formation of solid AgCl from $Ag^+(aq)$ and $Cl^-(aq)$ ions.

Note that the net ionic equation corresponding to the double-replacement reaction between $KCl(aq)$ and $AgClO_4(aq)$ is the same as that for the reaction between $NaCl(aq)$ and $AgNO_3(aq)$:

ionic equation:

$$\cancel{K^+(aq)} + Cl^-(aq) + Ag^+(aq) + \cancel{ClO_4^-(aq)} \rightarrow \cancel{K^+(aq)} + \cancel{ClO_4^-(aq)} + AgCl(s)$$

net ionic equation:

$$Ag^+(aq) + Cl^-(aq) \rightarrow AgCl(s)$$

The main advantage to writing net ionic equations is that they focus on the reactant ions that are involved directly in the reaction.

In order to predict if a precipitate forms in a reaction occurring in aqueous solution, it is necessary to know whether a possible reaction product is soluble in water or not. We shall see in Chapter 19 that each compound has a definite solubility (for example, so many grams per liter of water), but in this chapter we shall simply indicate whether a compound is insoluble or not.

Example 4-10: Write the net ionic equation for the reaction between $Pb(NO_3)_2$ and K_2CrO_4, given that lead chromate, $PbCrO_4$, is the only insoluble product.

Solution: The complete ionic equation is

$$Pb^{2+}(aq) + 2NO_3^-(aq) + 2K^+(aq) + CrO_4^{2-}(aq) \rightarrow 2K^+(aq) + 2NO_3^-(aq) + PbCrO_4(s)$$

We get the net ionic equation by canceling the $K^+(aq)$ and $NO_3^-(aq)$ ions on both sides:

$$Pb^{2+}(aq) + CrO_4^{2-}(aq) \rightarrow \underset{\text{yellow-orange}}{PbCrO_4(s)}$$

Problems 4-27 to 4-30 deal with net ionic equations.

4-11. NEUTRALIZATION IS THE REACTION BETWEEN AN ACID AND A BASE

We have seen that the driving force for double-replacement reactions can be the formation of a precipitate. It can also be the formation of an un-ionized (molecular) compound from ionic reactants. The most important example of such a reaction is one in which water is produced. For example,

$$HCl(aq) + NaOH(aq) \rightarrow NaCl(aq) + H_2O(l)$$

The complete ionic equation is

$$H^+(aq) + Cl^-(aq) + Na^+(aq) + OH^-(aq) \rightarrow Na^+(aq) + Cl^-(aq) + H_2O(l)$$

and the net ionic equation is

$$H^+(aq) + OH^-(aq) \rightarrow H_2O(l)$$

Because H_2O exists almost exclusively as neutral molecules, the $H^+(aq)$ and $OH^-(aq)$ are depleted from the reaction mixture by the formation of H_2O. Note the comparison between the two driving forces for double-replacement reactions: the formation of a precipitate and the formation of a molecular compound. In each case reactant ions are depleted from the reaction mixture.

The reaction between HCl and NaOH is amazing. We probably all know that acids like hydrochloric acid are corrosive. They react with many metals and many rocks, such as limestone, $CaCO_3$, and concentrated solutions of them damage flesh, causing painful burns and blisters. Less familiar, perhaps, are the chemical properties of a substance like NaOH. Recall that substances that produce $OH^-(aq)$ ions when dissolved in water are bases. The Group 1 hydroxides and the soluble Group 2 hydroxides are bases. Bases like sodium hydroxide are also very corrosive, causing painful burns and blisters on the skin, just as hydrochloric acid does. Sodium hydroxide is the principal ingredient in oven and drain cleaners. Thus the reaction between HCl(aq) and NaOH(aq) is one between two reactive and hazardous substances, but the products are sodium chloride and water, a harmless salt solution. We say that the acid (HCl) and the base (NaOH) have *neutralized* each other. The chemical reaction between an acid and a base is called a *neutralization reaction.* For example,

Sodium hydroxide is a key ingredient in Drāno and other cleaners.

$$\underset{\text{acid}}{HCl(aq)} + \underset{\text{base}}{NaOH(aq)} \rightarrow \underset{\text{salt}}{NaCl(aq)} + \underset{\text{water}}{H_2O(l)}$$

One definition of a *salt* is that it is an ionic compound formed in the reaction between an acid and a base.

Because acids and bases are so important in chemistry, we have summarized some of their properties in Table 4-7. As the table shows, acidic solutions taste sour. Vinegar tastes sour because it is a dilute solution of acetic acid, lemons taste sour because they contain citric

Table 4-7 Some properties of acids and bases

Acids	*Bases*
solutions taste sour (don't taste any, but recall the taste of vinegar)	solutions taste bitter and feel slippery to the touch (don't taste any)
produce hydrogen ions when dissolved in water	produce hydroxide ions when dissolved in water
neutralize bases to produce salts and water	neutralize acids to produce salts and water
solutions turn litmus paper red	solutions turn litmus paper blue
react with many metals to produce hydrogen gas	

Never attempt to determine whether a chemical solution is acidic or basic by tasting it.

acid, and rhubarb tastes sour because it contains oxalic acid. Basic solutions taste bitter and feel slippery; an example of a basic solution is soapy water. An interesting and important property of acidic and basic solutions is their effect on certain dyes and vegetable matter. *Litmus,* which is a vegetable substance obtained from lichens, is red in acidic solutions and blue in basic solutions. Paper impregnated with litmus, called *litmus paper,* serves as a quick test to see if a solution is acidic or basic.

Litmus is called an indicator because it indicates whether a solution is acidic or basic. There are many different indicators.

In summary, then, a chemical characteristic of an acid is that it produces $H^+(aq)$ in aqueous solutions and a chemical characteristic of a base is that it produces $OH^-(aq)$ in aqueous solutions. When an acid and a base neutralize each other, the $H^+(aq)$ and $OH^-(aq)$ react to produce $H_2O(l)$ and in doing so nullify the acidic and basic character of the separate solutions:

$H^+(aq)$	+	$OH^-(aq)$	$\rightarrow$	$H_2O(l)$
acid		base		water

Example 4-11: Complete and balance the equations for the following neutralization reactions and name the salt formed in each case:

Problem 4-33 is similar to Example 4-11.

$$H_2SO_4(aq) + KOH(aq) \rightarrow$$

$$HNO_3(aq) + Ca(OH)_2(aq) \rightarrow$$

Solution: Each of these reactions is between an acid and a base, and so a salt and water are produced in each case. The first reaction is between potassium hydroxide and sulfuric acid, and so the salt formed is potassium sulfate:

$H_2SO_4(aq)$ +	$2KOH(aq)$	$\rightarrow$	$K_2SO_4(aq)$	+ $2H_2O(l)$
sulfuric acid	potassium hydroxide		potassium sulfate	
an acid	a base		a salt	

The second reaction is between calcium hydroxide and nitric acid, and so the salt formed is calcium nitrate (Table 4-1):

$2HNO_3(aq)$ +	$Ca(OH)_2(aq)$	$\rightarrow$	$Ca(NO_3)_2(aq)$	+ $2H_2O(l)$
nitric acid	calcium hydroxide		calcium nitrate	
an acid	a base		a salt	

4-12. THE CONCENTRATION OF AN ACID OR A BASE CAN BE DETERMINED BY TITRATION

We can utilize neutralization reactions to determine the concentration of solutions of acids or bases. Suppose we have a basic solution whose concentration is not known. We measure out a certain volume of the basic solution and then slowly add an acidic solution of known concentration until the base is completely neutralized. Such a process is called a *titration* and can be carried out with the apparatus shown in Figures 4-11 and 4-12. Knowing the volume and concentration of the acidic solution required to neutralize the base is sufficient to determine the concentration of the basic solution. As an example, suppose we find that it requires 27.25 mL of 0.150 M HCl solution to neutralize 30.00 mL of a NaOH solution. The number of moles of HCl required to neutralize the NaOH(*aq*) is given by the equation (Section 3-10):

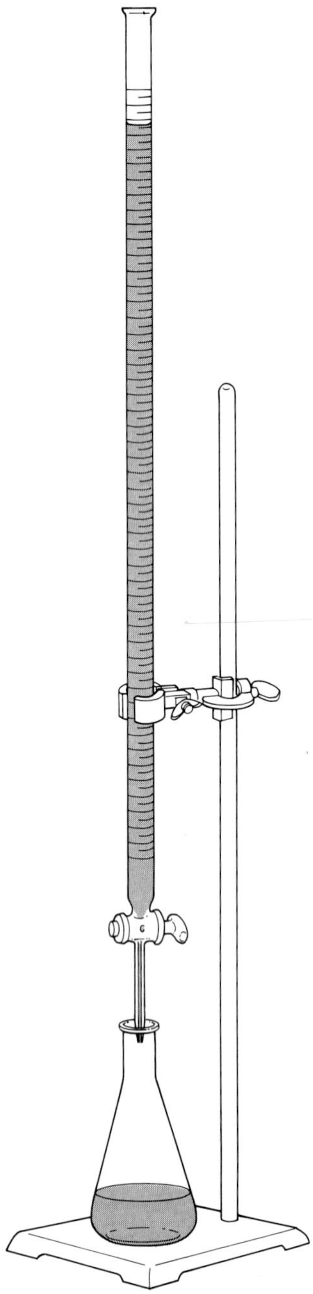

$$\text{number of moles} = (\text{molarity})\,(\text{volume of solution})$$
$$n = MV$$

Thus

$$\text{mol HCl} = (0.150\ \text{mol}\cdot\text{L}^{-1})(27.25 \times 10^{-3}\ \text{L}) = 4.09 \times 10^{-3}\ \text{mol}$$

The reaction between HCl(*aq*) and NaOH(*aq*) is

$$HCl(aq) + NaOH(aq) \rightarrow NaCl(aq) + H_2O(l)$$

which indicates that 1 mol of HCl is required to neutralize 1 mol of NaOH. Therefore

$$\text{mol NaOH} = \text{mol HCl} = 4.09 \times 10^{-3}\ \text{mol}$$

There is 4.09×10^{-3} mol of NaOH in the 30.00 mL of the NaOH solution, and so the concentration of the solution is

$$M = n/V = \frac{4.09 \times 10^{-3}\ \text{mol}}{30.00 \times 10^{-3}\ \text{L}} = 0.136\ \text{M}$$

Figure 4-11 A *buret* is a precision-made piece of glassware used to measure accurately the volume of a solution. Burets are particularly useful for delivering accurate volumes of one solution to another.

Example 4-12: By titration, it is found that 37.60 mL of 0.210 M NaOH is required to neutralize 25.05 mL of $H_2SO_4(aq)$. Calculate the concentration of the H_2SO_4 solution.

Solution: The number of moles of NaOH required to neutralize the H_2SO_4 is

$$\text{mol NaOH} = MV = (0.210\ \text{mol}\cdot\text{L}^{-1})(37.60 \times 10^{-3}\ \text{L}) = 7.90 \times 10^{-3}\ \text{mol}$$

According to the reaction

$$H_2SO_4(aq) + 2NaOH(aq) \rightarrow Na_2SO_4(aq) + 2H_2O(l)$$

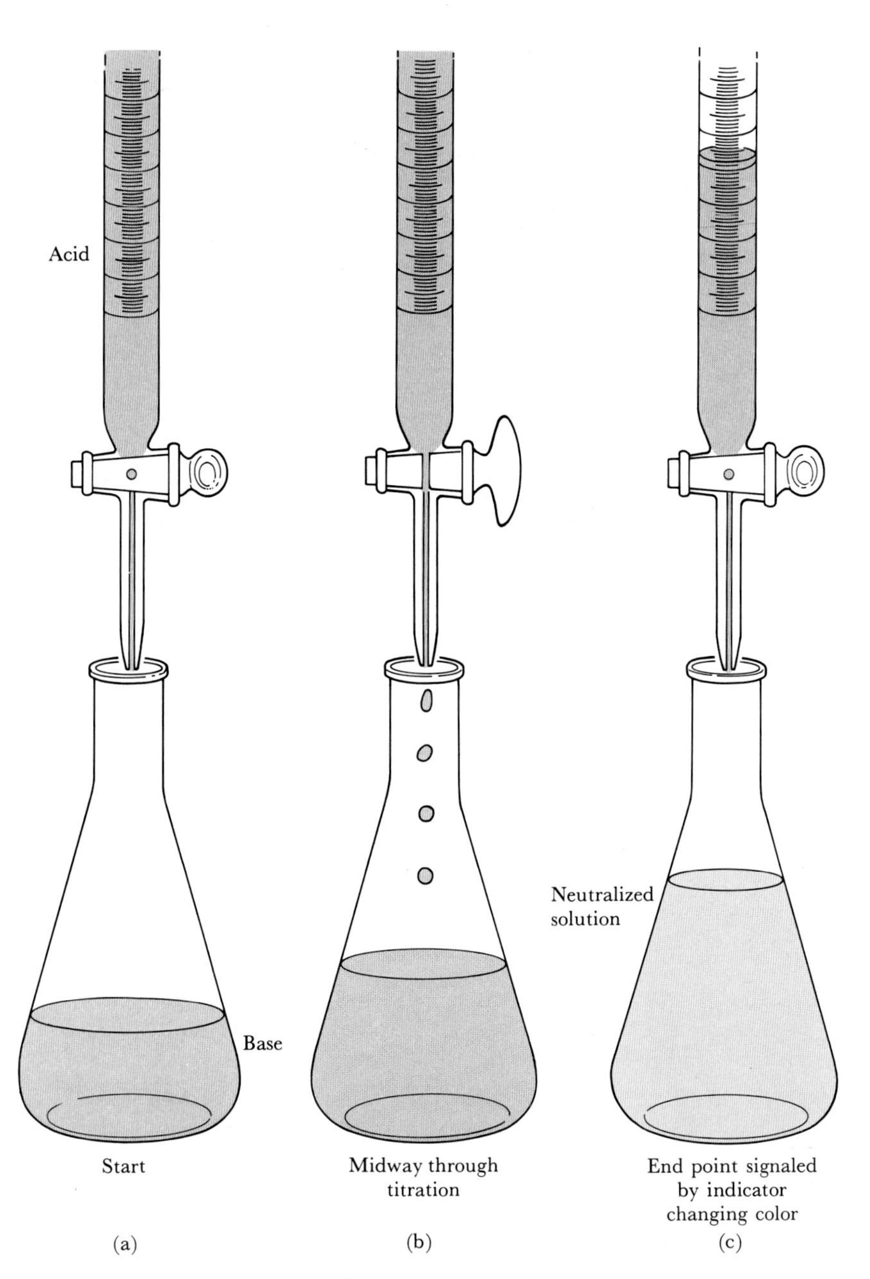

Figure 4-12 A titration experiment to determine the concentration of a basic solution. (a) A known volume of the basic solution has been measured into a flask and a few drops of litmus solution added. Because the solution in the flask is basic, the litmus turns blue. (b) A solution of acid of known concentration is added slowly to the flask until (c) the solution just barely turns pink. At this point, called the *end point* of the titration, the base has been completely neutralized and the last drop of acid added has caused the solution to turn acidic, changing the color of the litmus from blue to red. The litmus is used to indicate when the base has just been neutralized and is called an *indicator*.

it requires 2 mol of NaOH to neutralize 1 mol of H_2SO_4. Consequently, we have

$$\text{mol } H_2SO_4 = (7.90 \times 10^{-3} \text{ mol NaOH})\left(\frac{1 \text{ mol } H_2SO_4}{2 \text{ mol NaOH}}\right)$$
$$= 3.95 \times 10^{-3} \text{ mol}$$

The concentration of the H_2SO_4 solution is

$$M = \frac{n}{V} = \frac{3.95 \times 10^{-3} \text{ mol}}{25.05 \times 10^{-3} \text{ L}} = 0.158 \text{ M}$$

Example 4-13: A 1.50-g sample of an unknown acid is dissolved to make 100 mL of solution and neutralized with 0.200 M NaOH(*aq*). The volume of NaOH solution required to neutralize the acid is 75.0 mL. Assume that the acid has only one acidic proton per molecule and compute its formula mass.

Solution: The number of moles of base required to neutralize the acid is

$$\text{mol OH}^- = MV = (0.200 \text{ mol}\cdot\text{L}^{-1})(0.0750 \text{ L}) = 1.50 \times 10^{-2} \text{ mol}$$

Therefore, the number of moles of acid present in the original 100 mL of solution is 1.50×10^{-2} mol. Thus we see that

Recall that the symbol ≎ means "combines with" or "stoichiometrically equivalent to."

$$1.50 \text{ g acid} \approxeq 1.50 \times 10^{-2} \text{ mol acid}$$

Upon dividing by 1.50×10^{-2}, we obtain

$$100 \text{ g acid} \approxeq 1.00 \text{ mol acid}$$

Thus the formula mass of the unknown acid is 100.

Example 4-14: Suppose we wish to neutralize 20.0 g of solid NaOH by adding 1.00 M aqueous HCl. How many milliliters of HCl should be added?

Solution: The equation for the neutralization reaction is

$$\text{NaOH}(s) + \text{HCl}(aq) \rightarrow \text{NaCl}(aq) + H_2O(l)$$

Problems 4-43 to 4-54 deal with calculations involving titration.

It requires 1 mol of HCl to neutralize 1 mol of NaOH. The 20.0 g of NaOH corresponds to

$$(20.0 \text{ g NaOH})\left(\frac{1 \text{ mol NaOH}}{40.00 \text{ g NaOH}}\right) = 0.500 \text{ mol NaOH}$$

To neutralize the 0.500 mol of NaOH, we need to add 0.500 mol of HCl. Given the concentration (1.00 M) and the number of moles (0.500 mol), we can use the fact that $n = MV$ to calculate the volume to be used:

$$V_{HCl} = \frac{n}{M} = \frac{0.500\ \text{mol}}{1.00\ \text{mol}\cdot\text{L}^{-1}} = 0.500\ \text{L}$$

One of the most difficult questions that a beginning student of chemistry faces is to predict the products of a reaction when only the reactants are given. This is often a difficult question even for a chemist. The classification of reaction types presented in this chapter is helpful in this regard, but it still requires additional chemical experience to develop confidence. As you see more chemical reactions throughout this book, think about each one and try to classify it according to the scheme developed in this chapter.

SUMMARY

Many chemical reactions can be classified as being one of four types:

1. Combination: a reaction of two substances to form a single product.
2. Decomposition: a reaction in which a substance breaks down into two or more simpler substances.
3. Single-replacement: a reaction involving the substitution of one element in a compound by another (also called a substitution reaction).
4. Double-replacement: a reaction in which the cations of two ionic compounds exchange anionic partners. The driving force can be the formation of a precipitate or the formation of a molecular product such as water.

Single-replacement reactions can be used to order the metals in terms of relative reactivity. The resulting reactivity series of the metals can be used to predict whether or not a single-replacement reaction occurs. The nonmetals are too varied to allow a correspondingly simple activity series, but the relative reactivities of the halogens are $F_2 > Cl_2 > Br_2 > I_2$.

Acids are substances that produce the species $H^+(aq)$ when dissolved in water; bases are substances that produce hydroxide ions, $OH^-(aq)$, when dissolved in water. Two classes of acids are binary acids and oxyacids. Solutions of oxyacids yield relatively stable polyatomic anions, which remain intact in many chemical reactions. Acids and bases react with each other to produce a salt and water; such reactions are called neutralization reactions. The names of salts are based upon the names of the acids that produce them (Table 4-1). Acids and bases cause certain dyes to change color. For example, litmus turns red in acidic solutions and blue in basic solutions.

TERMS YOU SHOULD KNOW

combination reaction 123
combustion reaction 124
molecular compound 124
binary compound 125
polyatomic ion 126
base 129
basic anhydride 129
acid 130
hydrogen ion, $H^+(aq)$ 130
hydronium ion, $H_3O^+(aq)$ 130
acidic anhydride 130
acidic hydrogen atom 130
acidic proton 130
oxyacid 131
binary acid 131
decomposition reaction 132
thermal decomposition 133
single-replacement reaction 134
substitution reaction 134
reactivity series 136
thermite reaction 136
double-replacement reaction 139
precipitate 139
spectator ion 140
ionic equation 140
net ionic equation 140
neutralization reaction 142
salt 142
litmus paper 143
titration 144
buret 144
end point 145
indicator 145

PROBLEMS

NOMENCLATURE

4-1. Name the following compounds:

(a) $RbCN$
(b) $AgClO_4$
(c) $La_2(CrO_4)_3$
(d) $CsMnO_4$
(e) $Al(OH)_3$

4-2. Name the following compounds:

(a) $NaC_2H_3O_2$
(b) $Ca(ClO_3)_2$
(c) $(NH_4)_2CO_3$
(d) KOH
(e) $Ba(NO_3)_2$

4-3. Name the following compounds, which are used as fertilizers:

(a) $NaNO_3$
(b) $(NH_4)_2SO_4$
(c) $(NH_4)_3PO_4$
(d) $Ca_3(PO_4)_2$
(e) K_3PO_4

4-4. Name the following compounds, which are used in photography:

(a) $(NH_4)_2S_2O_3$ (fixer)
(b) Na_2CO_3 (preservative)
(c) Na_2SO_3 (preservative)
(d) K_2CO_3 (activator)
(e) $Na_2S_2O_3$ (fixer)

4-5. Using the names of the polyatomic ions given in Table 4-1, write the formula for each of the following:

(a) sodium thiosulfate
(b) aluminum hydrogen carbonate
(c) potassium perchlorate
(d) calcium carbonate
(e) barium acetate

4-6. Using the names of the polyatomic ions given in Table 4-1, write the formula for each of the following:

(a) acetic acid
(b) chloric acid
(c) carbonic acid
(d) perchloric acid
(e) permanganic acid

4-7. Give the chemical formula for each of the following:

(a) sodium chromate
(b) potassium phosphate
(c) rubidium hydroxide
(d) lithium perchlorate
(e) ammonium nitrate

4-8. Give the chemical formula for each of the following:

(a) sodium perchlorate
(b) potassium permanganate
(c) silver acetate
(d) calcium sulfite
(e) lithium cyanide

4-9. Using Table 4-2, give the systematic name for each of the following:

(a) HgO (d) $CoBr_2$
(b) SnO_2 (e) $Cr(CN)_3$
(c) CuI

4-10. Using Table 4-2, give the systematic name for each of the following:

(a) $CrSO_4$ (d) $Sn(NO_3)_2$
(b) Fe_2O_3 (e) Cu_2CO_3
(c) $Co(CN)_2$

4-11. Using Table 4-2, write the formula for each of the following:

(a) copper(II) oxide (d) iron(II) acetate
(b) gold(I) chloride (e) cobalt(III) sulfate
(c) tin(II) phosphate

4-12. Using Table 4-2, write the formula for each of the following:

(a) mercury(I) acetate (d) iron(II) hydrogen carbonate
(b) mercury(II) cyanide (e) chromium(II) sulfite
(c) cobalt(III) hydroxide

PREDICTION OF REACTIONS

4-13. Complete and balance the equations for the following combination reactions:

(a) $Mg(s) + N_2(g) \rightarrow$
(b) $H_2(g) + S(s) \rightarrow$
(c) $K(s) + Br_2(l) \rightarrow$
(d) $Al(s) + O_2(g) \rightarrow$
(e) $MgO(s) + SO_2(g) \rightarrow$

4-14. Complete and balance the equations for the following combination reactions:

(a) $Na(s) + Cl_2(g) \rightarrow$
(b) $K(s) + O_2(g) \rightarrow$
(c) $MgO(s) + CO_2(g) \rightarrow$
(d) $H_2(g) + O_2(g) \rightarrow$
(e) $N_2(g) + H_2(g) \rightarrow$

4-15. Complete and balance the following equations:

(a) $Li_2O(s) + H_2O(l) \xrightarrow[H_2O(l)]{}$
(b) $SO_3(g) + H_2O(l) \xrightarrow[H_2O(l)]{}$
(c) $H_2SO_4(l) \xrightarrow[H_2O(l)]{}$
(d) $HBr(g) \xrightarrow[H_2O(l)]{}$

4-16. Complete and balance the following equations:

(a) $SrO(s) + H_2O(l) \xrightarrow[H_2O(l)]{}$
(b) $HNO_3(l) \xrightarrow[H_2O(l)]{}$
(c) $Cs_2O(s) + H_2O(l) \xrightarrow[H_2O(l)]{}$
(d) $HI(g) \xrightarrow[H_2O(l)]{}$

4-17. Complete and balance the equations for the following decomposition reactions:

(a) $CaCO_3(s) \xrightarrow{\text{high T}}$
(b) $H_2O_2(aq) \xrightarrow{\text{light}}$
(c) $PbSO_3(s) \xrightarrow{\text{high T}}$
(d) $Ag_2O(s) \xrightarrow{\text{high T}}$

4-18. Complete and balance the equations for the following decomposition reactions:

(a) $MgCO_3(s) \xrightarrow{\text{high T}}$
(b) $NaClO_3(s) \xrightarrow{\text{high T}}$
(c) $NaN_3(s) \xrightarrow{\text{high T}}$
(d) $Au_2O_3(s) \xrightarrow{\text{high T}}$

4-19. Complete and balance the equations for the following replacement reactions (if no reaction, write N.R.):

(a) $Na(l) + AlCl_3(s) \rightarrow$
(b) $Cu(s) + Fe_2O_3(s) \rightarrow$
(c) $Zn(s) + K_2SO_4(aq) \rightarrow$
(d) $Fe(s) + SnCl_2(aq) \rightarrow$
(e) $Br_2(l) + NaI(aq) \rightarrow$

4-20. Complete and balance the equations for the following replacement reactions (if no reaction, write N.R.):

(a) $Ba(s) + H_2O(g) \rightarrow$
(b) $Fe(s) + H_2SO_4(aq) \rightarrow$
(c) $Ca(s) + HBr(aq) \rightarrow$
(d) $Pb(s) + HCl(aq) \rightarrow$
(e) $Zn(s) + CaCl_2(aq) \rightarrow$

4-21. Complete and balance the following equations (if no reaction, write N.R.):

(a) $Cu(NO_3)_2(aq) + Au(s) \rightarrow$
(b) $NaCl(aq) + I_2(aq) \rightarrow$
(c) $AgNO_3(aq) + Mg(s) \rightarrow$
(d) $Ag(s) + HCl(aq) \rightarrow$
(e) $Pb(s) + H_2O(l) \rightarrow$

4-22. Complete and balance the following equations (if no reaction, write N.R.):

(a) $Na(l) + NiO(s) \rightarrow$
(b) $Ca(s) + PbCl_2(s) \xrightarrow{\text{high T}}$
(c) $Mg(s) + Fe_3O_4(s) \xrightarrow{\text{high T}}$
(d) $Cu(s) + Al_2O_3(s) \rightarrow$
(e) $Ag(s) + Zn(NO_3)_2(s) \rightarrow$

RELATIVE REACTIVITIES

4-23. Arrange the following metals in order of decreasing reactivity:

Na Au Fe Sn

4-24. Arrange the following metals in order of decreasing reactivity:

Pt Ag K Zn

4-25. Lead production became important in Roman times because of its use in making pipes to carry water to the famous Roman baths. The manufacture of lead had been developed by the Greeks as a by-product of the silver mines outside Athens. Lead occurs in silver ores as the sulfide galena, PbS. The silver occurs as the oxide. The ore is first roasted, that is, heated in air to convert the lead to an oxide. The ore is then smelted, that is, heated with charcoal. Write the chemical equations for the reactions that occur in the roasting and smelting of the lead-silver ore.

4-26. The first metal to be prepared from its ore was copper, perhaps as early as 6000 BC in the Middle East. The ore, which contained copper(II) oxide, was heated with charcoal, which was prepared by the incomplete burning of wood and is mainly elemental carbon. Later, iron and tin were prepared in the same way. Bronze was made by mixing copper-containing ore with tin ore and heating the mixture in the presence of charcoal. Write the chemical equations for the preparation of these metals. Assume the ores to be CuO, SnO_2, and Fe_2O_3.

NET IONIC EQUATIONS

4-27. Write the net ionic equation corresponding to each of the following equations:

(a) $Na_2S(aq) + 2HCl(aq) \rightarrow 2NaCl(aq) + H_2S(g)$
(b) $PbCl_2(aq) + Na_2S(aq) \rightarrow 2NaCl(aq) + PbS(s)$
(c) $H_2SO_4(aq) + 2KOH(aq) \rightarrow K_2SO_4(aq) + 2H_2O(l)$
(d) $Na_2O(s) + 2HCl(aq) \rightarrow 2NaCl(aq) + H_2O(l)$
(e) $NH_3(g) + HCl(aq) \rightarrow NH_4Cl(aq)$

4-28. Write the net ionic equation corresponding to each of the following equations:

(a) $HClO_3(aq) + KOH(aq) \rightarrow KClO_3(aq) + H_2O(l)$
(b) $Pb(NO_3)_2(aq) + Na_2CO_3(aq) \rightarrow 2NaNO_3(aq) + PbCO_3(s)$
(c) $2AgClO_4(aq) + (NH_4)_2SO_4(aq) \rightarrow 2NH_4ClO_4(aq) + Ag_2SO_4(s)$
(d) $K_2S(aq) + Zn(NO_3)_2(aq) \rightarrow 2KNO_3(aq) + ZnS(s)$
(e) $Hg_2(ClO_3)_2(aq) + SrCl_2(aq) \rightarrow Sr(ClO_3)_2(aq) + Hg_2Cl_2(s)$

4-29. Given that the sulfides, carbonates, and hydroxides of the transition metals are insoluble in water and that transition-metal nitrates and perchlorates are soluble in water, complete and balance the following equations. In each case write out the corresponding net ionic equation:

(a) $Fe(NO_3)_2(aq) + NaOH(aq) \rightarrow$
(b) $Zn(ClO_4)_2(aq) + K_2S(aq) \rightarrow$
(c) $Pb(NO_3)_2(aq) + KOH(aq) \rightarrow$
(d) $Zn(NO_3)_2(aq) + Na_2CO_3(aq) \rightarrow$
(e) $Cu(ClO_4)_2(aq) + Na_2CO_3(aq) \rightarrow$

ACIDS AND BASES

4-31. Decide, based on your personal experience and on Table 4-7, which of the following solutions are acidic and which are basic:

(a) carbonated soft drinks
(b) apple cider
(c) milk of magnesia (suspension)
(d) tomatoes
(e) soap

4-33. Complete and balance the equation for each of the following acid-base reactions and name the salt produced in each case:

(a) $HClO_3(aq) + Ba(OH)_2(aq) \rightarrow$
(b) $HC_2H_3O_2(aq) + KOH(aq) \rightarrow$
(c) $HI(aq) + Mg(OH)_2(s) \rightarrow$
(d) $H_2SO_4(aq) + RbOH(aq) \rightarrow$
(e) $H_3PO_4(aq) + LiOH(aq) \rightarrow$

4-30. Given that most Ag^+, Hg^{2+}, and Pb^{2+} salts except those of NO_3^- and ClO_4^- are insoluble in water, complete and balance the following equations. In each case write out the corresponding net ionic equation:

(a) $AgNO_3(aq) + Na_2S(aq) \rightarrow$
(b) $H_2SO_4(aq) + Pb(NO_3)_2(aq) \rightarrow$
(c) $Hg(NO_3)_2(aq) + NaI(aq) \rightarrow$
(d) $CdCl_2(aq) + AgClO_4(aq) \rightarrow$
(e) $LiBr(aq) + Pb(ClO_4)_2(aq) \rightarrow$

4-32. Decide, based on your personal experience and on Table 4-7, which of the following solutions are acidic and which are basic:

(a) laundry detergent in water
(b) orange juice
(c) jam
(d) bicarbonate of soda in water
(e) household ammonia

4-34. Red ants contain an appreciable amount of formic acid, $\underline{H}CHO_2$, where only the underlined proton is acidic. It is observed that ants sprayed with window cleaner containing ammonia die quickly. Write the neutralization reaction between formic acid and ammonia.

CLASSIFICATION OF REACTIONS

4-35. Classify each of the following reactions as combination, decomposition, single-replacement, or double-replacement:

(a) $CaCO_3(s) \rightarrow CaO(s) + CO_2(g)$
(b) $NH_3(g) + HCl(g) \rightarrow NH_4Cl(s)$
(c) $2AgBr(s) + Cl_2(g) \rightarrow 2AgCl(s) + Br_2(l)$
(d) $Ag_2SO_4(s) + 2NaI(aq) \rightarrow 2AgI(s) + Na_2SO_4(aq)$

4-37. Complete and balance the following equations and classify each:

(a) $ZnBr_2(s) + Cl_2(g) \rightarrow$
(b) $HCl(aq) + Mg(OH)_2(s) \rightarrow$
(c) $BaO(s) + CO_2(g) \rightarrow$
(d) $Ag_2O(s) \xrightarrow{\text{heat}}$
(e) $Li(s) + H_2O(l) \rightarrow$

4-36. Classify each of the following reactions as combination, decomposition, single-replacement, or double-replacement:

(a) $Pb(NO_3)_2(aq) + 2NaI(aq) \rightarrow PbI_2(s) + 2NaNO_3(aq)$
(b) $2KClO_3(s) \rightarrow 2KCl(s) + 3O_2(g)$
(c) $2NaCl(s) + H_2SO_4(l) \rightarrow 2HCl(g) + Na_2SO_4(s)$
(d) $Fe(s) + 2HBr(aq) \rightarrow FeBr_2(aq) + H_2(g)$

4-38. Complete and balance the following equations and classify each:

(a) $SrCO_3(s) \xrightarrow{\text{heat}}$
(b) $Li(s) + O_2(g) \rightarrow$
(c) $Zn(s) + H_2SO_4(aq) \rightarrow$
(d) $Na_2O(s) + CO_2(g) \rightarrow$
(e) $XeF_4(s) \xrightarrow{\text{heat}}$

4-39. Write the balanced chemical equation for each of the following reactions:

(a) An aqueous solution of hydrochloric acid is added to an aqueous solution of potassium cyanide with the emission of toxic hydrogen cyanide gas. This reaction is extremely dangerous and must be done in a well-ventilated hood.
(b) Potassium metal dissolves in water with the evolution of hydrogen gas.
(c) Hydrogen peroxide decomposes to produce oxygen gas and a colorless liquid.
(d) Excess hydrogen gas is injected into a flask containing a small amount of liquid bromine. Upon heating, the bromine liquid disappears and a colorless gas is formed.

4-40. Write the balanced chemical equation for each of the following reactions:

(a) Zinc sulfide dissolves in aqueous hydrochloric acid with the emission of foul-smelling and highly toxic hydrogen sulfide gas.
(b) A solution containing calcium chloride is added to dilute aqueous phosphoric acid, and solid white calcium phosphate forms.
(c) Hydrochloric acid is spilled on a laboratory bench and solid sodium carbonate is thrown over the spill to neutralize the acid. Some of the solid dissolves, and carbon dioxide is given off.
(d) An aqueous solution of zinc chloride is added to an aqueous solution of sodium sulfide to produce a white precipitate of zinc sulfide.

4-41. The metal the ancient Greeks produced from lead-containing ores was an alloy of lead and silver. They were able to separate the two metals by melting the alloy and blowing air over the molten metal. Write the equation for the reaction that takes place.

4-42. Mercury has been known since early times, although we do not know when it was first discovered. Probably because it is a liquid metal, mercury has been thought to have mystical properties throughout history. It occurs in the ore cinnabar as the sulfide, HgS. The metal was prepared by heating the ore and condensing the mercury vapor. Write the equation for this reaction.

NEUTRALIZATION AND TITRATION

4-43. By titration, it is found that 17.3 mL of 0.250 M NaOH(*aq*) is required to neutralize 37.6 mL of HCl(*aq*). Calculate the concentration of the hydrochloric acid solution.

4-44. By titration, it is found that 24.6 mL of 0.300 M $H_2SO_4(aq)$ is required to neutralize 20.0 mL of NaOH(*aq*). Calculate the concentration of the NaOH solution.

4-45. (a) What volume of 0.250 M HNO_3 solution is required to neutralize 15.0 μL of 0.010 M $Ca(OH)_2$?
(b) What volume of 0.300 M H_2SO_4 solution is required to neutralize 25.0 mL of 0.100 M NaOH?

4-46. Commercial antacid tablets contain a base, often an insoluble metal hydroxide, that reacts with stomach acid. Two bases used for this are $Mg(OH)_2(s)$ and $Al(OH)_3(s)$. Given that stomach acid is about 0.10 M HCl(*aq*) compute the number of milliliters of stomach acid that can be neutralized by 500 mg of each of these bases.

4-47. A 20.0-g sample of KOH(*s*) is dissolved in water to a final volume of 0.200 L, and the resulting solution is added to 1.00 L of 0.125 M HCl(*aq*). Compute the molarity of KCl(*aq*) in the resulting solution.

4-48. A 500-mL sample of 0.200 M NaOH(*aq*) is added to 200 mL of 0.100 M HBr(*aq*). Compute the molarities of the various species after the reaction occurs.

4-49. A 0.365-g sample of a mixture of NaOH and NaCl requires 31.7 mL of 0.150 M HCl to react with all the NaOH. What is the mass percentage of NaOH in the mixture?

4-50. In order to test the purity of NaOH after its manufacture, 0.400 g is dissolved in enough water to make 100 mL of a 0.100 M solution. The solution is titrated with 0.100 M HCl to determine the actual concentration of NaOH. It is found that 25.0 mL of NaOH is neutralized by 23.2 mL of HCl. Calculate the purity of the solid NaOH. What assumption do you have to make?

4-51. A 1.00-g sample of an unknown acid is dissolved to make 100.0 mL of solution and neutralized with 0.250 M NaOH(*aq*). The volume of NaOH(*aq*) required to neutralize the acid was 66.6 mL. Assume that the acid has only one acidic proton per molecule and compute the formula mass of the acid.

0.250 M NaOH × 0.0666 L = 0.01665 mol NaOH

1 g acid / 0.01665 mol acid = 60.1 g/mol acid

4-52. A 1.00-g sample of an unknown acid is dissolved to make 100.0 mL of solution and neutralized with 0.250 M NaOH(*aq*). The volume of NaOH(*aq*) required to neutralize the acid was 86.9 mL. Assume that the acid has only one acidic proton per molecule and compute the formula mass of the acid.

4-53. A 1.00-g sample of an unknown acid is dissolved to make 100.0 mL of solution and neutralized with 0.250 M NaOH(*aq*). The volume of NaOH(*aq*) required to neutralize the acid was 88.8 mL. Assume that the acid has two acidic protons per molecule and compute the formula mass of the acid.

0.250 M NaOH × 0.0888 L = 0.0222 mol NaOH

0.0222 mol NaOH ⇔ 0.0111 mol acid

1 g / 0.0111 mol = 90.1 g/mol acid

4-54. A 1.00-g sample of an unknown acid is dissolved to make 100.0 mL of solution and neutralized with 0.250 M NaOH(*aq*). The volume of NaOH(*aq*) required to neutralize the acid was 67.7 mL. Assume that the acid has two acidic protons per molecule and compute the formula mass of the acid.

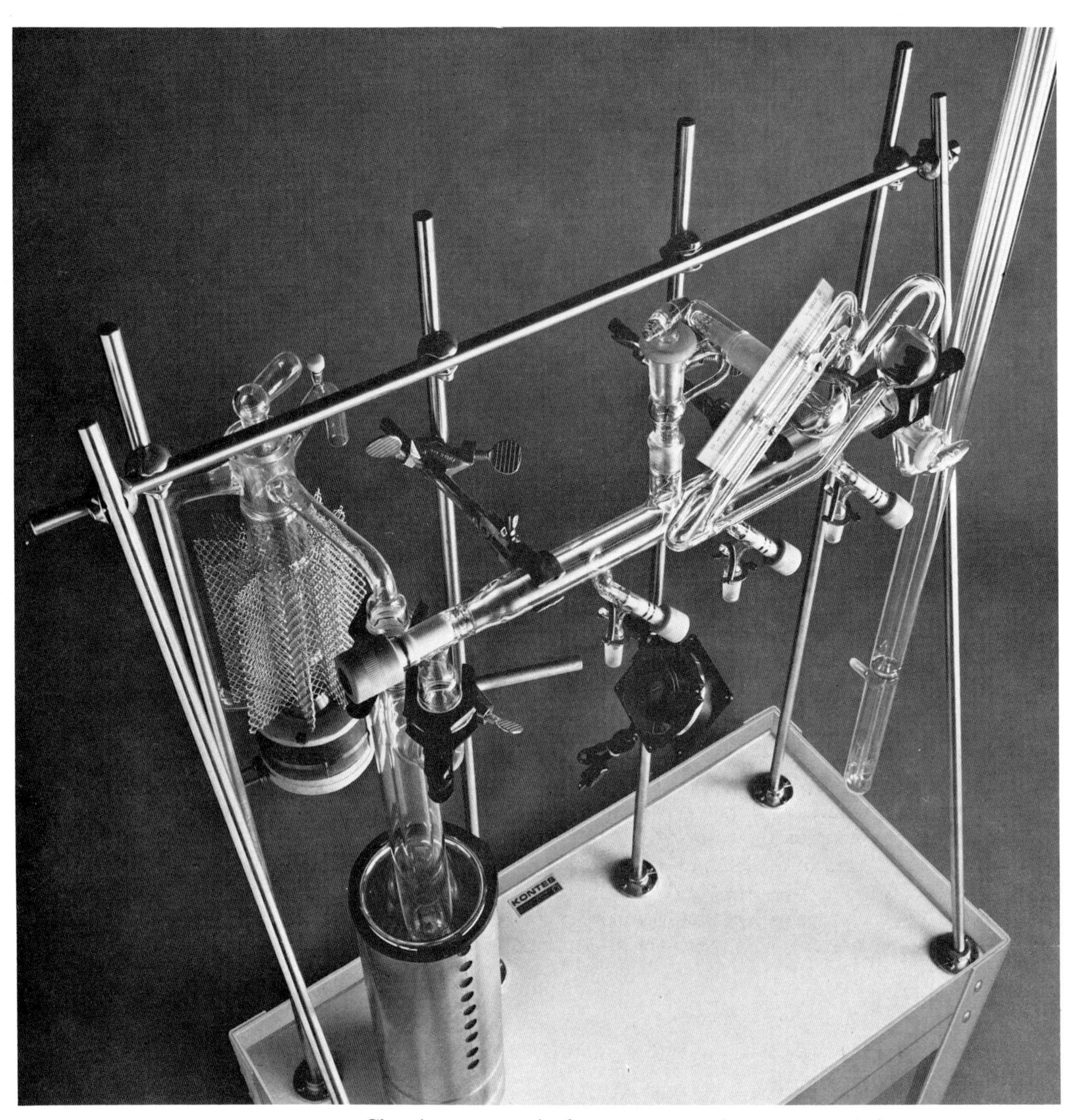

Chemists can manipulate gases at various pressures using an apparatus called a vacuum rack. The vacuum is produced by a pump (not shown). The device with a scale attached is a pressure gauge. The metal-covered container in the front is a cold trap, which is used to condense substances.

5 / The Properties of Gases

In this chapter, we study the properties of gases. Many chemical reactions involve gases as reactants or products or both, and so we must learn how the properties of gases depend upon conditions such as temperature, pressure, volume, and number of moles. We shall see how gases respond to changes in pressure and temperature and then discuss how the pressure, temperature, and volume of a gas are related to each other. After presenting a number of experimental observations concerning gases, we shall discuss the kinetic theory of gases, which gives a nice insight into the molecular nature of gases.

5-1. THE PHYSICAL STATES OF MATTER ARE SOLID, LIQUID, AND GAS

To understand the nature of gases, we must first discuss the three physical states of matter: *solid, liquid,* and *gas.* A *solid* has a fixed volume and shape. A *liquid,* on the other hand, has a fixed volume but assumes the shape of the container into which it is poured. A *gas* has neither a fixed volume nor a fixed shape; it always expands to occupy the entire volume of any closed container into which it is placed.

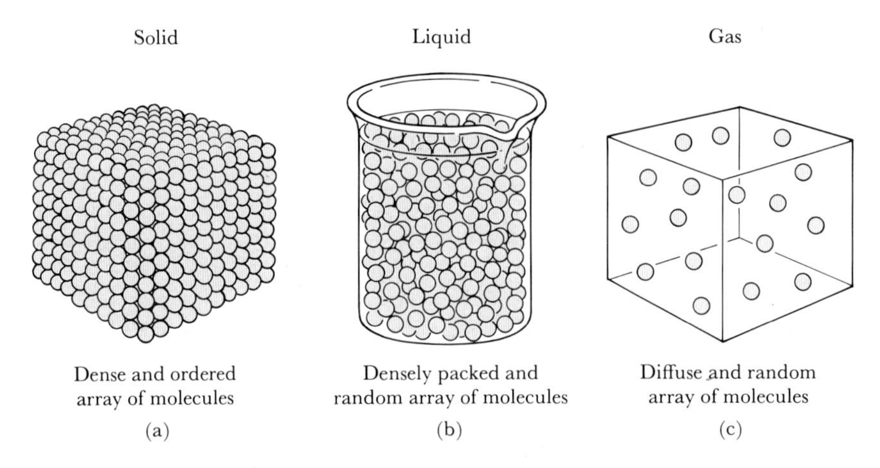

Figure 5-1 A molecular view of (a) a solid, (b) a liquid, and (c) a gas.

The molecular picture of a solid is that of a lattice (that is, an ordered network) of particles (atoms, molecules, or ions), as shown in Figure 5-1a. Each particle is essentially fixed at a lattice position and is not free to move about. The fixed lattice positions of the particles of a solid are reflected by the fixed volume and shape that characterize a solid.

A molecular view of a liquid (Figure 5-1b) is that the particles are in continuous contact with each other but are free to move about throughout the liquid. There is no orderly, fixed arrangement of particles in a liquid as there is in a solid. When a solid melts and becomes a liquid, the lattice array breaks down and the constituent particles are no longer held in fixed positions. The fact that the densities of the solid phase and the liquid phase of any substance do not differ greatly from each other indicates that the separation between the particles is similar in the two phases. Furthermore, the solid and liquid phases of a substance have similar, small *compressibilities*, meaning that their volume does not change appreciably with increasing pressure. The similar compressibilities of the solid and liquid phases of a substance is further evidence that the particles in the two phases have similar separations.

Compressibility is a measure of how readily the volume of a substance changes.

When a liquid is vaporized, there is a huge increase in volume. For example, 1 mol of liquid water occupies 17.3 mL at 100°C, whereas 1 mol of water vapor occupies over 30,000 mL under the same conditions. Upon vaporization, the molecules of a substance become widely separated, as indicated in Figure 5-1(c). The picture of a gas as being made up of widely separated particles accounts nicely for the relative ease with which gases can be compressed. The particles take up only a small fraction of the total space occupied by a gas; most of the volume of a gas is empty space. As we see in this chapter, the volume of a gas decreases markedly with increasing pressure.

5-2. THE PRESSURE OF A GAS CAN SUPPORT A COLUMN OF LIQUID

As we have seen, a gas is mostly empty space, with the molecules widely separated from each other. The molecules are in constant motion, traveling about at high speeds and colliding with each other and with the walls of the container. It is the force of these incessant, numerous collisions with the walls of the container that is responsible for the *pressure* exerted by a gas.

A common laboratory setup used to measure the pressure exerted by a gas is a *manometer*, which is a glass U-shaped tube partially filled with a liquid (Figure 5-2). Mercury is commonly used as the liquid because it has a high density and is fairly unreactive. Figure 5-2 illustrates the measurement of gas pressure with a manometer. The height *h* of the column of mercury that is supported by the gas in the flask is directly proportional to the pressure of the gas. Because of this direct proportionality, it is convenient to express pressure in terms of the height of a column of mercury that the gas will support. This height is usually expressed in millimeters, and so pressure is expressed in terms of millimeters of mercury (mm Hg). The pressure

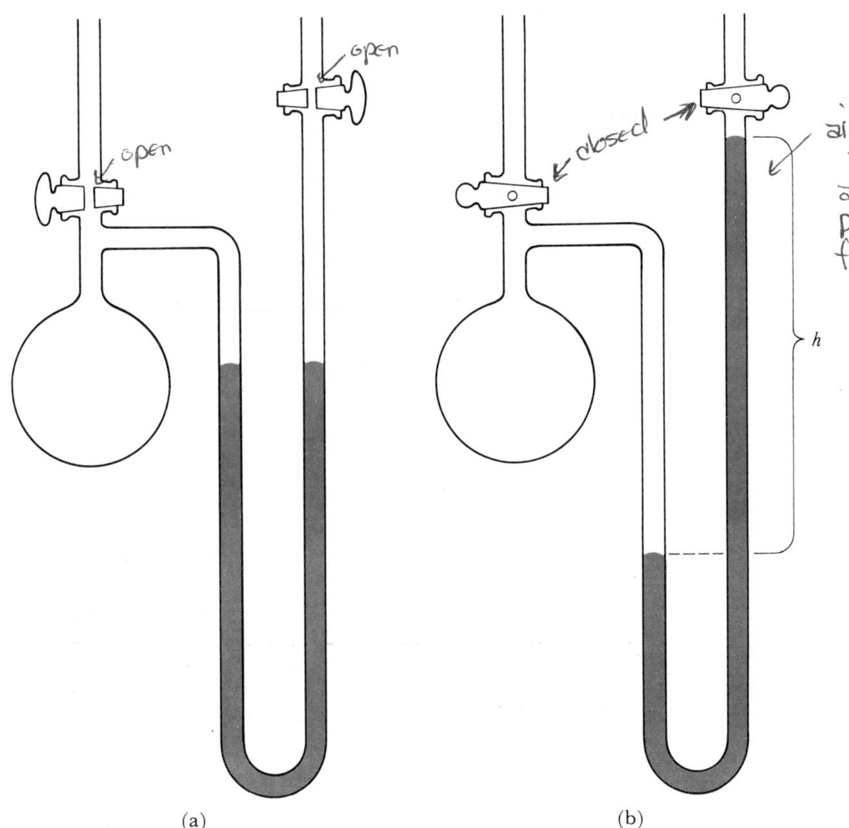

Figure 5-2 A mercury manometer. (a) Both stopcocks are open to the atmosphere, and so both columns are exposed to atmospheric pressure. Both columns are at the same height because the pressure is the same on both surfaces. (b) The two stopcocks are closed, and the air in the right-hand column has been evacuated so that there is essentially no pressure on the right-hand column of mercury. As a result, the heights of the columns are no longer the same. The difference in heights is a direct measure of the pressure of the gas in the flask.

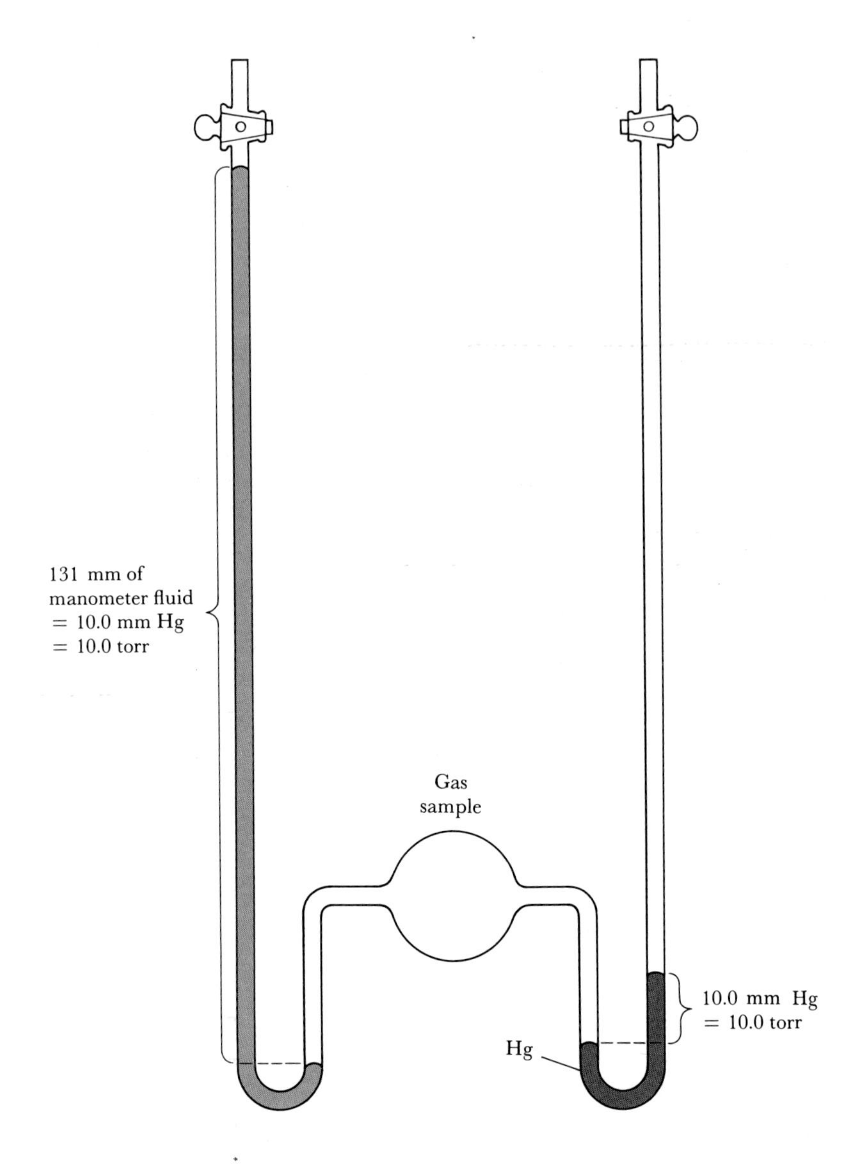

Figure 5-3 The height of a column of liquid that can be supported by a particular gas is inversely proportional to the density of the liquid. The denser the liquid, the shorter the column that can be supported by the gas. The liquid in the left column has a density of $1.04\ g\cdot mL^{-1}$.

unit mm Hg is called a *torr*, after the Italian scientist Evangelista Torricelli, who invented the barometer, which is described in the next section. Thus we say, for example, that the pressure of a gas is 600 torr.

Although mercury is most often used as the liquid in a manometer, many other liquids are suitable. The height of the column of liquid that can be supported by a gas is inversely proportional to the density of the liquid; that is, the less dense the liquid, the taller will be the column (Figure 5-3).

Example 5-1: Suppose the pressure of a gas is 760 torr. What is the height of a column of water that will be supported by the gas? The density of mercury is $13.6\ g\cdot mL^{-1}$, and that of water is $1.00\ g\cdot mL^{-1}$.

Solution: Mercury is 13.6 times as dense as water, and so the column of water supported will be 13.6 times higher than the column of mercury. The column of mercury supported is 760 mm, and so

$$\text{column of water} = (760 \text{ mm})\left(\frac{13.6 \text{ g}\cdot\text{mL}^{-1}}{1.00 \text{ g}\cdot\text{mL}^{-1}}\right)\left(\frac{1 \text{ m}}{1000 \text{ mm}}\right) = 10.3 \text{ m } (= 33.8 \text{ ft})$$

5-3. THE EARTH'S ATMOSPHERE EXERTS A PRESSURE

The atmosphere surrounding the earth is a gas that exerts a pressure. The manometer pictured in Figure 5-2 can be used to demonstrate this. If the flask is open to the atmosphere and the air in the right-hand side is evacuated, a column of mercury will be supported by the atmospheric pressure. The height of the mercury column depends upon elevation above sea level, temperature, and climatic conditions, but at sea level on a clear day it is about 760 mm. The pressure is the result of the mass of the atmosphere that presses down on the earth's surface. Figure 5-4 shows how this pressure is measured with a *barometer;* the measured atmospheric pressure is called the *barometric pressure.*

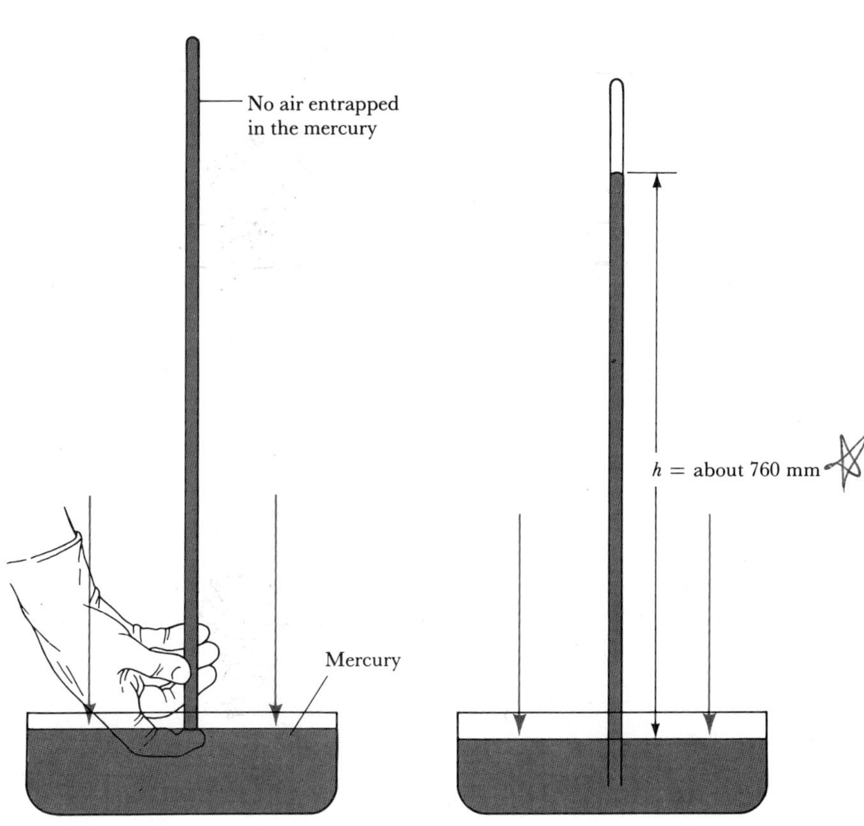

Figure 5-4 The measurement of atmospheric pressure by means of a barometer, as first done by Torricelli in the 1600's. (a) A long cylinder sealed at one end is completely filled with mercury and then inverted in a mercury bath. (b) A column of mercury about 760 mm high can be supported by the pressure exerted by the atmosphere, which is represented by the arrows.

5-4. PRESSURE IS EXPRESSED IN VARIOUS UNITS

Several units can be used to express pressure. A pressure of 760 torr is defined as one *standard atmosphere* (atm). It is common to express pressure in terms of standard atmospheres or, more simply, atmospheres.

Example 5-2: Given that the measured barometric pressure at Boulder, Colorado, is 680 torr, express this pressure in atmospheres.

Solution: The conversion between torr and atmospheres is

$$1 \text{ atm} = 760 \text{ torr}$$

Therefore,

$$P = (680 \text{ torr})\left(\frac{1 \text{ atm}}{760 \text{ torr}}\right) = 0.895 \text{ atm}$$

The barometric pressure decreases with increasing altitude.

The precise definition of a pascal requires a knowledge of elementary physics.

Strictly speaking, torr and atmosphere are not units of pressure because pressure is defined as a *force per unit area*. The SI unit of pressure is the *pascal* (Pa). The precise definition of a pascal is given in Appendix B; however, a simple, operational definition is that a pascal is the pressure exerted on a 1-m^2 surface by a mass of 102 g. More important for us is the relation between torr, atmosphere, and pascal:

$$760 \text{ torr} = 1 \text{ atm} = 1.013 \times 10^5 \text{ Pa}$$

Several common units for expressing pressure are summarized in Table 5-1. The units torr and atmosphere are so widely used by chemists that their replacement by the pascal will be extremely slow and painful. Consequently, in most cases, we shall use torr or atmosphere in this text, but a section of problems using SI pressure units is included at the end of this chapter.

Example 5-3: Convert 2280 torr to standard atmospheres and to kilopascals.

Solution: To convert from torr to standard atmospheres, we use the fact that 1 atm = 760 torr. Therefore,

$$(2280 \text{ torr})\left(\frac{1 \text{ atm}}{760 \text{ torr}}\right) = 3.00 \text{ atm}$$

To convert from torr to kilopascals, we use the conversion factor 760 torr = 101.3 kPa (Table 5-1):

It is useful to remember that one kilopascal is approximately equal to 0.01 atmosphere.

$$(2280\ \text{torr})\left(\frac{101.3\ \text{kPa}}{760\ \text{torr}}\right) = 304\ \text{kPa}$$

Note that, because there are about 100 kPa in 1 atm, 1 kPa is almost equal to 0.01 atm. Thus 304 kPa is approximately 3.0 atm.

Table 5-1 Various units for expressing pressure

SI unit
1 pascal (Pa) = pressure (force per unit area) exerted by a mass of 102 g on a 1-m^2 surface
(see Appendix B for precise definition)

"convenience" unit
height of a column of mercury supported by the pressure; commonly expressed as torr (1 torr = 1 mm Hg)

defined unit
1 standard atmosphere = 1.013×10^5 Pa
= 101.3 kPa
= 760 torr
= 14.7 lb·in.$^{-2}$

meteorological unit
1 bar = 10^5 Pa
1 atm = 1.013 bar = 1013 mbar (the bar is derived from *bar*ometer)

5-5. AT CONSTANT TEMPERATURE, THE VOLUME OF A GAS IS INVERSELY PROPORTIONAL TO ITS PRESSURE

The first systematic study of the behavior of gases under different applied pressures was carried out in the 1660's by the Irish scientist Robert Boyle, who trapped a sample of gas in the closed end of a J-shaped tube and varied the volume of the gas by adding mercury to the open arm (Figure 5-5). In this way he was able to measure the pressure of the gas as a function of its volume, where the pressure is determined by the height of the column of mercury that the gas supports. Boyle was able to show that, at constant temperature, the volume of a given sample of gas is inversely proportional to the pressure:

$$V \propto \frac{1}{P}$$

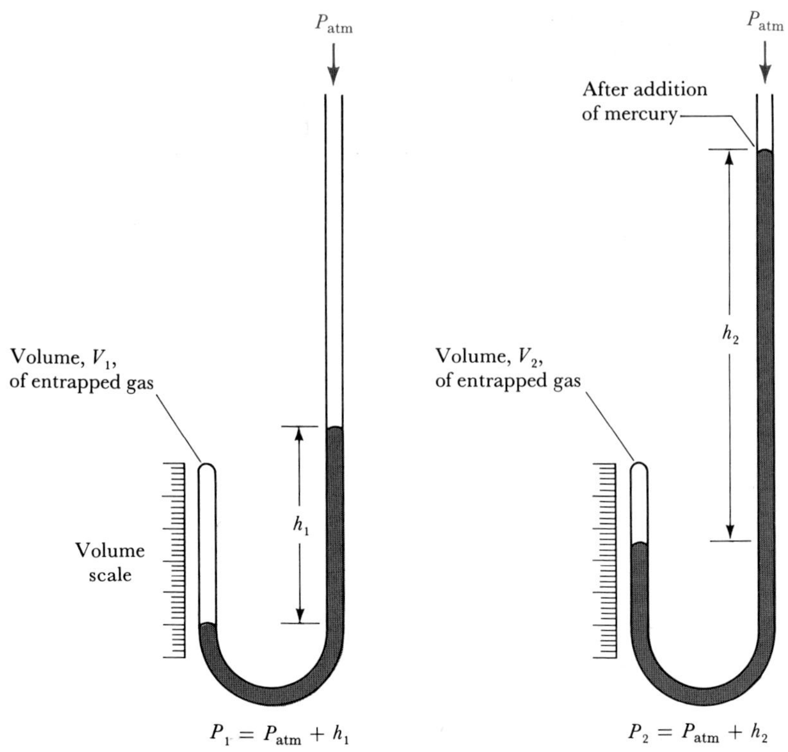

Figure 5-5 An apparatus similar to that used by Boyle to study the pressure-volume relationship of gases. A sample of gas is trapped in the sealed end of a J-shaped tube. The pressure on the gas can be varied by adding mercury to the right-hand column. Because the right-hand column is open to the atmosphere, the pressure (in torr) acting on the gas is given by $P_{gas} = P_{atm} + h$.

In terms of an equation, we have

$$V = \frac{c}{P} \qquad \text{(constant temperature)}$$

or

$$PV = c \qquad \text{(constant temperature)} \tag{5-1}$$

You can always convert a proportionality statement to an equality by inserting a proportionality constant.

where c is a proportionality constant. The relationship between pressure and volume expressed in Equation (5-1) is known as *Boyle's law.* The greater the pressure on a gas, the smaller the volume at constant temperature. If we double the pressure on a gas, then its volume decreases by a factor of two.

Table 5-2 Experimental data for the volume of 0.580 g of air as a function of temperature at three different pressures

	V/L		
$t/°C$	*0.500 atm*	*1.00 atm*	*2.00 atm*
0	0.90	0.45	0.22
100	1.22	0.61	0.30
200	1.55	0.78	0.39
300	1.88	0.94	0.47
400	2.21	1.10	0.55
500	2.54	1.27	0.63

5-6. AT CONSTANT PRESSURE, THE VOLUME OF A GAS IS PROPORTIONAL TO ITS TEMPERATURE

Jacques Charles, the French scientist and adventurer, was the first to show that there is a linear relationship between the volume of a gas and its temperature. Typical experimental data are given in Table 5-2, and these data are plotted as volume versus temperature in Figure 5-6.

Jacques Charles did many of his experiments on gases aloft in a hydrogen-filled balloon.

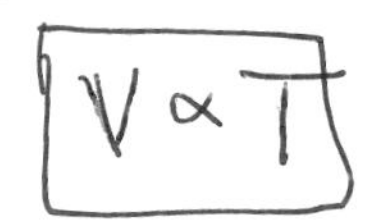

Not only does Figure 5-6 show that there is a linear relationship between volume and temperature, but, more important, it also suggests that we can define a new temperature scale by adding 273° to

Figure 5-6 The volume of 0.580 g of air plotted as a function of temperature at three different pressures. Note that all three curves extrapolate to $V = 0$ at $-273°C$. These plots suggest that we can define an absolute temperature scale by adding 273 to the Celsius scale to get $T = t(°C) + 273$. This new temperature scale is called the *absolute temperature scale* or *Kelvin scale,* and the unit is the kelvin (K). The style for designation of units in labels of axes follows SI convention; recall the explanatory marginal note on p. 18.

the Celsius scale. By doing so, all the curves in Figure 5-6 can be represented by the simple equation

$$V = kT \quad \text{(constant pressure)} \tag{5-2}$$

where the temperature (T) is

$$T\,(\text{in K}) = t\,(\text{in °C}) + 273$$

and where the value of k depends upon the pressure and quantity of the gas.

The relation between volume and absolute temperature given by Equation (5-2) is called *Charles's law:* the volume of a fixed mass of gas at a fixed pressure is directly proportional to its absolute temperature.

Note that there is no degree sign on the symbol for Kelvin.

The temperature scale introduced here is called the *absolute temperature scale* or the *Kelvin temperature scale;* the unit for this scale is the kelvin (K). The precise relation between the two scales is

$$T/\text{K} = t/°\text{C} + 273.15 \tag{5-3}$$

The Kelvin scale is the most fundamental temperature scale. Figure 5-7 shows the temperatures of a number of fixed points in three temperature scales. The lowest possible temperature on the Kelvin scale is 0 K, which corresponds to −273.15°C.

It is impossible to achieve a temperature of 0 K. The lowest temperature yet achieved is 10^{-6} K.

A capillary tube is a tube having a small inside diameter.

Example 5-4: A simple *gas thermometer* can be made by trapping a small sample of gas with a drop of mercury in a glass capillary tube that is sealed at one end and open at the other (Figure 5-8). Suppose that in such a thermometer, the gas occupies a volume of 0.180 mL at 0°C. The thermometer is then immersed in a liquid, and the final volume of the gas is 0.232 mL. What is the temperature of the liquid?

Solution: From Charles's law, we have $V = kT$, or

$$\frac{V}{T} = k$$

As long as the pressure on the gas remains the same, the value of k does not change. Consequently, we can write

$$\frac{V_i}{T_i} = k \quad \text{and} \quad \frac{V_f}{T_f} = k$$

or

$$\frac{V_f}{T_f} = \frac{V_i}{T_i} \tag{5-4}$$

Figure 5-7 The temperatures of fixed reference points in the Fahrenheit, Celsius, and Kelvin temperature scales. The lowest possible temperature is 0 K, which corresponds to $-273.15\,°C$ and to $-459.67\,°F$.

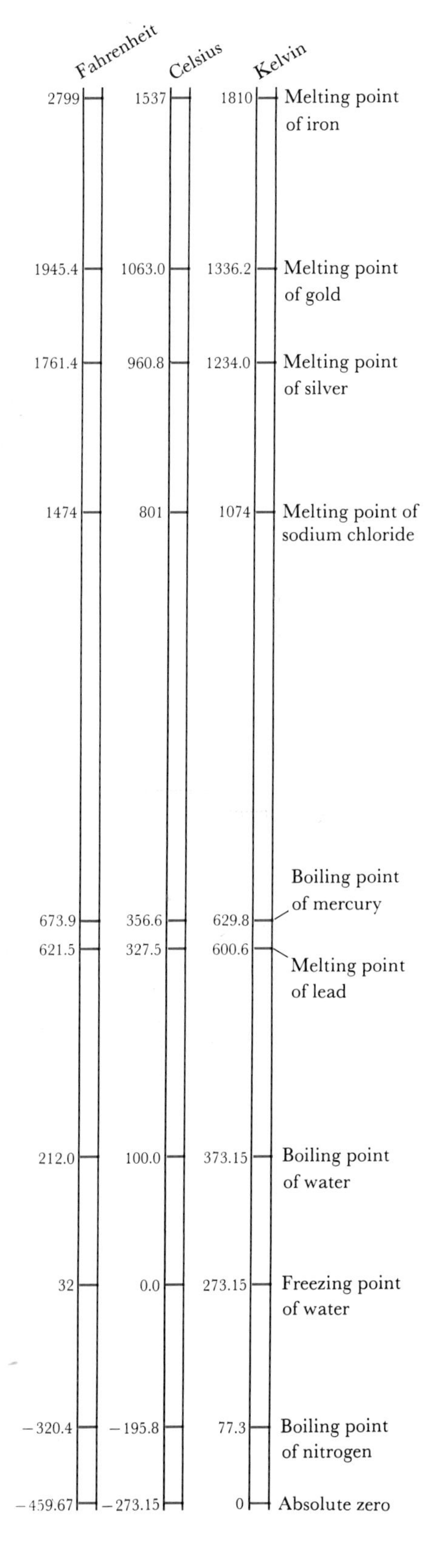

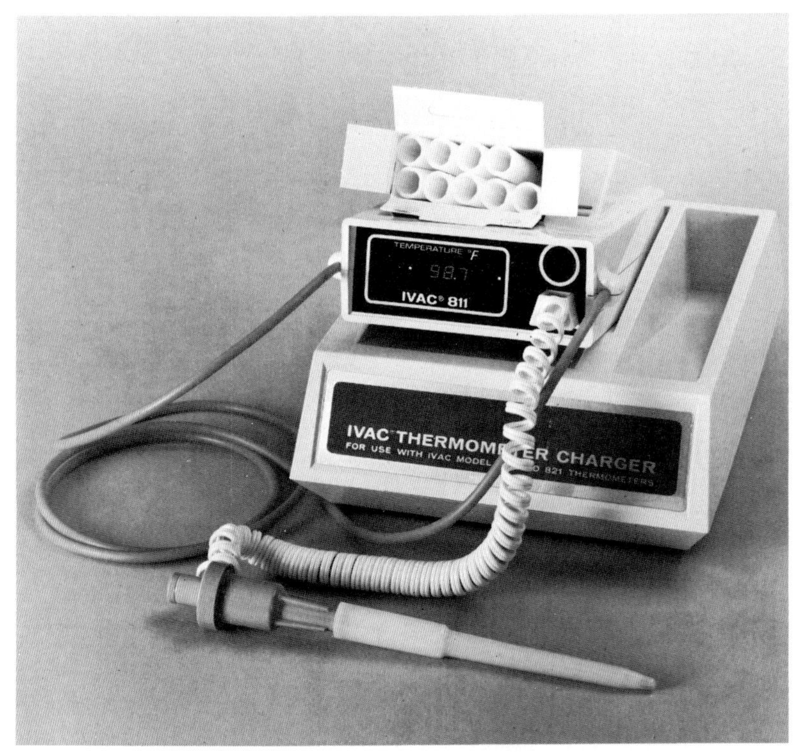

The clinical digital thermometer uses a thermistor to sense the temperature. A thermistor is a metal-oxide chip whose resistance depends strongly on the temperature.

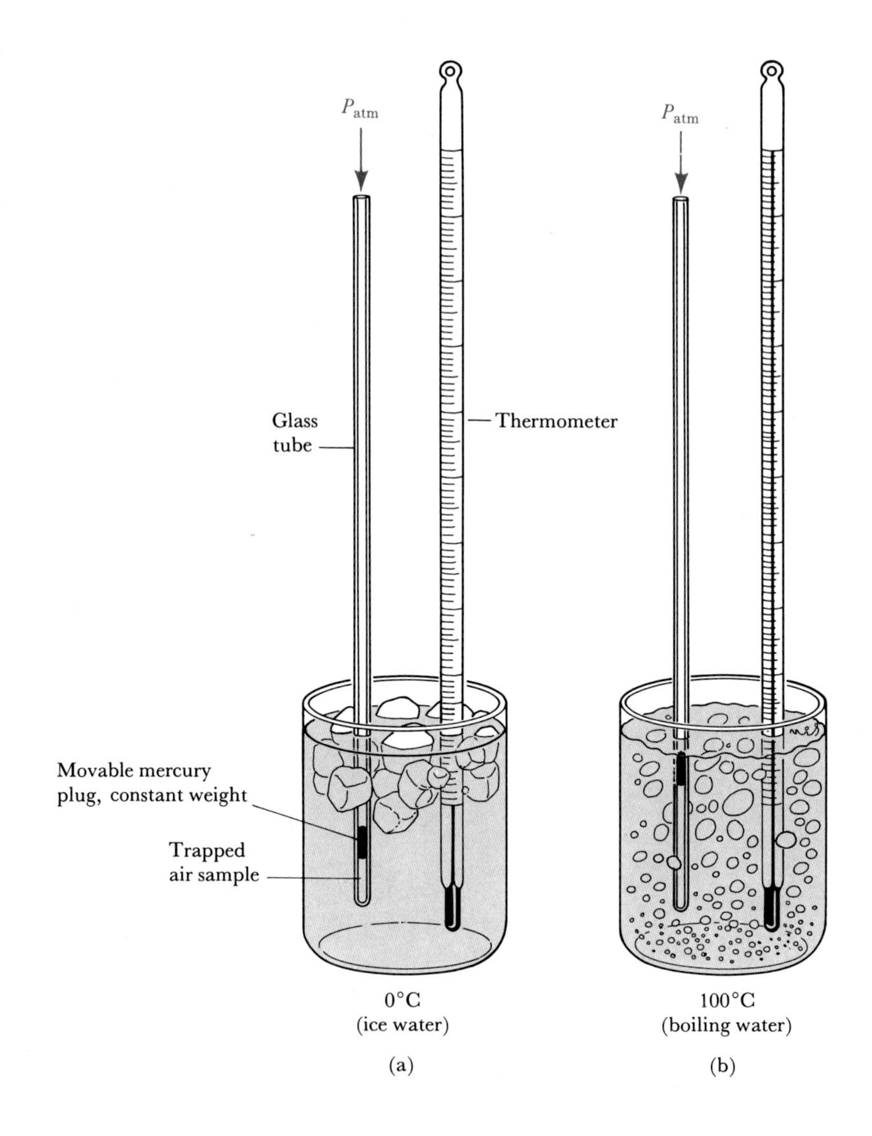

Figure 5-8 A gas thermometer. A sample of air is trapped by a drop of mercury in a capillary tube that is sealed at the bottom. According to Charles's law, the volume of the air is directly proportional to the Kelvin temperature. The atmosphere maintains a constant pressure on the air trapped below the mercury, which moves up or down to a position where the pressure of the trapped air equals the atmospheric pressure. As the temperature increases, the drop of mercury rises because the gas expands.

where the subscripts i and f stand for *i*nitial and *f*inal, respectively. We must always remember to use absolute temperatures in Charles's law:

$$T_i = 0° + 273 = 273\ \text{K}$$

We are seeking T_f, and so we solve Equation (5-4) for T_f to get

$$T_f = T_i \frac{V_f}{V_i} = (273\ \text{K})\left(\frac{0.232\ \text{mL}}{0.180\ \text{mL}}\right) = 352\ \text{K}$$

The corresponding temperature in degrees Celsius is

$$t_f = 352 - 273 = 79°\text{C}$$

5-7. EQUAL VOLUMES OF GASES AT THE SAME PRESSURE AND TEMPERATURE CONTAIN EQUAL NUMBERS OF MOLECULES

Early experiments with gaseous reactions showed a remarkable property. It was observed by Gay-Lussac in 1809 that if all volumes are measured at the same pressure and temperature, then the volumes in which gases combine in chemical reactions are related to each other by simple whole numbers. For example,

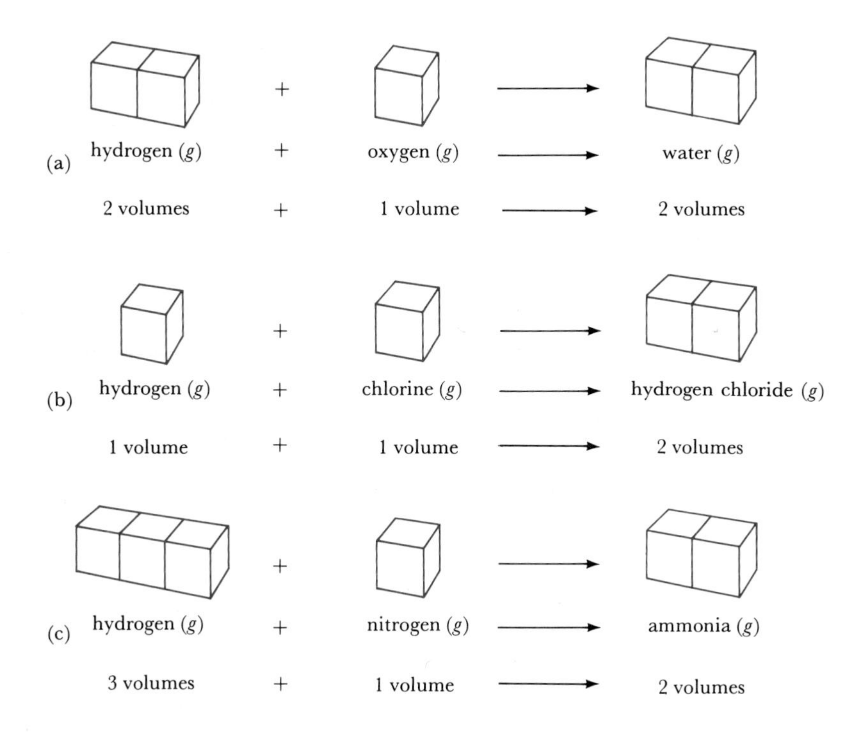

Note that in each of these cases the relative volumes of reactants and products are in the proportion of simple whole numbers. This observation is known as *Gay-Lussac's law of combining volumes* and was one of the earliest indications of the existence of atoms and molecules. The interpretation of Gay-Lussac's law by Avogadro in 1811 led to the realization that many of the common gaseous elements, such as hydrogen, oxygen, nitrogen, and chlorine, occur naturally as diatomic molecules (H_2, O_2, N_2 and Cl_2) rather than as single atoms. Let's review Avogadro's line of reasoning and see why this is so.

Following Gay-Lussac's observations, Avogadro postulated that equal volumes of gases at the same pressure and temperature contain equal numbers of molecules. This statement was known as *Avogadro's hypothesis* at the time, but now it is accepted as a law. Avogadro was the first to point out the distinction between atoms and molecules. Consider the reaction between hydrogen and chlorine to form hydro-

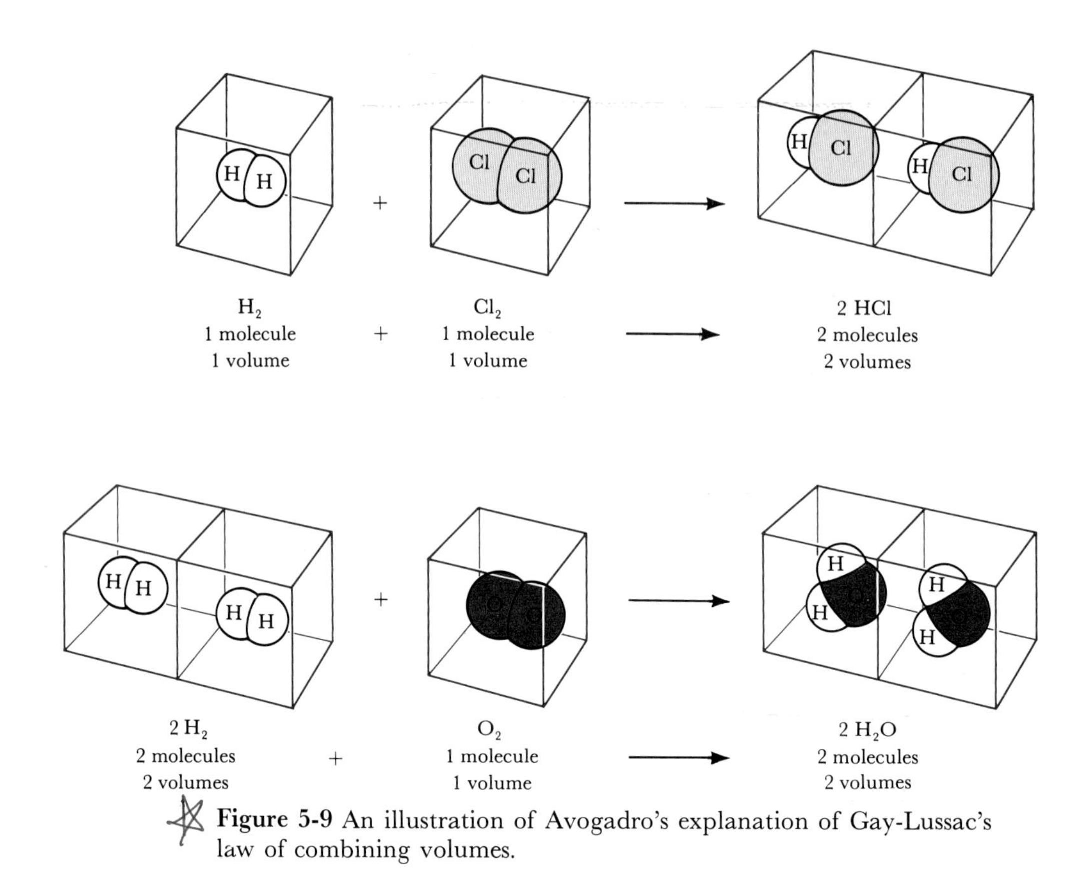

Figure 5-9 An illustration of Avogadro's explanation of Gay-Lussac's law of combining volumes.

gen chloride. Recall that *one volume* of hydrogen reacts with *one volume* of chlorine to produce *two volumes* of hydrogen chloride. According to Avogadro's reasoning, this means that *one molecule* of hydrogen reacts with *one molecule* of chlorine to produce *two molecules* of hydrogen chloride (Figure 5-9). If one molecule of hydrogen can form two molecules of hydrogen chloride, then a hydrogen molecule must consist of two atoms (at least) of hydrogen. Avogadro pictured both hydrogen and chlorine as diatomic gases and was able to represent the reaction between them as

$$
\begin{array}{rcl}
H_2(g) + & Cl_2(g) & \rightarrow 2HCl(g) \\
1 \text{ molecule} + & 1 \text{ molecule} & \rightarrow 2 \text{ molecules} \\
1 \text{ volume} + & 1 \text{ volume} & \rightarrow 2 \text{ volumes}
\end{array}
$$

Prior to Avogadro's explanation, it was difficult to see how one volume of hydrogen could produce two volumes of hydrogen chloride. If hydrogen occurred simply as atoms, there would be no way to explain Gay-Lussac's law of combining volumes. Two other examples of Avogadro's law applied to chemical reactions are

$$3H_2(g) + N_2(g) \rightarrow 2NH_3(g)$$
$$3 \text{ molecules} + 1 \text{ molecule} \rightarrow 2 \text{ molecules}$$
$$3 \text{ volumes} + 1 \text{ volume} \rightarrow 2 \text{ volumes}$$

$$2H_2(g) + O_2(g) \rightarrow 2H_2O(g)$$
$$2 \text{ molecules} + 1 \text{ molecule} \rightarrow 2 \text{ molecules}$$
$$2 \text{ volumes} + 1 \text{ volume} \rightarrow 2 \text{ volumes}$$

It is interesting to note that, in spite of the beautiful simplicity of Avogadro's explanation of these reactions, his work was largely ignored and chemists continued to confuse atoms and molecules and to use many incorrect chemical formulas. It wasn't until the mid-1800's that Avogadro's hypothesis was finally appreciated and generally accepted.

5-8. BOYLE'S, CHARLES'S, AND AVOGADRO'S LAWS CAN BE COMBINED INTO ONE EQUATION

Avogadro postulated that equal volumes of gases at the same pressure and temperature contain the same number of molecules. This implies that equal volumes of gases at the same pressure and temperature contain equal numbers of moles, n. Thus, we can write Avogadro's law as

$$V \propto n \qquad \text{(fixed } P \text{ and } T\text{)}$$

Boyle's law and Charles's law are, respectively,

$$V \propto \frac{1}{P} \qquad \text{(fixed } T \text{ and } n\text{)}$$

$$V \propto T \qquad \text{(fixed } P \text{ and } n\text{)}$$

We can combine these three proportionality statements for V into one by writing

$$V \propto \frac{nT}{P}$$

Note how the three individual statements for V are all included in the combined statement. For example, if P and n are fixed, then only T can vary and we see that $V \propto T$, which is Charles's law. If P and T are fixed, then only n can vary and we have Avogadro's law, $V \propto n$. Lastly, if T and n are fixed, we have Boyle's law, $V \propto 1/P$.

We can convert the combined proportionality statement for V to an equation by introducing a proportionality constant R:

$$V = \frac{RnT}{P}$$

This equation is equivalently, but more commonly, written as

$$PV = nRT \tag{5-5}$$

and is called the *ideal-gas law* or *ideal-gas equation.* It is based upon Boyle's law, Charles's law, and Avogadro's law. Boyle's law and Charles's law are valid only at low pressures (less than a few atmospheres, say) and so Equation (5-5) is valid only at low pressures. It turns out, however, that most gases obey Equation (5-5) within a few percent up to tens of atmospheres, and so Equation (5-5) is very useful. Gases that satisfy the ideal-gas equation are said to behave ideally, or to be *ideal gases.*

Before we can use Equation (5-5), we must determine the value of R, which is called the *gas constant.* It has been determined experimentally that 1 mol of an ideal gas at 0°C and 1.00 atm occupies 22.4 L. The volume 22.4 L, shown in Figure 5-10, is called the *molar volume* of an ideal gas at 0°C and 1.00 atm. If we solve Equation (5-5) for R and substitute this information into the resulting equation, then we find that

$R = 0.0821\ \text{L}\cdot\text{atm}\cdot\text{K}^{-1}\cdot\text{mol}^{-1}$

$$R = \frac{PV}{nT} = \frac{(1.00\ \text{atm})(22.4\ \text{L})}{(1.00\ \text{mol})(273\ \text{K})} = 0.0821\ \text{L}\cdot\text{atm}\cdot\text{mol}^{-1}\cdot\text{K}^{-1} \tag{5-6}$$

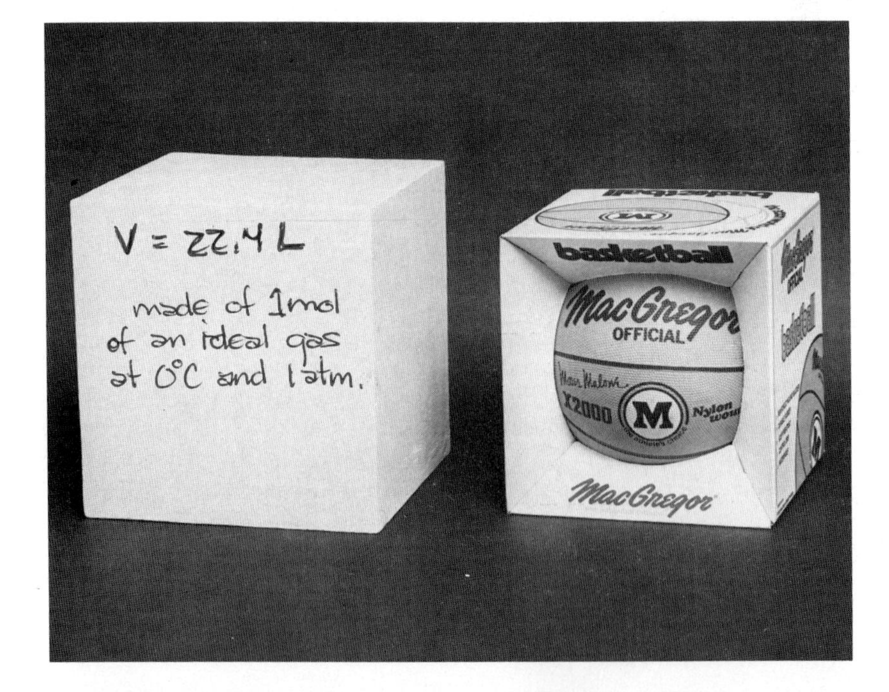

Figure 5-10 The volume of 1 mol of an ideal gas at 0°C and 1 atm. A volume of 22.4 L can be represented by a cube whose edges measure 28.2 cm. The carton that a basketball comes in is shown here for comparison.

You should pay careful attention to the units of R. When the value $R = 0.0821\ \mathrm{L \cdot atm \cdot mol^{-1} \cdot K^{-1}}$ is used in Equation (5-5), P must be expressed in atmospheres, V in liters, n in moles, and T in kelvins.

Notice that we say that T is in kelvins, not in degrees kelvin.

Now that we have determined the value of the gas constant, we can use Equation (5-5) in many ways.

Example 5-5: Calculate the pressure developed in a 0.80-L bicycle tire when 4.0 g of air is pumped in at 40°C. Take the average molecular mass of air as 29.

Because air is approximately 80% N_2 and 20% O_2, we take the effective mass of air to be (0.80)(28) + (0.20)(32) = 29.

Solution: Comparison with Equation (5-5), $PV = nRT$, shows that we are given three quantities (V, T, and the mass of air, from which we can compute n) and we wish to calculate the fourth (P). To calculate P from Equation (5-5), we must be sure that V, T, and n are in the correct units:

$$n = (4.0\ \mathrm{g})\left(\frac{1\ \mathrm{mol}}{29\ \mathrm{g}}\right) = 0.14\ \mathrm{mol}$$

$$T = 40 + 273 = 313\ \mathrm{K}$$

We now solve Equation (5-5) for P:

$$P = \frac{nRT}{V} = \frac{(0.14\ \mathrm{mol})(0.0821\ \mathrm{L \cdot atm \cdot mol^{-1} \cdot K^{-1}})(313\ \mathrm{K})}{(0.80\ \mathrm{L})}$$

$$= 4.5\ \mathrm{atm}$$

Since 1 atm equals 101 kPa, 4.5 atm corresponds to

$$4.5\ \mathrm{atm} \times 101 \frac{\mathrm{kPa}}{\mathrm{atm}} = 450\ \mathrm{kPa}$$

Since 1 atm also equals 14.7 pounds per square inch (psi), 4.5 atm corresponds to

$$4.5\ \mathrm{atm} \times 14.7 \frac{\mathrm{psi}}{\mathrm{atm}} = 66\ \mathrm{psi}$$

Example 5-6: The pressure of oxygen in inhaled air is 157 torr. The total volume of the average adult lungs when expanded is about 6.0 L, and body temperature is 37°C. Calculate the mass of O_2 required to occupy a volume of 6.0 L at a pressure of 157 torr and a temperature of 37°C.

Normal human body temperature (98.6°F) corresponds to 37.0°C.

Solution: Once again note that we are given three quantities (V, P, and T) and we wish to calculate the fourth (n). We must express P and T in the correct units:

$$P = (157\ \text{torr})\left(\frac{1\ \text{atm}}{760\ \text{torr}}\right) = 0.207\ \text{atm}$$

$$T = 37 + 273 = 310\ \text{K}$$

Solving Equation (5-5) for n, we get

Problems 5-7 to 5-26 deal with the ideal-gas equation.

$$n = \frac{PV}{RT} = \frac{(0.207\ \text{atm})(6.0\ \text{L})}{(0.0821\ \text{L}\cdot\text{atm}\cdot\text{mol}^{-1}\cdot\text{K}^{-1})(310\ \text{K})}$$
$$= 0.049\ \text{mol}$$

The number of grams of O_2 is

$$\text{grams of } O_2 = (0.049\ \text{mol})\left(\frac{32.0\ \text{g}}{1\ \text{mol}}\right)$$
$$= 1.6\ \text{g}$$

In Examples 5-5 and 5-6 we were given three quantities and had to calculate a fourth. Another type of application of the ideal-gas equation involves changes from one set of conditions to another.

Example 5-7: One mole of O_2 gas occupies 22.4 L at 0°C and 1.00 atm. What volume does it occupy at 100°C and 4.00 atm?

Solution: Note that in this problem we are given V at one set of conditions (T and P) and asked to calculate V under another set of conditions (that is, a different T and P). Because R is a constant, *and because n is a constant in this problem,* we can write the ideal-gas equation as

$$\frac{PV}{T} = nR = \text{constant}$$

This equation says that the ratio PV/T remains constant, and so we can write

$$\frac{P_iV_i}{T_i} = \frac{P_fV_f}{T_f} \tag{5-7}$$

Recall that we wish to calculate a final volume, and so we solve this equation for V_f:

$$V_f = V_i\left(\frac{P_i}{P_f}\right)\left(\frac{T_f}{T_i}\right)$$

If we substitute the given quantities into this equation, then we obtain

$$V_f = (22.4\text{ L})\left(\frac{1.00\text{ atm}}{4.00\text{ atm}}\right)\left(\frac{373\text{ K}}{273\text{ K}}\right)$$

$$= 7.65\text{ L}$$

Note that the increase in pressure (from 1.00 atm to 4.00 atm) decreases the gas volume (the gas is compressed), whereas the increase in temperature increases it. The pressure increases by a factor of 4.00, whereas the temperature increases only by a factor of 373/273 = 1.37; thus the *net* effect is a decrease in the volume of the gas.

Note that in Example 5-7 we multiplied the initial volume (22.4 L) by a pressure ratio and a temperature ratio. The pressure increased from 1.00 atm to 4.00 atm, and the pressure ratio used was 1.00/4.00, resulting in a smaller volume, as you would expect. Similarly, the temperature increased from 273 K to 373 K, and the temperature ratio used was 373/273, resulting in an increased volume. A "common sense" method of solving this problem is to write

An increase in pressure decreases the volume of a gas. An increase in temperature increases the volume of a gas.

$$V_f = V_i \times \text{pressure ratio} \times \text{temperature ratio}$$

and to decide by simple reasoning whether each ratio to be used is greater or less than unity.

5-9. THE IDEAL-GAS EQUATION CAN BE USED TO CALCULATE GAS DENSITIES AND MOLAR MASSES

Some of the most important applications of the ideal-gas equation involve the calculation of densities and molar masses. The *molar mass, M*, of a substance is simply the mass in grams of one mole of that substance. Molar masses are numerically equal to formula masses, but have units of $g \cdot mol^{-1}$. For example, the formula mass of H_2O is 18.0, and its molar mass is 18.0 $g \cdot mol^{-1}$.

Molar mass is numerically equal to molecular mass, but has units of $g \cdot mol^{-1}$.

If Equation (5-5) is solved for n/V, then we get

$$\frac{n}{V} = \frac{P}{RT}$$

The ratio n/V is equal to gas density in the units moles per liter. We can convert from moles per liter to grams per liter by multiplying both sides of the equation by the molar mass. If we denote the density in grams per liter by the symbol ρ (the Greek letter rho), then we can write

$$\rho = \frac{Mn}{V} = \frac{MP}{RT} \qquad (5\text{-}8)$$

Note that gas density increases as pressure increases and as temperature decreases.

Example 5-8: Calculate the density of ammonia gas at 0°C and 1.00 atm.

Solution: The molar mass of NH_3 is 17.0 $g \cdot mol^{-1}$. Using Equation (5-8) with appropriate units gives

$$\rho = \frac{MP}{RT} = \frac{(17.0\ g \cdot mol^{-1})(1.00\ atm)}{(0.0821\ L \cdot atm \cdot mol^{-1} \cdot K^{-1})(273\ K)}$$
$$= 0.758\ g \cdot L^{-1}$$

Another important application of the ideal-gas equation involves the calculation of molar masses.

Like liquids, gases have flow properties and can be poured from one container to another if they are denser than air.

Example 5-9: A 0.286-g sample of chlorine gas occupies 250 mL at 300 torr and 25°C. Determine the molar mass of chlorine.

Solution: We are given V, P, and T, and so we can use Equation (5-5) to calculate n:

$$n = \frac{PV}{RT} = \frac{(300\ torr)\left(\frac{1\ atm}{760\ torr}\right)(0.250\ L)}{(0.0821\ L \cdot atm \cdot mol^{-1} \cdot K^{-1})(298\ K)}$$
$$= 4.03 \times 10^{-3}\ mol$$

Thus 0.286 g of chlorine gas corresponds to 4.03×10^{-3} mol:

$$0.286\ g \approx 4.03 \times 10^{-3}\ mol$$

We would like to have this stoichiometric correspondence read

$$\text{a certain number of grams} \approx 1.00\ mol$$

We can achieve this by dividing both sides by 4.03×10^{-3}:

$$\frac{0.286\ g}{4.03 \times 10^{-3}} \approx \frac{4.03 \times 10^{-3}\ mol}{4.03 \times 10^{-3}}$$
$$71.0\ g \approx 1.00\ mol$$

Thus, we find that the molar mass of chlorine is 71.0 $g \cdot mol^{-1}$, or that the molecular mass of chlorine is 71.0. This implies that chlorine is a diatomic gas (Cl_2) because the atomic mass of chlorine is 35.45.

An alternate solution is to use the relation

$$\text{number of moles} = \frac{\text{mass in grams}}{\text{molar mass}}$$

$$n = \frac{m}{M}$$

Problems 5-33 to 5-40 ask you to do calculations involving gas density.

Solving for M we obtain

$$M = \frac{m}{n} = \frac{0.286\text{ g}}{4.03 \times 10^{-3}\text{ mol}} = 71.0\text{ g}\cdot\text{mol}^{-1}$$

We can combine a problem like Example 5-9 with a determination of the simplest formula of a compound from chemical analysis to determine the molecular formula.

Example 5-10: Acetylene gas is used in oxyacetylene welding torches. Chemical analysis shows that acetylene is 92.3 percent carbon and 7.70 percent hydrogen by mass. It has a density of $0.711\text{ g}\cdot\text{L}^{-1}$ at 20°C and 500 torr. Use these data to determine the molecular formula of acetylene.

Solution: The determination of the simplest formula from chemical analysis is explained in Section 3-3. Following the procedure given there, we write

$$92.3\text{ g C} \approx 7.70\text{ g H}$$

Dividing the left side by 12.01 g C/mol C and the right side by 1.008 g H/mol H gives

$$7.69\text{ mol C} \approx 7.64\text{ mol H}$$

Dividing both sides by 7.64 and rounding off yields

$$1\text{ mol C} \approx 1\text{ mol H}$$

and so the simplest formula of acetylene is CH.

We now use the density data to determine the molar mass of acetylene. Solving Equation (5-8) for M gives

$$M = \frac{\rho RT}{P} = \frac{(0.711\text{ g}\cdot\text{L}^{-1})(0.0821\text{ L}\cdot\text{atm}\cdot\text{mol}^{-1}\cdot\text{K}^{-1})(293\text{ K})}{\left(\frac{500}{760}\text{ atm}\right)}$$

$$= 26.0\text{ g}\cdot\text{mol}^{-1}$$

The formula mass of acetylene is 26.0 and its simplest formula is CH ($M = 13.0\text{ g}\cdot\text{mol}^{-1}$); therefore, the molecular formula of acetylene is C_2H_2 ($M = 26.0\text{ g}\cdot\text{mol}^{-1}$).

5-10. THE TOTAL PRESSURE OF A MIXTURE OF GASES IS THE SUM OF THE PARTIAL PRESSURES OF ALL THE GASES IN THE MIXTURE

Up to this point we have not considered explicitly mixtures of gases, and yet mixtures of gases are of great importance. For example, air is a mixture of nitrogen (78 percent), oxygen (20 percent), and argon (1 percent) with lesser amounts of other gases, such as carbon dioxide. Many industrial processes involve gaseous mixtures. For example, the commercial production of ammonia involves the reaction

$$3H_2(g) + N_2(g) \xrightarrow[300\text{ atm}]{500°\text{C}} 2NH_3(g)$$

and thus the reaction vessel contains a mixture of N_2, H_2, and NH_3. In a mixture of gases, each gas exerts a pressure as if it were present alone in the container:

$$P_{total} = P_1 + P_2 + P_3 \cdots \qquad (5\text{-}9)$$

The pressure exerted by each gas is called its *partial pressure*, and Equation (5-9) is known as *Dalton's law of partial pressures*. For a mixture of two gases, then, we have

$$P_{total} = P_1 + P_2$$

where each of the gases obeys the ideal-gas equation:

$$P_1 = \frac{n_1RT}{V} \qquad P_2 = \frac{n_2RT}{V}$$

Notice that the volume occupied by each gas is V, because each gas in a mixture occupies the entire container. If the partial pressures P_1 and P_2 are substituted into Equation (5-9), then we get for our two-gas mixture,

$$\begin{aligned} P_{total} &= \frac{n_1RT}{V} + \frac{n_2RT}{V} \\ &= (n_1 + n_2)\frac{RT}{V} \qquad (5\text{-}10) \\ &= n_{total}\frac{RT}{V} \end{aligned}$$

The total pressure exerted by a mixture of gases is determined by the total number of moles of gas in the mixture.

Practical applications of Dalton's law of partial pressures arise often in the laboratory. A standard method for determining the quantity of gas evolved in a chemical reaction is diagramed in Fig-

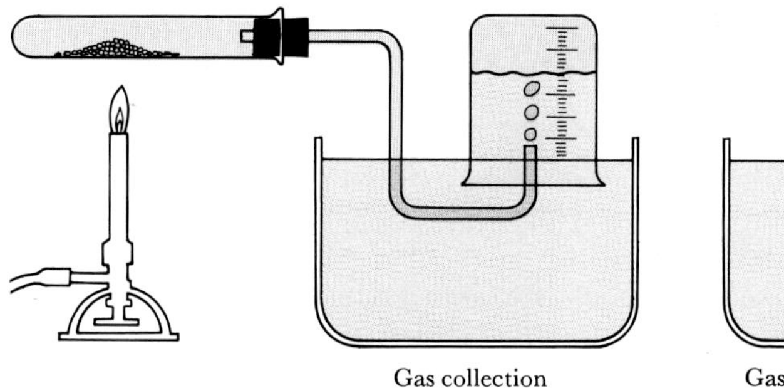
Gas collection   Gas volume measurement

Figure 5-11 The collection of a gas over water. When the water levels inside and outside the container are equal, the pressure inside the container and the atmospheric pressure must be equal.

ure 5-11. The gas displaces the water from an inverted beaker that is initially filled with water. When the reaction is completed, the beaker is raised or lowered until the water levels inside and outside are the same. When the two levels are the same, the pressure inside the beaker is equal to the atmospheric pressure. The pressure inside the beaker, however, is not due just to the gas collected; there is also water vapor present. Thus, the pressure inside the beaker is

We shall study the pressure due to water vapor more fully in Chapter 13.

$$P_{total} = P_{gas} + P_{H_2O} = P_{atmospheric}$$

The vapor pressure of H_2O depends only upon the temperature; Table 13-2 gives a table of the vapor pressure of H_2O at various temperatures.

Example 5-11: The reaction

$$2KClO_3(s) \rightarrow 2KCl(s) + 3O_2(g)$$

represents a common laboratory procedure for producing small quantities of pure oxygen. A 0.250-L flask is filled with oxygen that was collected over water when the atmospheric pressure was 730 torr (see Figure 5-11). The gas temperature is 14°C. Compute the volume of dry oxygen at 0°C and 760 torr. The vapor pressure of H_2O at 14°C is 12.0 torr.

Solution: To calculate the volume of dry oxygen, we must first determine its partial pressure. The atmospheric pressure is 730 torr, and the vapor pressure of H_2O at 14°C is 12.0 torr; therefore

$$P_{O_2} = P_{total} - P_{H_2O} = (730 - 12.0)\text{ torr} = 718\text{ torr}$$

We wish to calculate the volume of O_2 at 0°C and 760 torr, and so we use the equation from Example 5-7:

Problems 5-41 to 5-46 deal with partial pressures.

$$V_f = V_i\left(\frac{P_i}{P_f}\right)\left(\frac{T_f}{T_i}\right) = (0.250\text{ L})\left(\frac{718\text{ torr}}{760\text{ torr}}\right)\left(\frac{273\text{ K}}{287\text{ K}}\right)$$

$$= 0.225\text{ L}$$

5-11. THE KINETIC THEORY OF GASES VIEWS THE MOLECULES OF A GAS AS BEING IN CONSTANT MOTION

The fact that Boyle's, Charles's, and Dalton's laws and the ideal-gas equation are valid for all gases suggests that these laws reflect the fundamental nature of gases. As we have seen, the space occupied by a gas is mostly empty, with the molecules being widely separated from each other and in constant motion.

By applying the laws of physics to the motion of the molecules, it is possible to calculate the pressure exerted by gas molecules on the walls of a container and to show that the pressure is given by the ideal-gas equation. Because this theory focuses on the motion of the molecules, it is called the *kinetic theory of gases*. The kinetic theory predicts that, as long as the volume taken up by the molecules is much smaller than the volume of the container and as long as the molecules do not attract each other, then the gas behaves ideally.

The word kinetic is derived from the Greek word *kinetikos,* meaning motion or moving.

A body in motion has an energy by virtue of the fact that it is in motion. The energy associated with the motion of a body is called the *kinetic energy* E and is given by the formula

$$E = \frac{1}{2} mv^2 \qquad (5\text{-}11)$$

where m is the mass of the body and v is its speed. If m is expressed in kilograms and v in meters per second, then E has the units $\text{kg}\cdot\text{m}^2\cdot\text{s}^{-2}$. This combination of units is called a *joule* (J), which is the SI unit of energy: $1\ \text{J} = 1\ \text{kg}\cdot\text{m}^2\cdot\text{s}^{-2}$. (See Appendix B.)

The molecules in a gas do not all have the same speed. As the caption to Figure 5-12 explains, there is a distribution of molecular

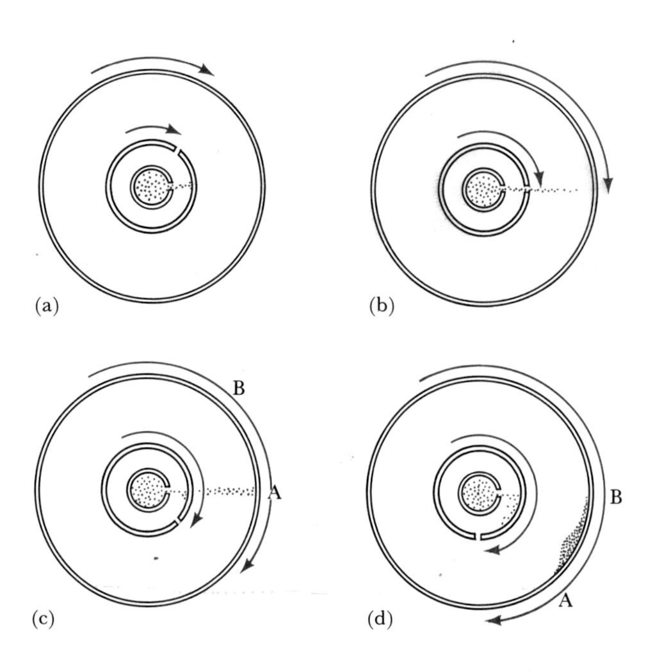

Figure 5-12 An experimental setup that can be used not only to demonstrate that there is a distribution of molecular speeds in a gas but also to measure that distribution. The apparatus consists of three concentric evacuated cylindrical drums. The two outermost drums rotate together at the same angular speed and the innermost drum is stationary. The innermost drum contains a gas or vapor, say, silver vapor. The two innermost drums have small slits, and when these momentarily line up (b), a beam of silver atoms is directed to the inner surface of the outermost drum. The silver atoms with the greatest speed reach the outermost drum first, at point A in (c). By the time the slower silver atoms reach the outer drum, it will have rotated some and so the deposit of silver atoms will be spread out. The thickness of the silver deposit is proportional to the number of silver atoms with a certain speed, and so the variation in thickness represents the actual distribution of speeds of the silver atoms in the vapor.

speeds in a gas. Figure 5-13 shows the distribution of molecular speeds for N_2 gas at two different temperatures. Notice that, as the temperature increases, more molecules travel at higher speeds.

A fundamental postulate of the kinetic theory of gases says that the average kinetic energy per mole of a gas $\bar{E}_{av}$ is directly proportional to the absolute temperature:

$$\bar{E}_{av} \propto T$$

The bar in $\bar{E}_{av}$ indicates that it is the *average kinetic energy per mole.* The proportionality constant between $\bar{E}_{av}$ and T is $\frac{3}{2}R$; that is

$$\bar{E}_{av} = \tfrac{3}{2}RT \qquad (5\text{-}12)$$

This R is the same R (the gas constant) that appears in the ideal-gas equation. Because $\bar{E}_{av}$ is expressed in $\mathrm{J \cdot mol^{-1}}$, we must express R in $\mathrm{J \cdot K^{-1} \cdot mol^{-1}}$ when we use Equation (5-12). It is shown in Appendix B that $R = 8.31\ \mathrm{J \cdot K^{-1} \cdot mol^{-1}}$.

We can *define* an average speed by

$$\bar{E}_{av} \equiv \tfrac{1}{2}M_{kg}v_{av}^2 \qquad (5\text{-}13)$$

where the $\equiv$ sign means that the relation is a definition of v_{av}. Because $\bar{E}_{av}$ is the average kinetic energy *in joules per mole,* the mass on the right-hand side, M_{kg}, is the mass in kilograms of a mole of molecules, which is equal to the molar mass of the gas divided by 1000 g/kg. If we substitute Equation (5-12) into Equation (5-13) and solve for v_{av}, then we get

$$v_{av}^2 = \frac{2\bar{E}_{av}}{M_{kg}} = \frac{2(\frac{3}{2}RT)}{M_{kg}} = \frac{3RT}{M_{kg}}$$

Taking the square root of both sides yields

$$v_{av} = \left(\frac{3RT}{M_{kg}}\right)^{1/2} \qquad (5\text{-}14)$$

We can use Equation (5-14) to calculate v_{av}, which is an *average speed* of the molecules in a gas.

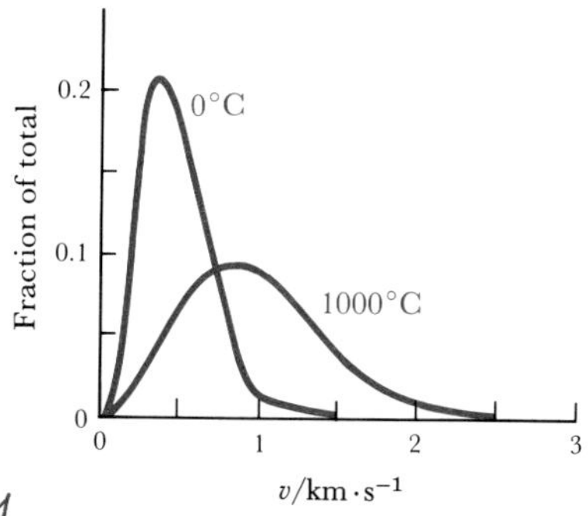

Figure 5-13 The distribution of speeds for nitrogen gas molecules at 0 and 1000°C. The distribution is represented by the fraction of nitrogen molecules that have speed v plotted versus that speed. Note, for example, that the fraction of molecules with a speed of 1 $\mathrm{km \cdot s^{-1}}$ is greater at 1000°C than at 0°C.

Several different definitions of average speeds occur in the kinetic theory of gases. The quantity v_{av} defined by Equation (5-13) is the only average speed that we shall use in this book.

Example 5-12: Calculate v_{av} for N_2 at 20°C.

Solution: The molar mass of N_2 in kilograms per mole is

$$M_{kg} = \frac{28.0\ \mathrm{g \cdot mol^{-1}}}{1000\ \mathrm{g \cdot kg^{-1}}} = 0.0280\ \mathrm{kg \cdot mol^{-1}}$$

Thus, using Equation (5-14) we compute

$1\ \mathrm{J} = 1\ \mathrm{kg \cdot m^2 \cdot s^{-2}}$

$$v_{av} = \left(\frac{3RT}{M_{kg}}\right)^{1/2}$$

$$= \left[\frac{(3)(8.31\ \mathrm{J \cdot mol^{-1} \cdot K^{-1}})(293\ \mathrm{K})}{0.028\ \mathrm{kg \cdot mol^{-1}}}\right]^{1/2}$$

$$= (2.61 \times 10^5\ \mathrm{J \cdot kg^{-1}})^{1/2}$$

$$= (2.61 \times 10^5\ \mathrm{m^2 \cdot s^{-2}})^{1/2} = 511\ \mathrm{m \cdot s^{-1}}$$

A speed of $511\ \mathrm{m \cdot s^{-1}}$ is equivalent to about 1100 mph.

Values of v_{av} for several gases are given in Table 5-3. Note that v_{av} decreases with increasing molecular mass at constant temperature, as is required by Equation (5-14).

A 6.35-mm rifle bullet exits the rifle barrel with a velocity of over $1200\ \mathrm{m \cdot s^{-1}}$.

A sound wave is a pressure wave that travels through a substance. The speed with which a sound wave travels through a gas depends upon the speeds of the molecules in the gas. It can be shown from the kinetic theory of gases that the speed of sound through a gas is about $0.7v_{av}$. The speed of sound in air at 20°C and 1 atm is about 760 mph, or $340\ \mathrm{m \cdot s^{-1}}$.

Table 5-3 Values of v_{av} for gases at 20°C and 1000°C

The molecules of a gas move at speeds of hundreds of meters per second.

		$v_{av}/m \cdot s^{-1}$	
Molecule	*Formula mass*	t = 20°C	t = 1000°C
H_2	2.0	1900	4000
N_2	28.0	510	1060
O_2	32.0	480	1000
CO_2	44.0	410	850

5-12. THE AVERAGE DISTANCE A MOLECULE TRAVELS BETWEEN COLLISIONS IS THE MEAN FREE PATH

Although the molecules in a gas at 1 atm and 20°C travel with speeds of hundreds of meters per second, they do not travel any appreciable distances that rapidly. We all have observed that it may take several minutes for an odor to spread through a room. The explanation for this lies in the fact that the molecules in a gas undergo many collisions, and so their actual path is a chaotic, zigzag path like that shown in Figure 5-14. Between collisions, gas molecules travel with speeds of hundreds of meters per second, but their net progress is quite slow. The average distance traveled between collisions is called the *mean free path* (l). According to the kinetic theory, the mean free path is given by

$$l = (3.1 \times 10^7 \text{ pm}^3 \cdot \text{atm} \cdot \text{K}^{-1}) \frac{T}{\sigma^2 P} \tag{5-15}$$

where T is the temperature in kelvins, σ (the Greek letter sigma) is the diameter of a molecule in picometers, and P is the pressure in atmospheres. Table 5-4 lists *molecular diameters* for various gas molecules.

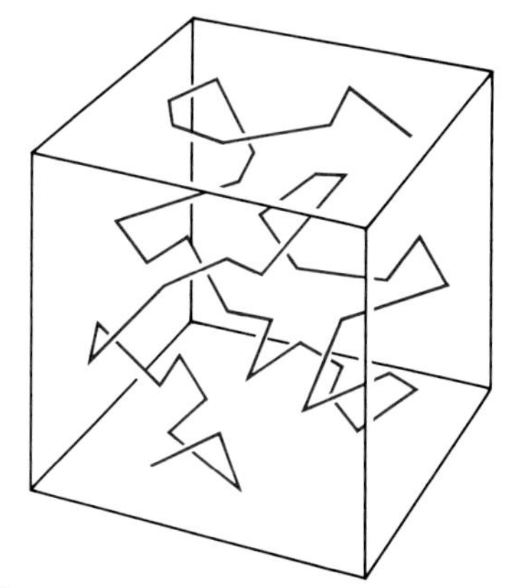

Figure 5-14 A typical path followed by a gas molecule. The molecule travels in a straight line until it collides with another molecule, at which point its direction is changed in an almost random manner. At 0°C and 1 atm, a molecule undergoes about 10^{10} collisions per second.

We can use Equation (5-15) to calculate the mean free path in N_2 at 1.00 atm and 20°C. For N_2, $\sigma = 370$ pm (Table 5-4), and so we compute

$$\begin{aligned} l &= (3.1 \times 10^7 \text{ pm}^3 \cdot \text{atm} \cdot \text{K}^{-1}) \left(\frac{293 \text{ K}}{(370 \text{ pm})^2 (1.00 \text{ atm})} \right) \\ &= 6.6 \times 10^4 \text{ pm} \\ &= 6.6 \times 10^{-8} \text{ m} \end{aligned}$$

where we have used the fact that 1 pm $= 10^{-12}$ m. A distance of 66,000 pm is over 175 times the diameter of a nitrogen molecule. Thus, we see that at 1.00 atm and 20°C a nitrogen molecule travels an average distance of over 175 molecular diameters between collisions.

If we divide v_{av} by the mean free path, we get an estimate of the number of collisions (z) that one molecule undergoes per second. This quantity is called the *collision frequency* and is given by

$$\begin{aligned} z &= \text{number of collisions per second per molecule} \\ &\approx \frac{\text{distance traveled per second}}{\text{distance traveled per collision}} = \frac{\text{collisions}}{\text{second}} = \frac{v_{av}}{l} \end{aligned} \tag{5-16}$$

Thus, for N_2 at 20°C and 1.00 atm we compute

$$\begin{aligned} z = \frac{v_{av}}{l} &= \frac{511 \text{ m} \cdot \text{s}^{-1}}{6.6 \times 10^{-8} \text{ m} \cdot \text{collision}^{-1}} \\ &= 7.7 \times 10^9 \text{ collisions/s} \end{aligned}$$

Thus we see that one nitrogen molecule undergoes about 8 billion collisions per second at 20°C and 1 atm.

Table 5-4 Molecular diameters for some common gases

Gas	*Diameter/pm*
He	210
Ne	250
Ar	360
Kr	410
Xe	470
H_2	280
N_2	370
O_2	370

5-13. DIFFERENT GASES LEAK THROUGH A SMALL HOLE IN A CONTAINER AT DIFFERENT RATES

We can use Equation (5-14) to derive a formula for the relative rates at which gases leak from a container through a small hole, a process called *effusion*. For two gases at the same pressure and temperature, the rate of effusion is directly proportional to the average speed of the molecules. We let $v_{av,A}$ and $v_{av,B}$ be the average speeds of two gases A and B, and we use Equation (5-14) to write

$$v_{av,A} = \left(\frac{3RT}{M_{kg,A}}\right)^{1/2} \quad \text{and} \quad v_{av,B} = \left(\frac{3RT}{M_{kg,B}}\right)^{1/2}$$

The temperature does not have a subscript because both gases are at the same temperature. If we divide $v_{av,A}$ by $v_{av,B}$, we can obtain

$$\frac{v_{av,A}}{v_{av,B}} = \left(\frac{M_{kg,B}}{M_{kg,A}}\right)^{1/2} = \left(\frac{M_B}{M_A}\right)^{1/2}$$

where M_A and M_B are the formula masses of A and B. The rate of effusion is directly proportional to v_{av}, and so

$$\frac{\text{rate}_A}{\text{rate}_B} = \left(\frac{M_B}{M_A}\right)^{1/2} \tag{5-17}$$

This relation was observed experimentally by Graham in the 1840's and is called *Graham's law of effusion.*

Example 5-13: Use Graham's law of effusion to calculate the time it will take 1 μmol of hydrogen to leak out of a certain container if it is observed that it takes 72 s for 1 μmol of nitrogen to leak out at the same temperature and pressure.

Solution: The time it takes for a given quantity of gas to leak from a container is inversely proportional to the rate at which it effuses. Thus we write

$$\frac{t_B}{t_A} = \frac{\text{rate}_A}{\text{rate}_B} = \left(\frac{M_B}{M_A}\right)^{1/2} \tag{5-18}$$

If we let gas B be H_2 and gas A be N_2, then

$$t_{H_2} = t_{N_2}\left(\frac{M_{H_2}}{M_{N_2}}\right)^{1/2} = (72\text{ s})\left(\frac{2.016}{28.02}\right)^{1/2} = 19\text{ s}$$

Notice that the lighter gas escapes more quickly, as you might have expected.

Problems 5-59 to 5-62 deal with Graham's law.

Effusion may be used to separate the components of a mixture of two gases. The mixture is placed in a porous container and the lighter gas effuses out more quickly than the heavier gas, leaving the mixture remaining in the container enriched in the heavier component. During the development of the atomic bomb, it was necessary to separate uranium-235 from uranium-238. The two isotopes have identical chemical properties and so cannot be separated chemically. They were separated by first making the gaseous compound uranium hexafluoride, UF_6. Naturally occurring fluorine consists of only one isotope, fluorine-19. Thus UF_6 consists of two isotopic species, $^{235}UF_6$ and $^{238}UF_6$. These two species were then separated by effusion, with the lighter $^{235}UF_6$ effusing through a porous barrier slightly more rapidly than the heavier $^{238}UF_6$. Although the ratio of the rates of effusion is only

$$\frac{\text{rate}_{235}}{\text{rate}_{238}} = \left(\frac{238 + 6 \times 19}{235 + 6 \times 19}\right)^{1/2} = 1.004$$

almost complete separation is achieved by repeating the process over and over.

The kinetic theory of gases gives us a detailed picture of the molecular nature of gases. From it we can calculate the properties of gases in terms of molecular quantities. The kinetic theory has numerous applications throughout chemistry, and we shall refer to it frequently in later chapters.

5-14. THE IDEAL-GAS EQUATION IS NOT VALID AT HIGH PRESSURES

The ideal-gas equation is valid for all gases at sufficiently low densities and sufficiently high temperatures. As the pressure on a given quantity of gas is increased, however, deviations from the ideal-gas equation appear. These deviations can be displayed graphically by plotting PV/RT as a function of pressure, as shown in Figure 5-15. For 1 mol of an ideal gas, PV/RT is equal to unity for any value of P, and so deviations from ideal-gas behavior occur as deviations of the ratio PV/RT from unity. The extent of *deviation from ideality* at a given pressure depends upon the temperature and upon the nature of the gas. The closer the gas is to the point at which it liquefies, the larger will be the deviation from ideal behavior. The kinetic theory of gases assumes that the molecules of a gas are simply point masses and that they have no attraction for each other. The deviations from ideality shown in Figure 5-15 are due to the inaccuracy of these two assumptions.

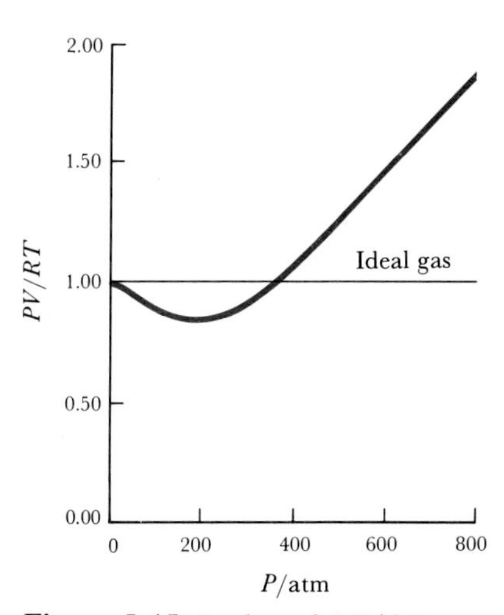

Figure 5-15 A plot of PV/RT versus P for 1 mol of methane at 300 K. This figure shows that the ideal-gas equation is not valid at high pressures.

The behavior shown in Figure 5-15 for methane is typical for all gases. As the pressure is increased, the deviation from ideal-gas behavior first lies below the ideal-gas prediction. This negative deviation from ideality can be explained by recognizing that there is an intrinsic attraction between molecules. This attraction is responsible for a gas condensing to a liquid as the temperature is lowered. We discuss the attraction between molecules in more detail in Chapter 13, but here we need to realize only that molecules do attract each other. When the molecules in a gas collide with each other, their mutual attraction causes them to stay together somewhat longer than if they did not attract each other. Consequently, the number of molecules, or moles, is effectively reduced. There are fewer collisions with the walls of the container than there would be if the gas were ideal (that is, if there were no attraction between the molecules), and so the pressure and the product PV are less than predicted by the ideal-gas equation.

The fact that gases condense to liquids establishes that molecules attract each other.

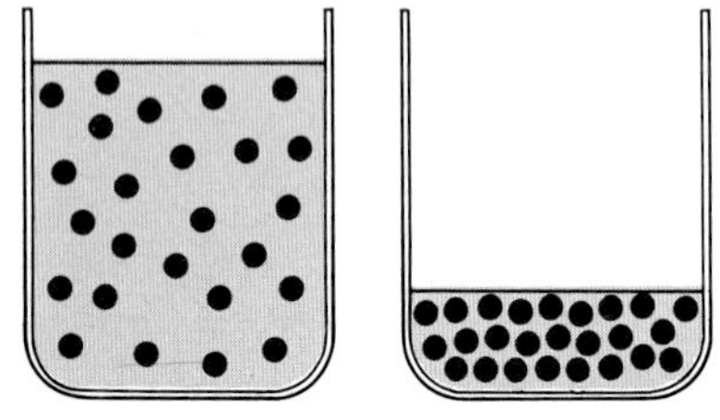

Figure 5-16 At high pressures, the volume of the molecules of a gas is no longer negligible relative to the volume of the container.

Figure 5-15 also shows that at higher pressures the deviation from ideal-gas behavior lies above the ideal-gas prediction. This positive deviation from ideality can be understood by recognizing that molecules have a finite size. At high pressures, the volume of the gas molecules is not negligible relative to the volume of the container (Figure 5-16). Consequently, the volume that is *available* to any given molecule is less than the total volume of the container. If we let b be the volume of the molecules, then $V - b$ is the volume available to any molecule. This free volume, $V - b$, should be used in calculations instead of simply V. If we replace V by $V - b$ in the ideal-gas equation, then we obtain

$$P(V - b) = RT$$

If we divide by RT and then by $V - b$, then we obtain

$$\frac{P}{RT} = \frac{1}{V - b}$$

We now multiply by V to get an expression for PV/RT:

$$\frac{PV}{RT} = \frac{V}{V - b} > 1$$

The ratio $V/(V - b)$ is greater than unity, and so PV/RT is greater than unity, as in Figure 5-15. Thus, the positive deviations from ideal-gas behavior at high pressures are due to the finite size of the gas molecules.

Do you see that the ratio $V/(V - b)$ is always greater than unity because b is positive?

There are many equations that modify the ideal-gas equation to account for the attraction between molecules and for their finite size. The best known is called the *van der Waals equation:*

An equation that connects P, V, T, and n is called an equation of state. The van der Waals equation is one of many equations of state.

$$\left(P + \frac{n^2 a}{V^2}\right)(V - nb) = nRT \tag{5-19}$$

where a and b are constants, called *van der Waals constants,* whose values depend upon the particular gas. Each gas has its own values of a and b, as shown in Table 5-5. The quantity b is proportional to the volume of a gas molecule. The quantity a is related to the attraction between the molecules; the more strongly the molecules attract each other, the larger a is.

Let's use the van der Waals equation to calculate the pressure exerted at 300 K by 1.00 mol of methane occupying a 250-mL container. From Table 5-5 we find that $a = 2.253\ \mathrm{L^2 \cdot atm \cdot mol^{-2}}$ and $b = 0.0428\ \mathrm{L \cdot mol^{-1}}$. If we divide Equation (5-19) by $V - nb$ and solve for P, then we obtain

$$P = \frac{nRT}{V - nb} - \frac{n^2 a}{V^2} \tag{5-20}$$

Table 5-5 The van der Waals constants of some gases

Name	*Formula*	*$a/\mathrm{L^2 \cdot atm \cdot mol^{-2}}$*	*$b/\mathrm{L \cdot mol^{-1}}$*
ammonia	NH_3	4.170	0.0371
argon	Ar	1.345	0.0322
carbon dioxide	CO_2	3.592	0.0427
chlorine	Cl_2	6.493	0.0562
helium	He	0.0341	0.0237
hydrogen	H_2	0.244	0.0266
hydrogen chloride	HCl	3.667	0.0408
krypton	Kr	2.318	0.0398
methane	CH_4	2.253	0.0428
neon	Ne	0.211	0.0171
nitrogen	N_2	1.390	0.0391
oxygen	O_2	1.360	0.0380
propane	C_3H_8	8.664	0.0844
sulfur dioxide	SO_2	6.714	0.0564
xenon	Xe	4.194	0.0510

Substituting $n = 1.00$ mol, $R = 0.0821\ \mathrm{L \cdot atm \cdot mol^{-1} \cdot K^{-1}}$, $T =$ 300 K, $V = 0.0250$ L, and the values of a and b into Equation (5-20), we obtain

$$P = \frac{(1.00\ \mathrm{mol})(0.0821\ \mathrm{L \cdot atm \cdot mol^{-1} \cdot K^{-1}})(300\ \mathrm{K})}{0.250\ \mathrm{L} - (1.00\ \mathrm{mol})(0.0428\ \mathrm{L \cdot mol^{-1}})} - \frac{(1.00\ \mathrm{mol})^2(2.253\ \mathrm{L^2 \cdot atm \cdot mol^{-2}})}{(0.250\ \mathrm{L})^2} = 82.8\ \mathrm{atm}$$

By comparison, the ideal-gas equation predicts that

$$P = \frac{nRT}{V} = \frac{(1.00\ \mathrm{mol})(0.0821\ \mathrm{L \cdot atm \cdot mol^{-1} \cdot K^{-1}})(300\ \mathrm{K})}{0.250\ \mathrm{L}} = 98.5\ \mathrm{atm}$$

The prediction of the van der Waals equation is in excellent agreement with the experimental value.

SUMMARY

In a gas, the particles are widely separated and travel throughout the entire volume of their container in a chaotic manner, colliding with each other and with the walls of the container. The pressure exerted by a gas is due to the incessant collisions of the gas molecules on the walls of the container. At constant temperature, the volume and the pressure of a gas are related by Boyle's law, which says that volume and pressure are inversely related. The relation between the volume of a gas and its temperature is given by Charles's law, which serves also to define the fundamental temperature scale, the Kelvin scale.

The experimental study of the combining volumes of reacting gases led to Gay-Lussac's law of combining volumes, which states that, if all volumes are measured at the same temperature and pressure, then the volumes in which gases combine in chemical reactions are related by simple whole numbers. Gay-Lussac's law leads to Avogadro's law, which states that equal volumes of gases at the same pressure and temperature contain equal numbers of molecules. Boyle's, Charles's, and Avogadro's laws can be combined into one law, the ideal-gas law, which is the relation among the pressure, volume, temperature, and number of moles of a gas.

All the experimental gas laws can be explained by the kinetic theory of gases. One of the central postulates of the kinetic theory of gases is that the average kinetic energy of a gas is directly proportional to its absolute temperature. The kinetic theory provides equations that can be used to calculate the average speed of a molecule, the mean free path, the rate of effusion, and other molecular quantities.

The ideal-gas equation is valid for all gases at sufficiently low densities and sufficiently high temperatures. As the pressure on a given quantity of gas is increased, however, deviations from the ideal-gas equation are observed. The van der Waals equation, which takes into account that the molecules of a gas attract each other and have a finite size, is valid at higher densities and lower temperatures than the ideal-gas equation.

TERMS YOU SHOULD KNOW

solid 155
liquid 155
gas 155
compressibility 156
pressure 157
manometer 157
torr (mm Hg) 158
barometer 159
barometric pressure 159
standard atmosphere (atm) 160
pascal (Pa) 160
Boyle's law 162
absolute (Kelvin) temperature scale 164
Charles's law 164
gas thermometer 164
Gay-Lussac's law of combining volumes 167
Avogadro's law 167
ideal-gas law 170
ideal-gas equation 170
ideal gas 170
gas constant, R 170
molar volume 170
molar mass 173
partial pressure 176
Dalton's law of partial pressures 176
kinetic theory of gases 178
kinetic energy 178
joule, J 178
average kinetic energy 179
average speed, v_{av} 179
mean free path, l 181
molecular diameter, σ 181
collision frequency, z 181
effusion 182
Graham's law of effusion 182
deviations from ideality 183
van der Waals equation 185
van der Waals constants 185

EQUATIONS YOU SHOULD KNOW HOW TO USE

$PV = nRT$ (5-5) (ideal-gas equation)

$\rho = \frac{MP}{RT}$ (5-8) (density of an ideal gas)

$P_{total} = P_1 + P_2 + P_3 + \dots$ (5-9) (Dalton's law of partial pressures)

$\bar{E}_{av} = \frac{3}{2}RT$ (5-12) (average kinetic energy per mole of a gas)

$v_{av} = \left(\frac{3RT}{M}\right)^{1/2}$ (5-14) (average speed of gas molecules)

$l = (3.1 \times 10^7\ \text{pm}^3\cdot\text{atm}\cdot\text{K}^{-1})\frac{T}{\sigma^2 P}$ (5-15) (mean free path of gas molecules)

$$z = \frac{v_{av}}{l} \quad \text{(5-16)} \quad \text{(collision frequency of gas molecules)}$$

$$\frac{\text{rate}_A}{\text{rate}_B} = \frac{t_B}{t_A} = \left(\frac{M_B}{M_A}\right)^{1/2} \quad \text{(5-18)} \quad \text{(Graham's law of effusion)}$$

$$\left(P + \frac{n^2 a}{V^2}\right)(V - nb) = nRT \quad \text{(5-19)} \quad \text{(van der Waals equation)}$$

0.0821

$$P = \frac{nRT}{V - nb} - \frac{n^2 a}{V^2}$$

Important numbers:
760 torr/atm
1.013 bar/atm
0.0821 L × atm/mol × K
1.013 × 10^5 Pa/atm
22.4 L/mol
8.31 J/mol × K
3.1 × 10^7 pm^3 × atm/K
J = 1 kg × m^2/sec^2 or 1 N·m
1 pm = 10^-12 m
1 Pa = 1 N/m^2

PROBLEMS

PRESSURE

5-1. The atmospheric pressure at the surface of Venus is about 100 atm. Convert 100 atm to torr and to bars.

100 atm × 760 torr/atm = 76000 torr
100 atm × 1.013 bar/atm = 101 bar

5-2. The partial pressure of water vapor in air saturated with water at 25°C is 24 torr. Convert 24 torr to atmospheres and to millibars.

5-3. The atmospheric pressure in Mexico City is about 580 torr. Convert this pressure to atmospheres and to bars.

580 torr / 760 torr/atm = 0.76 atm
580 torr / 760 torr/atm × 1.013 bar/atm = 0.77 bar

5-4. A barometer used by meteorologists measures atmospheric pressure in millibars. Because of a low-pressure front, the atmospheric pressure was 990 mbar. Convert this reading to atmospheres and to torr.

5-5. The organic compound di-*n*-butylphthalate, $C_{16}H_{22}O_4$, is sometimes used as a low-density ($1.043\ \text{g}\cdot\text{mL}^{-1}$) manometer fluid. Compute the pressure in torr of a gas that supports a 500-mm column of di-*n*-butylphthalate.

500 mm (1.043 g/mL / 13.6 g/mL) = 38.3 mm Hg
38.3 torr
in a way, you are saying $P_P = P_1 \rho_1$

5-6. Gallium metal melts at 29.8°C and boils at 2403°C. The density of liquid gallium at 30°C is $6.095\ \text{g}\cdot\text{mL}^{-1}$. Because of its wide liquid range (30 to 2403°C), gallium could be used as a manometer or barometer fluid at high temperatures. Compute the height of the liquid gallium column in a gallium barometer on a day (air temperature above 30°C) when a mercury barometer reads 740 torr.

5-7. The apparatus shown in Figure 5-2b is called a closed-end manometer. Sketch a closed-end manometer with a gas pressure of 500 torr.

500 mm Hg

5-8. The apparatus shown in Figure 5-5 is called an open-end manometer. Sketch an open-end manometer in which the gas sample exerts a pressure of 300 torr. Assume that P_{atm} is 1 atm.

BOYLE'S LAW

5-9. A gas bubble has a volume of 0.650 mL at the bottom of a lake, where the pressure is 3.46 atm. What is the volume of the bubble at the surface of the lake, where the pressure is 1.00 atm? Assume that the temperature is constant.

PV = P₁V₁ → $PV = P_1V_1$
3.46 atm · 0.650 mL = 1.00 atm · V_1
V_1 = 2.25 mL

5-10. Suppose we wish to inflate a weather balloon with helium. The balloon has a volume of 100 m³, and we wish to inflate it to a pressure of 0.10 atm. If we use 50-L cylinders of compressed helium gas at a pressure of 100 atm, how many cylinders do we need? Assume that the temperature remains constant.

5-11. The volume of one cylinder in a particular automobile engine is 0.44 L. The cylinder is filled with a mixture of gasoline and air at 1.0 atm. The cylinder is compressed to 0.073 L prior to ignition of the combustible mixture. What pressure must be applied to produce this compression? Assume that the temperature remains constant.

$PV = P_1V_1$
1 atm · 0.44 L = P_1 · 0.073 L
P_1 = 6.0 atm

5-12. A human adult breathes in approximately 0.50 L of air at 1.00 atm with each breath. If a tank holds 50 L of air at 200 atm, how many breaths will the tank supply? Assume a temperature of 37°C.

CHARLES'S LAW

5-13. Suppose that in a gas thermometer the gas occupies 14.7 mL at 0°C. The thermometer is immersed in boiling water (100°C). What is the volume of the gas at 100°C?

5-14. A balloon filled with air has a volume of 3.25 L at 30°C and is placed in a freezer at −10°C. What is the volume of the balloon at −10°C?

GAY-LUSSAC'S LAW

5-15. Methane burns according to the reaction

$$CH_4(g) + 2O_2(g) \rightarrow CO_2(g) + 2H_2O(g)$$

What volume of air, which is 20 percent oxygen by volume, is required to burn 1.0 L of methane when both are the same temperature and pressure?

5-16. Hydrogen and oxygen react violently with each other once the reaction is initiated. For example, a spark can set off the reaction and cause the mixture to explode. What volume of oxygen will react with 0.55 L of hydrogen if both are at 300°C and 1 atm? What volume of water will be produced at 300°C and 1 atm?

IDEAL-GAS LAW

5-17. Calculate the volume that 1.00 mol of ammonia gas occupies at 37°C and 600 torr.

5-18. Calculate the number of grams of chlorine in a 1.5-L container at a pressure of 9.6 atm and a temperature of 25°C.

5-19. Calculate the pressure exerted by 18 g of steam (H_2O) confined to a volume of 18 L at 100°C. What volume would the water occupy if the steam were condensed to liquid water at 25°C? Hint: The density of liquid water is $1.00\ g \cdot mL^{-1}$.

5-20. Calculate the volume in liters occupied by 100 g of propane gas (C_3H_8) at 870 torr and 27°C.

5-21. Calculate the number of molecules of helium in 1.0 L at −200°C and 0.0010 atm. Compare this value with the number in 1.0 L at 0°C and 1.0 atm.

5-22. Calculate the number of molecules of SO_3 in 100 L at 100°C and 1.00 atm.

5-23. The ozone molecules in the stratosphere absorb much of the ultraviolet radiation from the sun. The temperature of the stratosphere is −23°C, and the pressure due to the ozone is 1.4×10^{-7} atm. Calculate the number of ozone molecules present in 1.0 L.

5-24. A low pressure of 1.0×10^{-3} torr is readily obtained in the laboratory by means of a vacuum pump. Calculate the number of molecules in 1.00 mL of gas at this pressure and 20°C.

5-25. A 0.500-L container is occupied by nitrogen at a pressure of 800 torr and a temperature of 0°C. If the highest pressure the container can withstand is 3.0 atm, what is the highest temperature the gas should be heated to?

5-26. A weather balloon is partially filled with helium at 20°C to a volume of 31.5 L and a pressure of 1.3 atm. The balloon rises to the stratosphere, where the temperature and pressure are −23.0°C and 3.00×10^{-3} atm. Calculate the volume of the balloon in the stratosphere.

IDEAL-GAS LAW AND CHEMICAL REACTIONS

5-27. Acetylene is prepared by the reaction of calcium carbide with water:

$$CaC_2(s) + 2H_2O(l) \rightarrow Ca(OH)_2(s) + C_2H_2(g)$$

What volume of acetylene can be obtained from 100 g of calcium carbide at 0°C and 1.00 atm? What volume results when the temperature is 120°C and the pressure is 1.00 atm?

5-28. A sheet of pure aluminum is placed under pure oxygen at 1.00 atm pressure in a sealed 1.00-L container at 25°C. One hour later, the pressure has dropped to 0.91 atm. Calculate the number of grams of oxygen that have reacted with the aluminum.

5-29. Cellular respiration occurs according to the equation

$$\underset{\text{glucose}}{C_6H_{12}O_6(s)} + 6O_2(g) \rightarrow 6CO_2(g) + 6H_2O(l)$$

Calculate the volume of $CO_2(g)$ produced at 37°C (body temperature) and 1.00 atm when 1.00 g of glucose is metabolized.

5-30. Several television commercials state that it requires 10,000 gallons of air to burn 1 gal of gasoline. Using octane, C_8H_{18}, as the chemical formula of gasoline and using the fact that air is 20 percent oxygen by volume, calculate the volume of air at 0°C and 1.0 atm that is required to burn 1 gal of gasoline. The density of octane is 0.70 $g \cdot mL^{-1}$. Hint: 1 gal = 4 qt and 1 qt = 0.946 L.

5-31. Chlorine is produced by the electrolysis of a solution of sodium chloride:

$$2NaCl(aq) + 2H_2O(l) \rightarrow 2NaOH(aq) + H_2(g) + Cl_2(g)$$

The hydrogen gas and chlorine gas are collected separately at 10.0 atm and 25°C. What volume of each can be obtained from 2.50 kg of sodium chloride?

5-32. Chlorine gas can be prepared in the laboratory by the reaction

$$MnO_2(s) + 4HCl(aq) \rightarrow MnCl_2(aq) + 2H_2O(l) + Cl_2(g)$$

How much MnO_2 should be added to excess HCl to obtain 500 mL of chlorine gas at 25°C and 750 torr?

GAS DENSITY

5-33. Calculate the density of water in the gas phase at 100°C and 1.00 atm. Compare this value with the density of liquid water at 100°C and 1.00 atm (0.958 $g \cdot mL^{-1}$).

5-34. Calculate the density of the gas CF_2Cl_2 at 0°C and 1.00 atm.

5-35. A 1.21-g sample of ether was vaporized in a sealed 250-mL container at 50°C. The pressure due to the ether gas was 1.73 atm. Determine the molar mass of ether.

5-36. A 2.97-g sample of a major component of petroleum was vaporized in a sealed 500-mL container at 95°C. The pressure due to the gaseous compound was 2.13 atm. Determine the molar mass of this compound.

5-37. Ethylene is a gas produced in petroleum-cracking and is used to synthesize a variety of important chemicals, such as polyethylene and polyvinylchloride. Chemical analysis shows that ethylene is 85.60 percent carbon and 14.40 percent hydrogen by mass. It has a density of 0.9588 $g \cdot L^{-1}$ at 25°C and 635 torr. Use these data to determine the molecular formula of ethylene.

5-38. Benzene is the fifteenth most widely used chemical in the United States, and its principal source is petroleum. Benzene has a wide range of uses, including its use as a solvent and in the synthesis of nylon and detergents. Chemical analysis shows that benzene is 92.24 percent carbon and 7.76 percent hydrogen by mass. A 2.334-g sample of benzene was vaporized in a sealed 500-mL container at 100°C, producing a pressure of 1.83 atm. Use these data to determine the molecular formula of benzene.

5-39. Lactic acid is produced by the muscles when insufficient oxygen is available and is responsible for muscle cramps during vigorous exercising. It also provides the acidity found in dairy products. Chemical analysis shows that lactic acid is 39.99 percent carbon, 6.73 percent hydrogen, and 53.28 percent oxygen by mass. A 0.3338-g sample of lactic acid was vaporized in a sealed 300-mL container at 150°C, producing a pressure of 326 torr. Use these data to determine the molecular formula of lactic acid.

5-40. Ethyl acetate is synthesized from acetic acid and ethanol and is used as a synthetic fruit essence. Chemical analysis shows that ethyl acetate is 54.52 percent carbon, 9.17 percent hydrogen, and 36.31 percent oxygen by mass. A 1.203-g sample of ethyl acetate was vaporized in a sealed 250-mL container at 95°C, producing a pressure of 1.65 atm. Use these data to determine the molecular formula of ethyl acetate.

PARTIAL PRESSURES

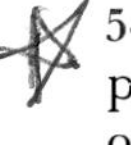

5-41. A mixture of O_2 and N_2 is reacted with white phosphorus, which removes the oxygen. If the volume of the mixture decreases from 50.0 mL to 35.0 mL, calculate the partial pressures of O_2 and N_2 in the mixture. Assume that the total pressure remains constant at 740 torr.

5-43. Nitroglycerin decomposes according to the equation

$$4C_3H_5(NO_3)_3(s) \rightarrow 12CO_2(g) + 10H_2O(g) + 6N_2(g) + O_2(g)$$

What is the total volume of gases produced when collected at 1.0 atm and 25 °C from 10 g of nitroglycerin? What pressure is produced if the reaction is confined to a volume of 0.50 L at 25 °C? Assume that you can use the ideal-gas equation.

5-45. A gaseous mixture of three volumes of carbon dioxide and one volume of water vapor at 200 °C and 2.00 atm is cooled to 50 °C, thereby condensing the water vapor. If the total volume remains the same, what is the pressure of the carbon dioxide at 50 °C?

5-42. Scuba gear delivers to the lungs air that is at the same pressure as the external pressure on the diver. As a diver goes deeper into the water, the pressure on the lungs increases. If the pressure of the air were not increased, the diver's lungs would be compressed. Air is 20 percent oxygen and 80 percent nitrogen by volume. What is the pressure of O_2 in air at a depth where the total pressure is 4.0 atm? The body cannot handle oxygen properly at a high pressure. How can you reduce the partial pressure of the O_2?

5-44. Explosions occur when a substance decomposes very rapidly with the production of a large volume of gases. When detonated, TNT (trinitrotoluene) decomposes according to the equation

$$2C_7H_5(NO_2)_3(s) \rightarrow 12CO(g) + 2C(s) + 5H_2(g) + 3N_2(g)$$

What is the total volume of gases produced from 1.00 kg of TNT if collected at 0 °C and 1.0 atm? What pressure is produced if the reaction is confined to a 50-L container at 500 °C? Assume that you can use the ideal-gas equation.

5-46. Hydrogen gas is prepared in the laboratory by adding zinc to an acid:

$$Zn(s) + 2H^+(aq) \rightarrow Zn^{2+}(aq) + H_2(g)$$

A 0.200-L flask is filled with hydrogen collected over water at 752 torr and 24 °C. What is the volume of dry hydrogen at 0 °C and 760 torr? How many moles of hydrogen are collected? The partial pressure of water at 24 °C is 22.4 torr.

MOLECULAR SPEEDS

5-47. Calculate the average speed, v_{av}, of a chlorine molecule at 25 °C.

5-49. If the temperature of a gas is doubled, how much is the average speed of the molecules increased?

5-48. Calculate the average speed, v_{av}, of a neon atom at 25 °C.

5-50. The speed of sound in air at sea level at 20 °C is about 760 mph. Compare this value with the average speed of N_2 and O_2 gas molecules at 20 °C (see Table 5-3).

5-51. Two gases at the same temperature have the same average kinetic energy. Prove that if $v_{av,A}$ and $v_{av,B}$ are the average speeds of the gases, then

$$\frac{v_{av,A}}{v_{av,B}} = \left(\frac{M_B}{M_A}\right)^{1/2}$$

where M_A and M_B are the formula masses. Use this result to calculate the ratio of the average speeds of an oxygen molecule and a helium atom at the same temperature.

5-52. Consider a mixture of $H_2(g)$ and $I_2(g)$. Compute the ratio of the average speeds of H_2 and I_2 molecules in the reaction mixture.

MEAN FREE PATH

5-53. Calculate the mean free path and the collision frequency experienced by one molecule of hydrogen at 20°C and 1.0 atm.

5-54. Using the molecular diameters given in Table 5-4, calculate the mean free path of helium and krypton at 1.00 torr and at 760 torr when the temperature is 20°C.

5-55. Interstellar space has an average temperature of about 10 K and an average density of hydrogen atoms of about one hydrogen atom per cubic meter. Compute the mean free path of hydrogen atoms in interstellar space. Take $\sigma = 100$ pm.

5-56. Calculate the pressures at which the mean free path of a hydrogen molecule will be 1.00 μm, 1.00 mm, and 1.00 m at 20°C.

5-57. Calculate the number of collisions per second of one oxygen molecule at 20°C and 1.0 atm.

5-58. Calculate the number of collisions that one molecule of nitrogen undergoes at 20°C and 1.0×10^{-3} torr.

GRAHAM'S LAW OF EFFUSION

5-59. Two identical balloons are filled, one with helium and one with nitrogen, at the same temperature and pressure. If the nitrogen leaks out from its balloon at the rate of 75 mL/h, then what will be the rate of leakage of the helium-filled balloon?

5-60. Two identical porous containers are filled, one with hydrogen and one with carbon dioxide, at the same temperature and pressure. After one day, 1.50 mL of carbon dioxide had leaked out of its container. How much hydrogen has leaked out in one day?

5-61. It takes 145 s for 1.00 mL of N_2 to effuse from a certain porous container. Given that it takes 230 s for 1.00 mL of an unknown gas to effuse under the same temperature and pressure, calculate the molecular mass of the unknown gas.

5-62. Suppose that it takes 175 s for 1.00 mL of N_2 to effuse from a porous container under a certain temperature and pressure and that it takes 200 s for 1.00 mL of a CO-CO_2 mixture to effuse under the same conditions. What is the volume percentage of CO in the mixture?

VAN DER WAALS EQUATION

5-63. Use the van der Waals equation to calculate the pressure exerted by 30.0 g of N_2 confined to a 155-mL container at 400 K. Compare your answer with the pressure calculated using the ideal-gas equation.

5-64. Show that the van der Waals equation reduces to the ideal-gas equation at low densities, where the molecules are very far apart from each other.

5-65. Calculate the number of Cl_2 gas molecules in a volume of 5.00 mL at 40°C and 2.15×10^4 Pa.

5-66. Calculate the pressure in pascals that is exerted by 6.15 mg of CO_2 occupying 2.10 mL at 75°C.

5-67. A sample of radon occupies 7.12 μL at 22°C and 8.72×10^4 Pa. Calculate the volume at 0°C and 1.013×10^5 Pa. What is the mass of the radon?

5-68. Calculate the mass of $N_2O(g)$ that occupies 2.10 L at a pressure of 4.50×10^4 Pa and a temperature of 15°C.

5-69. Calculate the density of $ND_3(g)$ at 0°C and 2.00×10^3 Pa. Take $M = 20.06\ \text{g}\cdot\text{mol}^{-1}$.

5-70. Using the fact that 1 atm = 1.013×10^5 Pa, calculate the gas constant in SI units, $\text{J}\cdot\text{mol}^{-1}\cdot\text{K}^{-1}$.

$\rho = \frac{MP}{RT}$

$\rho = \frac{20.06\ \text{g/mol} \times 2.00 \times 10^{3}\ \text{Pa}}{8.31\ \text{N}\cdot\text{m/K}\cdot\text{mol} \times 273\ \text{K}}$

$\rho = 17.7\ \text{g/m}^3$

When using Pa as pressure, V must be in m^3 and R must be 8.31.

INTERCHAPTER C

Nitrogen

An ammonia plant. This plant produces 750 tons of ammonia per day from hydrogen gas and nitrogen gas. The nitrogen comes from air and the hydrogen is obtained from the reaction between methane and steam.

The most significant property of elemental nitrogen, N_2, is its lack of chemical reactivity. Nitrogen, as N_2, does not take part in many chemical reactions. Although nitrogen compounds are essential nutrients for animals and plants, only a few microorganisms are able to utilize elemental nitrogen directly by converting it to water-soluble compounds of nitrogen. The conversion of nitrogen from the free element to nitrogen compounds is one of the most important problems of modern chemistry and is called *nitrogen fixation.*

C-1. OVER 37 BILLION POUNDS OF NITROGEN IS PRODUCED ANNUALLY

Nitrogen is a colorless, odorless gas that exists as a diatomic molecule, N_2. The principal source of nitrogen is the atmosphere, which is about 78 percent N_2 by volume. Pure nitrogen is produced by the fractional distillation of liquid air. Nitrogen boils at $-196°C$, whereas oxygen, the other principal component of air, boils at $-183°C$.

In terms of U.S. industrial production, nitrogen is the third leading chemical. Over 37 billion pounds of pure nitrogen is produced from air each year. Nitrogen is also found in potassium nitrate, KNO_3 (saltpeter), and in sodium nitrate, $NaNO_3$ (Chile saltpeter). Vast deposits of these two nitrates are found in the arid northern region of Chile, where there is insufficient rainfall to wash away these soluble compounds. The Chilean nitrate deposits are about 200 miles long, 20 miles wide, and many feet thick. At one time the economy of Chile was based primarily upon the sale of nitrates for use as fertilizers.

Large quantities of nitrogen are stored and shipped as the liquid in insulated metal cylinders. Smaller quantities are shipped as the gas in heavy-walled steel cylinders. The most convenient source of nitrogen gas in the laboratory is a steel cylinder charged with compressed N_2 gas. An alternative source is to heat an aqueous solution of ammonium nitrite, which thermally decomposes according to the equation

$$NH_4NO_2(aq) \rightarrow N_2(g) + 2H_2O(l)$$

Ammonium nitrite is a potentially explosive solid, and so the aqueous ammonium nitrite solution is made by adding ammonium chloride and sodium nitrite, both stable compounds, to water. Even so, the solution must be heated carefully to avoid an explosion.

Nitrogen also occurs in all living organisms, both animal and vegetable. Proteins and nucleic acids, such as DNA and RNA, contain significant quantities of nitrogen.

C-2. MOST NITROGEN IS CONVERTED TO AMMONIA BY THE HABER PROCESS

The inertness of nitrogen toward most other chemical substances makes reactions in which nitrogen combines with other elements economically important. Nitrogen fixation occurs both industrially and in nature. The most important industrial nitrogen-fixation reaction is the *Haber process,* in which nitrogen reacts directly with hydrogen at high pressure and high temperature to form ammonia:

$$N_2(g) + 3H_2(g) \xrightarrow[500°C]{300\text{ atm}} 2NH_3(g)$$

Ammonia is a colorless gas with a sharp, irritating odor. It is the effective agent in some forms of "smelling salts." Unlike nitrogen, ammonia is extremely soluble in water. Household ammonia is about a 2 M solution of NH_3 in water together with a detergent. Ammonia was the first complex molecule to be identified in interstellar space. Ammonia occurs in galactic dust clouds in the Milky Way and, in the solid form, constitutes the rings of Saturn. Over 38 billion pounds of ammonia is produced annually in the United States by the Haber process. Rated in terms of pounds produced per year, ammonia is the second ranked industrial chemical in the United States, being surpassed only by sulfuric acid.

C-3. NITROGEN IS A KEY INGREDIENT IN FERTILIZERS

Ammonia is readily soluble in water, binds to many components of soil, and is easily converted to usable plant food. Concentrated aqueous solutions of ammonia or pure liquid ammonia can be sprayed directly into the soil (Figure C-1). Ammonia is inexpensive and high in nitrogen. The increased growth of plants when fertilized by ammonia is spectacular. Liquid ammonia is toxic and injurious to living tissue, however, and must be handled carefully.

For some purposes it is more convenient to use a solid fertilizer instead of ammonia solutions. For example, ammonia combines directly with sulfuric acid to produce ammonium sulfate:

$$2NH_3(aq) + H_2SO_4(aq) \rightarrow (NH_4)_2SO_4(aq)$$

Figure C-1 This photo demonstrates the method of spraying ammonia into the soil. Liquid ammonia, called anhydrous ammonia, is used extensively as a fertilizer because it is cheap, high in nitrogen, and easy to apply.

Table C-1 World fertilizer consumption

	Consumption/10^6 metric tons			*Annual growth/%*
	1972	1980		
North America	16.5	24.3		5.0
Western Europe	17.5	22.6		3.3
Eastern Europe and USSR	18.9	32.0		6.8
Asia (excluding Japan)	10.3	19.9		8.6
Latin America	3.2	7.2		10.7
Africa	1.3	2.2		6.8
Japan, Israel, South Africa, Oceania	4.3	5.5		3.1
World total	72.3	113.7	Average	6.3

Ammonium sulfate is the most important solid fertilizer in the world. Its annual U.S. production exceeds 4 billion pounds.

The primary fertilizer nutrients are nitrogen, phosphorus, and potassium, and fertilizers are rated by how much of each they contain. For example, a 5-10-5 fertilizer has 5 percent by mass total available nitrogen, 10 percent by mass phosphorus (equivalent to the form P_2O_5), 5 percent by mass potassium (equivalent to the form K_2O), and 80 percent inert ingredients. The production of fertilizers is one of the largest and most important industries in the world. Table C-1 lists world fertilizer usage figures for 1972 and 1980.

C-4. NITRIC ACID IS PRODUCED BY THE OSTWALD PROCESS

About half of all the ammonia produced is converted to nitric acid by the *Ostwald process*. The first step in this process is the conversion of ammonia to nitrogen oxide:

$$(1)\quad 4NH_3(g) + 5O_2(g) \xrightarrow[825\,^\circ C]{Pt} 4NO(g) + 6H_2O(g)$$

The second step in the Ostwald process involves the oxidation of NO to nitrogen dioxide by reaction with oxygen:

$$(2)\quad 2NO(g) + O_2(g) \rightarrow 2NO_2(g)$$

Figure C-2 Label from a bottle of concentrated nitric acid. Notice that the label contains information on the hazardous properties of the substance.

In the final step, the NO_2 is dissolved in water to yield nitric acid:

$$(3)\quad 3NO_2(g) + H_2O(l) \rightarrow 2HNO_3(aq) + NO(g)$$

The NO(g) evolved is recycled back to step (2).

Laboratory grade nitric acid is approximately 70 percent HNO_3 by weight with a density of $1.42\ g\cdot mL^{-1}$ and a concentration of 16 M (Figure C-2). The U.S. annual production of nitric acid is over 18 billion pounds, which makes it the tenth-ranked industrial chemical. Nitric acid is used in many important chemical processes, including the production of explosives such as trinitrotoluene (TNT), nitroglycerine, and nitrocellulose (gun cotton). It is also used in etching and photoengraving processes to produce grooves in metal surfaces. For example, nitric acid readily reacts with copper metal:

$$3Cu(s) + 8HNO_3(aq) \rightarrow 3Cu(NO_3)_2(aq) + 2NO(g) + 4H_2O(l)$$

Copper metal does not react directly with hydrochloric acid and sulfuric acid. Note that the above reaction does not involve the liberation of H_2 gas, as in the reaction of, say, Zn(s) with HCl(aq).

C-5. CERTAIN BACTERIA CAN FIX NITROGEN

Nitrogen fixation by microorganisms is an important source of plant nutrients. The most common of these nitrogen-fixing bacteria is the *Rhizobium* bacterium, which invades the roots of leguminous plants, such as alfalfa, clover, beans, and peas. The *Rhizobium* forms nodules on the roots of these legumes and has a symbiotic (mutually beneficial) relationship with the plant (Figure C-3). The plant produces carbohydrates through photosynthesis, and the *Rhizobium* uses the carbohydrate as fuel for fixing the nitrogen, which is incorporated into plant protein. Alfalfa is the most potent nitrogen-fixer, followed by clover, soybeans, other beans, peas, and peanuts. In modern agriculture, crops are rotated, meaning that a nonleguminous crop and a leguminous crop are alternated on one piece of land. The leguminous crop is either harvested, leaving behind nitrogen-rich roots, or plowed into the soil, adding both nitrogen and organic matter. A plowed-back crop of alfalfa may add as much as 400 lb of fixed nitrogen to the soil per acre.

Figure C-3 Nitrogen-fixing nodules on the roots of a leguminous plant. The nodules contain *Rhizobium,* a soil bacterium that converts atmospheric elemental nitrogen to water-soluble nitrogen compounds.

C-6. NITROGEN FORMS SEVERAL IMPORTANT COMPOUNDS WITH HYDROGEN AND OXYGEN

The most important nitrogen-hydrogen compounds are ammonia, NH_3, hydrazine, N_2H_4, and hydrazoic acid, HN_3. Hydrazine is a colorless, fuming, reactive liquid. It is produced by the *Raschig synthesis,* in which ammonia is reacted with hypochlorite ion (household bleach is sodium hypochlorite in water) in basic solution:

$$2NH_3(aq) + ClO^-(aq) \xrightarrow{OH^-(aq)} N_2H_4(aq) + H_2O(l) + Cl^-(aq)$$

Household bleach should *never* be mixed with household ammonia because extremely toxic and explosive chloramines, such as H_2NCl and $HNCl_2$, are produced as by-products. The reaction of hydrazine with oxygen,

$$N_2H_4(l) + O_2(g) \rightarrow N_2(g) + 2H_2O(g)$$

is accompanied by the release of a large amount of energy, and hydrazine and some of its derivatives are used as rocket fuels (Chapter 6).

Nitrous acid, HNO_2, is prepared by reacting an equimolar mixture of nitrogen oxide and nitrogen dioxide with a basic solution (for example, NaOH):

$$NO(g) + NO_2(g) + 2NaOH(aq) \rightarrow 2NaNO_2(aq) + H_2O(l)$$

Addition of acid to the resulting solution yields nitrous acid:

$$NO_2^-(aq) + H^+(aq) \rightarrow HNO_2(aq)$$

Salts of nitrous acid are called nitrites. Sodium nitrite, $NaNO_2$, is used as a meat preservative. The nitrite ion combines with the hemoglobin in meat to produce a deep red color. The main problem with the extensive use of nitrites in foods is that the nitrite ion reacts with amines in the body's gastric juices to produce compounds called nitrosamines, such as $(CH_3)_2NNO$, dimethylnitrosamine, which are carcinogenic.

The reaction of nitrous acid with hydrazine in acidic solution yields hydrazoic acid:

$$N_2H_4(aq) + HNO_2(aq) \rightarrow HN_3(aq) + 2H_2O(l)$$

Hydrazoic acid is a colorless, toxic liquid and a dangerous explosive. Its lead and mercury salts, $Pb(N_3)_2$ and $Hg(N_3)_2$, which are called *azides,* are used in detonation caps; both compounds are dangerously explosive. Sodium azide, NaN_3, is used as the gas source in automobile air safety bags, which inflate quickly on rapid deceleration.

Dinitrogen oxide, which is more commonly called nitrous oxide, is prepared by gently heating liquid ammonium nitrate:

$$NH_4NO_3(l) \rightarrow N_2O(g) + 2H_2O(g)$$

The reaction is potentially explosive, with the evolution of nitrogen, oxygen, and water. Nitrous oxide, which is fairly unreactive, is also called *laughing gas* and was once widely used as a general anesthetic in dentistry. Today it is used as a propellant in canned "whipped cream" products.

Nitrogen reacts directly with a few highly reactive metals, such as lithium, magnesium, and aluminum, to form metal *nitrides.* For example, it reacts with lithium at room temperature to produce black lithium nitride:

$$6Li(s) + N_2(g) \rightarrow 2Li_3N(s)$$

Hot magnesium reacts with nitrogen to form yellow magnesium nitride, Mg_3N_2. When magnesium is burned in air, most of it is converted to white magnesium oxide, but small yellow flecks of magnesium nitride can be seen in the white powder.

Nitrides react with water to produce ammonia gas. For example,

$$Li_3N(s) + 3H_2O(l) \rightarrow 3LiOH(aq) + NH_3(g)$$

The reaction of lithium nitride with heavy water, D_2O, provides a convenient preparation of deuterated ammonia:

$$Li_3N(s) + 3D_2O(l) \rightarrow 3LiOD(aq) + ND_3(g)$$

The production of ammonia via the formation of lithium nitride and the subsequent reaction of $Li_3N(s)$ with water is not an economical way of fixing nitrogen because of the high cost of producing lithium metal. The search for nitrogen-fixation processes that are more economical than the Haber process is an active area of contemporary chemical research.

QUESTIONS

C-1. Describe what is meant by nitrogen fixation.

C-2. Briefly describe the Haber process.

C-3. Using balanced chemical equations, outline the Ostwald process.

C-4. What are azides? How are they made?

C-5. What are nitrides? How are they made?

C-6. Give the chemical formula for each of the following compounds:

(a) ammonia
(b) nitric acid
(c) sodium nitrite
(d) nitrous oxide
(e) nitrogen oxide
(f) nitrogen dioxide
(g) sodium azide
(h) lithium nitride
(i) ammonium nitrate
(j) hydrazine

C-7. Describe the Raschig synthesis of hydrazine.

C-8. Why should you never mix household ammonia and bleach?

C-9. What percentage of air (by volume) is N_2?

C-10. Outline a method for the preparation of DNO_3 in $D_2O(l)$.

The energy evolved in chemical reactions is used to power rockets and spaceships. This photo shows lift-off of a reusable space shuttle.

6 / Chemical Reactions and Energy

In this chapter we begin our discussion of thermodynamics, which is that part of chemistry that deals with energy changes. Most chemical reactions are accompanied by a change in energy. These changes may occur in the form of the absorption or the evolution of heat. Energy evolved in the burning of petroleum, natural gas, and coal supplies over 90 percent of the energy used annually in the United States. All living organisms store energy in the form of certain chemicals. Rockets, missiles, and explosives all derive their power from the energy of chemical reactions. Our primary objective in this chapter is to develop an understanding of the energy changes involved in chemical reactions, including the factors that determine the value of the energy change.

6-1. ENERGY IS CONSERVED

When chemical reactions take place, they almost always involve a transfer of energy between the system of interest and its surroundings. The energy is most often transferred in the form of heat. *Thermodynamics* (thermo = heat, dynamics = changes) is the study of these energy transfers. One of the fundamental principles of thermodynamics is the *first law of thermodynamics*, which states that energy is neither created nor destroyed but is simply converted from one form to another. The first law of thermodynamics is the law of *conservation of energy*.

In thermodynamics energy is denoted by the symbol U. We denote the *energy change of a reaction* by ΔU_{rxn} (read delta U of the reaction).

The symbol Δ (delta) denotes a change in a quantity. It is determined by subtracting the initial value of the quantity from the final value. Thus, $\Delta T = T_f - T_i$ denotes a change in temperature, $\Delta V = V_f - V_i$ denotes a change in volume, and $\Delta U_{rxn} = U_f - U_i$ denotes a change in energy for a reaction.

The energy of a system can change as a result of a transfer of energy as *heat* or as *work*. Application of the first law of thermodynamics to a chemical reaction yields

$$\Delta U_{rxn} = q + w \tag{6-1}$$

where q is the energy transferred as heat and w is the energy transferred as work. A common form of work is the energy transferred when a volume change is produced as a result of a difference in pressures between the reaction system and its surroundings. Heat is energy that is transferred as a result of a difference in temperature between the reaction system and its surroundings. Heat flows spontaneously from a higher- to a lower-temperature system.

If a reaction occurs and there is no change in volume, then no energy is transferred as work and $w = 0$ in Equation (6-1). In this case we have

$$\Delta U_{rxn} = q_V = \begin{array}{l}\text{heat evolved or absorbed}\\ \text{at constant volume}\end{array} \tag{6-2}$$

The V subscript on q emphasizes that q_V is the heat transferred when the volume of the reaction system is constant.

Most chemical reactions take place at constant pressure because most are run in reaction vessels open to the atmosphere and the atmospheric pressure is constant during the course of the reaction (Figure 6-1). For reactions that occur at constant pressure, it is convenient to introduce a thermodynamic function H, called the *enthalpy:*

$$H = U + PV \tag{6-3}$$

where U is the energy of the reaction, P is the pressure, and V is the volume. Enthalpy has the convenient property that, for a reaction taking place at constant pressure,

$$\Delta H_{rxn} = q_P = \begin{array}{l}\text{heat evolved or absorbed}\\ \text{at constant pressure}\end{array} \tag{6-4}$$

The P subscript on q_P emphasizes that q_P is the heat transferred when the reaction takes place at constant pressure. Note that Equation (6-4) is the constant-pressure analog of Equation (6-2). Because most reactions occur at constant pressure, ΔH_{rxn} is often called the *heat of reaction.* Equation (6-4) is a key equation in *thermochemistry,* which deals with the heat evolved or absorbed in chemical reactions.

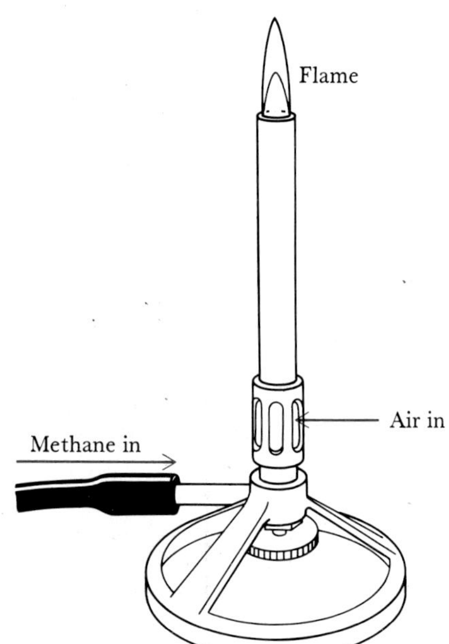

Figure 6-1 Combustion of methane gas in air (in a Bunsen burner). Because the reaction is open to the atmosphere, it is a constant-pressure reaction. For a reaction run at constant pressure, $\Delta H_{rxn} = q_P$.

6-2. CHEMICAL REACTIONS EVOLVE OR ABSORB ENERGY AS HEAT

We are all familiar with the utilization of chemical reactions as energy sources. For example, there are many types of *fuels* that burn in oxygen to provide heat. The burning of a substance in oxygen is called *combustion*. For example, the equation for the combustion of methane is

$$CH_4(g) + 2O_2(g) \rightarrow CO_2(g) + 2H_2O(g)$$

This reaction releases energy as heat and thus is called an *exothermic reaction* (exo = out). All combustion reactions are highly exothermic.

Reactions that absorb energy as heat are called *endothermic reactions* (endo = in). An exothermic and an endothermic reaction are illustrated schematically in Figure 6-2. In Figure 6-2(a), the exothermic reaction, the enthalpy of the reactants is greater than the enthalpy of the products and so energy is released as the reaction proceeds. In Figure 6-2(b), the endothermic reaction, the enthalpy of the reactants is less than the enthalpy of the products and so heat must be supplied in order for the reaction to proceed. In a sense, the heat must be supplied to push the reaction "uphill" with respect to enthalpy.

The *enthalpy change* for a chemical reaction is the total enthalpy of the products minus the total enthalpy of the reactants:

$$\Delta H_{rxn} = H_{prod} - H_{react} \quad (6\text{-}5)$$

For an exothermic reaction, H_{prod} is less than H_{react} and so $\Delta H_{rxn} < 0$. For an endothermic reaction, H_{prod} is greater than H_{react} and so $\Delta H_{rxn} > 0$. We have noted that the combustion of methane is exothermic:

$$CH_4(g) + 2O_2(g) \rightarrow CO_2(g) + 2H_2O(g)$$

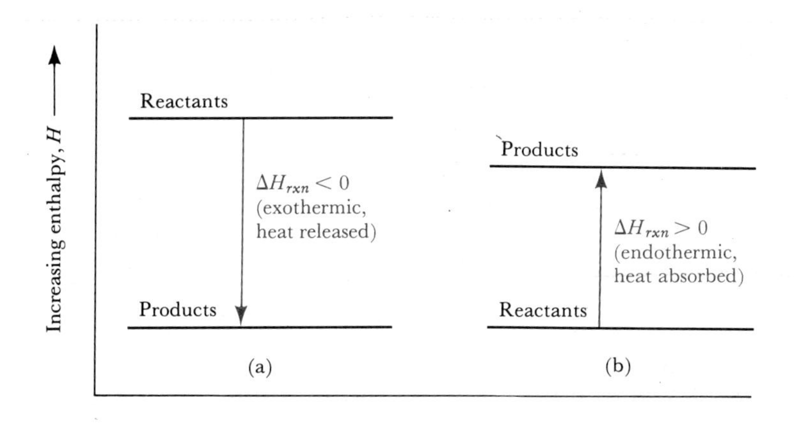

Figure 6-2 An enthalpy diagram for (a) an exothermic reaction and (b) an endothermic reaction. In an exothermic reaction the enthalpy of the products is less than that of the reactants. The enthalpy difference is released as heat during the reaction at constant pressure. In an endothermic reaction, the enthalpy of the products is greater than that of the reactants. The enthalpy difference must be supplied as heat in order for the reaction to occur at constant pressure.

The value of ΔH°_{rxn} for this reaction is -802 kJ at 25°C. The superscript degree sign on ΔH°_{rxn} tells us that the enthalpy change is the *standard enthalpy change* for the reaction. The standard enthalpy change refers to reactants and products at 1 atm pressure.

An example of an endothermic reaction is the *water-gas reaction* in which steam is passed over hot carbon to produce a mixture of carbon monoxide and hydrogen:

$$C(s) + H_2O(g) \rightarrow CO(g) + H_2(g)$$

For this reaction, $\Delta H^\circ_{rxn} = +131$ kJ at 25°C and so heat must be supplied. Both CO and H_2 are combustible gases, and thus the water-gas reaction can be used to produce from coal a combustible, gaseous mixture that is readily transported through pipelines.

The standard enthalpy change for the combustion of methane can be expressed as follows:

$$\Delta H^\circ_{rxn} = \overline{H}^\circ[CO_2(g)] + 2\overline{H}^\circ[H_2O(g)] - \overline{H}^\circ[CH_4(g)] - 2\overline{H}^\circ[O_2(g)] = -802 \text{ kJ}$$

where $\overline{H}^\circ[CO_2(g)]$ denotes the enthalpy of *1 mol* of $CO_2(g)$ at 1 atm, $\overline{H}^\circ[H_2O(g)]$ denotes the enthalpy of *1 mol* of $H_2O(g)$ at 1 atm, and so on. The coefficient 2 occurs in front of $\overline{H}^\circ[H_2O(g)]$ and $\overline{H}^\circ[O_2(g)]$ because there are 2 mol of each of these compounds in the balanced chemical equation. The negative value of ΔH°_{rxn} tells us that the reaction gives off heat and is therefore exothermic.

For the water-gas reaction we have

$$\Delta H^\circ_{rxn} = \overline{H}^\circ[CO(g)] + \overline{H}^\circ[H_2(g)] - \overline{H}^\circ[C(s)] - \overline{H}^\circ[H_2O(g)] = +131 \text{ kJ}$$

where the positive value of ΔH°_{rxn} tells us that the reaction absorbs energy as heat and is therefore endothermic.

You may have noticed that we specified the temperature (25°C) when we gave the values of ΔH°_{rxn} for the combustion of methane and for the water-gas reaction. We did this because the value of ΔH°_{rxn} usually depends upon the temperature at which a reaction occurs. The variation in ΔH°_{rxn} with temperature usually is not great, but we should specify the reaction temperature. In this book we do not study the variation in ΔH°_{rxn} with temperature; instead we neglect the dependence of ΔH°_{rxn} on temperature, which is a satisfactory approximation for our purposes.

We have seen that the combustion of methane is an exothermic reaction and that the water-gas reaction is an endothermic reaction. The key question is, what molecular property or properties of the reactants and products give rise to the observed value of ΔH°_{rxn}? What determines the sign and magnitude of ΔH°_{rxn}? We shall answer these questions in Section 6-5, but first we must learn a little more about enthalpies (heats) of reaction.

6-3. HEATS OF REACTION CAN BE CALCULATED FROM TABULATED HEATS OF FORMATION

The value of ΔH°_{rxn} for the reaction of carbon with oxygen is

$$C(s) + O_2(g) \rightarrow CO_2(g) \qquad \Delta H^\circ_{rxn} = -393.5\ \text{kJ}$$

Because 1 mol of $CO_2(g)$ is formed directly from carbon and oxygen, the value of ΔH°_{rxn} for this reaction is *defined* to be the *standard molar enthalpy of formation* of $CO_2(g)$ from its elements. We denote the molar enthalpy of formation of a substance by $\Delta \bar{H}^\circ_f$. For the formation of carbon dioxide from its elements

$$\begin{aligned}\Delta H^\circ_{rxn} &= \Delta \bar{H}^\circ_f[CO_2(g)] = -393.5\ \text{kJ}\cdot\text{mol}^{-1}\\ &= \bar{H}^\circ[CO_2(g)] - \bar{H}^\circ[C(s)] - \bar{H}^\circ[O_2(g)]\end{aligned}$$

The bar in $\bar{H}^\circ$ indicates that the value is for 1 mol of $CO_2(g)$, and the subscript f stands for *f*ormation from the elements. A $\Delta \bar{H}^\circ_f[CO_2(g)]$ value of -393.5 kJ tells us that 1 mol of $CO_2(g)$ lies 393.5 kJ "downhill" on the enthalpy scale relative to its elements (Figure 6-3). The molar enthalpies of formation of water vapor and acetylene from their elements are equal to the ΔH°_{rxn} values for the reactions in which these compounds are formed from their elements (Figures 6-4 and 6-5):

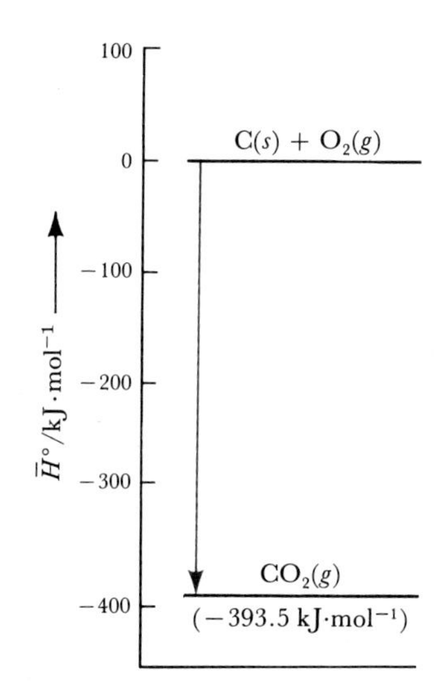

Figure 6-3 An enthalpy diagram showing that the molar enthalpy of $CO_2(g)$ lies 393.5 kJ below the molar enthalpies of $C(s)$ and $O_2(g)$.

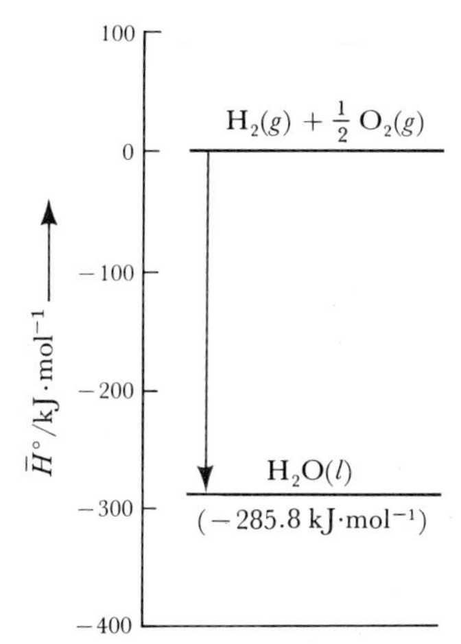

Figure 6-4 An enthalpy diagram showing that the molar enthalpy of $H_2O(l)$ lies 285.8 kJ below the enthalpy of a mixture of 1 mol of $H_2(g)$ and 0.5 mol of $O_2(g)$.

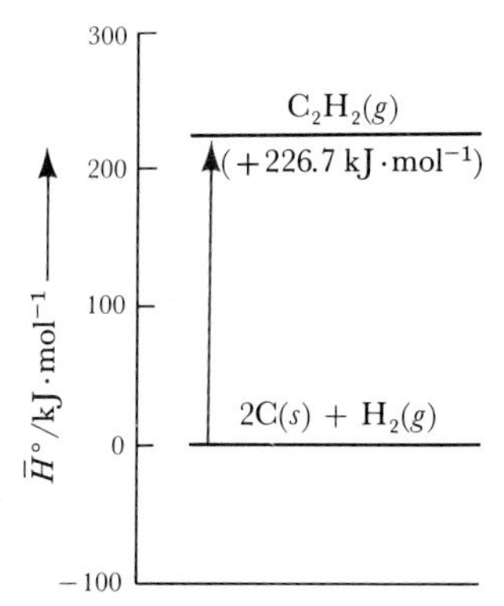

Figure 6-5 An enthalpy diagram showing that the molar enthalpy of acetylene lies 226.7 kJ above the enthalpy of a mixture of 2 mol of $C(s)$ and 1 mol of $H_2(g)$.

$$H_2(g) + \tfrac{1}{2}O_2(g) \rightarrow H_2O(l) \qquad \Delta H^\circ_{rxn} = \Delta \bar{H}^\circ_f[H_2O(l)] = -285.8\ \text{kJ}\cdot\text{mol}^{-1}$$

$$2C(s) + H_2(g) \rightarrow C_2H_2(g) \qquad \Delta H^\circ_{rxn} = \Delta \bar{H}^\circ_f[C_2H_2(g)] = +226.7\ \text{kJ}\cdot\text{mol}^{-1}$$

Note that the first of these equations has a $\tfrac{1}{2}$ as a balancing coefficient. This simply means that $\tfrac{1}{2}$ mol of $O_2(g)$ reacts with 1 mol of $H_2(g)$. We shall use fractional balancing coefficients occasionally.

Figures 6-3, 6-4, and 6-5 are combined onto one graph in Figure 6-6. As suggested by Figure 6-6, we can set up a table of $\Delta \bar{H}^\circ_f$ values for *compounds* by setting the $\Delta \bar{H}^\circ_f$ values for the *elements* equal to zero. That is, for each element in its normal physical state at 25°C and 1 atm, we set $\Delta \bar{H}^\circ_f$ equal to zero. Thus, for example,

The form of an element for which we take $\Delta \bar{H}^\circ_f = 0$ is the natural (common) form of the element at that temperature.

$$\Delta \bar{H}^\circ_f[Fe(s)] = 0 \qquad \Delta \bar{H}^\circ_f[S(s)] = 0$$
$$\Delta \bar{H}^\circ_f[Hg(l)] = 0 \qquad \Delta \bar{H}^\circ_f[Cl_2(g)] = 0$$
$$\Delta \bar{H}^\circ_f[O_2(g)] = 0 \qquad \Delta \bar{H}^\circ_f[C(s), \text{graphite}] = 0$$

Realize, however, that $\Delta \bar{H}^\circ_f$ for gaseous oxygen atoms, $O(g)$, for example, is not equal to zero but is equal to +247 kJ. This is because pure oxygen normally occurs as diatomic molecules and energy must be supplied to produce $O(g)$ from $O_2(g)$:

$$O_2(g) \rightarrow 2O(g) \qquad \Delta H^\circ_{rxn} = +494\ \text{kJ}$$

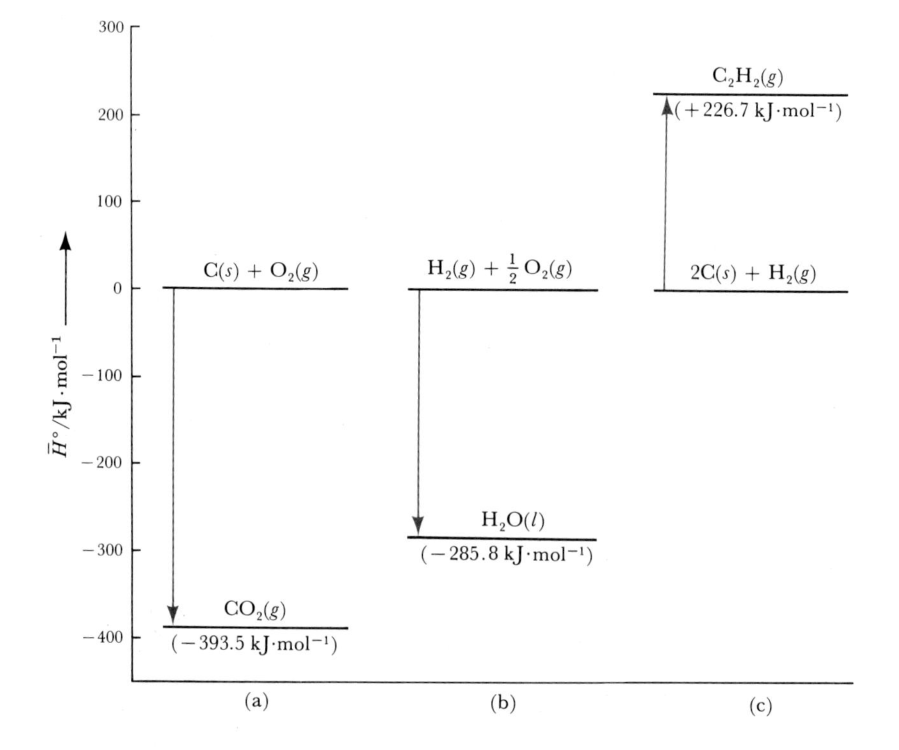

Figure 6-6 Enthalpy changes involved in the formation of (a) $CO_2(g)$, (b) $H_2O(l)$, and (c) $C_2H_2(g)$ from their elements.

Thus

$$\Delta H^\circ_{rxn} = 2\Delta \bar{H}^\circ_f[O(g)] - \Delta \bar{H}^\circ_f[O_2(g)]$$
$$= 2\Delta \bar{H}^\circ_f[O(g)] - 0 = +494 \text{ kJ}$$

$$\Delta \bar{H}^\circ_f[O(g)] = \frac{494 \text{ kJ}}{2 \text{ mol}} = 247 \text{ kJ}\cdot\text{mol}^{-1}$$

Example 6-1: Given that

$$N_2(g) \rightarrow 2N(g) \qquad \Delta H^\circ_{rxn} = +945.2 \text{ kJ}$$

at 25°C, calculate $\Delta \bar{H}^\circ_f[N(g)]$ at 25°C.

Solution: First we use

$$\Delta H^\circ_{rxn} = 2\Delta \bar{H}^\circ_f[N(g)] - \Delta \bar{H}^\circ_f[N_2(g)] = +945.2 \text{ kJ}$$

Now, using the convention that $\Delta \bar{H}^\circ_f[N_2(g)] = 0$ because $N_2(g)$ is the normal form of nitrogen at 25°C, we write

$$\Delta \bar{H}^\circ_f[N(g)] = \frac{945.2 \text{ kJ}}{2 \text{ mol}} = 472.6 \text{ kJ}\cdot\text{mol}^{-1}$$

We can write a general chemical equation,

$$aA + bB \rightarrow yY + zZ$$

where the lower-case letters are the balancing coefficients and the capital letters represent the formulas of the reactants and products. The value of ΔH°_{rxn} for this general equation is

$$\Delta H^\circ_{rxn} = y\Delta \bar{H}^\circ_f[Y] + z\Delta \bar{H}^\circ_f[Z] - a\Delta \bar{H}^\circ_f[A] - b\Delta \bar{H}^\circ_f[B] \quad (6\text{-}6)$$

The normal form of bromine at 25°C is $Br_2(l)$, and this is the form of bromine at 25°C for which we take $\Delta \bar{H}^\circ_f[Br_2(l)] = 0$. Heat must be applied to convert $Br_2(l)$ to $Br_2(g)$, and thus $\Delta \bar{H}^\circ_f[Br_2(g)]$ is equal to ΔH°_{rxn} for the reaction $Br_2(l) \rightarrow Br_2(g)$.

In using Equation (6-6), it is necessary to specify whether each substance is a gas, liquid, or solid because the value of $\Delta \bar{H}^\circ_f$ depends on the physical state of the substance. Using Equation (6-6), the value of ΔH°_{rxn} for the reaction

$$2C_2H_2(g) + 5O_2(g) \rightarrow 4CO_2(g) + 2H_2O(l)$$

is

$$\Delta H^\circ_{rxn} = 4\Delta \bar{H}^\circ_f[CO_2(g)] + 2\Delta \bar{H}^\circ_f[H_2O(l)] - 2\Delta \bar{H}^\circ_f[C_2H_2(g)] - 5\Delta \bar{H}^\circ_f[O_2(g)]$$

Using data in Figure 6-6, we obtain

$$\Delta H^\circ_{rxn} = (4 \text{ mol})(-393.5 \text{ kJ}\cdot\text{mol}^{-1}) + (2 \text{ mol})(-285.8 \text{ kJ}\cdot\text{mol}^{-1}) - (2 \text{ mol})(+226.7 \text{ kJ}\cdot\text{mol}^{-1}) - (5 \text{ mol})(0 \text{ kJ}\cdot\text{mol}^{-1})$$
$$= -2599.0 \text{ kJ}$$

Note that $\Delta\bar{H}_f^\circ[O_2(g)] = 0$ because the $\Delta\bar{H}_f^\circ$ value of any element in its normal state at 25°C is zero. Table 6-1 lists $\Delta\bar{H}_f^\circ$ values for a variety of substances at 25°C.

Example 6-2: Use the $\Delta\bar{H}_f^\circ$ data in Table 6-1 to compute ΔH_{rxn}° for the combustion of ethyl alcohol:

$$C_2H_5OH(l) + 3O_2(g) \rightarrow 2CO_2(g) + 3H_2O(g)$$

Solution: Referring to Table 6-1, we find

$$\Delta\bar{H}_f^\circ[CO_2(g)] = -393.5\ \text{kJ}\cdot\text{mol}^{-1} \quad \Delta\bar{H}_f^\circ[H_2O(g)] = -241.8\ \text{kJ}\cdot\text{mol}^{-1}$$

$$\Delta\bar{H}_f^\circ[O_2(g)] = 0 \quad \Delta\bar{H}_f^\circ[C_2H_5OH(l)] = -277.7\ \text{kJ}\cdot\text{mol}^{-1}$$

Problems 6-5 to 6-16 ask you to use the data in Table 6-1 to calculate standard heats of reactions.

and thus

$$\begin{aligned}\Delta H_{rxn}^\circ &= 2\Delta\bar{H}_f^\circ[CO_2(g)] + 3\Delta\bar{H}_f^\circ[H_2O(g)] \\ &\quad - \Delta\bar{H}_f^\circ[C_2H_5OH(l)] - 3\Delta\bar{H}_f^\circ[O_2(g)] \\ &= (2\ \text{mol})(-393.5\ \text{kJ}\cdot\text{mol}^{-1}) + (3\ \text{mol})(-241.8\ \text{kJ}\cdot\text{mol}^{-1}) \\ &\quad - (1\ \text{mol})(-277.7\ \text{kJ}\cdot\text{mol}^{-1}) - (3\ \text{mol})(0) \\ &= -1234.7\ \text{kJ}\end{aligned}$$

The reaction is highly exothermic.

The combustion reaction of ethyl alcohol was used to power the *Redstone* missiles. Ethyl alcohol burns in air at a much lower temperature than, say, kerosene or methane, and for this reason it has been used for the burners in chemistry sets.

We shall have further use for Table 6-1 in the next section.

6-4. HESS'S LAW SAYS THAT ENTHALPY CHANGES FOR CHEMICAL REACTIONS ARE ADDITIVE

One of the most useful properties of ΔH_{rxn}° values is that we can add the ΔH_{rxn}° values for two or more reactions to obtain the ΔH_{rxn}° value for another reaction. This property is best illustrated by example. Consider the two reactions

Chemical equations can be added and subtracted like algebraic equations.

(1) $C(s) + \frac{1}{2}O_2(g) \rightarrow CO(g)$ $\quad \Delta H_{rxn}^\circ(1) = -110.5\ \text{kJ}$

(2) $CO(g) + \frac{1}{2}O_2(g) \rightarrow CO_2(g)$ $\quad \Delta H_{rxn}^\circ(2) = -283.0\ \text{kJ}$

If we add these two chemical equations as if they were algebraic equations, then we get

(3) $C(s) + CO(g) + \frac{1}{2}O_2(g) + \frac{1}{2}O_2(g) \rightarrow CO(g) + CO_2(g)$

If we cancel $CO(g)$ from both sides, then we get

(3) $C(s) + O_2(g) \rightarrow CO_2(g)$

Table 6-1 Standard molar enthalpies of formation, $\Delta \bar{H}_f^\circ$, for various substances at 25°C

Substance	*Formula*	$\Delta \bar{H}_f^\circ / kJ \cdot mol^{-1}$
acetylene	$C_2H_2(g)$	+226.7
ammonia	$NH_3(g)$	−46.19
benzene	$C_6H_6(l)$	+49.03
bromine vapor	$Br_2(g)$	+30.91
carbon dioxide	$CO_2(g)$	−393.5
carbon monoxide	$CO(g)$	−110.5
carbon tetrachloride	$CCl_4(l)$	−135.4
	$CCl_4(g)$	−103.0
diamond	$C(s)$	+1.897
ethane	$C_2H_6(g)$	−84.68
ethanol (ethyl alcohol)	$C_2H_5OH(l)$	−277.7
ethylene	$C_2H_4(g)$	+52.28
glucose	$C_6H_{12}O_6(s)$	−1260
graphite	$C(s)$	0
hydrazine	$N_2H_4(l)$	+50.6
hydrogen bromide	$HBr(g)$	−36.4
hydrogen chloride	$HCl(g)$	−92.31
hydrogen fluoride	$HF(g)$	−271.1
hydrogen iodide	$HI(g)$	+26.1
hydrogen peroxide	$H_2O_2(l)$	−187.8
iodine vapor	$I_2(g)$	+62.4
magnesium carbonate	$MgCO_3(s)$	−1096
magnesium oxide	$MgO(s)$	−601.7
methane	$CH_4(g)$	−74.86
methanol (methyl alcohol)	$CH_3OH(l)$	−238.7
	$CH_3OH(g)$	−200.7
nitrogen oxide	$NO(g)$	+90.37
nitrogen dioxide	$NO_2(g)$	+33.85
dinitrogen tetroxide	$N_2O_4(g)$	+9.66
	$N_2O_4(l)$	−19.5
propane	$C_3H_8(g)$	−103.8
sodium carbonate	$Na_2CO_3(s)$	−1131
sodium oxide	$Na_2O(s)$	−418.0
sucrose	$C_{12}H_{22}O_{11}(s)$	−2220
sulfur dioxide	$SO_2(g)$	−296.8
sulfur trioxide	$SO_3(g)$	−395.7
water	$H_2O(l)$	−285.8
	$H_2O(g)$	−241.8

Elemental carbon occurs both as diamond and as graphite at 25°C. We take $\Delta \bar{H}_f^\circ = 0$ for graphite because it is the more common form. Note that energy is required to convert graphite to diamond.

Elemental forms for which we take $\Delta \bar{H}_f^\circ = 0$ at 25°C

$H_2(g)$	$Na(s)$
$O_2(g)$	$Br_2(l)$
$N_2(g)$	$Cl_2(g)$
$C(s, \textit{graphite})$	$F_2(g)$
$S(s, \textit{rhombic})$	$I_2(s)$
$Mg(s)$	

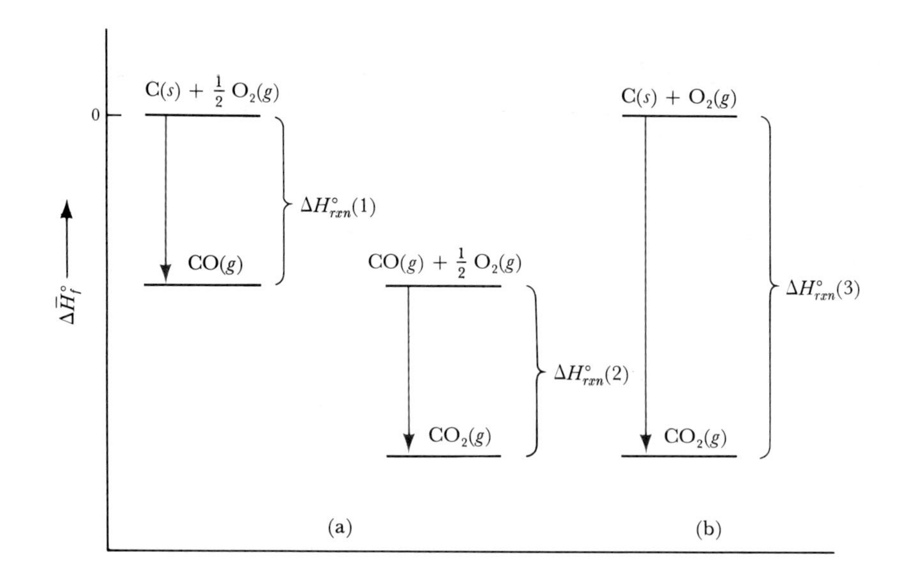

Figure 6-7 An illustration of Hess's law. The value of ΔH°_{rxn} for the reaction $C(s) + O_2(g) \rightarrow CO_2(g)$ is independent of whether the reaction occurs in two steps (a) or in one step (b).

The additive property of ΔH°_{rxn} says that ΔH°_{rxn} for Equation (3) is simply

$$\begin{aligned}\Delta H^\circ_{rxn}(3) &= \Delta H^\circ_{rxn}(1) + \Delta H^\circ_{rxn}(2) \\ &= -110.5 \text{ kJ} + (-283.0 \text{ kJ}) = -393.5 \text{ kJ}\end{aligned} \tag{6-7}$$

From the ΔH°_f data in Table 6-1, we can compute for Equation (3)

$$\begin{aligned}\Delta H^\circ_{rxn}(3) &= \Delta \bar{H}^\circ_f[CO_2(g)] - \Delta \bar{H}^\circ_f[C(s)] - \Delta \bar{H}^\circ_f[O_2(g)] \\ &= (1 \text{ mol})(-393.5 \text{ kJ}\cdot\text{mol}^{-1}) - (1 \text{ mol})(0) - (1 \text{ mol})(0) \\ &= -393.5 \text{ kJ}\end{aligned}$$

which agrees with the value for $\Delta H^\circ_{rxn}(3)$ obtained by summing $\Delta H^\circ_{rxn}(1)$ and $\Delta H^\circ_{rxn}(2)$.

The additivity property of ΔH°_{rxn} values is known as *Hess's law:* if two or more chemical equations are added together, then the value of ΔH°_{rxn} for the resulting equation is equal to the sum of the ΔH°_{rxn} values for the separate equations. Hess's law is shown graphically in Figure 6-7 for Equations (1) through (3).

Example 6-3: Given the following ΔH°_{rxn} values:

(1) $SO_2(g) \rightarrow S(s) + O_2(g)$ $\qquad \Delta H^\circ_{rxn}(1) = +296.8 \text{ kJ}$

(2) $2S(s) + 3O_2(g) \rightarrow 2SO_3(g)$ $\qquad \Delta H^\circ_{rxn}(2) = -791.4 \text{ kJ}$

Compute ΔH°_{rxn} for the reaction

(3) $2SO_2(g) + O_2(g) \rightarrow 2SO_3(g)$

Solution: To obtain Equation (3) from Equations (1) and (2), it is first necessary to multiply Equation (1) through by 2 because Equation (3) involves 2 mol of $SO_2(g)$ as a reactant. If we multiply an equation through by 2, then its ΔH°_{rxn} must also be multiplied by 2 because twice as many moles of reactants are consumed and twice as many moles of products are produced. Thus the amount of heat evolved or absorbed is doubled. For the equation

(4) $$2SO_2(g) \rightarrow 2S(s) + 2O_2(g)$$

we have

$$\Delta H^\circ_{rxn} = 2\Delta H^\circ_{rxn}(1) = 2 \times 296.8\ \text{kJ} = +593.6\ \text{kJ}$$

Addition of Equations (2) and (4) yields

$$2S(s) + 3O_2(g) + 2SO_2(g) \rightarrow 2S(s) + 2O_2(g) + 2SO_3(g)$$

Problems 6-17 to 6-26 deal with Hess's law.

If we cancel $2S(s)$ and $2O_2(g)$ from both sides, then we get Equation (3):

$$2SO_2(g) + O_2(g) \rightarrow 2SO_3(g)$$

The corresponding value of ΔH°_{rxn} is

$$\begin{aligned}\Delta H^\circ_{rxn}(3) &= 2\Delta H^\circ_{rxn}(1) + \Delta H^\circ_{rxn}(2)\\ &= +593.6\ \text{kJ} - 791.4\ \text{kJ}\\ &= -197.8\ \text{kJ}\end{aligned}$$

Note that the conversion of sulfur dioxide to sulfur trioxide is an exothermic reaction. This is the reaction that takes place in the catalytic converters of automobiles that run on unleaded gasoline and in the manufacture of sulfuric acid (Interchapter D).

Another useful property of ΔH°_{rxn} values follows from Equation (6-6). If we reverse a chemical equation, then the reactants become the products and the products become the reactants. Thus

$$\Delta H^\circ_{rxn}(\text{reverse rxn}) = -\Delta H^\circ_{rxn}(\text{forward rxn}) \qquad (6\text{-}8)$$

The value of ΔH°_{rxn} for the reaction

(1) $$CO_2(g) \rightarrow C(s) + O_2(g)$$

which is the reverse of the reaction

(2) $$C(s) + O_2(g) \rightarrow CO_2(g)$$

is

$$\begin{aligned}\Delta H^\circ_{rxn}(1) &= -\Delta H^\circ_{rxn}(2)\\ &= -(-393.5\ \text{kJ}) = +393.5\ \text{kJ}\end{aligned}$$

The arrangement of the atoms in ethyl alcohol and methyl ether are

```
  H H             H     H
  | |             |     |
H-C-C-O-H       H-C--O--C-H
  | |             |     |
  H H             H     H
ethyl alcohol    methyl ether
```

The utility of Hess's law can be seen from the following example. Two compounds that have the same molecular formula but different arrangements of atoms are said to be *isomers*. The compounds ethyl alcohol and methyl ether are isomers. Both have the molecular formula C_2H_6O. We shall write the formula for ethyl alcohol as $C_2H_5OH(l)$ and that for methyl ether as $CH_3OCH_3(l)$. The heats of combustion of the isomeric compounds ethyl alcohol, $C_2H_5OH(l)$, and methyl ether, $CH_3OCH_3(l)$, are

$$(1)\quad C_2H_5OH(l) + 3O_2(g) \rightarrow 2CO_2(g) + 3H_2O(g) \qquad \Delta H^\circ_{rxn}(1) = -1234.7\ \text{kJ}$$

$$(2)\quad CH_3OCH_3(l) + 3O_2(g) \rightarrow 2CO_2(g) + 3H_2O(g) \qquad \Delta H^\circ_{rxn}(2) = -1328.3\ \text{kJ}$$

Consider now the isomerization reaction in which ethyl alcohol is converted to methyl ether:

$$(3)\qquad C_2H_5OH(l) \rightarrow CH_3OCH_3(l)$$

The value of ΔH°_{rxn} for Equation (3) cannot be determined directly because the reaction occurs very slowly. However, it is not difficult to carry out the separate combustion reactions (1) and (2) and then combine the experimental results and use Hess's law to obtain ΔH°_{rxn} for the isomerization reaction. Equation (3) is obtained by adding the reverse of Equation (2) to Equation (1). By adding these two equations and canceling $3O_2(g)$, $2CO_2(g)$, and $3H_2O(g)$ from both sides of the resulting equation, we obtain Equation (3):

$$C_2H_5OH(l) \rightarrow CH_3OCH_3(l)$$

From Hess's law the value of $\Delta H^\circ_{rxn}(3)$, that is, the enthalpy change for the isomerization reaction, is

$$\begin{aligned}\Delta H^\circ_{rxn}(3) &= \Delta H^\circ_{rxn}(1) + [-\Delta H^\circ_{rxn}(2)]\\ &= -1234.7\ \text{kJ} + [+1328.3\ \text{kJ}] = +93.6\ \text{kJ}\end{aligned}$$

We can use Hess's law to understand how the molar heats of formation in Table 6-1 are used to calculate enthalpy changes. Consider the general reaction

$$\text{aA} + \text{bB} \rightarrow \text{yY} + \text{zZ}$$

We can calculate ΔH°_{rxn} in two steps, as shown in the following diagram:

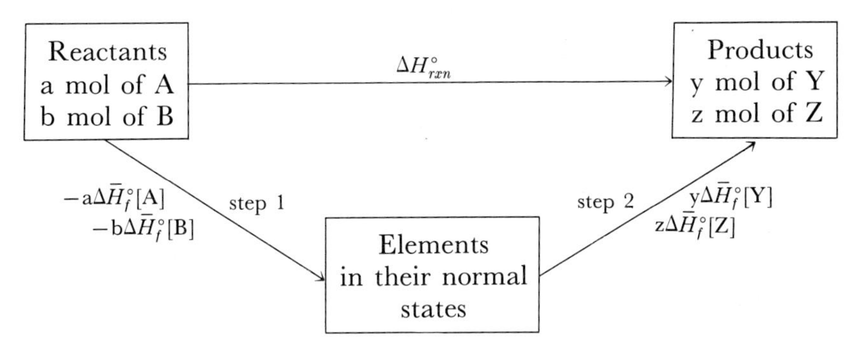

First we decompose compounds A and B into their constituent elements (step 1) and then combine the elements to form the compounds Y and Z (step 2). In the first step, we have

$$\Delta H^\circ_{rxn}(1) = -a\Delta\bar{H}^\circ_f[A] - b\Delta\bar{H}^\circ_f[B]$$

The minus signs occur here because the reaction is the reverse of the formation of the compounds from their elements; we are forming the elements from the compounds. In the second step, we have

$$\Delta H^\circ_{rxn}(2) = y\Delta\bar{H}^\circ_f[Y] + z\Delta\bar{H}^\circ_f[Z]$$

The sum of $\Delta H^\circ_{rxn}(1)$ and $\Delta H^\circ_{rxn}(2)$ gives ΔH°_{rxn}:

$$\begin{aligned}\Delta H^\circ_{rxn} &= y\Delta\bar{H}^\circ_f[Y] + z\Delta\bar{H}^\circ_f[Z] - a\Delta\bar{H}^\circ_f[A] - b\Delta\bar{H}^\circ_f[B] \\ &= \Delta H^\circ_f(\text{products}) - \Delta H^\circ_f(\text{reactants})\end{aligned}$$

Example 6-4: The combustion of acetylene has the stoichiometry

$$C_2H_2(g) + \tfrac{5}{2}O_2(g) \rightarrow 2CO_2(g) + H_2O(g)$$

According to Equation (6-6), the value of ΔH°_{rxn} is

$$\Delta H^\circ_{rxn} = 2\Delta\bar{H}^\circ_f[CO_2(g)] + \Delta\bar{H}^\circ_f[H_2O(g)] - \Delta\bar{H}^\circ_f[C_2H_2(g)]$$

where we have taken $\Delta\bar{H}^\circ_f[O_2(g)] = 0$. Use Hess's law and the diagram at the top of the page to derive this expression for ΔH°_{rxn}.

Solution: We start with the following equations, which represent the formation of the compounds from the elements:

(1) $2C(s) + H_2(g) \rightarrow C_2H_2(g)$ $\quad \Delta H^\circ_{rxn}(1) = \Delta\bar{H}^\circ_f[C_2H_2(g)]$

(2) $C(s) + O_2(g) \rightarrow CO_2(g)$ $\quad \Delta H^\circ_{rxn}(2) = \Delta\bar{H}^\circ_f[CO_2(g)]$

(3) $H_2(g) + \tfrac{1}{2}O_2(g) \rightarrow H_2O(g)$ $\quad \Delta H^\circ_{rxn}(3) = \Delta\bar{H}^\circ_f[H_2O(g)]$

We now reverse Equation (1), multiply Equation (2) by 2, and add the result to Equation (3) to get

$$C_2H_2(g) + 2C(s) + 2O_2(g) + H_2(g) + \tfrac{1}{2}O_2(g) \rightarrow 2C(s) + H_2(g) + 2CO_2(g) + H_2O(g)$$

After canceling terms on both sides, we obtain

$$C_2H_2(g) + \tfrac{5}{2}O_2(g) \rightarrow 2CO_2(g) + H_2O(g)$$

By adding the ΔH°_{rxn} values, we obtain

$$\begin{aligned}\Delta H^\circ_{rxn} &= 2\Delta H^\circ_{rxn}(2) + \Delta H^\circ_{rxn}(3) - \Delta H^\circ_{rxn}(1)\\ &= 2\Delta \bar{H}^\circ_f[CO_2(g)] + \Delta \bar{H}^\circ_f[H_2O(g)] - \Delta \bar{H}^\circ_f[C_2H_2(g)]\end{aligned}$$

in nice accord with the result obtained using Equation (6-6).

6-5. THE VALUE OF ΔH°_{rxn} IS DETERMINED PRIMARILY BY THE DIFFERENCE IN THE BOND ENTHALPIES OF THE REACTANT AND PRODUCT MOLECULES

In this section we learn what molecular properties of reactants and products give rise to the observed value of ΔH°_{rxn}. The enthalpy change for the reaction

$$H_2O(g) \rightarrow O(g) + 2H(g)$$

is $\Delta H^\circ_{rxn} = +925$ kJ. Although we have not studied bonding yet, we shall learn in Chapter 11 that a water molecule has the structure

```
  O
 / \
H   H
```

An input of energy as heat is necessary to break a chemical bond.

There are two O—H bonds, represented by the lines joining the H atoms to the O atom, in a water molecule. At constant pressure an input of 925 kJ of energy as heat is required to break the two O—H bonds in 1 mol of water molecules. Let's denote the average O—H *bond enthalpy per mole* of O—H bonds in water by $\bar{H}$(O—H). The average molar bond enthalpy $\bar{H}$(O—H) is equal to one half the total energy input required to break the two O—H bonds:

$$\bar{H}(\text{O—H}) = \frac{925\text{ kJ}}{2} = 463\text{ kJ}$$

The values of the O—H molar bond enthalpy in a variety of compounds are approximately the same as the value for water. For example, the enthalpy change for the reaction

```
   H                H
   |                |
H—C—O—H  →  H—C—O + H
   |                |
   H                H
```

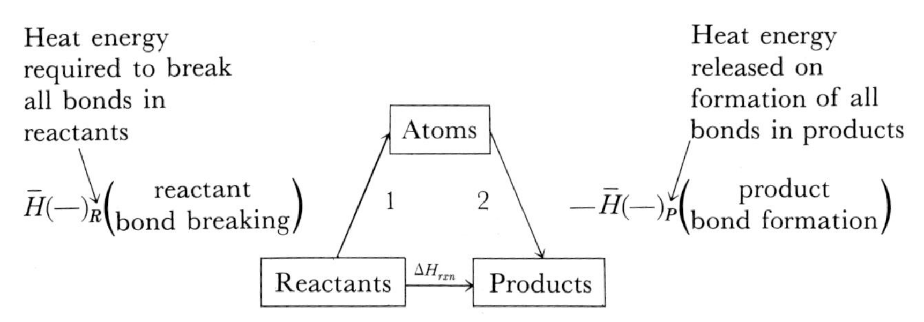

Figure 6-8. The enthalpy change of a reaction is given approximately by the difference between the heat energy required to break all the chemical bonds in the reactants and the energy released as heat on the formation of all the chemical bonds in the products.

in which a single O—H bond is broken, is $\Delta H^\circ_{rxn} = 464$ kJ. Molar bond enthalpies are often referred to simply as *bond energies.*

In Figure 6-8 we picture a chemical reaction as taking place in two steps. In step 1 the reactant molecules are broken down into their constituent atoms, and in step 2 the atoms are rejoined to form the product molecules. Step 1 requires an input of energy to break all the bonds in the reactant molecules, whereas step 2 evolves energy as the bonds in the product molecules are formed. The total enthalpy change for the reaction is

The formation of a chemical bond releases energy as heat.

$$\Delta H^\circ_{rxn} \approx \begin{pmatrix}\text{heat energy input to}\\ \text{break all bonds in}\\ \text{reactants}\end{pmatrix} - \begin{pmatrix}\text{energy evolved as heat}\\ \text{on formation of all}\\ \text{bonds in products}\end{pmatrix}$$

$$\approx \bar{H}(\text{bond})_R - \bar{H}(\text{bond})_P \qquad (6\text{-}9)$$

where $\bar{H}(\text{bond})_R$ represents the sum of the molar bond enthalpies for all the reactant bonds and $\bar{H}(\text{bond})_P$ represents the sum of the molar bond enthalpies for all the product bonds in the chemical equation for the reaction. (The reason for the approximately equals sign in Equation (6-9) is discussed later in this section.)

If more energy is released on formation of the product bonds than is required to break the reactant bonds, then the value of ΔH°_{rxn} is negative (energy released). If less energy is released on the formation of the product bonds than is required to break the reactant bonds, then the value of ΔH°_{rxn} is positive (energy consumed). The value of ΔH°_{rxn} is determined primarily by the difference in the bond enthalpies of the reactants and products. Values of average bond enthalpies for a variety of chemical bonds are given in Table 6-2. These values are the heat energies required to break Avogadro's number of bonds (in other words, they are molar bond enthalpies).

Example 6-5: Chemical reactions can be used to produce flames for heating. The nonnuclear reaction with the highest attainable flame temperature (approximately 6000°C, about the surface temperature of the sun) is

$$\text{H—H}(g) + \text{F—F}(g) \rightarrow 2\text{H—F}(g)$$

Use the molar bond enthalpies in Table 6-2 to estimate ΔH°_{rxn} for this reaction.

Table 6-2 Average molar bond enthalpies ("bond energies")

Bond	*Molar bond enthalpy,* $\bar{H}(bond)/kJ\cdot mol^{-1}$	*Bond*	*Molar bond enthalpy,* $\bar{H}(bond)/kJ\cdot mol^{-1}$
O—H	464	N—H	390
C—O	351	F—F	155
C—C	347	Cl—Cl	243
C—H	414	Br—Br	192
C—F	439	H—H	435
C—Cl	331	H—F	565
C—Br	276	H—Cl	431
C—N	293	H—Br	368
		H—S	364

The value of a given bond energy varies slightly from molecule to molecule.

Solution: The reaction involves the rupture of one H—H and one F—F bond and the formation of two H—F bonds; thus:

$$\Delta H^\circ_{rxn} \approx \bar{H}(\text{H—H}) + \bar{H}(\text{F—F}) - 2\bar{H}(\text{H—F})$$

and

$$\begin{aligned}\Delta H^\circ_{rxn} &\approx (1\text{ mol})(435\text{ kJ}\cdot\text{mol}^{-1}) + (1\text{ mol})(155\text{ kJ}\cdot\text{mol}^{-1}) \\ &\quad - (2\text{ mol})(565\text{ kJ}\cdot\text{mol}^{-1}) \\ &\approx -540\text{ kJ}\end{aligned}$$

Problems 6-27 to 6-34 deal with bond energies.

The value of ΔH°_{rxn} we would obtain using data from Table 6-1 is -542.2 kJ, which is in nice agreement with what we have found here using bond energies.

Although the value of ΔH°_{rxn} is determined primarily by the difference in bond energies of the reactants and the products, interactions (attractive forces) *between* chemical species in the liquid and in solid phases (Chapter 13) can also make significant contributions to the value of ΔH°_{rxn}. As an example, consider the vaporization of liquid water at 25°C,

$$H_2O(l) \rightarrow H_2O(g) \qquad \Delta H^\circ_{rxn} = +44.0\text{ kJ}$$

Note that no internal O—H bonds are broken in this process. In the vaporization of water, it is the attractive forces *between* the water molecules that must be overcome. This is one of the reasons that ΔH°_{rxn} values obtained from bond enthalpies, rather than from $\Delta \bar{H}^\circ_f$ values, are only approximate. Another reason is that the bond energies given in Table 6-2 are *average* values.

6-6. HEAT CAPACITY MEASURES THE ABILITY OF A SUBSTANCE TO TAKE UP ENERGY AS HEAT

The *heat capacity* of a substance is defined as the heat required to raise the temperature of the substance by one kelvin. If the substance is heated at constant pressure, then the heat added is equal to ΔH and the heat capacity is denoted by C_P, where the subscript P denotes constant pressure. In terms of an equation, we can write

$$C_P = \frac{\Delta H}{\Delta T} \qquad (6\text{-}10)$$

where $\Delta H = q_P$ and ΔT is the increase in temperature of the substance arising from the heat input. All that is necessary to determine the heat capacity of a substance is to add a known quantity of energy as heat and then measure the resulting increase in temperature. The following example illustrates the use of Equation (6-10).

The heat capacity of a substance is always positive.

Example 6-6: When 421.2 J of heat is added to 36.0 g of liquid water, the temperature of the water increases from 10.000°C to 12.800°C. Compute the heat capacity of the 36.0 g of $H_2O(l)$.

Solution: From Equation (6-10) we have

$$C_P = \frac{\Delta H}{\Delta T}$$

ΔT is equal to 12.800°C − 10.000°C = 2.800°C. If we use the fact that a kelvin degree and a Celsius degree are the same size, then we have

$$C_P = \frac{421.2\ \text{J}}{2.800\ \text{K}} = 150.4\ \text{J}\cdot\text{K}^{-1}$$

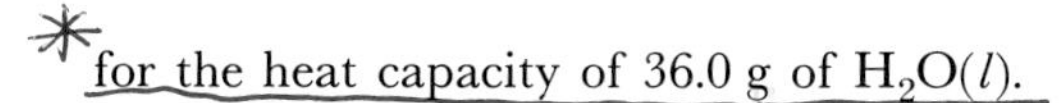

for the heat capacity of 36.0 g of $H_2O(l)$.

We shall denote the *heat capacity per mole*, or the *molar heat capacity*, of any substance by $\bar{C}_P$, where the bar denotes a molar quantity. The heat capacity per mole of water can be computed from the C_P value of 150.4 J·K^{-1} calculated in Example 6-6 for 36.0 g of water. Since a mole of water has a mass of 18.0 g, 36.0 g contains 2.00 moles of water. Thus the value of $\bar{C}_P$ for liquid water is

$$\bar{C}_P = \frac{150.4\ \text{J}\cdot\text{K}^{-1}}{2.00\ \text{mol}} = 75.2\ \text{J}\cdot\text{K}^{-1}\cdot\text{mol}^{-1}$$

It requires an input of 75.2 joules of heat to raise the temperature of 18.0 grams of liquid water 1.00 kelvin.

The *specific heat of a substance* is the heat capacity per gram of the substance. The heat capacity per mole of liquid water is

75.2 J·K^{-1}·mol^{-1}. There are 18.0 g in a mole of water, and thus the specific heat, c_{sp}, of liquid water is

$$c_{sp} = \frac{75.2\ \text{J}\cdot\text{K}^{-1}\cdot\text{mol}^{-1}}{18.0\ \text{g}\cdot\text{mol}^{-1}} = 4.18\ \text{J}\cdot\text{K}^{-1}\cdot\text{g}^{-1}$$

The specific heat of ice is 2.09 J·K^{-1}·g^{-1}, which is only half the specific heat of liquid water. Specific heats are especially useful in dealing with substances that cannot be represented by a chemical formula, for example, wood or stone.

Example 6-7: Compute the energy as heat required to raise the temperature of 151 kg of water (that is, 40 gal, the volume of a typical home water heater) from 18.0°C to 60.0°C, assuming no loss of energy to the surroundings. Take the specific heat of water to be 4.18 J·K^{-1}·g^{-1}.

The total heat capacity, C_P, is related to the specific heat, C_{sp}, by

$$C_P = C_{sp} \times (\text{mass in grams})$$

Solution: The heat energy required can be computed from Equation (6-10):

$$\Delta H = C_P\Delta T = (4.18\ \text{J}\cdot\text{K}^{-1}\cdot\text{g}^{-1})(151\ \text{kg})\left(\frac{1000\ \text{g}}{1\ \text{kg}}\right)(333 - 291)\ \text{K}$$

$$= 2.65 \times 10^7\ \text{J} = 26.5\ \text{MJ}$$

where 1 MJ = 1 megajoule = 1×10^6 J. Most U.S. natural gas suppliers charge about 40 cents per 100 MJ of heat energy. It thus costs about 10 cents to heat 40 gal of water from 18°C to 60°C, assuming that all the heat goes into the water. In actual practice, only about half the heat is absorbed by the water; the remainder is lost to the surroundings, and so the actual cost is about 20 cents.

The molar heat capacities and specific heats for a variety of substances are given in Table 6-3. Especially noteworthy is the high specific heat of liquid water. Only helium and hydrogen gases and liquid lithium have greater specific heats than $H_2O(l)$. However, $He(g)$ and $H_2(g)$ at 1 atm pressure have much lower densities than $H_2O(l)$; the heat capacities per liter at atmospheric pressure are

Substance	$C_P/J\cdot K^{-1}\cdot L^{-1}$
$He(g)$	0.8
$H_2(g)$	1.2
$Na(l)$	1300
$Li(l)$	3100
$H_2O(l)$	4180

Table 6-3 Molar heat capacities and specific heats at constant pressure for some common substances

Substance	*Formula*	$\bar{C}_P/J \cdot K^{-1} \cdot mol^{-1}$	$c_{sp}/J \cdot K^{-1} \cdot g^{-1}$
ammonia	$NH_3(g)$	35.1	2.06
carbon dioxide	$CO_2(g)$	37.1	0.84
copper	$Cu(s)$	24.5	0.39
ethylene	$C_2H_4(g)$	42.0	1.50
helium	$He(g)$	20.8	5.2
hydrogen	$H_2(g)$	28.8	14.4
ice	$H_2O(s)$	37.7	2.09
lithium (liquid)	$Li(l)$	40.0	5.7
mercury	$Hg(l)$	28.0	0.14
nitrogen	$N_2(g)$	29.2	1.04
oxygen	$O_2(g)$	29.4	0.92
sodium (liquid)	$Na(l)$	30.8	1.34
steel	—	—	0.46
water	$H_2O(l)$	75.2	4.18
water vapor	$H_2O(g)$	35.5	1.97
wood	—	—	1.8
glass, rock, dirt, fiberglass, asbestos, cement	—	—	0.84

On a volume basis, liquid water has a higher heat capacity than any other liquid, and this fact, together with the fact that water is the cheapest chemical substance, makes water the most widely used coolant fluid. It is the primary coolant for most power plants (both conventional and nuclear) and combustion engines.

6-7. A CALORIMETER IS A DEVICE USED TO MEASURE THE AMOUNT OF HEAT EVOLVED OR ABSORBED IN A REACTION

The value of ΔH°_{rxn} for a chemical reaction can be measured in a device called a *calorimeter*. A simple calorimeter consisting of a Dewar flask ("thermos bottle") equipped with a high-precision thermometer is shown in Figure 6-9. A calorimeter works on the principle of

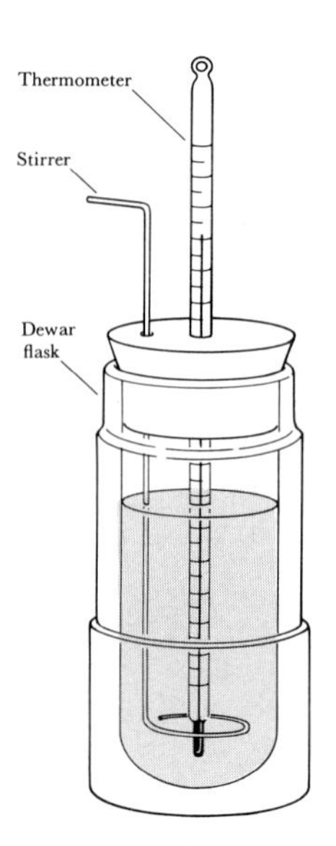

Figure 6-9 A simple calorimeter, consisting of a Dewar flask with cover, which prevents a significant loss or gain of heat from the surrounding air; a high-precision thermometer, which gives the temperature to within ± 0.001 K; and a simple ring-type stirrer. One reactant is placed in the Dewar flask, and then the other reactant, at the same temperature, is added. The reaction mixture is stirred and the change in temperature is measured.

energy conservation, which requires that the total energy is always conserved. Recall that the principle of energy conservation (first law of thermodynamics) states that energy cannot be created or destroyed but only transferred from one system to another.

For a chemical reaction occurring at fixed pressure, the value of ΔH_{rxn} is equal to the heat evolved or absorbed in the process

$$\Delta H_{rxn} = q_P \tag{6-4}$$

Consider an exothermic reaction run in a Dewar flask. The heat evolved by the reaction cannot escape from the flask and thus is absorbed by the calorimeter contents (reaction mixture, thermometer, stirrer, and so on). The heat absorption leads to an increase in the temperature of the calorimeter contents. Because all of the heat evolved by the reaction is absorbed by the calorimeter contents, we can write

$$\Delta H_{rxn} = -\Delta H_{calorimeter} \tag{6-11}$$

From the definition of heat capacity, Equation (6-10), we have

$$\Delta H_{calorimeter} = C_{P,\, calorimeter}\, \Delta T \tag{6-12}$$

where ΔT is the observed temperature change. Substitution of Equation (6-12) into Equation (6-11) yields

$$\Delta H_{rxn} = -C_{P,\, calorimeter}\, \Delta T \tag{6-13}$$

Equation (6-13) tells us that, if we run a chemical reaction in a calorimeter with a known heat capacity ($C_{P,\, calorimeter}$) and if we determine the temperature change, then the value of ΔH_{rxn} can be computed. The value of $C_{P,\,calorimeter}$ can be determined by electrical heating or by running a reaction with a known value of ΔH_{rxn} and determining the associated ΔT value. In the latter case we have

$$C_{P,\,calorimeter} = \frac{-\Delta H_{rxn}}{\Delta T} \qquad (6\text{-}14)$$

As an example of the application of Equation (6-14), consider the reaction between HCl(aq) and NaOH(aq):

$$\mathrm{HCl}(aq) + \mathrm{NaOH}(aq) \rightarrow \mathrm{H_2O}(l) + \mathrm{NaCl}(aq)$$

for which $\Delta H^\circ_{rxn} = -55.70\ \mathrm{kJ \cdot mol^{-1}}$. The superscript zero on ΔH°_{rxn} for a reaction that involves solution-phase species refers to dilute solutions. The value of ΔH_{rxn} depends on concentration, but this effect is usually not large. We shall ignore the effect of concentration on ΔH_{rxn} values and assume that $\Delta H_{rxn} = \Delta H^\circ_{rxn}$. Suppose that we add 0.500 L of 0.200 M HCl(aq) to 0.500 L of 0.200 M NaOH(aq) in the calorimeter shown in Figure 6-9 and that the observed ΔT is 1.210 K. Thc number of moles of HCl(aq) available before reaction is

$$\begin{matrix}\text{mol HCl}(aq) \\ \text{available}\end{matrix} = MV = (0.200\ \mathrm{mol \cdot L^{-1}})(0.500\ \mathrm{L}) = 0.100\ \mathrm{mol}$$

The number of moles of NaOH(aq) available before reaction is

$$\text{mol NaOH}(aq)\text{ available} = (0.200\ \mathrm{mol \cdot L^{-1}})(0.500\ \mathrm{L}) = 0.100\ \mathrm{mol}$$

The reaction stoichiometry tells us that 1 mol of NaOH(aq) reacts with 1 mol of HCl(aq), and thus a total of 0.100 mol of NaOH(aq) and 0.100 mol of HCl(aq) react to form 0.100 mol of H_2O(l) and 0.100 mol of NaCl(aq). The value of ΔH°_{rxn} for 0.100 mol of HCl(aq) or NaOH(aq) reacted is

$$\Delta H^\circ_{rxn} = -(55.7\ \mathrm{kJ \cdot mol^{-1}})(0.100\ \mathrm{mol}) = -5.57\ \mathrm{kJ}$$

The value of $C_{P,\,calorimeter}$ can be computed from Equation (6-14):

$$C_{P,\,calorimeter} = \frac{-(-5570\ \mathrm{J})}{1.210\ \mathrm{K}} = 4.60 \times 10^3\ \mathrm{J \cdot K^{-1}}$$

This value of $C_{P,\,calorimeter}$ can now be used to determine other ΔH°_{rxn} values using this calorimeter and Equation (6-13).

Example 6-8: A 0.500-L sample of 0.200 M $NaCl(aq)$ is added to 0.500 L of 0.200 M $AgNO_3(aq)$ in a calorimeter with $C_P = 4.60 \times 10^3\ J \cdot K^{-1}$. The observed ΔT is +1.423 K. Compute the value of ΔH°_{rxn} for the reaction

Problems 6-43 to 6-46 are similar to Example 6-8.

$$AgNO_3(aq) + NaCl(aq) \rightarrow AgCl(s) + NaNO_3(aq)$$

Solution: The observed increase in temperature arises from the formation of the precipitate $AgCl(s)$. A total of 0.100 mol of $AgCl(s)$ is formed in the reaction:

$$\text{mol } AgCl(s) \text{ formed} = (0.500\ L)(0.200\ mol \cdot L^{-1}) = 0.100\ mol$$

The value of ΔH°_{rxn} for the formation of 0.100 mol of $AgCl(s)$ is

$$\Delta H^\circ_{rxn} = -C_{P,\,calorimeter}\,\Delta T$$
$$= -(4.60 \times 10^3\ J \cdot K^{-1})(1.423\ K) = -6550\ J$$

(for 0.1 mol)

For the formation of 1.00 mol of $AgCl(s)$, the value of ΔH°_{rxn} is 10 times the value for 0.100 mol; thus

$$\Delta H^\circ_{rxn} = -6550\ J \times 10 = -65500\ J = -65.5\ kJ$$

6-8. COMBUSTION REACTIONS CAN BE USED AS ENERGY SOURCES

An important measure of the quality of a fuel is the quantity of heat evolved per gram when the fuel is burned. The more energy evolved as heat per unit mass, the greater the utility of the fuel as an energy source. Other important criteria for fuels are cost, ease of transport, and utilization hazards.

The heat of combustion of a substance, which is a special case of the heat of reaction, can be determined in a *bomb calorimeter* like that shown in Figure 6-10. A known mass of the substance whose heat of combustion is to be determined is loaded into the bomb calorimeter along with an ignition wire. The calorimeter is then pressurized with excess oxygen gas at about 30 atm. The combustion reaction is then started by passing a short burst of high-voltage current through the ignition wire. The heat of combustion is determined by measuring the temperature increase of the calorimeter and of the water in which it is immersed. From the known heat capacity of the calorimeter assembly and the observed ΔT, the heat of combustion can be computed in a manner analogous to that described for the Dewar-flask calorimeter.

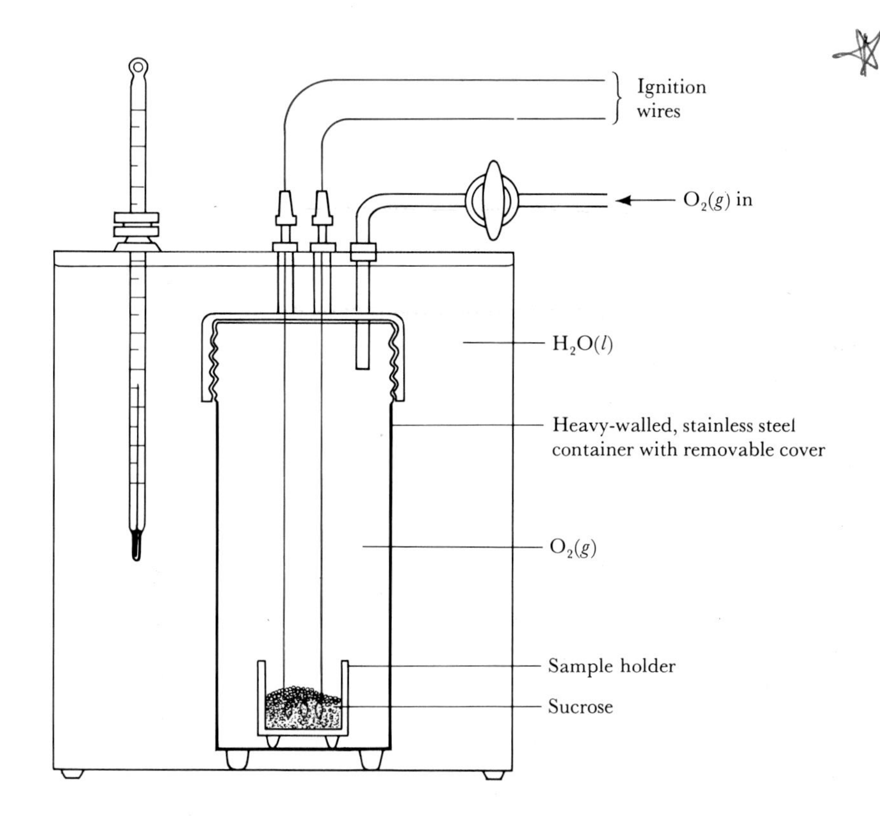

Figure 6-10 Combustion of sucrose in a bomb calorimeter. The sample is placed in the sample holder together with a coil of ignition wire. The cover is then screwed into place, and the vessel is pressurized with excess oxygen gas to about 30 atm. The calorimeter is then placed in an insulated water bath equipped with a thermometer. A short burst of current heats the ignition wire to red-hot and thereby initiates the combustion reaction.

A bomb calorimeter like that shown in Figure 6-10 has a fixed total volume, and thus the reaction takes place at constant volume. When the volume is constant, no work is done. Substitution of $w = 0$ into Equation (6-1) yields Equation (6-2):

$$\Delta U_{rxn} = q_V = \begin{array}{l}\text{heat evolved or absorbed} \\ \text{at constant volume}\end{array} \qquad (6\text{-}2)$$

Thus the energy evolved or absorbed as heat in a reaction run in a sealed, rigid container is equal to ΔU_{rxn}, the energy change for the reaction.

The difference between a constant-pressure process and a constant-volume process can be seen by considering a reaction in which the volume of the reaction mixture increases during the course of the reaction. In this case, the mixture has to expend some energy to push back the surrounding atmosphere as it expands. Consequently, the energy evolved or absorbed as heat in a constant-pressure reaction is not the same as that for the same reaction run at constant volume. The difference between ΔU_{rxn} and ΔH_{rxn} is just the energy that is required for the system to expand against a constant atmospheric pressure. It turns out that the numerical difference between ΔH_{rxn} and ΔU_{rxn} is usually small, and we shall assume that $\Delta H_{rxn} \approx \Delta U_{rxn}$.

Example 6-9: A 1.00-g sample of octane, $C_8H_{18}(l)$, is burned in a calorimeter like that shown in Figure 6-10, and the observed temperature increase is 1.681 K. The heat capacity of the calorimeter assembly is $C_{calorimeter} = 28.46\ \text{kJ}\cdot\text{K}^{-1}$. Compute the heat of combustion per gram and per mole of $C_8H_{18}(l)$. Assume that $\Delta H^\circ_{rxn} \approx \Delta U^\circ_{rxn}$.

Solution: The combustion reaction is

$$C_8H_{18}(l) + {}^{25}\!/_2 O_2(g) \rightarrow 8CO_2(g) + 9H_2O(g)$$

The heat of combustion is

$$\Delta H^\circ_{rxn} \approx \Delta U^\circ_{rxn} = -C_{calorimeter}\,\Delta T$$

Using this equation, we find that the value of ΔH°_{rxn} for 1.00 g of $C_8H_{18}(l)$ is

$$\Delta H^\circ_{rxn} \text{ per g } C_8H_{18}(l) = -(28.46\ \text{kJ}\cdot\text{K}^{-1})(1.681\ \text{K}\cdot\text{g}^{-1}) = -47.84\ \text{kJ}\cdot\text{g}^{-1}$$

because the observed ΔT is for the combustion of 1.00 g of octane. The value of ΔH°_{rxn} per mole of octane (molecular mass = 114.2) is

$$\begin{aligned}\Delta H^\circ_{rxn} \text{ per mol } C_8H_{18}(l) &= -(47.84\ \text{kJ}\cdot\text{g}^{-1})(114.2\ \text{g}\cdot\text{mol}^{-1}) \\ &= -5463\ \text{kJ}\cdot\text{mol}^{-1}\end{aligned}$$

The heats of combustion of a variety of fuels are given in Table 6-4. Combustible fuels are used extensively as thermal energy sources. Note that in every case in Table 6-4 the combustion reaction is highly exothermic. A large negative value of ΔH°_{rxn} for a combustion reaction is a prerequisite for an effective fuel.

Example 6-10: Using Table 6-1, calculate the value of the heat of combustion of acetylene given in Table 6-4.

Solution: The chemical equation for the combustion of acetylene is

$$C_2H_2(g) + {}^{5}\!/_2 O_2(g) \rightarrow 2CO_2(g) + H_2O(g)$$

The heat of reaction is

$$\begin{aligned}\Delta H^\circ_{rxn} = {} & 2\Delta \bar{H}^\circ_f[CO_2(g)] + \Delta \bar{H}^\circ_f[H_2O(g)] \\ & - \Delta \bar{H}^\circ_f[C_2H_2(g)] - \tfrac{5}{2}\Delta \bar{H}^\circ_f[O_2(g)]\end{aligned}$$

Using molar heat of formation values from Table 6-1, we have

Table 6-4 Heats of combustion of some fuels

Fuel	*Reaction*	$\Delta H^\circ_{rxn}/kJ\cdot mol^{-1}$	$\Delta H^\circ_{rxn}/kJ\cdot g^{-1}$
hydrogen	$H_2(g) + \frac{1}{2}O_2(g) \rightarrow H_2O(g)$	−242	−121
methane	$CH_4(g) + 2O_2(g) \rightarrow CO_2(g) + 2H_2O(g)$	−802	−50
carbon	$C(s) + O_2(g) \rightarrow CO_2(g)$	−394	−33
acetylene	$C_2H_2(g) + \frac{5}{2}O_2(g) \rightarrow 2CO_2(g) + H_2O(g)$	−1256	−48
methyl alcohol	$CH_3OH(g) + \frac{3}{2}O_2(g) \rightarrow CO_2(g) + 2H_2O(g)$	−676	−21
benzene	$C_6H_6(l) + \frac{15}{2}O_2(g) \rightarrow 6CO_2(g) + 3H_2O(g)$	−3135	−40
isooctane	$C_8H_{18}(l) + \frac{25}{2}O_2(g) \rightarrow 8CO_2(g) + 9H_2O(g)$	−5460	−48
refined heating oil[a]	—	—	−44
gasoline, kerosene, or diesel fuel[a]	—	—	−48
coal[a]	—	—	−28
dry, seasoned wood[a]	—	—	−25

[a]Complex mixtures for which it is not possible to give a chemical formula.

$$\begin{aligned}\Delta H^\circ_{rxn} &= (2\text{ mol})(-393.5\text{ kJ}\cdot\text{mol}^{-1}) + (1\text{ mol})(-241.8\text{ kJ}\cdot\text{mol}^{-1}) \\ &\quad - (1\text{ mol})(226.7\text{ kJ}\cdot\text{mol}^{-1}) - 0 \\ &= -1255.5\text{ kJ}\end{aligned}$$

This value is the heat of combustion of 1 mol of acetylene (Table 6-4). The molecular mass of acetylene is 26.0, and so the heat of combustion per gram is

$$\Delta H^\circ_{rxn} = (-1255.5\text{ kJ}\cdot\text{mol}^{-1})\left(\frac{1\text{ mol}}{26.0\text{ g}}\right) = -48.3\text{ kJ}\cdot\text{g}^{-1}$$

in agreement with the value in Table 6-4.

Methane, CH_4, propane, C_3H_8, and butane, C_4H_{10}, are also hydrocarbon fuels. Propane and butane are stored as liquids in tanks, and the vapor over the liquid has a sufficiently high pressure to flow out of the tank to where combustion is to take place. Disposable cigarette lighters are charged with liquid butane, which expands out the lighter valve and is ignited by a spark when the wheel is rotated over the flint:

$$\underset{\text{butane}}{C_4H_{10}(g)} + \tfrac{13}{2}O_2(g) \rightarrow 4CO_2(g) + 5H_2O(g)$$

$$\Delta H^\circ_{rxn} = -2658\text{ kJ}$$

Gasohol is a mixture of gasoline and ethyl alcohol. The most common blend is 90 percent conventional gasoline plus 10 percent ethyl alcohol. The alcohol is produced by fermenting sugars from plants, with yeasts as the fermentation agent; for example,

$$\underset{\text{glucose, a sugar}}{C_6H_{12}O_6(aq)} \xrightarrow{\text{yeast}} \underset{\text{ethyl alcohol}}{2C_2H_5OH(aq)} + 2CO_2(g)$$

One critical point to be made regarding the use of ethanol as a gasoline additive is that the heat of combustion per gram of ethyl alcohol is significantly less than for a hydrocarbon. An automobile running on pure C_2H_5OH would use about 70 percent more ethyl alcohol than gasoline per mile.

6-9. FOOD IS FUEL

The food we eat constitutes the fuel necessary to maintain our body temperature and accomplish other physiological functions and to provide the energy we need to move about. To maintain body weight, a normally active, healthy adult must take in about 130 kJ of food energy per kilogram of body weight per day. The common unit for the energy content of food is the *calorie*, a term you undoubtedly have heard of. The calorie used to be the unit of energy for scientific work, but it is slowly being replaced by the joule. Nevertheless, nutritionists, physicians, and the popular press still use the term calorie, and so it is necessary to be able to convert from one unit to another. There are 4.184 J in 1 cal, and the unit conversion factor is

$$\frac{4.184\ \text{J}}{1\ \text{cal}} = 1$$

In this section we express all energy values in kilojoules and kilocalories. Incidentally, the popular term calorie is actually a kilocalorie and is sometimes written Calorie. The total daily energy intakes required to maintain various body weights are

Body weight	*Required daily energy input*	
110 lb (50 kg)	7500 kJ	1800 kcal
175 lb (80 kg)	10,600 kJ	2500 kcal
250 lb (114 kg)	15,000 kJ	3600 kcal

About 100 kJ per kilogram of body weight per day is required to keep the body functioning at a minimal level. If we consume more food than we require for our normal activity level, then the excess that is not eliminated is stored in the body as fat. One kilogram of body fat contains about 39,000 kJ of stored energy. The per-gram chemical energy values for fats, proteins, and carbohydrates are

Food	*Energy value*	
fats	$39\ kJ\cdot g^{-1}$	$9.3\ kcal\cdot g^{-1}$
proteins	$17\ kJ\cdot g^{-1}$	$4.1\ kcal\cdot g^{-1}$
carbohydrates	$16\ kJ\cdot g^{-1}$	$3.8\ kcal\cdot g^{-1}$

Note that carbohydrates and proteins have less than one half the energy content of fats.

Exercise is great for improving muscle tone and thereby firming up sagging tissue, but exercise is not an effective way to lose weight. A 1-h brisk walk over average terrain consumes only about 700 kJ of stored energy, which corresponds to about one fourth of the energy content of a quarter-pound hamburger on a roll. The energy values of some common foods are given in Table 6-5. The message is that calories (or better, kilojoules) count.

Table 6-5 Approximate energy values of some common foods

Food	*kJ/100 g*	*kcal/4 oz*[b]
green vegetables[a]	115	31
beer	200	54
fruits[a]	250	67
milk	300	80
seafood (steamed)[a]	400	107
cottage cheese, low-fat yogurt	450	120
chicken (broiled, meat only)	600	160
steak (broiled, no fat)	1400	350
yogurt	1000	270
liquor (80 proof)	1000	270
ice cream (bulk)	1100	295
bread, cheese[a]	1200	320
hamburger, hot dogs, popcorn, sugar	1600	430
potato chips, nuts (roasted)	2400	640
butter, cream, margarine, mayonnaise	3000	800
fat	3900	1045

[a] Average values for the more common varieties; variations within categories are as much as 10 to 15 percent.
[b] 1 kcal = 4.184 kJ; 1 oz = 28 g.

6-10. ROCKETS AND EXPLOSIVES UTILIZE HIGHLY EXOTHERMIC REACTIONS WITH GASEOUS PRODUCTS

A study of the ΔH°_{rxn} data in Table 6-4 leads to the conclusion that the most energy-rich fuel on a mass basis is hydrogen, which has an energy content per gram of well over twice that of the next best fuel. This is the major reason hydrogen was used in the second stage of the Apollo spaceships that traveled to the moon. The main disadvantages of liquid hydrogen fuel are that it can be maintained as a liquid only at very low temperatures (about 20 K at 1 atm) and that hydrogen forms explosive mixtures with air. The first and third stages of the Apollo spaceships were powered by the reaction between kerosene and liquid oxygen ("LOX").

The lunar landers were powered to and from the lunar surface by the energy released from the reaction between the liquid fuel *N,N*-dimethylhydrazine, $H_2NN(CH_3)_2(l)$, and the liquid oxidizer dinitrogen tetroxide, $N_2O_4(l)$:

$$H_2NN(CH_3)_2(l) + 2N_2O_4(l) \rightarrow 3N_2(g) + 2CO_2(g) + 4H_2O(g)$$
$$\Delta H^\circ_{rxn} = -29\ \mathrm{kJ/g\ fuel}$$

This reaction is especially suitable for a lunar escape vehicle because it starts spontaneously on mixing. No battery or spark plugs and associated electrical circuitry are required.

The powerful and dangerous explosive nitroglycerin is made by the reaction of a mixture of nitric acid with glycerin:

$$\begin{array}{c} \mathrm{H} \\ | \\ \mathrm{H{-}C{-}OH} \\ | \\ \mathrm{H{-}C{-}OH} \\ | \\ \mathrm{H{-}C{-}OH} \\ | \\ \mathrm{H} \end{array} + 3HNO_3 \rightarrow \begin{array}{c} \mathrm{H} \\ | \\ \mathrm{H{-}C{-}O{-}NO_2} \\ | \\ \mathrm{H{-}C{-}O{-}NO_2} \\ | \\ \mathrm{H{-}C{-}O{-}NO_2} \\ | \\ \mathrm{H} \end{array} + 3H_2O$$

glycerin nitroglycerin

Nitroglycerin can detonate on even the slightest disturbance. In 1867, the Swedish chemist Alfred Nobel tamed nitroglycerin by adding the solid absorbent material diatomaceous earth (a soft, bulky material composed of the skeletal remains of certain algae) to produce *dynamite*. The Nobel prizes are funded by the interest that accrues on a fund established by Nobel using part of the money he made on the licensing of his numerous patents on explosives.

SUMMARY

Chemical reactions involve transfers of energy between the system of interest and its surroundings. Commonly, the energy is either absorbed or released as heat. Reactions that give off energy as heat are called exothermic reactions, and reactions that take in energy as heat are called endothermic reactions. The total energy involved is always conserved; that is, energy is neither created nor destroyed but is simply transferred from one system to another. This principle of energy conservation is known as the first law of thermodynamics, and the study of energy changes is called thermodynamics.

For a constant-pressure process, the heat evolved or absorbed, q_P, is equal to the enthalpy change; thus $\Delta H_{rxn} = q_P$ for a reaction run at constant pressure. For a reaction run at constant volume, the heat evolved or absorbed, q_V, is equal to the change in energy for the reaction; thus $\Delta U_{rxn} = q_V$.

The standard molar enthalpy of formation of a compound from its elements, $\Delta \bar{H}_f^\circ$, is determined by assigning a value of zero to the standard enthalpies of formation of the normal state of each element at 25°C and 1 atm. A table of $\Delta \bar{H}_f^\circ$ values can be used to compute ΔH_{rxn}° values. The value of ΔH_{rxn}° for a particular reaction is determined primarily by the difference in the bond energies of the reactant and the product molecules. The breaking of chemical bonds in the reactant molecules consumes energy, and the formation of chemical bonds in the product molecules releases energy. The difference between these values is ΔH_{rxn}°.

The heat capacity of a substance is a measure of its capacity to take up energy as heat. The higher the heat capacity of a substance, the smaller the resulting temperature increase for a given amount of heat energy added. A reactant used to provide heat energy is called a fuel. Combustion reactions are those in which a fuel reacts with oxygen. The amount of energy absorbed or evolved as heat by a chemical reaction can be measured in a calorimeter.

TERMS YOU SHOULD KNOW

thermodynamics 203
first law of thermodynamics 203
conservation of energy 203
energy change, ΔU_{rxn} 203
heat, q 204
work, w 204
enthalpy 204
heat of reaction 204
thermochemistry 204
fuel 205
combustion 205
exothermic reaction 205
endothermic reaction 205
enthalpy change, ΔH_{rxn} 205
standard enthalpy change, ΔH_{rxn}° 206
water-gas reaction 206
standard molar enthalpy of formation, $\Delta \bar{H}_f^\circ$ 207
Hess's law 212
isomer 214
molar bond enthalpy, $\bar{H}$(bond) 216
bond energy 217
heat capacity, C_P 219
molar heat capacity, $\bar{C}_P$ 219
specific heat, c_{sp} 219
calorimeter 221
Dewar flask 222
bomb calorimeter 224
calorie 228
dynamite 230

EQUATIONS YOU SHOULD KNOW HOW TO USE

$$\Delta U_{rxn} = q_V \quad (6\text{-}2)$$

$$\Delta H_{rxn} = q_P \quad (6\text{-}4)$$

For the reaction aA + bB → yY + zZ,

$$\Delta H^\circ_{rxn} = y\Delta \bar{H}^\circ_f[\mathrm{Y}] + z\Delta \bar{H}^\circ_f[\mathrm{Z}] - a\Delta \bar{H}^\circ_f[\mathrm{A}] - b\Delta \bar{H}^\circ_f[\mathrm{B}] \quad (6\text{-}6)$$

$$\Delta H^\circ_{rxn} \approx \bar{H}(\mathrm{bond})_R - \bar{H}(\mathrm{bond})_P \quad (6\text{-}9)$$

$$C_P = \frac{\Delta H}{\Delta T} = \frac{q_P}{\Delta T} \quad (6\text{-}10)$$

$$\Delta H_{rxn} = -C_{P,calorimeter}\Delta T \quad (6\text{-}13)$$

PROBLEMS

HEAT AND ENERGY

6-1. When 10.0 g of methane burns in oxygen, 501 kJ of heat is evolved. Calculate the amount of heat (in kilojoules) evolved when 1 mol of methane burns.

6-2. When 2.46 g of barium reacts with chlorine, 15.4 kJ is evolved. Calculate the heat evolved (in kilojoules) when 1.00 mol of barium chloride is formed from barium and chlorine.

6-3. When 0.320 g of carbon reacts with sulfur to give carbon disulfide, 2.38 kJ is absorbed. Calculate the heat absorbed (in kilojoules) when 1 mol of carbon disulfide is formed from carbon and sulfur.

6-4. When 0.165 g of magnesium is burned in oxygen, 4.08 kJ is evolved. Calculate the heat evolved (in kilojoules) when 1.00 mol of magnesium oxide is formed from magnesium and oxygen.

HEATS OF FORMATION

6-5. Use the $\Delta \bar{H}^\circ_f$ data in Table 6-1 to compute ΔH°_{rxn} for the following reactions:

(a) $C_2H_4(g) + H_2O(l) \rightarrow C_2H_5OH(l)$
(b) $C_2H_4(g) + 3O_2(g) \rightarrow 2CO_2(g) + 2H_2O(g)$
(c) $C_2H_4(g) + H_2(g) \rightarrow C_2H_6(g)$

In each case state whether the reaction is endothermic or exothermic.

6-6. Use the $\Delta \bar{H}^\circ_f$ data in Table 6-1 to compute ΔH°_{rxn} for the following reactions:

(a) $2H_2O_2(l) \rightarrow 2H_2O(l) + O_2(g)$
(b) $MgO(s) + CO_2(g) \rightarrow MgCO_3(s)$
(c) $4NH_3(g) + 5O_2(g) \rightarrow 4NO(g) + 6H_2O(g)$

In each case state whether the reaction is endothermic or exothermic.

6-7. Use the $\Delta \bar{H}^\circ_f$ data in Table 6-1 to compute ΔH°_{rxn} for the following combustion reactions:

(a) $C_2H_5OH(l) + 3O_2(g) \rightarrow 2CO_2(g) + 3H_2O(g)$
(b) $C_2H_6(g) + \frac{7}{2}O_2(g) \rightarrow 2CO_2(g) + 3H_2O(g)$

Compare the heat of combustion per gram of the fuels $C_2H_5OH(l)$ and $C_2H_6(g)$.

6-8. Use the $\Delta \bar{H}^\circ_f$ data in Table 6-1 to compute ΔH°_{rxn} for the following combustion reactions:

(a) $CH_3OH(l) + \frac{3}{2}O_2(g) \rightarrow CO_2(g) + 2H_2O(g)$
(b) $N_2H_4(l) + O_2(g) \rightarrow N_2(g) + 2H_2O(g)$

Compare the heat of combustion per gram of the fuels $CH_3OH(l)$ and $N_2H_4(l)$.

6-9. Given that $\Delta H^\circ_{rxn} = -2815.8$ kJ for the combustion of 1.00 mol of glucose:

$$C_6H_{12}O_6(s) + 6O_2(g) \rightarrow 6CO_2(g) + 6H_2O(l)$$

use the $\Delta \bar{H}^\circ_f$ data in Table 6-1 together with the given ΔH°_{rxn} value to compute the value of $\Delta \bar{H}^\circ_f$ for glucose.

6-10. Use the fact that $\Delta H^\circ_{rxn} = -5646.7$ kJ for the combustion of 1.00 mol of sucrose,

$$C_{12}H_{22}O_{11}(s) + 12O_2(g) \rightarrow 12CO_2(g) + 11H_2O(l)$$

plus the $\Delta \bar{H}^\circ_f$ data in Table 6-1 to compute $\Delta \bar{H}^\circ_f$ for sucrose.

6-11. Given that $\Delta \bar{H}_f^\circ = 142\ \mathrm{kJ \cdot mol^{-1}}$ for $O_3(g)$ and $\Delta \bar{H}_f^\circ = 247.5\ \mathrm{kJ \cdot mol^{-1}}$ for $O(g)$, calculate ΔH°_{rxn} for the reaction

$$O_2(g) + O(g) \rightarrow O_3(g)$$

This is one of the reactions that produce ozone in the atmosphere.

$\Delta H^\circ_{rxn} = \Delta \bar{H}_f^\circ[O_3] - \Delta \bar{H}_f^\circ[O_2] - \Delta \bar{H}_f^\circ[O]$
$= (1\ \mathrm{mol})(142\ \mathrm{kJ/mol}) - (1\ \mathrm{mol})(247.5\ \mathrm{kJ/mol})$
$= -105.5\ \mathrm{kJ}$

6-12. The $\Delta \bar{H}_f^\circ$ values for $Cu_2O(s)$ and $CuO(s)$ are $-169.0\ \mathrm{kJ \cdot mol^{-1}}$ and $-157.3\ \mathrm{kJ \cdot mol^{-1}}$, respectively. Compute ΔH°_{rxn} for the reaction

$$CuO(s) + Cu(s) \rightarrow Cu_2O(s)$$

$\Delta H_{rxn} = -11.7\ \mathrm{kJ}$

6-13. Calculate $\Delta \bar{H}_f^\circ$ for the atomic species for each of the following:

$945.2\ \mathrm{kJ} = (2\ \mathrm{mol})\Delta \bar{H}_f^\circ[N] - 0$

(a) $N_2(g) \rightarrow 2N(g)$ $\quad \Delta H^\circ_{rxn} = +945.2\ \mathrm{kJ}$ $\quad$ 472.6 kJ/mol
(b) $F_2(g) \rightarrow 2F(g)$ $\quad \Delta H^\circ_{rxn} = +158.0\ \mathrm{kJ}$ $\quad$ 79.0 kJ/mol
(c) $H_2(g) \rightarrow 2H(g)$ $\quad \Delta H^\circ_{rxn} = +436.0\ \mathrm{kJ}$ $\quad$ 218.0 kJ/mol
(d) $Cl_2(g) \rightarrow 2Cl(g)$ $\quad \Delta H^\circ_{rxn} = +243.4\ \mathrm{kJ}$ $\quad$ 121.7 kJ/mol

Which of these diatomic molecules has the greatest bond strength?

N_2

6-14. Using Table 6-1, calculate ΔH°_{rxn} for

(a) $H_2(g) + F_2(g) \rightarrow 2HF(g)$ $\quad \Delta H^\circ_{rxn} = -542.2\ \mathrm{kJ}$ exo
(b) $2CO(g) + O_2(g) \rightarrow 2CO_2(g)$ $\quad -566.0\ \mathrm{kJ}$ exo
(c) $3H_2(g) + N_2(g) \rightarrow 2NH_3(g)$ $\quad -92.38\ \mathrm{kJ}$ exo
(d) $2NO(g) + O_2(g) \rightarrow 2NO_2(g)$ $\quad -113.04\ \mathrm{kJ}$ exo

State whether each reaction is endothermic or exothermic.

6-15. Using Table 6-1, calculate the heat required to vaporize 1.00 mol of iodine at 25°C.

$I_2(s) \rightarrow I_2(g)$ $\quad \Delta H^\circ_{rxn} = \Delta \bar{H}_f^\circ[I_2(g)] - \Delta \bar{H}_f^\circ[I_2(s)]$
$= (1\ \mathrm{mol})(62.4\ \mathrm{kJ/mol}) - 0$
$= 62.4\ \mathrm{kJ}$

6-16. Using Table 6-1, calculate the heat required to vaporize 1.00 mol of water at 25°C.

HESS'S LAW

6-17. The ΔH°_{rxn} value for the reaction

$$ZnS(s) + \tfrac{1}{2}O_2(g) \rightarrow ZnO(s) + S(s)$$

is $-145.4\ \mathrm{kJ}$. Compute the value of ΔH°_{rxn} for the reaction

$$ZnO(s) + S(s) \rightarrow ZnS(s) + \tfrac{1}{2}O_2(g)$$

$\Delta H^\circ_{rxn}(2) = -\Delta H^\circ_{rxn}(1)$ $\quad$ $+145.4\ \mathrm{kJ}$

6-18. The ΔH°_{rxn} value for the reaction

$$CaO(s) + H_2O(l) \rightarrow Ca(OH)_2(s)$$

is $-56.27\ \mathrm{kJ}$. Compute the amount of heat (in kilojoules) required to convert 1.00 g of $Ca(OH)_2(s)$ to $CaO(s)$.

6-19. Given that

$$2P(s) + 3Cl_2(g) \rightarrow 2PCl_3(g) \qquad \Delta H^\circ_{rxn} = -613\ \mathrm{kJ}$$
$$2P(s) + 5Cl_2(g) \rightarrow 2PCl_5(g) \qquad \Delta H^\circ_{rxn} = -790\ \mathrm{kJ}$$

calculate ΔH°_{rxn} for the reaction

$$PCl_3(g) + Cl_2(g) \rightarrow PCl_5(g)$$

$2PCl_3 \rightarrow 2P + 3Cl_2$ $\quad +613\ \mathrm{kJ}$
$2P + 5Cl_2 \rightarrow 2PCl_5$ $\quad -790\ \mathrm{kJ}$
$2PCl_3 + 2Cl_2 \rightarrow 2PCl_5$
$\Delta H^\circ_{rxn} = 613\ \mathrm{kJ} + (-790\ \mathrm{kJ})$
$\Delta H^\circ_{rxn} = -177\ \mathrm{kJ}$
$PCl_3 + Cl_2 \rightarrow PCl_5$ $\quad \Delta H_{rxn} = \frac{-177\ \mathrm{kJ}}{2} = -88.5\ \mathrm{kJ}$

6-20. Use the values of ΔH°_{rxn} given for these reactions,

$$Cu(s) + Cl_2(g) \rightarrow CuCl_2(s) \qquad \Delta H^\circ_{rxn} = -206\ \mathrm{kJ}$$
$$2Cu(s) + Cl_2(g) \rightarrow 2CuCl(s) \qquad \Delta H^\circ_{rxn} = -136\ \mathrm{kJ}$$

to calculate ΔH°_{rxn} for the reaction

$$CuCl_2(s) + Cu(s) \rightarrow 2CuCl(s)$$

6-21. The ΔH°_{rxn} values for the following reactions are

$$2Fe(s) + \tfrac{3}{2}O_2(g) \rightarrow Fe_2O_3(s) \qquad \Delta H^\circ_{rxn} = -823.41\ \mathrm{kJ}$$
$$3Fe(s) + 2O_2(g) \rightarrow Fe_3O_4(s) \qquad \Delta H^\circ_{rxn} = -1120.48\ \mathrm{kJ}$$

Use the above data to compute ΔH°_{rxn} for the reaction

$$3Fe_2O_3(s) \rightarrow 2Fe_3O_4(s) + \tfrac{1}{2}O_2(g)$$

$3Fe_2O_3 \rightarrow 6Fe + \tfrac{9}{2}O_2$ $\quad \Delta H^\circ_{rxn} = +2470.23\ \mathrm{kJ}$
$6Fe + 4O_2 \rightarrow 2Fe_3O_4$ $\quad \Delta H^\circ_{rxn} = -2240.96\ \mathrm{kJ}$
$3Fe_2O_3 \rightarrow 2Fe_3O_4 + \tfrac{1}{2}O_2$
$\Delta H^\circ_{rxn} = 2470.23\ \mathrm{kJ} - 2240.96\ \mathrm{kJ}$
$\Delta H_{rxn} = +229.27\ \mathrm{kJ}$

6-22. Given that

$$H_2(g) + F_2(g) \rightarrow 2HF(g) \qquad \Delta H^\circ_{rxn} = -542.2\ \mathrm{kJ}$$
$$2H_2(g) + O_2(g) \rightarrow 2H_2O(l) \qquad \Delta H^\circ_{rxn} = +571.6\ \mathrm{kJ}$$

$2H_2O \rightarrow 2H_2 + O_2$ $\quad$ -1084.4

calculate ΔH°_{rxn} for

$$2F_2(g) + 2H_2O(l) \rightarrow 4HF(g) + O_2(g)$$

$\Delta H^\circ_{rxn} = -512.8\ \mathrm{kJ}$

6-23. The standard molar heats of combustion at 25°C for sucrose, glucose, and fructose are

Compound	$\Delta\bar{H}^{\circ}_{rxn}/kJ\cdot mol^{-1}$
$C_{12}H_{22}O_{11}$, sucrose	−5646.7
$C_6H_{12}O_6$, glucose	−2815.8
$C_6H_{12}O_6$, fructose	−2826.7

Use Hess's law to compute ΔH°_{rxn} at 25°C for the reaction

$$C_{12}H_{22}O_{11}(s) + H_2O(l) \rightarrow C_6H_{12}O_6(s) + C_6H_{12}O_6(s)$$

sucrose → glucose + fructose

$C_{12}H_{22}O_{11} + 12O_2 \rightarrow 12CO_2 + 11H_2O \quad \Delta H^{\circ}_{rxn} = -5646.7\ kJ$ (sucrose)

$C_6H_{12}O_6 + 6O_2 \rightarrow 6CO_2 + 6H_2O \quad \Delta H^{\circ}_{rxn} = -2815.8\ kJ$ (glucose)

$C_6H_{12}O_6 + 6O_2 \rightarrow 6CO_2 + 6H_2O \quad \Delta H^{\circ}_{rxn} = -2826.7\ kJ$ (fructose)

$C_{12}H_{22}O_{11} + 12O_2 \rightarrow 12CO_2 + 11H_2O \quad -5646.7$

$6CO_2 + 6H_2O \rightarrow C_6H_{12}O_6 + 6O_2 \quad +2815.8$

$6CO_2 + 6H_2O \rightarrow C_6H_{12}O_6 + 6O_2 \quad +2826.7$

$C_{12}H_{22}O_{11} + H_2O \rightarrow C_6H_{12}O_6 + C_6H_{12}O_6$

$\Delta H^{\circ}_{rxn} = -5646.7 + 2815.8 + 2826.7 = -4.2\ kJ$

6-24. The standard molar heats of combustion of the isomers *m*-xylene and *p*-xylene, both $(CH_3)_2C_6H_4$, are $-4553.9\ kJ\cdot mol^{-1}$ and $-4556.8\ kJ\cdot mol^{-1}$, respectively. Use these data, together with Hess's law, to compute ΔH°_{rxn} for the reaction

$$m\text{-xylene} \rightarrow p\text{-xylene}$$

6-25. Given that

$$4NH_3(g) + 5O_2(g) \rightarrow 4NO(g) + 6H_2O(l) \qquad \Delta H^{\circ}_{rxn} = -1170\ kJ$$

$$4NH_3(g) + 3O_2(g) \rightarrow 2N_2(g) + 6H_2O(l) \qquad \Delta H^{\circ}_{rxn} = -1530\ kJ$$

calculate $\Delta\bar{H}^{\circ}_f$ for NO(*g*).

$4NH_3 + 5O_2 \rightarrow 4NO + 6H_2O \quad -1170\ kJ$

$2N_2 + 6H_2O \rightarrow 4NH_3 + 3O_2 \quad +1530\ kJ$

$2N_2 + 2O_2 \rightarrow 4NO \quad 360\ kJ$

$\frac{1}{2}N_2 + \frac{1}{2}O_2 \rightarrow NO \quad 90\ kJ$

$\Delta H^{\circ}_{rxn} = 90\ kJ$

6-26. Given that

$$Xe(g) + F_2(g) \rightarrow XeF_2(s) \qquad \Delta H^{\circ}_{rxn} = -164\ kJ$$

$$Xe(g) + 2F_2(g) \rightarrow XeF_4(s) \qquad \Delta H^{\circ}_{rxn} = -262\ kJ$$

calculate ΔH°_{rxn} for the reaction

$$XeF_2(s) + F_2(g) \rightarrow XeF_4(s)$$

BOND ENTHALPIES

6-27. The enthalpy change for the reaction

$$ClF_3(g) \rightarrow Cl(g) + 3F(g)$$

is 514 kJ. Calculate the average Cl—F bond energy in ClF_3.

The bonding in ClF_3 is

F—Cl—F
   |
   F

$\Delta H^{\circ}_{rxn} = 3\bar{H}(Cl-F)$

$514\ kJ = (3\ mol)\bar{H}(Cl-F)$

$\bar{H}(Cl-F) = \frac{514\ kJ}{3\ mol} = 171\ kJ/mol$

6-28. The enthalpy change for the reaction

$$OF_2(g) \rightarrow O(g) + 2F(g)$$

is 368 kJ. Calculate the average O—F bond energy of OF_2.

The bonding in OF_2 is

F—O—F

$\Delta H_{rxn} = 2\bar{H}^{\circ}[OF]$

$368\ kJ = 2\bar{H}^{\circ}(OF)$

$\bar{H}^{\circ} = 184\ kJ$

6-29. Use the bond enthalpy data in Table 6-2 to estimate the ΔH°_{rxn} for the reaction

$$CH_4(g) + 4Cl_2(g) \rightarrow CCl_4(g) + 4HCl(g)$$

The bonding in CH_4 and CCl_4 is

```
    H            Cl
    |            |
 H—C—H       Cl—C—Cl
    |            |
    H            Cl
```

$\Delta H^{\circ}_{rxn} \approx \bar{H}(\text{bond})_R - \bar{H}(\text{bond})_P$

$\approx 4\bar{H}(C-H) + 4\bar{H}(Cl-Cl) - 4\bar{H}(C-Cl) - 4\bar{H}(H-Cl)$

$\approx (4\ mol)(414\ kJ/mol) + (4\ mol)(243\ kJ/mol) - (4\ mol)(331\ kJ/mol) - (4\ mol)(431\ kJ/mol)$

$\approx -420\ kJ$

6-30. Use the bond enthalpy data in Table 6-2 to estimate the ΔH°_{rxn} for the reaction

$$CH_3Cl(g) + F_2(g) \rightarrow CH_2FCl(g) + HF(g)$$

The bonding in CH_3Cl and CH_2FCl is

```
    Cl           Cl
    |            |
 H—C—H        H—C—H
    |            |
    H            F
```

6-31. The formation of water from oxygen and hydrogen involves the reaction

$$2H_2(g) + O_2(g) \rightarrow 2H_2O(g)$$

Use the bond energies given in Table 6-2 and the $\Delta\bar{H}_f^\circ$ value for $H_2O(g)$ given in Table 6-1 to compute the oxygen-oxygen bond energy in $O_2(g)$.

The bonding in H_2O is

H—O—H

6-32. The formation of ammonia from hydrogen and nitrogen involves the reaction

$$N_2(g) + 3H_2(g) \rightarrow 2NH_3(g)$$

Use the bond energies given in Table 6-2 and the $\Delta\bar{H}_f^\circ$ value for $NH_3(g)$ given in Table 6-1 to compute the bond energy in $N_2(g)$.

The bonding in NH_3 is

```
H—N—H
  |
  H
```

6-33. Given that $\Delta\bar{H}_f^\circ[H(g)] = 218\ kJ\cdot mol^{-1}$, $\Delta\bar{H}_f^\circ[C(g)] = 709\ kJ\cdot mol^{-1}$, and $\Delta\bar{H}_f^\circ[CH_4(g)] = -74.86\ kJ\cdot mol^{-1}$, calculate the average C—H bond energy in CH_4.

The bonding in CH_4 is

```
   H
   |
H—C—H
   |
   H
```

6-34. Given that $\Delta\bar{H}_f^\circ[Cl(g)] = 128\ kJ\cdot mol^{-1}$, $\Delta\bar{H}_f^\circ[C(g)] = 709\ kJ\cdot mol^{-1}$, and $\Delta\bar{H}_f^\circ[CCl_4(g)] = -103\ kJ\cdot mol^{-1}$, calculate the average C—Cl bond energy in CCl_4.

The bonding in CCl_4 is

```
    Cl
    |
Cl—C—Cl
    |
    Cl
```

HEAT CAPACITY

6-35. When 2210 J of heat is added to 73.0 g of ethyl alcohol, C_2H_5OH, the temperature increases by 12.3 C°. Compute the total heat capacity of the 73.0 g, the molar heat capacity, and the specific heat of ethyl alcohol.

6-36. When 285 J of heat is added to 33.6 g of hexane, C_6H_{14}, a component of gasoline, the temperature rises from 25.00°C to 28.74°C. Calculate the specific heat and molar heat capacity of hexane.

6-37. A 50-kg sample of liquid water is used to cool an engine. Calculate the heat removed (in joules) from the engine when the temperature of the water is raised from 25.0°C to 49.7°C. Take $\bar{C}_P = 75.2\ J\cdot K^{-1}\cdot mol^{-1}$ for $H_2O(l)$.

6-38. Liquid sodium is being considered as an engine coolant. How many grams of sodium are needed to absorb 1.00 MJ of heat if the temperature of the sodium is not to increase by more than 10 C°? Take $\bar{C}_P = 30.8\ J\cdot K^{-1}\cdot mol^{-1}$ for Na(*l*).

6-39. A 25.0-g sample of copper at 25.0°C is placed in 100.0 g of water at 40.0°C. The copper and water quickly come to the same temperature by the process of heat transfer from water to copper. Calculate the final temperature of the water in °C. The specific heat of copper is $0.385\ J\cdot K^{-1}\cdot g^{-1}$.

6-40. If a 50.0-g piece of copper is heated to 100°C and then put into a vessel containing 250 mL of water at 0°C, what will be the final temperature (in °C) of the water? The specific heat of copper is $0.385\ J\cdot K^{-1}\cdot g^{-1}$.

6-41. A 1.00-kg block of aluminum ($\bar{C}_P = 24.2\ J\cdot K^{-1}\cdot mol^{-1}$) at 600°C is placed in contact with a 1.00-kg block of copper ($\bar{C}_P = 24.5\ J\cdot K^{-1}\cdot mol^{-1}$) at 100°C. What will be the final temperature (in °C) of the two blocks? Assume that no heat is lost to the surroundings.

6-42. A 50.0-g sample of aluminum at 25.0°C is placed in 99.9 g of water at 55.0°C. Heat is transferred from the water to the aluminum. The final temperature of the water and aluminum is 52.1°C. Calculate the specific heat of aluminum in $J\cdot K^{-1}\cdot g^{-1}$.

CALORIMETRY

6-43. A 0.0500-L sample of 0.200 M aqueous hydrochloric acid is added to 0.0500 L of 0.200 M aqueous ammonia in a calorimeter whose heat capacity is 480 $J \cdot K^{-1}$. The temperature increase is 1.17 K. Calculate ΔH°_{rxn} (in kJ) for the reaction

$$HCl(aq) + NH_3(aq) \rightarrow NH_4Cl(aq)$$

which occurs when the two solutions are mixed.

6-44. A 0.0500-L sample of 0.500 M barium nitrate is added to 0.0500 L of 0.500 M magnesium sulfate in a calorimeter with a heat capacity of 455 $J \cdot K^{-1}$. The observed increase in temperature is 1.43 K. Calculate ΔH°_{rxn} (in kJ) for the reaction

$$Ba(NO_3)_2(aq) + MgSO_4(aq) \rightarrow BaSO_4(s) + Mg(NO_3)_2(aq)$$

which occurs when the two solutions are mixed.

6-45. Under the right conditions, such as high temperature, ammonium nitrate is an explosive, decomposing according to the reaction

$$2NH_4NO_3(s) \rightarrow 2N_2(g) + 4H_2O(g) + O_2(g)$$

A 2.00-g sample of NH_4NO_3 is detonated in a calorimeter whose heat capacity is 4.92 $kJ \cdot K^{-1}$. The temperature increase is 3.06 K. Calculate the heat of reaction for the decomposition of 1.00 mol of ammonium nitrate.

6-46. Calcium hydroxide is prepared by adding calcium oxide (lime) to water. It is important to know how much heat is evolved in order to provide for adequate cooling. A 10.0-g sample of calcium oxide is added to 1.00 L of water in a calorimeter whose heat capacity is 4.37 $kJ \cdot K^{-1}$. The observed increase in temperature is 2.70 K. Calculate ΔH°_{rxn} for the formation of 1.00 mol of $Ca(OH)_2$. The reaction is

$$CaO(s) + H_2O(l) \rightarrow Ca(OH)_2(s)$$

6-47. A 5.00-g sample of potassium chloride is dissolved in 1.00 L of water in a calorimeter whose heat capacity is 4.51 $kJ \cdot K^{-1}$. The temperature decreases 0.256 K. Calculate the molar heat of solution of potassium chloride.

6-48. A 5.00-g sample of nitric acid is dissolved in 1.00 L of water in a calorimeter whose heat capacity is 5.16 $kJ \cdot K^{-1}$. The temperature increases 0.511 K. Calculate the molar heat of solution of nitric acid in water.

COMBUSTION OF FUELS

6-49. Propane is often used as a home fuel in areas where natural gas is not available. When 3.00 g of propane, C_3H_8, is burned in a calorimeter whose heat capacity is 32.7 $kJ \cdot K^{-1}$, the temperature of the calorimeter increases by 4.25 K. Calculate the heat of combustion in kJ per gram and kJ per mole of propane.

6-50. Fructose, a sugar found in fruits, is a source of energy for the body. The combustion of fructose takes place according to the reaction

$$C_6H_{12}O_6(s) + 6O_2(g) \rightarrow 6CO_2(g) + 6H_2O(l)$$

When 5.00 g of fructose is burned in a calorimeter with a heat capacity of 29.7 $kJ \cdot K^{-1}$, the temperature of the calorimeter increases by 2.635 K. Calculate the heat of combustion per gram and per mole of fructose. How much heat (in kJ) is released when 1.00 g of fructose is converted to $CO_2(s)$ and $H_2O(l)$ in the body?

6-51. When 1.00 g of oxalic acid, $H_2C_2O_4$, is burned in a calorimeter whose heat capacity is 8.75 $kJ \cdot K^{-1}$, the temperature increases 0.312 K. Calculate the heat of combustion in kilojoules per mole of oxalic acid. Using the $\Delta \overline{H}^\circ_f$ values for $CO_2(g)$ and $H_2O(l)$ given in Table 6-1, calculate the heat of formation of oxalic acid.

6-52. When 2.62 g of lactic acid, $C_3H_6O_3$, is burned in a calorimeter whose heat capacity is 21.7 $kJ \cdot K^{-1}$, the temperature of the calorimeter increases 1.800 K. Calculate the heat of combustion (in kJ) per mole of lactic acid. Using the $\Delta \overline{H}^\circ_f$ values for $CO_2(g)$ and $H_2O(l)$ given in Table 6-1, calculate $\Delta \overline{H}^\circ_f$ for lactic acid. Lactic acid is produced in muscle when there is a shortage of oxygen, such as during vigorous exercise. A buildup of lactic acid is responsible for muscle cramps.

MISCELLANEOUS PROBLEMS

6-53. Consider a carefully insulated 40-gal water heater that is heated by combustion of natural gas, $CH_4(g)$. Compute the number of moles of $CH_4(g)$ that must be burned to raise the temperature of 40 gal of water from 46°F to 135°F. Assume that 20 percent of the evolved heat is lost to the surroundings, take the specific heat of water as $4.2\ J\cdot K^{-1}\cdot g^{-1}$, and use the data in Table 6-4. See Appendix B for the Fahrenheit temperature scale to the Celsius temperature scale.

6-54. Bicycle riding at 13 mph (a moderate pace) consumes 2800 kJ/h for a 150-lb person. How many miles must this person ride in order to lose 1 lb of body fat?

6-55. One proposal for an effortless method of losing weight is to drink large amounts of cold water. Water has no food value. The body must provide heat in order to bring the temperature of the water to body temperature, 37°C. This heat is provided by the burning of stored carbohydrates or fat. How much heat must the body provide to warm 1.0 L of water at 0°C to 37°C? How many kilograms of body fat must be burned to provide this heat?

6-56. Glucose is used as fuel in the body according to the reaction

$$C_6H_{12}O_6(aq) + 6O_2(g) \rightarrow 6CO_2(g) + 6H_2O(l) \qquad \Delta H^\circ_{rxn} = -2820\ kJ\cdot mol^{-1}$$

How many grams of glucose must be burned to raise the temperature of the body from 34°C to 37°C for an 82-kg person? Assume that all the heat of combustion is used to heat the body. Assume also that the specific heat of the body is that of water.

6-57. The French chemists Pierre L. Dulong and Alexis T. Petit noted in 1819 that the molar heat capacity of many solids at ordinary temperatures is proportional to the number of atoms per formula unit of the solid. They quantified their observations in what is known as Dulong and Petit's rule, which says that the molar heat capacity of a solid can be expressed as

$$\bar{C}_p \simeq N \times 25\ J\cdot K^{-1}\cdot mol^{-1}$$

where N is the number of atoms per formula unit. The observed specific heat of a compound containing thallium and chlorine is $0.208\ J\cdot K^{-1}\cdot g^{-1}$. Use Dulong and Petit's rule to determine the formula of the compound.

6-58. The specific heat of an oxide of rubidium is $0.64\ J\cdot K^{-1}\cdot g^{-1}$. Use Dulong and Petit's rule to determine the formula of the compound.

6-59. The mineral, stilleite, contains zinc and selenium. The observed specific heat is $0.348\ J\cdot K^{-1}\cdot g^{-1}$. Use Dulong and Petit's rule to determine the formula of stilleite.

6-60. The mineral, matlockite, contains lead, fluorine, and chlorine. The observed specific heat of matlockite is $0.290\ J\cdot K^{-1}\cdot g^{-1}$. Use Dulong and Petit's rule to determine the formula of matlockite.

INTERCHAPTER D

Sulfur

Many of the hot springs and geysers in the western United States have yellow deposits of sulfur around them.

Sulfur (atomic number 16, atomic mass 32.06) is a yellow, tasteless, odorless solid that is often found in nature as the free element. Sulfur is essentially insoluble in water but dissolves readily in carbon disulfide, CS_2. It does not react with dilute acids or bases, but it does react with many metals at elevated temperatures to form metal sulfides.

Sulfur, which constitutes only 0.05 percent of the earth's crust, is not one of the most prevalent elements. Yet it is one of the most commercially important ones because it is the starting material for the most important industrial chemical, sulfuric acid.

Figure D-1 Io, one of the moons of Jupiter, appears yellow because of large deposits of sulfur from volcanic activity.

D-1. SULFUR IS RECOVERED FROM LARGE UNDERGROUND DEPOSITS BY THE FRASCH PROCESS

Prior to 1900, most of the world's supply of sulfur came from Sicily, where sulfur occurs at the surfaces around hot springs and volcanoes. In the early 1900's, however, large subsurface deposits of sulfur were found along the Gulf Coast of the United States. The sulfur occurs in limestone caves, over 1000 feet beneath layers of rock, clay, and quicksand. The recovery of the sulfur from these deposits posed a great technological problem, which was solved by the engineer Herman Frasch. The *Frasch process* (Figure D-2) uses an arrangement of three concentric pipes (diameters of 1 in., 3 in., and 6 in.) placed in a bore hole that penetrates to the base of the sulfur-bearing calcite

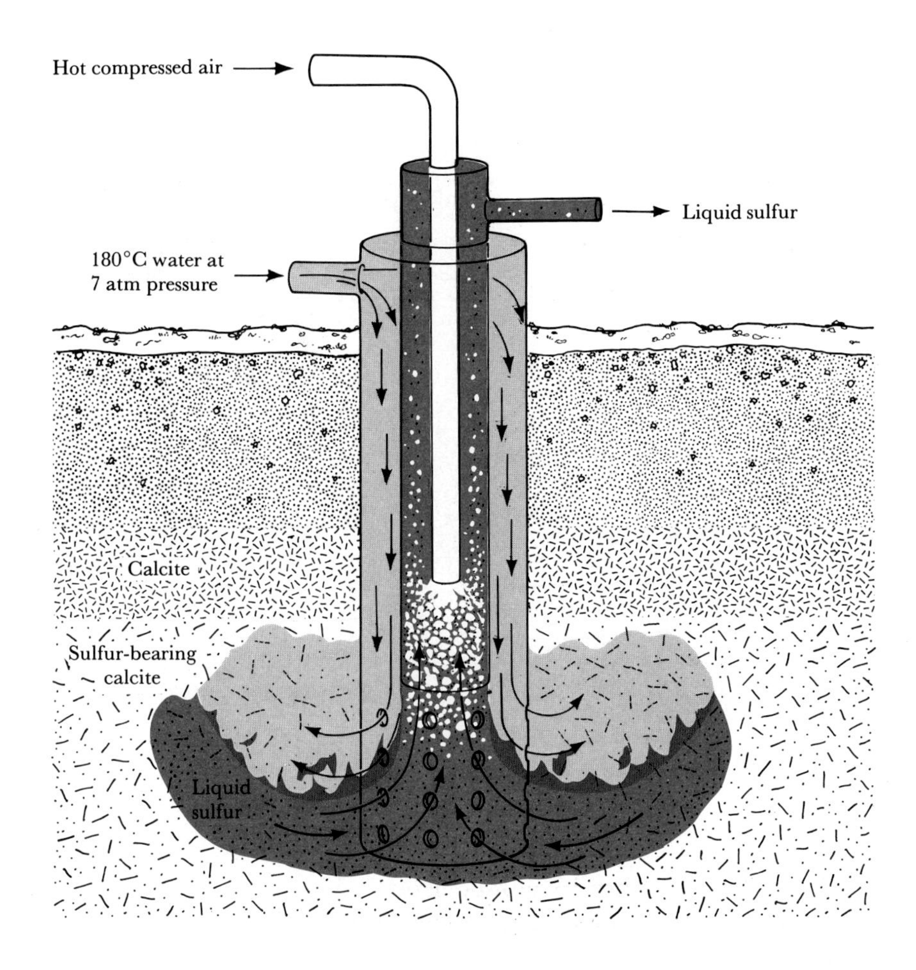

Figure D-2 The Frasch process for sulfur extraction. Three concentric pipes are sunk into sulfur-bearing calcite rock. Water at 180°C is forced down the outermost pipe to melt the sulfur. Hot compressed air is forced down the innermost pipe and mixes with the molten sulfur, forming a foam of water, air, and sulfur. The mixture rises to the surface through the center pipe. The resulting dried sulfur has a purity of 99.5 percent.

($CaCO_3$) rock formation. Pressurized hot water (180°C) is forced down the space between the 6-in. and 3-in. pipes to melt the sulfur (melting point 119°C). The molten sulfur, which is twice as dense as water, sinks to the bottom of the deposit and is then forced up the space between the 3-in. and 1-in. pipes as a foam by the action of compressed air injected through the innermost pipe. The molten sulfur rises to the surface, where it is pumped into tank cars for shipment or into storage areas. About 80 percent of the U.S. annual sulfur production of over 10 million metric tons is obtained by the Frasch process from the region around the Gulf of Mexico in Louisiana and Texas (Figure D-3).

Sulfur is also obtained in increasingly large quantities from the hydrogen sulfide (H_2S) in so-called sour natural gas and from the H_2S produced when sulfur is removed from petroleum. Hydrogen sulfide is burned in air to produce sulfur dioxide gas, which is then reacted with additional hydrogen sulfide to produce sulfur:

$$2H_2S(g) + 3O_2(g) \rightarrow 2SO_2(g) + 2H_2O(g)$$

$$SO_2(g) + 2H_2S(g) \rightarrow 3S(l) + 2H_2O(g)$$

Figure D-3 Sulfur is mined in huge quantities by the Frasch process. This is a mountain of pure sulfur ready for shipment in Newhall, Texas.

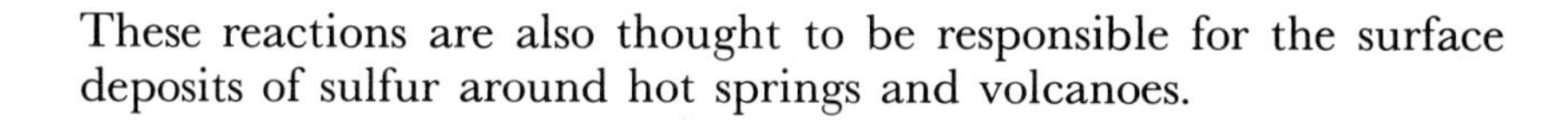

These reactions are also thought to be responsible for the surface deposits of sulfur around hot springs and volcanoes.

D-2. SULFIDE ORES ARE IMPORTANT SOURCES OF SEVERAL METALS

Deposits of metal sulfides are found in many regions and are valuable ores of the respective metals. Galena (PbS), cinnabar (HgS), and antimony sulfide (Sb_2S_3) are examples of metal sulfides that are ores (Figure D-4). In obtaining metals from sulfide ores, the ores are usu-

(a)

(b)

(c)

Figure D-4 Many metal sulfides are valuable ores of the respective metals. Shown here are (a) galena (PbS), (b) cinnabar (HgS), and (c) stibnite (Sb_2S_3).

ally *roasted,* meaning that they are heated in an oxygen atmosphere. The chemical equation for the roasting of galena is

$$2PbS(s) + 3O_2(g) \rightarrow 2PbO(s) + 2SO_2(g)$$

The lead oxide then is reduced by heating with carbon in the form of coke:

$$PbO(s) + C(s) \rightarrow Pb(l) + CO(g)$$

The sulfur dioxide produced by roasting is a serious atmospheric pollutant and should be recovered.

Iron pyrite, also known as fool's gold (Figure D-5), is a famous metal sulfide that has little commercial value. Sulfur is also found in nature in a few insoluble sulfates, such as *gypsum,* $CaSO_4 \cdot 2H_2O$ (calcium sulfate dihydrate) (Figure D-6), and *barite,* $BaSO_4$.

Sulfur also occurs in many proteins. Hair protein is fairly rich in sulfur. In fact, the formation of a "permanent" wave in hair involves the breaking and remaking of sulfur bonds (Chapter 26).

Figure D-5 Iron pyrite, FeS_2, is known as fool's gold. Novice gold miners often mistake iron pyrite for gold because the two look so much alike. Gold is much denser and softer than iron pyrite.

Figure D-6 Large deposits of gypsum, $CaSO_4 \cdot 2H_2O$, an insoluble mineral, are found in many areas. The notation $\cdot 2H_2O$ denotes 2 mol of water of crystallization per mole of $CaSO_4$. Left, the dunes of White Sands National Monument in New Mexico are composed of gypsum. Above is a 3-inch cluster of gypsum crystals.

(a)

(b)

Figure D-7 Sulfur occurs as (a) rhombic and (b) monoclinic crystals. Rhombic sulfur is the stable form below 96°C. From 96°C to 119°C (the normal melting point) monoclinic sulfur is the stable form. The terms rhombic and monoclinic are derived from the shape of the crystals.

D-3. SULFUR EXISTS AS RINGS OF EIGHT SULFUR ATOMS

Sulfur exhibits a property called *allotropy,* which means that there is more than one possible molecular form of sulfur. Below 96°C sulfur exists as yellow, transparent, *rhombic* crystals, shown in Figure D-7(a). If rhombic sulfur is heated above 96°C, then it becomes opaque and the crystals expand into *monoclinic* crystals (Figure D-7b). The molecular units of the rhombic form are rings containing eight sulfur atoms, S_8 (Figure D-8). Monoclinic sulfur is the stable form from 96°C to the melting point. The molecular units of monoclinic sulfur are also S_8 rings, but the rings themselves are arranged differently in rhombic and monoclinic sulfur, which are allotropic forms of solid sulfur.

Cold rhombic sulfur is colorless.

Monoclinic sulfur melts at 119°C to a thin, pale yellow liquid consisting of S_8 rings. Upon heating to about 150°C there is little

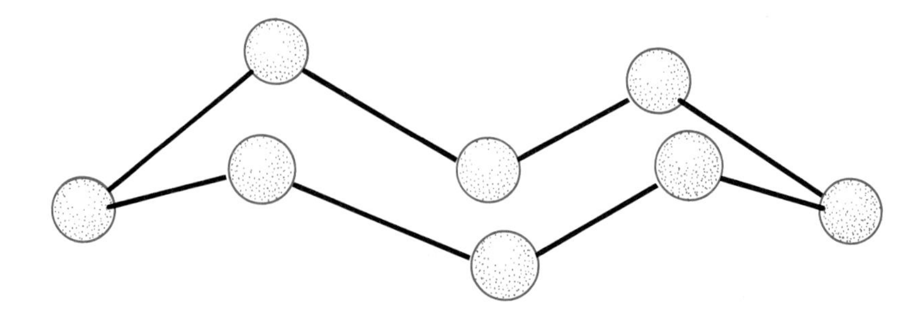

Figure D-8 Under most conditions, sulfur exists as eight-membered rings, S_8. The ring is not flat but puckered in such a way that four of the atoms lie in one plane and the other four lie in another plane.

20°C

120°C

200°C

400°C

Figure D-9 Molten sulfur at various temperatures. The change in color and physical properties of liquid sulfur with increasing temperature (120° to about 250°C) is a result of the conversion of eight-membered rings to long chains of sulfur atoms. Above 250°C, the long chains begin to break up into smaller segments and the sulfur once again pours freely.

change, but beyond 150°C the liquid sulfur begins to thicken and turns reddish brown. By 200°C, the liquid is so thick that it hardly pours (Figure D-9). The molecular explanation for this behavior is simple. At about 150°C, thermal agitation causes the S_8 rings to begin to break apart and form chains of sulfur atoms:

$$S_8 \text{ (ring)} \longrightarrow \text{S—S—S—S—S—S—S—S}$$

Figure D-10 If liquid sulfur at about 200°C is cooled quickly by pouring it into cold water, a rubbery substance called plastic sulfur is obtained.

These chains can then join together to form longer chains, which become entangled in each other and cause the liquid to thicken. Above 250°C, the liquid begins to flow more easily because the thermal agitation is sufficient to begin to break the chains of sulfur atoms. At the boiling point (445°C), liquid sulfur pours freely and the vapor molecules consist mostly of S_8 rings.

If liquid sulfur at about 200°C is placed quickly in cold water (this process is called *quenching*), then a rubbery substance known as *plastic sulfur* is formed (Figure D-10). The material is rubbery because the long, coiled chains of sulfur atoms can straighten out some if they are pulled. This molecular explanation of rubbery character, or *elasticity,* is shown schematically in Figure N-4, in Interchapter N. As plastic sulfur cools, it slowly becomes hard again as it rearranges itself into the rhombic form.

D-4. SULFURIC ACID IS THE LEADING INDUSTRIAL CHEMICAL

By far the most important use of sulfur is in the manufacture of sulfuric acid. Most sulfuric acid is made by a process called the *contact process.* The sulfur is first burned in oxygen to produce sulfur dioxide:

$$S(s) + O_2(g) \rightarrow SO_2(g)$$

The sulfur dioxide is then converted to sulfur trioxide:

$$2SO_2(g) + O_2(g) \xrightarrow{V_2O_5(s)} 2SO_3(g)$$

The V_2O_5 over the arrow in this equation means that V_2O_5 (vanadium pentoxide) is a catalyst for the reaction. The sulfur trioxide is then absorbed into nearly pure liquid sulfuric acid (99 percent sulfuric acid plus 1 percent water) to form *fuming sulfuric acid* (*oleum*):

$$H_2SO_4(l) + SO_3(g) \rightarrow \underset{\text{oleum}}{H_2S_2O_7} \text{ (35\% in } H_2SO_4)$$

The oleum is then added to water or aqueous sulfuric acid to produce the desired final concentration of aqueous sulfuric acid. Sulfur trioxide cannot be absorbed directly in water because the acid mist of H_2SO_4 that forms is very difficult to condense.

Over 80 billion pounds of sulfuric acid are produced annually in the United States. Commercial-grade sulfuric acid is one of the least expensive chemicals, costing less than 10 cents per pound in bulk quantities. Very large quantities of sulfuric acid are used in the production of fertilizers and numerous industrial chemicals, the petroleum industry, metallurgical processes, synthetic fiber production,

Table D-1 Annual U.S. industrial use of sulfuric acid (1980)

Use	*Quantity used/ millions of tons*
manufacture of ammonium sulfate and phosphate fertilizers	20
manufacture of HNO_3, HCl, H_3PO_4, HF, explosives	10
purification of petroleum products	4
manufacture of paints and pigments	1
cleaning metal surfaces, metallurgy	1
other	4

and paints, pigments, and explosives manufacture. Table D-1 lists the quantities of sulfuric acid used in its various industrial applications.

Pure, anhydrous sulfuric acid is a colorless, syrupy liquid that freezes at 10°C and boils at 290°C. The standard laboratory acid is 98 percent H_2SO_4 and 18 M in H_2SO_4 (Figure D-11). Concentrated sulfuric acid is a powerful dehydrating agent. Gases are sometimes bubbled through it to remove traces of water vapor—provided, of course, that the gases do not react with the acid.

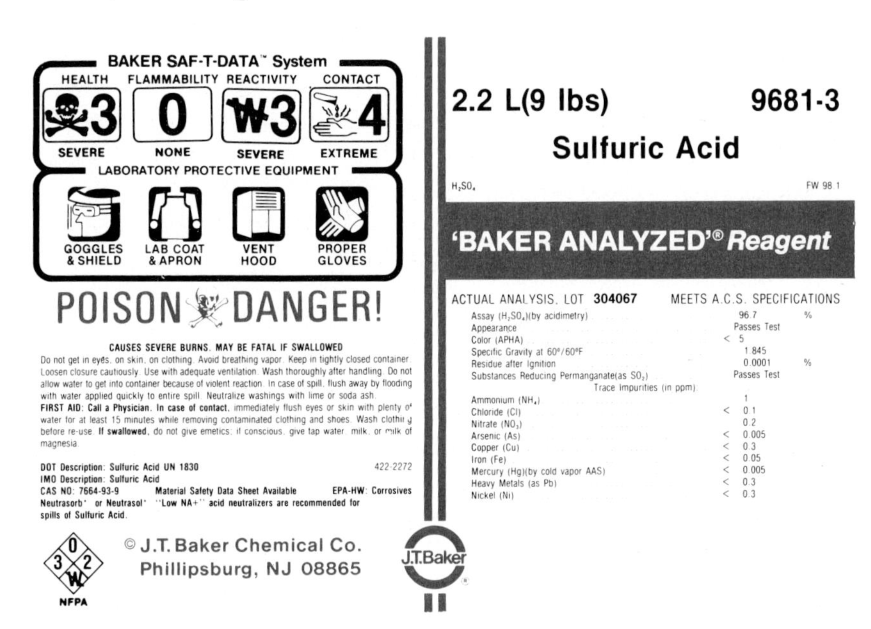

Figure D-11 Sulfuric acid is sold for laboratory use as an 18 M solution that is 98 percent sulfuric acid and 2 percent water. When diluting concentrated sulfuric acid with water, the heavier sulfuric acid (density = $1.84\ g \cdot mL^{-1}$) should always be added to the water so that thorough mixing can occur rapidly and the large amount of heat evolved can be distributed throughout the liquid volume. If water is added to sulfuric acid, then the lighter water forms a layer on top of the acid. The heat evolved is localized to the region between the two layers and can be great enough to explosively boil off water containing sulfuric acid and may even crack the container. Never add water to any acid; *always add acid to water*. For the reaction $H_2SO_4(l) + H_2O(l) \rightarrow H_2SO_4\ (\text{in}\ H_2O), \Delta H_{rxn} = -96\ kJ \cdot mol^{-1}$.

Sulfuric acid is such a strong dehydrating agent that it can remove water from carbohydrates, such as cellulose and sugar, even though these substances contain no free water. If concentrated sulfuric acid is poured over sucrose, $C_{12}H_{22}O_{11}$, then we have the reaction

$$C_{12}H_{22}O_{11}(s) \xrightarrow[H_2SO_4\ (98\%)]{} 12C(s) + 11H_2O \text{ (in } H_2SO_4)$$

This impressive reaction is shown in Figure D-12. Similar reactions are responsible for the destructive action of concentrated sulfuric acid on wood, paper, and skin.

Figure D-12 Concentrated (98%) sulfuric acid is a powerful dehydrating agent capable of converting sucrose to carbon.

The high boiling point and strength of sulfuric acid are the basis of its use in the production of other acids. For example, dry hydrogen chloride gas is produced by the reaction of sodium chloride with sulfuric acid:

$$2NaCl(s) + H_2SO_4(l) \rightarrow Na_2SO_4(s) + 2HCl(g)$$

The high boiling point of the sulfuric acid allows the $HCl(g)$ to be driven off by heating. The $HCl(g)$ is then added to water to produce hydrochloric acid. Note that this reaction is a double replacement reaction driven by the removal of a gaseous product from the reaction mixture.

D-5. SULFUR FORMS SEVERAL WIDELY USED COMPOUNDS

Sulfur burns in oxygen to form sulfur dioxide, a colorless gas with a characteristic choking odor. A pressure of 3 atm is sufficient to liquefy sulfur dioxide at 20°C. For this reason, SO_2 was once used in industrial refrigeration units, but the unpleasant odor and toxicity brought on its replacement by Freons.

Sulfur dioxide is very soluble in water; over 200 g of sulfur dioxide dissolve in one liter of water. Some of the sulfur dioxide reacts with the water to form sulfurous acid:

$$SO_2(g) + H_2O(l) \rightarrow H_2SO_3(aq)$$

but most of it exists in solution as $SO_2(aq)$.

Most sulfur dioxide is used to make sulfuric acid, but some is used as a bleaching agent in the manufacture of paper products, oils and starch, and as a food additive to inhibit browning. Large quantities are used in the wine industry as a fungicide for grapevines.

The salts of sulfurous acid are called sulfites. For example, if sodium hydroxide is added to an aqueous solution of sulfur dioxide, then sodium sulfite is formed according to the equation

$$2NaOH(aq) + H_2SO_3(aq) \rightarrow \underset{\text{sodium sulfite}}{Na_2SO_3(aq)} + 2H_2O(l)$$

Sodium sulfite is used occasionally as a preservative, especially for dehydrated fruits. The sulfite ion acts as a fungicide; however, it imparts a characteristic sulfur dioxide odor and taste to the food.

The thiosulfate ion is produced when an aqueous solution of a metal sulfite, such as $Na_2SO_3(aq)$, is boiled in the presence of solid sulfur:

$$S(s) + SO_3^{2-}(aq) \rightarrow \underset{\text{thiosulfate}}{S_2O_3^{2-}(aq)}$$

The designation thio denotes the replacement of an oxygen atom by a sulfur atom. Sodium thiosulfate is used extensively as "hypo" ($Na_2S_2O_3 \cdot 5H_2O$) in black-and-white photography.

Another compound of sulfur that is widely used is hydrogen sulfide, H_2S, a colorless gas with an offensive odor suggestive of rotten eggs. It is also very poisonous. Collapse, coma, and death from respiratory failure can occur within minutes of heavy exposure to H_2S. Although the odor of hydrogen sulfide usually gives an emphatic warning, the sense of smell can become fatigued at high H_2S concentrations.

Trace amounts of hydrogen sulfide occur naturally in the atmosphere due to volcanic activity and the decay of organic matter. In fact, the presence of hydrogen sulfide in the atmosphere is demonstrated by the tarnishing of silver. In the presence of oxygen, silver reacts with hydrogen sulfide according to the reaction

$$4Ag(s) + 2H_2S(g) + O_2(g) \rightarrow 2Ag_2S(s) + 2H_2O(l)$$

The silver sulfide formed by this reaction is a black, insoluble solid that appears as a dark tarnish on the surface of the silver (Figure D-13). Hydrogen sulfide is an important reagent in various qualitative analysis schemes in which metal ions are selectively removed from solution as insoluble metal sulfides (Section 19-13).

Organic compounds that contain an —SH group are called *mercaptans* and are notoriously foul-smelling. For example, the Guinness *Book of World Records* says that ethyl mercaptan, CH_3CH_2SH, has the worst odor of any substance; many other mercaptans have comparably obnoxious odors. The odor of a skunk's spray is due to a mixture of mercaptans.

Figure D-13 Trace amounts of hydrogen sulfide in the atmosphere arising from volcanic activity and the decay of organic matter, which contains sulfur compounds, are responsible for the tarnishing of silver.

QUESTIONS

D-1. Describe the Frasch process.

D-2. Use balanced chemical equations to show how zinc can be obtained from the ore zinc blende.

D-3. Give the chemical formula for each of the following substances:

(a) sulfur dioxide
(b) sulfur trioxide
(c) sulfuric acid
(d) sulfurous acid
(e) hydrogen sulfide
(f) iron pyrite
(g) gypsum
(h) cinnabar
(i) sodium thiosulfate pentahydrate

D-4. Describe what happens at various stages when sulfur (initially in the rhombic form) is heated slowly from 90°C to 450°C.

D-5. Describe, using balanced chemical equations, the contact process for the manufacture of sulfuric acid.

D-6. When gypsum, $CaSO_4 \cdot 2H_2O(s)$, is heated, part of the water of crystallization is driven off and plaster of Paris, $CaSO_4 \cdot \frac{1}{2}H_2O(s)$, is formed. Write a balanced chemical equation for this process. Addition of water to plaster of Paris gives back gypsum. Commercial plaster contains plaster of Paris mixed with fibrous material, such as animal hair, to provide structural strength.

D-7. Salts of sulfuric acid are called ______. Salts of sulfurous acid are called ______.

D-8. Write a balanced chemical equation for the formation of the fertilizer ammonium sulfate from ammonia and sulfuric acid.

D-9. Why would it be unwise to attempt to increase the acidity of the soil around plants by adding concentrated sulfuric acid?

D-10. Epsom salt is magnesium sulfate heptahydrate. Given that hepta is Greek for seven, write the formula for Epsom salt.

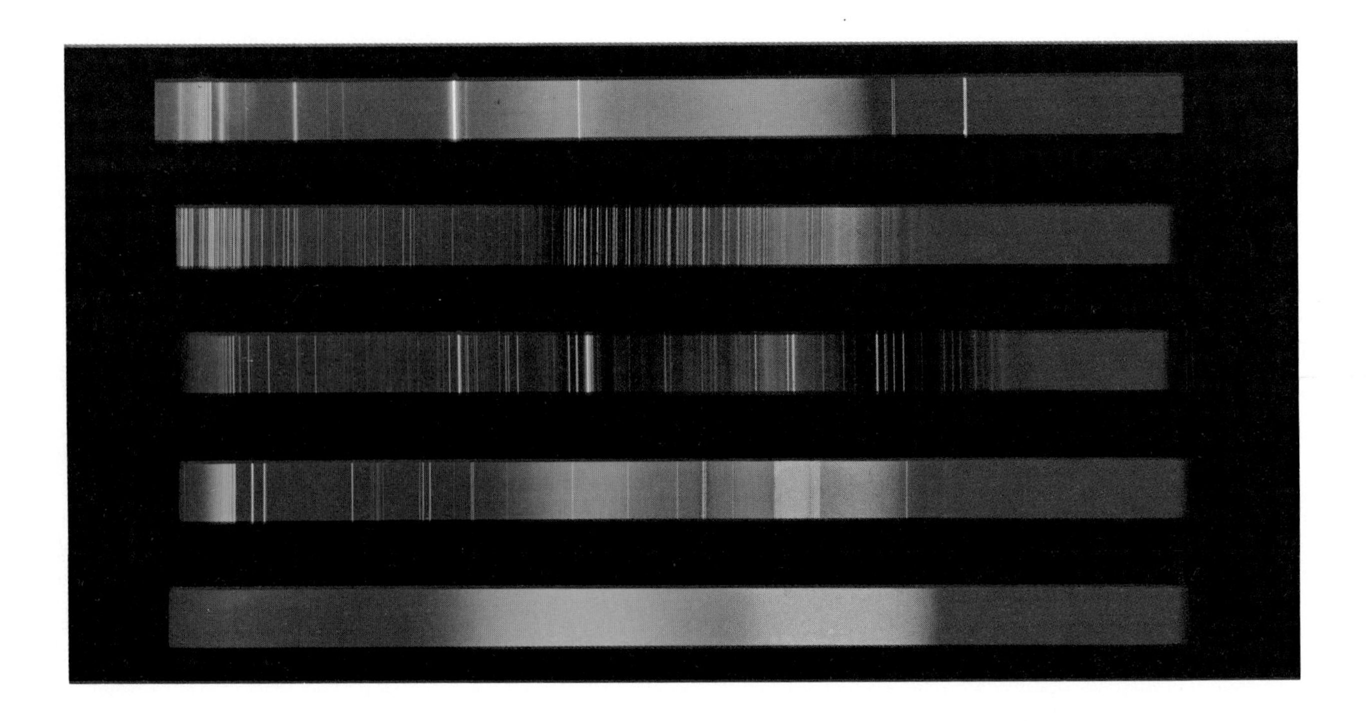

When an electric spark is allowed to pass through a gas, the atoms of the gas emit radiation of certain characteristic wavelengths, or frequencies. The above photo shows the emission spectra of lithium, iron, barium, and calcium and a continuous spectrum obtained by passing white light through a prism.

7 / The Quantum Theory and the Hydrogen Atom

We learned in Chapter 1 that atoms consist of protons, neutrons, and electrons, with the protons and neutrons making up the dense, central nucleus and the electrons arranged in some unspecified manner around the nucleus. Our model pictures an atom as mostly empty space, with the diameter of the nucleus being roughly 10^{-5} times that of the whole atom. We shall learn in this chapter that a description of the arrangement of electrons around a nucleus requires a new and unexpected way of looking at nature. This new perspective is given by what is called the quantum theory. One of the principal results of the quantum theory is that the electronic energies of atoms are quantized, meaning that they can take on only certain discrete values. In this chapter we trace the development of the quantum theory and apply it to the hydrogen atom.

7-1. FIRST IONIZATION ENERGY IS A PERIODIC PROPERTY

The periodic table offers a great deal of insight into the *electronic structure* of the atom. For example, elements in the same column in the periodic table are similar chemically, and so we might expect that their electronic arrangements, particularly those of the outermost and hence most chemically important electrons, are similar. A direct indication of the arrangement of electrons about a nucleus is its *ionization energies*. The ionization energy of an atom or an ion is the energy required to remove an electron completely from the gaseous atom or ion; this energy can be determined experimentally.

The *first ionization energy* of an atom is the minimum energy required to remove an electron from the neutral atom, A, to produce a positively charged ion, A^+:

$$A(g) \rightarrow A^+(g) + e^-(g)$$

The *second ionization energy* is the minimum energy required to remove an electron from the A^+ ion to produce an A^{2+} ion:

$$A^+(g) \rightarrow A^{2+}(g) + e^-(g)$$

We can go on to define and measure third, fourth, and successive ionization energies. We denote ionization energies by I_1, I_2, I_3, and so forth. We expect the second and higher ionization energies to be greater than the first ionization energy because, in removing successive electrons from an atom, we must overcome an increasingly greater electrical attraction between the positively charged ion and the electron that is being removed. Thus we find that $I_1 < I_2 < I_3 < I_4$, and so forth, for any given atom.

If we plot the first ionization energies of the elements against atomic number (Figure 7-1), then we find that there is a periodic pattern in these data. Note that the noble gases have relatively large first ionization energies. This means that it is relatively difficult to remove electrons from noble-gas atoms and hence suggests that the

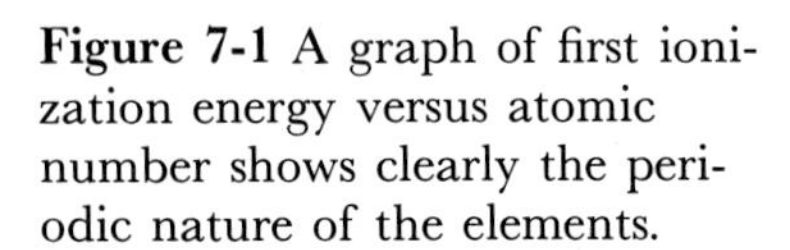

Figure 7-1 A graph of first ionization energy versus atomic number shows clearly the periodic nature of the elements.

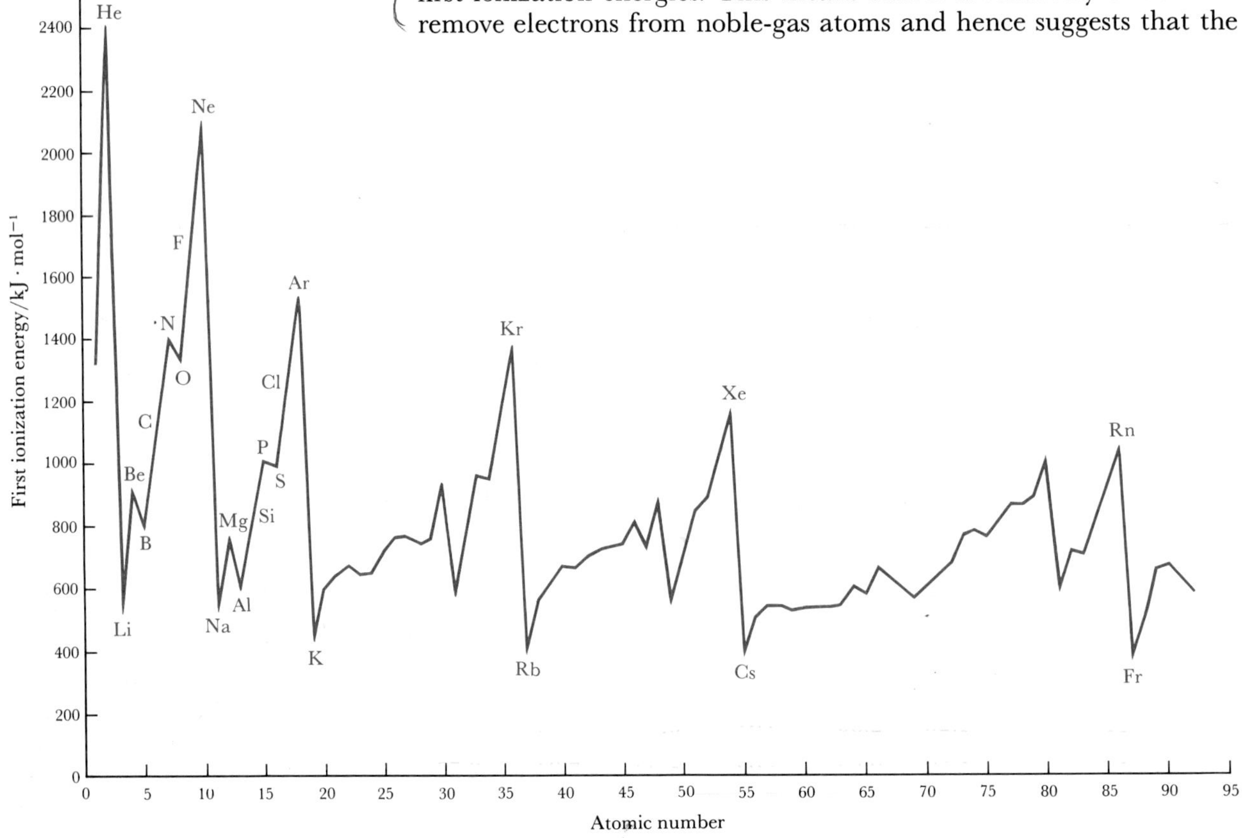

electronic structures of these atoms are more stable than that of the elements that precede and follow them in the periodic table. Furthermore, the alkali metals have relatively low ionization energies (Figure 7-1), in accord with their extremely reactive nature. Thus we see that ionization energies as well as chemical properties display a periodic character because both depend upon electronic structure. Figure 7-2 illustrates the trend in first ionization energies in the periodic table.

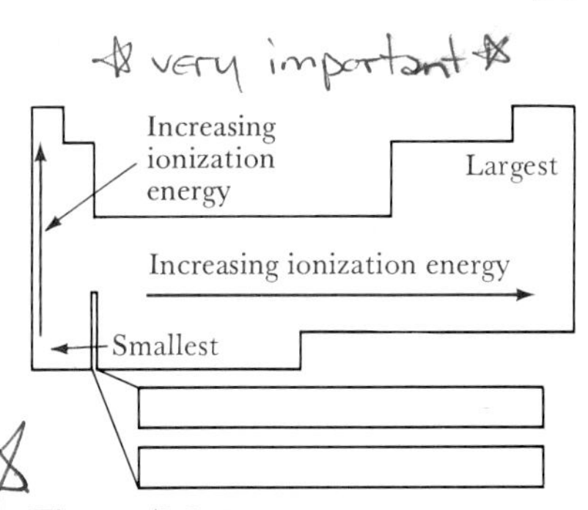

Figure 7-2 The trend in first ionization energy in the periodic table. Ionization energy increases as we go from left to right across the rows and as we go up the columns.

7-2. THE VALUES OF SUCCESSIVE IONIZATION ENERGIES SUGGEST A SHELL STRUCTURE

We can gain more insight into electronic structure by listing not just the first ionization energies but successive ionization energies, as in Table 7-1, where I_1 through I_{10} are tabulated for the elements hydrogen through argon.

Table 7-1 Successive ionization energies of the elements hydrogen through argon

		Ionization energy/MJ·mol^{-1}									
Z	*Element*	I_1	I_2	I_3	I_4	I_5	I_6	I_7	I_8	I_9	I_{10}
1	H	1.31									
2	He	2.37	5.25								
3	Li	0.52	7.30	11.81							
4	Be	0.90	1.76	14.85	21.01						
5	B	0.80	2.42	3.66	25.02	32.82					
6	C	1.09	2.35	4.62	6.22	37.83	47.28				
7	N	1.40	2.86	4.58	7.48	9.44	53.27	64.36			
8	O	1.31	3.39	5.30	7.47	10.98	13.33	71.33	84.08		
9	F	1.68	3.37	6.05	8.41	11.02	15.16	17.87	92.04	106.43	
10	Ne	2.08	3.95	6.12	9.37	12.18	15.24	20.00	23.07	115.38	131.43
11	Na	0.50	4.56	6.91	9.54	13.35	16.61	20.11	25.49	28.93	141.37
12	Mg	0.74	1.45	7.73	10.54	13.62	17.99	21.70	25.66	31.64	35.46
13	Al	0.58	1.82	2.74	11.58	14.83	18.38	23.30	27.46	31.86	38.46
14	Si	0.79	1.58	3.23	4.36	16.09	19.78	23.79	29.25	33.87	38.73
15	P	1.06	1.90	2.91	4.96	6.27	21.27	25.40	29.85	35.87	40.96
16	S	1.00	2.25	3.36	4.56	7.01	8.49	27.11	31.67	36.58	43.14
17	Cl	1.26	2.30	3.82	5.16	6.54	9.36	11.02	33.60	38.60	43.96
18	Ar	1.52	2.67	3.93	5.77	7.24	8.78	11.99	13.84	40.76	46.19

Note: The lines separate regions of relatively low and relatively high ionization energies.

Let's look at helium first. The first ionization energy is quite large, once again indicating the extraordinary stability of the helium atom. The second ionization energy is even higher, being more than twice as large as the first. Realize, however, that here we are removing an electron from a positively charged He^+ ion, and so we should expect I_2 to be greater than I_1 because of the attraction between the positively charged ion and the negatively charged electron we are removing.

The case of lithium is more interesting than helium. The first ionization energy is $0.52\ MJ \cdot mol^{-1}$, and the second is $7.30\ MJ \cdot mol^{-1}$. If we assume that the difference between I_2 and I_1 for helium ($5.25 - 2.37 = 2.88\ MJ \cdot mol^{-1}$) accounts for the attraction between a positive ion and a negative electron, then the value for I_2 in lithium is far greater than can be accounted for by the electrical attraction and indicates clearly the extraordinary stability of Li^+. Once Li^+ is ionized to Li^{2+}, the next ionization energy is regular, as with helium. This pattern of ionization energies suggests that the lithium atom has one chemically important electron; when it loses that electron, the result is a Li^+ ion with two electrons and helium-like stability. The Li^+ ion usually takes part in chemical reactions as a spectator ion.

Ionization energies of beryllium in $MJ \cdot mol^{-1}$

I_1	I_2	I_3	I_4
0.90	1.76	14.85	21.01

Table 7-1 shows that for beryllium there is a large jump from I_2 to I_3. This jump suggests that the four electrons in beryllium are arranged such that two of them are readily detached, and hence chemically active, whereas the other two constitute a very stable helium-like inner core.

The elements sodium through argon show a pattern quite similar to the elements lithium through neon, but now it appears that the inner-core structure is like neon rather than like helium. For example, we can picture a sodium atom as a neon-like core with a loosely bound electron outside this core. The *electronic structure* of a sodium atom can be seen by plotting successive ionization energies against number of electrons removed, as in Figure 7-3. This figure suggests that the electrons in the sodium atom are arranged in three separate groups, called *shells*. The first electron is relatively easily removed to give Na^+, which has a neon-like stable core. This stability is indicated by a large jump between I_1 and I_2. After nine electrons have been removed, we have only two electrons left, which are arranged in a tightly bound helium-like core. There is a large jump in energy in going from I_9 to I_{10}, which means that the last two electrons to be removed from the sodium atom constitute a third shell.

Electrons are arranged about nuclei in shells.

Table 7-1 suggests that the electrons in atoms are arranged in shells consisting of noble-gas-like cores and outer, chemically active electrons. We can summarize these ideas by presenting the atoms as in Table 7-2. In the second column we indicate the noble-gas-like core by using the symbol for the gas enclosed in brackets; the outer-shell electrons are indicated by the dots. Thus, for example, we represent beryllium by a helium-like inner core with two electrons outside this core. The placement of the dots is arbitrary at this point.

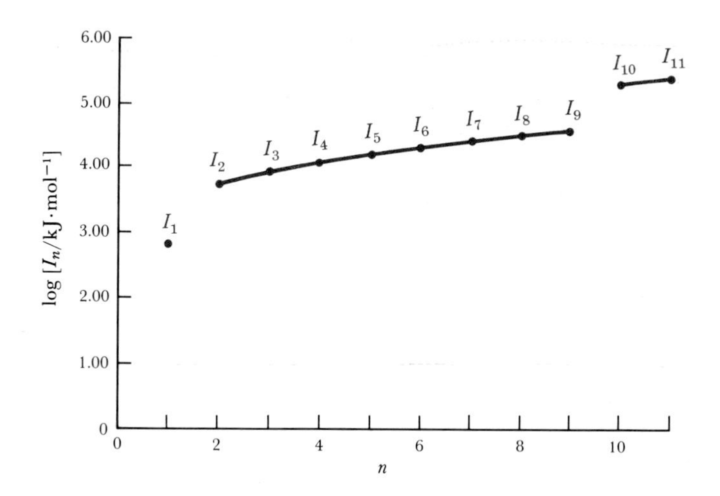

Figure 7-3 The logarithms of the 11 ionization energies of the sodium atom versus the number of electrons removed (n). This graph suggests that the electrons in a sodium atom are arranged in three shells. (Logarithms of the ionization energies are plotted simply to compress the vertical scale.)

The third column in Table 7-2 is an abbreviated version of the second column. Only the outer electrons are indicated, and the appropriate noble-gas core is understood. Note that the number of outer electrons increases from one to eight as we go across a row in the periodic table from alkali metal to noble gas. This pattern repeats itself from row to row. It is the outer electrons that play the key role in the chemical properties of the elements; thus, fluorine and

Table 7-2 A simple representation of the first 18 elements, indicating their noble-gas-like inner core and their outer electrons

Symbol	*Inner-core representation*	*Lewis electron-dot formula*	*Symbol*	*Inner-core representation*	*Lewis electron-dot formula*
H	H·	H·			
He	[He]	·He·			
Li	[He]·	Li·	Na	[Ne]·	Na·
Be	·[He]·	·Be·	Mg	·[Ne]·	·Mg·
B	·[Hẹ]·	·Ḅ·	Al	·[Ṅe]·	·Ȧl·
C	·[Ḥ̇e]·	·Ċ̣·	Si	·[Ṇ̇e]·	·Ṩi·
N	·[Ḥ̇e]:	·Ṇ̈·	P	·[Ṇ̈e]·	·P̣̈·
O	·[Ḧ̤e]·	·Ö̤·	S	·[N̤̈e]·	·S̤̈·
F	:[Ḥ̈e]:	:F̣̈:	Cl	:[Ṇ̈e]:	:C̣̈l:
Ne	[Ne]	:N̤̈e:	Ar	[Ar]	:Ä̤r:

chlorine, for example, are chemically similar because they both have seven outer electrons. This representation, called a *Lewis electron-dot formula,* was introduced in 1916 by G. N. Lewis, one of the greatest American chemists. Lewis formulas show only the outer and hence chemically important electrons.

One final feature of Table 7-1 we should discuss is that the ionization energy decreases in going from beryllium to boron and from magnesium to aluminum. This decrease shows that a structure with two electrons outside a noble-gas core is relatively stable, suggesting that the eight electrons around a core can be placed into *subshells,* with the first subshell containing two electrons and the second subshell containing six electrons. Table 7-3 summarizes our observations of ionization energies and their implications for the electronic structure of the elements hydrogen through neon.

Table 7-3 An interpretation of the ionization energies of the first 10 elements in terms of a subshell electronic structure

Lewis electron-dot formula	*Arrangement of electrons*	*First ionization energy/MJ·mol⁻¹*
H·	one electron	1.31
He:	one completed shell	2.37
Li·	one electron outside a completed shell	0.52
Be:	two electrons outside a completed shell; these two electrons constitute a completed *subshell*	0.90
·Ḃ·	one electron outside a completed subshell	0.80
·C· (4 dots)	two electrons outside a completed subshell	1.09
·N· (5 dots)	three electrons outside a completed subshell	1.40
·O· (6 dots)	four electrons outside a completed subshell	1.31
:F: (7 dots)	five electrons outside a completed subshell	1.68
:Ne: (8 dots)	six electrons outside a completed subshell, constituting a second completed subshell; the two subshells, one containing two electrons and the other containing six electrons, constitute a completed shell of eight electrons	2.08

7-3. ELECTROMAGNETIC RADIATION CAN BE DESCRIBED AS WAVES

In order to describe better the electronic structure of atoms, we must first briefly discuss electromagnetic radiation. Visible light, ultraviolet and infrared radiation, radio waves, and X-rays are all forms of electromagnetic radiation. For many years scientists disagreed over whether electromagnetic radiation exists as beams of particles or as waves. Many experiments supported one viewpoint or the other, but toward the end of the nineteenth century most evidence favored a wave picture.

Figure 7-4 depicts typical waves, whose motion can be visualized as moving across the page. The distance between successive crests or troughs is called the *wavelength* and is denoted by the Greek letter lambda, λ. If we picture each wave as moving across the page, the number of crests that pass a given point per second is called the *frequency* and is denoted by the Greek letter nu, ν. Although the units of wavelength are meters per cycle and the units of frequency are cycles per second, the term cycle is understood and so omitted from both units. Thus the units of wavelength are meters and the units of frequency are reciprocal seconds. The product of the wavelength and frequency, $\lambda\nu$, is the speed with which the wave travels. All forms of electromagnetic radiation travel at a speed of $3.00 \times 10^8\ \text{m}\cdot\text{s}^{-1}$, which is usually called simply the *speed of light*. If we denote the speed of light by c, we can write

The unit of reciprocal second is 1/second, or s^{-1}.

You may know the speed of light as 186,000 miles per second.

$$\lambda\nu = c \tag{7-1}$$

The various forms of electromagnetic radiation differ only in their frequency or their wavelength. For example, a look at a radio dial shows that 100 MHz (megahertz) is in the middle of the FM dial. The

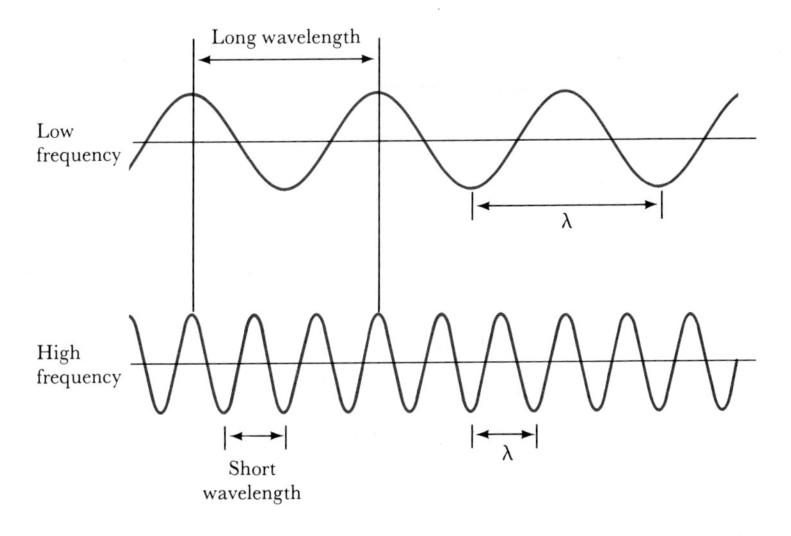

Figure 7-4 Two wave forms, one with a wavelength three times as large as the other. If both waves moved across the page with the same speed, three crests of the bottom wave would pass by a given point for every one crest of the top wave. Thus the frequency of the bottom wave is three times greater than the frequency of the top wave.

unit *hertz* is the same as the unit reciprocal second, and so 100 MHz is equal to $100 \times 10^6\ s^{-1}$. The wavelength of a 100 MHz signal is

$$\lambda = \frac{c}{\nu} = \frac{3.00 \times 10^8\ m \cdot s^{-1}}{1.00 \times 10^8\ s^{-1}} = 3.00\ m$$

Problems 7-9 and 7-10 are similar to Example 7-1.

Example 7-1: X-rays are produced by bombarding a metal surface with a beam of energetic electrons. If copper is the metal used, the wavelength of the X-radiation is 0.154 nm. Calculate the frequency of this radiation.

Solution: Recall that 1 nm is 10^{-9} m. Because $\lambda = 0.154 \times 10^{-9}$ m,

$$\nu = \frac{c}{\lambda} = \frac{3.00 \times 10^8\ m \cdot s^{-1}}{0.154 \times 10^{-9}\ m}$$

$$= 1.95 \times 10^{18}\ s^{-1} = 1.95 \times 10^{18}\ Hz$$

1nm = 10^{-9}m

We see from these calculations that the range of wavelengths and frequencies of electromagnetic radiation, called the *electromagnetic spectrum,* is truly enormous. Table 7-4 illustrates the wavelengths and frequencies associated with certain regions in the electromagnetic spectrum.

7-4. THE SPECTRA EMITTED BY ATOMS ARE LINE SPECTRA

When white light is passed through a prism, we see that the light is made up of many colors. The same effect can be seen in a rainbow, where white light from the sun passes through water droplets in the atmosphere and is separated into its component colors. The wavelengths of the radiation in white light vary from about 400 nm to 700 nm. This is the region of the electromagnetic spectrum to which the human eye is sensitive and is called the *visible region.* The short-wavelength end of the visible region (400 nm) is violet and the long-wavelength end (700 nm) is red.

The spectrum of white light has no gaps in it and is called a *continuous spectrum.* Yet, if we examine radiation emitted from a glass tube containing a gas through which an electric spark is passed, we find that the resultant spectrum is not continuous but consists of several separate lines (Figure 7-5). This type of spectrum is called a *line spectrum* and is characteristic of the particular gas used in the discharge tube. Thus *atomic spectra* are line spectra. The simplest atomic spectrum is that of the hydrogen atom. Part of this spectrum

Table 7-4 The regions of the electromagnetic spectrum

Region	*Approximate wavelength/m*	*Comments*
radio waves	1000 – 1	AM band 190–560 m shortwave band 14– 75 m FM band 2.8– 3.4 m television bands 5.6–4.2 m (chs. 2, 3, 4) 4.0–3.4 m (chs. 5, 6) 1.7–1.4 m (chs. 7–13)
microwaves	1 – 10^{-4}	includes radar; used to probe rotational motion of molecules and in microwave ovens
infrared light	10^{-4} – 7×10^{-7}	can be felt as heat; used to probe vibrational motions of molecules
visible light	7×10^{-7} – 4×10^{-7}	consists of the colors red, orange, yellow, green, blue, indigo, and violet
ultraviolet light	4×10^{-7} – 10^{-8}	causes sunburn; kills bacteria
X-rays	10^{-8} – 10^{-11}	penetrate human tissue and other matter; have well-known medical applications
gamma rays	10^{-11} – 10^{-13}	emitted by energetic nuclei; very penetrating radiation; used to kill cancer cells
cosmic rays	10^{-13} – 10^{-14}	very high-energy, penetrating radiation of cosmic origin

is shown in Figure 7-6, and the frontispiece of this chapter shows spectra for lithium, iron, barium, and calcium atoms.

Each atom in a mixture of atoms emits its own atomic spectrum, and so the spectrum of a mixture is a superposition of each of the constituent atomic spectra. Atomic spectra are readily obtained in the laboratory and so provide a convenient way to determine the atomic composition of an unknown substance.

Figure 7-5 A schematic drawing of an atomic spectrometer. When a spark is passed through the discharge tube, hydrogen molecules are dissociated into hydrogen atoms, which emit radiation of only certain frequencies. This radiation is collimated by the slits, passed through a prism, and projected onto a photographic plate. The result is a line spectrum. The wavelengths observed in a line spectrum depend upon the particular gas in the discharge tube.

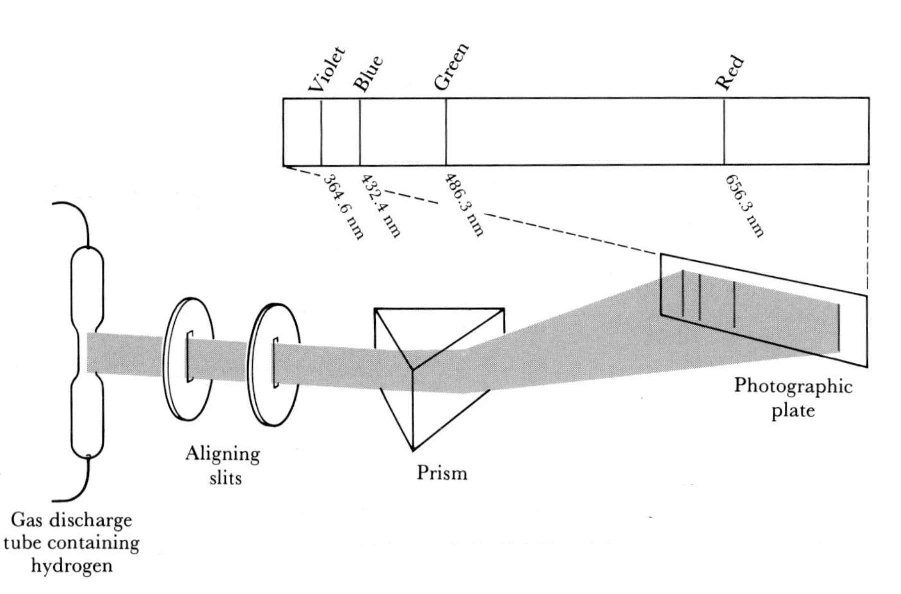

The line spectrum associated with an element serves as a fingerprint for that element. The study and analysis of the spectral lines emitted by atoms, or, more generally, the study of the interaction of electromagnetic radiation and atoms, is called *atomic spectroscopy*. One can compare an observed spectrum with those in a handbook of atomic spectra and identify each type of atom present by comparison. Atomic spectroscopy is a standard technique of analytical chemistry, which is that part of chemistry principally involved with chemical analysis. Spectroscopy is used also in agriculture, archaeology, art, criminal investigation, and many other fields. For example, we have probably all seen a movie or television show in which the crime lab analyzes soil taken from a suspect's shoes and compares it with soil collected from the scene of a crime.

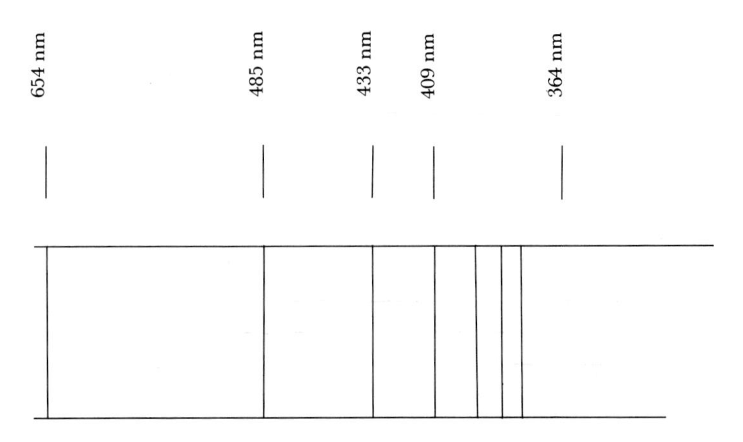

Figure 7-6 The emission spectrum of the hydrogen atom in the visible and near-ultraviolet regions.

7-5. A STUDY OF BLACKBODY RADIATION LED PLANCK TO THE FIRST QUANTUM HYPOTHESIS, $E = h\nu$

The study of atomic spectra was an important contribution to our present-day picture of the electronic structure of atoms, but the most important experiments, and the ones that led directly to what we now know as the quantum theory, had to do with the radiation given off by solid bodies when they are heated. We all know, for instance, that when the burner of an electric stove is heated, it first turns a dull red and progressively becomes redder as its temperature increases. If it were possible to heat the burner to even higher temperatures, it would turn white and eventually dull blue. Thus we see that a heated body continuously changes color from red through white and then blue at higher and higher temperatures. The frequency of the radiation emitted shifts from a lower frequency (red) to a higher frequency (blue). The distribution of frequencies emitted depends on what material is being heated, but an ideal body, one which absorbs and emits all frequencies, is called a *blackbody* and serves as an excellent idealization for any heated body that is radiating energy.

The radiation emitted by a blackbody is called *blackbody radiation.* Plots of the intensity of blackbody radiation versus wavelength are given for several temperatures in Figure 7-7. Note that, as the temperature is increased, both the total amount of radiation and the fraction of the radiation lying in the visible region increase. In the late 1800's many physicists tried to derive theoretical equations that could reproduce the curves in Figure 7-7, but they were all unsuccessful.

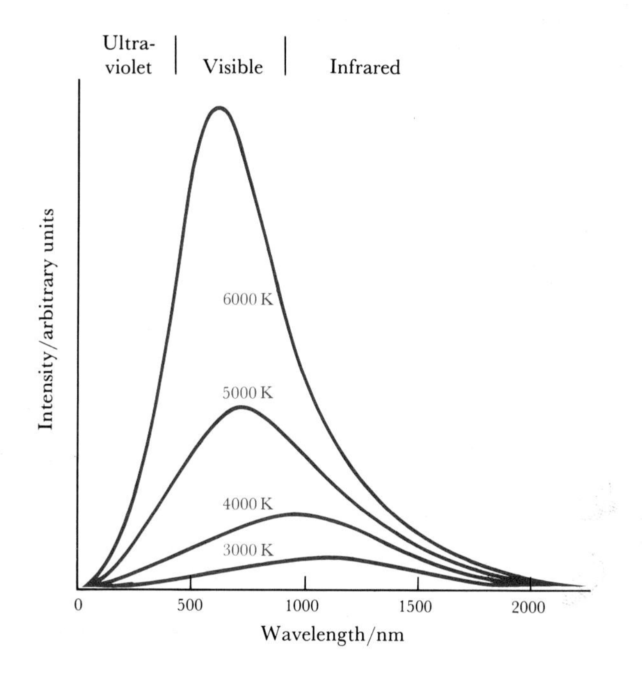

Figure 7-7 Distribution of the intensity of the radiation emitted by a blackbody. As the temperature increases, the total radiation emitted (the area under each curve) increases greatly. In addition, the major portion of each curve shifts from the infrared into the visible and toward the ultraviolet region. This is in accord with the observed fact that heated bodies first become red hot and then white hot as their temperature is increased. It was Planck's explanation of these curves in 1900 that led to the quantum theory.

The first person to offer a successful explanation of blackbody radiation was the German physicist Max Planck. Like those before him, Planck assumed that the radiation emitted by a blackbody was due to the oscillations of the electrons in the constituent particles of the material of the blackbody. These electrons were pictured as oscillating within the atom much like electrons oscillate in an antenna to give off radio waves. With these "atomic antennae," however, the oscillations occur at a high frequency, and hence we find radiation in the infrared, visible, and ultraviolet regions rather than in the radio-wave region of the electromagnetic spectrum. Everyone before Planck assumed that the energies of the electronic oscillators responsible for the emission of the radiation could have any value whatsoever. In 1900, Planck had the great revolutionary insight to break away from this thinking. Instead, he assumed that the radiation could be emitted only in little packets, called *quanta*, and that the energy associated with these quanta is proportional to the frequency of the radiation. In terms of an equation, Planck assumed that

$$E = h\nu \qquad (7\text{-}2)$$

$h = 6.626 \times 10^{-34}$ J·s

where E is the energy, h is a proportionality constant now called *Planck's constant,* and ν is the frequency of the radiation. Planck was able to reproduce all the data on blackbody radiation by choosing h to be 6.626×10^{-34} J·s. Planck's constant is one of the most famous and fundamental constants of science.

Example 7-2: The region of the electromagnetic spectrum that lies outside the violet end of the visible region is called the ultraviolet (UV) region. The region that lies outside the red end of the visible region is called the infrared (IR) region. Typical frequencies in these two regions are

$$\nu_{UV} = 3 \times 10^{16}\ \text{s}^{-1} \qquad \nu_{IR} = 1 \times 10^{14}\ \text{s}^{-1}$$

Calculate the energies associated with the radiation of these frequencies.

Solution: We can use Equation (7-2) directly:

$$E_{UV} = h\nu_{UV} = (6.626 \times 10^{-34}\ \text{J}\cdot\text{s})(3 \times 10^{16}\ \text{s}^{-1}) = 2 \times 10^{-17}\ \text{J}$$

$$E_{IR} = h\nu_{IR} = (6.626 \times 10^{-34}\ \text{J}\cdot\text{s})(1 \times 10^{14}\ \text{s}^{-1}) = 7 \times 10^{-20}\ \text{J}$$

Note that the energy increases going from the infrared to the ultraviolet region. Ultraviolet radiation is more energetic than infrared radiation.

Example 7-3: Cosmic rays are penetrating electromagnetic radiation produced by certain catastrophic cosmic events. Calculate the energy associated with cosmic radiation and compare your result with the energy of radio waves. Use Table 7-4 for the necessary information.

Solution: From Table 7-4, we see that the average wavelength of a radio wave is 500 m and that the average wavelength of a cosmic ray is about 10^{-14} m. To use Equation (7-2), we must first convert wavelength to frequency by means of Equation (7-1):

Problems 7-11 to 7-20 deal with the energy of electromagnetic radiation.

$$\nu_{cosmic} = \frac{c}{\lambda_{cosmic}} = \frac{3.00 \times 10^{8}\ \mathrm{m \cdot s^{-1}}}{10^{-14}\ \mathrm{m}} = 3 \times 10^{22}\ \mathrm{s^{-1}}$$

$$\nu_{radio} = \frac{c}{\lambda_{radio}} = \frac{3.00 \times 10^{8}\ \mathrm{m \cdot s^{-1}}}{500\ \mathrm{m}} = 6 \times 10^{5}\ \mathrm{s^{-1}}$$

We now use Equation (7-2) to calculate the energy associated with each frequency:

$$E_{cosmic} = h\nu_{cosmic} = (6.626 \times 10^{-34}\ \mathrm{J \cdot s})(3 \times 10^{22}\ \mathrm{s^{-1}}) = 2 \times 10^{-11}\ \mathrm{J}$$

$$E_{radio} = h\nu_{radio} = (6.626 \times 10^{-34}\ \mathrm{J \cdot s})(6 \times 10^{5}\ \mathrm{s^{-1}}) = 4 \times 10^{-28}\ \mathrm{J}$$

Cosmic radiation is about 10^{17} times more energetic than radio waves. The energy increases as one goes from the radio and microwave regions to the X-ray and gamma-ray regions. This is certainly in accord with the fact that X-rays are more penetrating, and hence more energetic, than, for example, radio waves.

Planck's theory of blackbody radiation, with its unconventional assumption that a body can emit radiation only in small, quantized packets, was not widely accepted in spite of its success in predicting curves like those in Figure 7-7. Most scientists considered Planck's theory to be somewhat of a curiosity and believed that in time a more satisfactory theory of blackbody radiation would emerge. Just a few years later, however, in 1905, Albert Einstein used the very same idea to explain the results of some experiments in which electrons were ejected from a metal surface when ultraviolet radiation was shone on it. Following Planck, Einstein proposed that electromagnetic radiation consists of little packets of energy, $E = h\nu$. These little packets, or quanta, are now called *photons*, and it is often convenient to picture electromagnetic radiation as a beam of photons of energy $h\nu$. Blackbody radiation now is used in astronomy to determine the surface temperatures of stars (Figure 7-8).

Figure 7-8 The electromagnetic spectrum of the sun as measured in the upper atmosphere of the earth. A comparison of this figure with Figure 7-7 shows that the sun's surface radiates as a blackbody at about 6000 K. The theory of blackbody radiation is used regularly in astronomy to estimate the surface temperatures of stars. Sirius, the brightest star in the sky, has a surface temperature of 11,000 K.

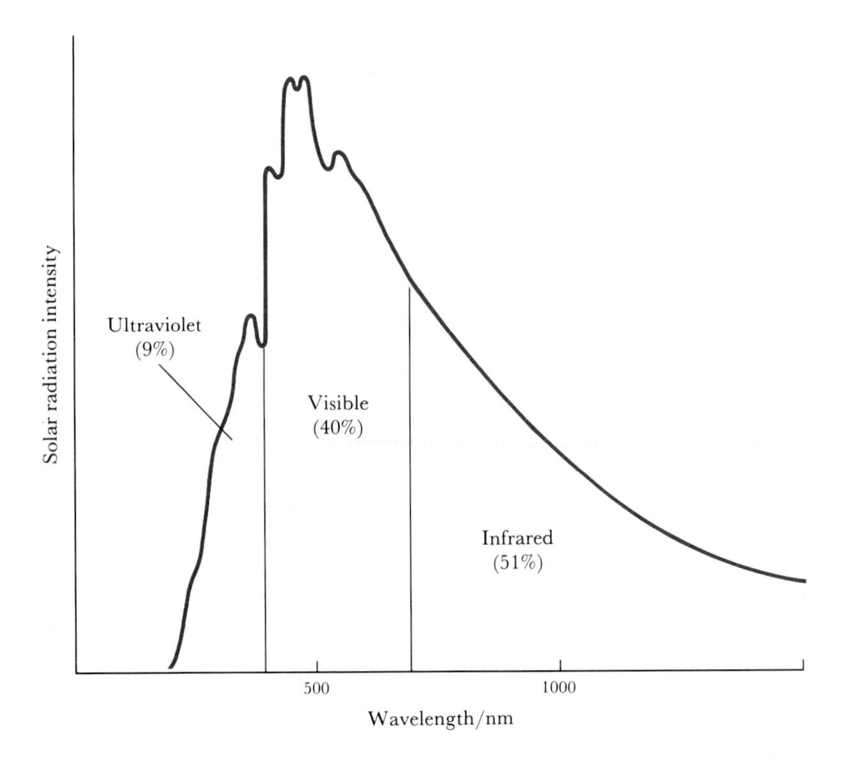

7-6. DE BROGLIE POSTULATED THAT MATTER HAS WAVELIKE PROPERTIES

Scientists have always had difficulty describing the nature of light. In many experiments light exhibits a definite wavelike character, but in many others it seems to behave as a stream of little particles. Because light appears sometimes to be wavelike and sometimes to be particle-like, we talk of the *wave-particle duality* of light. In 1924, Louis de Broglie, a young French physicist, proposed for his doctoral thesis that, if light can display this wave-particle duality, then matter, which certainly appears to be particle-like, might also display wavelike properties under certain conditions. This is a rather strange proposal at first sight, but it does suggest a nice symmetry in nature. Certainly if light can appear to be particle-like at times, why shouldn't matter appear to be wavelike at times?

De Broglie put this idea into a quantitative scheme by proposing that both light *and* matter obey the equation

$$\lambda = \frac{h}{mv} \tag{7-3}$$

where h is Planck's constant, m is the mass of a particle, and v is its speed. Equation (7-3) predicts that a particle having mass m and moving with a speed v has a *de Broglie wavelength* $\lambda = h/mv$.

Example 7-4: Calculate the de Broglie wavelength of an electron traveling at 1.00 percent of the speed of light.

Problems 7-25 to 7-30 ask you to calculate the de Broglie wavelengths of moving particles.

Solution: The mass of an electron is 9.11×10^{-31} kg (inside back cover). Its speed is

$$v = (0.0100)(3.00 \times 10^{8}\ \mathrm{m \cdot s^{-1}}) = 3.00 \times 10^{6}\ \mathrm{m \cdot s^{-1}}$$

and so

$$mv = (9.11 \times 10^{-31}\ \mathrm{kg})(3.00 \times 10^{6}\ \mathrm{m \cdot s^{-1}}) = 2.73 \times 10^{-24}\ \mathrm{kg \cdot m \cdot s^{-1}}$$

The de Broglie wavelength of this electron is

$$\lambda = \frac{h}{mv} = \frac{6.626 \times 10^{-34}\ \mathrm{J \cdot s}}{2.73 \times 10^{-24}\ \mathrm{kg \cdot m \cdot s^{-1}}} = 2.43 \times 10^{-10}\ \mathrm{m} = 0.243\ \mathrm{nm}$$

We have used the fact that $\mathrm{J} = \mathrm{kg \cdot m^2 \cdot s^{-2}}$ (Chapter 5). By referring to Table 7-4, we see that the wavelength of the electron in this example corresponds to the wavelength of X-rays.

$1\ \mathrm{J} = 1\ \mathrm{kg \cdot m^2 \cdot s^{-2}}$

Problem 7-26 asks you to show that the wavelength for a baseball traveling at 90 mph is 1.2×10^{-34} m, which is a ridiculously small wavelength. Thus, although Equation (7-3) is of trivial consequence for a macroscopic object like a baseball, it predicts that electrons can act like X-rays.

7-7. DE BROGLIE WAVES ARE OBSERVED EXPERIMENTALLY

When a beam of X-rays is directed at a thin foil of a crystalline substance, the beam is scattered in a definite manner characteristic of the atomic structure of the crystalline substance. This phenomenon, called *X-ray diffraction,* happens because the length of the interatomic spacings in the crystal is close to the wavelength of the X-rays. A similar pattern, called an *electron diffraction pattern,* occurs when a beam of electrons is used (Figure 7-9). The similarity of the two patterns demonstrates that both X-rays and electrons do indeed behave similarly in these experiments.

The wavelike property of electrons is used in *electron microscopes.* Electron wavelengths can be controlled by an applied voltage; the small deBroglie wavelengths attainable provide a more precise probe than that of an ordinary light microscope. In addition, in contrast to electromagnetic radiation of similar wavelengths (X-rays and UV), electron beams can be readily focused by using electric and magnetic fields. Electron microscopes are now routine tools in chemical and biological investigations of molecular structures (Figure 7-10).

(a)

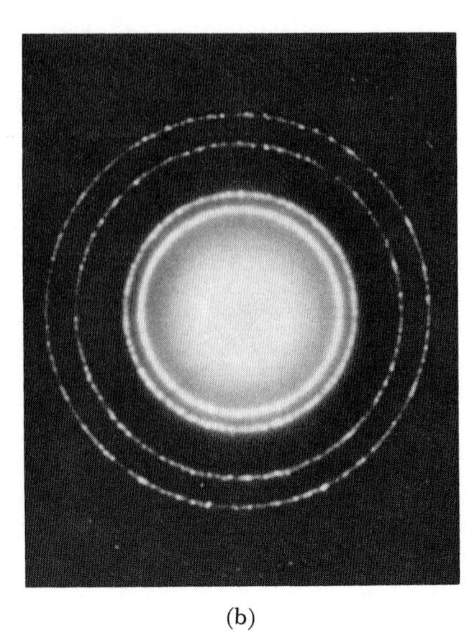
(b)

Figure 7-9 (a) The X-ray diffraction pattern from aluminum foil. (b) The electron diffraction pattern from aluminum foil. The similarity of these two patterns shows that electrons can behave like X-rays and display wavelike properties.

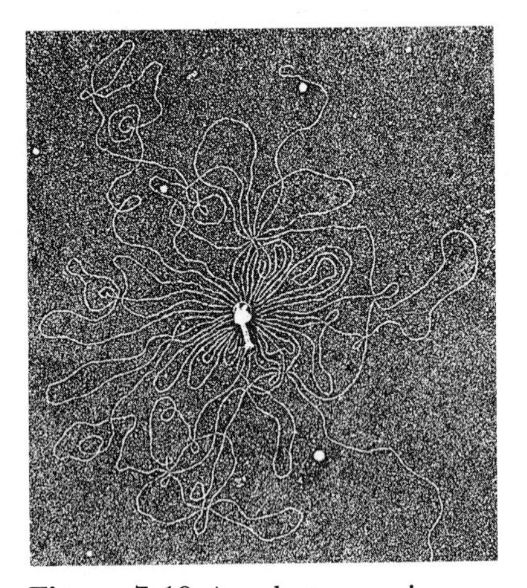

Figure 7-10 An electron micrograph of a virus particle that has burst and released strands of DNA, the long, cylindrical molecule revealed beautifully in the figure.

An interesting aside in the concept of the wave-particle duality of matter is that it was J. J. Thomson who first showed, in 1895, that the electron is a subatomic particle and G. P. Thomson who was one of the first to show experimentally, in 1926, that the electron could act as a wave. These two Thomsons were father and son. The father received a Nobel Prize in 1906 for showing that the electron is a particle, and the son received a Nobel Prize in 1937 for showing that it is a wave.

7-8. THE ENERGY OF THE ELECTRON IN A HYDROGEN ATOM IS QUANTIZED

According to Planck and Einstein, light can exist only in little energy packets, called quanta or photons. According to de Broglie, light and matter exhibit a wave-particle duality; that is, light and matter have both wavelike and particle-like properties. Because light exists only in discrete energy packets, it may not be surprising to learn, then, that the energies of particles can have only certain fixed values. We say that the energies are *quantized,* and the theory that predicts these energies is called the *quantum theory*. Just as the de Broglie wavelength of a moving particle is important only for very small particles, such as electrons and atoms, the results of the quantum theory are especially important only for small particles.

In 1926 Erwin Schrödinger (Figure 7-11) first presented what has become one of the most famous and important equations in science: the *Schrödinger equation,* the central equation of quantum theory. This equation takes into account the wave nature of particles as de Broglie proposed in 1924. It is too complicated to present here, but we can examine some of its consequences. As a specific example, let's consider the hydrogen atom, where we have only one electron moving under the electrical attraction of one proton. When we solve the Schrödinger equation for the hydrogen atom, we obtain a discrete set of energies that the electron is allowed to have. It turns out that these energies are given by the equation

$$E_n = -\frac{2.18 \times 10^{-18}\,\mathrm{J}}{n^2} \qquad n = 1, 2, 3, \ldots \qquad (7\text{-}4)$$

Note that the value of n is restricted to integer values. When $n = 1$, $E_1 = -2.18 \times 10^{-18}$ J; when $n = 2$, $E_2 = -0.545 \times 10^{-18}$ J; and so on. The energy is quantized because n can take on only integral values. These allowed *energy states* are presented schematically in Figure 7-12.

Figure 7-11 Erwin Schrödinger, a Viennese physicist, formulated the modern quantum theory of atoms and molecules in 1925. His theory is summarized by what is called the Schrödinger equation, which describes the motion of extremely small particles, such as electrons, atoms, and molecules.

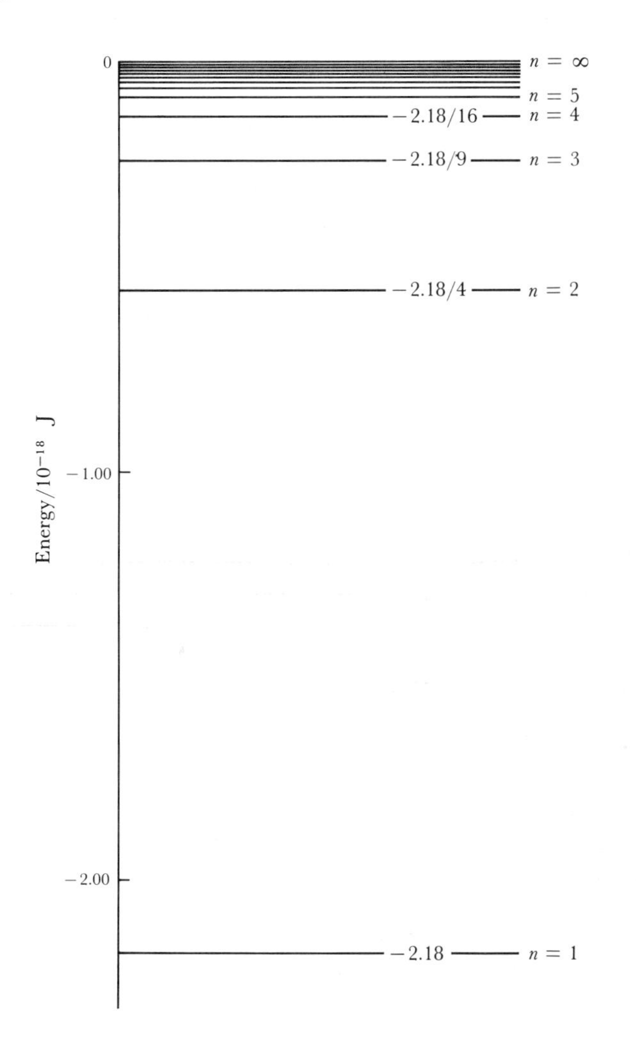

Figure 7-12 The energy states of the electron in the hydrogen atom according to the quantum theory. The electron can have only the energies given by $E_n = (-2.18 \times 10^{-18}\,\mathrm{J})/n^2$, $n = 1, 2, 3, \ldots$, and no others.

Because of the minus sign, all the energies given by Equation (7-4) are negative. The lowest energy is E_1, and $E_1 < E_2 < E_3$, and so on. The state of zero energy occurs when $n = \infty$. In the state of zero energy, the proton and electron are so far apart that they do not attract each other at all, and so we take their interaction energy to be zero. At closer distances, the proton and electron attract each other because they have opposite charges. A negative energy state is more stable than a state of zero energy.

Allowed energy states are called *stationary states* in the quantum theory. The stationary state of lowest energy is called the *ground state*, and the states of higher energies are called *excited states*. The state with $n = 1$ is the ground state; the state with $n = 2$ is the *first excited state;* the state with $n = 3$ is the *second excited state;* and so on.

7-9. ATOMS EMIT OR ABSORB ELECTROMAGNETIC RADIATION WHEN THEY UNDERGO TRANSITIONS FROM ONE STATIONARY STATE TO ANOTHER

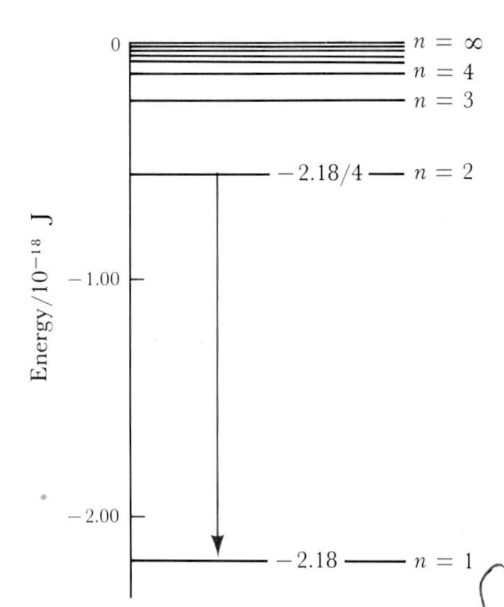

Figure 7-13 The electron in a hydrogen atom undergoing a transition from the $n = 2$ state to the $n = 1$ state. The energy of the $n = 1$ state is less than that in the $n = 2$ state, and so this transition is accompanied by the emission of electromagnetic radiation.

When an atom is in a stationary state, it does not absorb or emit electromagnetic radiation. When it undergoes a transition from one stationary state to another, however, it emits or absorbs electromagnetic radiation. Consider a hydrogen atom that undergoes a transition from the $n = 2$ state to the $n = 1$ state (Figure 7-13). In this case the electron goes from a higher energy state to a lower energy state. The energy released is given by $E_2 - E_1$ and is emitted as electromagnetic radiation. The frequency of the emitted radiation satisfies the equation

$$\Delta E = E_2 - E_1 = h\nu_{2\to1} \tag{7-5}$$

Equation (7-5) is identical to Planck's original quantum hypothesis (Equation 7-2).

Equation (7-5) can be written in a form that emphasizes the law of conservation of energy. Initially, the hydrogen atom is in state 2, with energy E_2. It then makes a transition to state 1, with energy E_1, and emits a photon of energy $h\nu_{2\to1}$. The total energy after the transition is $E_1 + h\nu_{2\to1}$, and the total energy before the transition is E_2. According to the law of conservation of energy,

$$E_2 = E_1 + h\nu_{2\to1} \tag{7-6}$$

Equation (7-6) is exactly the same as Equation (7-5). We can obtain the frequency of the electromagnetic radiation emitted by solving Equation (7-5) or (7-6) for $\nu_{2\to1}$:

$$\nu_{2\to1} = \frac{E_2 - E_1}{h}$$

If we examine the transition from an arbitrary excited state ($n = 2, 3, 4, \ldots$) to the ground state ($n = 1$), then we find a series of lines whose frequencies are

$$\nu_{n\to 1} = \frac{E_n - E_1}{h} \qquad n = 2, 3, 4, \ldots \tag{7-7}$$

If we substitute Equation (7-4) into Equation (7-7), then we get

$$\nu_{n\to 1} = \left(\frac{2.18 \times 10^{-18}\,\text{J}}{6.626 \times 10^{-34}\,\text{J}\cdot\text{s}}\right)\left(\frac{1}{1^2} - \frac{1}{n^2}\right) \qquad n = 2, 3, 4, \ldots$$

$$= (3.29 \times 10^{15}\,\text{s}^{-1})\left(\frac{1}{1^2} - \frac{1}{n^2}\right) \qquad n = 2, 3, 4, \ldots \tag{7-8}$$

Table 7-5 Frequencies and wavelengths of the lines in the Lyman series for hydrogen

n	$\nu_{n\to 1}/10^{15}\,s^{-1}$	$\lambda_{n\to 1}/nm$
2	2.47	122
3	2.92	103
4	3.08	97.3
5	3.16	95.0
6	3.20	93.7
⋮	⋮	⋮
∞	3.29	91.2

Equation (7-8) predicts that there is a series of lines in the hydrogen atom emission spectrum that correspond to transitions from state n to state 1. The frequencies of these lines are given by Equation (7-8) with $n = 2, 3, 4$, and so on. The values of these frequencies are given in Table 7-5. This series of lines occurs in the ultraviolet region and is called the *Lyman series*. The agreement between the frequencies or wavelengths calculated from Equation (7-8) and those observed experimentally is excellent.

Equation (7-8) indicates that the lines of the Lyman series emission spectrum are caused by transitions from excited states ($n > 1$) to the ground state ($n = 1$). Figure 7-14 shows that there is also a series due to transitions from higher excited states to the $n = 2$ state. In this case Equation (7-7) becomes

$$\nu_{n\to 2} = \frac{E_n - E_2}{h} \qquad n = 3, 4, 5, \ldots$$

and Equation (7-8) becomes

$$\nu_{n\to 2} = (3.29 \times 10^{15}\,\text{s}^{-1})\left(\frac{1}{2^2} - \frac{1}{n^2}\right) \qquad n = 3, 4, 5, \ldots \tag{7-9}$$

Equation (7-9) predicts the series of lines due to transitions from state n to state 2. This series of lines is called the *Balmer series.*

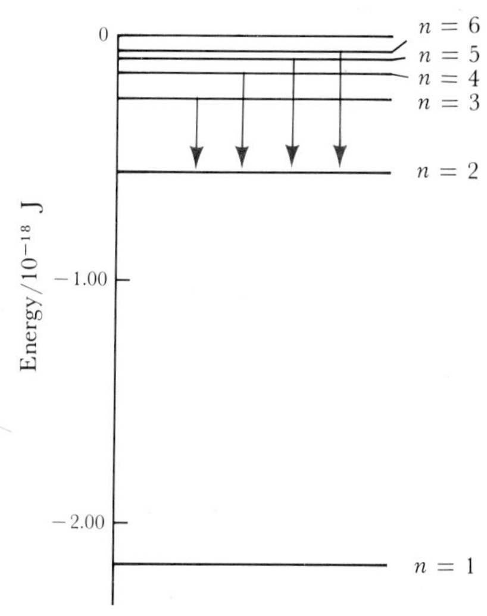

Figure 7-14 Transitions from higher energy states to the $n = 2$ state for a hydrogen atom. Each transition is accompanied by the emission of a photon. This series of lines is called the Balmer series.

Example 7-5: Calculate the frequencies and wavelengths of the first few lines in the Balmer series.

Solution: We use Equation (7-9) and set up a table like Table 7-5. We let $n = 3, 4, 5$, and so on in Equation (7-9). For example, with $n = 3$ we have

$$\nu_{3\to 2} = (3.29 \times 10^{15}\,\text{s}^{-1})\left(\frac{1}{4} - \frac{1}{9}\right) = 4.57 \times 10^{14}$$

The other values of $\nu_{n\to 2}$ are

n	$\nu_{n\to 2}/10^{14}\ s^{-1}$
3	4.57
4	6.17
5	6.91
6	7.31
⋮	⋮
∞	8.23

To calculate the corresponding wavelengths, we use Equation (7-1). For example,

$$\lambda_{3\to 2} = \frac{c}{\nu_{3\to 2}} = \frac{3.00 \times 10^{8}\ \text{m}\cdot\text{s}^{-1}}{4.57 \times 10^{14}\ \text{s}^{-1}} = 6.56 \times 10^{-7}\ \text{m} = 656\ \text{nm}$$

The other values of $\lambda_{n\to 2}$ are

n	$\lambda_{n\to 2}/nm$
3	656
4	486
5	434
6	410
⋮	⋮
∞	365

The agreement of these predictions with experimental data is excellent (Figure 7-6).

As we said before, *emission spectra* are obtained when a gas is placed in a discharge tube and a spark is discharged through the gas. The discharge is a pulse of energy that promotes the atoms into excited states. As the atoms return to their ground state, the electrons fall down through the allowed energy states and produce the observed emission spectrum.

Absorption spectra are observed experimentally when an atomic gas is irradiated with electromagnetic radiation containing all frequencies (Figure 7-15). Only radiation of certain frequencies is absorbed by the gas.

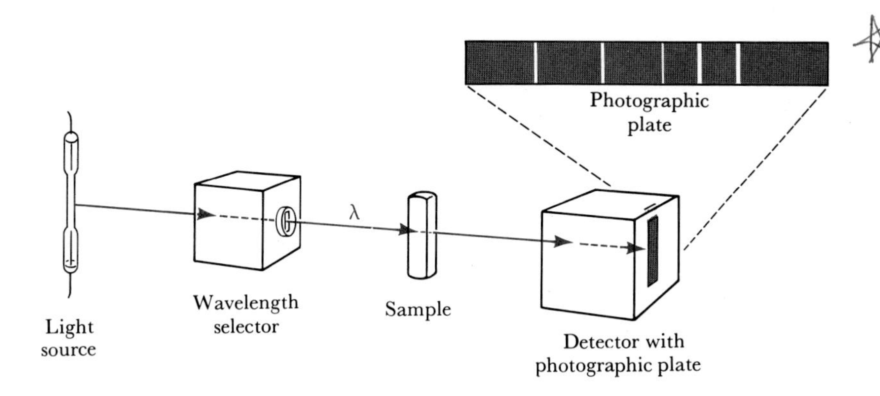

Figure 7-15 A schematic diagram of an atomic absorption experiment. Electromagnetic radiation is passed through a gas sample and then through a wavelength selector (a prism) and then is passed through the sample. The transmitted beam is then directed onto a detector, such as a photographic plate. Only certain wavelengths are absorbed by the sample gas, and the photographic plate is exposed at all wavelengths except those absorbed by the sample. The wavelengths absorbed by the sample appear as unexposed lines on the plate.

Let's consider transitions from the ground state ($n = 1$) of a hydrogen atom to some excited state ($n > 1$). Before the transition, we have a hydrogen atom in its ground state with energy E_1 and a photon of energy $h\nu_{1\to n}$. After the transition, we have a hydrogen atom in state n with energy E_n. Conservation of energy requires that the total energy before the transition ($E_1 + h\nu_{1\to n}$) be equal to the total energy after the transition (E_n):

$$h\nu_{1\to n} + E_1 = E_n \qquad n = 2, 3, 4, \ldots \qquad (7\text{-}10)$$

If we solve Equation (7-10) for $\nu_{1\to n}$, we find that

$$\nu_{1\to n} = \frac{E_n - E_1}{h} \qquad n = 2, 3, 4, \ldots \qquad (7\text{-}11)$$

Note that the frequency of absorption is the same as the frequency of emission given by Equation (7-7). Equation (7-11) represents the Lyman series in the absorption spectrum of a hydrogen atom. Figure 7-16 shows the transitions that correspond to the Lyman series in an absorption spectrum.

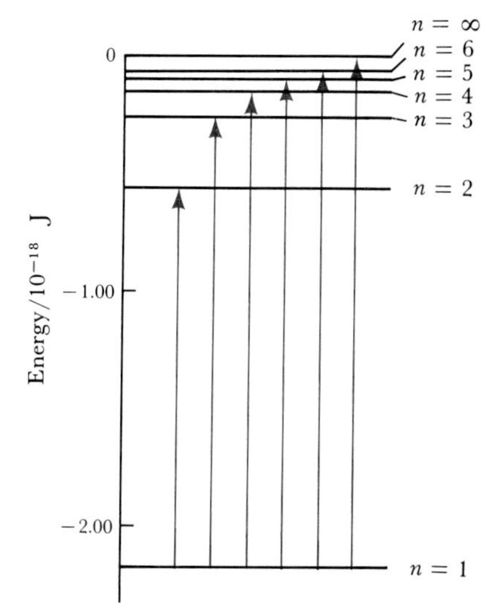

Figure 7-16 A hydrogen atom in its ground electronic state ($n = 1$) can absorb electromagnetic radiation. When this occurs, the electron is promoted to an excited state.

Example 7-6: Use Equation (7-4) to calculate the ionization energy of the hydrogen atom.

Solution: The ionization energy is the energy required to completely remove the electron from the ground state of the atom. Therefore, the ionization must correspond to the transition from $n = 1$ to $n = \infty$. The energy associated with this transition is $E_\infty - E_1$. If we use Equation (7-4) for E_n, then we obtain for the ionization energy

$$E_\infty - E_1 = 0 - \left(\frac{-2.18 \times 10^{-18}\,\text{J}}{1^2}\right) = I = 2.18 \times 10^{-18}\,\text{J}$$

This is the ionization energy per atom. If we multiply this result by Avogadro's number, then we obtain the ionization energy per mole:

$$I = \left(2.18 \times 10^{-18}\frac{\text{J}}{\text{atom}}\right)\left(\frac{6.02 \times 10^{23}\ \text{atoms}}{1\ \text{mol}}\right) = 1310\ \text{kJ}\cdot\text{mol}^{-1}$$

which is in excellent agreement with the value given in Table 7-1.

7-10. A WAVE FUNCTION IS USED TO CALCULATE THE PROBABILITY THAT AN ELECTRON WILL BE FOUND IN SOME REGION OF SPACE

Besides giving a discrete set of energies for atoms, the Schrödinger equation also provides an associated set of functions, called *wave functions* or *orbitals*. Wave functions are customarily denoted by the Greek letter psi, ψ, and are functions of the position of the electron. We emphasize this dependence by writing $\psi = \psi(x, y, z)$, where x, y, and z are coordinates that are used to denote the position of the electron. The square of a wave function, ψ^2, has a direct physical interpretation. The value of the square of a wave function, $\psi^2(x, y, z)$, is a *probability density* in the sense that $\psi^2 \Delta V$ is the probability that the electron will be found in a small volume ΔV surrounding the point (x, y, z). This is a profound statement because it says that we cannot locate the electron precisely; we only can assign a *probability* that the electron is in a certain region.

When we solve the Schrödinger equation for a particular atom, molecule, or ion, we find two things: (1) a set of discrete energies and (2) a set of wave functions, or probabilities. The allowed energies of the electron in a hydrogen atom are given by Equation (7-4). Any particular energy is specified by the integer n, called a *quantum number*. Although only one quantum number is needed to specify the energy of an electron, three quantum numbers are required to specify the wave functions. These three quantum numbers are denoted by n, ℓ, and m_ℓ. Let's look at the significance of each of these quantum numbers in turn and see how they apply to the hydrogen atom.

The quantum number n is called the *principal quantum number*. It alone determines the energy of the electron in a hydrogen atom, and we have already seen that n can take on the values $n = 1, 2, 3$, and so on.

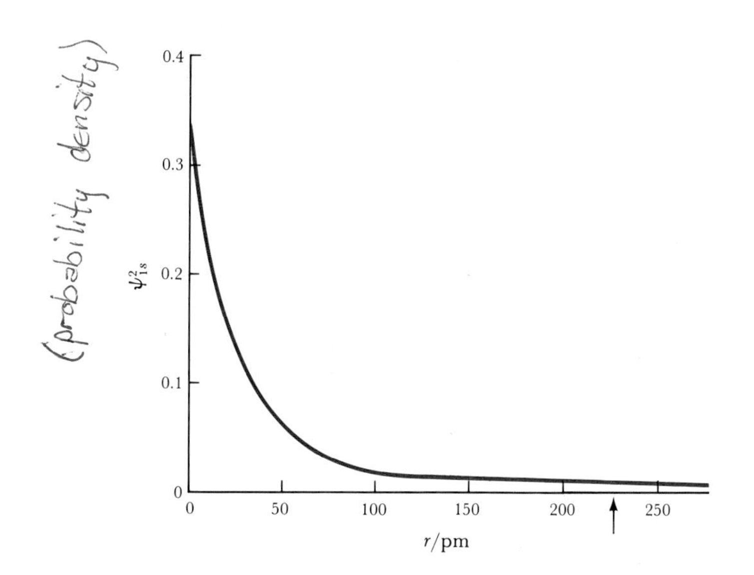

Figure 7-17 A graph of ψ_{1s}^2 versus r. Even though the electron is most likely to be found near the nucleus, the curve never quite falls to zero. Thus there is a nonzero probability, however small, of finding the electron at *any* distance from the nucleus. The arrow indicates the distance beyond which there is only a 1 percent chance of finding the electron.

When $n = 1$, the energy is the lowest allowed value. This is the ground state of the hydrogen atom. The wave function that describes the ground state of the hydrogen atom depends upon the distance of the electron from the proton and can be written $\psi(r)$, where r is the distance of the electron from the nucleus. For reasons that we'll soon know, the ground-state wave function of the hydrogen atom is denoted by ψ_{1s} rather than by just ψ_1. The probability density, $\psi_{1s}^2(r)$, is plotted in Figure 7-17. Note that probability density falls off rapidly with distance. Because ψ_{1s}^2 depends upon only the magnitude of r and not upon the direction of r in space, ψ_{1s}^2 is said to be *spherically symmetric*.

Several other representations of wave functions, or orbitals, are more lucid than simply plotting the square of the wave function. For example, we can represent the 1*s* orbital by the stippled diagram in Figure 7-18(a). The number of dots in a volume ΔV is proportional to the probability of finding the electron in that volume. The relation

Figure 7-18 Three different representations of a hydrogen 1*s* orbital, or wave function. (a) The density of the stippled dots in any small region is proportional to the probability of finding the electron in that region. (b) The curves are electron probability contours and enclose regions within which there is a certain probability of finding the electrons. Each mark along the axes represents 100 pm. (c) The sphere encloses a 99 percent probability of finding the electron. Realize that the 1*s* orbital is spherically symmetric and that (a) and (b) represent cross sections through a sphere.

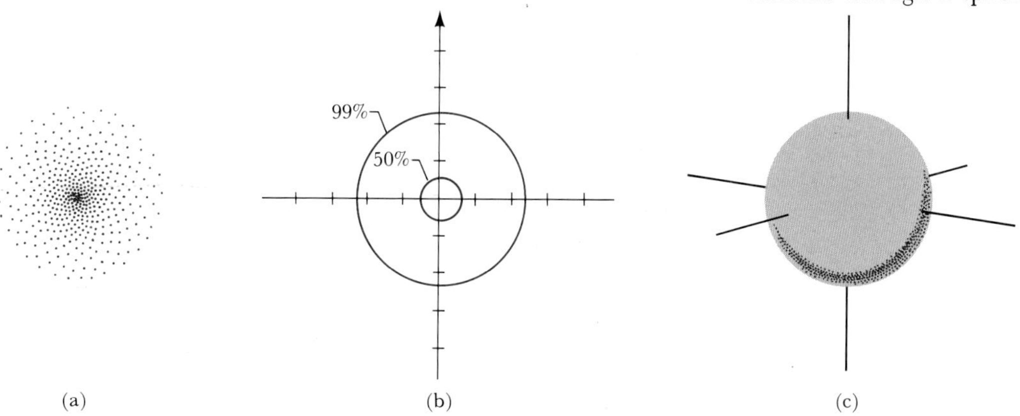

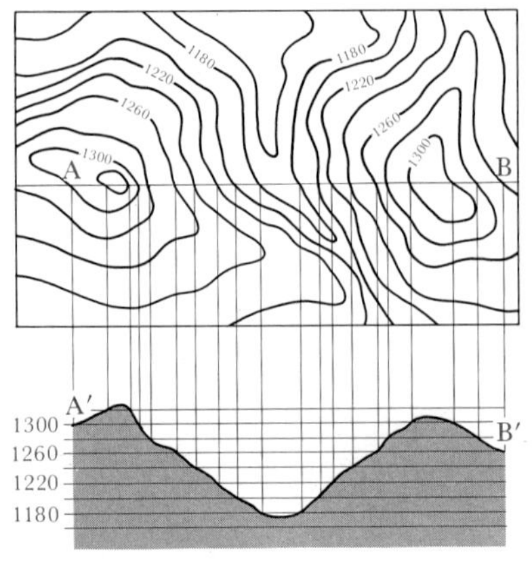

Figure 7-20 Side views of some hills, along with the contour map describing them. The lines in the contour map connect points at the same elevation. Closely spaced lines mean steep slopes and widely separated lines represent more level areas.

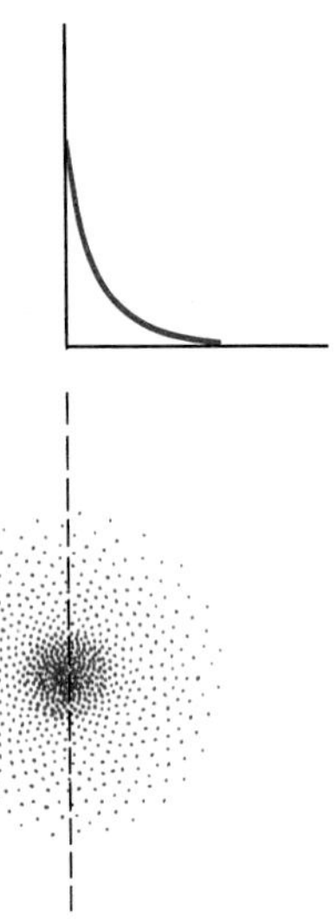

Figure 7-19 The relation between a plot of ψ_{1s}^2 versus r and the stippled representation of a 1*s* orbital. Both show that the probability of finding the electron around some point decays rapidly with distance from the nucleus.

between a plot of ψ_{1s}^2 versus r and the stippled representation of Figure 7-18 is shown in Figure 7-19.

We also can represent a 1*s* orbital by a *contour diagram*. Contour maps are maps that indicate land elevation. Figure 7-20 presents a top and side view of some hills plus the associated contour map. In the contour diagram of a 1*s* orbital (Figure 7-18b), the contour lines define the probability of finding the electron within a given region: Figure 7-18(b) shows contours for 50 percent and 99 percent probability. The contours are circular because they represent a cross section of the spherically symmetric 1*s* orbital.

A third representation of the 1*s* orbital shows the three-dimensional surface that constitutes the space where the electron has a certain chance of being found. The sphere in Figure 7-18(c) represents the 99 percent contour and is simply the three-dimensional representation of the outer contour in Figure 7-18(b). The representation in Figure 7-18(c) has the advantage of portraying clearly the three-dimensional shape of the orbital.

7-11. THE AZIMUTHAL QUANTUM NUMBER DETERMINES THE SHAPE OF AN ORBITAL

The principal quantum number, n, specifies the size, or the extent, of an orbital. The quantum number ℓ specifies the shape of an orbital. Orbitals with different values of ℓ have different shapes. This second quantum number is called the *azimuthal quantum number*, although we could just as well call it the shape quantum number. A direct result of solving the Schrödinger equation is that ℓ is restricted to the values $0, 1, \ldots, n-1$. The allowed values of ℓ depend upon the value of n according to

n	ℓ
1	0
2	0, 1
3	0, 1, 2
4	0, 1, 2, 3
⋮	⋮

For historical reasons, the values of ℓ are designated by letters:

ℓ	*Designation*
0	*s*
1	*p*
2	*d*
3	*f*
⋮	⋮

The letters *s*, *p*, *d*, and *f* stand for *s*harp, *p*rincipal, *d*iffuse, and *f*undamental, which are the designations of the series in the atomic emission spectra of the alkali metals. For $\ell = 4$ and greater, the letters follow alphabetical order after *f*.

Orbitals are denoted by first writing the numerical value of n (1, 2, 3, . . .) and then following this by the letter designation for the value of ℓ (*s*, *p*, *d*, *f*, . . .). For example, an orbital for which $n = 1$ and $\ell = 0$ is called a 1*s* orbital. An orbital for which $n = 3$ and $\ell = 2$ is called a 3*d* orbital. Table 7-6 lists the orbitals for $n = 1$ through 4.

Table 7-6 The designation of orbitals by letters

n	ℓ	*Designation*
1	0	1*s*
2	0	2*s*
	1	2*p*
3	0	3*s*
	1	3*p*
	2	3*d*
4	0	4*s*
	1	4*p*
	2	4*d*
	3	4*f*

Example 7-7: Why is there no 1*p* or 3*f* orbital listed in Table 7-6?

Solution: When $n = 1$, ℓ can have only the value 0 because $\ell = 0, 1, 2, \ldots, n - 1$ and $n - 1 = 0$ when $n = 1$. Thus there is no such orbital as a 1*p* orbital. Similarly, when $n = 3$, ℓ can have only the values 0, 1, and 2. A 3*f* orbital would require that $n = 3$ and $\ell = 3$, and so there is no 3*f* orbital.

An electron described by a 1*s* orbital has an energy that is obtained from Equation (7-4) by setting $n = 1$. When $n = 2$, ℓ can be 0 or 1, and so we have two possibilities, a 2*s* and a 2*p* orbital. Both of these orbitals have a principal quantum number $n = 2$, and so an electron described by either of these orbitals has an energy E_2 in Equation (7-4). These two orbitals have different shapes, however,

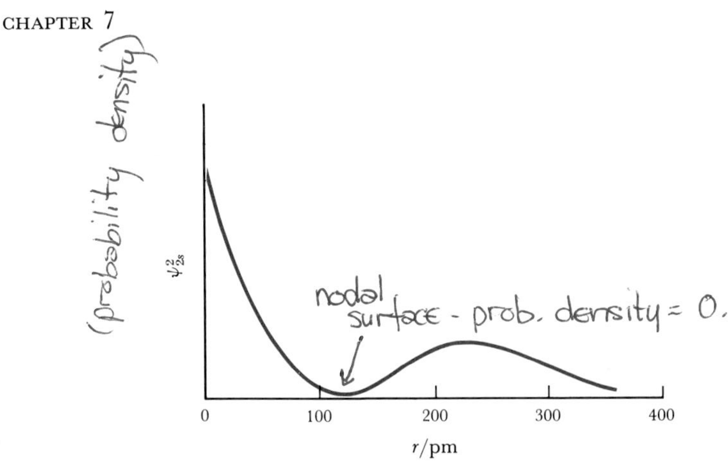

Figure 7-21 A graph of ψ^2_{2s} versus r. If we compare this curve with Figure 7-17, we see that a $2s$ orbital extends farther from the nucleus than does a $1s$ orbital. The radius of a sphere that encloses a 90 percent probability of finding the electron in a $2s$ orbital is almost 500 pm; the corresponding radius for a $1s$ orbital is about 200 pm. Because an s orbital is spherically symmetric, there is a spherical nodal surface in a $2s$ orbital (see Figure 7-25).

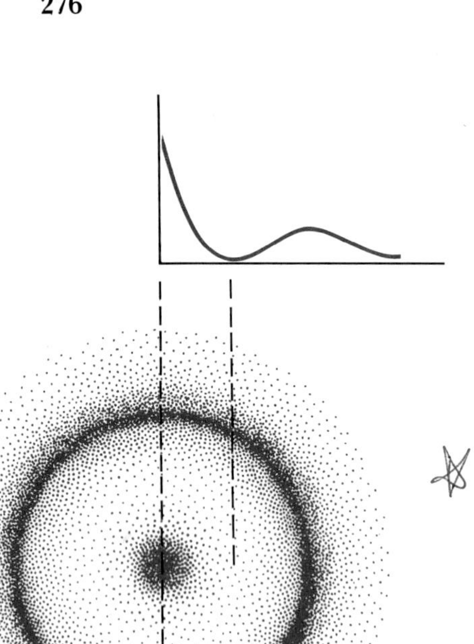

Figure 7-22 The relationship between a plot of ψ^2_{2s} versus r and a stippled representation of a $2s$ orbital. The distance at which the plot of ψ^2_{2s} touches zero indicates a spherical nodal surface in the orbital. Remember that s orbitals are spherically symmetric and that this is a cross section of a three-dimensional diagram.

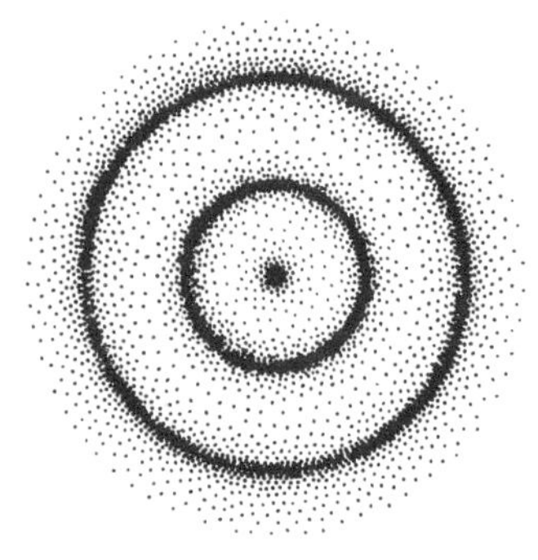

Figure 7-23 A stippled representation of a $3s$ orbital, showing that it has two nodal surfaces.

because they are associated with different values of l. All s orbitals are spherically symmetric. In Figure 7-21, ψ^2_{2s} is plotted versus r. By comparing this graph with Figure 7-17, we see that a $2s$ orbital is larger than a $1s$ orbital. Figure 7-22 illustrates the relation between a plot of ψ^2_{2s} versus r and a stippled diagram representing the $2s$ orbital probability density. Figure 7-21 shows that the probability density is zero over a spherical surface whose radius is 106 pm. It is common for orbitals to have surfaces on which the probability density is zero. Such surfaces are called *nodal surfaces.*

The representations shown in Figure 7-18(b) and (c) for a $1s$ orbital would look the same, only larger, for a $2s$ orbital. A $3s$ orbital has two nodal surfaces, as depicted in Figure 7-23. Its contour diagram and its surface of 99 percent probability look the same as those of a $2s$ orbital, simply larger. Figure 7-24 shows the 99 percent contour surfaces for $1s$, $2s$, and $3s$ orbitals. We see from this figure that n determines the size, or spatial extent, of an orbital.

We also have a $2p$ orbital to consider when $n = 2$. Figure 7-25(a) shows a stippled diagram of a $2p$ orbital. The most obvious feature of a $2p$ orbital is that it is *not* spherically symmetric. The representation shown in Figure 7-25(c) shows the three-dimensional shape of a $2p$ orbital. When viewed along the z axis, the $2p$ orbital appears to be circular. We say that the $2p$ orbital is *cylindrically symmetric* about its long axis (the z axis in Figure 7-25). Note that the xy plane that bisects the $2p$ orbital is a nodal surface; the $2p$ orbital vanishes everywhere on that surface. Just as all s orbitals are spherically symmetric, all p orbitals are cylindrically symmetric about their long axis. Figure 7-26 shows two representations of a $3p$ orbital. Although a $3p$ orbital looks more complicated than a $2p$ orbital, it nevertheless is

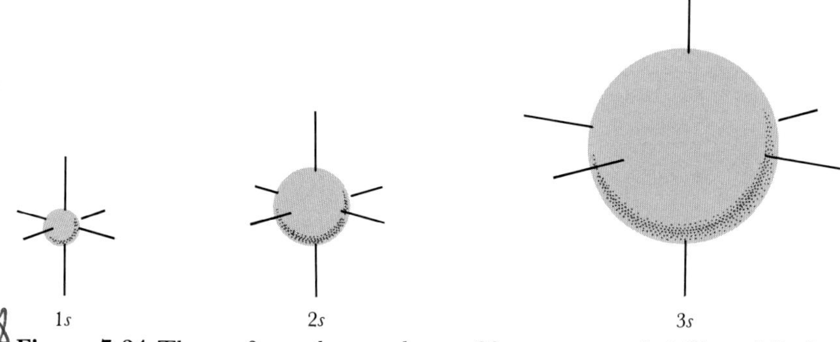

Figure 7-24 The surfaces that enclose a 99 percent probability of finding the electron in a 1*s*, 2*s*, and 3*s* orbital. Because *s* orbitals are spherically symmetric, these surfaces are spherical. The radii of the spheres shown above are in a ratio of about 1:2:5.

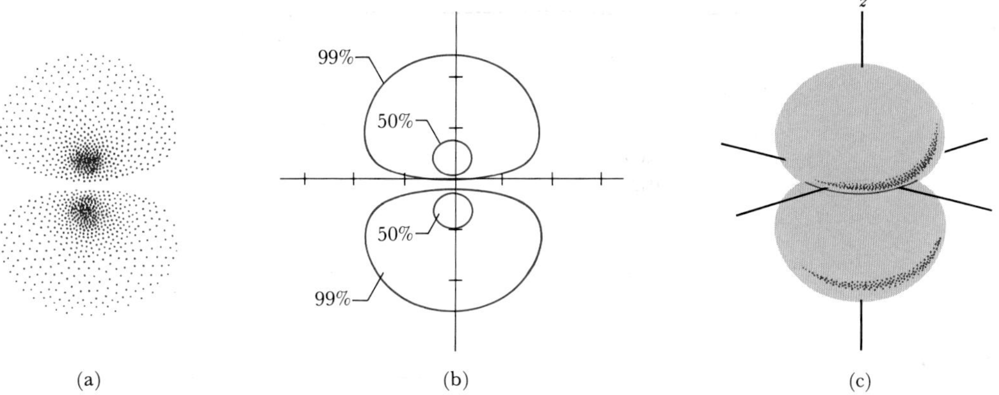

Figure 7-25 Three representations of a 2*p* orbital: (a) a stippled representation, (b) the 50 percent and 99 percent probability contours, and (c) the surfaces that enclose a 99 percent probability of finding the 2*p* electron within them. The representation in (c) clearly depicts the shape of a 2*p* orbital, which is *not* spherically symmetric. The first two representations are cross sections, and the three-dimensional representation is obtained by rotating each figure around the *z* axis. The resulting orbital is said to be cylindrically symmetric. The marks on the axes in (b) indicate intervals of 50 pm.

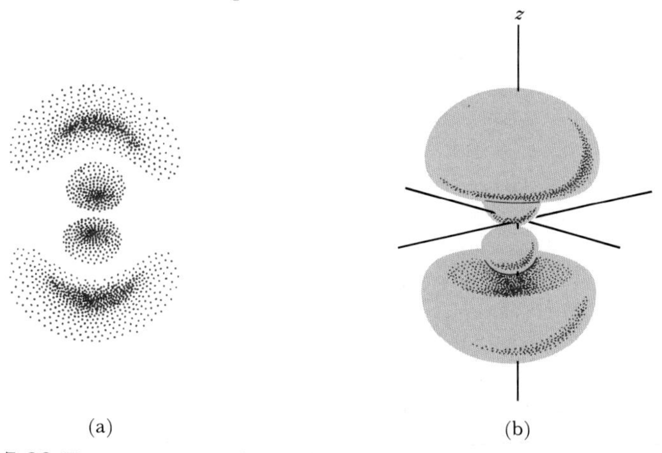

Figure 7-26 Two representations of a 3*p* orbital: (a) a stippled representation and (b) the surfaces that enclose a 99 percent probability of finding the 3*p* electron within them.

cylindrically symmetric about its long axis. A $3p$ orbital differs from a $2p$ orbital in being larger (because n is larger) and in having more nodal surfaces. The most important property of p orbitals for our purposes is that they are directed along an axis, as shown in Figures 7-25 and 7-26.

7-12. THE MAGNETIC QUANTUM NUMBER DETERMINES THE SPATIAL ORIENTATION OF AN ORBITAL

The third quantum number, m_ℓ, called the *magnetic quantum number*, determines the spatial orientation of an orbital. It turns out that the magnetic quantum number can assume only the values ℓ, $\ell - 1$, $\ell - 2, \ldots, 0, -1, -2, \ldots, -\ell$. The allowed values of m_ℓ depend upon the value of ℓ according to

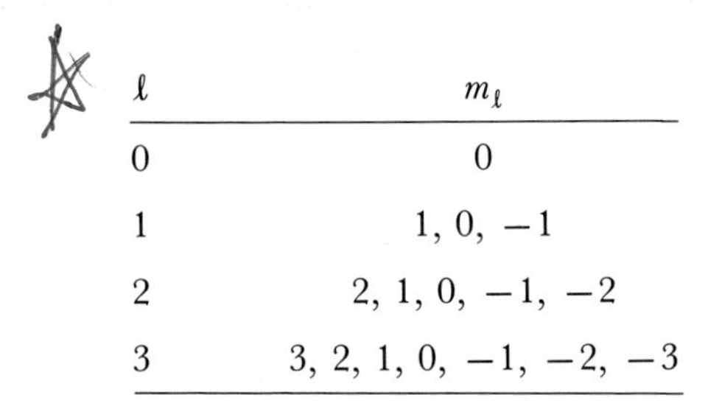

ℓ	m_ℓ
0	0
1	1, 0, −1
2	2, 1, 0, −1, −2
3	3, 2, 1, 0, −1, −2, −3

For an s orbital, $\ell = 0$, and so the only value that m_ℓ can have is 0. For a p orbital, $\ell = 1$ and so m_ℓ can have the values +1, 0, and −1. Table 7-7 summarizes the allowed values of ℓ and m_ℓ for $n = 1$ through $n = 4$.

Table 7-7 The allowed values of ℓ and m_ℓ for $n = 1$ through 4

n	ℓ	m_ℓ	*Orbital*	*Number of orbitals*
1	0	0	1*s*	1
2	0	0	2*s*	1
	1	1, 0, −1	2*p*	3
3	0	0	3*s*	1
	1	1, 0, −1	3*p*	3
	2	2, 1, 0, −1, −2	3*d*	5
4	0	0	4*s*	1
	1	1, 0, −1	4*p*	3
	2	2, 1, 0, −1, −2	4*d*	5
	3	3, 2, 1, 0, −1, −2, −3	4*f*	7

Example 7-8: Without referring to Table 7-7, list all the values of ℓ and m_ℓ that are allowed for $n = 3$.

Solution: When $n = 3$, ℓ can have the values 0, 1, and 2. Thus we have a $3s$ orbital ($\ell = 0$), a $3p$ orbital ($\ell = 1$), and a $3d$ orbital ($\ell = 2$). For the $3s$ orbital, $\ell = 0$ and so m_ℓ must also equal 0. For the $3p$ orbital, $\ell = 1$ and so m_ℓ can be $+1$, 0, or -1. For the $3d$ orbital, $\ell = 2$ and so m_ℓ can be $+2$, $+1$, 0, -1, or -2.

Table 7-7 shows that there is only one s orbital for each value of n, three p orbitals for $n \geq 2$, five d orbitals for $n \geq 3$, and seven f orbitals for $n \geq 4$. Let np denote a $2p$ orbital, a $3p$ orbital, and so on. The three np orbitals differ by the value of the magnetic quantum number. All np orbitals have the same shape because they all have the same value of ℓ $(=1)$, but they have different orientations because they all have different values of m_ℓ. The value of m_ℓ determines the spatial orientation of an orbital. The three $2p$ orbitals are shown in Figure 7-27. One $2p$ orbital is directed along the z axis as in Figure 7-25. The other two have the same shape as the one directed along the z axis but are directed along the x axis and the y axis. The p orbitals are designated by p_x, p_y, and p_z, with the subscripts indicating the axis along which the orbital is directed.

We could go on to discuss d orbitals and f orbitals, but for most of the topics we shall discuss, s orbitals and p orbitals will be sufficient (although d orbitals are used in Chapter 23).

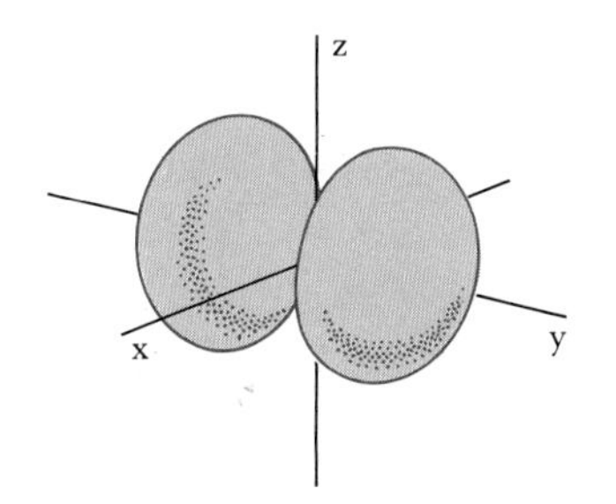

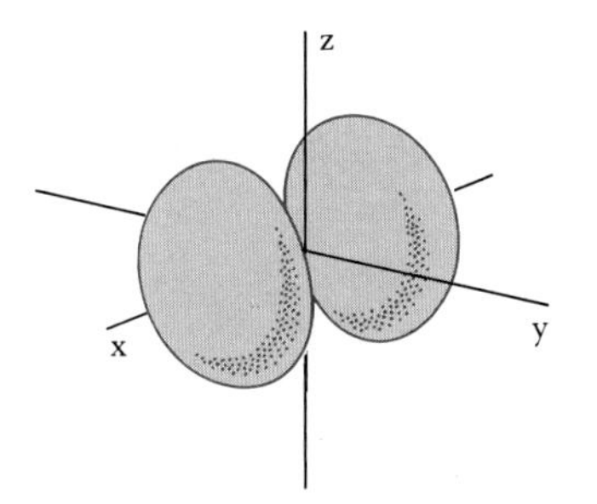

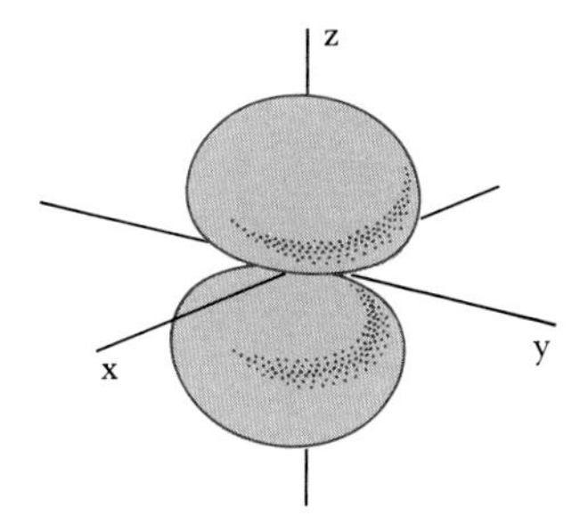

Figure 7-27 The three $2p$ orbitals. Each has the same shape, but their spatial orientations differ. This is because they have the same azimuthal quantum number ($\ell = 1$) but different magnetic quantum numbers ($m_\ell = +1, 0, -1$). Recall that ℓ determines the shape of an orbital and that m_ℓ determines its orientation. The three p orbitals are directed along the x, y, and z axes and are sometimes designated by p_x, p_y, and p_z.

7-13. AN ELECTRON POSSESSES AN INTRINSIC SPIN

As we know, the Schrödinger equation yields three quantum numbers, n, ℓ, and m_ℓ. When first presented, this equation explained a great deal of experimental data, but some scattered observations still did not fit into the picture. For example, close examination of some atomic spectral lines shows that they actually consist of two closely spaced lines. Even though a fine detail, the splitting of spectral lines was perplexing. In 1926, Wolfgang Pauli, a young German physicist, argued that this splitting could be explained if the electron exists in two different states. Shortly after this, two Dutch scientists, George Uhlenbeck and Samuel Goudsmit, identified these two different states with a property called the *intrinsic electron spin*. The electron spins like a top in one of two directions about its axis. The intrinsic spin of an electron introduces a fourth quantum number, called the *spin quantum number*, denoted by m_s. It designates the spin state of the electron and takes on one of two possible values: $+\frac{1}{2}$ or $-\frac{1}{2}$.

Example 7-9: Deduce the possible sets of the four quantum numbers (n, ℓ, m_ℓ, and m_s) when $n = 2$.

Solution: When $n = 2$, ℓ can be 0 or 1. Let's consider the case $\ell = 0$ first. If $\ell = 0$, then $m_\ell = 0$. Thus, we have so far that $n = 2$, $\ell = 0$, and $m_\ell = 0$. The spin quantum number can have the value $+\frac{1}{2}$ or $-\frac{1}{2}$, regardless of the values of the other three quantum numbers. We thus have two possible sets of quantum numbers;

n	ℓ	m_ℓ	m_s
2	0	0	$+\frac{1}{2}$
2	0	0	$-\frac{1}{2}$

Now consider the case $n = 2$ and $\ell = 1$. When $\ell = 1$, m_ℓ can be $+1$, 0, or -1. Thus we have

n	ℓ	m_ℓ
2	1	1
2	1	0
2	1	-1

Each of these sets of three quantum numbers can have $m_s = +\frac{1}{2}$ or $-\frac{1}{2}$, and so

n	ℓ	m_ℓ	m_s
2	1	1	$+\frac{1}{2}$
2	1	1	$-\frac{1}{2}$
2	1	0	$+\frac{1}{2}$
2	1	0	$-\frac{1}{2}$
2	1	-1	$+\frac{1}{2}$
2	1	-1	$-\frac{1}{2}$

There are eight possible sets of the four quantum numbers when $n = 2$.

The introduction of the spin quantum number implies that it takes four quantum numbers to specify the state of the electron in a hydrogen atom. These values are

$$n = 1, 2, 3, \ldots$$
$$\ell = 0, 1, 2, \ldots, n - 1$$
$$m_\ell = \ell, \ell - 1, \ldots, 0, -1, \ldots, -\ell$$
$$m_s = +\tfrac{1}{2} \text{ or } -\tfrac{1}{2}$$

Table 7-8 The allowed combinations of the four quantum numbers for $n = 1$ through 3

n	ℓ	m_ℓ	m_s
1	0	0	$+\frac{1}{2}$ or $-\frac{1}{2}$
2	0	0	$+\frac{1}{2}$ or $-\frac{1}{2}$
	1	1	$+\frac{1}{2}$ or $-\frac{1}{2}$
		0	$+\frac{1}{2}$ or $-\frac{1}{2}$
		-1	$+\frac{1}{2}$ or $-\frac{1}{2}$
3	0	0	$+\frac{1}{2}$ or $-\frac{1}{2}$
	1	1	$+\frac{1}{2}$ or $-\frac{1}{2}$
		0	$+\frac{1}{2}$ or $-\frac{1}{2}$
		-1	$+\frac{1}{2}$ or $-\frac{1}{2}$
	2	2	$+\frac{1}{2}$ or $-\frac{1}{2}$
		1	$+\frac{1}{2}$ or $-\frac{1}{2}$
		0	$+\frac{1}{2}$ or $-\frac{1}{2}$
		-1	$+\frac{1}{2}$ or $-\frac{1}{2}$
		-2	$+\frac{1}{2}$ or $-\frac{1}{2}$

Table 7-8 summarizes the allowed combinations of the four quantum numbers for $n = 1$ through $n = 3$.

We have dwelt on the quantum theory of the hydrogen atom not only because this is the simplest atom but also because it serves as a prototype for other atoms. The application of the quantum theory to other atoms yields a similar set of orbitals that can be used to build up their electronic structure. We shall do this in the next chapter and find a beautiful correlation between electronic structure and the periodic table.

SUMMARY

First ionization energy is a periodic property. The values of successive ionization energies suggest that electrons in atoms are arranged in a shell structure (Figure 7-3). The shell structures deduced from a study of successive ionization energies are depicted in Lewis electron-dot formulas given in Table 7-2.

Atoms emit line spectra, which provide a detailed indication of their electronic structure. The emission spectrum of atomic hydrogen is the simplest atomic spectrum because the hydrogen atom is the simplest atom, having only one electron. The spectrum consists of several series of lines, such as the Lyman series and the Balmer series, all accounted for by the quantum theory.

The first quantum theory was developed by Planck in 1900 to describe the distribution of the intensity of blackbody radiation. He assumed that electromagnetic radiation could be emitted only in little packets of energy, given by $E = h\nu$. This idea was used five years later by Einstein to describe electromagnetic radiation as a beam of photons.

In 1924 de Broglie proposed that matter has wavelike properties in addition to particle-like properties and presented a simple formula for calculating the wavelength of matter waves. A year later Schrödinger proposed what is now called the Schrödinger equation, which is the central equation of the quantum theory and governs the motion of small particles, such as electrons, atoms, and molecules. One consequence of the Schrödinger equation is that the electrons in atoms and molecules can have only certain discrete, or quantized, energies. In addition, Schrödinger showed that an electron in an atom or molecule must be described by a wave function, or orbital, which is obtained by solving the Schrödinger equation. The square of a wave function gives the probability density associated with finding the electron in some region of space. The hydrogen atom wave functions serve as the prototype for all other atoms. The wave functions are specified by three quantum numbers: n, the principal quantum number; ℓ, the azimuthal quantum number; and m_ℓ, the magnetic quantum number.

Orbitals with $\ell = 0$ are called *s* orbitals, orbitals with $\ell = 1$ are called *p* orbitals, orbitals with $\ell = 2$ are called *d* orbitals, and orbitals with $\ell = 3$ are called *f* orbitals. For a given value of n, there is one *s* orbital and there may be three *p* orbitals, five *d* orbitals, and seven *f* orbitals.

To explain certain fine details in atomic spectra, Uhlenbeck, Goudsmit, and Pauli introduced a fourth quantum number, the spin quantum number, m_s, which specifies the intrinsic spin of an electron. The spin quantum number can have the value $+\frac{1}{2}$ or $-\frac{1}{2}$.

TERMS YOU SHOULD KNOW

electronic structure 251
ionization energy 251
first ionization energy 252
second ionization energy 252
shell 254
outer electron 254
Lewis electron-dot formula 256
subshell 256
wavelength, λ 257
frequency, ν 257
speed of light 257
hertz, Hz 258
electromagnetic spectrum 258
visible region 258
continuous spectrum 258
line spectrum 258

atomic spectrum 258
atomic spectroscopy 260
blackbody 261
blackbody radiation 261
quanta 262
Planck's constant, h 262
photon 263
wave-particle duality 264
de Broglie wavelength 264
X-ray diffraction 265
electron diffraction 265
electron microscope 265
quantized 266
quantum theory 266
Schrödinger equation 266
energy state 267
stationary state 268
ground state 268
excited state 268
first excited state 268
second excited state 268
Lyman series 269
Balmer series 269
emission spectrum 270
absorption spectrum 270
wave function, ψ 272
orbital 272
probability density 272
quantum number 272
principal quantum number, n 272
spherically symmetric 273
contour diagram 274
azimuthal quantum number, ℓ 274
s orbital 275
p orbital 275
d orbital 275
f orbital 275
nodal surface 276
cylindrically symmetric 276
magnetic quantum number, m_ℓ 278
intrinsic electron spin 280
spin quantum number, m_s 280

EQUATIONS YOU SHOULD KNOW HOW TO USE

$\lambda\nu = c$ (7-1) (relation between wavelength and frequency)

$E = h\nu$ (7-2) (energy of a photon)

$\lambda = \dfrac{h}{mv}$ (7-3) (de Broglie wavelength)

↑in kg

$E_n = -\dfrac{2.18 \times 10^{-18}\,\text{J}}{n^2}$ $n = 1, 2, 3, \ldots$ (7-4) (energies of the electron in a hydrogen atom)

$\Delta E = E_2 - E_1$ $\Delta E = h\Delta\nu$

$\nu_{n\to1} = (3.29 \times 10^{15}\,\text{s}^{-1})\left(\dfrac{1}{1^2} - \dfrac{1}{n^2}\right)$ $n = 2, 3, 4, \ldots$ (7-8) (Lyman series frequencies)

$\nu_{n\to2} = (3.29 \times 10^{15}\,\text{s}^{-1})\left(\dfrac{1}{2^2} - \dfrac{1}{n^2}\right)$ $n = 3, 4, 5, \ldots$ (7-9) (Balmer series frequencies)

$n = 1, 2, \ldots$

$\ell = 0, 1, 2, \ldots, n - 1$

$m_\ell = \ell, \ell - 1, \ell - 2, \ldots, 0, -1, -2, \ldots, -\ell$

$m_s = +\frac{1}{2}$ or $-\frac{1}{2}$

$h = 6.626 \times 10^{-34}$ J·s

1 pm = 10^{-12} m 1nm = 10^{-9} m

PROBLEMS

IONIZATION ENERGIES

7-1. Arrange the following species in order of increasing ionization energy:

He Ar Ne Li^+

7-2. Arrange the following species in order of increasing ionization energy:

F Ne Be^{2+} Li^+

7-3. Use the data in Table 7-1 to plot the logarithms of the ionization energies of the lithium atom versus the number of electrons removed. What does the plot suggest about the electronic structure of lithium?

7-4. Use the data in Table 7-1 to plot the logarithms of the ionization energies of beryllium versus the number of electrons removed. Compare your plot to Figure 7-3.

LEWIS ELECTRON-DOT FORMULAS

7-5. Draw Lewis electron-dot formulas for all the alkali metal atoms and for all the halogen atoms. What is the similarity in all the alkali metal atom formulas? In all the halogen formulas?

7-6. Draw Lewis electron-dot formulas for all the noble gases. What is the similarity in all these formulas?

7-7. Write the Lewis electron-dot formula for each of the following species:

B^+ N^{3-} F^- O^{2-} Na^+

7-8. Write the Lewis electron-dot formula for each of the following species:

Ar S S^{2-} Al^{3+} Cl^-

ELECTROMAGNETIC RADIATION

7-9. A helium-neon laser produces light of wavelength 633 nm. What is the frequency of this light?

7-10. The radiation given off by a sodium lamp, which is used in streetlights, has a wavelength of 589.2 nm. What is the frequency of this radiation?

7-11. The first ionization energy of sodium is 495.8 $kJ \cdot mol^{-1}$. What is the wavelength of light that is just sufficient to ionize one sodium atom?

7-12. The first ionization energy of neon is 2.08 $MJ \cdot mol^{-1}$. Do X-rays with a wavelength of 80 pm have sufficient energy to ionize neon?

7-13. Ultraviolet radiation is capable of destroying a variety of bacteria by rupturing chemical bonds in certain critical molecules of the bacteria. Given that the wavelength of UV light is about 100 nm, estimate the energy of the critical chemical bonds in the bacteria.

7-14. X-rays are known to cause cancer by rupturing chemical bonds in DNA molecules in human cells. Given that the average chemical bond energy in DNA molecules is about 300 $kJ \cdot mol^{-1}$, estimate the number of chemical bonds that can be ruptured by a 1-nm X-ray photon.

7-15. Calculate the energy of 1.00 mol of X-ray photons of wavelength 210 pm. The energy of 1 mol of photons is called an *einstein;* its value depends upon the energy of the photons.

7-16. A photographic plate produces an image when the incoming light dissociates a AgBr molecule. The energy required to dissociate a AgBr molecule is 1×10^{-19} J. What is the (a) frequency and (b) wavelength of the incoming light?

7-17. Sunlight reaches the earth's surface at Madison, WI, with an average power of about 1.0 $kJ \cdot s^{-1} \cdot m^{-2}$. If the sunlight consists of photons with an average wavelength of 510 nm, how many photons strike a 1-cm^2 area per second?

7-18. The human eye can detect as little as 2.35×10^{-18} J of green light of wavelength 510 nm. Calculate the minimum number of photons that can be detected by the human eye.

7-19. A nitrogen laser "lases" at 337.1 nm. If the laser produces 10 mJ per pulse, how many photons are produced per pulse?

7-20. A carbon dioxide laser produces radiation of wavelength 10.6 μm. Calculate the energy of one photon produced by this laser. If the laser produces about 1 J of energy per pulse, how many photons are produced per pulse?

BLACKBODY RADIATION

7-21. A detailed mathematical discussion of blackbody radiation would allow us to derive a famous formula connecting λ_{max}, the wavelength at which the blackbody radiation intensity is a maximum, and the temperature:

$$\lambda_{max} T = 2.89 \times 10^{-3}\ \mathrm{m \cdot K}$$

Use this equation and Figure 7-8 to estimate the surface temperature of the sun.

7-22. The color of a hot metal object can be used to estimate its temperature. Using the formula given in Problem 7-21, estimate the temperature if a metal object is

(a) dull red, with $\lambda_{max} = 800$ nm
(b) red hot, with $\lambda_{max} = 680$ nm
(c) yellow hot, with $\lambda_{max} = 600$ nm
(d) white hot, with $\lambda_{max} = 510$ nm

7-23. Sirius, one of the brightest stars, has a surface temperature of 11,000 K. What color does Sirius appear to be?

7-24. The stars called red giants are, as their name implies, red. What is their surface temperature?

DE BROGLIE WAVELENGTH

7-25. Calculate the de Broglie wavelength of a proton traveling at a speed of $1.00 \times 10^5\ \mathrm{m \cdot s^{-1}}$. The mass of a proton is 1.67×10^{-27} kg.

7-26. A baseball has a mass of 5.0 oz. Calculate the de Broglie wavelength of a baseball traveling at 90 mph.

7-27. Calculate the de Broglie wavelength of a nitrogen molecule traveling with a speed of $500\ \mathrm{m \cdot s^{-1}}$.

7-28. Neutrons that are at equilibrium at a temperature T are called *thermal neutrons*. First, use Equation (5-14) to calculate the average speed of thermal neutrons at 1000 K. (The mass of a neutron is 1.67×10^{-27} kg.) Now calculate the de Broglie wavelength of a thermal neutron at 1000 K.

7-29. With what speed would an electron have to be traveling to have a de Broglie wavelength of 1.0×10^{-11} m?

7-30. Calculate the speed of an electron that has a wavelength of 100 pm.

HYDROGEN ATOMIC SPECTRUM

7-31. Give three examples of "quantized" quantities encountered in the macroscopic world.

7-32. Why do the same lines appear in both absorption spectra and emission spectra?

7-33. How much energy is required for an electron in a hydrogen atom to make a transition from the $n = 2$ state to the $n = 3$ state? What is the wavelength of a photon having this energy?

7-34. A line in the Lyman series of hydrogen has a wavelength of 1.03×10^{-7} m. Find the original energy level of the electron.

7-35. Compute the wavelength of a photon capable of ionizing a hydrogen atom in its ground state.

7-36. Use Equation (7-4) to compute the ionization energy of a hydrogen atom in which the electron is in the first excited state.

7-37. The energy levels of one-electron ions, such as He^+ and Li^{2+}, are given by the equation

$$E_n = -\frac{2.18 \times 10^{-18} Z^2}{n^2} \text{J}$$

where Z is the atomic number. Compare the measured ionization energies (Table 7-1) for He^+, Li^{2+}, and Be^{3+} ions with the values calculated from this equation.

7-38. A helium ion is called hydrogen-like because it consists of one electron and one nucleus. The Schrödinger equation can be applied to He^+, and the result that corresponds to Equation (7-4) is

$$E_n = -\frac{8.72 \times 10^{-18}}{n^2} \text{J}$$

Show that the spectrum of He^+ consists of a number of separate series, just as the spectrum of atomic hydrogen does.

QUANTUM NUMBERS AND ORBITALS

7-39. Explain what information is given by a dot-density graph of an electron orbital.

7-40. What does it mean when the contours in a contour diagram are closely spaced? When they are widely separated?

7-41. What is the total number of orbitals associated with the $n = 4$ level?

7-42. How many different orbitals are there in a $4f$ subshell?

7-43. Indicate which of the following atomic orbital designations are impossible:

(a) $7s$ (b) $1p$ (c) $5d$ (d) $2d$ (e) $4f$

7-44. Indicate which of the following atomic orbital designations are impossible:

(a) $3f$ (b) $6p$ (c) $6g$ (d) $5f$ (e) $4d$

7-45. Give all the possible sets of four quantum numbers for an electron in a $4f$ orbital.

7-46. Give all the possible sets of four quantum numbers for an electron in a $4d$ orbital.

7-47. Give the corresponding atomic orbital designations (that is, $1s$, $3p$, and so on) for electrons with the following sets of quantum numbers:

	n	l	m_l	m_s
(a)	2	1	0	$-\frac{1}{2}$
(b)	4	0	0	$+\frac{1}{2}$
(c)	5	3	-1	$-\frac{1}{2}$
(d)	3	2	$+1$	$-\frac{1}{2}$

7-48. Give the corresponding atomic orbital designations for electrons with the following sets of quantum numbers:

	n	l	m_l	m_s
(a)	3	2	2	$+\frac{1}{2}$
(b)	3	2	-2	$+\frac{1}{2}$
(c)	2	1	0	$+\frac{1}{2}$
(d)	4	3	-2	$+\frac{1}{2}$

7-49. If $l = 2$, what can you deduce about n? If $m_l = 3$, what can you say about l?

7-50. Which subshells in the following list exist?

$4p$ $2d$ $3s$ $1f$ $2p$

7-51. Indicate which of the following sets of quantum numbers are allowed (that is, possible) for an electron in an atom:

	n	l	m_l	m_s	
(a)	2	1	0	$+\frac{1}{2}$	poss.
(b)	3	0	(+1)	$-\frac{1}{2}$	not poss.
(c)	3	2	-2	$-\frac{1}{2}$	poss.
(d)	1	(1)	0	$+\frac{1}{2}$	not poss.
(e)	2	1	0	(0)	not poss.

7-52. Indicate which of the following sets of quantum numbers are allowed for an electron in an atom:

	n	l	m_l	m_s
(a)	2	0	-1	$+\frac{1}{2}$
(b)	4	2	$+2$	$-\frac{1}{2}$
(c)	5	3	0	$-\frac{1}{2}$
(d)	2	2	0	$-\frac{1}{2}$
(e)	3	1	-1	0

IONIZATION ENERGIES

7-53. Estimate the value of ΔH_{rxn} for the following reactions using the data given in Table 7-1.

(a) $Li(g) + Na^+(g) \rightarrow Li^+(g) + Na(g)$
(b) $Mg^{2+}(g) + Mg(g) \rightarrow 2Mg^+(g)$
(c) $Al^{3+}(g) + 3e^- \rightarrow Al(g)$

7-54. Estimate the value of ΔH_{rxn} for the following reactions using the data given in Table 7-1.

(a) $B(g) \rightarrow B^{3+}(g) + 3e^-(g)$
(b) $2Li(g) + Mg^{2+}(g) \rightarrow 2Li^+(g) + Mg(g)$
(c) $O^{2+}(g) + 2e^-(g) \rightarrow O(g)$

a. $Li \rightarrow Li^+ + e^-$ $\quad \Delta H^\circ_{rxn} = I_1 = 0.52\text{ MJ}$
$Na^+ + e^- \rightarrow Na$ $\quad \Delta H^\circ_{rxn} = -I_1 = -0.50\text{ MJ}$

$Li + Na^+ \rightarrow Li^+ + Na$ $\quad \Delta H^\circ_{rxn} = 0.02\text{ MJ} = \boxed{20\text{ kJ}}$

b. $Mg^{2+} + e^- \rightarrow Mg^+$ $\quad \Delta H^\circ_{rxn} = -I_2 = -1.45\text{ MJ}$
$Mg \rightarrow Mg^+ + e^-$ $\quad \Delta H^\circ_{rxn} = I_1 = 0.74\text{ MJ}$

$Mg^{2+} + Mg \rightarrow 2Mg^+$ $\quad \Delta H^\circ_{rxn} = -0.71\text{ MJ} = \boxed{-710\text{ kJ}}$

c. $Al^{3+} + e^- \rightarrow Al^{2+}$ $\quad \Delta H^\circ_{rxn} = -I_3 = -2.74\text{ MJ}$
$Al^{2+} + e^- \rightarrow Al^+$ $\quad \Delta H^\circ_{rxn} = -I_2 = -1.82\text{ MJ}$
$Al^+ + e^- \rightarrow Al$ $\quad \Delta H^\circ_{rxn} = -I_1 = -0.58\text{ MJ}$

$Al^{3+} + 3e^- \rightarrow Al$ $\quad \Delta H^\circ_{rxn} = -5.14\text{ MJ}$
$= \boxed{-5140\text{ kJ}}$

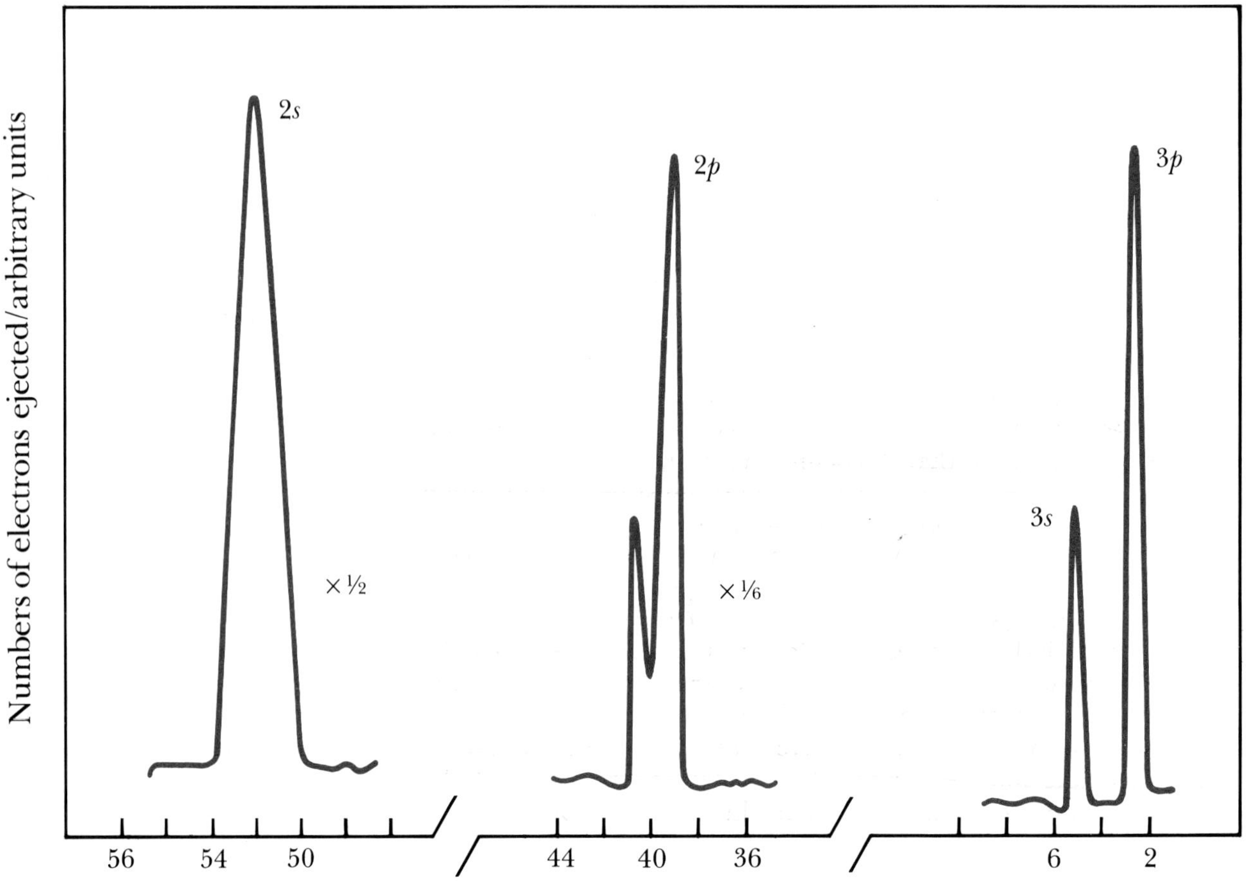

2s

2p

3p

3s

× ½

× ⅙

Numbers of electrons ejected/arbitrary units

56 54 50 44 40 36 6 2

Energy/10^{-18} J

When gas-phase atoms are irradiated with X-rays, electrons are ejected from the atoms. From a knowledge of the energy of the X-radiation and the energies of the ejected electrons, it is possible to deduce the energies of the electrons in the atoms. The peaks in the above spectrum are due to electrons ejected from the various atomic orbitals of argon. The small shoulders on the 2*p* and 3*p* peaks are a fine detail arising from the spin of the electron.

8 / Electronic Structure and Periodic Properties of Atoms

In Chapter 7 we discussed the application of the quantum theory to the hydrogen atom. Because this atom has only one electron around its nucleus, it turns out that the Schrödinger equation can be solved fairly easily for it. Although large computers must be used to obtain the wave functions and energies for atoms with two or more electrons, the resulting wave functions and energies have the same general features as those of the hydrogen atom. The energy states of all atoms can be associated with orbitals that are similar to the hydrogen atomic orbitals. Detailed calculations show, however, that the energies of these orbitals are not in exactly the same order as in the hydrogen atom. In this chapter we discuss some of the results for atoms with two or more electrons. We then introduce the Pauli exclusion principle and use it to correlate electronic structures with the chemical properties of the elements in the periodic table.

8-1. THE ENERGY STATES OF ATOMS WITH TWO OR MORE ELECTRONS DEPEND UPON BOTH n AND ℓ

Recall from Chapter 7 that the electronic energy of the hydrogen atom is given by

$$E_n = -\frac{2.18 \times 10^{-18}\,\text{J}}{n^2} \qquad n = 1, 2, 3, \ldots \qquad (8\text{-}1)$$

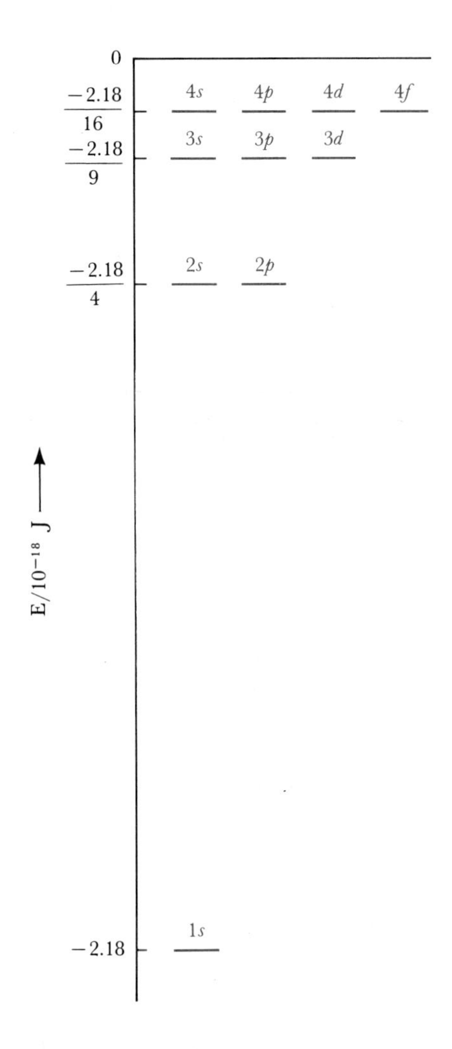

Figure 8-1 The relative energies of the hydrogen atomic orbitals. Because the energy depends upon only the principal quantum number, orbitals with the same value of n have the same energy.

Notice in Equation (8-1) that the energy depends upon only the principal quantum number n. Figure 8-1 shows the energies of the various hydrogen atomic orbitals. Because the energies depend upon only the principal quantum number, the $2s$ and $2p$ orbitals, for example, have the same energy. This is *not* the case for atoms containing more than one electron. In multielectron atoms there are not only electron-nucleus interactions but also electron-electron interactions. Because of these electron-electron interactions, the relationship between energy and quantum numbers is more complicated for multielectron atoms than that given by Equation (8-1). The electronic energies of multielectron atoms depend in a complicated way on the azimuthal quantum number l as well as on the principal quantum number n. Thus, for example, the $2s$ and $2p$ orbitals for atoms other than hydrogen have different energies. The ordering of the orbital energies, shown in Figure 8-2, is $1s < 2s < 2p < 3s < 3p < 4s <$

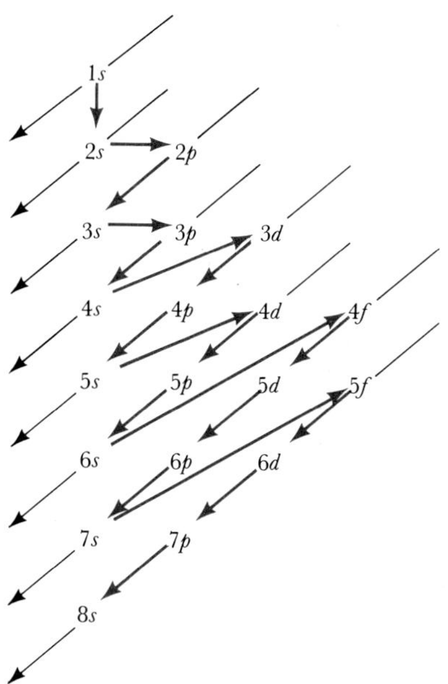
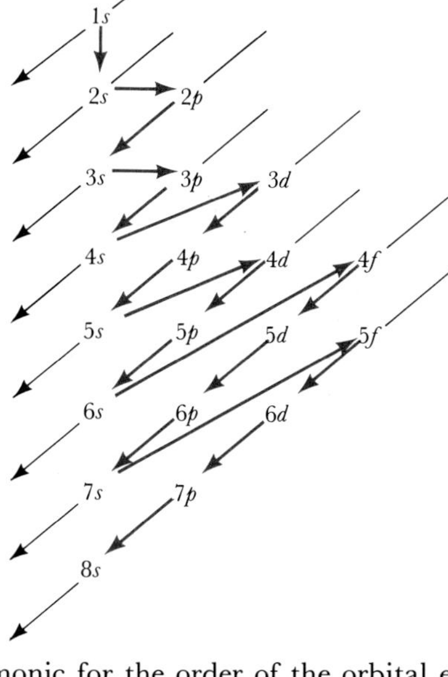

Figure 8-3 A mnemonic for the order of the orbital energies of atoms containing more than one electron. The orbitals are arranged as shown, and then diagonal black lines are drawn through them. The correct order of the orbital energies of atoms beyond hydrogen is obtained by going down a black line as far as possible and then jumping to the top of the next line.

Figure 8-2 The relative energies of the orbitals of atoms containing more than one electron. For multielectron atoms, the orbital energies depend upon both the principal quantum number n and the azimuthal quantum number ℓ; orbitals with the same value of n but different values of ℓ have different energies.

$3d <$ . . . . Note that, as n increases, the dependence of the energy upon ℓ becomes so pronounced that the energy of the $4s$ orbital is less than that of the $3d$ orbital. As in the case of the hydrogen atom, the orbital energies bunch together as n increases, and so this type of "reversal" becomes even more pronounced at higher energies. Fortunately, there is a simple mnemonic for remembering the order of the orbitals, which is shown in Figure 8-3.

8-2. NO TWO ELECTRONS IN THE SAME ATOM CAN HAVE THE SAME SET OF FOUR QUANTUM NUMBERS

Before we can correlate electronic structure with the periodic table, we must learn how to assign the electrons to the various orbitals. It was Wolfgang Pauli, in 1926, who first determined how to make this assignment. In what is now called the *Pauli exclusion principle*, he proposed that *no two electrons in the same atom can have the same set of four quantum numbers*. The Pauli exclusion principle is a fundamental principle of physics and can be used to understand the periodic table and the other periodic properties of the elements.

Table 8-1 lists the allowed sets of four quantum numbers (n, ℓ, m_ℓ, m_s). Note that there are only two allowed combinations for $n = 1$: $(1, 0, 0, +\frac{1}{2})$ and $(1, 0, 0, -\frac{1}{2})$. Both combinations have $n = 1$ and $\ell = 0$ and so correspond to two electrons in a 1*s* orbital. The two electrons differ only in their spin quantum numbers. We can represent this pictorially by a circle enclosing two vertical arrows:

$$(\uparrow\downarrow)$$

The circle represents the orbital, and the two arrows represent the two electrons with different spin quantum numbers. The arrow pointing upward represents $m_s = +\frac{1}{2}$, and the arrow pointing downward represents $m_s = -\frac{1}{2}$. This pictorial representation is so ingrained that chemists often use the terms *spin up* and *spin down* to refer to electrons with $m_s = +\frac{1}{2}$ and $m_s = -\frac{1}{2}$, respectively. The Pauli exclusion principle states that it is not possible for the spin quantum numbers of the electrons in a given orbital to be the same; if they were, the electrons would have the same set of four quantum numbers. Thus the representations $(\uparrow\uparrow)$ and $(\downarrow\downarrow)$ are not allowed; that is, they are forbidden.

No orbital can be occupied by more than two electrons.

The $n = 1$ level is complete with two electrons because there are only two possible sets of four quantum numbers with $n = 1$. When $n = 2$, there are two possible values of ℓ, namely, 0 and 1. The $\ell = 0$ value corresponds to a 2*s* orbital, which can hold two electrons of opposite spins. The $\ell = 1$ value corresponds to the three 2*p* orbitals ($m_\ell = 1$, $m_\ell = 0$, $m_\ell = -1$), each of which can hold two electrons of opposite spins, giving in all six electrons in the *p* orbitals. The $n = 2$ level, then, can hold eight electrons (two in the 2*s* orbital and six in the 2*p* orbitals):

$$(\uparrow\downarrow) \quad (\uparrow\downarrow)(\uparrow\downarrow)(\uparrow\downarrow)$$

2*s* 2*p*

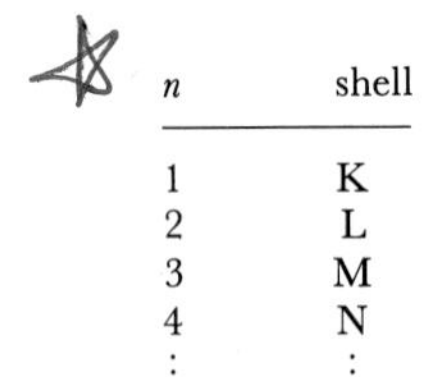

n	shell
1	K
2	L
3	M
4	N
⋮	⋮

For historical reasons, the levels designated by n are called *shells*. The $n = 1$ shell is called the K shell, the $n = 2$ shell is called the L shell, the $n = 3$ shell is called the M shell, and so forth. The groups of orbitals designated by ℓ values within these shells are called *subshells*. For $n = 2$, there are two subshells: the *s* subshell, which can contain a maximum of two electrons, and the *p* subshell, which can contain a maximum of six electrons.

For $n = 3$, we have the 3*s*, 3*p*, and 3*d* orbitals. The only new feature here is the *d* subshell. Because there are five *d* orbitals and each one can contain 2 electrons with opposite spins, the *d* subshell can contain up to 10 electrons. Thus, as Table 8-1 shows, the $n = 3$ level, or M shell, can contain 18 electrons. The only new feature for

Table 8-1 The sets of four quantum numbers allowed by the Pauli exclusion principle

n	ℓ	m_ℓ	m_s	*Number of electrons*
1 (K shell) (2 electrons)	0 (s subshell) (2 electrons)	0	$+\frac{1}{2}$ or $-\frac{1}{2}$	2
2 (L shell) (8 electrons)	0 (s subshell) (2 electrons)	0	$+\frac{1}{2}$ or $-\frac{1}{2}$	2
	1 (p subshell) (6 electrons)	+1	$+\frac{1}{2}$ or $-\frac{1}{2}$	
		0	$+\frac{1}{2}$ or $-\frac{1}{2}$	6
		−1	$+\frac{1}{2}$ or $-\frac{1}{2}$	= 8
3 (M shell) (18 electrons)	0 (s subshell) (2 electrons)	0	$+\frac{1}{2}$ or $-\frac{1}{2}$	2
	1 (p subshell) (6 electrons)	+1	$+\frac{1}{2}$ or $-\frac{1}{2}$	
		0	$+\frac{1}{2}$ or $-\frac{1}{2}$	6
		−1	$+\frac{1}{2}$ or $-\frac{1}{2}$	
	2 (d subshell) (10 electrons)	+2	$+\frac{1}{2}$ or $-\frac{1}{2}$	
		+1	$+\frac{1}{2}$ or $-\frac{1}{2}$	
		0	$+\frac{1}{2}$ or $-\frac{1}{2}$	10
		−1	$+\frac{1}{2}$ or $-\frac{1}{2}$	
		−2	$+\frac{1}{2}$ or $-\frac{1}{2}$	= 18
4 (N shell) (32 electrons)	0 (s subshell) (2 electrons)	0	$+\frac{1}{2}$ or $-\frac{1}{2}$	2
	1 (p subshell) (6 electrons)	+1	$+\frac{1}{2}$ or $-\frac{1}{2}$	
		0	$+\frac{1}{2}$ or $-\frac{1}{2}$	6
		−1	$+\frac{1}{2}$ or $-\frac{1}{2}$	
	2 (d subshell) (10 electrons)	+2	$+\frac{1}{2}$ or $-\frac{1}{2}$	
		+1	$+\frac{1}{2}$ or $-\frac{1}{2}$	
		0	$+\frac{1}{2}$ or $-\frac{1}{2}$	10
		−1	$+\frac{1}{2}$ or $-\frac{1}{2}$	
		−2	$+\frac{1}{2}$ or $-\frac{1}{2}$	
	3 (f subshell) (14 electrons)	+3	$+\frac{1}{2}$ or $-\frac{1}{2}$	
		+2	$+\frac{1}{2}$ or $-\frac{1}{2}$	
		+1	$+\frac{1}{2}$ or $-\frac{1}{2}$	
		0	$+\frac{1}{2}$ or $-\frac{1}{2}$	14
		−1	$+\frac{1}{2}$ or $-\frac{1}{2}$	
		−2	$+\frac{1}{2}$ or $-\frac{1}{2}$	
		−3	$+\frac{1}{2}$ or $-\frac{1}{2}$	= 32

$n = 4$ is the f subshell. Because there are seven f orbitals and each one can contain 2 electrons with opposite spins, the f subshell can contain up to 14 electrons, giving a total capacity of 32 $(2 + 6 + 10 + 14)$ electrons for the $n = 4$ level.

8-3. ELECTRONIC CONFIGURATIONS ARE OBTAINED BY PLACING ELECTRONS INTO THE AVAILABLE ORBITALS OF LOWEST ENERGY

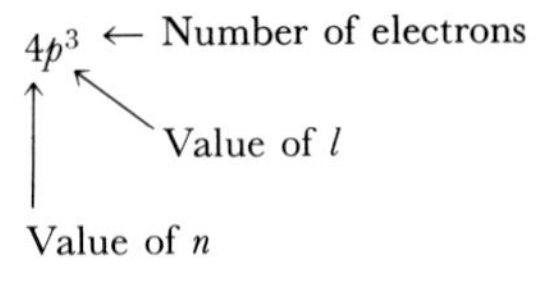

We are now ready to use Table 8-1 to understand some of the principal features of the periodic table in terms of electronic structure. We first consider the helium atom. The lowest energy state of the helium atom is achieved by placing both electrons in the 1s orbital because this is the orbital with the lowest energy. Thus we can represent the *ground electronic state* in helium by (⇅) or by $1s^2$. The latter notation is standard. The 1s means that we are considering a 1s orbital, and the exponent denotes that there are two electrons in the orbital. It is understood that the electrons have different spin quantum numbers, or opposite spins. If we are depicting five electrons in the 3p orbitals, then we write $3p^5$. The arrangement of electrons in the orbitals is called the *electron configuration* of the atom. We say that the electron configuration of the ground state of helium is $1s^2$.

Example 8-1: What is the meaning of the symbol $3d^4$?

Solution: The 3d signifies the 3d orbital, and the exponent 4 means that there are four electrons in the 3d orbital. The 3 tells us that $n = 3$, and the d tells us that $\ell = 2$.

Example 8-2: What is the meaning of $1s^2 2s^1$? If this describes the orbital assignment of the electrons in a neutral atom, what element is it?

Solution: There are two orbitals, 1s and 2s, indicated in $1s^2 2s^1$. The exponent 2 indicates that there are two electrons in the 1s orbital. The exponent 1 indicates that there is one electron in the 2s orbital. Because there is a total of three electrons and the atom is neutral, the atom described is lithium.

Let's go on now and consider the case of lithium with its three electrons. It is not possible for three electrons in a 1s orbital to have different sets of the four quantum numbers. The 1s orbital is completely filled by two of the electrons, and so the third electron must be assigned to the next available orbital, the 2*s* orbital. The electron in the 2*s* orbital can have $m_s = +\frac{1}{2}$ or $-\frac{1}{2}$, and so we can represent the lithium atom by

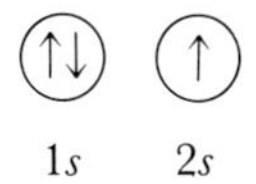

or by

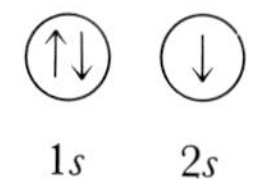

The direction of the arrow in the 2*s* orbital is not important here, and it is customary to use the spin up picture. The more standard notation is $1s^22s^1$.

Note that as we add an electron in going from helium to lithium, we must place this electron in a new orbital. Two of the electrons are in the filled 1*s* orbital, and the other is in the 2*s* orbital. We should compare this with Table 7-2, where it is argued from an examination of the experimental values of ionization energies that lithium can be represented as a helium core with one outer electron. In Table 7-2, we represent the lithium atom by the electron-dot formula ·[He]. We see that this same conclusion follows naturally from quantum theory.

The ground state of beryllium ($Z = 4$) is obtained by placing the fourth electron in the 2*s* orbital such that the two electrons there have opposite spins. Pictorially we have for the ground state of the beryllium atom

beryllium (↑↓) (↑↓)
1*s* 2*s*

and the standard notation for this ground-state electron configuration is $1s^22s^2$.

In boron ($Z = 5$) both the 1*s* and 2*s* orbitals are filled, and so we must use the 2*p* orbitals. Thus we have for boron

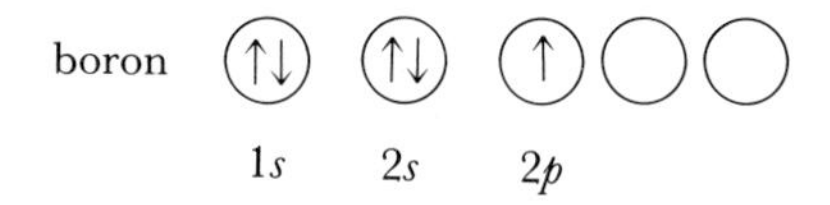

Recall from Chapter 7 that the three *p* orbitals for the hydrogen atom have the same energy in the absence of any external electric or magnetic fields. This is also true for multielectron atoms, as is indicated in Figure 8-2. Thus, it does not matter into which of the three p orbitals we place the electron. In addition, recall from above that the direction of the arrow in the $2p$ orbital is not important and that it is customary to draw such unpaired electrons as spin up. The ground-state electron configuration of boron is written $1s^2 2s^2 2p^1$.

Example 8-3: Given the representation

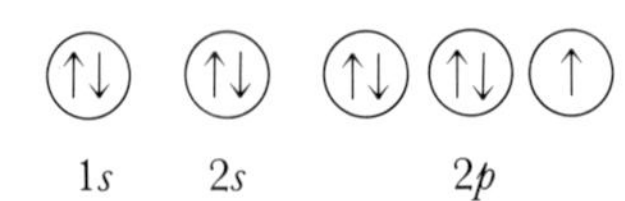

write the corresponding electron configuration. Given that the electron configuration describes a neutral atom, which atom is it?

Solution: Because there are two electrons in the $1s$ orbital, two in the $2s$ orbital, and five in the $2p$ orbitals, the electron configuration is $1s^2 2s^2 2p^5$. A total of nine electrons in the neutral atom indicates that it is a fluorine atom ($Z = 9$).

Example 8-4: The ground-state electron configuration of ions can be described by the same notation that we have discussed for atoms. What is the ground-state electron configuration of Li^+?

Solution: Li^+ has two electrons ($Z = 3$ for Li, and $Z - 1 = 2$ electrons for Li^+). The ground electronic state is obtained by placing both of these electrons in the $1s$ orbital:

Li^+ (↑↓)

or simply $1s^2$.

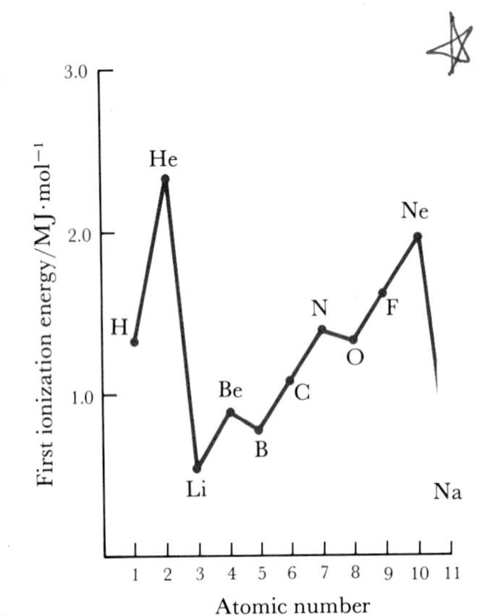

Figure 8-4 The first ionization energy I_1 of the elements H through Na. Notice that I_1 is relatively high for the noble gases (He and Ne), indicating that a completed shell is extraordinarily stable. Also notice that Be and N show small maxima, indicating some degree of stability for a filled or half-filled subshell.

We might guess that the unpaired electron in the boron atom is relatively easier to detach than is a paired $2s$ electron in beryllium. Figure 8-4 is a plot of the first ionization energies of the elements hydrogen through sodium. Note that these data confirm this guess.

8-4. HUND'S RULE IS USED TO PREDICT GROUND-STATE ELECTRON CONFIGURATIONS

For a carbon atom ($Z = 6$) we have three distinct choices for the placement of the two $2p$ electrons. The three configurations that obey the Pauli exclusion principle are

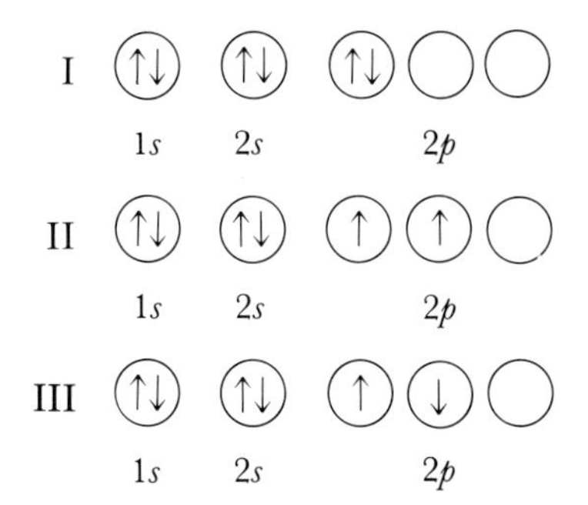

There are, however, small differences in the energies of these three configurations. In configuration I, both electrons are in the same p orbital and hence are restricted, on the average, to the same region in space. In the other two cases, the two electrons are in different p orbitals and so are, on the average, in different regions. Because electrons have the same charge and so repel each other, the placement of the two electrons into different p orbitals and hence different regions allows repulsion between electrons to be minimized. Thus we conclude that configurations II and III have lower energies and so are favored over configuration I.

We can argue a choice between configurations II and III as well. In II the electrons have the same spin, and in III they have opposite spins. If the electrons were in the same orbital, configuration II with its parallel spins would not be allowed but configuration III would be. We can surmise, then, that there is an effective exclusion between electrons with the same spin even though they are in different, but nevertheless closely related, orbitals. The two electrons in configuration II effectively exclude each other more than they do in configuration III. Thus they remain on the average farther away from each other and so decrease their electrical repulsion. This leads to a lower energy for configuration II. Of the three possible configurations for the carbon atom, the one in which the two p electrons are placed in different p orbitals with *parallel spins* leads to the lowest-energy, or ground-state, configuration. The ground-state configuration of the carbon atom is

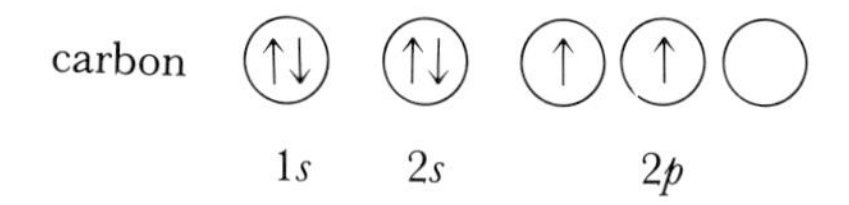

These arguments can be generalized to give what is called *Hund's rule,* which states that, for any set of orbitals of the same energy, that is, for any subshell, the ground-state electron configuration is obtained by placing the electrons in different orbitals of this set with parallel spins. No orbital in the subshell contains two electrons until each one contains one electron. Using Hund's rule, we write for nitrogen ($Z = 7$)

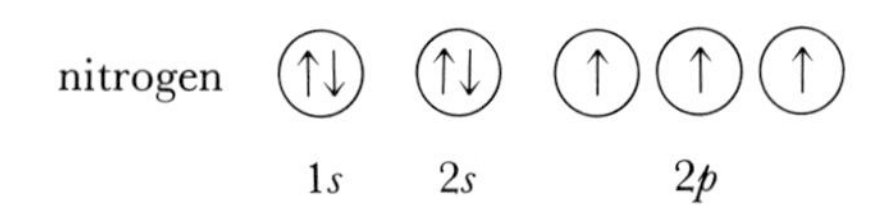

The more standard notation is $1s^22s^22p_x^12p_y^12p_z^1$. This often is condensed to $1s^22s^22p^3$. In both cases the reader is assumed to know that the three $2p$ electrons have parallel spins.

For an oxygen atom ($Z = 8$) we begin to pair up the p electrons and have

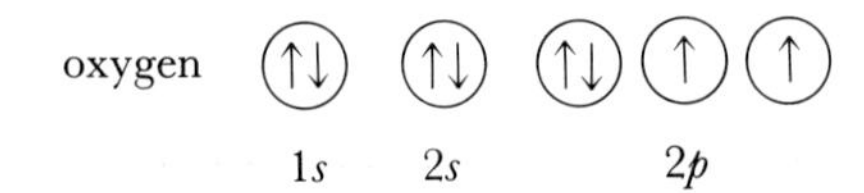

or $1s^22s^22p_x^22p_y^12p_z^1$, or simply $1s^22s^22p^4$. Realize that it does not matter into which p orbital we place the paired electrons. The electron configurations $1s^22s^22p_x^12p_y^22p_z^1$ and $1s^22s^22p_x^12p_y^12p_z^2$ are equivalent to each other and to $1s^22s^22p_x^22p_y^12p_z^1$.

Example 8-5: What is the ground-state electron configuration of O^+?

Solution: The O^+ ion has seven electrons (for O, $Z = 8$; for O^+ we have $8 - 1 = 7$ electrons). Four of the electrons are in the $1s$ and $2s$ orbitals. The other three are in the $2p$ orbitals. According to Hund's rule, the three $2p$ electrons are in different $2p$ orbitals and all have the same spin:

O^+ (↑↓) (↑↓) (↑)(↑)(↑)

1s 2s 2p

The electron configuration is $1s^22s^22p_x^12p_y^12p_z^1$, or simply $1s^22s^22p^3$.

8-5. THE SECOND AND THIRD ROWS OF THE PERIODIC TABLE HAVE SIMILAR OUTER ELECTRON CONFIGURATIONS

The ground-state electron configurations of the first 10 elements are summarized in Table 8-2. Note that helium has a filled $n = 1$ shell and that neon has a filled $n = 2$ shell. These electron configurations are for the ground electronic state of the atom, that is, the state of lowest energy. The ground-state electron configuration is obtained by filling up the atomic orbitals of lowest energy in accord with the Pauli exclusion principle and Hund's rule.

We saw in Chapter 7 that an atom can absorb electromagnetic radiation. In this process an electron is promoted to an orbital of higher energy, and the atom is said to be in an *excited state*. For example, a sodium atom absorbs electromagnetic radiation of wavelength 589 nm according to

Table 8-2 Ground-state electron configurations for the first 10 elements

Element	*Electron configuration*	*Pictorial representation* 1s	2s	2p		
H	$1s^1$	(↑)				
He	$1s^2$	(↑↓)				
Li	$1s^22s^1$	(↑↓)	(↑)			
Be	$1s^22s^2$	(↑↓)	(↑↓)			
B	$1s^22s^22p^1$	(↑↓)	(↑↓)	(↑)		
C	$1s^22s^22p^2$	(↑↓)	(↑↓)	(↑)	(↑)	
N	$1s^22s^22p^3$	(↑↓)	(↑↓)	(↑)	(↑)	(↑)
O	$1s^22s^22p^4$	(↑↓)	(↑↓)	(↑↓)	(↑)	(↑)
F	$1s^22s^22p^5$	(↑↓)	(↑↓)	(↑↓)	(↑↓)	(↑)
Ne	$1s^22s^22p^6$	(↑↓)	(↑↓)	(↑↓)	(↑↓)	(↑↓)

$$\text{Na}(1s^22s^22p^63s^1) + h\nu \rightarrow \text{Na}(1s^22s^22p^63p^1)$$

We see that the electron in the $3s$ orbital is promoted to a $3p$ orbital in the process. The resulting sodium atom is in an excited state, and its electron configuration is $1s^22s^22p^63p^1$. We are interested primarily in ground electronic states, but we should realize that the ground state is just the lowest of a set of allowed atomic energy states.

Example 8-6: What is the electron configuration of the first excited state of neon?

Solution: The first excited state is obtained by promoting the electron of highest energy in the ground state to the next available orbital. The ground-state electron configuration of neon is $1s^22s^22p^6$. The electron of highest energy is any one of the $2p$ electrons. The next available orbital is the $3s$ orbital, and so

$$\text{Ne}^* \text{ (first excited state)} \qquad 1s^22s^22p^53s^1$$

The asterisk indicates an excited state.

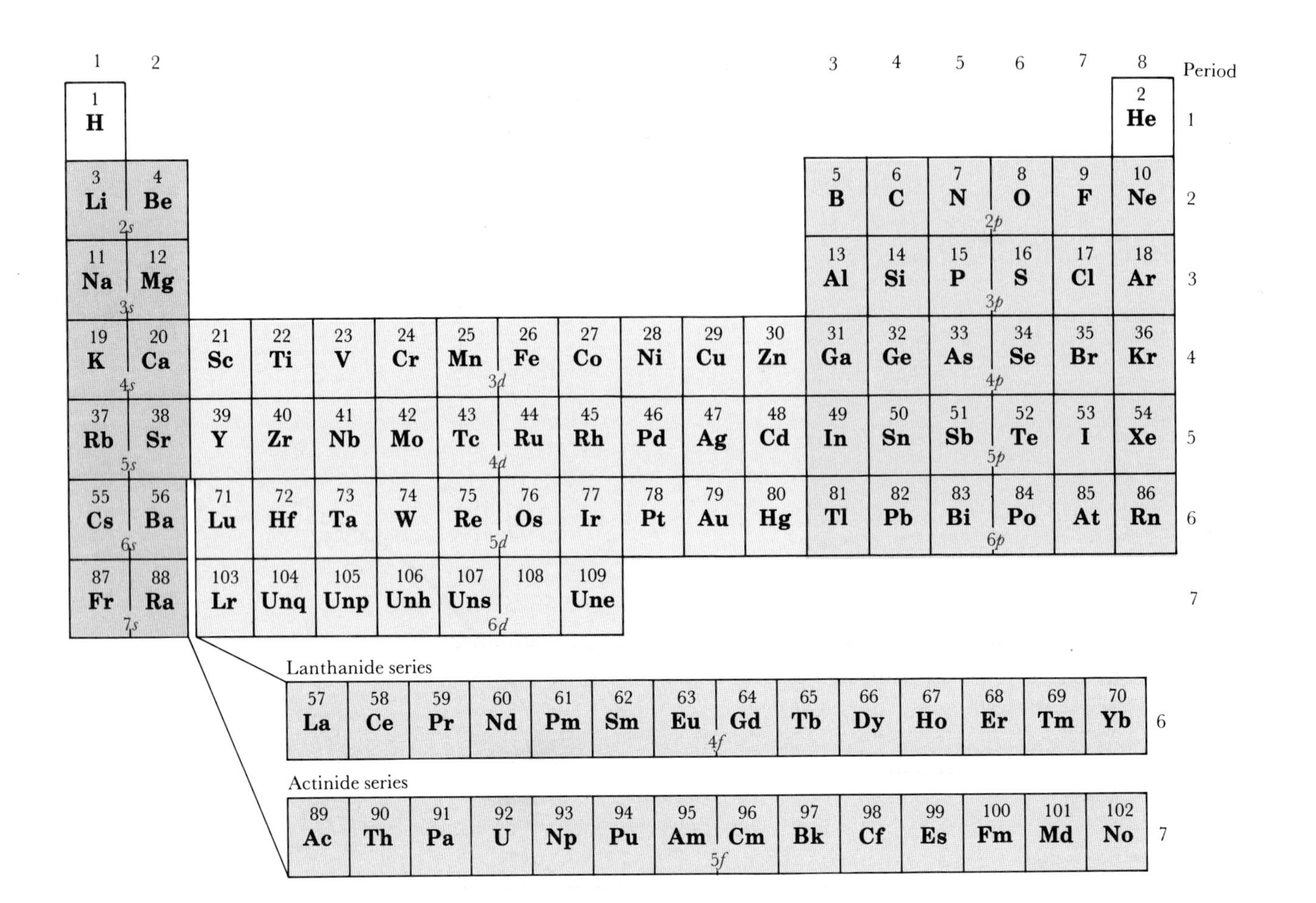

Figure 8-5. A periodic table indicating which orbitals are occupied by the valence electrons of each element. Blocks of elements with the same shading have the same valence electron subshells.

Figure 8-5 is a periodic table indicating the orbitals that are used in building up the electron configuration of each element. Note that after neon we must go to the 3*s* and 3*p* orbitals to obtain the next row of the periodic table:

Element	*Ground-state configuration*	*Abbreviated form of ground-state configuration*
sodium	$1s^22s^22p^63s^1$	$[Ne]3s^1$
magnesium	$1s^22s^22p^63s^2$	$[Ne]3s^2$
aluminum	$1s^22s^22p^63s^23p^1$	$[Ne]3s^23p^1$
silicon	$1s^22s^22p^63s^23p^2$	$[Ne]3s^23p^2$
phosphorus	$1s^22s^22p^63s^23p^3$	$[Ne]3s^23p^3$
sulfur	$1s^22s^22p^63s^23p^4$	$[Ne]3s^23p^4$
chlorine	$1s^22s^22p^63s^23p^5$	$[Ne]3s^23p^5$
argon	$1s^22s^22p^63s^23p^6$	$[Ne]3s^23p^6$

In this series of elements we are filling up the 3*s* and 3*p* orbitals outside a neon inner-shell structure. It is therefore common practice to use the abbreviated form of the electron configurations shown in the right-hand column. If we compare the Na through Ar electron configurations to the Li through Ne electron configurations (Table 8-2), then we see why these two series of elements have a periodic correlation in chemical properties. Their outer electron configurations vary from ns^1 to ns^2np^6 ($n = 2$ and $n = 3$) in the same manner. The outer electrons often are called *valence shell electrons*. Reference to Figure 8-5 shows that the number at the top of each column in the periodic table is equal to the number of valence electrons.

8-6. THE PRINCIPAL QUANTUM NUMBER OF THE OUTER *s* ELECTRONS IS EQUAL TO THE NUMBER OF THE ROW IN THE PERIODIC TABLE

Figure 8-2 shows that, after argon, the next available orbital is the 4*s* orbital. Thus the electron configurations of the next two elements after argon, namely, potassium and calcium, are

potassium	$[Ar]4s^1$
calcium	$[Ar]4s^2$

where [Ar] denotes the ground-state electron configuration of an argon atom. If we consider the ground-state electron configurations of lithium, sodium, and potassium, then we can see why they fall naturally into the same column of the periodic table. Each has an ns^1 configuration outside a noble-gas configuration, that is,

lithium	$[He]2s^1$
sodium	$[Ne]3s^1$
potassium	$[Ar]4s^1$

Also note that the principal quantum number of the outer *s* orbital coincides with the number of the row of the periodic table (Figure 8-5). Each row starts off with an alkali metal, whose electron configuration is [noble gas]ns^1. For example, rubidium, which follows krypton and begins the fifth row of the table, has the electron configuration

rubidium $[Kr]5s^1$

The ground-state electron configurations of the other alkali metals are

cesium	$[Xe]6s^1$
francium	$[Rn]7s^1$

Alkali metals
[noble gas]ns^1

Alkaline earth metals
[noble gas]ns^2

The same type of observation can be used to explain why the alkaline earths all occur in the second column in the periodic table. The electron configuration of an alkaline earth metal is [noble gas]ns^2:

beryllium	[He]$2s^2$
magnesium	[Ne]$3s^2$
calcium	[Ar]$4s^2$
strontium	[Kr]$5s^2$
barium	[Xe]$6s^2$
radium	[Rn]$7s^2$

Once again note that the principal quantum number of the outer s orbital corresponds to the number of the row in the periodic table.

8-7. IN THE FIRST SET OF TRANSITION METALS THE FIVE 3*d* ORBITALS ARE FILLED SEQUENTIALLY

Once we reach calcium ($Z = 20$), the $4s$ orbital is completely filled. Figure 8-2 shows that the next available orbitals are the five $3d$ orbitals. Each of these can be occupied by 2 electrons of opposite spins, giving a total of 10 electrons in all. Note that this corresponds perfectly with the 10 transition metals that occur between calcium and gallium in the periodic table. Thus we see that in the first set of transition metals there is the sequential filling of the five $3d$ orbitals. Because of this, the first set of transition metals is called the *3d transition-metal series*.

You may have thought that the ground-state electron configurations of these 10 elements go smoothly from [Ar]$4s^23d^1$ to [Ar]$4s^23d^{10}$, but this is not so. The ground-state electron configurations of the $3d$ transition metals are

scandium	[Ar]$4s^23d^1$
titanium	[Ar]$4s^23d^2$
vanadium	[Ar]$4s^23d^3$
chromium	[Ar]$4s^13d^5$
manganese	[Ar]$4s^23d^5$
iron	[Ar]$4s^23d^6$
cobalt	[Ar]$4s^23d^7$
nickel	[Ar]$4s^23d^8$
copper	[Ar]$4s^13d^{10}$
zinc	[Ar]$4s^23d^{10}$

We see that chromium and copper have only one $4s$ electron. Note that in each case an electron has been taken from the $4s$ orbital in order to either half-fill or completely fill all of the $3d$ orbitals. This happens because an extra stability is realized by the electron configurations

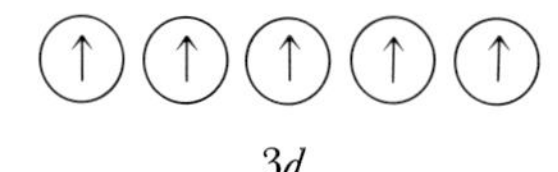

3d

and

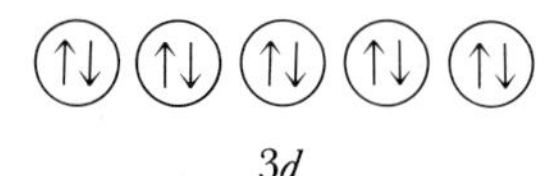

3d

relative to the *incorrect* $4s^2 3d^4$ and $4s^2 3d^9$ ground-state configurations for these elements. It so happens that the energies of the 4*s* and 3*d* electrons are rather close to each other (Figure 8-2), and the exchange of an electron between these two types of orbitals occurs easily. This is one reason that the transition metals exhibit ions with different charges, such as Fe^{2+} and Fe^{3+}.

After the 3*d* orbitals are filled, the next available orbitals are the 4*p* orbitals, which fill up as follows:

gallium	$[Ar]4s^2 3d^{10} 4p^1$
germanium	$[Ar]4s^2 3d^{10} 4p^2$
arsenic	$[Ar]4s^2 3d^{10} 4p^3$
selenium	$[Ar]4s^2 3d^{10} 4p^4$
bromine	$[Ar]4s^2 3d^{10} 4p^5$
krypton	$[Ar]4s^2 3d^{10} 4p^6$

For these six elements the 4*p* orbitals are sequentially filled, and these elements fall naturally into the fourth row of the periodic table under the sequence of elements B through Ne and Al through Ar, which fill the 2*p* and 3*p* orbitals, respectively (Figure 8-5).

Krypton, like all the noble gases, has a completely filled set of orbitals whose principal quantum number corresponds to the row in the periodic table. Figure 8-3 shows that the 5*s* orbital follows the 4*p* orbital, and so we are back to the left-hand column of the periodic table with the alkali metal rubidium and the alkaline earth metal strontium. These two metals have the ground-state electron configurations $[Kr]5s^1$ and $[Kr]5s^2$, respectively. The next available orbitals are the 4*d* orbitals, which lead to the second transition-metal series, or the 4*d* transition-metal series, Y through Cd. Figure 8-6 gives the ground-state electron configurations for these 10 metals and shows irregularities like those found in the 3*d* transition-metal series. After cadmium, $[Kr]5s^2 4d^{10}$, the 5*p* orbitals are filled to give the six elements indium through the noble gas xenon, which has the ground-state electron configuration $[Kr]5s^2 4d^{10} 5p^6$. As before, the comple-

1	2											3	4	5	6	7	8
ns^1	ns^2											ns^2np^1	ns^2np^2	ns^2np^3	ns^2np^4	ns^2np^5	ns^2np^6
1 **H** $1s^1$																	2 **He** $1s^2$
3 **Li** $2s^1$	4 **Be** $2s^2$											5 **B** $2s^22p^1$	6 **C** $2s^22p^2$	7 **N** $2s^22p^3$	8 **O** $2s^22p^4$	9 **F** $2s^22p^5$	10 **Ne** $2s^22p^6$
11 **Na** $3s^1$	12 **Mg** $3s^2$											13 **Al** $3s^23p^1$	14 **Si** $3s^23p^2$	15 **P** $3s^23p^3$	16 **S** $3s^23p^4$	17 **Cl** $3s^23p^5$	18 **Ar** $3s^23p^6$
19 **K** $4s^1$	20 **Ca** $4s^2$	21 **Sc** $3d^14s^2$	22 **Ti** $3d^24s^2$	23 **V** $3d^34s^2$	24 **Cr** $3d^54s^1$	25 **Mn** $3d^54s^2$	26 **Fe** $3d^64s^2$	27 **Co** $3d^74s^2$	28 **Ni** $3d^84s^2$	29 **Cu** $3d^{10}4s^1$	30 **Zn** $3d^{10}4s^2$	31 **Ga** $4s^24p^1$	32 **Ge** $4s^24p^2$	33 **As** $4s^24p^3$	34 **Se** $4s^24p^4$	35 **Br** $4s^24p^5$	36 **Kr** $4s^24p^6$
37 **Rb** $5s^1$	38 **Sr** $5s^2$	39 **Y** $4d^15s^2$	40 **Zr** $4d^25s^2$	41 **Nb** $4d^45s^1$	42 **Mo** $4d^55s^1$	43 **Tc** $4d^65s^1$	44 **Ru** $4d^75s^1$	45 **Rh** $4d^85s^1$	46 **Pd** $4d^{10}$	47 **Ag** $4d^{10}5s^1$	48 **Cd** $4d^{10}5s^2$	49 **In** $5s^25p^1$	50 **Sn** $5s^25p^2$	51 **Sb** $5s^25p^3$	52 **Te** $5s^25p^4$	53 **I** $5s^25p^5$	54 **Xe** $5s^25p^6$
55 **Cs** $6s^1$	56 **Ba** $6s^2$	71 **Lu** $5d^16s^2$	72 **Hf** $5d^26s^2$	73 **Ta** $5d^36s^2$	74 **W** $5d^46s^2$	75 **Re** $5d^56s^2$	76 **Os** $5d^66s^2$	77 **Ir** $5d^76s^2$	78 **Pt** $5d^96s^1$	79 **Au** $5d^{10}6s^1$	80 **Hg** $5d^{10}6s^2$	81 **Tl** $6s^26p^1$	82 **Pb** $6s^26p^2$	83 **Bi** $6s^26p^3$	84 **Po** $6s^26p^4$	85 **At** $6s^26p^5$	86 **Rn** $6s^26p^6$
87 **Fr** $7s^1$	88 **Ra** $7s^2$	103 **Lr** $6d^17s^2$	104 **Unq** $6d^27s^2$	105 **Unp** $6d^37s^2$	106 **Unh** $6d^47s^2$	107 **Uns** $6d^57s^2$	108	109 **Une** $6d^77s^2$									

Lanthanide series

57 **La** $5d^16s^2$	58 **Ce** $4f^26s^2$	59 **Pr** $4f^36s^2$	60 **Nd** $4f^46s^2$	61 **Pm** $4f^56s^2$	62 **Sm** $4f^66s^2$	63 **Eu** $4f^76s^2$	64 **Gd** $4f^75d^16s^2$	65 **Tb** $4f^96s^2$	66 **Dy** $4f^{10}6s^2$	67 **Ho** $4f^{11}6s^2$	68 **Er** $4f^{12}6s^2$	69 **Tm** $4f^{13}6s^2$	70 **Yb** $4f^{14}6s^2$

Actinide series

89 **Ac** $6d^17s^2$	90 **Th** $6d^27s^2$	91 **Pa** $5f^26d^17s^2$	92 **U** $5f^36d^17s^2$	93 **Np** $5f^46d^17s^2$	94 **Pu** $5f^66d^07s^2$	95 **Am** $5f^76d^07s^2$	96 **Cm** $5f^76d^17s^2$	97 **Bk** $5f^96d^07s^2$	98 **Cf** $5f^{10}7s^2$	99 **Es** $5f^{11}7s^2$	100 **Fm** $5f^{12}7s^2$	101 **Md** $5f^{13}7s^2$	102 **No** $5f^{14}7s^2$

Figure 8-6 A periodic table showing the ground-state electron configurations of the outer electrons of the elements. The general outer electron configurations of the various groups are given above each group. Thus, the alkali metals have the outer electron configuration ns^1, the alkaline earths ns^2, and so on.

tion of a set of *p* orbitals leads to a noble gas located in the right-hand column of the periodic table. The two reactive metals cesium and barium follow xenon by filling the 6*s* orbital to give the ground-state electron configurations [Xe]$6s^1$ and [Xe]$6s^2$, respectively.

8-8. THE LANTHANIDES HAVE SIMILAR CHEMICAL PROPERTIES

After filling the 6*s* orbital, we go to the seven 4*f* orbitals. Because each of these 7 orbitals can hold 2 electrons of opposite spin, we expect that the next 14 elements should involve the filling of the 4*f* orbitals. The elements lanthanum ($Z = 57$) through ytterbium ($Z = 70$) constitute what are called the *lanthanides* because they begin with the element lanthanum in the periodic table. The chemistry of these elements is so similar that for many years it proved very difficult to separate them. However, separations are possible using modern chromatographic methods (Interchapter B).

Example 8-7: By referring to Figure 8-5, predict the ground-state electron configuration of a neodymium atom ($Z = 60$).

Solution: Neodymium occurs in the sixth row of the periodic table. The noble gas preceding this row is xenon. The ground-state electron configuration of barium, the element that precedes the lanthanides, is [Xe]$6s^2$. Neodymium is the fourth member of the lanthanides, and so we predict that it has four $4f$ electrons. The predicted ground-state electron configuration is

$$\text{Nd} \quad [\text{Xe}]6s^2 4f^4$$

Figure 8-6 shows that this result is correct. Notice, however, that there are several irregularities in the electron configurations of the lanthanides.

If we consider that the lanthanides differ only in the number of electrons in the $4f$ subshells, with the $6s$ and $5p$ subshells already filled, the reason for their chemical similarity becomes clear. According to the quantum theory, the average distance of an electron from a nucleus depends upon both the principal quantum number n and the azimuthal quantum number ℓ. Although the average distance from the nucleus increases with n, it decreases as ℓ increases, and so electrons with large values of ℓ are on the average closer to the nucleus than are electrons with smaller values of ℓ. Therefore, we can conclude that the average distance of $4f$ electrons from the nucleus is smaller than that of $6s$ or $5p$ electrons. For the $4f$ electrons, not only is n smaller but ℓ is larger than for $6s$ or $5p$ electrons. The $4f$ electrons, then, tend to lie deeper in the interior of the atom and so have little effect on the chemical activity of the atom, which is dominated by the outer (valence) electrons. For this reason, the lanthanides are also called *inner transition metals*. The outer electron configuration, which plays a principal role in determining chemical activity, is the same for all the lanthanides ($5p^6 6s^2$) and accounts for their similar chemical properties.

On the average, $4f$ electrons are closer to the nucleus than are $6s$ or $5p$ electrons.

Following the lanthanides is a third transition metal series (the $5d$ transition-metal series) consisting of the elements lutetium ($Z = 71$) through mercury ($Z = 80$). This series, in which the $5d$ orbitals are filled, is followed by the six elements thallium ($Z = 81$) through radon ($Z = 86$). Radon, a radioactive noble gas with the ground-state electron configuration [Xe]$6s^2 4f^{14} 5d^{10} 6p^6$, finishes the sixth row of the periodic table.

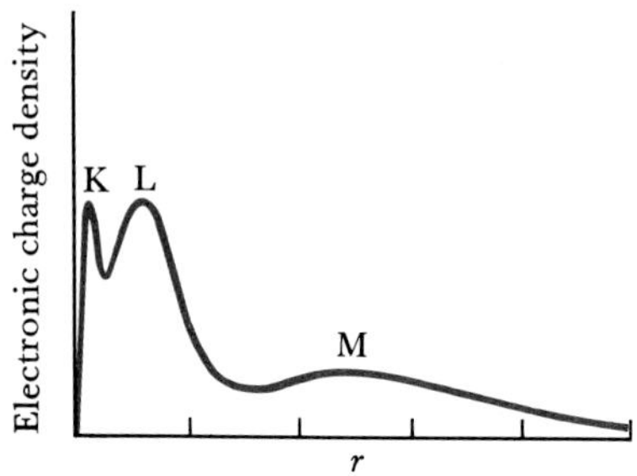

Figure 8-7 The distribution of electronic charge density versus distance from the nucleus for an argon atom. Note that there appear to be three shells. Two of these are well defined and close to the nucleus (the K and L shells). The third, outermost shell (the M shell) is more diffuse.

The next two elements, the radioactive metals francium, [Rn]$7s^1$, and radium, [Rn]$7s^2$, are followed by another inner transition-metal series in which the $5f$ orbitals are filled. This series begins with actinium ($Z = 89$) and ends with nobelium ($Z = 102$) and is called the *actinide series*. All the elements in this series occur only in radioactive form, and in fact, with the exception of plutonium, the elements beyond uranium ($Z = 92$) have not been found in nature. They are synthesized in nuclear reactors and are called the *transuranium elements* (Chapter 24).

The ground-state outer electron configurations of all the elements are given in Figure 8-6. It is a good exercise to go through the periodic table and predict the electron configuration of each element. A few irregularities occur as n increases, but the general features should be apparent.

8-9. ATOMIC RADIUS IS A PERIODIC PROPERTY

In Chapter 7 we learned that the electrons distribute themselves about a nucleus in a diffuse, cloud-like manner so that there is no sharp boundary at the "edge" of an atom. Although the Schrödinger equation is complicated for multielectron atoms, it can be solved with a computer. The results of such a calculation for argon are sketched in Figure 8-7. One can clearly discern three shells; the inner two are well defined and the outermost is more diffuse.

Even though atoms do not have well-defined radii, we can propose operational definitions for *atomic radii* that are based on models. For example, the atoms in a crystal of an element are arranged in ordered arrays. A simple version of such an ordered array is shown in Figure 8-8. The atoms are arranged in a simple cubic array, and we can propose that one half of the distance between adjacent nuclei be used as an effective atomic radius. Real crystals usually exist in more complicated geometric patterns than simple cubic, but effective atomic radii can still be deduced. Atomic radii obtained in this manner are called *crystallographic radii*. The crystallographic radii of the elements are plotted versus atomic number in Figure 8-9, indicating the periodic dependence of the radii on atomic number.

The crystallographic radii of the elements lithium through fluorine decrease uniformly from left to right across this row of the periodic table because the nuclear charge increases and attracts the electrons more strongly. Both the K shell and L shell contract, giving a smaller effective radius as the atomic number increases. This same trend is seen in Figure 8-9 for the other rows of the periodic table. Atomic radii usually decrease as one goes from left to right in a row across the periodic table.

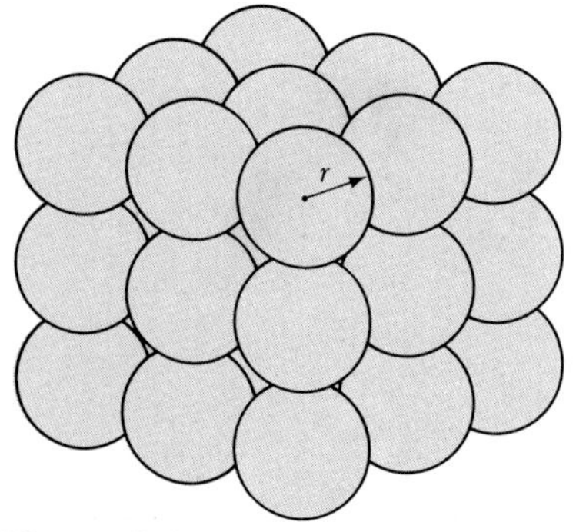

Figure 8-8 A simple cubic arrangement of atoms in a crystal.

The crystallographic radii of the alkali metal group increase as one goes down the periodic table from lithium to cesium. Although the

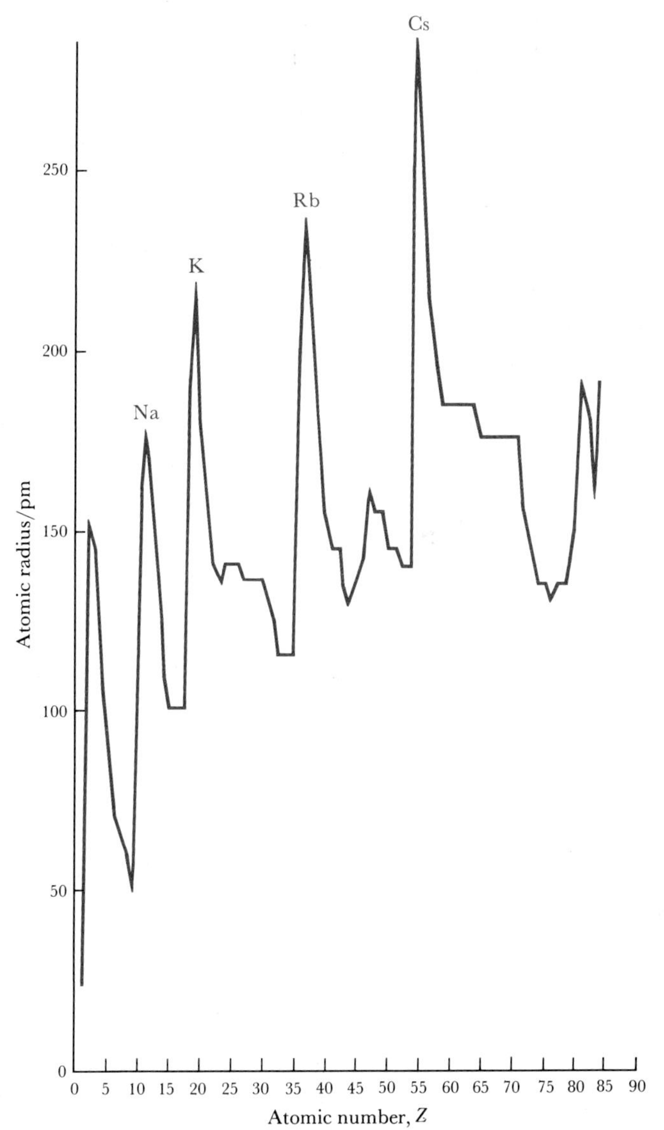

Figure 8-9 Crystallographic radii of the elements versus atomic number. Note that atomic radius is a periodic property and that the atomic radii of the alkali metals are larger than those of the preceding noble-gas radii.

nuclear charge increases, the outermost electrons begin new shells, and this effect outweighs the increased nuclear attraction. Similar behavior is found for other groups in the periodic table. Atomic radii usually increase as one goes down the periodic table within a group, as shown in Figure 8-10. Figure 8-11 gives the numerical values of atomic radii.

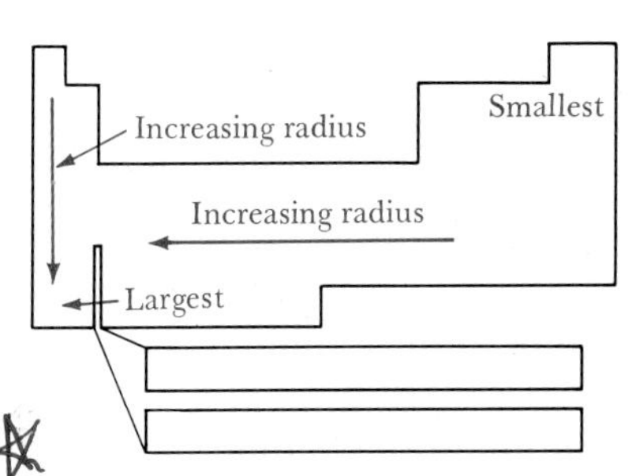

Figure 8-10 The trend of atomic radii in the periodic table.

1	2											3	4	5	6	7	8
1 **H** 25																	2 **He** 50
3 **Li** 145	4 **Be** 105											5 **B** 85	6 **C** 70	7 **N** 65	8 **O** 60	9 **F** 50	10 **Ne** 70
11 **Na** 180	12 **Mg** 150											13 **Al** 125	14 **Si** 110	15 **P** 100	16 **S** 100	17 **Cl** 100	18 **Ar** 95
19 **K** 220	20 **Ca** 180	21 **Sc** 160	22 **Ti** 140	23 **V** 135	24 **Cr** 140	25 **Mn** 140	26 **Fe** 140	27 **Co** 135	28 **Ni** 135	29 **Cu** 135	30 **Zn** 135	31 **Ga** 130	32 **Ge** 125	33 **As** 115	34 **Se** 115	35 **Br** 115	36 **Kr** 110
37 **Rb** 235	38 **Sr** 200	39 **Y** 180	40 **Zr** 155	41 **Nb** 145	42 **Mo** 145	43 **Tc** 135	44 **Ru** 130	45 **Rh** 135	46 **Pd** 140	47 **Ag** 160	48 **Cd** 155	49 **In** 155	50 **Sn** 145	51 **Sb** 145	52 **Te** 140	53 **I** 140	54 **Xe** 130
55 **Cs** 266	56 **Ba** 215	71 **Lu** 175	72 **Hf** 155	73 **Ta** 145	74 **W** 135	75 **Re** 135	76 **Os** 130	77 **Ir** 135	78 **Pt** 135	79 **Au** 135	80 **Hg** 150	81 **Tl** 190	82 **Pb** 180	83 **Bi** 160	84 **Po** 190	85 **At**	86 **Rn**
87 **Fr**	88 **Ra** 215	103 **Lr**	104 **Unq**	105 **Unp**	106 **Unh**	107 **Uns**	108	109 **Une**									

Lanthanide series

57 **La** 195	58 **Ce** 185	59 **Pr** 185	60 **Nd** 185	61 **Pm** 185	62 **Sm** 185	63 **Eu** 185	64 **Gd** 180	65 **Tb** 175	66 **Dy** 175	67 **Ho** 175	68 **Er** 175	69 **Tm** 175	70 **Yb** 175

Actinide series

89 **Ac** 195	90 **Th** 180	91 **Pa** 180	92 **U** 175	93 **Np** 175	94 **Pu** 175	95 **Am**	96 **Cm**	97 **Bk**	98 **Cf**	99 **Es**	100 **Fm**	101 **Md**	102 **No**

Figure 8-11 The crystallographic radii of the elements. The units are picometers.

The reasoning we have just used to explain the variation of atomic radii in the periodic table can be used to explain variations in *first ionization energies*. The steady decrease with increasing atomic number within a group is due to the increase in atomic radius as one goes down the periodic table. The farther the electron is from the nucleus, the less the nuclear attraction, and so the electron is more easily removed. Therefore, ionization energies decrease as one goes down the periodic table within a group. Figure 8-12 shows the first ionization energies of the elements.

1	2											3	4	5	6	7	8
1 **H** 1310																	2 **He** 2370
3 **Li** 520	4 **Be** 899											5 **B** 800	6 **C** 1090	7 **N** 1400	8 **O** 1310	9 **F** 1680	10 **Ne** 2080
11 **Na** 496	12 **Mg** 738											13 **Al** 577	14 **Si** 786	15 **P** 1060	16 **S** 999	17 **Cl** 1260	18 **Ar** 1520
19 **K** 419	20 **Ca** 590	21 **Sc** 633	22 **Ti** 659	23 **V** 650	24 **Cr** 652	25 **Mn** 717	26 **Fe** 762	27 **Co** 758	28 **Ni** 737	29 **Cu** 745	30 **Zn** 906	31 **Ga** 579	32 **Ge** 785	33 **As** 965	34 **Se** 941	35 **Br** 1140	36 **Kr** 1350
37 **Rb** 403	38 **Sr** 549	39 **Y** 637	40 **Zr** 671	41 **Nb** 653	42 **Mo** 693	43 **Tc**	44 **Ru** 724	45 **Rh** 743	46 **Pd** 804	47 **Ag** 731	48 **Cd** 868	49 **In** 558	50 **Sn** 708	51 **Sb** 834	52 **Te** 869	53 **I** 1007	54 **Xe** 1170
55 **Cs** 376	56 **Ba** 503	71 **Lu** 482	72 **Hf** 531	73 **Ta** 760	74 **W** 770	75 **Re** 759	76 **Os** 840	77 **Ir** 888	78 **Pt** 865	79 **Au** 890	80 **Hg** 1010	81 **Tl** 589	82 **Pb** 716	83 **Bi** 703	84 **Po** 813	85 **At**	86 **Rn** 1040
87 **Fr**	88 **Ra** 509	103 **Lr**	104 **Unq**	105 **Unp**	106 **Unh**	107 **Uns**	108	109 **Une**									

Lanthanide series

57 **La** 541	58 **Ce** 667	59 **Pr** 556	60 **Nd** 609	61 **Pm**	62 **Sm** 540	63 **Eu** 547	64 **Gd** 594	65 **Tb** 650	66 **Dy** 658	67 **Ho**	68 **Er**	69 **Tm**	70 **Yb** 598

Actinide series

89 **Ac** 666	90 **Th**	91 **Pa**	92 **U**	93 **Np**	94 **Pu**	95 **Am**	96 **Cm**	97 **Bk**	98 **Cf**	99 **Es**	100 **Fm**	101 **Md**	102 **No**

Figure 8-12 The first ionization energies of the elements. The units are $kJ \cdot mol^{-1}$.

SUMMARY

The energy states of the hydrogen atom depend upon only the principal quantum number n, but for all other atoms the energy states depend upon both n and the azimuthal quantum number ℓ. According to the Pauli exclusion principle, no two electrons in an atom can have the same set of four quantum numbers (n, ℓ, m_ℓ, m_s). Using this principle and the order of the energy states given in Figures 8-2 and 8-3, we are able to write ground-state electron configurations and to correlate these with the periodic table. Electron configurations enable us to discuss atomic radii and the trend of atomic radii within the periodic table. The filling of the atomic orbitals that result from the solution of the Schrödinger equation, subject to the constraint of the Pauli exclusion principle, provides a beautiful rationale for the structure and organization of the periodic table.

TERMS YOU SHOULD KNOW

Pauli exclusion principle 291	excited state 298
spin up 292	valence shell electron 301
spin down 292	*3d* transition-metal series 302
opposite spins 292	lanthanide series 304
shell 292	inner transition metal 305
subshell 292	actinide series 306
ground electronic state 294	transuranium elements 306
electron configuration 294	atomic radius 306
parallel spins 297	crystallographic radius 306
Hund's rule 297	

PROBLEMS

ORBITALS AND ELECTRONS

8-1. For each pair, indicate which set of orbitals has the higher energy in neutral, multielectron gaseous atoms:

(a) $3p$ or $4p$
(b) $3d$ or $4p$
(c) $5p$ or $6s$
(d) $4d$ or $5p$
(e) $4f$ or $6s$

8-2. Which subshell in a neutral gaseous atom is filled next after the following subshells are filled:

(a) $4f$
(b) $3d$
(c) $6s$
(d) $4p$
(e) $5d$

8-3. Give all the possible sets of four quantum numbers for an electron in a $3d$ orbital.

8-4. Give all the possible sets of four quantum numbers for an electron in a $4f$ orbital.

8-5. Without referring to the text, deduce the maximum number of electrons that can occupy an s orbital, a subshell of p orbitals, a subshell of d orbitals, and a subshell of f orbitals.

8-6. Without referring to the text, deduce the maximum number of electrons that can occupy a K shell, an L shell, an M shell, and an N shell.

8-7. Explain why there are 10 members of each d transition series.

8-8. Explain why there are 14 members of each f transition series.

THE PAULI EXCLUSION PRINCIPLE

8-9. Indicate which of the following electron configurations are ruled out by the Pauli exclusion principle:

(a) $1s^22s^22p^5$
(b) $1s^22s^22p^63s^3$
(c) $1s^22s^22p^63s^23p^64s^23d^{12}$
(d) $1s^22s^22p^63s^23p^8$

8-10. Indicate which of the following electron configurations are ruled out by the Pauli exclusion principle:

(a) $1s^22s^22p^7$
(b) $1s^22p^1$
(c) $1s^22s^22p^63s^23p^64s^23d^{14}4p^6$
(d) $1s^22s^22p^63s^23p^64s^23d^{10}4p^6$

8-11. For each of the following pictorial representations for the electrons in a carbon atom, determine if the representation corresponds to an excited state, a ground state, or a forbidden state:

	1s	2s	2p			3s	3p		
(a)	↑↓	↑↓	↑		↑				
(b)	↑↓	↑	↑	↑				↑	
(c)	↑↓	↑↓	↑	↑					
(d)	↑↓	↑↓	↓	↓					

8-12. Explain why the following ground-state electron configurations are not possible:

(a) $1s^22s^32p^1$
(b) $1s^22s^22p^33s^3$
(c) $1s^22s^22p^73s^23p^3$
(d) $1s^22s^22p^63s^23p^14s^23d^{14}$

8-13. Write the corresponding electron configuration for each of the following pictorial representations. Name the element, assuming that the configuration describes a neutral atom:

	1s	2s	2p	3s	3p	3d	4s	4p
(a)	(↑↓)	(↑↓)	(↑↓)(↑↓)(↑↓)	(↑↓)	(↑)(↑)( )			
(b)	(↑↓)	(↑↓)	(↑↓)(↑↓)(↑↓)	(↑↓)	(↑↓)(↑↓)(↑↓)	(↑)(↑)(↑)(↑)(↑)	(↑)	
(c)	(↑↓)	(↑↓)	(↑↓)(↑↓)(↑↓)	(↑↓)	(↑↓)(↑↓)(↑↓)	(↑↓)(↑↓)(↑↓)(↑↓)(↑↓)	(↑↓)	(↑)(↑)( )
(d)	(↑↓)	(↑↓)	(↑↓)(↑↓)(↑↓)	(↑↓)	(↑↓)(↑↓)(↑↓)	(↑↓)(↑↓)(↑↓)(↑↓)(↑↓)	(↑↓)	(↑↓)(↑↓)(↑)
(e)	(↑↓)	(↑↓)	(↑)					

8-14. Write the corresponding electron configuration for each of the following pictorial representations. Name the element that each represents, assuming neutral atoms:

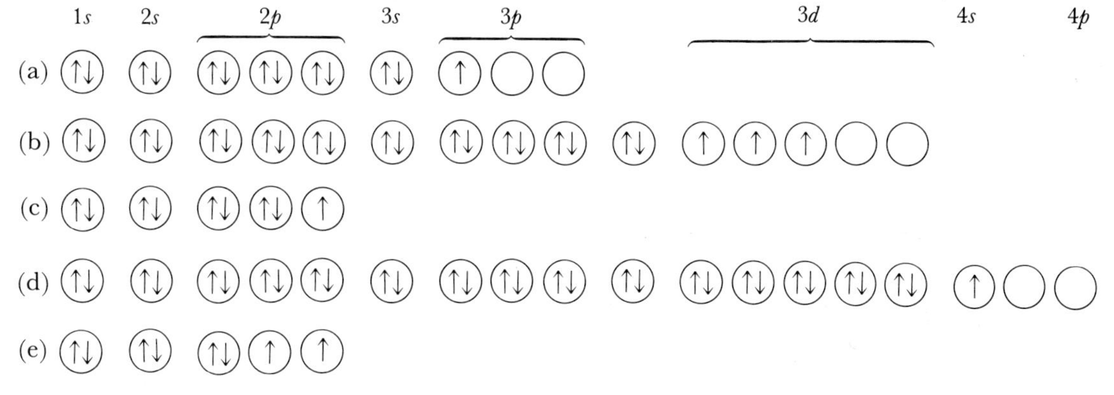

8-15. Write the ground-state electron configurations for the following elements:

(a) Ti (b) Mo (c) Cu (d) Zn (e) Ag

8-16. Write the ground-state electron configurations for the following elements:

(a) Fe (b) W (c) As (d) Ne (e) K

8-17. Write the ground-state electron configurations for the following neutral atoms:

(a) P (b) Ni (c) I (d) Cd (e) Nd

8-18. Write the ground-state electron configuration for the following neutral atoms:

(a) Se (b) Mn (c) Sn (d) Au (e) Eu

8-19. Without counting the total number of electrons, determine the neutral atom whose ground-state electron configuration is

(a) $1s^22s^22p^63s^23p^64s^23d^8$
(b) $1s^22s^22p^63s^23p^64s^23d^{10}4p^65s^14d^{10}$
(c) $1s^22s^22p^63s^23p^4$
(d) $1s^22s^22p^63s^23p^64s^23d^{10}4p^65s^24d^{10}5p^66s^24f^{14}5d^{10}6p^2$

8-20. Without counting the total number of electrons, determine the neutral atom whose ground-state electron configuration is

(a) $1s^22s^22p^63s^23p^1$
(b) $1s^22s^22p^63s^23p^64s^23d^{10}4p^65s^14d^5$
(c) $1s^22s^22p^63s^23p^64s^23d^6$
(d) $1s^22s^22p^63s^23p^64s^23d^{10}4p^65s^24d^{10}5p^66s^24f^{14}5d^{10}$

8-21. Referring only to the periodic table on the inside front cover, write the ground-state electron configuration of

(a) Mg [Ne]$3s^2$
(b) Br [Ar]$4s^2 3d^{10} 4p^5$
(c) Au [Xe]$6s^2 4f^{14} 5d^{10}$
(d) Cs [Xe]$6s^1$
(e) Ga [Ar]$4s^2 3d^{10} 4p^1$

8-22. Referring only to the periodic table on the inside front cover, give two examples of

(a) an atom with a half-filled subshell
(b) an atom with a completed outer shell
(c) an atom with its outer electrons occupying a half-filled subshell and a filled subshell

8-23. Without looking at a periodic table, deduce the atomic numbers of the other elements that are in the same family as the element with atomic number (a) 16 and (b) 11.

a. 8, 16, 34, 52, 84
b. 1, 3, 11, 19, 37, 55, 87

8-24. Referring only to the periodic table on the inside front cover, determine the element of lowest atomic number whose ground state contains

(a) an *f* electron
(b) three *d* electrons
(c) a complete *d* subshell
(d) ten *p* electrons
(e) four complete *s* subshells

8-25. Referring only to the periodic table on the inside front cover, indicate which elements have an outer

(a) *s* electron configuration Groups 1 + 2
(b) *p* electron configuration Groups 3, 4, 5, 6, 7, 8
(c) *d* electron configuration transition metals
(d) *f* electron configuration lanthanides + actinides

Some chemists call these various elements the *s*-block, *p*-block, *d*-block, and *f*-block elements.

8-26. Determine the atomic numbers of the series of elements that you think would use $5g$ orbitals as their highest occupied orbitals. How many elements would there be in this series?

8-27. Referring only to the periodic table on the inside front cover, write Lewis electron-dot formulas for (see Table 7-2)

(a) Si
(b) P
(c) Se
(d) I
(e) Sr

8-28. Which columns in the periodic table correspond to the following electron-dot formulas:

(a) [ ]·
(b) ·[ ]:
(c) ·[ ]·
(d) ·[ ]·
(e) ·[ ]·

HUND'S RULE

8-29. Use Hund's rule to write ground-state electron configurations for

(a) O^+
(b) C^-
(c) F^+
(d) O^{2+}
(e) N^-

8-30. According to Hund's rule, which of the following electronic configurations are ground states:

	1s	2s	2p	3s	3p
(a)	(↑↓)	(↑↓)	( ) (↑↓) (↑↓)	( )	( ) ( ) ( )
(b)	(↑↓)	(↑↓)	(↑↓) (↑↓) (↑↓)	(↑↓)	(↑↓) ( ) ( )
(c)	(↑↓)	(↑↓)	(↑↓) (↑) (↑)	( )	( ) ( ) ( )
(d)	(↑↓)	(↑↓)	(↑↓) (↑↓) (↑↓)	(↑↓)	(↑↓) (↑↓) ( )
(e)	(↑↓)	(↑↓)	(↑↓) (↑↓) (↑↓)	(↑↓)	(↑↓) (↑↓) (↑)

8-31. How many unpaired electrons are there in the ground state of each of the following atoms:

(a) Si 2
(b) P 3
(c) Ti 2
(d) Se 2
(e) Zr 2

8-32. How many unpaired electrons are there in the ground state of each of the following ions:

(a) Cl^-
(b) O^+
(c) Al^{3+}
(d) Xe^+
(e) N^-

LEWIS ELECTRON-DOT FORMULAS

8-33. Write the Lewis electron-dot formulas for the Group 6 elements.

8-34. Write the Lewis electron-dot formulas for the Group 2 elements.

8-35. Nonmetals add electrons under certain conditions in order to attain a noble-gas electron configuration. How many electrons must be gained in this process by the following elements? Write the Lewis electron-dot formula for each ion that is formed. What noble-gas electron configuration is attained in each case:

(a) H (c) C (e) Cl
(b) O (d) S

8-36. Metals lose electrons to attain a noble-gas electron configuration. How many electrons are lost by the following elements when they attain such a configuration? Write the Lewis electron-dot formula for the ion thus formed. What is the corresponding noble-gas-like inner core in each case:

(a) Ca (c) Na (e) Al
(b) Li (d) Mg

GROUND-STATE ELECTRON CONFIGURATIONS OF IONS

8-37. Write the ground-state electron configuration for the following ions:

(a) K^+ (c) S^{2-} (e) Al^{3+}
(b) Br^- (d) Ba^{2+}

What do these electron configurations have in common?

8-38. Write the ground-state electron configuration for the following ions:

(a) Se^{2-} (c) Ga^+ (e) P^{3-}
(b) Be^+ (d) Pb^{2+}

8-39. Determine the number of unpaired electrons in the ground state of the following species:

(a) F^- (b) Sn^{2+} (c) Bi^{3+} (d) Ar^+ (e) S^-

8-40. Determine the number of unpaired electrons in the ground state of the following species:

(a) K^+ (b) Sc^{3+} (c) F^+ (d) Cl^- (e) O^-

8-41. Arrange the following species into groups of isoelectronic species (see Example 1-9):

F^-	Sc^{3+}	Be^{2+}	Rb^+	O^{2-}	Na^+
Ti^{4+}	Ar	B^{3+}	He	Se^{2-}	Y^{3+}

8-42. Arrange the following species into groups of isoelectronic species (see Example 1-9):

Mg^{2+}	K^+	Ca^{2+}	Li^+	H^-	Sr^{2+}
Ne	N^{3-}	P^{3-}	C^{4+}	Br^-	Kr

8-43. Describe the following processes in terms of the electronic configurations of the species involved:

(a) $O(g) + 2e^- \rightarrow O^{2-}(g)$
(b) $Ca(g) + Sr^{2+}(g) \rightarrow Sr(g) + Ca^{2+}(g)$

8-44. Describe the following processes in terms of the electronic configurations of the species involved:

(a) $I(g) + e^- \rightarrow I^-(g)$
(b) $K(g) + F(g) \rightarrow K^+(g) + F^-(g)$

EXCITED-STATE ELECTRON CONFIGURATIONS

8-45. Write the electron configuration for the first excited state of each of the following ions:

(a) Na^+ (c) F^-
(b) He^+ (d) O^+

8-46. Write the electron configuration of the first excited state for each of the following ions:

(a) Be^{2+} (c) Al^{3+}
(b) Mg^{2+} (d) O^{2-}

8-47. Which of the following electron configurations of neutral atoms represent excited states:

(a) $1s^22s^22p^53s^1$
(b) $1s^22s^22p^63s^23p^64s^13d^4$
(c) $1s^12s^1$
(d) $1s^22s^22p^33s^2$

8-48. Which of the following electron configurations of neutral atoms represent excited states:

(a) $1s^22s^22p^63s^1$
(b) $1s^22s^22p^53s^2$
(c) $1s^22s^22p^63s^13p^1$
(d) $1s^22s^22p^63s^23p^63d^1$

ATOMIC AND IONIC RADII

8-49. Determine the member of each of the following pairs of atoms that has the larger radius (do not use any references except the periodic table):

(a) N and P
(b) P and S
(c) S and Ar
(d) Ar and Kr

8-50. Arrange the following sets of atoms in order of increasing atomic radius:

(a) Kr, He, Ar, Ne
(b) K, Na, Rb, Li
(c) Be, Ne, F, N, B

8-51. Without using any references except the periodic table, arrange the members of the following groups in order of increasing size:

(a) Li, Na, Cs, Rb — Li, Na, Rb, Cs
(b) Al, Na, Mg, P — P, Al, Mg, Na
(c) Ca, Ba, Sr — Ca, Sr, Ba

8-52. The radii of lithium and its ions are

Species	*Radius/pm*
Li	135
Li^+	60
Li^{2+}	18

Explain why the radii decrease from Li to Li^{2+}.

IONIZATION ENERGIES

8-53. The first ionization energies of the alkaline earth metals are

Metal	*Ionization energy/$MJ \cdot mol^{-1}$*
Be	0.899
Mg	0.738
Ca	0.590
Sr	0.549
Ba	0.503

Explain why the ionization energies decrease from beryllium to barium. Because the atomic radius increases from beryllium to barium. The outermost electrons are further away in successive element of the group, and removing the is easier. So ↓ ionization energy

8-54. Arrange the following sets of atoms in order of increasing ionization energy:

(a) B, O, Ne, F
(b) Te, I, Sb, Xe
(c) K, Ca, Rb, Cs
(d) Ar, Na, S, Al

8-55. Arrange the following species in terms of increasing first ionization energy (do not use any references):

F Cs Br Cl K

Cs < K < Br < Cl < F

8-56. Arrange the following species in terms of increasing first ionization energy (do not use any references):

O Ca Ba S Se Sr

INTERCHAPTER E

The Chemistry of the Atmosphere

The earth from space. The cloud cover is clearly visible and gives the earth a bright appearance.

The atmosphere is the sea of gas that envelops the earth. We live at the bottom of this gaseous sea and owe our existence to its chemical properties. The major emphasis of this interchapter is on the chemical composition and reactions of the atmosphere and their influence on life on earth.

E-1. THE EARTH'S ATMOSPHERE CAN BE DIVIDED INTO FOUR DISTINCT REGIONS

Because of the effect of the earth's gravitational field, the pressure of the atmosphere decreases with increasing altitude, from a maximum of about 760 torr (1 atm) at sea level to effectively zero at several

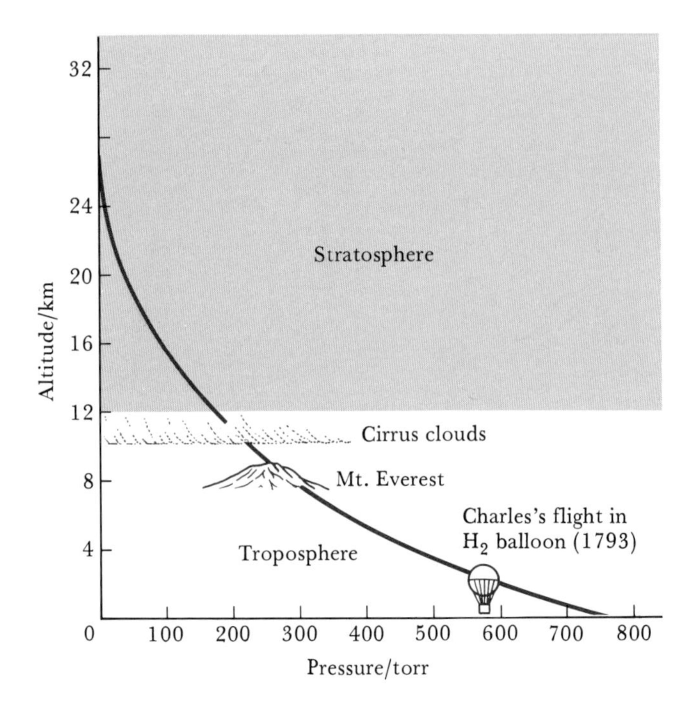

Figure E-1 The pressure of the atmosphere as a function of height above sea level. Note that pressure decreases rapidly with increasing altitude and drops to 0.5 atm (380 torr) at about 6 km.

hundred kilometers (Figure E-1). Most properties of the atmosphere change gradually with altitude, but it is nonetheless convenient to divide the atmosphere into four separate regions, which can be defined by considering the temperature as a function of altitude (Figure E-2).

The lowest region, in which the temperature decreases steadily with increasing altitude, is the *troposphere.* At the top of the troposphere (about 10 km), the temperature is about −55°C. The troposphere accounts for more than 80 percent of the mass and virtually all the water vapor, clouds, and precipitation in the earth's atmosphere. It is characterized by strong vertical mixing. For example, in clear air, a molecule can traverse the entire depth of the troposphere in a few days; during severe thunderstorms, the traversal may occur in minutes. All our weather takes place in the troposphere.

Above the troposphere lies the *stratosphere.* In this region, the temperature increases with altitude until it reaches about −10°C at roughly 50 km. The troposphere and stratosphere together account for 99.9 percent of the mass of the atmosphere. Compared with the troposphere, the stratosphere is relatively calm. It is characterized by very little vertical mixing. Debris from nuclear explosions and dust from volcanic eruptions remain in the stratosphere for years before becoming mixed with the troposphere.

Beyond the stratosphere is the region called the *mesosphere* (literally, the middle sphere). Like the troposphere, it is a region in which temperature decreases with altitude.

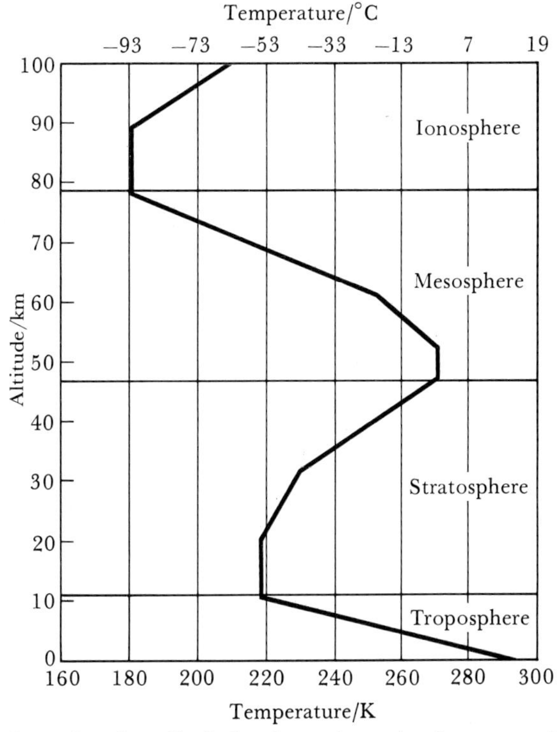

Figure E-2 The temperature of the earth's atmosphere as a function of height above sea level. The average temperature over the surface of the earth is about 15 °C. The temperature drops at the rate of 7 C°/km for the first 10 km or so, and then stays constant for a few kilometers. It increases with altitude from 20 km up to about 50 km, after which it decreases until 80 km and then increases again. This temperature profile delineates four separate regions of the atmosphere.

The fourth region is called the *ionosphere*. As the name implies, the ionosphere contains ions and electrons, which are produced by high-energy solar radiation.

Geologists believe that much of the earth's atmosphere was formed from gases that were discharged from the earth's interior through volcanic activity. Its present composition is fairly uniform up to about 100 km (Table E-1) and is roughly 78 percent nitrogen and 21 percent oxygen, whereas the gaseous emissions from volcanos are a mixture of about 85 percent water vapor, 10 percent carbon dioxide, and a few percent nitrogen (as N_2) and sulfur compounds. Elemental oxygen is notably absent from volcanic emissions; the oxygen in the atmosphere is a result of photosynthesis. Nitrogen is a fairly unreactive molecule, and so most of the nitrogen released by volcanic activity remains in the atmosphere. Consequently, nitrogen has become the dominant constituent of the earth's atmosphere.

The atmosphere was capable of holding only a small fraction of the water vapor that resulted from volcanic eruptions. Eventually the accumulated water vapor gave rise to clouds and rain and subsequently to the bodies of water on the earth's surface. The main source of atmospheric water vapor today is evaporation from the earth's surface. The evaporated water is incorporated into clouds and then returned by precipitation. The average time a water molecule spends in the atmosphere is about one week. The concentration of water vapor is highest near the ground and drops to very low values above 10 km.

Table E-1 The composition of the earth's atmosphere below 100 km

Constituent	*Content in fraction of total molecules or ppm*[a]
nitrogen (N_2)	0.7808 (75.51% by mass)
oxygen (O_2)	0.2095 (23.14% by mass)
argon	0.0093 (1.28% by mass)
water vapor	0–0.04
carbon dioxide	325 ppm
neon	18 ppm
helium	5 ppm
methane	2 ppm
krypton	1 ppm
hydrogen (H_2)	0.5 ppm
dinitrogen oxide	0.5 ppm
xenon	0.1 ppm

[a]Parts per million; for example, 325 ppm of carbon dioxide means that of each 1 million molecules 325 are CO_2.

E-2. THE CONCENTRATION OF CARBON DIOXIDE IN THE EARTH'S ATMOSPHERE IS SLOWLY INCREASING

Carbon dioxide is produced not only by volcanic eruptions but also by respiration, the decay of organic matter, and the combustion of fossil fuels. It is removed from the atmosphere by photosynthesis, dissolution in the oceans, and the formation of shales and carbonate rocks (primarily as $CaCO_3$ and $MgCO_3$). There is evidence that the rate of removal of carbon dioxide from the earth's atmosphere is not high enough to keep pace with the ever-increasing rate at which it is added to the atmosphere as a result of the combustion of fossil fuels. The concentration of carbon dioxide in the atmosphere has increased by almost 8 percent since the beginning of this century. Between 1958 and 1975 the increase was 5 percent (Figure E-3). The present worldwide rate of increase of carbon dioxide concentration is about half the rate at which it is produced by the combustion of fossil fuels.

This steady increase in the concentration of carbon dioxide in the atmosphere could have serious consequences. The solar radiation that penetrates the earth's atmosphere is mostly visible radiation, with a little ultraviolet and infrared. This radiation is absorbed by the ground and the oceans. The earth's surface is thus warmed and as a result emits heat (infrared) radiation. Carbon dioxide, as well as

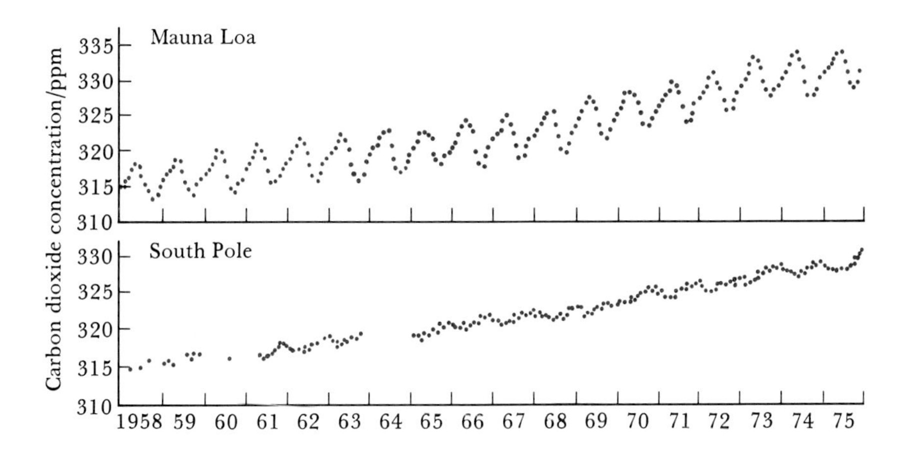

Figure E-3 Atmospheric carbon dioxide concentrations measured at Mauna Loa, Hawaii, and at the South Pole from 1958 through 1975. The oscillations in the Hawaii data reflect the depletion of CO_2 from the air by photosynthesis during the growing season and its subsequent buildup between growing seasons.

water vapor, absorbs infrared radiation strongly. Hence, instead of passing through the atmosphere, the radiated energy is absorbed by the atmosphere, and the temperature of the atmosphere is increased. This phenomenon is called the *greenhouse effect* because it is similar to the trapping of the sun's heat in a greenhouse. (This is really a misnomer, however, because greenhouses attain higher temperatures than the outside air primarily because the glass cover restricts the vertical movement of the air that is heated by solar radiation.) It has been estimated that the greenhouse effect is responsible for an average increase of about 45 C° in the temperature at the earth's surface. Without this effect, the average world temperature would be −30°C instead of 15°C. The planet Venus, which has a very dense atmosphere of carbon dioxide, has a surface temperature of about 500°C.

Calculations indicate that a doubling of the concentration of carbon dioxide in the atmosphere would cause an increase in the average world temperature of several degrees Celsius. (Recall that the concentration of carbon dioxide has increased by almost 8 percent in this century.) An increase in the average world temperature by a few degrees is not trivial; such an increase could cause a large part of the polar ice caps to melt, raising the level of the oceans and flooding coastal cities. The calculations that lead to this dire forecast, however, are controversial.

The greenhouse effect due to water vapor in the atmosphere is a major factor in the weather of deserts and other regions. Although desert days may be very hot, there is little water vapor in the atmosphere to absorb the reradiated infrared radiation at night and so the nights are very cool.

E-3. THE NOBLE GASES WERE NOT DISCOVERED UNTIL 1893

In 1893, the English physicist Lord Rayleigh noticed a small discrepancy between the density of nitrogen obtained by the removal of oxygen, water vapor, and carbon dioxide from air and the density of nitrogen prepared by chemical reaction, such as the thermal decomposition of ammonium nitrite:

$$NH_4NO_2(s) \rightarrow N_2(g) + 2H_2O(g)$$

One liter of nitrogen at 0°C and 1 atm obtained by the removal of all the other known gases from air (Figure E-4) has a mass of 1.2572 g, whereas one liter of dry nitrogen obtained from ammonium nitrite has a mass of 1.2505 g under the same conditions. This slight difference led Lord Rayleigh to suspect that some other gas was present in the sample of nitrogen obtained from air.

The English chemist William Ramsay found that, if hot calcium metal is placed in a sample of nitrogen obtained from air, about 1 percent of the gas fails to react. Pure nitrogen would react completely. Because of the inertness of the residual gas, Ramsay gave it the name argon (Greek, idle). He then liquefied the residual gas and, upon measuring its boiling point, discovered that it consisted of five components, each with its own characteristic boiling point (Table E-2). The component present in the greatest amount retained the name argon. The others were named helium (sun), neon (new), krypton (hidden), and xenon (stranger). Helium was named after the Greek

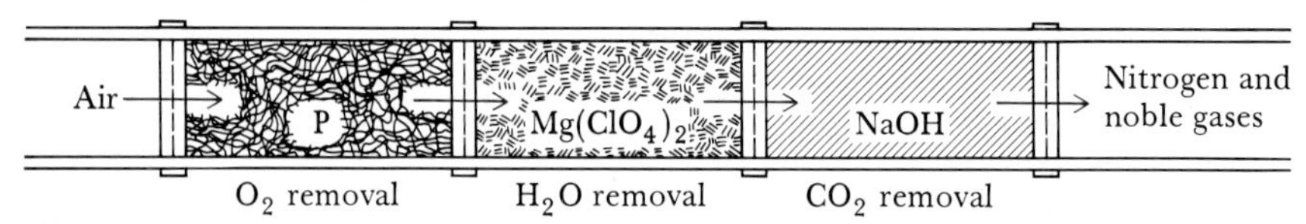

Figure E-4 A schematic illustration of the removal of O_2, H_2O, and CO_2 from air. First the oxygen is removed by allowing the air to pass over phosphorus:

$$P_4(s) + 5O_2(g) \rightarrow P_4O_{10}(s)$$

The residual air is passed through anhydrous magnesium perchlorate to remove the water vapor:

$$Mg(ClO_4)_2(s) + 6H_2O(g) \rightarrow Mg(ClO_4)_2 \cdot 6H_2O(s)$$

and then through sodium hydroxide to remove the CO_2:

$$NaOH(s) + CO_2(g) \rightarrow NaHCO_3(s)$$

The gas that remains is primarily nitrogen with about 1 percent noble gases.

Table E-2 Properties of the noble gases

Gas	*ppm in the air*	*Density*[a]$/g \cdot L^{-1}$	*Melting point/°C*	*Boiling point/°C*	*Cost*[b]
helium	5.2	0.179	−272.2	−268.9	$115
neon	18.2	0.900	−248.6	−245.9	$100
argon	9340	1.78	−189.3	−185.8	$115
krypton	1.1	3.75	−157.1	−152.9	$400
xenon	0.08	5.90	−112.0	−108.1	$1400

[a]At 0°C and 1 atm.
[b]For 100 L at about 40 atm, research grade (1983).

word for sun (helios) because its presence in the sun had been detected earlier by spectroscopic methods.

The noble gases in the atmosphere are thought to have arisen as by-products of the decay of radioactive elements in the earth's crust (Chapter 24). For their work in discovering and characterizing an entire new family of elements, Rayleigh received the 1904 Nobel Prize in physics and Ramsay received the 1904 Nobel Prize in chemistry.

All the noble gases are colorless, odorless, and relatively inert. Helium is used in lighter-than-air craft, despite the fact that it is denser and hence has less lifting power than hydrogen, because it is nonflammable. Helium is also used in welding to provide an inert atmosphere around the welding flame and thus reduce corrosion of the heated metal. Neon is used in neon signs, which are essentially discharge tubes (Section 7-4) filled with neon or a neon-argon mixture. When placed in a discharge tube, neon emits an orange-yellow glow that penetrates fog very well. Argon, the most plentiful and least expensive noble gas, often is used in place of nitrogen in incandescent light bulbs of high candlepower because it does not react with the hot filament. Krypton and xenon are too scarce and costly to find much application, although they are used in lasers and flashtubes for high-speed photography.

Helium has 93 percent of the lifting power of hydrogen.

E-4. OXIDES OF SULFUR ARE MAJOR POLLUTANTS OF THE ATMOSPHERE

Except for the carbon dioxide produced from the combustion of fossil fuels, all the atmospheric constituents discussed up to this point arise from natural sources, meaning sources not due to human activities. Many machines and industries introduce harmful and irritating substances into the atmosphere, which collectively are referred to as pollutants. Some common pollutants are the oxides of sulfur (denoted by SO_x) that result from the combustion of sulfur-containing coal and oil, oxides of nitrogen (NO_x) and carbon monoxide (CO) produced in the internal combustion engine, and hydrocarbons, which arise from the incomplete combustion of gasoline.

Two oxides of sulfur, SO_2 and SO_3, are major pollutants in industrial and urban areas. Most coal and petroleum contain some sulfur, which becomes SO_2 when burned. Concentrations of SO_2 as low as 0.1 to 0.2 ppm can be incapacitating to persons suffering from respiratory conditions such as emphysema and asthma. Although SO_2 is not easily oxidized to SO_3, the presence of dust particles and other particulate matter or ultraviolet radiation facilitates the conversion. The SO_3 then reacts with water vapor to form a very fine sulfuric acid mist. Such a mist is also produced in automobile catalytic converters. Both sulfuric acid and sulfurous acid, H_2SO_3, which arises from the reaction

$$SO_2(g) + H_2O(g) \rightarrow H_2SO_3(\text{mist})$$

produce acid rain, which is rain that is up to 1000 times more acidic than normal rain. Acid rain occurs commonly in northern Europe and in the northeastern United States. Many lakes in these regions are so acidic that the fish life is disappearing.

Acid rain has a devastating effect on limestone and marble, both of which contain $CaCO_3$. The reaction that occurs is

$$CaCO_3(s) + H_2SO_4(aq) \rightarrow CaSO_4(s) + H_2O(l) + CO_2(g)$$

The formation of powdered calcium sulfate breaks down the limestone or marble structure. The decomposition of carbonates by acid rain is a major cause of the deterioration of the ancient buildings and monuments of Europe (Figure E-5).

Figure E-5 Many of the limestone and marble monuments of Europe, such as this one in Rome, are being damaged by acid rain.

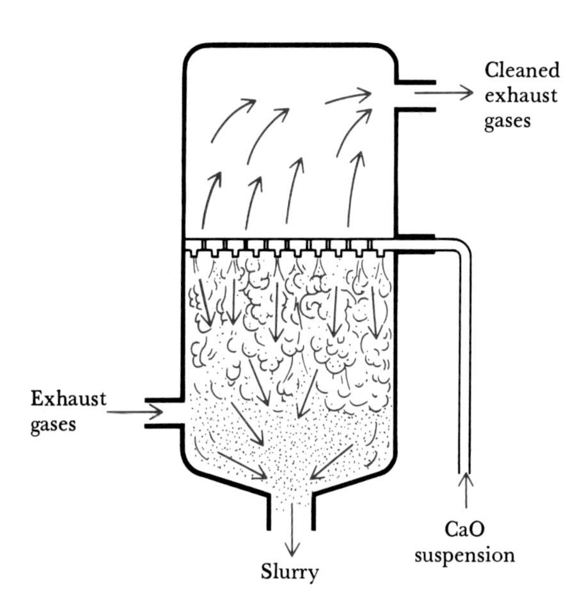

Figure E-6 In a scrubber, exhaust gases are passed through a spray of an aqueous suspension of calcium oxide, which removes the SO_2 from the gas. The chemical equation for the reaction that occurs is

$$CaO(s) + SO_2(g) \rightarrow CaSO_3(s)$$

There have been three major disasters attributed to air polluted with oxides of sulfur. In 1952, a gray fog highly polluted with oxides of sulfur settled over London for several days and was reportedly responsible for 4000 deaths. Such a *London fog,* as it is now called, also caused hundreds of deaths in Donora, Pennsylvania, in 1948 and along the Meuse Valley in Belgium in 1930.

Several methods can be used to control the amount of SO_2 introduced into the atmosphere. One obvious way is to burn low-sulfur coal and petroleum. Nigerian oil and some Middle East oil is low in sulfur, whereas Venezuelan oil is high in sulfur. In general, coal from east of the Mississippi River is higher in sulfur than western coal. One method for removing SO_2 from fossil fuel combustion products involves passing the effluent gases through a device called a *scrubber* (Figure E-6), where the gases are sprayed with an aqueous suspension of calcium oxide (lime). The scrubbing eliminates most of the SO_2 but produces large amounts of $CaSO_3$ and $CaSO_4$ that must be disposed of.

E-5. HYDROCARBONS AND OXIDES OF NITROGEN ARE THE PRIMARY INGREDIENTS OF PHOTOCHEMICAL SMOG

Under ordinary conditions, nitrogen and oxygen do not react with each other. When combined at high pressure and temperature, however, as in the cylinders of an automobile engine, they react to form nitrogen oxide, NO, which then reacts with O_2 to produce nitrogen dioxide. Ordinarily this reaction occurs too slowly at the low concentrations of NO in the atmosphere to account for any significant concentration of $NO_2(g)$, but, for reasons that are not yet understood, the reaction occurs rapidly in sunny, urban atmospheres. Nitrogen dioxide is a red-brown noxious gas that is responsible for the yellow-brown color of smog, first made famous in Los Angeles but now all too common in many urban areas. A concentration of 500 ppm of NO_2 in air is usually fatal; there is some disagreement concerning tolerable levels of NO_2, but they are not higher than 3 to 5 ppm. Levels of NO_2 reach 0.9 ppm in Los Angeles on particularly bad days.

The problem of NO_2 is not so much its primary toxicity but the fact that it is dissociated by radiation to produce atomic oxygen:

$$NO_2(g) \xrightarrow{\text{392-nm light}} NO(g) + O(g)$$

Because the dissociation of the NO_2 is caused by radiation (light), it is called *photodissociation*. The atomic oxygen then reacts with molecular oxygen to produce ozone. These two reactions account for the fact that ozone levels are higher on sunny days than on cloudy days. Ozone in the atmosphere makes up about 90 percent of the general category of pollutants called oxidants, which are now measured continually in many cities. Los Angeles has air pollution alerts when the level of oxidants exceeds 0.35 ppm.

The atomic oxygen produced by the photodissociation of NO_2 also attacks the hydrocarbons introduced into the atmosphere by the incomplete combustion of gasoline and diesel fuel. The reaction of atomic oxygen with hydrocarbons initiates a complicated sequence of chemical reactions. The end products of these reactions are a number of substances that attack living tissue and lead to great discomfort, if not serious disorders. These substances make up what is called *photochemical smog,* so called because the entire process is initiated by the photodissociation of NO_2. A particularly irritating product is PAN (peroxyacetylnitrate), which causes eyes to tear and smart, something that people who live in smoggy cities experience often.

The control of photochemical smog requires controlling the emission of its two principal ingredients, NO and hydrocarbons from automobile exhausts. The Congressional Clean Air Act of 1967, with its amendments in 1970 and 1977, imposed limitations on exhaust emissions. Although there are indications that smog has lessened in some cities, in many others smog and other types of pollution problems are still increasing.

E-6. OZONE IS A LIFE-SAVING CONSTITUENT OF THE STRATOSPHERE

Ozone in the stratosphere (Figure E-2) is essential to life as we know it. Oxygen molecules in the stratosphere and higher regions undergo photodissociation by absorption of solar radiation of wavelengths less than 240 nm:

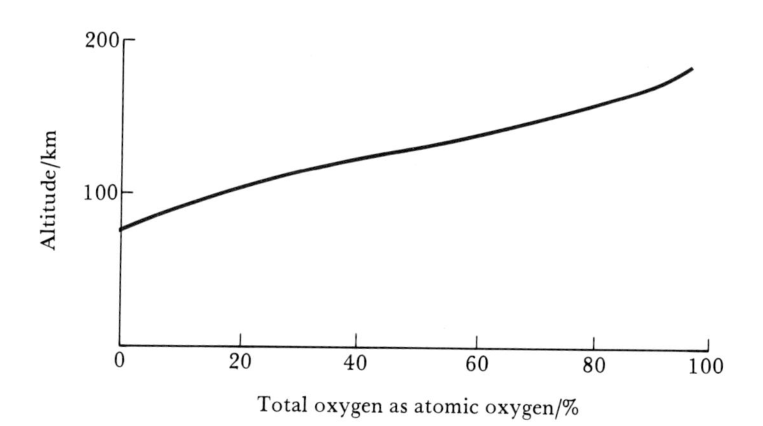

Figure E-7 The concentration of atomic oxygen as a function of altitude. Atomic oxygen is produced by the photodissociation of molecular oxygen.

$$O_2(g) \xrightarrow{\lambda < 240\ \text{nm}} 2O(g)$$

Virtually all the solar radiation in the region between 100 and 200 nm is absorbed by O_2. The atomic oxygen produced by this reaction is a major atmospheric constituent above 100 km; the total fraction of oxygen that exists as atomic oxygen is plotted versus altitude in Figure E-7.

Direct evidence for the presence of atomic oxygen in the upper atmosphere was obtained in 1956 when 8 kg of NO was released from a rocket at an altitude of 110 km. Immediately a yellow-red glow appeared in the sky as a result of the reaction of NO and O to produce NO_2. Since that time, small mass spectrometers have been part of experiments sent up in satellites and have provided data like those in Figure E-7. Because the gas density is so low above 100 km, oxygen atoms produced at greater altitudes can exist for a long time. At lower altitudes, where the total gas density is higher, the oxygen atoms are able to react not only with each other but also, and more important, with oxygen molecules to produce ozone. Below 70 km or so, atomic oxygen produced by photodissociation of O_2 reacts almost immediately with O_2 to produce O_3. Figure E-8 shows the concentration of ozone in the atmosphere versus altitude. Note that most of the ozone is found between 15 and 30 km; this region is called the *ozone layer.*

Ultraviolet radiation with wavelengths longer than 240 nm is not absorbed until it encounters the ozone layer, where it is then absorbed by photodissociation of O_3:

$$O_3(g) \xrightarrow{240\ \text{nm} < \lambda < 310\ \text{nm}} O_2(g) + O(g)$$

The oxygen atom produced in this reaction quickly recombines with another oxygen molecule:

$$O(g) + O_2(g) \rightarrow O_3(g)$$

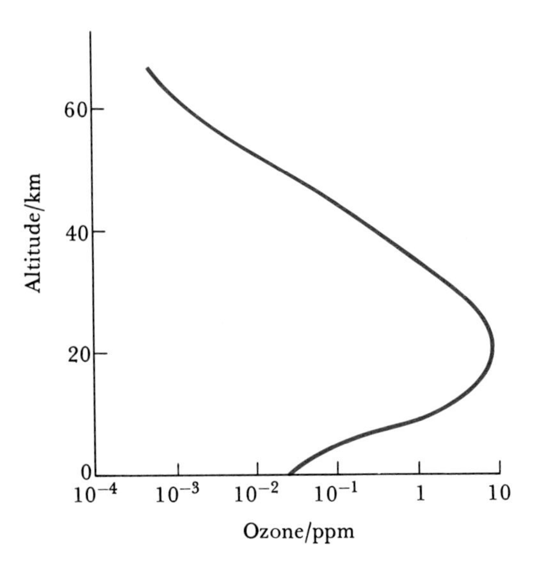

Figure E-8 The concentration of ozone in the earth's atmosphere as a function of altitude. Note that even in the ozone layer the concentration of ozone does not exceed 10 ppm.

By adding these two equations, we see that there is no net chemical change but there is an absorption of radiation. Through this pair of reactions repeated many times, a single ozone molecule ultimately leads to the absorption of many photons of radiation. Trace amounts of ozone in the stratosphere can absorb virtually all the solar ultraviolet radiation in the region 240 nm $< \lambda <$ 310 nm. When the solar spectrum is measured at the far reaches of the earth's atmosphere and at the earth's surface, it is seen that the ozone in the stratosphere absorbs nearly all the sun's ultraviolet radiation (Figure E-9). Were it not for the formation of ozone in the upper atmosphere, life as it currently exists on earth would not be possible because of the deleterious effect of ultraviolet radiation on most living organisms.

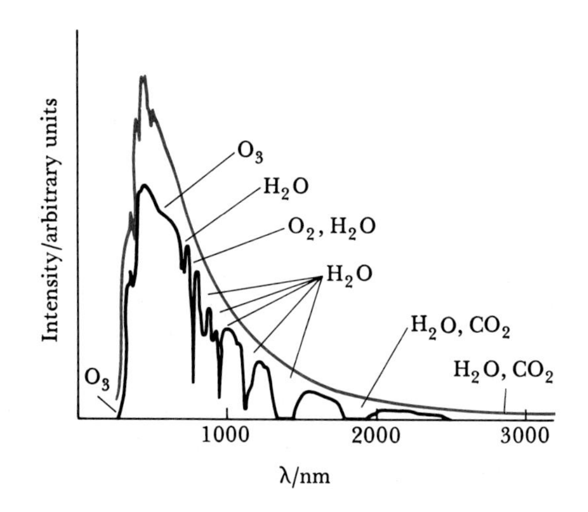

Figure E-9 A comparison of the intensity of the solar radiation incident upon the earth's outer atmosphere (red) with that reaching the surface of the earth (black). The molecular species responsible for filtering in various regions of the radiation spectrum are indicated. The visible region is between 400 and 700 nm, the region of greatest intensity reaching the earth's surface. The absorption of radiation with $\lambda >$ 1000 nm by CO_2 and H_2O is responsible for the greenhouse effect.

In the 1970's, it was suggested that compounds called chlorofluorocarbons, which are the principal components of many refrigerants and propellants in aerosol sprays, may diffuse up to the ozone layer and undergo a series of reactions that lead to the depletion of ozone. If we take the molecule CF_2Cl_2 as a typical chlorofluorocarbon, then the reaction that initiates the reactions with ozone is the photodissociation of CF_2Cl_2:

$$CF_2Cl_2(g) \xrightarrow{\lambda \approx 200 \text{ nm}} CF_2Cl(g) + Cl(g)$$

The atomic chlorine generated in this reaction reacts with ozone according to the two-step sequence

$$Cl(g) + O_3(g) \rightarrow ClO(g) + O_2(g)$$
$$ClO(g) + O(g) \rightarrow Cl(g) + O_2(g)$$

The net reaction is

$$O_3(g) + O(g) \rightarrow 2O_2(g)$$

indicating a depletion of ozone. The atomic chlorine that initiates this reaction is regenerated in the second step and so is able to react with another ozone molecule, repeating the process. In this way, one molecule of a chlorofluorocarbon may be responsible for the depletion of many ozone molecules. Whether or not these reactions take place to any significant extent in the stratosphere or pose a serious threat to the ozone layer is controversial. Nervertheless, the use of chlorofluorocarbons in most aerosol spray cans has been discontinued.

It should be emphasized that the concentration of atmospheric ozone is indeed small, never exceeding 10 ppm (Figure E-9). If the entire layer of ozone were compressed to a pressure of 1 atm at 0°C, its thickness would be only 3 mm.

E-7. SOLAR RADIATION OF WAVELENGTHS SHORTER THAN 100 nm IS SCREENED BY PHOTOIONIZATION IN THE IONOSPHERE

In the ionosphere (Figure E-2), solar radiation of wavelengths shorter than 100 nm causes the *photoionization* of oxygen atoms, oxygen molecules, and nitrogen molecules:

$$O(g) \xrightarrow{\lambda < 100\ \text{nm}} O^+(g) + e^-(g)$$

$$O_2(g) \xrightarrow{\lambda < 100\ \text{nm}} O_2^+(g) + e^-(g)$$

$$N_2(g) \xrightarrow{\lambda < 100\ \text{nm}} N_2^+(g) + e^-(g)$$

All the solar radiation with $\lambda < 100$ nm is absorbed by these photoionization reactions. Table E-3 summarizes the processes that are responsible for the absorption of the sun's ultraviolet radiation. Photoionization reactions in the ionosphere and photodissociation reactions in the mesosphere and stratosphere absorb most of the harmful solar radiation ($\lambda < 310$ nm) before it reaches the earth's surface.

The products formed in photoionization are very reactive, but the gas density in the ionosphere is so low that it takes hours for the recombination reactions to occur. Consequently, at any instant of time there are significant densities of charged particles in the ionosphere. We are all familiar with the effect of these particles on AM radio transmission. During the day, the ions in the ionosphere form a thick, almost continuous layer (called a Heaviside layer), depicted in a simplified manner in Figure E-10(a). The AM radio waves bounce off the lower region of this ionic layer and are picked up by a receiver at point *A*. During the night, two well-defined ionic layers form, which are depicted in Figure E-10(b). During night transmission, some of the AM signal passes through the lower layer, bounces off the upper layer, and is picked up at point *B*. This is why we are able to pick up AM radio stations from long distances after the sun goes down and the photoionization processes in the atmosphere cease.

Table E-3 The absorption of solar radiation by the earth's atmosphere

Wavelength/nm	*Fraction of total solar energy*	*Altitude where absorbed in earth's atmosphere/km*	*Primary absorption mechanism*	*Fraction absorbed by the atmosphere*
<100	3 parts in 10^6	90–200	ionization of O, O_2, and N_2	all
100–240	1 part in 10^4	50–110	dissociation of O_2	all
240–310	1.75%	30–60	dissociation of O_3	all
>310	98%+	0–10	absorption by water vapor	0.17

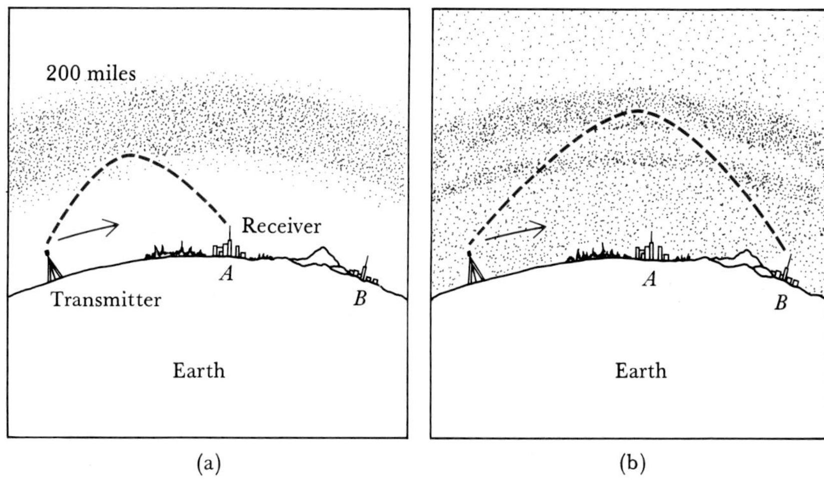

Figure E-10 (a) The fairly continuous region of relatively high electron density that occurs during daylight. Radio waves bounce off the bottom of this region and are reflected back to earth and picked up at a receiver located at point *A*. (b) After sundown, the region of relatively high electron density separates into two regions. Some radio waves penetrate the lower region and are reflected by the upper region and picked up by a receiver located at point *B*.

QUESTIONS

E-1. Name the four regions into which the earth's atmosphere can be divided.

E-2. In which region of the atmosphere does all our weather take place?

E-3. What is the origin of the nitrogen in our atmosphere?

E-4. What is the potential problem concerning the continuous increase of the concentration of carbon dioxide in the atmosphere?

E-5. Why are clear nights colder than cloudy nights?

E-6. Explain how the noble gases were discovered.

E-7. What is photochemical smog? Why is it so called?

E-8. What is acid rain?

E-9. Write a chemical equation for the reaction of the deterioration of limestone and marble structures by acid rain.

E-10. Why do we say that ozone is a life-saving constituent of the stratosphere?

E-11. What is meant by the ozone layer?

E-12. What is photoionization? Discuss the life-saving role of photoionization in the ionosphere.

An electrolytic solution in which NaCl has been added to distilled water. The light bulb shines brightly, indicating that an aqueous solution of NaCl is a strong conductor of electricity.

9 / Ionic Compounds

In Chapter 8 we learned about the relationship of the electronic structure of atoms and the periodic table. It seems reasonable to suppose that an understanding of atomic structure should lead to an understanding of the chemical bonding that occurs between atoms. For example, it is possible to use the electron configurations of the sodium atom and the chlorine atom to understand why one atom of sodium combines with just one atom of chlorine to form sodium chloride. Why is the result NaCl, instead of $NaCl_2$ or Na_2Cl? Why is NaCl an ionic compound, capable of conducting an electric current when it is melted or dissolved in water? Why do carbon and hydrogen combine to form the stable molecule methane, whose formula is CH_4 instead of CH or CH_2? Why is nitrogen a diatomic gas, N_2, at room temperature, whereas sodium chloride is a solid with a melting point of 800°C? Why doesn't a solution of sucrose dissolved in water conduct an electric current? These are the kind of questions that we answer in this and the next chapter.

9-1. THE ELECTROSTATIC ATTRACTION THAT BINDS OPPOSITELY CHARGED IONS TOGETHER IS CALLED AN IONIC BOND

To understand ionic bonds, let's first consider the reaction between a sodium atom and a chlorine atom. The electron configurations of the sodium and chlorine atoms are

$$\text{Na} \quad [\text{Ne}]3s^1 \qquad \text{Cl} \quad [\text{Ne}]3s^23p^5$$

The Lewis electron-dot formulas for the sodium and chlorine atoms are Na · and $:\ddot{\underset{..}{Cl}}\cdot$. Note that the electron configuration of a sodium atom consists of a neon-like inner core with a 3*s* electron outside the core. If the sodium atom loses the 3*s* electron, then the result is a sodium ion, with an electron configuration like that of the noble gas neon. We can describe the ionization process by the equation

$$Na([Ne]3s^1) \rightarrow Na^+([Ne]) + e^-$$

Once a sodium atom loses its 3*s* electron, the resultant sodium ion has a neon-like electron configuration and is relatively stable to further ionization.

If a chlorine atom accepts an electron, then the result is a chloride ion, with an electron configuration like that of the noble gas argon. We write this as

$$Cl([Ne]3s^23p^5) + e^- \rightarrow Cl^-([Ar])$$

Noble-gas electron configurations are relatively stable.

Thus we see that both a sodium atom and a chlorine atom can achieve a noble-gas electron configuration through the transfer of an electron from the sodium atom to the chlorine atom. We can describe the electron transfer by the equation

$$Na([Ne]3s^1) + Cl([Ne]3s^23p^5) \rightarrow Na^+([Ne]) + Cl^-([Ar])$$

or, in terms of Lewis electron-dot formulas,

$$Na\cdot + \cdot\ddot{\underset{..}{Cl}}: \rightarrow \underbrace{Na^+ + :\ddot{\underset{..}{Cl}}:^-}_{Na^+Cl^-}$$

The sodium ion and the chloride ion have opposite charges and so attract each other. The electrostatic attraction binds the ions together and is called an *ionic bond*.

We have seen that noble-gas electron configurations are relatively stable to the gain or loss of additional electrons. Because both Na^+ and Cl^- have achieved a noble-gas electron configuration, the above reaction will occur easily and there is no tendency for additional electron transfer. We already can answer several of the questions posed in the introduction to this chapter. We have seen that a sodium atom has one and only one electron that it loses readily and that a chlorine atom readily gains one and only one electron. Therefore, when a sodium atom reacts with a chlorine atom, the transfer of an electron from the sodium atom to the chlorine atom occurs readily and the result is one sodium ion and one chloride ion. The chemical formula of sodium chloride, then, is NaCl and not $NaCl_2$, Na_2Cl, or anything other than NaCl. Furthermore, NaCl is an *ionic compound*, that is, a compound composed of ions.

9-2. REACTIVE METALS COMBINE WITH REACTIVE NONMETALS TO PRODUCE IONIC COMPOUNDS

We shall discuss the energetics of the reaction between a sodium atom and a chlorine atom later, but first let's discuss some other reactions in which the products are ions. Consider the reaction

$$\text{Ba}(g) + 2\text{Cl}(g) \rightarrow \text{BaCl}_2(g)$$

The electron configurations of the barium atom and the chlorine atom are

$$\text{Ba} \quad [\text{Xe}]6s^2 \qquad \text{Cl} \quad [\text{Ne}]3s^23p^5$$

In this case, the barium atom gives up two outer electrons to achieve the xenon electron configuration and each of the two chlorine atoms accepts one electron to achieve an argon electron configuration. We can write this as

$$\text{Ba}([\text{Xe}]6s^2) + 2\text{Cl}([\text{Ne}]3s^23p^5) \rightarrow \text{Ba}^{2+}([\text{Xe}]) + 2\text{Cl}^-([\text{Ar}])$$

This reaction occurs readily because all the product species achieve a noble-gas electron configuration. The formula BaCl_2 represents the neutral set of one doubly charged barium ion and two singly charged chloride ions. In terms of Lewis electron-dot formulas,

$$\cdot\text{Ba}\cdot + 2\ \cdot\underset{\cdot\cdot}{\overset{\cdot\cdot}{\text{Cl}}}\colon \rightarrow \underbrace{\colon\underset{\cdot\cdot}{\overset{\cdot\cdot}{\text{Cl}}}\colon^- + \text{Ba}^{2+} + \colon\underset{\cdot\cdot}{\overset{\cdot\cdot}{\text{Cl}}}\colon^-}_{\text{BaCl}_2}$$

The right-hand side of this equation indicates that the products are one barium ion and two chloride ions. The barium ion attracts both chloride ions, and so there are two ionic bonds in BaCl_2. Note that the net charge on BaCl_2 must be zero because it is a compound. The reaction between barium and chlorine atoms can be written as

$$\text{Ba}(g) + 2\text{Cl}(g) \rightarrow \text{BaCl}_2(g)$$

Barium chloride is a solid at room temperature.

The compound formed, BaCl_2, is an ionic compound.

Example 9-1: Predict the product of the reaction between sodium and sulfur.

Solution: The Lewis electron-dot formulas for the sodium and sulfur atoms are $\text{Na}\cdot$ and $\cdot\underset{\cdot\cdot}{\overset{\cdot\cdot}{\text{S}}}\cdot$. The sulfur atom can complete its outer shell and achieve an argon electron configuration by accepting two electrons. A sodium atom contributes one electron, and so two sodium atoms are required for each sulfur atom. The reaction can be written

$$2\text{Na}\cdot + \cdot\ddot{\underset{\cdot\cdot}{\text{S}}}\cdot \rightarrow \underbrace{\text{Na}^{+}:\ddot{\underset{\cdot\cdot}{\text{S}}}:^{2-}\text{Na}^{+}}_{\text{Na}_2\text{S}}$$

Problems 9-1 to 9-8 deal with electron configurations and chemical reactions.

The ionic product in this reaction is sodium sulfide, whose chemical formula is Na_2S. An understanding of the electron transfers in this reaction allows us to understand why the equation for the reaction is

$$2\text{Na}(s) + \text{S}(s) \rightarrow \text{Na}_2\text{S}(s)$$

Example 9-2: Use Lewis electron-dot formulas to predict the chemical formula of aluminum oxide.

Solution: Aluminum oxide results from the reaction of aluminum with oxygen. The Lewis electron-dot formulas of the aluminum and oxygen atoms are $\cdot\dot{\text{Al}}\cdot$ and $\cdot\ddot{\underset{\cdot\cdot}{\text{O}}}\cdot$. The aluminum atom loses three electrons to achieve a noble-gas (Ne) electron configuration. The oxygen atom needs two electrons to complete its shell, and so there must be a combination of two aluminum atoms (six electrons lost) and three oxygen atoms (six electrons gained) for the final product. The chemical formula of the product is Al_2O_3.

The examples that we have discussed so far involve reactions between reactive metals and reactive nonmetals. These atoms can lose or gain electrons relatively easily to achieve a noble-gas electron configuration. Figure 9-1 shows some of the common atoms that lose or gain electrons to achieve a noble-gas electron configuration. All the ions in Figure 9-1 have an outer electron configuration of ns^2np^6. We shall see often that eight electrons in an outer shell is a particularly stable arrangement and that there is a strong tendency for it to occur. This is especially true for the elements in the first two rows of the periodic table. Note that metallic elements lose electrons to become positively charged ions (*cations*) and nonmetallic elements gain electrons to become negatively charged ions (*anions*), and that the charges on these ions correspond exactly to the ionic charges discussed in Chapter 2. In fact, the rules developed there for writing correct formulas for simple chemical compounds reflect the fact that the group of ions indicated by the chemical formula must have no net electrical charge.

Example 9-3: Referring only to the periodic table, write the chemical formula of scandium bromide.

Solution: Scandium has three valence electrons ($3d^14s^2$), and so the scandium ion is Sc^{3+}; bromine is a Group 7 element, and the bromide ion is Br^-. For overall electrical neutrality, we must have three Br^- for each Sc^{3+}, and so the formula for scandium bromide is $ScBr_3$.

9-3. MANY IONS DO NOT HAVE A NOBLE-GAS ELECTRON CONFIGURATION

There are many metals not listed in Figure 9-1. For example, let's consider silver ($Z = 47$), which has the electron configuration $[Kr]5s^14d^{10}$. The silver atom would have to lose 11 electrons or gain 7 to achieve a noble-gas electron configuration. Table 7-1 shows that the first of these alternatives would require an enormous amount of energy. The energy required for the addition of seven electrons is also prohibitively large. Each successive electron would have to overcome a larger and larger repulsion as the negative charge on the ion is increased. Consequently, ions with charge greater than three are uncommon.

Although a silver atom cannot achieve a noble-gas configuration, its outer electron configuration will be $4s^24p^64d^{10}$ if it loses its $5s$ electron. This configuration, with 18 electrons in the outer shell, is relatively stable and is sometimes called a *pseudo-noble-gas electron configuration.* Thus, silver forms a unipositive ion according to the reaction

$$Ag([Kr]5s^14d^{10}) \rightarrow Ag^+([Kr]4d^{10}) + e^-$$

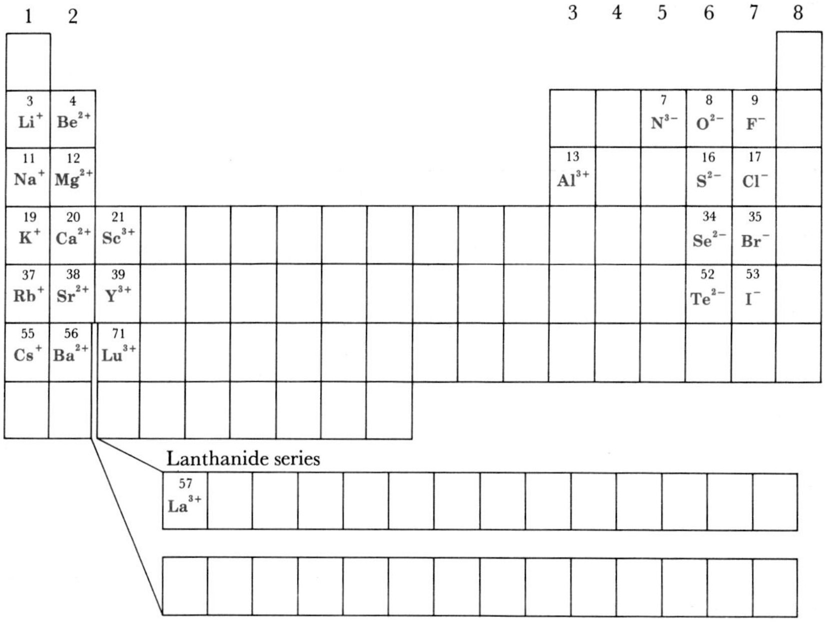

1	2		3	4	5	6	7	8
3 Li^+	4 Be^{2+}				7 N^{3-}	8 O^{2-}	9 F^-	
11 Na^+	12 Mg^{2+}		13 Al^{3+}			16 S^{2-}	17 Cl^-	
19 K^+	20 Ca^{2+}	21 Sc^{3+}				34 Se^{2-}	35 Br^-	
37 Rb^+	38 Sr^{2+}	39 Y^{3+}				52 Te^{2-}	53 I^-	
55 Cs^+	56 Ba^{2+}	71 Lu^{3+}						

Lanthanide series

57 La^{3+}

Figure 9-1 Some common ions with noble-gas outer electron configuration ns^2np^6

Example 9-4: Predict the electron configuration and the charge of a zinc ion.

The 18-electron outer electron configuration is $ns^2np^6nd^{10}$.

Solution: A zinc atom has a $1s^22s^22p^63s^23p^64s^23d^{10}$ configuration. The zinc atom can achieve the 18-electron outer configuration $3s^23p^63d^{10}$ by losing its two 4*s* electrons. Thus the electron configuration of the zinc ion is

$$Zn^{2+}(1s^22s^22p^63s^23p^63d^{10}) \quad \text{or} \quad Zn^{2+}([Ar]3d^{10})$$

and its charge is +2. Note that a Zn^{2+} ion has a completely filled M shell.

Other elements that behave similarly to zinc are shown in Figure 9-2. Note that these elements occur near the end of the *d* transition series.

Another outer electron configuration that is often found in ions can be illustrated by the element indium ($Z = 49$). The electron configuration of In is $[Kr]5s^24d^{10}5p^1$. Loss of the 5*p* electron yields the electron configuration $[Kr]5s^24d^{10}$ for the In^+ ion. Although In^+ does not have a noble-gas electron configuration, it does have all its subshells completely filled. As we discussed in Chapter 8, this is a relatively stable electron configuration. Other elements that behave like indium are shown in Figure 9-3. Note that these elements are in Groups 3, 4, and 5. The unipositive ions in Figure 9-3 can lose two more electrons to achieve the $ns^2np^6nd^{10}$ outer configuration. For example, if In^+ loses its two 5*s* electrons, then the resulting configura-

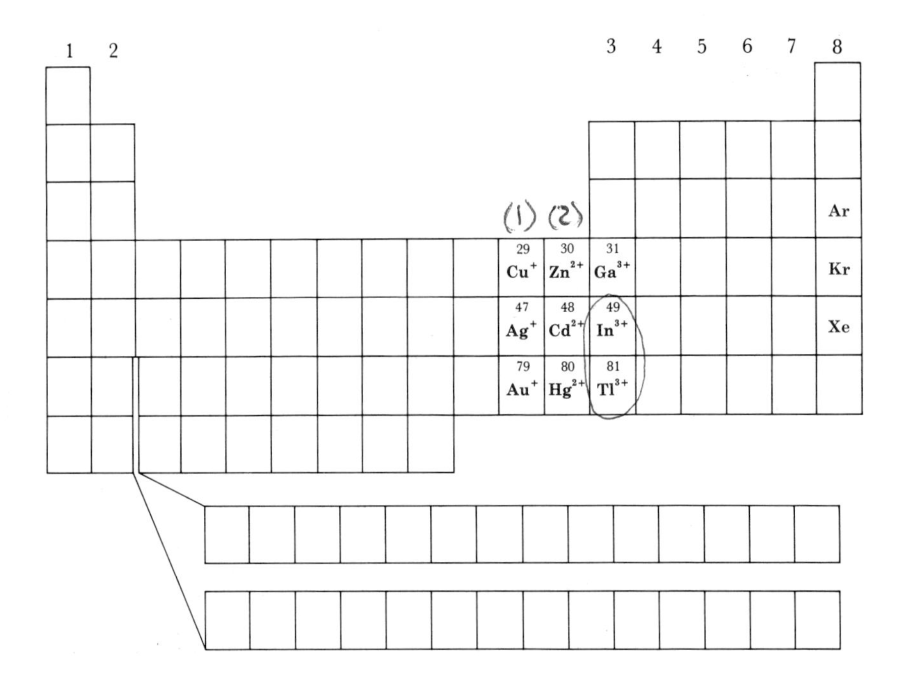

Figure 9-2 Some ions with a pseudo-noble-gas outer electron configuration $ns^2np^6nd^{10}$

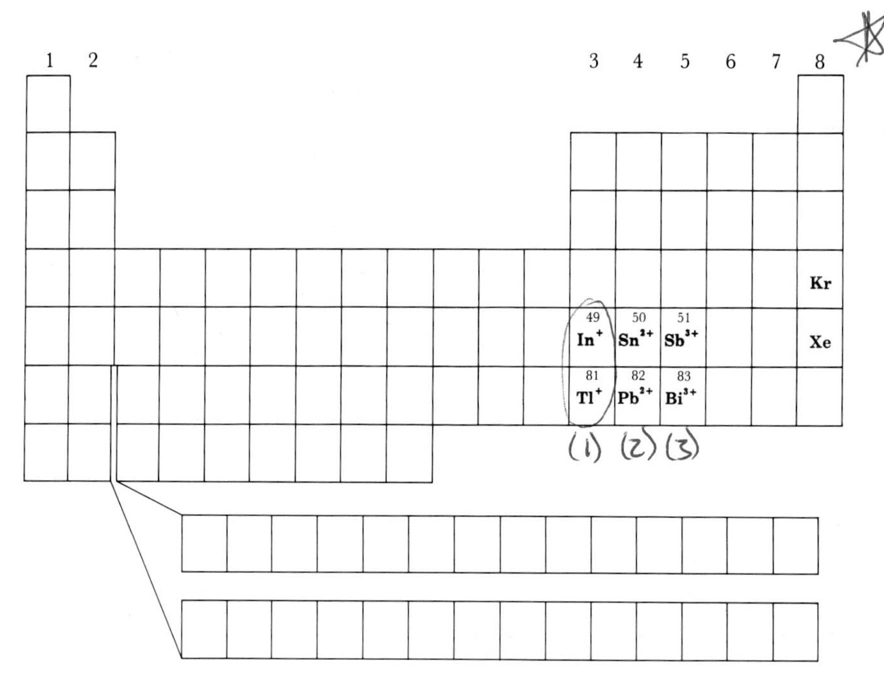

Figure 9-3 Some ions with the outer electron configuration [noble gas]$(n+1)s^2nd^{10}$.

tion for In^{3+} is [Kr]$4d^{10}$, which appears in Figure 9-2. Thus we see that indium and thallium have two possible ionic charges, +1 and +3. Gallium has only one ionic charge, +3.

Although these rules help to understand the ionic charges of many metals, there are many other metals that do not obey any simple rules for predicting the relatively stable ions. Figure 9-4 gives the outer electron configurations of some common transition-metal ions.

21	22	23	24 Cr^{2+} [Ar]$3d^4$ Cr^{3+} [Ar]$3d^3$	25 Mn^{2+} [Ar]$3d^5$ Mn^{3+} [Ar]$3d^4$	26 Fe^{2+} [Ar]$3d^6$ Fe^{3+} [Ar]$3d^5$	27 Co^{2+} [Ar]$3d^7$ Co^{3+} [Ar]$3d^6$	28 Ni^{2+} [Ar]$3d^8$	29	30
39	40	41	42 Mo^{2+} [Kr]$4d^4$	43	44 Ru^{2+} [Kr]$4d^6$ Ru^{3+} [Kr]$4d^5$	45	46 Pd^{2+} [Kr]$4d^8$	47	48
71	72	73	74	75	76	77	78 Pt^{2+} [Xe]$5d^8$ Pt^{4+} [Xe]$5d^6$	79	80

Figure 9-4 The outer electron configurations of some common transition-metal ions.

Although there is no simple pattern apparent in Figure 9-4, there is one interesting feature. We learned in Chapter 8 that the $3d$ orbitals are filled after the $4s$ orbital. When the transition-metal atoms lose electrons, we might expect that the $3d$ electrons are lost first, but this is not the case. For example, the ground-state electron configuration of Ni^{2+} is $[Ar]3d^8$ and *not* $[Ar]4s^23d^6$. The reason for this is that the energy levels shown in Figures 8-2 and 8-3 are for the *neutral* atoms. The charges on ions alter the order of the orbital energies so that, in Ni^{2+} and the other transition-metal ions, the energy of the $4s$ orbital is greater than that of the $3d$ orbitals.

Example 9-5: Predict the ground-state electron configuration of Ti^{3+}.

Solution: The ground-state electron configuration of the titanium atom is

$$[Ar]4s^23d^2$$

Although the $4s$ orbital is filled before the $3d$ orbitals in neutral atoms, this is not true for ions. In Ti^{3+} the $3d$ orbitals are used first. Therefore, the ground-state electron configuration of Ti^{3+} is

$$[Ar]3d^1$$

Note that the electron configuration of Ti^{3+} is regular in the sense that it is

$$1s^22s^22p^63s^23p^63d^1$$

The orbitals of the transition-metal ions are filled in the regular order.

Example 9-6: Referring only to the periodic table, predict the electron configurations of the ions Fe^{3+} and Pt^{4+}.

Problems 9-9 to 9-12 deal with ground-state electron configurations of transition-metal ions.

Solution: According to Example 9-5, the electron configurations of the transition-metal ions are regular. The atomic number of iron is 26, and so Fe^{3+} has 23 electrons. The electron configuration of Fe^{3+} is

$$1s^22s^22p^63s^23p^63d^5$$

or simply $[Ar]3d^5$.

The atomic number of platinum is 78, and so Pt^{4+} has 74 electrons and its electron configuration is

$$1s^22s^22p^63s^23p^63d^{10}4s^24p^64d^{10}4f^{14}5s^25p^65d^6$$

or simply $[Xe]4f^{14}5d^6$.

Note that the results here are the same as those given in Figure 9-4.

9-4. IONIC RADII ARE NOT THE SAME AS ATOMIC RADII

Because atoms and ions are different species, we should expect atomic radii and *ionic radii* to have different values. The 3*s* electron in a sodium atom is, on the average, farther away from the nucleus than the 1*s*, 2*s*, and 2*p* electrons because the 3*s* electron is in the M shell. When a sodium atom loses its 3*s* electron, only the K and L shells remain, and so the resulting sodium ion is smaller than a sodium atom. In addition, the excess positive charge draws the remaining electrons toward the nucleus and the K and L shells contract. The relative sizes of the alkali metal atoms and ions are shown in Figure 9-5. The Group 2 metals lose two outer *s* electrons in becoming ions. The excess positive charge of +2 contracts the remaining shells even more than in the case of the Group 1 metals. The relative sizes of the Group 2 atoms and ions are shown in Figure 9-6.

The atoms of nonmetals gain electrons in becoming ions. The excess negative charge results in an enlargement of the shells, and so negative ions are larger than their parent atoms. The relative atomic and ionic sizes of the halogen atoms are shown in Figure 9-7.

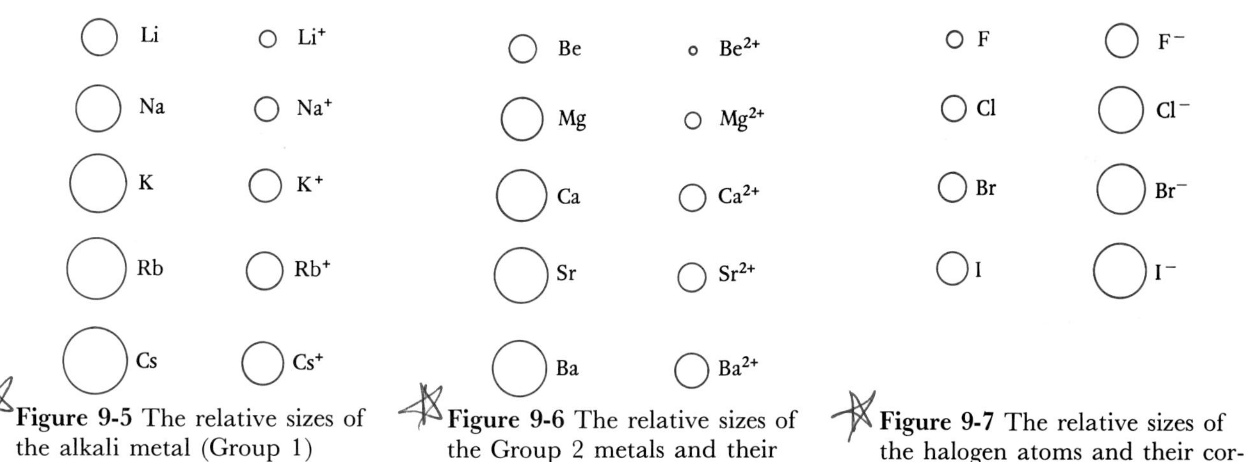

Figure 9-5 The relative sizes of the alkali metal (Group 1) atoms and ions. The positive charge from the loss of an electron draws the electrons in toward the nucleus. Consequently, positive ions are always smaller than their corresponding atoms.

Figure 9-6 The relative sizes of the Group 2 metals and their corresponding ions.

Figure 9-7 The relative sizes of the halogen atoms and their corresponding halide ions. The addition of an extra electron increases the electron-electron repulsion and causes the electron cloud to expand. Negative ions are always larger than their corresponding atoms for this reason.

Species that have the same number of electrons are called isoelectronic.

Example 9-7: Predict which of the isoelectronic species is the larger ion, Na^+ or F^-.

Solution: The electron configuration of both Na^+ and F^- is $1s^22s^22p^6$. The excess positive charge of Na^+ contracts the K and L shells, and the excess negative charge of F^- leads to an enlargement of the shells, and so we predict that F^- is larger than Na^+. Note also that both ions have the same number of electrons (10) but sodium has a nuclear charge of +11 and fluorine has a nuclear charge of only +9. The crystallographic radius of Na^+ is 95 pm and that of F^- is 136 pm.

Problems 9-21 to 9-26 involve ionic radii.

Figure 9-8 shows the relative sizes of a number of atoms and ions arranged in the form of a periodic table, and Table 9-1 gives the radii of some ions. We can use the ionic radii in Table 9-1 and the atomic

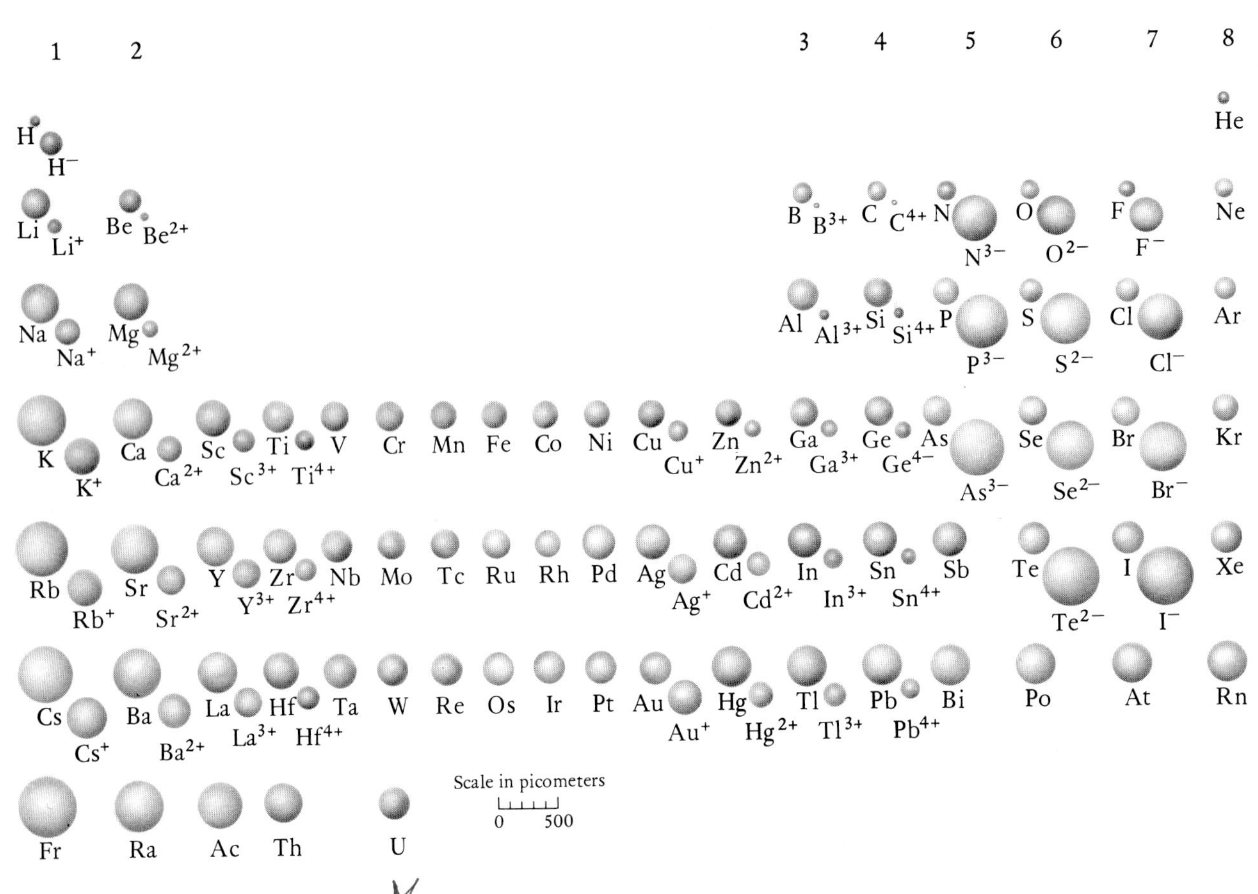

Figure 9-8 The relative sizes of atoms and ions arranged in the form of a periodic table.

Table 9-1 Some ionic radii

Ion	*Radius/pm*	*Ion*	*Radius/pm*	*Ion*	*Radius/pm*	*Ion*	*Radius/pm*
			Cations				
Ag^+	126	Ba^{2+}	135	Al^{3+}	50	Ce^{4+}	101
Cs^+	169	Be^{2+}	31	B^{3+}	20	Ti^{4+}	68
Cu^+	96	Ca^{2+}	99	Bi^{3+}	74	U^{4+}	97
K^+	133	Cd^{2+}	97	Cr^{3+}	65	Zr^{4+}	80
Li^+	60	Co^{2+}	82	Fe^{3+}	67		
Na^+	95	Cu^{2+}	70	Ga^{3+}	62		
NH_4^+	148	Fe^{2+}	78	In^{3+}	81		
Rb^+	148	Hg^{2+}	110	La^{3+}	115		
Tl^+	144	Mg^{2+}	65	Tl^{3+}	95		
		Mn^{2+}	80	Y^{3+}	93		
		Ni^{2+}	69				
		Pb^{2+}	116				
		Sr^{2+}	113				
		Zn^{2+}	74				
			Anions				
Br^-	195	O^{2-}	140	N^{3-}	171		
Cl^-	181	S^{2-}	184	P^{3-}	212		
F^-	136	Se^{2-}	196				
H^-	154	Te^{2-}	221				
I^-	216						

radii in Figure 8-11 to illustrate the changes that occur in some simple reactions. For example, the reaction between a sodium atom and a fluorine atom can be pictured as

$$Na + F \longrightarrow Na^+ + F^-$$

9-5. SOLUTIONS THAT CONTAIN IONS CONDUCT AN ELECTRIC CURRENT

Most ionic compounds yield ions when they dissolve in water. For example, an aqueous solution of NaCl consists of $Na^+(aq)$ and $Cl^-(aq)$ ions that move throughout the water. If electrodes connected to the poles of a battery are dipped into a solution containing ions (Figure 9-9), then the positive ions are attracted to the negative electrode and the negative ions are attracted to the positive electrode. The motion of the ions through the solution constitutes an electric current. The presence of ions in a solution is readily demonstrated by an apparatus like that shown in Figure 9-10. If there are ions in the solution, then an electric current can be carried by them and the light bulb will light up. If there are no ions in the solution, then no electric current can flow through the solution and the light bulb will not light up. An example of a solution that does not conduct an electric current is sucrose, $C_{12}H_{22}O_{11}$, in water. Substances like NaCl, which dissolve in water to produce solutions that conduct an electric current, are called *strong electrolytes*. Electrolytes exist as ions when they are dissolved in water. A substance such as sucrose, which

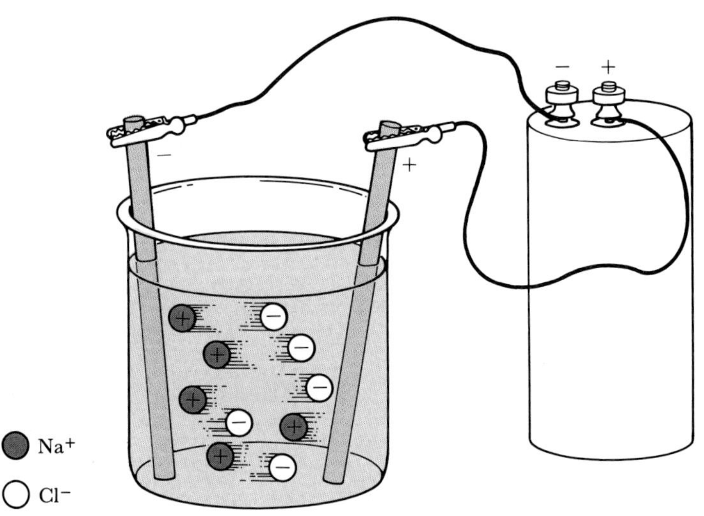

Figure 9-9 The conduction of an electric current by an aqueous solution of NaCl. An electric voltage is applied by dipping metal strips (electrodes) attached to the poles of a battery, into the solution. Like the poles of a battery, one of the electrodes is positive and the other is negative. The positively charged sodium ions are attracted to the negative electrode, and the negatively charged chloride ions are attracted to the positive electrode. Thus the Na^+ ions migrate to the left in the figure and the Cl^- ions migrate to the right. The migration of the ions constitutes an electric current through the solution.

Figure 9-10 A simple test for the presence of ions in a solution. If ions are present (NaCl), then an electric current can be carried by the solution, forming a closed electric circuit and lighting the light bulb. If there are no ions in the solution (sucrose), then no electric current is carried; the circuit is effectively open and the light bulb does not light up.

does not exist as ions when it is dissolved in water, is called a *nonelectrolyte.*

Not all electrolytes conduct an electric current to the same degree. For example, a 0.10 M $HgCl_2$ solution conducts an electric current only to the extent that the bulb in Figure 9-11 glows dimly. A substance such as $HgCl_2$ is called a *weak electrolyte*. Not all the $HgCl_2$ in aqueous solution exists as ions; much of it exists as molecular $HgCl_2$ units. Because only a fraction of the $HgCl_2$ exists as ions, there are fewer ions available to carry an electric current, and so a $HgCl_2$ aqueous solution does not conduct electricity as well as a NaCl solution of the same concentration. Figure 9-11 summarizes strong electrolytes, weak electrolytes, and nonelectrolytes.

Some helpful guidelines for determining whether a substance is a strong electrolyte, a weak electrolyte, or a nonelectrolyte are

1. The acids HCl, HBr, HI, HNO_3, H_2SO_4, and $HClO_4$ are strong electrolytes. Most other acids are weak electrolytes.
2. The soluble hydroxides (hydroxides of the Group 1 and 2 metals except beryllium) are strong electrolytes. Most other bases, and particularly ammonia, are weak electrolytes.

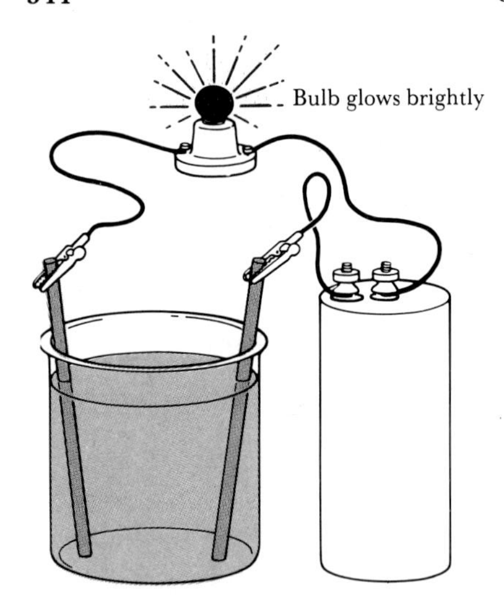

0.10 M NaCl (*aq*)
No NaCl molecules present, only ions present

(a) Strong electrolyte

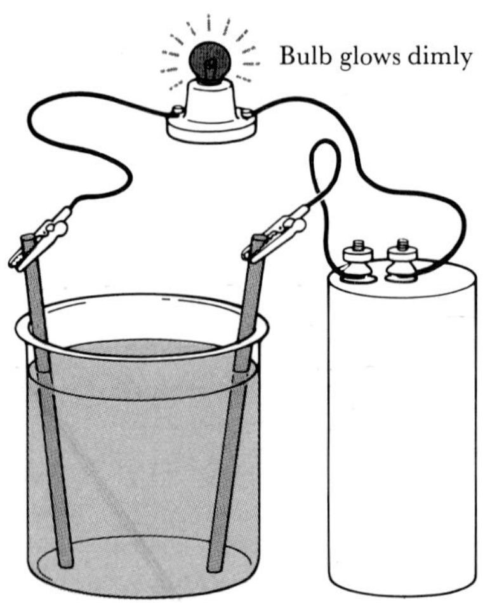

0.10 M $HgCl_2$ (*aq*)
Mostly molecules present, with some ions

(b) Weak electrolyte

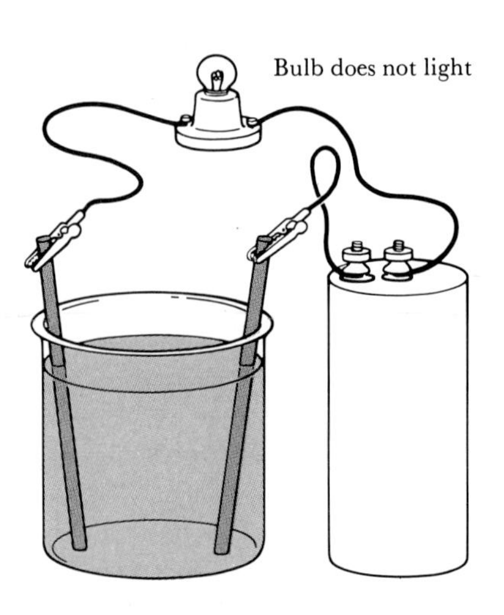

0.10 M sucrose (*aq*)
Only molecules present

(c) Nonelectrolyte

Figure 9-11 A summary of the electrical properties of strong electrolytes, weak electrolytes, and nonelectrolytes. (a) Strong electrolytes are substances that ionize completely in water and so are good conductors of electricity. (b) Weak electrolytes are substances that ionize only partially. They produce fewer ions per mole than a strong electrolyte and thus are poorer conductors of electricity. (c) Nonelectrolytes produce no ions in solution and so are nonconductors.

3. Most soluble salts are strong electrolytes. The halides and cyanides of metals with high atomic numbers (for example, mercury and lead) are often weak electrolytes.
4. Most organic compounds are nonelectrolytes. Notable exceptions are organic acids and bases, which are usually weak electrolytes.
5. Water is a very weak electrolyte.

Example 9-8: Classify each of the following compounds as a strong electrolyte, a weak electrolyte, or a nonelectrolyte: $KClO_3$, C_6H_6 (benzene), $Al(OH)_3$, H_2SO_3.

Solution: Since $KClO_3$ is a soluble salt, it is a strong electrolyte. Benzene, C_6H_6, is an organic compound and so is a nonelectrolyte. The compound $Al(OH)_3$ is a hydroxide but not one of a Group 1 or 2 metal, and so it is a weak electrolyte. Since H_2SO_3 is not among the acids classified as strong electrolytes, it is a weak electrolyte.

Problems 9-29 and 9-30 ask you to classify compounds as strong electrolytes, weak electrolytes, and nonelectrolytes.

9-6. IONIC COMPOUNDS CONDUCT AN ELECTRIC CURRENT WHEN THEY ARE MELTED

We learned in the previous section that aqueous solutions of ionic compounds conduct an electric current, that is, that ionic compounds are electrolytes. Ionic compounds also conduct an electric current when they are melted. *Ionic crystals* consist of ordered, three-dimensional arrays of positive and negative ions. The crystalline structure of NaCl is shown in Figure 9-12. Each Na^+ ion has six nearest-neighbor Cl^- ions, and each Cl^- ion has six nearest-neighbor Na^+ ions. This ordered arrangement of Na^+ and Cl^- ions extends throughout the NaCl crystal.

An ionic bond results from electrostatic attraction between two oppositely charged ions. When NaCl is melted (its melting point is 800°C), the ordered crystalline structure shown in Figure 9-12 breaks down and each ion is free to move throughout the fluid. If an electric voltage is applied across the NaCl melt, then the Na^+ ions migrate toward one electrode and the Cl^- ions migrate toward the other electrode. This migration of the ions under the applied voltage constitutes an electric current between the two electrodes. It is a characteristic property of ionic substances that they can conduct an electric current when melted. Figure 9-13 is an illustration of the solid, liquid, and gaseous states of an ionic compound such as sodium chloride.

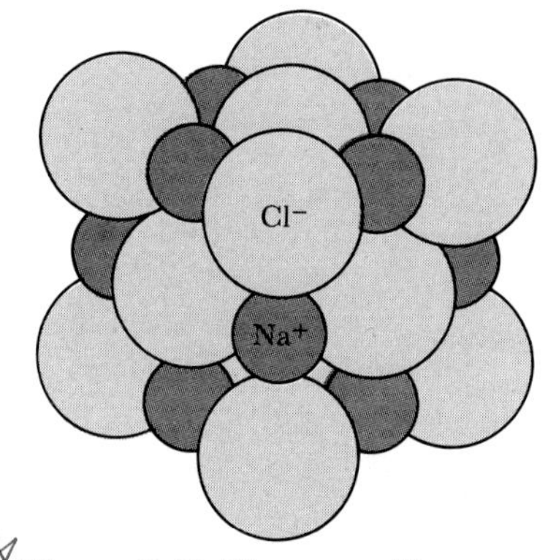

Figure 9-12 The crystalline structure of NaCl. Each Na^+ ion is surrounded by six Cl^- ions, and each Cl^- ion is surrounded by six Na^+ ions. This drawing shows the ions as spheres drawn to scale and illustrates the packing that occurs in the crystal.

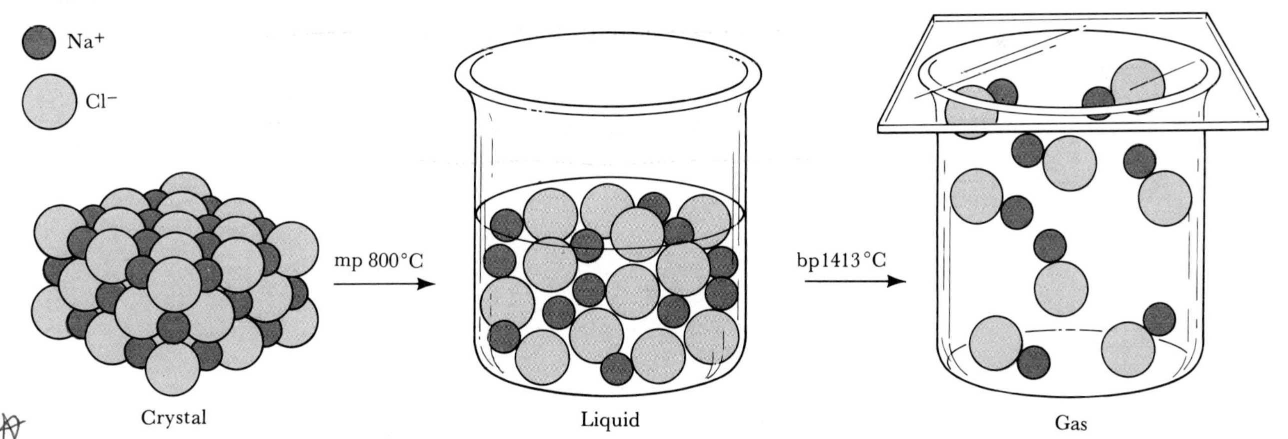

Figure 9-13 The three states of an ionic compound. In the crystal, the ions are arranged in an ordered array, forming a lattice. When the crystal melts, the lattice structure breaks down and the ions are free to move throughout the liquid. The gaseous phase consists of NaCl ion-pairs.

9-7. ELECTRON AFFINITY IS THE ENERGY RELEASED WHEN AN ATOM GAINS AN ELECTRON

Up to now our discussion of ionic bonds has been qualitative. We have argued that ionic bonds are formed when the ions have outer electron configurations that are ns^2np^6, like the noble gases, or variations of a noble-gas electron configuration, such as the pseudo-noble-gas configuration $ns^2np^6nd^{10}$. We now show by calculation that, when an ionic bond is formed, the energy of the ionic products is lower than that of the atomic reactants.

Let's consider the reaction

$$Na(g) + Cl(g) \rightarrow Na^+Cl^-(g)$$

The net energy change for this reaction can be calculated by breaking the reaction down into three separate steps:

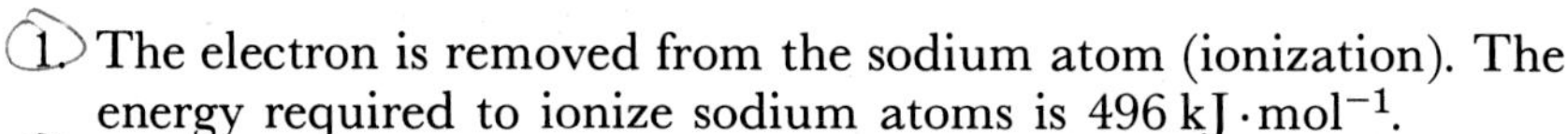

1. The electron is removed from the sodium atom (ionization). The energy required to ionize sodium atoms is $496 \text{ kJ} \cdot \text{mol}^{-1}$.
2. The electron is added to the chlorine atom. Energy is released in the process, and this energy is called the *electron affinity* of chlorine. We discuss the idea of electron affinity below. The electron affinity of atomic chlorine is $348 \text{ kJ} \cdot \text{mol}^{-1}$.
3. The sodium ion and chloride ion are brought together as shown in Figure 9-14. From Table 9-1 we see that the radius of Na^+ is 95 pm and that of Cl^- is 181 pm. Thus the centers of the two ions are 181 + 95 = 276 pm apart when the two ions are just touching.

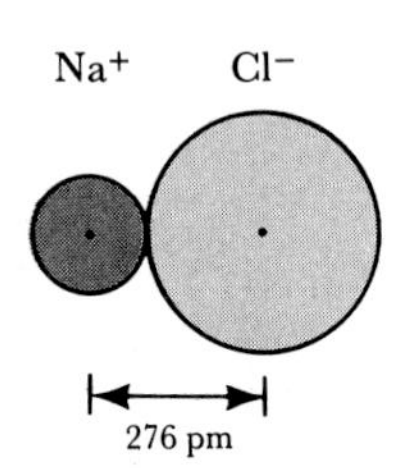

Figure 9-14 A solid-sphere representation of the ion-pair NaCl. According to Table 9-1, the Na^+ can be represented as a sphere of radius 95 pm and the chloride ion can be represented as a sphere of radius 181 pm. Because the two ions have opposite charges, they draw together until they touch, at a distance of 95 + 181 = 276 pm. They are bound together at this distance in an ionic bond.

We discussed ionization energies in Chapter 7. We now need to discuss the energy involved in steps 2 and 3. Many atoms, especially the atoms of the nonmetals, readily gain extra electrons to become ions. As mentioned above, the energy released in the process is called the electron affinity (EA) of the atom. We can define the electron affinity of any atom through the equation

$$\text{atom}(g) + \text{electron} \rightarrow \text{negative ion}(g) + \text{EA}$$

In this equation it is understood that both the atom and the ion are isolated from any others, as is the case in the gas phase. A positive EA indicates that energy is released upon formation of the ion. The more positive the EA, the greater the affinity of the neutral atom for an additional electron. The most important EA values for our purposes are those for the reactive nonmetals, which are given in Table 9-2.

Table 9-2 shows that the electron affinity of a chlorine atom is $348 \text{ kJ} \cdot \text{mol}^{-1}$. So far, then, we can write

$$\mathrm{Na}(g) + 496\ \mathrm{kJ} \rightarrow \mathrm{Na}^+(g) + \mathrm{e}^-$$
$$\mathrm{Cl}(g) + \mathrm{e}^- \rightarrow \mathrm{Cl}^-(g) + 348\ \mathrm{kJ}$$

If we add these two equations, we find that

$$\mathrm{Na}(g) + \mathrm{Cl}(g) + 148\ \mathrm{kJ} \rightarrow \mathrm{Na}^+(g) + \mathrm{Cl}^-(g) \qquad (9\text{-}1)$$

Notice that a net energy input of 148 kJ is required for this reaction.

Table 9-2 Electron affinities of some atoms

Element	*Electron affinity* $/kJ \cdot mol^{-1}$
H	72
C	121
O	136
S	200
F	333
Cl	348
Br	324
I	295

9-8. COULOMB'S LAW IS USED TO CALCULATE THE ENERGY OF AN ION-PAIR

Remember that each of the species in Equation (9-1) is in the gaseous phase, which means in particular that Na^+ and Cl^- are so far apart from each other that they are effectively isolated entities. We now must calculate the energy required to bring these two ions to their equilibrium separation of 276 pm (Figure 9-14). We use *Coulomb's law* to calculate this energy. Coulomb's law states that the energy of two ions whose centers are separated by a distance d is

$$E = (2.31 \times 10^{-16}\ \mathrm{J \cdot pm})\frac{Z_1 Z_2}{d} \qquad (9\text{-}2)$$

where E is in joules, d is in picometers, and Z_1 and Z_2 are the charges of the two ions (Figure 9-15). If the charges of the ions have the same sign, then E in Equation (9-2) is positive. If the ions are oppositely charged, E is negative. In our case, $Z_1 = +1\ (Na^+)$, $Z_2 = -1\ (Cl^-)$, and $d = 276$ pm, so that

$$E = \frac{(2.31 \times 10^{-16}\ \mathrm{J \cdot pm})(+1)(-1)}{276\ \mathrm{pm}}$$
$$= -8.37 \times 10^{-19}\ \mathrm{J}$$

The negative sign means that the ions attract each other, so that the energy at 276 pm is less than it is when the ions are isolated from each other.

Energy is released when Na^+ and Cl^- ions are brought together. The quantity -8.37×10^{-19} J is for one ion-pair. For 1 mol of NaCl ion-pairs, the energy released is

$$E = \left(\frac{-8.37 \times 10^{-19}\ \mathrm{J}}{\text{ion-pair}}\right)\left(6.02 \times 10^{23}\ \frac{\text{ion-pairs}}{\text{mol}}\right)$$
$$= -504\ \mathrm{kJ \cdot mol^{-1}}$$

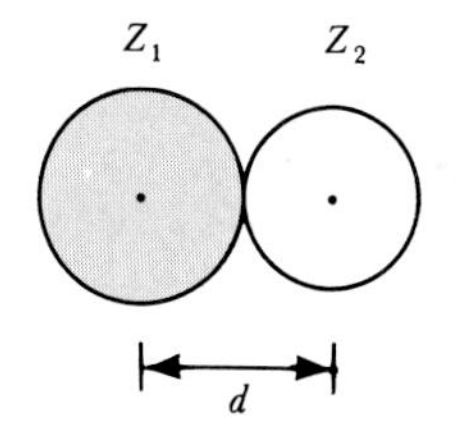

$$E = (2.31 \times 10^{-16}\ \mathrm{J \cdot pm})\frac{Z_1 Z_2}{d}$$

Figure 9-15 Two ions separated by a distance d. The charges of the ions are Z_1 and Z_2. The energy of interaction of two ions is given by Coulomb's law, shown in Equation (9-2).

In terms of a chemical equation, we have

$$Na^+(g) + Cl^-(g) \rightarrow Na^+Cl^-(g) + 504\ kJ \quad d = 276\ pm$$

If we add this equation to the equation

$$Na(g) + Cl(g) + 148\ kJ \rightarrow Na^+(g) + Cl^-(g)$$

we get

$$Na(g) + Cl(g) \rightarrow Na^+Cl^-(g) + 356\ kJ \quad d = 276\ pm$$

Thus we calculate that 356 kJ is released in the formation of 1 mol of NaCl ion-pairs from 1 mol of sodium atoms and 1 mol of chlorine atoms. The fact that energy is released in the process means that the energy of the ion-pair is lower than that of the two separated atoms. Because the ion-pair has a lower energy, it is stable with respect to the atoms. The overall process is illustrated in Figure 9-16.

Realize that this calculation is for gaseous NaCl, which exists as widely separated ion-pairs. The energy released in the formation of crystalline NaCl is greater than for gaseous NaCl because each ion in crystalline NaCl is surrounded by six ions of opposite charge, thus giving an additional stability. We shall discuss the energy released upon crystal formation in a later section.

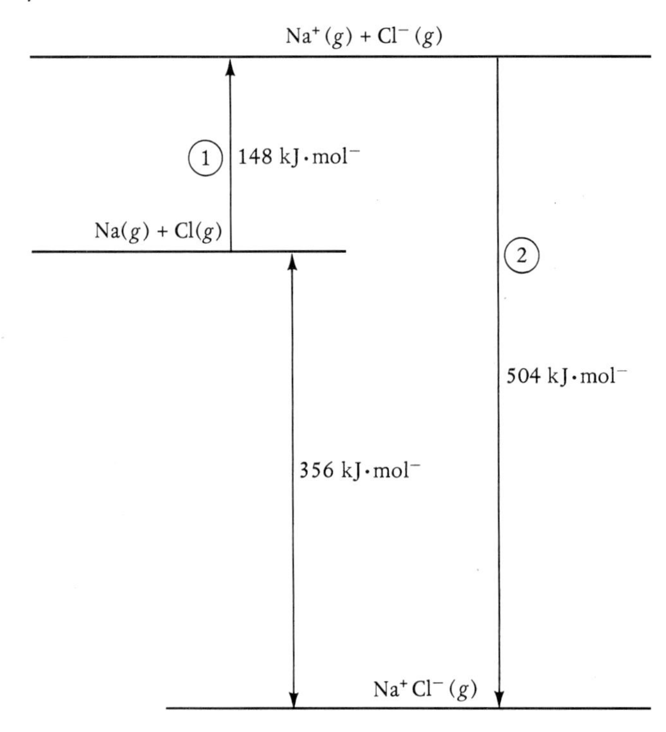

Figure 9-16 A summary of the steps used to calculate the energy evolved in the process $Na(g) + Cl(g) \rightarrow Na^+Cl^-(g)$. The process is broken down into two steps. First the atoms are converted to ions, and then the two ions are brought together to a distance equal to the sum of their ionic radii. The first step uses the ionization energy of sodium and the electron affinity of chlorine. The second step uses Coulomb's law to calculate the energy involved in bringing the two isolated ions together.

Example 9-9: Calculate the energy released in the reaction

$$Cs(g) + Cl(g) \rightarrow Cs^+Cl^-(g)$$

Solution: We must break this reaction down into three steps: (1) the ionization of Cs, (2) the addition of an electron to Cl, and (3) the bringing together of Cs^+ and Cl^- to their ion-pair separation.

1. From Figure 8-12, we see that the first ionization energy of Cs is $376\ \text{kJ} \cdot \text{mol}^{-1}$. Thus

$$Cs(g) + 376\ \text{kJ} \rightarrow Cs^+(g) + e^-$$

2. From Table 9-2, we see that the electron affinity of Cl is $348\ \text{kJ} \cdot \text{mol}^{-1}$, and so

$$Cl(g) + e^- \rightarrow Cl^-(g) + 348\ \text{kJ}$$

If we add the results of steps 1 and 2, we get

$$Cs(g) + Cl(g) + 28\ \text{kJ} \rightarrow Cs^+(g) + Cl^-(g)$$

3. We now must calculate the energy required to bring Cs^+ and Cl^- to their separation as an ion-pair. According to Table 9-1, the radius of Cs^+ is 169 pm and that of Cl^- is 181 pm. Their separation as an ion-pair, then, is 350 pm. We now use Equation (9-2):

$$\begin{aligned} E &= (2.31 \times 10^{-16}\ \text{J} \cdot \text{pm}) \frac{Z_1 Z_2}{d} \\ &= \frac{(2.31 \times 10^{-16}\ \text{J} \cdot \text{pm})(+1)(-1)}{350\ \text{pm}} \\ &= -6.60 \times 10^{-19}\ \text{J} \end{aligned}$$

This is the energy released by the formation of one ion-pair. For 1 mol of CsCl, we multiply this result by Avogadro's number:

$$\begin{aligned} E &= \left(\frac{-6.60 \times 10^{-19}\ \text{J}}{\text{ion-pair}}\right)\left(6.02 \times 10^{23}\ \frac{\text{ion-pairs}}{\text{mol}}\right) \\ &= -397\ \text{kJ} \cdot \text{mol}^{-1} \end{aligned}$$

The negative sign indicates that energy is released in the process. We can express this result in the form of an equation:

$$Cs^+(g) + Cl^-(g) \rightarrow Cs^+Cl^-(g) + 397\ \text{kJ}$$
$$d = 350\ \text{pm}$$

If we combine this with the net result of steps 1 and 2, that is, with

$$Cs(g) + Cl(g) + 28\ kJ \rightarrow Cs^+(g) + Cl^-(g)$$

we get

$$Cs(g) + Cl(g) \rightarrow Cs^+Cl^-(g) + 369\ kJ$$
$$d = 350\ pm$$

Thus 369 kJ is evolved in the formation of 1 mol of CsCl ion-pairs from 1 mol of cesium atoms and 1 mol of chlorine atoms.

Problems 9-41 to 9-44 are similar to Example 9-9.

9-9. THE FORMATION OF IONIC SOLIDS FROM THE ELEMENTS IS AN EXOTHERMIC PROCESS

Many of the reactions that we have considered so far in this chapter have been simplified in the sense that we have discussed only reactions between atoms. Chlorine, for example, exists as a diatomic molecule at room temperature. In addition, sodium is a solid at room temperature and so the reaction we should consider is

$$Na(s) + \tfrac{1}{2}Cl_2(g) \rightarrow NaCl(s)$$

instead of

$$Na(g) + Cl(g) \rightarrow NaCl(g)$$

We can break the first of these reactions down into five steps:

1. Vaporize 1 mol of sodium metal so that the sodium atoms are far apart and effectively isolated from each other. The energy required for this step is the energy of vaporization of sodium, which is $93\ kJ \cdot mol^{-1}$ at room temperature. We can write this process as

$$Na(s) + 93\ kJ \rightarrow Na(g)$$

2. Dissociate 0.5 mol of $Cl_2(g)$ into 1 mol of chlorine atoms. The energy required for this process is $122\ kJ \cdot mol^{-1}$, and so we write

$$\tfrac{1}{2}Cl_2(g) + 122\ kJ \rightarrow Cl(g)$$

3. Ionize the mole of $Na(g)$. The energy required is the first ionization energy of sodium, which is $496\ kJ \cdot mol^{-1}$ (Figure 8-12). Therefore, we write

$$Na(g) + 496\ kJ \rightarrow Na^+(g) + e^-$$

4. Add the mole of electrons generated in step 3 to 1 mol of chlorine atoms. This is step 2 in Example 9-9 and is

$$Cl(g) + e^- \rightarrow Cl^-(g) + 348\ \text{kJ}$$

5. Bring the mole of isolated sodium ions and the mole of isolated chloride ions together to form the NaCl crystal shown in Figure 9-12. Energy is released in this step. This energy, called the *lattice energy,* is known to be 780 $\text{kJ}\cdot\text{mol}^{-1}$ for NaCl(*s*). The equation describing this process is

$$Na^+(g) + Cl^-(g) \rightarrow NaCl(s) + 780\ \text{kJ}$$

Lattice energies are tabulated for many crystalline substances.

Notice that the lattice energy of NaCl is greater than the energy released upon gaseous ion-pair formation. As we said above, each Na^+ and Cl^- in the crystal is surrounded by six oppositely charged ions and so the crystal has a lower energy than the gas of ion-pairs.

We can summarize the five steps by the equations

1. $Na(s) + 93\ \text{kJ} \rightarrow Na(g)$
2. $\frac{1}{2}Cl_2(g) + 122\ \text{kJ} \rightarrow Cl(g)$
3. $Na(g) + 496\ \text{kJ} \rightarrow Na^+(g) + e^-$
4. $Cl(g) + e^- \rightarrow Cl^-(g) + 348\ \text{kJ}$
5. $Na^+(g) + Cl^-(g) \rightarrow NaCl(s) + 780\ \text{kJ}$

If we add these five equations, then the net result is

$$Na(s) + \tfrac{1}{2}Cl_2(g) \rightarrow NaCl(s) + 417\ \text{kJ}$$

Thus we see that the formation of NaCl(*s*) from Na(*s*) and $Cl_2(g)$ is accompanied by the evolution of a large amount of energy. Notice that it is the large value of the lattice energy (step 5) that is primarily responsible for the strongly exothermic character of the overall reaction. Generally, the lattice energies of ionic crystals are large enough to produce exothermic reactions for the formation of ionic solids from their elements. The various steps in the process are displayed in Figure 9-17.

Example 9-10: Calculate the energy released in the process

$$Rb(s) + \tfrac{1}{2}Br_2(l) \rightarrow RbBr(s)$$

The heat of vaporization of Rb(*s*) is 75.8 $\text{kJ}\cdot\text{mol}^{-1}$, the heat of vaporization of $Br_2(l)$ is 30.9 $\text{kJ}\cdot\text{mol}^{-1}$, the energy required to dissociate $Br_2(g)$ is 192 $\text{kJ}\cdot\text{mol}^{-1}$, and the lattice energy of RbBr(*s*) is 670 $\text{kJ}\cdot\text{mol}^{-1}$.

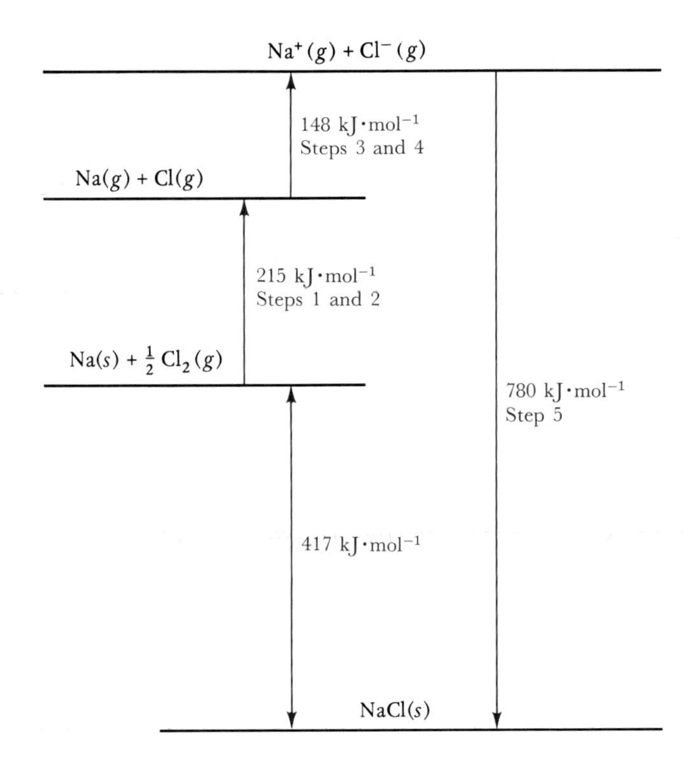

Figure 9-17 A summary of the steps used to calculate the energy evolved in the process $Na(s) + \frac{1}{2}Cl_2(g) \rightarrow NaCl(s)$. The process is broken down into three stages. First the sodium is vaporized and the chlorine is dissociated to produce Na(g) and Cl(g) (steps 1 and 2). Then these atoms are converted to ions (steps 3 and 4). Lastly, the $Na^+(g)$ and $Cl^-(g)$ are brought together to form the NaCl crystal lattice shown in Figure 9-12. The energy involved in this last step is called the lattice energy of sodium chloride. The complete cycle of steps shown in this figure is called The Born-Haber cycle.

Solution: We break the reaction down into six steps, one of which (#2) involves the vaporization of bromine because bromine is a liquid at room temperature.

1. Vaporize 1 mol of rubidium metal. The heat of vaporization of rubidium is 75.8 kJ·mol^{-1}, and so

$$Rb(s) + 75.8\ kJ \rightarrow Rb(g)$$

2. Vaporize 0.5 mol of $Br_2(l)$. The energy required for this is (30.9/2) kJ, and so

$$\tfrac{1}{2}Br_2(l) + 15.5\ kJ \rightarrow \tfrac{1}{2}Br_2(g)$$

3. Dissociate 0.5 mol of $Br_2(g)$ into 1 mol of Br(g). The energy required in this step is (192/2) kJ, and so

$$\tfrac{1}{2}Br_2(g) + 96\ kJ \rightarrow Br(g)$$

4. Ionize 1 mol of Rb(g). According to Figure 8-12, the ionization energy of rubidium is 403 kJ·mol^{-1}:

$$Rb(g) + 403\ kJ \rightarrow Rb^+(g)$$

5. Attach 1 mol of electrons to 1 mol of bromine atoms. The energy released in this process is the electron affinity of bromine. According to Table 9-2, the electron affinity of a bromine atom is 324 kJ. Therefore, we have

$$Br(g) + e^- \rightarrow Br^-(g) + 324 \text{ kJ}$$

6. Bring 1 mol of $Rb^+(g)$ and 1 mol of $Br^-(g)$ together to form crystalline RbBr. The energy evolved in this process is the lattice energy of rubidium bromide, which is 670 $kJ \cdot mol^{-1}$. We can write this process as

$$Rb^+(g) + Br^-(g) \rightarrow RbBr(s) + 670 \text{ kJ}$$

The energy released in the process

$$Rb(s) + \tfrac{1}{2}Br_2(l) \rightarrow RbBr(s)$$

is obtained by adding the reactions in steps 1 through 6:

$$Rb(s) + \tfrac{1}{2}Br_2(l) \rightarrow RbBr(s) + 404 \text{ kJ}$$

The formation of 1 mol of crystalline RbBr from the elements is a highly exothermic process.

Problems 9-47 to 9-51 ask you to calculate the energy released for the formation of crystalline compounds.

SUMMARY

An ionic bond is the electrostatic attraction that binds oppositely charged ions together. An ionic compound is formed between two reactants if one of them has a relatively low ionization energy and the other has a relatively high electron affinity. This is the case when a reactive metal reacts with a reactive nonmetal. When this occurs, one or more electrons are transferred completely from the metal atom to the nonmetal atom, resulting in an ionic bond. Substances that dissolve to produce solutions that conduct an electric current are called electrolytes.

Many ions are stable because they achieve a noble-gas electron configuration, although there are stable ions that have other types of electron configurations as well, called pseudo-noble-gas configurations. Ions are completely different species from atoms, and so ionic radii are not the same as atomic radii. The relative sizes of ions and their trend in the periodic table can be understood in terms of electron configurations. The chapter concludes with a quantitative discussion of the energetics of the formation of ionic compounds from their respective elements. These calculations use the concepts of ionization energy and electron affinity and are summarized in Figures 9-16 and 9-17.

TERMS YOU SHOULD KNOW

ionic bond 332
ionic compound 332
cation 334
anion 334
pseudo-noble-gas electron configuration 335
ionic radius 339
strong electrolyte 342
nonelectrolyte 343
weak electrolyte 343
ionic crystal 345
electron affinity, EA 346
Coulomb's law 347
lattice energy 351
Born-Haber cycle 352

EQUATIONS YOU SHOULD KNOW HOW TO USE

$$E = (2.31 \times 10^{-16}\ \mathrm{J \cdot pm}) \frac{Z_1 Z_2}{d} \qquad \text{(Coulomb's law)} \qquad (9\text{-}2)$$

(E in joules, d in picometers)

PROBLEMS

ELECTRON CONFIGURATIONS AND CHEMICAL REACTIONS

9-1. Describe the following reactions in terms of the electron configurations of the various species involved:

(a) $K(g) + Cl(g) \rightarrow KCl(g)$
(b) $Sr(g) + 2Br(g) \rightarrow SrBr_2(g)$
(c) $2Al(g) + 3O(g) \rightarrow Al_2O_3(g)$

9-2. Describe the following reactions in terms of the electron configurations of the various species involved:

(a) $Ca(g) + 2F(g) \rightarrow CaF_2(g)$
(b) $2Na(g) + O(g) \rightarrow Na_2O(g)$
(c) $Mg(g) + S(g) \rightarrow MgS(s)$

9-3. Use electron configurations to describe the formation of the following ionic compounds from the atoms:

(a) $GaF_3(g)$
(b) $CaS(g)$
(c) $Li_3N(g)$

9-4. Using electron configurations, predict the product of the reactions between the following atoms:

(a) calcium and bromine
(b) magnesium and nitrogen
(c) cesium and selenium

9-5. Describe the following reactions in terms of Lewis electron-dot formulas for the various species involved:

(a) $3Li(g) + N(g) \rightarrow Li_3N(g)$
(b) $Na(g) + H(g) \rightarrow NaH(g)$
(c) $Al(g) + 3I(g) \rightarrow AlI_3(g)$

9-6. Describe the following reactions in terms of Lewis electron-dot formulas for the various species involved:

(a) $K(g) + I(g) \rightarrow KI(g)$
(b) $Ba(g) + O(g) \rightarrow BaO(g)$
(c) $Be(g) + Se(g) \rightarrow BeSe(g)$

9-7. Predict the products of the following reactions from a consideration of the Lewis electron-dot formulas of the reactants and the achievement of noble-gas electron configurations in the product ions:

(a) calcium and sulfur (as S)
(b) aluminum and chlorine (as Cl)
(c) beryllium and oxygen (as O)

9-8. Predict the products of the following reactions from a consideration of the Lewis electron-dot formulas of the reactants and the achievement of noble-gas electron configurations in the product ions:

(a) sodium and oxygen (as O)
(b) scandium and chlorine (as Cl)
(c) aluminum and sulfur (as S)

ELECTRON CONFIGURATIONS OF IONS

9-9. Predict the ground-state electron configuration of

(a) Cr^{2+}
(b) Cu^{2+}
(c) Co^{3+}
(d) Mn^{2+}
(e) Ni^{3+}

9-10. Predict the ground-state electron configuration of

(a) Ru^{2+}
(b) W^{3+}
(c) Pd^{2+}
(d) Ag^{2+}
(e) Ir^{3+}

9-11. Which *d* transition-metal ions with a +2 charge have

(a) six *d* electrons
(b) ten *d* electrons
(c) one *d* electron
(d) five *d* electrons
(e) eight *d* electrons

Do you see a connection between the number of *d* electrons in the +2 ion and the position of the ion in its transition-metal series?

9-12. How many *d* electrons are there in

(a) Fe^{2+}
(b) Zn^{2+}
(c) V^{2+}
(d) Ni^{2+}
(e) Ti^{2+}

Can you see a pattern between the number of *d* electrons and the position of these ions in the first transition-metal series?

9-13. Using only a periodic table, predict the ground-state pseudo-noble-gas electron configuration and charge of the following ions:

(a) cadmium ion
(b) indium(III) ion
(c) thallium(III) ion
(d) zinc ion

9-14. Predict the ground-state pseudo-noble-gas electron configuration and the charge of the following ions:

(a) copper ion
(b) gallium ion
(c) mercury ion
(d) gold ion

9-15. Determine which of the following salts are composed of isoelectronic cations and anions:

(a) LiF
(b) NaF
(c) KBr
(d) KCl
(e) BaI_2
(f) AlF_3

9-16. Determine which of the following salts are composed of isoelectronic cations and anions:

(a) NaCl
(b) RbBr
(c) $SrCl_2$
(d) $SrBr_2$
(e) MgF_2
(f) KI

CHEMICAL FORMULAS OF IONIC COMPOUNDS

9-17. Write the chemical formula for

(a) yttrium sulfide
(b) lanthanum bromide
(c) magnesium telluride
(d) rubidium nitride
(e) aluminum selenide

9-18. Write the chemical formula for each of the following ionic compounds:

(a) aluminum iodide
(b) sodium fluoride
(c) calcium oxide
(d) barium bromide
(e) potassium sulfide

9-19. Write the chemical formula for each of the following compounds:

(a) silver chloride
(b) cadmium sulfide
(c) zinc nitride
(d) copper(I) bromide
(e) gallium oxide

9-20. Write the chemical formula for each of the following compounds:

(a) lead(II) chloride
(b) bismuth sulfide
(c) iron(II) oxide
(d) molybdenum(III) fluoride
(e) cobalt(III) oxide

IONIC RADII

9-21. The following pairs of ions are isoelectronic. Predict which is the larger ion in each pair:

(a) K^+ and Cl^-
(b) Ag^+ and Cd^{2+}
(c) Cu^+ and Zn^{2+}
(d) F^- and O^{2-}
(e) S^{2-} and P^{3-}

9-22. List the following ions in order of increasing radius:

Na^+ O^{2-} Mg^{2+} F^- Al^{3+}

9-23. Use data in Table 9-1 to estimate the distances between the centers of the ions in the following species:

(a) AgCl
(b) $HgCl_2$ (linear)
(c) ZnO

9-24. Use data in Table 9-1 to estimate the distances between the centers of the ions in the following species:

(a) $BeCl_2$ (linear)
(b) FeO
(c) NaH

9-25. Use the data in Figure 8-11 and Table 9-1 to estimate the percent increase in volume that occurs when Cl(g) is converted to Cl^-(g) (note: the volume of a sphere is $V = \frac{4}{3}\pi r^3$).

9-26. Use the data in Figure 8-11 and Table 9-1 to estimate the percent increase in volume that occurs when I(g) is converted to I^-(g) (note: the volume of a sphere is $V = \frac{4}{3}\pi r^3$).

ELECTROLYTES

9-27. Predict the ions produced when the following salts are dissolved in water:

(a) NaF
(b) $ZnCl_2$
(c) AgF
(d) SrI_2
(e) ScI_3

9-28. Predict the ions produced when the following substances are dissolved in water:

(a) KI
(b) $CdBr_2$
(c) $Ca(OH)_2$
(d) $NiCl_2$
(e) LiBr

9-29. Classify each of the following compounds as a strong electrolyte, a weak electrolyte, or a nonelectrolyte:

(a) $CaCl_2$
(b) $PbCl_2$
(c) $HC_2H_3O_2$ (acetic acid)
(d) CH_3COCH_3 (acetone)

9-30. Classify each of the following compounds as a strong electrolyte, a weak electrolyte, or a nonelectrolyte:

(a) C_3H_7OH (rubbing alcohol)
(b) KNO_3
(c) HClO
(d) Na_2SO_3

IONIZATION ENERGIES AND ELECTRON AFFINITIES

9-31. List the following atoms in order of the ease with which they gain electron(s) to form anions:

Br I H Cl

9-32. List the following atoms in order of the ease with which they lose electron(s) to form cations:

Ca K Na Al Li

9-33. Use the electron affinity data for Cl, Br, and I in Table 9-2 together with your knowledge of atomic periodicity trends to estimate the electron affinity of astatine.

9-34. Explain in qualitative terms why the electron affinities of all the metals are negative quantities, that is, why $\Delta H^\circ_{rxn} > 0$ for the reaction

$$M(g) + e^- \rightarrow M^-(g)$$

9-35. Using Figure 8-12 and Table 9-2, calculate ΔH°_{rxn} for the following reactions:

(a) $Rb(g) + Br(g) \rightarrow Rb^+(g) + Br^-(g)$
(b) $Li(g) + Cl(g) \rightarrow Li^+(g) + Cl^-(g)$
(c) $K(g) + F(g) \rightarrow K^+(g) + F^-(g)$

9-36. Using Figure 8-12 and Table 9-2, calculate ΔH°_{rxn} for the following reactions:

(a) $K(g) + Cl(g) \rightarrow K^+(g) + Cl^-(g)$
(b) $Na(g) + I(g) \rightarrow Na^+(g) + I^-(g)$
(c) $Li(g) + Br(g) \rightarrow Li^+(g) + Br^-(g)$

9-37. Use the ionization energy and electron affinity data in Figure 8-12 and Table 9-2 to estimate ΔH°_{rxn} for the following reactions:

(a) $Cu(g) + Cl(g) \rightarrow Cu^+(g) + Cl^-(g)$
(b) $Cl(g) + I^-(g) \rightarrow Cl^-(g) + I(g)$
(c) $Ag(g) + F(g) \rightarrow Ag^+(g) + F^-(g)$

9-38. Use the ionization energy and electron affinity data in Figure 8-12 and Table 9-2 to estimate ΔH°_{rxn} values for the following reactions:

(a) $Au(g) + I(g) \rightarrow Au^+(g) + I^-(g)$
(b) $F^-(g) + Cl(g) \rightarrow F(g) + Cl^-(g)$
(c) $In(g) + Br(g) \rightarrow In^+(g) + Br^-(g)$

CALCULATIONS INVOLVING COULOMB'S LAW

9-39. Use Coulomb's law to calculate the energy of a potassium ion and a chloride ion that are just touching.

9-40. Use Coulomb's law to calculate the energy of a sodium ion and a fluoride ion that are just touching.

9-41. Calculate the energy released in kilojoules per mole in the reaction

$$K(g) + Br(g) \rightarrow K^+Br^-(g)$$

9-42. Calculate the energy released in kilojoules per mole in the reaction

$$Ag(g) + Cl(g) \rightarrow Ag^+Cl^-(g)$$

9-43. Calculate the energy released in kilojoules per mole in the reaction

$$Na(g) + H(g) \rightarrow Na^+H^-(g)$$

9-44. Calculate the energy released in kilojoules per mole in the reaction

$$Cu(g) + I(g) \rightarrow CuI(g)$$

9-45. Construct a diagram like that shown in Figure 9-16 for $Li^+F^-(g)$ (see Figure 8-12 and Tables 9-1 and 9-2 for the necessary data).

9-46. Construct a diagram like that shown in Figure 9-17 for $Rb^+H^-(g)$ (see Figure 8-12 and Tables 9-1 and 9-2 for the necessary data).

ENERGETICS OF LATTICE FORMATION

9-47. Calculate the energy released in kilojoules per mole in the reaction

$$Na(s) + \tfrac{1}{2}F_2(g) \rightarrow NaF(s)$$

The heat of vaporization of Na is $93\ kJ \cdot mol^{-1}$, the dissociation of F_2 requires $155\ kJ \cdot mol^{-1}$, and the lattice energy of NaF is $919\ kJ \cdot mol^{-1}$.

9-48. Calculate the energy released in kilojoules per mole in the reaction

$$K(s) + \tfrac{1}{2}Br_2(l) \rightarrow KBr(s)$$

The heat of vaporization of potassium is $89\ kJ \cdot mol^{-1}$, the sum of the dissociation and vaporization energies for $Br_2(l)$ is $223\ kJ \cdot mol^{-1}$, and the lattice energy of KBr is $688\ kJ \cdot mol^{-1}$.

9-49. Calculate the energy released in kilojoules per mole when NaI(*s*) is formed in the reaction

$$Na(s) + \tfrac{1}{2}I_2(s) \rightarrow NaI(s)$$

The energy of vaporization of Na(*s*) is $93\ kJ \cdot mol^{-1}$. The sum of the enthalpies of dissociation and vaporization of $I_2(s)$ is $214\ kJ \cdot mol^{-1}$, and the lattice energy of NaI is $704\ kJ \cdot mol^{-1}$.

93 + ~~214~~ 107 + 496 − 295 − 704 = −303 kJ/mol

9-50. Calculate the energy released in kilojoules per mole when LiH(*s*) is formed in the reaction

$$Li(s) + \tfrac{1}{2}H_2(g) \rightarrow LiH(s)$$

The heat of vaporization of lithium is $161\ kJ \cdot mol^{-1}$, the dissociation energy of H_2 is $436\ kJ \cdot mol^{-1}$, and the lattice energy of LiH is $917\ kJ \cdot mol^{-1}$.

9-51. Calculate the energy released in the reaction

$$Ca(s) + Cl_2(g) \rightarrow CaCl_2(s)$$

The energy of vaporization of Ca(*s*) is $179\ kJ \cdot mol^{-1}$, the dissociation energy required for Cl_2 is $244\ kJ \cdot mol^{-1}$, the lattice energy of $CaCl_2(s)$ is $2256\ kJ \cdot mol^{-1}$, and the second ionization energy of Ca(*g*) is $1140\ kJ \cdot mol^{-1}$.

Ca(s) + 179 kJ/mol → Ca(g)
Cl₂(g) + 244 kJ/mol → 2Cl(g)
Ca(g) + 590 kJ/mol → Ca⁺(g) + e⁻
Ca⁺(g) + 1140 kJ/mol → Ca²⁺(g)
2Cl(g) + 2e⁻ → 2Cl⁻(g) + 696 kJ
Ca²⁺(g) + 2Cl⁻(g) → CaCl₂(s) + 2256 kJ/mol
(179 + 244 + 590 + 1140 − 696 − 2256 = −799)

Ca(s) + Cl₂(g) → CaCl₂(s) + 799 kJ/mol

ΔH°rxn = −799 kJ

9-52. For the reaction

$$K(s) + \tfrac{1}{2}Br_2(l) \rightarrow KBr(s)$$

$\Delta H^\circ_{rxn} = -392\ kJ \cdot mol^{-1}$. For the reaction

$$K(g) + Br(g) \rightarrow KBr(g)$$

$\Delta H^\circ_{rxn} = -329\ kJ \cdot mol^{-1}$. From these two reactions, calculate the enthalpy of vaporization for KBr(*s*) (see Problem 9-48 for additional data).

INTERCHAPTER F

The Alkali Metals

The alkali metals are soft. Here we see sodium being cut with a knife.

The alkali metals are lithium, sodium, potassium, rubidium, cesium, and francium. They occur in Group 1 of the periodic table and so have a [noble gas]ns^1 electron configuration and an ionic charge of +1. All the alkali metals have low ionization energies and therefore are very reactive. None occur as the free metal in nature. They must be stored under an inert substance, such as kerosene, because they react spontaneously with the oxygen and water vapor in the air.

F-1. THE HYDROXIDES OF THE ALKALI METALS ARE STRONG BASES

The alkali metals are all fairly soft and can be cut with a sharp knife. When freshly cut they are bright and shiny, but they soon take on a dull finish because of the reactions with air. The alkali metals are so called because their hydroxides, MOH, are all soluble, strong bases in water (alkaline means basic). The physical properties of the alkali metals are given in Table F-1, and their primary sources and commercial uses are given in Table F-2.

Table F-1 The physical properties of the alkali metals

Property	*Lithium*	*Sodium*	*Potassium*	*Rubidium*	*Cesium*
chemical symbol	Li	Na	K	Rb	Cs
atomic mass	6.941	22.98977	39.0983	85.4678	132.9054
electron configuration	[He]$2s^1$	[Ne]$3s^1$	[Ar]$4s^1$	[Kr]$5s^1$	[Xe]$6s^1$
melting point/°C	181	98	64	39	29
boiling point/°C	1347	892	774	696	670
density at 20°C/g·cm^{-3}	0.53	0.97	0.87	1.53	1.88
first ionization energy of M(g)/kJ·mol^{-1}	520	496	419	403	376
ionic radius/pm	60	95	133	148	169

Lithium is the least dense of all the elements that are solid or liquid at 20°C.

Sodium metal is obtained by electrolysis of molten mixtures of sodium chloride and calcium chloride:

$$2NaCl\ [in\ CaCl_2(l)] \xrightarrow[600°C]{electrolysis} 2Na(l) + Cl_2(g)$$

Chlorine gas is a useful by-product of the electrolysis. The $CaCl_2$ is added to the NaCl to lower the temperature necessary for the operation of the electrolysis cell. Pure NaCl melts at 800°C.

Sodium is the least expensive metal per unit volume.

Potassium can be obtained by electrolysis, but it can also be produced by the replacement reaction of molten potassium chloride with gaseous sodium in the absence of air:

$$KCl(l) + Na(g) \xrightarrow{780°C} NaCl(s) + K(g)$$

The success of the process is based on the fact that potassium is much more volatile than sodium. The boiling point of potassium is 118°C lower than that of sodium (Table F-1). Rubidium and cesium can be produced in an analogous manner. There are no stable isotopes of francium; all of them are radioactive.

F-2. THE ALKALI METALS ARE VERY REACTIVE

The alkali metals react directly with all the nonmetals except the noble gases (Table F-3). The increasing reactivity of the alkali metals with increasing atomic number is demonstrated in a spectacular manner by their reaction with water. When metallic lithium reacts with water, hydrogen gas is slowly evolved, whereas sodium reacts vigorously with water. The reaction of potassium with water pro-

Table F-2 The sources and commercial uses of the alkali metals

Metal	*Sources*	*Commercial uses*
lithium	certain mineral springs and salt lakes	lithium batteries; medicine; alloys (increases corrosion resistance and tensile strength); hydrogen bomb
sodium	salt waters, $NaCl(s)$, $NaNO_3(s)$	numerous important industrial compounds, such as $NaCl$, Na_2CO_3 (soda ash), $NaOH$, $NaHCO_3$ (baking soda), $Na_2S_2O_3 \cdot 5H_2O$ (hypo), $Na_2B_4O_7 \cdot 10H_2O$ (borax); sodium phosphates (detergents)
potassium	ancient ocean and salt lake beds; occurs in numerous mineral deposits at low levels, $KNO_3(s)$	important industrial compounds: KCl (salt substitute), KBr (sedatives), KI (iodized salts), KOH (detergents), KNO_3 (gun powder), $KClO_3$, KCN (gold mining), K_2SO_4 (glass, cathartic), $K_2Cr_2O_7$ (pigments)
rubidium	mineral springs (Searles Lake, Calif.; Manitoba; Michigan brines), certain rare minerals found in Elba	special types of glass; "getter" (O_2 remover) in vacuum tubes
cesium	water from certain mineral springs (Bernje Lake, Manitoba),	"getter" (O_2 remover) in vacuum tubes; photoelectric cells; atomic clocks

duces a fire because the heat generated by the reaction is sufficient to ignite the hydrogen gas evolved. Rubidium and cesium react with water with explosive violence.

Molten lithium is an exceedingly reactive substance. The only known substances that do not react with molten lithium are tungsten, molybdenum, and low-carbon stainless steels. If a piece of lithium metal is melted in a glass tube, then the molten lithium rapidly eats a hole through the glass. The reaction is accompanied by a brilliant green-yellow flame and considerable evolution of heat.

The alkali metals react directly with oxygen. Molten lithium ignites in oxygen to form $Li_2O(s)$; the reaction is accompanied by a bright red flame. The reactions of the other alkali metals do not yield the oxides M_2O. With sodium the *peroxide* Na_2O_2 is formed, and with potassium, rubidium, and cesium the *superoxides* KO_2, RbO_2, and CsO_2 are formed.

The alkali metals react directly with hydrogen at high temperatures to form hydrides. For example,

$$2Na(l) + H_2(g) \xrightarrow{500^\circ C} 2NaH(s)$$

The alkali metal hydrides are ionic compounds that contain the hydride ion, H^-. The hydrides react with water to liberate hydrogen,

Table F-3 Some of the more common reactions of the alkali metals

reaction with oxygen	reaction with water
$4Li(s) + O_2(g) \rightarrow 2Li_2O(s)$	$2M(s) + 2H_2O(l) \rightarrow 2MOH(s) + H_2(g)$
$2Na(s) + O_2(g) \rightarrow Na_2O_2(s)$	reaction with sulfur
$K(s) + O_2(g) \rightarrow KO_2(s)$	$2M(s) + S(s) \rightarrow M_2S(s)$
$Cs(s) + O_2(g) \rightarrow CsO_2(s)$	reaction with hydrogen
$Rb(s) + O_2(g) \rightarrow RbO_2(s)$	$2M(s) + H_2(g) \rightarrow 2MH(s)$
reaction[a] with halogens (denoted by X_2)	reaction with nitrogen
$2M(s) + X_2 \rightarrow 2MX(s)$	$6Li(s) + N_2(g) \rightarrow 2Li_3N(s)$

[a] $M(s)$ denotes any one of the alkali metals.

$$NaH(s) + H_2O(l) \rightarrow NaOH(aq) + H_2(g)$$

and are used to remove traces of water from organic solvents. In such cases, the metal hydroxide precipitates from the solution.

Lithium is the only element that reacts directly with nitrogen at room temperature:

$$6Li(s) + N_2(g) \rightarrow \underset{\text{lithium nitride}}{2Li_3N(s)}$$

The reddish black lithium nitride reacts directly with water to form ammonia:

$$Li_3N(s) + 3H_2O(l) \rightarrow 3LiOH(aq) + NH_3(g)$$

This reaction can be used to prepare deuterated ammonia, ND_3:

$$Li_3N(s) + 3D_2O(l) \rightarrow 3LiOD(aq) + ND_3(g)$$

Some of the more common reactions of the alkali metals are summarized in Table F-3.

Compounds of the alkali metals are for the most part white, high-melting ionic solids. With very few exceptions, alkali metal salts are soluble in water and the resulting solutions are electrolytic as a result of the dissociation of the salt into its constituent ions.

Not all the properties of lithium are analogous to those of the other members of the alkali metal family. For example, in contrast to the analogous salts of the other alkali metals, LiF and Li_2CO_3 are insoluble in water and LiCl is soluble in alcohols and ethers. The anomalous behavior of lithium is ascribed to the much smaller size (Table F-1) of the Li^+ ion, which gives the ion a much greater electron affinity [higher ionization energy of Li(*g*)] than the other alkali metal ions.

The alkali metals have the unusual property of dissolving in liquid ammonia to yield a blue electrolytic solution. The properties of such a solution are interpreted in terms of *solvated electrons* and alkali metal ions:

$$M(s) \xrightarrow[NH_3(l)]{} M^+(amm) + e^-(amm)$$

When the blue solutions are concentrated by evaporation, they become bronze in color and behave like liquid metals.

F-3. MANY ALKALI METAL COMPOUNDS ARE IMPORTANT COMMERCIALLY

Sodium hydroxide is the seventh ranked industrial chemical. Over 20 billion pounds of it is produced annually in the United States. Sodium hydroxide sometimes is called caustic soda and is prepared by the electrolysis of concentrated aqueous sodium chloride solutions:

$$2NaCl(aq) + 2H_2O(l) \xrightarrow{\text{electrolysis}} 2NaOH(aq) + H_2(g) + Cl_2(g)$$

or by the reaction between calcium hydroxide (called slaked lime) and sodium carbonate:

$$Na_2CO_3(aq) + Ca(OH)_2(aq) \rightarrow 2NaOH(aq) + CaCO_3(s)$$

The formation of the insoluble $CaCO_3$ is a driving force for this second reaction. The alkali metal hydroxides are white, translucent, corrosive solids that are extremely soluble in water; at 20°C the solubility of NaOH is 15 M and that of KOH is 13 M.

Sodium carbonate, which is called soda ash, is the twelfth ranked industrial chemical. The annual United States production of sodium carbonate exceeds 16 billion pounds. It is prepared from sodium chloride by the Solvay process, which was devised by the Belgian brothers Ernest and Edward Solvay in 1861. In this process, carbon dioxide is bubbled through a cooled solution of sodium chloride and ammonia. The reactions are

$$NH_3(aq) + CO_2(aq) + H_2O(l) \rightarrow NH_4^+(aq) + HCO_3^-(aq)$$

$$NaCl(aq) + NH_4^+(aq) + HCO_3^-(aq) \xrightarrow{15°C} NaHCO_3(s) + NH_4Cl(aq)$$

At 15°C the sodium hydrogen carbonate precipitates from the solution. Part of the sodium hydrogen carbonate is converted to sodium carbonate by heating:

$$2NaHCO_3(s) \xrightarrow{80°C} Na_2CO_3(s) + H_2O(l) + CO_2(g)$$

The carbon dioxide produced in this reaction is used again in the first reaction.

Table F-4 Some commercially important alkali metal compounds and their uses

Compound	*Uses*
lithium aluminum hydride, $LiAlH_4$	production of many pharmaceuticals, perfumes, and organic chemicals
sodium hydrogen carbonate (sodium bicarbonate), $NaHCO_3$	manufacture of effervescent salts and beverages, baking powder, gold plating
sodium carbonate, Na_2CO_3	manufacture of glass, pulp and paper, soaps and detergents, textiles
sodium hydroxide, NaOH	production of rayon, cellulose, paper, soaps, detergents, textiles, oven cleaner
sodium sulfate decahydrate (Glauber's salt), $Na_2SO_4 \cdot 10H_2O$	solar heating storage, air conditioning
potassium carbonate (potash), K_2CO_3	manufacture of special glass for optical instruments and electronic devices, soft soaps
potassium nitrate, KNO_3	pyrotechnics, explosives, matches

The commercial success of the Solvay process requires the recovery of the ammonia, which is relatively expensive. The ammonia is recovered from the NH_4Cl by the reaction

$$2NH_4Cl(aq) + Ca(OH)_2(s) \rightarrow 2NH_3(g) + CaCl_2(aq) + 2H_2O(l)$$

The calcium hydroxide and the carbon dioxide used in the process are obtained by heating limestone (primarily $CaCO_3$).

The raw materials of the Solvay process are sodium chloride, limestone, and water, all of which are inexpensive. The principal use of sodium carbonate is in the manufacture of glass (Interchapter N).

Some other important alkali metal compounds and their uses are given in Table F-4.

QUESTIONS

F-1. Why must the alkali metals be stored under kerosene?

F-2. How is sodium metal produced commercially?

F-3. Why, do you think, is sodium metal the least expensive metal per unit volume?

F-4. Explain why the reactivity of the alkali metals increases with atomic number.

F-5. What is the only element that reacts directly with nitrogen at room temperature?

F-6. What are the raw materials in the Solvay process?

F-7. Write chemical equations for the reactions in the Solvay process.

F-8. Complete and balance the following equations.

(a) $K(s) + O_2(g) \rightarrow$
(b) $Na(s) + H_2O(l) \rightarrow$
(c) $K(s) + N_2(g) \rightarrow$
(d) $NaH(s) + H_2O(l) \rightarrow$
(e) $Li_3N(s) + H_2O(l) \rightarrow$
(f) $Na(s) + F_2(g) \rightarrow$

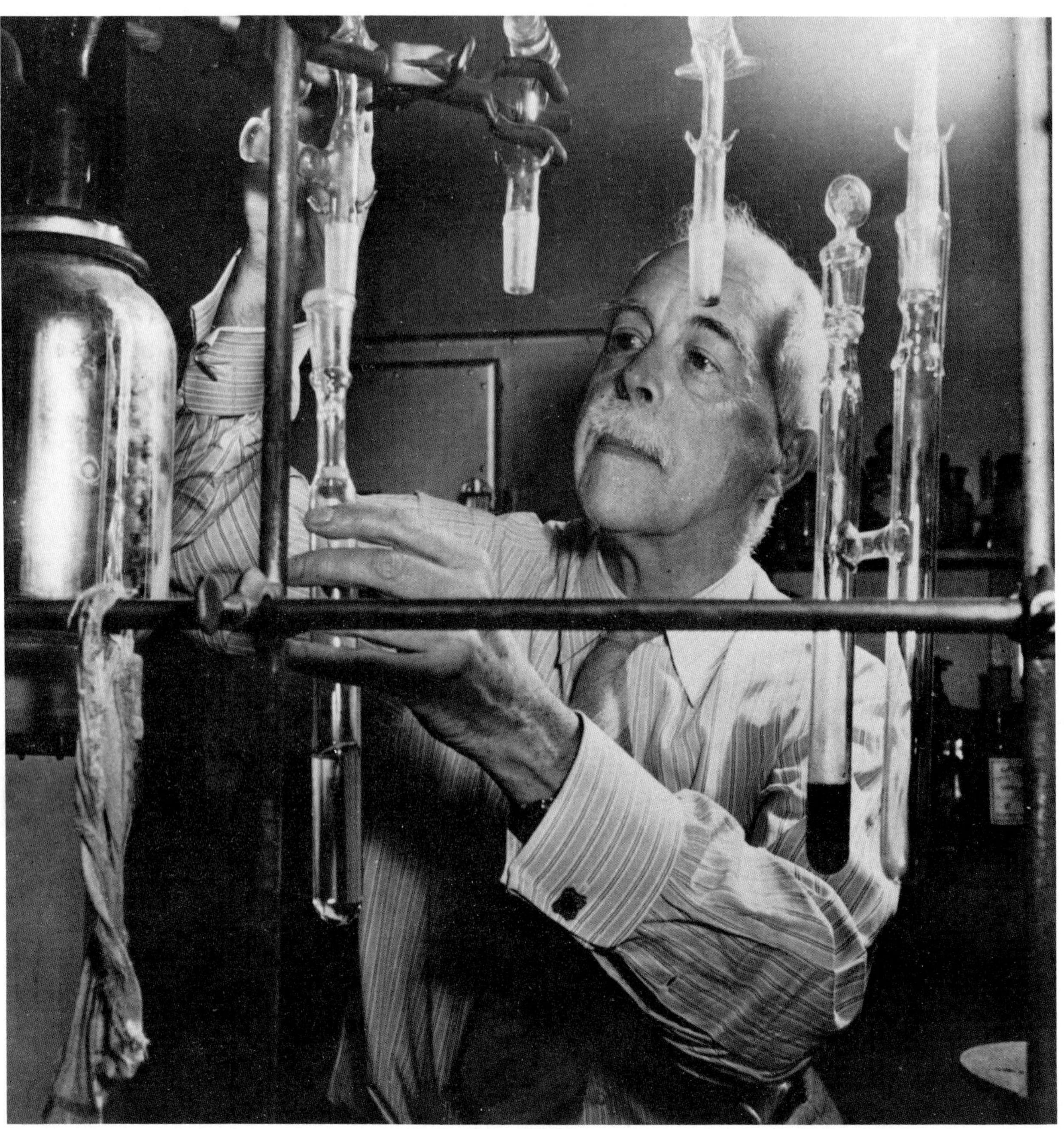

Gilbert N. Lewis was one of America's greatest chemists. He not only introduced the Lewis method of writing molecular formulas that is the subject of this chapter but also was one of the pioneers of chemical thermodynamics. He spent most of his career at the University of California at Berkeley, where he built up one of the strongest chemistry departments in the world.

10 / Lewis Formulas

In Chapter 9 we saw that, when metals react with nonmetals, *valence electrons* (that is, outer-shell electrons) of the metallic atoms are transferred completely to the nonmetallic atoms, resulting in ionic compounds. There is an enormous class of compounds that are not ionic. Many of these compounds are gases or liquids at room temperature, and they are poor conductors of an electric current. The bonding in these compounds differs from that in ionic compounds and is called *covalent bonding*. Although an understanding of covalent bonding requires a quantum theoretical description, it is possible to gain a qualitative and intuitive insight into covalent bonding in molecules by studying a method of writing molecular formulas that was introduced by the American chemist G. N. Lewis. A full decade before the quantum theory was formulated by Schrödinger in 1925, Lewis postulated that a covalent bond can be described as a pair of electrons that are shared between two atoms. The *electron-pair bond* idea was later given a firm theoretical basis by the quantum theory of chemical bonding, which we discuss in Chapter 12. The molecular formulas introduced by Lewis are called *Lewis formulas* and are one of the most useful and important concepts of chemistry.

10-1. A COVALENT BOND CAN BE DESCRIBED AS A PAIR OF ELECTRONS SHARED BY TWO ATOMS

Consider the molecule Cl_2. The Lewis electron-dot formula for a chlorine atom is

$$:\ddot{\underset{\cdot\cdot}{Cl}}\cdot$$

Recall that a Lewis electron-dot formula for an atom shows only the outer-shell (valence) electrons. The chlorine atom is one electron short of having eight electrons in its outer shell and achieving an argon-like electron configuration. A chlorine atom in Cl_2 could get this electron from the other chlorine atom, but certainly that chlorine atom does not wish to give up an electron. In a sense there is a stalemate with respect to electron transfer because both atoms have the same driving force to gain an electron. From a more quantitative point of view, although a chlorine atom has a high electron affinity ($348\ \mathrm{kJ \cdot mol^{-1}}$), it has a much higher ionization energy ($1260\ \mathrm{kJ \cdot mol^{-1}}$) and does not lose an electron very easily. Ionic bonds result only when one atomic reactant is a metal (relatively low ionization energy) and the other is a nonmetal (relatively high electron affinity).

Although we have ruled out the formation of an ionic bond in Cl_2, there is a way for the two chlorine atoms to achieve an argon-like electron configuration *simultaneously*. If the two *share* a pair of electrons between them, then the resulting distribution of valence electrons can be pictured as

$$:\ddot{\underset{\cdot\cdot}{\mathrm{Cl}}}\cdot + \cdot\ddot{\underset{\cdot\cdot}{\mathrm{Cl}}}: \rightarrow :\ddot{\underset{\cdot\cdot}{\mathrm{Cl}}}:\ddot{\underset{\cdot\cdot}{\mathrm{Cl}}}:$$

Notice that *each* chlorine atom has eight electrons in its outer shell:

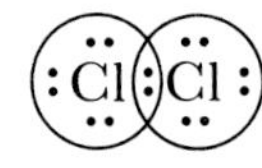

Thus by sharing a pair of electrons, each chlorine atom is able to achieve the stable argon-like outer electron configuration of eight electrons. According to Lewis's picture, the shared electron pair is responsible for holding the two chlorine atoms together as a chlorine molecule. The bond formed between two atoms by a shared electron pair is called a *covalent bond.*

The above electron-dot formula for Cl_2 is called a *Lewis formula.* It is conventional to indicate the electron-pair bond as a line joining the two atoms and the other electrons as pairs of dots surrounding the atoms:

$$:\ddot{\underset{\cdot\cdot}{\mathrm{Cl}}}\text{—}\ddot{\underset{\cdot\cdot}{\mathrm{Cl}}}:$$

The pairs of electrons that are not shared between the chlorine atoms are called *lone electron pairs,* or simply *lone pairs.* A Lewis formula correctly depicts a covalent bond as a pair of electrons shared between two atoms.

When Cl_2 is solidified (its freezing point is $-101\,^{\circ}\mathrm{C}$), it forms a *molecular crystal* (Figure 10-1). In contrast to an ionic crystal, the constituent particles of a molecular crystal are molecules, in this case Cl_2

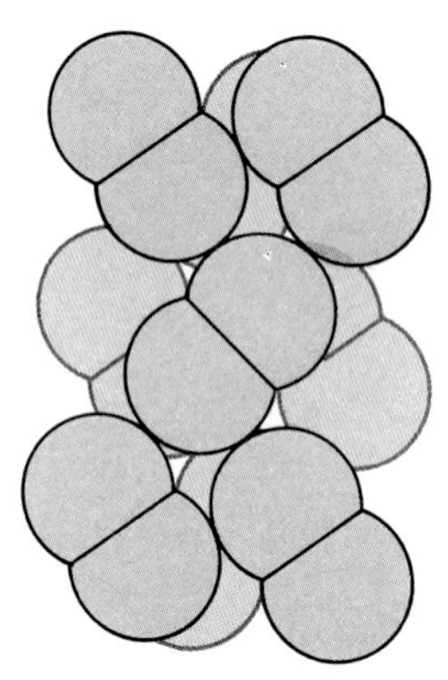

Figure 10-1 The regular arrangement of the chlorine molecules in crystalline Cl_2. This pattern is repeated throughout the crystal. The chlorine molecules are neutral and so do not attract each other as strongly as neighboring ions in an ionic lattice. Consequently, molecular crystals like Cl_2 usually have lower melting points than ionic crystals. The melting point of chlorine is $-101\,^{\circ}\mathrm{C}$.

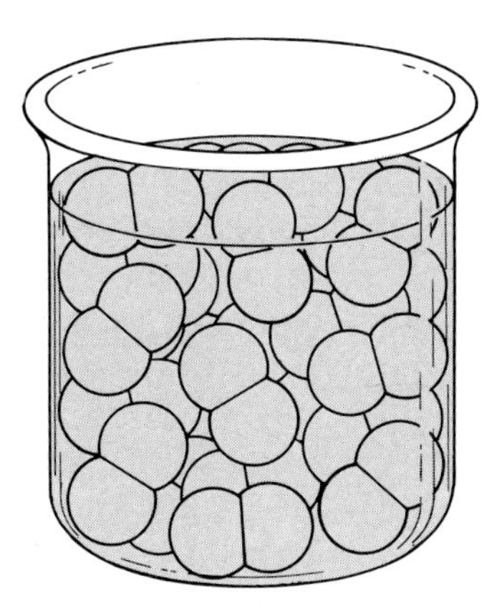

Figure 10-2 The arrangement of the chlorine molecules in liquid Cl_2. There is no orderly pattern in the liquid as there is in crystalline Cl_2. The chlorine molecules in the liquid phase move throughout the container and are randomly oriented.

molecules. The low melting point of chlorine indicates that the attraction between the molecules is weak relative to the attraction between ions in a crystal. The Cl_2 molecules are neutral, and thus there is no net electrostatic attraction between them in the crystal. When solid Cl_2 is melted, the Cl_2 molecules remain intact and move about the fluid as neutral molecules (Figure 10-2). Consequently, liquid Cl_2 does not conduct an electric current.

There are no ions in solid chlorine or liquid chlorine.

Example 10-1: Write the Lewis formula for a bromine molecule.

Solution: The Lewis electron-dot formula for bromine is

$$:\!\underset{\cdot\cdot}{\ddot{\mathrm{Br}}}\cdot$$

In an acceptable Lewis formula, each atom has a noble-gas-like electron configuration. Pictorially, this is achieved by each atom's having eight electrons, represented by dots, surrounding it. As in the case of Cl_2, this can be done for Br_2 by writing

$$:\!\underset{\cdot\cdot}{\ddot{\mathrm{Br}}}\cdot \;+\; \cdot\underset{\cdot\cdot}{\ddot{\mathrm{Br}}}\!: \;\rightarrow\; :\!\underset{\cdot\cdot}{\ddot{\mathrm{Br}}}\!:\!\underset{\cdot\cdot}{\ddot{\mathrm{Br}}}\!:$$

or

$$:\!\underset{\cdot\cdot}{\ddot{\mathrm{Br}}}\text{—}\underset{\cdot\cdot}{\ddot{\mathrm{Br}}}\!:$$

All the elemental halogens exist as diatomic molecules. The Lewis formulas for F_2 and I_2 are similar to the Lewis formulas for Cl_2 and Br_2. Note that each of the individual atoms has 7 valence electrons, giving a total of 14 valence electrons. There are seven electron pairs in the molecule, and so all the valence electrons are accounted for and each atom has eight electrons around its nucleus.

The halogens have the Lewis formula $:\!\underset{\cdot\cdot}{\ddot{\mathrm{X}}}\text{—}\underset{\cdot\cdot}{\ddot{\mathrm{X}}}\!:$, where X is F, Cl, Br, or I.

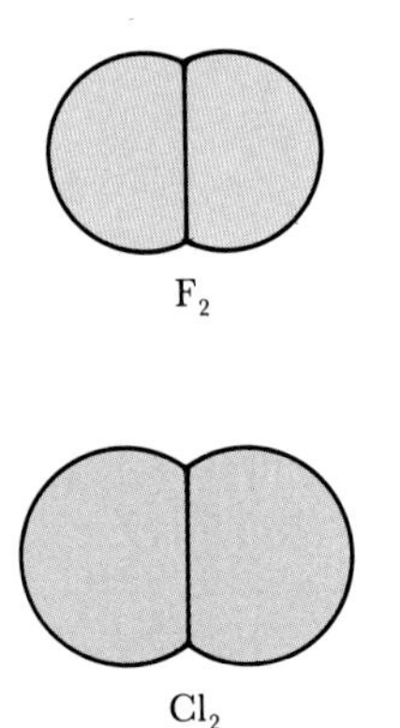

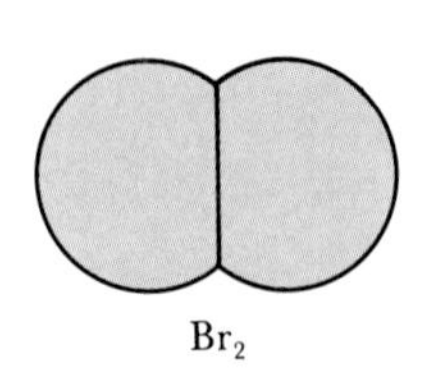

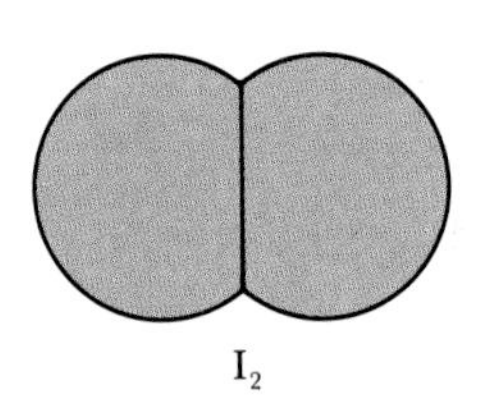

Figure 10-3 Molecular models of the halogen molecules, drawn to scale to indicate the relative sizes of the atoms. Note that the halogens become larger as one goes down the group in the periodic table.

Figure 10-3, which shows molecular models of the halogen molecules, can be used to help define *bond length*. The drawings suggest that the nuclei of the two halogen atoms are held a fixed distance apart. In a real molecule the atoms vibrate about these positions, but they vibrate about a well-defined average bond distance. The average bond lengths of the halogens are given in Table 10-1. Note that the bond lengths in the diatomic halogen molecules increase as one goes down the periodic table.

10-2. WE ALWAYS TRY TO SATISFY THE OCTET RULE WHEN WRITING LEWIS FORMULAS

When F_2 is bubbled through an aqueous solution of NaOH, the pale yellow gas oxygen difluoride, OF_2, is formed. Oxygen difluoride reacts explosively with the other halogens. We can deduce the Lewis formula for OF_2 by first writing the Lewis electron-dot formulas for the atoms:

$$:\underset{\cdot\cdot}{\overset{\cdot\cdot}{F}}\cdot \qquad \cdot\underset{\cdot\cdot}{\overset{\cdot\cdot}{O}}\cdot \qquad \cdot\underset{\cdot\cdot}{\overset{\cdot\cdot}{F}}:$$

We wish to join these three atoms such that each has eight electrons in its outer shell. Pictorially, we wish to join these atoms such that each one can be written with eight valence electrons surrounding the nucleus. By bringing the fluorine atoms in toward the oxygen atom, we see that the electron-dot formula

$$:\underset{\cdot\cdot}{\overset{\cdot\cdot}{F}}:\underset{\cdot\cdot}{\overset{\cdot\cdot}{O}}:\underset{\cdot\cdot}{\overset{\cdot\cdot}{F}}:$$

allows all three atoms to be surrounded simultaneously by eight electrons. Thus we conclude that a satisfactory Lewis formula for OF_2 is

$$:\underset{\cdot\cdot}{\overset{\cdot\cdot}{F}}-\underset{\cdot\cdot}{\overset{\cdot\cdot}{O}}-\underset{\cdot\cdot}{\overset{\cdot\cdot}{F}}:$$

As a final check of this formula, note that there are 20 valence electrons indicated in the Lewis formula for the molecule and that there is a total of 20 valence electrons $[(2 \times 7) + 6]$ in the Lewis electron-dot formulas for the individual atoms.

The fact that the Lewis formula for OF_2 depicts the two fluorine atoms attached to a central oxygen atom suggests that this is the actual bonding in the molecule. The great utility of Lewis formulas is that they suggest which atoms are bonded to which in a molecule.

We can formalize the writing of Lewis formulas by introducing the *octet rule*. The octet rule states that many elements form covalent bonds so as to end up with eight electrons in their outer shells. We shall see that, although there are exceptions to the octet rule, it is still useful because of the large number of compounds that do obey it. In general, we do not violate the octet rule in writing a Lewis formula

Table 10-1 The bond lengths of the halogen molecules

Molecule	*Bond length/pm*
F_2	128
Cl_2	198
Br_2	228
I_2	266

unless it is impossible to avoid doing so. The octet rule has its origin in the special stability of the noble-gas electron configuration. Thus, for example, carbon, nitrogen, oxygen, and fluorine achieve a neon-like electron configuration when they are surrounded by eight valence electrons.

When we discussed OF_2, we purposely placed the oxygen atom between the two fluorine atoms because we anticipated the correct answer. With a little experience, you can become adept at arranging atoms into satisfactory Lewis formulas. As another example, consider the molecule NF_3. Nitrogen forms a variety of halogen compounds, all of which are poisonous and many of which are dangerously explosive. The electron-dot formulas for the constituent atoms in NF_3 are

```
 ..     ..     ..
:F·    ·N·    ·F:
 ..     ·      ..
        ·
       :F:
        ..
```

Note that there is a total of 26 $[(3 \times 7) + 5 = 26]$ valence electrons in these four atoms. By joining the three fluorine atoms to the nitrogen atom, we have

```
 ..  ..  ..            ..  ..  ..
:F : N : F:    or     :F — N — F:
 ..  ..  ..            ..  |   ..
    :F:                   :F:
     ..                    ..
```

Each atom in

```
 .. .. ..
:F:N:F:
 .. .. ..
   :F:
    ..
```

satisfies the octet rule.

as the Lewis formula for NF_3. This Lewis formula shows that NF_3 has three N—F bonds. Note that there are 13 electron pairs, or 26 valence electrons, in NF_3, which agrees, as it must, with the total number of valence electrons in the individual atoms.

In constructing the Lewis formula for NF_3, we arranged the atoms correctly in anticipation of the final result, but there is really a pattern here. Notice that nitrogen needs three electrons to satisfy the octet rule and that it forms three bonds. In the case of OF_2, oxygen needs two electrons and it forms two bonds. In both cases, fluorine needs only one more valence electron to complete its octet and so forms one bond. In these examples, the number of bonds that each atom forms is equal to the number of electrons it needs to satisfy the octet rule. In addition, if there is only one of a particular atom, it is a good first try to assume that this atom is the central atom and that the other atoms are attached to it.

An alternative method for writing Lewis formulas is as follows:

1. Count the total number of valence electrons in the individual atoms.
2. Divide the total number of electrons by two to obtain the total number of electron pairs in the Lewis formula of the molecule.
3. Join the atoms by arranging the electron pairs such that each atom is surrounded simultaneously by eight electrons, or four electron pairs.

Let's apply this method to OF_2. There is a total of $7 + 7 + 6 = 20$ valence electrons in the individual atoms. Thus there are 10 electron pairs in the Lewis formula for OF_2. We place the unique atom (O) in the center and join the atoms together by arranging the 10 electron pairs such that each of the 3 atoms is surrounded simultaneously by 8 electrons, giving

$$:\ddot{\underset{..}{F}}-\ddot{\underset{..}{O}}-\ddot{\underset{..}{F}}:$$

With practice you will find this method effective in writing Lewis formulas.

Example 10-2: Draw the Lewis formula for carbon tetrafluoride, CF_4.

Solution: The electron-dot formulas for carbon and fluorine are

$$\cdot\dot{\underset{\cdot}{C}}\cdot \qquad \text{and} \qquad \cdot\ddot{\underset{..}{F}}:$$

We expect carbon to form four bonds because it needs four electrons to satisfy the octet rule. In addition, there is only one carbon atom, and so we assume that it is central and that the four fluorine atoms are attached to it. The Lewis formula for CF_4 is

Carbon, nitrogen, oxygen, and fluorine always satisfy the octet rule.

$$\begin{matrix} & :\ddot{F}: & \\ :\ddot{\underset{..}{F}}: & \ddot{\underset{..}{C}} & :\ddot{\underset{..}{F}}: \\ & :\underset{..}{F}: & \end{matrix} \qquad \text{or} \qquad \begin{matrix} & :\ddot{F}: & \\ & | & \\ :\ddot{\underset{..}{F}}- & C & -\ddot{\underset{..}{F}}: \\ & | & \\ & :\underset{..}{F}: & \end{matrix}$$

The alternative method is to see that there is a total of 32 valence electrons in the individual atoms and that there are 16 electron pairs in the molecule: 3 lone pairs about each of the 4 fluorine atoms (12 pairs) and 4 covalent bonds (4 pairs).

10-3. THE HYDROGEN ATOM IS AN EXCEPTION TO THE OCTET RULE

We have said that there are a number of exceptions to the octet rule. One important exception is the hydrogen atom. The noble gas closest to hydrogen in the periodic table is helium. We might expect, then, that hydrogen needs only two electrons in order to attain a noble-gas electron configuration. For example, let's consider H_2 itself. The electron-dot formula for hydrogen is $\cdot H$. Each hydrogen atom can be surrounded by two electrons if the two hydrogen atoms share electrons:

$$H\cdot + \cdot H \rightarrow H:H \qquad \text{or} \qquad H-H$$

In this way each hydrogen atom achieves a helium-like electron configuration.

The Lewis formulas for the hydrogen halides are obtained directly from the electron-dot formulas of the individual atoms. If we let X be F, Cl, Br, or I, we can write

$$\text{H}\cdot + \cdot\ddot{\underset{\cdot\cdot}{\text{X}}}: \rightarrow \text{H—}\ddot{\underset{\cdot\cdot}{\text{X}}}:$$

The hydrogen atom has two electrons surrounding it, and the halogen atom has eight. Molecular models of the hydrogen halides are shown in Figure 10-4. The bond lengths of the hydrogen halides are given in Table 10-2.

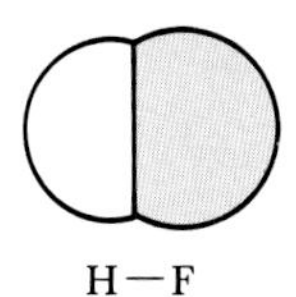

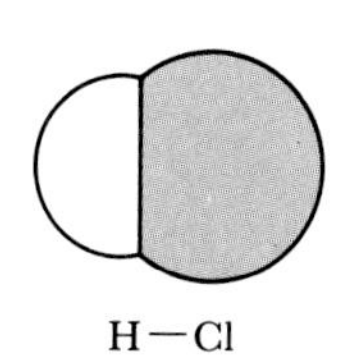

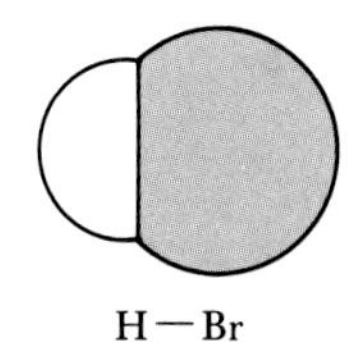

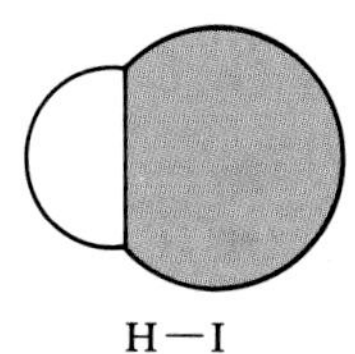

Figure 10-4 Molecular models of the hydrogen halides.

Example 10-3: Draw the Lewis formula for H_2O.

Solution: The electron-dot formulas for oxygen and hydrogen are

$$\text{H}\cdot \qquad \text{and} \qquad \cdot\ddot{\underset{\cdot\cdot}{\text{O}}}\cdot$$

Oxygen needs two additional electrons to be surrounded by eight electrons and so forms two bonds. Each hydrogen needs one electron to be surrounded by two electrons and so forms one bond. In addition, there are two hydrogen atoms and only one oxygen atom, and so we guess that the oxygen atom is central and that the two hydrogen atoms are attached to it. The Lewis formula for H_2O is

$$\text{H—}\ddot{\underset{\cdot\cdot}{\text{O}}}\text{—H}$$

Note that there are two lone pairs of electrons on the oxygen atom in the water molecule. The oxygen atom has eight electrons surrounding it, and each hydrogen atom has two.

Example 10-4: Draw the Lewis formula for chloroform, $HCCl_3$.

Solution: The chloroform molecule has three different types of atoms, and thus the first step is to decide how to arrange them in the Lewis formula. A key point to recognize is that, because hydrogen completes its valence shell with only two valence electrons, it is almost always a *terminal* atom in the Lewis formula; that is, it is bonded to only one other atom. Of the remaining four atoms (CCl_3), the carbon atom is unique, and therefore we guess that it is the central atom. Thus we have the postulated arrangement

$$\begin{array}{ccc} & \text{H} & \\ & \cdot & \\ :\ddot{\underset{\cdot\cdot}{\text{Cl}}}\cdot & \cdot\dot{\underset{\cdot}{\text{C}}}\cdot & \cdot\ddot{\underset{\cdot\cdot}{\text{Cl}}}: \\ & \cdot & \\ & :\underset{\cdot\cdot}{\text{Cl}}: & \end{array}$$

Table 10-2 The bond lengths of the hydrogen halides

Compound	*Bond length/pm*
HF	92
HCl	127
HBr	141
HI	161

There is a total of $4 + (3 \times 7) + 1 = 26$ valence electrons to be accommodated in the Lewis formula. If we use 8 of these electrons to form 4 bonds

$$\begin{array}{c} \text{H} \\ | \\ \text{Cl—C—Cl} \\ | \\ \text{Cl} \end{array}$$

then the remaining 18 (26 − 8) can be accommodated as 9 lone pairs (3 lone pairs on each of the 3 Cl atoms). Thus the Lewis formula of $HCCl_3$ is

$$\begin{array}{c} \text{H} \\ | \\ :\ddot{\underset{\cdot\cdot}{\text{Cl}}}\text{—C—}\ddot{\underset{\cdot\cdot}{\text{Cl}}}: \\ | \\ :\underset{\cdot\cdot}{\text{Cl}}: \end{array}$$

Note that the octet rule is satisfied for the carbon atom and for each chlorine atom.

At one time chloroform was used extensively as an inhalation anesthetic and analgesic, but these uses have been abandoned because of undesirable side effects, primarily nausea and intense headaches.

Hydrazine is a colorless, oily liquid that fumes in air. It has a penetrating odor resembling that of ammonia and is a violent poison.

Up to this juncture we have considered only molecules for which the arrangement of atoms in the Lewis formula is based on placing the unique atom (other than hydrogen, which is always a terminal atom) in the central position. However, consider the problem of writing a Lewis formula for hydrazine, N_2H_4. For this compound, there is no unique atom to place in the central position, and so we assume that the two nitrogen atoms must be bonded to each other with the four hydrogen atoms in terminal positions. Thus we write

$$\begin{array}{c} \text{H—N—N—H} \\ \;\;|\;\;\;\;| \\ \;\;\text{H}\;\;\text{H} \end{array}$$

There is a total of $(4 \times 1) + (2 \times 5) = 14$ valence electrons. The 5 bond pairs require a total of 10 valence electrons, and the remaining 4 valence electrons are placed as lone pairs on the nitrogens in order to complete the octets on the nitrogen atoms and simultaneously accommodate the 14 valence electrons.

$$\begin{array}{c} \text{H—}\ddot{\text{N}}\text{—}\ddot{\text{N}}\text{—H} \\ \;\;|\;\;\;\;| \\ \;\;\text{H}\;\;\text{H} \end{array}$$

Example 10-5: Draw the Lewis formula for methyl alcohol, CH_3OH.

Solution: We first assume that the hydrogen atoms are terminal. We next assume that, because carbon and oxygen cannot both be central atoms, they must be bonded to each other:

C—O

An oxygen atom has six valence electrons and thus usually completes its octet by forming two bonds, whereas a carbon atom has only four valence electrons and usually completes its octet by forming four bonds. Thus we write

```
     H
     |
  H—C—O—H
     |
     H
```

The CH_3OH molecule contains $(4 \times 1) + 4 + 6 = 14$ valence electrons. The 5 bonds require a total of 10 valence electrons. The remaining four valence electrons are accommodated in two lone pairs on the oxygen atom, which are also necessary to complete the octet on the oxygen atom. Thus the Lewis formula for methyl alcohol is

```
     H
     |   ..
  H—C—O—H
     |   ..
     H
```

Methyl alcohol is also called wood alcohol because it can be obtained by the destructive distillation of wood. It is a severe poison, causing convulsions, blindness, and death.

These simple procedures enable us to arrange the atoms correctly in the Lewis formula for most cases that involve at most two different types of atoms other than hydrogen. For more complex cases, a more extensive knowledge of molecular structure is required to predict correctly the arrangement of atoms in the Lewis formula. However, if we are given the arrangement of the atoms, then we can use the procedures developed in this chapter to write an acceptable Lewis formula.

10-4. MANY MOLECULES CONTAIN DOUBLE OR TRIPLE COVALENT BONDS

In all the molecules we have discussed so far, there is only one electron pair connecting two atoms. However, in many molecules, atoms are joined by two electron pairs (a *double bond*) or three electron pairs (a *triple bond*). For example, let's consider carbon dioxide, CO_2.

We first note that CO_2 has a total of $(2 \times 6) + 4 = 16$ valence electrons. We place the carbon atom between the two oxygen atoms and form electron-pair bonds between the carbon and the oxygen atoms. Using the remaining electrons to form lone pairs yields

$$\cdot\ddot{\underset{\cdot\cdot}{O}}—\dot{\underset{\cdot}{C}}—\ddot{\underset{\cdot\cdot}{O}}\cdot \qquad \text{violates the octet rule}$$

This Lewis formula violates the octet rule because there are only six valence electrons around the carbon atom and seven around each oxygen atom. However, if we combine the two unpaired electrons on carbon with the unpaired electrons on the oxygens and then move the pairs into the bonding regions between the oxygen atoms and the carbon atom, we obtain

$$:\!O{=}C{=}O\!: \qquad \text{satisfies the octet rule}$$

This Lewis formula satisfies the octet rule because there are now eight electrons around each atom:

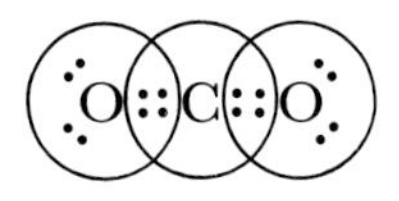

The Lewis formula for the CO_2 molecule predicts two carbon-oxygen double bonds. A double bond between two atoms is shorter and stronger than a single bond between the same two atoms. The carbon-oxygen bonds in CO_2 are indeed shorter and stronger than, for example, the carbon-oxygen single bond in methyl alcohol, CH_3OH. Table 10-3 gives typical bond lengths and bond energies for some single and double bonds.

Table 10-3 Average bond lengths and bond energies of single and double bonds

Bond	*Average bond length/pm*	*Average bond energy/kJ·mol^{-1}*
C—O	143	360
C=O	120	800
C—C	154	350
C=C	134	600
N—N	145	170
N=N	125	420

Example 10-6: Draw a Lewis formula for the ethylene molecule, C_2H_4.

Solution: The hydrogen atoms must be terminal atoms; consequently the two carbon atoms must be joined together. There is a total of $(4 \times 1) + (2 \times 4) = 12$ valence electrons in an ethylene molecule. Formation of the 5 electron-pair bonds requires a total of 10 valence electrons and leaves 1 unpaired electron on each carbon atom:

```
    H  H
    |  |
H—C—C—H      violates the octet rule
    •  •
```

The only way to satisfy the octet rule is to use the two unpaired electrons to form a double bond between the two carbon atoms:

```
    H  H
    |  |
H—C=C—H      satisfies the octet rule
```

An example of a molecule that has a triple bond is N_2. The electron-dot formula for a nitrogen atom is

```
 ••
•N•
 •
```

Each nitrogen atom needs three electrons to satisfy the octet rule. The only way this can happen is for the two nitrogen atoms to share *three* pairs of electrons between them:

```
 •        •
:N•  +  •N:  →  :N:::N:     or     :N≡N:
 •        •
```

The three shared electron pairs in N_2 constitute a triple bond. Another molecule that contains a triple bond is the poisonous gas hydrogen cyanide, HCN:

H—C≡N:

Note that 2 of the 10 valence electrons are used to form a C—H single bond, 6 are used to form a C≡N triple bond, and 2 are used to form a lone pair on the nitrogen atom.

Example 10-7: Write the Lewis formula for acetylene, C_2H_2.

Solution: The electron-dot formulas for hydrogen and carbon are

```
           •
H•   and  •C•
           •
```

Each hydrogen atom can form only one bond with a carbon atom, giving

$$\text{H}:\dot{\underset{\cdot}{\text{C}}}\cdot \qquad \cdot\dot{\underset{\cdot}{\text{C}}}:\text{H}$$

These two fragments can be joined just as two $:\dot{\underset{\cdot}{\text{N}}}\cdot$ can be joined, giving

Problems 10-21 to 10-32 ask you to draw Lewis formulas of molecules with double or triple bonds.

$$\text{H}:\text{C}:::\text{C}:\text{H} \qquad \text{or} \qquad \text{H—C}\equiv\text{C—H}$$

The acetylene molecule involves a carbon-carbon triple bond.

10-5. MOST POLYATOMIC IONS CONTAIN COVALENT BONDS

Many ions exist as a group of atoms bonded together by covalent bonds. For example, consider the hydroxide ion, OH^-. The electron-dot formulas for O and H are

$$\cdot\ddot{\underset{\cdot\cdot}{\text{O}}}\cdot \qquad \text{and} \qquad \cdot\text{H}$$

If we bond these two atoms covalently, then we have

$$\cdot\ddot{\underset{\cdot\cdot}{\text{O}}}:\text{H}$$

The OH^- ion has one more electron than OH because of the negative charge. This additional electron is placed on the oxygen atom to satisfy the octet rule, giving

$$\text{H—}\ddot{\underset{\cdot\cdot}{\text{O}}}:^{\ominus}$$

A negative charge often is circled in a Lewis formula so that it is not confused with an electron pair bond. The compound NaOH is an ionic compound and exists as Na^+ and OH^-, but the bonding within the OH^- ion is covalent.

Example 10-8: Write the Lewis formula for the ammonium ion, NH_4^+.

Solution: The electron-dot formulas for the atoms are $\cdot\ddot{\underset{\cdot}{\text{N}}}\cdot$ and $\cdot$H. The total number of valence electrons in NH_4^+ is $5 + (4 \times 1) - 1 = 8$, where we subtract 1 because NH_4^+ has a charge of $+1$. As usual, we guess that nitrogen, being the unique atom, is central and write

$$\left[\begin{array}{c} \text{H} \\ | \\ \text{H—N—H} \\ | \\ \text{H} \end{array}\right]^+$$

as the Lewis formula for NH_4^+.

10-6. FORMAL CHARGES CAN BE ASSIGNED TO ATOMS IN LEWIS FORMULAS

It is often convenient to assign a charge to each atom in a molecule or ion. Such charges are called *formal charges* because they are assigned by a set of arbitrary rules and do not necessarily represent the actual charges on the atoms. To assign formal charges, we *assume* that each pair of shared electrons is shared *equally* between the two atoms and assign one of these electrons to each atom. Lone-pair electrons are assigned to the atom on which they are located. The formal charge is the assigned net charge on the atom. The formal charge on an atom in a Lewis formula is calculated using the equation

$$\begin{pmatrix}\text{formal charge}\\\text{on an atom in}\\\text{a Lewis formula}\end{pmatrix} = \begin{pmatrix}\text{total number of}\\\text{valence electrons}\\\text{in the free atom}\end{pmatrix} - \begin{pmatrix}\text{total number}\\\text{of lone-}\\\text{pair electrons}\end{pmatrix} - \frac{1}{2}\begin{pmatrix}\text{total number}\\\text{of shared}\\\text{electrons}\end{pmatrix} \quad (10\text{-}1)$$

Consider NH_4^+. A hydrogen atom has one valence electron, and there are no lone-pair electrons in NH_4^+. Each hydrogen atom shares two electrons, and so the formal charge assigned to each hydrogen atom is, from Equation (10-1),

$$\text{formal charge on H in } NH_4^+ = 1 - 0 - \frac{1}{2}(2) = 0$$

A nitrogen atom, $\cdot\ddot{\text{N}}\cdot$, has five valence electrons. The nitrogen atom in NH_4^+ shares eight electrons, and so

$$\text{formal charge on N in } NH_4^+ = 5 - 0 - \frac{1}{2}(8) = +1$$

Thus NH_4^+ is written

```
     H
     |
  H—N⊕—H
     |
     H
```

where the ⊕ charge on N denotes a formal charge of +1 on the nitrogen atom. Note that the sum of the formal charges on the various atoms is equal to the net charge on the molecule or molecular ion.

Example 10-9: Write the Lewis formula for CO and assign formal charges.

Solution: The electron-dot formulas for the carbon and oxygen atoms are

$$\cdot\dot{\underset{\cdot}{C}}\cdot \quad \text{and} \quad \cdot\ddot{\underset{\cdot}{O}}:$$

There is a total of 10 valence electrons. The Lewis formula for CO is

$$:C\equiv O:$$

which involves a triple bond between the carbon and the oxygen. The formal charges on carbon and oxygen in CO are calculated using Equation (10-1):

$$\text{formal charge on C} = 4 - 2 - \frac{1}{2}(6) = -1$$

$$\text{formal charge on O} = 6 - 2 - \frac{1}{2}(6) = +1$$

Thus the Lewis formula for CO with the formal charges indicated is

$$\ominus:C\equiv O:\oplus$$

Once again, notice that the sum of the formal charges on the various atoms is equal to the net charge on the species, which is zero for CO.

10-7. RESONANCE CAN BE USED TO PREDICT RELATIVE BOND LENGTHS IN MOLECULES

There are a number of molecules and ions for which it is possible to write two or more satisfactory Lewis formulas. For example, let's consider the nitrite ion, NO_2^-. One Lewis formula for NO_2^- is

$$\ddot{N}\text{ with } :\ddot{O}{=}N \text{ (left) and } N{-}\ddot{\underset{..}{O}}\!:\ominus \text{ (right)}$$

One of the oxygen atoms, the right-hand one as written, has a formal charge of −1, having three lone pairs and one bond. Another equally acceptable Lewis formula for NO_2^- is

$$\ddot{N}\text{ with } \ominus:\ddot{\underset{..}{O}}{-}N \text{ (left) and } N{=}\ddot{O}: \text{ (right)}$$

In this case the negative formal charge is on the other oxygen atom. Both of these Lewis formulas satisfy the octet rule. When it is possible to write two or more satisfactory Lewis formulas *without altering the positions of the nuclei,* the actual formula is viewed as an average or as a superposition of the individual formulas. Each of the individual Lewis formulas is said to be a *resonance form* and the phenomenon itself is called *resonance.* We indicate resonance forms by means of a two-headed arrow, as in

Neither of the individual Lewis formulas taken separately accurately reflects the actual formula. Two separate Lewis formulas are necessary to describe the bonding in NO_2^-.

There is no generally accepted way to represent resonance pictorially, but one way is to write the formula as

Each of the NO bonds in NO_2^- can be thought of as the *average* of a single bond and a double bond. The two possible Lewis formulas for NO_2^- suggest that the two NO bonds in NO_2^- are equivalent, and this is in accord with experimental observation: the two do have exactly the same length, 113 pm. Notice that the individual Lewis formulas suggest that the two NO bonds are not equivalent, one being a single bond and the other being a double bond.

Another example of resonance occurs in the nitrate ion, NO_3^-. Three equally satisfactory Lewis formulas for NO_3^- are

Because each of these Lewis formulas is equally satisfactory, the actual structure is a superposition or an average of the three and can be represented pictorially as

A nitrate ion is planar, with each NO bond pointing to a vertex of an equilateral triangle.

In this case each of the N—O bonds is the average of a double bond and two single bonds. As the superimposed representation suggests, the three N—O bonds are equivalent, which is in agreement with experimental observations (each N—O bond is 122 pm in length). Furthermore, there are no known chemical reactions that can be used to distinguish one oxygen atom from another in a nitrate ion.

The need for resonance forms arises from the fact that Lewis formulas involve electron-pair bonds. If the species involves a bond intermediate between a single and a double bond, then we need to write two or more Lewis formulas to describe the bonding in the molecule.

Example 10-10: Draw Lewis formulas for the two resonance forms of SO_2. Indicate formal charges and discuss the bonding in this molecule. This example is particularly important to understand because it involves several of the important concepts that we have discussed in this chapter.

Solution: The electron-dot formulas for S and O are

$$\cdot\ddot{\underset{\cdot\cdot}{O}}\cdot \quad \text{and} \quad \cdot\ddot{\underset{\cdot\cdot}{S}}\cdot$$

There is a total of 18 valence electrons in the 3 atoms in SO_2, and thus, there are nine electron pairs. As usual, we place the unique atom (S) in the center and join the atoms together. The two resonance forms are

$$:\!\underset{\cdot\cdot}{O}\!=\!\ddot{S}^{\oplus}\!-\!\ddot{\underset{\cdot\cdot}{O}}\!:^{\ominus} \quad \longleftrightarrow \quad {}^{\ominus}\!:\!\ddot{\underset{\cdot\cdot}{O}}\!-\!\ddot{S}^{\oplus}\!=\!\underset{\cdot\cdot}{O}\!:$$

The formal charges indicated are calculated according to Equation (10-1):

$$\text{formal charge on S in } SO_2 = 6 - 2 - \tfrac{1}{2}(6) = +1$$

$$\text{formal charge on singly bonded O in } SO_2 = 6 - 6 - \tfrac{1}{2}(2) = -1$$

$$\text{formal charge on doubly bonded O in } SO_2 = 6 - 4 - \tfrac{1}{2}(4) = 0$$

These two Lewis formulas constitute resonance forms, and so the actual formula is an average of the two, which can be represented by

$$:\!\underset{\cdot\cdot}{O}\overset{\cdot\cdot}{\underset{}{\cdots S \cdots}}\underset{\cdot\cdot}{O}\!:$$

This formula suggests that the two S—O bonds in SO_2 are equivalent, or that the two S—O bond lengths in SO_2 are equal. This is in agreement with experiment.

Problems 10-33 to 10-40 involve resonance and formal charges.

10-8. FORMAL CHARGE CAN BE USED TO CHOOSE A PREFERRED LEWIS FORMULA

We discussed the molecule oxygen difluoride, OF_2, in Section 10-2 and wrote its Lewis formula as

$$:\ddot{\underset{..}{F}}—\ddot{\underset{..}{O}}—\ddot{\underset{..}{F}}: \qquad (1)$$

The OF_2 molecule was one of the examples we used to illustrate that it is a good first try to place the unique atom in the center. Note that the formal charge on each atom in formula (1) is zero. There is another Lewis formula for OF_2, however, that satisfies the octet rule:

$$:\ddot{\underset{..}{F}}—\overset{\oplus}{\ddot{\underset{..}{F}}}—\overset{\ominus}{\ddot{\underset{..}{O}}}: \qquad (2)$$

These two Lewis formulas predict entirely different bonding in OF_2. The first predicts that the oxygen atom is in the center of the molecule and that there are two O—F bonds. The second predicts that one of the fluorine atoms is in the center and that there is one F—F bond and one O—F bond.

We can use formal charges to select one of these Lewis formulas for OF_2 over the other. Although formal charges do not represent the actual charges on the atoms in a molecule, it is sometimes convenient to consider them as if they were real. For example, consider formula (2). Although it satisfies the octet rule, the central fluorine atom is assigned a formal charge of +1. Fluorine is the most reactive nonmetal and so gains electrons instead of losing them. Consequently, the formal charge of +1 on the fluorine atom is not chemically reasonable, and so we reject formula (2). We predict correctly, then, that the actual structure of OF_2 is represented by formula (1), with the oxygen atom being central. In general, the Lewis formula with the lower formal charges represents the preferred (lowest energy) Lewis formula.

Problems 10-37 and 10-38 ask you to use formal charges to choose between Lewis formulas.

10-9. THE OCTET RULE FAILS FOR SPECIES WITH AN ODD NUMBER OF ELECTRONS

As useful as the octet rule is, there are numerous cases in which it is not valid. For example, consider the stable molecule nitrogen oxide, NO. The electron-dot formulas of nitrogen and oxygen are

$$\cdot\ddot{\underset{\cdot}{N}}\cdot \qquad \text{and} \qquad \cdot\ddot{\underset{..}{O}}\cdot$$

If we try to write a Lewis formula for NO, we find that it is not possible to satisfy the octet rule. The best that we can do is

$$\ddot{\underset{\cdot}{\text{N}}}\!:\,:\ddot{\underset{\cdot\cdot}{\text{O}}} \qquad \text{or} \qquad \overset{\ominus}{\ddot{\underset{\cdot\cdot}{\text{N}}}}\!:\,:\overset{\oplus}{\ddot{\underset{\cdot}{\text{O}}}}$$

The difficulty here is that the total number of valence electrons is an odd number (11), and so it is not possible to pair up all the electrons as we have been doing. A species in which it is not possible to pair up all the electrons is called a *free radical*. Because of the unpaired electron, free radicals are usually very reactive.

A free radical is a species with an odd number of electrons.

Another example of a free radical is chlorine dioxide, ClO_2. The chlorine atom has seven valence electrons, and each oxygen atom has six. Therefore, ClO_2 has an odd number (19) of valence electrons. A Lewis formula for ClO_2 is

$$\ominus\!:\ddot{\underset{\cdot\cdot}{\text{O}}}:\overset{\oplus}{\ddot{\underset{\cdot\cdot}{\text{Cl}}}}:\ddot{\underset{\cdot\cdot}{\text{O}}}\cdot \qquad (1)$$

An equally satisfactory Lewis formula is obtained by writing the odd electron on the other oxygen atom:

$$\cdot\ddot{\underset{\cdot\cdot}{\text{O}}}:\overset{\oplus}{\ddot{\underset{\cdot\cdot}{\text{Cl}}}}:\ddot{\underset{\cdot\cdot}{\text{O}}}\!:\ominus \qquad (2)$$

Another resonance form of ClO_2 is

Chlorine dioxide is a yellow to reddish yellow gas with an unpleasant odor similar to that of chlorine. It reacts explosively with many substances.

$$\ominus\!:\ddot{\underset{\cdot\cdot}{\text{O}}}\text{—}\overset{(+2)}{\dot{\underset{\cdot\cdot}{\text{Cl}}}}\text{—}\ddot{\underset{\cdot\cdot}{\text{O}}}\!:\ominus \qquad (3)$$

We judge resonance form (3) to be less important than the other two because of the higher formal charges. Formulas (1) and (2) for ClO_2 are the major contributing resonance formulas, and the ClO_2 free radical is viewed as a superposition of the two. As expected, the Cl—O bonds in ClO_2 are equivalent.

The molecules NO and ClO_2 are free radicals. They have an odd number of electrons and so cannot satisfy the octet rule. There is also a class of compounds that have an even number of outer electrons but do not have enough electrons to form octets about each atom. Compounds of beryllium and boron serve as particularly important examples of *electron deficiency*. Consider the molecule BeH_2. The electron-dot formulas for Be and H are

$$\text{H}\cdot \qquad \text{and} \qquad \cdot\text{Be}\cdot$$

A Lewis formula for BeH_2 is

$$\text{H—Be—H}$$

The beryllium atom is four electrons short of satisfying the octet rule. Electron-deficient compounds are usually highly reactive species. For example, the electron-deficient compound BF_3 readily reacts with NH_3 to form H_3NBF_3:

$$H_3N: + BF_3 \rightarrow H_3\overset{\oplus}{N}-\overset{\ominus}{B}F_3$$

The lone electron pair in NH_3 can be shared between the nitrogen atom and the boron atom so that the octet rule is satisfied. A covalent bond that is formed when one atom contributes both electrons is called a *coordinate-covalent bond.* The product of the above reaction, H_3NBF_3, is called a *donor-acceptor complex.*

Example 10-11: Suggest a Lewis formula for BF_3 that satisfies the octet rule. Give a reason it can be rejected.

Solution: Each fluorine atom has 7 valence electrons and the boron atom has 3, for a total of 24. A Lewis formula for BF_3 using 12 electron pairs is

$$:\ddot{F}-\overset{\ominus}{B}(=\overset{\oplus}{F})-\ddot{F}:$$

The formal charge of +1 on the very reactive nonmetallic fluorine atom can be used to decide that this Lewis formula is less favorable than the formula

$$:\ddot{F}-B(-\ddot{F}:)-\ddot{F}:$$

for which the formal charges are all zero.

10-10. ATOMS IN THE THIRD ROW OF THE PERIODIC TABLE CAN ACCOMMODATE MORE THAN EIGHT ELECTRONS IN THEIR VALENCE SHELLS

The most common exceptions to the octet rule occur in molecules consisting of atoms found below the second row of the periodic table. For example, the atoms in the third row of the periodic table satisfy the octet rule by simulating the argon electron configuration, $3s^23p^6$. A typical molecule in this case is phosphorus trichloride, PCl_3, whose Lewis formula is

The phosphorus atom has a set of empty 3*d* orbitals, and these orbitals can be used for sharing electrons with other atoms. The 3*d* orbitals allow phosphorus to have more than eight electrons surrounding it, and thus phosphorus can form more than four bonds. The gaseous compound phosphorus pentachloride, PCl_5, is formed by the reaction

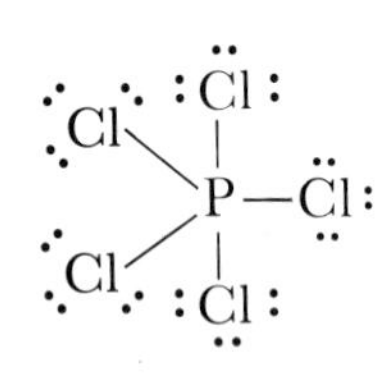

$$PCl_3(g) + Cl_2(g) \xrightarrow{300^\circ C} PCl_5(g)$$

The Lewis formula for PCl_5 is shown in the margin.

In a Lewis formula such as that for PCl_5, we say that the phosphorus atom has an *expanded valence shell.* Phosphorus expands its valence shell by using its *d* orbitals. It is not possible for the atoms in the second row of the periodic table to expand their valence shells beyond eight electrons because second-row elements complete the L shell when they satisfy the octet rule. Second-row elements would have to use orbitals in the M shell to accommodate more electrons, and the energies of the orbitals in the M shell are much higher than those of the orbitals in the L shell. Thus, although PF_5 has been synthesized, NF_5 has never been observed.

NCl_3 is a yellow, thick oily liquid; NCl_5 has never been made.

Sulfur is able to expand its shell to form compounds such as sulfur tetrafluoride, SF_4, and sulfur hexafluoride, SF_6. The Lewis formulas for these molecules are

Sulfur hexafluoride is one of the most dense gases, having a density about five times as great as air.

Example 10-12: Two Lewis formulas for SO_2Cl_2 are

(1) (2)

Can you present an argument for selecting one over the other?

SO_2Cl_2 is called sulfuryl chloride. It is a colorless liquid with a very pungent odor. Its vapor is corrosive to human skin and mucous membranes.

Solution: Lewis formula (1) involves an expanded octet on sulfur and zero formal charges on all the atoms. Lewis formula (2) obeys the octet rule, but the electron distribution gives rise to nonzero formal charges on the sulfur atom and the two oxygen atoms. Therefore, we can argue that the formula (2) represents an electronic distribution that requires an additional energy to separate the formal charges and so is less favorable than the formula (1), in which all the formal charges are zero.

10-11. MOST CHEMICAL BONDS ARE INTERMEDIATE BETWEEN PURELY IONIC AND PURELY COVALENT

Although we have discussed ionic bonds and covalent bonds in this and the previous chapter as distinct cases, most chemical bonds are neither purely ionic nor purely covalent, but are intermediate between the two. The bond in HCl serves as a good example to discuss this point.

When we introduced the concept of formal charge, we *arbitrarily* assigned one of the electrons in the covalent bond to each atom. The formal charges of H and Cl in HCl are zero. We emphasize here that this is a formal and arbitrary, but useful, procedure. We are tacitly assuming that the electrons in the covalent bond are shared equally by the hydrogen atom and the chlorine atom. We know, however, that different isolated atoms have different ionization energies and different electron affinities. It seems reasonable that different atoms attract electrons differently even if they are bonded to each other by a covalent bond. We can make this discussion quantitative by introducing the concept of *electronegativity*. Electronegativity is a measure of the force with which an atom attracts the electrons in its covalent bonds to other atoms.

Figure 10-5 gives the electronegativities of the elements according to a procedure devised by Linus Pauling. The larger the electronegativity, the greater the attraction of the atom for the electrons in its covalent bonds. Note that electronegativities increase from left to right across the short (second and third) rows of the periodic table. This left-to-right increase in electronegativities reflects the increasingly nonmetallic nature of the elements toward the right-hand side of the table. Note also that electronegativities decrease as one goes down a column. The reason for the top-to-bottom decrease in electronegativities is that the nuclear attraction of the outer electrons decreases as the size of the atom increases. The periodic trend of electronegativity is sketched in Figure 10-6. Note that fluorine is the most electronegative atom and that cesium and francium are the least electronegative. The order for the atoms with the greatest electronegativities is $F > O > N \approx Cl > C \approx S > H \approx P$.

1 H 2.1																	2 He -
3 Li 1.0	4 Be 1.5											5 B 1.9	6 C 2.5	7 N 3.0	8 O 3.5	9 F 4.0	10 Ne -
11 Na 0.9	12 Mg 1.2											13 Al 1.5	14 Si 1.8	15 P 2.1	16 S 2.5	17 Cl 3.0	18 Ar -
19 K 0.8	20 Ca 1.0	21 Sc 1.3	22 Ti 1.5	23 V 1.6	24 Cr 1.6	25 Mn 1.5	26 Fe 1.8	27 Co 1.8	28 Ni 1.8	29 Cu 1.9	30 Zn 1.5	31 Ga 1.6	32 Ge 1.8	33 As 2.0	34 Se 2.4	35 Br 2.8	36 Kr -
37 Rb 0.8	38 Sr 1.0	39 Y 1.2	40 Zr 1.4	41 Nb 1.6	42 Mo 1.8	43 Tc 1.9	44 Ru 2.2	45 Rh 2.2	46 Pd 2.2	47 Ag 1.7	48 Cd 1.4	49 In 1.7	50 Sn 1.8	51 Sb 1.9	52 Te 2.1	53 I 2.5	54 Xe -
55 Cs 0.7	56 Ba 0.9	57-71 1.1-1.2	72 Hf 1.3	73 Ta 1.5	74 W 1.7	75 Re 1.9	76 Os 2.2	77 Ir 2.2	78 Pt 2.2	79 Au 2.4	80 Hg 1.9	81 Tl 1.8	82 Pb 1.8	83 Bi 1.8	84 Po 2.0	85 At 2.2	86 Rn -
87 Fr 0.7	88 Ra 0.9	89 Ac 1.1	90 Th 1.3	91 Pa 1.5	92 U 1.7	93-103 Np-Lr 1.3											

Figure 10-5 Electronegativities of the elements (Pauling). Note that the electronegativities of the elements in the second and third rows increase from left to right, whereas they generally decrease from top to bottom in a given column.

It is the *difference* in electronegativities of the two atoms in a covalent bond that determines how the electrons in the bond are shared. If the electronegativities are the same, then the bond is shared equally and it is a *pure covalent bond*, or a *nonpolar bond*. Equal sharing of bonding electrons occurs in homonuclear diatomic molecules. If the electronegativities of the two atoms differ, then the bond is not shared equally and is said to be a *polar bond*. The extreme case of a polar bond occurs when the difference in electronegativities is large, say greater than about 1.7. For such a case, the electron pair ends up completely on the more electronegative atom, giving a *pure ionic bond*.

A polar bond can be illustrated by HCl. The electronegativity of H is 2.1 and that of Cl is 3.0, and the difference between them is 0.9. Thus the electrons in the bond are not shared equally. Because the electronegativity of Cl is greater than that of H, the chlorine atom attracts the electron pair more strongly than does the hydrogen atom. The bonding electrons are shifted a little toward the chlorine atom and so it acquires a *partial* negative charge, which leaves the hydrogen with a partial positive charge. We indicate partial charges by the Greek letter delta (δ) and we write

$$\overset{\delta+}{\text{H}}-\overset{\delta-}{\text{Cl}}$$

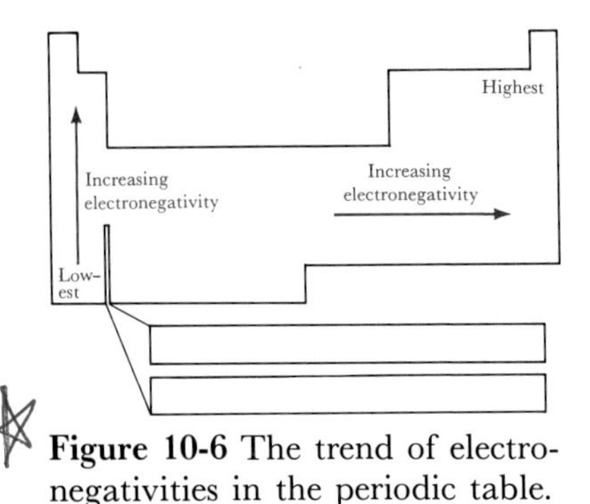

Figure 10-6 The trend of electronegativities in the periodic table.

It is important to understand that $\delta+$ or $\delta-$ represents only the *fraction* of an electronic charge that results from the unequal sharing of the electrons in the covalent bond. The numerical value of δ is of no importance to us. It denotes only that the hydrogen atom is slightly positively charged and that the chlorine atom is slightly negatively charged. From a quantum theoretical point of view, δ represents the fact that the two electrons in the covalent bonds are more likely to be found near the chlorine atom than near the hydrogen atom.

Example 10-13: Describe the charge distribution in the interhalogen compound ClF.

Solution: According to Figure 10-5, the electronegativity of F is 4.0 and that of Cl is 3.0, and so the Cl—F bond is a polar bond. The electron pair is somewhat more likely to be found near the fluorine atom than near the chlorine atom. Thus the fluorine atom has a slightly negative charge, $\delta-$, and the chlorine atom has a slightly positive charge, $\delta+$. We can represent this polar bond by writing

$$\overset{\delta+}{\text{Cl}}-\overset{\delta-}{\text{F}}$$

10-12. LEWIS FORMULAS DO NOT INDICATE THE SHAPES OF MOLECULES

A quantity that is a measure of the polarity of a bond is its *dipole moment.* The dipole moment of a bond customarily is represented as an arrow ($+\!\!\!\longrightarrow$) pointing along the bond from $\delta+$ to $\delta-$. The cross on the tail of the arrow indicates the positive charge. For HCl and ClF we have

$$\overset{+\!\!\!\longrightarrow}{\text{H—Cl}} \quad \text{and} \quad \overset{+\!\!\!\longrightarrow}{\text{Cl—F}}$$

This notation indicates the direction of the dipole moment. The magnitude of the dipole moment is the product of the length of the bond and the amount of net electric charge that is separated. The dipole moments of the hydrogen halides are given in Table 10-4. Note the dependence of the dipole moment on the electronegativity difference. The larger the difference in electronegativity, the larger the dipole moment.

Dipole moments are vector quantities.

A dipole moment is represented by an arrow because it is a quantity that has both magnitude *and* direction. A quantity that has magnitude *and* direction is called a *vector*. A familiar example of a vector is a force. A force must be described by both the direction in which it is applied *and* its magnitude. Vectors have the important property that they cancel if applied with the same magnitude in opposite directions. In a stalemated tug of war, both teams are pulling with

Table 10-4 The dipole moments of the hydrogen halides

Molecule	*Electronegativity difference*	*Dipole moment*[a]$/10^{-30}$ *C·m*
HF	1.9	6.36
HCl	0.9	3.43
HBr	0.7	2.63
HI	0.4	1.27

[a]The units of dipole moment are charge times distance, or coulombs times meters (C·m) in SI units.

the same magnitude of force but in opposing directions. The net result is an effective cancellation. This can be illustrated pictorially by

$$\longleftarrow\!\bullet\!\longrightarrow \;=\; \text{no net force}$$

If the forces are not applied in opposing directions, then there is a resultant net force:

It is not necessary for us to be able to calculate the magnitude of the net force, but only to see pictorially that the direction of the net force is as shown above.

Dipole moments have the same properties as forces. In CO_2, :O=C=O:, each C—O bond is polar but the bond dipole moments are directed in opposite directions:

$$\overset{\longleftarrow\!+\;\;+\!\longrightarrow}{:\!O\!=\!C\!=\!O\!:}$$

The bond dipole moments cancel exactly, and the CO_2 molecule has no dipole moment; the molecule is said to be nonpolar. Dipole moments can be measured experimentally, and it is found that CO_2 has zero dipole moment; therefore CO_2 must be a linear molecule (Figure 10-7).

Let's consider H_2O as another example. The O—H bonds are polar, and we can display this by

$$\overset{+\!\longrightarrow\;\;\longleftarrow\!+}{H\!-\!\ddot{O}\!-\!H}$$

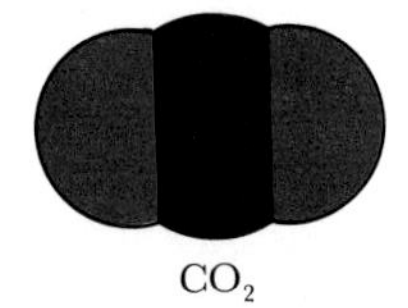

Figure 10-7 A molecular model of carbon dioxide, a linear molecule. The O—C—O bond angle is 180°.

This diagram implies that water is a nonpolar molecule. In fact, H_2O has a large dipole moment. The discrepancy is due to our assumption that H_2O is a linear molecule. We see in the next chapter that the water molecule is bent. The H—O—H bond angle is not 180°; rather it is about 104°. A molecular model of H_2O is shown in Figure 10-8. We can illustrate the bent structure by

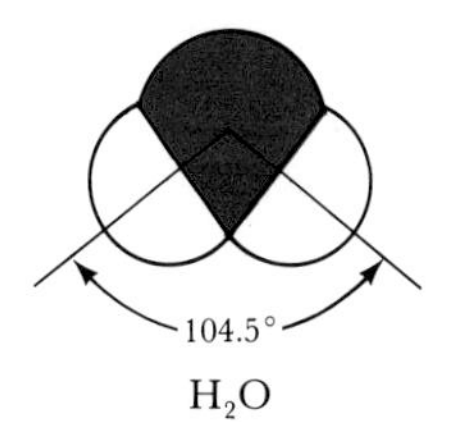

Figure 10-8 A molecular model of water, a nonlinear molecule. The H—O—H bond angle is 104.5°.

O
H 104.5° H

Thus the net dipole moment in H_2O has the orientation

The dipole moment of a molecule is the net result of its bond dipole moments.

O
H H

There is an important lesson for us here. The Lewis formulas that we have learned to write in this chapter suggest which atoms are bonded to which in the molecule. Although the formulas are very useful in this regard, they are *not* meant to indicate or suggest the three-dimensional arrangement of the atoms in a molecule. In the next chapter we learn some simple, useful rules for predicting the shapes of molecules using Lewis formulas.

SUMMARY

In a Lewis formula, a covalent bond is represented as a pair of electrons shared between two atoms. A Lewis formula shows the arrangement of the valence electrons in a molecule and suggests which atoms are bonded to which. The octet rule is a useful rule for writing Lewis formulas, especially for compounds involving the elements carbon, nitrogen, oxygen, and fluorine.

Although there are a number of exceptions to the octet rule, it is still a very useful guide for writing Lewis formulas for compounds. Another aid for writing Lewis formulas is the idea of formal charge. Formal charges can be assigned to the various atoms in a Lewis formula and used to select preferred Lewis formulas. Sometimes it is possible to write two or more Lewis formulas for a molecule without altering the position of the atoms. When this is possible, each individual formula is said to be a resonance form and the accepted Lewis formula is an average or a superposition of the individual formulas.

Most chemical bonds are neither purely ionic nor purely covalent. The degree of ionic character in a bond can be estimated by means of the electronegativity. When the electronegativities of the two atoms joined by a covalent bond are different, the bond is said to be polar.

Lewis formulas indicate not the shape of a molecule but only the bonding within a molecule.

TERMS YOU SHOULD KNOW

valence electron 365
covalent bond 366
Lewis formula 366
lone electron pair 366
lone pair 366
molecular crystal 366
bond length 368
octet rule 368
double bond 373
triple bond 373
formal charge 377
resonance form 379
resonance 379
free radical 382
electron deficiency 382
coordinate-covalent bond 383
donor-acceptor complex 383
expanded valence shell 384
electronegativity 385
pure covalent bond 386
nonpolar bond 386
polar bond 386
pure ionic bond 386
dipole moment 387
vector 387

EQUATIONS YOU SHOULD KNOW HOW TO USE

$$\begin{pmatrix}\text{formal charge}\\ \text{on an atom in}\\ \text{a Lewis formula}\end{pmatrix} = \begin{pmatrix}\text{total number of}\\ \text{valence electrons}\\ \text{in the free atom}\end{pmatrix} - \begin{pmatrix}\text{total number}\\ \text{of lone-}\\ \text{pair electrons}\end{pmatrix} - \frac{1}{2}\begin{pmatrix}\text{total number}\\ \text{of shared}\\ \text{electrons}\end{pmatrix} \quad (10\text{-}1)$$

PROBLEMS

INORGANIC COMPOUNDS INVOLVING SINGLE BONDS

10-1. Draw the Lewis formula for F_2 and I_2.

10-2. Draw the Lewis formula for

(a) SCl_2
(b) $GeCl_4$
(c) $AsBr_3$

10-3. Draw the Lewis formula for

(a) PCl_3
(b) SiF_4
(c) NI_3

10-4. Tetrafluorohydrazine, N_2F_4, is a colorless liquid that is used as rocket fuel. Draw the Lewis formula for tetrafluorohydrazine.

10-5. Phosphorus, arsenic, and antimony form compounds similar to NH_3:

Formula	*Name*	*Characteristics*
PH_3	phosphine	very poisonous gas with the odor of decaying fish
AsH_3	arsine	very poisonous gas with the odor of strong garlic
SbH_3	stibine	intensely poisonous gas with its own characteristic, disagreeable odor

Draw the Lewis formula for each of these compounds.

10-6. Draw the Lewis formula for

(a) HBr
(b) SiH_4
(c) NH_2OH

10-7. The Group 6 elements sulfur, selenium, and tellurium form the compounds H_2S, H_2Se, and H_2Te. Each of these compounds is a poisonous gas with a strong, offensive odor. Draw the Lewis formula for each.

10-8. Titanium tetrachloride, $TiCl_4$, is a colorless liquid that fumes profusely when exposed to moist air. Draw a Lewis formula for titanium tetrachloride.

10-9. Pure hydrogen peroxide, H_2O_2, is a colorless liquid that is caustic to the skin, but a 3% aqueous solution is a bleaching agent. Draw the Lewis formula for H_2O_2.

10-10. A hydroxide ion, OH^-, results when a proton is removed from H_2O. The ion that results when a proton is removed from NH_3 is called an amide ion, NH_2^-. Draw the Lewis formula for an amide ion. Name the following ionic compounds: $NaNH_2$ and $Ba(NH_2)_2$.

10-11. Nitrogen trichloride is a yellow, thick, oily liquid that has a pungent odor and is very unstable. Draw the Lewis formula for nitrogen trichloride.

10-12. Solid sulfur consists of eight-membered rings of sulfur atoms. Draw the Lewis formula for S_8.

ORGANIC COMPOUNDS INVOLVING SINGLE BONDS

10-13. Draw the Lewis formula for

(a) methane, CH_4
(b) fluoromethane, CH_3F
(c) aminomethane, CH_3NH_2

10-14. Draw the Lewis formula for

(a) methyl mercaptan, CH_3SH
(b) dimethyl ether, CH_3OCH_3
(c) trimethyl amine, $N(CH_3)_3$

10-15. Hydrocarbons are compounds that contain only hydrogen and carbon. The simplest hydrocarbon is methane, CH_4, but there are a great many others. In one type of hydrocarbon, the carbon atoms are connected to each other in a chain. Draw the Lewis formulas for the following "straight-chain" hydrocarbons:

(a) propane, C_3H_8
(b) butane, C_4H_{10}
(c) octane, C_8H_{18}

Have you heard of any of these hydrocarbons?

10-16. Another type of hydrocarbon is the branched-chain hydrocarbon, in which the carbon atoms are not connected to each other in a straight chain. Draw the Lewis formula for the following branched-chain hydrocarbons:

(a) isobutane, $(CH_3)_3CH$
(b) neopentane, $(CH_3)_4C$
(c) isopentane, $C_2H_5CH(CH_3)_2$

10-17. An alcohol is an organic compound containing an OH group. Draw the Lewis formula for the following alcohols:

(a) ethyl alcohol, CH_3CH_2OH
(b) *n*-propyl alcohol, $CH_3CH_2CH_2OH$
(c) isopropyl alcohol, $(CH_3)_2CHOH$

Rubbing alcohol is a 70% by volume aqueous solution of isopropyl alcohol.

10-18. Carbon, hydrogen, and chlorine form a set of compounds:

CCl_4	carbon tetrachloride (solvent)
$CHCl_3$	chloroform (cleaning agent)
CH_2Cl_2	methylene chloride (degreasing and cleaning fluid
CH_3Cl	methyl chloride (refrigerant)

Draw the Lewis formula for each of these compounds.

10-19. Draw all possible Lewis formulas for $C_2H_4Cl_2$.

10-20. A methoxide ion, CH_3O^-, results when a proton is removed from methanol, CH_3OH. Draw the Lewis formula for CH_3O^-. Name the compounds $KOCH_3$ and $Al(OCH_3)_3$.

MULTIPLE BONDS

10-21. Which of the following species have a triple bond:

(a) H_2CO (c) C_2H_4
(b) C_2^{2-}

10-23. Draw the Lewis formula for

(a) acrylonitrile, CH_2CHCN
(b) nitrous acid, HNO_2
(c) silicon dioxide, SiO_2

10-25. Formic acid is a colorless liquid with a penetrating odor. It is the irritating ingredient in the bite of ants. Its chemical formula is HCOOH. Draw the Lewis formula for formic acid.

10-27. Acetone is an organic compound widely used in the chemical industry as a solvent, for example, in paints and varnishes. You may be familiar with its sweet odor because it is used as a fingernail-polish remover. Its chemical formula is CH_3COCH_3. Draw the Lewis formula for acetone.

10-29. Propylene is the thirteenth ranked industrial chemical produced in the United States and is used in the synthesis of a wide range of compounds, such as polypropylene, glycerol, and isopropyl alcohol (rubbing alcohol). Its chemical formula is CH_3CHCH_2. Draw the Lewis formula for propylene.

10-31. Diazine, N_2H_2, is a yellow compound that is unstable above $-180°C$, decomposing to N_2, H_2, and N_2H_4. Draw the Lewis formula for diazine.

10-22. Draw the Lewis formula for

(a) Cl_2CO (b) HCN (c) HOOCCOOH

10-24. Vinyl chloride is an important industrial chemical used in the manufacture of polyvinyl chloride and other polymers. Its chemical formula is C_2H_3Cl. Draw the Lewis formula for vinyl chloride.

10-26. Acetic acid is a familiar organic acid; vinegar is a dilute aqueous solution of acetic acid. Its chemical formula is often written CH_3COOH. Draw the Lewis formula for acetic acid.

10-28. Acetaldehyde is a member of the group of organic compounds called aldehydes. Formaldehyde, the simplest aldehyde, is a well-known preservative for biological specimens. The chemical formula for acetaldehyde is CH_3CHO. Draw its Lewis formula.

10-30. Draw all the possible Lewis formulas for butadiene, C_4H_6, which contains two double bonds.

10-32. Hydrazoic acid, HN_3, is a dangerously explosive, colorless liquid. Azides of heavy metals explode when struck sharply and are used in detonation caps. Draw the Lewis formula for hydrazoic acid (HN_3) and for the azide ion (N_3^-).

RESONANCE AND FORMAL CHARGES

10-33. Draw Lewis formulas for the resonance forms of the formate ion, $HCOO^-$. Indicate formal charges and discuss the bonding of this ion.

10-35. Draw Lewis formulas for three resonance forms of the carbonate ion, CO_3^{2-}. Indicate formal charges and discuss the bonding in this ion.

10-37. Laughing gas, an anesthetic and propellant in whipped cream dispensing cans, has the composition N_2O. Use Lewis formulas and formal charge considerations to predict which of the structures, NNO or NON, is more likely.

10-34. Draw Lewis formulas for the resonance forms of the acetate ion, CH_3COO^-. Indicate formal charges and discuss the bonding in this ion.

10-36. Draw Lewis formulas for two resonance forms of ozone, O_3. Indicate formal charges and discuss the bonding in this molecule.

10-38. Use formal charges to predict the arrangement of the atoms in NOCl.

10-39. Draw the Lewis formula or formulas for each of the following species (give resonance structures where appropriate):

(a) CS_3^{2-}
(b) $C_2O_4^{2-}$

10-40. Nitrogen and oxygen form a number of compounds:

Formula	*Name*	*Color*	*Physical state at 25°C, 1 atm*
N_2O	dinitrogen oxide	colorless	gas
NO	nitrogen oxide	colorless	gas
N_2O_3	dinitrogen trioxide	dark blue	gas
NO_2	nitrogen dioxide	brown	gas
N_2O_4	dinitrogen tetroxide	colorless	gas
N_2O_5	dinitrogen pentoxide	white	solid

Draw the Lewis formula for each of these nitrogen oxides.

OCTET RULE VIOLATIONS

10-41. Which of the following species contain an odd number of electrons:

(a) NO_2
(b) CO
(c) O_3^-
(d) O_2^-
(e) SO_2

Write a Lewis formula for each of these species.

10-42. Which of the following species contain an odd number of electrons:

(a) BrO_3
(b) SO_3
(c) HNO
(d) HO_2
(e) SO_4^-

Write a Lewis formula for each of these species.

10-43. Nitrosamines are carcinogens that are found in tobacco smoke. They can also be formed in the body from the nitrites and nitrates used to preserve processed meats, especially bacon and sausage. The simplest nitrosamine is methylnitrosamine, H_3CNNO. Draw the Lewis formula for this molecule. Is it a free radical?

10-44. The oxides of nitrogen, often designated as NO_x, are formed from N_2 and O_2 (from the air) in the internal combustion engine. The oxides NO, NO_2, and nitric acid, HNO_3, which is formed by the reaction

$$HO + NO_2 \rightarrow HNO_3$$

are important species in the formation of smog (Interchapter E). Draw the Lewis formulas for NO, NO_2, HO, and HNO_3. Identify which of these species are free radicals.

10-45. Many free radicals combine to form molecules that do not contain any unpaired electrons. The driving force for the radical-radical combination reaction is the formation of a new electron-pair bond. Draw Lewis formulas for the reactant and product species in the following reactions:

(a) $CH_3(g) + CH_3(g) \rightarrow H_3CCH_3(g)$
(b) $N(g) + NO(g) \rightarrow NNO(g)$

10-46. Many free radicals combine to form molecules that do not contain any unpaired electrons. The driving force for the radical-radical combination reaction is the formation of a new electron-pair bond. Draw Lewis formulas for the reactant and product species in the following reactions:

(a) $2HO(g) \rightarrow H_2O_2(g)$
(b) $Cl(g) + CH_3(g) \rightarrow CH_3Cl(g)$

GENERAL LEWIS FORMULAS

10-47. Draw the Lewis formula for

(a) PCl_6^-
(b) I_3^-
(c) SiF_6^{2-}

10-48. Draw the Lewis formula for

(a) sulfate ion, SO_4^{2-}
(b) phosphate ion, PO_4^{3-}
(c) perchlorate ion, ClO_4^-

Show the formal charges and discuss the bonding in each ion.

10-49. Draw a Lewis formula for each of the following compounds of xenon:

Compound	*Form at 25°C*	*Melting point/°C*
XeF_2	colorless crystals	129
XeF_4	colorless crystals	117
XeF_6	colorless crystals	50
$XeOF_4$	colorless liquid	−46
XeO_2F_2	colorless crystals	31

10-50. The halogens form a number of interhalogen compounds. For example, chlorine pentafluoride, ClF_5, can be prepared by the reaction

$$KCl(s) + 3F_2(g) \rightarrow \underset{\text{colorless}}{ClF_5(g)} + KF(s)$$

The halogen fluorides are very reactive, combining explosively with water, for example. Draw the Lewis formula for each of the following halogen fluoride species:

(a) ClF_5
(b) IF_3
(c) IF_7
(d) IF_4^+
(e) BrF_6^+

10-51. Draw the Lewis formula for

(a) tetrafluoroammonium ion, NF_4^+
(b) tetrafluorochlorinium ion, ClF_4^+
(c) phosphonium ion, PH_4^+
(d) hexafluoroarsenate ion, AsF_6^-
(e) tetrafluorobromate ion, BrF_4^-

10-52. Draw the Lewis formula for each of the following oxychlorine species:

(a) perchlorate ion, ClO_4^-
(b) chlorine oxide, ClO
(c) chlorate ion, ClO_3^-
(d) chlorine dioxide, ClO_2
(e) hypochlorite ion, ClO^-

10-53. Draw the Lewis formula for the following acids:

(a) $HClO_3$
(b) HNO_2
(c) HIO_4
(d) $HBrO_2$
(e) H_2SO_3

10-54. Draw the Lewis formula for each of the following oxyacids of sulfur:

(a) sulfuric acid, H_2SO_4
(b) thiosulfuric acid, $H_2S_2O_3$
(c) disulfuric acid, $H_2S_2O_7$
(d) dithionic acid, $H_2S_2O_6$ (has an S—S bond)
(e) peroxydisulfuric acid, $H_2S_2O_8$ (has an O—O bond)

10-55. The Group 6 elements form a number of halides. Some of them are

(a) SCl_2 (d) S_2F_4
(b) SCl_4 (e) Se_2Br_2
(c) SeF_6

Draw a Lewis formula for each of these halides.

10-56. Phosphorus forms a number of oxohalides, X_3PO, in which X may be F, Cl, or Br. The commonest, phosphoryl chloride, is obtained by the reaction

$$2PCl_3(g) + O_2(g) \rightarrow 2Cl_3PO(g)$$

Draw the Lewis formula for the phosphoryl halides.

10-57. The dichromate ion, $Cr_2O_7^{2-}$, has no metal-metal and no oxygen-oxygen bonds. Draw a Lewis formula for $Cr_2O_7^{2-}$ (consider Cr to have six valence electrons in this ion).

10-58. In the P_4O_6 molecule each phosphorus atom is bonded to three oxygen atoms and each oxygen atom is bonded to two phosphorus atoms. Draw a Lewis formula for P_4O_6.

DIPOLE MOMENTS

10-59. Draw the Lewis formula for bromine chloride and indicate its dipole moment.

10-60. Arrange the following groups of molecules in order of increasing dipole moment:

(a) HCl HF HI HBr
(b) PH_3 NH_3 AsH_3 (tripod-shaped molecules)
(c) Cl_2O F_2O H_2O (bent molecules)
(d) ClF_3 BrF_3 IF_3 (T-shaped molecules)
(e) H_2O H_2S H_2Te H_2Se (bent molecules)

10-61. Describe the charge distribution in

(a) nitrogen trifluoride, NF_3
(b) oxygen difluoride, OF_2
(c) oxygen dibromide, OBr_2

10-62. Describe the charge distribution in

(a) hydrogen fluoride, HF
(b) phosphine, PH_3
(c) hydrogen sulfide, H_2S

Regular geometric shapes, like this crystal of the mineral franklinite, occur often in nature. In this chapter we shall learn that many molecules have regular geometric shapes.

11 / The Shapes of Molecules

Molecules come in a variety of shapes. In this chapter we devise a set of simple, systematic rules that allow us to predict the shapes of thousands of molecules. These rules are based upon the Lewis formulas that we developed in Chapter 10 and are collectively called the valence-shell electron-pair repulsion (VSEPR) theory. In spite of its rather imposing name, VSEPR theory is easy to understand, easy to apply, and reliable.

11-1. LEWIS FORMULAS DO NOT REPRESENT THE SHAPES OF MOLECULES

Lewis formulas indicate which atoms are bonded to which in a molecule but do not indicate the molecule's actual shape. Consider the molecule dichloromethane, CH_2Cl_2. One Lewis formula for dichloromethane is

```
         H
  ..     |     ..
 :Cl — C — Cl:        (1)
  ..     |     ..
         H
```

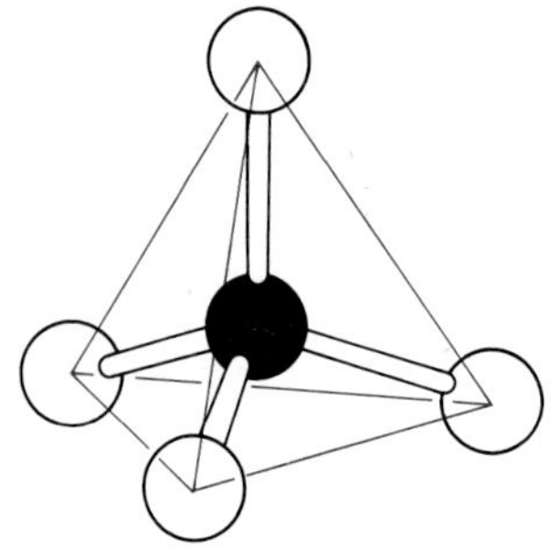

Figure 11-1 Each carbon-hydrogen bond in a methane molecule points toward the vertex of a regular tetrahedron. The positions of all four hydrogen atoms in CH_4 are equivalent by symmetry. All the H—C—H bond angles are the same, 109.5°.

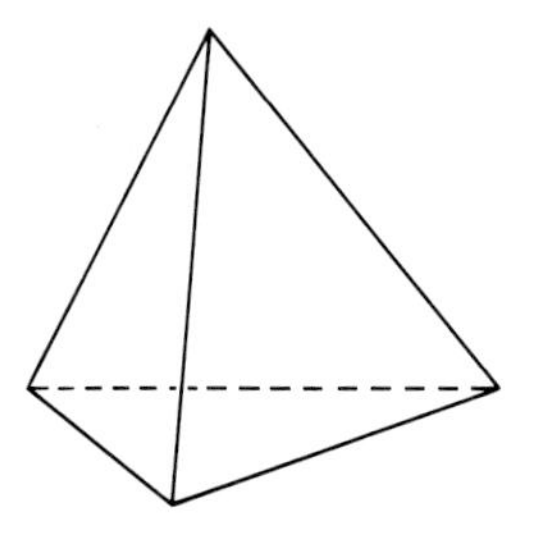

Figure 11-2 A regular tetrahedron is a symmetric body consisting of four equivalent vertices and four equivalent faces. Each face is an equilateral triangle. Note that a tetrahedron differs from the more familiar pyramid, which has a square base.

If we infer from this Lewis formula that dichloromethane is flat, or *planar,* then we must conclude that the Lewis formula

$$\begin{array}{ccc} & :\ddot{\text{Cl}}: & \\ & | & \\ \text{H}— & \text{C} & —\ddot{\text{Cl}}: \\ & | & \\ & \text{H} & \end{array} \qquad (2)$$

represents a different geometry for dichloromethane. In formula (1), the two chlorine atoms lie 180° apart, whereas in formula (2) they lie 90° apart. Molecules that have the same chemical formula, CH_2Cl_2 in this case, but different geometric arrangements, are called *geometric isomers.* Geometric isomers are different molecular species and so have different chemical and physical properties. For example, their boiling points differ, and so they can be separated by distillation.

Two isomers of dichloromethane have never been observed; therefore, our assumption that dichloromethane is a planar molecule is incorrect. The fact that there is only one kind of dichloromethane molecule suggests that the four bonds around the central atom are oriented such that there is only one distinct way of bonding two hydrogen atoms and two chlorine atoms to a central carbon atom. A geometric arrangement that explains why there are no geometric isomers of dichloromethane, as well as many other similar observations, was proposed by the Dutch chemist Jacobus van't Hoff and the French chemist Joseph Le Bel independently in 1874. They proposed that the four bonds about a central carbon atom are directed toward the corners of a regular *tetrahedron* (Figure 11-1). A regular tetrahedron is a four-sided figure that has four equivalent vertices and four identical faces, each of which is an equilateral triangle (Figure 11-2). Appendix D gives instructions for building a regular tetrahedron out of cardboard.

11-2. ALL FOUR VERTICES OF A REGULAR TETRAHEDRON ARE EQUIVALENT

You can see from a model or from Figures 11-1 and 11-2 that the four vertices of a tetrahedron are equivalent. Because of this there is only one way of bonding two hydrogen atoms and two chlorine atoms directly to a central carbon atom, in accord with the experimental fact that dichloromethane has no isomers.

The model in Figure 11-3 is called a *space-filling molecular model* and gives a fairly accurate representation of the angles between bonds and of the relative sizes of the atoms in a molecule. A less realistic model, but perhaps one in which the geometry is easier to see, is the *ball-and-stick molecular model* shown in Figure 11-4 (see also Figure 11-1).

In a tetrahedral molecule like methane, all the H—C—H bond angles are equal to 109.5°, which is called the *tetrahedral bond angle*. A carbon atom that is bonded to four other atoms is called a *tetravalent* carbon atom. The hypothesis of van't Hoff and Le Bel that the bonds of a tetravalent carbon atom are tetrahedrally oriented was the beginning of what is called *structural chemistry*, the area of chemistry in which the shapes and sizes of molecules are studied. Many experimental methods have been developed to determine molecular geometries. Most of the methods involve the interaction of electromagnetic radiation or electrons with molecules. Using such methods, it is possible to measure bond lengths and bond angles in molecules. It turns out that there is a great variety of molecular shapes. We saw in Chapter 10 that CO_2 is a linear molecule and that H_2O is bent. Methane is an example of a tetrahedral molecule. Some examples of other molecular geometries that have been determined experimentally are shown in Figure 11-5.

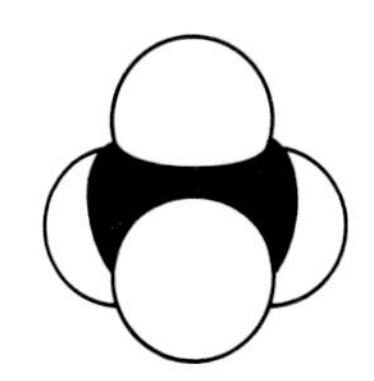

Figure 11-3 A space-filling molecular model of methane.

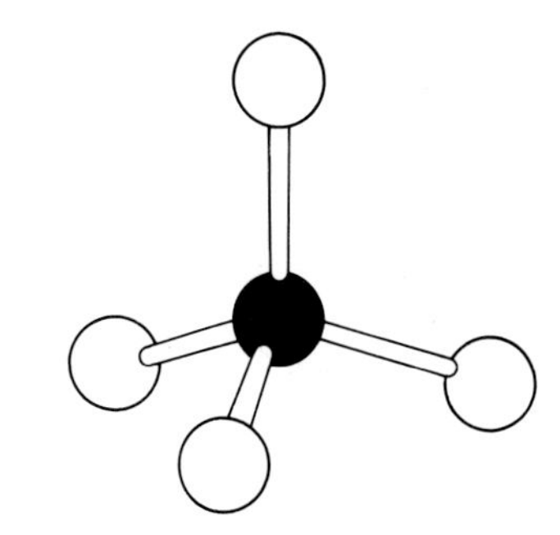

Figure 11-4 A ball-and-stick molecular model of methane.

11-3. VALENCE-SHELL ELECTRON-PAIR REPULSION THEORY IS USED TO PREDICT THE SHAPES OF MOLECULES

A set of rules has been devised that allows us to predict the shapes of the molecules shown in Figure 11-5 and the shapes of other molecules as well. The method is based upon the total number of bonds and

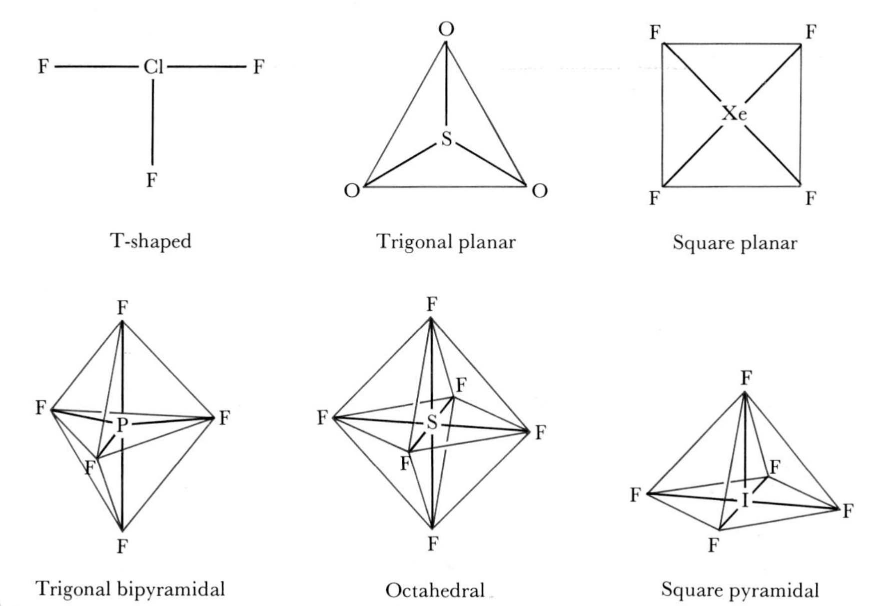

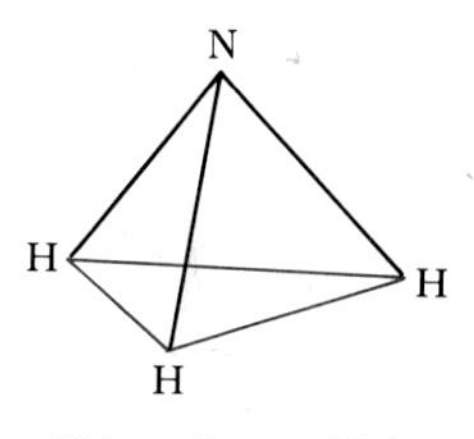

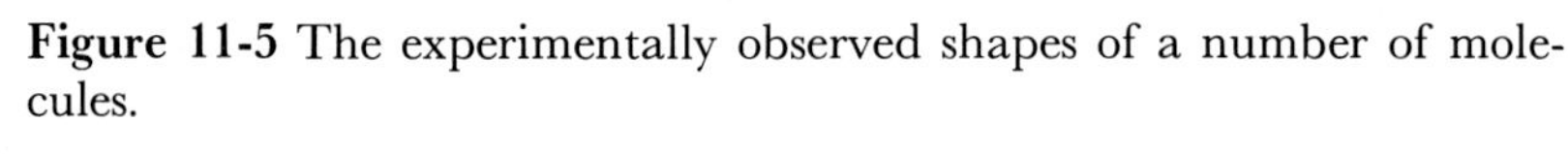

Figure 11-5 The experimentally observed shapes of a number of molecules.

valence electron pairs in the valence shell of the central atom in a molecule. The postulate is that the shape of a molecule is determined by the mutual repulsion of the electron pairs in the valence shell of the central atom, and so the method that we are about to describe is called the *valence-shell electron-pair repulsion theory,* or the *VSEPR theory.*

Consider the electron-deficient molecule $BeCl_2$. The Lewis formula for this molecule is

$$:\ddot{\underset{\cdot\cdot}{Cl}}—Be—\ddot{\underset{\cdot\cdot}{Cl}}:$$

Although the central beryllium atom has no lone electron pairs, it does have two covalent bonds and thus has two electron pairs in its valence shell. These valence-shell electron pairs repel each other and can minimize their mutual repulsion by being as far apart as possible. If we visualize the central beryllium atom as a sphere and the two valence-shell electron pairs as being on the surface of the sphere, then the two bonds minimize their mutual repulsion by being at opposite poles of the sphere. Thus the two bonds are on opposite sides of the central beryllium atom and the Cl—Be—Cl bond angle is 180°. The shape of a molecule is determined by the positions of the atomic nuclei in the molecule, and so we say that $BeCl_2$ is a linear molecule. This is in accord with experimental studies of $BeCl_2$. The positioning of the two valence-shell electron pairs on opposite sides of the central atom is shown in Figure 11-6(a).

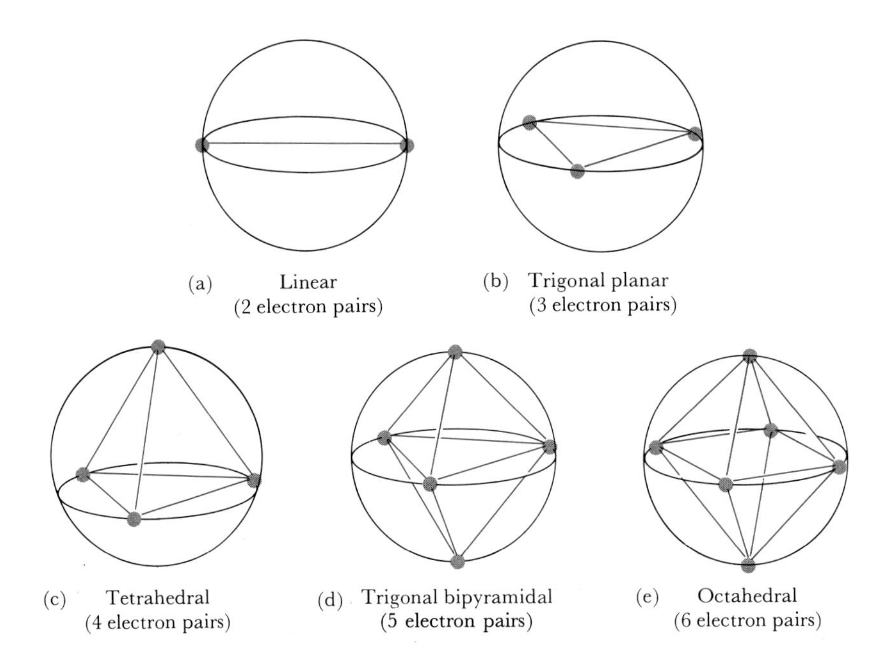

Figure 11-6 Sets of electron pairs (●) arranged on the surfaces of spheres so as to minimize the mutual repulsion between them. (a) Two electron pairs lie at opposite poles of a sphere. (b) Three electron pairs lie on an equator and enclose an equilateral triangle. (c) Four electron pairs enclose a regular tetrahedron. (d) Five electron pairs are arranged such that two lie at poles and the other three lie on the equator and enclose an equilateral triangle. (e) Six electron pairs lie at the vertices of a regular octahedron. (See Frontispiece.)

Example 11-1: Predict the shape of the covalent molecule zinc chloride, $ZnCl_2$, which is used in antiperspirants and as a wood preservative.

Solution: The outer electron configuration of Zn is $3s^2 3p^6 4s^2 3d^{10}$. The $n = 3$ shell is complete, and the two $4s$ electrons are valence electrons. The Lewis electron-dot formula for Zn is :Zn. The electron-dot formula for Cl is ·C̤̈l:, and so the Lewis formula for $ZnCl_2$ is

:C̤̈l—Zn—C̤̈l:

VSEPR theory predicts that the two electron pairs in the valence shell of the central zinc atom arrange themselves as in Figure 11-6(a), and therefore $ZnCl_2$ is a linear molecule.

Zinc chloride is an ionic solid at room temperature (20°C), but the vapor consists of individual zinc chloride molecules with the Lewis formula :C̤̈l—Zn—C̤̈l:. VSEPR theory applies only to individual molecules, as in the vapor state. Throughout this chapter we always refer to individual molecules in the gas phase even though the substance may be an ionic solid at room temperature.

Consider now a molecule with three electron pairs in the valence shell of the central atom. An example is the electron-deficient compound boron trifluoride, whose Lewis formula is

```
      ..
     :F:
      |
      B
     / \
   :F:  :F:
```

The three valence-shell electron pairs surrounding the boron atom can minimize their mutual repulsion by maximizing their mutual separation. This results in a planar triangular, or *trigonal planar,* structure (Figure 11-6b). We predict, then, that BF_3 is a planar, symmetric molecule with F—B—F bond angles equal to 120°. This is in complete agreement with experiment.

Boron trifluoride is a colorless gas with a pungent, suffocating odor. It acts as a catalyst for a number of reactions.

11-4. THE NUMBER OF VALENCE-SHELL ELECTRON PAIRS DETERMINES THE SHAPE OF A MOLECULE

An example of a molecule with four covalent bonds surrounding a central atom is methane. These four electron pairs minimize their mutual repulsion by pointing toward the vertices of a regular tetrahedron (Figure 11-6c). We see, then, that the tetrahedral geometry of methane is a result of the mutual repulsion of the four electron pairs making up its four covalent bonds. All the H—C—H bond angles in methane are equal to the tetrahedral bond angle, 109.5°.

Silane is a gas with a repulsive odor and is used in the preparation of extremely pure silicon for semiconductors (Interchapter H).

Example 11-2: Silicon lies below carbon in the periodic table and is also tetravalent. Predict the geometry of silane, SiH_4.

Solution: The Lewis formula for silane is

```
     H
     |
  H—Si—H
     |
     H
```

There are four valence-shell electron pairs about the central silicon atom, and so we predict correctly that silane is tetrahedral and that the H—Si—H bond angles are 109.5°. This is in excellent agreement with experiment.

The next example shows that VSEPR theory can be applied to molecular ions.

Example 11-3: Predict the geometry of the ammonium ion, NH_4^+.

Solution: The Lewis formula for NH_4^+ is

```
     H
     | ⊕
  H—N—H
     |
     H
```

Because the valence shell of the nitrogen atom contains a total of four electron pairs (four covalent bonds), we predict that the ammonium ion is tetrahedral. This is the observed structure of NH_4^+.

Many molecules have five electron pairs in the form of five covalent bonds in the valence shell of the central atom. An example is phosphorus pentachloride, PCl_5, whose Lewis formula is

```
        ..
       :Cl:
        |  .Cl:
  ..    | /
 :Cl—P
  ..    | \
        |  .Cl:
       :Cl:
        ..
```

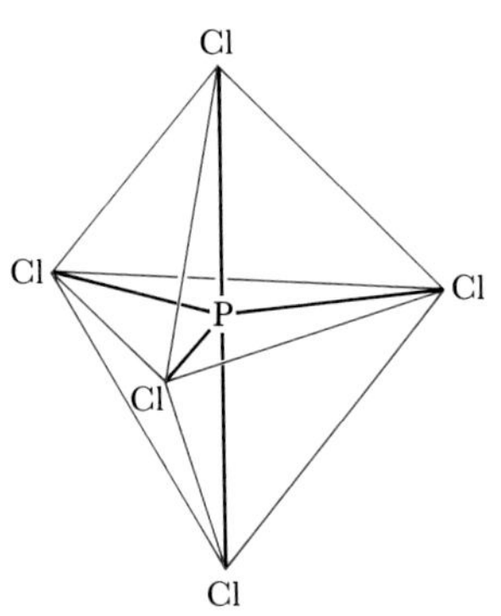

Figure 11-7 The shape of the phosphorus pentachloride molecule.

The five electron pairs in the valence shell of the phosphorus atom are positioned to minimize their mutual repulsion, outlining a *trigonal bipyramid* (Figures 11-6(d) and 11-7). Notice that the three vertices on the equator of the sphere in Figure 11-6(d) form an equilateral triangle and that the vertices lying at the poles lie above and below the center of the equilateral triangle. The five vertices of a trigonal bipyramid are *not* equivalent. The three vertices lying on the equator in Figure 11-6(d) are equivalent and are called *equatorial vertices;* the two vertices lying at the poles are equivalent and are called *axial vertices.*

Some other examples of trigonal bipyramidal molecules are antimony pentachloride, $SbCl_5$, and arsenic pentafluoride, AsF_5.

11-5. MOLECULES WITH SIX COVALENT BONDS ABOUT A CENTRAL ATOM ARE OCTAHEDRAL

Consider the molecule sulfur hexafluoride, SF_6, whose Lewis formula is

```
        ..
   .. .:F:. ..
   :F    |   F:
     \   |  /
        S
     /   |  \
   :F    |   F:
   ..  .:F:.  ..
        ..
```

This Lewis formula shows that there are six electron pairs in the valence shell of the central sulfur atom. These six electron pairs mutually repel each other. The repulsion is minimized if the six electron pairs, or covalent bonds in this case, point toward the corners of a regular *octahedron.* An octahedron (Figure 11-8) has six vertices and eight faces. All eight faces are identical equilateral triangles. An important property of a regular octahedron is that all six vertices are equivalent. This can be seen easily by building an octahedron using the directions given in Appendix D. We see, then, that SF_6 is octahedral and that the six fluorine atoms are equivalent. There is no way, by either chemical or physical methods, to distinguish among the six sulfur-fluorine bonds in SF_6.

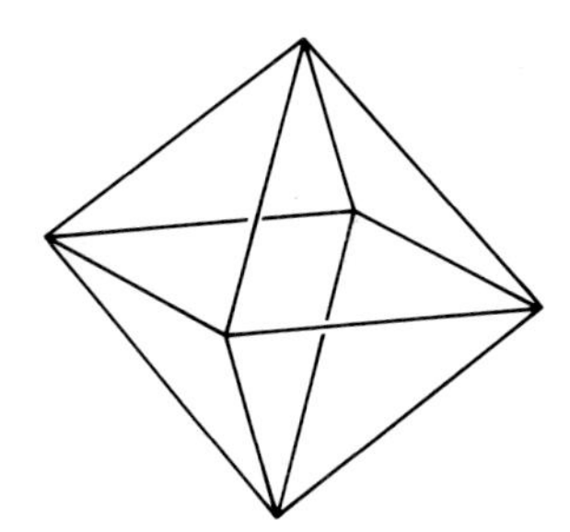

Figure 11-8 A regular octahedron is a symmetric body consisting of six equivalent vertices and eight identical faces that are equilateral triangles.

Example 11-4: Predict the shape of the hexachlorophosphate ion, PCl_6^-.

Solution: The Lewis formula for PCl_6^- is

```
 [       ..         ]-
 [ .. .:Cl:. ..     ]
 [ :Cl    |   Cl:   ]
 [    \   |  /      ]
 [        P         ]
 [    /   |  \      ]
 [ :Cl    |   Cl:   ]
 [ ..  .:Cl:.  ..   ]
 [       ..         ]
```

The six covalent bonds are directed toward the vertices of a regular octahedron, and we predict correctly that the PCl_6^- ion is octahedral.

Problems 11-1 to 11-4 are similar to Examples 11-1 to 11-4.

Table 11-1 shows the bond angles associated with the molecular shapes we have discussed thus far.

Table 11-1 The bond angles associated with the shapes shown in Figure 11-6

Example	*Shape*	*Structure*
$BeCl_2$	linear	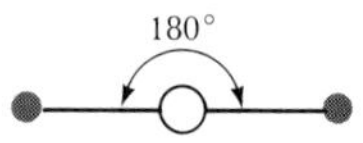 180°
BF_3	trigonal planar	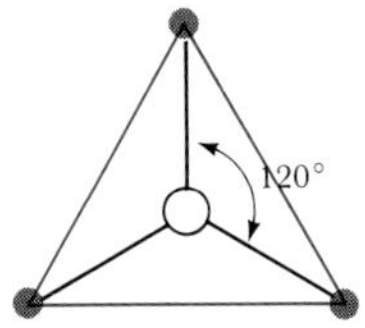 120°
CH_4	tetrahedral	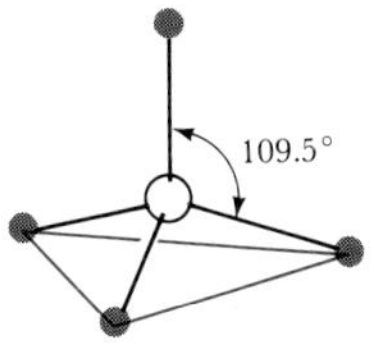 109.5°
PCl_5	trigonal bipyramidal	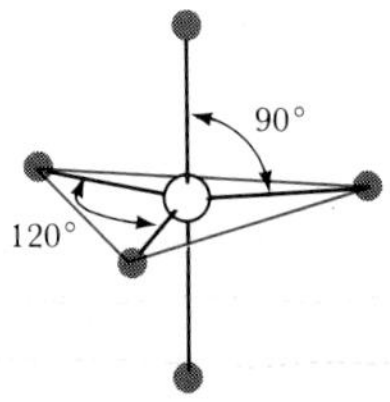 90°, 120°
SF_6	octahedral	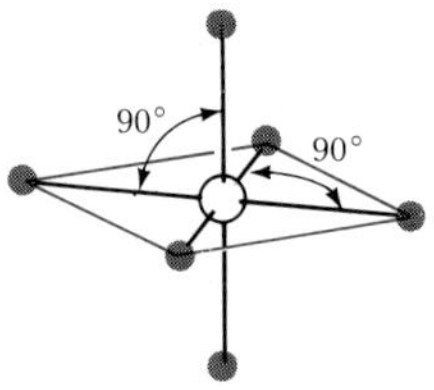 90°, 90°

11-6. LONE ELECTRON PAIRS IN THE VALENCE SHELL AFFECT THE SHAPES OF MOLECULES

In each case we have discussed so far, all the electron pairs in the valence shell of the central atom have been in covalent bonds. Let's now consider cases in which there are lone pairs of electrons as well as covalent bonds in the valence shell of the central atom. As an example, consider the ammonia molecule. The Lewis formula for NH_3 is

$$\mathrm{H{-}\ddot{N}{-}H}$$
$$\mathrm{\;\;|\;\;}$$
$$\mathrm{H}$$

There are four electron pairs in the valence shell of the nitrogen atom. Three of them are in covalent bonds, and one is a lone pair. These four valence-shell electron pairs mutually repel each other and so are directed toward the corners of a tetrahedron (Figure 11-9). The three hydrogen atoms form an equilateral triangle, and the nitrogen atom sits in the center of and above the plane of the triangle. Such a structure is called a triangular pyramid or *trigonal pyramid*. The ammonia molecule is shaped like a tripod, with the three N—H bonds forming the legs of the tripod. A space-filling molecular model of NH_3 is shown in Figure 11-10. It is important to keep in mind that the shape of a molecule is given by the positions of the nuclei in the molecule. The electrons are diffuse and cloud-like and are spread over the molecule. The much more massive nuclei are relatively fixed, and it is the position of the nuclei that defines what we mean by the shape or geometry of a molecule.

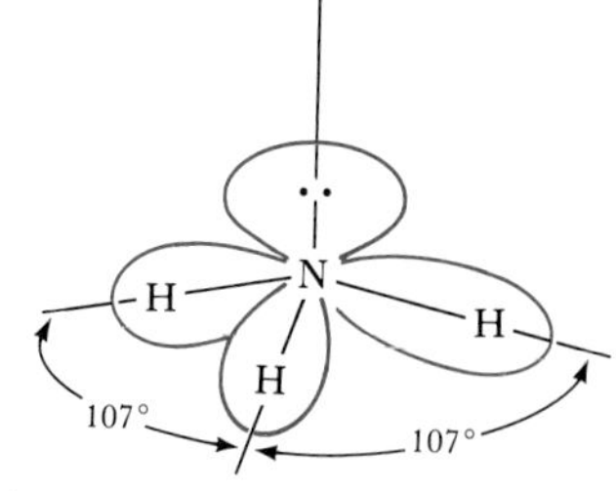

Figure 11-9 A schematic view of the electron pairs in an ammonia molecule. The pairs that constitute the covalent bonds are localized between the nitrogen and hydrogen atoms. The lone pair on the nitrogen atom is not localized between any two atoms.

Figure 11-10 A space-filling molecular model of ammonia.

If the four electron pairs in NH_3 pointed to the corners of a regular tetrahedron, then the H—N—H bond angles would be 109.5°. The four electron pairs in this case are not all of the same type, however. Three of them occur in covalent bonds, and the fourth is a lone pair. Thus we might expect some distortion from purely regular tetrahedral behavior. Electron pairs in covalent bonds are shared between two atoms and are localized between them. A lone pair of electrons, on the other hand, is associated with only the central atom and so is not as localized as the pair of electrons in a covalent bond. Thus a lone pair of electrons is more spread out and is bulkier; that is, it takes up more room around a central atom than does a covalent bond. This means that the repulsion between a lone pair of electrons and the electron pair in a covalent bond is greater than the repulsion between the electron pairs in two covalent bonds. This causes the H—N—H bond angles in NH_3 to decrease slightly from the 109.5° regular tetrahedral angle to 107.3°. Although VSEPR theory is not able to give actual numerical values of bond angles, it can be used to make qualitative statements. We cannot predict that the H—N—H bond angles in NH_3 are 107.3°, but we can predict that they are slightly less than the ideal tetrahedral angle of 109.5°. We say that VSEPR is a qualitative rather than a quantitative theory. There is no simple, reliable quantitative theory of chemical bonding and molecular structure.

11-7. THE VALUES OF m AND n IN THE GENERAL FORMULA AX_mE_n CAN BE USED TO PREDICT THE SHAPE OF A MOLECULE

The example of NH_3 shows that it is the *total* number of electron pairs in the valence shell of the central atom that determines the shape of a molecule. For example, the Lewis formula for H_2O is

$$\mathrm{H{-}\ddot{\underset{..}{O}}{-}H}$$

The four valence-shell electron pairs are directed toward the corners of a tetrahedron (Figure 11-11c). The lone pairs take up more room around the oxygen atom than the two pairs in the covalent bonds, and so we expect the repulsion between the lone electron pairs to be greater than either that between a lone pair and a covalent bond or that between two covalent bonds. We can summarize this by writing

$$\text{lone-pair–lone-pair repulsion} > \text{lone-pair–bond-pair repulsion} > \text{bond-pair–bond-pair repulsion} \quad (11\text{-}1)$$

We predict, then, that the H—O—H bond angle in H_2O is less than the regular tetrahedral angle of 109.5° and that it is even smaller than the H—N—H bond angle (107.3°) in NH_3. The experimentally measured bond angle in H_2O is 104.5°.

Each of the molecules CH_4, NH_3, and H_2O has four electron pairs in the valence shell of the central atom. The four electron pairs are directed toward the corners of a tetrahedron as shown in Figure 11-11. The shape of a molecule is described by giving the positions of the *nuclei*. Thus, CH_4 is tetrahedral, NH_3 is trigonal pyramidal, and H_2O is bent.

Figure 11-11 The role of bonding and nonbonding electron pairs in determining molecular geometry.

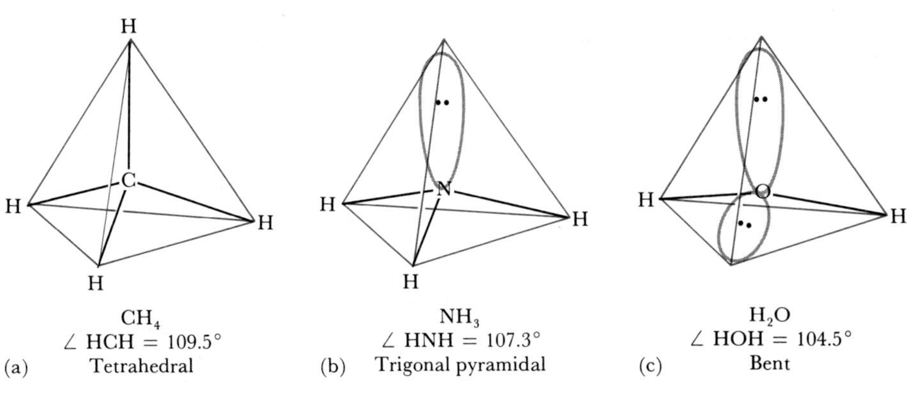

We can classify molecules by introducing the following terminology. Let A represent a central atom, X an atom bonded to the central atom, and E a lone pair of electrons. We call an atom bonded to a central atom a *ligand*. The molecules that we are discussing in this chapter can be classified as AX_mE_n, where m is the number of ligands and n is the number of lone electron pairs in the valence shell of the central atom A. Therefore, the methane molecule belongs to the class AX_4, the ammonia molecule belongs to the class AX_3E, and the water molecule belongs to the class AX_2E_2. The classes of molecules that we discuss in this chapter are given in Figure 11-12.

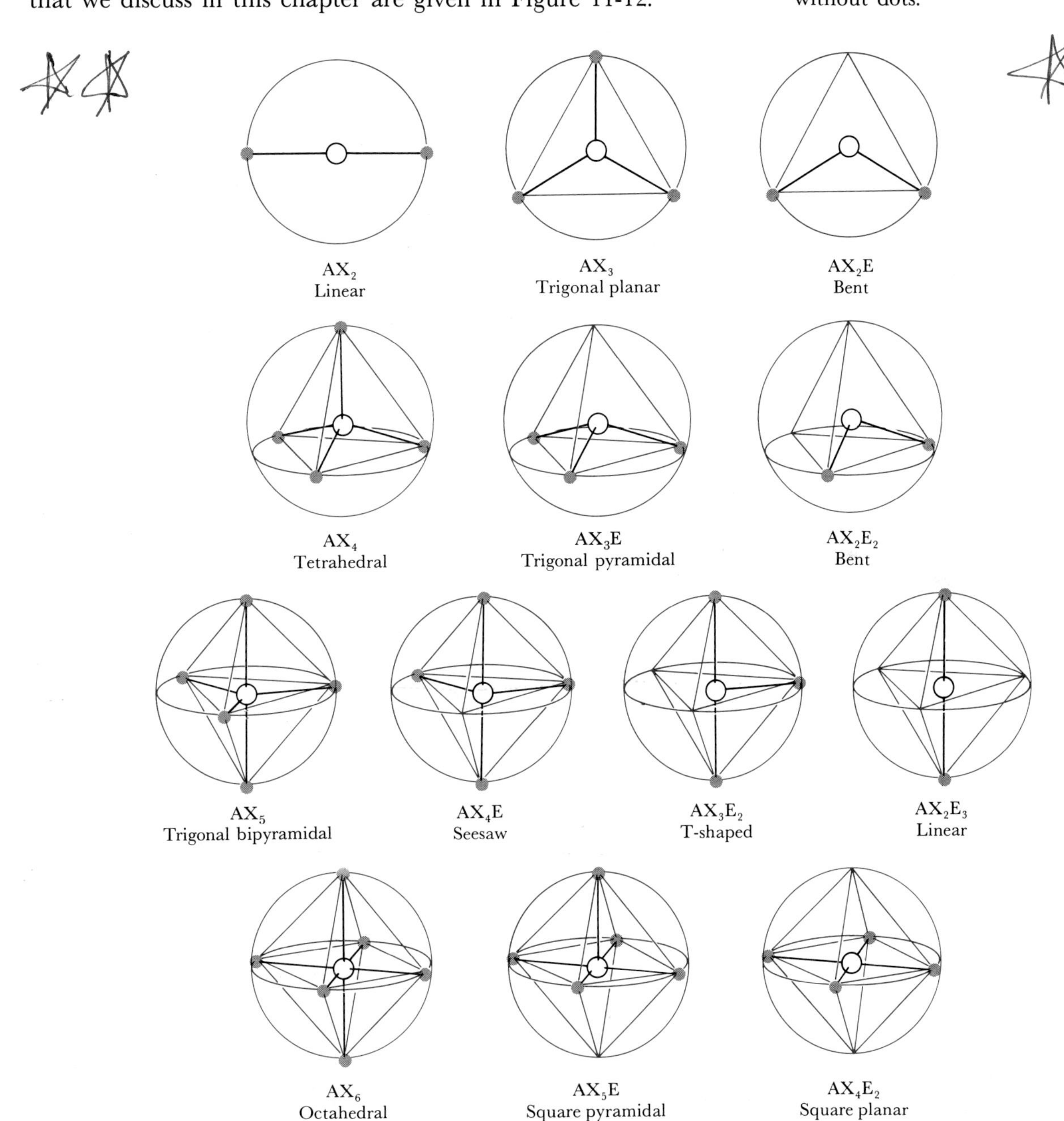

Figure 11-12 A summary of the various molecular shapes that result when m ligands (X) and n lone electron pairs (E) surround a central atom (A) to form an AX_mE_n molecule. The open circle at the center of each sphere represents the central atom, and the gray dots on the surfaces of the spheres represent the attached ligands. The lone electron pairs are at the vertices without dots.

11-8. VSEPR THEORY IS APPLICABLE TO MOLECULES THAT CONTAIN DOUBLE OR TRIPLE BONDS

In predicting the shapes of molecules by VSEPR theory, a double or triple bond is counted simply as a bond connecting the ligand X to the central atom A. For example, the Lewis formula for CO_2 is

$$:\ddot{O}=C=\ddot{O}:$$

Thus CO_2 is classified as an AX_2 molecule, and we predict that it is linear. Another example of an AX_2 molecule that contains a multiple bond is HCN, the Lewis formula for which is

$$H—C\equiv N:$$

There are two groups of electrons around the central carbon atom, and so we predict that HCN is a linear molecule.

Example 11-5: Predict the shape of a formaldehyde molecule, H_2CO.

Solution: The Lewis formula for H_2CO is

H
  \
   C=O:
  /
H

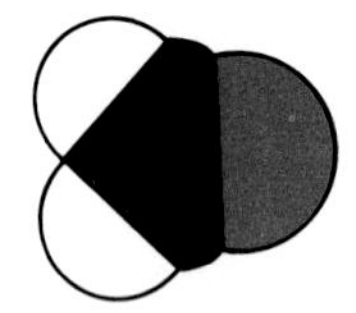

Figure 11-13 A space-filling molecular model of formaldehyde.

We treat the double bond as any other bond and classify H_2CO as an AX_3 molecule. We predict correctly that H_2CO is a trigonal planar molecule, meaning that all four atoms lie in a single plane. Figure 11-13 shows a model of the formaldehyde molecule.

Although a double or triple bond is treated as one covalent bond when a molecule is placed in an AX_mE_n class, we must recognize that multiple bonds contain a higher electronic charge density than single covalent bonds. Consequently, multiple bonds repel single bonds more strongly than single bonds repel other single bonds. In this regard, multiple bonds act like lone electron pairs. Because of this, we predict that the H—C—H bond angle in H_2CO (Example 11-5) is slightly less than 120° and that the H—C—O bond angles are equal but slightly larger than 120°. The actual experimental values are 116° and 122°, respectively:

H (122°)
116° C=O:
H (122°)

Example 11-6: Compare the shapes of the molecules $COCl_2$ (phosgene) and $SOCl_2$ (thionyl chloride).

Phosgene is a colorless, highly toxic gas. When diluted with air, it has an odor resembling moldy hay. It was used as a war gas in World War I.

Solution: The Lewis formulas for these two molecules are

$$\ddot{O}\!=\!C(\ddot{Cl}:)_2 \quad \text{and} \quad :\ddot{Cl}\!-\!\ddot{S}(\!=\!\ddot{O})\!-\!\ddot{Cl}:$$

Phosgene belongs to the class AX_3 and so is planar, like formaldehyde. The lone pair on the sulfur atom in thionyl chloride puts this molecule in the AX_3E class. Thus, thionyl chloride is trigonal pyramidal, with the two chlorine atoms and the oxygen atom lying in a plane and the sulfur atom lying above the plane.

The different shapes of $COCl_2$ and $SOCl_2$ result from the presence of a lone pair.

Notice that even though the chemical formulas for phosgene and thionyl chloride are similar, the shapes of the two molecules are different.

11-9. VSEPR THEORY CAN BE APPLIED TO MOLECULES THAT ARE DESCRIBED BY RESONANCE

VSEPR theory applies to molecules that are described by a superposition of resonance forms as well as it does to molecules that can be represented by just one Lewis formula, as illustrated in the following examples.

Example 11-7: Predict the shape of the ozone molecule, O_3.

Solution: There are two contributing resonance forms for O_3:

$$\ominus:\ddot{\underset{..}{O}}\!-\!\underset{\oplus}{\ddot{O}}\!=\!\ddot{O}\!: \leftrightarrow \ :\ddot{O}\!=\!\underset{\oplus}{\ddot{O}}\!-\!\ddot{\underset{..}{O}}:\ominus$$

The superposition of these two resonance forms is

$$:\ddot{O}\!\overset{\text{---}}{=}\!\ddot{O}\!\overset{\text{---}}{=}\!\ddot{O}:$$

which is called a resonance hybrid and shows that O_3 belongs to the class AX_2E. Thus we predict that the ozone molecule is bent. In addition, we predict that the lone pair on the central oxygen atom causes the O—O—O bond angle to be less than 120°. The observed bond angle in O_3 is 116.7°.

Example 11-8: Predict the shape of the nitrate ion, NO_3^-.

Solution: There are three resonance forms of NO_3^-:

The acceptable Lewis formula is a superposition of these three formulas:

This shows that NO_3^- is of the class AX_3, and so we predict that the nitrate ion is trigonal planar and that the three O—N—O bond angles are equal to 120°.

11-10. LONE-PAIR ELECTRONS OCCUPY THE EQUATORIAL VERTICES OF A TRIGONAL BIPYRAMID

Recall (Figure 11-7) that the five vertices of a trigonal bipyramid are not equivalent to each other. They form a set of three equivalent equatorial vertices and two equivalent axial vertices. Consequently, in considering the class of molecules designated by AX_4E, for example, we have two nonequivalent choices for the position of the lone pair. We can place it at an axial vertex or at an equatorial vertex.

Figure 11-14(a) shows that an equatorial pair has only two nearest neighbors at 90° and Figure 11-14(b) shows that an axial electron pair has three nearest neighbors at 90°. The two other neighbors in Figure 11-14(a) lie at 120° from the position labeled E and thus are far enough away that their interaction with E can be ignored. Conse-

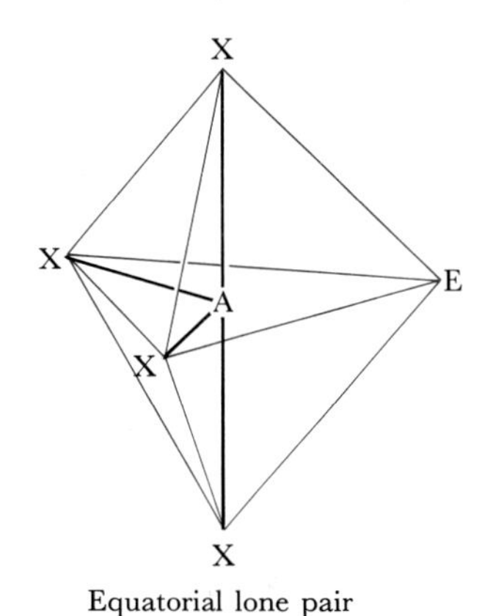

(a) Equatorial lone pair
Two neighbor atoms at 90°

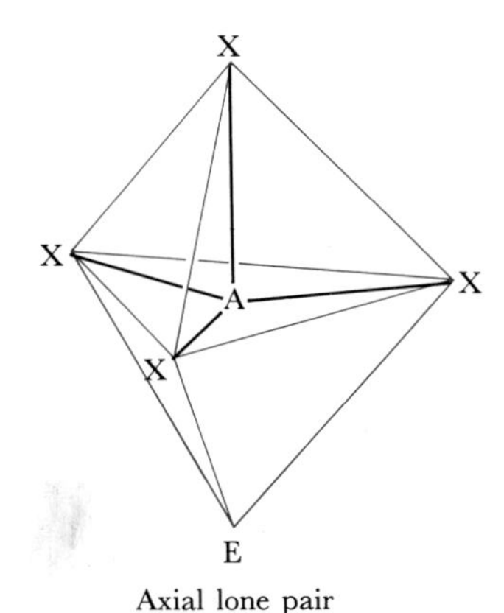

(b) Axial lone pair
Three neighbor atoms at 90°

Figure 11-14 Because the five vertices of a trigonal bipyramid fall into two distinct classes, there are two nonequivalent positions available for the lone electron pair in a molecule of the class AX_4E. A lone pair at an equatorial position (a) has only two nearest neighbors at 90°, whereas a lone pair at an axial position (b) has three nearest neighbors at 90°. Consequently, the repulsion due to a lone electron pair is minimized by placing it at an equatorial vertex, and molecules with the general formula AX_4E are shaped like a seesaw (a).

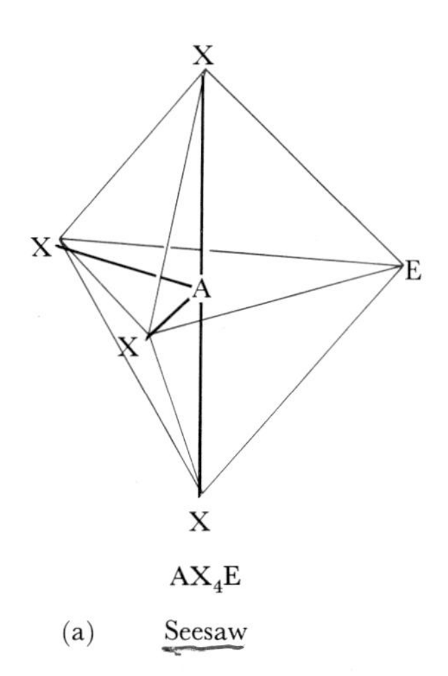

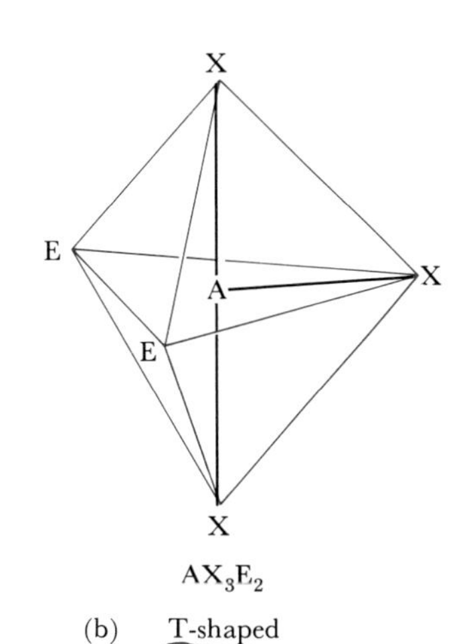

(c) Linear

Figure 11-15 The shapes of molecules that belong to the classes AX_4E, AX_3E_2, and AX_2E_3. The lone pairs occupy the equatorial positions in each case.

quently, the repulsion due to a lone electron pair is minimized by placing it at an equatorial vertex rather than at an axial vertex. Thus, molecules that belong to the classes AX_4E, AX_3E_2, and AX_2E_3 have the lone electron pairs at the equatorial positions, as shown in Figure 11-15. Because the shape of a molecule is determined by the positions of only the atomic nuclei, we see from Figure 11-15 that an AX_4E molecule is shaped like a seesaw, an AX_3E_2 molecule is T-shaped, and an AX_2E_3 molecule is linear.

The Lewis formula for sulfur tetrafluoride

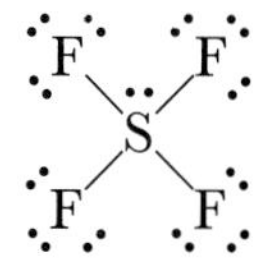

shows that this molecule belongs to the class AX_4E. The lone pair is placed at one of the equatorial positions of a trigonal bipyramid, and so the ideal shape of SF_4 is as shown in Figure 11-16(a). This ideal shape predicts that the F(axial)—S—F(axial) bond angle is 180° and

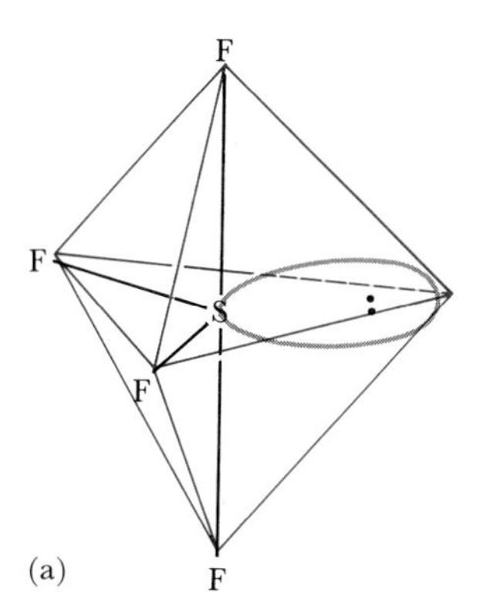

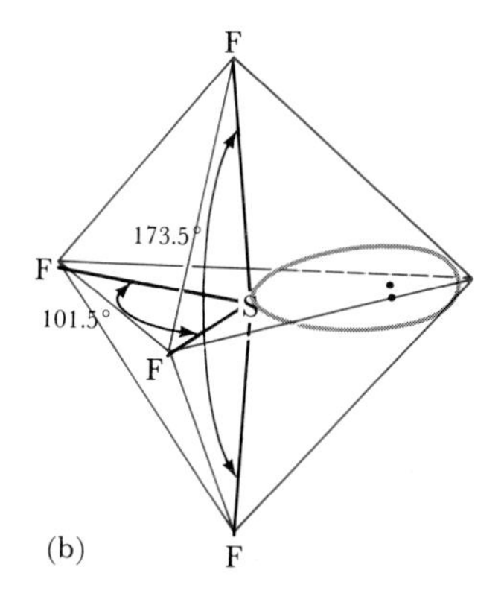

Figure 11-16 (a) The ideal shape of a sulfur tetraflouride molecule, which belongs to the class AX_4E. (b) The lone electron pair at the equatorial position repels the four covalent sulfur-fluorine bonds and distorts the molecule away from ideal geometry.

Table 11-2 Some interhalogen compounds

AX	AX_3E_2	AX_5E
IF	IF_3	IF_5
BrF	BrF_3	BrF_5
ClF	ClF_3	ClF_5
ICl		
BrCl		
IBr		

that the F(equatorial)—S—F(equatorial) bond angle is 120°. The lone pair at the equatorial position causes a small distortion from ideal behavior, however, and the actual shape of SF_4 is that shown in Figure 11-16(b). The experimentally observed bond angles in SF_4, which are indicated in Figure 11-16(b), are in accord with the rule that says lone pairs take up more space than single covalent bonds.

There are a number of compounds formed between halogen atoms in which a less electronegative central halogen atom is bonded to more electronegative halogen atoms. Some of these *interhalogen compounds* are listed in Table 11-2. All known interhalogen molecules obey the predictions of VSEPR theory. Consider the molecule chlorine trifluoride, ClF_3, whose Lewis formula is

:F̤̈—C̈l̈—F̤̈:
   |
  :F̤̈:

This Lewis formula shows that ClF_3 belongs to the class AX_3E_2. The ideal shape of ClF_3 is shown in Figure 11-17(a). Notice that the ClF_3 molecule is T-shaped. The ideal shape predicts that the F(axial)—Cl—F(equatorial) bond angles are 90°. The two lone pairs cause small distortions, however, and so these bond angles are somewhat less than 90°, as shown in Figure 11-17(b).

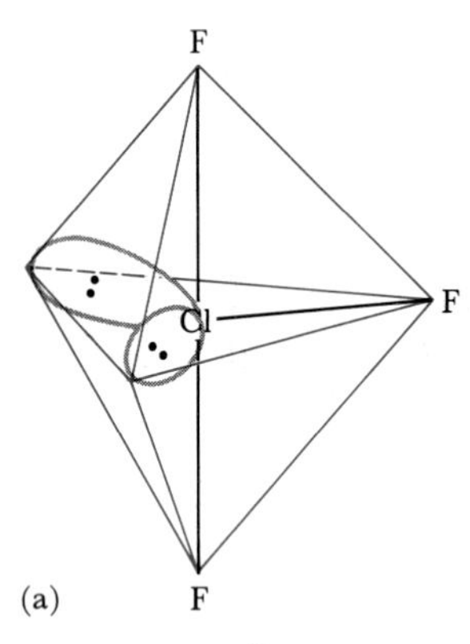

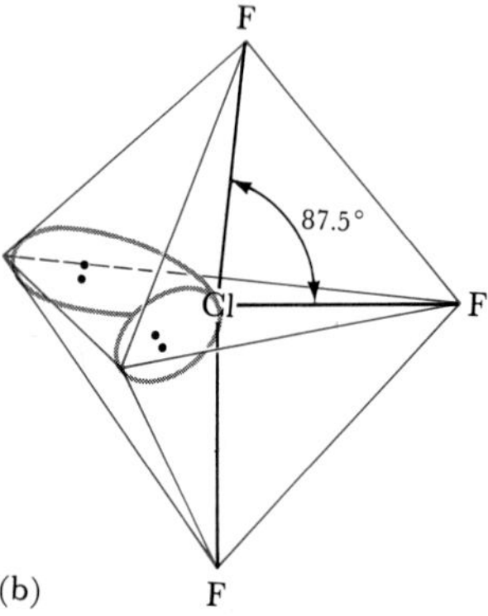

Figure 11-17 (a) The ideal shape of a chlorine trifluoride molecule, which belongs to the class AX_3E_2. (b) The two lone electron pairs at the equatorial positions repel the chlorine-fluorine bonds and distort the molecule away from ideal geometry.

Example 11-9: Although elemental iodine, I_2, is not very soluble in water, it is quite soluble in aqueous solutions of potassium iodide. The increased solubility is due to the formation of the triiodide ion, I_3^-, through the reaction

$$I_2(aq) + I^-(aq) \rightarrow I_3^-(aq)$$

Predict the geometry of the triiodide ion.

Solution: The Lewis formula for I_3^- is

      ⊖
:Ï—Ï—Ï:

This Lewis formula shows that I_3^- belongs to the class AX_2E_3. Figure 11-15(c) indicates that the three lone pairs occupy equatorial positions in a trigonal bipyramid, and thus I_3^- is a linear ion.

One of the impressive successes of VSEPR theory is its correct prediction of the structures of noble-gas compounds. Some noble-gas compounds that have been synthesized are the xenon fluorides (XeF_2, XeF_4, and XeF_6), xenon oxyfluorides ($XeOF_4$, XeO_2F_2), xenon oxides (XeO_3, XeO_4), and krypton fluorides (KrF_2, KrF_4).

Example 11-10: Predict the shape of xenon difluoride, XeF_2.

Solution: The Lewis formula for XeF_2 is

$$:\ddot{\underset{..}{F}}-\ddot{\underset{..}{Xe}}-\ddot{\underset{..}{F}}:$$

This shows that XeF_2 belongs to the class AX_2E_3. Figure 11-18 shows the linear shape of XeF_2.

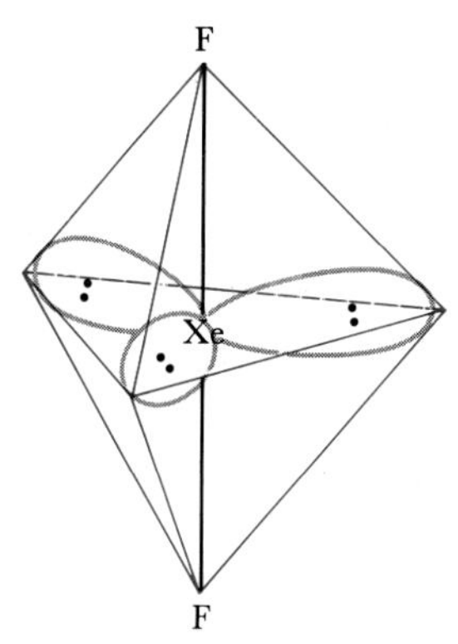

Figure 11-18 The compound XeF_2 belongs to the class AX_2E_3. The three lone pairs occupy the equatorial vertices of a trigonal bipyramid. The two fluorine atoms occupy the axial positions, and so XeF_2 is a linear molecule.

11-11. TWO LONE ELECTRON PAIRS OCCUPY OPPOSITE VERTICES OF AN OCTAHEDRON

The octahedral classes AX_6, AX_5E, AX_4E_2, and AX_3E_3 are shown in Figure 11-19. Because all six vertices of a regular octahedron are equivalent, all six possible positions of the lone pair in AX_5E are equivalent. In order to minimize the lone-pair–lone-pair repulsion in an AX_4E_2 molecule, however, the two lone pairs are placed at opposite vertices, as shown in Figure 11-19(c). There are no known AX_3E_3 molecules, but Figure 11-19(d) predicts that they would be T-shaped.

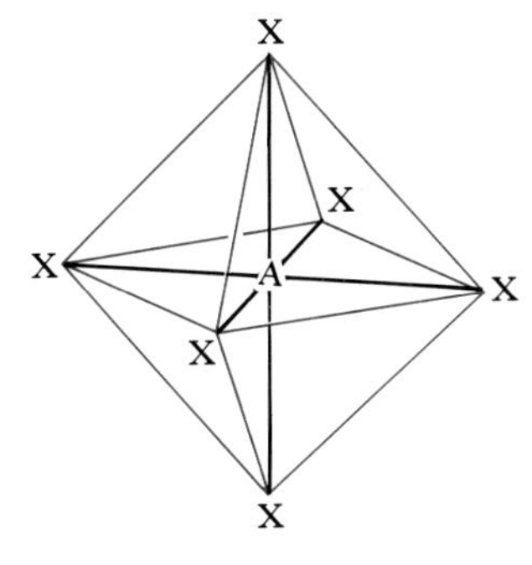

(a) AX_6 Octahedral

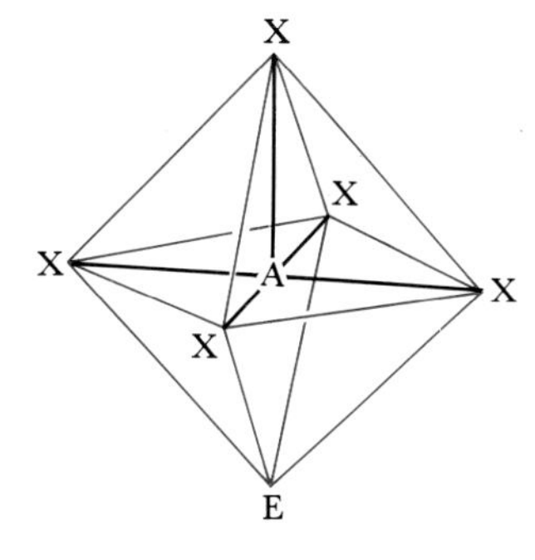

(b) AX_5E Square pyramidal

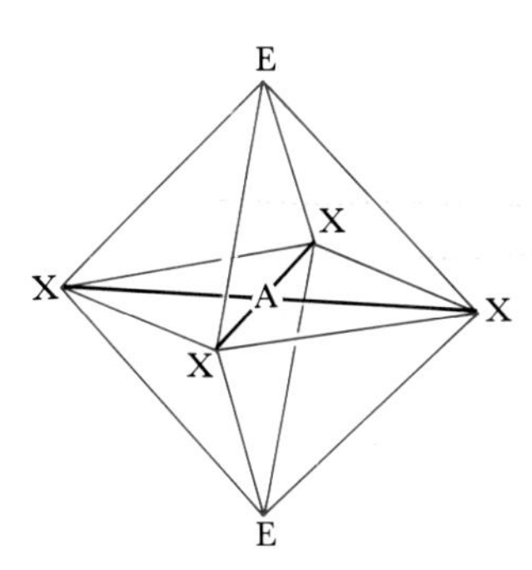

(c) AX_4E_2 Square planar

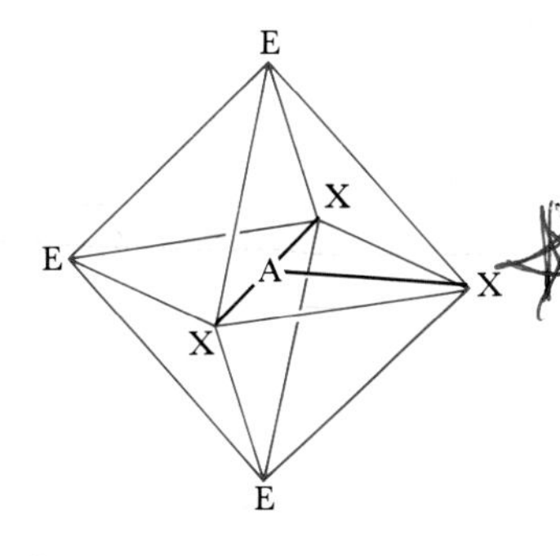

(d) AX_3E_3 T-shaped

Figure 11-19 The ideal shapes associated with the classes AX_6, AX_5E, AX_4E_2, and AX_3E_3. In (c) and (d), two of the lone electron pairs occupy opposite vertices because this placement minimizes the relatively strong lone-pair–lone-pair repulsion.

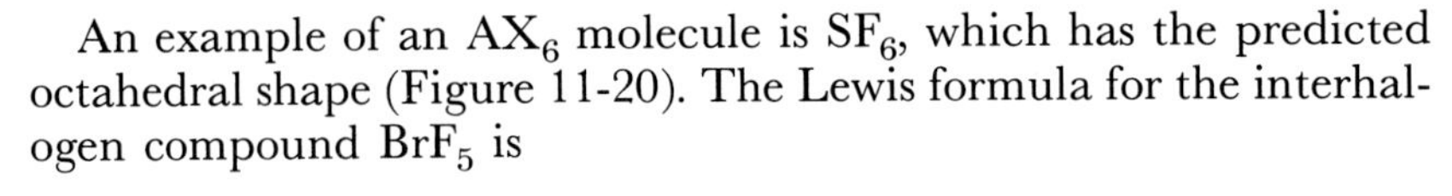

An example of an AX_6 molecule is SF_6, which has the predicted octahedral shape (Figure 11-20). The Lewis formula for the interhalogen compound BrF_5 is

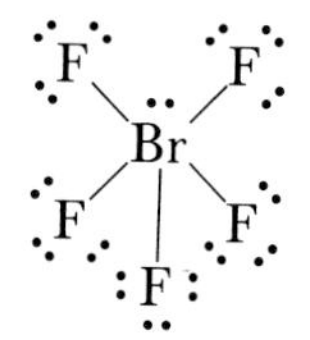

This shows that BrF_5 belongs to the class AX_5E. According to Figure 11-19(b), then, we predict that BrF_5 has a square pyramidal shape. The shape of BrF_5 is shown in Figure 11-21. The F—Br—F bond angles are slightly less than the ideal 90° because of the lone pair sitting at one vertex. The next example shows that XeF_4 is an AX_4E_2 molecule and so is square planar.

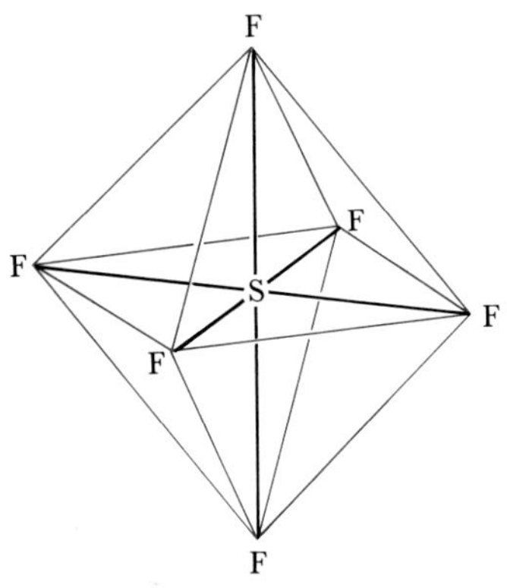

Figure 11-20 The octahedral sulfur hexafluoride molecule. The six sulfur-fluorine bonds point toward the vertices of a regular octahedron. The six fluorine atoms in SF_6 are equivalent by symmetry. There is no chemical or physical way to distinguish among them.

Example 11-11: Xenon tetrafluoride is prepared by heating Xe and F_2 at 400°C at a pressure of 6 atm in a nickel container. The reaction is

$$Xe(g) + 2F_2(g) \rightarrow XeF_4(s)$$

Predict the shape of XeF_4.

Solution: The Lewis formula for XeF_4 is

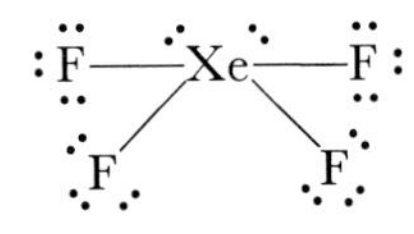

This molecule belongs to the class AX_4E_2, and so according to Figure 11-19(c), we predict that XeF_4 is a square planar molecule. This is indeed the observed structure of xenon tetrafluoride, as shown in Figure 11-22.

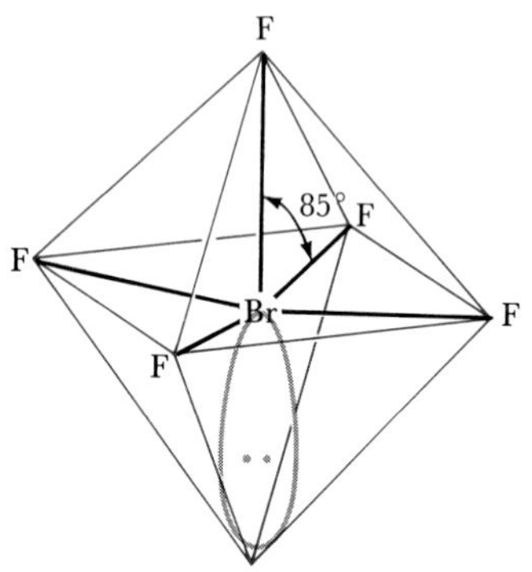

Figure 11-21 The shape of the interhalogen molecule bromine pentafluoride, BrF_5. The lone electron pair repels the bromine-fluorine bonds, causing the bromine atom to lie below the plane formed by four of the fluorine atoms. The BrF_5 molecule has a shape somewhat like an opened umbrella with its handle pointing upward.

The VSEPR theory is simple and useful, and the shapes of thousands of molecules and ions can be predicted from it. As we mentioned before, VSEPR is a qualitative theory, in that it does not predict precise numerical values of bond angles and bond lengths, but it does predict *relative* bond angles. For example, although the theory cannot predict that the H—N—H bond angles in NH_3 are 107.3°, it can predict that they are slightly less than the ideal tetrahedral angle of 109.5°.

VSEPR theory, however, does not give us any insight into the nature of chemical bonds. For example, it does not even tell us why two atoms bond together in the first place. In Chapter 12 we study chemical bonding from a more fundamental point of view.

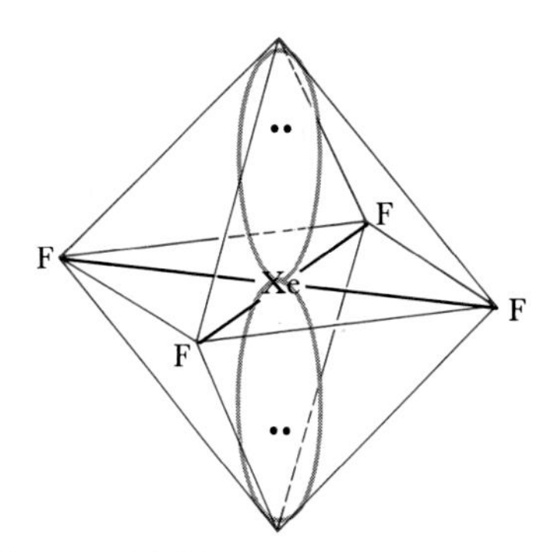

Figure 11-22 The geometry of the noble-gas compound xenon tetrafluoride, XeF_4, which belongs to the class AX_4E_2. The two lone electron pairs occupy opposite vertices of the octahedron, and so XeF_4 is a planar molecule.

SUMMARY

The valence-shell electron-pair repulsion (VSEPR) theory is used to predict the shapes of molecules. It is applicable primarily to molecules in which there is a central atom bonded to ligands.

The procedure for using VSEPR theory to predict molecular shapes can be summarized as follows:

1. Write the Lewis formula for the molecule or ion. If there are two or more resonance forms, the Lewis formula is their superposition.
2. Use the Lewis formula to determine the class AX_mE_n to which the molecule belongs.
3. Given the class AX_mE_n to which the molecule or ion belongs, use Table 11-3 (p. 416) to predict its shape.

The shapes in Table 11-3 can be rationalized as follows:

1. The valence-shell electron pairs around a central atom are arranged so as to minimize their mutual repulsion. The resultant ideal geometries are

Number of valence-shell pairs	*Ideal geometry*
2	linear
3	trigonal
4	tetrahedral
5	trigonal bipyramidal
6	octahedral

In determining the ideal geometry, a multiple bond is counted as one valence-shell pair.

2. Repulsions between valence-shell pairs are in the order lone-pair–lone-pair > lone-pair–bond-pair > bond-pair–bond-pair. This causes small distortions from ideal geometry. Multiple bonds behave like lone pairs in this regard.
3. Lone pairs and multiple bonds are bulkier than single bonds and so occupy a site that results in the least repulsion with neighbors, for example, an equatorial site in the trigonal bipyramid. If sites are equivalent, the bulkier pairs are as far apart as possible, for example, at opposite vertices of an octahedron.

Table 11-3 A summary of molecular shapes

Molecular class	Ideal shape	Examples	Molecular class	Ideal shape	Examples
AX_2	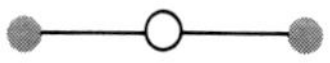 Linear	CO_2, HCN, $BeCl_2$	AX_4E	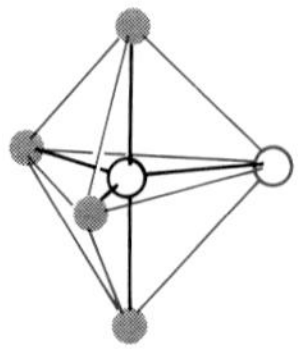 Seesaw	SF_4, XeO_2F_2, IF_4^+, $IO_2F_2^-$
AX_3	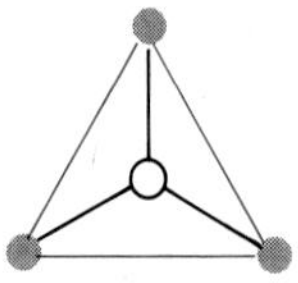 Trigonal planar	SO_3, BF_3, NO_3^-, CO_3^{2-}	AX_3E_2	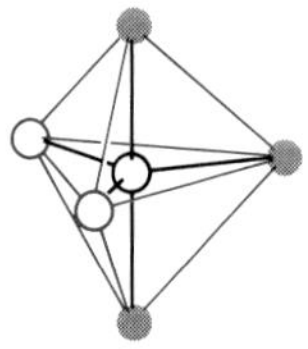 T-shaped	ClF_3, BrF_3
AX_2E	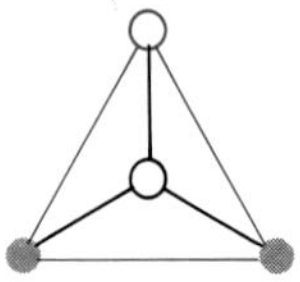 Bent	SO_2, O_3, PbX_2, SnX_2	AX_2E_3	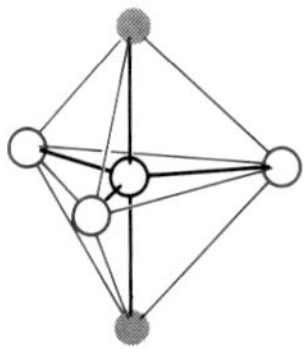 Linear	XeF_2, I_3^-, IF_2^-
AX_4	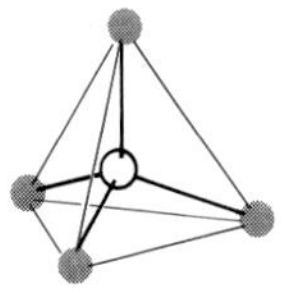 Tetrahedral	SiH_4, CH_4, SO_4^{2-}, ClO_4^-, PO_4^{3-}, XeO_4	AX_6	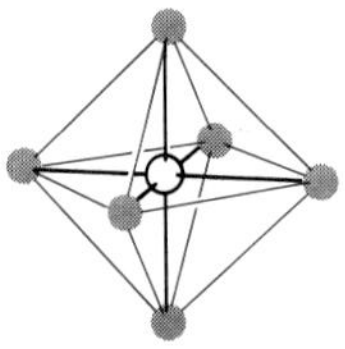 Octahedral	SF_6, IOF_5, $Te(OH)_6$
AX_3E	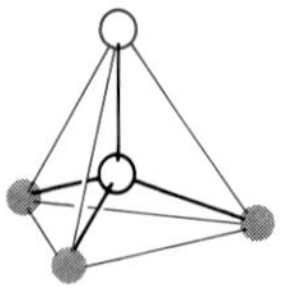 Trigonal pyramidal	NH_3, PF_3, $AsCl_3$, ClO_3^-, H_3O^+, XeO_3	AX_5E	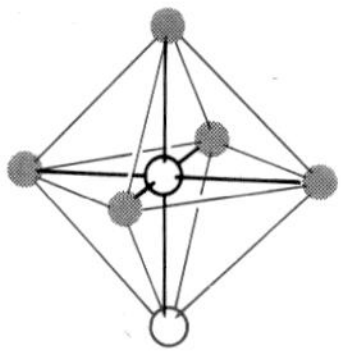 Square pyramidal	IF_5, TeF_5^-, $XeOF_4$
AX_2E_2	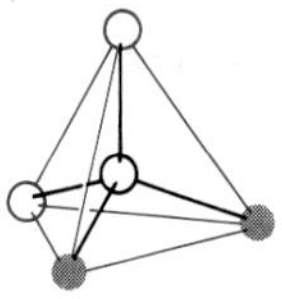 Bent	H_2O, OF_2, SF_2	AX_4E_2	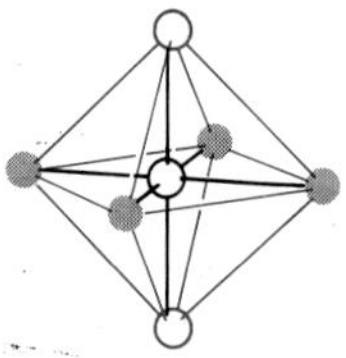 Square planar	XeF_4, ICl_4^-
AX_5	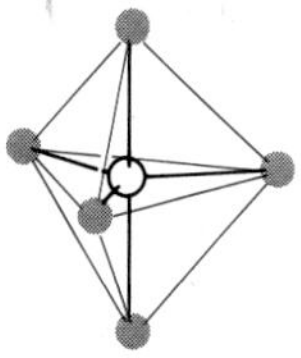 Trigonal bipyramidal	PCl_5, AsF_5, SOF_4			

TERMS YOU SHOULD KNOW

planar 398
geometric isomer 398
tetrahedron 398
space-filling molecular model 398
ball-and-stick molecular model 398
tetrahedral bond angle (109.5°) 399
tetravalent 399
structural chemistry 399
valence-shell electron-pair repulsion (VSEPR) theory 400
trigonal planar 401
trigonal bipyramid 402
equatorial vertex 403
axial vertex 403
octahedron 403
trigonal pyramid 405
ligand 407
AX_mE_n 407
interhalogen compound 412

EQUATIONS YOU SHOULD KNOW HOW TO USE

lone-pair–lone-pair repulsion > lone-pair–bond-pair repulsion > bond-pair–bond-pair repulsion (11-1)

PROBLEMS

MOLECULES AND IONS INVOLVING NO LONE PAIRS

11-1. Which of the following molecules and ions have bond angles of 90°:

(a) TeF_6 (b) $ZnCl_4^{2-}$ (c) $AsBr_5$ (d) $HgCl_2$ (e) GaI_3

11-2. Which of the following molecules and ions have bond angles of 90°:

(a) NH_4^+ (b) $AlCl_3$ (c) AlF_6^{3-} (d) $SiCl_4$ (e) PCl_5

11-3. Which of the following molecules and ions have 120° bond angles:

(a) $HgCl_4^{2-}$ (b) $SbBr_6^-$ (c) $SbCl_5$ (d) $InCl_3$ (e) $GeCl_4$

11-4. Which of the following molecules have 180° bond angles:

(a) SeF_6 (b) CdI_2 (c) $AsCl_5$ (d) $SiCl_4$ (e) ZnI_2

MOLECULES AND IONS INVOLVING ONLY SINGLE BONDS

11-5. Which of the following triatomic molecules are linear and which are bent:

(a) CdI_2 (b) $SnBr_2$ (c) KrF_2 (d) OF_2 (e) $TeCl_2$

11-6. Which of the following triatomic ions are linear and which are bent:

(a) NH_2^- (b) PF_2^+ (c) IF_2^+ (d) Br_3^- (e) ICl_2^-

11-7. Which of the following molecules and ions are tetrahedral:

(a) XeF_4 (b) $CdCl_4^{2-}$ (c) SiH_4 (d) SeF_4 (e) $TiCl_4$

11-8. Which of the following molecules and ions are trigonal pyramidal:

(a) OH_3^+ (b) ClF_3 (c) PF_3 (d) BF_3 (e) NH_3

11-9. Which of the following molecules are trigonal bipyramidal:

(a) BrF_5 (b) $SbCl_5$ (c) PF_5 (d) IF_5

11-10. Which of the following molecules and ions are square planar:

(a) SF_4 (b) ICl_4^- (c) $SiCl_4$ (d) $TeCl_4$

11-11. Which geometry (tetrahedral, seesaw, or square planar) describes the shape of each of the following molecules:

(a) SF_4
(b) KrF_4
(c) CF_4
(d) $GeCl_4$
(e) $TeCl_4$

11-12. Which geometry (tetrahedral, seesaw, or square planar) describes the shape of each of the following ions:

(a) IBr_4^-
(b) PCl_4^+
(c) BF_4^-
(d) IF_4^+
(e) $AlCl_4^-$

11-13. In which of the following molecules and ions

(a) TeF_6
(b) $SbCl_5$
(c) ICl_4^-
(d) $InBr_3$

do the following bond angles occur:

(1) 90° (2) 109.5° (3) 120°

11-14. In which of the following molecules and ions

(a) GeH_4
(b) PF_5
(c) BCl_3
(d) $SnCl_6^{2-}$

do the following bond angles occur:

(1) 90° (2) 109.5° (3) 120°

11-15. In which of the following molecules and ions

(a) SiH_4
(b) SbF_6^-
(c) BrF_4^-
(d) $AsCl_4^+$

do the following bond angles occur:

(1) 90° (2) 109.5° (3) 120°

11-16. Of the following molecules and ions

(a) SeF_6
(b) $SnCl_3^-$
(c) BrF_3
(d) IF_5

which have a bond angle that is

(1) exactly 90°
(2) slightly less than 90°
(3) exactly 109.5°
(4) slightly less than 109.5°
(5) exactly 120°
(6) slightly less than 120°
(7) slightly greater than 120°
(8) exactly 180°
(9) slightly less than 180°

11-17. Of the following molecules and ions

(a) $SbCl_3$
(b) ICl_2^-
(c) TeF_6
(d) SF_4
(e) BrF_5

which have a bond angle that is

(1) exactly 90°
(2) slightly less than 90°
(3) exactly 109.5°
(4) slightly less than 109.5°
(5) exactly 120°
(6) slightly less than 120°
(7) slightly greater than 120°
(8) exactly 180°
(9) slightly less than 180°

11-18. Of the following molecules

(a) ClF_3
(b) $SiCl_4$
(c) PCl_3
(d) XeF_4
(e) SF_6

which have a bond angle that is

(1) exactly 90°
(2) slightly less than 90°
(3) exactly 109.5°
(4) slightly less than 109.5°
(5) exactly 120°
(6) slightly less than 120°
(7) slightly greater than 120°
(8) exactly 180°
(9) slightly less than 180°

11-19. VSEPR theory has been very successful in predicting the geometry of interhalogen molecules and ions. Predict the shapes of the interhalogen compounds given in Table 11-2.

11-20. Predict the shapes of the following iodofluorine ions:

(a) IF_2^+
(b) IF_2^-
(c) IF_4^+
(d) IF_4^-
(e) IF_6^+

11-21. Give one example of each of the following:

(a) bent molecule
(b) bent ion
(c) tetrahedral ion
(d) octahedral molecule
(e) trigonal pyramidal ion

11-22. Give one example of each of the following:

(a) trigonal planar molecule
(b) trigonal pyramidal molecule
(c) T-shaped molecule
(d) octahedral ion
(e) linear ion

MOLECULAR SHAPES AND DIPOLE MOMENTS

11-23. Draw a Lewis formula for each of the following molecules. Use VSEPR theory to predict the shape of each. Indicate which ones have a dipole moment (discussed in Section 10-12):

(a) $HgCl_2$
(b) AsF_5
(c) SCl_2
(d) Cl_2O
(e) $GaCl_3$

11-24. Draw a Lewis formula for each of the following molecules. Use VSEPR theory to predict the shape of each. Indicate which ones have a dipole moment (discussed in Section 10-12):

(a) $GeCl_4$
(b) $TeCl_4$
(c) PoF_6
(d) XeF_2
(e) BrF_3

11-25. Draw a Lewis formula for each of the following molecules. Use VSEPR theory to predict the shape of each. Indicate which ones have a dipole moment (discussed in Section 10-12):

(a) $TeBr_4$
(b) CdI_2
(c) SeF_4
(d) $SbCl_5$
(e) IF_5

11-26. Draw a Lewis formula for each of the following molecules. Use VSEPR theory to predict the shape of each. Indicate which ones have a dipole moment (discussed in Section 10-12):

(a) TeF_6
(b) ClF_5
(c) $HgCl_2$
(d) $SeCl_2$
(e) ClF_3

11-27. Predict which of the following molecules are polar (dipole moments are discussed in Section 10-12):

(a) CF_4
(b) PF_5
(c) XeF_4
(d) BF_3
(e) SeF_4

11-28. Predict which of the following molecules are polar (dipole moments are discussed in Section 10-12):

(a) $TeCl_4$
(b) BCl_3
(c) $CdCl_2$
(d) PCl_5
(e) $PbCl_2$

MOLECULES OR IONS THAT MAY INVOLVE MULTIPLE BONDS

11-29. Draw a Lewis formula for each of the following molecules and use VSEPR theory to predict the shapes:

(a) $SOCl_2$
(b) SO_2Cl_2
(c) ClO_2F
(d) ClO_3F
(e) $POCl_3$

11-30. Draw a Lewis formula for each of the following and use VSEPR theory to predict the shapes:

(a) NOCl
(b) IOF_5
(c) $PO_2F_2^-$
(d) PO_3F^{2-}
(e) SO_3

11-31. Draw a Lewis formula for each of the following and use VSEPR theory to predict the shapes:

(a) CCl_2O
(b) NSF_3
(c) N_3^-
(d) NO_2Cl
(e) $PSCl_3$

11-32. Draw a Lewis formula for each of the following and use VSEPR theory to predict the shapes:

(a) $IO_2F_2^-$
(b) ClO_2^-
(c) $XeOF_4$
(d) $SeOCl_2$
(e) ClO_3F

11-33. Predict the shapes of the following molecules and ions:

(a) BrO_2^-
(b) SiO_2
(c) SO_3Cl^-
(d) IO_4^-
(e) PO_3I^{2-}

11-34. Predict the shapes of the following molecules and ions:

(a) SOF_4
(b) SO_2Cl_2
(c) $AsO_2F_2^-$
(d) XeO_2
(e) $AsOCl_3$

11-35. VSEPR theory has also been very successful in predicting the molecular geometry of noble-gas compounds. Predict the molecular geometry of each of the following:

(a) XeO_3
(b) XeO_4
(c) KrF_2
(d) XeO_6^{4-}
(e) XeF_5^+

11-36. Predict the geometry of each of the following phosphorus-containing species:

(a) POF_3
(b) PCl_6^-
(c) PCl_5
(d) PCl_4^+
(e) PO_4^{3-}

11-37. Compare the shapes of the following oxysulfur ions:

(a) sulfoxylate ion, SO_2^{2-}
(b) sulfite ion, SO_3^{2-}
(c) sulfate ion, SO_4^{2-}

11-38. Compare the shapes of the following oxychloro ions:

(a) chlorite, ClO_2^-
(b) chlorate, ClO_3^-
(c) perchlorate, ClO_4^-

11-39. Draw Lewis formulas for the following molecules and predict their shapes:

(a) XeO_2F_4
(b) $XeOF_4$
(c) IO_2F
(d) IO_3F
(e) IO_2F_3

11-40. Draw Lewis formulas for the following molecules and ions and predict their shapes:

(a) CO_3^{2-}
(b) IO_2^-
(c) CS_3^{2-}
(d) SO_4^{2-}
(e) NO_2Cl

11-41. The species NO_2^+ and NO_2^- have O—N—O bond angles of 180° and 115°, respectively. Use VSEPR theory to explain the difference in bond angles.

11-42. Compare the shapes of the oxynitrogen ions

(a) NO_2^-
(b) NO_3^-
(c) NO_2^+
(d) NO_4^{3-}

PREDICTING THE NUMBER OF GEOMETRIC ISOMERS

11-43. Describe the possible geometric isomers that might exist for

(a) a tetrahedral molecule AX_3Y
(b) a tetrahedral molecule AX_2YZ
(c) a square planar molecule AX_3Y
(d) a square planar molecule AX_2Y_2

11-44. The molecule $Pt(NH_3)_2Cl_2$ is square planar with the platinum atom in the center of the square. How many isomers of $Pt(NH_3)_2Cl_2$ exist?

11-45. Describe the possible geometric isomers that might exist for an octahedral molecule whose formula is

(a) AX_5Y
(b) AX_4Y_2
(c) AX_3Y_3

11-46. (a) The ion $Co(NH_3)_4Cl_2^+$ is octahedral with the cobalt in the center. How many isomers of $Co(NH_3)_4Cl_2^+$ are there? (b) The molecule $Co(NH_3)_3Cl_3$ is also octahedral with the cobalt in the center. How many isomers of $Co(NH_3)_3Cl_3$ are there?

11-47. The ions in the following series are octahedral with the platinum in the center. How many isomers of each are there:

(a) $Pt(NH_3)_6^{4+}$
(b) $Pt(NH_3)_5Cl^{3+}$
(c) $Pt(NH_3)_4Cl_2^{2+}$
(d) $Pt(NH_3)_3Cl_3^+$

11-48. Describe the possible geometric isomers that might exist for a trigonal bipyramidal molecule whose formula is

(a) AX_4Y
(b) AX_3Y_2
(c) AX_2Y_3

INTERCHAPTER G

Spectroscopy

Modern spectroscopic methods involve complex instrumentation. This photo shows a mass spectrometer (Section G-3).

Chemists are continually faced with the problem of identifying substances of unknown composition. Such problems arise in many areas and range from very simple to extremely complex. Fortunately, there are numerous powerful analytical methods available for determining the chemical composition of unknown substances. Many of these analytical methods establish the identity of chemical compounds by showing that certain characteristic properties of the unknown compound are identical to those of a known compound. The analytical methods we discuss in this interchapter are infrared spectroscopy, nuclear magnetic resonance spectroscopy, mass spectrometry, and X-ray fluorescence. In addition to these physical methods, there are innumerable analytical tests in which a compound is identified by means of characteristic chemical reactions. Such chemical tests often are structured to produce a colored reaction product that can be identified visually or spectrophotometrically.

The branch of chemistry that deals with the identification of unknown samples is called analytical chemistry.

G-1. THE INFRARED SPECTRUM OF A COMPOUND GIVES CONSIDERABLE INFORMATION ABOUT ITS STRUCTURE

Many methods that are used to identify compounds involve the interaction of electromagnetic radiation with matter. In this section we discuss the interaction of infrared radiation with matter.

The infrared region of the electromagnetic spectrum lies between the visible and radio wave regions, that is, in the wavelength region between 10^{-6} and 10^{-4} m (Table 7-4). When a molecule is irradiated with infrared radiation of various wavelengths, some of the radiation is absorbed by the molecule. The energy of the absorbed radiation is stored in the molecule as *molecular vibrations,* that is, as vibrations of the various nuclei in the molecule. A plot of the amount of radiation absorbed versus the frequency or wavelength of the radiation is called a *spectrum.* If the radiation is in the infrared region, then the spectrum is called an *infrared spectrum.*

The infrared spectrum of an organic molecule gives a great deal of information about the structure of the molecule. Figure G-1 shows the infrared spectra of carbon tetrachloride and chloroform. The Lewis formulas for these two compounds are

```
       ..                        ..
      :Cl:                      :Cl:
  ..   |   ..                    |   ..
 :Cl—C—Cl:       and        H—C—Cl:
  ..   |   ..                    |   ..
      :Cl:                      :Cl:
       ..                        ..
carbon tetrachloride        chloroform
```

Note from Figure G-1 that the substitution of one chlorine atom in CCl_4 by a hydrogen atom causes a dramatic change in the infrared spectrum. The absorption maximum (transmission minimum) that occurs around 13 μm in both spectra is due to the back-and-forth stretching motion of a C—Cl bond. The absorption around 3.3 μm in the chloroform spectrum is due to the back-and-forth stretching motion of the C—H bond. The various vibrational motions of groups of atoms in molecules lead to characteristic absorption of infrared radiation and can be used to verify the presence of certain bonds. For example, molecules that contain one or more C—H bonds absorb at around 3.3 μm.

The infrared spectra of three different compounds with the same chemical formula (C_4H_8O) are shown in Figure G-2; note that the spectra are very different. Each compound has a characteristic infrared spectrum that is determined by the types and arrangements of the bonds in the molecules. In effect, the infrared spectrum of a compound constitutes a "fingerprint" of the molecule. If the infrared spectrum of an unknown compound is known, then the compound can be identified by comparing its spectrum with the spectra of various known compounds. The infrared spectrum provides a variety of important clues to the structure of the molecule. Analytical methods that involve spectra are called *spectroscopic methods.*

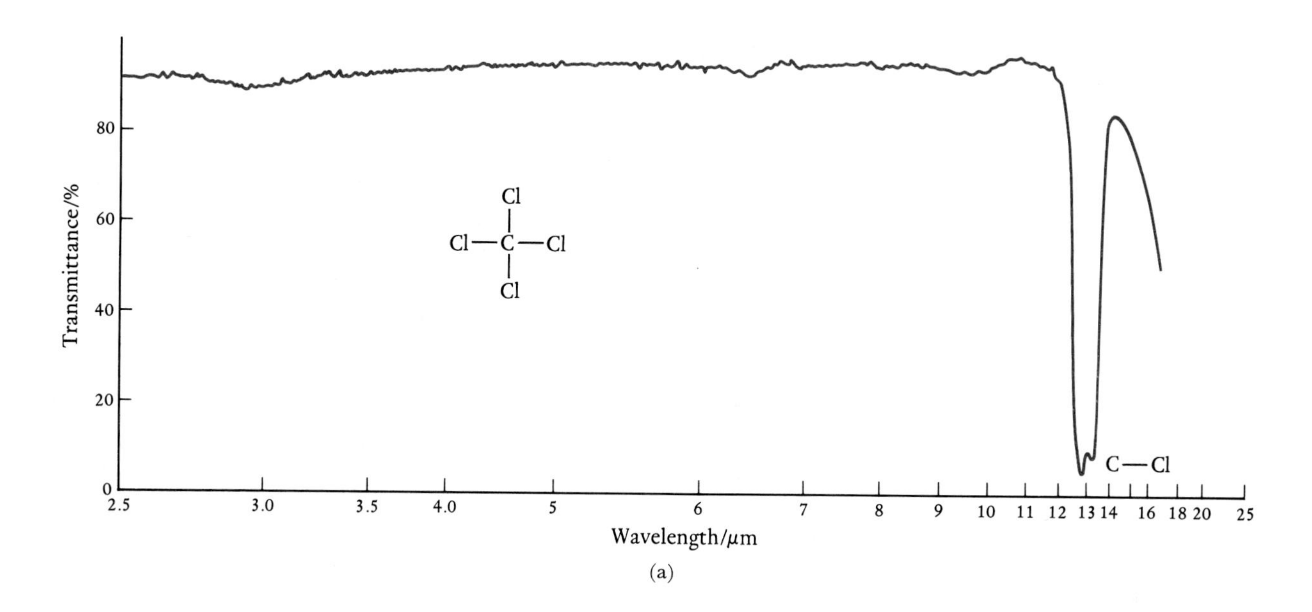

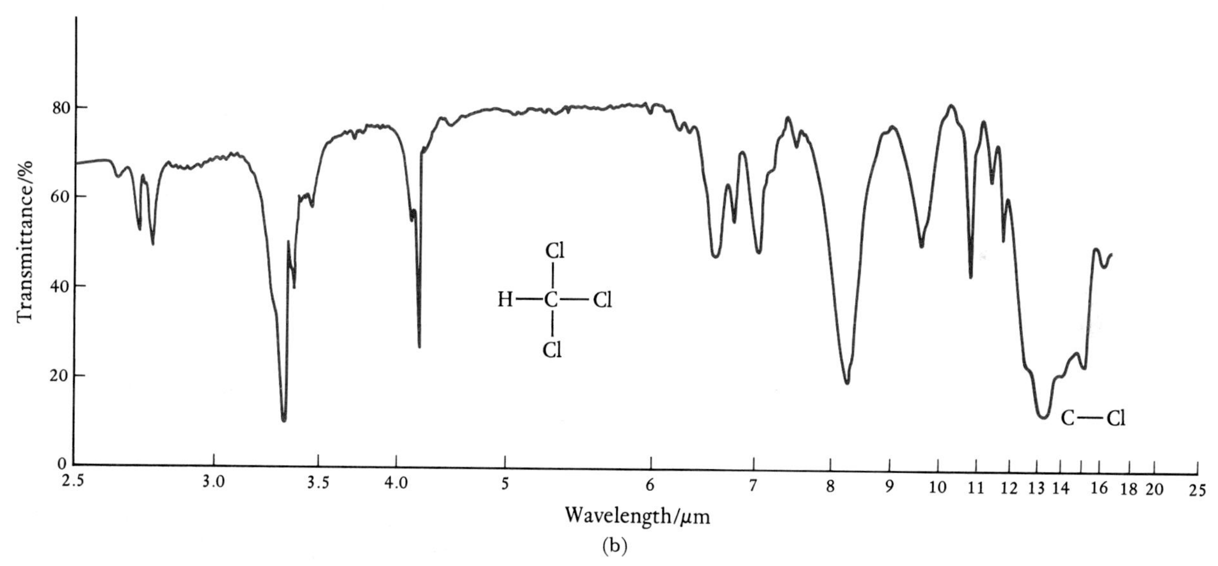

Figure G-1 The infrared spectra of (a) carbon tetrachloride and (b) chloroform. The absorption of the infrared radiation at a given wavelength appears as a downward-pointing peak at that wavelength. The absorption is due to vibrational motion of the atoms in the molecule. The substitution of one chlorine atom in carbon tetrachloride by a hydrogen atom leads to a completely different infrared spectrum.

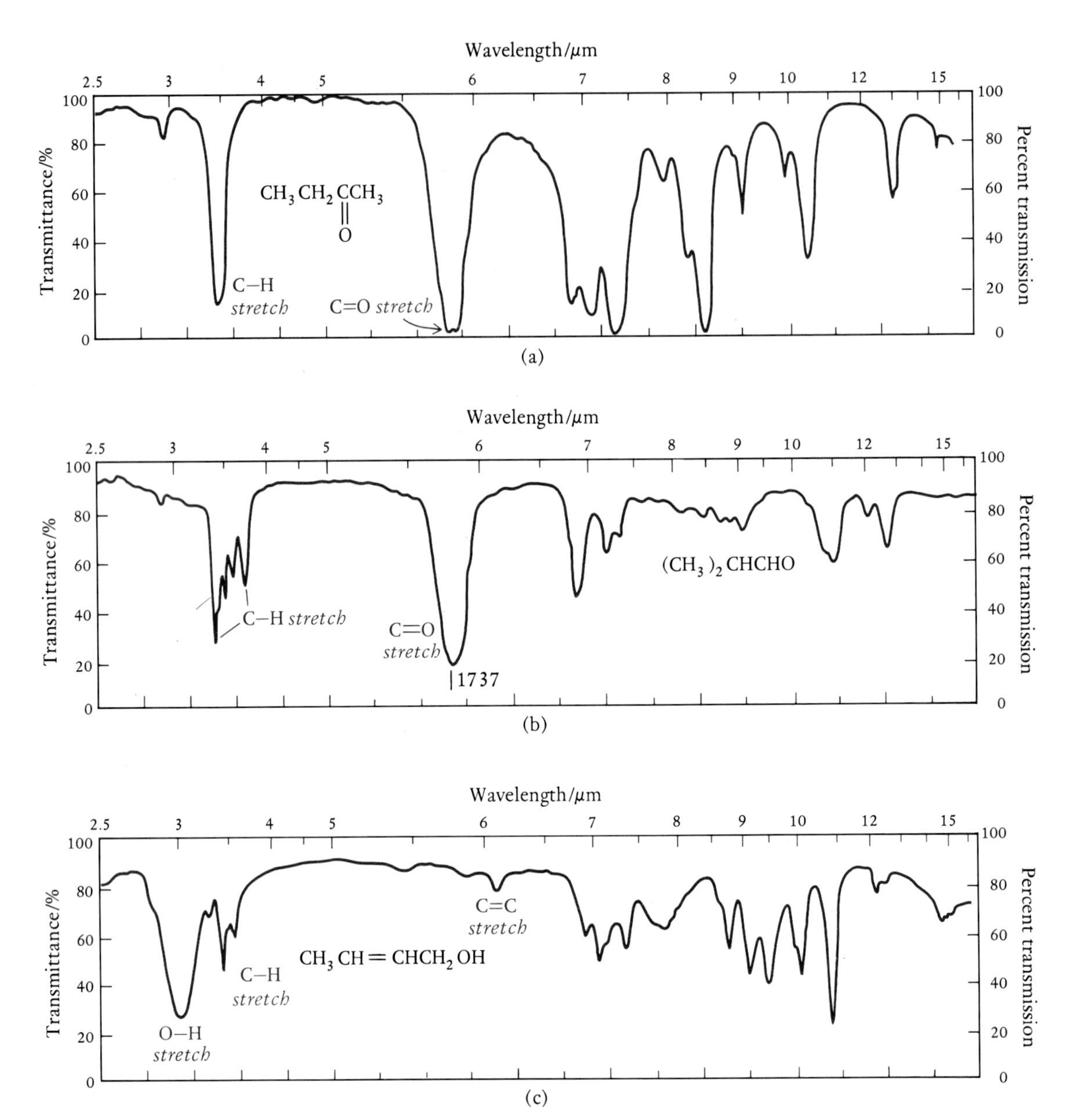

Figure G-2 The infrared spectra of three different compounds with the same chemical formula, C_4H_8O. Although the three compounds have the same chemical formula, their infrared spectra are easily distinguished, reflecting the different arrangements of the atoms in the molecules.

G-2. NUCLEAR MAGNETIC RESONANCE SPECTRA INDICATE THE POSITIONS OF HYDROGEN ATOMS IN A MOLECULE

When a sample of a compound containing hydrogen atoms is placed in a magnetic field and irradiated with electromagnetic energy in the radio wave region, the hydrogen nuclei absorb part of the radiation. This phenomenon is called *nuclear magnetic resonance* (NMR), and the resulting spectrum is called an *NMR spectrum*. The wavelength at which a hydrogen atom absorbs is strongly dependent on its neighboring atoms. A hydrogen atom in $CHCl_3$, for example, absorbs at a different wavelength than a hydrogen atom in $CHBr_3$. The absorption wavelength can be correlated with the position of a hydrogen atom in a molecule, so that an observed absorption can be identified with a particular hydrogen atom. Figure G-3 shows the NMR spectra of two molecules with the same chemical formula, $C_3H_6O_2$. One of these molecules is methyl acetate, whose Lewis formula is

In practice, the wavelength of the electromagnetic radiation is fixed, and the strength of the magnetic field is varied in an NMR spectrometer.

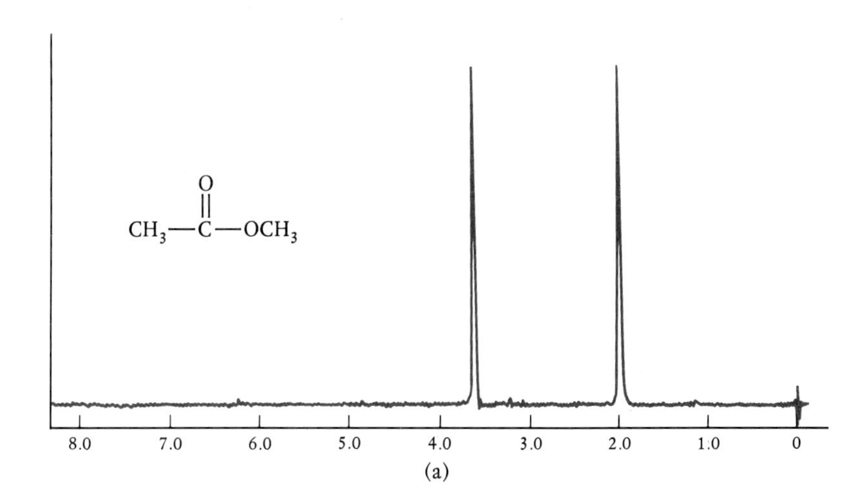

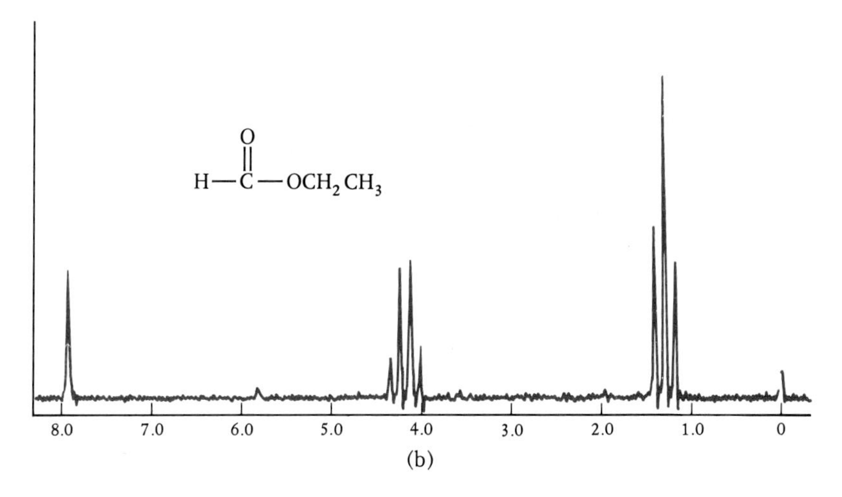

Figure G-3 The nuclear magnetic resonance spectra of (a) methyl acetate and (b) ethyl formate. The empirical formula for both these compounds is $C_3H_6O_2$, but their NMR spectra are quite different.

```
     H  :O:      H
     |  ||       |
  H—C—C—Ö—C—H
     |           |
     H           H
```

and the other is ethyl formate, whose Lewis formula is

```
    :O:     H  H
    ||      |  |
  H—C—Ö—C—C—H
            |  |
            H  H
```

Note that the two spectra, and hence the two compounds, are easily distinguished from each other.

Figure G-4 shows the NMR spectra of two isomeric compounds, whose Lewis formulas are

```
    :C̈l: H                     H  H
     |   |                     |  |
  H—C—C—H        and     :C̈l—C—C—C̈l:
     |   |                     |  |
    :C̤l: H                     H  H
```

These two molecules have the same chemical formula ($C_2H_4Cl_2$) and differ only in whether the two chlorine atoms are attached to the same carbon atom or to different carbon atoms. Yet their NMR spectra are easily distinguished from each other. With just a little experience, it is possible to interpret NMR spectra and deduce the positions of hydrogen atoms (and even other atoms) in molecules, and so learn a great deal about molecular structure. NMR spectroscopy is one of the most widely used analytical techniques in organic chemistry.

G-3. COMPOUNDS CAN BE IDENTIFIED BY THEIR CHARACTERISTIC MASS FRAGMENTATION PATTERNS

If electrons of sufficiently high energy are directed at molecules in the gas phase, then the molecules are ionized to positive ions. For example,

$$\underset{C_5H_{12}}{CH_3-\overset{\displaystyle CH_3}{\underset{\displaystyle CH_3}{\overset{|}{\underset{|}{C}}}}-CH_3} + e^- \text{ (high-energy)} \rightarrow \underset{C_5H_{12}^+}{\left[CH_3-\overset{\displaystyle CH_3}{\underset{\displaystyle CH_3}{\overset{|}{\underset{|}{C}}}}-CH_3\right]^{\oplus}} + 2e^-$$

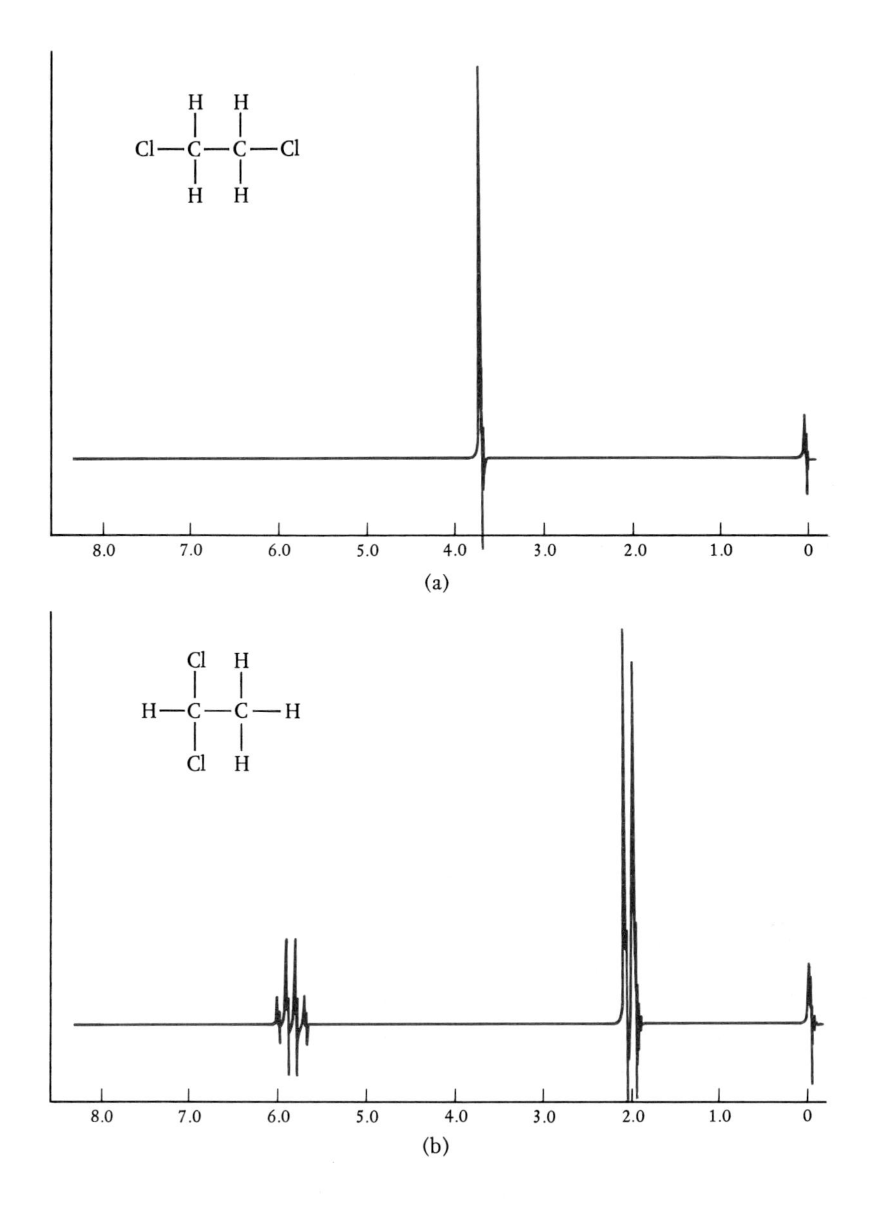

Figure G-4 The nuclear magnetic resonance spectra of two isomers. These two very different spectra suggest that NMR is a sensitive probe of the structure of a molecule.

The molecular ion $C_5H_{12}^+$ is unstable and fragments into several other ions:

$$C_5H_{12}^+ \rightarrow C_4H_9^+,\quad C_3H_5^+,\quad C_2H_5^+,\quad C_2H_3^+$$

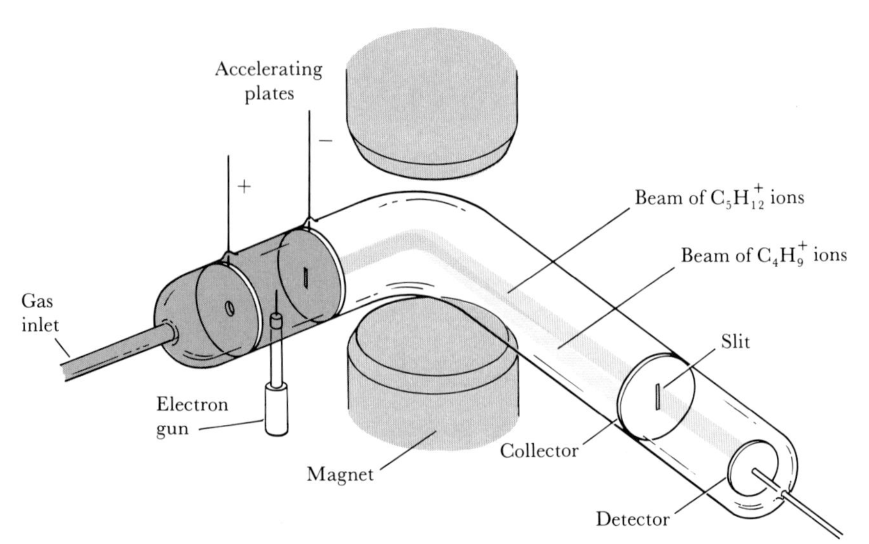

Figure G-5 A schematic drawing of a mass spectrometer. Gas molecules are ionized by electron bombardment, and the ions are accelerated by an electric field. The ion beam is passed through a magnetic field, where it is resolved into component beams of ions of equal mass. Light ions are deflected more strongly than heavy ions by the magnetic field. In a beam containing $C_5H_{12}^+$ and $C_4H_9^+$ ions, the lighter $C_4H_9^+$ ions are deflected more than the heavier $C_5H_{12}^+$ ions. The mass spectrometer depicted here is adjusted to detect the $C_5H_{12}^+$ ions. By changing the magnitude of the magnetic or electric field, the beam of $C_4H_9^+$ can be moved to strike the collector at the slit, where it would then pass through to the detector and be measured as a current. This portion of a mass spectrometer can be seen clearly in the front part of the apparatus shown in the Frontispiece.

The various molecular ions produced in the electron bombardment are accelerated by an electric field (positive ions move toward the negative plate) into a magnetic field (Figure G-5). The ion beam, which constitutes an electric current, is deflected by the magnetic field. The greater the mass of an ion, the less its deflection in the magnetic field. By varying the magnitude of the electric or the magnetic field, the relative proportions of ions of a particular mass number are measured as a current by a detector.

A mass spectrometer fragmentation pattern, that is, a *mass spectrum,* is characteristic of a particular compound. An unknown compound often can be identified by comparison of its mass spectrum with tabulated mass spectra of known compounds. If two mass spectra match peak for peak (mass numbers and relative peak heights), then it is essentially certain that the two compounds are identical. Matching mass spectra of compounds is another example of molecular identification by a "fingerprint" method.

Figure G-6 shows the mass spectra of two compounds with the chemical formula C_8H_{18}, whose structures are

```
                                                    H          H
                                                    |          |
                                                  H—C—H      H—C—H
   H  H  H  H  H  H  H  H                      H    |      H   |     H
   |  |  |  |  |  |  |  |                      |    |      |   |     |
 H—C—C—C—C—C—C—C—C—H         and           H—C—C————C————C—C—H
   |  |  |  |  |  |  |  |                      |    |      |   |  |
   H  H  H  H  H  H  H  H                      H    |      H   H  H
                                                  H—C—H
                                                    |
                                                    H
```

Note how different the two spectra are even though the two compounds have the same total mass. The different molecular structures cause the two compounds to fragment in different ways.

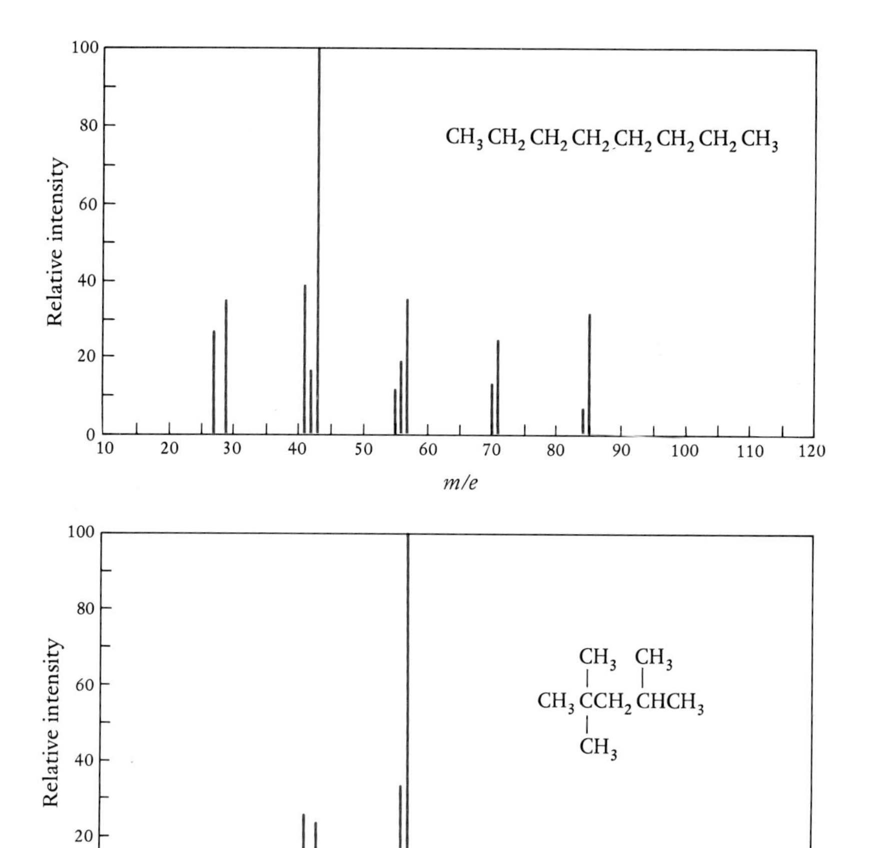

Figure G-6 Mass spectra of two molecules that have the same empirical formula, C_8H_{18}. The horizontal axis is the mass-to-charge ratio of a fragment.

If a compound is unknown, then its mass spectrum is useful for obtaining the molecular mass (which would be the largest observed mass value) and details about the molecular structure. This structural information is obtained by comparing the fragmentation pattern with accumulated knowledge about how various types of molecules tend to fragment.

Mass spectrometry has many practical applications. For example, it has been used to identify compounds in the blood of people found unconscious as a result of ingestion of toxic substances.

G-4. THE ELEMENTS IN A SAMPLE CAN BE IDENTIFIED BY THEIR CHARACTERISTIC X-RAY FLUORESCENCE SPECTRA

Consider the problem of characterizing a solid sample of unknown composition, given that we do not even know what elements are present in the sample. One powerful new tool available for the determination of the nature and the relative amounts of the various elements present in a sample is *X-ray fluorescence* (XRF) *spectroscopy*.

If a sample is subjected to an X-ray beam of sufficiently high energy, then the X-rays impinging on the sample eject electrons from the inner shells of the atoms in the sample. The ejection of these electrons is followed by the emission of X-radiation from the atom at a set of wavelengths that is characteristic of the particular element. Thus each element gives rise to a characteristic X-ray emission (fluorescence) spectrum, as shown in Figure G-7. The relative amount of an element in a sample is determined by measuring the intensity of the observed X-ray fluorescence arising from the sample and comparing the result with that obtained using standard samples for calibration.

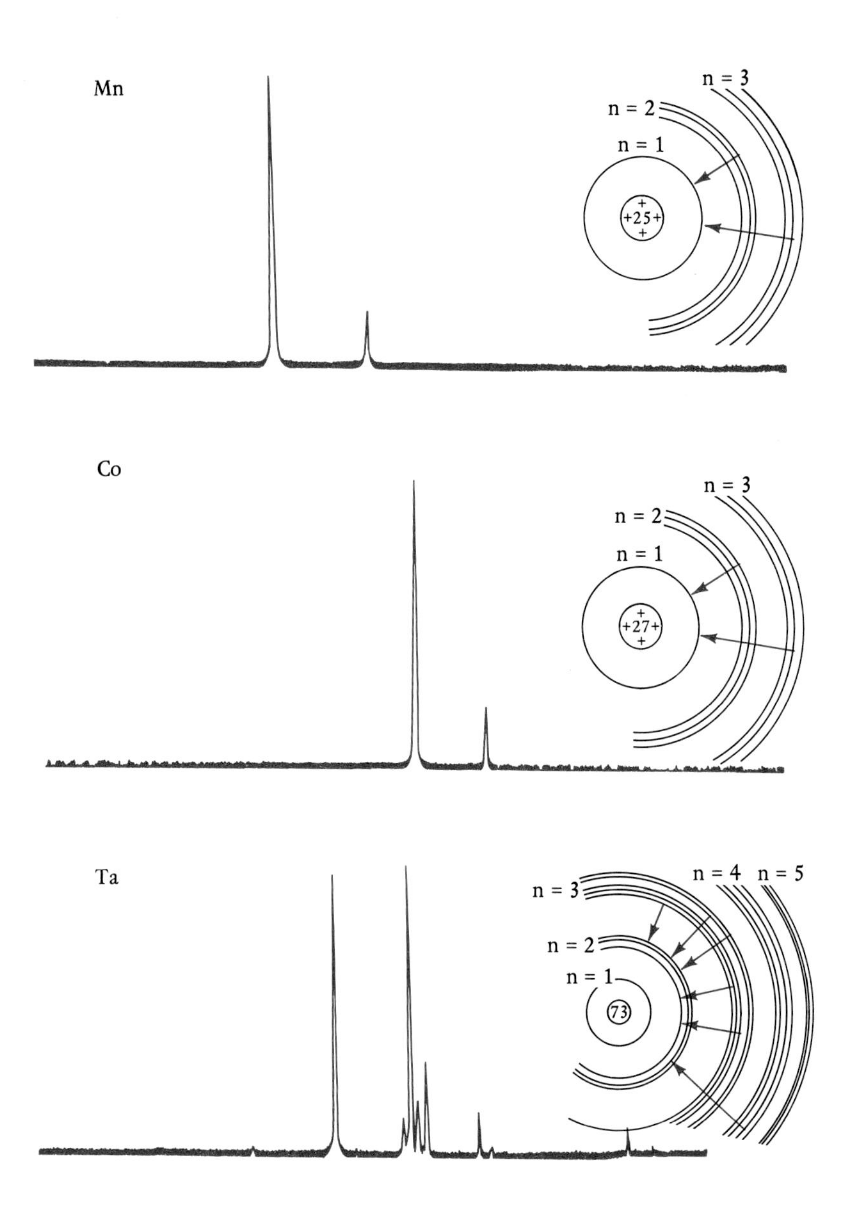

Figure G-7 When an atom is bombarded with X-rays of sufficiently high energy, electrons are ejected from the inner shells, such as the $n = 1$ shell and the $n = 2$ shell. Electrons from higher shells then jump into the partially filled inner shells and emit X-rays in the process. The frequencies of the X-rays emitted are characteristic of the atom. This figure shows the X-ray emission spectra of manganese, cobalt, and tantalum.

X-ray fluorescence spectroscopy is readily adaptable to the nondestructive analysis of minute samples and thus is used extensively in environmental studies, archaeology, geochemistry, and criminal investigations. For example, it is easy to determine whether a dirt particle on, say, a shoe sole is the same as the dirt at a crime location, whether airborne particulates are coming from a certain smokestack, or whether a reputed Rembrandt was actually painted by Rembrandt.

X-ray fluorescence spectroscopy is one of several extremely sensitive analytical techniques that are used for elemental or atomic analysis. We shall learn about two other such techniques, neutron activation analysis and particle-induced X-ray emission (PIXIE), in Chapter 24. These techniques are extremely sensitive; samples that are much too small to be seen by the naked eye can be analyzed.

QUESTIONS

G-1 What is an infrared spectrum?

G-2 Describe how infrared spectroscopy can be used to distinguish two substances from each other.

G-3 What type of molecular motion leads to the absorption of infrared radiation?

G-4 What do the initials NMR stand for?

G-5 Why do you think that the nuclear magnetic resonance described in this interchapter is sometimes called proton magnetic resonance (PMR)?

G-6 What is a mass spectrum?

G-7 Describe how mass spectrometry can be used to identify an unknown compound.

G-8 Describe how a mass spectrometer works (see Figure G-7).

G-9 What is the origin of the spectrum in X-ray fluorescence spectroscopy?

G-10 Describe how X-ray fluorescence spectroscopy can be used to identify unknown compounds.

Linus Pauling, one of America's greatest chemists, was a pioneer in the theory and understanding of chemical bonding. His book, *The Nature of the Chemical Bond,* first published in 1939, is one of the most influential chemistry texts of the twentieth century. During the 1950s, Pauling was in the forefront of the fight against nuclear bomb testing. He was awarded the Nobel Prize for chemistry in 1954 and the Nobel Peace Prize in 1963.

12 / Bonding in Molecules

We learned in Chapter 8 that the electrons in atoms are described in terms of atomic orbitals. In this chapter we shall learn that the electrons in molecules are described in terms of molecular orbitals. Polyatomic molecules can be pictured as a group of atoms held together by covalent bonds. When covalent bonds are localized between pairs of atoms, they are called localized covalent bonds. We shall learn how to describe localized covalent bonds in terms of localized bond orbitals, which are formed by combining atomic orbitals. In a polyatomic molecule, the localized bond orbitals are occupied by the valence electrons and account for the bonding in the molecule. The atomic orbitals we use to construct bond orbitals allow us to describe the geometry of molecules in terms of their bond orbitals. Chemical bonding and molecular geometry are closely related.

In the final sections of the chapter we present a theory of bonding that serves as a basis for localized bond orbitals. We apply this theory, called molecular orbital theory, to the homonuclear diatomic molecules H_2 through Ne_2, learning how to write electron configurations for these molecules and how to predict relative bond lengths and bond energies.

12-1. THE ELECTRONS IN MOLECULES ARE DESCRIBED BY MOLECULAR ORBITALS

The simplest neutral molecule is H_2, which has only two electrons. The Schrödinger equation that applies to H_2 can be solved with a computer. The results for H_2 are interesting because they are similar

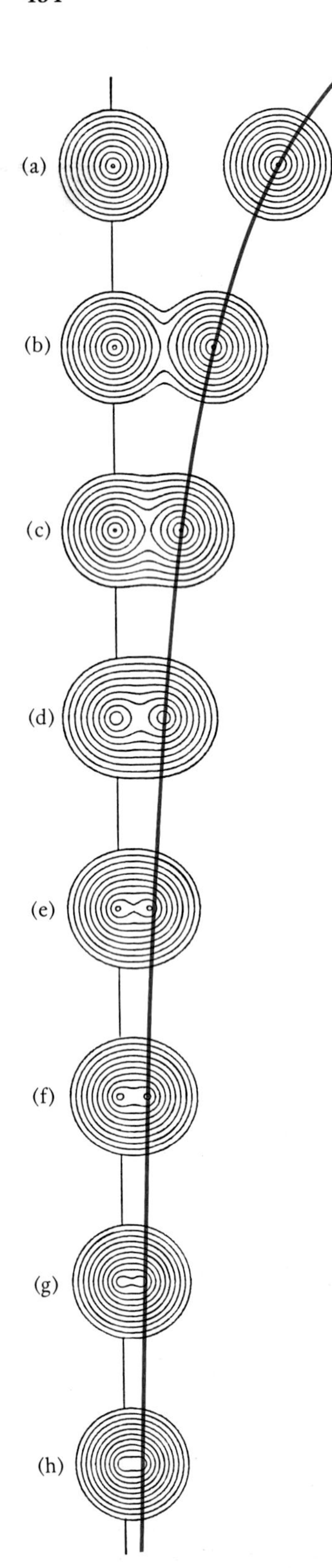

Figure 12-1 Electron-density contour maps of two hydrogen atoms as a function of their separation. At large separations (a), the two orbitals appear simply as those of two separate atoms. As the atoms come together, the two separate atomic orbitals combine into one molecular orbital (through h) encompassing both nuclei.

to the results for more complicated molecules. As a first step in setting up the Schrödinger equation for H_2, the two nuclei are fixed at some given separation. Then the two electrons are included, and the equation is solved to give the wave functions and energies that describe the two electrons. The wave function that corresponds to the lowest energy is the *ground-state wave function* and can be used to compute contour diagrams that show the distribution of the electron density around the two nuclei.

Figure 12-1 shows ground-state electron-density contour diagrams as a function of the internuclear separation of two hydrogen atoms. Note that at large separations the two atoms hardly interact with each other and so the electron density is just that of two electrons, each in a 1*s* orbital about one of the hydrogen atoms. As the separation decreases, however, the two 1*s* orbitals combine into one orbital that is distributed around both nuclei. Such an orbital is called a *molecular orbital* because it extends over both nuclei in the molecule. The buildup of electron density between the nuclei results in a covalent bond that attracts the two nuclei together. Note how the detailed quantum theoretical results shown in Figure 12-1 correspond to our notion that a covalent bond results when an electron pair is shared between two nuclei.

Figure 12-2 shows the energies that correspond to the electron densities of Figure 12-1. Notice that interaction energies have negative values for any separations at which the atoms attract each other. These negative values mean that energy is released when the H_2 bond is formed. Figure 12-2 shows that, for H_2, the interaction energy versus separation R has a minimum at $R = 74$ pm. The internuclear separation at the minimum energy is the predicted length of a bond. The experimentally measured value of the bond length in H_2 is 74 pm, in perfect agreement with the calculated value.

12-2. THE BONDING IN POLYATOMIC MOLECULES CAN BE DESCRIBED IN TERMS OF BOND ORBITALS

Most molecules can be viewed as a group of atoms that are held together by covalent bonds and thus are called *polyatomic molecules*. As a first approximation, the bonding in polyatomic molecules can be analyzed in terms of localized bond orbitals between pairs of bonded atoms. The approach is similar to that described for H_2 in the pre-

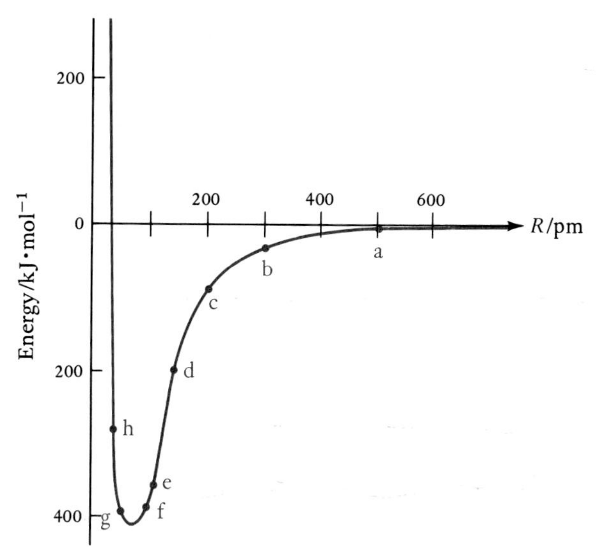

Figure 12-2 The energy of two hydrogen atoms as a function of their separation R. The labels (a) through (h) correspond to those in Figure 12-1. At large distances (a), the two hydrogen atoms do not interact and so their interaction energy is zero. As the two come together, they attract each other and so their interaction energy becomes negative. When they come less than 74 pm apart, the interaction energy increases and they repel each other. The bond length of H_2 is the distance at which the energy is a minimum, that is, 74 pm. The energy at this distance is $-436\ kJ\cdot mol^{-1}$, which is the energy required to dissociate H_2 into two separate hydrogen atoms.

ceding section. For example, consider the methane molecule, CH_4. Each hydrogen atom is joined to the central carbon atom by a covalent bond formed from an atomic orbital on carbon and the 1*s* orbital on hydrogen. As Figure 12-3 suggests, the bonding electrons, and hence the orbitals that describe them, are localized along the carbon-hydrogen bonds. The orbitals that describe the electrons in localized covalent bonds, such as those in methane, are called *localized bond orbitals* and are concentrated primarily in the region between the two atoms that are joined by the bond. The two electrons that occupy a localized bond orbital constitute a localized covalent bond. Note the similarity between the bonding picture in CH_4 and the four bonding electron pairs in the Lewis formula for CH_4:

```
     H
     |
  H—C—H
     |
     H
```

We shall often use Lewis formulas as a guide to constructing bond orbitals.

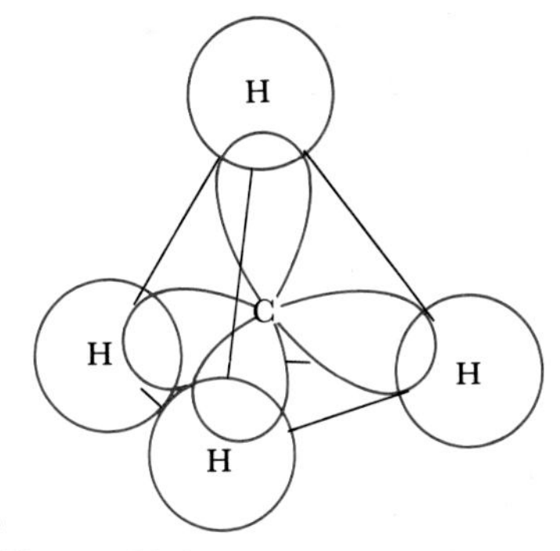

Figure 12-3 The atomic orbitals used to construct localized bond orbitals. The bond orbitals in a methane molecule can be pictured as four carbon-hydrogen bond orbitals, directed toward the vertices of a tetrahedron. A localized bond orbital that is occupied by two electrons with opposite spins constitutes a covalent bond localized between two atoms.

12-3. A SET OF sp^3 ORBITALS POINT TOWARD THE VERTICES OF A TETRAHEDRON

A methane molecule is tetrahedral, and its four carbon-hydrogen bonds are equivalent. Thus we must construct four equivalent localized bond orbitals to describe the bonding in methane. Using a procedure of quantum mechanics that need not concern us here, we can combine the 2*s* orbital and all three 2*p* orbitals on the carbon atom to get four equivalent *hybrid atomic orbitals* (Figure 12-4), so called because they are formed by combining different atomic orbitals on the same atom. Because these four equivalent hybrid orbitals result from combining the 2*s* and all three 2*p* orbitals on the carbon atom, they are called sp^3 ("s-p-three") *orbitals*. Figure 12-4 shows that sp^3 hybrid orbitals point to the vertices of a tetrahedron. The four equivalent localized bond orbitals in CH_4 are formed by combining each one of the four sp^3 orbitals with a different hydrogen 1*s* orbital. The combination of a carbon sp^3 orbital and a hydrogen 1*s* orbital is shown in Figure 12-5. There are four valence electrons from the carbon atom and one valence electron from each of the four hydrogen atoms for a total of eight valence electrons in CH_4. Two valence electrons of opposite spin occupy each of the four equivalent localized bond orbitals and so form the four equivalent covalent bonds in CH_4.

Figure 12-4 The 2*s* and three 2*p* orbitals on an atom can be combined to give four sp^3 hybrid orbitals, which are all equivalent and point toward the vertices of a tetrahedron. The angle between sp^3 orbitals is the tetrahedral bond angle, 109.5°.

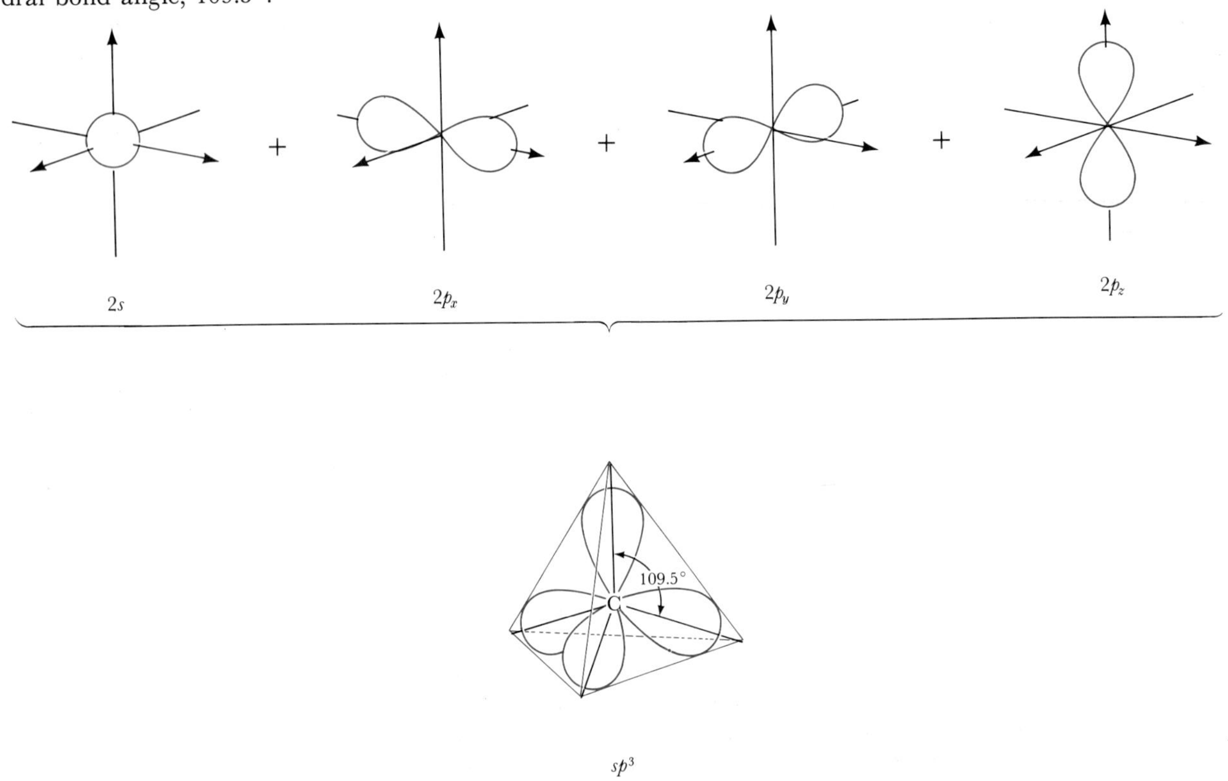

2*s* + $2p_x$ + $2p_y$ + $2p_z$

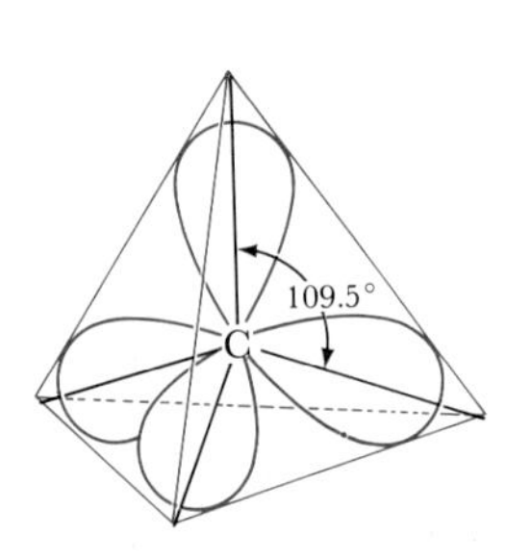

sp^3

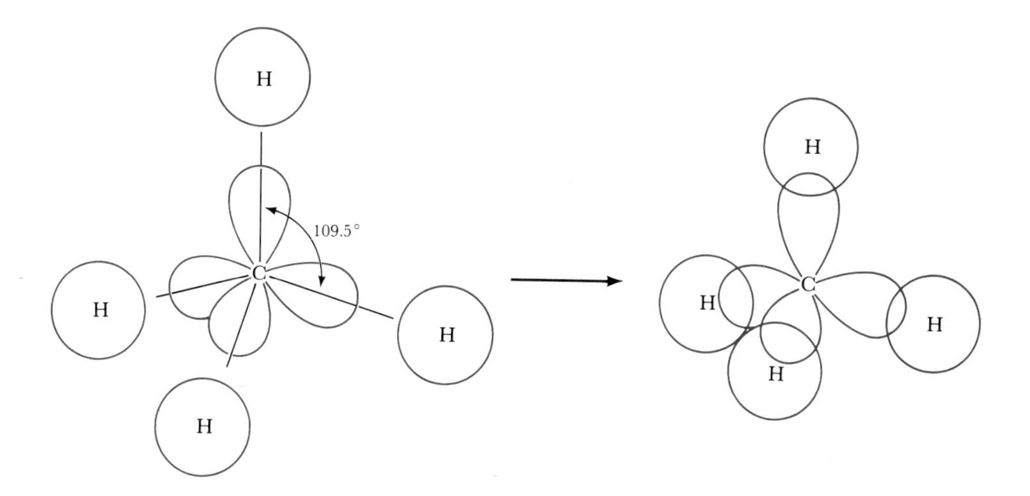

Each carbon-hydrogen bond orbital in Figure 12-5 is cylindrically symmetric, that is, has a circular cross section when viewed along a line between the hydrogen and carbon nuclei. Atomic orbitals that are spherically symmetric, and hence have a circular cross section, are called *s* orbitals. Bond orbitals or molecular orbitals that have circular cross sections when viewed along an internuclear axis are called *σ orbitals* ("sigma orbitals"; σ is the Greek letter corresponding to s). The carbon-hydrogen bond orbitals in methane are σ orbitals. When a σ orbital is occupied by two electrons of opposite spins, the result is called a *σ bond.*

Although we have used the carbon atom in a methane molecule to illustrate the use of sp^3 hybrid orbitals, these orbitals also can be used on other atoms. The following example shows the use of sp^3 hybrid orbitals on a nitrogen atom to describe the bonding in the ammonium ion.

Figure 12-5 Four equivalent localized bond orbitals in CH_4 are formed by combining the four carbon sp^3 orbitals and the hydrogen 1*s* orbitals. There are eight valence electrons in CH_4 (four from the carbon atom and one from each hydrogen atom). Each of the four localized bond orbitals is occupied by a pair of electrons of opposite spin, accounting for the four localized carbon-hydrogen bonds in CH_4.

Example 12-1: Describe the bonding in the ammonium ion, NH_4^+, whose Lewis formula is

```
⎡     H    ⎤+
⎢     |    ⎥
⎢ H — N — H⎥
⎢     |    ⎥
⎣     H    ⎦
```

Solution: We learned in Chapter 11 that NH_4^+ is tetrahedral. Thus we wish to form four localized bond orbitals that point toward the vertices of a tetrahedron. We can do this by forming sp^3 hybrid orbitals on the nitrogen atom by combining the nitrogen 2*s* orbital and all three of the nitrogen 2*p* orbitals. The sp^3 hybrid orbitals on a nitrogen atom are similar to the carbon atom sp^3 orbitals shown in Figure 12-4. We can form four equivalent localized bond orbitals by combining each sp^3 orbital on the nitrogen atom with a hydrogen 1*s* orbital. There are eight valence electrons in NH_4^+. Two valence electrons of opposite spins occupy each of the four bond orbitals, accounting for the four covalent bonds in NH_4^+. The bonding and the shape of the ammonium ion are similar to what is shown for methane in Figure 12-5.

Ethane is called a hydrocarbon because it consists of hydrogen and carbon. Ethane is one of many hydrocarbons.

We use sp^3 orbitals to describe the bonding in many molecules. For example, consider the hydrocarbon ethane, C_2H_6, whose Lewis formula is

```
    H   H
    |   |
H — C — C — H
    |   |
    H   H
```

Ball-and-stick and space-filling models of ethane are shown in Figure 12-6. The shape of an ethane molecule can be described in terms of sp^3 hybrid orbitals on the carbon atoms. The carbon-carbon bond orbital in ethane is formed by the combination of two sp^3 orbitals, one from each carbon atom; the six carbon-hydrogen bond orbitals in ethane result from the combination of the three remaining sp^3 orbitals on each carbon atom with the hydrogen 1*s* atomic orbitals. Note that there are seven bond orbitals in ethane. Each carbon atom has 4 valence electrons and each hydrogen atom has 1, giving a total of 14 valence electrons in ethane. The 14 valence electrons occupy the 7 bond orbitals in ethane such that each bond orbital has 2 electrons of opposite spins. The resulting bonding in ethane is shown

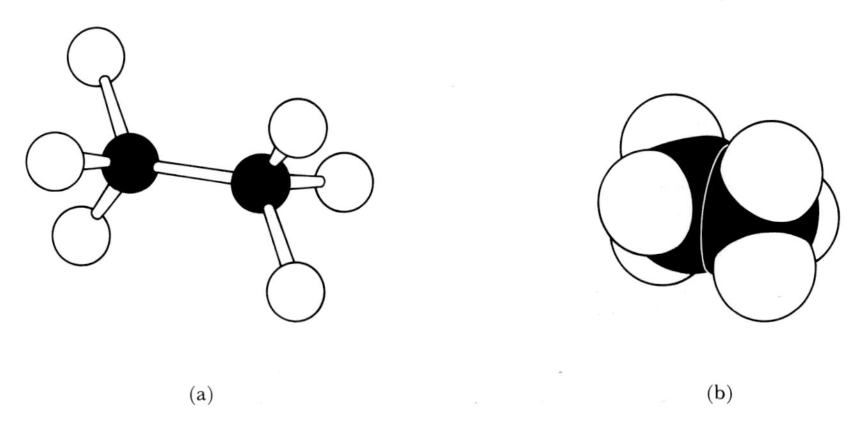

Figure 12-6 (a) Ball-and-stick and (b) space-filling models of ethane. Note that the bonds about each carbon atom are tetrahedrally oriented.

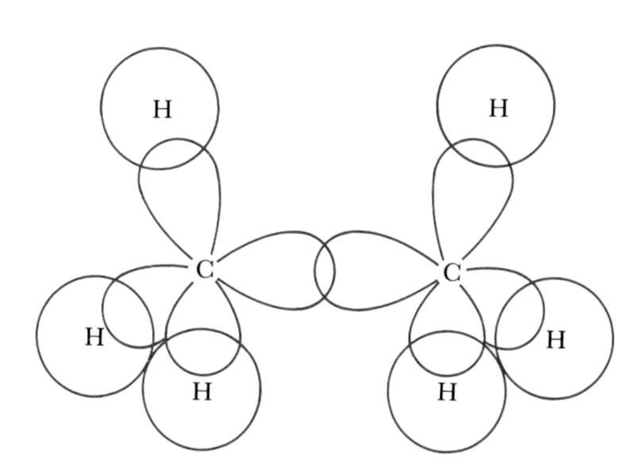

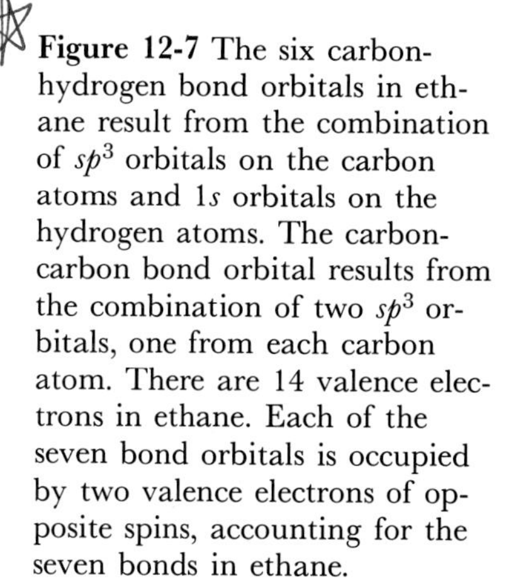

Figure 12-7 The six carbon-hydrogen bond orbitals in ethane result from the combination of sp^3 orbitals on the carbon atoms and 1s orbitals on the hydrogen atoms. The carbon-carbon bond orbital results from the combination of two sp^3 orbitals, one from each carbon atom. There are 14 valence electrons in ethane. Each of the seven bond orbitals is occupied by two valence electrons of opposite spins, accounting for the seven bonds in ethane.

in Figure 12-7. The two sp^3 orbitals are combined such that the resulting carbon-carbon bond orbital is circular when viewed along a line joining the two carbon atoms. Thus the carbon-carbon bond orbital in ethane is cylindrically symmetric around the bond axis and is a σ orbital, and the carbon-carbon bond is a σ bond. The ethane molecule has seven σ bonds.

12-4. WE USE sp^3 ORBITALS TO DESCRIBE THE BONDING IN MOLECULES CONTAINING OXYGEN ATOMS

We can use sp^3 orbitals to describe the bonding about an oxygen atom as well as about a carbon atom. For example, consider H_2O,

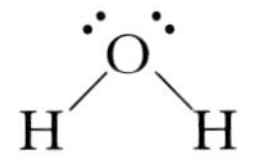

The two bond orbitals in a water molecule can be pictured as a combination of each hydrogen 1s orbital with one of the sp^3 orbitals on the oxygen atom to produce two O—H σ bond orbitals (Figure 12-8). There are eight valence electrons in H_2O, six from the oxygen atom and one from each hydrogen atom. Four of the valence electrons occupy the two O—H σ bond orbitals. The other four occupy the two nonbonded sp^3 orbitals and constitute the two lone electron pairs on the oxygen atom. We would predict on the basis of this bonding that the H—O—H bond angle is 109.5°, whereas the experimental value is 104.5°. The 5° difference can be attributed to the fact that the four orbitals about the oxygen atom are not equivalent. Two are used to form bonds with the hydrogen atoms, and two are used for the lone-pair electrons. Recall from our discussion of VSEPR theory that we predicted the H—O—H bond angle in H_2O to be somewhat less than the tetrahedral value of 109.5° because the lone electron pairs repel the two hydrogen-oxygen bonds.

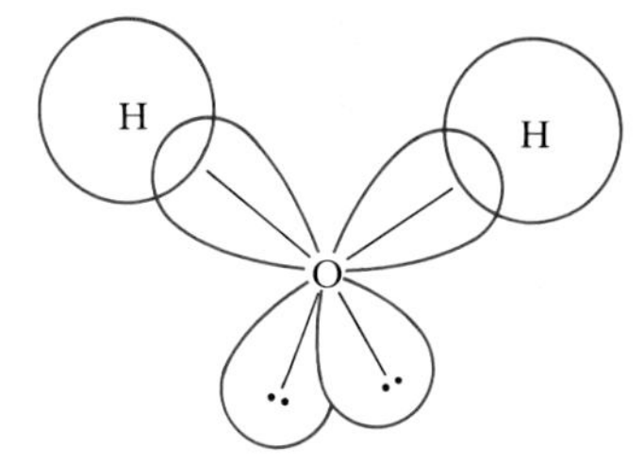

Figure 12-8 Bonding in the water molecule. The two hydrogen-oxygen bond orbitals are formed by the combination of two sp^3 orbitals on the oxygen atom with the hydrogen 1s orbitals. Of the eight valence electrons in the molecule, four occupy the two bond orbitals and four occupy the two nonbonded sp^3 orbitals on the oxygen atom. The latter are lone electron pairs.

Example 12-2: The ammonia molecule is trigonal pyramidal,

with H—N—H bond angles of 107°. Use hybrid orbitals to describe the bonding in NH_3.

Solution: Ammonia has three covalent bonds and one lone pair of electrons. We know from VSEPR theory that the four electron pairs in the valence shell of the nitrogen atom point toward the vertices of a tetrahedron. We can describe the bonding in NH_3 by assuming that the $2s$ and three $2p$ orbitals of the nitrogen atom form four sp^3 orbitals. Three of these sp^3 orbitals form localized bond orbitals by combining with the hydrogen $1s$ orbitals. Thus we can describe the bonding in an ammonia molecule in terms of three localized bond orbitals and a lone-pair (nonbonded) sp^3 orbital on the nitrogen atom.

There are eight valence electrons in NH_3. Six of them occupy the three localized bond orbitals and two occupy the nonbonded sp^3 orbital. The three fully occupied bond orbitals describe the three covalent bonds in NH_3; the fully occupied nonbonded sp^3 orbital on the nitrogen atom describes the lone pair of electrons in NH_3 (Figure 12-9). The use of sp^3 orbitals predicts that the H—N—H bond angles are 109.5°. The four valence orbitals in NH_3 are not equivalent (one describes a lone pair), however, and we should expect to find small deviations from a regular tetrahedral shape. The observed H—N—H bond angles in NH_3 are 107°.

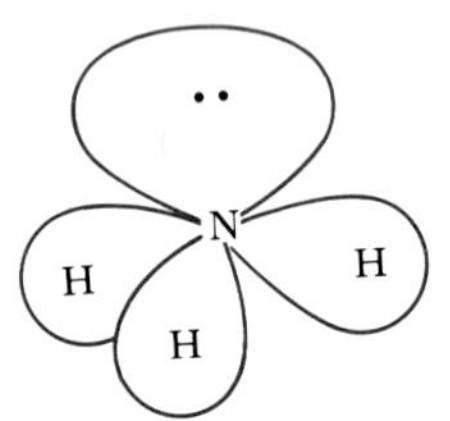

Figure 12-9 The use of sp^3 hybrid orbitals on nitrogen to describe the bonding in NH_3. Three of the nitrogen sp^3 orbitals are combined with hydrogen $1s$ orbitals to form three equivalent localized bond orbitals. The fourth nitrogen sp^3 orbital is a nonbonded orbital and is occupied by the lone pair of electrons in ammonia.

12-5 A DOUBLE BOND CONSISTS OF A σ BOND AND A π BOND

One of the simplest molecules in which there is a double bond is ethene, C_2H_4, which is more commonly known as ethylene. Its Lewis formula is

Ethylene is another example of a hydrocarbon.

Ethylene is a gas at room temperature. It has the unusual property of causing fruit to ripen.

The geometry of an ethylene molecule is quite different from that of ethane or the other alkanes. All six atoms in an ethylene molecule lie in one plane, as shown in Figure 12-10. The bonding in ethylene cannot be described in terms of sp^3 hybrid orbitals because the bonds to carbon are not arranged tetrahedrally.

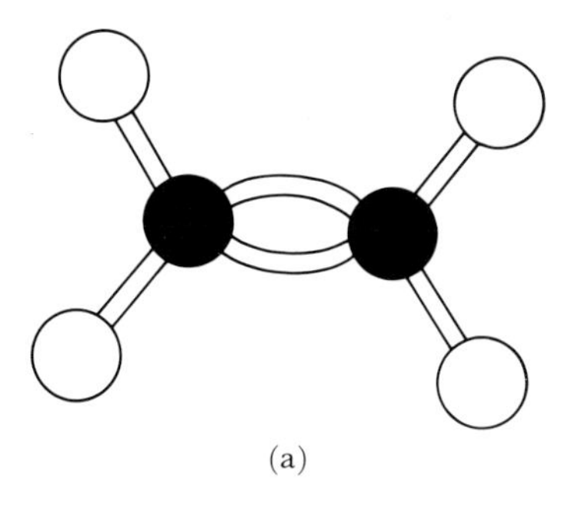

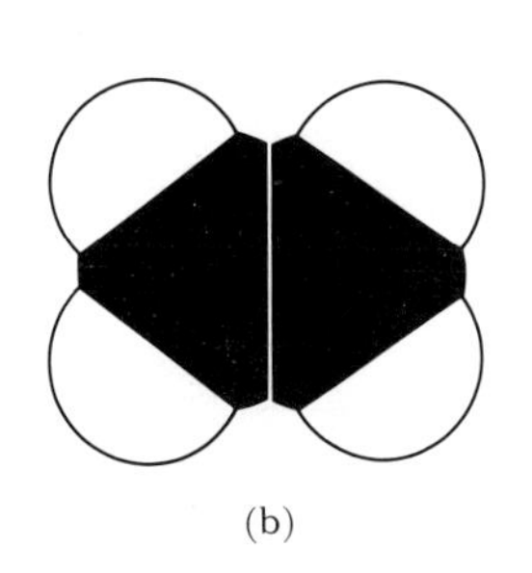

Figure 12-10 (a) Ball-and-stick and (b) space-filling models of ethylene. The important structural feature of an ethylene model is that it is planar; all six atoms lie in a single plane. This is due to the geometry associated with a carbon-carbon double bond.

If instead of combining a carbon $2s$ orbital and three $2p$ orbitals, we combine the $2s$ orbital with only two $2p$ orbitals, then the result is three equivalent hybrid orbitals called *sp^2 ("s-p-two") orbitals.* The three sp^2 orbitals lie in a plane and are directed 120° from each other, as shown in Figure 12-11. Note that the three sp^2 orbitals point toward the vertices of an equilateral triangle. Three sp^2 hybrid orbitals are formed by combining the $2s$ and two $2p$ orbitals on an atom. The remaining p orbital is perpendicular to the plane formed by the three sp^2 orbitals (Figure 12-11b). The sp^2 orbitals are used to describe the bonding to an atom about which the geometry is trigonal planar.

The first step in describing the bonding in ethylene is to join the two carbon atoms by combining an sp^2 orbital from each, as shown in Figure 12-12. The resulting carbon-carbon bond is a σ bond. The four hydrogen atoms are bonded, two to each carbon atom, by combining the hydrogen $1s$ orbitals with the four remaining sp^2 orbitals

Figure 12-11 The formation of sp^2 hybrid orbitals by combining the $2s$ orbital and two $2p$ orbitals on an atom. The three sp^2 orbitals formed are equivalent, lie in a plane, and are directed 120° from each other. In (a), for simplicity, only the two $2p$ orbitals that are combined with the $2s$ orbital are shown. In (b), all three $2p$ orbitals are shown. The $2p$ orbital that is not combined with the $2s$ orbital is perpendicular to the plane formed by the three sp^2 orbitals.

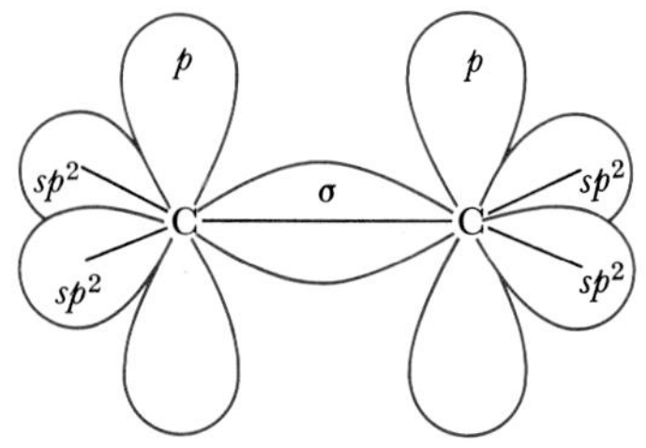

Figure 12-12 Two carbon atoms joined by the combination of an sp^2 orbital from each. The resulting bond orbital is cylindrically symmetric around the carbon-carbon axis and so is a σ bond orbital. The carbon-carbon σ bond orbital constitutes part of the double bond in ethylene.

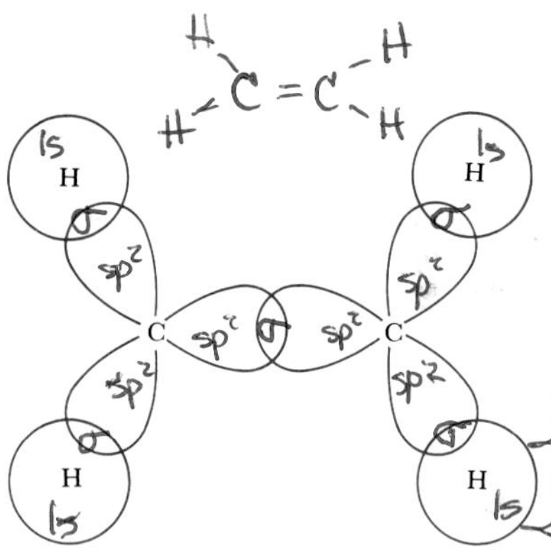

Figure 12-13 The σ-bond framework in ethylene. The carbon-carbon bond orbital results from the combination of two sp^2 orbitals, one from each carbon atom. The four carbon-hydrogen bond orbitals result from the combination of carbon sp^2 orbitals and hydrogen 1*s* orbitals. The remaining *p* orbitals on the carbon atoms are not shown but are perpendicular to the page.

on the carbon atoms, as shown in Figure 12-13. All five bonds formed so far are σ bonds, and Figure 12-13 shows the *σ-bond framework* in ethylene.

Recall that there is an unused 2*p* orbital on each carbon atom and that both are perpendicular to the H—C—H plane (Figure 12-12). The two *p* orbitals are combined to give the orbital denoted by the symbol π in Figure 12-14. The double bond in ethylene is described by the σ orbital *and* the π orbital in Figure 12-14. The orbital that we denote by π in Figure 12-14 is not cylindrically symmetric along the bond axis. It does not have a circular cross section. Instead, its cross section is similar to that of an atomic *p* orbital, and so it is called a *π orbital* ("pi orbital"; π is the Greek letter that corresponds to p). When two electrons occupy a π orbital, the result is a *π bond.* A σ bond and a π bond do not have the same energy, and so a double bond, although stronger than a single bond, is *not* twice as strong as a single bond. Single carbon-carbon bond energies are about $350\ kJ \cdot mol^{-1}$, whereas double carbon-carbon bond energies are about $600\ kJ \cdot mol^{-1}$.

Figures 12-13 and 12-14 show the six bond orbitals in an ethylene molecule. Five are σ bond orbitals (Figure 12-13) and one is a π bond orbital (Figure 12-14). There are 12 valence electrons in ethylene (4 from each carbon atom and 1 from each hydrogen atom), and they occupy the 6 bond orbitals in pairs to give four single bonds and one double bond.

Example 12-3: Describe the bonding in the formaldehyde molecule, which has the Lewis formula

$$\begin{array}{l} H \\ \quad \diagdown \\ \quad\ C{=}\ddot{\underset{..}{O}}: \\ \quad \diagup \\ H \end{array}$$

Solution: From VSEPR theory we conclude that the formaldehyde molecule is planar with a trigonal geometry around the carbon atom. Because the bond angles are about 120°, we use sp^2 hybrid orbitals on carbon. Further, because there are also three groups of electrons around the oxygen atom (the double bond and two lone pairs), we also use sp^2 hybrid orbitals on the oxygen. An sp^2 orbital on carbon is combined with an sp^2 orbital on oxygen to form a carbon-oxygen σ bond. The remaining two sp^2 orbitals on carbon are combined with the 1*s* orbitals on H to form the two C—H σ bonds. The remaining two sp^2 orbitals on oxygen are used to house the two lone pairs on oxygen. The remaining *p* orbital on carbon and the remaining *p* orbital on oxygen, both of which are perpendicular to the plane of the molecule, are combined to form a carbon-oxygen π bond. Thus, the carbon-oxygen double bond is composed of a σ bond and a π bond. The bonding in formaldehyde is shown in Figure 12-15.

12-6. THERE IS NO ROTATION ABOUT CARBON-CARBON DOUBLE BONDS

The double bond in ethylene consists of a σ bond and a π bond, each containing two electrons. The π bond locks the molecule into a planar shape, and so ethylene is a flat, or planar, molecule. *There is no rotation about carbon-carbon double bonds.* Consider the molecule 1,2-dichloroethene, ClCH=CHCl. Because there is no rotation about the carbon-carbon double bond, there are two distinct forms of 1,2-dichloroethene:

:Cl: and H on C=C, H and :Cl: — *trans* and :Cl: and :Cl: on C=C, H and H — *cis*

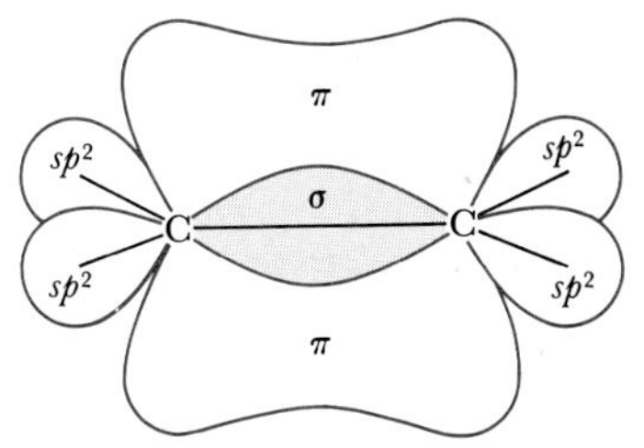

Figure 12-14 A double bond consists of a σ bond and a π bond. The σ bond results from the combination of two sp^2 orbitals, one from each atom. The π bond results from the combination of two p orbitals, one from each atom. The π orbital maintains the σ-bond framework in a planar shape and prevents rotation about the double bond.

The first of these is called *trans*-1,2-dichloroethene because the chlorine atoms lie across (trans means across) the double bond from each other. The other is called *cis*-1,2-dichloroethene because the chlorine atoms lie on the same side of the double bond. The 1,2- notation tells us that the chlorine atoms are attached to different carbon atoms. Molecules with the same atom-to-atom bonding but different spatial arrangements are called *stereoisomers*. The particular type of stereoisomerism that is displayed by 1,2-dichloroethene is called *cis-trans isomerism*. Stereoisomers, and *cis-trans* isomers in particular, have different physical properties. For example, the boiling point of the *trans* isomer of 1,2-dichloroethene is 48 °C and that of the *cis* isomer is 60 °C.

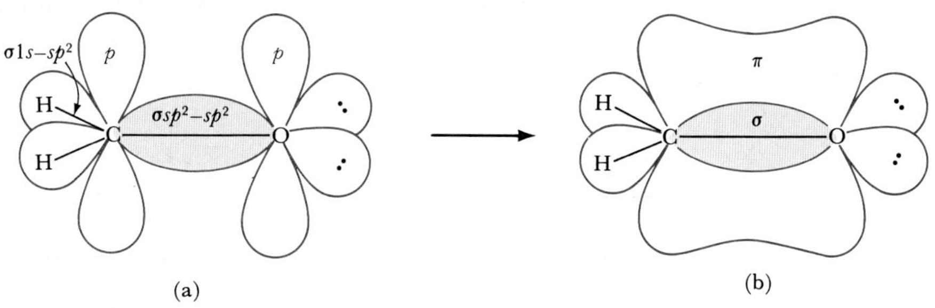

Figure 12-15 The bonding in formaldehyde, HCHO. (a) The σ-bond framework, showing the unused $2p$ orbitals that are perpendicular to the plane formed by the four atoms. These two $2p$ orbitals combine to form a π bond. (b) The carbon-oxygen double bond consists of one σ bond and one π bond.

12-7. PHYSIOLOGICAL ACTIVITY CAN BE AFFECTED DRAMATICALLY BY MOLECULAR SHAPE

An example of the importance of *cis-trans* isomerism occurs in the chemistry of vision. Although we have stated that there is no rotation allowed about double bonds, *cis* and *trans* isomers can interconvert if the molecule is supplied with sufficient energy in the form of heat or light:

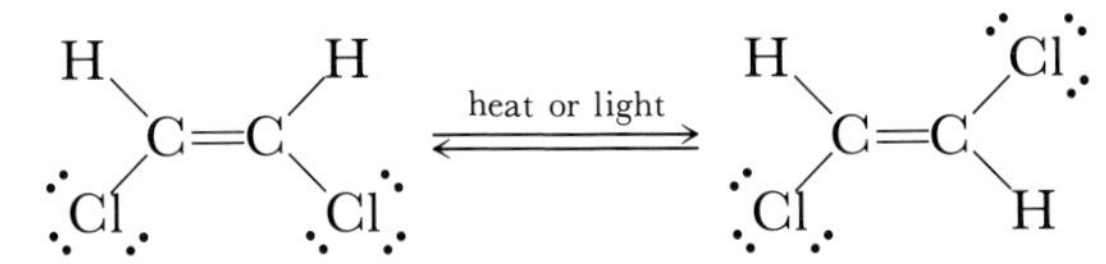

cis $\underset{}{\overset{\text{heat or light}}{\rightleftharpoons}}$ *trans*

It was determined in the 1950's that the chemistry of vision involves *cis*-to-*trans* isomerization. The retina of the eye contains a substance called rhodopsin, which consists of a molecule called 11-*cis*-retinal combined with a protein called opsin. When 11-*cis*-retinal is struck by a photon of visible light, it isomerizes at the *cis* double bond to give 11-*trans*-retinal:

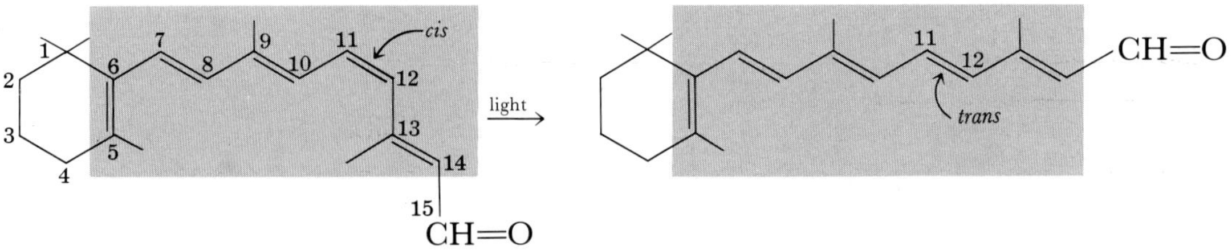

11-*cis*-retinal $\xrightarrow{\text{light}}$ 11-*trans*-retinal

(The numbers in this drawing refer to carbon atoms, which are not shown.) The shading in these formulas represents a planar region in the molecule. The shapes of the *cis* and *trans* isomers are significantly different, and the light-induced change in shape triggers a response in the optic nerve cells that is transmitted to the brain and perceived as vision. The vision response occurs through a sequence of processes that has been investigated thoroughly. The primary event, however, is the conversion of the *cis* to the *trans* isomer of retinal.

12-8. A TRIPLE BOND CONSISTS OF ONE σ BOND AND TWO π BONDS

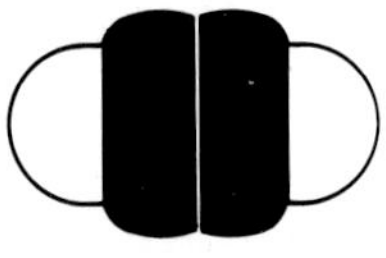

Let us next consider a molecule that contains a triple bond. A good example is ethyne, C_2H_2, which is more commonly called acetylene (Figure 12-16). The Lewis formula for acetylene is

$$H{-}C{\equiv}C{-}H$$

Figure 12-16 A space-filling model of acetylene, C_2H_2, a linear molecule.

2s 2p sp

(a)

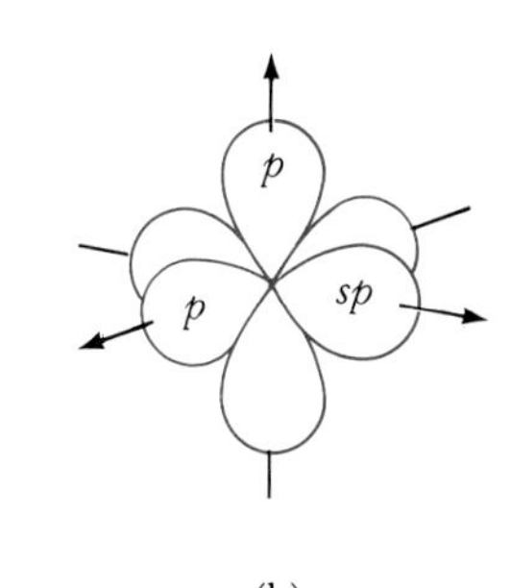

(b)

Figure 12-17 The formation of *sp* hybrid orbitals results from combining the 2*s* orbital and one 2*p* orbital on an atom. The two *sp* orbitals are equivalent and are directed 180° from each other. In (a), for simplicity, only the 2*p* orbital that is combined with the 2*s* orbital is shown. In (b), all three 2*p* orbitals are shown. The two 2*p* orbitals that are not combined with the 1*s* orbital are perpendicular to each other and to the line formed by the *sp* orbitals.

Acetylene is a colorless gas with a penetrating odor. One of its most important uses is in oxyacetylene torches.

If a 2*s* orbital and only one 2*p* orbital on an atom are combined, the result is a set of two equivalent hybrid orbitals, called *sp orbitals,* that point 180° from each other (Figure 12-17). When the *sp* orbitals are combined with atomic orbitals from other atoms, the resulting geometry is *linear.* Figure 12-17 also shows that the two remaining *p* orbitals are perpendicular to the two *sp* orbitals. The bonding in acetylene is described in terms of *sp* hybrid orbitals. We can build a σ-bond framework for acetylene in two steps. We first form the carbon-carbon bond orbital by combining two *sp* orbitals, one from each carbon atom. Then we form the carbon-hydrogen bond orbitals by combining a hydrogen 1*s* orbital with the remaining *sp* orbital on each carbon atom. The σ-bond framework of acetylene is shown in Figure 12-18. Because the *sp* orbitals on the carbon atoms are directed 180° from each other, the σ-bond framework of acetylene makes it a linear molecule.

The remaining carbon 2*p* orbitals are perpendicular to the H—C—C—H axis, as shown in Figure 12-19. These orbitals can combine to produce two π bond orbitals. There are five bond orbitals in acetylene, three σ and two π. There are 10 valence electrons in acetylene [(2 × 4) + (2 × 1)], and they occupy the five bonding orbitals. *The carbon-carbon triple bond consists of one σ bond and two π bonds.* The shape of *p* orbitals is such that the two π bond orbitals in acetylene together form a cylindrical shell along the carbon-carbon bond, producing a distribution of π electrons like that shown in Figure 12-20.

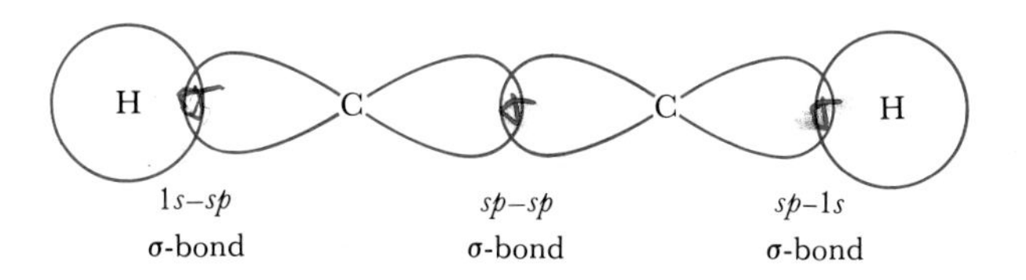

Figure 12-18 The σ-bond framework of acetylene. The carbon-carbon σ bond orbital results from combining two *sp* orbitals, one from each carbon atom. Each two carbon-hydrogen bond orbital results from combining a carbon *sp* orbital and a hydrogen 1*s* orbital.

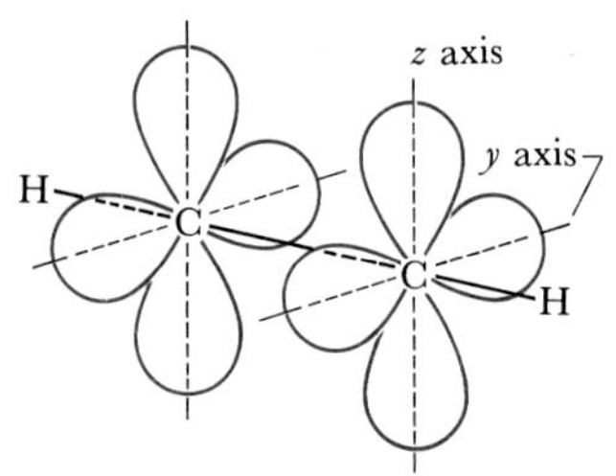

Figure 12-19 The $2p$ orbitals on the carbon atoms in acetylene. The $2p$ orbitals that are directed along the z axis combine to form one π bond orbital, and the $2p$ orbitals directed along the y axis overlap to form another π bond orbital.

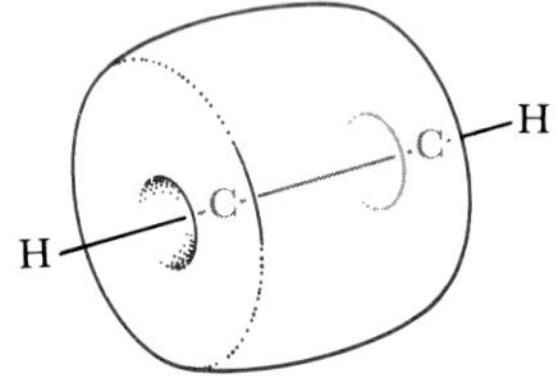

Figure 12-20 The π electron distribution in a triple bond, which consists of one σ bond and two π bonds. Although neither π bond orbital is separately cylindrically symmetric about the carbon-carbon axis, the two together are cylindrically symmetric.

Example 12-4: Compare the bonding in acetylene with that in hydrogen cyanide, HCN.

Solution: The Lewis formula for HCN is

$$\mathrm{H{-}C{\equiv}N:}$$

Because a triple bond consists of one σ bond and two π bonds, we must use *sp* orbitals on both the carbon atom and the nitrogen atom in HCN. The σ-bond framework of HCN can be written as

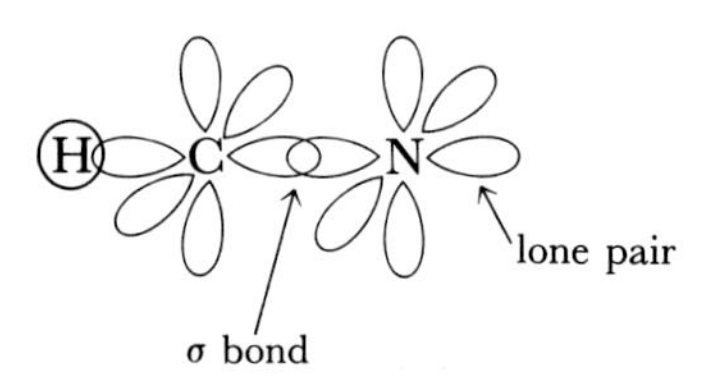

We see that HCN is a linear molecule and that its σ-bond framework is similar to that of C_2H_2 (Figures 12-18 and 12-19). The unused $2p$ orbitals of the carbon and nitrogen atoms combine to form the two π bond orbitals. There are four bond orbitals in HCN; two are σ bond orbitals and two are π bond orbitals. There are 10 valence electrons in HCN: 8 occupy the 4 bond orbitals and 2 occupy the nitrogen *sp* orbital and constitute a lone electron pair on the nitrogen atom. The π electron density in HCN is cylindrically symmetric, as it is in C_2H_2.

Table 12-1 summarizes the properties of sp^3, sp^2, and *sp* hybrid orbitals.

12-9. THE π ELECTRONS IN BENZENE ARE DELOCALIZED

There are many molecules that have π orbitals that extend over more than two adjacent atoms. One of the simplest examples of such a molecule is benzene, C_6H_6, a clear, colorless, poisonous, highly flammable liquid (bp 80.1°C) with a characteristic odor. Benzene is ob-

Table 12-1 Properties of hybrid orbitals

Hybrid orbital	*Geometry*	*Number formed*	*Bond angle*	*Example*
sp	linear	2	180°	acetylene
sp^2	trigonal planar	3	120°	ethylene
sp^3	tetrahedral	4	109.5°	methane, ethane

tained from petroleum and coal tar and has a great many chemical uses. It has the two resonance Lewis formulas

Benzene is a hydrocarbon.

```
        H                      H
        |                      |
  H    C     H          H     C     H
    \ //  \ /             \ /  \\ /
     C      C              C      C
     |      ||     ↔       ||     |
     C      C              C      C
    / \\  / \             / \  // \
  H    C     H          H    C     H
       |                      |
       H                      H
```

Using the concept of superposing resonance structures, we can write the Lewis formula for benzene as

```
        H
        |
  H    C     H
    \ /  \ /
     C ( ) C
     | ( ) |
     C ( ) C
    / \  / \
  H    C     H
       |
       H
```

The bonds in this formula are superpositions of single and double bonds and may be considered to be intermediate between single and double bonds. The carbon-carbon bond distance in benzene is 140 pm, intermediate between the usual single-bond (154 pm) and double-bond (134 pm) carbon-carbon bond lengths.

We can describe the bonding in benzene in terms of σ and π bonds. The bonding about each carbon atom in benzene is trigonal, and so we assign sp^2 orbitals to each. This leads directly to the σ-bond framework shown in Figure 12-21. The angles in a regular hexagon

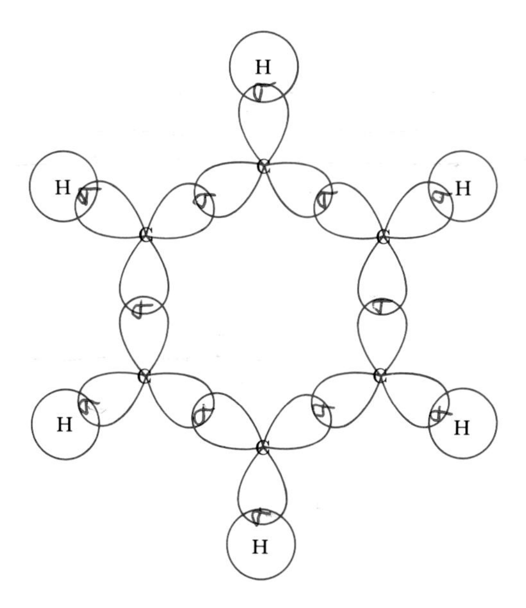

Figure 12-21 The σ-bond framework in a benzene molecule. Each carbon-carbon bond orbital results from the combination of sp^2 orbitals, and each carbon-hydrogen bond orbital results from the combination of a carbon sp^2 orbital and a hydrogen $1s$ orbital. All 12 atoms lie in a single plane, so benzene is a planar molecule. The six carbon atoms form a regular hexagon.

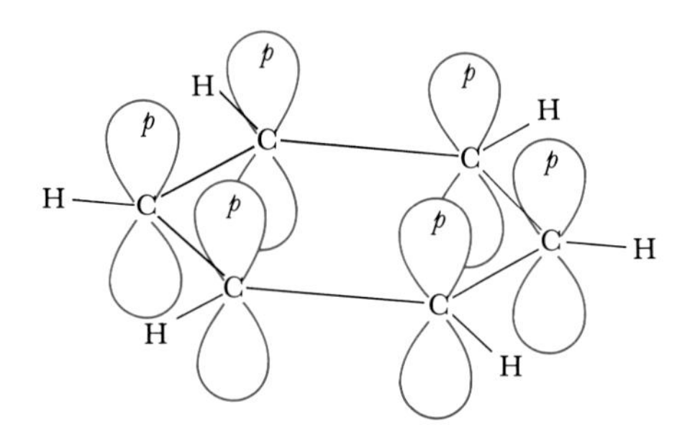

Individual p orbitals of benzene

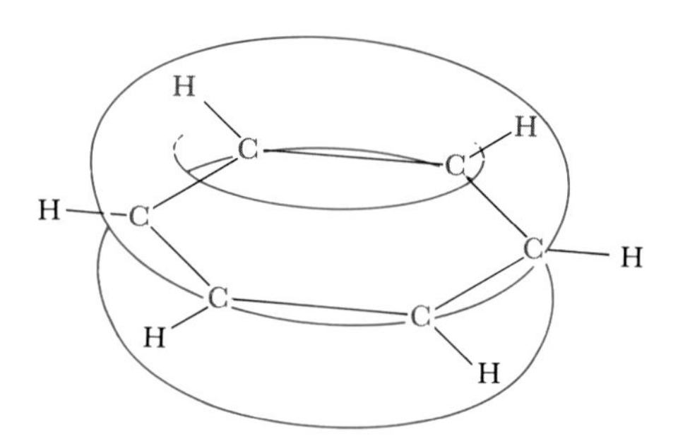

Nonlocalized bonding between benzene p orbitals

Figure 12-22 Each carbon atom in a benzene ring has a $2p$ orbital perpendicular to the ring. These six $2p$ orbitals combine to form three π orbitals that are spread uniformly over the entire ring.

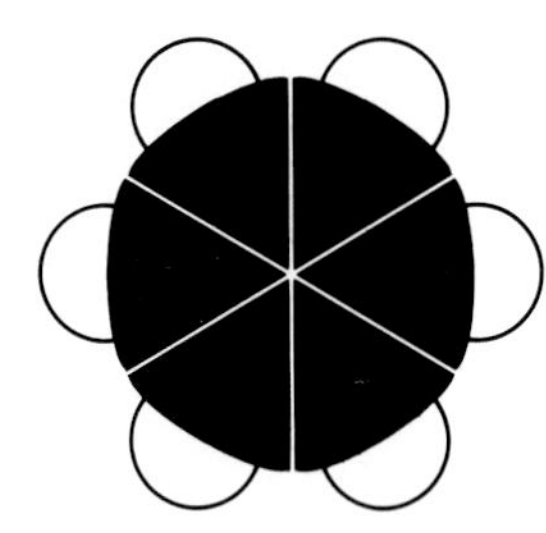

Figure 12-23 A space-filling model of a benzene molecule. Notice the planar hexagonal geometry.

are 120°, and this fits nicely with the 120° bond angles associated with sp^2 orbitals if the σ-bond framework in benzene is planar and if the carbon atoms form a regular hexagon (Figure 12-21). Note that there are 12 σ bond orbitals in benzene. Each carbon atom has a $2p$ orbital that is perpendicular to the hexagonal plane. These six p orbitals can combine into three π orbitals spread uniformly over the entire ring (Figure 12-22). A molecular model of benzene is shown in Figure 12-23.

Example 12-5: Determine the total number of valence electrons in benzene and compare this result with the total number of bond orbitals in the molecule.

Solution: Each of the 6 carbon atoms has 4 valence electrons, and each of the 6 hydrogen atoms has 1, giving a total of 30 valence electrons in benzene. Figure 12-21 shows that there are 12 σ bond orbitals. In addition to these, there are three π orbitals. Thus, there is a total of 15 bond orbitals in benzene. The 30 valence electrons occupy the 15 bond orbitals such that each orbital is occupied by 2 electrons of opposite spins.

The π orbitals in benzene are not associated with any particular pair of carbon atoms. Instead they are spread over the entire ring (Figure 12-22). Thus the π orbitals in benzene, and the electrons that occupy these orbitals, are said to be *delocalized*. The electrons in π orbitals are referred to as π electrons. The delocalization of the π electrons around the benzene ring is an example of electron *charge delocalization*, a phenomenon that confers an extra degree of stability on a molecule and accounts for the relative resistance of the benzene ring to many chemical reagents.

The delocalization of the π electrons in benzene is in accord with the superposition of its resonance formulas. It is common to write the Lewis formula for benzene as where the circle represents the even distribution of π electrons over all six carbon atoms. A common abbreviation for this Lewis formula is simply

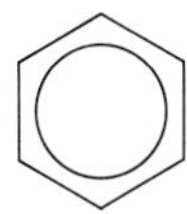

in which each vertex represents a carbon atom and the hydrogen atoms are understood.

The benzene ring is part of the chemical formulas of a great many important chemicals.

12-10. MOLECULAR ORBITALS CAN BE ORDERED ACCORDING TO THEIR ENERGIES

Although we have been able to describe the bonding in a variety of molecules, we still have no insight into such questions as why, for example, two hydrogen atoms join to form a stable molecule whereas two helium atoms do not. In these sections we discuss a theory of bonding, called *molecular orbital theory,* that can answer such questions. Although molecular orbital theory can be applied to all molecules, for simplicity, we shall consider only *homonuclear diatomic molecules,* that is, molecules consisting of two identical atoms.

Recall that we describe the electronic structure of atoms in terms of atomic orbitals, which are based on the set of orbitals that were given for a hydrogen atom. Because a hydrogen atom has only one electron, its atomic orbitals are relatively simple and serve as approximate orbitals for more complicated atoms. A one-electron system that applies to homonuclear diatomic molecules is the *hydrogen molecular ion,* H_2^+, which consists of two protons and one electron. The H_2^+ ion is stable relative to a separated H and H^+; its bond length is 106 pm and its dissociation energy is 255 $kJ \cdot mol^{-1}$.

The H_2^+ molecular orbitals form the basis of diatomic molecular orbitals, just as hydrogen atomic orbitals form the basis for the atomic orbitals of other atoms.

When we solve the Schrödinger equation for H_2^+, we obtain a set of wave functions, or orbitals, and a corresponding set of energies. Because these orbitals extend over both nuclei in H_2^+, they are molecular orbitals. Just as we discussed the shapes of the various hydrogen atomic orbitals and then used them to build up more complicated atoms, we discuss the various H_2^+ molecular orbitals and then use them to build up more complicated diatomic molecules. Figure 12-24 shows the shapes of the first few molecular orbitals of H_2^+. Each shape represents the three-dimensional surface that encloses a certain probability of finding the electron within the volume enclosed by that surface.

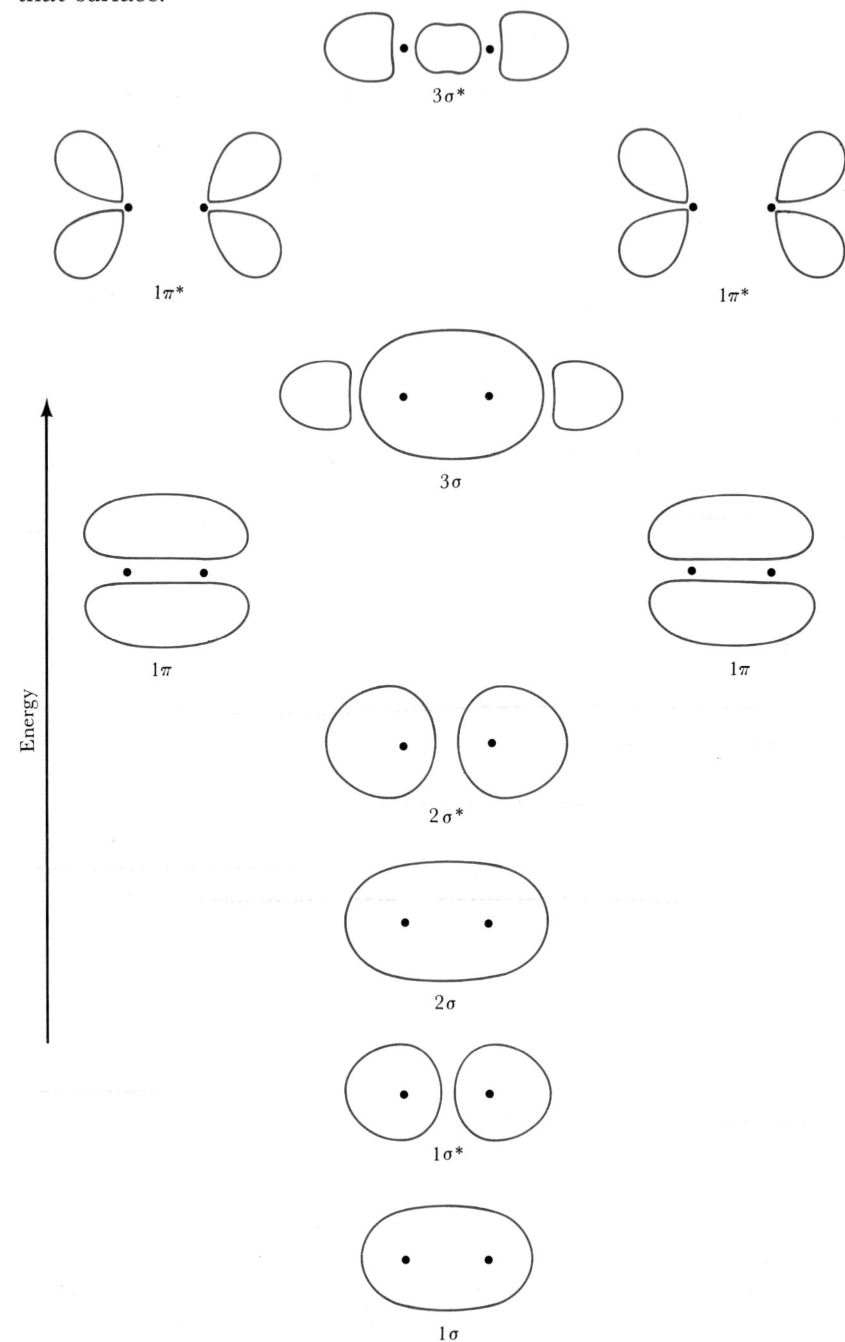

Figure 12-24 The three-dimensional surfaces that depict the shapes, but not the relative sizes of the first few H_2^+ molecular orbitals. The orbitals are listed in order of increasing energy. Note that some molecular orbitals have zero values in the region between the two nuclei, which are shown as heavy dots. Also note that the two molecular orbitals designated by 1π have the same energy and that the two designated by 1π* have the same energy.

12-11. MOLECULAR ORBITALS ARE BONDING OR ANTIBONDING

The first two H_2^+ molecular orbitals shown in Figure 12-24 are σ orbitals and so have circular cross sections when viewed along the internuclear axis. Figure 12-24 shows, however, that these two orbitals are different when viewed perpendicular to the internuclear axis. The first molecular orbital, the one that corresponds to the lowest energy in H_2^+, is concentrated between the two nuclei. Electrons in this molecular orbital are likely to be found in the region between the two nuclei and so draw the two nuclei toward each other. This effective attraction results in a covalent bond and provides a molecular orbital interpretation of the electron-pair bond in Lewis formulas. A molecular orbital that is concentrated in a region between two nuclei is called a *bonding orbital* because electrons in such an orbital act to bond the two nuclei together.

The H_2^+ molecular orbital that corresponds to the second lowest energy, that is, the second lowest one in Figure 12-24, is zero in the region between the nuclei. It is concentrated more on the far sides of the two nuclei. Electrons in this orbital are likely to be found not in the region between the two nuclei but on the far sides of the nuclei. Thus, electrons in this orbital tend to draw the nuclei apart. A molecular orbital that is zero in the region between two nuclei and concentrated on the far sides of the nuclei is called an *antibonding orbital*. A bonding orbital is designated simply by its Greek letter, and an antibonding orbital is designated by its Greek letter followed by an asterisk. The σ and σ* orbitals that have the lowest energy are called the 1σ and 1σ* orbitals. They are the first σ orbital and the first σ* orbital in order of increasing energy. The orbitals designated by 2σ and 2σ* are the second σ orbital and the second σ* orbital. Note that molecular orbitals come in bonding-antibonding pairs. Thus we have both a 1π (first π orbital) and a 1π* (first π* orbital) orbital, a 3σ and a 3σ* orbital, and so on.

Figure 12-25 The electron configuration of H_2. The two electrons occupy the molecular orbital corresponding to the lowest energy and have opposite spins in accord with the Pauli exclusion principle.

A σ* orbital is called a "sigma-star-orbital."

The 1σ orbital has the lowest energy of all the H_2^+ molecular orbitals, and the 1σ* has the next lowest energy. We now construct electron configurations of homonuclear diatomic molecules by placing electrons in these orbitals in accord with the Pauli exclusion principle. The hydrogen molecule, H_2, has two electrons. According to the Pauli exclusion principle, we place two electrons of opposite spins in the 1σ orbital and write the electron configuration of H_2 as $(1\sigma)^2$. The ground-state electron configuration of H_2 is illustrated in Figure 12-25, where, for simplicity, only the first two molecular orbitals are shown. The two electrons with opposite spins occupy the 1σ orbital. The two electrons in the bonding orbital constitute a bonding pair of electrons and account for the single-bond character of H_2.

Figure 12-26 The electron configuration of the hypothetical molecule He_2. There are two electrons in a bonding orbital and two in an antibonding orbital, and so He_2 has no net bonding. The molecule He_2 has never been detected experimentally.

Now let's consider the species He_2, which has four electrons. Two of the electrons of He_2 occupy the 1σ orbital, and two occupy the 1σ* orbital (Figure 12-26). Thus, there are two electrons in a bonding orbital and two in an antibonding orbital. Electrons in a bonding

Table 12-2 Molecular properties of H_2^+, H_2, He_2^+, and He_2

Species	*Number of electrons*	*Ground-state electron configuration*	*Bond order*	*Bond length/pm*	*Bond energy/kJ·mol^{-1}*
H_2^+	1	$(1\sigma)^1$	$\frac{1}{2}$	106	255
H_2	2	$(1\sigma)^2$	1	74	436
He_2^+	3	$(1\sigma)^2(1\sigma^*)^1$	$\frac{1}{2}$	108	251
He_2	4	$(1\sigma)^2(1\sigma^*)^2$	0	not observed	not observed

orbital tend to draw the nuclei together, whereas electrons in an antibonding orbital tend to draw the nuclei apart. The result is that the effect of the antibonding electrons cancels the effect of the bonding electrons and there is no net bonding. This is in accord with the fact that the species He_2 has never been observed experimentally.

Table 12-2 summarizes the properties of the molecular species, H_2^+, H_2, He_2^+ and He_2. The property *bond order* in Table 12-2 is defined as

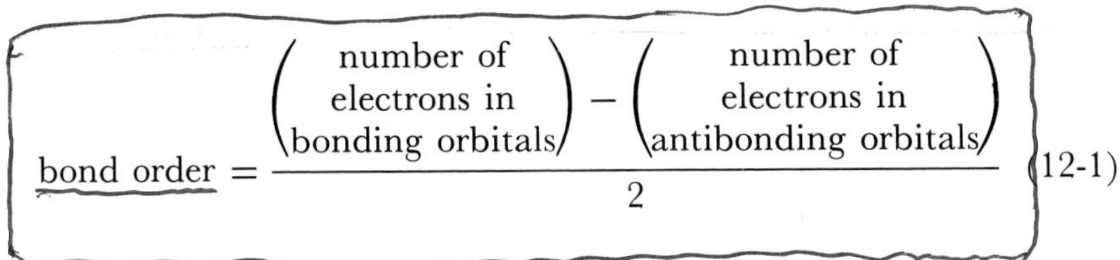

$$\text{bond order} = \frac{\begin{pmatrix}\text{number of}\\ \text{electrons in}\\ \text{bonding orbitals}\end{pmatrix} - \begin{pmatrix}\text{number of}\\ \text{electrons in}\\ \text{antibonding orbitals}\end{pmatrix}}{2} \tag{12-1}$$

A bond order of 1 indicates a single bond, a bond order of 2 indicates a double bond, and so on. The bond order of 0 for He_2 indicates that there is no helium-helium covalent bond. Note from Table 12-2 that bond orders do not necessarily have to be whole numbers.

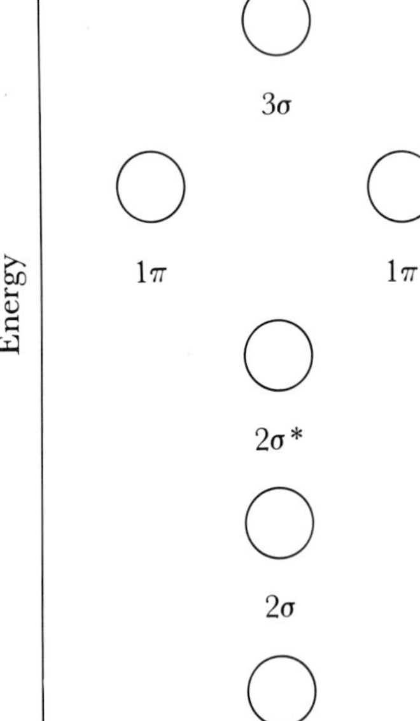

Figure 12-27 An energy-level diagram to be used for the homonuclear diatomic molecules H_2 through Ne_2. The orbitals are listed in order of increasing energy, $1\sigma < 1\sigma^* < 2\sigma < 2\sigma^* < 1\pi < 3\sigma < 1\pi^* < 3\sigma^*$. Electrons occupy these orbitals in accord with the Pauli exclusion principle.

12-12. MOLECULAR ORBITAL THEORY PREDICTS THAT DIATOMIC NEON DOES NOT EXIST

Figure 12-27 shows an energy-level diagram for the molecular orbitals 1σ to $3\sigma^*$. We can use Figure 12-27 to write ground-state electron configurations for the homonuclear diatomic molecules Li_2 through Ne_2.

Lithium vapor contains diatomic lithium molecules, Li_2. A lithium atom has three electrons, and so Li_2 has a total of six electrons. In the ground state of Li_2, the six electrons occupy the lowest three molecular orbitals in Figure 12-27 in accord with the Pauli exclusion principle. The ground-state electron configuration of Li_2 is $(1\sigma)^2(1\sigma^*)^2(2\sigma)^2$. There is a net of two bonding electrons, and so the bond order is 1. Thus we predict that Li_2 is stable relative to two separated lithium atoms. Table 12-3 shows that Li_2 has a bond length of 267 pm and a bond energy of 101 kJ·mol^{-1}. The process

$$Li_2(g) \rightarrow 2Li(g) \qquad \Delta H^\circ_{rxn} = 101 \text{ kJ}$$

is endothermic. The ground-state electron configurations of the homonuclear diatomic molecules Li_2 through Ne_2 are given in Table 12-3.

Example 12-6: Use Figure 12-27 to write the ground-state electron configuration of N_2. Calculate the bond order of N_2 and compare your result with the Lewis formula for N_2.

Solution: There are 14 electrons in N_2, and, using Figure 12-27, we see that its ground-state electron configuration is $(1\sigma)^2(1\sigma^*)^2(2\sigma)^2(2\sigma^*)^2(1\pi)^4(3\sigma)^2$. According to Equation (12-1), the bond order in N_2 is

$$\text{bond order} = \frac{10-4}{2} = 3$$

The Lewis formula for N_2, :N≡N:, is thus in agreement with the more fundamental molecular orbital theory result. The triple bond in N_2 accounts for its short bond length (110 pm) and its large bond energy (941 kJ·mol^{-1}). The bond in N_2 is the second strongest known bond.

Table 12-3 Properties of the homonuclear diatomic molecules of elements in the second row of the periodic table

Species	*Ground-state electron configuration*	*Bond order*	*Bond length/pm*	*Bond energy/kJ·mol^{-1}*
Li_2	$(1\sigma)^2(1\sigma^*)^2(2\sigma)^2$	1	267	101
Be_2	$(1\sigma)^2(1\sigma^*)^2(2\sigma)^2(2\sigma^*)^2$	0	not observed	not observed
B_2	$(1\sigma)^2(1\sigma^*)^2(2\sigma)^2(2\sigma^*)^2(1\pi)^2$	1	159	289
C_2	$(1\sigma)^2(1\sigma^*)^2(2\sigma)^2(2\sigma^*)^2(1\pi)^4$	2	124	599
N_2	$(1\sigma)^2(1\sigma^*)^2(2\sigma)^2(2\sigma^*)^2(1\pi)^4(3\sigma)^2$	3	110	941
O_2	$(1\sigma)^2(1\sigma^*)^2(2\sigma)^2(2\sigma^*)^2(1\pi)^4(3\sigma)^2(1\pi^*)^2$	2	121	494
F_2	$(1\sigma)^2(1\sigma^*)^2(2\sigma)^2(2\sigma^*)^2(1\pi)^4(3\sigma)^2(1\pi^*)^4$	1	142	154
Ne_2	$(1\sigma)^2(1\sigma^*)^2(2\sigma)^2(2\sigma^*)^2(1\pi)^4(3\sigma)^2(1\pi^*)^4(3\sigma^*)^2$	0	not observed	not observed

Example 12-7: Use molecular orbital theory to argue that neon does not form a stable diatomic molecule.

Solution: Neon has 10 electrons, and so Ne_2 would have 20 electrons. According to Figure 12-27, its ground-state electron configuration would be $(1\sigma)^2(1\sigma^*)^2(2\sigma)^2(2\sigma^*)^2(1\pi)^4(3\sigma)^2(1\pi^*)^4(3\sigma^*)^2$. The bond order associated with this electron configuration is

$$\text{bond order} = \frac{10 - 10}{2} = 0$$

indicating that there is no neon-neon bond and that neon does not form a diatomic molecule.

12-13. A PHOTOELECTRON SPECTRUM DEMONSTRATES THE EXISTENCE OF MOLECULAR ORBITALS

To many students beginning their study of chemistry, the idea of atomic orbitals and molecular orbitals is rather abstract and sometimes appears to be far removed from "reality." It so happens, however, that the electron configurations of molecules can be verified experimentally. If high-energy electromagnetic radiation, such as X-radiation, is directed into a gas, then electrons are ejected from the molecules of the gas. The energy required to eject an electron from a molecule in the gas is a direct measure of how strongly the electron is bound in the molecule. The energy with which an electron is bound in a molecule is called the *binding energy* of that electron. The binding energy of an electron in a molecule depends upon which molecular orbital the electron occupies. The lower the energy of the molecular orbital that the electron occupies, the more energy it takes to remove the electron from the molecule.

The measurement of the energies of the electrons ejected by radiation incident to gaseous molecules is called *photoelectron spectroscopy*. A *photoelectron spectrum* of N_2 is shown in Figure 12-28. The peaks in this figure correspond to the energies of the electrons in N_2. Note how they correspond to the indicated molecular orbital energies. Also note that the 1*s* electrons in N_2 are much more tightly bound than the other electrons. Photoelectron spectra provide a striking experimental verification of the molecular orbital picture that we are developing here.

The frontispiece for Chapter 8 shows a photoelectron spectrum of argon.

12-14. OXYGEN MOLECULES ARE PARAMAGNETIC

The prediction of the distribution of the electrons in an oxygen molecule is one of the most impressive successes of molecular orbital the-

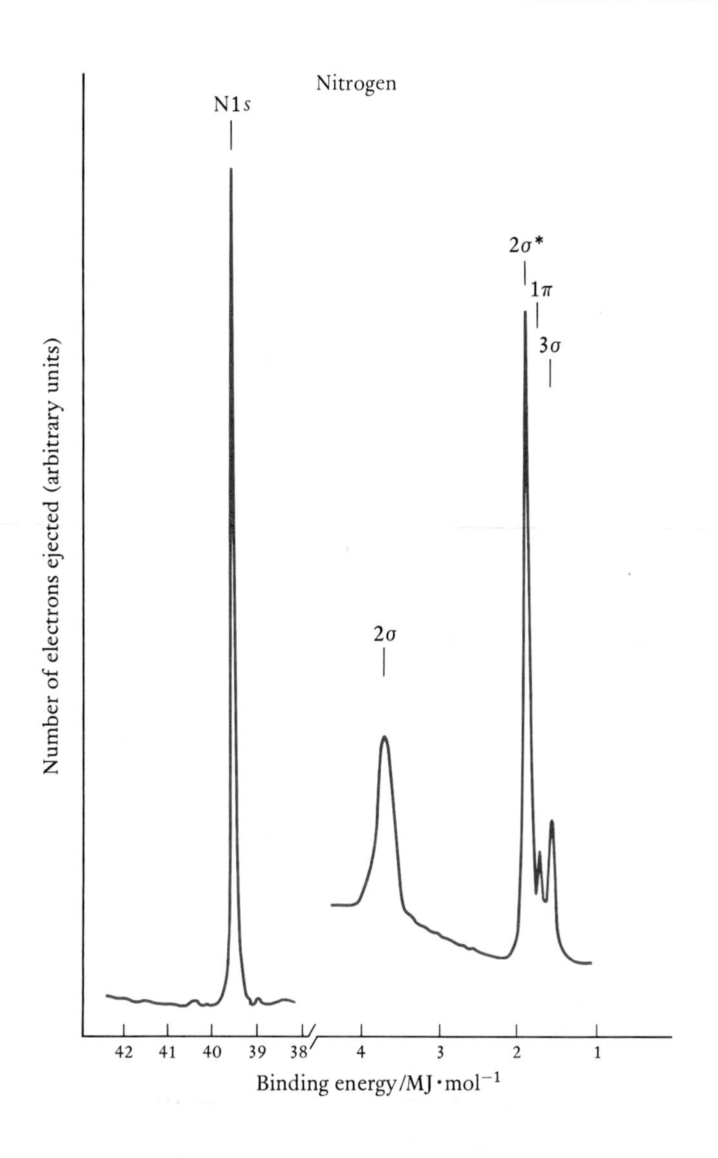

Figure 12-28 The photoelectron spectrum of N_2. When gas-phase molecules are irradiated with high-energy radiation, electrons are ejected from the molecules. By measuring the energies of the ejected electrons, we can deduce the energies of the electrons in the molecules. The peaks in this plot are due to electrons being ejected from various molecular orbitals.

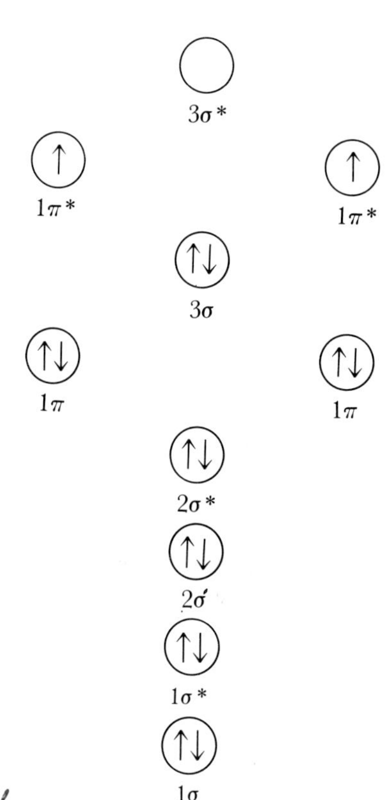

Figure 12-29 The ground-state electron configuration of O_2. There are 16 electrons in O_2, and they occupy the molecular orbitals as shown. Note that two of the electrons occupy the $1\pi^*$ orbitals in accord with Hund's rule, being placed in separate orbitals with unpaired spins. The molecule itself has a net electronic spin and so acts as a tiny magnet.

ory. Oxygen molecules are *paramagnetic*. This means that oxygen is a magnetic substance and so is attracted to a region between the poles of a magnet.

Each oxygen atom has 8 electrons; thus, O_2 has a total of 16 electrons. When the 16 electrons are placed according to the molecular orbital diagram given in Figure 12-27, the last two go into the $1\pi^*$ orbitals. As in the atomic case, we apply *Hund's rule* and place one electron in each $1\pi^*$ orbital such that the two electrons have unpaired spins. This is shown in Figure 12-29. The ground-state electron configuration of O_2 is $(1\sigma)^2(1\sigma^*)^2(2\sigma)^2(2\sigma^*)^2(1\pi)^4(3\sigma)^2(1\pi^*)^2$. Each $1\pi^*$ orbital is occupied by one electron such that the spins are unpaired. Thus, an oxygen molecule has a net electron spin and so acts as a tiny magnet. These tiny magnets cause O_2 to be drawn between the poles of a magnet.

The amount of oxygen in air can be monitored by measuring the paramagnetism of a sample of air. Because oxygen is the only major component in air that is paramagnetic, the measured paramagnetism of air is directly proportional to the amount of oxygen present. Linus Pauling developed this method, which was used to monitor oxygen levels in submarines and airplanes in World War II. It is still used by physicians to monitor the oxygen content in blood during anesthesia.

The Lewis formula of O_2 does not account for the unpaired electrons and hence does not account for the paramagnetism of O_2. According to the octet rule, we should write $\ddot{\underset{\cdot\cdot}{O}}=\ddot{\underset{\cdot\cdot}{O}}$ for the Lewis formula of O_2, but this implies that all the electrons are paired, which is in disagreement with the paramagnetism of O_2. The oxygen molecule is an exception to the utility of Lewis formulas, whereas the more fundamental molecular orbital theory is able to account successfully for the distribution of the electrons in O_2.

We also can use molecular orbital theory to discuss the relative bond lengths and bond energies of O_2^+, O_2, O_2^-, and O_2^{2-}. Because O_2 has 16 electrons, O_2^+ has 15 electrons, O_2^- has 17 electrons, and O_2^{2-} has 18 electrons. Using Figure 12-27, we obtain for the ground-state electron configurations and bond orders of these species

Species	*Ground-state electron configuration*	*Bond order*
O_2^+	$(1\sigma)^2(1\sigma^*)^2(2\sigma)^2(2\sigma^*)^2(1\pi)^4(3\sigma)^2(1\pi^*)^1$	$2\frac{1}{2}$
O_2	$(1\sigma)^2(1\sigma^*)^2(2\sigma)^2(2\sigma^*)^2(1\pi)^4(3\sigma)^2(1\pi^*)^2$	2
O_2^-	$(1\sigma)^2(1\sigma^*)^2(2\sigma)^2(2\sigma^*)^2(1\pi)^4(3\sigma)^2(1\pi^*)^3$	$1\frac{1}{2}$
O_2^{2-}	$(1\sigma)^2(1\sigma^*)^2(2\sigma)^2(2\sigma^*)^2(1\pi)^4(3\sigma)^2(1\pi^*)^4$	1

We predict that the bond lengths decrease and the bond energies increase with increasing bond order. This is in nice agreement with the experimental values, which are

Species	*Bond length/pm*	*Bond energy/kJ·mol⁻¹*
O_2^+	112	643
O_2	121	494
O_2^-	133	395
O_2^{2-}	149	$\simeq$250

Example 12-8: Use molecular orbital theory to discuss the relative bond energies and bond lengths of N_2^+ and N_2.

Solution: Using Figure 12-27, we find that the ground-state electron configurations and bond orders of N_2^+ and N_2 are

Species	*Ground-state electron configuration*	*Bond order*
N_2^+	$(1\sigma)^2(1\sigma^*)^2(2\sigma)^2(2\sigma^*)^2(1\pi)^4(3\sigma)^1$	$2\frac{1}{2}$
N_2	$(1\sigma)^2(1\sigma^*)^2(2\sigma)^2(2\sigma^*)^2(1\pi)^4(3\sigma)^2$	3

Because the bond order of N_2 is greater than that of N_2^+, we predict that N_2 has a larger bond energy and a shorter bond length than N_2^+. The experimental values are

Species	*Bond energy/kJ·mol⁻¹*	*Bond length/pm*
N_2	941.7	109.8
N_2^+	842.2	111.6

Molecular orbital theory can be applied to *heteronuclear diatomic molecules* (that is, diatomic molecules in which the two atoms are different) and to polyatomic molecules; however, we do not discuss these extensions here.

Problems 12-52 to 12-54 ask you to apply molecular orbital theory to heteronuclear diatomic molecules.

SUMMARY

Electrons in molecules are described by molecular orbitals, which are orbitals that are spread over two or more atoms. The bonding in polyatomic molecules can be described in terms of localized bond orbitals, which are formed by combining atomic orbitals from atoms that are bonded together.

The bond orbitals of all the polyatomic molecules discussed in this chapter can be described in terms of hybrid orbitals. Different types of hybrid orbitals are used to describe different molecular geometries. Single bonds can be described in terms of sp^3 hybrid orbitals. The bonds formed using sp^3 orbitals are tetrahedrally oriented and are called σ bonds.

Double bonds can be described by using sp^2 hybrid orbitals. Double bonds consist of one σ bond and one π bond. The π bond is formed by the combination of p orbitals on adjacent atoms. Double bonds have a significant effect on the shape of a molecule. The atoms bonded directly to double-bonded atoms all lie in one plane. In addition, there is no rotation about double bonds, which leads to the possibility of *cis* and *trans* isomers. Many biological and physiological functions of molecules are dramatically sensitive to the structures and shapes of the molecules.

Triple bonds consist of one σ bond and two π bonds. The shape of two π bonds between two atoms is that of a cylinder, and so triple bonds are cylindrically symmetric. In some molecules, such as benzene, the π bond orbitals are spread uniformly over many atoms and are said to be delocalized. The electrons that occupy delocalized orbitals are also delocalized, giving what is called charge delocalization. Charge delocalization confers an extra degree of stability on a molecule or ion and accounts for the relative stability of benzene.

A set of diatomic molecular orbitals can be obtained for the hydrogen molecular ion, H_2^+, and these can be used to write electron configurations of diatomic molecules. This is in analogy with the atomic case, where we used hydrogen atomic orbitals to write electron configurations of other atoms. There are two types of molecular orbitals, bonding and antibonding. The bonding properties of molecules depend upon the number of electrons in bonding and antibonding orbitals. The scheme used in this chapter is called molecular orbital theory. Molecular orbital theory correctly predicts that the diatomic molecules He_2 and Ne_2 do not exist and that O_2 is paramagnetic.

TERMS YOU SHOULD KNOW

ground-state wave function 434
molecular orbital 434
polyatomic molecule 434
localized bond orbital 435
hybrid atomic orbital 436
sp^3 orbital 436
σ orbital 437
σ bond 437
sp^2 orbital 441
σ-bond framework 442
π orbital 442
π bond 442
stereoisomers 443
cis-trans isomerism 443
sp orbital 445
delocalized orbital 448
charge delocalization 448
molecular orbital theory 449
homonuclear diatomic molecule 449
hydrogen molecular ion, H_2^+ 449
bonding orbital 451
antibonding orbital 451
σ^* orbital 451
π^* orbital 451
bond order 452
binding energy 454
photoelectron spectroscopy 454
photoelectron spectrum 454
paramagnetic 455
Hund's rule 455
heteronuclear diatomic molecule 457

EQUATIONS YOU SHOULD KNOW HOW TO USE

$$\text{bond order} = \frac{\begin{pmatrix}\text{number of}\\ \text{electrons in}\\ \text{bonding orbitals}\end{pmatrix} - \begin{pmatrix}\text{number of electrons}\\ \text{in antibonding}\\ \text{orbitals}\end{pmatrix}}{2} \qquad (12\text{-}1)$$

PROBLEMS

BOND FORMATION

12-1. Explain why the interaction energy of two hydrogen atoms 74 pm apart is negative.

12-2. In Figure 12-2, the interaction energy between two hydrogen atoms is plotted as a function of their separation. Suggest a reason for the enormous increase in the energy when the separation distance is less than 74 pm.

12-3. How many localized bonds and lone pairs are there in

(a) PCl_3
(b) SF_4
(c) SiH_4
(d) OF_2
(e) HOCl

12-4. How many localized bonds and lone pairs are there in

(a) PH_3
(b) H_2O_2
(c) H_3O^+
(d) Cl_2O
(e) CH_3OH

12-5. How many bonds are there in propane, whose Lewis formula is

```
   H  H  H
   |  |  |
H—C—C—C—H
   |  |  |
   H  H  H
```

How many valence electrons are there in three carbon atoms and eight hydrogen atoms?

12-6. How many bonds are there in butane, whose Lewis formula is

```
   H  H  H  H
   |  |  |  |
H—C—C—C—C—H
   |  |  |  |
   H  H  H  H
```

How many valence electrons are there in 4 carbon atoms and 10 hydrogen atoms?

sp^3 HYBRID ORBITALS

12-7. How many valence electrons are there in carbon tetrafluoride, CF_4? Describe the bonding in this molecule.

12-8. The formula for chloroform is $HCCl_3$. How many valence electrons are there in chloroform? Describe the bonding in the chloroform molecule.

12-9. Describe the bonding in carbon tetrachloride, CCl_4. Which atomic orbitals on the chlorine atoms are used to form the localized bond orbitals?

12-10. Describe the bonding in OF_2. Which atomic orbitals on the fluorine atoms are used to form localized bond orbitals?

12-11. The hydronium ion, H_3O^+, is trigonal pyramidal with H—O—H bond angles of 110°. Use hybrid orbitals to describe the bonding in H_3O^+.

12-12. Discuss the bonding in NF_3. Which atomic orbitals are used to form the localized bond orbitals in this molecule? How many valence electrons are there?

12-13. Describe the bonding in hydrazine, N_2H_4, in terms of sp^3 hybrid orbitals on the nitrogen atoms.

12-14. How many valence electrons are there in hydrogen peroxide, H_2O_2? How many localized bonds are there? Discuss the bonding in H_2O_2.

12-15. A class of organic compounds called *amines* may be viewed as derived from NH_3 with one or more hydrogen atoms replaced by alkyl groups, which are hydrocarbon groups, such as $-CH_3$ (methyl) and $-CH_2CH_3$ (ethyl). Examples of amines are

CH_3NH_2	$(CH_3)_2NH$	$(CH_3)_3N$
methylamine	dimethylamine	trimethylamine

Discuss the bonding and the shape of methylamine.

12-16. Discuss the bonding and shape of dimethylamine (Problem 12-15). How many σ bonds are there? How many lone pairs of electrons? How many valence shell electrons are there in the constituent atoms?

12-17. A class of organic compounds called *alcohols* may be viewed as derived from HOH by replacing one of the hydrogen atoms by an alkyl group (Problem 12-15). Two simple alcohols are

```
    H                 H  H
    |   ..            |  |   ..
  H—C—O—H          H—C—C—O—H
    |   ..            |  |   ..
    H                 H  H
methyl alcohol     ethyl alcohol
```

Discuss the bonding and shape of methyl alcohol.

12-18. Discuss the bonding and the shape of ethyl alcohol (Problem 12-17). How many σ bonds are there? How many lone pairs of electrons? How many valence shell electrons are there in the constituent atoms?

12-19. If both hydrogen atoms in HOH are replaced by alkyl groups, the result is an *ether,* ROR′, where R and R′ are alkyl groups that may or may not be different. The simplest ether is

```
    H     H
    |  .. |
  H—C—O—C—H
    |  .. |
    H     H
 dimethyl ether
```

Discuss the bonding in dimethyl ether.

12-20. Discuss the bonding in ethyl methyl ether, whose Lewis formula is

```
    H  H     H
    |  |  .. |
  H—C—C—O—C—H
    |  |  .. |
    H  H     H
 ethyl methyl ether
```

12-21. The H—As—H bond angles in AsH_3 are about 90°. What atomic orbitals would you use to form the localized bond orbitals in this molecule?

12-22. The bond angle in H_2Te is about 90°. What atomic orbitals would you use to form the localized bond orbitals in this molecule?

*sp*² HYBRID ORBITALS

12-23. Boron hydride, BH_3, is a planar, symmetric molecule with 120° H—B—H bond angles. Discuss the bonding in BH_3. What type of hybrid orbital best describes the bonding in this molecule?

12-24. Aluminum chloride, $AlCl_3$, is a planar symmetric molecule with 120° Cl—Al—Cl bond angles. Discuss the bonding in $AlCl_3$. What type of hybrid orbital best describes the bonding in this molecule?

12-25. Propene is a hydrocarbon with the Lewis formula

$$\begin{array}{ccc} H & & CH_3 \\ & C{=}C & \\ H & & H \end{array}$$

How many σ bonds are there in propene? How many π bonds? How many valence electrons?

12-26. The compound 2-butene is a hydrocarbon with the Lewis formula

$$\begin{array}{ccc} H & & CH_3 \\ & C{=}C & \\ H_3C & & H \end{array}$$

How many σ bonds are there in 2-butene? How many π bonds? How many valence electrons?

12-27. How many σ bonds and π bonds are there in each of the following molecules:

(a) $Cl_2C{=}CH_2$

(b) $H_2C{=}CHCH{=}CH_2$

(c) $H_2C{=}C{=}CH_2$

(d) a six-membered ring: CH_2, $CH{=}CH$, CH_2, $CH{=}CH$ (cyclohexa-1,4-diene: HC=CH and CH=CH joined by two CH_2 groups)

12-28. How many σ bonds and π bonds are there in each of the following molecules:

(a) $F_2C{=}CF_2$

(b) $H_2C{=}CHCH_3$

(c) $H_2C{=}C{=}CCl_2$

(d) a four-membered ring of C=C and C=C joined by single bonds, with an H on each carbon

12-29. Use hybrid orbitals to describe the bonding in propene:

$$\begin{array}{ccc} H & & H \\ & C{=}C & \\ H & & CH_3 \end{array}$$

12-30. Use hybrid orbitals to describe the bonding in vinyl chloride:

$$\begin{array}{ccc} H & & H \\ & C{=}C & \\ H & & :\!\ddot{Cl}\!: \end{array}$$

Vinyl chloride is used to make the plastic polyvinyl chloride.

12-31. Aldehydes are organic compounds that have the general Lewis formula

$$\begin{array}{cc} R & \\ & C{=}\ddot{O}: \\ H & \end{array}$$

where R is either a hydrogen atom (giving formaldehyde) or an alkyl group such as $-CH_3$ (methyl) or $-CH_2CH_3$ (ethyl). Discuss the bonding in acetaldehyde, CH_3CHO.

12-32. Ketones are organic compounds with the general Lewis formula

$$\begin{array}{cc} R' & \\ & C{=}\ddot{O}: \\ R & \end{array}$$

where R and R′ are alkyl groups that may or may not be different. The simplest ketone is acetone, $(CH_3)_2CO$, one of the most important solvents. Discuss the bonding and shape of acetone.

12-33. The Lewis formula of formamide, a slightly viscous, colorless, odorless liquid that is used as a softener for paper, is

$$\begin{array}{ccccc} H & - & \ddot{N} & - & C & - & H \\ & & | & & \| & & \\ & & H & & :\!O\!: & & \end{array}$$

Discuss the bonding in formamide.

12-34. Urea is the end product of the body's metabolism of nitrogen-containing compounds. An average adult secretes over 25 g of urea in urine daily. Discuss the bonding in urea, whose formula is

$$\begin{array}{c} H_2NCNH_2 \\ \| \\ :\!O\!: \end{array}$$

sp HYBRID ORBITALS

12-35. How many σ bonds are there in ethylacetylene, $CH_3CH_2C{\equiv}CH$? How many π bonds? How many valence electrons?

12-36. How many σ bonds and π bonds are there in methyl cyanide, CH_3CN? How many valence electrons?

12-37. Use *sp* orbitals to describe the bonding and the lone pairs in carbon monoxide, CO.

12-38. Use *sp* orbitals to describe the bonding and the lone pairs in the acetylide ion, C_2^{2-}.

12-39. Discuss the bonding and shape of methylacetylene, $CH_3C{\equiv}CH$. How many σ and π bonds are there? How many valence electrons?

12-40. Discuss the bonding and shape of dimethylacetylene, $CH_3C{\equiv}CCH_3$. How many σ and π bonds are there? How many valence electrons?

DELOCALIZED BONDS

12-41. Phenol (carbolic acid), a derivative of benzene in which one hydrogen atom is replaced by —OH, has the formula C_6H_5OH and is used as an antiseptic and disinfectant. Draw a Lewis formula for phenol and discuss its bonding.

12-42. Aniline, $C_6H_5NH_2$, a derivative of benzene in which one hydrogen atom is replaced by $—NH_2$, is used in the manufacture of numerous dyes. Draw a Lewis formula for aniline and discuss its bonding.

12-43. Draw the complete Lewis formula for and discuss the bonding in naphthalene,

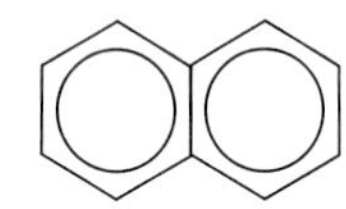

naphthalene

a white, crystalline solid with an odor characteristic of mothballs.

12-44. Draw the complete Lewis formula for and discuss the bonding in anthracene,

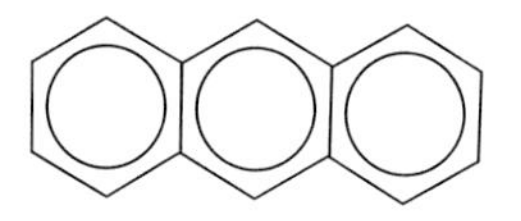

anthracene

a yellow, crystalline solid found in coal tar.

12-45. Discuss the bonding and the geometry of hexamethylbenzene, in which each hydrogen atom in benzene is replaced by a methyl group ($—CH_3$). How many σ and π orbitals are there? How many valence electrons?

12-46. There are many molecules that are similar to benzene but have one or more of the carbon atoms replaced by other atoms. Such molecules are called *heterocyclic molecules*. One of the best known nitrogen heterocyclic molecules is the foul-smelling substance pyridine, C_5H_5N. Draw the Lewis formula for and discuss the bonding in pyridine.

MOLECULAR ORBITAL THEORY

12-47. Use molecular orbital theory to explain why diatomic beryllium does not exist.

12-48. Use molecular orbital theory to predict whether or not diatomic boron is paramagnetic.

12-49. Use molecular orbital theory to explain why the bond energy of N_2 is greater than that of N_2^+, but the bond energy of O_2 is less than that of O_2^+.

12-50. Use molecular orbital theory to predict the relative bond energies and bond lengths of F_2 and F_2^+.

12-51. Use molecular orbital theory to predict the relative bond energies and bond lengths of diatomic carbon, C_2, and the acetylide ion, C_2^{2-}.

12-52. Molecular orbital theory can be applied to heteronuclear diatomic molecules. The energy-level scheme in Figure 12-27 can be used if the atomic numbers of the two atoms in the heteronuclear diatomic molecule differ by only one or two atomic numbers. Use Figure 12-27 to determine the ground-state electron configurations and bond orders of NF, NF^+, and NF^-. Which of these species do you predict to be paramagnetic?

12-53. Molecular orbital theory can be applied to heteronuclear diatomic molecules. The energy-level scheme in Figure 12-27 can be used if the atomic numbers of the two atoms in the heteronuclear diatomic molecule differ by only one or two atomic numbers. Use Figure 12-27 to determine the bond order of carbon monoxide and compare your result with the Lewis formula for CO.

12-54. Molecular orbital theory can be applied to heteronuclear diatomic molecules. The energy-level scheme in Figure 12-27 can be used if the atomic numbers of the two atoms in the heteronuclear diatomic molecule differ by only one or two atomic numbers. Use Figure 12-27 to determine the bond order of the cyanide ion, CN^-, and compare your result with the Lewis formula for this ion.

Crystals organize in one of these six crystal systems. The six systems are (starting at top and moving clockwise) orthorhombic, hexagonal, monoclinic, tetragonal, triclinic, and (center) isometric.

13 / Liquids and Solids

The molecules in a gas are widely separated from each other and travel distances equal to many molecular diameters between collisions. The molecules in solids and liquids are in contact with each other, and so the properties of solids and liquids depend upon how strongly the molecules interact with, or attract, each other. The attractions between molecules hold solids and liquids in the condensed state. If molecules did not attract each other, gases would not condense to form liquids. In this chapter we study molecular interactions and how they affect the properties of liquids and solids. Some of the properties we shall study are the enthalpies of fusion and vaporization, vapor pressure, and surface tension. We also shall study crystals, which are characterized by a repetitive, ordered arrangement of atoms. We learn that the underlying orderly arrangement of the atoms or molecules in a crystal can be determined by the patterns that result when X-rays are passed through the crystal. We conclude the chapter with a study of phase diagrams, which are diagrams that include detailed information about the solid, liquid, and gas phases of a substance on one graph.

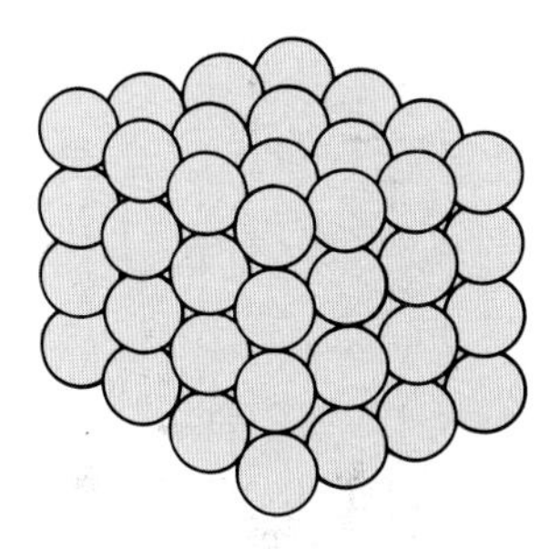

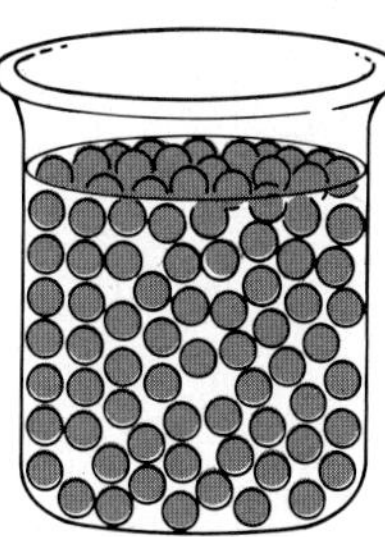

Figure 13-1 A molecular view of the difference between a solid and a liquid. The molecules in a solid are situated at fixed lattice positions, whereas the molecules in a liquid move throughout the entire volume of the liquid.

13-1. INTERACTIONS BETWEEN MOLECULES ARE MUCH GREATER IN THE SOLID AND LIQUID PHASES THAN IN THE GAS PHASE

We learned in Chapter 5 that a solid retains its shape, has a definite volume, and does not flow. A liquid takes the shape of the container that it occupies, has a definite volume, and flows readily (Figure 13-1). A gas occupies the entire volume of its container.

(a)

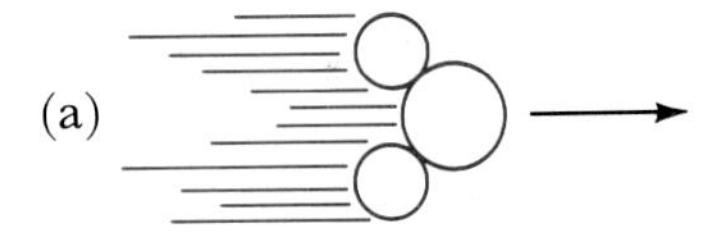

(b)

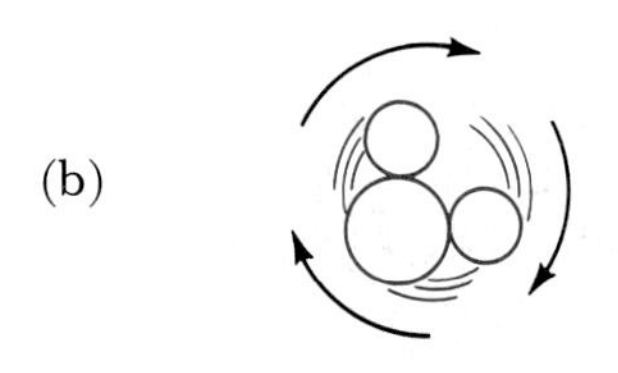

(c)

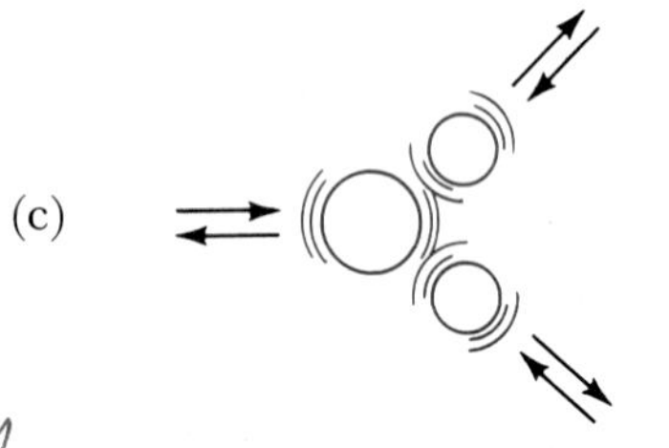

Figure 13-2 Molecular motion may be *translational*, *rotational*, or *vibrational*. (a) Translation is movement of the entire molecule through space. (b) Rotation is the spinning of a molecule in space. (c) Vibration is the back-and-forth movement of nuclei about fixed relative positions.

The molecules of a solid are close together and restricted to fixed positions in space. The individual atoms of the molecules in a solid vibrate about fixed positions in space, but the molecules themselves usually do not rotate or move easily from site to site (Figure 13-2). The molecules in a liquid are close together, but they are not restricted to fixed positions in space; they rotate and move throughout the fluid as well as vibrate. Molecules in a liquid are essentially in constant interaction with one another, and the distance that a molecule travels between collisions is about one molecular diameter or less (Figure 13-3). The translational and rotational motion of liquid molecules is hindered by frequent collisions between the molecules. Recall from Chapter 5 that the average distance between the molecules in a gas at 1 atm pressure is large relative to that in a liquid or a solid. The molecules in a gas can rotate freely, and the distance that a molecule travels between collisions is large relative to the molecular diameter, being hundreds of molecular diameters at 1 atm. The distinctions between a solid, a liquid, and a gas can be summarized as follows:

Phase	*Translation*	*Rotation*	*Vibration*	*Average distance between particles*
solid	none	usually none	about fixed positions in space	less than 1 molecular diameter
liquid	hindered	hindered	free	less than 1 molecular diameter
gas	free	free	free	about 10 molecular diameters at 1 atm

The molar volume of a substance is a direct indication of the average separation between the molecules. The molar volume of the liquid phase of a compound is approximately equal to the molar volume of the solid phase of the same compound, yet the molar volume of a gas is much larger than that of a liquid or a solid at the same pressure and temperature. The molar volumes for solid, liquid, and gaseous iodine and water at 0°C and 1 atm and for water at 100°C and 1 atm are

Substance	*Solid volume/ $mL \cdot mol^{-1}$*	*Liquid volume/ $mL \cdot mol^{-1}$*	*Gas volume/ $mL \cdot mol^{-1}$*
I_2 at 0°C	51.4	65.1	22,400
H_2O at 0°C	19.8	18.0	22,400
H_2O at 100°C	—	18.8	30,600

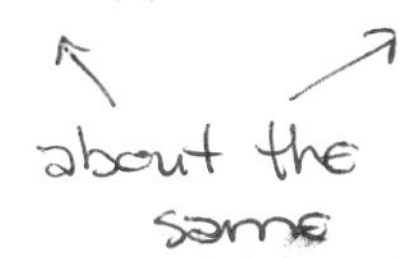

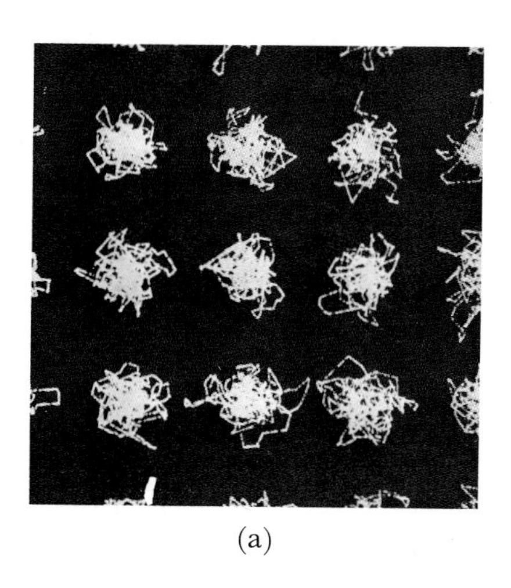
(a)

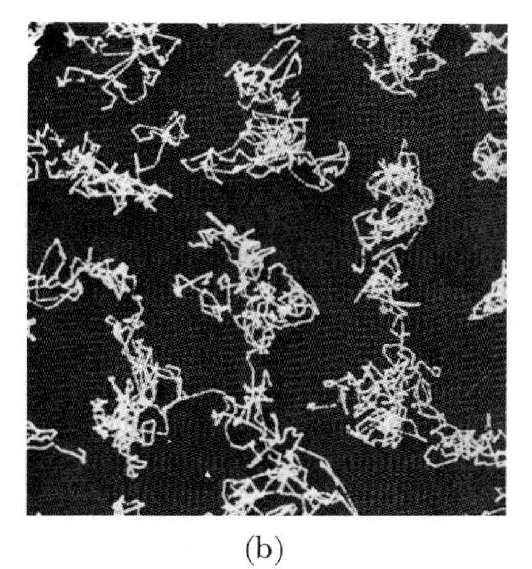
(b)

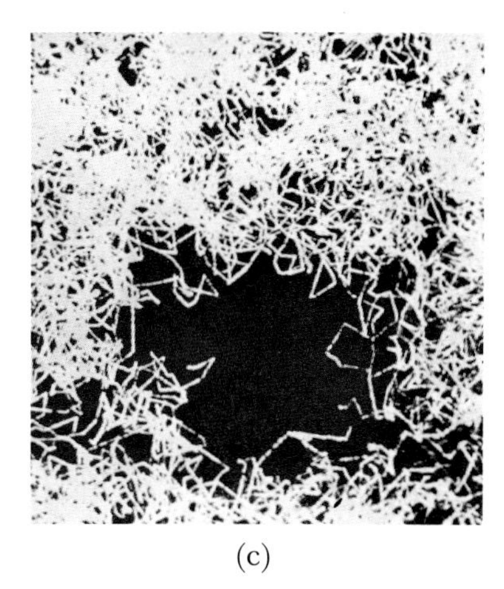
(c)

Figure 13-3 Computer-calculated paths of particles appear as bright lines on the face of a cathode-ray tube coupled to the computer. (a) Motion of atoms in an atomic crystal (note that the atoms move only about fixed positions). (b) A crystal in the process of melting (note the breakdown of the ordered array). (c) A liquid and its vapor (the dark area represents a gas bubble surrounded by particles whose motions characterize a liquid).

Note that the molar volume of water increases by a factor of 1630 (30,600/18.8) upon vaporization at 100°C. The average separation between the molecules on going from the liquid to the gas phase at 1 atm increases approximately 10-fold.

13-2. THE PROCESSES OF MELTING AND BOILING APPEAR AS HORIZONTAL LINES ON A HEATING CURVE

Figure 13-4 shows how the temperature of a substance (water in this case) varies with time if it is heated at a constant rate. Such a plot is called a *heating curve*. Figure 13-4 illustrates several important quantities. Initially the water is at a temperature of −10°C and so is in the form of ice. As heat is added, the temperature of the ice increases until 0°C, the melting point of ice, is reached. At 0°C, the temperature remains constant for 60 min even though the ice is being heated at a constant rate of $100\ \mathrm{J \cdot min^{-1}}$. The heat being added at 0°C melts the ice to liquid water. The temperature of the ice-water mixture remains at 0°C until all the ice is melted. The energy absorbed as heat that is required to melt 1 mol of any substance is called the *molar enthalpy of fusion* and is denoted by $\Delta\bar{H}_{fus}$. The experiment illustrated in Figure 13-4 shows that it requires 60 min to melt 1 mol of ice when heat is added at the rate of $100\ \mathrm{J \cdot min^{-1}}$, and so

$$\begin{aligned}\Delta\bar{H}_{fus} &= (60\ \mathrm{min \cdot mol^{-1}})(100\ \mathrm{J \cdot min^{-1}}) \\ &= 6000\ \mathrm{J \cdot mol^{-1}} \\ &= 6.0\ \mathrm{kJ \cdot mol^{-1}}\end{aligned}$$

After all the ice is melted, and not until then, the temperature increases from 0°C. The temperature continues to increase until 100°C, the boiling point of water, is reached. At 100°C, the temperature remains constant for 407 min even though heat is being added

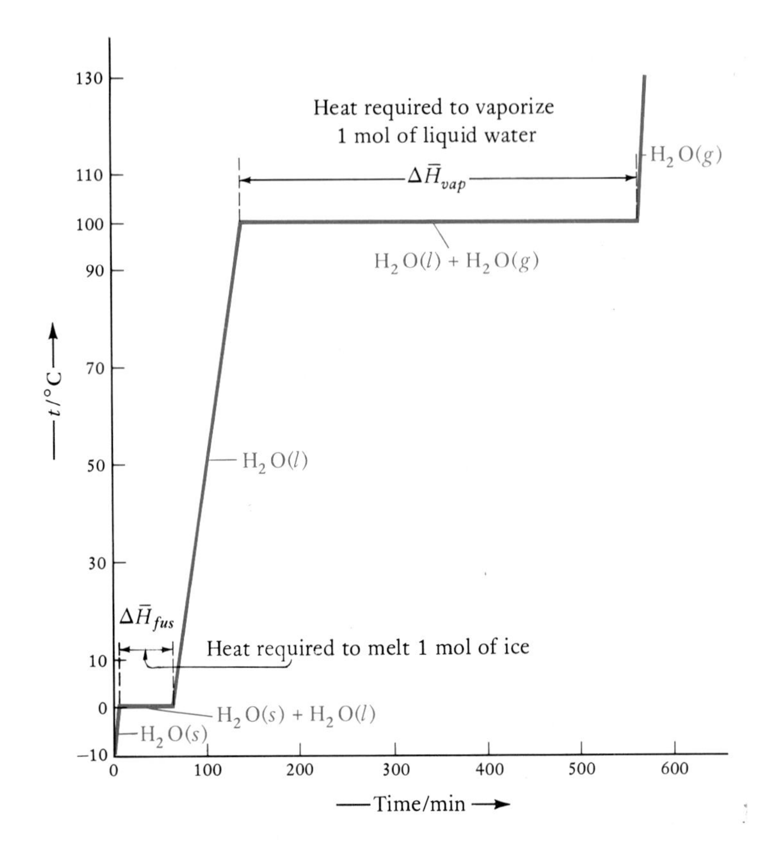

Figure 13-4 The heating curve for 1 mol of water starting with ice at $-10\,°C$. The energy is added as heat at a constant rate of $100\ J \cdot min^{-1}$. The most noteworthy features of the heating curve are the horizontal portions, which represent the heat of fusion, $\Delta\bar{H}_{fus}$, and the heat of vaporization, $\Delta\bar{H}_{vap}$. Note that the heat of vaporization is much larger than the heat of fusion.

at a rate of $100\ J \cdot min^{-1}$. The heat being absorbed at $100\,°C$ is vaporizing the liquid water to water vapor (steam). The temperature of the liquid-vapor mixture remains at $100\,°C$ until all the water is vaporized. The energy absorbed as heat that is required to vaporize 1 mol of any substance is called the *molar enthalpy of vaporization* and is denoted by $\Delta\bar{H}_{vap}$. According to Figure 13-4, $\Delta\bar{H}_{vap}$ for water is

The enthalpy of vaporization is often called the heat of vaporization.

$$\begin{aligned}\Delta\bar{H}_{vap} &= (407\ min \cdot mol^{-1})(100\ J \cdot min^{-1}) \\ &= 40{,}700\ J \cdot mol^{-1} \\ &= 40.7\ kJ \cdot mol^{-1}\end{aligned}$$

Once all the water is vaporized, the temperature increases from $100\,°C$, as shown in the figure.

The horizontal portions of the heating curve represent the heat of fusion and the heat of vaporization. The other regions represent pure phases being heated at a constant rate. We learned in Chapter 6 that it requires energy to raise the temperature of a substance. The heat absorbed in raising the temperature of a substance from T_1 to T_2 without a change in phase is given by

Heat capacity is discussed in Chapter 6.

$$q_P = n\bar{C}_P(T_2 - T_1) \tag{13-1}$$

where $\bar{C}_P$ is the molar heat capacity at constant pressure and n is the number of moles. The heat capacity is the measure of the ability of a substance to take up energy as heat. The different rates of temperature increase for $H_2O(s)$, $H_2O(l)$, and $H_2O(g)$ that appear in Figure 13-4 (steep segments of the heating curve) are a result of the fact that

$$\bar{C}_P[H_2O(l)] > \bar{C}_P[H_2O(s)] > \bar{C}_P[H_2O(g)]$$

Example 13-1: Construct a heating curve for water if energy is added as heat at a rate of $200\ \text{J}\cdot\text{min}^{-1}$. Use the following data:

$$\Delta\bar{H}_{fus} = 6.0\ \text{kJ}\cdot\text{mol}^{-1} \qquad \Delta\bar{H}_{vap} = 40.7\ \text{kJ}\cdot\text{mol}^{-1}$$

$$\bar{C}_P[H_2O(s)] = 37.7\ \text{J}\cdot\text{K}^{-1}\cdot\text{mol}^{-1}$$

$$\bar{C}_P[H_2O(g)] = 35.2\ \text{J}\cdot\text{K}^{-1}\cdot\text{mol}^{-1}$$

$$\bar{C}_P[H_2O(l)] = 75.3\ \text{J}\cdot\text{K}^{-1}\cdot\text{mol}^{-1}$$

Solution: Let's start at $-10°\text{C}$ as in Figure 13-4. It requires

$$\begin{aligned} q_P &= n\bar{C}_P(T_2 - T_1) \\ &= (1\ \text{mol})(37.7\ \text{J}\cdot\text{K}^{-1}\cdot\text{mol}^{-1})(10\ \text{K}) \\ &= 377\ \text{J} \end{aligned}$$

to raise the temperature of the ice from $-10°\text{C}$ to $0°\text{C}$. The sample is heated at a rate of $200\ \text{J}\cdot\text{min}^{-1}$, and so it takes

$$t = \frac{377\ \text{J}}{200\ \text{J}\cdot\text{min}^{-1}} = 1.89\ \text{min}$$

The molar enthalpy of fusion of ice is 6.0 kJ, and so it requires

$$t = \frac{(6.0\ \text{kJ})\left(\dfrac{1000\ \text{J}}{\text{kJ}}\right)}{200\ \text{J}\cdot\text{min}^{-1}} = 30\ \text{min}$$

to melt the ice.

To raise the temperature of water from $0°\text{C}$ to $100°\text{C}$ requires

$$\begin{aligned} q_P &= (1\ \text{mol})(75.3\ \text{J}\cdot\text{K}^{-1}\cdot\text{mol}^{-1})(100\ \text{K}) \\ &= 7530\ \text{J} \end{aligned}$$

The time required is

$$t = \frac{7530\ \text{J}}{200\ \text{J}\cdot\text{min}^{-1}} = 37.7\ \text{min}$$

The molar enthalpy of vaporization is $40.7\ \text{kJ}\cdot\text{mol}^{-1}$, and so the time it takes to vaporize the water is

$$t = \frac{(40.7\ \text{kJ})\left(\frac{1000\ \text{J}}{\text{kJ}}\right)}{200\ \text{J}\cdot\text{min}^{-1}} = 204\ \text{min}$$

Finally, it requires

$$\begin{aligned} q_P &= (1\ \text{mol})(35.2\ \text{J}\cdot\text{K}^{-1}\cdot\text{mol}^{-1})(20\ \text{K}) \\ &= 704\ \text{J} \end{aligned}$$

to heat the water vapor from 100° to 120°C, and it takes

$$t = \frac{704\ \text{J}}{200\ \text{J}\cdot\text{min}^{-1}} = 3.52\ \text{min}$$

Problems 13-11 to 13-14 involve heating curves.

to do so.

Putting all these calculations together, we obtain the heating curve shown in Figure 13-5.

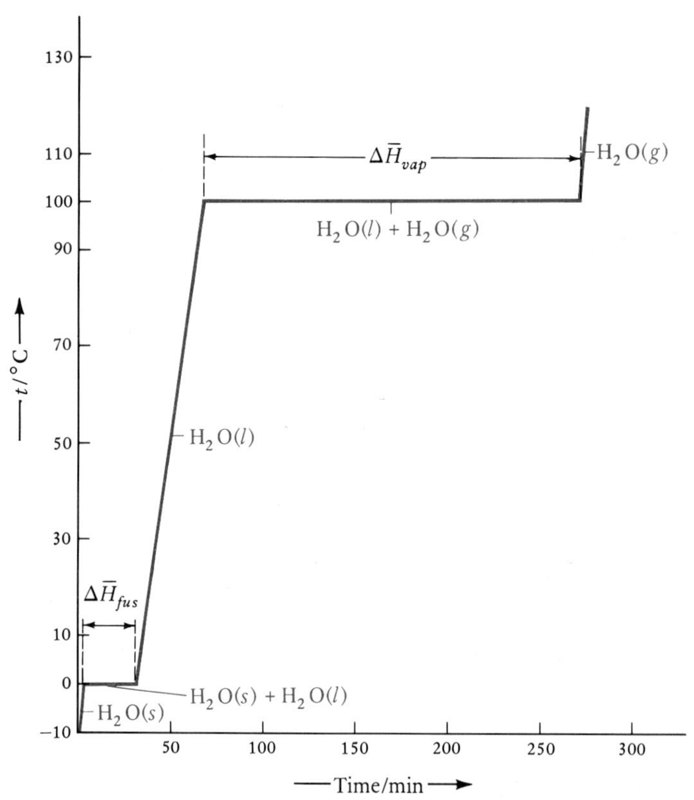

Figure 13-5 The heating curve for 1 mol of water heated at a rate of $200\ \text{J}\cdot\text{min}^{-1}$.

13-3. IT REQUIRES ENERGY TO MELT A SOLID AND TO VAPORIZE A LIQUID

The melting, or fusion, of ice can be represented by

$$H_2O(s) \rightarrow H_2O(l) \qquad \Delta\bar{H}_{fus} = 6.0 \text{ kJ}$$

or, in general, we write

$$X(s) \rightarrow X(l) \qquad \Delta\bar{H}_{fus}$$

where X stands for any element or compound. The enthalpy of fusion is necessarily positive because it requires energy to break up the crystal lattice. Recall that a positive value of ΔH means that heat is absorbed in the process. The vaporization of water can be represented by

$\Delta\bar{H}_{fus} > 0$

$$H_2O(l) \rightarrow H_2O(g) \qquad \Delta\bar{H}_{vap} = 40.7 \text{ kJ}$$

or, in general,

$$X(l) \rightarrow X(g) \qquad \Delta\bar{H}_{vap}$$

The value of $\Delta\bar{H}_{vap}$ is always positive because it requires energy to separate the molecules in a liquid from each other. Gas-phase molecules are so far apart from each other that they interact only very weakly relative to liquid-phase molecules. Essentially all the energy put in as heat of vaporization is required to separate the molecules of the liquid from one another. Because the temperature does not change during vaporization, the average kinetic energy of the molecules does not change. The molar enthalpies of vaporization of several compounds are given in Table 13-1.

$\Delta\bar{H}_{vap} > 0$

The consumption of heat in the vaporization of a liquid is one of the mechanisms used by the human body to regulate its temperature at about 310 K. When the body becomes overheated, sweating begins; the evaporation of water from the surface of the body consumes 2.3 kJ per gram of sweat that is evaporated. The rate of evaporation can be enhanced greatly by continuously sweeping away the water-saturated air near the skin surface, as occurs in a breeze.

Example 13-2: Ammonia is used as a refrigerant in some industrial refrigeration units. The molar enthalpy of vaporization of liquid ammonia is $23.4 \text{ kJ} \cdot \text{mol}^{-1}$. Compute the amount of heat absorbed in the vaporization of 5.00 kg of $NH_3(l)$.

Table 13-1 Melting points, boiling points, and molar enthalpies of vaporization and fusion

Compound	*Chemical formula*	*mp/K*	*bp/K*	$\Delta\overline{H}_{fus}/kJ\cdot mol^{-1}$	$\Delta\overline{H}_{vap}/kJ\cdot mol^{-1}$
ammonia	NH_3	195	240	5.65	23.4
argon	Ar	84	87	1.17	6.52
bromine	Br_2	266	332	10.6	29.5
carbon dioxide	CO_2	217	(195) sublimes	8.33	(25.2) sublimes
chlorine	Cl_2	172	239	6.41	20.4
formaldehyde	H_2CO	181	252	—	24.5
helium	He	—	4.2	0.014	0.081
hydrogen	H_2	14	20	0.12	0.90
iodine	I_2	387	458	15.5	41.9
krypton	Kr	116	121	1.63	9.03
lithium bromide	LiBr	823	1583	17.6	148.1
mercury	Hg	234	630	2.30	59.1
methane	CH_4	91	112	0.94	8.17
neon	Ne	24	27	0.33	1.76
nitrogen	N_2	63	77	0.72	5.58
oxygen	O_2	54	90	0.44	6.82
water	H_2O	273	373	6.01	40.7
xenon	Xe	160	166	2.30	12.63

Solution: Each mole of $NH_3(l)$ that vaporizes consumes 23.4 kJ of energy as heat. The molar mass of NH_3 is 17.0 g·mol^{-1}; thus the amount of heat, q_P, absorbed in vaporizing 5.00 kg of $NH_3(l)$ is

$$q_P = (5.00\text{ kg})\left(\frac{1000\text{ g}}{\text{kg}}\right)\left(\frac{1\text{ mol}}{17.0\text{ g}}\right)(23.4\text{ kJ}\cdot\text{mol}^{-1}) = 6.88 \times 10^3\text{ kJ}$$

Example 13-3: Compute the energy released as heat when 28 g of liquid water at 18°C is converted to ice at 0°C. (An ice cube contains about 1 oz of water, and 1 oz is equivalent to 28 g.) The molar heat capacity of $H_2O(l)$ is $\overline{C}_P = 75.3\text{ J}\cdot\text{K}^{-1}\cdot\text{mol}^{-1}$, and $\Delta\overline{H}_{fus} = 6.0\text{ kJ}\cdot\text{mol}^{-1}$ for ice.

454 g = 1 lb
16 oz = 1 lb
1 oz = 28 g

Solution: The overall process must be broken down into two steps. We must first bring the $H_2O(l)$ from 18°C to 0°C (the freezing point of water) and then consider the process $H_2O(l) \rightarrow H_2O(s)$ at 0°C:

28 g $H_2O(l)$ at 18°C	$\xrightarrow{\text{step 1}}$	28 g $H_2O(l)$ at 0°C	$\xrightarrow{\text{step 2}}$	28 g $H_2O(s)$ at 0°C

For step 1 we have, from Equation (13-1), where n is the number of moles of water,

$$\begin{aligned} q_P &= n\bar{C}_P(T_2 - T_1) \\ &= (28\text{ g})\left(\frac{1\text{ mol } H_2O}{18\text{ g } H_2O}\right)(75.3\text{ J}\cdot\text{K}^{-1}\cdot\text{mol}^{-1})\left(\frac{1\text{ kJ}}{1000\text{ J}}\right)(-18\text{ K}) \\ &= -2.1\text{ kJ} \end{aligned}$$

The negative sign for q_P reflects the fact that energy must be removed to lower the temperature of the water. For step 2, where n is the number of moles of H_2O,

$$q_P = n(-\Delta\bar{H}_{fus}) = (28\text{ g})\left(\frac{1\text{ mol } H_2O}{18\text{ g } H_2O}\right)(-6.0\text{ kJ}\cdot\text{mol}^{-1}) = -9.3\text{ kJ}$$

where the minus sign in front of $\Delta\bar{H}_{fus}$ arises because freezing is the reverse of fusion, that is, $\Delta\bar{H}_{freezing} = -\Delta\bar{H}_{fus}$. The total amount of energy that must be *removed* as heat from the 28 g of water is 2.1 kJ + 9.3 kJ = 11.4 kJ.

13-4. VAN DER WAALS FORCES ARE ATTRACTIVE FORCES BETWEEN MOLECULES

In the process of vaporization (Figure 13-6), the molecules of the liquid, which are in contact with each other, become separated from each other and widely dispersed. The value of the enthalpy of vaporization reflects how strongly the molecules attract each other in the liquid. The more strongly the molecules attract each other, the greater the value of $\Delta\bar{H}_{vap}$.

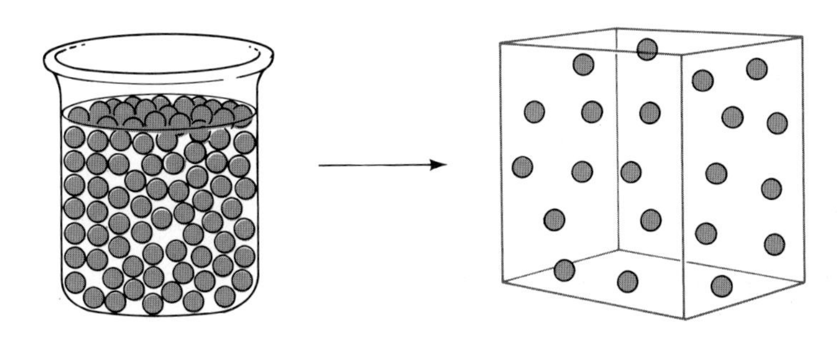

Figure 13-6 When a liquid is vaporized, the molecules become widely separated from each other.

Ionic compounds have relatively high boiling points and large molar enthalpies of vaporization.

The simplest force at the atomic or molecular level is the force between ions. We studied this force in Chapter 9 when we discussed ionic compounds. Two ions with opposite charges attract each other; ions with like charges repel each other. The force between ions is a relatively strong force; it requires a relatively large amount of energy to separate ions of opposite charge. The enthalpies of vaporization of ionic compounds are much larger than those of nonionic compounds. Enthalpies of vaporization of ionic compounds are typically at least 100 $kJ \cdot mol^{-1}$. In addition, the boiling points of ionic compounds are higher than those of nonionic compounds.

Most of the compounds listed in Table 13-1 are molecular compounds. The fact that all substances can be liquefied means that even neutral molecules attract each other. In Chapter 10 we learned that some molecules have dipole moments, or are polar. An example of a polar molecule is formaldehyde, whose Lewis formula is

$$\begin{array}{l} H \\ \quad \searrow \\ \quad\quad C{=}\ddot{\underset{..}{O}}: \\ \quad \nearrow \\ H \end{array}$$

Because the electronegativity of an oxygen atom is greater than that of a carbon atom, the oxygen atom has a small negative charge and the carbon atom has a small positive charge. The electronegativities of a carbon atom and a hydrogen atom are almost equal, and so carbon-hydrogen bonds are not polar. A formaldehyde molecule has partial charges on the carbon and oxygen atoms, and we write

$$\begin{array}{l} H \\ \quad \searrow \\ \quad\quad \overset{\delta+}{C}{=}\overset{\delta-}{\ddot{\underset{..}{O}}}: \\ \quad \nearrow \\ H \end{array}$$

Even though a formaldehyde molecule is electrically neutral overall, it has a positively charged end and a negatively charged end and thus is a polar molecule.

Figure 13-7 Even though polar molecules are electrically neutral overall, they attract each other by a dipole-dipole force. The molecules orient themselves as shown because the positive end of one attracts the negative end of another. The dipoles are said to be oriented head to tail.

Polar molecules attract each other (Figure 13-7), and the attraction between them is called a *dipole-dipole attraction.* Charges in polar molecules are considerably smaller than the full electronic charges on ions, and so dipole-dipole forces are smaller than ion-ion forces. Enthalpies of vaporization for many polar compounds are around 20 $kJ \cdot mol^{-1}$.

A particularly important dipole-dipole attraction occurs when one or more hydrogen atoms are bonded to an electronegative atom, as in the case of water and ammonia. Let's consider the molar enthalpies of vaporization of water, ammonia, and methane:

H_2O	NH_3	CH_4
40.7 $kJ \cdot mol^{-1}$	23.4 $kJ \cdot mol^{-1}$	8.2 $kJ \cdot mol^{-1}$

These three compounds have approximately the same molecular mass (18, 17, 16, respectively), but the amounts of energy required to separate the molecules of the liquids are very different.

We can represent a water molecule by

$$\underset{\delta+}{\mathrm{H}}\diagup\overset{\mathrm{O}^{2\delta-}}{}\diagdown\mathrm{H}_{\delta+}$$

Water molecules in liquid water attract each other through the electrostatic interaction between a hydrogen atom and the oxygen atom on a different molecule (Figure 13-8):

$$\mathrm{H}_{\delta+}\!-\!\overset{2\delta-}{\mathrm{O}}\!-\!\underset{\delta+}{\mathrm{H}}\cdots\overset{2\delta-}{\mathrm{O}}(\mathrm{H}^{\delta+})\!-\!\mathrm{H}_{\delta+}$$

hydrogen bond

The attraction that occurs between molecules when a hydrogen atom is covalently bonded to a highly electronegative atom, such as O, N, or F, is called *hydrogen bonding*. Because a hydrogen atom is so small, the charge on it is highly concentrated, and so it strongly attracts electronegative atoms in neighboring molecules.

Hydrogen bonds are a particularly strong form of dipole-dipole attraction. The pattern shown in Figure 13-8 extends throughout liquid water and gives water its large value of $\Delta\bar{H}_{vap}$ (40.7 $\mathrm{kJ\cdot mol^{-1}}$) and its high boiling point. We shall see that hydrogen bonding gives water many special properties.

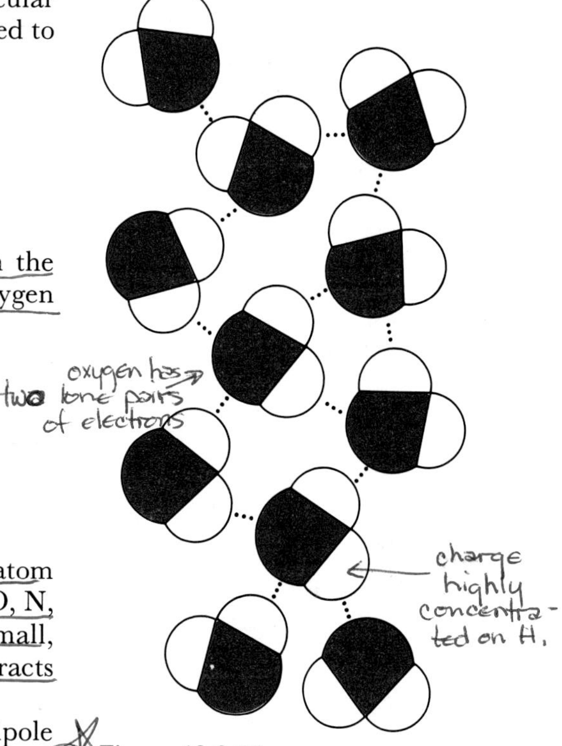

Figure 13-8 There are many hydrogen bonds in liquid water; each oxygen atom can form two hydrogen bonds because each oxygen atom has two lone pairs of electrons (see Figure 11-11c).

Hydrogen bonding greatly affects the structure of ice (Figure 13-9), which is described as an *open structure* because of the significant fraction that is unoccupied. The open structure of ice is a direct consequence of the fact that each hydrogen atom is hydrogen-bonded to an oxygen atom of an adjacent molecule. Note the tetrahedral arrangement of the oxygen atoms in ice. Every oxygen atom sits in the center of a tetrahedron formed by four other oxygen atoms.

The structure of liquid water is less open than the structure of ice because when ice melts the total number of hydrogen bonds decreases. Unlike most other substances, the water *increases* in density on going from solid to liquid because of a partial breakdown of the hydrogen-bonded structure. The extent of hydrogen bonding in liquid water is only about 80 percent, whereas in ice 100 percent of the oxygen atoms are hydrogen-bonded. The extent of hydrogen bonding in water decreases as the temperature increases.

Hydrogen bonding effects the shapes of proteins and many other biochemical molecules.

Hydrogen bonding occurs also in liquid ammonia, but the individual hydrogen bonds are weaker than in water because nitrogen is less electronegative than oxygen and the fractional charges on the nitrogen and hydrogen atoms in NH_3 are less than those on the oxygen and hydrogen atoms in H_2O. Furthermore, there are fewer hydrogen bonds in NH_3 because each nitrogen atom can form only

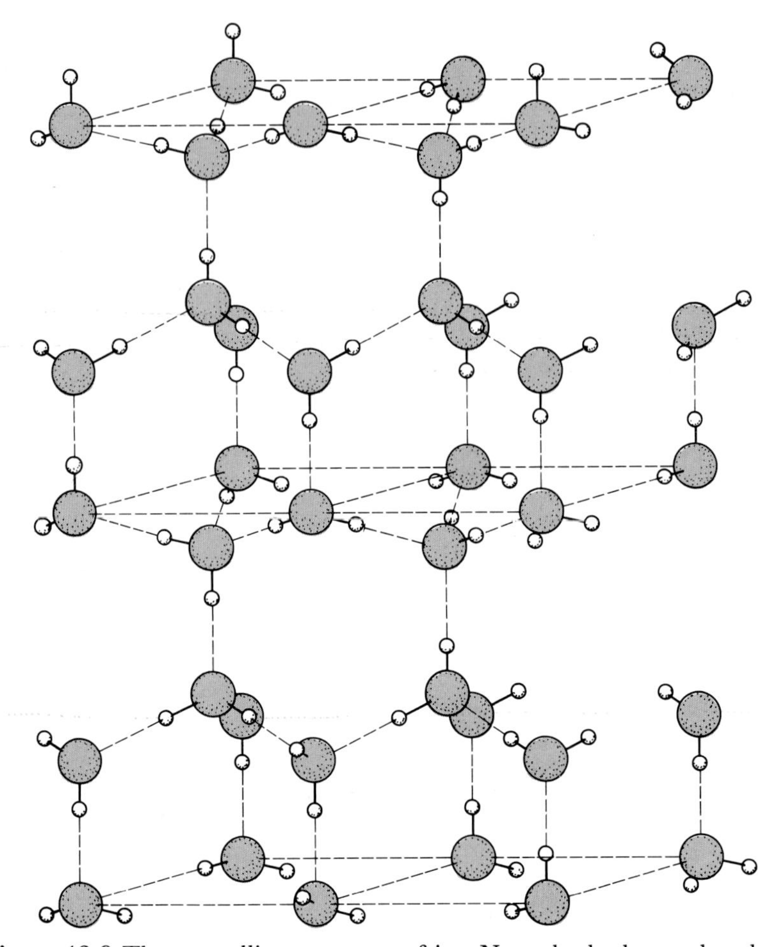

Figure 13-9 The crystalline structure of ice. Note the hydrogen bonds and the tetrahedral arrangement of the oxygen atoms. Each oxygen atom is located in the center of a tetrahedron formed by four other oxygen atoms. The entire structure is held together by hydrogen bonds.

Figure 13-10 There are fewer hydrogen bonds in liquid NH_3 than in H_2O because each nitrogen atom can form only one hydrogen bond. Because NH_3 has only one lone pair, Figure 11-11, there is only one vertex at which to form a hydrogen bond.

one hydrogen bond (Figure 13-10). Consequently, $\Delta \bar{H}_{vap}$ for ammonia is only about half that of water.

Methane is a nonpolar molecule because it is tetrahedral and because the electronegativities of a carbon atom and a hydrogen atom are almost the same. There is no hydrogen bonding or dipole-dipole attraction in methane, and thus, of H_2O, NH_3, and CH_4, CH_4 has the lowest value of $\Delta \bar{H}_{vap}$. There is one other attractive force that we must consider in this case, however. Methane liquefies at 91 K and has an enthalpy of vaporization of 8.2 $kJ \cdot mol^{-1}$; therefore, methane molecules must attract each other. Even the noble gases, which consist of single, spherical atoms, can be liquefied.

How neutral, nonpolar molecules attract each other was not understood until the quantum theory was developed. Let's consider two argon atoms. As electrons move about in each atom, an instantaneous dipole moment is set up in each atom (Figure 13-11). When the atoms approach each other, the electrons on one atom influence the electrons on the other atom in such a way that the instantaneous dipole moments on each atom are head to tail (Figure 13-12). The electronic distribution then has a lower energy than if the instantaneous dipole moments were head to head. The instantaneous dipole-dipole attraction depicted in Figure 13-12 accounts for the attractive force between nonpolar molecules as well as atoms. This force was first explained by the German physicist Fritz London in 1930 and is now called a *London force.*

Figure 13-11 If it were possible to take an instantaneous view of an atom, it might look like this drawing. The instantaneous position of the electrons leads to an instantaneous dipole moment. The negative charge is due to a greater-than-average electronic-charge density, and the positive charge is due to a less-than-average electronic-charge density. As the electrons move around, the dipole moment points in all directions equally and averages out to zero.

Because London forces are due to the motion of electrons, their strength depends upon the number of electrons. The more electrons there are in the two interacting molecules, the stronger their attraction for each other. Therefore, we expect the value of $\Delta \bar{H}_{vap}$ to increase with the number of electrons, or even with the size of the molecules. The following data support this prediction:

Substance	$\Delta \bar{H}_{vap}/kJ\cdot mol^{-1}$	*Substance*	$\Delta \bar{H}_{vap}/kJ\cdot mol^{-1}$
He	0.08	F_2	6.5
Ne	1.76	Cl_2	20.4
Ar	6.52	Br_2	29.5
Kr	9.03	I_2	41.9
Xe	12.63		

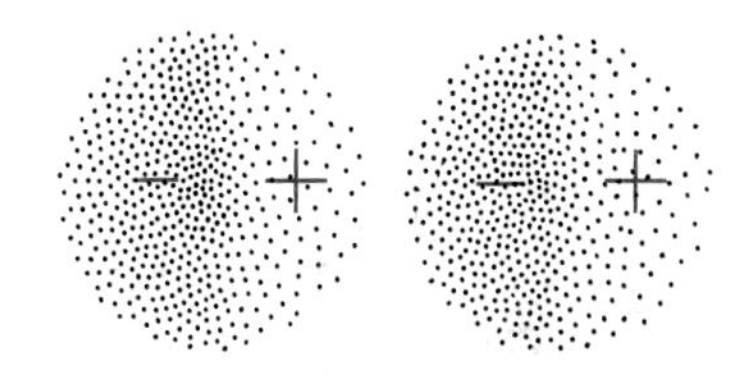

Figure 13-12 When two atoms are near each other, the motion of the electrons in the two atoms affect each other so that the instantaneous dipole moments are head to tail. This leads to an instantaneous dipole-dipole attraction between the two atoms.

Note that, within each group, the value of $\Delta \bar{H}_{vap}$ increases with the number of electrons.

Figure 13-13 plots the melting points and boiling points of the noble gases and the hydrides of the nonmetallic elements. The hydrogen-bonded compounds (H_2O, NH_3, and HF) have unusually high melting points and boiling points. Except for the hydrogen-bonded compounds, there is a general increase of melting point and boiling point with increasing molecular mass. This increase is due to the increase in London forces with an increase in the number of electrons in a molecule. The attractive forces between molecules, be they dipole-dipole forces or London forces, are collectively called *van der Waals forces.*

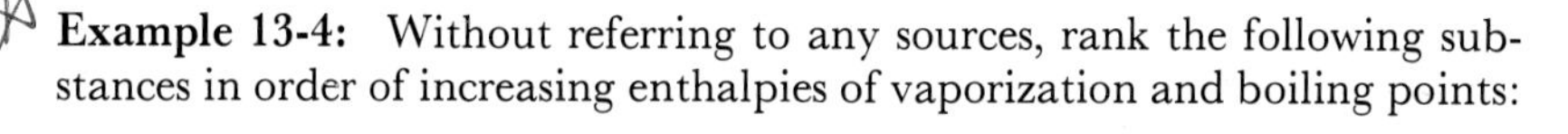

Example 13-4: Without referring to any sources, rank the following substances in order of increasing enthalpies of vaporization and boiling points:

NaCl H_2 CH_3OH

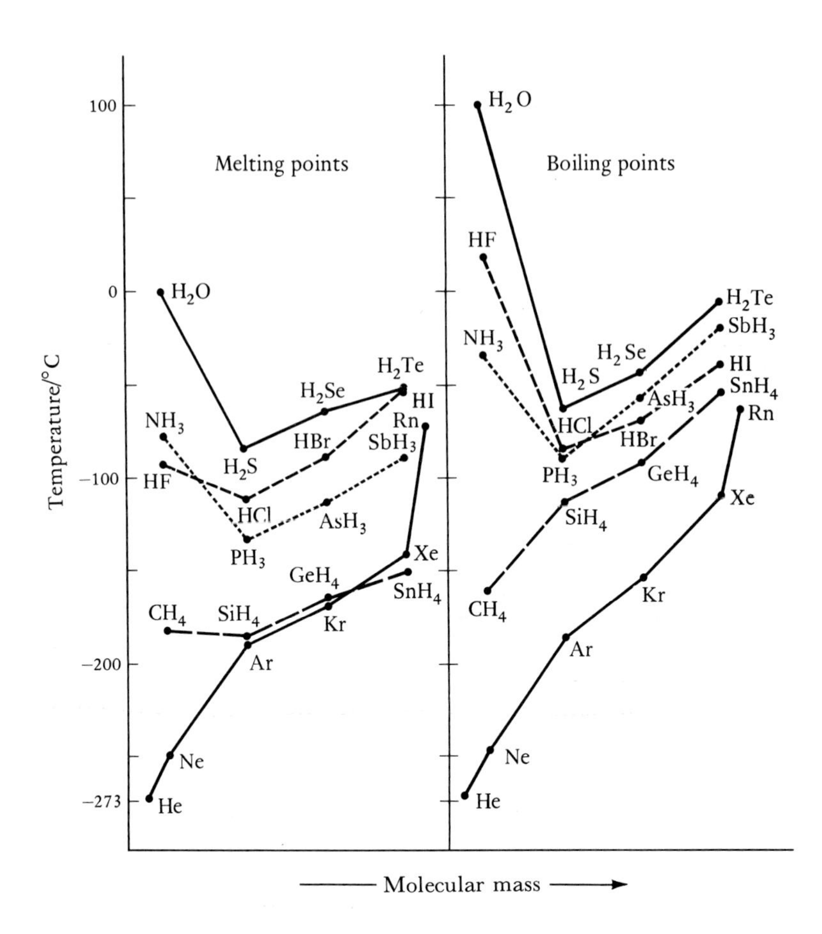

Figure 13-13 Melting and boiling points of the noble gases and hydrides of the nonmetallic elements. Note the abnormally high values for hydrogen fluoride, water, and ammonia, which are the result of hydrogen bonds in the liquid and solid phases.

Solution: The only ionic compound listed is NaCl. Thus we predict that NaCl has the largest value of $\Delta\bar{H}_{vap}$ and the highest boiling point. The Lewis formula for CH_3OH is

```
     H
     |   ..
  H—C—O—H
     |   ..
     H
```

and thus CH_3OH is hydrogen-bonded. H_2 is a nonpolar molecule, and so we predict that

$$\Delta\bar{H}_{vap}[\mathrm{NaCl}] > \Delta\bar{H}_{vap}[\mathrm{CH_3OH}] > \Delta\bar{H}_{vap}[\mathrm{H_2}]$$

The actual values are $170\ \mathrm{kJ\cdot mol^{-1}}$, $35.3\ \mathrm{kJ\cdot mol^{-1}}$, and $0.90\ \mathrm{kJ\cdot mol^{-1}}$, respectively. The boiling points are 1690 K for NaCl, 337 K for CH_3OH, and 20.4 K for H_2.

Problems 13-15 to 13-20 are similar to Example 13-4.

13-5. THE ENTHALPY OF SUBLIMATION IS A MEASURE OF THE ENERGY OF INTERACTION BETWEEN THE MOLECULES IN A SOLID

Sublimation is the process whereby a solid is converted directly to a gas without passing through the liquid phase. The *molar enthalpy of sublimation*, $\Delta \bar{H}_{sub}$, is the energy absorbed as heat when 1 mol of a solid is sublimed at constant pressure. Essentially all the energy put in as heat in the sublimation process is used to separate the molecules in the solid from one another. The larger the value of $\Delta \bar{H}_{sub}$, the stronger the intermolecular attractions in the solid. The best-known example of sublimation is the conversion of dry ice, $CO_2(s)$, to carbon dioxide gas:

The enthalpy of sublimation is often called the heat of sublimation.

$$CO_2(s) \rightarrow CO_2(g) \qquad \Delta \bar{H}_{sub} = 25.2 \text{ kJ} \cdot \text{mol}^{-1}$$

The name dry ice is used because the CO_2 does not become a liquid at 1 atm pressure. Dry ice at 1 atm has a temperature of $-78\,^\circ\text{C}$ and is widely used as a one-time low-temperature refrigerant. The sublimation of 44 g of dry ice carries off 25.2 kJ of heat.

Like carbon dioxide, iodine and moth balls pass directly from the solid to the vapor phase at 1 atm.

Ice sublimes at temperatures below its melting point (0°C). For the sublimation of ice,

$$H_2O(s) \rightarrow H_2O(g) \qquad \Delta \bar{H}_{sub} = 46.7 \text{ kJ} \cdot \text{mol}^{-1}$$

Snow often sublimes, and so does the ice in the freezer compartment of your refrigerator. Ice cubes left in the freezer get smaller over a period of a few weeks.

The various terms involved in the different types of phase changes that we have discussed up to this point are summarized in the following diagram:

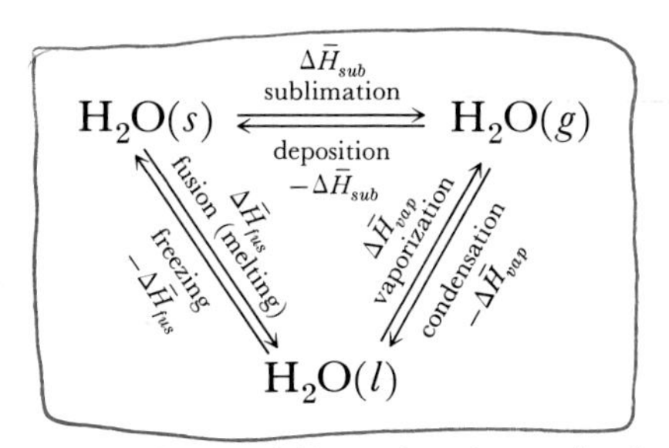

Note from the diagram that sublimation is equivalent to fusion plus vaporization, from which we conclude that

$$\Delta \bar{H}_{sub} = \Delta \bar{H}_{fus} + \Delta \bar{H}_{vap} \qquad (13\text{-}2)$$

From Equation (13-2) and Table 13-1, we can compute the molar heat of sublimation of water:

$$\Delta \bar{H}_{sub} = 6.0 \text{ kJ} \cdot \text{mol}^{-1} + 40.7 \text{ kJ} \cdot \text{mol}^{-1} = 46.7 \text{ kJ} \cdot \text{mol}^{-1}$$

Example 13-5: Prove that $\Delta \bar{H}_{sub}$ is always greater than $\Delta \bar{H}_{vap}$.

Solution: According to Equation (13-2)

$$\Delta \bar{H}_{sub} = \Delta \bar{H}_{fus} + \Delta \bar{H}_{vap}$$

We can rewrite this equation as

$$\Delta \bar{H}_{sub} - \Delta \bar{H}_{vap} = \Delta \bar{H}_{fus}$$

The value of $\Delta \bar{H}_{fus}$ is always positive, however, and so

$$\Delta \bar{H}_{sub} - \Delta \bar{H}_{vap} > 0$$

or

$$\Delta \bar{H}_{sub} > \Delta \bar{H}_{vap}$$

The reason that $\Delta \bar{H}_{sub} > \Delta \bar{H}_{vap}$ is that $\Delta \bar{H}_{sub}$ involves the energy required to break up the crystal lattice in addition to vaporizing the liquid.

Figure 13-14 When a liquid is placed in a closed container that has been evacuated, eventually the pressure of the vapor above the liquid reaches a constant value that depends upon the particular liquid and the temperature.

13-6. THE EQUILIBRIUM BETWEEN A LIQUID AND ITS VAPOR IS A DYNAMIC EQUILIBRIUM

Suppose that a liquid is placed in an evacuated container with a vapor space and that the container is closed (Figure 13-14). Then it is observed that the pressure of the vapor over the liquid increases rapidly at first and then progressively more slowly until a constant pressure is reached. Initially, molecules in the liquid phase escape from the surface of the liquid and go into the vapor phase. The number of molecules that leave the surface of the liquid is proportional to the surface area of the liquid. Because the surface area is constant, the rate of evaporation is constant (Figure 13-15). There

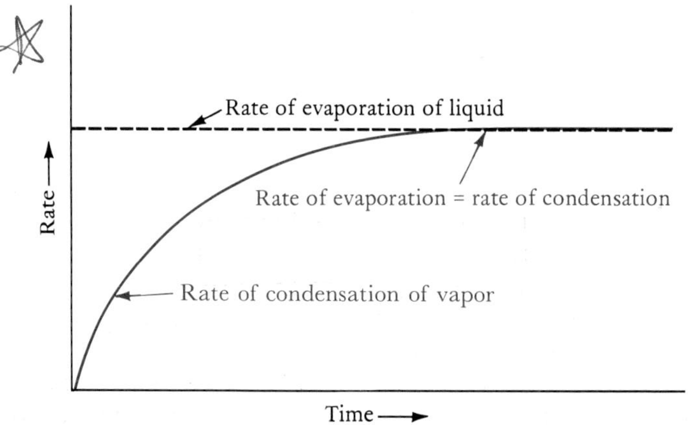

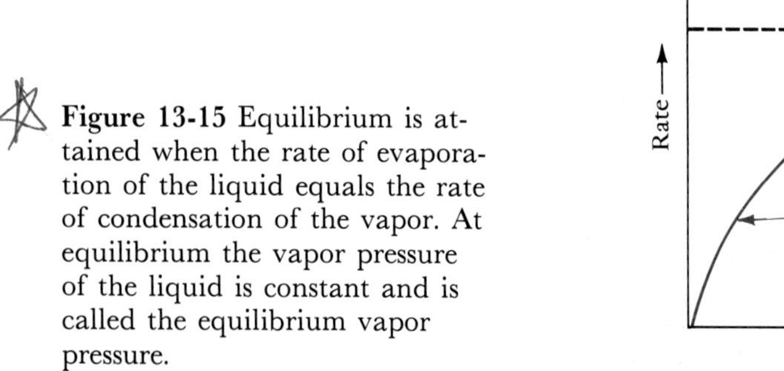

Figure 13-15 Equilibrium is attained when the rate of evaporation of the liquid equals the rate of condensation of the vapor. At equilibrium the vapor pressure of the liquid is constant and is called the equilibrium vapor pressure.

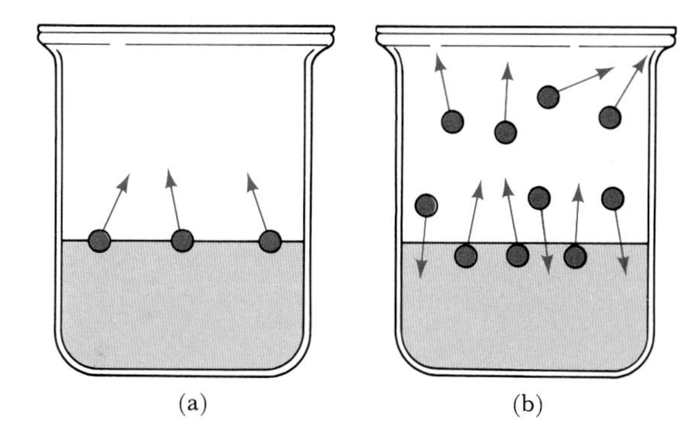

Figure 13-16 When a liquid is placed in a closed container, the rate at which molecules escape from the surface is constant, but the rate at which molecules enter the liquid from the vapor is proportional to the number of molecules in the vapor. When the number of molecules in the vapor is such that the rate of escape from the surface is equal to the rate of condensation from the vapor, the liquid and vapor are in equilibrium with each other (b).

are no molecules in the vapor phase initially, and so there is no condensation from the vapor phase to the liquid phase. As the concentration of molecules in the vapor phase increases, the pressure of the vapor increases and the number of vapor-phase molecules that collide with the liquid surface increases. As a result, the rate of condensation of the vapor increases. Eventually a point is reached where the rate of evaporation from the liquid surface is equal to the rate of condensation from the vapor phase. The pressure of the vapor no longer increases but takes on a constant value. The evaporation-condensation process appears to have stopped, and we say that the system is at *equilibrium*, meaning that there is no apparent change taking place (Figure 13-16).

The equilibrium between the liquid and the vapor is a *dynamic equilibrium*. That is, the liquid continues to evaporate and the vapor continues to condense, but the rate of evaporation is exactly equal to the rate of condensation and thus there is no *net* change. In an equation, we have

We shall see that the concept of dynamic equilibium occurs frequently in chemistry.

$$\text{rate of evaporation} = \text{rate of condensation}$$

The pressure of the vapor at equilibrium is called the *equilibrium vapor pressure*. We shall see that the value of the equilibrium vapor pressure depends upon the particular liquid and the temperature.

13-7. A LIQUID HAS A UNIQUE EQUILIBRIUM VAPOR PRESSURE AT EACH TEMPERATURE

Let's consider the approach to a dynamic liquid-vapor equilibrium at two different temperatures. The higher the temperature is, the more rapidly the molecules in the liquid phase are moving and so the higher the rate of evaporation. Figure 13-17 shows that, because the rate of evaporation at T_2 is greater than the rate of evaporation at T_1 (given that $T_2 > T_1$), the equilibrium vapor pressure at T_2 is greater than that at T_1. Thus we see that the value of the equilibrium vapor pressure of a liquid increases with increasing temperature. At each

Figure 13-17 The change in vapor pressure with time for a liquid as it approaches equilibrium. Because the rate of evaporation increases with increasing temperature, the equilibrium vapor pressure increases with increasing temperature ($T_2 > T_1$).

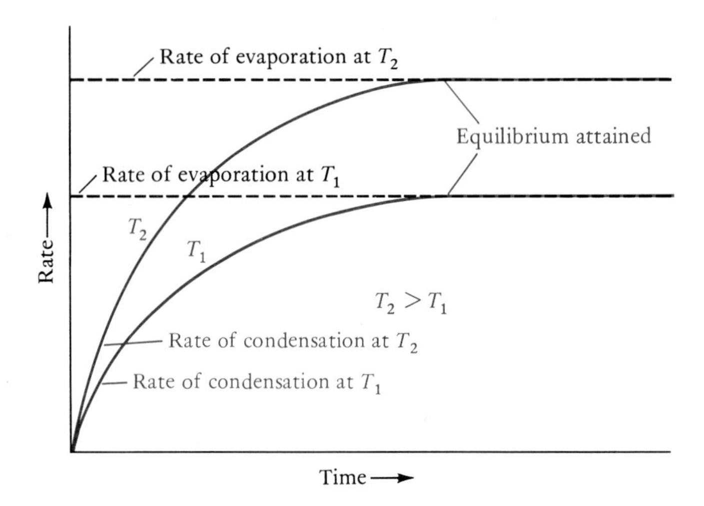

temperature, a liquid has a definite equilibrium vapor pressure. The equilibrium vapor pressure of water as a function of the absolute temperature, which is called its *vapor pressure curve,* is given in Figure 13-18 and in Table 13-2. Figure 13-18 also shows the equilibrium vapor pressure curve for ethanol, CH_3CH_2OH, or ethyl alcohol.

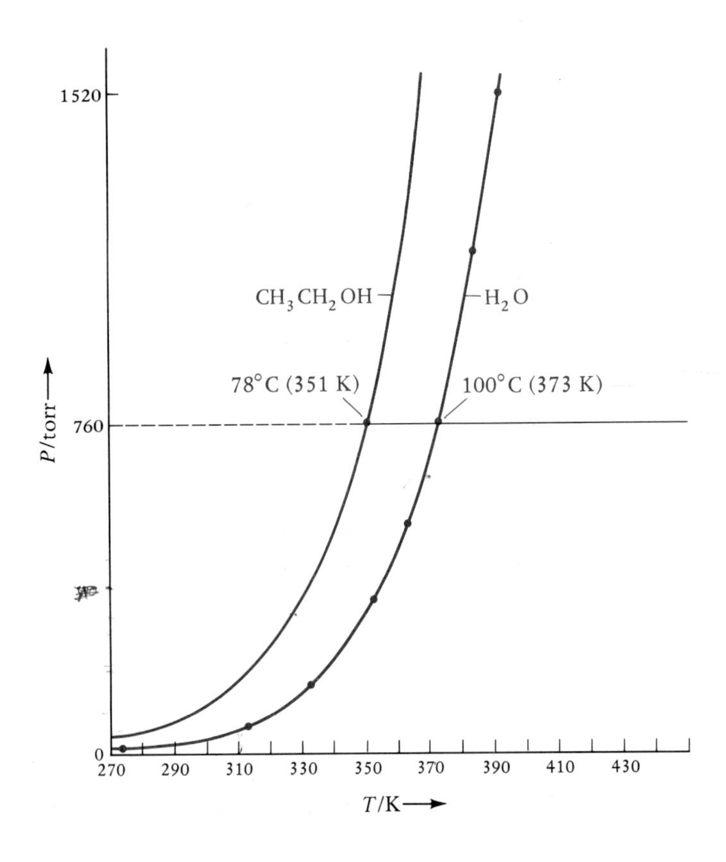

Figure 13-18 The equilibrium vapor pressure curves for water and ethanol over the temperature range 270 to 430 K (−3 to 157°C). Note the rapid increase in vapor pressure with increasing temperature. The equilibrium vapor pressure curve for ethanol lies above that for water because ethanol has a higher equilibrium vapor pressure than water at the same temperature.

The boiling point of a liquid is the temperature at which its vapor pressure equals atmospheric pressure. The *normal boiling point* (that at exactly 1 atm) of water is 373 K (100°C). If the atmospheric pressure is less than 1.00 atm, then the temperature at which the vapor pressure of liquid water equals atmospheric pressure is less than 100°C. For example, at 8000 feet elevation, where the atmospheric pressure is about 0.75 atm, water boils at 92°C, and at 2.0 atm, as in a pressure cooker, it boils at 120°C. Figure 13-19 shows the effect of elevation on the boiling point of water. Because atmospheric pressure decreases with increasing elevation and because the rate at which food cooks depends strongly on the temperature, it requires a significantly longer time to cook food by boiling at high elevations than at sea level. An egg must be boiled for about 5 min at 9200 ft in order to be cooked to the same extent as one boiled for 3 min at sea level.

Elevation/ft	Pressure/torr	Boiling temperature of water °F	°C
30,000	240	158	71
Mt. Everest			
Mt. McKinley (Alaska)			
20,000	355	176	80
Mt. Whitney (California)			
10,000	530	194	90
Vail, Colorado			
5000	630	203	95
4000	655	205	96
3000	680	206	97
2000	705	208	98
1000	730	210	99
Sea level 0	760	212	100

Figure 13-19 The effect of elevation on the boiling point of water.

Table 13-2 Equilibrium vapor pressure of water as a function of temperature

$t/°C$	P/atm	$P/torr$
0	0.0060	4.6
5	0.0086	6.5
10	0.0121	9.2
15	0.0168	12.8
20	0.0230	17.4
25	0.0313	23.8
30	0.0418	31.6
35	0.0555	42.2
40	0.0728	55.3
45	0.0946	71.9
50	0.122	92.5
55	0.155	118.0
60	0.197	149.4
65	0.247	187.5
70	0.308	233.7
75	0.380	289.1
80	0.467	355.1
85	0.571	433.6
90	0.692	525.8
95	0.834	633.9
100	1.000	760.0
105	1.192	906.1
110	1.414	1074.6

Example 13-6: The following figure gives the vapor pressure of a certain liquid as a function of temperature:

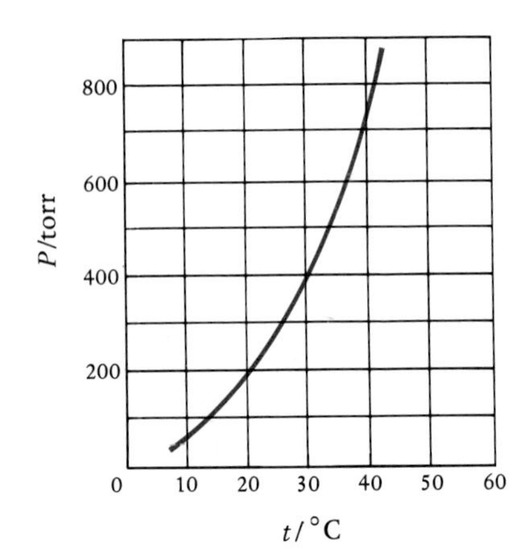

(a) Estimate the normal boiling point of the liquid. (b) If the liquid boils at 35°C, then what is the pressure of the atmosphere?

Solution: (a) When the pressure is 760 torr, the temperature is about 40°C. Thus, the normal boiling point is 40°C. (b) The vapor pressure at 35°C is about 500 torr. Because the liquid is boiling at 35°C, the pressure of the atmosphere must be about 500 torr.

13-8. RELATIVE HUMIDITY IS BASED UPON THE VAPOR PRESSURE OF WATER

The amount of water vapor in the atmosphere is expressed in terms of *relative humidity*. Relative humidity is the ratio of the partial pressure of the water vapor in the atmosphere to the equilibrium vapor pressure of water at the same temperature times 100. In terms of an equation we have

We define relative humidity as a percentage quantity.

$$\text{relative humidity} = \frac{P_{H_2O}}{P^{\circ}_{H_2O}} \times 100 \qquad (13\text{-}3)$$

where P_{H_2O} is the partial pressure of the water vapor in the air and $P^{\circ}_{H_2O}$ is the equilibrium vapor pressure of water at the same temperature. At 20°C, the equilibrium vapor pressure of water is 17.4 torr. If the partial pressure of the water vapor in the air is 11.2 torr, then the relative humidity is

$$\text{relative humidity} = \frac{11.2\ \text{torr}}{17.4\ \text{torr}} \times 100 = 64.4\%$$

If the temperature of the air is lowered to 13°C, where the equilibrium vapor pressure of water is 11.2 torr, then the relative humidity is

$$\text{relative humidity} = \frac{11.2\ \text{torr}}{11.2\ \text{torr}} \times 100 = 100\%$$

At 13°C, air that contains water vapor at a partial pressure of 11.2 torr is saturated with water vapor. Below this temperature, the water vapor begins to condense as dew or fog, which consists of small droplets of water. The air temperature at which the relative humidity reaches 100 percent is called the *dew point.* Most people begin to feel uncomfortable when the dew point rises above 20°C, and air with a dew point above 24°C is generally regarded as extremely humid or muggy.

Dew forms when the night temperature drops below the dew point. Frost is frozen dew.

Example 13-7: Calculate the relative humidity and the dew point if the partial pressure of water vapor in the air is 22.2 torr and the temperature of the air is 30°C. The equilibrium vapor pressure of water at 30°C is 31.6 torr.

Problems 13-35 to 13-38 involve relative humidity and dew point.

Solution: The relative humidity, given by Equation (13-3), is

$$\text{relative humidity} = \frac{P_{H_2O}}{P^{\circ}_{H_2O}} \times 100 = \frac{22.2\ \text{torr}}{31.6\ \text{torr}} \times 100 = 70.3\%$$

The dew point is the temperature at which the equilibrium vapor pressure of water is equal to 22.2 torr. According to Table 13-2, this is about 24°C. Such a day would be considered very uncomfortable.

The dew point is a better indicator of comfort than is the relative humidity, as is illustrated by the following comparison of two days with the same relative humidity:

Air temperature/°C	*Relative humidity/%*	*Dew point/°C*
20	70	14
30	70	24

Although the relative humidity is the same for both days, the day with the 24°C dew point is uncomfortable whereas the day with the 14°C dew point is pleasant. In warm weather, the higher the dew point, the more uncomfortable we feel.

Figure 13-20 Molecules in the interior of a liquid are attracted in all directions, but molecules at the surface experience a net inward attraction that minimizes the surface area of the liquid and results in a surface tension.

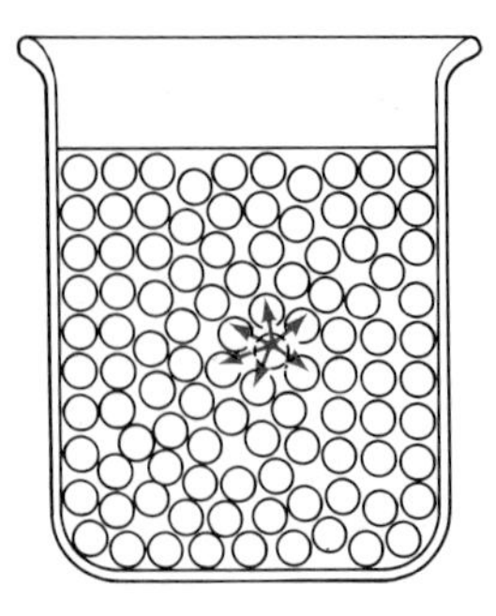

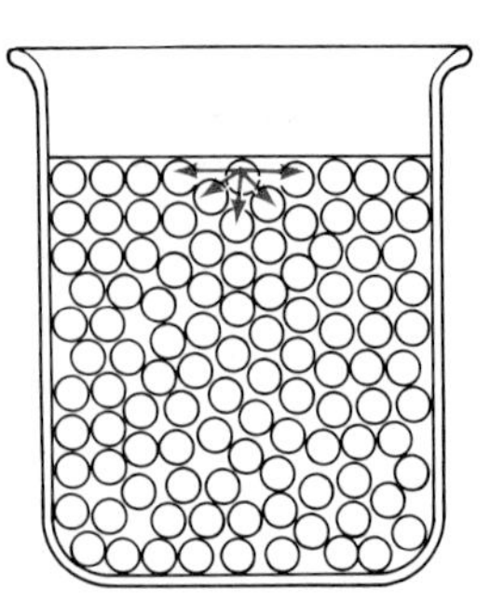

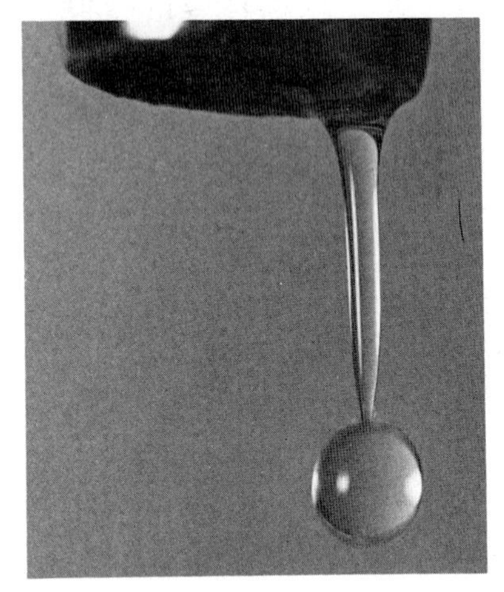

Surface tension causes freely falling drops of liquids to assume a spherical shape.

13-9. SURFACE TENSION MINIMIZES THE SURFACE AREA OF A LIQUID

A molecule in the body of a liquid is subject to attractive forces in all directions, but a molecule at the surface experiences an attractive force toward the interior of the liquid (Figure 13-20). Thus, molecules at the surface of a liquid experience a net inward force. This force tends to minimize the number of molecules at the surface and so minimize the surface area of the liquid. We call this force *surface tension*. The surface tension of a liquid tends to hold a drop of liquid in a spherical shape because a sphere is the shape that has the smallest surface area for a given volume. The higher the surface tension, the more nearly spherical the drop (Figure 13-21).

The fewer number of attractive interactions at the surface causes the molecules at the surface of a liquid to have a higher energy than molecules in the interior. Thus it requires energy to create a liquid surface, and surface tension can be thought of as the energy that is required to create a unit area of surface. The dimensions of surface tension are energy per unit area.

If a paper clip is carefully placed on a water surface, it floats even though the density of the paper clip is greater than the density of water (Figure 13-22). The clip is held up by the surface tension. Water striders and some other insects can walk on water, being supported by the surface tension of the water, which resists penetration of the surface.

Figure 13-21 Shapes of equal volumes of Hg, H_2O, $(CH_3)_2SO$, and CH_3COCH_3 (acetone), from left to right. Surface tension holds the drops in a spherical shape and gravity flattens them. The effect of gravity is the same for all the drops, and thus the higher the surface tension, the more nearly spherical the drop. The surface tensions are in units of $mJ \cdot m^{-2}$.

Certain compounds, such as sodium dodecylsulfate (SDS), $NaC_{12}H_{25}SO_3$, lower the surface tension of a liquid by concentrating at the liquid surface. Such molecules are called *surfactants* (surface active agents). A 0.1% SDS solution has a surface tension of 20 $mJ \cdot m^{-2}$, whereas pure water has a surface tension of 72 $mJ \cdot m^{-2}$. The surface tension reduction of water by surfactants is the basis of detergent action. The reduction of the surface tension of water by the detergent makes the surface tension of the water comparable to that of the grime (oils) and allows the solution to wet the grime.

The rise of a liquid in a thin tube is called *capillary action*. Capillary action occurs when the adhesive forces between the molecules of the capillary wall surface and the molecules of the liquid are sufficiently great that the liquid adheres to the solid surface. The adhesive force pulls the liquid up into the capillary. The liquid column rises until the upward adhesive force is balanced by the downward gravitational force. Capillary action plays a major role in the movement of water in plants, animals, and soil. The water is pulled by capillary action up into and through living structures (Figure 13–23).

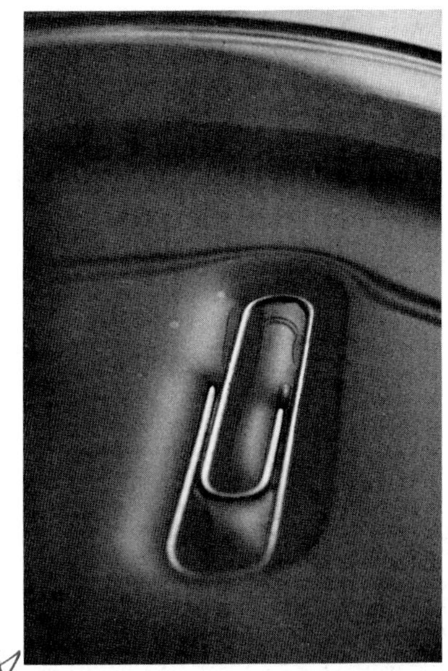

Figure 13-22 A metal paper clip floats on water because of the distribution of its weight and the surface tension of the water. Water has a high surface tension because the surface molecules form fewer hydrogen bonds than the interior molecules. The surface of water is like an elastic skin that resists penetration.

Surfactants concentrate at the surface of a liquid.

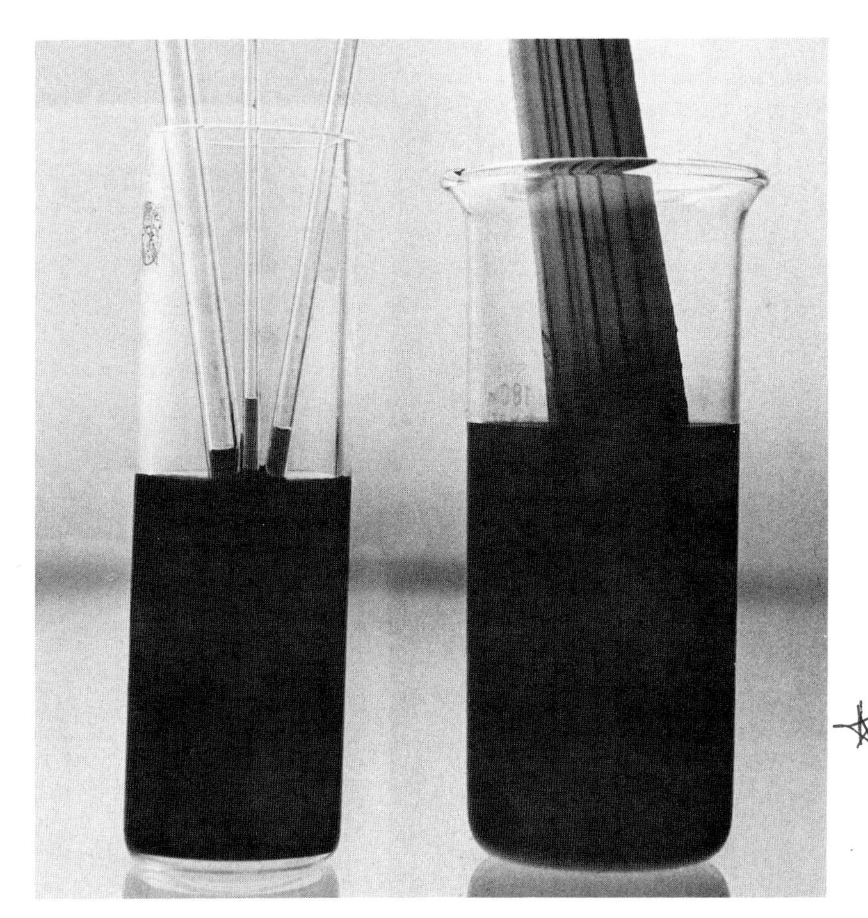

Figure 13-23 Capillary action is shown here as colored water rises in these glass tubes and a celery stalk. The cells in a celery stalk impart a capillary structure.

Figure 13-24 The X-ray diffraction pattern produced by a crystal of sodium chloride. The symmetry and spacing of the dots carry detailed information regarding the arrangement of atoms in the crystals.

13-10. X-RAY DIFFRACTION PATTERNS YIELD INFORMATION ABOUT THE STRUCTURE OF CRYSTALS

A distinguishing characteristic of crystals is the ordered nature of the crystal lattice. When X-rays with a wavelength comparable to the nearest-neighbor distance between atoms in a crystal pass through the crystal, an *X-ray diffraction pattern* results (Figure 13-24). The presence of a definite ordered array of atoms in the crystal produces a definite diffraction pattern of the X-rays, which can be recorded as an array of spots on an X-ray film. X-ray diffraction patterns are used to determine the arrangement of the atoms and molecules in a crystal.

To get a feel for the origin of X-ray diffraction patterns from crystals, let's examine the *optical diffraction patterns* formed by light passing through holes in opaque sheets (Figure 13-25). Notice that each size and arrangement of holes yields a particular diffraction pattern that can be used to determine the arrangement of the holes that produced it. Therefore, the optical diffraction patterns carry information regarding the relative positions and the spacings of the holes, which is similar to the type of information carried by the X-ray diffraction pattern regarding the atoms in the crystal.

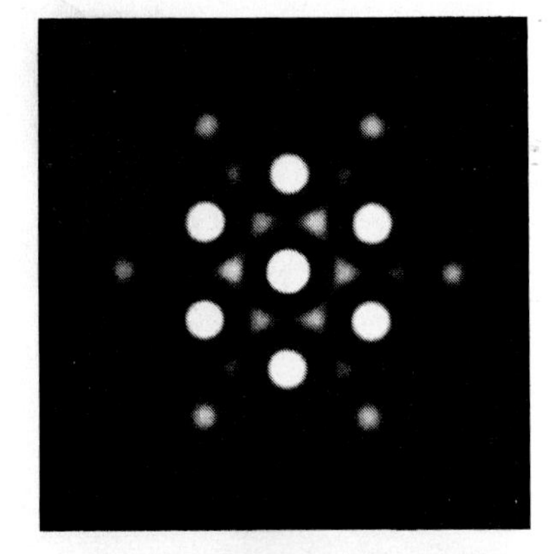

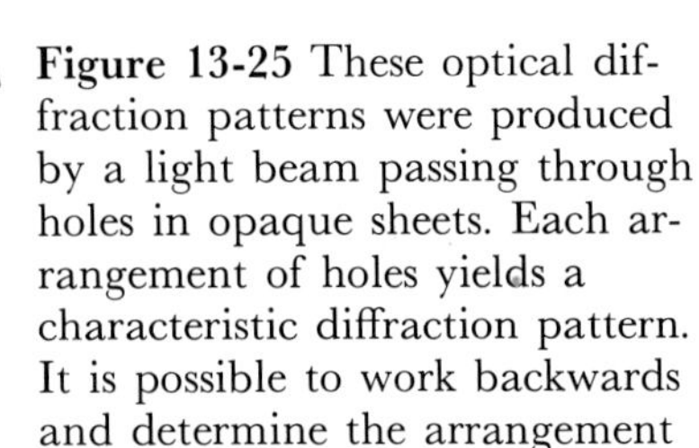

Figure 13-25 These optical diffraction patterns were produced by a light beam passing through holes in opaque sheets. Each arrangement of holes yields a characteristic diffraction pattern. It is possible to work backwards and determine the arrangement of holes that gives a certain diffraction pattern.

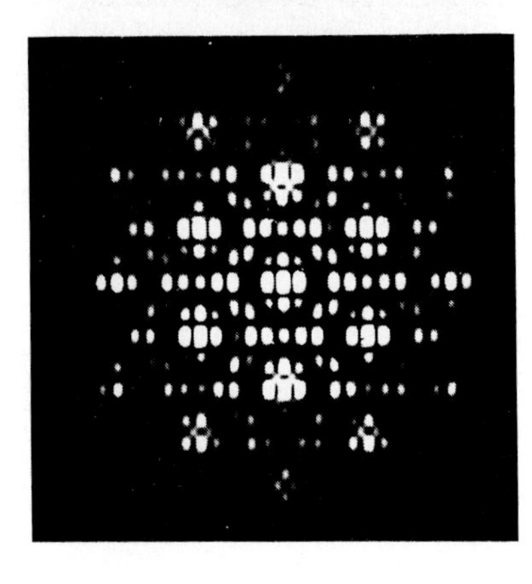

Table 13-3 The seven simple unit cells

Type	*Length of axes*[a]	*Angle*[a]
cubic	$a = b = c$	$\alpha = \beta = \gamma = 90°$
tetragonal	$a = b \neq c$	$\alpha = \beta = \gamma = 90°$
orthorhombic	$a \neq b \neq c$	$\alpha = \beta = \gamma = 90°$
monoclinic	$a \neq b \neq c$	$\alpha = \gamma = 90°, \beta \neq 90°$
triclinic	$a \neq b \neq c$	$\alpha \neq \beta \neq \gamma \neq 90°$
hexagonal	$a = b \neq c$	$\alpha = \beta = 90°, \gamma = 120°$
rhombohedral	$a = b = c$	$\alpha = \beta = \gamma \neq 90°$

[a]The three axes and three angles are shown in Figure 13-26.

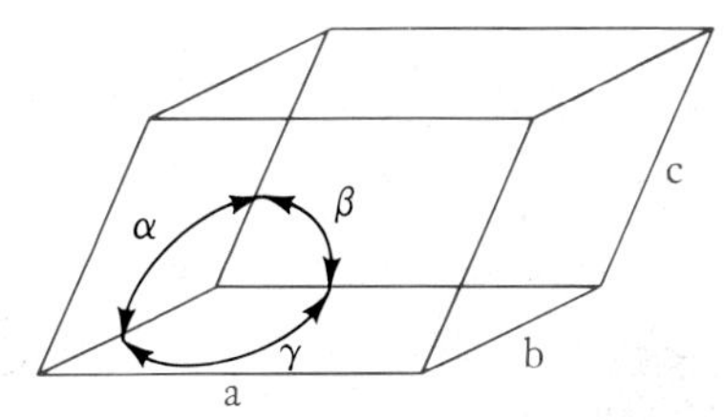

Figure 13-26 The axes and angles used to describe the seven simple unit cells.

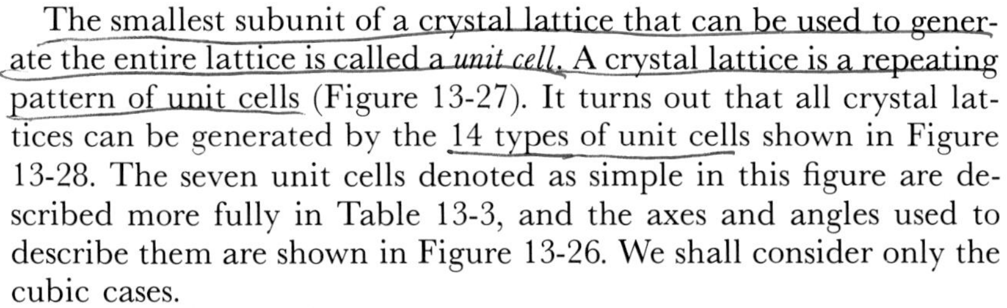
The smallest subunit of a crystal lattice that can be used to generate the entire lattice is called a *unit cell*. A crystal lattice is a repeating pattern of unit cells (Figure 13-27). It turns out that all crystal lattices can be generated by the 14 types of unit cells shown in Figure 13-28. The seven unit cells denoted as simple in this figure are described more fully in Table 13-3, and the axes and angles used to describe them are shown in Figure 13-26. We shall consider only the cubic cases.

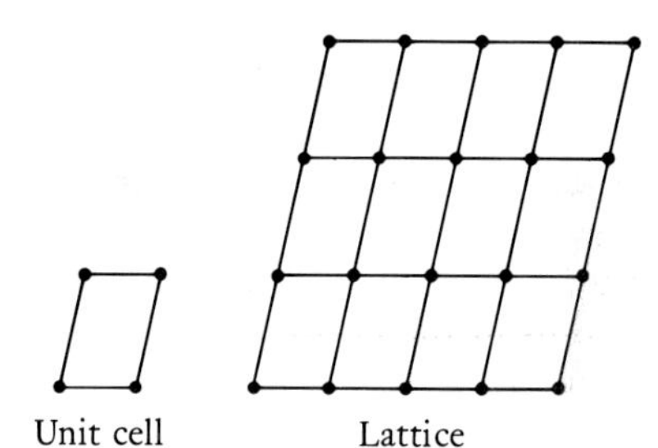

Figure 13-27 A two-dimensional illustration of the generation of a crystal lattice by a unit cell.

The three cubic unit cells and the lattices they generate are shown in Figures 13-29 and 13-30. The crystalline structures of many metals fall into one of these classes. Only one metal, polonium, has a *simple cubic* lattice. Some examples of metals that have *body-centered cubic* lattices are Ba, Cs, K, Li, Mo, Na, Ta, U, and V. *Face-centered cubic* lattices are found in Ag, Al, Au, Cu, Sr, Ni, Pb, and Pt.

Example 13-8: Copper exists as a face-centered cubic lattice. How many copper atoms are there in a unit cell?

Solution: Reference to Figure 13-30(c) shows that each of the eight copper atoms at the corners of the unit cell are shared by eight unit cells, and so we assign one copper atom ($8 \times \frac{1}{8} = 1$) to each unit cell. Each of the six atoms at the faces is shared by two unit cells, and so we assign three more copper atoms ($6 \times \frac{1}{2} = 3$) to each unit cell, giving a total of four copper atoms in a unit cell. Figure 13-30 also illustrates the counting process for the simple cubic and body-centered cubic unit cells.

All the particles in metallic and noble-gas crystals are spherical and identical, and it is common to picture these lattices in terms of packed spheres, as in Figure 13-30. The face-centered cubic arrangement represents the closest possible packing of layers of identical spheres and is sometimes called the *cubic closest-packed structure*. We can build up a face-centered cubic lattice of spheres in the following

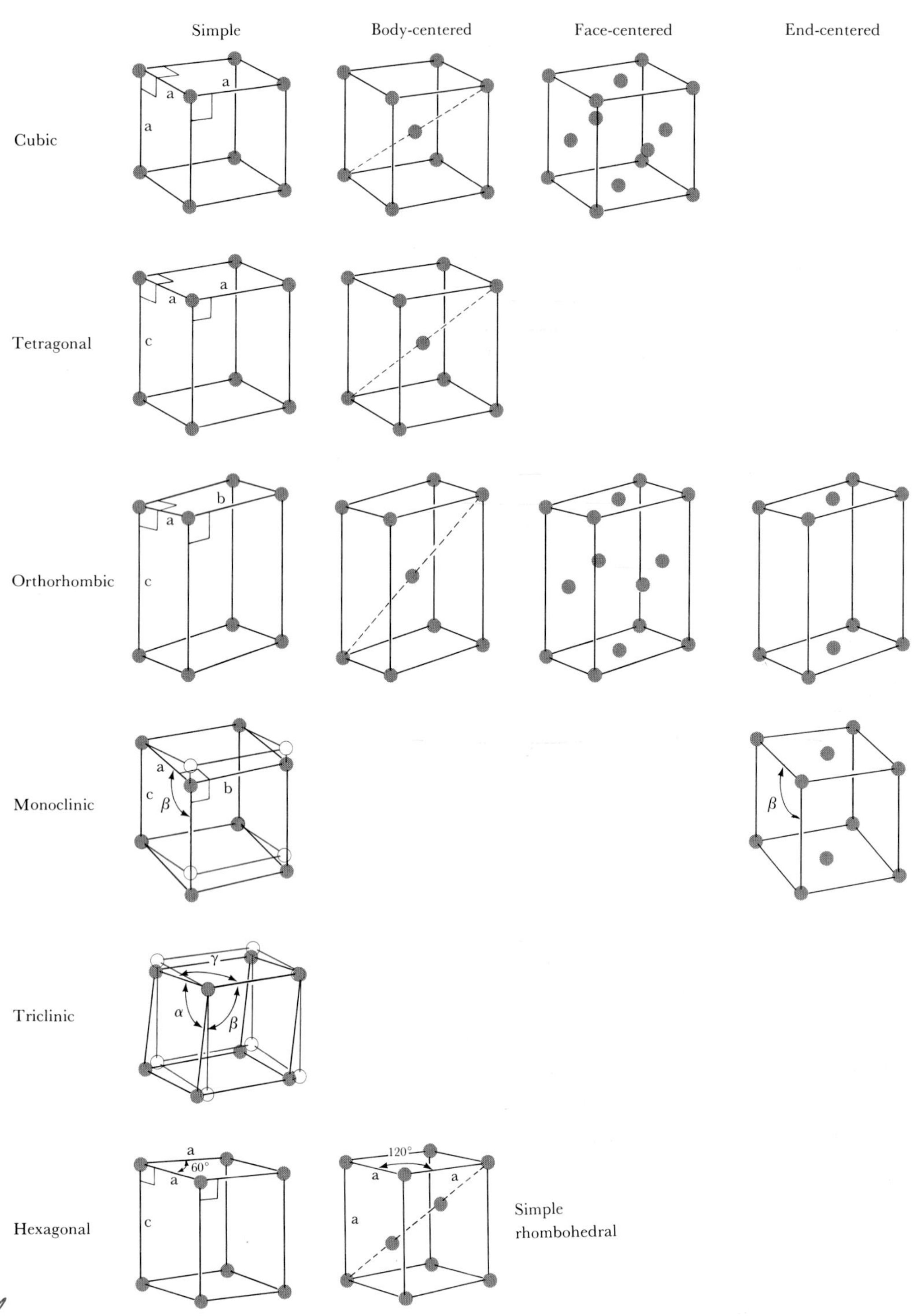

Figure 13-28 The 14 unit cells that generate all the possible three-dimensional crystal lattices. The terms cubic, tetragonal, orthorhombic, and so on are described more fully in Table 13-3. We shall deal with only the cubic lattices in this chapter.

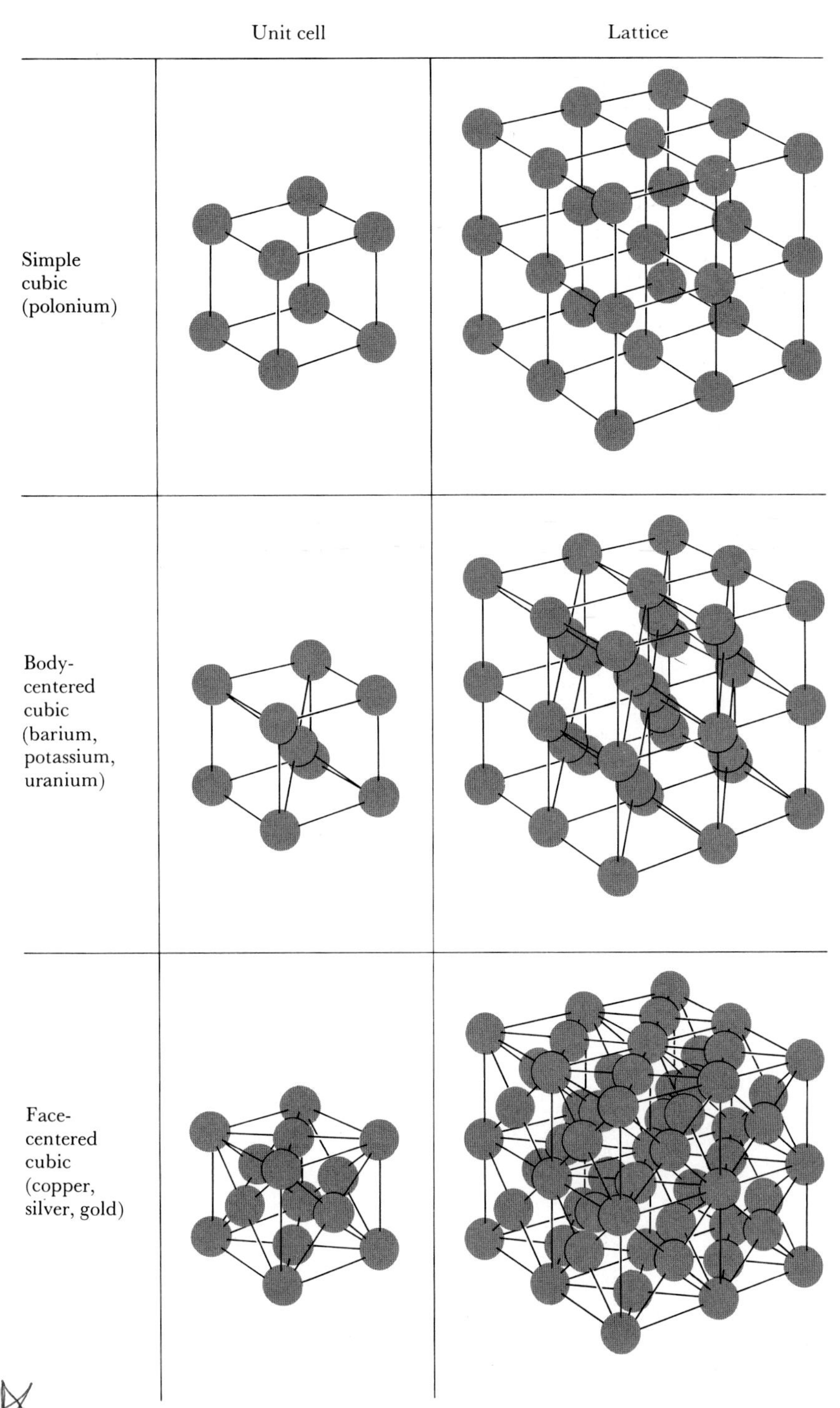

Figure 13-29 The three cubic unit cells are called simple cubic, body-centered cubic, and face-centered cubic. Note that the body-centered cubic unit cell has an atom at the center of the unit cell and that a face-centered unit cell has an atom at the center of each face.

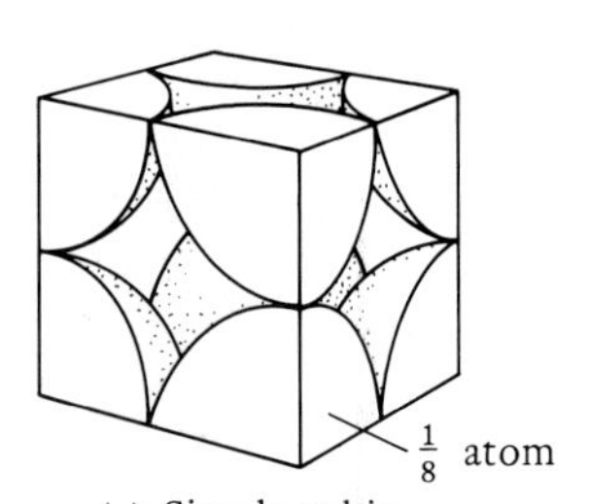

(a) Simple cubic

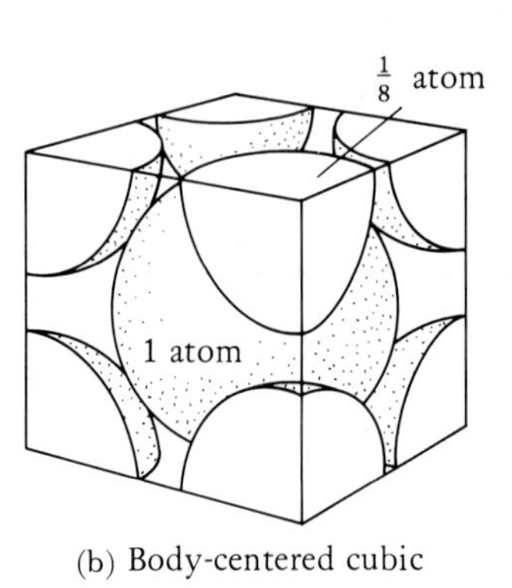

(b) Body-centered cubic

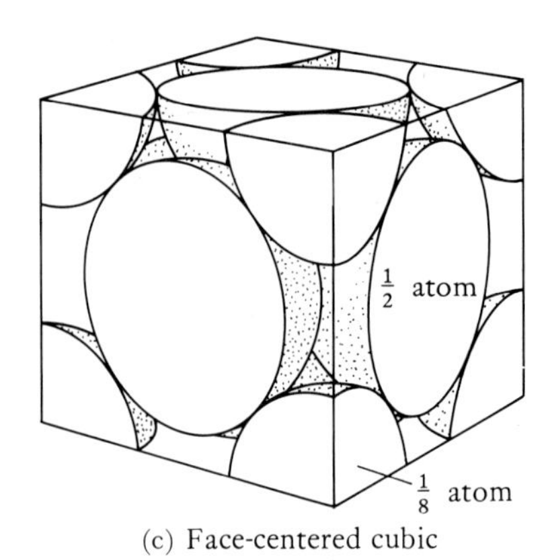

(c) Face-centered cubic

Figure 13-30 The three cubic unit cells, illustrating how the atoms are shared by adjacent unit cells. Figure 13-29 shows a more open perspective of these unit cells.

manner. First form a closest-packed single layer of spheres. Figure 13-31 shows that this layer forms not a square array but rather a hexagonal array. Now form a second closest-packed layer over the hexagonal array by placing each sphere in a well of the first layer (Figure 13-32b). Now form a third layer by placing spheres in the wells of the second layer (Figure 13-32c). By continuing in this manner, we generate the face-centered cubic arrangement shown in Figure 13-29. Notice that the spheres in the face-centered cubic structure actually touch each other. This fact can be used to calculate the radii of atoms.

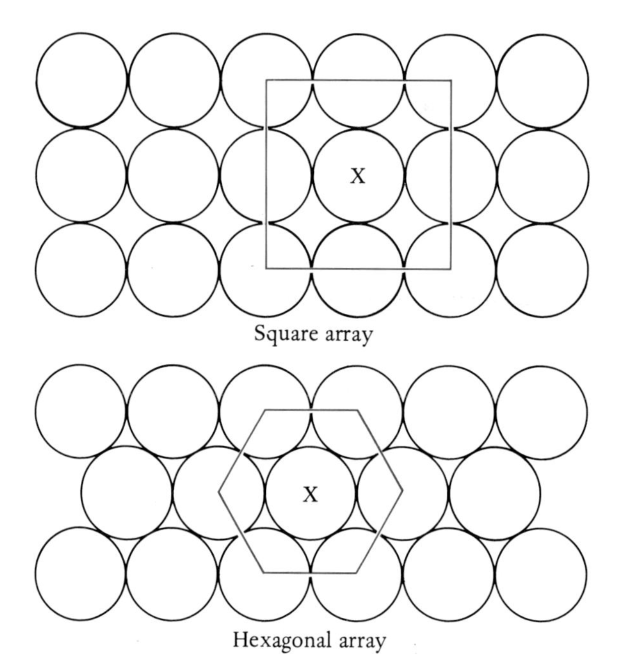

Figure 13-31 A layer of identical spheres packed in a hexagonal array is more closely packed than a layer of identical spheres packed in a square array.

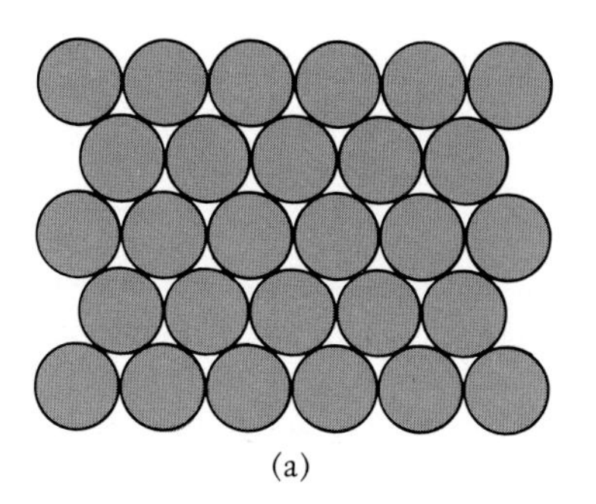

(a)

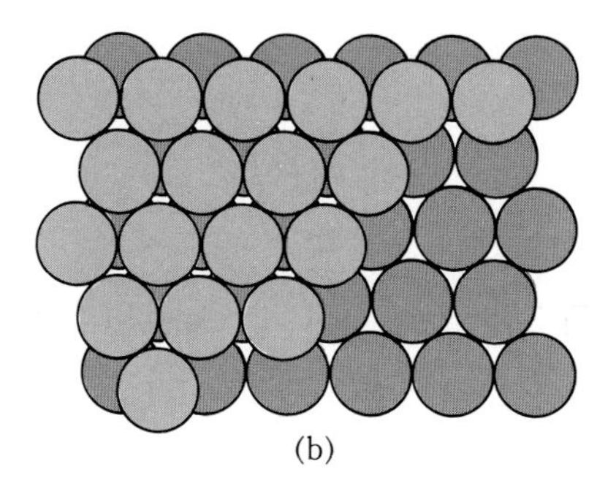

(b)

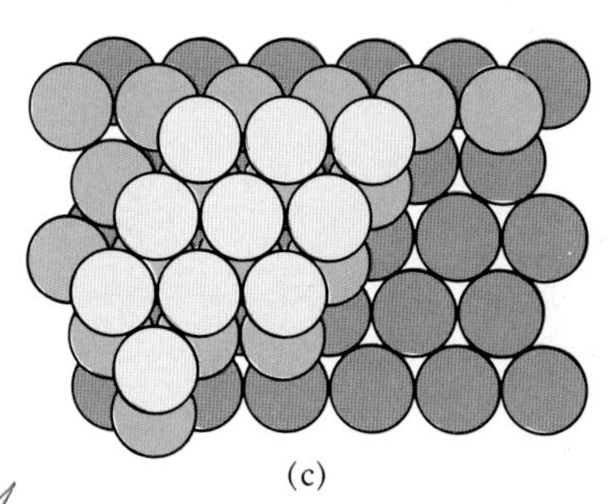

(c)

Figure 13-32 An illustration of how a face-centered cubic array can be constructed from identical spheres. The resulting unit cells are face-centered cubic, and the packing is maximal, that is, there is a minimum of void space in this arrangement.

Example 13-9: Copper, which crystallizes as a face-centered cubic lattice, has a density of $8.930\ \mathrm{g \cdot cm^{-3}}$ at 20°C. Calculate the radius of a copper atom.

Solution: The molar volume of copper is given by

$$\bar{V} = \frac{\text{molar mass}}{\text{density}} = \frac{63.55\ \mathrm{g \cdot mol^{-1}}}{8.930\ \mathrm{g \cdot cm^{-3}}} = 7.116\ \mathrm{cm^3 \cdot mol^{-1}}$$

In Example 13-8, we found that the unit cell of copper contains four copper atoms. The volume occupied by the unit cell, then, is given by

$$v = \frac{\left(\dfrac{4\ \text{atoms}}{\text{unit cell}}\right)(7.116\ \mathrm{cm^3 \cdot mol^{-1}})}{6.022 \times 10^{23}\ \mathrm{atom \cdot mol^{-1}}}$$

$$= 4.727 \times 10^{-23}\ \frac{\mathrm{cm^3}}{\text{unit cell}}$$

Because the unit cell is cubic, the length of one edge is

$$l = (4.727 \times 10^{-23}\ \mathrm{cm^3})^{1/3} = 3.616 \times 10^{-8}\ \mathrm{cm} = 361.6\ \mathrm{pm}$$

The length of a diagonal of any face of the unit cell is

$$\text{diagonal} = \sqrt{2}(361.6\ \mathrm{pm}) = 511.4\ \mathrm{pm}$$

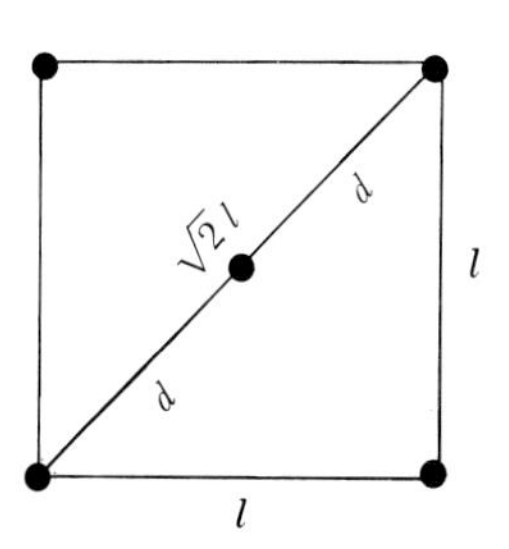

$\sqrt{2}l = 2$ copper atom diameters.

According to the face-centered cubic unit cell in Figure 13-30(c), the diameter of a copper atom is

$$\text{diameter} = \frac{511.4\ \mathrm{pm}}{2} = 255.7\ \mathrm{pm}$$

and thus the radius is 127.9 pm. The result is in good agreement with the value given in Figure 8-11.

Problems 13-43 to 13-54 ask you to do calculations involving crystal structures.

If the density and the dimensions of a unit cell are known from X-ray analysis, then it is possible to determine the value of Avogadro's number.

Example 13-10: Potassium crystallizes in a body-centered cubic lattice, and the length of a unit cell is 533.3 pm. Given that the density of potassium is 0.8560 $g\cdot cm^{-3}$, calculate Avogadro's number.

Solution: The molar volume of potassium is

$$\bar{V} = \frac{\text{molar mass}}{\text{density}} = \frac{39.10\ g\cdot mol^{-1}}{0.8560\ g\cdot cm^{-3}} = 45.68\ cm^3\cdot mol^{-1}$$

The volume of a unit cell is

$$v = (533.3\ pm)^3 = (5.333 \times 10^{-8}\ cm)^3 = 1.517 \times 10^{-22}\ cm^3$$

The number of unit cells per mole is

$$\text{unit cells per mole} = \frac{\bar{V}}{v} = \frac{45.68\ cm^3\cdot mol^{-1}}{1.517 \times 10^{-22}\ cm^3/\text{unit cell}}$$

$$= 3.011 \times 10^{23}\ \frac{\text{unit cells}}{\text{mol}}$$

Problems 13-47 and 13-48 are similar to Example 13-10.

There are two atoms per unit cell in this case [Figure 13-30(b)], however, and so the number of atoms per mole is

$$\left(3.011 \times 10^{23}\ \frac{\text{unit cells}}{\text{mol}}\right)\left(\frac{2\ \text{atoms}}{\text{unit cell}}\right) = 6.022 \times 10^{23}\ \text{atom}\cdot mol^{-1}$$

In 1976, scientists at the National Bureau of Standards used very precise X-ray measurements on ultra-pure silicon to obtain a value of $(6.0220976 \pm 63) \times 10^{23}$ for Avogadro's number. The ± 63 indicates the uncertainty in the last two digits.

The crystalline structure of ionic solids often depends upon the difference in size of the cations and anions. For example, the unit cells of sodium chloride and cesium chloride crystals are shown in Figure 13-33. The different packing arrangements for NaCl and CsCl are a direct consequence of the fact that cesium ions are larger than sodium ions.

Example 13-11: Refer to Figure 13-33 and determine the number of sodium ions and chloride ions in a unit cell of sodium chloride.

Solution: There are 12 sodium ions at the edges of the cube shown in Figure 13-33(a). Each of these sodium ions is shared by four unit cells,

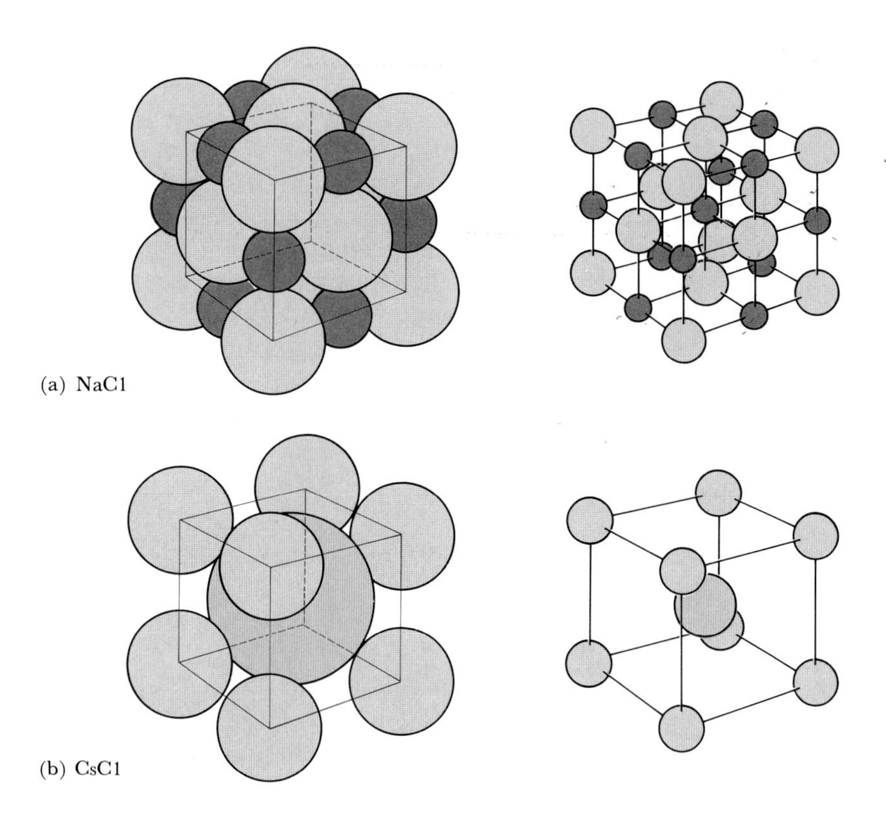

Figure 13-33 Different representations of the unit cells of (a) NaCl and (b) CsCl. The different crystalline structures in the two cases are a direct consequence of the relative sizes of the cations and the anions. Recall that cations are positively charged ions and that anions are negatively charged ions.

giving us three ($12 \times \frac{1}{4} = 3$) sodium ions to be assigned to the unit cell. These three plus the sodium ion at the center of the cell give a total of four sodium ions in the unit cell. There must be four chloride ions also. Each of the eight chloride ions at the corners is shared by eight unit cells, giving one chloride ion to assign to the unit cell. Each of the chloride ions at the six faces of the cube is shared by two unit cells, giving three more chloride ions, for a total of four. Thus, there are four sodium chloride ion pairs in the unit cell of sodium chloride, as shown in Figure 13-34.

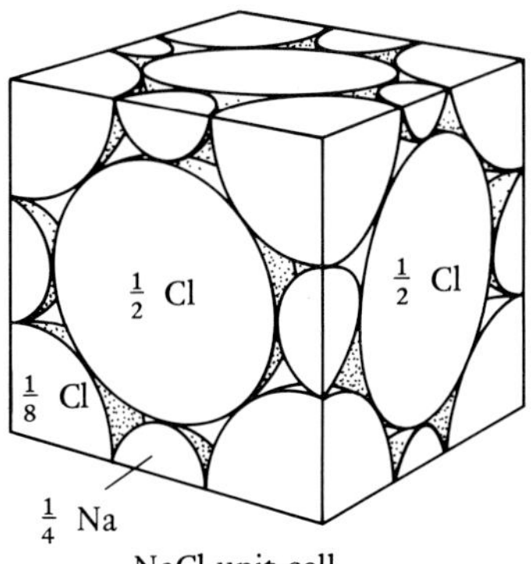

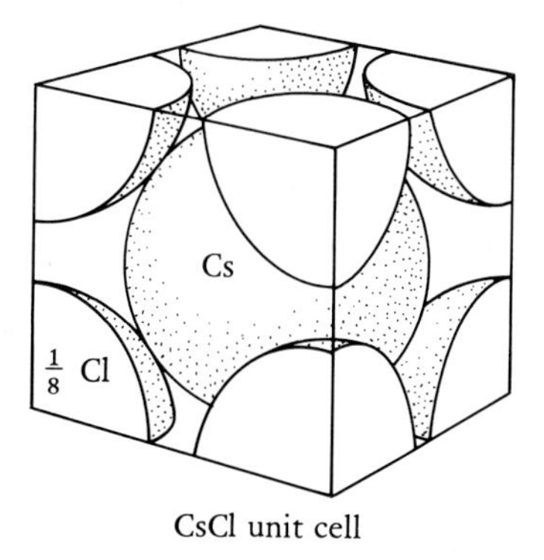

Figure 13-34 The unit cells of NaCl and CsCl, illustrating how the ions are shared by adjacent unit cells.

The crystalline structures of salts that consist of more than two ions per formula unit are more complicated than those for simple salts such as NaCl and CsCl. For example, Figure 13-35 shows the structure of CaF_2, called the fluorite structure. Other crystals with the fluorite structure are SrF_2, BaF_2, K_2O, UO_2, CeO_2, and Na_2S.

Molecular crystals come in a great variety of structures. The noble gases neon through xenon and methane have face-centered cubic structures. Carbon dioxide crystallizes in a cubic lattice (Figure 13-36), and the halogens chlorine, bromine, and iodine crystallize in a face-centered orthorhombic structure (Figure 13-37). As a more involved example, the unit cell of the crystalline structure of anthracene

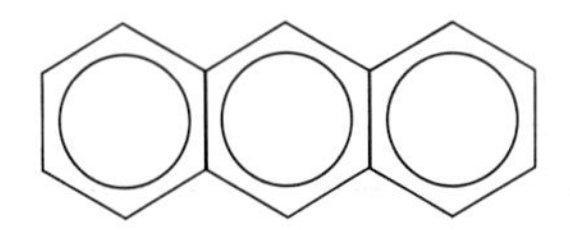

is shown in Figure 13-38. The unit cell contains two anthracene molecules per unit cell. An important point to notice in Figure 13-38 is that the position of the atoms in the unit cell of a molecular crystal specifies the position of *all* the atoms in the molecule. Thus, determination of the crystalline structure of a molecular solid is equivalent to a determination of the structure of one molecule. In fact, X-ray crystallography is used extensively to determine the structure of molecules and is one of the most powerful methods available for the determination of molecular structure.

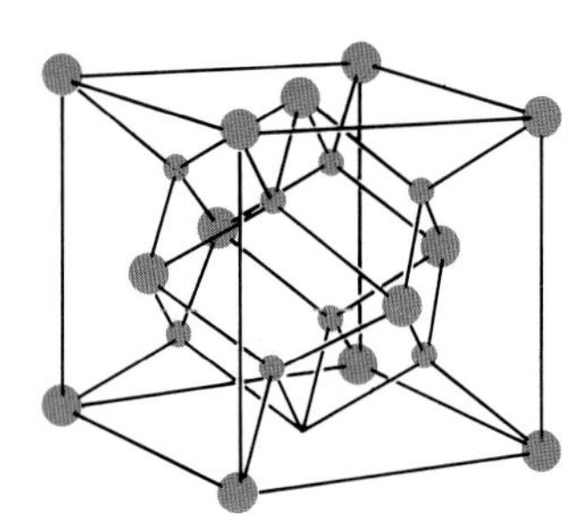

Figure 13-35 The unit cell of CaF_2.

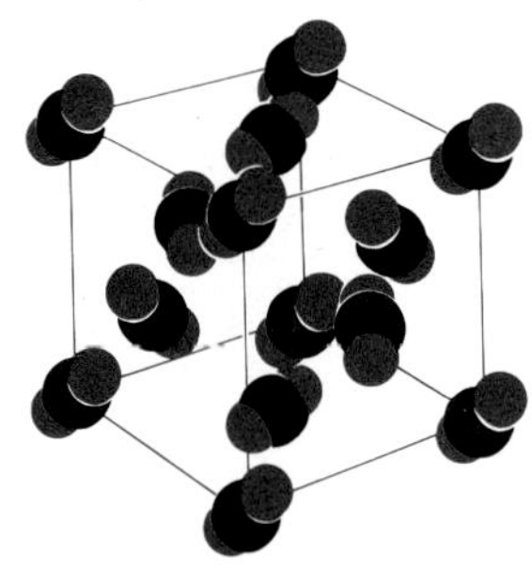

Figure 13-36 The unit cell of CO_2. The molecules have been reduced in size for clarity.

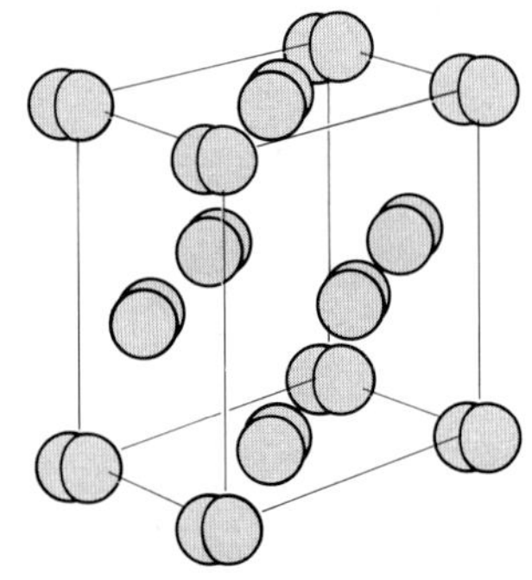

Figure 13-37 The unit cell of Cl_2, Br_2, and I_2. The molecules have been reduced in size for clarity. The unit cell is face-centered orthorhombic.

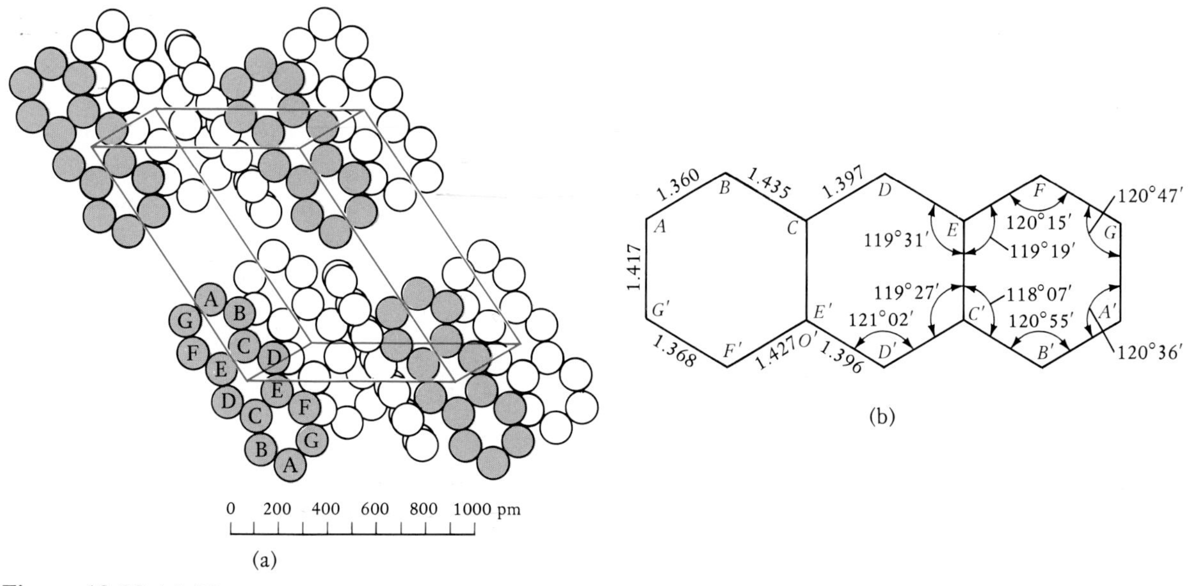

Figure 13-38 (a) The arrangement of anthracene molecules in the unit cell of crystalline anthracene. (b) The dimensions of the individual anthracene molecules. This is the type of information that can be obtained from X-ray diffraction experiments.

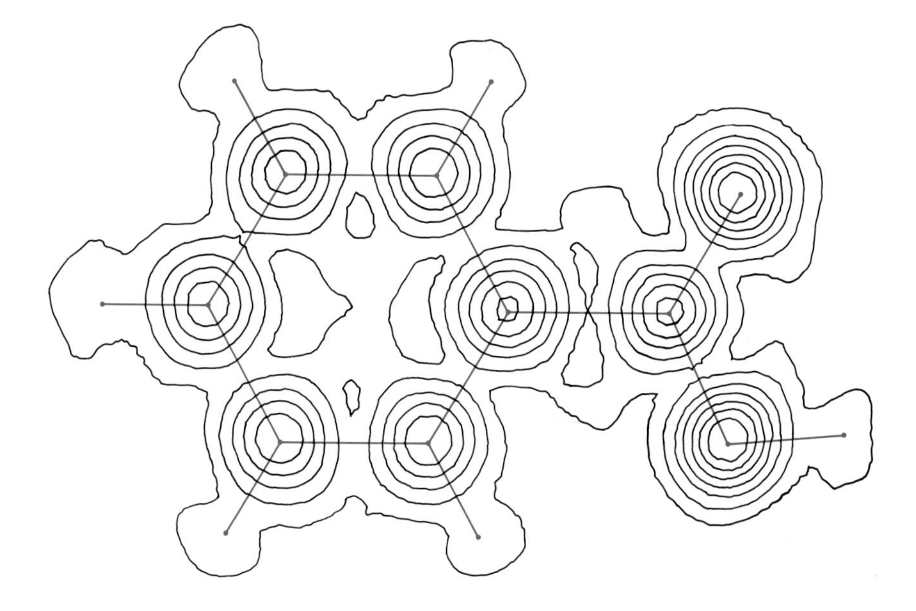

An electron density contour map for benzoic acid obtained from x-ray diffraction measurements. The positions of the nuclei in the molecule are readily deduced from this contour map. Notice the hexagonal array of carbon atoms. Also notice the COOH group that is attached to the benzene ring. The overall molecule is planar.

13-11. A PHASE DIAGRAM DISPLAYS THE PHASE EQUILIBRIUM LINES SIMULTANEOUSLY

The vapor pressure curve, the sublimation pressure curve, and the solid-liquid curve (melting point curve) can be combined into a single diagram called a *phase diagram*. A phase diagram tells us the state of a substance at any pressure and temperature. For example, Figure 13-39 shows that water exists as ice at 1 atm and 263 K (−10°C) and as a liquid at 200 atm and 323 K (50°C) and also at 200 atm and 500 K. The lines that separate the various phases are the sublimation pressure curve (OT), the melting point curve (TM), and the vapor pressure curve (TC). The *sublimation pressure curve* is analogous to the vapor pressure curve (Section 13-7) except that the condensed phase in this case is the solid phase.

The *melting point curve* gives the temperature at which the solid and liquid phases are in equilibrium at a particular pressure. Melting points are only weakly dependent on pressure, and so a melting point curve is an almost vertical line (Figure 13-39). For almost all substances, the melting point increases with increasing pressure at a rate of 0.01 to 0.03 $K \cdot atm^{-1}$. Water is anomalous in that its melting point decreases with increasing pressure. The melting point of ice decreases by 0.01 K per atmosphere of applied pressure. At 2050 atm, ice melts at −22°C. Between 0°C and −22°C, ice can be melted by the application of pressure. At any temperature lower than −22°C, ice cannot be melted by the application of pressure because of the existence of other high-pressure forms of ice.

Along the vapor pressure curve, liquid and vapor exist together at equilibrium. The equilibrium vapor pressure of a liquid increases with temperature up to the *critical point* (point C in Figure 13-39), where the vapor pressure curve terminates abruptly. The critical point for water occurs at 218 atm and 647 K. Liquid does not exist at a temperature greater than the *critical temperature*. A gas above its critical temperature cannot be liquefied no matter how high a pressure is applied.

The *triple point* (point T in Figure 13-39) is the point at which the vapor pressure curve, the sublimation pressure curve, and the melting point curve all intersect. At the triple point, and *only at the triple point,* all three phases—solid, liquid, and gas—coexist in equilibrium. The triple point for water occurs at 4.58 torr and 273.16 K.

Very clean liquid water can be *supercooled;* that is, its temperature can be reduced below the triple point (curve TS in Figure 13-39). Supercooled liquid water converts spontaneously to ice if an ice crystal is added to it because the ice crystal acts as a nucleation center for crystallization.

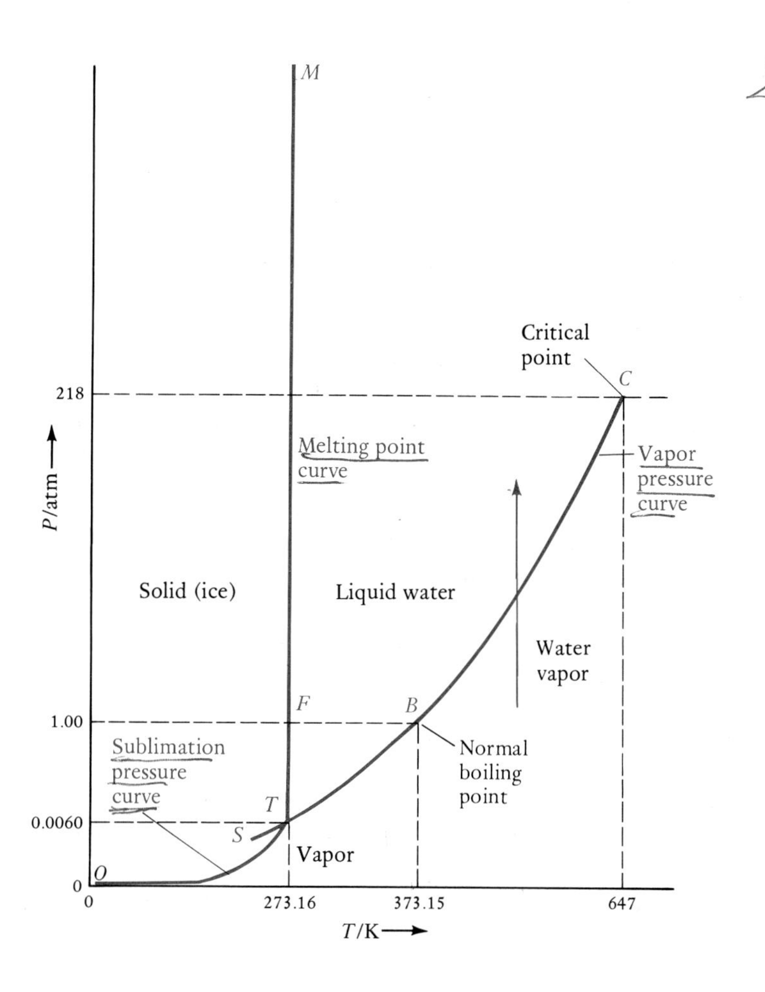

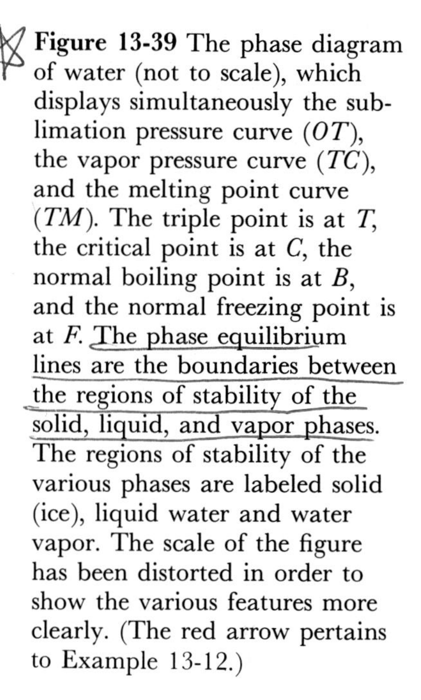
Figure 13-39 The phase diagram of water (not to scale), which displays simultaneously the sublimation pressure curve (OT), the vapor pressure curve (TC), and the melting point curve (TM). The triple point is at T, the critical point is at C, the normal boiling point is at B, and the normal freezing point is at F. The phase equilibrium lines are the boundaries between the regions of stability of the solid, liquid, and vapor phases. The regions of stability of the various phases are labeled solid (ice), liquid water and water vapor. The scale of the figure has been distorted in order to show the various features more clearly. (The red arrow pertains to Example 13-12.)

Example 13-12: Use the phase diagram of water given in Figure 13-39 to predict the result of increasing the pressure of water vapor initially at 1 atm and 500 K, keeping the temperature constant.

Solution: This process is indicated by the red vertical arrow in Figure 13-39. At 1 atm and 500 K water exists as a vapor. As the pressure is increased, we cross the liquid-vapor curve at a pressure of about 30 atm and the vapor condenses to a liquid.

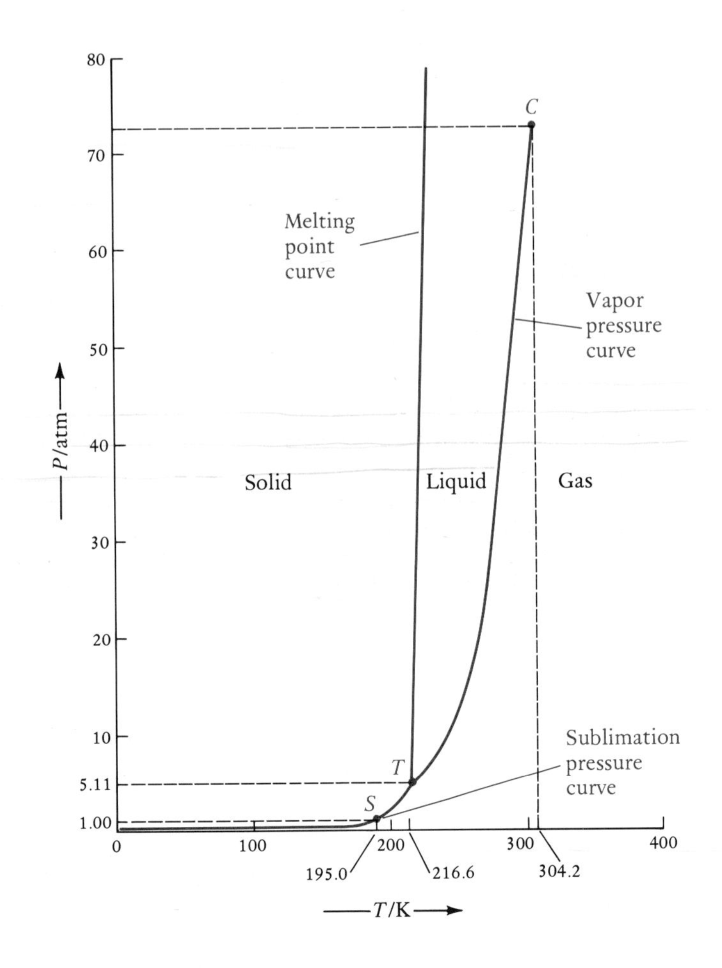

Figure 13-40 The phase diagram for CO_2. The point C is the critical point, and the point T is the triple point. Note that the triple point lies above 1 atm and thus $CO_2(s)$ at 1 atm does not melt—it sublimes. The point S is the normal sublimation point of CO_2.

The phase diagram of carbon dioxide is shown in Figure 13-40. Although it looks similar to that of H_2O, there are several important differences. The melting point curve of CO_2 goes up and to the right, indicating that the melting point of CO_2 increases with increasing pressure. Recall that the melting point curve of water points up and slightly to the left, indicating that the melting point of H_2O decreases with increasing pressure.

Another difference between Figures 13-39 and 13-40 is in the triple point. The triple point for CO_2 occurs at 5.11 atm and 216.6 K, and therefore CO_2 does not have a normal (1 atm) boiling point. However, CO_2 does have a normal sublimation point, which occurs at $-78.2°C$: this is the temperature of dry ice at 1 atm. Liquid CO_2 can be obtained by compressing $CO_2(g)$ at a temperature below its critical point (30°C). A pressure of about 60 atm is required to liquefy CO_2 at 25°C. A carbon dioxide–filled fire extinguisher at 25°C contains liquid CO_2 at a pressure of about 60 atm. A CO_2 fire extinguisher works by displacing O_2 from the combustion zone. Carbon dioxide will not support combustion, and in the absence of oxygen the fire suffocates.

Two allotropes of carbon are graphite and diamond.

13-12. DIAMOND CAN BE PRODUCED FROM GRAPHITE BY APPLYING VERY HIGH PRESSURE

Numerous solids exist in more than one form. The different forms differ in crystalline structure and therefore in physical properties, including density. For example, solid carbon can exist as graphite or as diamond. Diamond has an extended, covalently bonded tetrahedral structure. Each carbon atom lies at the center of a tetrahedron formed by four other carbon atoms (Figure 13-41). The C—C bond distance is 154 pm, which is the same as the C—C bond distance in ethane. The diamond crystal is, in effect, a gigantic molecule. The hardness of diamond is due to the fact that each carbon atom is covalently bonded to four others and thus many covalent bonds must be broken in order to cleave a diamond. Graphite has the unusual layered structure shown in Figure 13-42. The C—C bond distance within a layer is 139 pm, which is close to the C—C bond distance in benzene. The distance between layers is about 340 pm. The bonding within a layer is strong, but the interaction between layers is weak. Therefore, the layers easily slip past each other. This is the molecular basis of the lubricating action of graphite. The "lead" of lead pencils is actually graphite. Layers of the graphite rub off from the pencil onto the paper.

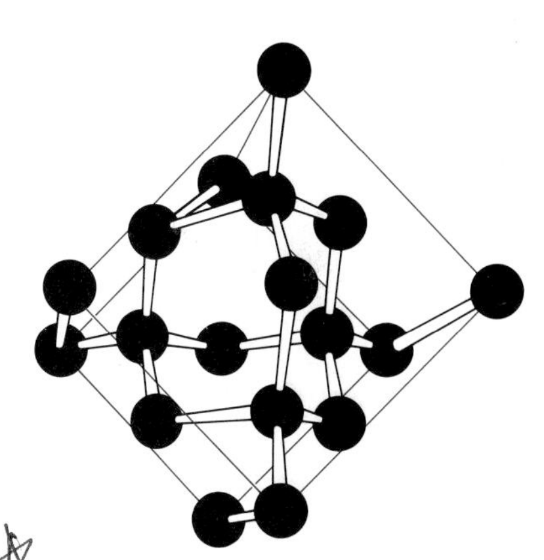

Figure 13-41 The crystalline structure of diamond. Each carbon atom is covalently bonded to four other carbon atoms, forming a tetrahedral network. A diamond crystal is essentially one gigantic molecule.

The diamond-graphite equilibrium curve of the phase diagram of carbon shows that graphite is the stable form at ordinary temperatures and pressures and that diamond is the stable form at high pressures (Figure 13-43). There is a nice, intuitive argument for the relative stability of diamond at high pressure. The density of graph-

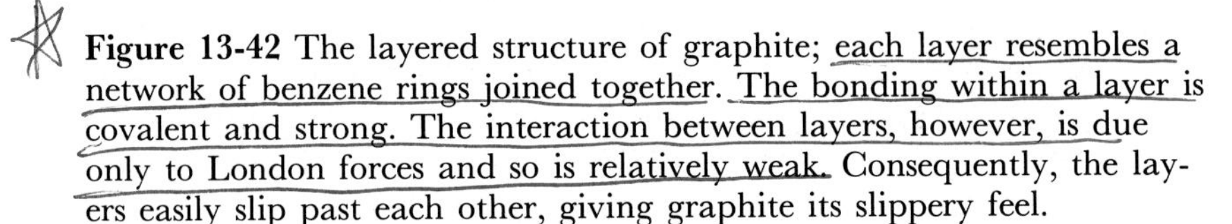

Graphite is used as a lubricant in railroad freight-car wheel assemblies.

Figure 13-42 The layered structure of graphite; each layer resembles a network of benzene rings joined together. The bonding within a layer is covalent and strong. The interaction between layers, however, is due only to London forces and so is relatively weak. Consequently, the layers easily slip past each other, giving graphite its slippery feel.

ite is 2.2 $g \cdot cm^{-3}$, and that of diamond is 3.5 $g \cdot cm^{-3}$. Consequently, the molar volume of graphite is 5.5 $cm^3 \cdot mol^{-1}$ and the molar volume of diamond is 3.4 $cm^3 \cdot mol^{-1}$. When a high pressure is applied to graphite, it can relieve some of the effect of the pressure by converting to diamond, which has a smaller volume. Any substance that can exist in two different forms with different densities exists as the higher-density, or smaller-volume, form under high pressure. This is why ice melts under pressure; the liquid water has a higher density, or a smaller volume, than ice. Recall that ice floats on water.

The graphite-diamond phase equilibrium curve in Figure 13-43

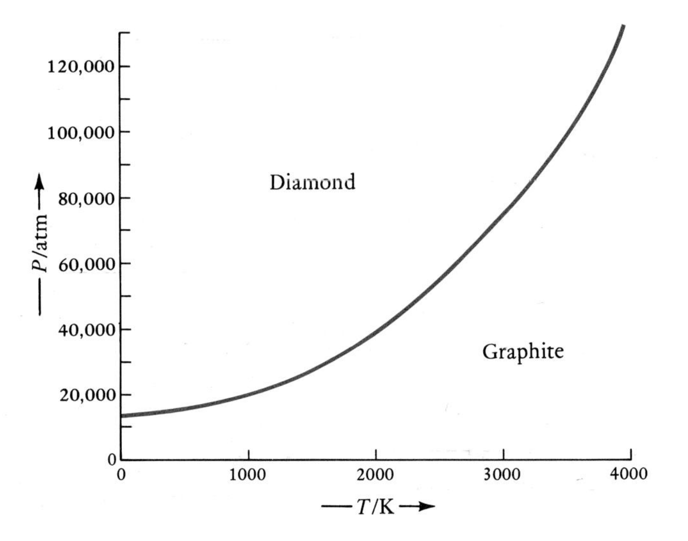

Figure 13-43 The graphite-diamond phase equilibrium line. Above the curve diamond is the stable form of solid carbon, and below the curve graphite is the stable form. At 25°C, the equilibrium transition pressure for the conversion of graphite to diamond is 15,000 atm.

defines the minimum conditions necessary for the production of diamond from graphite. Diamond is the hardest substance known and consequently is extensively used as an abrasive and cutting material where very high resistance to wear is required. The industrial market for diamonds is much greater than the market for diamonds as jewels.

To produce diamonds from graphite it is necessary to subject the graphite to a pressure and temperature above the graphite-diamond equilibrium line in Figure 13-43. For example, at 300 K a pressure greater than 15,000 atm is required. However, under these conditions the rate of conversion of graphite to diamond is extremely slow. Higher pressures and temperatures are required to achieve the conversion in a reasonable time. At 2500 K and 150,000 atm, essentially complete conversion occurs in a few minutes. Rapid reduction of the pressure and temperature traps the carbon in the diamond form. A tetrahedral anvil is used to generate the high pressures necessary for the production of diamonds from graphite (Figure 13-44).

Figure 13-44 A tetrahedral anvil used to synthesize diamonds by applying pressures above 100,000 atm to graphite.

The rate of conversion of diamond to graphite at ordinary temperatures is negligible. However, diamonds are not forever. If they are heated to 1500°C, then the conversion to graphite occurs in minutes. Synthetic diamonds are not artificial diamonds; they are *real* diamonds in every scientific sense. They have exactly the same chemical composition, structure, and physical properties as diamonds formed by natural forces.

SUMMARY

Molecular interactions in solids and liquids are much greater than those in gases. The values of $\Delta \bar{H}_{vap}$ and $\Delta \bar{H}_{sub}$ are quantitative measures of the strength of molecular interactions in liquids and solids, respectively. All molecules attract each other. If the molecules are polar, then they attract each other by dipole-dipole forces. The polar molecules orient themselves so that their dipole moments are head to tail; this is the orientation that minimizes the energy of the interacting molecules. Two nonpolar molecules attract each other because the electrons in the molecules redistribute such that the instantaneous dipole moments on each molecule are head to tail. The resulting attraction is called a London force. All the attractions between molecules are collectively called van der Waals forces.

Two important properties of liquids are vapor pressure and surface tension. Pure liquids have a unique equilibrium vapor pressure at each temperature; the equilibrium vapor pressure increases with increasing temperature. A liquid boils when its vapor pressure equals the pressure of the atmosphere. The normal boiling point of a liquid is the temperature at which the vapor pressure of the liquid is equal to 1 atm. The surface tension of a liquid is a force that minimizes the surface area of the liquid. Surface tension is a consequence of the fact that surface molecules have a higher energy than molecules in the bulk liquid.

A crystalline structure can be determined from the diffraction pattern of X-rays passed through the crystal. The diffraction pattern yields information regarding the position of the atoms in the crystal. The smallest segment of a crystal lattice that contains all of the structural information about the crystal is called the unit cell.

The various phase equilibrium lines of a substance can be displayed simultaneously on a phase diagram. The normal melting point is the temperature at which the pure solid and liquid phases are in equilibrium at 1 atm. The normal boiling point is the temperature at which the pure liquid and gas phases are in equilibrium at 1 atm. The triple point is the only point at which the solid, liquid, and gas phases are simultaneously in equilibrium with one another. The critical point is the temperature above which a gas cannot be liquefied by the application of pressure. Some solids can exist in more than one form; the form that has the higher density, or the smaller molar volume, is the more stable form at high pressure.

TERMS YOU SHOULD KNOW

vibrational motion 466
rotational motion 466
translational motion 466
heating curve 467
molar enthalpy of fusion, $\Delta\bar{H}_{fus}$ 467
molar enthalpy of vaporization, $\Delta\bar{H}_{vap}$ 468
dipole-dipole attraction 474
hydrogen bonding 475
London force 477
van der Waals force 477
sublimation 479
molar enthalpy of sublimation, $\Delta\bar{H}_{sub}$ 479
equilibrium 481
dynamic equilibrium 481
equilibrium vapor pressure 481
vapor pressure curve 482
normal boiling point 483
relative humidity 484
dew point 485
surface tension 486
surfactant 487
capillary action 487
X-ray diffraction pattern 488
optical diffraction pattern 488
unit cell 489
simple cubic 489
body-centered cubic 489
face-centered cubic 489
cubic closest-packed structure 489
phase diagram 498
sublimation pressure curve 498
melting point curve 498
critical point 498
critical temperature 498
triple point 498
supercooled 498

EQUATIONS YOU SHOULD KNOW HOW TO USE

$$q_P = n\bar{C}_P(T_2 - T_1) \quad (13\text{-}1)$$

$$\Delta \bar{H}_{sub} = \Delta \bar{H}_{fus} + \Delta \bar{H}_{vap} \quad (13\text{-}2)$$

$$\text{relative humidity} = \frac{P_{H_2O}}{P^{\circ}_{H_2O}} \times 100 \quad (13\text{-}3)$$

PROBLEMS

HEATS OF VAPORIZATION AND FUSION

13-1. Rubidium boils at 686°C, and its heat of vaporization is 69 $kJ \cdot mol^{-1}$. Calculate the quantity of heat evolved when 1.00 kg of rubidium condenses.

13-2. The heat of vaporization of einsteinium was determined to be 128 $kJ \cdot mol^{-1}$ using only a 100-μg sample. How much heat is required to vaporize 100 μg of einsteinium?

13-3. Given that 26.2 kJ of heat is required to completely vaporize 60.0 g of benzene, C_6H_6, at 80.1°C, calculate the molar heat of vaporization, $\Delta \bar{H}_{vap}$, of benzene.

13-4. Given that 87.9 kJ of heat is required to completely vaporize 100.0 g of ethyl alcohol, C_2H_5OH, at 78.5°C, calculate the molar enthalpy of vaporization, $\Delta \bar{H}_{vap}$, of ethyl alcohol.

13-5. Calculate the heat required to raise the temperature of 0.500 kg of ice at 0°C to water at 25°C.

13-6. The heat of vaporization of the refrigerant Freon-12, CCl_2F_2, is 155 $J \cdot g^{-1}$. Estimate the number of grams of Freon-12 that must be evaporated to freeze a tray of 16 one-ounce (1 oz = 28 g) ice cubes with the water initially at 18°C.

13-7. Calculate the energy released as heat when 20.1 g of liquid mercury at 25°C is converted to solid mercury at its melting point. The heat capacity of Hg(l) is 28.0 $J \cdot K^{-1} \cdot mol^{-1}$.

13-8. Calculate the heat required to convert 7.0 g of ice at −10°C to water at 52°C.

13-9. A method for lowering the temperature of a person suffering from a fever or overheating is to bathe the person with rubbing (isopropyl) alcohol, C_3H_7OH. The evaporation of the alcohol removes heat from the patient, thereby reducing body temperature. How much heat is removed when 100.0 g of rubbing alcohol evaporates if its $\Delta \bar{H}_{vap}$ is 42.1 $kJ \cdot mol^{-1}$?

13-10. Suppose you are stranded in a mountain cabin by a snowstorm. You have some food and fuel, but you wish to conserve them as long as possible. You remember reading that you should melt snow to get water to drink and not eat the snow directly because the body expends energy when it has to melt the snow. Explain why and estimate how much energy is used per gram of snow.

HEATING CURVES

13-11. Sketch a heating curve for 10.0 g of mercury from 300 to 800 K using a heat input rate of 100 $J \cdot min^{-1}$. Refer to Table 13-1 for some of the necessary data for mercury. The molar heat capacities of liquid and gaseous mercury are 28.0 $J \cdot K^{-1} \cdot mol^{-1}$ and 20.8 $J \cdot K^{-1} \cdot mol^{-1}$, respectively.

13-12. What would take longer, heating 10.0 g of water at 50.0°C to 100.0°C or vaporizing the 10.0 g if the rate of heating in both cases is 5 $J \cdot s^{-1}$?

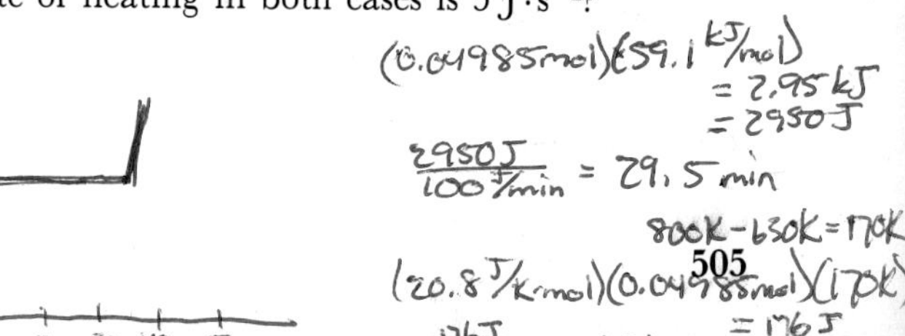

13-13. Heat was added to 10.00 g of solid lead at the rate of 100 $J \cdot min^{-1}$. At 327°C, the temperature remained constant for 2.30 min. Calculate $\Delta \bar{H}_{fus}$ for lead.

13-14. Heat was added to 50.0 g of liquid acetone, CH_3COCH_3, at the rate of 500.0 $J \cdot min^{-1}$. The temperature remained constant at 56.2°C for 55.0 min. Calculate $\Delta \bar{H}_{vap}$ for acetone.

VAN DER WAALS FORCES

13-15. Arrange the following molecules in order of increasing boiling point:

KBr C_2H_5OH C_2H_6 He

13-16. Arrange the following molecules in order of increasing boiling point:

Ar NH_3 Kr NaCl

13-17. Which of the following molecules have polar interactions:

NaF ClF NF_3 F_2

13-18. Which of the following molecules involve hydrogen bonding:

H_2 HF CH_4 CH_3OH

13-19. Arrange the following molecules in order of increasing molar enthalpy of vaporization:

CH_4 C_2H_6 CH_3OH C_2H_5OH

13-20. Arrange the following molecules in order of increasing molar enthalpy of vaporization:

CCl_4 $SiCl_4$ CH_4 $SiBr_4$

13-21. Trouton's rule states that the molar enthalpy of vaporization of a liquid that does not involve strong molecular interactions such as hydrogen bonding or ion-ion attractions is given by

$$\Delta \bar{H}_{vap} = (85\ J \cdot K^{-1} \cdot mol^{-1})T_b$$

where T_b is the normal boiling point of the liquid in kelvins. Use Trouton's rule to estimate $\Delta \bar{H}_{vap}$ for argon, given that its T_b is 87 K.

13-22. Apply Trouton's rule, given in Problem 13-21, to water and suggest a molecular explanation for any discrepancy with the value of $\Delta \bar{H}_{vap}$ given in Table 13-1.

HEAT OF SUBLIMATION

13-23. Calculate the heat absorbed by the sublimation of 100.0 g of solid carbon dioxide.

13-24. Compute the number of moles of water at 0°C that can be frozen by 1.00 mol of solid carbon dioxide. See Table 13-1 for necessary data.

13-25. Using the data in Table 13-1, calculate the enthalpy of sublimation of (a) ammonia and (b) methane.

13-26. Although the temperature may not exceed 0°C, the amount of ice on a sidewalk decreases due to sublimation. A source of heat for the sublimation is solar radiation. The average daily solar radiation in February for Boston is 8.1 $MJ \cdot m^{-2}$. Calculate how much ice will disappear from a 1.0-m^2 area in one day assuming that all the radiation is used to sublime the ice. The density of ice is 0.917 $g \cdot cm^{-3}$.

VAPOR PRESSURE

13-27. The temperature of the human body is about 37°C. Use data in Table 13-2 and Figure 13-18 to estimate the vapor pressure of water in exhaled air. Assume that air in the lungs is saturated with water vapor.

13-28. Using Figure 13-18, estimate the boiling point of ethyl alcohol at 2 atm.

13-29. A 0.50-g sample of ethyl alcohol is placed in a sealed 250-mL container. Is there any liquid present when the temperature is held at 60°C?

13-30. Calculate the concentration in $mol \cdot L^{-1}$ of water in air saturated with water vapor at 25°C.

13-31. At what temperature would water boil on top of an 18,000-ft mountain where the atmospheric pressure is one half that at sea level?

13-32. Mexico City lies at an elevation of 7400 ft (2300 m). Water boils at 93°C in Mexico City. What is the normal atmospheric pressure there?

13-33. Atmospheric pressure decreases with altitude. Plot the following data to obtain the relationship between pressure and altitude:

Altitude/ft	*Atmospheric pressure/atm*
5000	0.83
10,000	0.70
15,000	0.58
20,000	0.47

Using your plot and the vapor pressure curve of water (Figure 13-18), estimate the boiling point of water at the following locations:

Location	*Altitude/ft*
Denver	5280
Mount Kilimanjaro	19,340
Mt. Washington	6290
The Matterhorn	14,690

13-34. British surveyors were prevented from extending their survey of India into the Himalayas because entry into Tibet was banned by the Chinese emperor. In 1865, the Indian Nain Singh secretly entered Lhasa, the capital city of Tibet, and determined its correct location for map placement. Singh was not able to bring instruments for measuring altitude with him and so was forced to estimate the altitude from the boiling point of water. He estimated that Lhasa was 3420 m above sea level. Its true elevation is 3540 m. Describe how altitude can be estimated from a measurement of the boiling point of water.

RELATIVE HUMIDITY AND DEW POINT

13-35. The relative humidity in a greenhouse at 40°C is 92 percent. Calculate the vapor pressure of water vapor in the greenhouse.

13-36. The relative humidity of air at 10°C is 50 percent. Calculate the relative humidity of the air heated to 25°C.

13-37. The relative humidity is 56 percent on a certain day on which the temperature is 30°C. As the air cools during the night, what will be the dew point?

13-38. Moisture often forms on the outside of a glass containing a mixture of ice and water. Use the principles developed in this chapter to explain this phenomenon.

SURFACE TENSION

13-39. Suggest a simple explanation based on hydrogen bonding for the unusually high surface tension of liquid water relative to other nonmetallic liquids.

13-40. The advertised "sheeting action" of various dishwasher detergents is based on the presence of a surfactant in the detergent. Explain sheeting action in terms of surface tension changes. The surface area of a sphere of radius a is $4\pi a^2$.

13-41. The surface tension of water is 72 mJ·m^{-2}. What is the energy required to change a spherical drop of water with a diameter of 2 mm to two smaller spherical drops of equal size? The surface area of a sphere of radius a is $4\pi a^2$.

13-42. The surface tension of water is 72 mJ·m^{-2}. Calculate the energy required to disperse one spherical drop of radius 3.0 mm into spherical drops of radius 3.0×10^{-3} mm. The surface area of a sphere of radius a is $4\pi a^2$.

CRYSTAL STRUCTURES

13-43. The density of platinum is 21.45 g·cm^{-3} at 20°C. Given that the unit cell of platinum is face-centered cubic, calculate the length of an edge of a unit cell.

13-44. The density of tantalum is 16.69 g·cm^{-3} at 20°C. Given that the unit cell of tantalum is body-centered cubic, calculate the length of an edge of a unit cell.

13-45. The unit cell of aluminum is face-centered cubic, and the length of an edge of a unit cell is 405 pm at 25°C. Calculate the density of aluminum at 25°C.

13-46. The unit cell of lithium is body-centered cubic, and the length of an edge of a unit cell is 351 pm at 20°C. Calculate the density of lithium at 20°C.

13-47. Copper crystallizes in a face-centered cubic lattice with a density of 8.93 g·cm^{-3}. Given that the length of an edge of a unit cell is 361.5 pm, calculate Avogadro's number.

13-48. Chromium crystallizes in a body-centered cubic lattice with a density of 7.20 g·cm^{-3}. Given that the length of an edge of a unit cell is 288.4 pm, calculate Avogadro's number.

13-49. Crystalline potassium fluoride has the NaCl-type structure shown in Figure 13-33. Given that the density of KF is 2.481 g·cm^{-3} at 20°C, calculate the unit cell length and the nearest-neighbor distance in KF. (The nearest-neighbor distance is the shortest distance between any two ions in the lattice.)

13-50. Crystalline cesium bromide has the CsCl-type structure shown in Figure 13-33. Given that the density of CsBr is 4.44 g·cm^{-3} at 20°C, calculate the unit cell length and the nearest-neighbor distance in CsBr.

13-51. The unit cell of calcium oxide is one of the three cubic crystalline structures. Given that the density of CaO is 3.25 g·cm^{-3} and that the length of an edge of a unit cell is 481 pm, determine how many formula units of CaO there are in a unit cell. Which type of cubic structure is the unit cell?

13-52. The unit cell of potassium bromide is one of the three cubic crystalline structures. Given that the density of KBr is 2.75 g·cm^{-3} and that the length of an edge of a unit cell is 654 pm, determine how many formula units of KBr there are in a unit cell. Which type of cubic structure is the unit cell?

13-53. Cesium chloride has the crystal structure shown in Figure 13-33. The length of a side of a unit cell is determined by X-ray diffraction to be 412.1 pm. What is the density of cesium chloride?

13-54. Sodium chloride has the crystal structure shown in Figure 13-33. By X-ray diffraction, it is determined that the distance between a sodium ion and a chloride ion is 282 pm. Using the fact that the density of sodium chloride is 2.163 g·cm^{-3}, calculate Avogadro's number.

PHASE DIAGRAMS

13-55. Determine whether water is a solid, liquid, or gas at the following pressure and temperature combinations (use Figure 13-39):

(a) 373 K, 0.90 atm (c) 400 K, 100 atm
(b) 0°C, 0.005 atm (d) 0°C, 200 atm

13-56. Referring to Figure 13-40, state the phase of CO_2 under the following conditions:

(a) 127°C, 8 atm (c) 50°C, 1 atm
(b) −50°C, 40 atm (d) −80°C, 5 atm

13-57. The phase diagram for sulfur is

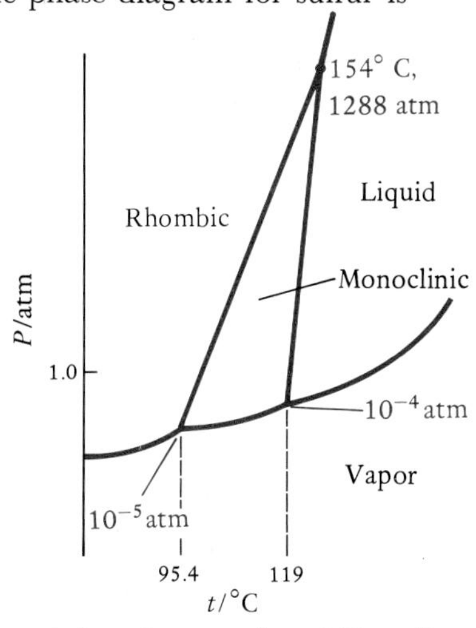

How many triple points are there? Describe what happens if sulfur is heated from 40°C at 1 atm to 200°C at 1 atm. Below what pressure will sublimation occur?

3 Triple points. Sulfer goes from rhombic to monoclinic at 96°C and melts at 119°C. Sublimation occurs at a pressure less than 10^{-5} atm.

13-58. Sketch the phase diagram for oxygen using the following data:

	Triple point	*Critical point*
temperature/K	54.3	154.6
pressure/torr	1.14	37,823

The normal melting point and normal boiling point of oxygen are −218.4°C and −182.9°C. Does oxygen melt under an applied pressure as water does?

CLAUSIUS-CLAPEYRON EQUATION

13-59. Two pressures on the equilibrium vapor pressure curve are related by an equation called the Clausius-Clapeyron equation, which is

$$\log \frac{P_2}{P_1} = \frac{\Delta \bar{H}_{vap}}{2.30R}\left(\frac{1}{T_1} - \frac{1}{T_2}\right)$$

where P_2 is the vapor pressure at the kelvin temperature T_2, P_1 is the vapor pressure at the kelvin temperature T_1, and R is the molar gas constant, $8.314\ \mathrm{J \cdot K^{-1} \cdot mol^{-1}}$. Given that $\Delta \bar{H}_{vap}$ for water is $40.7\ \mathrm{kJ \cdot mol^{-1}}$ and that the normal boiling point of water is 100°C, estimate the vapor pressure of water in a pressure cooker at 110°C.

13-60. The Clausius-Clapeyron equation is presented in Problem 13-59. The vapor pressure of water is 31.8 torr at 30°C and 92.5 torr at 50°C. Calculate $\Delta \bar{H}_{vap}$ for water.

$$\log \frac{P_2}{1.00\ \mathrm{atm}} = \frac{40.7 \times 10^3\ \mathrm{J/mol}}{(2.3)(8.314\ \mathrm{J/K \cdot mol})}\left(\frac{1}{373\ \mathrm{K}} - \frac{1}{383\ \mathrm{K}}\right)$$

$P_2 = \boxed{1.4\ \mathrm{atm}}$

13-61. The vapor pressures (in torr) of solid and liquid uranium hexafluoride are given by

$$\log P_s = 10.646 - \frac{2559.1\ \mathrm{K}}{T} \qquad \text{(solid)}$$

$$\log P_l = 7.538 - \frac{1511\ \mathrm{K}}{T} \qquad \text{(liquid)}$$

where T is the absolute temperature. Calculate the temperature and pressure at the triple point of UF_6.

$$10.646 - \frac{2559.1\ \mathrm{K}}{T_t} = 7.538 - \frac{1511\ \mathrm{K}}{T_t}$$

$T_t = \boxed{337.2\ \mathrm{K}}$

$$\log P = 7.538 - \frac{1511\ \mathrm{K}}{337.2\ \mathrm{K}}$$

$P_t = \boxed{1140\ \mathrm{torr}}$

13-62. The vapor pressures (in torr) of solid and liquid chlorine are given by

$$\log P_s = 10.560 - \frac{1640\ \mathrm{K}}{T} \qquad \text{(solid)}$$

$$\log P_l = 7.769 - \frac{1159\ \mathrm{K}}{T} \qquad \text{(liquid)}$$

where T is the absolute temperature. Calculate the temperature and pressure at the triple point of chlorine.

INTERCHAPTER H

Silicon: A Semimetal

Ultrapure (99.9999 percent) silicon is produced in the form of a cylinder and is then sliced into thin wafers for semiconductor manufacture.

Silicon constitutes 28 percent of the mass of the earth's mantle and is the second most abundant element in the mantle, being exceeded only by oxygen. Silicon does not occur as the free element in nature. It occurs primarily as the oxide and in various silicates. Elemental silicon has a gray, metallic luster. It is fairly inert and reacts at normal temperatures only with the halogens and with dilute aqueous alkalis. The major use of silicon is in the manufacture of glass and transistors.

H-1. ELEMENTAL SILICON IS USED TO MAKE SEMICONDUCTORS

The elements that lie on the border between the metals and the nonmetals in the periodic table are called semimetals. A comparison of the properties of metals, semimetals, and nonmetals is given in

Table H-1 Comparison of physical properties of metals, semimetals, and nonmetals

Metals	*Semimetals*	*Nonmetals*
basic oxides	weakly acidic oxides	acidic oxides
high electrical and thermal conductance	intermediate electrical and thermal conductance	insulators
high reflectance (bright, shiny)	intermediate reflectance	low reflectance (dull)
electrical resistance increases with increasing temperature	electrical resistance decreases with increasing temperature	resistance insensitive to temperature
malleable and ductile	brittle	not malleable, not ductile
nonvolatile and high-melting oxides, halides, and hydrides	volatile and low-melting oxides, halides and hydrides	volatile and low-melting oxides, halides, and hydrides

Table H-1. Silicon, which lies between aluminum and phosphorus in the periodic table, is the most important industrial semimetal. Elemental silicon is made by the high-temperature reduction of silicon dioxide (the major constituent of numerous sands) with carbon:

$$SiO_2(l) + C(s) \xrightarrow{3000^\circ C} Si(l) + CO_2(g)$$

The 98 percent pure silicon prepared by this reaction must be further purified before it can be used to make transistors. Therefore, it is converted to the liquid silicon tetrachloride by reaction with chlorine:

$$Si(s) + 2Cl_2(g) \rightarrow SiCl_4(l)$$

The silicon tetrachloride is further purified by repeated distillation and then converted to silicon by reaction with magnesium:

$$SiCl_4(g) + 2Mg(s) \rightarrow 2MgCl_2(s) + Si(l)$$

The resulting silicon is purified still further by a special method of recrystallization called *zone refining*. In this process, solid silicon is packed in a tube that is mounted in a vertical position (Figure H-1) with an electric heating loop around the base of the tube. The solid near the heating loop is melted by passing a current through the loop, and the tube is then lowered very slowly through the loop. As the melted solid cools slowly in the region of the tube below the heating loop, pure crystals separate out, leaving most of the impurities behind in the moving molten zone. The process can be repeated as often as necessary to achieve the desired purity of the recrystallized solid. Purities up to 99.9999 percent are possible with zone refining.

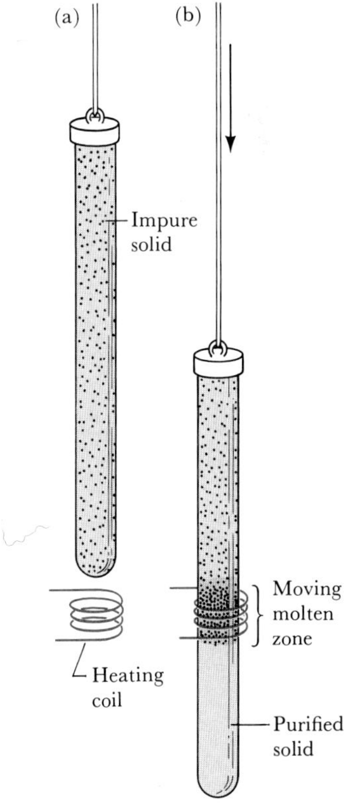

Figure H-1 Zone refining. An impure solid is packed tightly in a glass tube, and the tube is lowered slowly through a heating coil that melts the solid. Pure solid crystallizes out from the bottom of the melted zone, and the impurities concentrate in the moving molten zone.

A germanium crystal is being produced by the zone-refining method. This single crystal of ultrapure germanium was grown from a melt.

H-2. THERE ARE TWO TYPES OF SEMICONDUCTORS, *n*-TYPE AND *p*-TYPE

In a crystal, there are two sets of energy levels because of the combination of the valence orbitals of all the atoms. These two sets of energy levels are analogous to the bonding and antibonding orbitals that occur when orbitals from just two atoms are combined. The lower set of energy levels is called the *valence band* and is occupied by the valence electrons of the atoms. The higher set corresponds to antibonding orbitals and is called the *conduction band.* Electrons in the conduction band can move readily throughout the crystal (Figure H-2).

An electric current is carried in a solid by the electrons in the conduction band, which are called the conduction electrons. In an insulator (such as a nonmetal), there are essentially no electrons in the conduction band because the energies there are much higher than the energies in the valence band. Metals are excellent electrical conductors because there is no energy gap between the conduction

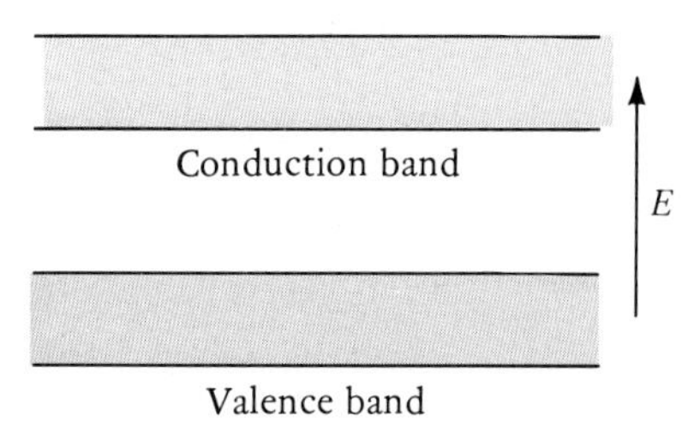

Figure H-2 When the atoms of a crystal are brought together to form the crystal lattice, the valence orbitals of the atoms combine to form two sets of energy levels, called the valence band and the conduction band.

band and the valence band. The valence electrons in a metal are conduction electrons. In a semiconductor the energy separation between the conduction band and the valence band is comparable to thermal energies, and thus some of the valence electrons can be thermally excited into the conduction band. Thus a semiconductor has electrical properties intermediate between those of metals and insulators. Figure H-3 illustrates the difference between a metal, an insulator, and a semiconductor.

An insulator like silicon can be converted to a semiconductor by addition of selected impurity atoms. For example, an *n*-type (*n* for negative) silicon semiconductor is produced when trace amounts of atoms with five valence electrons, such as phosphorus or antimony, are added to silicon, which has four valence electrons (Figure H-4a). The excess valence electrons on the impurity atoms, which substitute

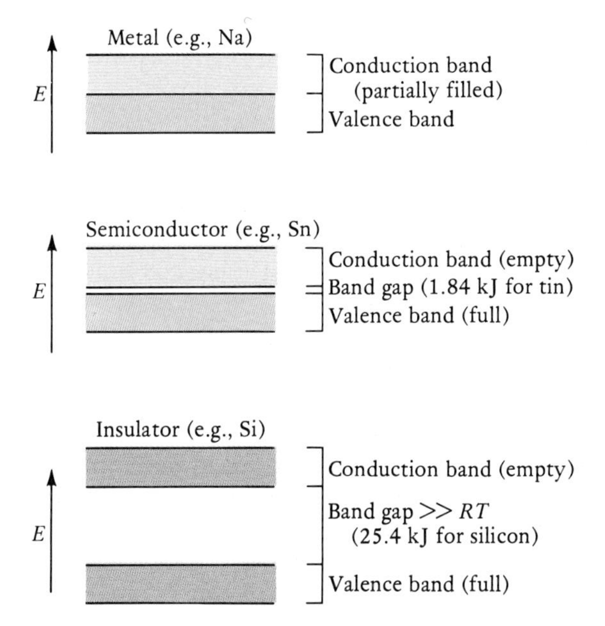

Figure H-3 A comparison of the energy separations between the valence bands (bonding electron energy levels) and conduction bands (accessible energy levels for mobile electrons) of metals, semiconductors, and insulators. Metals have no band gap, semiconductors have a small band gap, and insulators have a large band gap. Thermal energy at 300 K is $RT = (8.314\ \mathrm{J \cdot K^{-1} \cdot mol^{-1}})(300\ \mathrm{K}) = 2.5\ \mathrm{kJ \cdot mol^{-1}}$. Note that the band gap energy separation for a semiconductor is comparable to RT in magnitude.

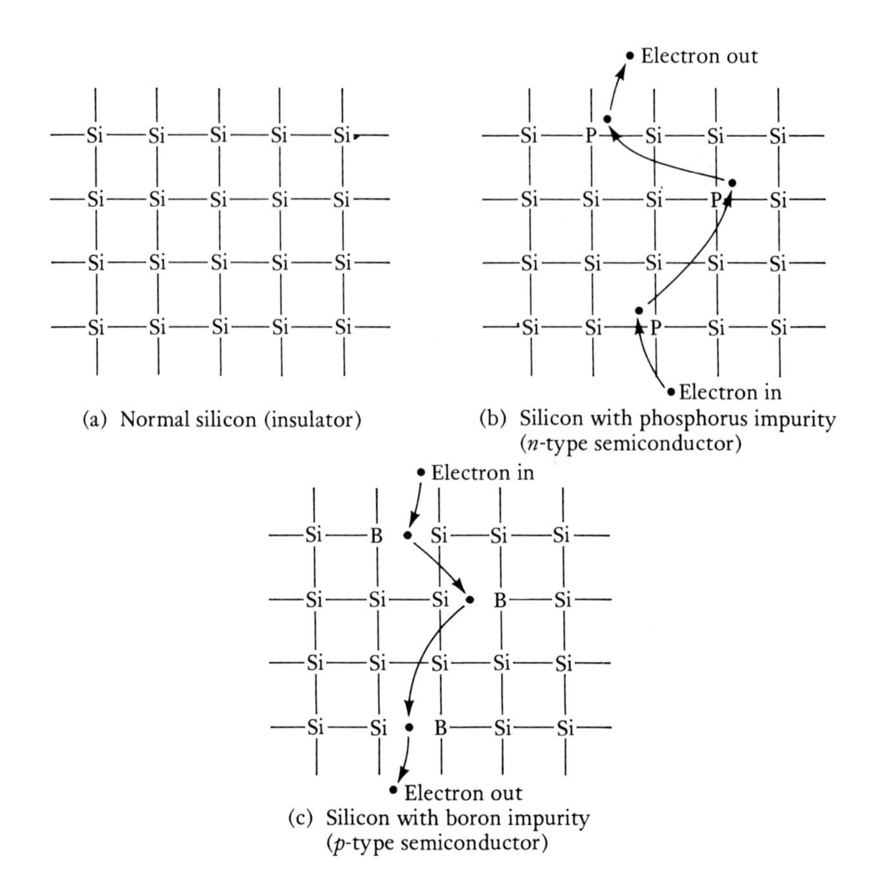

Figure H-4 Comparison of normal, *n*-type, and *p*-type silicon. (a) Silicon has four valence electrons, and each silicon atom forms four 2-electron bonds to other silicon atoms. (b) Phosphorus has five valence electrons, and thus when a phosphorus atom substitutes for an silicon atom in a silicon crystal, there is an unused valence electron on each phosphorus atom that can become a conduction electron. (c) Boron has only three valence electrons, and thus when a boron atom substitutes for a silicon atom in a silicon crystal, there results an electron vacancy (a "hole"). Electrons from the silicon valence bond can move through the crystal by hopping from one vacancy site to another.

for some of the silicon atoms in the crystal, become the current carriers in the crystal (Figure H-4b). A *p*-type (*p* for positive) semiconductor is produced when trace amounts of atoms with three valence electrons, such as boron or indium, are added to silicon. The deficiency of valence electrons on the impurity atoms functions as "holes" by means of which electrons can "hop" through the silicon crystal (Figure H-4c). Because impurity atoms have a major effect on the electrical properties of semiconductors, it is necessary to use extremely pure ($\geqslant$99.9999 percent) silicon and to add precise amounts of impurities of carefully controlled composition to the crystal in order to obtain the desired electrical properties.

The South Bay area of San Francisco is called the Silicon Valley because of the large number of companies that produce semi-conductors, transistors, and computer chips.

It would be difficult to exaggerate the impact of semiconductor devices on modern technology. With their minute size and very low power requirements, these devices have made possible computers with incredible computing, storage, and retrieval capabilities. It is possible to make a computer memory chip with over a million transistors (the solid-state equivalent of the now-obsolete vacuum tube) in a space of only 1 mm^2 (Figure H-5). Hand-held computers have made the slide rule obsolete, and the personal computer industry is expanding at an amazing rate.

Figure H-5 Computer chips in a contact lens. Each chip may consist of over a million transistors.

H-3. SILICATES OCCUR WIDELY IN NATURE

Silicates occur in numerous minerals and in asbestos, mica, and clays. Cement, bricks, tiles, porcelains, glass, and pottery are all made from silicates. All silicates involve silicon-oxygen single bonds, of which there are two types. Terminal —Si—O bonds involve oxygen bonded to silicon and no other atoms, and bridging —Si—O—Si— bonds involve oxygen linking two silicon atoms.

The simplest silicate anion is the tetrahedral *orthosilicate* ion, SiO_4^{4-}:

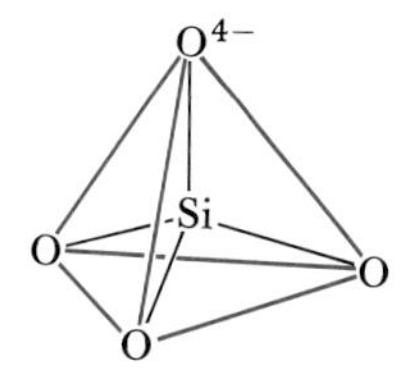

The SiO_4^{4-} ion is found in the minerals zircon, $ZrSiO_4$, and willemite, Zn_2SiO_4, and also in sodium silicate, which, when dissolved in water, is called water glass, $Na_4SiO_4(aq)$. The minerals enstatite, $MgSiO_3$, and spondumene, $LiAl(SiO_3)_2$, are silicates that contain long, straight-chain silicate polyanions involving the SiO_3^{2-} chain unit:

```
      O⊖      O⊖      O⊖      O⊖      O⊖
      |       |       |       |       |
 —O—Si—O—Si—O—Si—O—Si—O—Si—
      |       |       |       |       |
      O⊖      O⊖      O⊖      O⊖      O⊖
   SiO3(2-) SiO3(2-) SiO3(2-) SiO3(2-) SiO3(2-)
```

Figure H-6 shows the tetrahedral shape of each SiO_4^{4-} unit. Structures that result from joining many smaller units together are called polymers (Interchapter N). The straight-chain silicate anions shown above are called polyanions because they result from joining together

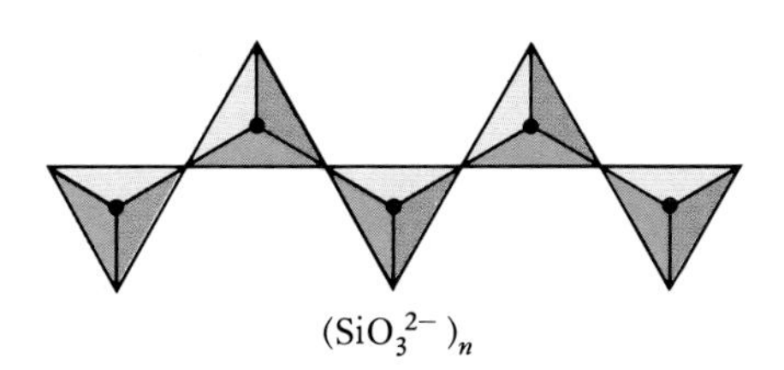

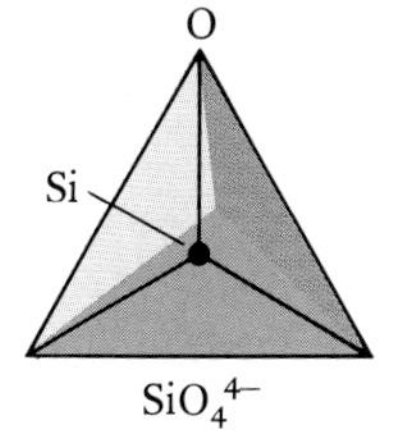

Figure H-6 Tetrahedral SiO_4^{4-} units are linked together through oxygen atoms that are shared by tetrahedra to form straight-chain silicate polyanions.

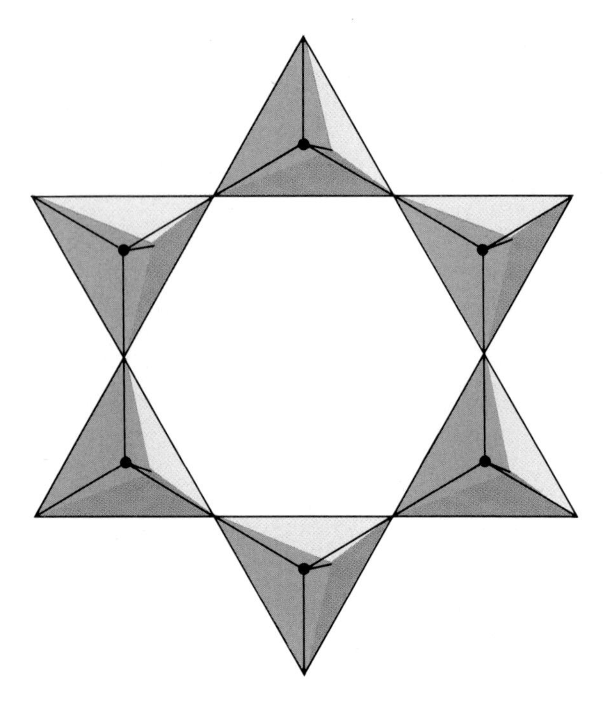

Figure H-7 The cyclic polysilicate ion $Si_6O_{18}^{12-}$, which occurs in the mineral beryl. Six SiO_4^{4-} tetrahedral units are joined in a ring with the tetrahedra linked by shared oxygen atoms.

many silicate anions. The mineral beryl, $Be_3Al_2Si_6O_{18}$, contains the cyclic polysilicate anion $Si_6O_{18}^{12-}$ (Figure H-7). These cyclic polysilicate anions can themselves be joined together to form polymeric, cyclic polysilicate anions with the composition $(Si_4O_{11}^{6-})_n$ and the structure shown in Figure H-8. The best example of a mineral containing polymeric, cyclic polysilicate chains is asbestos (Figure H-9). The fibrous character of asbestos is a direct consequence of the molecular structure of the $(Si_4O_{11}^{6-})_n$ polymeric chains.

Can you identify the two $Si_4O_{11}^{6-}$ subunits in Figure H-8?

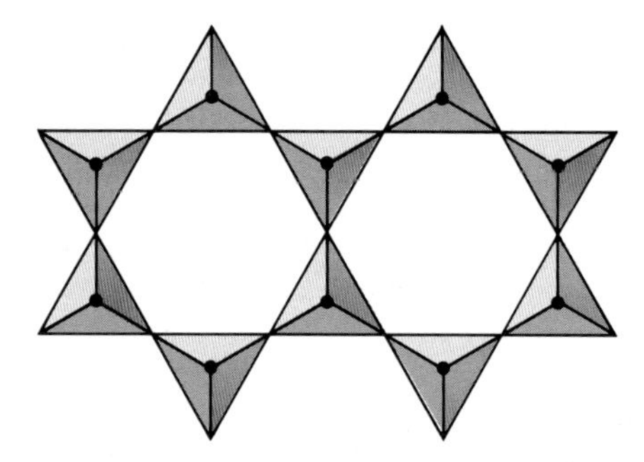

Figure H-8 The cyclic polysilicate ion $Si_6O_{18}^{12-}$ can form a polymeric cyclic network like that shown here. The composition of the cyclic network is $(Si_4O_{11}^{6-})_n$. Asbestos has this structure.

Figure H-9 Asbestos. The fibrous character of this mineral is a direct consequence of its $(Si_4O_{11}^{6-})_n$ polymeric chains.

The silicate minerals mica and talc contain two-dimensional, polymeric silicate sheets with the overall silicate composition $Si_2O_5^{2-}$. The structure of these sheets is illustrated in Figure H-10, and Figure H-11 shows how mica can easily be fractured into thin sheets. Talc has the composition $Mg_3(OH)_2(Si_2O_5)_2$, whereas micas have a variety of compositions, one example of which is lepidolite, $KLi_2Al(Si_2O_5)_2(OH)$. The ease with which mica can be separated into thin sheets and the slippery feel of talcum powder arise from the layered structure of the silicates in these minerals.

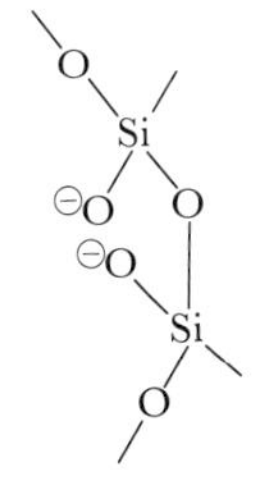

The $Si_2O_5^{2-}$ subunit

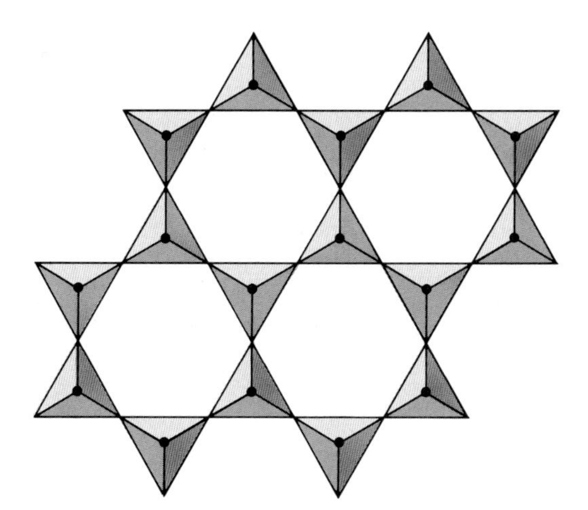

Figure H-10 Structure of polysilicate sheets composed of $(Si_2O_5^{2-})_n$ subunits. Mica has this structure.

Because of its high thermal stability, mica was used as a window material in high-temperature ovens; however, special borate glasses are now used for this purpose.

Figure H-11 The ease with which the mineral mica can be separated into thin sheets is a direct consequence of the existence of polymeric silicate sheets with the composition $(Si_2O_5^{2-})_n$. (See Figure H-10.)

The so-called framework minerals contain SiO_4^{4-} and AlO_4^{5-} tetrahedra in which each oxygen atom is shared by two tetrahedra to form a three-dimensional network (Figure H-12). Zeolites, feldspars, and ultramarines are examples of minerals with this type of structure. The designation framework minerals denotes the fact that such solids contain cavities of various sizes into which small molecules can enter. Zeolites also possess the property of being able to exchange cations in the crystal with external cations, as a result of the presence of loosely held cations in the cavities of the crystal.

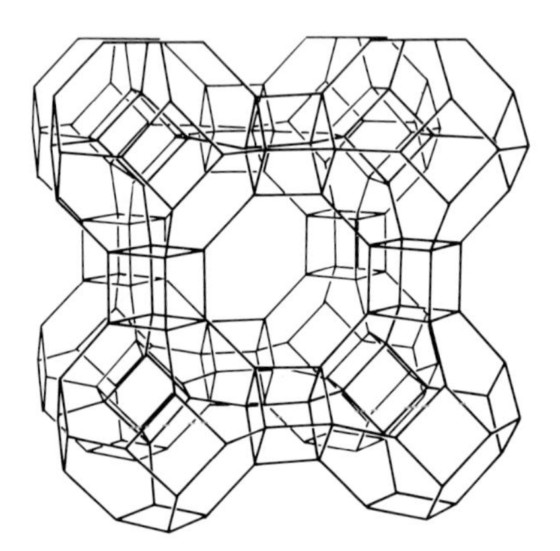

Figure H-12 Zeolite structure. Note the presence of channels and cavities. These open regions are capable of housing small molecules such as water.

Molecular sieves are synthetic zeolites, the most common of which has the composition $Na_{12}(AlO_2)_{12}(SiO_2)_{12} \cdot 27H_2O$. The waters of hydration are lost on heating (200°C), and the resulting anhydrous zeolite has a very high affinity for water. Anhydrous zeolites are used to remove traces of water from organic solvents. Absolute ethanol is obtained from 95% ethanol by the use of molecular sieves. Molecular sieves are also used to separate straight-chain from branched-chain hydrocarbons; the straight-chain hydrocarbons move into the channels in the sieves, but the branched-chain compounds are too bulky to fit into the sieve channels.

H-4. MOST GLASSES ARE SILICATES

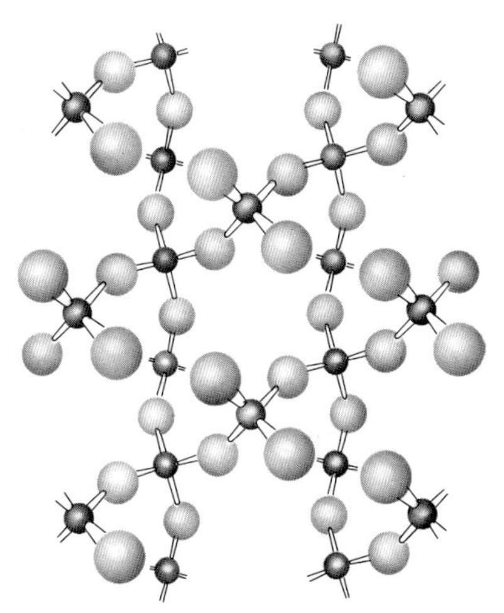

Figure H-13 The crystalline structure of quartz. Note that each silicon atom is surrounded by 4 oxygen atoms. The silicon atoms are linked by the oxygen atoms.

Quartz is a crystalline material with the composition SiO_2 and the crystalline structure shown in Figure H-13. When crystalline quartz is melted and then cooled quickly to prevent the formation of crystals, there is formed a disordered three-dimensional array of polymeric chains, sheets, and other three-dimensional clusters. The resulting material is called quartz glass. All glass consists of a random array of these clusters.

Glass manufacturing is a 10-billion-dollar-per-year industry in the United States. The major component in glass is almost pure quartz sand. Among the other components of glass, soda (Na_2O) comes from soda ash (Na_2CO_3), lime (CaO) comes from limestone ($CaCO_3$), and aluminum oxide comes from feldspars, which have the general formula $M_2O \cdot Al_2O_3 \cdot 6SiO_2$, where M is K or Na. All the components of glass are fairly inexpensive chemicals.

A wide variety of glass properties can be produced by varying the glass composition (Table H-2). For example, partial replacement of CaO and Na_2O by B_2O_3 gives a glass that does not expand on heating or contract on cooling and is thus used in making glass utensils meant to be heated. Colored glass is made by adding a few percent of a colored transition metal oxide, such as CoO to make blue "cobalt" glass and Cr_2O_3 to make orange glass. Lead glass, which contains PbO, has attractive optical properties and is used to make decorative, cut-glass articles.

Quartz often forms large, beautiful crystals.

Table H-2 Oxide composition of some common glasses

	Oxide/%					
	SiO_2	CaO	Na_2O	B_2O_3	Al_2O_3	Other
quartz	100	0	0	0	0	0
window glass (soda lime)	72	11	13	0	0.3	3.8 K_2O
aluminosilicate (oven cookware)	55	15	0	0	20	15 MgO
heat-resistant glass	76	3	5	13	2	0.5 K_2O
lead glass (decorative "crystal")	67	1	10	0	0	15 PbO, 7 K_2O
bioglass (surgical implants)	45	24	25	0	0	6 P_2O_5
optical	69	12	6	0.3	0	12 K_2O

The addition of K_2O increases the hardness of glass and makes it easier to grind to precise shapes. Optical glass contains about 12 percent K_2O. Photochromic eyeglasses have a small amount of added silver chloride dispersed throughout and trapped in the glass. When sunlight strikes this type of glass, the tiny AgCl grains decompose into opaque clusters of silver atoms and chlorine atoms:

$$\underset{\text{(clear)}}{\text{AgCl}} \underset{\text{dark}}{\overset{\text{sunlight}}{\rightleftharpoons}} \underset{\text{(opaque)}}{\text{Ag} + \text{Cl}}$$

The chlorine atoms are trapped in the crystal lattice, and the silver and chlorine atoms recombine in the dark to form silver chloride, which causes the glass to become clear.

The etching of glass by hydrofluoric acid, HF(*aq*), is a result of the reaction

$$SiO_2(s) + 6HF(aq) \rightarrow H_2SiF_6(s) + 2H_2O(l)$$

and this reaction is used to "frost" the inside surface of lightbulbs.

Porcelain has a much higher percentage of Al_2O_3 than glass and as a result is a heterogeneous substance. Porcelain is stronger than glass because of this hetereogeneity and is also more chemically resistant than glass. Earthenware is similar in composition to porcelain but is more porous because it is fired at a lower temperature.

QUESTIONS

H-1 Describe how silicon is produced.

H-2 Describe the process of zone refining. What purity of silicon is obtainable by zone refining?

H-3 Discuss the difference between a conductor, an insulator, and a semiconductor in terms of the gap between the conduction band and the valence band.

H-4 Discuss the difference between *p*-type and *n*-type semiconductors.

H-5 Use VSEPR theory to predict the shape of an orthosilicate ion, SiO_4^{4-}.

H-6 Describe the various silicate structures of mica and asbestos.

H-7 Sketch a straight-chain silicate polyanion. How are the SiO_4^{4-} units linked to each other?

H-8 Sketch the cyclic polysilicate ion, $Si_6O_{18}^{12-}$. In which mineral does this structure occur?

H-9 Discuss the silicate polyanion structure of mica.

H-10 What are the three principal components of glass?

H-11 Describe how photochromic eyeglasses work.

H-12 Describe the reaction by which HF attacks glass.

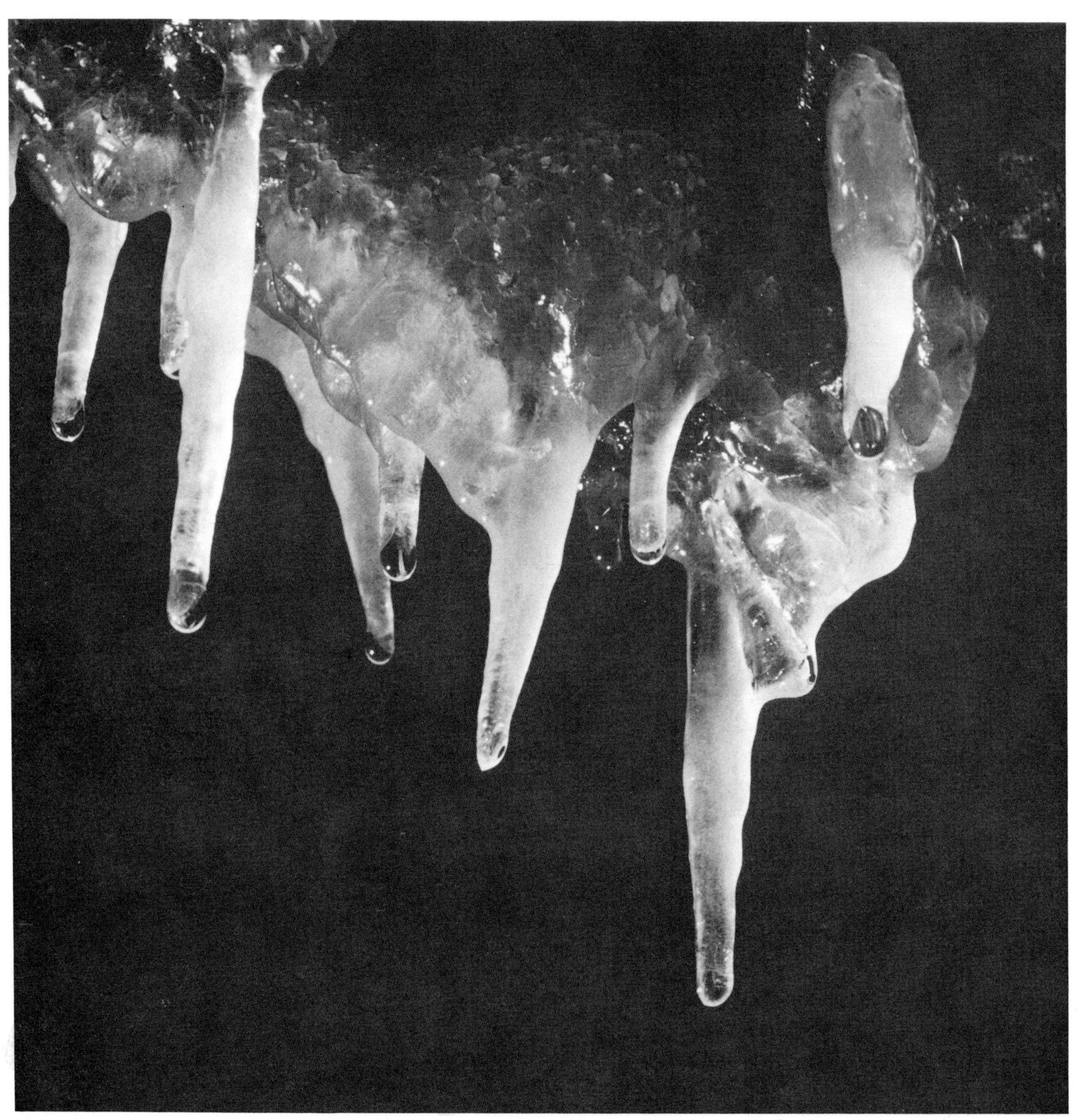

Stalactites are deposits of $CaCO_3(s)$ that form from water containing $Ca^{2+}(aq)$ and $HCO_3^-(aq)$ that slowly evaporates from the drops at the tips of stalactites.

14 / Solutions

In this chapter we discuss some of the elementary properties of solutions from a molecular viewpoint. We introduce the concept of solubility, which tells us how much of one substance can be dissolved in another substance. The major emphasis of this chapter is on the colligative properties of solutions, which are properties that depend primarily on the ratio of the number of solute particles to the number of solvent particles in the solution. The colligative properties of solutions are vapor pressure lowering, boiling point elevation, freezing point depression, and osmotic pressure.

We learned in Chapter 13 that pure water has a unique equilibrium vapor pressure at each temperature. We learn in this chapter that the equilibrium vapor pressure of pure water always decreases when a substance is dissolved in it. The key to understanding the colligative properties of solutions is understanding why the equilibrium vapor pressure of a solvent decreases when a substance is dissolved in it.

14-1. A SOLUTION IS A HOMOGENEOUS MIXTURE OF TWO OR MORE SUBSTANCES

Most chemical and biological processes take place in solution, particularly in aqueous solution. A *solution* is a mixture of two or more substances that is homogeneous at the molecular level. A solution must be *homogeneous,* meaning that it must have the same properties from one region to another. The most common examples of solutions involve a solid, such as NaCl, dissolved in water. The resulting solu-

Table 14-1 Types and examples of solutions

State of component 1	*State of component 2*	*State of resulting solution*	*Examples*
gas	gas	gas	air; vaporized gasoline-air mixture in the combustion chambers of a car
gas	liquid	liquid	oxygen in water; carbon dioxide in carbonated beverages
gas	solid	solid	hydrogen in palladium and platinum
liquid	liquid	liquid	water and alcohol
liquid	solid	solid	mercury in gold and silver
solid	liquid	liquid	sodium chloride and water
solid	solid	solid	metal alloys

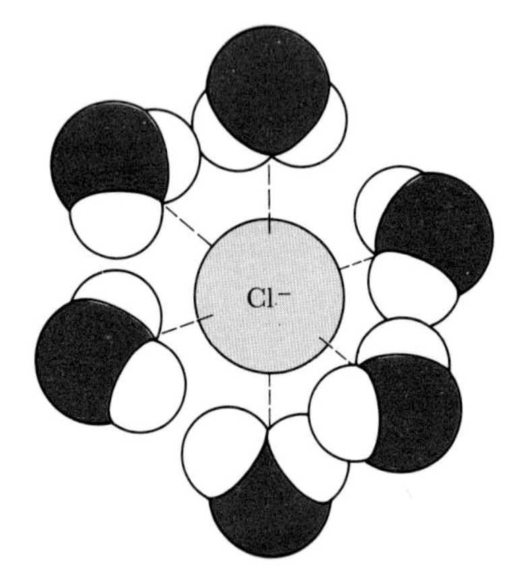

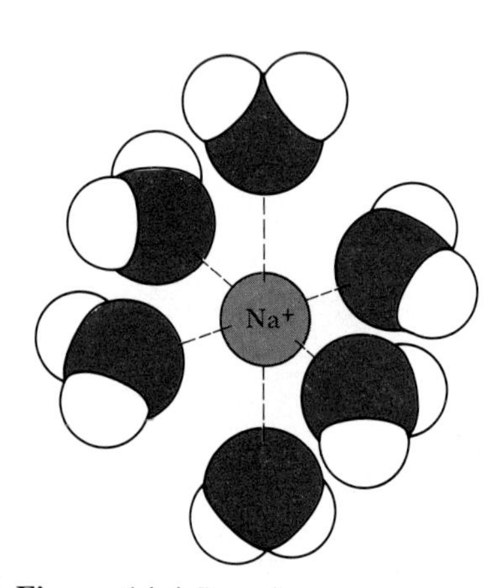

Figure 14-1 Ions in aqueous solutions are stabilized by the interaction with water molecules. Such ions are said to be solvated.

tion is clear and homogeneous. From a molecular point of view, the sodium ions and chloride ions are uniformly dispersed among the water molecules. The *components* of a solution are the pure substances that are mixed to form the solution. The components do not have to be a solid and a liquid. There are many other types of solutions (Table 14-1).

The component of a solution that has the same state (gas, liquid, or solid) as the resulting solution is called the *solvent*. If two or more components have the same state as the resulting solution, then the component present in excess is the solvent. The other components of a solution are called *solutes*. A solute is often considered to be a substance that is dissolved in a solvent. The terms solvent and solute are merely terms of convenience; all components of a solution are uniformly dispersed throughout the solution.

14-2. SOLUBILITY INVOLVES A DYNAMIC EQUILIBRIUM

Consider a solid, such as crystals of NaCl, at the bottom of a beaker of water. As we know from Chapter 13, the molecular picture of a solid is that of an ordered lattice, with the individual particles restricted to lattice sites. The molecules of a liquid, on the other hand, move about continuously, colliding very frequently with each other. The water molecules in our example collide not only with each other but also with the surfaces of the sodium chloride crystals at the bottom of the beaker. In doing so, they jar sodium ions and chloride ions loose, allowing them to enter into solution (Figure 14-1). The sodium and chloride ions interact strongly with water molecules, and this interaction facilitates the solution process.

The ions that dissolve are free to move about throughout the solution, colliding with water molecules, other ions, the walls of the beaker, and the surfaces of the NaCl crystals remaining at the bottom of the beaker. Some of these ion-crystal collisions result in sodium and chloride ions sticking to the crystal surfaces and being incorporated back into the crystalline structure. We can represent the solution process by

$$\mathrm{NaCl}(s) \underset{\text{crystallization}}{\overset{\text{solution, } \mathrm{H_2O}(l)}{\rightleftharpoons}} \mathrm{Na}^+(aq) + \mathrm{Cl}^-(aq)$$

Notice that the reaction proceeds both forward (solution) and backward (crystallization). According to our molecular picture, both processes are occurring constantly. As the number of ions in solution increases, the number of collisions with the crystal surfaces increases, and so more dissolved ions are incorporated back into the solid phase. Eventually a balance is reached, where at any instant the number of ions going into solution is equal to the number of ions being incorporated into the crystal.

Let's call the number of ions that leave the crystals and enter the solution per second the *rate of solution* and the number of ions that leave the solution and attach to the crystals per second the *rate of crystallization*. When the rate of solution is equal to the rate of crystallization, then the quantity of NaCl at the bottom of the beaker no longer changes and it appears to the naked eye that the solution process has ceased. From a molecular point of view, however, some ions are still entering the solution and others are being deposited onto the crystal surfaces. The rates of solution and crystallization are equal, and so no *net* change is observed. To emphasize that there is still a great deal of molecular activity even though there is no net progress, we say that the solution and crystallization processes are in *dynamic equilibrium* (Figure 14-2).

Equilibrium vapor pressure is pictured as a dynamic equilibrium.

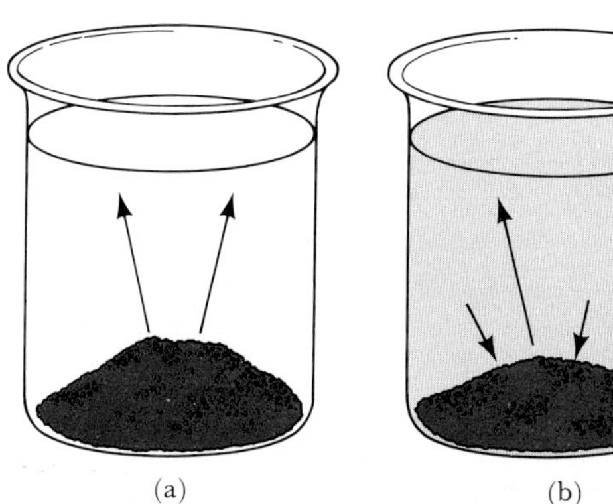
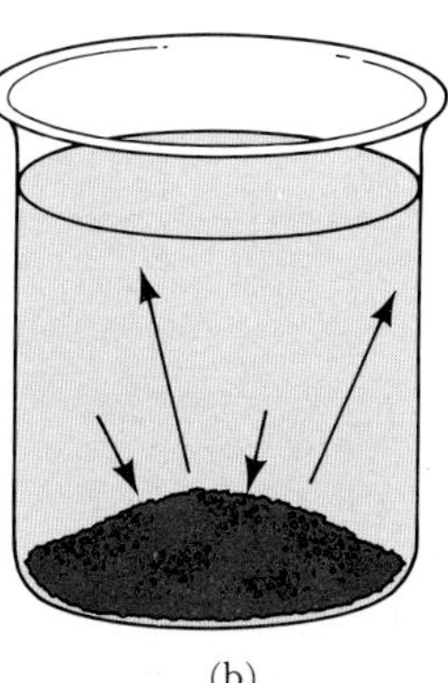
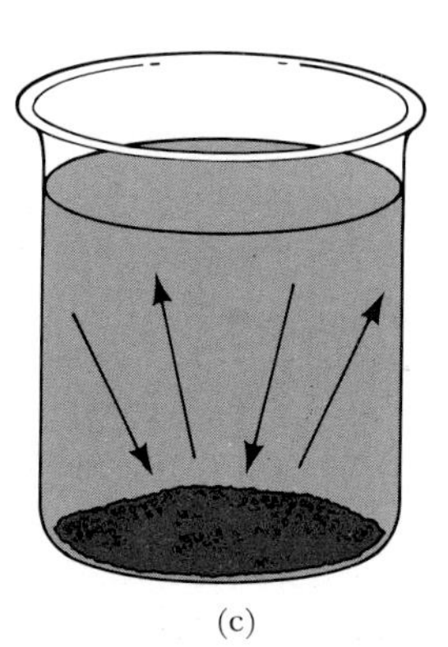

Figure 14-2 Dynamic equilibrium between ions going into solution and ions being incorporated back into the crystal lattice. When the rates of these two processes are the same, then the system is said to be in dynamic equilibrium (c).

Dynamic equilibrium occurs when the number of ions being deposited on the surface of the crystals is equal to the number of ions entering the solution at any instant. At this equilibrium point, no more ions can be accommodated by the solution; more ions in solution result only in more collisions with the crystal surfaces and thus more ions deposited. The solution is then said to be *saturated,* and the quantity of solute dissolved is called the *solubility* of that solute. Solubility can be expressed as grams of solute per 100 g of solvent. For example, the solubility of KBr in water at 20°C is about 65 g per 100 g of H_2O.

It is important to realize that the solubility of a substance is the maximum quantity that can be dissolved in a given quantity of solvent. The solubility of NaCl at 20°C is about 35 g per 100 g of H_2O. If we add 50 g of NaCl to 100 g of H_2O at 20°C, then 35 g dissolves and 15 g is left as undissolved NaCl(*s*). The solution is saturated. If we add 25 g of NaCl to 100 g of H_2O, then all of the NaCl dissolves to form what is called an *unsaturated solution,* which is a solution that is still able to dissolve more solute.

In most cases the solubility of a substance depends upon temperature. The effect of temperature on the solubility of several salts in water is shown in Figure 14-3. Almost all salts become more soluble in water as the temperature increases. For example, potassium nitrate is about five times more soluble in water at 40°C than in water at 0°C. At higher temperatures both the ions in the crystal and the

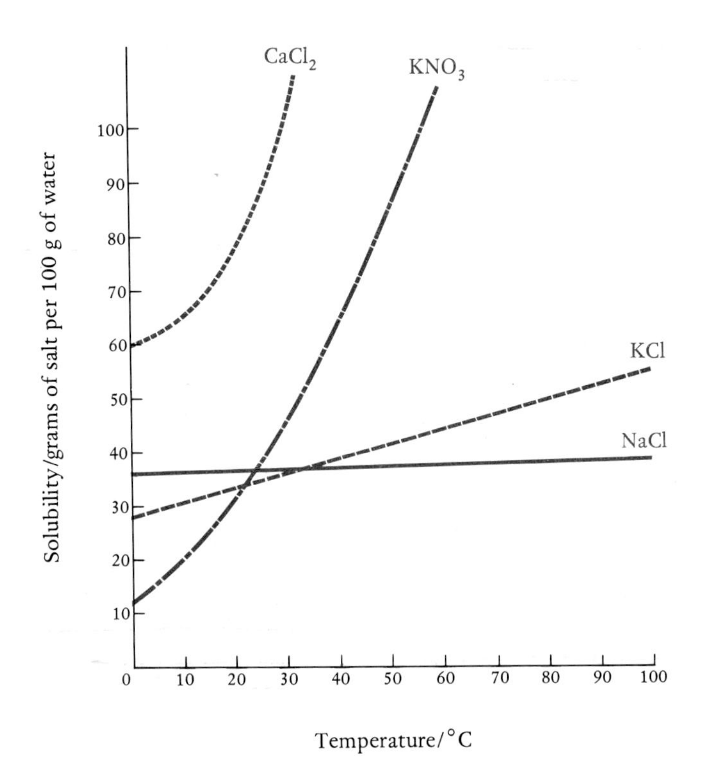

Figure 14-3 The solubility of several salts as a function of temperature. These curves show that solubility depends upon temperature and usually increases with increasing temperature.

Table 14-2 Solubilities in water at 0°C and 100°C

	Solubility/g per 100 g of H_2O	
Substance	0°C	100°C
$NaCl(s)$	35	39
$KCl(s)$	29	57
$KNO_3(s)$	13	247
$NaOH(s)$	42	347
$NH_4NO_3(s)$	118	871
$CaCl_2(s)$	60	159

Salts can be purified by forming a saturated solution at a high temperature and then cooling the solution. The decreased solubility as the temperature decreases results in the precipitation of the salt. The soluble impurities tend to remain in solution.

water molecules in the solvent have higher average energies, and thus the breakdown of the crystal lattice that must occur in order for solution to take place is more easily accomplished. Table 14-2 gives solubility data for some common substances.

14-3. THE EQUILIBRIUM VAPOR PRESSURE OF A PURE LIQUID IS ALWAYS DECREASED WHEN A SUBSTANCE IS DISSOLVED IN THE LIQUID

How the solvent vapor pressure is decreased when a solute is dissolved in a solvent is shown in Figure 14-4 for the case of a nonvolatile solute such as glucose or NaCl, which both have a totally negligible vapor pressure. The presence of solute molecules at or near the surface of the liquid decreases the rate of evaporation of the solvent molecules from the surface simply because there are fewer solvent molecules per unit area of surface. Note that in the simple picture

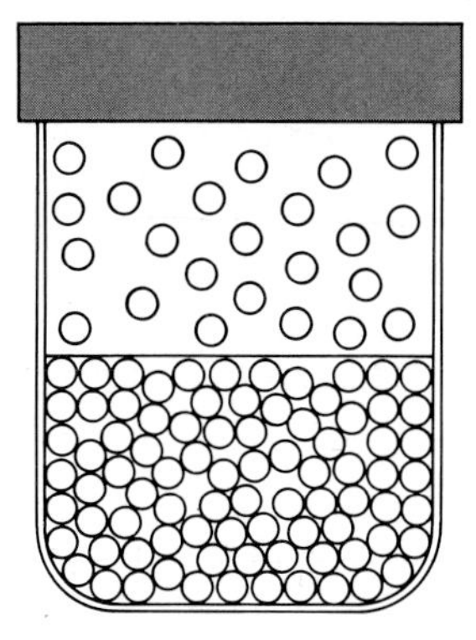

Pure solvent

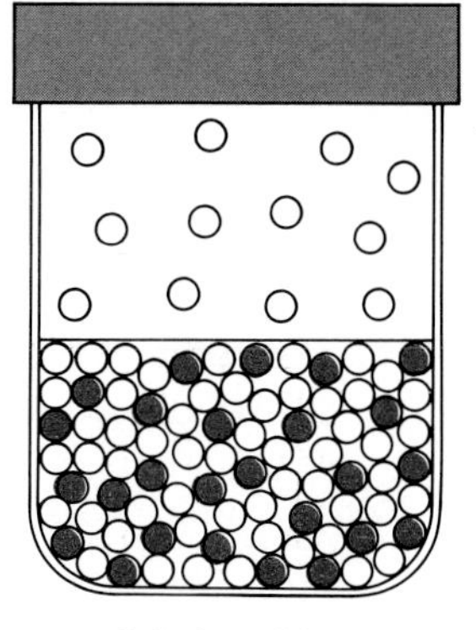

Solution with a nonvolatile solute

Figure 14-4 The effect of a nonvolatile solute on the equilibrium vapor pressure of a solvent at a fixed temperature. The solute molecules lower the equilibrium vapor pressure of the solvent relative to that of the pure solvent by partially blocking the escape of solvent molecules from the surface of the solution.

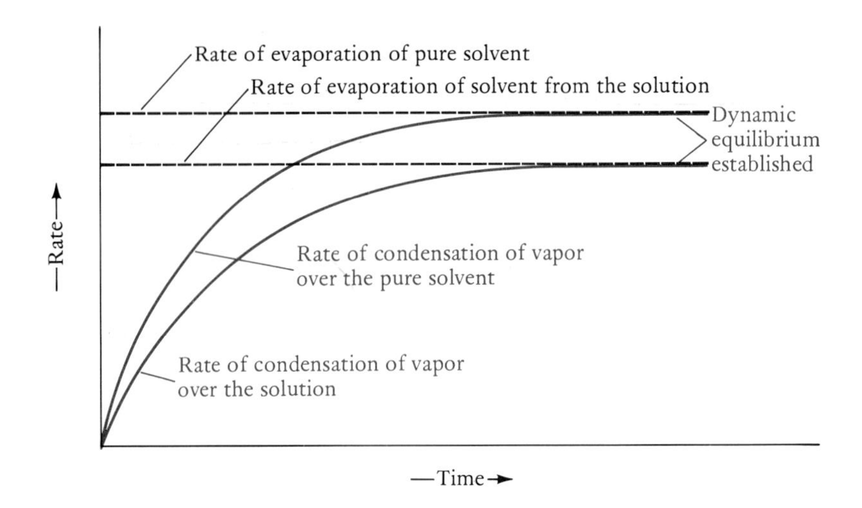

Figure 14-5 The effect of a nonvolatile solute on solvent evaporation rate. The decreased solvent evaporation rate leads to a lower equilibrium vapor pressure for the solvent.

presented in Figure 14-4, it is the ratio of the number of solute particles to the number of solvent particles that is important. The *nature* of the solute is unimportant in this simplified view of solutions.

Recall from Chapter 13 that as the concentration of molecules in a vapor phase increases, the rate of condensation of the vapor increases because the rate of condensation is directly proportional to the number of molecules in the vapor phase. The vapor pressure reaches a constant, or equilibrium, value when the rate of condensation equals the rate of evaporation. Consequently, the lower evaporation rate from a solution gives rise to a lower equilibrium vapor pressure of the solvent over the solution than over pure solvent at the same temperature (Figure 14-5).

14-4. THE EQUILIBRIUM VAPOR PRESSURE OF THE SOLVENT OVER A SOLUTION IS PROPORTIONAL TO THE MOLE FRACTION OF THE SOLVENT IN THE SOLUTION

Consider a solution composed of molecules of compound A and molecules of compound B. If the interaction between an A and a B molecule is the same as that between two A molecules or two B molecules, then the A and B molecules will be randomly distributed throughout the solution, including the region near the surface (Figure 14-6). Such a solution is called an *ideal solution*. For an ideal solution the equilibrium vapor pressure of the solvent is directly proportional to a quantity called the *mole fraction* of the solvent. In a solution containing n_1 moles of solvent and n_2 moles of solute, the mole fraction of the solvent, X_1, is defined as

In formulas involving solutions, the solvent is usually designated as component 1, and a quantity referring to the solvent carries a subscript 1 as in Equation (14-1).

$$X_1 = \frac{n_1}{n_1 + n_2} \tag{14-1}$$

moles solvent
mole fraction
moles solvent and solute

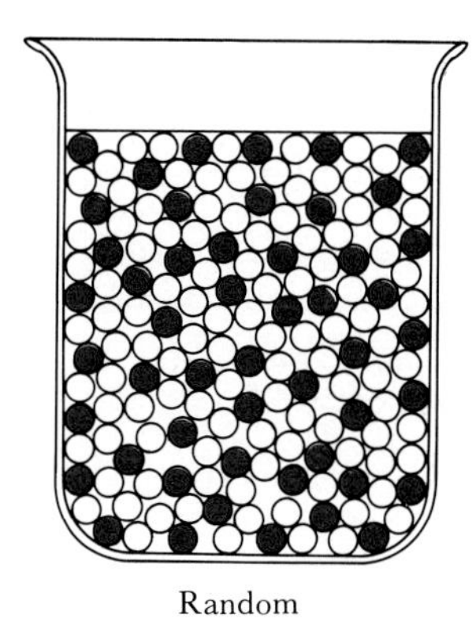

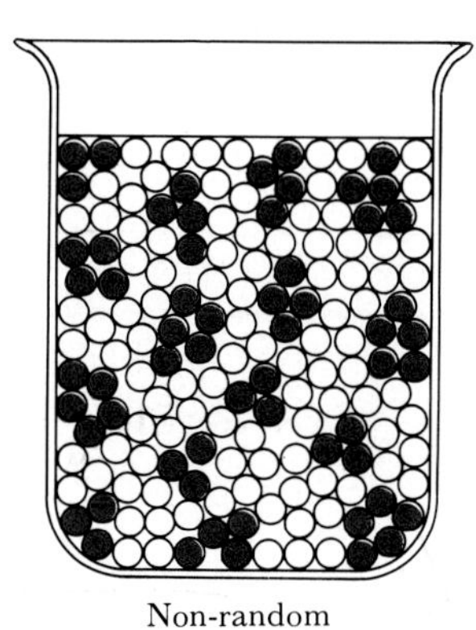

Figure 14-6 Random and non-random distribution of two types of molecules, A (open circles) and B (solid circles), in solution. Note in the non-random case that AAA... and BBB... clusters form in the solution. The random distribution represents an ideal solution.

Example 14-1: Suppose that we dissolve 50.0 g of sucrose, $C_{12}H_{22}O_{11}$, in 200 g of water. Compute the mole fraction of water in the resulting solution.

Solution: The formula mass of sucrose is 342.3, and the formula mass of water is 18.02. Thus the mole fraction of water in the solution is

A mole fraction may also be thought of as a molecular fraction. It represents the fraction of molecules of a species in a solution.

$$X_{H_2O} = \frac{n_{H_2O}}{n_{H_2O} + n_{sucrose}}$$

$$= \frac{(200\ g)\left(\dfrac{1\ mol\ H_2O}{18.02\ g\ H_2O}\right)}{(200\ g)\left(\dfrac{1\ mol\ H_2O}{18.02\ g\ H_2O}\right) + (50.0\ g)\left(\dfrac{1\ mol\ sucrose}{342.3\ g\ sucrose}\right)}$$

$$= 0.987$$

Note that mole fraction is a dimensionless quantity and that it cannot exceed unity.

The relation between the vapor pressure of the solvent and the mole fraction of solvent is

$$P_1 \propto X_1$$

where P_1 is the solvent equilibrium vapor pressure and X_1 is the mole fraction of solvent. Mathematically, a proportion such as this can be written as an equation by introducing a proportionality constant. Therefore, we can write

A proportion may be converted to an equation by inserting a proportionality constant.

$$P_1 = kX_1 \tag{14-2}$$

Raoult's law tells us that the vapor pressure of a solvent is directly proportional to the mole fraction of the solvent in the solution.

where k is a proportionality constant that is determined as follows. When the mole fraction of the solvent is unity, that is, when we have pure solvent, then P_1 is equal to the vapor pressure of the pure solvent, which we denote by P_1°. Thus when $X_1 = 1$ in Equation (14-2), we have $P_1 = P_1^\circ$. If we substitute $X_1 = 1$ and $P_1 = P_1^\circ$ into Equation (14-2), then we see that $k = P_1^\circ$. Thus we can rewrite Equation (14-2) as

$$\boxed{P_1 = X_1 P_1^\circ} \tag{14-3}$$

which is known as *Raoult's law*.

The amount by which the vapor pressure of a solution is less than the vapor pressure of the pure solvent, that is, $P_1^\circ - P_1$, is called the *vapor pressure lowering*.

Example 14-2: The vapor pressure of pure water at 100°C is 1.00 atm (760 torr). Use Raoult's law to estimate the vapor pressure of an aqueous solution at 100°C containing 30.0 g of glucose, $C_6H_{12}O_6$, dissolved in 200 g of water. Also compute the vapor pressure lowering.

Solution: The mole fraction of water in the solution is

$$X_{H_2O} = \frac{n_{H_2O}}{n_{H_2O} + n_{glucose}}$$

$$= \frac{(200\ \text{g})\left(\dfrac{1\ \text{mol H}_2\text{O}}{18.02\ \text{g H}_2\text{O}}\right)}{(200\ \text{g})\left(\dfrac{1\ \text{mol H}_2\text{O}}{18.02\ \text{g H}_2\text{O}}\right) + (30.0\ \text{g})\left(\dfrac{1\ \text{mol glucose}}{180.2\ \text{g glucose}}\right)}$$

$$= 0.985$$

We compute the equilibrium vapor pressure of water over the solution from Raoult's law (Equation 14-3):

$$P_{H_2O} = X_{H_2O} P_{H_2O}^\circ = (0.985)(760\ \text{torr}) = 749\ \text{torr}$$

The vapor pressure lowering of water produced by the dissolved glucose is

$$P_{H_2O}^\circ - P_{H_2O} = (760 - 749)\ \text{torr} = 11\ \text{torr}$$

The lowering of the solvent vapor pressure by a solute is small for solutions of low to moderate solute concentration.

14-5. COLLIGATIVE PROPERTIES OF SOLUTIONS DEPEND ONLY ON THE SOLUTE PARTICLE CONCENTRATION

There are several properties of solutions that are known as *colligative properties*. Colligative properties depend primarily upon the ratio of the number of solute particles to the number of solvent particles and not upon the chemical nature of the solute. The major colligative properties of solutions are *vapor pressure lowering, boiling point elevation, freezing-point depression,* and *osmotic pressure*.

It is convenient for the analysis of colligative properties to introduce the molality concentration scale for solutes. We define the *molality m* of a solute as the number of moles of solute per 1000 g of solvent:

Molality is proportional to the mole fraction of the solute in a dilute solution.

$$\text{molality} = \frac{\text{moles of solute}}{1000\text{ g of solvent}} = \frac{\text{moles of solute}}{\text{kilogram of solvent}}$$

For example, a solution prepared by dissolving 1.00 mol of sodium chloride (58.44 g) in 1.00 kg of water is 1.00 molal (1.00 m) in NaCl.

Example 14-3: Compute the molality of a solution prepared by dissolving 10.0 g of NaCl(*s*) in 500 g of water.

Solution: The number of moles of NaCl in 10.0 g is

$$(10.0\text{ g})\left(\frac{1\text{ mol NaCl}}{58.44\text{ g NaCl}}\right) = 0.171\text{ mol}$$

When 0.171 mol of NaCl is dissolved in 0.500 kg of water, the molality of NaCl in the resulting solution is

$$m = \frac{0.171\text{ mol}}{0.500\text{ kg}} = 0.342\,\frac{\text{mol}}{\text{kg}} = 0.342\text{ m}$$

The molality concentration scale, which is denoted by *m*, is not the same as the molarity concentration scale, denoted by *M* (Section 3-10). To prepare 500 mL of a 0.342 M solution of NaCl in water, we dissolve 0.171 mol of NaCl in less than 500 mL of water and dilute the resulting solution with enough water to yield exactly 500 mL *of solution*. Compare this procedure with that described in Example 14-3 for the preparation of a 0.342 m solution.

It is essential for the understanding of colligative properties to recognize that it is the ratio of the number of solute particles to the number of solvent particles that determines the magnitude of a colligative effect. A 0.10 m aqueous NaCl solution has *twice as many* solute particles per mole of water as a 0.10 m aqueous glucose solution because NaCl is an electrolyte and dissociates completely in water to $Na^+(aq)$ and $Cl^-(aq)$, whereas glucose exists in solution as intact $C_6H_{12}O_6$ molecules. A 0.10 m aqueous acetic acid solution exists primarily as undissociated acetic acid, and thus the number of solute particles per unit volume in the 0.10 m solution is roughly the same as in a 0.10 m glucose solution. In contrast to the situation in water, acetic acid in benzene exists predominantly as hydrogen-bonded *dimers* (pairs of identical molecules bonded together) of the type

$$\begin{array}{c} \quad O\cdots H—O \\ CH_3—C \qquad\qquad C—CH_3 \\ \quad O—H\cdots O \end{array}$$

Therefore, a 0.10 m solution of $HC_2H_3O_2$ in benzene has only half as many solute particles as a 0.10 m solution of, say, acetone, which forms no dimers, in benzene. We thus distinguish between the molality, denoted by m, and the *colligative molality*, denoted by m_c, of a solute. The distinction is illustrated numerically in Table 14-3.

Table 14-3 Comparison of molality and colligative molality

Solute	*Solvent*	*Solute molality, m*	*Solute particles per formula unit*	*Colligative molality, m_c*
$C_6H_{12}O_6$	H_2O	0.10	1	0.10
NaCl	H_2O	0.10	2	0.20
$CaCl_2$	H_2O	0.10	3	0.30
$HC_2H_3O_2$	H_2O	0.10	1	0.10
$HC_2H_3O_2$	C_6H_6	0.10	$\frac{1}{2}$	0.050

Example 14-4: A 1.0-mol sample of each of the following substances is dissolved in 1000 g of water. Determine the colligative molality of each of the resulting solutions:

$$CH_3OH \qquad AgNO_3 \qquad K_2SO_4$$

Solution: Molality is the number of moles of solute per kilogram of solvent, and so the molality of each of these substances is 1.0 m. Methanol is a nonelectrolyte, and thus the colligative molality is the same as the molality. Silver nitrate is a strong electrolyte in water, yielding $Ag^+(aq)$ and $NO_3^-(aq)$, and thus $m_c = 2$ m $= 2.0$ m_c. Potassium sulfate is a strong electrolyte in water, yielding $2K^+(aq) + SO_4^{2-}(aq)$, and thus $m_c = 3$ m $= 3.0$ m_c.

14-6. NONVOLATILE SOLUTES INCREASE THE BOILING POINT OF A LIQUID

Recall that the boiling point is the temperature at which the equilibrium vapor pressure equals the atmospheric pressure. We know from Section 14-3 that the equilibrium vapor pressure of the solvent over a solution containing a nonvolatile solute is less than that for the pure solvent at the same temperature. Therefore, the temperature at which the equilibrium vapor pressure reaches atmospheric pressure is higher for the solution than for the pure solvent (Figure 14-7). In other words, the boiling point of the solution, T_b, is higher than the boiling point of the pure solvent, T_b°. Figure 14-7 suggests that the magnitude of the boiling point elevation, $T_b - T_b^\circ$, is proportional to the magnitude of the vapor pressure lowering, $P_1^\circ - P_1$:

$$(T_b - T_b^\circ) \propto (P_1^\circ - P_1)$$

The vapor pressure lowering is proportional to the mole fraction of solute or to the colligative molality of the solute:

$$(P_1^\circ - P_1) \propto m_c$$

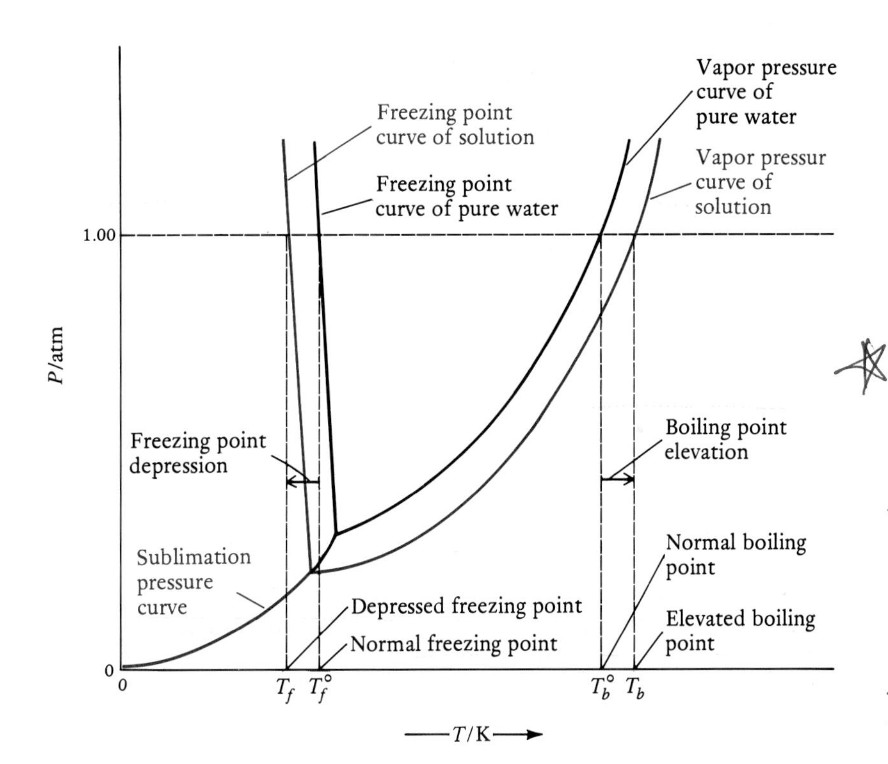

Figure 14-7 Phase diagrams for pure water (black lines) and for water containing a nonvolatile solute (red lines). The presence of the solute lowers the vapor pressure of the solvent. The reduced vapor pressure of the solvent results in an *increase* in the boiling point of the solution relative to that of the pure solvent and a *decrease* in the freezing point of the solution relative to that of the pure solvent.

Table 14-4 Boiling point elevation constants (K_b) and freezing point depression constants (K_f) for various solvents

Solvent	*Boiling point/°C*	$K_b/K \cdot m_c^{-1}$	*Freezing point/°C*	$K_f/K \cdot m_c^{-1}$
water	100.00	0.52	0.00	1.86
acetic acid	117.9	2.93	16.6	3.90
benzene	80.0	2.53	5.50	5.10
chloroform	61.2	3.63	−63.5	4.68
cyclohexane	80.7	2.79	6.5	20.2
nitrobenzene	210.8	5.24	5.7	6.87
camphor	208.0	5.95	179.8	40.0

Because the boiling point elevation is proportional to the vapor pressure lowering, it is also proportional to the colligative molality. Thus we can write

$$T_b - T_b^\circ = K_b m_c \tag{14-4}$$

where K_b is a proportionality constant. Equation (14-4) tells us that the increase in the boiling point of a solution containing a nonvolatile solute is directly proportional to the solute (particle) concentration, that is, directly proportional to the colligative molality. The value of K_b depends only upon the solvent and is called the *boiling point elevation constant.* Values of K_b for several solvents are given in Table 14-4.

The magnitude of the boiling point elevation is small unless a solution is concentrated. For example, we see from Equation (14-4) and Table 14-4 that $T_b - T_b^\circ$ for 1.0 m solution of glucose in water is

Note that the boiling point elevation is a small effect for a 1.0 m_c solution.

$$T_b - T_b^\circ = [0.52\ \mathrm{K \cdot m_c^{-1}}](1.0\ \mathrm{m_c}) = 0.52\ \mathrm{K}$$

14-7. SOLUTES DECREASE THE FREEZING POINT OF A LIQUID

How the freezing point of a liquid is depressed by a solute can be understood with the aid of Figure 14-7. When the temperature of an aqueous solution is lowered to the freezing point, the solid that separates out is almost always pure ice. For example, the ice that forms from seawater is pure; impurities may be trapped in occlusions, which are air or liquid pockets, but the solid phase is pure. The lowering of the vapor pressure of water by the solute causes the equilibrium vapor pressure curve to intersect the equilibrium sublimation pressure curve at a lower temperature than in the case of pure water (Figure 14-7). The freezing point curve begins at the intersection of the sublimation pressure and vapor pressure curves in Figure 14-7, that is, at the triple point. Therefore, the freezing point curve for the solution lies at lower temperatures than the freezing point curve of the pure liquid.

Solutes increase the boiling point and decrease the melting point of the solvent.

By arguments analogous to those used to obtain the boiling point elevation equation, the magnitude of the freezing point depression produced by a solute with colligative molality m_c is found to be proportional to m_c:

freezing pt. depression constant

$$T_f^\circ - T_f = K_f m_c \qquad \Delta T_f = K_f m_c \tag{14-5}$$

where T_f° is the freezing point of the pure solvent, T_f is the freezing point of the solution ($T_f^\circ > T_f$), and K_f is the *freezing point depression constant*. The value of K_f depends only upon the solvent.

The value of the freezing point depression constant of water is $1.86\ \mathrm{K \cdot m_c^{-1}}$ (Table 14-4). Thus we predict that an aqueous solution with a colligative molality of $0.50\ \mathrm{m}_c$ has a freezing point depression of

$$T_f^\circ - T_f = K_f m_c = (1.86\ \mathrm{K \cdot m_c^{-1}})(0.50\ \mathrm{m_c}) = 0.93\ \mathrm{K}$$

and the freezing point is $-0.93°\mathrm{C}$.

Example 14-5: One method used to melt ice on streets and sidewalks is to spread rock salt (NaCl) crystals on the ice. The solubility of NaCl in liquid water around 0 °C is 4.8 m. Explain why NaCl can be used to melt ice and estimate the lowest temperature at which NaCl can melt ice.

Solution: Sodium chloride in water completely dissociates into $Na^+(aq)$ and $Cl^-(aq)$, and so the colligative molality is twice the molality. Thus the freezing point depression of a saturated aqueous NaCl solution is

$$T_f^\circ - T_f = K_f m_c = (1.86\ \mathrm{K \cdot m_c^{-1}})(9.6\ \mathrm{m_c}) = 18\ \mathrm{K}$$

The freezing point of the solution, therefore, is $-18°\mathrm{C}$. On contact with ice, salt dissolves to form a saturated aqueous solution with a freezing point of $-18°\mathrm{C}$. Above $-18°\mathrm{C}$ spreading salt on ice causes the ice to melt as it forms a concentrated aqueous salt solution. When the ice temperature is below $-18°\mathrm{C}$, spreading salt on it will not melt the ice because ice freezes out from the saturated $NaCl(aq)$ solution.

An 18 °C temperature decrease can be produced by the addition of salt to ice. This effect is used in the preparation of homemade ice cream to obtain a low enough temperature to solidify cream.

14-8. THE FREEZING POINT DEPRESSION EFFECT IS THE BASIS OF THE ACTION OF ANTIFREEZE

The most commonly used *antifreeze* is ethylene glycol,

```
          H   H
   ··     |   |   ··
H—O—C—C—O—H
   ··     |   |   ··
          H   H
```

whose boiling point is 197°C and whose freezing point is −17.4°C. The addition of ethylene glycol to water depresses the freezing point and elevates the boiling point of the water. A 50% by volume ethylene glycol in water solution has a freezing point of −36°C.

Example 14-6: Estimate the freezing point of a 5.0 m_c solution of ethylene glycol in water.

Solution: The freezing point of the solution is computed from Equation (14-5):

$$T_f^\circ - T_f = K_f m_c$$
$$= (1.86\ \mathrm{K \cdot m_c^{-1}})(5.0\ \mathrm{m_c}) = 9.3\ \mathrm{K}$$

The freezing point of the solution is 263.9 K, or −9.3°C.

The freezing point of aqueous ethylene glycol solutions as a function of the volume percent of ethylene glycol is given in Figure 14-8.

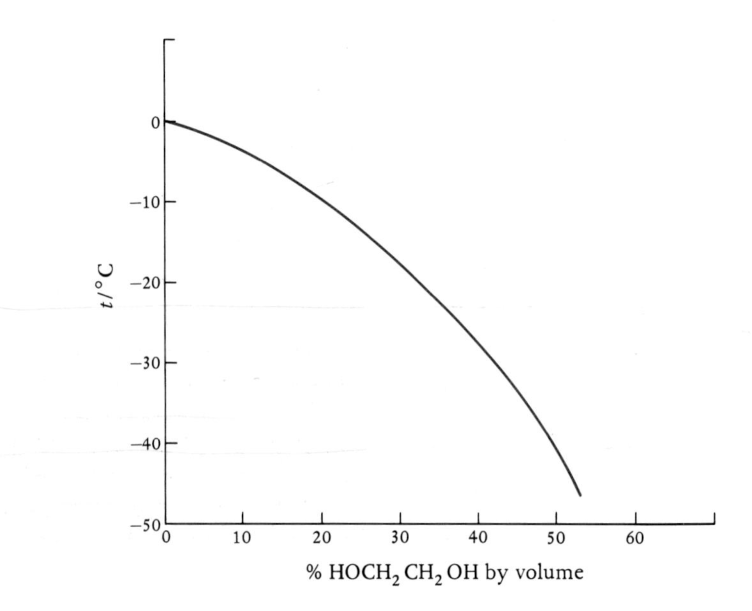

Figure 14-8 The freezing-point curve for aqueous ethylene glycol (antifreeze) solutions.

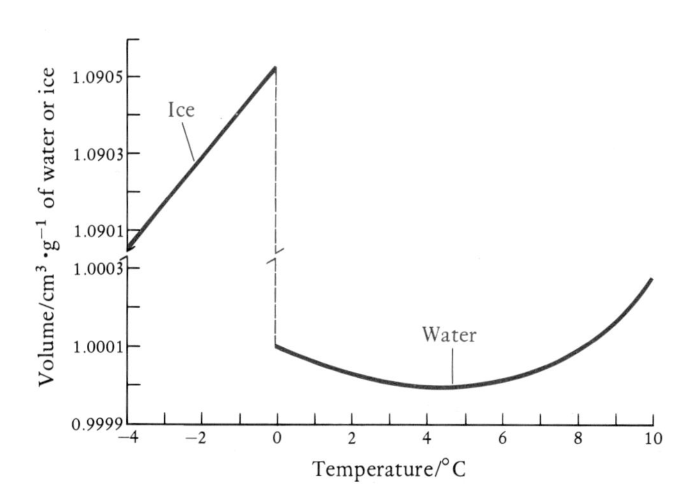

Figure 14-9 Plot of the volume of 1 g of water as a function of temperature. Note the 9 percent increase in volume when liquid water freezes.

The effectiveness of ethylene glycol as an antifreeze is a result of several factors, among which are its high boiling point and chemical stability and the tendency of the ice that freezes out of the solution to form a slushy mass rather than a solid block. In the absence of a sufficient amount of antifreeze, the 9 percent volume expansion of water on freezing generates a force of 30,000 lb/in^2 at $-22\,^\circ$C, which is sufficient to rupture a radiator or even a metal engine block (Figure 14-9).

The freezing point of seawater is about $-1.85\,^\circ$C, and this is the temperature of the seawater surrounding an iceberg (Figure 14-10). There are two species of fish that live in the cold waters of the Ross Sea of Antarctica near the sea ice. On the basis of the total concentration of solutes dissolved in the blood serum of these fish, the freez-

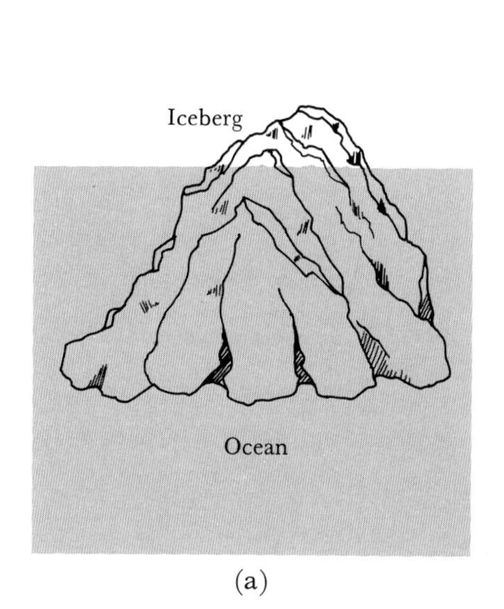

(a)

(b)

Figure 14-10 Over 90 percent of an iceberg is below the surface of the water. (a) Shows a diagram of this distribution of mass.

Photo taken in Antarctica of one of the species of fish that has "antifreeze" proteins in its blood. (Photo courtesy Prof. Robert Feeney.)

ing point of the serum should be $-1.46°C$. This means that the blood should freeze in the $-1.85°C$ water, but the fish is protected from freezing by "antifreeze" proteins, which have an enhanced capacity to lower the freezing point of water. The freezing point depression of solutions containing the antifreeze proteins is much greater than that predicted by Equation (14-5). The freezing point of the fish blood serum after removal of all the salts (but not of the antifreeze proteins) is $-0.60°C$. The concentration of antifreeze proteins in the fish blood is about $3 \times 10^{-4}\ m_c$. According to Equation (14-5) the predicted freezing point depression of a $3 \times 10^{-4}\ m_c$ solution of the antifreeze protein is

$$T_f^\circ - T_f = (1.86\ \mathrm{K \cdot m_c^{-1}})(3 \times 10^{-4}\ \mathrm{m_c}) = 0.0006\ \mathrm{K}$$

and the solution should freeze at $-0.0006°C$. The observed freezing point of the antifreeze protein solution is thus 1000 times $(-0.60°C/-0.0006°C)$ greater than that predicted by Equation (14-5). How the antifreeze proteins work is not entirely clear. A hypothesis is that they are adsorbed onto the surfaces of ice crystal nuclei, thereby stopping further growth of the crystals.

14-9. ONLY SOLVENT MOLECULES CAN PASS THROUGH A SEMIPERMEABLE MEMBRANE

Suppose we place pure water in one beaker and an equal volume of seawater in another beaker and then place both beakers under a bell jar, as shown in Figure 14-11(a). We observe that as time passes the

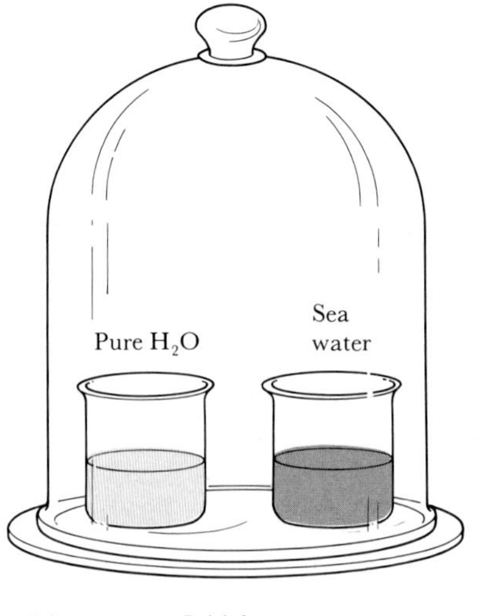

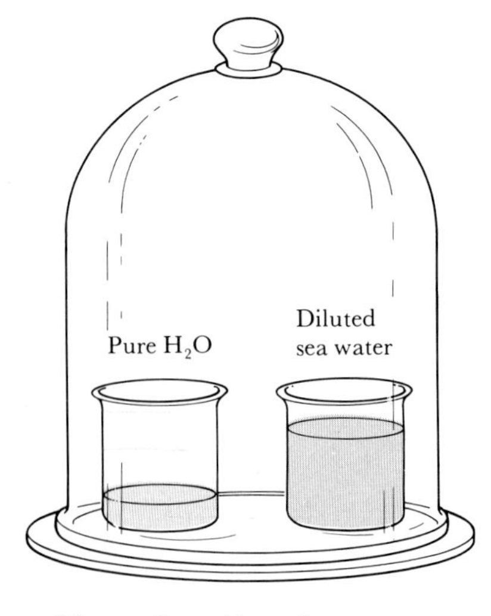

Figure 14-11 (a) One beaker contains pure water, and the other contains seawater. The equilibrium vapor pressure of the pure water is greater than that of the water over the seawater solution. (b) As time passes, pure water is transferred via the vapor phase from the beaker containing pure water to the beaker containing seawater, thereby diluting the seawater. If we wait long enough, all the pure liquid water will transfer to the seawater beaker and the final concentration of salts in the seawater will be half the initial value.

volume of pure water decreases and the volume of seawater increases (Figure 14-11b). The pure water has a higher equilibrium vapor pressure than the seawater, and thus the rate of condensation of the water into the seawater is greater than the rate of evaporation of the water from the seawater. The net effect is the transfer of water, via the vapor phase, from the beaker with pure water to the beaker with seawater. This transfer continues until no pure liquid water remains, and the seawater ends up diluted.

If pure water and seawater are separated by a membrane that is permeable to water but not to the ions in seawater, then the water passes from the pure water side of the membrane to the seawater side (Figure 14-12). Such a membrane is called a *semipermeable membrane,* and the tendency of the water to pass through the membrane is called *escaping tendency*. The escaping tendency of water from pure water is greater than the escaping tendency of water from seawater. As water passes through the membrane to the seawater side, the escaping tendency of the water in the seawater increases, not only because the seawater is being diluted but also because of the increased pressure on the seawater side of the membrane. This pressure increase arises from the hydrostatic pressure head of the solution. The column of seawater rises until the escaping tendency of the pure water is equal to the escaping tendency of the water in the seawater. At this point, equilibrium is reached and the column no longer rises. The hydrostatic pressure head produced in this process is called the *osmotic pressure* (Figure 14-12). The spontaneous passage of solvent from one solution to a more concentrated solution is called *osmosis.*

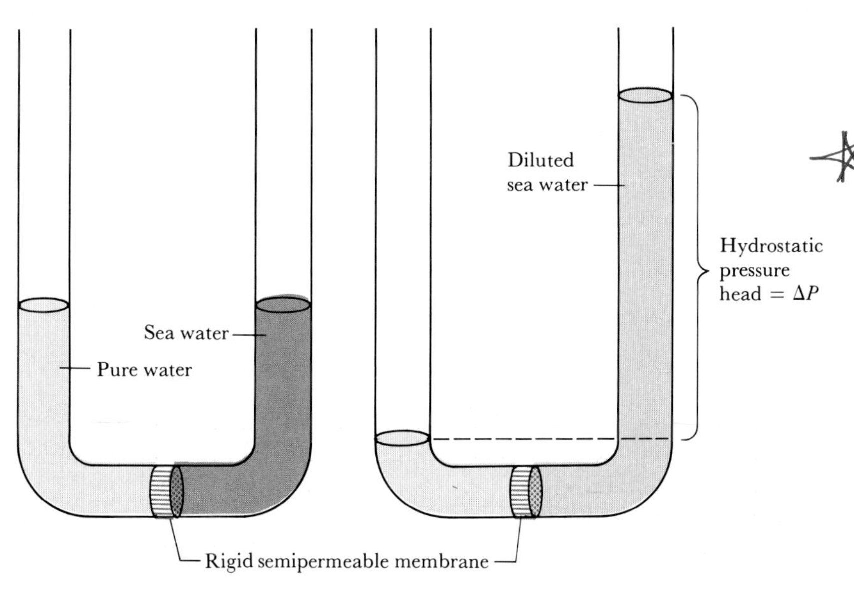

Figure 14-12 Passage of water through a rigid semipermeable membrane separating pure water from seawater. The water passes through the membrane until the escaping tendency of the pure water equals the escaping tendency of the water from the seawater. The escaping tendency of water from the seawater side of the membrane increases as the seawater is diluted and as a result of the increased hydrostatic pressure head on the seawater, which results from the increase in the seawater column height.

Osmotic pressure has important consequences in biology as well as chemistry. The osmotic pressure π of a solution is given by the equation

$$\pi = RTM_c \tag{14-6}$$

where R is the gas constant, $0.0821\ \text{L}\cdot\text{atm}\cdot\text{K}^{-1}\cdot\text{mol}^{-1}$; T the absolute temperature; and M_c the *colligative molarity,* that is, the molarity multiplied by the number of particles per formula unit.

Example 14-7: Estimate the osmotic pressure of a 0.100 M solution of glucose in water at 37°C.

Solution: Using Equation (14-6) we have

$$\begin{aligned}\pi &= RTM_c \\ &= (0.0821\ \text{L}\cdot\text{atm}\cdot\text{K}^{-1}\cdot\text{mol}^{-1})(310\ \text{K})(0.100\ \text{mol}\cdot\text{L}^{-1}) \\ &= 2.55\ \text{atm} = 1930\ \text{torr}\end{aligned}$$

Note that the osmotic pressure is large compared to atmospheric pressure.

Example 14-8: As a rough approximation, seawater can be regarded as 0.55 M in NaCl. Estimate the osmotic pressure of seawater at 15°C.

Solution: The colligative molarity of seawater is approximately 2×0.55 M because NaCl dissociates in water to yield two ions per formula unit. Thus the osmotic pressure of seawater is

$$\begin{aligned}\pi &= RTM_c \\ &= (0.0821\ \text{L}\cdot\text{atm}\cdot\text{K}^{-1}\cdot\text{mol}^{-1})(288\ \text{K})(2 \times 0.55\ \text{mol}\cdot\text{L}^{-1}) \\ &= 26\ \text{atm}\end{aligned}$$

Note that the osmotic pressure of seawater is 26 times higher than atmospheric pressure.

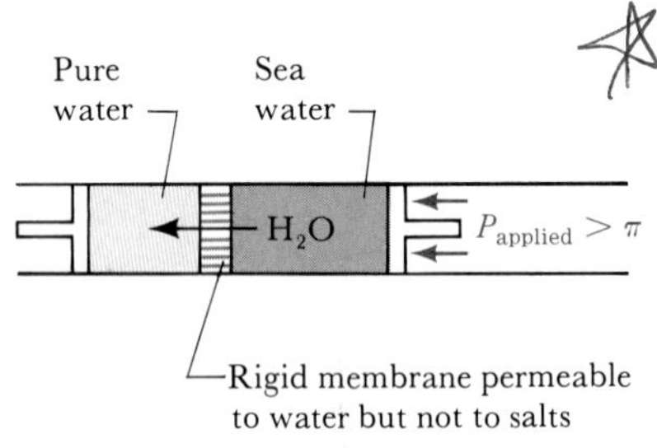

Figure 14-13 Reverse osmosis. A rigid semipermeable membrane separates pure water from seawater. A pressure in excess of the osmotic pressure of seawater (26 atm) is applied to the seawater, and this increases the escaping tendency of water from the seawater to a value above that of pure water. Under these conditions, the net flow of water is from the seawater side through the semipermeable membrane to the pure water side. The net effect is the production of fresh water from seawater.

If a pressure in excess of 26 atm is applied to seawater at 15°C, then the escaping tendency of the water in the seawater will exceed that of pure water. Thus pure water can be obtained from seawater by using a rigid semipermeable membrane and an applied pressure in excess of the osmotic pressure. This process is known as *reverse osmosis* (Figure 14-13). Reverse osmosis units are commercially available and are used to obtain fresh water from brine.

One unit of a reverse osmosis desalination system produces about 2770 m^3 (600,000 gal) per day of high-quality drinking water from deep, brackish wells.

14-10. THE MOLECULAR MASS OF PROTEINS CAN BE DETERMINED BY OSMOTIC PRESSURE MEASUREMENTS

Osmotic pressure values are relatively high, and this fact makes osmotic pressure measurement an especially good method for determining the molecular mass of proteins. Proteins have large molecular masses and therefore yield a relatively small number of solute particles for a given dissolved mass. The only colligative effect sufficiently sensitive to provide useful molecular information on proteins is osmotic pressure.

Example 14-9: A 4.00-g sample of human hemoglobin was dissolved to make 0.100 L of solution, and the osmotic pressure of the solution at 7°C was found to be 0.0132 atm. Estimate the molecular mass of the hemoglobin.

Solution: From Equation (14-6), the concentration of the hemoglobin in the aqueous solution is

$$M_c = \frac{\pi}{RT} = \frac{0.0132\ \text{atm}}{(0.0821\ \text{L}\cdot\text{atm}\cdot\text{K}^{-1}\cdot\text{mol}^{-1})(280\ \text{K})}$$

$$= 5.74 \times 10^{-4}\ \text{mol}\cdot\text{L}^{-1}$$

The molecular mass can be computed from the concentration of the protein because the dissolved mass is known. We have that

$$5.74 \times 10^{-4}\ \text{mol}\cdot\text{L}^{-1} \Bumpeq \frac{4.00\ \text{g}}{0.100\ \text{L}} = 40.0\ \text{g}\cdot\text{L}^{-1}$$

and therefore

$$5.74 \times 10^{-4}\ \text{mol} \Bumpeq 40.0\ \text{g}$$

By dividing both sides of this stoichiometric correspondence by 5.74×10^{-4}, we find that

$$1\ \text{mol} \Bumpeq 69{,}700\ \text{g}$$

The molecular mass of the hemoglobin is 69,700. Protein molecular masses can be as large as 1,000,000.

Example 14-10: A sample of human hemoglobin with a molecular mass of 69.7×10^3 (Example 14-9) was found to contain 0.321 percent iron by mass. Use this result to compute the number of iron atoms in a hemoglobin molecule.

Solution: The percent by mass of an element in a compound is given by the formula

$$\begin{pmatrix}\text{\% by mass of}\\ \text{an element}\\ \text{in a compound}\end{pmatrix} = \frac{\begin{pmatrix}\text{atomic mass}\\ \text{of the element}\end{pmatrix}}{\begin{pmatrix}\text{molecular mass}\\ \text{of the compound}\end{pmatrix}} \times \begin{pmatrix}\text{number of}\\ \text{atoms of the}\\ \text{element in}\\ \text{the compound}\end{pmatrix} \times 100$$

We know all the quantities in this equation except the number of atoms of the element in one molecule of the compound. If we call this number n, then we have

$$0.321 = \left(\frac{55.85}{6.97 \times 10^4}\right)(n)(100)$$

$$n = \frac{(0.321)(6.97 \times 10^4)}{(55.85)(100)} = 4.00$$

There are four iron atoms per human hemoglobin molecule.

14-11. OSMOTIC PRESSURE PLAYS A MAJOR ROLE IN LIVING SYSTEMS BY KEEPING CELLS INFLATED

The colligative concentration of the solution inside a typical biological cell is approximately 0.3 M_c. Most animal cells have about the same internal colligative molarity as the extracellular fluid in which the cells exist. Both plant cells and animal cells have membranes that are permeable to water but not, for example, to sucrose.

Water passes spontaneously through a cell membrane from the side with the lower colligative molarity (higher water escaping tendency) to the side with the higher colligative molarity (lower water escaping tendency). The entry of water into an animal cell causes the cell to expand, and the exit of water from the cell causes the cell to contract. The cell assumes its normal volume when it is placed in a solution with a colligative molarity of 0.3 M_c. More concentrated solutions cause the cell to contract, and less concentrated solutions cause it to expand (Figure 14-14). When cells are placed in distilled water at 27°C, equilibrium is obtained at an internal cell pressure equal to the osmotic pressure of a 0.30 M_c solution, that is,

The greater the colligative molarity is, the less the escaping tendency of the water from the solution.

$$\pi = RTM_c = (0.0821\ \text{L}\cdot\text{atm}\cdot\text{K}^{-1}\cdot\text{mol}^{-1})(300\ \text{K})(0.30\ \text{mol}\cdot\text{L}^{-1})$$
$$= 7.4\ \text{atm}$$

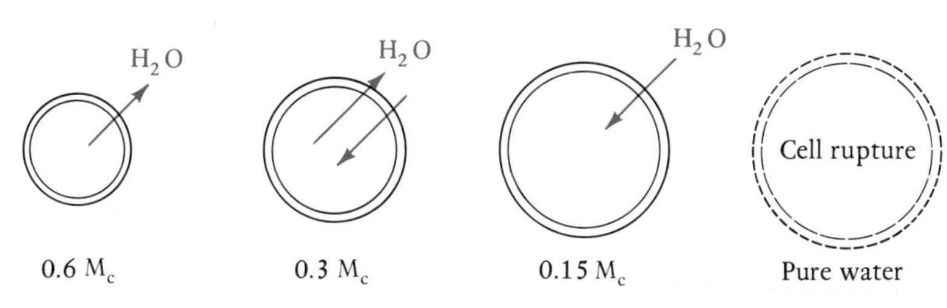

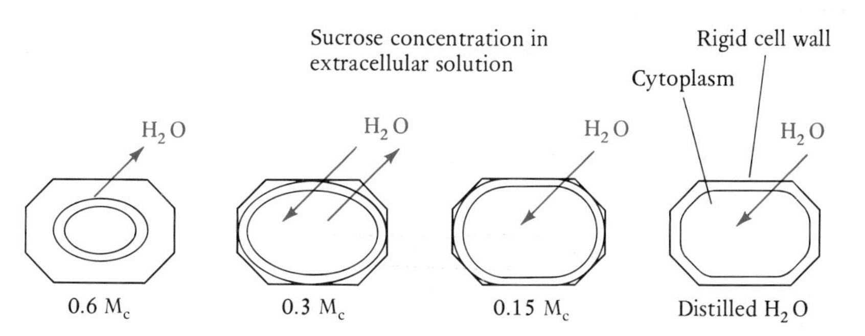

Figure 14-14 Osmotic phenomena in living cells. Animal cells contract as a result of water loss when placed in a solution with a colligative molarity greater than 0.3 M_c and expand when placed in a solution with a colligative molarity less than 0.3 M_c because they take up water. Plant cells act like automobile tires—they become rigid when inflated but still retain their basic shape when deflated—whereas animal cells behave more like balloons.

Osmosis enables non-woody plants to stand erect. These are mermaid's cups.

Because a pressure of 7.4 atm cannot be maintained by animal cell membranes, the cells burst. Plant cell walls are rigid, and they can tolerate a pressure of 7.4 atm. Thus the entry of water into plant cells gives nonwoody plants the rigidity required to stand erect.

14-12. THE EQUILIBRIUM CONCENTRATION OF A DISSOLVED GAS IS PROPORTIONAL TO THE PRESSURE OF THE GAS OVER THE SOLUTION

Henry's law states that the solubility of a gas in a liquid is directly proportional to the equilibrium pressure of the gas over the solution:

$$P_{gas} = k_h M_{gas} \tag{14-7}$$

where P_{gas} is the equilibrium partial pressure of the gas over the solution, k_h is the *Henry's law constant* for the gas, and M_{gas} is the molarity of the gas in the liquid. Henry's law follows from the fact that, at equilibrium, the rate of dissolution of a gas in a solvent is equal to the rate of escape of the dissolved gas from the solution. If the gas pressure is doubled, then the number of gas molecules that strike the solution surface per unit time is doubled and thus the concentration of dissolved gas is doubled (Figure 14-15).

The dissolution of gas in a liquid is a dynamic equilibrium.

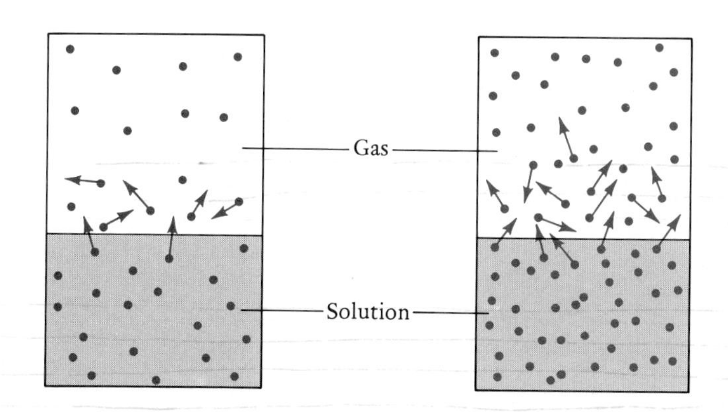

Figure 14-15 The solubility of a gas in a liquid is proportional to the pressure of the gas over the solution. If the gas pressure is doubled, then at equilibrium the number of molecules entering and leaving the solution is doubled and the concentration of dissolved gas is doubled.

When the equilibrium partial pressure of oxygen over water is plotted as a function of the concentration of dissolved oxygen, the result is a straight line (Figure 14-16), as predicted by Henry's law.

Example 14-11: Compute the concentration of O_2 in water that is in equilibrium with air at 25°C. The Henry's law constant for O_2 in water at 25°C is 780 atm·M^{-1}.

Solution: The partial pressure of O_2 in the atmosphere is 0.20 atm, and thus, from Equation (14-7),

$$M_{O_2} = \frac{0.20\text{ atm}}{780\text{ atm}\cdot\text{M}^{-1}} = 2.6 \times 10^{-4}\text{ M}$$

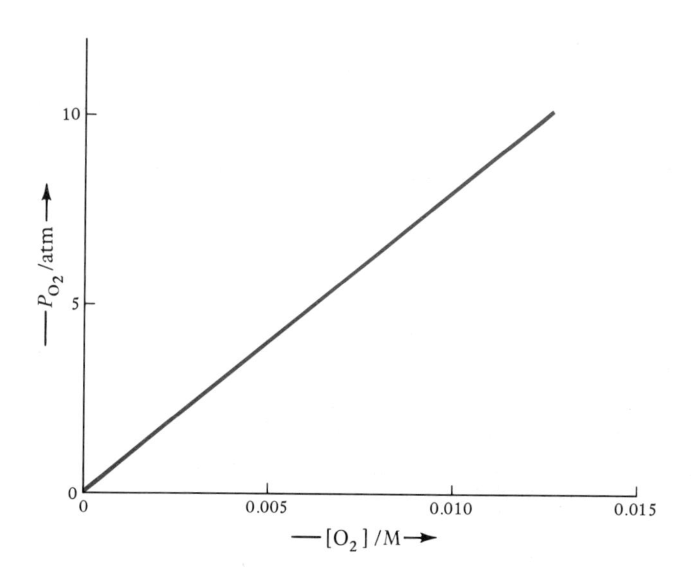

Figure 14-16 The equilibrium pressure of oxygen over water at 25°C plotted against the concentration of oxygen in the water The linear plot confirms that Henry's law holds for oxygen in water over the range 0 to 10 atm $O_2(g)$.

Table 14-5 Henry's law constants for gases in water at 25°C

Gas	$k_h/atm \cdot M^{-1}$
He	2.7×10^3
N_2	1.6×10^3
O_2	7.8×10^2
CO_2	29
H_2S	10

Henry's law constants for several common gases are given in Table 14-5. The smaller the value of this constant for a gas, the greater the solubility of the gas because $M_{gas} = P_{gas}/k_h$.

Carbonated beverages are pressurized with CO_2 gas at a pressure above 1 atm; champagne is pressurized at 4 to 5 atm. The CO_2 pressure is responsible for the rush of escaping gas that causes the "pop" when the carbonated drink container is opened. The loss of CO_2 from the solution begins because the average atmospheric partial pressure of CO_2 is only 3×10^{-4} atm. The bubbles that form in the liquid are mostly CO_2 plus some water vapor at about 1 atm total pressure.

The air that is breathed by a diver under water is significantly above atmospheric pressure because the diver must exhale the air into an environment that has a pressure greater than atmospheric pressure. For example, at a depth of 90 ft the pressure is about 3.7 atm, and so the diver must breathe air at 3.7 atm. At this pressure the solubilities of N_2 and O_2 in the blood are 3.7 times as great as at sea level. If a diver ascends too rapidly, then the sudden pressure drop causes the dissolved nitrogen to form numerous small gas bubbles in the blood. This phenomenon, which is extremely painful and can result in death, is called the bends because it causes the afflicted person to bend over in pain. Oxygen, which is readily metabolized, does not accumulate as bubbles in the blood. However, pure oxygen cannot be used for breathing because at high oxygen pressures the need to breathe is greatly reduced and so CO_2 accumulates in the bloodstream and leads to CO_2 asphyxiation. The solution to this

Active fish prefer cooler water because of the higher concentration of dissolved oxygen and nutrients.

problem was proposed by the chemist Joel Hildebrand and consists of substituting helium for nitrogen. Helium is only about half as soluble in blood as nitrogen, and thus the magnitude of the problem is cut in half. Divers' "air" tanks contain a mixture of He and O_2 adjusted so that the pressure of O_2 is about 0.20 atm at maximum dive depth.

The solubility of gases in liquids decreases with increasing temperature. This effect is depicted in Figure 14-17 for O_2 dissolved in water that is in equilibrium with air. Cold water in equilibrium with air has a higher concentration of dissolved oxygen than warm water. The decreased solubility of oxygen with increased temperature is the reason why most fish, especially the more active ones, prefer cooler water. Tuna are especially sensitive to water temperature; they greatly prefer seawater at 14 to 15°C. Tuna fishermen use this knowledge to locate schools of tuna. If the water temperature is below 14°C or above 15°C, then it is unlikely that tuna will be found.

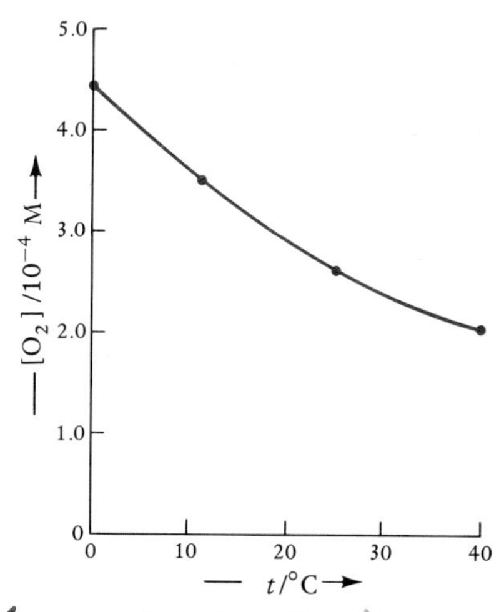

Figure 14-17 Solubility of oxygen in water as a function of temperature. The partial pressure of O_2 in each case is 0.20 atm, which is its partial pressure in air. Note that the solubility of O_2 *decreases* as the temperature of the solution increases. This result is generally observed for the solubility of a gas in a liquid. The decreasing solubility of O_2 in water with increasing temperature has important ramifications for plant and animal life.

SUMMARY

The key point in understanding the colligative properties of solutions is that the equilibrium vapor pressure of a solvent is reduced when a solute is dissolved in it. If a solution is ideal, then the equilibrium vapor pressure of any component is given by Raoult's law. Ideal solutions result when the interactions between unlike molecules in the solution are the same as the interactions between like molecules. If this is not the case, then the solution is not ideal and deviations from Raoult's law are observed.

The equilibrium vapor pressure of water over an aqueous solution is a direct measure of the escaping tendency of water from the solution. The higher the vapor pressure, the greater the escaping tendency. The escaping tendency of water from the liquid phase can be decreased by dissolving a solute in the water or by decreasing the temperature of the solution. The escaping tendency of water from the liquid phase can be increased by decreasing the solute concentration, by increasing the temperature, or by increasing the total pressure on the liquid phase.

Colligative properties of solutions (vapor pressure lowering, boiling point elevation, freezing point depression, and osmotic pressure) depend only on the solute particle concentration and are independent of the chemical nature of the solute.

Osmotic pressure is the largest of the colligative effects and is used to determine the molecular mass of proteins. Osmotic pressure effects are important in biological systems in that osmotic pressure keeps living cells inflated.

Henry's law states that the solubility of a gas in a liquid is directly proportional to the pressure of the gas over the solution. Gas solubility decreases with increasing temperature.

TERMS YOU SHOULD KNOW

solution 523
homogeneous 523
components 524
solvent 524
solute 524
rate of solution 525
rate of crystallization 525
dynamic equilibrium 525
saturated solution 526
solubility 526
unsaturated solution 526
ideal solution 528
mole fraction, X 528
Raoult's law 530
vapor pressure lowering 530
colligative property 531
boiling point elevation 531
freezing point depression 531
molality, m 531
dimer 532
colligative molality, m_c 532
boiling point elevation constant, K_b 534
freezing point depression constant, K_f 535
antifreeze 536
semipermeable membrane 539
escaping tendency 539
osmotic pressure, π 539
osmosis 539
colligative molarity, M_c 540
reverse osmosis 540
Henry's law 544
Henry's law constant, k_h 544

EQUATIONS YOU SHOULD KNOW HOW TO USE

$$X_1 = \frac{n_1}{n_1 + n_2} \qquad (14\text{-}1) \qquad \text{(mole fraction)}$$

$$P_1 = X_1 P_1^\circ \qquad (14\text{-}3) \qquad \text{(Raoult's law)}$$

$$T_b - T_b^\circ = K_b m_c \qquad (14\text{-}4) \qquad \text{(boiling point elevation)}$$

$$T_f^\circ - T_f = K_f m_c \qquad (14\text{-}5) \qquad \text{(freezing point depression)}$$

$$\pi = RTM_c \qquad (14\text{-}6) \qquad \text{(osmotic pressure)}$$

$$P_{gas} = k_h M_{gas} \qquad (14\text{-}7) \qquad \text{(Henry's law)}$$

PROBLEMS

MOLE FRACTION

14-1. Calculate the mole fractions in a solution that is made up of 25.0 g of ethyl alcohol, C_2H_5OH, and 40.0 g of water.

14-2. A solution of 40% formaldehyde, H_2CO, 10% methyl alcohol, CH_3OH, and 50% water by mass is called formalin. Calculate the mole fractions of formaldehyde, methyl alcohol, and water in formalin. Formalin is used to disinfect dwellings, ships, storage houses, and so forth.

14-3. Describe how you would prepare an aqueous solution of acetone, CH_3COCH_3, in which the mole fraction of acetone is 0.19.

14-4. In a solution of formaldehyde, H_2CO, in water, the mole fraction of formaldehyde is 0.186. Calculate the mole fraction of water in the mixture.

RAOULT'S LAW

14-5. The vapor pressure of pure water at 37°C is 47.1 torr. Use Raoult's law to estimate the vapor pressure of an aqueous solution at 37°C containing 10.0 g of glucose, $C_6H_{12}O_6$, dissolved in 100.0 g of water. Also compute the vapor pressure lowering.

14-6. Water at 37°C has a vapor pressure of 47.1 torr. Calculate the vapor pressure lowering if 50.0 g of glycerin, $C_3H_8O_3$, is added to 100.0 mL of water. The density of water at 37°C is 0.993 $g \cdot mL^{-1}$.

14-7. The vapor pressure of pure water at 25°C is 23.76 torr. Use Raoult's law to estimate the vapor pressure of an aqueous solution at 25°C containing 30.0 g of sucrose, $C_{12}H_{22}O_{11}$, dissolved in 195 g of water. Also calculate the vapor pressure lowering.

14-8. The vapor pressure of pure water at 100°C is 1.00 atm. Use Raoult's law to estimate the vapor pressure of water over an aqueous solution at 100°C containing 50.0 g of ethylene glycol, $C_2H_6O_2$, dissolved in 100.0 g of water. Also calculate the vapor pressure lowering for water.

14-9. We have presented Raoult's law for the case in which the solute is nonvolatile. How do you think Raoult's law is extended to the case in which both the solute *and* the solvent are volatile, with vapor pressures P_2° and P_1°, respectively?

14-10. The vapor pressure lowering of a solution, $P_1^\circ - P_1$, is proportional to the mole fraction of nonvolatile solute:

$$P_1^\circ - P_1 \propto X_2$$

This proportionality can be written as an equation:

$$P_1^\circ - P_1 = cX_2$$

where c is a proportionality constant. Determine c.

14-11. If both the solute *and* the solvent in an ideal solution are volatile, then each component obeys Raoult's law, so that

$$P_1 = X_1P_1^\circ$$

$$P_2 = X_2P_2^\circ$$

Show that the total pressure over such a solution is

$$P_{total} = P_2^\circ + X_1(P_1^\circ - P_2^\circ)$$

Plot P_{total} versus X_1 for $P_1^\circ = 348$ torr and $P_2^\circ = 270$ torr.

14-12. The vapor pressures of benzene, C_6H_6, and toluene, C_7H_8, at 60°C are 380 torr and 140 torr, respectively. Calculate the partial pressures of benzene and toluene and the total vapor pressure of a solution made up of 50.0 g of benzene and 50.0 g of toluene at 60°C (see Problem 14-11).

MOLALITY

14-13. The solubility of iodine in carbon tetrachloride is 2.603 g per 100.0 g of CCl_4 at 35°C. Calculate the molality of iodine in a saturated CCl_4 solution.

14-14. Describe how you would prepare an aqueous solution that is 1.75 m in $Ba(NO_3)_2$.

14-15. Estimate the molality of sucrose, $C_{12}H_{22}O_{11}$, in a solution prepared by dissolving 2.00 teaspoons of sucrose in 1 cup of water (3 teaspoons = 1 tablespoon = 0.50 oz = 14 g and 4 cups = 1 qt = 0.946 L). Take the density of water as 1.00 $g \cdot mL^{-1}$.

14-16. How many kilograms of water would have to be added to 18.0 g of oxalic acid, $H_2C_2O_4$, to prepare a 0.050 m solution?

14-17. A 1.0-mol sample of each of the following substances is dissolved in 1000 g of water. Determine the colligative molality of the substance in each case:

(a) $MgSO_4$
(b) $CuCl_2$
(c) C_2H_5OH

14-18. A 1.0-mol sample of each of the following substances is dissolved in 1000 g of water. Determine the colligative molality of the substance in each case:

(a) CH_3OH
(b) $Al(NO_3)_3$
(c) $Fe(NO_3)_2$

14-19. Given that the vapor pressure of water is 17.54 torr at 20°C, calculate the vapor pressure lowering of aqueous solutions that are 0.25 m in

(a) NaCl
(b) $CaCl_2$
(c) sucrose, $C_{12}H_{22}O_{11}$

14-20. The colligative molality of the contents of a typical human cell is about 0.30 m_c. Compute the equilibrium vapor pressure of water at 37°C for the cell solution. Take $P^\circ_{H_2O} = 47.1$ torr at 37°C.

14-21. It is possible to convert from molality to molarity if the density of a solution is known. The density of a 2.00 m NaOH aqueous solution is 1.22 $g\cdot mL^{-1}$. Calculate the molarity of this solution.

14-22. In many fields outside chemistry, solution concentrations are expressed in mass percent. Calculate the molality of an aqueous solution that is 24.0% potassium chromate, K_2CrO_4, by mass. Given that the density of the solution is 1.21 $g\cdot mL^{-1}$, calculate the molarity.

BOILING POINT ELEVATION

14-23. Estimate the boiling point of a 0.75 m aqueous solution of $Ca(NO_3)_2$.

14-24. How much NaCl would have to be dissolved in 1000 g of water in order to raise the boiling point 1.0°C?

14-25. Estimate the boiling point of a solution of 5.0 g of picric acid, $C_6H_2(OH)(NO_2)_3$, dissolved in 100.0 g of cyclohexane. Assume that the colligative molality and the molality are the same for picric acid.

14-26. The colligative molality of seawater is about 1.10 m_c. Estimate the boiling point of seawater at 1.00 atm and its vapor pressure at 15°C. The vapor pressure of pure water at 15°C is 12.79 torr.

14-27. The boiling point of ethylene glycol is 197°C, whereas the boiling point of ethyl alcohol is 78°C. Ethylene glycol is called a "permanent" antifreeze and ethyl alcohol a "temporary" one. Explain the difference between "permanent" and "temporary" antifreezes.

14-28. It is claimed that radiator antifreeze also provides "antiboiling" protection for automobile cooling systems. Estimate the boiling point of a solution composed of 50.0 g of water and 50.0 g of ethylene glycol. Assume that the vapor pressure of ethylene glycol is negligible at 100°C. The formula of ethylene glycol is $C_2H_6O_2$.

FREEZING POINT DEPRESSION

(K_f values given in Table 14-4)

14-29. Estimate the freezing point of an aqueous solution of 30.0 g of glucose, $C_6H_{12}O_6$, dissolved in 500.0 g of water.

14-30. Estimate the freezing point of a 0.15 m aqueous solution of NaCl.

14-31. Estimate the freezing point of a solution of 12.0 g of carbon tetrachloride dissolved in 750 g of benzene.

14-32. Quinine is a natural product extracted from the bark of the cinchona tree, which is native to South America. Quinine is used as an antimalarial agent. When 1.00 g of quinine is dissolved in 10.0 g of cyclohexane, the freezing point is lowered 6.23 K. Calculate the molecular mass of quinine.

14-33. Vitamin K is involved in normal blood clotting. When 0.500 g of vitamin K is dissolved in 10.0 g of camphor, the freezing point of the solution is lowered by 4.43 K. Calculate the molecular mass of vitamin K.

$\Delta T_f = K_f m_c$
$4.43\text{K} = (40.0\text{ K}/m_c) m_c$
$m_c = 0.111 m_c$
$\frac{0.111 \text{ mol vit. K}}{1 \text{ kg solvent}} \times 0.01 \text{ kg camphor} = 0.00111 \text{ mol K}$
$\frac{0.5\text{ g}}{0.00111 \text{ mol}} = \boxed{450 \text{ g/mol}}$

14-34. The Rast method was used to determine the molecular mass of a class of compounds called polychlorinated biphenyls (PCBs), which are highly resistant to decomposition and have been used as coolants in transformers. However, PCBs are carcinogenic and are being phased out of use. A 0.100-g sample of a PCB dissolved in 10.0 g of camphor depressed the freezing point 1.22 K. Calculate the molecular mass of the compound.

14-35. Don Juan Pond in the Wright Valley of Antarctica freezes at $-57\,°\text{C}$. The major solute in the pond is $CaCl_2$. Estimate the concentration of $CaCl_2$ in the pond water.

$\Delta T_f = K_f m_c$
$57°\text{C} = (1.86\text{ K}/m_c) m_c$
$m_c = 30.6 m_c$
$m = \frac{30.6 m_c}{3} = \boxed{10 m}$

14-36. Most wines are about 12% ethyl alcohol by volume, and many hard liquors are about 80 proof (200 proof is 100% ethyl alcohol; an 80 proof liquor is 40% by volume ethyl alcohol). Assuming that the only major nonaqueous constituent of wine and vodka is ethyl alcohol, estimate the freezing points of wine and vodka. Take the density of ethyl alcohol as $0.79\text{ g}\cdot\text{mL}^{-1}$ and the density of water as $1.00\text{ g}\cdot\text{mL}^{-1}$. The formula of ethyl alcohol is C_2H_5OH.

14-37. Use Figure 14-8 to determine the percent by volume of ethylene glycol in water that is necessary to give antifreeze protection down to $-40\,°\text{C}$. Compare the result obtained from Figure 14-8 with that calculated from the freezing point depression equation. Take the density of ethylene glycol to be $1.116\text{ g}\cdot\text{mL}^{-1}$ and the density of water to be $1.00\text{ g}\cdot\text{mL}^{-1}$. The formula of ethylene glycol is $HOCH_2CH_2OH$.

52%

$\Delta T_f = K_f m_c$
$40°\text{C} = (1.86\text{ K}/m_c) m_c$
$m_c = 21.5 m_c$
$\frac{21.5 \text{ mol} \times 62.07 \text{ g/mol}}{1.116 \text{ g/mL}} = 1200 \text{ mL}$
$\frac{1200 \text{ mL}}{1200 + 1000} \times 100 = \boxed{55\%}$

14-38. Menthol is a crystalline substance with a peppermint taste and odor. A solution of 6.54 g of menthol per 100.0 g of cyclohexane freezes at $-1.95\,°\text{C}$. Determine the formula mass of menthol.

14-39. A solution of mercury(II) chloride, $HgCl_2$, is a poor conductor of electricity. A 40.7-g sample of $HgCl_2$ is dissolved in 100.0 g of water, and the freezing point of the solution is found to be $-2.83\,°\text{C}$. Explain why $HgCl_2$ in solution is a poor conductor of electricity.

$\Delta T_f = K_f m_c$
$2.83\text{K} = 1.86\text{ K}/m_c \times m_c$
$m_c = 1.52 m_c$
$\frac{40.7\text{ g}}{271.5 \text{ g/mol}} = 0.1499 \text{ mol}$
$\frac{0.1499 \text{ mol } HgCl_2}{0.1 \text{ kg } H_2O} = 1.50 m$
$\boxed{HgCl_2 \text{ is essentially undissociated}}$

14-40. Mayer's reagent, K_2HgI_4, is used in analytical chemistry. In order to determine its extent of dissociation in water, its effect on the freezing point of water is investigated. A 0.25 m aqueous solution is prepared, and its freezing point is found to be $-1.41\,°\text{C}$. Suggest a possible dissociation reaction that takes place when K_2HgI_4 is dissolved in water.

OSMOTIC PRESSURE

14-41. Calculate the osmotic pressure of a 0.50 M aqueous solution of sucrose, $C_{12}H_{22}O_{11}$, at 50°C.

$\pi = RTM_c$ $\quad \pi = (0.0821)(323\text{ K})(0.50 \text{ mol/L})$ $\quad \pi = \boxed{13 \text{ atm}}$

14-42. Calculate the osmotic pressure of seawater at 37°C. Take $M_c = 1.10\text{ mol}\cdot\text{L}^{-1}$ for seawater.

14-43. Insulin is a small protein hormone that regulates carbohydrate metabolism by decreasing blood glucose levels. A deficiency of insulin leads to diabetes. A 2.00-g sample of insulin is dissolved in enough water to make 10.0 mL of solution, and the osmotic pressure of the solution at 25°C is found to be 0.853 atm. Estimate the molecular mass of insulin.

$\pi = RTM_c$
$0.853 = (0.0821)(298\text{ K})M_c$
$M_c = 0.0349 \text{ mol/L}$
$0.0349 \text{ mol/L} \times 0.01 \text{ L} = 3.49 \times 10^{-4} \text{ mol}$
$\frac{2.00\text{ g}}{3.49 \times 10^{-4} \text{ mol}} = \boxed{5730 \text{ g/mol}}$

14-44. Pepsin is the principal digestive enzyme of gastric juice. A 3.00-g sample of pepsin is dissolved in enough water to make 10.0 mL of solution, and the osmotic pressure of the solution at 25°C is found to be 0.213 atm. Estimate the molecular mass of pepsin.

14-45. Immunoglobulin G, formerly called gamma globulin, is the principal antibody in blood serum. A 5.00-g sample of immunoglobulin G is dissolved in enough water to make 0.100 L of solution, and the osmotic pressure of the solution at 25°C is found to be 6.19 torr. Estimate the molecular mass of immunoglobulin G.

$\pi = RTM_c$

$8.145\times10^{-3}\ \text{atm} = 0.0821 \times 298\text{K} \times M_c$

$M_c = 3.33\times10^{-4}\ \text{mol/L}$

$3.33\times10^{-4}\ \text{mol/L} \times 0.1\ \text{L} = 3.33\times10^{-5}\ \text{mol}$

$\frac{5.00\ \text{g}}{3.33\times10^{-5}\ \text{mol}} = \boxed{1.50\times10^{5}\ \text{g/mol}}$

14-46. A 2.0-g sample of the polymer polyisobutylene, $[CH_2C(CH_3)_2]_x$, is dissolved in enough cyclohexane to make 10.0 mL of solution at 20°C and produces an osmotic pressure of 2.0×10^{-2} atm. Determine the formula mass and the number of units (x) in the polymer.

14-47. In reverse osmosis, water flows out of a salt solution until the osmotic pressure of the solution equals the applied pressure. If a pressure of 100 atm is applied to seawater, what will be the final concentration of the seawater at 20°C when reverse osmosis stops? Given that seawater is a 1.1 M_c solution of NaCl(aq), calculate how many liters of seawater are required to produce 10 L of fresh water at 20°C with an applied pressure of 100 atm.

$\pi = RTM_c$

$100\ \text{atm} = (0.0821)(293\ \text{K})M_c$

$M_c = 4.16\ \text{mol/L}$

$M_B V_B = M_A V_A$

$(1.1\ M_c)V_B = (4.16\ M_c)V_a$

$V_b - V_a = 10\ \text{L}$

$V_a = V_b - 10\ \text{L}$

$(1.1 M_c)V_B = (4.16 M_c)(V_B - 10\ \text{L})$

$V_B = \boxed{13.6\ \text{L}}$

14-48. What is the minimum pressure that must be applied at 25°C to obtain pure water by reverse osmosis from water that is 0.15 M in NaCl and 0.015 M in $MgSO_4$?

HENRY'S LAW

14-49. Calculate the concentration of nitrogen in water at a nitrogen gas pressure of 1.0 atm and 25°C.

$P_{gas} = k_h M_{gas}$

$1.0\ \text{atm} = 1.6\times10^{3}\ \text{atm/M} \cdot M \qquad M = \boxed{6.3\times10^{-4}\ M}$

14-50. Which gas—N_2, O_2, or CO_2—has the highest concentration in water at 25°C when each has a pressure of 1.0 atm?

14-51. The Henry's law constant for CO_2 in water at 25°C is 29 atm·M^{-1}. Estimate the concentration of dissolved CO_2 in a carbonated soft drink pressurized with 2.0 atm of CO_2 gas.

$P_{gas} = k_h M_{gas} \qquad 2.0\ \text{atm} = 29\ \text{atm/M} \cdot M \qquad M = 6.9\times10^{-2}\ M = \boxed{0.069\ M}$

14-52. Dry air consists of 20 percent O_2 and 79 percent N_2 by volume. If the air pressure above a pool of water is 1.00 atm, then calculate the concentrations of dissolved O_2 and N_2 at 25°C.

14-53. Every 33 ft under water the pressure increases by 1 atm. How many feet would a diver have to descend before the blood concentration of oxygen reaches 1.28×10^{-3} M, assuming that compressed air is used (20 percent O_2 by volume)?

$P_{gas} = k_h M_{gas}$

$P_{gas} = (7.8\times10^{2}\ \text{atm/M})(1.28\times10^{-3}\ \text{M})$

$P_{gas} = 1.0\ \text{atm}$

$(0.20)P_{air} = P_{O_2}$

$P_{air} = \frac{1.0\ \text{atm}}{0.20} = 5.0\ \text{atm}$

increase in pressure = 4.0 atm

$(33\ \text{ft/atm})(4.0\ \text{atm}) = \boxed{130\ \text{ft}}$

14-54. What volume percent of oxygen should be used in a diver's air tanks to make the partial pressure of oxygen in the air supply 0.20 atm at a water depth of 90 ft (see Problem 14-53)?

INTERCHAPTER I

Main-Group Metals

When magnesium vapor is condensed to a solid, dendrites of magnesium are formed. Dendrites are delicate, lacelike crystals consisting of thin branches that are interwoven.

The metals that occur in Groups 1 through 5 in the periodic table are called the main-group metals. The outer electrons of the main-group metals occupy *s* and *p* orbitals, and most of the compounds of these elements are ionic. The chemical reactivities of these metals vary greatly, from the very reactive alkali metals, which combine spontaneously with the oxygen and water vapor in the air, to the relatively unreactive tin, which is used to make tinned cans. We discussed the alkali metals in Interchapter F, and here we discuss the other main-group metals.

Figure I-1 The mineral beryl, $Be_3Al_2Si_6O_{18}$, occurs in hexagonal prisms. It is a light green color. Beryl is the chief source of beryllium and is used as a gem.

I-1. THE ALKALINE EARTH METALS FORM IONIC COMPOUNDS CONSISTING OF M^{2+} IONS

The alkaline earth metals are beryllium, magnesium, calcium, strontium, barium, and radium. They occur in Group 2 in the periodic table and all have a [noble gas]ns^2 electron configuration. Only the +2 ionic charge is important in the chemistry of the alkaline earth metals. The term alkaline earth metal originally applied only to calcium, strontium, and barium because the chemical properties of beryllium and magnesium differ somewhat from those of calcium, strontium, and barium. Usually the chemical properties of the first members of a group are somewhat different from those of later elements in the group. Some of the physical properties of the alkaline earth metals are given in Table I-1, and their sources and commercial uses are given in Table I-2.

The alkaline earth metals are too reactive to occur as the free metals in nature and must be prepared by electrolysis. For example, calcium, strontium, and barium are prepared by high-temperature electrolysis of the molten chloride:

$$CaCl_2(l) \xrightarrow[\text{high T}]{\text{electrolysis}} Ca(l) + Cl_2(g)$$

The alkaline earth metals react rapidly with water, but the rates of these reactions are much lower than those for the alkali metals. Beryllium and magnesium react slowly with water at ordinary temperatures, although hot magnesium reacts violently with water.

The alkaline earth metals burn in oxygen to form the MO oxides, which are ionic solids. Magnesium is used as an incendiary in warfare because of its vigorous reaction with oxygen. It burns even more

Table I-1 Physical properties of the alkaline earth metals

	Beryllium	*Magnesium*	*Calcium*	*Strontium*	*Barium*
chemical symbol	Be	Mg	Ca	Sr	Ba
atomic mass	9.0218	24.305	40.08	87.62	137.33
electron configuration	[He]$2s^2$	[Ne]$3s^2$	[Ar]$4s^2$	[Kr]$5s^2$	[Xe]$6s^2$
melting point/°C	1278	651	845	769	725
boiling point/°C	2970	1107	1487	1384	1740
density at 20°C/$g \cdot cm^{-3}$	1.85	1.74	1.55	2.54	3.51
sum of first and second ionization energies of M(g)/$kJ \cdot mol^{-1}$	2656	2187	1734	1608	1462
ionic radius/pm	31	65	99	113	135

Table I-2 Major sources and commercial uses of the alkaline earth metals

Metal	*Sources*	*Method of preparation*	*Major commercial uses of metal or its compounds*
beryllium	beryllium aluminum silicates, including *beryl*, $Be_3Al_2Si_6O_{18}$	electrolysis of K_2BeF_4 melt	lightweight alloys (improves corrosion resistance and resistance to fatigue and temperature changes); rockets and satellites
magnesium	dolomite, $CaMg(CO_3)_2$; carbonates and silicates; seawater and well brines	electrolysis of $MgCl_2$ melt	alloys for airplanes; flashbulbs; pyrotechnics (white flame); medicine
calcium	limestone, $CaCO_3$; gypsum, $CaSO_4 \cdot 2H_2O$; fluorite, CaF_2; apatite, $Ca_{10}(OH)_2(PO_4)_6$ (major constituent of tooth enamel)	electrolysis of $CaCl_2$ melt	quicklime, CaO, and slaked lime, $Ca(OH)_2$; widely used as a base in the chemical industry; cements; bleaching powder
strontium	celestite, $SrSO_4$; strontianite, $SrCO_3$	electrolysis of $SrCl_2$ melt	salts used in pyrotechnics (red flame); beet sugar refining
barium	witherite, $BaCO_3$; barite, $BaSO_4$	electrolysis of $BaCl_2$ melt	$BaSO_4$; white paints; $BaCO_3$, rat poison; $Ba(NO_3)_2$, pyrotechnics (green flame)
radium	pitchblende and carnotite ores	electrolysis of $RaCl_2(aq)$	skin cancer treatments

rapidly when sprayed with water and reacts with carbon dioxide via the reaction

$$2Mg(s) + CO_2(g) \rightarrow 2MgO(s) + C(s)$$

Covering burning magnesium with sand slows the combustion, but the molten magnesium reacts with the silicon dioxide (the principal component of sand) to form magnesium oxide:

$$2Mg(l) + SiO_2(s) \rightarrow 2MgO(s) + Si(s)$$

Magnesium ribbon is used in flashbulbs. The brilliant flash is produced by the reaction of magnesium with oxygen.

Barium reacts with excess oxygen to form barium peroxide:

$$Ba(s) + O_2(g) \rightarrow BaO_2(s)$$

Except for beryllium, the alkaline earth metals react vigorously with dilute acids:

$$Mg(s) + 2HCl(aq) \rightarrow MgCl_2(aq) + H_2(g)$$

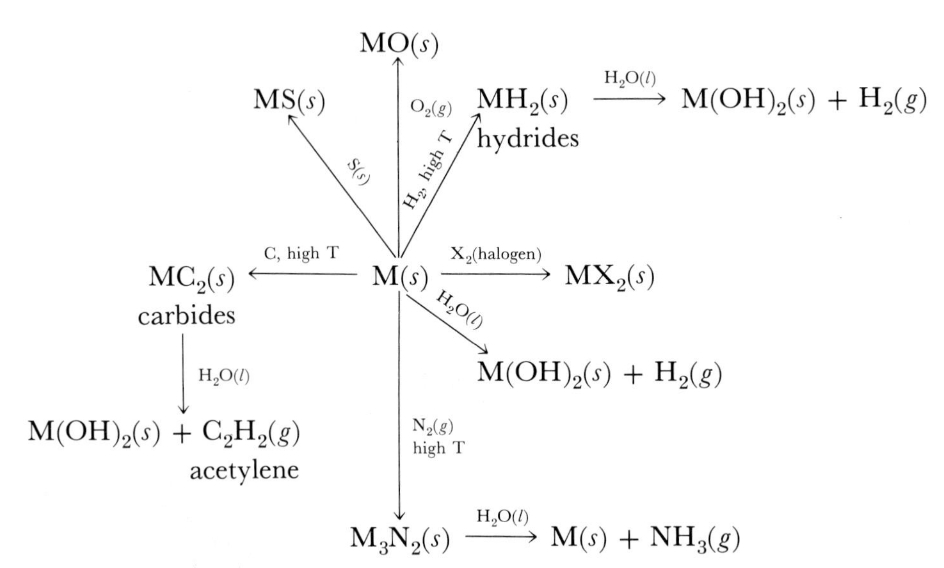

Figure I-2 Representative reactions of Group 2 metals.

The alkaline earth metals react with most of the nonmetals to form ionic binary compounds. The reactions of the alkaline earth metals are summarized in Figure I-2.

Many alkaline earth metal compounds are important commercially. Magnesium hydroxide is only slightly soluble in water, and suspensions of it are sold as the antacid Milk of Magnesia.

Magnesium sulfate heptahydrate, $MgSO_4 \cdot 7H_2O$, known as Epsom salt, is used as a cathartic, or purgative. The name Epsom comes from the place where the compound was first discovered in 1695, in a natural spring in Epsom, England.

Calcium (Figure I-3) is an essential constituent of bones and teeth, limestone, plants, and the shells of marine organisms. The Ca^{2+} ion plays a major role in muscle contraction, vision, and nerve excitation. Calcium oxide, or quicklime, is made by heating limestone:

$$CaCO_3(s) \rightarrow CaO(s) + CO_2(g)$$

Calcium oxide is the third ranked industrial chemical, over 35 billion pounds being produced annually in the United States. It is mixed with water to form calcium hydroxide, which is also called slaked lime:

$$CaO(s) + H_2O(l) \rightarrow Ca(OH)_2(aq)$$

Slaked lime is used to make cement, mortar, and plaster. Plaster of Paris is $CaSO_4 \cdot \frac{1}{2}H_2O$, which combines with water to form gypsum:

(a)

(b)

$$\underset{\text{plaster of Paris}}{CaSO_4 \cdot \tfrac{1}{2}H_2O(s)} + \tfrac{3}{2}H_2O(l) \rightarrow \underset{\text{gypsum}}{CaSO_4 \cdot 2H_2O(s)}$$

Asbestos is a calcium magnesium silicate with the approximate composition $CaMg_3(SiO_3)_4$. It can resist very high temperatures, but, because small asbestos fibers are a confirmed carcinogen, it is being phased out as a construction material.

Strontium salts produce a brilliant red flame and are used in signal flares and fireworks (Figure I-4). The radioactive isotope strontium-90, which is produced in atomic bomb explosions, is a major health hazard because it behaves like calcium and incorporates in bone marrow, causing various cancers.

Figure I-3 (a) Calcium metal is rarely seen in its pure form because it oxidizes rapidly. The calcium turnings shown here appear to be dull because of surface oxidation. (b) Fluorite (calcium fluoride) and calcite (calcium carbonate) are two common minerals containing calcium.

Figure I-4 Strontium compounds are used in pyrotechnics, signal flares, and tracer bullets because they emit a brilliant red color when heated to high temperatures.

Table I-3 Physical properties of the Group 3 metals

	Aluminum	*Gallium*	*Indium*	*Thallium*
chemical symbol	Al	Ga	In	Tl
atomic mass	26.98154	69.72	114.82	204.37
electron configuration	$[Ne]2s^2 3p^1$	$[Ar]4s^2 3d^{10} 4p^1$	$[Kr]5s^2 4d^{10} 5p^1$	$[Xe]6s^2 4f^{14} 5d^{10} 6p^1$
melting point/°C	660	30	157	304
boiling point/°C	2467	2250	2070	1457
density at 20°C/$g \cdot cm^{-3}$	2.70	5.90	7.30	11.85
sum of the first three ionization energies of $M(g)/kJ \cdot mol^{-1}$	5137	5520	5063	5415
ionic radius/pm	50	62	81	95 (Tl^{3+}) 144 (Tl^{+})

I-2. THE GROUP 3 METALS ARE ALUMINUM, GALLIUM, INDIUM, AND THALLIUM

Some of the physical properties of Group 3 metals are presented in Table I-3, and their major sources and commercial uses are given in Table I-4.

Aluminum is the most abundant metal in the earth's crust, where it occurs in various silicates and in vast deposits of bauxite, AlO(OH)

Table I-4 Major sources and commercial uses of the Group 3 metals

Metal	*Sources*	*Method of preparation*	*Major commercial uses*
aluminum	bauxite, AlO(OH); clays	electrolysis of Al_2O_3 in Na_3AlF_6	exterior building trim; high-voltage transmission lines; alloys for aircraft and ships; $KAl(SO_4)_2$ soil acidifier; explosives; catalysts
gallium	trace impurity in bauxite and zinc blendes	electrolysis of chloride	high-temperature heat transfer fluid
indium	trace impurity in zinc blendes	electrolysis of chloride or cyanide	corrosion-resistant plating; jewelry; dental alloys; metallic seals
thallium	impurity in pyrites; flue dust in H_2SO_4 manufacture	$TlI_3(l) + 3Na(l) \rightarrow 3NaI(l) + Tl(g)$	salts are rodent poisons; Tl(I) salts especially dangerous because toxic amounts can be absorbed through the skin

Figure I-5 Bauxite, a reddish-brown ore, is the principal source of aluminum. Bauxite is heated with coke (the black substance shown) to produce aluminum oxide, a white powder. This is further refined through electrolysis to produce aluminum.

(Figure I-5), from which it is obtained by electrolysis in the Hall process (Chapter 21). Although aluminum reacts with oxygen to form aluminum oxide, aluminum metal resists corrosion by forming a protective, adherent layer of aluminum oxide. Freshly polished aluminum has a bright, silvery appearance, but weathered aluminum has a dull tarnish because of the aluminum oxide coating. Aluminum is light, soft, and widely used in lightweight alloys with silicon, copper, and magnesium.

Gallium, indium, and thallium are soft, silvery-white metals. Gallium has the greatest liquid range (over 2000 C°) of any known substance and melts at 30°C (Figure 2-10).

Aluminum and gallium dissolve in *both* acids and bases. The reaction of aluminum with an acid is

$$2Al(s) + 6H^+(aq) \rightarrow 2Al^{3+}(aq) + 3H_2(g)$$

and the reaction of aluminum with a base is

$$2Al(s) + 6H_2O(l) + 2OH^-(aq) \rightarrow \underset{\text{aluminate ion}}{2Al(OH)_4^-(aq)} + 3H_2(g)$$

A scanning electron micrograph of pure recrystallized aluminum (×510).

The dissolution of aluminum in strong base is used as the basis of some drain cleaners, such as Drano, to unplug drains by a combination of the grease-dissolving action of the strong base (NaOH) and the agitation produced by the bubbles of evolved hydrogen gas. Note that explosive mixtures of hydrogen and oxygen gas may form in the sink and all sparks and flames should be avoided.

In contrast to the other Group 3 metals, thallium exhibits both +3 and +1 ionic charges in aqueous media. Thallium(I) compounds are very poisonous, and even trace amounts can cause complete loss of body hair. The aqueous-solution chemistry of $Tl^+(aq)$ is similar to that of $Ag^+(aq)$; for example, the Cl^-, Br^-, I^-, and S^{2-} salts are all insoluble and the halide salts darken on exposure to sunlight.

I-3. THE GROUP 4 METALS ARE TIN AND LEAD

The ground-state electron configurations of tin and lead are

$$\text{Sn}\ ([\text{Kr}]5s^2 4d^{10} 5p^2)$$
$$\text{Pb}\ ([\text{Xe}]6s^2 4f^{14} 5d^{10} 6p^2)$$

The principal compounds of both tin and lead involve M(II) and M(IV). Tin is found primarily in the mineral *cassiterite*, SnO_2, which occurs in rare but large deposits in Malaysia, China, the USSR, and the United States. Total U.S. natural reserves of tin are very small. It is used in plating (tin-plated food cans) and in various alloys, including solders, type metal, pewter, bronze, and gun metal. Table I-5 gives the compositions of some tin and lead alloys. Tin objects are subject to a condition called tin disease, which is the conversion of the *white* allotrope of tin to the *gray* allotrope. This conversion occurs slowly below 13°C and results in the brittle gray allotrope of tin.

Lead is obtained primarily from the ore *galena*, PbS, by roasting with carbon:

$$\text{PbS}(s) + \text{C}(s) + 2\text{O}_2(g) \rightarrow \text{Pb}(l) + \text{CO}_2(g) + \text{SO}_2(g)$$

Table I-5 Composition of tin and lead alloys[a]

solder	50 Sn	50 Pb		
aluminum solder	86 Sn	9 Zn	5 Al	
type metal	3 Sn	82 Pb	15 Sb	
pewter	85 Sn	7 Cu	6 Bi	2 Sb
bronze	15 Sn	80 Cu	5 Zn	
gun metal	10 Sn	90 Cu		
bell metal	22 Sn	78 Cu		
battery plate	—	94 Pb	6 Sb	

[a]The numbers are the respective percentages by mass of the indicated metals.

(a)

(b)

Figure I-6 (a) The mineral mimetite contains lead arsenate and chloride, $Pb_5Cl(AsO_4)_3$. (b) Crocoite is a rare reddish orange mineral of lead chromate, $PbCrO_4$.

Lead is resistant to corrosion and is used in a variety of alloys. Lead storage batteries constitute the major use of lead. The metal is also used in cable coverings, ammunition, and the synthesis of tetraethyl lead, $(CH_3CH_2)_4Pb$, which is used in leaded gasolines. Lead was once used in paints—$PbCrO_4$ is yellow and Pb_3O_4 is red—but lead salts constitute a serious health hazard, as they are cumulative poisons, and their use in paints has been discontinued. The Romans used lead vessels to store wine and other consumables and to conduct water in lead-lined aqueducts; thus lead poisoning may have had more to do with the collapse of the Roman Empire than any other factor. The use of lead-containing glazes on pottery for food use is now prohibited in the United States. A daily intake of more than 1 mg of lead is hazardous. Early symptoms of lead poisoning are constipation, anemia, loss of appetite, and pain in the joints; later stages are paralysis and brain damage.

I-4. BISMUTH IS THE ONLY GROUP 5 METAL

Bismuth has five valence electrons ($6s^26p^3$). The principal compounds of bismuth contain either Bi(III) or Bi(V). It is a pink-white metal that occurs rarely as the free metal (Figure I-7). The most common source of bismuth is the sulfide ore *bismuthinite,* Bi_2S_3 (Figure I-8).

Bismuth metal is obtained from the ore by roasting the ore with carbon in air. Bismuth is also obtained as a by-product in lead smelting. It burns in air with a bright blue flame, forming the yellow oxide Bi_2O_3, and is used in a variety of alloys, including pewter and low-

Figure I-7 Bismuth is the only Group 5 metal.

Figure I-8 Bismuthinite, Bi_2S_3, is a principal ore of bismuth.

melting alloys that are used in fire-extinguisher sprinkler-head plugs, electrical fuses, and relief valves for compressed-gas cylinders. Bismuth alloys contract on heating and thus find use in alloys that might otherwise crack because of thermal expansion when subjected to high temperatures.

The oxide Bi_2O_3 is soluble in strongly acidic aqueous solutions. The bismuthyl ion, $BiO^+(aq)$, and the bismuthate ion, $BiO_3^-(aq)$, are important in the aqueous-solution chemistry of bismuth. The bismuthyl ion forms insoluble compounds such as BiOCl and BiO(OH), whereas BiO_3^- is a powerful oxidizing agent. Bismuth pentafluoride, BiF_5, is a potent fluorinating agent that transfers fluorine to various compounds and is converted to the trifluoride, BiF_3.

The major sources and commercial uses of tin, lead, and bismuth are summarized in Table I-6.

Table I-6 Major sources and commercial uses of tin, lead, and bismuth

Metal	*Sources*	*Method of preparation*	*Major commercial uses*
tin	cassiterite, SnO_2	roasting of SnO_2	tin plating; soft solder, type metal, pewter, bronze; production of glass; superconductors
lead	galena, PbS; anglesite, $PbSO_4$; cerussite, $PbCO_3$	roasting of PbS	storage batteries; plumbing, cable covering; ammunition, tetraethyl lead; radiation shields; type metal
bismuth	bismuthinite, Bi_2S_3, and bismite, Bi_2O_3	roasting of ores	low-melting alloys for fuses, sprinkler systems, safety plugs, castings, type metal

QUESTIONS

I-1 How are the Group 2 metals obtained?

I-2 Why would you not throw water on a magnesium fire?

I-3 Write the balanced equation for the reaction that occurs when a flashbulb flashes.

I-4 Give the chemical formula for each of the following substances:

(a) Milk of Magnesia
(b) Epsom salt
(c) quicklime
(d) limestone
(e) slaked lime
(f) plaster of Paris
(g) gypsum
(h) asbestos
(i) bauxite
(j) tetraethyl lead

I-5 What substance has the greatest liquid range?

I-6 Write balanced chemical equations for the dissolution of gallium metal in HCl(*aq*) and NaOH(*aq*).

I-7 List the metals of Groups 3, 4, and 5.

I-8 Explain why both Sn and Pb form compounds involving M(II) and M(IV).

I-9 What is tin disease?

I-10 How is lead produced from galena?

Effect of temperature on the reaction equilibrium

$$N_2O_4(g) \rightleftharpoons 2NO_2(g)$$

colorless brown

An increase in temperature from 0°C (ice water) to 25°C converts some of the N_2O_4 to NO_2 and results in a darker color for the reaction mixture.

15 / Chemical Equilibrium

In this chapter we show how the concept of dynamic equilibrium, which was developed in Chapter 13 to describe the equilibrium vapor pressure of a liquid and in Chapter 14 to describe the dissolution of a solid in a liquid, can be extended to cover all types of chemical reactions. We learn how to describe a chemical system at equilibrium quantitatively. We shall see that a chemical reaction equilibrium is characterized by a quantity called an equilibrium constant. Two key questions arise in the analysis of reaction equilibria: (1) In which direction (left to right or right to left) does a reaction at equilibrium shift in response to a change in conditions that disturbs the equilibrium? (2) If we prepare a nonequilibrium mixture of reactants and products, then in what direction does the reaction proceed toward equilibrium and what are the equilibrium values of the concentrations? We learn in this chapter how to answer these questions and thereby how to predict the reaction conditions necessary to maximize the amount of a desired product of a chemical reaction.

15-1. A CHEMICAL EQUILIBRIUM IS A DYNAMIC EQUILIBRIUM

We learned in Chapter 13 that a pure liquid has a unique equilibrium vapor pressure at each temperature. We also learned that a dynamic equilibrium between a liquid and its vapor is attained when the rate of evaporation from the liquid phase equals the rate of condensation from the vapor phase. If the liquid is water, then we can express the equilibrium process by the equation

$$H_2O(l) \rightleftharpoons H_2O(g)$$

The rate of evaporation from the liquid phase is said to be the *forward rate*, and the rate of condensation from the vapor phase is said to be the *reverse rate*. The double arrows denote a reaction equilibrium. In general, a state of *chemical equilibrium* is attained when the rates of the forward and reverse processes are equal. Furthermore, a true equilibrium can be attained from either direction. For example, the equilibrium vapor pressure of water at a particular temperature is exactly the same whether we start with pure liquid water that evaporates until the vapor pressure is constant or with pure water vapor that is initially at a pressure in excess of the equilibrium value. In the latter case, water vapor condenses until the equilibrium vapor pressure is attained and then the *net* condensation of water vapor ceases.

Other chemical reactions can attain an equilibrium state in a manner analogous to that for the $H_2O(l) \rightleftharpoons H_2O(g)$ equilibrium. For example, consider the chemical reaction described by the equation

$$\underset{\text{colorless}}{N_2O_4(g)} \rightleftharpoons \underset{\text{brown}}{2NO_2(g)} \qquad (15\text{-}1)$$

in which the colorless gas nitrogen tetroxide, $N_2O_4(g)$, dissociates into the reddish-brown gas nitrogen dioxide, $NO_2(g)$. Reaction (15-1), like all chemical reactions, is really two opposing ones. The forward reaction is the dissociation of N_2O_4 molecules into NO_2 molecules:

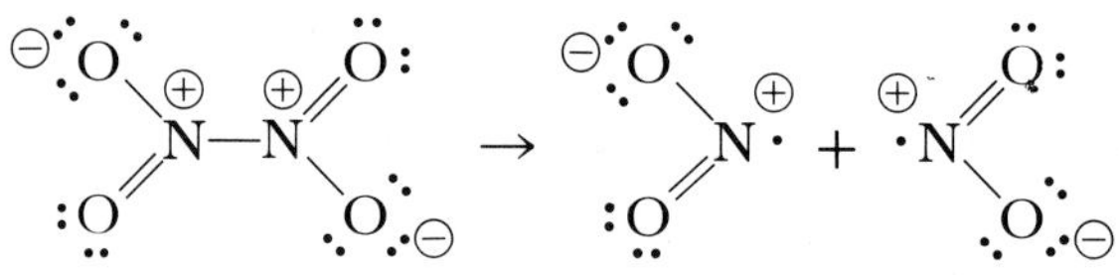

and the reverse reaction is the association of NO_2 molecules into N_2O_4 molecules:

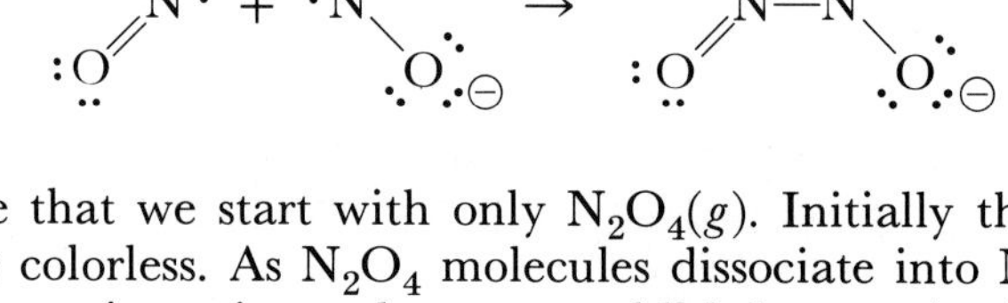

Suppose that we start with only $N_2O_4(g)$. Initially the reaction mixture is colorless. As N_2O_4 molecules dissociate into NO_2 molecules, the reaction mixture becomes reddish brown. As the concentration of NO_2 increases, more and more NO_2 molecules associate back into N_2O_4 molecules. Thus the reverse rate of Reaction (15-1) increases with time. Eventually, the forward rate and the reverse rate become equal and a state of equilibrium exists. The concentrations of N_2O_4 and NO_2 no longer change with time. The equilibrium is a *dynamic equilibrium* because N_2O_4 molecules are still dissociating into NO_2 molecules and NO_2 molecules are still associating into N_2O_4 molecules. The rates of these two processes are exactly the same, however, and so there is no net change in the concentrations of N_2O_4 and NO_2.

At equilibrium the forward reaction rate equals the reverse reaction rate.

15-2. A CHEMICAL EQUILIBRIUM IS APPROACHABLE FROM EITHER SIDE

We can study the approach to equilibrium of Reaction (15-1) quantitatively. If we determine the value of the NO_2 concentration, denoted as $[NO_2]$, in the reaction mixture and if we know the concentration of N_2O_4 that we started with, then it is a simple matter to determine the value of the N_2O_4 concentration, denoted as $[N_2O_4]$, in the reaction mixture. For example, suppose that we start with $1.00\ \text{mol}\cdot\text{L}^{-1}$ of N_2O_4 and no NO_2. We can denote these initial concentrations by

$$[N_2O_4]_0 = 1.00\ \text{M} \qquad \text{and} \qquad [NO_2]_0 = 0$$

where the subscript 0 indicates an initial concentration. Suppose also that at some later time we find that

$$[NO_2] = 0.40\ \text{M}$$

What is the corresponding value of $[N_2O_4]$? According to the equation $N_2O_4(g) \rightleftharpoons 2NO_2(g)$, each molecule of N_2O_4 that dissociates gives two molecules of NO_2. Thus in order for there to be a 0.40 M concentration of $NO_2(g)$, 0.20 M of $N_2O_4(g)$ must have dissociated, because from the reaction stoichiometry

$$\left(\begin{matrix}\text{moles per liter}\\ \text{of } N_2O_4 \text{ consumed}\end{matrix}\right) = \left(\begin{matrix}\text{moles per liter}\\ \text{of } NO_2 \text{ produced}\end{matrix}\right)\left(\frac{1\ \text{mol}\ N_2O_4}{2\ \text{mol}\ NO_2}\right)$$

Thus, if $[NO_2] = 0.40$ M, then

$$\left(\begin{matrix}\text{moles per liter}\\ \text{of } N_2O_4 \text{ consumed}\end{matrix}\right) = \left(\frac{0.40\ \text{mol}\ NO_2}{\text{L}}\right)\left(\frac{1\ \text{mol}\ N_2O_4}{2\ \text{mol}\ NO_2}\right)$$

$$= 0.20\ \text{mol}\cdot\text{L}^{-1}$$

Therefore, the value of $[N_2O_4]$ when $[NO_2] = 0.40$ M is

$$[N_2O_4] = \underset{\text{initial value}}{1.00\ M} - \underset{\text{amount consumed}}{0.20\ M} = 0.80\ M$$

The following example illustrates the calculation of *equilibrium concentrations* in reaction mixtures.

Example 15-1: Suppose that the ammonia synthesis reaction,

$$N_2(g) + 3H_2(g) \rightleftharpoons 2NH_3(g)$$

is carried out with the following initial concentrations:

$$[N_2]_0 = 2.00\ M \qquad [H_2]_0 = 0.50\ M \qquad [NH_3]_0 = 0$$

and that at equilibrium we find that $[NH_3] = 0.20$ M. Compute the equilibrium values of $[N_2]$ and $[H_2]$.

Solution: From the reaction stoichiometry we have

$$\begin{pmatrix}\text{moles per liter}\\ \text{of } N_2 \text{ consumed}\end{pmatrix} = \begin{pmatrix}\text{moles per liter}\\ \text{of } NH_3 \text{ produced}\end{pmatrix}\left(\frac{1\ \text{mol}\ N_2}{2\ \text{mol}\ NH_3}\right)$$

$$= \left(\frac{0.20\ \text{mol}\ NH_3}{L}\right)\left(\frac{1\ \text{mol}\ N_2}{2\ \text{mol}\ NH_3}\right) = 0.10\ M$$

Therefore, the value of $[N_2]$ when $[NH_3] = 0.20$ M is

$$[N_2] = \underset{\text{initial value}}{2.00\ M} - \underset{\text{amount consumed}}{0.10\ M} = 1.90\ M$$

Similarly, we compute for $[H_2]$ at equilibrium:

$$[H_2] = 0.50\ M - \left(\frac{0.20\ \text{mol}\ NH_3}{L}\right)\left(\frac{3\ \text{mol}\ H_2}{2\ \text{mol}\ NH_3}\right)$$

$$= \underset{\text{initial value}}{0.50\ M} - \underset{\text{amount consumed}}{0.30\ M} = 0.20\ M$$

Note that three times as much H_2 as N_2 is consumed in the reaction, as is required by the stoichiometry.

Let's continue our discussion of the N_2O_4–NO_2 reaction. If we introduce 1.00 mol of N_2O_4 into a 1.00-L reaction vessel held at 100°C and then determine the concentration of $[NO_2]$, and thus that

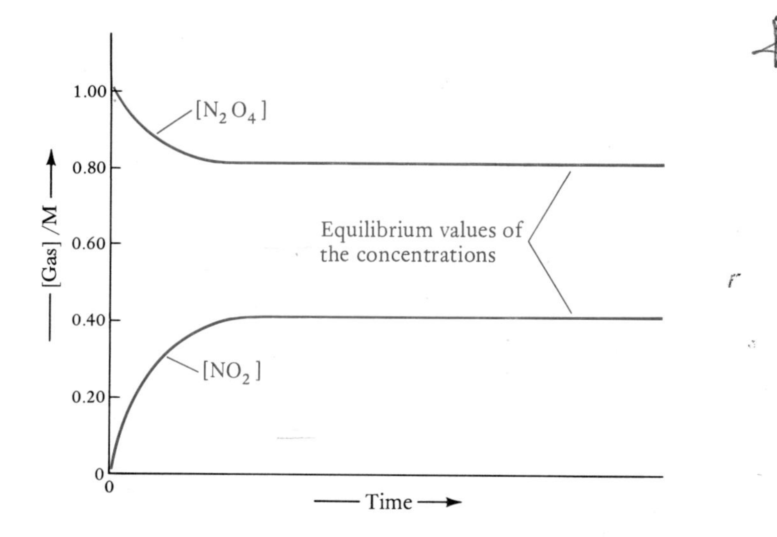

Figure 15-1 Plot of $[N_2O_4]$ and $[NO_2]$ as a function of time at 100°C for the reaction $N_2O_4(g) \rightleftharpoons 2NO_2(g)$. At time zero, $[N_2O_4]_0 = 1.00$ M and $[NO_2]_0 = 0$. As the reaction proceeds, $[N_2O_4]$ decreases from 1.00 M to 0.80 M and then remains constant, whereas $[NO_2]$ increases from zero to 0.40 M and then remains constant. The values $[N_2O_4] = 0.80$ M and $[NO_2] = 0.40$ M are equilibrium values. When the reaction mixture attains equilibrium, the concentrations of $N_2O_4(g)$ and $NO_2(g)$ remain constant.

of $[N_2O_4]$, as a function of time, then we obtain the results shown in Figure 15-1. Note that the value of $[N_2O_4]$, which starts out at 1.00 M, decreases to a constant value of 0.80 M. The value of $[NO_2]$, which starts out at zero, increases to a constant value of 0.40 M.

If we start with 2.00 mol of $NO_2(g)$ per liter and 0 mol of $N_2O_4(g)$, then we find that the concentrations $[NO_2]$ and $[N_2O_4]$ change with time in the manner shown in Figure 15-2. Note that the equilibrium values of $[N_2O_4]$ and $[NO_2]$ are the same in Figure 15-2 as they are in Figure 15-1.

At equilibrium the reactant and product concentrations are constant.

The equilibrium values of $[N_2O_4]$ and $[NO_2]$ for several sets of initial conditions are given in Table 15-1. The most remarkable feature of the data in Table 15-1 is shown in the last column. The value of the ratio $[NO_2]^2/[N_2O_4]$ at equilibrium for the reaction

$$N_2O_4(g) \rightleftharpoons 2NO_2(g)$$

is equal to a constant. From Table 15-1, we see that

$$\frac{[NO_2]^2}{[N_2O_4]} = 0.20 \text{ M (at equilibrium at 100°C)}$$

Table 15-1 shows that the constant value of $[NO_2]^2/[N_2O_4]$ at equilibrium *is independent of the initial values,* $[NO_2]_0$ *and* $[N_2O_4]_0$. The reason behind this result is given in the next section.

The data in Figures 15-1 and 15-2 illustrate two essential facets of chemical equilibrium:

1. At equilibrium, the reactant and product concentrations show no further change with time.
2. The same equilibrium state is attained starting either from the reactant side or the product side of the equation.

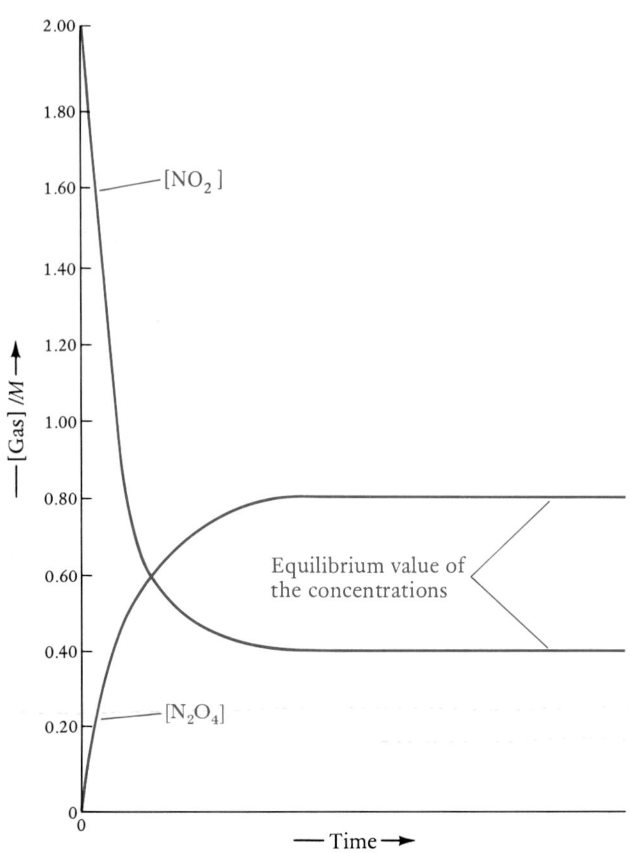

Figure 15-2 Plot of $[N_2O_4]$ and $[NO_2]$ as a function of time at 100°C for the reaction $N_2O_4(g) \rightleftharpoons 2NO_2(g)$. At time zero $[NO_2]_0 = 2.00$ M and $[N_2O_4]_0 = 0$. As the reaction proceeds, $[N_2O_4]$ increases from zero to 0.80 M and then remains constant, whereas $[NO_2]$ decreases from 2.00 M to 0.40 M and then remains constant. The values $[N_2O_4] =$ 0.80 M and $[NO_2] = 0.40$ M are equilibrium values. When the reaction mixture attains equilibrium, $[N_2O_4]$ and $[NO_2]$ remain constant.

Table 15-1 Initial and equilibrium values at 100°C of $[N_2O_4]$ and $[NO_2]$ for the reaction $N_2O_4(g) \rightleftharpoons 2NO_2(g)$

Initial concentration/M		*Equilibrium concentration*/M		*Value at equilibrium of the quantity*
$[N_2O_4]_0$	$[NO_2]_0$	$[N_2O_4]$	$[NO_2]$	$\{[NO_2]^2/[N_2O_4]\}/M$
1.00	0	0.80	0.40	0.20
2.00	0	1.71	0.58	0.20
0	2.00	0.80	0.40	0.20
0	1.00	0.36	0.27	0.20

Example 15-2: Suppose that it was determined that at 100°C $[NO_2] = 0.175$ M at equilibrium for the reaction

$$N_2O_4(g) \rightleftharpoons 2NO_2(g)$$

Calculate $[N_2O_4]$ at equilibrium at 100°C.

Solution: According to Table 15-1,

$$\frac{[NO_2]^2}{[N_2O_4]} = 0.20 \text{ M}$$

at equilibrium at 100°C. If $[NO_2] = 0.175$ M, then we have

$$[N_2O_4] = \frac{[NO_2]^2}{0.20 \text{ M}} = \frac{[0.175 \text{ M}]^2}{0.20 \text{ M}} = 0.15 \text{ M}$$

When given one equilibrium concentration & you have to find the other you need to know: K or in. [ ].

15-3. THE EQUILIBRIUM CONSTANT EXPRESSION FOR A CHEMICAL EQUATION IS EQUAL TO THE RATIO OF PRODUCT CONCENTRATION TERMS TO REACTANT CONCENTRATION TERMS

The general solution to the problem of formulating equilibrium constant expressions for chemical reactions was put forth by the Norwegian chemists Cato Guldberg and Peter Waage in 1864 and is referred to in this book as the *law of concentration action.*

The law of concentration action tells us how to write the equilibrium constant expression for a chemical equation.

Consider the balanced generalized chemical equation

$$aA(g) + bB(soln) + cC(s) \rightleftharpoons xX(g) + yY(soln) + zZ(l) \quad (15\text{-}2)$$

at equilibrium. Guldberg and Waage, on the basis of their experimental observations on a variety of chemical reactions, postulated that the *equilibrium constant* for Reaction (15-2) is given by the expression

$$K_c = \frac{[X]^x[Y]^y}{[A]^a[B]^b} \quad (15\text{-}3)$$

In other words, the *equilibrium constant expression* for a reaction is *the ratio of product concentrations to reactant concentrations, with each concentration factor raised to a power equal to the stoichiometric coefficient of that species in the balanced equation. Pure liquids and solids, whose concentration cannot be varied, do not appear in the equilibrium constant expression.* Because their concentrations cannot be varied, pure solids and liquids exert a constant effect on the reaction equilibrium and so are not included in Equation (15-3).

The subscript c in Equation (15-3) emphasizes that the equilibrium constant is expressed in terms of concentrations. We shall soon see that we can express equilibrium constants in terms of pressures also, in which case we write K_p instead of K_c. The crucial point here is that the value of K_c in Equation (15-3) is equal to a constant at a given temperature.

Application of the law of concentration action to the equilibrium

$$N_2O_4(g) \rightleftharpoons 2NO_2(g)$$

yields

$$K_c = \frac{[NO_2]^2}{[N_2O_4]} \tag{15-4}$$

From the results in the preceding section, we have for the $N_2O_4(g)$ dissociation reaction at 100°C that $K_c = 0.20$ M.

There is a simple way to see why $[NO_2]$ is squared in Equation (15-4). If we rewrite Reaction (15-1) as

$$N_2O_4(g) \rightleftharpoons NO_2(g) + NO_2(g)$$

then the origin of the $[NO_2]^2$ term is seen more clearly:

$$K_c = \frac{[NO_2][NO_2]}{[N_2O_4]} = \frac{[NO_2]^2}{[N_2O_4]}$$

In most books the Guldberg and Waage law of concentration action is referred to as the *law of mass action* or the *mass action law*. If we interpret Guldberg and Waage's "active masses" as concentrations, which was apparently their intent, then the designation law of concentration action is a more informative description of the principle involved in the law than is law of mass action.

15-4. EQUILIBRIUM CONSTANTS CAN BE EXPRESSED IN TERMS OF PRESSURES FOR GAS PHASE REACTIONS

We can express the equilibrium constant for Reaction (15-1) in terms of pressure as well as concentration. To see how to do this, recall that

$$PV = nRT \tag{15-5}$$

The concentration of a gas is equal to the number of moles of gas per liter, that is,

$$[\text{gas}] = \frac{n}{V} \tag{15-6}$$

where the square brackets denote *molar* concentration. Equation (15-5) can be rearranged to

$$P = \left(\frac{n}{V}\right)RT \qquad (15\text{-}7)$$

Substitution of Equation (15-6) into (15-7) yields

$$P = [\text{gas}]RT \qquad (15\text{-}8)$$

or

$$[\text{gas}] = \frac{P}{RT} \qquad (15\text{-}9)$$

The concentration of a gas is directly proportional to the pressure of the gas.

If we substitute Equation (15-9) into (15-4), then we find that

$$K_c = \frac{(P_{NO_2}/RT)^2}{P_{N_2O_4}/RT} = \frac{1}{RT}\frac{P^2_{NO_2}}{P_{N_2O_4}} \qquad (15\text{-}10)$$

We denote the ratio of the pressures in Equation (15-10) by K_p:

$$K_p = \frac{P^2_{NO_2}}{P_{N_2O_4}} \qquad (15\text{-}11)$$

where the subscript p indicates that the equilibrium constant is expressed in terms of pressures rather than concentrations. The relation between K_c and K_p for Reaction (15-1) is

The relation between K_c and K_p depends on the reaction stoichiometry.

$$K_c = \frac{1}{RT}K_p \qquad (15\text{-}12)$$

Note that K_c and K_p have the same form; K_c is a ratio of concentrations, and K_p is a similar ratio in terms of pressures. The important point is that if K_c is a constant at constant temperature, then so is K_p because they differ only by a factor RT, which is a constant at constant temperature.

Example 15-3: Derive the relation between K_c and K_p for the reaction

$$2SO_2(g) + O_2(g) \rightleftharpoons 2SO_3(g)$$

The reaction in Example 15-3 is an important step in the production of sulfuric acid (Interchapter D).

Solution: The expression for K_c is

$$K_c = \frac{[SO_3]^2}{[SO_2]^2[O_2]}$$

If we substitute Equation (15-9) into this expression for K_c, then we obtain

$$K_c = \frac{(P_{SO_3}/RT)^2}{(P_{SO_2}/RT)^2(P_{O_2}/RT)} = RT\frac{P_{SO_3}^2}{P_{SO_2}^2 P_{O_2}} = RT\,K_p$$

Note that the relation between K_c and K_p depends upon the stoichiometry of the reaction.

Some additional examples of the application of the law of concentration action (Equation 15-3) are

(1) $$C(s) + H_2O(g) \rightleftharpoons CO(g) + H_2(g)$$

$$K_c = \frac{[CO][H_2]}{[H_2O]} \qquad K_p = \frac{P_{CO}P_{H_2}}{P_{H_2O}}$$

Note that C(s) does not appear in the K_c and K_p expressions because it is a pure solid.

(2) $$HNO_2(aq) \rightleftharpoons H^+(aq) + NO_2^-(aq)$$

$$K_c = \frac{[H^+][NO_2^-]}{[HNO_2]}$$

There is no corresponding K_p expression in this case because none of the reactants and products are in the gas phase.

Example 15-4: Use the law of concentration action to write the equilibrium constant expressions for the following reactions:

(a) $N_2(g) + 3H_2(g) \rightleftharpoons 2NH_3(g)$

(b) $C(s) + CO_2(g) \rightleftharpoons 2CO(g)$

Solution: For Reaction (a), all the reactants and products are gases and so we have

$$K_c = \frac{[NH_3]^2}{[N_2][H_2]^3}$$

Problems 15-1 through 15-6 will give you practice in writing K_c expressions.

whereas for Reaction (b) we have

$$K_c = \frac{[CO]^2}{[CO_2]}$$

Note that C(s) does not appear in the K_c expression for Reaction (b) because carbon is a pure solid. The relations between K_c and K_p are

$$K_c = \frac{(P_{NH_3}/RT)^2}{(P_{N_2}/RT)(P_{H_2}/RT)^3} = (RT)^2 \frac{P^2_{NH_3}}{P_{N_2}P^3_{H_2}} = (RT)^2 K_p$$

and

$$K_c = \frac{(P_{CO}/RT)^2}{(P_{CO_2}/RT)} = \frac{1}{RT}\frac{P^2_{CO}}{P_{CO_2}} = \frac{1}{RT}K_p$$

Example 15-5: Suppose that a sample of HI(g) is injected into a reaction vessel and the following reaction equilibrium is established:

$$2HI(g) \rightleftharpoons H_2(g) + I_2(g)$$

Analysis of the equilibrium reaction mixture yields the following results:

$$[HI] = 0.27\ M \qquad [H_2] = [I_2] = 0.86\ M$$

Use these results to calculate K_c for the reaction.

The reaction equilibrium in Example 15-5 can be studied by observing the intensity of the purple color, which is due to $I_2(g)$ because $H_2(g)$ and HI(g) are colorless.

Solution: Application of the law of concentration action to the reaction yields

$$K_c = \frac{[H_2][I_2]}{[HI]^2}$$

Substituting the values of the equilibrium concentrations into the K_c expression yields

$$K_c = \frac{(0.86\ M)(0.86\ M)}{(0.27\ M)^2} = 10$$

15-5. EQUILIBRIUM CONSTANTS ARE USED IN A VARIETY OF CALCULATIONS

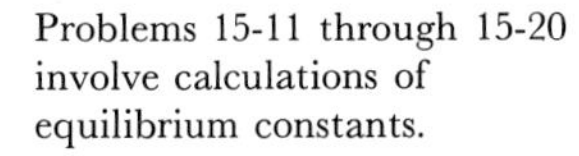

Problems 15-11 through 15-20 involve calculations of equilibrium constants.

The following three examples illustrate some types of equilibrium calculations.

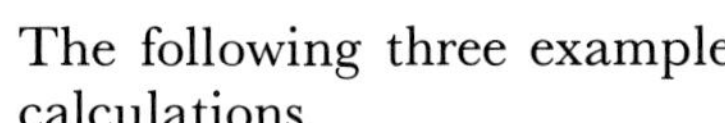

Example 15-6: Suppose that 2.00 mol of $N_2O_4(g)$ is injected into a 1.00-L reaction vessel held at 100°C and that the reaction

$$N_2O_4(g) \rightleftharpoons 2NO_2(g) \qquad K_c = 0.20\ M$$

proceeds to equilibrium. Compute the equilibrium values of $[N_2O_4]$ and $[NO_2]$.

Solution: From the law of concentration action and Table 15-1, we have

$$K_c = \frac{[NO_2]^2}{[N_2O_4]} = 0.20\ \text{M} \tag{15-13}$$

It is helpful in working equilibrium calculations to set the problem up in tabular form with the initial concentrations and equilibrium concentrations of each species placed in separate rows directly below the species in the chemical equation. Thus from the data given we have

	$N_2O_4(g)$ $\rightleftharpoons$	$2NO_2(g)$
initial concentration	2.00 M	0
equilibrium concentration		

We now use the reaction stoichiometry to obtain expressions for the equilibrium concentrations of N_2O_4 and NO_2. Let the number of moles per liter of $N_2O_4(g)$ that dissociates into $NO_2(g)$ be x; then the equilibrium concentration of N_2O_4 is

$$[N_2O_4] = 2.00\ \text{M} - x$$

Each mole of N_2O_4 that dissociates produces 2 mol of NO_2, and so the equilibrium concentration of $NO_2(g)$ expressed in terms of x is

$$[NO_2] = 2x$$

Thus our completed table for the calculation is

	$N_2O_4(g)$ $\rightleftharpoons$	$2NO_2(g)$
initial concentration	2.00 M	0
equilibrium concentration	2.00 M $- x$	$2x$

Substituting the equilibrium concentration data from this table into Equation (15-13) yields

$$\frac{(2x)^2}{(2.00\ \text{M} - x)} = 0.20\ \text{M}$$

or

$$4x^2 = (0.20\ \text{M})(2.00\ \text{M} - x) = 0.40\ \text{M}^2 - (0.20\ \text{M})x$$

Rearranging this equation to the standard form of the quadratic equation we obtain

$$4x^2 + (0.20\ \text{M})x - 0.40\ \text{M}^2 = 0$$

The two roots of this equation are given by the *quadratic formula** and thus

The quadratic formula is discussed in detail in Appendix A and in the Study Guide/Solutions Manual.

$$x = \frac{-0.20\ \text{M} \pm \sqrt{(0.20\ \text{M})^2 - (4)(4)(-0.40\text{M}^2)}}{(2)(4)} = 0.29\ \text{M}$$

We have rejected the negative value of x (-0.34 M) as physically unacceptable because only positive values of concentrations have physical meaning. From the value $x = 0.29$ M, we compute that at equilibrium

$$[NO_2] = 2x = (2)(0.29\ \text{M}) = 0.58\ \text{M}$$

$$[N_2O_4] = 2.00\ \text{M} - x = 2.00\ \text{M} - 0.29\ \text{M} = 1.71\ \text{M}$$

As a final check, we note that $[NO_2]^2/[N_2O_4]$ is equal to 0.20 M, the value of K_c given in Equation (15-13).

The quantity K_c defined in Equation (15-3) is an equilibrium constant in terms of concentrations and generally has units involving concentration. For example, Equation (15-13) shows that K_c has units of M^2/M, or M, for the reaction $N_2O_4(g) \rightleftharpoons 2NO_2(g)$. More advanced treatments of chemical equilibria use equilibrium constants that are defined in such a way that they are unitless. In recognition of this, some general chemistry textbooks do not include units for K_c, but this is inconsistent with the definition in Equation 15-3. In this text we always include the units of K_c. We do this not only because it is correct but because the units in calculations like those in Example 15-6 follow logically. A similar argument applies to K_p, which generally has units involving pressure.

Example 15-7: Suppose that we place 1.0 mol of each of the gases HI(g), $H_2(g)$, and $I_2(g)$ in a 2.0-L reaction vessel and that the following equilibrium is established:

$$2HI(g) \rightleftharpoons H_2(g) + I_2(g)$$

Given that $K_c = 10$ for this reaction at the temperature of the equilibrium reaction mixture, compute the equilibrium values of [HI], $[H_2]$, and $[I_2]$.

*The *quadratic equation* $ax^2 + bx + c = 0$ has the solutions

$$x = \frac{-b \pm \sqrt{b^2 - 4ac}}{2a} \qquad (15\text{-}14)$$

Only positive values of pressure or concentration have physical significance.

Solution: We proceed with the calculation by setting up a table of initial and equilibrium concentrations. The initial concentrations of [HI], $[H_2]$, and $[I_2]$ are (1.0 mol/2.0 L) = 0.50 M. If we let x be the number of moles per liter of H_2 and I_2 that are produced by the dissociation of HI, then we obtain

It takes $2x$ moles of HI to make x moles of H_2 or I_2.

	$2HI(g)$	$\rightleftharpoons H_2(g)$	$+\ I_2(g)$
initial concentration	0.50 M	0.50 M	0.50 M
equilibrium concentration	$0.50\text{ M} - 2x$	$0.50\text{ M} + x$	$0.50\text{ M} + x$

Application of the law of concentration action to the reaction equilibrium yields

$$K_c = \frac{[H_2][I_2]}{[HI]^2} = 10$$

Substitution of the equilibrium concentration expressions from the table into the K_c expression yields

$$\frac{(0.50\text{ M} + x)^2}{(0.50\text{ M} - 2x)^2} = 10$$

We note that the left side of this equation is a perfect square, and thus we take the square root of both sides and obtain

$$\frac{0.50\text{ M} + x}{0.50\text{ M} - 2x} = \sqrt{10} = 3.16$$

Thus

$$0.50\text{ M} + x = 3.16(0.50\text{ M} - 2x)$$
$$0.50\text{ M} + x = 1.58\text{ M} - 6.32x$$

Rearranging yields

$$7.32x = 1.08\text{ M}$$
$$x = 1.08\text{ M}/7.32 = 0.148\text{ M}$$

The equilibrium concentrations are

$$[H_2] = [I_2] = 0.50\text{ M} + x = 0.50\text{ M} + 0.15\text{ M} = 0.65\text{ M}$$
$$[HI] = 0.50\text{ M} - 2x = 0.50\text{ M} - 2(0.15\text{ M}) = 0.20\text{ M}$$

As a final check, we note that

$$\frac{[H_2][I_2]}{[HI]^2} = \frac{(0.65\text{ M})(0.65\text{ M})}{(0.20\text{ M})^2} = 10.6 \approx 10$$

Although 10.6 rounds off to 11 and $K_c = 10$, the difference is just a result of accumulated round-off error in the calculation.

Example 15-8: Given that $CO_2(g)$ reacts with $C(s)$ via the reaction

$$C(s) + CO_2(g) \rightleftharpoons 2CO(g) \qquad K_p = 1.90\text{ atm}$$

and that at equilibrium the total pressure in the reaction vessel is 2.00 atm, compute P_{CO_2} and P_{CO} at equilibrium. Note that this problem involves an equilibrium constant expressed in pressure units rather than concentration units.

Carbon monoxide (CO) is a very poisonous gas produced in the incomplete combustion of charcoal, wood, and other fuels.

Solution: We are given only the total pressure, which is equal to the sum of the partial pressures of $CO(g)$ and $CO_2(g)$:

$$P_{tot} = P_{CO} + P_{CO_2} = 2.00\text{ atm}$$

The reaction equilibrium fixes the ratio P^2_{CO}/P_{CO_2} at 1.90 atm, and so

$$K_p = \frac{P^2_{CO}}{P_{CO_2}} = \frac{P^2_{CO}}{2.00\text{ atm} - P_{CO}} = 1.90\text{ atm}$$

We can rearrange this equation to get

$$P^2_{CO} = (1.90\text{ atm})(2.00\text{ atm} - P_{CO}) = 3.80\text{ atm}^2 - (1.90\text{ atm})\,P_{CO}$$

or

$$P^2_{CO} + (1.90\text{ atm})\,P_{CO} - 3.80\text{ atm}^2 = 0$$

This is a quadratic equation in P_{CO}, and the two roots are obtained from the quadratic formula:

$$P_{CO} = \frac{-1.90\text{ atm} \pm \sqrt{(1.90\text{ atm})^2 - (4)(1)(-3.80\text{ atm}^2)}}{2}$$

$$= \frac{-1.90\text{ atm} \pm 4.34\text{ atm}}{2}$$

The two values that we find are 1.22 atm and -3.12 atm. Only positive values of pressure or concentration have physical significance, and so we reject the negative root and write

$$P_{CO} = 1.22\text{ atm}$$

The pressure of $CO_2(g)$ at equilibrium is thus

$$P_{CO_2} = 2.00\text{ atm} - P_{CO} = 2.00\text{ atm} - 1.22\text{ atm} = 0.78\text{ atm}$$

Problems 15-21 through 15-38 give more examples of calculations involving chemical equilibrium.

15-6. A CHEMICAL REACTION DISPLACED FROM EQUILIBRIUM PROCEEDS TOWARD A NEW EQUILIBRIUM STATE IN THE DIRECTION THAT AT LEAST PARTIALLY OFFSETS THE CHANGE IN CONDITIONS

Henri Le Châtelier, a French physical chemist, is most famous for postulating the principle that now bears his name. He also investigated gas combustion and platinum alloys.

The direction in which an equilibrium chemical reaction shifts (left to right or right to left) in response to a change in conditions can be predicted by using *Le Châtelier's principle* (pronounced luh shat′elyay). In the application of Le Châtelier's principle, we consider the chemical reaction to be initially at equilibrium and then subject it to a change in conditions that displaces the system from equilibrium. We do not need to know the numerical value of the equilibrium constant K in order to apply Le Châtelier's principle, which can be stated as follows:

If a chemical reaction at equilibrium is subjected to a change in conditions that displaces it from equilibrium, then the reaction proceeds toward a new equilibrium state in the direction that—at least partially—offsets the change in conditions.

The conditions that can affect a reaction equilibrium are

1. Concentration of a reactant or product
2. Reaction volume or applied pressure
3. Temperature

Although Le Châtelier's principle sounds imposing, it is actually simple to apply. A key point to recognize is that the equilibrium constant depends only on the temperature; it does not change when the reactant or product concentrations, the reaction volume, or the applied pressure is changed.

Consider the reaction equilibrium

$$C(s) + CO_2(g) \rightleftharpoons 2CO(g)$$

If we disturb the equilibrium by injecting some additional $CO_2(g)$ into the reaction vessel, then the concentration of $CO_2(g)$ is increased. In response to the change in conditions, the reaction equilibrium shifts from left to right because this is the direction in which $CO_2(g)$ is consumed and this leads to a partial decrease of the increase in the CO_2 concentration. In the new equilibrium state, the concentration of CO_2 and the concentration of CO are both greater than in the original equilibrium state, but the concentration of CO_2 in the new equilibrium state is less than it was immediately after the additional CO_2 was injected.

Le Châtelier's principle is used to make qualitative predictions of shifts in reaction equilibria.

If we disturb the reaction equilibrium by injecting some additional $CO(g)$ into the reaction vessel, then the concentration of CO is increased and the reaction equilibrium shifts from right to left because this is the direction that decreases the CO concentration.

If we inject some additional $C(s)$ into the reaction vessel, then there is *no* shift in the reaction equilibrium because the *concentration* of a solid is independent of the amount present. In other words, the injection or removal of some C(*s*) does not displace the reaction from equilibrium. Generally, the further addition or partial removal of any solid reactant or product does not shift the reaction equilibrium.

A dynamic equilibrium responds to a change in conditions that disturbs the equilibrium.

If we remove some CO(*g*) from the equilibrium reaction mixture, then the reaction equilibrium shifts from left to right to produce more CO(*g*) because this is the direction that partially offsets the change in conditions. The concentration of CO(*g*) in the new equilibrium state is less than that in the original equilibrium state but greater than the concentration of CO(*g*) in the reaction vessel immediately after the removal of CO(*g*).

Example 15-9: Consider the reaction equilibrium

$$N_2(g) + 3H_2(g) \rightleftharpoons 2NH_3(g)$$

Use Le Châtelier's principle to predict the effect on the equilibrium concentration of $NH_3(g)$ and of $N_2(g)$ produced by an increase in the concentration of $H_2(g)$.

Solution: An increase in the concentration of $H_2(g)$, which is a reactant, shifts the equilibrium from left to right, and this produces an *increase* in the equilibrium concentration of $NH_3(g)$. The equilibrium concentration of $N_2(g)$ decreases because N_2 is consumed in the production of $NH_3(g)$.

15-7. A DECREASE IN VOLUME OR INCREASE IN APPLIED PRESSURE SHIFTS A REACTION EQUILIBRIUM TOWARD THE SIDE WITH FEWER MOLES OF GAS

Consider the effect of a change in volume on the reaction equilibrium $N_2O_4(g) \rightleftharpoons 2NO_2(g)$. Suppose that we decrease the volume available to the reaction system by increasing the applied pressure twofold. The immediate effect is a decrease in the available volume by a factor of 2 and a doubling of the total number of gas molecules per unit volume (Figure 15-3). The reaction system can partially offset the increase in the number of gas molecules per unit volume by shifting from right to left. It requires two NO_2 molecules to make one N_2O_4 molecule, and thus a right-to-left shift decreases the total number of gas molecules per unit volume and thereby partially offsets the change in conditions.

In general, a decrease in volume shifts a reaction equilibrium toward the side with the smaller number of moles of gas. If the change in number of moles of gas for a reaction is positive, as in the $N_2O_4(g) \rightleftharpoons 2NO_2(g)$ reaction, for example, then a decrease in volume shifts the reaction equilibrium from right to left. If the change in number of moles of gas for a reaction is negative, as in $N_2(g) + 3H_2(g) \rightleftharpoons 2NH_3(g)$, then a decrease in volume shifts the reaction equilibrium from left to right.

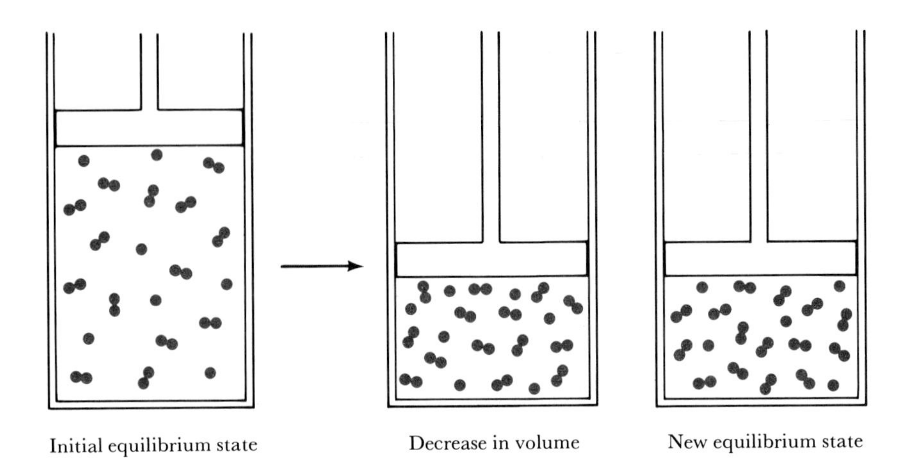

Figure 15-3 The immediate result of a decrease in total volume on the reaction

$$N_2O_4(g) \rightleftharpoons 2NO_2(g)$$
$$(\bullet\!\bullet \rightleftharpoons 2\bullet)$$

is a doubling of the total number of gas molecules per unit volume. The reaction shifts right to left to a new equilibrium state because that is the direction that decreases the total number of gas molecules per unit volume.

If the number of moles of gas is the same on both sides, as in the equation

$$H_2(g) + I_2(g) \rightleftharpoons 2HI(g)$$

then a change in volume has no effect on the reaction equilibrium because there is no direction in which the reaction can shift to change the total number of gas molecules per unit volume.

Example 15-10: Should the total volume of the equilibrium reaction mixture

$$C(s) + CO_2(g) \rightleftharpoons 2CO(g)$$

be increased or decreased in order to increase the extent of conversion of carbon and carbon dioxide to carbon monoxide?

Solution: An increase in total volume favors the side of the reaction with the greater number of moles of *gas*. Thus, an increase in volume would shift this reaction from left to right, thereby increasing the production of $CO(g)$ from $C(s)$ and $CO_2(g)$.

Note that, in assessing the effect of a volume change on a reaction equilibrium, we do not have to consider pure liquid and solid phases because the change in total gas volume does not affect the concentration of species in solid and liquid phases.

A change in reaction volume or applied pressure produces a change in the concentration of a gaseous species.

From Boyle's law we know that the volume of an ideal gas is inversely proportional to the applied pressure. Thus, the results that we have just deduced for changes in volume are the opposite of what we would predict for changes in applied pressure. For example, in Example 15-10, Le Châtelier's principle says that we should decrease the pressure on the reaction system in order to produce more $CO(g)$.

A decrease in the applied pressure is equivalent to an increase in the reaction volume.

Example 15-11: Consider the reaction equilibrium

$$N_2(g) + 3H_2(g) \rightleftharpoons 2NH_3(g)$$

Suppose that the applied pressure on the system is increased. In what direction does the reaction shift toward a new equilibrium state?

Solution: An increase in applied pressure causes the reaction to shift to the side with the fewer moles of gas because this is the direction that partially offsets the increase in pressure. Thus the reaction equilibrium shifts to the right when the applied pressure is increased.

15-8. AN INCREASE IN TEMPERATURE SHIFTS A REACTION EQUILIBRIUM IN THE DIRECTION IN WHICH HEAT IS ABSORBED

The value of ΔH°_{rxn} for the reaction

$$CaCO_3(s) \rightleftharpoons CaO(s) + CO_2(g)$$

is +158 kJ. In other words, the conversion of 1.00 mol of $CaCO_3(s)$ to 1.00 mol of $CaO(s)$ plus 1.00 mol of $CO_2(g)$ requires an input of energy of 158 kJ as heat, and the reverse reaction liberates 158 kJ of energy as heat. Thus we have

$$CaCO_3(s) + 158\text{ kJ} \rightleftharpoons CaO(s) + CO_2(g)$$

or

$$CaCO_3(s) + \text{heat} \rightleftharpoons CaO(s) + CO_2(g)$$

An increase in temperature increases the availability of energy to the reaction, and thus Le Châtelier's principle tells us that the equilibrium shifts in the direction that counteracts this change, that is, from left to right. The left-to-right shift produces an increase in the equilibrium concentration of $CO_2(g)$ in the system. When the temperature increases, the system absorbs energy as heat in an attempt to reduce its temperature to the original value. If the temperature increase is maintained by continuous heating, then the reaction continues to proceed from left to right until a new equilibrium state is achieved at an increased value of $[CO_2]$.

For an endothermic reaction ($\Delta H^\circ_{rxn} > 0$), a reaction equilibrium shifts to the right when the temperature is increased and to the left when the temperature is decreased. For an exothermic reaction ($\Delta H^\circ_{rxn} < 0$), the equilibrium shifts to the left when the temperature is increased and to the right when the temperature is decreased.

The effect of temperature on a reaction equilibrium has its origin in the dependence of the reaction equilibrium constant on temperature. If ΔH°_{rxn} for a reaction is positive, then the value of K increases with increasing temperature. If ΔH°_{rxn} for a reaction is negative, then K decreases with increasing temperature.

The various possible changes in reaction conditions and the resulting effects on reaction equilibria as deduced from Le Châtelier's principle are summarized in Table 15-2.

Example 15-12: For the reaction

$$N_2(g) + 3H_2(g) \rightleftharpoons 2NH_3(g)$$

$\Delta H^\circ_{rxn} = -92$ kJ. Will an increase in the temperature increase or decrease the extent of conversion of N_2 and H_2 to NH_3?

In Problems 15-39 through 15-50 you use Le Châtelier's principle to predict the effect on the equilibrium arising from changes in reaction conditions.

Solution: The value of ΔH°_{rxn} is negative, and thus the reaction evolves energy as heat:

$$N_2(g) + 3H_2(g) \rightleftharpoons 2NH_3(g) + 92 \text{ kJ}$$

An increase in temperature favors the absorption of energy as heat and thus shifts the equilibrium to the left, which decreases the yield of ammonia.

15-9. CHEMICAL REACTIONS ALWAYS PROCEED TOWARD EQUILIBRIUM

At 100°C the equilibrium constant expression for the reaction

$$N_2O_4(g) \rightleftharpoons 2NO_2(g)$$

is

$$K_c = \frac{[NO_2]^2}{[N_2O_4]} = 0.20 \text{ M} \qquad (15\text{-}15)$$

We now define the *reaction quotient*, Q_c, as the quantity that has exactly the same algebraic form as the equilibrium-constant expression for a reaction, but has *arbitrary* concentrations for the substances involved. Thus, for the N_2O_4–NO_2 reaction, we have the Q_c expression

$$Q_c = \frac{[NO_2]_0^2}{[N_2O_4]_0} \qquad (15\text{-}16)$$

Table 15-2 Effects on reaction equilibria arising from changes in reaction conditions, as predicted by Le Châtelier's principle

Change in conditions	*Direction of shift in reaction equilibrium*
Concentration	
increase in concentration of a reactant	$\longrightarrow$
increase in concentration of a product	$\longleftarrow$
decrease in concentration of a reactant	$\longleftarrow$
decrease in concentration of a product	$\longrightarrow$
Temperature	
endothermic reaction, $\Delta H^{\circ}_{rxn} > 0$	
increase in temperature	$\longrightarrow$
decrease in temperature	$\longleftarrow$
exothermic reaction, $\Delta H^{\circ}_{rxn} < 0$	
increase in temperature	$\longleftarrow$
decrease in temperature	$\longrightarrow$
Total Volume	
more moles of gas on right	
decrease in total volume	$\longleftarrow$
increase in total volume	$\longrightarrow$
more moles of gas on left	
decrease in total volume	$\longrightarrow$
increase in total volume	$\longleftarrow$

where the zero subscripts denote either arbitrary or initial concentrations. The difference between Equation (15-15) and Equation (15-16) is that only equilibrium values of the concentrations can be used in Equation (15-15), because Equation (15-15) applies only at equilibrium. However, arbitrary values of the concentrations can be used in Equation (15-16). The reaction quotient Q_c is not a constant; instead, it takes on whatever value results from the way we prepare the reaction system. For example, suppose we mix 2.00 mol of $N_2O_4(g)$ with 2.00 mol of $NO_2(g)$ in a 1.00-L reaction vessel at 100°C. The value of Q_c for the reaction

$$N_2O_4(g) \rightleftharpoons 2NO_2(g)$$

is then equal to

$$Q_c = \frac{[NO_2]_0^2}{[N_2O_4]_0} = \frac{(2.00\text{ M})^2}{(2.00\text{ M})} = 2.00\text{ M}$$

The value of K_c at 100°C for this reaction is 0.20 M; therefore, $Q_c \neq K_c$ and the reaction mixture is not at equilibrium. At equilibrium the ratio of $[NO_2]^2$ to $[N_2O_4]$ must equal 0.20 M:

$$K_c = \frac{[NO_2]^2}{[N_2O_4]} = 0.20\text{ M} \qquad \text{(at equilibrium at 100°C)}$$

In order for the reaction system with $Q_c = 2.00$ M to attain equilibrium, the value of Q_c must decrease from 2.00 M to 0.20 M. Because $Q_c = [NO_2]_0^2/[N_2O_4]_0$, we see that Q_c is decreased if $[NO_2]_0$ decreases and $[N_2O_4]_0$ increases. Consequently, the reaction in which $[N_2O_4]_0 = 2.00$ M and $[NO_2]_0 = 2.00$ M proceeds from right to left toward equilibrium because this is the direction in which the value of $[NO_2]$ decreases and the value of $[N_2O_4]$ increases. When a reaction system reaches equilibrium, the value of Q_c is equal to the value of K_c. That is, for any reaction at equilibrium,

$$\frac{Q_c}{K_c} = 1 \qquad \text{(at equilibrium)}$$

The *direction of reaction spontaneity* is always toward equilibrium. The numerical value of the ratio Q_c/K_c tells us the direction (left to right or right to left) in which a reaction system not at equilibrium spontaneously proceeds toward equilibrium. The various possibilities are as follows:

Value of Q/K	*Direction the reaction proceeds toward equilibrium*
$Q/K < 1$	$\longrightarrow$
$Q/K > 1$	$\longleftarrow$
$Q/K = 1$	no net change (equilibrium state)

In other words, a system that is not in equilibrium proceeds toward equilibrium in the direction in which Q approaches K in magnitude. If Q is greater than K, then the value of Q decreases toward K and the reaction proceeds from right to left as the reaction moves toward equilibrium. If Q is smaller than K, then the value of Q increases toward K and the reaction proceeds from left to right as the reaction moves toward equilibrium. If $Q = K$, then the reaction is at equilibrium and no further net change occurs.

Both Q and K must be expressed in the same units. Thus if we express K in concentration units (K_c), then Q also must be expressed in the same concentration units (Q_c).

Example 15-13: Suppose that $CO_2(g)$ and $CO(g)$ are brought into contact with $C(s)$ at 1000 K at $P_{CO_2} = 2.00$ atm and $P_{CO} = 0.50$ atm. Is the reaction

$$C(s) + CO_2(g) \rightleftharpoons 2CO(g) \qquad K_p = 1.90 \text{ atm}$$

at equilibrium? If not, in what direction will the reaction proceed toward equilibrium?

Solution: The value of Q_p for the reaction system as prepared is

$$Q_p = \frac{P_{CO}^2}{P_{CO_2}} = \frac{(0.50 \text{ atm})^2}{(2.00 \text{ atm})} = 0.13 \text{ atm}$$

The value of K_p is given as 1.90 atm, and so

$$\frac{Q_p}{K_p} = \frac{0.13}{1.90} < 1$$

In Problems 15-51 through 15-56 you calculate the value of Q and then use it to predict the direction in which the reaction proceeds.

Therefore, the reaction system is not at equilibrium. Because $Q/K < 1$, the reaction proceeds toward equilibrium from left to right with P_{CO} increasing and P_{CO_2} decreasing until $P_{CO}^2/P_{CO_2} = 1.90$ atm, that is, until equilibrium is attained.

Example 15-14: For the reaction

$$N_2O_4(g) \rightleftharpoons 2NO_2(g)$$

$K_c = 0.20$ M at 100°C. Suppose that this reaction is initially at equilibrium and that the reaction volume is then decreased by a factor of 2. Determine the value of Q_c in the initial equilibrium state and immediately following the twofold decrease in volume.

Solution: When a reaction is at equilibrium $Q_c = K_c$, and thus $Q_c = 0.20$ M in the initial equilibrium state. The twofold decrease in the reaction volume has no effect on K_c because the value of K_c is constant at a given temperature. However, the value of Q_c does change as a result of the change in volume because the twofold volume decrease doubles the concentrations of $NO_2(g)$ and $N_2O_4(g)$. Thus, denoting the equilibrium concentrations prior to the volume change by $[NO_2]_{eq}$ and $[N_2O_4]_{eq}$, we have for the value of Q_c immediately following the volume decrease

$$Q_c = \frac{[NO_2]_0^2}{[N_2O_4]_0} = \frac{\{2[NO_2]_{eq}\}^2}{2[N_2O_4]_{eq}} = \frac{4}{2}\frac{[NO_2]_{eq}^2}{[N_2O_4]_{eq}} = 2K_c = 0.40\text{ M}$$

Therefore

$$\frac{Q_c}{K_c} = 2 > 1$$

and the reaction shifts from right to left toward the new equilibrium state, just as we would predict using Le Châtelier's principle.

It is important to understand the distinction between Q and K. The expressions for Q and K for a particular reaction have exactly the same algebraic form. However, only equilibrium values of the concentrations can be substituted into the K expression, whereas nonequilibrium values of the concentrations can be substituted into the Q expression. For a given reaction equilibrium at a particular temperature, the value of K is fixed. The value of Q is not fixed, however, but rather is determined by how we prepare the reaction system. Only if $Q = K$ is an arbitrarily prepared reaction system at equilibrium; otherwise the composition of the reaction system, and the value of Q, change as the reaction proceeds toward equilibrium.

15-10. EQUILIBRIUM CONSTANTS FOR CHEMICAL REACTIONS CAN BE COMBINED TO OBTAIN EQUILIBRIUM CONSTANTS FOR OTHER REACTIONS

It is sometimes necessary to compute the equilibrium constant for a chemical reaction from the equilibrium constants for other, algebraically related chemical reactions. As a simple example, consider the pair of equations

$$(1)\quad N_2O_4(g) \rightleftharpoons 2NO_2(g) \qquad K_f = \frac{[NO_2]^2}{[N_2O_4]}$$

$$(2)\quad 2NO_2(g) \rightleftharpoons N_2O_4(g) \qquad K_r = \frac{[N_2O_4]}{[NO_2]^2}$$

Reaction (2) is simply the reverse of Reaction (1). Comparison of the equilibrium constant expressions for the *f*orward reaction, K_f, and the *r*everse reaction, K_r, leads to the conclusion that the equilibrium constant for the reverse reaction is equal to the reciprocal of the equilibrium constant for the forward reaction. Algebraically, we have

$$K_r = \frac{[N_2O_4]}{[NO_2]^2}$$

$$\frac{1}{K_f} = \frac{1}{[NO_2]^2/[N_2O_4]} = \frac{[N_2O_4]}{[NO_2]^2}$$

A comparison of these two expressions yields

$$\boxed{K_r = \frac{1}{K_f}} \qquad (15\text{-}17)$$

Equation (15-17) is a general result which is easily verified for any particular case.

Example 15-15: The value of K_c at 100°C for the reaction

$$(1) \quad N_2O_4(g) \rightleftharpoons 2NO_2(g)$$

is $K_c = 0.20$ M. Compute the value of K_c at 100°C for the reaction

$$(2) \quad 2NO_2(g) \rightleftharpoons N_2O_4(g)$$

Solution: Reaction (2) is the reverse of Reaction (1), and thus, using Equation (15-17), we have for Reaction (2)

$$K_r = \frac{1}{K_f} = \frac{1}{0.20\ \text{M}} = 5.0\ \text{M}^{-1}$$

Thus $K_c = 5.0\ \text{M}^{-1}$ for Reaction (2).

Consider the two reactions

$$(1) \quad C(s) + H_2O(g) \rightleftharpoons CO(g) + H_2(g)$$
$$(2) \quad CO_2(g) + 2H_2(g) \rightleftharpoons 2H_2O(g) + C(s)$$

If we add Equation (2) to Equation (1), then we obtain

$$C(s) + H_2O(g) + CO_2(g) + 2H_2(g) \rightleftharpoons CO(g) + H_2(g) + 2H_2O(g) + C(s)$$

Cancellation of like terms on the two sides of this equation yields

$$(3) \quad CO_2(g) + H_2(g) \rightleftharpoons CO(g) + H_2O(g)$$

Multiplication of the equilibrium constant expressions for Reactions (1) and (2) yields

$$K_1K_2 = \frac{[CO][H_2]}{[H_2O]} \times \frac{[H_2O]^2}{[CO_2][H_2]^2} = \frac{[CO][H_2O]}{[CO_2][H_2]}$$

The equilibrium constant expression for Reaction (3) is

$$K_3 = \frac{[CO][H_2O]}{[CO_2][H_2]}$$

Comparison of the K_3 expression with the expression for K_1K_2 shows

$$K_3 = K_1K_2 \qquad (15\text{-}18)$$

Equation (15-18) is a general result. If we *add two equations* to obtain a third, then the equilibrium constant of the third equation is equal to the product of the equilibrium constants of the two equations that are added together.

Example 15-15: Given the equilibrium constants

(1) $H^+(aq) + NO_2^-(aq) \rightleftharpoons HNO_2(aq)$ $\quad K_1 = 2.22 \times 10^3\ M^{-1}$

(2) $H_2O(l) \rightleftharpoons H^+(aq) + OH^-(aq)$ $\quad K_2 = 1.00 \times 10^{-14}\ M^2$

compute the value of K for the reaction

(3) $NO_2^-(aq) + H_2O(l) \rightleftharpoons HNO_2(aq) + OH^-(aq)$

Problems 15-57 through 15-60 involve calculations of equilibrium constants for combinations of reactions.

Solution: Equation (3) is obtained by adding together Equations (1) and (2) and cancelling like terms. Therefore, the equilibrium constant for Equation (3) is equal to the product K_1K_2 (Equation 15-18):

$$K_3 = K_1K_2 = (2.22 \times 10^3\ M^{-1})(1.00 \times 10^{-14}\ M^2) = 2.22 \times 10^{-11}\ M$$

where

$$K_3 = \frac{[HNO_2][OH^-]}{[NO_2^-]}$$

15-11. LE CHÂTELIER'S PRINCIPLE CAN BE USED TO SELECT THE CONDITIONS THAT MAXIMIZE THE EQUILIBRIUM YIELD OF A REACTION PRODUCT

Ammonia is produced commercially by the *Haber process,* which is described in Interchapter C. Because of the tremendous scale of NH_3 production, it is important that the reaction be run under the most favorable conditions, that is, conditions that maximize the *yield* of product.

The ammonia production reaction is

$$N_2(g) + 3H_2(g) \rightleftharpoons 2NH_3(g) \qquad \Delta H^\circ_{rxn} = -92 \text{ kJ}$$

Because there are more moles of gaseous species on the left of this equation, a decrease in the total reaction volume or, equivalently, an increase in the total reaction pressure favors the conversion of reactants to products. Thus, the percent conversion of N_2 and H_2 to NH_3 (the yield) increases as the total pressure increases. The value of ΔH°_{rxn} for the reaction is negative, and so we write

$$N_2(g) + 3H_2(g) \rightleftharpoons 2NH_3(g) + 92 \text{ kJ}$$

Thus an increase in temperature favors the conversion of products back to reactants; that is, the equilibrium shifts right to left when the temperature is increased. The equilibrium constant decreases with increasing temperature, as is shown in the following table:

The change in the equilibrium constant with temperature is given by the van't Hoff equation, which is described in Problem 15-61 and applied in Problems 15-61 through 15-64.

$t/°C$	$(K_p = P^2_{NH_3}/P_{N_2}P^3_{H_2})/atm^{-2}$
25	6.0×10^5
227	0.10
500	1.5×10^{-5}

Thus at a given total pressure the percent conversion of nitrogen plus hydrogen to ammonia *decreases* as the temperature increases.

The application of Le Châtelier's principle to the ammonia synthesis reaction leads to the prediction that *at equilibrium* the yield of ammonia is greater the higher the total pressure and the lower the temperature. However, the *rate* of the reaction at 25°C is negligibly slow. A high yield is of no commercial value if it takes forever to achieve the conversion. The rate of most reactions increases with increasing temperature. The ammonia production reaction is run at an elevated temperature (500°C), even though the equilibrium yield is not as favorable as at lower temperatures, in order to make the reaction proceed at an economically feasible rate. The low value of K at 500°C is offset by using a very high pressure (300 atm), as shown in the following table:

$$N_2(g) + 3H_2(g) \xrightleftharpoons{500°C} 2NH_3(g)$$

P_{tot}/atm	P_{NH_3}/atm	*% conversion of 3:1 ratio of $H_2:N_2$ to NH_3*
1.00	1.26×10^{-3}	0.25
300	152	30

The Haber process is thus based on a compromise between equilibrium (yield) and rate (speed of reaction) considerations. We shall study rates of reactions in the next chapter.

SUMMARY

A chemical reaction equilibrium is dynamic. At equilibrium the rates of the forward and reverse reactions are balanced and there is no net change in the system. A chemical reaction equilibrium is characterized quantitatively by the equilibrium constant expression for the reaction. The law of concentration action tells us how to formulate the equilibrium-constant expression for a chemical reaction.

The direction in which an established dynamic chemical reaction equilibrium shifts in response to a change in conditions, such as a change in reactant or product concentration, in temperature, or in volume, is predicted by Le Châtelier's principle. The direction in which a reaction mixture not at equilibrium proceeds toward equilibrium is predicted from the ratio of the reaction quotient, Q, to the reaction equilibrium constant, K.

The equilibrium constant for the reverse reaction is equal to the reciprocal of the equilibrium constant for the forward reaction. The equilibrium constant for a reaction obtained by algebraically adding two reactions is equal to the product of the equilibrium constants for the two reactions that are added together.

In selecting a set of reaction conditions to maximize the yield of a desired product, both equilibrium factors and reaction rate factors must be considered.

TERMS YOU SHOULD KNOW

forward rate 566
reverse rate 566
chemical equilibrium 566
dynamic equilibrium 567
equilibrium concentration 568
law of concentration action 571
equilibrium constant, K_c 571
equilibrium constant expression 571
K_p and K_c 573
quadratic formula 577
quadratic equation 577
Le Châtelier's principle 580
reaction quotient, Q 584
Q/K 586
direction of reaction spontaneity 586
addition of reactions 590
Haber process 590
yield 590

EQUATIONS YOU SHOULD KNOW HOW TO USE

$$K_c = \frac{[X]^x[Y]^y}{[A]^a[B]^b} \qquad (15\text{-}3)$$

Value of $\dfrac{Q}{K}$ compared to unity

$$K_r = \frac{1}{K_f} \qquad (15\text{-}17)$$

$$K_3 = K_1K_2 \qquad (15\text{-}18)$$

PROBLEMS

EQUILIBRIUM-CONSTANT EXPRESSION

15-1. Use the law of concentration action to write the equilibrium-constant expression (K_c) for the following reactions:

(a) $SnO_2(s) + 2H_2(g) \rightleftharpoons Sn(s) + 2H_2O(g)$
(b) $PCl_5(s) + H_2O(l) \rightleftharpoons 2HCl(g) + POCl_3(g)$

15-3. Use the law of concentration action to write the equilibrium-constant expression (K_c) for the following reactions:

(a) $2H_2S(g) \rightleftharpoons 2H_2(g) + S_2(g)$
(b) $C_2H_4(g) + H_2(g) \rightleftharpoons C_2H_6(g)$

15-5. Write the equilibrium-constant expression (K_c) for each of the following reactions:

(a) $2H_2(g) + O_2(g) \rightleftharpoons 2H_2O(g)$
(b) $2N_2O_5(g) \rightleftharpoons 4NO_2(g) + O_2(g)$
(c) $CuCl_2 \cdot 2H_2O(s) \rightleftharpoons CuCl_2(s) + 2H_2O(g)$

15-7. Write K_p expressions for the reactions in Problem 15-5.

15-9. Explain why the equilibrium constant for a reaction cannot be negative.

15-2. Use the law of concentration action to write the equilibrium-constant expression (K_c) for the following reactions:

(a) $CO(g) + 2H_2(g) \rightleftharpoons CH_3OH(g)$
(b) $2NaHCO_3(s) \rightleftharpoons Na_2CO_3(s) + CO_2(g) + H_2O(g)$

15-4. Use the law of concentration action to write the equilibrium-constant expression (K_c) for the following reactions:

(a) $N_2(g) + O_2(g) \rightleftharpoons 2NO(g)$
(b) $2HgO(s) \rightleftharpoons 2Hg(l) + O_2(g)$

15-6. Write the equilibrium-constant expression (K_c) for each of the following reactions:

(a) $NH_2COONH_4(s) \rightleftharpoons 2NH_3(g) + CO_2(g)$
(b) $CO_2(g) + C(s) \rightleftharpoons 2CO(g)$
(c) $N_2(g) + 2O_2(g) \rightleftharpoons N_2O_4(g)$

15-8. Write K_p expressions for the reactions in Problem 15-6.

15-10. One method by which hydrogen chloride gas is prepared commercially is the reaction of liquid sulfuric acid with sodium chloride at elevated temperatures. Write the net reaction for the process and give the equilibrium-constant expression for the reaction.

CALCULATION OF EQUILIBRIUM CONSTANTS

15-11. Phosgene, a toxic gas used in the synthesis of a variety of organic compounds, decomposes according to the equation

$$COCl_2(g) \rightleftharpoons CO(g) + Cl_2(g)$$

Phosgene is heated at 527°C in a reaction vessel. At equilibrium, the concentrations of the constituents are found to be [CO] = 0.0456 M, [Cl_2] = 0.0456 M, and [$COCl_2$] = 0.449 M. Calculate the equilibrium constant for the reaction at 527°C.

15-13. Consider the reaction

$$CuSO_4 \cdot 4NH_3(s) \rightleftharpoons CuSO_4 \cdot 2NH_3(s) + 2NH_3(g)$$

At 20°C, the equilibrium pressure of NH_3 is 62 torr. Compute K_p for this reaction.

15-12. The decomposition of phosphorus pentachloride occurs via the reaction

$$PCl_5(g) \rightleftharpoons PCl_3(g) + Cl_2(g)$$

A sample of PCl_5 is placed in a reaction vessel held at 250°C at an initial concentration of 1.10 M. When equilibrium is attained, the concentration of PCl_5 is 0.33 M. Calculate K_c for the reaction.

15-14. At 1000°C, methane and water react according to

$$CH_4(g) + H_2O(g) \rightleftharpoons CO(g) + 3H_2(g)$$

At equilibrium, it was found that $P_{CH_4} = 0.31$ atm, $P_{H_2O} = 0.83$ atm, $P_{CO} = 0.57$ atm, and $P_{H_2} = 2.26$ atm. Calculate K_p for this reaction.

15-15. A mixture of 1.00 mol of $H_2(g)$ and 1.00 mol of $I_2(g)$ is placed in a 1.00-L container held at a constant temperature. After equilibrium is attained, 1.56 mol of $HI(g)$ is found. Calculate K_c for the reaction

$$H_2(g) + I_2(g) \rightleftharpoons 2HI(g)$$

15-17. The equilibrium constant for the reaction

$$N_2(g) + 3H_2(g) \rightleftharpoons 2NH_3(g)$$

is $K_p = 0.10\ \text{atm}^{-2}$ at 227°C. Compute the value of K_c for the reaction at 227°C.

15-19. Synthesis gas is produced by passing steam over coal according to the reaction

$$C(s) + H_2O(g) \rightleftharpoons CO(g) + H_2(g)$$

The equilibrium concentrations at 800°C were found to be $[CO] = 6.75 \times 10^{-2}$ M, $[H_2] = 6.75 \times 10^{-2}$ M, and $[H_2O] = 5.70 \times 10^{-2}$ M. Calculate K_c and K_p.

15-16. Nitrogen dioxide decomposes at high temperatures according to the reaction

$$2NO_2(g) \rightleftharpoons 2NO(g) + O_2(g)$$

Suppose initially we have pure $NO_2(g)$ at 1000 K and 0.500 atm. If the *total* pressure is 0.732 atm when equilibrium is reached, what is the value of K_p?

15-18. According to Table 15-1, $K_c = 0.20$ M at 100°C for the reaction

$$N_2O_4(g) \rightleftharpoons 2NO_2(g)$$

Calculate K_p at the same temperature.

15-20. Hydrogen sulfide decomposes by the reaction

$$2H_2S(g) \rightleftharpoons 2H_2(g) + S_2(g)$$

Chemical analysis gives that $[H_2S] = 0.36$ M, $[H_2] = 0.027$ M, and $[S_2] = 0.075$ M at 1200°C. Calculate K_c and K_p for this reaction.

EQUILIBRIUM CALCULATIONS

15-21. Given that $[N_2O_4] = 0.730$ M at equilibrium for the reaction

$$N_2O_4(g) \rightleftharpoons 2NO_2(g) \qquad K_c = 0.20\ \text{M}$$

calculate the concentration of $NO_2(g)$ at equilibrium.

15-23. Phosphorus pentachloride decomposes according to the reaction

$$PCl_5(g) \rightleftharpoons PCl_3(g) + Cl_2(g) \quad K_c = 1.8\ \text{M at } 250°\text{C}$$

A 1.50-mol sample of PCl_5 is injected into a 0.50-L reaction vessel held at 250°C. Calculate the concentrations of PCl_5 and PCl_3 at equilibrium.

15-25. At 1200°C, $K_c = 2.51 \times 10^4$ for the reaction

$$H_2(g) + Cl_2(g) \rightleftharpoons 2HCl(g)$$

If 0.250 mol each of H_2 and Cl_2 are introduced initially into a reaction vessel, how many moles of HCl are there at equilibrium?

15-22. The equilibrium constant for the reaction

$$C(s) + CO_2(g) \rightleftharpoons 2CO(g)$$

at 1000 K is 1.90 atm. If the equilibrium pressure of CO is 1.50 atm, what is the equilibrium pressure of CO_2?

15-24. Carbon disulfide is prepared by heating sulfur and charcoal via the reaction

$$S_2(g) + C(s) \rightleftharpoons CS_2(g) \qquad K_c = 9.40 \text{ at } 900\ \text{K}$$

How much CS_2 can be prepared by heating 10.0 mol of sulfur with excess carbon in a 5.00-L reaction vessel held at 900 K until equilibrium is attained?

15-26. At 1000°C, $K_p = 0.263\ \text{atm}^{-1}$ for the reaction

$$C(s) + 2H_2(g) \rightleftharpoons CH_4(g)$$

Calculate the equilibrium pressure of $CH_4(g)$ if 0.250 mol of CH_4 is placed in a 4.00-L container at 1000°C.

15-27. Ammonium hydrogen sulfide decomposes according to the reaction

$$NH_4HS(s) \rightleftharpoons NH_3(g) + H_2S(g)$$

The equilibrium constant, K_c, is $1.81 \times 10^{-4}\ M^2$ at 25°C. If $NH_4HS(s)$ is placed in an evacuated reaction vessel at 25°C, what is the total gas pressure in the vessel when equilibrium is attained?

15-28. Sodium hydrogen carbonate, commonly called sodium bicarbonate, is used in baking soda and in fire extinguishers as a source of CO_2. It decomposes according to the reaction

$$2NaHCO_3(s) \rightleftharpoons Na_2CO_3(s) + CO_2(g) + H_2O(g)$$

Given that $K_p = 0.25\ atm^2$ at 125°C, calculate the partial pressures of CO_2 and H_2O at equilibrium when $NaHCO_3$ is heated to 125°C in a closed vessel.

15-29. The equilibrium constant for the reaction

$$2ICl(g) \rightleftharpoons I_2(g) + Cl_2(g)$$

is $K_c = 0.11$. Calculate the equilibrium concentrations of ICl, I_2, and Cl_2 when 0.33 mol of I_2 and 0.33 mol of Cl_2 are added to a 1.5-L reaction vessel.

15-30. Suppose that 5.00 mol of CO(*g*) is mixed with 2.50 mol of $Cl_2(g)$ in a 10.0-L reaction vessel and the following reaction attains equilibrium:

$$CO(g) + Cl_2(g) \rightleftharpoons COCl_2(g)$$

Given that $K_c = 4.0\ M^{-1}$, compute the equilibrium values of [CO], [Cl_2], and [$COCl_2$].

15-31. At 500°C, hydrogen iodide decomposes according to the reaction

$$2HI(g) \rightleftharpoons H_2(g) + I_2(g)$$

For HI heated to 500°C in a 1.00-L reaction vessel, chemical analysis gave the following concentrations at equilibrium: $[H_2] = 0.42$ M, $[I_2] = 0.42$ M, and $[HI] = 3.52$ M. If an additional mol of HI is introduced to the reaction vessel, what are the equilibrium concentrations after the new equilibrium has been reached?

15-32. Tin can be prepared by heating SnO_2 ore with hydrogen gas:

$$SnO_2(s) + 2H_2(g) \rightleftharpoons Sn(s) + 2H_2O(g)$$

When the reactants are heated to 500°C in a closed vessel, $[H_2O] = [H_2] = 0.25$ M at equilibrium. If more hydrogen is added so that its initial concentration becomes 0.50 M, what are the concentrations of H_2 and H_2O when equilibrium is restored?

EQUILIBRIUM CALCULATIONS INVOLVING TOTAL PRESSURE

15-33. Suppose that N_2O_4 and NO_2 are mixed together in a reaction vessel and that the total pressure at equilibrium is 2.40 atm. Calculate $P_{N_2O_4}$ and P_{NO_2} at equilibrium when the value of K_p is 4.90 atm for the reaction

$$N_2O_4(g) \rightleftharpoons 2NO_2(g)$$

15-34. Given that $H_2(g)$ reacts with $I_2(s)$ via the reaction

$$H_2(g) + I_2(s) \rightleftharpoons 2HI(g) \qquad K_p = 8.6\ atm$$

and that at equilibrium the total pressure in the reaction vessel is 4.5 atm, calculate P_{HI} and P_{H_2} at equilibrium. (Neglect the vapor pressure of $I_2(s)$.)

15-35. Zinc metal is produced by the reaction of its oxide with carbon monoxide at high temperature via the reaction

$$ZnO(s) + CO(g) \rightleftharpoons Zn(s) + CO_2(g) \qquad K_p = 600$$

At equilibrium the total pressure in the reaction vessel is 11.8 atm. Calculate P_{CO_2} and P_{CO} at equilibrium.

15-36. At equilibrium, the total pressure in the reaction vessel for the reaction

$$C(s) + 2H_2(g) \rightleftharpoons CH_4(g)$$

is 2.11 atm. Given that $K_p = 0.263\ atm^{-1}$ at 1000°C, compute P_{H_2} and P_{CH_4}.

15-37. The equilibrium constant for the methanol synthesis reaction,

$$2H_2(g) + CO(g) \rightleftharpoons CH_3OH(g)$$

is $K_p = 2.25 \times 10^4\ \text{atm}^{-2}$ at 25°C.

(a) Compute the value of P_{CH_3OH} at equilibrium when $P_{H_2} = 0.020$ atm and $P_{CO} = 0.010$ atm.
(b) Given that at equilibrium $P_{tot} = 10.0$ atm and $P_{H_2} = 0.020$ atm, compute P_{CO} and P_{CH_3OH}.

15-38. The equilibrium constant for the reaction

$$N_2(g) + 3H_2(g) \rightleftharpoons 2NH_3(g)$$

is $K_p = 0.10\ \text{atm}^{-2}$ at 227°C.

(a) Given that at equilibrium $P_{N_2} = 1.00$ atm and $P_{H_2} = 3.00$ atm, compute P_{NH_3} at equilibrium.
(b) Given that at equilibrium the total pressure is 2.00 atm and also that the mole fraction of H_2, X_{H_2}, is 0.20, compute X_{NH_3}. (Note that $X_{N_2} + X_{H_2} + X_{NH_3} = 1$.)

LE CHATELIER'S PRINCIPLE

15-39. Consider the reaction equilibrium

$$H_2(g) + CO_2(g) \rightleftharpoons H_2O(g) + CO(g)$$

Use Le Châtelier's principle to predict the effect on the equilibrium pressure of CO_2 and of CO produced by an increase in the pressure of H_2O.

15-40. Predict the effect of increasing the reaction volume of the equilibrium reaction mixture

$$H_2(g) + CO_2(g) \rightleftharpoons H_2O(g) + CO(g)$$

15-41. Consider the reaction equilibrium

$$2NO(g) + Br_2(g) \rightleftharpoons 2NOBr(g)$$

Use Le Châtelier's principle to predict the effect on the equilibrium concentration of NOBr and of NO produced by an increase in the concentration of Br_2.

15-42. For the reaction

$$2NO(g) + Br_2(g) \rightleftharpoons 2NOBr(g)$$

$\Delta H^\circ_{rxn} = -46.4$ kJ. Should the temperature be increased or decreased in order to increase the extent of conversion to NOBr?

15-43. Consider the reaction

$$C(s) + 2H_2(g) \rightleftharpoons CH_4(g) \qquad \Delta H^\circ_{rxn} = -75\ \text{kJ}$$

Which way will the equilibrium shift in response to each of the following changes in conditions (if the equilibrium is unaffected by the change, then write *no change*):

(a) increase in temperature
(b) increase in total reaction volume
(c) increase in P_{H_2}
(d) decrease in P_{CH_4}
(e) addition of C(*s*)

15-44. For the reaction

$$Ni(s) + 4CO(g) \rightleftharpoons Ni(CO)_4(g) \qquad \Delta H^\circ_{rxn} < 0$$

which way will the equilibrium shift in response to each of the following changes in conditions (if the equilibrium is unaffected by the change, then write *no change*):

(a) increase in temperature
(b) increase in reaction volume
(c) removal of $Ni(CO)_4(g)$
(d) addition of Ni(*s*)

15-45. Predict the direction in which the equilibrium

$$2SO_2(g) + O_2(g) \rightleftharpoons 2SO_3(g) \qquad \Delta H^\circ_{rxn} = -198\ \text{kJ}$$

will shift in response to each of the following changes in conditions:

(a) decrease in temperature
(b) increase in reaction volume
(c) addition of $O_2(g)$
(d) removal of $SO_2(g)$

15-46. In which direction will the equilibrium shift in response to the following changes in conditions:

$$N_2(aq) \rightleftharpoons N_2(g) \qquad \Delta H^\circ_{rxn} > 0$$

(a) increase in temperature
(b) increase in vapor space volume over the solution
(c) addition of $H_2O(l)$
(d) addition of $N_2(g)$

15-47. Use the following reaction to fill out the table given below:

$$SO_2(g) + NO_2(g) \rightleftharpoons SO_3(g) + NO(g) \qquad \Delta H^\circ_{rxn} = -42 \text{ kJ}$$

Change	*Effect on equilibrium*
(a) decrease in total volume	no change
(b) increase in temperature	←
(c) increase in partial pressure of $NO_2(g)$	→
(d) decrease in partial pressure of products	→

15-48. An important synthetic gas is *water gas*, an approximately 50/50 mixture of carbon monoxide and hydrogen made by passing steam over hot coal at about 650°C:

$$C(s) + H_2O(g) \rightleftharpoons CO(g) + H_2(g) \quad \Delta H^\circ_{rxn} = 131 \text{ kJ}$$

The equilibrium constant for this reaction at 650°C is 0.39 atm.

(a) How will the percent conversion of H_2O to CO and H_2 be affected by an increase in the total reaction volume?
(b) How will the percent conversion of H_2O to CO and H_2 be affected by a decrease in the temperature at fixed reaction volume?

15-49. Several key reactions in coal gasification are
1. the synthesis gas reaction:

$$C(s) + H_2O(g) \rightleftharpoons CO(g) + H_2(g) \qquad \Delta H^\circ_{rxn} = +131 \text{ kJ}$$

2. the water-gas-shift reaction:

$$CO(g) + H_2O(g) \rightleftharpoons CO_2(g) + H_2(g) \qquad \Delta H^\circ_{rxn} = -41 \text{ kJ}$$

3. the catalytic methanation reaction:

$$CO(g) + 3H_2(g) \rightleftharpoons H_2O(g) + CH_4(g) \qquad \Delta H^\circ_{rxn} = -206 \text{ kJ}$$

(a) Write the equilibrium constant expressions in terms of concentrations, K_c, for each of these reactions.
(b) Predict the direction in which each equilibrium shifts in response to (i) an increase in temperature and (ii) a decrease in reaction volume.

15-50. An important modern chemical problem is the liquefication of coal because it is still relatively abundant whereas oil is a dwindling resource. The first step is heating the coal with steam to produce synthesis gas:

$$C(s) + H_2O(g) \rightleftharpoons CO(g) + H_2(g) \qquad \Delta H^\circ_{rxn} = 131 \text{ kJ}$$

Carbon monoxide can be hydrogenated to form the important chemical methyl alcohol:

$$CO(g) + 2H_2(g) \rightleftharpoons CH_3OH(g) \qquad \Delta H^\circ_{rxn} = -128 \text{ kJ}$$

Use Le Châtelier's principle to suggest conditions that maximize the yield of CH_3OH from $CO(g)$ and $H_2(g)$.

REACTION QUOTIENT CALCULATIONS

15-51. At 900 K the equilibrium constant for the reaction

$$2SO_2(g) + O_2(g) \rightleftharpoons 2SO_3(g)$$

is 13 M^{-1}. If we mix the following concentrations of the three gases, predict in which direction the reaction will proceed toward equilibrium:

	$[SO_2]/M$	$[O_2]/M$	$[SO_3]/M$
(a)	0.20	0.20	0.20
(b)	0.10	0.20	0.20

15-52. Suppose that $H_2(g)$ and $CH_4(g)$ are brought into contact with $C(s)$ at 500°C with $P_{H_2} = 0.20$ atm and $P_{CH_4} = 3.0$ atm. Is the reaction

$$C(s) + 2H_2(g) \rightleftharpoons CH_4(g) \qquad K_p = 2.69 \times 10^3 \text{ atm}^{-1}$$

at equilibrium? If not, in what direction will the reaction proceed toward equilibrium?

15-53. Suppose we have a mixture of the gases H_2, CO_2, CO, and H_2O at 1260 K, with $P_{H_2} = 0.50$ atm, $P_{CO_2} = 0.25$ atm, $P_{CO} = 1.25$ atm, and $P_{H_2O} = 0.50$ atm. Is the reaction

$$H_2(g) + CO_2(g) \rightleftharpoons CO(g) + H_2O(g) \qquad K_p = 1.59$$

at equilibrium? If not, in what direction will the reaction proceed toward equilibrium?

15-54. Suppose $S_2(g)$ and $CS_2(g)$ are brought into contact with solid carbon at 900 K with $P_{S_2} = 1.78$ atm and $P_{CS_2} = 0.794$ atm. Is the reaction

$$S_2(g) + C(s) \rightleftharpoons CS_2(g) \qquad K_p = 9.40$$

at equilibrium? If not, in what direction will the reaction proceed toward equilibrium?

15-55. The equilibrium constant for the reaction

$$2SO_2(g) + O_2(g) \rightleftharpoons 2SO_3(g)$$

is $K_p = 0.14\ \text{atm}^{-1}$ at 900 K and $\Delta H^\circ_{rxn} = -198$ kJ. Suppose the reaction system is prepared at 900 K with the initial pressures $P_{O_2} = 0.20$ atm, $P_{SO_2} = 0.10$ atm, and $P_{SO_3} = 0.20$ atm.

(a) Compute the value of Q for the reaction with these pressures.
(b) Indicate the direction in which the reaction proceeds toward equilibrium.

15-56. Given that $K_p = 2.25 \times 10^4\ \text{atm}^{-2}$ at 25°C for the reaction

$$2H_2(g) + CO(g) \rightleftharpoons CH_3OH(g)$$

predict the direction in which a reaction mixture for which $P_{CH_3OH} = 10.0$ atm, $P_{H_2} = 0.010$ atm, and $P_{CO} = 0.0050$ atm proceeds toward equilibrium.

EQUILIBRIUM CONSTANTS FOR COMBINATIONS OF REACTIONS

15-57. Given that

$$CO(g) + H_2O(g) \rightleftharpoons CO_2(g) + H_2(g) \qquad K_1 = 1.44$$

$$CH_4(g) + H_2O(g) \rightleftharpoons CO(g) + 3H_2(g) \qquad K_2 = 25.6\ \text{atm}^2$$

calculate K_p for the reaction

$$CH_4(g) + 2H_2O(g) \rightleftharpoons CO_2(g) + 4H_2(g)$$

15-58. Given that $K_c = 0.20$ M at 100°C for the reaction

$$N_2O_4(g) \rightleftharpoons 2NO_2(g)$$

calculate K_c and K_p for the reaction

$$NO_2(g) \rightleftharpoons \tfrac{1}{2}N_2O_4(g)$$

15-59. Given that for the reactions at 973 K

$$MgCl_2(s) + \tfrac{1}{2}O_2(g) \rightleftharpoons MgO(s) + Cl_2(g) \qquad K_p = 2.95\ \text{atm}^{1/2}$$

$$MgCl_2(s) + H_2O(g) \rightleftharpoons MgO(s) + 2HCl(g) \qquad K_p = 8.40\ \text{atm}$$

determine the equilibrium constant at 973 K for the reaction

$$Cl_2(g) + H_2O(g) \rightleftharpoons 2HCl(g) + \tfrac{1}{2}O_2(g)$$

15-60. Given the following equilibrium constants at 1000 K:

$$CaCO_3(s) \rightleftharpoons CaO(s) + CO_2(g) \qquad K_1 = 0.039\ \text{atm}$$

$$C(s) + CO_2(g) \rightleftharpoons 2CO(g) \qquad K_2 = 1.9\ \text{atm}$$

determine the equilibrium constant at 1000 K for the reaction

$$CaCO_3(s) + C(s) \rightleftharpoons CaO(s) + 2CO(g)$$

TEMPERATURE DEPENDENCE OF K_p

15-61. The value of the equilibrium constant changes with temperature according to the *van't Hoff equation:*

$$\log \frac{K_2}{K_1} = \frac{\Delta H^\circ_{rxn}}{2.30R}\left(\frac{T_2 - T_1}{T_1 T_2}\right)$$

where K_2 is the equilibrium constant at the absolute temperature T_2, K_1 is the equilibrium constant at temperature T_1, ΔH°_{rxn} is the enthalpy of reaction, and R is the molar gas constant ($8.31\ \mathrm{J \cdot mol^{-1} \cdot K^{-1}}$). The enthalpy of reaction for

$$\log \frac{K_2}{1.78} = \frac{(92.9)}{2.30(8.31\times10^{-3})}\left(\frac{1}{523} - \frac{1}{673}\right)$$

$$PCl_5(g) \rightleftharpoons PCl_3(g) + Cl_2(g)$$

is $+92.9\ \mathrm{kJ \cdot mol^{-1}}$. The value of K_p is 1.78 atm at 250°C. Calculate K_p at 400°C.

$K_2 = 210\ \mathrm{atm}$

15-62. The equilibrium constant for the reaction

$$H_2(g) + I_2(g) \rightleftharpoons 2HI(g)$$

is 617 at 25°C and $\Delta H^\circ_{rxn} = -10.2\ \mathrm{kJ \cdot mol^{-1}}$. Use the van't Hoff equation to calculate K_c for the reaction at 100°C (see Problem 15-61).

$$\log \frac{K_2}{617} = \frac{(-10.2)}{2.30(8.31\times10^{-3})}\left(\frac{1}{298} - \frac{1}{373}\right)$$

$$\log \frac{K_2}{617} = \frac{(-10.2)}{1.91\times10^{-2}}\left(3.356\times10^{-3} - 2.68\times10^{-3}\right)$$

$\log K_2 - \log 617 = -0.361$

$\log K_2 - 2.79 = -0.361$

$\log K_2 = 2.429$

$K_2 = 269$

15-63. The reaction

$$2SO_2(g) + O_2(g) \rightleftharpoons 2SO_3(g) \qquad \Delta H^\circ_{rxn} = -198\ \mathrm{kJ}$$

occurs in the catalytic converter of an automobile. For this reaction $K_p = 0.14\ \mathrm{atm^{-1}}$ at 627°C. Estimate the value of K_p at 1000°C (see Problem 15-61).

$$\log \frac{K_2}{0.14} = \frac{(-198)}{(2.30)(8.31\times10^{-3})}\left(\frac{1}{900} - \frac{1}{1273}\right)$$

$K_2 = 5.9\times10^{-5}/\mathrm{atm}$

15-64. A key component of photochemical smog is NO, which is produced by the reaction between N_2 and O_2 at the high temperatures that occur in the internal combustion engine. At 2000°C the equilibrium constant for the reaction $N_2(g) + O_2(g) \rightleftharpoons 2NO(g)$ is 4×10^{-4}. The reaction takes place to a much lesser extent at low temperatures. Estimate the equilibrium constant at 25°C, using the value $\Delta H_{rxn} = 181\ \mathrm{kJ}$ at 25°C (see Problem 15-61).

15-65. Given the following equilibrium constant data for the Deacon process,

$$Cl_2(g) + H_2O(g) \rightleftharpoons 2HCl(g) + \tfrac{1}{2}O_2(g)$$

T/K	$\log K$
723	−0.706
873	−0.002

estimate the value of ΔH°_{rxn} (see Problem 15-61).

$$(-0.706) - (0.002) = \frac{\Delta H_{rxn}}{(2.30)(8.31\times10^{-3})}\left(\frac{1}{723} - \frac{1}{873}\right)$$

$\Delta H_{rxn} = 56.6\ \mathrm{kJ/mol}$

15-66. For the reaction

$$CO(g) + 2H_2(g) \rightleftharpoons CH_3OH(g)$$

$\Delta H^\circ_{rxn} = -128\ \mathrm{kJ}$ and $K_p = 2.25 \times 10^4\ \mathrm{atm^{-2}}$ at 25°C. The reaction is normally run at 300°C. What is the value of the equilibrium constant at 300°C (see Problem 15-61)?

The Belousov-Zhabotinskii reaction involves the oxidation of citric acid by potassium bromate. The spirals of chemical activity were induced by touching the surface of the solution with a hot filament. The bubbles are carbon dioxide, a product of the chemical reaction. The shading variations are due to an iron-complex ion that is orange in the Fe (III) form and light blue in the Fe (II) form.

16 / Rates and Mechanisms of Chemical Reactions

Chemists study the rates of chemical reactions to determine the conditions under which a particular reaction can be made to proceed at a favorable rate. The reaction rate law, which must be determined by experiment, tells us how the rate of a reaction depends on the concentration of the reactants and of other added substances, such as catalysts. The reaction rate law provides the most important clue to the reaction mechanism, which is the sequence of steps by which the reactants are converted to products. An understanding of reaction mechanisms enables us to adjust reaction conditions in order to produce a desired reaction rate and increases our understanding of how chemical reactions occur at the molecular level.

The attainment of chemical equilibrium is the exception rather than the rule. Most chemical systems of interest are not at chemical equilibrium, and thus a study of chemical reaction rates (chemical kinetics) is an essential prerequisite to understanding the chemistry of such systems.

16-1. A RATE TELLS US HOW FAST A QUANTITY IS CHANGING WITH TIME

We are all familiar with the concept of rate. For example, we talk about the rate of inflation, interest rates, annual percentage rates, growth rates, mortgage rates, and birth and death rates. A *rate* tells us how fast a quantity is changing with time.

Let's consider the rate of a chemical reaction. Suppose that the reaction is

$$A \rightarrow P$$

where A represents the reactant and P represents the product. We define the rate at which the product P is formed as

The units of reaction rate are moles per liter per second ($mol \cdot L^{-1} \cdot s^{-1}$, or $M \cdot s^{-1}$).

$$\underset{\text{units of } mol \cdot L^{-1} \cdot s^{-1}}{\text{rate}} = \frac{\Delta[P]}{\Delta t} = \frac{[P]_2 - [P]_1}{t_2 - t_1} \tag{16-1}$$

where $[P]_1$ is the concentration of P at time t_1 and $[P]_2$ is the concentration of P at some later time t_2. Note that the value of $t_2 - t_1$ is positive because $t_2 > t_1$ and that the value of $[P]_2 - [P]_1$ is positive because the product concentration increases as the reaction takes place. Thus $\Delta[P]/\Delta t$ is a positive quantity, and so the rate of a reaction is a positive quantity. Note that the units of rate are $mol \cdot L^{-1} \cdot s^{-1}$.

We also can express the rate in terms of the concentrations of reactants. The rate at which the reactant A is consumed in the reaction is defined as

$$\underset{\text{units of } mol \cdot L^{-1} \cdot s^{-1}}{\text{rate}} = -\frac{\Delta[A]}{\Delta t} = -\left(\frac{[A]_2 - [A]_1}{t_2 - t_1}\right) \tag{16-2}$$

where $[A]_1$ is the concentration of A at time t_1 and $[A]_2$ is the concentration of A at some later time t_2. The value of $t_2 - t_1$ is positive because $t_2 > t_1$, but $[A]_2$ is less than $[A]_1$ because A is being consumed by the reaction. Thus $[A]_2 - [A]_1 < 0$. A minus sign is inserted in front of $\Delta[A]/\Delta t$, which is negative, to make $-\Delta[A]/\Delta t$ (= rate) a positive quantity. A *reaction rate* is the rate at which a reactant is consumed (moles per liter reacted per second) or the rate at which a product is formed (moles per liter produced per second). The relation between Equations (16-1) and (16-2) is

The rate of a chemical reaction is always a positive quantity.

$$\text{rate} = -\frac{\Delta[A]}{\Delta t} = \frac{\Delta[P]}{\Delta t}$$

The rate of a chemical reaction can be studied in numerous ways. For example, suppose that the reaction is

$$\underset{\text{brown}}{2NO_2(g)} \rightarrow \underset{\text{colorless}}{2NO(g) + O_2(g)}$$

The species NO_2 is brown, whereas the species NO and O_2 are colorless. Thus we could determine the reaction rate by measuring the decrease in the intensity of the brown color in the reaction mixture as a function of time. The intensity of the brown color is directly proportional to the concentration of NO_2. Alternatively, we could determine the increase in pressure in the reaction mixture as a function of time. Note that the reaction produces 3 mol of gaseous products, $2NO(g) + O_2(g)$, for each 2 mol of gaseous reactant, $2NO_2(g)$, consumed. The increase in reaction-mixture pressure with time can be related to the reaction rate.

The rate at which NO_2 is consumed is equal to the rate at which NO is formed but is twice as great as the rate at which O_2 is formed. Two NO_2 molecules react for each O_2 molecule produced. Thus, we have

$$\frac{\Delta[NO]}{\Delta t} = -\frac{\Delta[NO_2]}{\Delta t}$$

$$\frac{\Delta[O_2]}{\Delta t} = -\left(\frac{1}{2}\right)\frac{\Delta[NO_2]}{\Delta t}$$

Example 16-1: Let the rate of the reaction

$$2N_2O_5(g) \rightarrow 4NO_2(g) + O_2(g)$$

be expressed as $\Delta[O_2]/\Delta t$. Express the rate of the reaction in terms of the other species involved.

Solution: The rate of loss of N_2O_5 is twice as great as the rate of production of O_2 because two N_2O_5 molecules are consumed for each O_2 molecule produced. Thus, we have

$$\text{rate} = \frac{\Delta[O_2]}{\Delta t} = -\left(\frac{1}{2}\right)\frac{\Delta[N_2O_5]}{\Delta t}$$

The rate of production of O_2 is one fourth as great as the production of NO_2 because only one O_2 molecule is formed for every four NO_2 molecules produced. Thus,

$$\text{rate} = \frac{\Delta[O_2]}{\Delta t} = -\frac{1}{2}\frac{\Delta[N_2O_5]}{\Delta t} = \frac{1}{4}\frac{\Delta[NO_2]}{\Delta t}$$

Notice that the fractions here are related to the balancing coefficients in the chemical equation.

The equation for the decomposition of gaseous nitrogen pentoxide to gaseous nitrogen dioxide and oxygen is

$$2N_2O_5(g) \rightarrow 4NO_2(g) + O_2(g) \qquad (16\text{-}3)$$

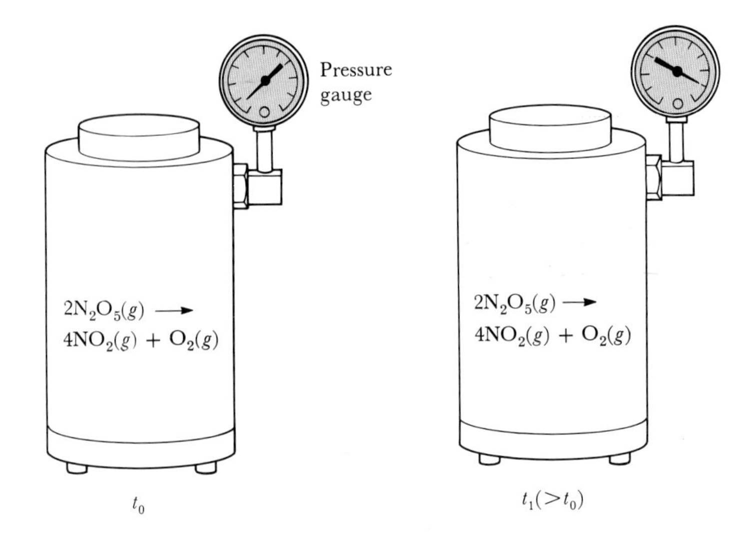

Figure 16-1 Measurement of the rate of the reaction $2N_2O_5(g) \rightarrow 4NO_2(g) + O_2(g)$ by the determination of the pressure of the reaction mixture as a function of time. Because there are more moles of gaseous products than there are of gaseous reactants, the pressure in the reaction vessel increases as the reaction proceeds.

The rate of this reaction can be determined by measuring the increase in the pressure of the reaction mixture as a function of time (Figure 16-1). The concentrations of $N_2O_5(g)$, $NO_2(g)$, and $O_2(g)$ in a reaction mixture with the initial concentrations $[N_2O_5]_0 = 1.24 \times 10^{-2}$ M, $[NO_2]_0 = 0$, and $[O_2]_0 = 0$ are given in Table 16-1. The data in Table 16-1 are plotted in Figure 16-2, which

Table 16-1 Concentration of $N_2O_5(g)$, $NO_2(g)$, and $O_2(g)$ as a function of time at 45°C for an initial concentration of $[N_2O_5]_0 = 1.24 \times 10^{-2}$ M in the reaction $2N_2O_5(g) \rightarrow 4NO_2(g) + O_2(g)$

t/min	$[N_2O_5]$/M	$[NO_2]$/M	$[O_2]$/M
0	1.24×10^{-2}	0	0
10	0.92×10^{-2}	0.64×10^{-2}	0.16×10^{-2}
20	0.68×10^{-2}	1.12×10^{-2}	0.28×10^{-2}
30	0.50×10^{-2}	1.48×10^{-2}	0.37×10^{-2}
40	0.37×10^{-2}	1.74×10^{-2}	0.44×10^{-2}
50	0.28×10^{-2}	1.92×10^{-2}	0.48×10^{-2}
60	0.20×10^{-2}	2.08×10^{-2}	0.52×10^{-2}
70	0.15×10^{-2}	2.18×10^{-2}	0.55×10^{-2}
80	0.11×10^{-2}	2.26×10^{-2}	0.57×10^{-2}
90	0.08×10^{-2}	2.32×10^{-2}	0.58×10^{-2}
100	0.06×10^{-2}	2.36×10^{-2}	0.59×10^{-2}

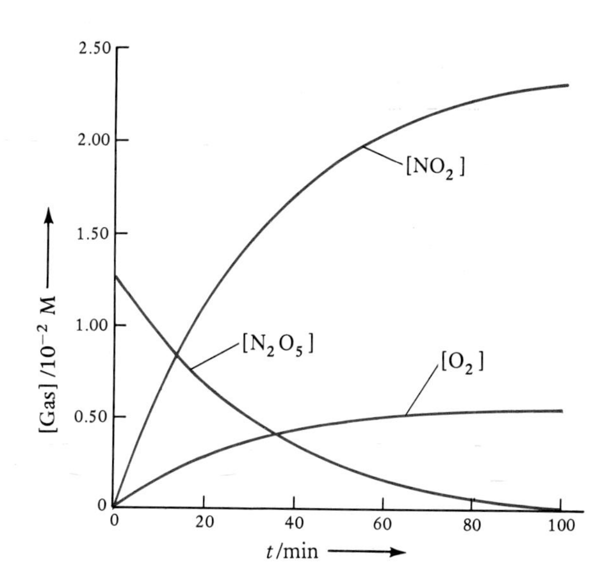

Figure 16-2 The change in concentration of $N_2O_5(g)$, $NO_2(g)$, and $O_2(g)$ as a function of time for the reaction $2N_2O_5(g) \rightarrow 4NO_2(g) + O_2(g)$ at 45°C. The initial concentrations are $[N_2O_5]_0 = 1.24 \times 10^{-2}$ M, $[NO_2]_0 = 0$, and $[O_2]_0 = 0$. Note that both $[NO_2]$ and $[O_2]$ increase with time because NO_2 and O_2 are reaction products.

illustrates both the decrease with time of the reactant concentration, $[N_2O_5]$, and the increase with time of the product concentrations, $[NO_2]$ and $[O_2]$. Note that $[NO_2]$ increases twice as fast as $[N_2O_5]$ decreases because 2 mol of NO_2 are produced for each mole of N_2O_5 that decomposes. The value of $[O_2]$ increases only half as fast as $[N_2O_5]$ decreases because only 0.5 mol of O_2 molecule is produced for each mole of N_2O_5 that decomposes.

16-2. THE RATE LAW EXPRESSES THE RATE OF A REACTION IN TERMS OF THE CONCENTRATIONS OF THE REACTANTS

An inspection of the data in Table 16-1 shows that the reaction rate decreases as $[N_2O_5]$ decreases. For example, the rate of decomposition of N_2O_5 over the first 10 min of the reaction is

$$\text{rate} = -\frac{\Delta[N_2O_5]}{\Delta t} = \frac{-(0.92 \times 10^{-2} - 1.24 \times 10^{-2})\text{ M}}{(10 - 0)\text{ min}}$$
$$= 3.2 \times 10^{-4}\text{ mol}\cdot\text{L}^{-1}\cdot\text{min}^{-1}$$

whereas the rate over the period 10 min to 20 min is

$$\text{rate} = \frac{-(0.68 \times 10^{-2} - 0.92 \times 10^{-2})\text{ M}}{(20 - 10)\text{ min}}$$
$$= 2.4 \times 10^{-4}\text{ mol}\cdot\text{L}^{-1}\cdot\text{min}^{-1}$$

Example 16-2: Use the data in Table 16-1 to calculate the rate of production of NO_2 over the first 10 min of the reaction

$$2N_2O_5(g) \rightarrow 4NO_2(g) + O_2(g)$$

Solution: Because NO_2 is a product species, the rate of production of NO_2 is given by $\Delta[NO_2]/\Delta t$. For the first 10 min of the reaction

Problems 16-9 through 16-12 ask you to calculate the rates of some other reactions.

$$\frac{\Delta[NO_2]}{\Delta t} = \frac{(0.64 \times 10^{-2} - 0)\ M}{(10 - 0)\ min} = 6.4 \times 10^{-4}\ M \cdot min^{-1}$$

Note that the rate of production of NO_2 over the first 10 min is twice as great as the rate of consumption of N_2O_5 over the same time period.

The rates of chemical reactions usually depend upon the concentrations of the reactants. There are both practical and theoretical reasons to assume that the rate of a reaction is proportional to the concentrations of the reactants raised to small integer powers. For the case of the thermal decomposition of N_2O_5, we assume that

$$\text{rate} \propto [N_2O_5]^x \quad (16\text{-}4)$$

We can write this proportionality as an equation by inserting a proportionality constant:

$$\text{rate} = k[N_2O_5]^x \quad (16\text{-}5)$$

The value of x in Equation (16-5) must be determined experimentally; it is *not* necessarily the same as the balancing coefficient of N_2O_5 in the chemical equation. Table 16-2 lists some experimentally determined rates for Reaction (16-3) as a function of $[N_2O_5]$. Notice

Table 16-2 Some experimentally observed rate values for the reaction $2N_2O_5(g) \rightarrow 4NO_2(g) + O_2(g)$ as a function of $[N_2O_5]$

$Rate/mol \cdot L^{-1} \cdot h^{-1}$	$[N_2O_5]/mol \cdot L^{-1}$
0.018	0.010
0.036	0.020
0.072	0.040
0.108	0.060
0.181	0.100

Notice that when the concentration of $N_2O_5(g)$ is doubled, the rate doubles.

from Table 16-2 that the reaction rate is directly proportional to $[N_2O_5]$. Thus $x = 1$ in Equation (16-5) and we have

$$\text{rate} = k[N_2O_5] \qquad (16\text{-}6)$$

Equation (16-6) expresses the rate of Reaction (16-3) as a function of $[N_2O_5]$. This equation is called the *rate law* for Equation (16-3), and k is called the *rate constant* of the reaction. Because $[N_2O_5]$ is raised to the first power in Equation (16-6), the rate law is said to be a *first-order* rate law and the reaction is said to be a *first-order* reaction. We can determine the numerical value of the rate constant k by using the data in Table 16-2. If we substitute the fact that the rate $= 0.072\ \text{mol}\cdot\text{L}^{-1}\cdot\text{h}^{-1}$ when $[N_2O_5] = 0.040\ \text{mol}\cdot\text{L}^{-1}$, then we obtain

A rate law tells us how a reaction rate depends on concentration.

$$k = \frac{\text{rate}}{[N_2O_5]} = \frac{0.072\ \text{mol}\cdot\text{L}^{-1}\cdot\text{h}^{-1}}{0.040\ \text{mol}\cdot\text{L}^{-1}} = 1.8\ \text{h}^{-1}$$

Notice that the units of the rate constant for a first-order reaction are time^{-1}.

Example 16-3: Hydrogen peroxide decomposes in the presence of iodide ion. The reaction is

$$2H_2O_2(aq) \xrightarrow{I^-(aq)} 2H_2O(l) + O_2(g)$$

Using the following data, determine the rate law and the value of the rate constant for the reaction:

Rate/mol·L^{-1}·min^{-1}	*$[H_2O_2]$/mol·L^{-1}*
1.01×10^{-3}	0.100
2.02×10^{-3}	0.200
4.04×10^{-3}	0.400
6.06×10^{-3}	0.600

Solution: The data indicate that the rate is directly proportional to $[H_2O_2]$, and so we have a first-order rate law:

$$\text{rate} = k[H_2O_2]$$

The value of the rate constant can be obtained from any pair of data in the table. For example,

$$k = \frac{\text{rate}}{[H_2O_2]} = \frac{4.04 \times 10^{-3}\ \text{mol}\cdot\text{L}^{-1}\cdot\text{min}^{-1}}{0.400\ \text{mol}\cdot\text{L}^{-1}} = 1.01 \times 10^{-2}\ \text{min}^{-1}$$

Problems 16-13 and 16-14 are similar to Example 16-3.

The rate law is

$$\text{rate} = (1.01 \times 10^{-2}\text{mol}\cdot\text{L}^{-1}\cdot\text{min}^{-1})[\text{H}_2\text{O}_2]$$

The next example shows that not all reaction rate laws are first order.

Example 16-4: At 325°C, $NO_2(g)$ reacts with $CO(g)$ to yield $NO(g)$ and $CO_2(g)$:

$$\text{NO}_2(g) + \text{CO}(g) \rightarrow \text{NO}(g) + \text{CO}_2(g)$$

A reactant that does not appear in the rate law has no effect on the reaction rate.

It is observed that the reaction rate does not depend upon the concentration of $CO(g)$, and so the reaction rate law does not involve [CO]. The rate does depend upon $[NO_2]$, however, and the following data are obtained:

Rate/mol·L⁻¹·s⁻¹	*[NO₂]/mol·L⁻¹*
0.011	0.15
0.045	0.30
0.18	0.60
0.40	0.90

Determine the rate law and the value of the rate constant.

The rate of the reaction in Example 16-4 can be determined by measuring the intensity of the brown color of NO_2 with time.

Solution: The rate is *not* directly proportional to $[NO_2]$. As we double $[NO_2]$ in going from 0.15 M to 0.30 M, we quadruple the rate. In going from 0.15 M to 0.60 M, we increase the rate by a factor of 16. Thus, we see that the rate is proportional to the square of $[NO_2]$, and so the rate law is

$$\text{rate} = k[\text{NO}_2]^2$$

Because the exponent in the rate law is 2, the rate law is a *second-order* rate law. We can evaluate k by using

Second order reaction:
k = /M·s

$$k = \frac{\text{rate}}{[\text{NO}_2]^2} = \frac{0.045\ \text{mol}\cdot\text{L}^{-1}\cdot\text{s}^{-1}}{(0.30\ \text{mol}\cdot\text{L}^{-1})^2} = 0.50\ \text{mol}^{-1}\cdot\text{L}\cdot\text{s}^{-1}$$

Notice that the units of k for a second-order reaction are $M^{-1}\cdot s^{-1}$.

In Example 16-4 we saw that a reaction rate law that is of the form

$$\text{rate} = k[\text{reactant}]^2$$

is called a *second-order rate law*. We learn in the next section that the rate law for the reaction

$$2NO_2(g) + F_2(g) \rightarrow 2NO_2F(g)$$

is

$$\text{rate} = (1.0 \times 10^{-4}\ M^{-1}\cdot s^{-1})[NO_2][F_2]$$

This rate law is first order in both $[NO_2]$ and $[F_2]$ and is said to be second order overall. More generally, if a reaction rate law is of the form

$$\text{rate} = k[A]^x[B]^y[C]^z$$

then the rate law is x order in [A], y order in [B], and z order in [C] with an overall order of $x + y + z$.

Example 16-5: Given the following reactions and rate laws, determine the order with respect to each reactant, the overall order, and the units of k.

(1)
$$2NOBr(g) \rightarrow 2NO(g) + Br_2(g)$$
$$\text{rate} = k[NOBr]^2$$

(2)
$$\text{cyclopropane } (H_2C{-}CH_2 \text{ ring with } CH_2) \rightarrow \text{propene } (H_2C{=}CH{-}CH_3)$$

cyclopropane → propene

$$\text{rate} = k[\text{cyclopropane}]$$

(3)
$$2H_2(g) + 2NO(g) \rightarrow 2H_2O(g) + N_2(g)$$
$$\text{rate} = k[H_2][NO]^2$$

(4)
$$I^-(aq) + OCl^-(aq) \rightarrow IO^-(aq) + Cl^-(aq)$$
$$\text{rate} = k[I^-][OCl^-]$$

Solution:

(1) The rate law is second order in [NOBr] and second order overall. The expression for k is

$$k = \frac{\text{rate}}{[NOBr]^2}$$

and the units of k are

$$\text{units of } k = \frac{M\cdot s^{-1}}{M^2} = M^{-1}\cdot s^{-1}$$

The units of a rate constant depend on the order of the reaction.

(2) The rate law is first order in [cyclopropane] and first order overall. The expression for k is

$$k = \frac{\text{rate}}{[\text{cyclopropane}]}$$

and the units of k are

$$\text{units of } k = \frac{M\cdot s^{-1}}{M} = s^{-1}$$

(3) The rate law is first order in $[H_2]$, second order in [NO], and third order overall. The expression for k is

$$k = \frac{\text{rate}}{[H_2][NO]^2}$$

and the units of k are

$$\text{units of } k = \frac{M \cdot s^{-1}}{M^3} = M^{-2} \cdot s^{-1}$$

(4) The rate law is first order in $[I^-]$, first order in $[OCl^-]$, and second order overall. The expression for k is

$$k = \frac{\text{rate}}{[I^-][OCl^-]}$$

and the units of k are

$$\text{units of } k = \frac{M \cdot s^{-1}}{M^2} = M^{-1} \cdot s^{-1}$$

Rate laws must be determined experimentally. There is no relation between the balancing coefficients in a chemical equation and the order of the reaction.

16-3. THE METHOD OF INITIAL RATES IS USED TO DETERMINE REACTION RATE LAWS

The initial rate is the rate at the start of the reaction.

Up to now, we have determined rate laws by referring to tables of reaction rates and concentrations of reactants. In practice, however, rate laws often are determined from data involving reaction rates at the early stage of reactions. Using the *method of initial rates,* we measure the rate of a reaction over an initial time interval that is short enough so that the concentrations of the reactants do not vary appreciably from their initial values. If we use zeros as subscripts to denote initial values, then we can write

$$(\text{rate})_0 = k[A]_0^x[B]_0^y[C]_0^z$$

We can then determine x, y, and z as we did in the previous section.

If the initial concentration of a reactant is doubled and the initial reaction rate doubles, then the rate law is first order in the concentration of that reactant. For example, consider the following initial-rate data for the reaction

$$2NO_2(g) + F_2(g) \rightarrow 2NO_2F(g)$$

	Initial concentration		
Run	$[NO_2]_0$/M	$[F_2]_0$/M	*Initial rate*/$M \cdot s^{-1}$
1	1.00	1.00	1.00×10^{-4}
2	2.00	1.00	2.00×10^{-4}
3	1.00	2.00	2.00×10^{-4}
4	2.00	2.00	4.00×10^{-4}

A comparison of the data for runs 1 and 2 shows that a twofold increase in the initial concentration of NO_2 (with $[F_2]_0$ held fixed) increases the initial rate by a factor of 2, and thus the rate must be proportional to the first power of $[NO_2]_0$:

$$\text{initial rate} \propto [NO_2]_0$$

A comparison of the data for runs 1 and 3 shows that a twofold increase in the initial concentration of F_2 (with $[NO_2]_0$ held fixed) increases the initial rate by a factor of 2, and thus the rate must also be proportional to $[F_2]_0$, that is,

$$\text{initial rate} \propto [F_2]_0$$

Combination of the above two results yields the reaction rate law:

In the method of initial rates we usually vary the concentration of only one species at a time.

$$\text{initial rate} = k[NO_2]_0[F_2]_0$$

Comparison of runs 1 and 4 shows that simultaneous twofold increases in $[NO_2]_0$ and $[F_2]_0$ increases the initial rate by a factor of 4, which is consistent with the above rate law. The reaction rate constant is (from run 1)

$$k = \frac{\text{initial rate}}{[NO_2]_0[F_2]_0} = \frac{1.00 \times 10^{-4}\ \text{M}\cdot\text{s}^{-1}}{[1.00\ \text{M}][1.00\ \text{M}]} = 1.00 \times 10^{-4}\ \text{M}^{-1}\cdot\text{s}^{-1}$$

The following two examples illustrate other determinations of reaction rate laws using the method of initial rates.

Example 16-6: Sucrose (table sugar) decomposes in aqueous solution into a mixture of two 6-carbon sugars, glucose and fructose:

The decomposition of sucrose shown in Example 16-6 is the first step in a series of reactions whereby sucrose is used as an energy source by animals.

$$\underset{\text{sucrose}}{C_{12}H_{22}O_{11}(aq)} + H_2O(l) \rightarrow \underset{\text{glucose}}{C_6H_{12}O_6(aq)} + \underset{\text{fructose}}{C_6H_{12}O_6(aq)}$$

The observed initial rates of decomposition of sucrose at two different initial sucrose concentrations are (at 25°C)

Run	$[sucrose]_0$/M	*Initial rate*/$\text{M}\cdot s^{-1}$
1	0.10	8.55×10^{-3}
2	0.20	1.71×10^{-2}

Find the dependence of the reaction rate on the concentration of sucrose and compute the rate constant for the reaction.

Solution: We note from the rate data that a twofold increase in $[\text{sucrose}]_0$ produces a twofold increase in rate, and thus the rate is proportional to the first power of the sucrose concentration:

$$\text{initial rate} = k[\text{sucrose}]_0$$

where k is the reaction rate constant, given by

$$k = \frac{\text{initial rate}}{[\text{sucrose}]_0}$$

Thus from run 1

$$k = \frac{8.55 \times 10^{-3}\,\text{M}\cdot\text{s}^{-1}}{0.10\,\text{M}} = 8.6 \times 10^{-2}\,\text{s}^{-1}$$

and from run 2

$$k = \frac{1.71 \times 10^{-2}\,\text{M}\cdot\text{s}^{-1}}{0.20\,\text{M}} = 8.6 \times 10^{-2}\,\text{s}^{-1}$$

The agreement between the two values of k confirms the postulated rate law. If the two calculated values of k differed significantly, then the reaction rate law could not be first order in the sucrose concentration.

Example 16-7: The following data were obtained for the reaction

$$2\text{NO}(g) + \text{Br}_2(g) \rightarrow 2\text{NOBr}(g)$$

Run	$[NO]_0/\text{M}$	$[Br_2]_0/\text{M}$	*Initial rate*/$\text{M}\cdot min^{-1}$
1	1.0	1.0	1.30×10^{-3}
2	2.0	1.0	5.20×10^{-3}
3	4.0	2.0	4.16×10^{-2}

Determine the reaction rate law and the rate constant.

Problems 16-15 through 16-26 involve the method of initial rates.

Solution: From run 1 to run 2, $[\text{NO}]_0$ doubles and $[\text{Br}_2]_0$ remains the same. The initial rate quadruples, and so the rate is second order in [NO]:

$$\text{rate} = k[\text{NO}]^2[\text{Br}_2]^x$$

From run 2 to run 3, both $[\text{NO}]_0$ and $[\text{Br}_2]_0$ double and the initial rate increases eightfold. We already know that the rate law is second order in [NO], and so the initial rate increases by a factor of 4 as a result of the increase in $[\text{NO}]_0$ alone. The rest of the increase in the initial rate, a factor of 2, is due to the increase in $[\text{Br}_2]_0$. Thus, assuming that the rate law does not change during the course of the reaction, we have

$$\text{rate} = k[\text{NO}]^2[\text{Br}_2]$$

The value of the rate constant is (run 2)

$$k = \frac{\text{initial rate}}{[\text{NO}]_0^2[\text{Br}_2]_0} = \frac{5.20 \times 10^{-3}\ \text{M}\cdot\text{min}^{-1}}{(2.0\ \text{M})^2(1.0\ \text{M})} = 1.3 \times 10^{-3}\ \text{M}^{-2}\cdot\text{min}^{-1}$$

16-4. THE HALF-LIFE FOR A FIRST-ORDER RATE LAW IS GIVEN BY $t_{1/2} = 0.693/k$

Using calculus, it is possible to show that, if a rate law is first order in the concentration of a reactant A, then the dependence of the concentration of A on time is given by

$$\log[\text{A}] = \log[\text{A}]_0 - \frac{kt}{2.30} \quad (16\text{-}7)$$

where $[\text{A}]_0$ is the concentration of A at $t = 0$.

It is convenient to plot Equation (16-7) as a straight line. You may have learned in algebra that the mathematical equation of a straight line is

$$y = mx + b$$

$y = mx + b$ (↑ slope, ↑ intercept)

where m and b are constants. The quantity b is the *intercept* of the line with the y axis and m is the *slope* of the line. The slope of a straight line is a measure of its steepness. Equation (16-7) may not look like it is of the form $y = mx + b$, but if we let $y = \log[\text{A}]$ and $x = t$, then Equation (16-7) is a straight line whose intercept is $\log[\text{A}]_0$ and whose slope is $-k/2.30$. Therefore, a plot of $\log[\text{A}]$ versus t is a straight line with a slope of $-k/2.30$. Further, if the rate law is not first order in [A], then a plot of $\log[\text{A}]$ versus t will not be linear.

Plotting data in the form of a straight line is discussed in Appendix A and in the Study Guide/Solutions Manual.

The $[\text{N}_2\text{O}_5]$ values given in Table 16-1 are converted to $\log[\text{N}_2\text{O}_5]$ data in Table 16-3. The data in Table 16-3 are plotted in the form $\log[\text{N}_2\text{O}_5]$ versus t in Figure 16-3. The fact that the plot of $\log[\text{N}_2\text{O}_5]$ versus t is linear confirms that the rate law is first order in $[\text{N}_2\text{O}_5]$, that is,

$$\text{rate} = -\frac{\Delta[\text{N}_2\text{O}_5]}{\Delta t} = k[\text{N}_2\text{O}_5]$$

Equation (16-7) can be rearranged to the form

$$\log\frac{[\text{A}]}{[\text{A}]_0} = -\frac{kt}{2.30} \quad (16\text{-}8)$$

Table 16-3 $[N_2O_5]$ and $\log[N_2O_5]$ as a function of time for the reaction $2N_2O_5(g) \rightarrow 4NO_2(g) + O_2(g)$

t/min	$[N_2O_5]$/M	$\log([N_2O_5]$/M)
0	1.24×10^{-2}	−1.91
10	0.92×10^{-2}	−2.04
20	0.68×10^{-2}	−2.17
30	0.50×10^{-2}	−2.30
40	0.37×10^{-2}	−2.43
50	0.28×10^{-2}	−2.55
60	0.20×10^{-2}	−2.70
70	0.15×10^{-2}	−2.82
80	0.11×10^{-2}	−2.96
90	0.08×10^{-2}	−3.10
100	0.06×10^{-2}	−3.22

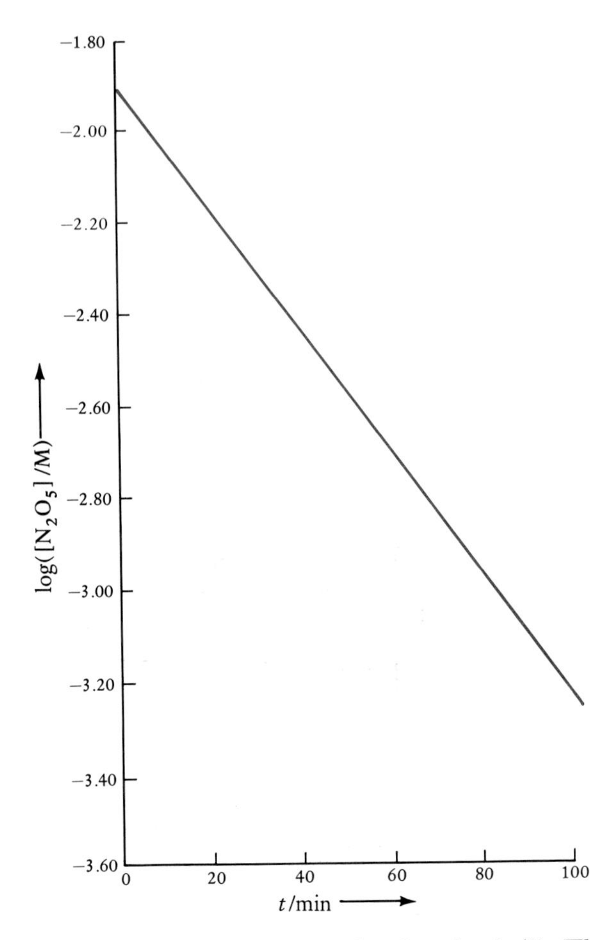

Figure 16-3 Plot of $\log[N_2O_5]$ versus t for the reaction $2N_2O_5(g) \rightarrow 4NO_2(g) + O_2(g)$.

where we have used the fact that $\log a - \log b = \log(a/b)$. The *half-life*, $t_{1/2}$, of a reactant is defined as the time that it takes for the concentration of the reactant to decrease by a factor of 2. Thus when $t = t_{1/2}$, $[A] = [A]_0/2$ in Equation (16-8). Substituting $t = t_{1/2}$ and $[A] = [A]_0/2$ in Equation (16-8) gives

$$\log\left(\frac{1}{2}\right) = -0.301 = -\frac{kt_{1/2}}{2.30}$$

Solving this equation for $t_{1/2}$ yields

$$t_{1/2} = \frac{0.693}{k} \tag{16-9}$$

Equation (16-9) says that the half-life for a first order decomposition reaction is *independent of the initial concentration of the reactant*. Conversely, if it is observed that $t_{1/2}$ for a reactant species A is independent of $[A]_0$, then the rate expression must contain [A] to the first power.

Let's return to the data in Table 16-1 on the $N_2O_5(g)$ decomposition reaction. Figure 16-4 shows a plot of $[N_2O_5]$ versus time on a

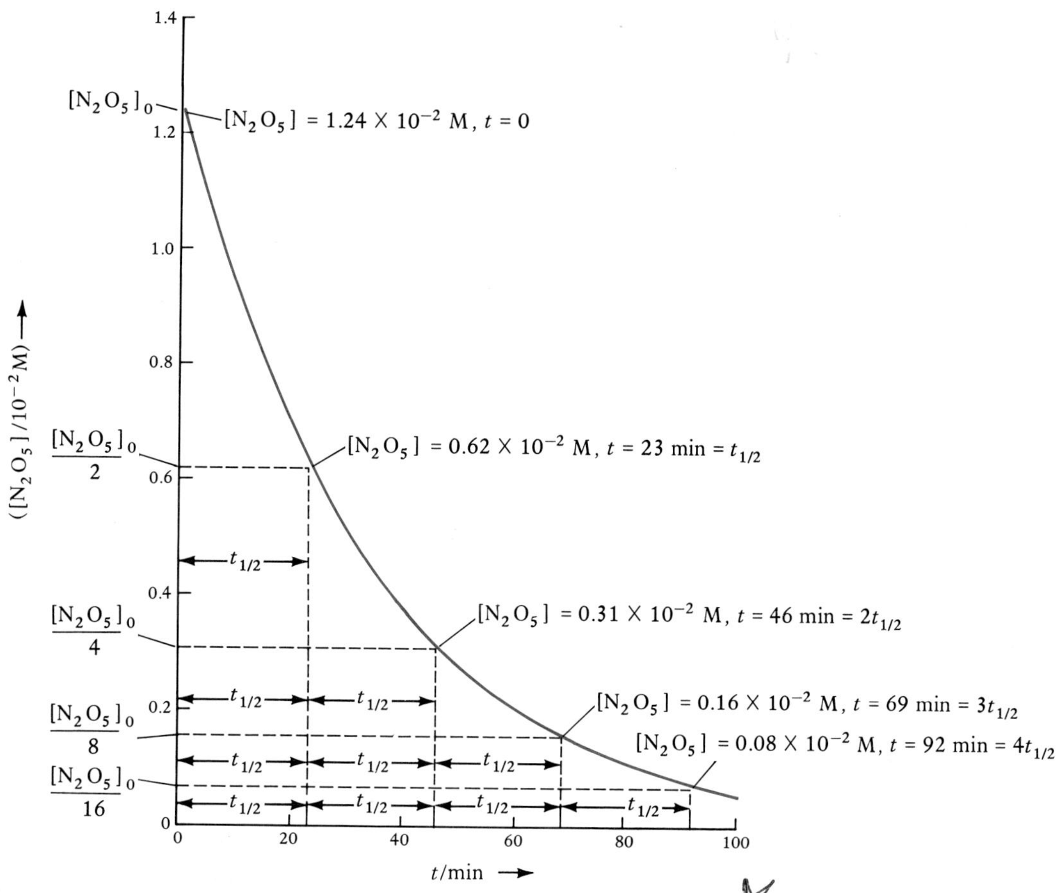

Figure 16-4 Plot of $[N_2O_5]$ versus time, illustrating the half-life for $[N_2O_5]$. The initial $N_2O_5(g)$ concentration is $[N_2O_5]_0 = 1.24 \times 10^{-2}$ M.

larger scale than that shown in Figure 16-2. The initial concentration of $N_2O_5(g)$ is $[N_2O_5]_0 = 1.24 \times 10^{-2}$ M. The time required for $[N_2O_5]$ to decrease from 1.24×10^{-2} M to $\frac{1}{2}(1.24 \times 10^{-2}\text{ M}) = 0.62 \times 10^{-2}$ M can be read off the plot in Figure 16-4 as $t_{1/2} = 23$ min. Note also in Figure 16-4 that it requires another 23 min for $[N_2O_5]$ to decrease from 0.62×10^{-2} M to $0.31 \times 10^{-2}\text{ M} = \frac{1}{2}(0.62 \times 10^{-2}\text{ M})$. The half-life for the first-order N_2O_5 decomposition reaction is *independent* of the concentration of unreacted $N_2O_5(g)$. This result is shown in Table 16-4.

Table 16-4 Half-life values for the N_2O_5 decomposition reaction

Twofold decrease in $[N_2O_5]$	$t_{1/2}$/*min*	*Total time/min*
1.24×10^{-2} M $\rightarrow$ 0.62×10^{-2} M	23	23
0.62×10^{-2} M $\rightarrow$ 0.31×10^{-2} M	23	46 (2×23)
0.31×10^{-2} M $\rightarrow$ 0.16×10^{-2} M	23	69 (3×23)
0.16×10^{-2} M $\rightarrow$ 0.08×10^{-2} M	23	92 (4×23)

Example 16-8: The rate law for the decomposition of aqueous hydrogen peroxide at 70°C,

$$2H_2O_2(aq) \rightarrow 2H_2O(l) + O_2(g)$$

is rate = $k[H_2O_2]$. The half-life for the $H_2O_2(aq)$ decomposition at 25°C is $t_{1/2} = 20$ min. Given that the initial concentration is $[H_2O_2]_0 = 0.30$ M, compute the value of $[H_2O_2]$ 60 min after the 0.30 M $H_2O_2(aq)$ solution is prepared.

Solution: A reaction time of 60 min corresponds to three half-lives:

$$\frac{60 \text{ min}}{20 \text{ min/half-life}} = 3 \text{ half-lives}$$

The concentration of a reactant A that remains unreacted after n half-lives is

$$[A] = [A_0]\left(\frac{1}{2}\right)^n \tag{16-10}$$

If $[A]/[A_0]$ is known, then Equation (16-10) can be solved for n, the number of half-lives. Also, noninteger values of n can be used in Equation (16-10).

where $[A]_0$ is the initial reactant concentration. Using Equation (16-10), we compute

$$[H_2O_2] = [H_2O_2]_0\left(\frac{1}{2}\right)^n = (0.30 \text{ M})\left(\frac{1}{2}\right)^3 = \frac{0.30 \text{ M}}{8} = 0.038 \text{ M}$$

16-5. THE HALF-LIFE OF A SECOND-ORDER RATE LAW DEPENDS UPON THE INITIAL CONCENTRATION

If a reaction rate is second order in the concentration of a reactant A, then it can be shown from calculus that the concentration of A as a function of time is given by

$$\frac{1}{[A]} = \frac{1}{[A]_0} + kt \tag{16-11}$$

We can plot Equation (16-11) as a straight line by letting $y = 1/[A]$ and $x = t$ in the equation $y = mx + b$. The intercept is $1/[A]_0$, and the slope is k. Thus a plot of $1/[A]$ versus t is a straight line whose slope is k. An example of a second-order reaction is

$$NO_2(g) + CO(g) \rightarrow NO(g) + CO_2(g)$$

Example 16-4 shows that the rate law for this reaction at 325°C is

$$\text{rate} = (0.50 \text{ M}^{-1}\cdot\text{s}^{-1})[NO_2]^2$$

Table 16-5 Kinetic data for the reaction $NO_2(g) + CO(g) \rightarrow NO(g) + CO_2(g)$ at 325°C

t/s	$[NO_2]/M$	$(1/[NO_2])/M^{-1}$
0	0.100	10.0
5.0	0.080	12.5
10.0	0.067	15.0
15.0	0.057	17.5
20.0	0.050	20.0
30.0	0.040	25.0
40.0	0.033	30.3
50.0	0.029	34.5
60.0	0.025	40.0

Table 16-5 and Figure 16-5 give values of $[NO_2]$ versus time for this reaction. The data in Table 16-5 are plotted in the form $1/[NO_2]$ versus t in Figure 16-6. The fact that $1/[NO_2]$ versus t is linear estab-

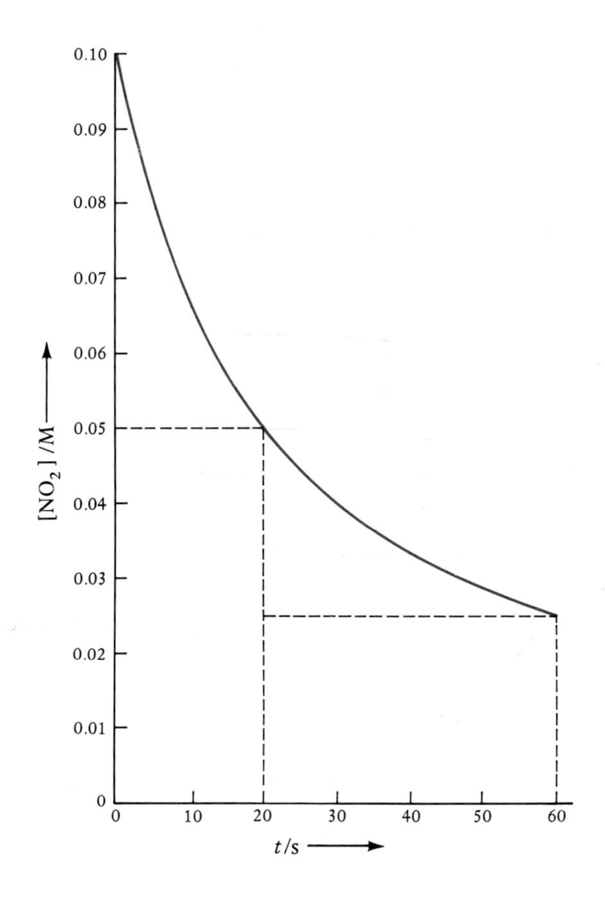

Figure 16-5 Plot of $[NO_2]$ versus time for the reaction $NO_2(g) + CO(g) \rightarrow NO(g) + CO_2(g)$ at 325°C.

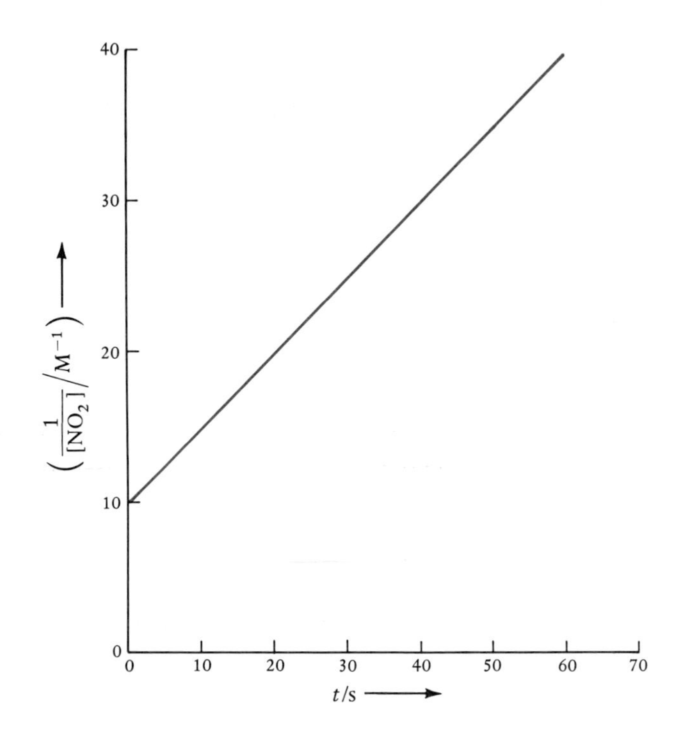

Figure 16-6 Plot of $1/[NO_2]$ versus time for the reaction $NO_2(g) + CO(g) \rightarrow NO(g) + CO_2(g)$.

lishes that the reaction rate law is second order in $[NO_2]$, that is,

$$\text{rate} = k[NO_2]^2$$

The half-life of a first-order reaction is independent of the initial concentration. The half-life of a second-order reaction is inversely proportional to the initial concentration.

We can use Equation (16-11) to derive an equation for the half-life of a second-order reaction. If we substitute $t = t_{1/2}$ and $[A] = [A]_0/2$ into Equation (16-11), then we obtain

$$\frac{2}{[A]_0} = \frac{1}{[A]_0} + kt_{1/2}$$

$$t_{1/2} = \frac{1}{k[A]_0} \qquad (16\text{-}12)$$

Note that the half-life of a second-order reaction is *not* independent of the initial concentration $[A]_0$, as it is for a first-order reaction.

Example 16-9: Use the data in Table 16-5 to verify that the half-life of a second-order reactant is not independent of the initial concentration of that reactant.

Solution: Referring to Table 16-5, we see that it takes 20 s for $[NO_2]$ to decrease from 0.100 M to 0.050 M and 40 s for $[NO_2]$ to decrease from 0.050 M to 0.025 M. Thus, the time that it takes for $[NO_2]$ to decrease by one half depends upon the value of $[NO_2]$ at the starting point.

Table 16-6 Kinetic properties of first-order and second-order reaction rate laws

Rate law	*Order*	*Units of k*	*Dependence of [A] on time*	*Half-life*	*Test plot*[a]
rate = $k[A]$	1	s^{-1}	$\log[A] = \log[A]_0 - \frac{kt}{2.30}$	$\frac{0.69}{k}$	$\log[A]$ vs t
rate = $k[A]^2$	2	$M^{-1} \cdot s^{-1}$	$\frac{1}{[A]} = \frac{1}{[A]_0} + kt$	$\frac{1}{k[A]_0}$	$\frac{1}{[A]}$ vs t

[a]If the indicated plot is linear, then the postulated reaction order is confirmed. That is, (a) if log[A] versus t is linear, then the rate law is first order in [A] as shown in the following plots:

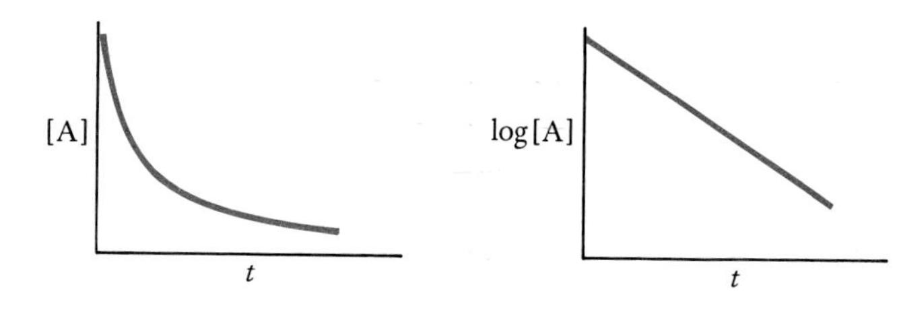

(b) if 1/[A] versus t is linear, then the rate law is second order in [A] as shown in the following plots:

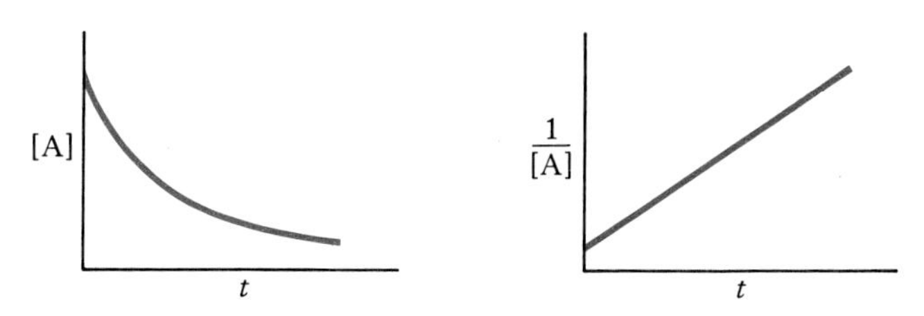

Table 16-6 summarizes the properties of first-order and second-order reactions.

16-6. A RATE LAW CANNOT BE DEDUCED FROM THE REACTION STOICHIOMETRY

A balanced chemical equation gives us the relative numbers of moles of reactants consumed and moles of products formed. If a chemical reaction occurs in a single step, then the reaction is called an *elementary process*. Two examples of elementary processes are the reaction of gaseous nitrogen dioxide to produce nitrogen monoxide and nitrogen trioxide,

$$NO_2(g) + NO_2(g) \rightarrow NO(g) + NO_3(g)$$

for which

$$\text{rate} = k[NO_2]^2$$

and the reaction of $NO_3(g)$ with carbon monoxide to yield carbon dioxide and NO_2,

$$NO_3(g) + CO(g) \rightarrow CO_2(g) + NO_2(g)$$

for which

$$\text{rate} = k[NO_3][CO]$$

An elementary process is a chemical reaction that occurs in a single step.

When we say that this second reaction is an elementary process, we mean that the conversion of NO_3 and CO to CO_2 and NO_2 can occur directly when an NO_3 molecule collides with a CO molecule; there are no other detectable chemical species involved in the reaction. The rate law for an elementary process can be written directly from the stoichiometry of the process; the rate is proportional to the product of the reactant concentrations (see the two examples presented above).

Only for elementary processes can the rate law be deduced from the reaction stoichiometry.

Most chemical reactions, however, are not elementary processes. In most chemical reactions, the conversion of reactants to products requires more than one step. For example, the reaction

$$NO_2(g) + CO(g) \rightarrow CO_2(g) + NO(g)$$

is known to involve at least two steps; the first step is slow and the second step is fast:

$$NO_2(g) + NO_2(g) \rightarrow NO(g) + NO_3(g) \qquad \text{slow}$$
$$NO_3(g) + CO(g) \rightarrow CO_2(g) + NO_2(g) \qquad \text{fast}$$

A series of elementary processes that add up to give an overall reaction is called a *reaction mechanism*. A reaction mechanism is a detailed description of the pathway of a chemical reaction. One of the goals of chemical kinetics is to deduce the mechanisms of reactions. As we shall see, the idea is to construct a mechanism that is consistent with the observed rate law. It is relatively easy to show that a certain rate law is consistent with a given mechanism. It takes more experience to construct a mechanism that agrees with a given rate law.

A rate-determining step acts as a bottleneck that controls the reaction rate.

If one step in a reaction mechanism is much slower than any other steps, then the slow step controls the overall reaction rate and is called the *rate-determining step*. The rate of the slow step in the $NO_2 + CO$ reaction is

$$\text{rate} = k[NO_2]^2$$

Because the rate of this step controls the rate for the overall reaction, this is also the rate law for the overall reaction. In effect, the CO molecules have to wait around for NO_3 molecules to be formed in the slow step. The NO_3 molecules, once formed, are consumed very rapidly by reaction with CO.

Note that the overall reaction stoichiometry does not disclose the involvement of the intermediate species NO_3. An *intermediate* is a species formed from the reactants that is involved in the conversion of reactants to products, but which does not appear as either reactant or product in the overall reaction. It is the possible involvement of intermediates in the reaction process that makes it impossible to deduce the reaction mechanism solely from the overall reaction stoichiometry.

Intermediates are often present only at low concentrations and may be difficult to detect experimentally.

As a second example, consider the reaction

$$H_2(g) + Br_2(g) \rightarrow 2HBr(g)$$

The rate law for the early stages of this reaction is

$$\text{rate} = k[H_2][Br_2]^{1/2}$$

A mechanism that is consistent with the observed rate law is

$$Br_2(g) \rightleftharpoons 2Br(g) \qquad \text{rapid equilibrium, } (K = [Br]^2/[Br_2])$$

$$Br(g) + H_2(g) \xrightarrow{k_2} HBr(g) + H(g) \qquad \text{slow}$$

$$H(g) + Br_2(g) \xrightarrow{k_3} HBr(g) + Br(g) \qquad \text{fast}$$

The rate-determining step is the slow elementary reaction, the rate law for which is

$$\text{rate} = k_2[H_2][Br]$$

Bromine atoms are an intermediate in this mechanism, and so [Br] is not easily measurable. We can eliminate [Br] from the rate law by using the rapid equilibrium of the first elementary reaction in the mechanism. By rapid equilibrium we mean that the equilibrium between bromine atoms and bromine molecules occurs so rapidly that the relation

$$K = \frac{[Br]^2}{[Br_2]}$$

holds at any instant. If we solve this expression for [Br], then we obtain

$$[Br] = K^{1/2}[Br_2]^{1/2}$$

Substitution of this expression for [Br] into the above rate law gives

$$\text{rate} = k_2K^{1/2}[H_2][Br_2]^{1/2} = k[H_2][Br_2]^{1/2}$$

where we have denoted $k_2K^{1/2}$ by k.

Example 16-10: The reaction

$$2NO_2(g) + F_2(g) \rightarrow 2NO_2F$$

A reaction mechanism is a postulated sequence of elementary processes by which the reactants are converted to products.

is thought to proceed via the following two-step mechanism:

$$NO_2 + F_2 \xrightarrow{k_1} NO_2F + F \qquad \text{slow}$$

$$F + NO_2 \xrightarrow{k_2} NO_2F \qquad \text{fast}$$

Identify any species that are intermediates in the reaction mechanism and derive the rate law for this mechanism.

Solution: The fluorine atom is an intermediate in the reaction mechanism. Note that fluorine atoms are produced in the first step and consumed in the second. Also note that the sum of the two elementary processes yields the overall reaction.

The rate law is given by the slow elementary step, and so

$$\text{rate} = k_1[NO_2][F_2]$$

Once again, notice that this rate law could not be deduced from the overall reaction.

Example 16-11: A possible mechanism for the decomposition of ozone to oxygen,

$$2O_3(g) \rightarrow 3O_2(g)$$

consists of the two steps

$$O_3(g) \rightleftharpoons O_2(g) + O(g) \qquad \text{rapid equilibrium}$$

$$O_3(g) + O(g) \xrightarrow{k_2} 2O_2(g) \qquad \text{slow}$$

Deduce a rate law consistent with this mechanism.

It is not possible to prove that a postulated mechanism is the actual mechanism; however, postulated mechanisms can be disproved.

Solution: The rate-determining step is the (slow) second reaction. The rate law for this elementary process is

$$\text{rate} = k_2[O_3][O]$$

The oxygen atoms are an intermediate species in the mechanism, and so their concentration is not easily measurable. We can eliminate [O] from the rate law by utilizing the fact that the first reaction is in rapid equilibrium, which means that the equilibrium adjusts essentially instantaneously. Thus, at any instant we have

$$K = \frac{[O_2][O]}{[O_3]}$$

If we solve this expression for [O], we obtain

$$[O] = \frac{K[O_3]}{[O_2]}$$

If we substitute this expression for [O] into the rate law, we obtain

Problems 16-47 through 16-58 involve reaction mechanisms.

$$\text{rate} = k_2 K \frac{[O_3]^2}{[O_2]} = k \frac{[O_3]^2}{[O_2]}$$

as the rate law for the decomposition of ozone to oxygen.

16-7. MOLECULES MUST COLLIDE BEFORE THEY CAN REACT

A basic postulate of the *collision theory* of reaction rates is that a chemical reaction takes place only if the energy of the collision between two molecules is sufficient to break chemical bonds. In a mixture of the gases A and B with $P_A = P_B = 1.0$ atm, the collision frequency between A and B molecules is about 10^{33} collisions$\cdot s^{-1} \cdot mL^{-1}$. If every collision led to a reactive event, then the initial reaction rate would be about 10^{14} $M \cdot s^{-1}$. Although there are a few reactions that occur at this very high rate, most occur at a much lower rate. The inescapable conclusion is that most collisions do not lead to reaction. Rather, the colliding molecules simply bounce off one another and remain unchanged. Only the more energetic collisions that occur with the necessary relative orientations lead to reaction.

In the collision theory of reaction rates, the rate constant k for the elementary process

$$A + B \xrightarrow{k} C + D$$

which has the rate law

$$\text{rate} = k[A][B]$$

is given by

$$k = \begin{pmatrix} \text{fraction of collisions} \\ \text{in which molecules have} \\ \text{the required relative} \\ \text{orientations} \end{pmatrix} \times \begin{pmatrix} \text{collision} \\ \text{frequency} \end{pmatrix} \times \begin{pmatrix} \text{fraction of} \\ \text{collisions with} \\ \text{the required} \\ \text{energy} \end{pmatrix}$$

Figure 16-7 Plot of the number of collisions between A and B molecules per unit time versus the kinetic energy of the colliding molecules. The plot is made for two different temperatures, T_1 and T_2 ($T_2 > T_1$). Note that the hatched area under the curve beyond E_a is greater for the T_2 curve than for the T_1 curve. The number of collisions with an energy E_a or greater increases with increasing temperature.

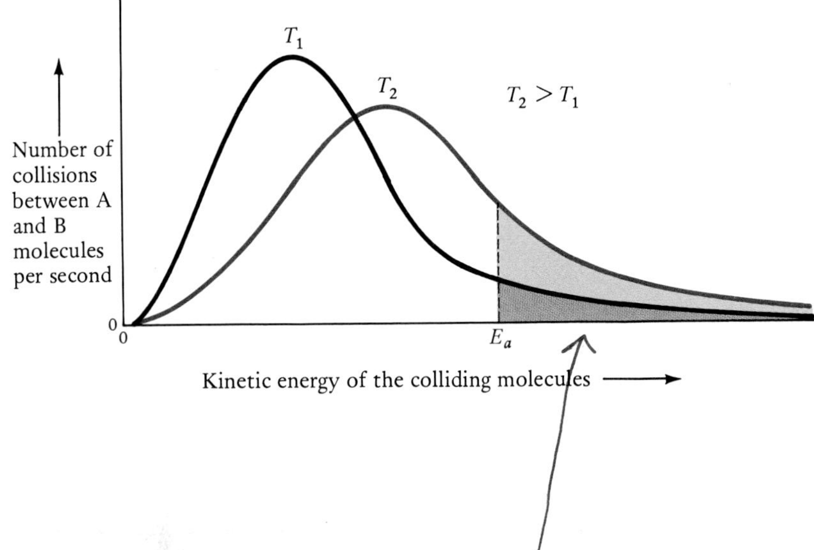

A plot of the number of collisions per second of the A and B molecules versus the kinetic energy of the colliding molecules has the form shown in Figure 16-7. The *activation energy*, E_a, is the minimum energy necessary to cause a reaction between colliding molecules. Molecules that collide with a kinetic energy less than E_a simply bounce off one another. Molecules that collide with a kinetic energy greater than E_a can react, provided they collide with the required relative orientation (Figure 16-8). The fraction of the collisions that involve an energy equal to or greater than the activation energy increases with increasing temperature because the average molecular speed increases with increasing temperature.

Figure 16-8 Nonreactive and reactive collisions of the molecules NO_2 and F_2 in the reaction $NO_2(g) + F_2(g) \rightarrow FNO_2(g) + F(g)$. (a) Nonreactive collision. Molecules bounce off one another without reacting because the kinetic energy of the colliding particles is less than the activation energy. (b) Reactive collision. Molecules collide with a kinetic energy greater than E_a and react. (c) Nonreactive collision. Molecules collide with a kinetic energy greater than E_a but do not react because they do not have the correct orientation for reaction.

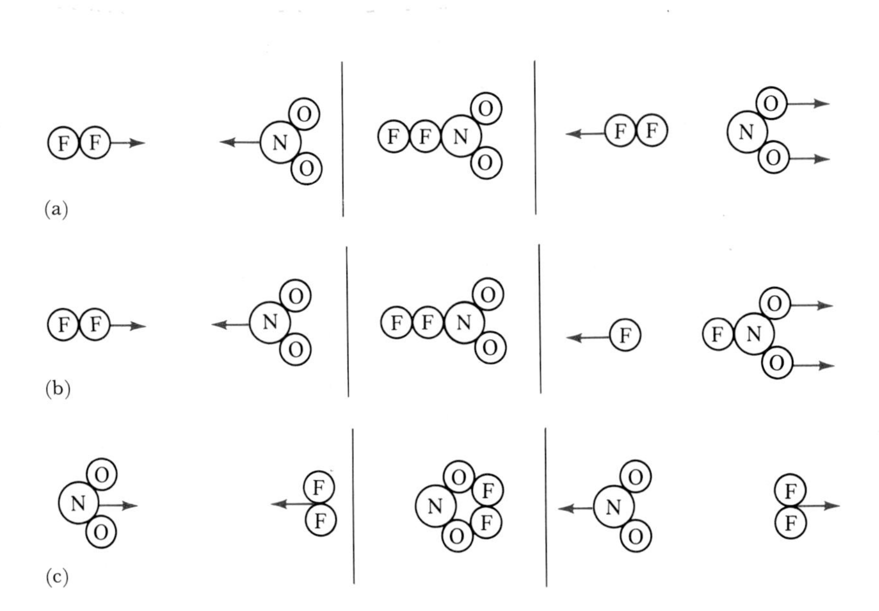

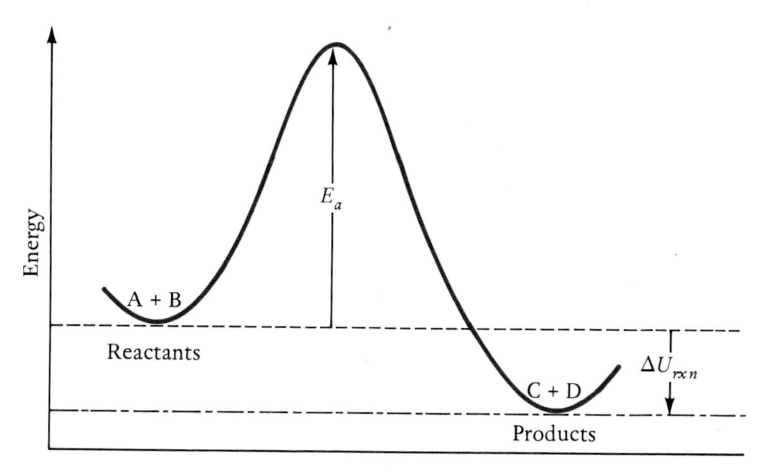

Figure 16-9 The energy of the molecules as the reaction A + B → C + D proceeds. The value of E_a has no relationship to the value of the overall energy change of the reaction, $\Delta U_{rxn} = U(\mathrm{C + D}) - U(\mathrm{A + B})$.

16-8. THE ACTIVATION ENERGY IS AN ENERGY BARRIER THAT THE REACTANTS MUST SURMOUNT IN ORDER TO REACT

The activation energy for the general reaction A + B → C + D is shown in Figure 16-9. If the colliding molecules do not collide with sufficient kinetic energy to go over the activation energy "hump," then the reaction does not take place. The temperature dependence of a reaction rate constant depends upon the activation energy of the reaction. The equation that describes the temperature dependence of a rate constant is called the *Arrhenius equation:*

Svante Arrhenius, a native of Sweden, was awarded the Nobel Prize in chemistry in 1903.

$$\log\left(\frac{k_2}{k_1}\right) = \frac{E_a}{2.30R}\left(\frac{T_2 - T_1}{T_1 T_2}\right) \tag{16-13}$$

where k_1 and k_2 are the rate constants at the absolute temperatures T_1 and T_2, respectively, E_a is the activation energy, and R is the molar gas constant ($8.31\ \mathrm{J \cdot K^{-1} \cdot mol^{-1}}$).

The activation energy, E_a, is a positive quantity.

Example 16-12: The activation energy for the reaction

$$2NO_2(g) + F_2(g) \rightarrow 2NO_2F(g)$$

is $E_a = 43.5\ \mathrm{kJ \cdot mol^{-1}}$. Estimate the increase in the rate constant of the reaction for an increase in temperature from 300 K to 310 K.

Solution: Using Equation (16-13) we have

Logarithms are discussed in Appendix A2.

$$\log\left(\frac{k_2}{k_1}\right) = \frac{E_a}{2.30R}\left(\frac{T_2 - T_1}{T_1 T_2}\right)$$

Inserting the quantities $E_a = 43.5\ \mathrm{kJ \cdot mol^{-1}}$, $T_1 = 300$ K, and $T_2 = 310$ K into this equation yields

$$\log\left(\frac{k_2}{k_1}\right) = \frac{43.5 \times 10^3\ \mathrm{J \cdot mol^{-1}}}{2.30 \times 8.31\ \mathrm{J \cdot K^{-1} \cdot mol^{-1}}}\left[\frac{310\ \mathrm{K} - 300\ \mathrm{K}}{(310 \times 300)\ \mathrm{K}}\right] = 0.24$$

Rough rule of thumb: around room temperature a 10°C increase in temperature leads to a twofold increase in reaction rate.

$$\frac{k_2}{k_1} = 10^{0.24} = 1.7$$

Thus the reaction rate constant increases by about a factor of 2 for the 10 K temperature increase.

Rate constants always increase with temperature because the activation energy is necessarily a positive quantity. The proof goes as follows. If $T_2 > T_1$ in Equation (16-13), then the right-hand side is necessarily positive. If the right-hand side is positive, then k_2 must be greater than k_1, in order that the left-hand side also be positive. Thus we see that $k_2 > k_1$ if $T_2 > T_1$ or, in other words, that the reaction rate constant increases with increasing temperature.

Example 16-13: It is observed that it takes about 4.5 min to cook a boiled egg in Denver, Colorado, to the same extent as a boiled egg cooked for 3.0 min at sea level. The boiling point of water in Denver is about 92°C. Estimate the activation energy for the coagulation of egg albumin, which is the reaction that occurs when an egg is hard-boiled.

Solution: At sea level, where the boiling point of water is 100°C, a boiled egg cooks about 1.5 times faster than at 92°C (4.5 min/3.0 min = 1.5). Thus the rate constant for the coagulation process of egg albumin is 1.5 times greater at 373 K than it is at 365 K. Substitution of these quantities into the Arrhenius equation yields

Problems 16-61 through 16-68 involve applications of the Arrhenius equation.

$$\log\left(\frac{k_2}{k_1}\right) = \frac{E_a}{2.30R}\left(\frac{T_2 - T_1}{T_1 T_2}\right)$$

$$\log 1.5 = \left(\frac{E_a}{2.30 \times 8.31\ \mathrm{J \cdot K^{-1} \cdot mol^{-1}}}\right)\left[\frac{373\ \mathrm{K} - 365\ \mathrm{K}}{(373 \times 365)\ \mathrm{K^2}}\right]$$

$$0.18 = \frac{E_a}{3.25 \times 10^5\ \mathrm{J \cdot mol^{-1}}}$$

$$E_a = 59 \times 10^3\ \mathrm{J \cdot mol^{-1}} = 59\ \mathrm{kJ \cdot mol^{-1}}$$

16-9. A CATALYST IS A SUBSTANCE THAT INCREASES THE REACTION RATE BUT IS NOT CONSUMED AS A REACTANT

The rates of many reactions are increased by catalysts. A *catalyst* is a reaction facilitator that acts by providing a different and faster reaction mechanism than is possible in the absence of the catalyst. For example, the reaction rate law for the decomposition of peroxydisulfate ion in water,

$$2S_2O_8^{2-}(aq) + 2H_2O(l) \rightarrow 4SO_4^{2-}(aq) + O_2(g) + 4H^+(aq)$$

is

$$\text{rate} = k[S_2O_8^{2-}][H^+]$$

The solvated hydrogen ion, $H^+(aq)$, does not appear as a reactant in this reaction, but nevertheless the reaction rate is proportional to $[H^+]$. The $H^+(aq)$ ion facilitates the decomposition of peroxydisulfate ion by attaching to one of the two oxygen atoms that link the two sulfate groups in the peroxydisulfate ion (Figure 16-10), thereby weakening the —O—O— bond in the $S_2O_8^{2-}(aq)$ ion. The weakened —O—O— bond in $-\underset{\substack{| \\ H}}{O}-O-$ ruptures more readily than the unprotonated —O—O— bond.

A catalyst acts by providing a new reaction pathway with a lower activation energy, and thus a larger rate constant. Because the rate of a reaction is proportional to the rate constant, the fact that the rate constant is larger means that the reaction goes faster. The role of a catalyst is illustrated in Figure 16-11. It increases the rates of both the forward and reverse reactions and hence increases the rate at which the reaction proceeds toward equilibrium. Nevertheless, *a catalyst does not affect the position of equilibrium.* In effect, a catalyst helps to get the job done faster, but the final result is the same.

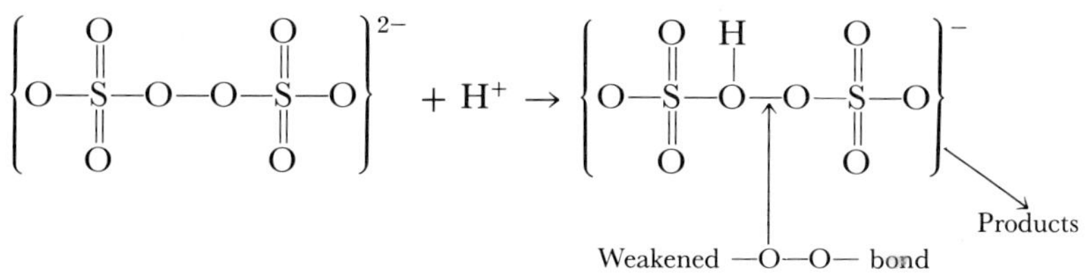

$$\left[O-\underset{\substack{\| \\ O}}{\overset{\substack{O \\ \|}}{S}}-O-O-\underset{\substack{\| \\ O}}{\overset{\substack{O \\ \|}}{S}}-O \right]^{2-} + H^+ \rightarrow \left[O-\underset{\substack{\| \\ O}}{\overset{\substack{O \\ \|}}{S}}-\overset{\substack{H \\ |}}{O}-O-\underset{\substack{\| \\ O}}{\overset{\substack{O \\ \|}}{S}}-O \right]^{-} \rightarrow \text{Products}$$

Weakened —O—O— bond

Products

Figure 16-10 The $H^+(aq)$ catalysis of the decomposition of peroxydisulfate ion in water. Protonation of the peroxide bond, —O—O—, weakens it and facilitates breakage.

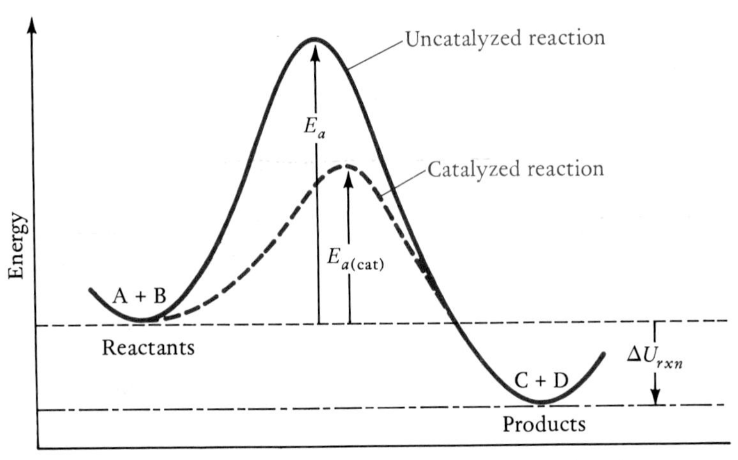

Figure 16-11 A comparison of the activation energies for the uncatalyzed, E_a, and catalyzed, E_a(cat), reaction A + B $\rightarrow$ C + D. The catalyst lowers the activation energy barrier to the reaction and thereby increases the reaction rate.

The following reaction is another example of how a catalyst works. Aqueous cerium(IV) ions, $Ce^{4+}(aq)$, react with aqueous thallium(I) ions, $Tl^{+}(aq)$, via the reaction

$$2Ce^{4+}(aq) + Tl^{+}(aq) \rightarrow 2Ce^{3+}(aq) + Tl^{3+}(aq)$$

The equilibrium constant for this reaction is large ($K \approx 10^{12}$ at 25°C), but the reaction is very slow. The low reaction rate is thought to be a consequence of the requirement that the reactive event, that is, the simultaneous transfer of two electrons from a $Tl^{+}(aq)$ ion to two different $Ce^{4+}(aq)$ ions, requires that two $Ce^{4+}(aq)$ ions be present simultaneously near a $Tl^{+}(aq)$ ion. The rate law is

$$\text{rate} = k[Tl^{+}][Ce^{4+}]^2$$

Three-body encounters are only about 10^{-5} times as likely to occur as two-body encounters, and thus the reaction rate is low. The $2Ce^{4+}(aq) + Tl^{+}(aq)$ reaction is catalyzed by $Mn^{2+}(aq)$. The catalytic action of Mn^{2+} has been attributed to the availability of the Mn^{3+} and Mn^{4+} oxidation states, which provide a new reaction pathway involving a sequence of three 2-body elementary processes:

$$Ce^{4+} + Mn^{2+} \xrightarrow{\text{slow}} Mn^{3+} + Ce^{3+} \qquad \text{rate-determining}$$

$$Ce^{4+} + Mn^{3+} \xrightarrow{\text{fast}} Mn^{4+} + Ce^{3+}$$

$$Tl^{+} + Mn^{4+} \xrightarrow{\text{fast}} Mn^{2+} + Tl^{3+} \qquad \text{two electrons transferred}$$

The sum of these three steps corresponds to the overall reaction stoichiometry. The rate law for the $Mn^{2+}(aq)$-catalyzed reaction is determined by the slowest step in the mechanism, and thus the rate law is

$$\text{rate} = k_{cat}[Ce^{4+}][Mn^{2+}]$$

where k_{cat} is the rate constant for the catalyzed reaction.

Innumerable industrial- and laboratory-scale chemical reactions are carried out in the presence of catalysts to increase the rate of production of the desired products. Some major industrial examples are the Haber process for NH_3 production, the manufacture of HNO_3 and H_2SO_4, and catalytic hydrocarbon-cracking processes in petroleum refining.

16-10. MANY METAL SURFACES ACT AS HETEROGENEOUS CATALYSTS

Platinum and palladium are surface catalysts for a variety of reactions, such as the hydrogenation of double bonds:

$$H_2C{=}CH_2 + H_2(g) \xrightarrow{Pt(s)} H{-}CH_2{-}CH_2{-}H(g)$$

The first step in this hydrogenation involves the adsorption of hydrogen onto the platinum surface, and this is followed by dissociation of the adsorbed H_2 into adsorbed hydrogen atoms:

$$H_2(\textit{surface}) \rightarrow 2H(\textit{surface})$$

The adsorbed hydrogen atoms can move around on the platinum surface and eventually react stepwise with adsorbed ethylene:

$$H_2C{=}CH_2(\textit{surface}) + H(\textit{surface}) \rightarrow H_2\dot{C}{-}CH_3(\textit{surface})$$

$$H_2\dot{C}{-}CH_3(\textit{surface}) + H(\textit{surface}) \rightarrow H_3C{-}CH_3(g)$$

The ethane, H_3CCH_3, produced does not interact strongly with the platinum surface and thus leaves the surface immediately after it is formed.

Platinum metal also catalyzes a wide variety of oxygenation reactions, including the oxidation of SO_2 to SO_3 in the production of sulfuric acid (Interchapter D). The first step is the adsorption of O_2 onto the platinum surface. The adsorbed O_2 dissociates into O atoms to form a surface layer of reactive O atoms:

$$O_2(g) \rightarrow O_2(surface)$$
$$O_2(surface) \rightarrow 2O(surface)$$

The final step involves the rapid reaction of SO_2 with surface O atoms (Figure 16-12):

$$SO_2 + O(surface) \rightarrow SO_3$$

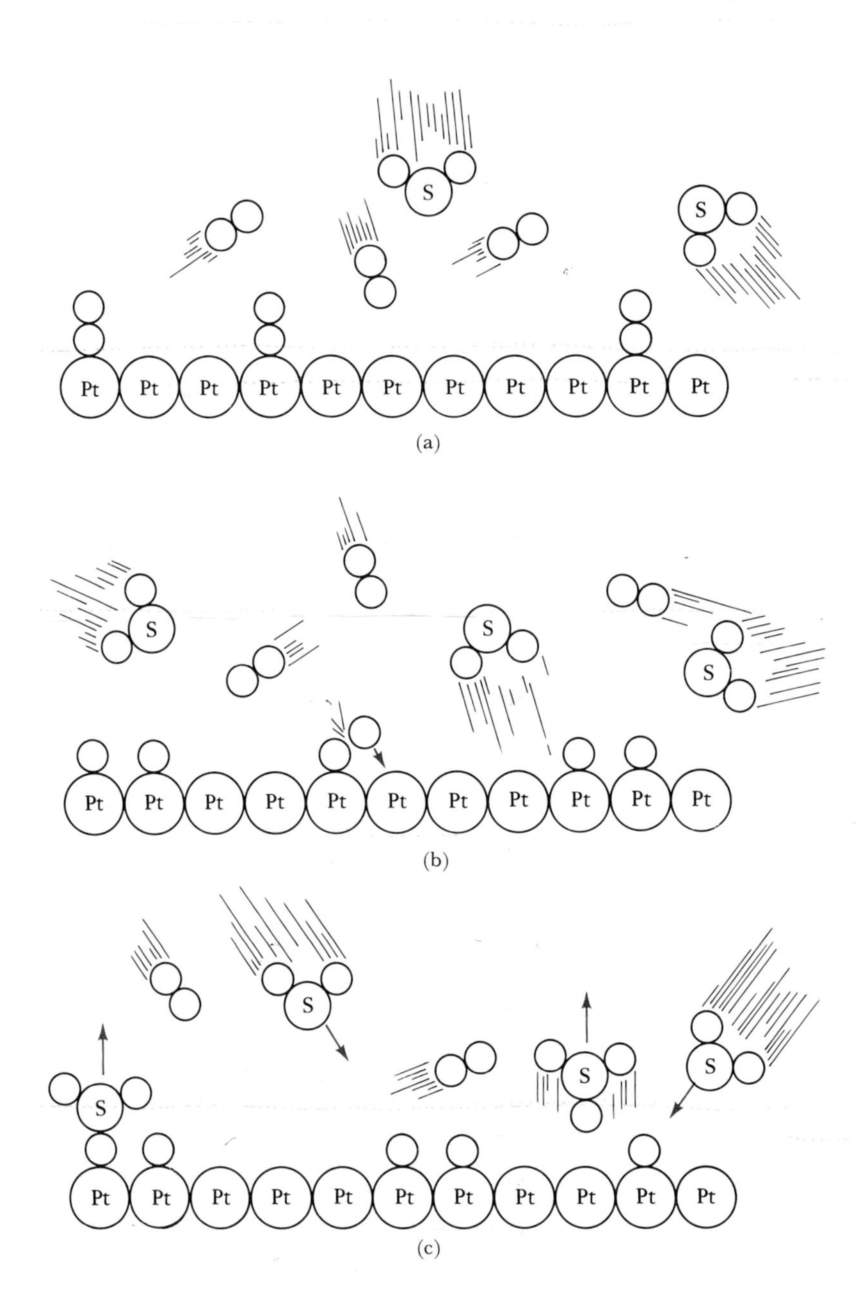

Figure 16-12 Heterogeneous (contact) catalysis by a platinum metal surface. The catalyzed reaction is $2SO_2(g) + O_2(g) \xrightarrow{Pt(s)} 2SO_3(g)$. The platinum surface catalyzes the reaction by causing the dissociation of adsorbed O_2 molecules into O atoms. The surface-bound O atoms then react with SO_2 to give SO_3. (a) Adsorption of O_2 onto the platinum surface. (b) Dissociation of O_2 to O atoms on the platinum surface. (c) Reaction of SO_2 with surface O atoms and desorption of SO_3 from the surface.

Such reactions are examples of *heterogeneous catalysis* because the catalyst and the reactants are in different phases.

16-11. AT EQUILIBRIUM, THE FORWARD REACTION RATE EQUALS THE REVERSE REACTION RATE

Consider the progression toward equilibrium of the reaction

$$2NOBr(g) \rightleftharpoons 2NO(g) + Br_2(g)$$

Recall from Chapter 15 that the equilibrium state can be approached from either side. When the equilibrium is approached from the left (forward direction), the observed rate law is

$$\text{rate} = k_f[NOBr]^2$$

The subscript f indicates that k_f is the rate constant of the forward reaction. When the equilibrium is approached from the right (reverse direction), the observed rate law is

$$\text{rate} = k_r[NO]^2[Br_2]$$

The subscript r indicates that k_r is the rate constant of the reverse reaction.

The *net reaction rate* is the difference between the forward rate and the reverse rate:

$$\text{net reaction rate} = \text{forward reaction rate} - \text{reverse reaction rate}$$
$$= k_f[NOBr]^2 - k_r[NO]^2[Br_2] \qquad (16\text{-}14)$$

At equilibrium, the net reaction rate is zero, the forward and reverse rates are equal, and Equation (16-14) becomes

$$k_f[NOBr]^2_{eq} = k_r[NO]^2_{eq}[Br_2]_{eq}$$

$$\frac{k_f}{k_r} = \frac{[NO]^2_{eq}[Br_2]_{eq}}{[NOBr]^2_{eq}} \qquad (16\text{-}15)$$

At equilibrium the forward reaction rate equals the reverse reaction rate.

Because the right-hand side of this equation is the equilibrium constant expression for the NOBr dissociation reaction, it is equal to the equilibrium constant K_c, and so we have

$$K_c = \frac{k_f}{k_r} \qquad (16\text{-}16)$$

Equation (16-16) shows that the equilibrium constant is equal to the ratio of the forward rate constant to the reverse rate constant for the reaction and illustrates the connection between thermodynamics (equilibrium) and kinetics (rates). An equilibrium state is a *dynamic* state in which the forward and reverse reactions continue to occur, but at equal rates. Thus there is no *net* change in the concentrations of the various species at equilibrium.

To derive Equation (16-15) we assumed that both forward and reverse reactions are elementary processes. Let's consider the more general case in which the reaction occurs in more than one step. For example, we saw earlier that the reaction

$$NO_2(g) + CO(g) \rightleftarrows CO_2(g) + NO(g)$$

involves two elementary steps:

$$(1) \quad NO_2(g) + NO_2(g) \rightleftarrows NO(g) + NO_3(g)$$

$$(2) \quad NO_3(g) + CO(g) \rightleftarrows CO_2(g) + NO_2(g)$$

For each elementary step, we can write

$$K_1 = \frac{k_{f1}}{k_{r1}} = \frac{[NO][NO_3]}{[NO_2]^2}$$

$$K_2 = \frac{k_{f2}}{k_{r2}} = \frac{[CO_2][NO_2]}{[NO_3][CO]}$$

The equilibrium constant of the overall reaction is given by the product of the equilibrium constants for each step (Section 15-10):

$$K = K_1K_2 = \frac{k_{f1}k_{f2}}{k_{r1}k_{r2}} = \frac{[NO][NO_3]}{[NO_2]^2} \times \frac{[CO_2][NO_2]}{[NO_3][CO]} = \frac{[NO][CO_2]}{[NO_2][CO]}$$

Thus, we see that, even though the reaction may proceed via a multistep mechanism, the equilibrium properties of the reaction are independent of the details of the pathway from reactants to products.

SUMMARY

A reaction rate tells us how fast a reactant is being consumed or how fast a product is being formed in a chemical reaction. The reaction rate law describes the dependence of the reaction rate on the concentrations of the various species involved in the reaction. A first-order rate law is proportional to the first power of the reactant concentration; a second-order rate law is proportional to the second power of the reactant concentration. The reaction rate law cannot be deduced solely from the reaction stoichiometry.

An elementary process is a chemical reaction that occurs in a single step. In this case, the rate law can be deduced from the stoichiometry. Most chemical reactions are not elementary processes, and so their rate laws *cannot* be deduced from the reaction stoichiometry. The reaction mechanism is the sequence of elementary processes by which reactants are converted to products. A reaction mechanism that involves more than one elementary process necessarily involves reaction intermediates. An intermediate is a species that is formed from the reactants and is involved in the conversion of reactants to products but does not appear in the overall reaction stoichiometry.

In order to react, molecules have to collide. However, only those collisions in which the molecules have the correct orientation and have a combined kinetic energy equal to or greater than the activation energy for the reaction lead to products. The temperature dependence of a rate constant is given by the Arrhenius equation.

A catalyst is a species that increases the rate of a reaction but is not consumed as a reactant. A catalyst provides a different and faster reaction pathway than is possible in its absence. A catalyst lowers the activation energy of a reaction but has no effect on the equilibrium concentration of a species.

TERMS YOU SHOULD KNOW

rate 601
reaction rate 602
rate law 607
rate constant 607
first-order rate law 607
second-order rate law 608
method of initial rates 610
slope 613
half-life, $t_{1/2}$ 614
elementary process 619
reaction mechanism 620
rate-determining step 620
intermediate 621
collision theory 623
activation energy, E_a 624
Arrhenius equation 625
catalyst 627
heterogeneous catalysis 631
net reaction rate 631

EQUATIONS YOU SHOULD KNOW HOW TO USE

rate $= k[A]$ (first order)

$$\log[A] = \log[A]_0 - \frac{kt}{2.30} \quad \text{(first order)} \quad (16\text{-}7)$$

$\ln[A] = \ln[A]_0 - kt$

$$t_{1/2} = \frac{0.693}{k} \quad \text{(first order)} \quad (16\text{-}9)$$

$\ln 0.5 = -kt$

$$\text{rate} = k[\text{A}]^2 \qquad \text{(second order)}$$

$$\frac{1}{[\text{A}]} = \frac{1}{[\text{A}]_0} + kt \qquad \text{(second order)} \qquad (16\text{-}11)$$

$$t_{1/2} = \frac{1}{k[\text{A}]_0} \qquad \text{(second order)} \qquad (16\text{-}12)$$

$$\log\left(\frac{k_2}{k_1}\right) = \frac{E_a}{2.30R}\left(\frac{T_2 - T_1}{T_1 T_2}\right) \qquad \text{(Arrhenius equation)} \qquad (16\text{-}13)$$

PROBLEMS

REACTION RATES

16-1. What are the units, in terms of M and s, of the rate constant for

(a) a second-order reaction
(b) a reaction that is 3/2 order overall

16-2. What are the units, in terms of M and s, of the rate constant for a

(a) zero-order reaction
(b) third-order reaction

16-3. Suggest a method for measuring the rate of the following reactions:

(a) $3I^-(aq) + S_2O_8^{2-}(aq) \rightarrow I_3^-(aq) + 2SO_4^{2-}(aq)$
colorless, colorless, blue with starch, colorless

(b) $(CaSO_4)_2 \cdot H_2O(s) + 3H_2O(g) \rightarrow 2CaSO_4 \cdot 2H_2O(s)$

16-4. Suggest a method for measuring the rate of the following reactions:

(a) $2H_2O_2(aq) \rightarrow 2H_2O(l) + O_2(g)$

(b) $2HBr(g) \rightarrow H_2(g) + Br_2(g)$
colorless, colorless, dark red-brown

16-5. For the reaction

$$O_3(g) + NO(g) \rightarrow O_2(g) + NO_2(g)$$

at 310 K

$$\text{rate} = (3.0 \times 10^6\ \text{M}^{-1}\cdot\text{s}^{-1})[O_3][NO]$$

Given that $[O_3] = [NO] = 2.0 \times 10^{-6}$ M at $t = 0$, compute the rate of the reaction at $t = 0$.

16-6. Given the rate law in Problem 16-5, calculate the initial rate of the reaction if $[O_3] = 3.0 \times 10^{-4}$ M and $[NO] = 6.3 \times 10^{-3}$ M.

16-7. For the reaction

$$2O_3(g) \rightarrow 3O_2(g)$$

$-\Delta[O_3]/\Delta t$ was found to be $6.3 \times 10^{-3}\ \text{atm}\cdot\text{s}^{-1}$. Determine the value of $\Delta[O_2]/\Delta t$ during this period of time. Use pressure instead of concentration for O_2.

16-8. For the reaction

$$2C_5H_6(g) \rightarrow C_{10}H_{12}(g)$$

$-\Delta[C_5H_6]/\Delta t$ was found to be $2.3\ \text{torr}\cdot\text{s}^{-1}$. Determine the value of $\Delta[C_{10}H_{12}]/\Delta t$ during this period of time. Use pressure instead of concentration for $C_{10}H_{12}$.

16-9. Given the following data on the decomposition of N_2O_5,

$$2N_2O_5(soln) \rightarrow 4NO_2(soln) + O_2(g)$$

at 45 °C in carbon tetrachloride, calculate the rate of decomposition of N_2O_5 over the first 175 s and also over the interval 845 s to 1202 s:

t/s	$[N_2O_5]$/M
0	1.48
175	1.32
506	1.07
845	0.87
1202	0.69

16-10. The value of $[H^+]$ in the reaction

$$CH_3OH(aq) + H^+(aq) + Cl^-(aq) \rightarrow CH_3Cl(aq) + H_2O(l)$$

was measured over a period of time:

$[H^+]$/M	t/min
2.12	0
1.90	31
1.78	61
1.61	121

Find the average rate of disappearance of $H^+(aq)$ for the time interval between each measurement. What is the average rate of disappearance of CH_3OH for the same time intervals? The average rate of appearance of CH_3Cl?

16-11. The rate of the reaction

$$2CO(g) \rightarrow CO_2(g) + C(s)$$

was studied by injecting some $CO(g)$ into a reaction vessel and measuring the total pressure while maintaining a constant reaction volume:

P_{tot}/torr	t/s
250	0
238	398
224	1002
210	1801

Calculate the average rate of disappearance of CO for the time interval between each measurement. Find the average rate of appearance of CO_2 for the same time intervals. Use torr instead of concentration.

16-12. The rate of the reaction

$$SO_2Cl_2(g) \rightarrow SO_2(g) + Cl_2(g)$$

is first-order in $P_{SO_2Cl_2}$. A study of the rate of the reaction in the gas phase gives the following data:

$P_{SO_2Cl_2}$/torr	t/s
760	0
680	5000
610	10,000

Calculate the reaction rate over the time intervals 0 to 5000 s and 5000 s to 10,000 s for this reaction. Use the partial pressure of SO_2Cl_2 in place of concentration.

16-13. Sulfuryl chloride decomposes according to the reaction

$$SO_2Cl_2(g) \rightarrow SO_2(g) + Cl_2(g)$$

Using the following data, determine the order of the reaction with respect to SO_2Cl_2:

$Rate/mol \cdot L^{-1} \cdot s^{-1}$	$[SO_2Cl_2]/mol \cdot L^{-1}$
2.2×10^{-6}	0.10
4.4×10^{-6}	0.20
6.6×10^{-6}	0.30
8.8×10^{-6}	0.40

rate = $k[SO_2Cl_2]$
first-order reaction.

16-14. Nitrosyl bromide decomposes according to the reaction

$$2NOBr(g) \rightarrow 2NO(g) + Br_2(g)$$

Using the following data, determine the order of the reaction with respect to NOBr(g):

$Rate/mol \cdot L^{-1} \cdot s^{-1}$	$[NOBr]/mol \cdot L^{-1}$
0.80	0.20
3.20	0.40
7.20	0.60
12.80	0.80

INITIAL RATES

16-15. The reaction

$$C_2H_5Cl(g) \rightarrow C_2H_4(g) + HCl(g)$$

has been studied at 300 K and the following data collected:

Run	*Reactant concentration,* $[C_2H_5Cl]_0/M$	*Initial rate of formation of* $C_2H_4/M \cdot s^{-1}$
1	0.33	2.40×10^{-30}
2	0.66	4.80×10^{-30}
3	1.32	9.60×10^{-30}

Determine the rate law, the order, and the rate constant for the reaction. rate = $k[C_2H_5Cl]$ 1st order

$\frac{2.40 \times 10^{-30}\ M/sec}{0.33\ M} = 7.3 \times 10^{-30}/sec$ k

16-16. The reaction

$$2C_5H_6(g) \rightarrow C_{10}H_{12}(g)$$

has been studied at 373 K and the following data collected:

Run	*Initial reactant pressure,* $[P_{C_5H_6}]_0/torr$	*Initial rate of formation of* $C_{10}H_{12}/torr \cdot s^{-1}$
1	200	5.76
2	400	23.04
3	800	92.2

Determine the rate law in terms of pressure rather than concentration, and compute the rate constant for the reaction.

16-17. The reaction

$$2NOCl(g) \rightarrow 2NO(g) + Cl_2(g)$$

has been studied at 400 K and the following data collected:

Run	*Reactant concentration,* $[NOCl]_0/M$	*Initial rate of formation of* $NO/M \cdot s^{-1}$
1	0.25	1.75×10^{-6}
2	0.50	7.00×10^{-6}
3	0.75	1.57×10^{-5}

Determine the rate law, the order, and the rate constant for the reaction. rate = $k[NOCl]^2$ 2nd order

$k = \frac{rate}{[NOCl]^2}$

$k = \frac{1.75 \times 10^{-6}\ M/sec}{(0.25M)^2}$

$k = 2.8 \times 10^{-5}/M\text{-sec}$

16-18. The following data were obtained for the decomposition of N_2O_3:

$$N_2O_3(g) \rightarrow NO(g) + NO_2(g)$$

$P_{N_2O_3}/torr$	*Initial rate/torr* $\cdot s^{-1}$
0.91	5.5
1.4	8.4
2.1	13

Determine the rate law for the reaction, expressed in terms of $P_{N_2O_3}$ rather than $[N_2O_3]$. Calculate the rate constant for the reaction.

16-19. Consider the reaction

$$Cr(H_2O)_6^{3+}(aq) + SCN^-(aq) \rightarrow Cr(H_2O)_5NCS^{2+}(aq) + H_2O(l)$$

for which the following kinetic data were obtained at 25°C:

$[Cr(H_2O)_6^{3+}]_0$/M	$[SCN^-]_0$/M	*Initial rate*/M·s^{-1}
1.0×10^{-4}	0.10	2.0×10^{-11}
1.0×10^{-3}	0.10	2.0×10^{-10}
1.5×10^{-3}	0.20	6.0×10^{-10}
1.5×10^{-3}	0.50	1.5×10^{-9}

Determine the rate law and the rate constant for the reaction.

rate = $k[Cr(H_2O)_6^{3+}][SCN^-]$

$k = \frac{\text{rate}}{[Cr(H_2O)_6^{3+}][SCN^-]}$ $k = \frac{2.0\times10^{-11} M/sec}{(1.0\times10^{-4} M)(0.10)} = 2.0\times10^{-6}/M\cdot sec$

16-20. The reaction

$$CoBr(NH_3)_5^{2+}(aq) + OH^-(aq) \rightarrow Co(NH_3)_5OH^{2+}(aq) + Br^-(aq)$$

has been studied at 25°C and the following data collected:

$[CoBr(NH_3)_5^{2+}]_0$/M	$[OH^-]_0$/M	*Initial rate* /M·s^{-1}
0.030	0.030	1.37×10^{-3}
0.060	0.030	2.74×10^{-3}
0.030	0.090	4.11×10^{-3}
0.090	0.090	1.23×10^{-2}

Determine the rate law, the overall order of the rate law, and the rate constant for the reaction.

16-21. Given the following initial-rate data for the reaction

$$ClO_3^-(aq) + 9I^-(aq) + 6H^+(aq) \rightarrow 3I_3^-(aq) + Cl^-(aq) + 3H_2O(l)$$

determine the reaction rate law:

$[I^-]_0$/M	$[ClO_3^-]_0$/M	$[H^+]_0$/M	*Initial rate/M*·s^{-1}
0.10	0.10	0.10	*x*
0.10	0.20	0.10	2*x*
0.20	0.20	0.10	4*x*
0.20	0.20	0.20	16*x*

rate = $k[ClO_3^-][I^-][H^+]^2$

16-22. Given the following initial-rate data for the reaction

$$BrO_3^-(aq) + 9I^-(aq) + 6H^+(aq) \rightarrow 3I_3^-(aq) + Br^-(aq) + 3H_2O(l)$$

determine the reaction rate law:

$[I^-]_0$/M	$[BrO_3^-]_0$/M	$[H^+]_0$/M	*Initial rate*/M·s^{-1}
0.10	0.10	0.10	*x*
0.20	0.20	0.10	4*x*
0.10	0.20	0.10	2*x*
0.20	0.20	0.20	4*x*

16-23. A reaction of importance in the formation of smog is that between ozone and nitrogen monoxide:

$$O_3(g) + NO(g) \rightarrow O_2(g) + NO_2(g)$$

The rate law for this reaction is

$$\text{rate} = k[O_3][NO]$$

(a) Given that $k = 2.99 \times 10^6\ M^{-1}\cdot s^{-1}$ at 310 K, compute the initial reaction rate when $[O_3]_0 = 2.0 \times 10^{-6}$ M and $[NO]_0 = 6.0 \times 10^{-5}$ M. (b) Given that $[O_3]$ and $[NO]$ remain essentially constant at the values given in part (a) owing to continuous production from separate sources, compute the number of moles of $NO_2(g)$ produced per hour per liter of air.

(handwritten) a. rate = $(2.99 \times 10^6/M\cdot sec)(2.0 \times 10^{-6} M)(6.0 \times 10^{-5} M)$
rate = 3.6×10^{-4} M/sec
b. $[NO_2]$ = rate × time
= $(3.6 \times 10^{-4}$ M/sec$)(3600$ sec/hr$)$ = 1.3 mol/L·hr

16-24. The rate law for the reaction

$$C_2H_4Br_2(aq) + 3I^-(aq) \rightarrow C_2H_4(g) + 2Br^-(aq) + I_3^-(aq)$$

at 300 K is

$$\text{rate} = (5.0 \times 10^{-3}\ M^{-1}\cdot s^{-1})[C_2H_4Br_2][I^-]$$

Fill in the missing entries in the following table:

Run	$[C_2H_4Br_2]_0$/M	$[I^-]_0$/M	*Initial rate of* C_2H_4/M·s^{-1}
1	0.20	0.20	*(handwritten)* 2.0×10^{-4}
2	0.20	*(handwritten)* 0.40	4.0×10^{-4}
3	*(handwritten)* 0.80	0.20	8.0×10^{-4}

16-25. Given the following initial-rate data at 300 K for the reaction

$$2NO_2(g) + O_3(g) \rightarrow N_2O_5(g) + O_2(g)$$

determine the reaction rate law and rate constant:

$[NO_2]_0$/M	$[O_3]_0$/M	*Initial rate*/M·s^{-1}
1.00	1.00	5.0×10^4
2.00	1.00	1.0×10^5
2.00	2.00	2.0×10^5

(handwritten) rate = $k[NO_2][O_3]$

16-26. Given the following initial-rate data for the reaction

$$CH_3COCH_3(aq) + Br_2(aq) \xrightarrow{H^+(aq)} CH_3COCH_2Br(aq) + H^+(aq) + Br^-(aq)$$

determine the reaction rate law and rate constant:

$[CH_3COCH_3]_0$/M	$[Br_2]_0$/M	$[H^+]_0$/M	*Initial rate*/M·s^{-1}
1.00	1.00	1.00	4.0×10^{-3}
2.00	1.00	1.00	8.0×10^{-3}
2.00	2.00	1.00	8.0×10^{-3}
1.00	2.00	2.00	8.0×10^{-3}

FIRST-ORDER REACTIONS

16-27. Calculate the time required for the concentration to decrease to 1/10 of its initial value for a first-order reaction with a rate constant $k = 10\ s^{-1}$.

(handwritten)
$\log[A] = \log[A_0] - \frac{kt}{2.30}$
$\log\frac{[A]}{[A_0]} = -\frac{kt}{2.30}$
$\log 0.1 = -(10/sec)\,t\,/(2.30)$
$t = 0.23$ sec

16-28. Calculate the time required for the concentration to decrease by 10 percent of its initial value for a first-order reaction with $k = 10\ s^{-1}$.

16-29. The reaction

$$SO_2Cl_2(g) \rightarrow SO_2(g) + Cl_2(g)$$

is first order with a rate constant of $2.2 \times 10^{-5}\ s^{-1}$ at 320°C. What fraction of a sample of SO_2Cl_2 will remain if it is heated for 5.0 h at 320°C?

16-30. The rate constant for the first-order reaction

$$\underset{\text{cyclopropane}}{C_3H_6(g)} \rightarrow \underset{\text{propene}}{CH_3CH{=}CH_2(g)}$$

at 500°C is $5.5 \times 10^{-4}\ s^{-1}$. Compute the half-life of cyclopropane at 500°C. Given an initial cyclopropane concentration of 1.00×10^{-3} M at 500°C, compute the concentration of cyclopropane that remains after 2.0 h.

16-31. Suppose that you place 100 bacteria into a flask containing nutrients for the bacteria and that you find the following data at 37°C:

t/min	*Number of bacteria*
0	100
15	200
30	400
45	800
60	1600

What is the order of the rate of production of the bacteria? How many bacteria do you predict there will be after 2 h? What is the rate constant for the process?

16-32. The U.S. Public Health Service requires that milk fresh from a pasteurizer may contain no more than 20,000 bacteria per milliliter. It has been reported that bacteria in milk stored at 40°F may double in 39 h. If a milk sample had 20,000 bacteria per milliliter after pasteurization, what is the bacteria count per milliliter after 10 days?

16-33. Azomethane, $CH_3N_2CH_3$, decomposes according to the equation

$$CH_3N_2CH_3(g) \rightarrow CH_3CH_3(g) + N_2(g)$$

Given that the decomposition is a first-order process with $k = 4.0 \times 10^{-4}\ s^{-1}$ at 300°C, calculate the fraction of azomethane that remains after one millisecond.

16-34. Methyl iodide, CH_3I, decomposes according to the equation

$$2CH_3I(g) \rightarrow C_2H_6(g) + I_2(g)$$

Given that the decomposition is a first-order process with $k = 1.5 \times 10^{-4}\ s^{-1}$ at 300°C, calculate the fraction of methyl iodide that remains after one millisecond.

16-35. Show that for a first-order reaction the time required for 99.9 percent of the reaction to take place is ten times that required for 50 percent of the reaction to take place.

16-36. Show that for a first-order reaction, the time required for 99.99 percent of the reaction to take place is twice as long as the time required for 99.0 percent of the reaction to take place

16-37. Peroxydisulfate ion, $S_2O_8^{2-}$, decomposes in aqueous solution according to the reaction

$$S_2O_8^{2-}(aq) + H_2O(l) \rightarrow 2SO_4^{2-}(aq) + \tfrac{1}{2}O_2(g) + 2H^+(aq)$$

Given the following data from an experiment with $[S_2O_8^{2-}]_0 = 0.100$ M in a solution with $[H^+]$ fixed at 0.100 M, determine the reaction rate law and calculate the rate constant:

t/min	$[S_2O_8^{2-}]$/M
0	0.100
17	0.050
34	0.025
51	0.012

plot a graph to see if it is first order.
rate = $k[S_2O_8^{2-}]$
$t_{1/2} = \frac{0.693}{k}$
$k = 0.041$/min
or use formula $\log[A] = \log[A_0] - \frac{kt}{2.30}$

16-38. At 400 K oxalic acid decomposes by the reaction

$$H_2C_2O_4(g) \rightarrow CO_2(g) + HCHO_2(g)$$

The rate of this reaction can be studied by measurement of the total pressure. Determine the rate law and the rate constant of this reaction from the following measurements, which give the total pressure reached after 2.00×10^4 s from the given starting pressure of oxalic acid:

$P_{H_2C_2O_4}$/torr (at $t = 0$)	P_{tot}/torr (at $t = 2.00 \times 10^4$ s)
5.0	7.2
7.0	10
8.4	12

16-39. Assuming that the loss of ability to recall learned material is a first-order process with a half-life of 70 days, compute the number of days required to forget 90 percent of the material that you have learned in preparation for an exam. (Assume constant temperature and no further reference to the learned material during the decay period.)

$t_{1/2} = \frac{0.693}{k}$ $k = 9.9 \times 10^{-3}$/day
$\log 0.1 = -\frac{kt}{2.30}$
$\log 0.1 = -\frac{9.9 \times 10^{-3}\, t}{2.30}$ $t = 230$ days

16-40. A rough estimate for the number of nerve cells in the average human brain is 20 billion (2×10^{10}). This number reaches a maximum around 30 years of age and then decreases at a constant rate of 2×10^5 cells per day.

(a) Determine the rate law for the disappearance of nerve cells after 30 years of age.
(b) Estimate the age in years at which the number of cells has dropped to one half of the maximum value.

SECOND-ORDER REACTIONS

16-41. The decomposition of NOBr(g),

$$2NOBr(g) \rightarrow 2NO(g) + Br_2(g)$$

is second order in [NOBr]. The rate constant for the reaction at 10°C is $0.80\ M^{-1} \cdot s^{-1}$. If $[NOBr]_0 = 0.052$ M, what is [NOBr] after the reaction has run for 1 min?

$\frac{1}{[A]} = \frac{1}{[A_0]} + kt$ $\frac{1}{[A]} = \frac{1}{0.052M} + (0.80)(60)$ $A = 0.015M$

16-42. The decomposition of $NO_2(g)$,

$$2NO_2(g) \rightarrow 2NO(g) + O_2(g)$$

is second order in $NO_2(g)$. The rate constant for the reaction at 300°C is $0.54\ M^{-1} \cdot s^{-1}$. If $[NO_2]_0 = 1.25$ M, what is $[NO_2]$ after the reaction has run for 2 min?

$\frac{1}{[A]} = \frac{1}{[A_0]} + kt$ $\frac{1}{[A]} = \frac{1}{1.25M} + (0.54)(120)$ $A = 0.015M$

16-43. The rate law for the reaction

$$2N_2O(g) \rightarrow 2N_2(g) + O_2(g)$$

is second order in $[N_2O]$. The reaction was carried out at 900 K with an initial concentration of N_2O of 2.0×10^{-2} M. It took 4500 s for N_2O to fall to half its initial concentration. What is the rate constant for this reaction?

$t_{1/2} = \frac{1}{k[A_0]}$ $4500\text{sec} = \frac{1}{k(2.0 \times 10^{-2} M)}$
$k = 1.1 \times 10^{-2}$/M·s

16-44. The rate constant for the reaction

$$H^+(aq) + OH^-(aq) \rightarrow H_2O(l)$$

has been determined to be $1.3 \times 10^{11}\ M^{-1} \cdot s^{-1}$. Calculate the half-life of the neutralization of HCl by NaOH when both are originally at a concentration of 1.0×10^{-3} M. When $[H^+]_0 = [OH^-]_0$, the rate law can be written as

$$\text{rate} = k[H^+]^2 = k[OH^-]^2$$

16-45. Identify in each of the following cases the order of the reaction rate law with respect to the reactant A, where A $\rightarrow$ products:

(a) The half-life of A is independent of the initial concentration of A.
(b) The rate of decrease of A is a constant.
(c) A twofold increase in the initial concentration of A leads to a 1.41-fold increase in the initial rate.

REACTION MECHANISMS

16-47. Write the rate law for each of the following elementary reactions:

(a) $N_2O(g) + O(g) \rightarrow 2NO(g)$
(b) $O(g) + O_3(g) \rightarrow 2O_2(g)$
(c) $ClCO(g) + Cl_2(g) \rightarrow Cl_2CO(g) + Cl(g)$

16-49. What is the overall reaction for the following reaction mechanism:

$$O_3(g) \rightleftharpoons O_2(g) + O(g) \qquad \text{fast}$$
$$O_3(g) + O(g) \rightarrow 2O_2(g) \qquad \text{slow}$$

16-51. The reaction of CO_2 with hydroxide ion in aqueous solution is postulated to occur according to the mechanism

$$CO_2(aq) + OH^-(aq) \rightarrow HCO_3^- \qquad \text{slow}$$
$$HCO_3^-(aq) + OH^-(aq) \rightarrow CO_3^{2-}(aq) + H_2O(l) \qquad \text{fast}$$

The rate law for the disappearance of $CO_2(aq)$ was found experimentally to be

$$\text{rate} = k[CO_2][OH^-]$$

Is this mechanism consistent with the observed rate law? Explain your answer.

16-46. Identify in each of the following cases the order of the reaction rate law with respect to the reactant A, where A $\rightarrow$ products:

(a) A twofold increase in the initial concentration of A leads to a fourfold increase in the initial rate.
(b) The half-life for the disappearance of A is inversely proportional to the initial concentration of A.
(c) The time required for $[A]_0$ to decrease to $[A]_0/2$ is equal to the time required for [A] to decrease from $[A]_0/2$ to $[A]_0/4$.

16-48. Write the rate law for each of the following elementary reactions:

(a) $K(g) + HCl(g) \rightarrow KCl(g) + H(g)$
(b) $Cl(g) + ICl(g) \rightarrow Cl_2(g) + I(g)$
(c) $NO_3(g) + CO(g) \rightarrow NO_2(g) + CO_2(g)$

16-50. What is the overall reaction for the following reaction mechanism:

$$NO_2(g) + NO_2(g) \rightarrow NO(g) + NO_3(g) \qquad \text{slow}$$
$$NO_3(g) \rightarrow NO(g) + O_2(g) \qquad \text{fast}$$

16-52. The reaction

$$2N_2O_5(g) \rightarrow 4NO_2(g) + O_2(g)$$

occurs according to the rate law

$$\text{rate} = k[N_2O_5]$$

Which of the following mechanisms is consistent with the observed rate law:

(1) $N_2O_5(g) \rightarrow NO_2(g) + NO_3(g)$ slow
$NO_3(g) + N_2O_5(g) \rightarrow 3NO_2(g) + O_2(g)$ fast

(2) $N_2O_5(g) \rightleftharpoons NO_2(g) + NO_3(g)$ fast
$NO_3(g) + N_2O_5(g) \rightarrow N_2O_4(g) + NO_2(g) + O_2(g)$ slow
$N_2O_4(g) \rightarrow 2NO_2(g)$ fast

16-53. What is the overall reaction for the following reaction mechanism:

$$2N_2O_5(g) \rightleftharpoons 2NO_2(g) + 2NO_3(g) \qquad \text{fast}$$

$$NO_2(g) + NO_3(g) \rightarrow NO(g) + O_2(g) + NO_2(g) \qquad \text{slow}$$

$$NO(g) + NO_3(g) \rightarrow 2NO_2(g) \qquad \text{fast}$$

$$2N_2O_5 \rightleftharpoons 4NO_2 + O_2$$

Write the rate law for the disappearance of $N_2O_5(g)$.

$$\text{rate} = k[NO_2][NO_3]$$

16-54. The proposed mechanism for a reaction is given by the following sequence:

$$H_2O_2(aq) + I^-(aq) \rightarrow HOI(aq) + OH^-(aq) \qquad \text{slow}$$

$$HOI(aq) + I^-(aq) \rightarrow I_2(aq) + OH^-(aq) \qquad \text{fast}$$

$$OH^-(aq) + H^+(aq) \rightarrow H_2O(l) \qquad \text{fast}$$

Determine the overall reaction and write the rate law for the disappearance of H_2O_2.

16-55. A proposed mechanism for the rate of formation of phosgene is

$$Cl_2(g) \rightleftharpoons 2Cl(g) \qquad \text{fast}$$

$$Cl(g) + CO(g) \rightleftharpoons ClCO(g) \qquad \text{fast}$$

$$ClCO(g) + Cl_2(g) \rightarrow Cl_2CO(g) + Cl(g) \qquad \text{slow}$$

$$2Cl_2 + CO \rightleftharpoons Cl_2CO + 2Cl$$

Show that a rate law that is consistent with this mechanism is

$$\text{rate} = k[Cl_2]^{3/2}[CO]$$

$$\text{rate} = k[ClCO][Cl_2]$$

$$K_2 = \frac{[ClCO]}{[Cl][CO]} \qquad [ClCO] = K_2[Cl][CO]$$

$$K_1 = \frac{[Cl]^2}{[Cl_2]} \qquad [Cl] = K_1^{1/2}[Cl_2]^{1/2}$$

$$[ClCO] = K_2K_1^{1/2}[Cl_2]^{1/2}[CO]$$

$$\text{rate} = kK_2K_1^{1/2}[Cl_2]^{1/2}[CO][Cl_2]$$

$$\text{rate} = k[Cl_2]^{3/2}[CO]$$

16-56. The rate law for the decomposition of phosgene

$$COCl_2(g) \rightarrow CO(g) + Cl_2(g)$$

is known to be

$$\text{rate} = k[COCl_2][Cl_2]^{1/2}$$

Show that the following mechanism is consistent with this rate law:

$$Cl_2(g) \rightleftharpoons 2Cl(g) \qquad \text{fast}$$

$$COCl_2(g) + Cl(g) \rightarrow COCl(g) + Cl_2(g) \qquad \text{slow}$$

$$COCl(g) \rightleftharpoons CO(g) + Cl(g) \qquad \text{fast}$$

$$\text{rate} = k[COCl_2][Cl] \qquad K_1 = \frac{[Cl]^2}{[Cl_2]} \qquad [Cl] = K_1^{1/2}[Cl_2]^{1/2}$$

$$\text{rate} = k[COCl_2][Cl_2]^{1/2}$$

16-57. The available kinetic data for the reaction

$$2NO(g) + O_2(g) \rightarrow 2NO_2(g)$$

is consistent with the following reaction mechanism:

$$NO(g) + O_2(g) \rightleftharpoons NO_3(g) \qquad \text{fast}$$

$$NO_3(g) + NO(g) \rightarrow 2NO_2(g) \qquad \text{slow}$$

Obtain the rate law for the reaction from the above mechanism. Express your rate law in terms of the reactant concentrations in the reaction.

$$\text{rate} = k[NO_3][NO]$$

$$K_1 = \frac{[NO_3]}{[NO][O_2]} \qquad [NO_3] = K_1[NO][O_2]$$

$$\text{rate} = kK_1[NO][O_2][NO]$$

$$\text{rate} = k[NO]^2[O_2]$$

16-58. The rate law for the reaction

$$Hg_2^{2+}(aq) + Tl^{3+}(aq) \rightarrow 2Hg^{2+}(aq) + Tl^+(aq)$$

is

$$\text{rate} = k\frac{[Hg_2^{2+}][Tl^{3+}]}{[Hg^{2+}]}$$

Show that the following mechanism is consistent with the observed rate law:

$$Hg_2^{2+}(aq) \rightleftharpoons Hg(aq) + Hg^{2+}(aq) \qquad \text{fast}$$

$$Hg(aq) + Tl^{3+}(aq) \rightarrow Tl^+(aq) + Hg^{2+}(aq) \qquad \text{slow}$$

$$\text{rate} = k[Hg][Tl^{3+}]$$

$$K_1 = \frac{[Hg][Hg^{2+}]}{[Hg_2^{2+}]} \qquad [Hg] = \frac{K_1[Hg_2^{2+}]}{[Hg^{2+}]}$$

$$\text{rate} = k\frac{[Hg_2^{2+}][Tl^{3+}]}{[Hg^{2+}]}$$

COLLISION THEORY

16-59. Many radical-radical recombination reactions, such as

$$2CH_3(g) \rightarrow C_2H_6(g)$$

proceed at the diffusion-controlled limit; that is, essentially every collision leads to a reactive event. Why do you think this is so? (Explain in terms of chemical bonds.)

16-60. Consider the following mechanism for the decomposition of $N_2O_5(g)$ to $NO_2(g)$ and $O_2(g)$:

$$N_2O_5(g) + N_2O_5(g) \rightleftharpoons N_2O_5^*(g) + N_2O_5(g) \qquad \text{fast}$$

$$N_2O_5^*(g) \rightarrow NO_2(g) + NO_3(g) \qquad \text{slow}$$

$$NO_3(g) + N_2O_5(g) \rightarrow 3NO_2(g) + O_2(g) \qquad \text{fast}$$

where $N_2O_5^*$ denotes a vibrationally excited molecule. Show that this mechanism gives rise to a rate law first order in N_2O_5 even though the mechanism involves collision of two N_2O_5 molecules.

ARRHENIUS EQUATION

16-61. The reaction

$$2N_2O_5(soln) \rightarrow 2N_2O_4(soln) + O_2(g)$$

takes place in carbon tetrachloride at room temperature. The rate constant is $2.35 \times 10^{-4}\ s^{-1}$ at 293 K and $9.15 \times 10^{-4}\ s^{-1}$ at 303 K. Calculate E_a for this reaction.

16-62. The rate constant for the reaction

$$H_2(g) + I_2(g) \rightarrow 2HI(g)$$

was determined to be $0.0234\ M^{-1}\cdot s^{-1}$ at 400°C and $0.750\ M^{-1}\cdot s^{-1}$ at 500°C. Calculate E_a for this reaction.

16-63. The activation energy for the reaction

$$C_4H_8(g) \rightarrow 2C_2H_4(g)$$

is $262\ kJ\cdot mol^{-1}$. At 600 K the rate constant is $6.07 \times 10^{-8}\ s^{-1}$. What is the rate constant at 800 K?

16-64. The activation energy for the reaction

$$2N_2O_5(g) \rightarrow 4NO_2(g) + O_2(g)$$

is $102\ kJ\cdot mol^{-1}$. At 45°C the rate constant is $5.0 \times 10^{-4}\ s^{-1}$. What is the rate constant at 65°C?

16-65. The denaturation of a certain virus is a first-order process with an activation energy of $586\ kJ\cdot mol^{-1}$. The half-life of the reaction at 29.6°C is 4.5 h. Compute the half-life at 37.0°C.

16-66. Cryosurgical procedures involve lowering the body temperature of the patient prior to surgery. Given that the activation energy for the beating of the heart muscle is about 30 kJ, estimate the pulse rate at 72.0°F. Assume the pulse rate at 98.6°F (310 K) to be 65 beats·min^{-1}.

16-67. Given the following rate-constant data on the gas-phase decomposition of $NO_2[2NO_2(g) \rightarrow 2NO(g) + O_2(g)]$, plot log k vs. $1/T$ and calculate E_a for the reaction.

T/K	k/$M^{-1}\cdot s^{-1}$
600	0.70
625	1.83
650	4.46
700	21.8

Use your plot to estimate k at 500 K.

16-68. The decomposition of dinitrogen pentoxide is a first-order process. The temperature-dependence of the observed rate constant is given by

t/°C	k/$10^{-5}\ s^{-1}$
0	0.0787
25	3.46
45	49.8
65	487

Plot log k vs $1/T$ and calculate E_a for the reaction. Use your plot to estimate k at 50°C.

16-69. Can a catalyst affect the position of equilibrium in a chemical reaction? Explain.

16-70. Explain why a catalyst for a forward reaction must also be a catalyst for the reverse reaction.

16-71. The aqueous decomposition of hydrogen peroxide in the presence of $Br^-(aq)$ and $H^+(aq)$,

$$2H_2O_2(aq) \rightarrow 2H_2O(l) + O_2(g)$$

has the rate law

$$\text{rate} = k[H_2O_2][H^+][Br^-]$$

(a) Identify the catalysts for the reaction.
(b) What is the overall order of the reaction?
(c) Suppose $[H_2O_2]_0 = 0.10$ M, $[H^+]_0 = 1.00 \times 10^{-3}$ M, and $[Br^-]_0 = 1.00 \times 10^{-3}$ M. Sketch the concentrations of these three species as a function of time given that $k = 1.0 \times 10^3\ M^{-2}\cdot s^{-1}$.

16-72. Given the following initial-rate data for the decomposition of $H_2O_2(aq)$,

$$2H_2O_2(aq) \rightarrow 2H_2O(l) + O_2(g)$$

$[H_2O_2]_0$/M	$[I^-]_0$/M	$[H^+]_0$/M	Initial rate of formation of O_2/$M\cdot s^{-1}$
0.20	0.010	0.010	2.0×10^{-3}
0.40	0.010	0.010	4.0×10^{-3}
0.40	0.020	0.010	8.0×10^{-3}
0.20	0.020	0.020	1.6×10^{-2}

(a) Determine the rate law.
(b) Pick out the catalyst(s), if any.

16-73. It is suspected that a reaction is catalyzed by the wall of the reaction vessel. What experiments would you perform to check out this possibility?

16-74. It is suspected that a reaction rate is influenced by light. What experiments would you perform to check out this possibility?

16-75. Figure 16-12 outlines the platinum-catalyzed mechanism for the reaction

$$2SO_2(g) + O_2(g) \rightleftharpoons 2SO_3(g)$$

It is observed that, except for very low pressures, the rate of the catalyzed reaction is independent of the pressures of $SO_2(g)$ and $O_2(g)$. That is, the rate law is

$$\text{rate} = k$$

Explain how the mechanism outlined in Figure 16-12 leads to this rate law.

16-76. The rate of decomposition of gases on hot metal surfaces often is found to be independent of the concentration of the gas in the gas phase; that is, the rate law is zero order in the reactant gas:

$$\text{rate} = k$$

Such a situation is found for the catalytic decomposition of ammonia on tungsten,

$$2NH_3(g) \xrightarrow{W(s)} N_2(g) + 3H_2(g)$$

and for the catalytic decomposition of nitrous oxide on platinum,

$$2N_2O(g) \xrightarrow{Pt(s)} 2N_2(g) + O_2(g)$$

How do you explain these observations in mechanistic terms?

16-77. The elementary reaction

$$C_2H_5Br(g) \underset{k_r}{\overset{k_f}{\rightleftharpoons}} C_2H_4(g) + HBr(g)$$

has a rate constant $k_f = 6.7 \times 10^{-7}\ s^{-1}$ and an equilibrium constant $K = 0.14$ M at 600 K. Evaluate k_r.

$K = \frac{k_f}{k_r}$ $0.14\,M = \frac{6.7 \times 10^{-7}\ 1/sec}{k_r}$ $k_r = 4.8 \times 10^{-6}$ /M·sec

16-78. For the elementary reaction

$$CO_2(aq) + OH^-(aq) \underset{k_r}{\overset{k_f}{\rightleftharpoons}} HCO_3^-(aq)$$

$K = 8.4 \times 10^7\ M^{-1}$ and $k_f = 8.4 \times 10^3\ M^{-1}\cdot s^{-1}$. Evaluate k_r.

16-79. Consider the reaction

$$2HI(g) \rightarrow H_2(g) + I_2(g)$$

Given that the rate law for the forward reaction is

$$\text{rate} = k_f[HI]^2$$

deduce the rate law for the reverse reaction.

rate $= k_r[H_2][I_2]$

16-80. Consider the reaction

$$HNO_2(aq) + OH^-(aq) \rightarrow NO_2^-(aq) + H_2O(l)$$

Given that the rate law for the forward reaction is

$$\text{rate} = k_f[HNO_2][OH^-]$$

deduce the rate law for the reverse reaction.

16-81. The reaction

$$2HCrO_4^-(aq) + 3HSO_3^-(aq) + 5H^+(aq) \rightleftharpoons 2Cr^{3+}(aq) + 3SO_4^{2-}(aq) + 5H_2O(l)$$

has the forward reaction rate law

$$\text{rate} = k_f[HCrO_4^-][HSO_3^-]^2[H^+]$$

Derive the rate law for the reverse reaction.

$K = \frac{k_f}{k_r}$ $k_f = K k_r$

$K = \frac{[Cr^{3+}]^2[SO_4^{2-}]^3}{[HCrO_4^-]^2[HSO_3^-]^3[H^+]^5}$

$k_f = k_r \cdot \frac{[Cr^{3+}]^2[SO_4^{2-}]^3}{[HCrO_4^-]^2[HSO_3^-]^3[H^+]^5}$

rate $= k_r \cdot \frac{[Cr^{3+}]^2[SO_4^{2-}]^3[HCrO_4^-][HSO_3^-]^2[H^+]}{[HCrO_4^-]^2[HSO_3^-]^3[H^+]^5}$

rate $= k_r \frac{[Cr^{3+}]^2[SO_4^{2-}]^3}{[HCrO_4^-][HSO_3^-][H^+]^4}$

16-82. The reaction

$$3HNO_2(aq) \rightleftharpoons H^+(aq) + NO_3^-(aq) + 2NO(g) + H_2O(l)$$

has the forward reaction rate law

$$\text{rate} = \frac{k[HNO_2]^4}{[NO]^2}$$

Derive the rate law for the reverse reaction.

Firefighters use an airport snowblower to blow sodium carbonate onto 20,000 gal of nitric acid spilled from a tankcar in Denver on April 3, 1983. The spill was brought under control (neutralized) in a few hours, and no serious injuries were caused by the accident. The neutralization reaction is $2HNO_3(aq) + Na_2CO_3(s) \rightarrow 2NaNO_3(aq) + H_2O(l) + CO_2(g)$.

17 / Acids and Bases

A quantitative understanding of the chemistry of acids and bases is essential to an understanding of many chemical reactions and most biochemical reactions. In this chapter we first present a definition of acids and bases that incorporates the key role that water plays in acid-base chemistry. Then we introduce the concept of pH, which is a convenient measure of the strength of an acidic or basic solution. We shall see that acids and bases can be classified as either strong or weak. A central, quantitative theme of this chapter is the calculation of the acidity, or pH, of an acidic or basic solution as a function of the concentration of the acid or base. We shall learn that solutions of salts can be acidic or basic.

17-1. AN ACID IS A PROTON DONOR AND A BASE IS A PROTON ACCEPTOR

In Chapter 4 we defined an acid as a substance that produces $H^+(aq)$ in aqueous solution and a base as a substance that produces $OH^-(aq)$ in aqueous solution. Acidic solutions taste sour (for example, vinegar, lemons, and grapefruit), and basic solutions feel slippery and taste bitter (for example, soap).

You should never taste chemicals if you are not absolutely certain that they are harmless.

The definition of acids and bases that we have used up to now is due to Arrhenius, and thus such substances are called *Arrhenius acids and bases.* In this chapter we use a more general definition of acids and bases, one due to Johannes Brønsted and Thomas Lowry. In the *Brønsted-Lowry acid-base* classification scheme, an acid is defined as a *proton donor* and a base as a *proton acceptor*. For example, when $HCl(g)$ is dissolved in water, the acid HCl donates a proton to the base H_2O to produce a *hydronium ion,* $H_3O^+(aq)$, which is a hydrated proton:

$$HCl(g) + H_2O(l) \rightarrow H_3O^+(aq) + Cl^-(aq)$$

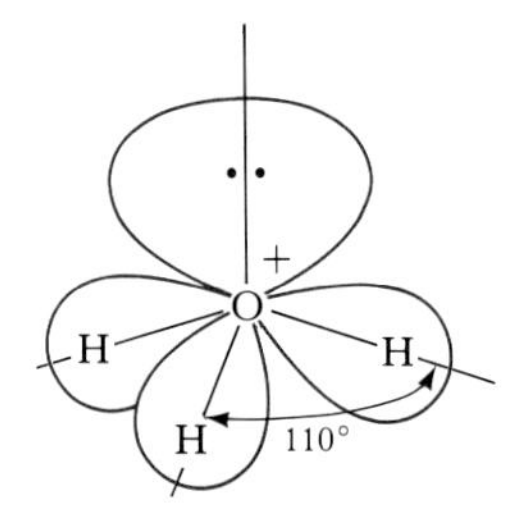

Figure 17-1 Side view of the hydronium ion, which has a tripod shape. All the O—H bond lengths are identical (106 pm), and all the H–O–H angles are identical (110°). The Lewis formula is $H—\ddot{O}^{\oplus}—H$ with H bonded below O, and thus the ion has one lone pair of electrons.

In water, a proton associates very strongly with a water molecule to form a hydronium ion, which has a trigonal pyramidal structure (Figure 17-1). The $H_3O^+(aq)$ notation tells us that the hydronium ion is further solvated by other water molecules, which are represented by (aq).

Up to now we have designated a hydrated proton by $H^+(aq)$, but it is more convenient to use the designation $H_3O^+(aq)$ when using the Brønsted-Lowry definition of acids and bases. For the dissociation of HCl, we write

$$HCl(aq) \rightarrow H^+(aq) + Cl^-(aq) \qquad \text{Arrhenius}$$

$$HCl(aq) + H_2O(l) \rightarrow H_3O^+(aq) + Cl^-(aq) \qquad \text{Brønsted-Lowry}$$

Note that the Brønsted-Lowry notation for the dissociation of HCl requires that we include $H_2O(l)$ explicitly to indicate the species to which HCl donates a proton.

A reaction involving the transfer of a proton from one molecule to another is called a *proton-transfer reaction* or a *protonation reaction.*

17-2. IN ANY AQUEOUS SOLUTION THE ION CONCENTRATION PRODUCT $[H_3O^+][OH^-]$ IS A CONSTANT

Acid-base reactions are proton-transfer reactions.

Pure water contains a small number of $H_3O^+(aq)$ and $OH^-(aq)$ ions that arise from the reaction

$$H_2O(l) + H_2O(l) \rightleftharpoons H_3O^+(aq) + OH^-(aq) \qquad (17\text{-}1)$$

In this reaction water molecules transfer protons to other water molecules. Note that water acts as both an acid (proton donor) and a base (proton acceptor).

The equilibrium constant expression for Reaction (17-1) is

$$K_w = [H_3O^+][OH^-] \qquad (17\text{-}2)$$

Note that the concentration of water does not apear in the K_w expression because $[H_2O]$ is effectively constant at 55.5 M in aqueous solutions (Chapter 15). The quantity K_w is called the *ion product constant of water.* At 25°C the experimental value of K_w is

$$K_w = [H_3O^+][OH^-] = 1.00 \times 10^{-14}\ M^2 \qquad (17\text{-}3)$$

Equation (17-3) tells us when $[H_3O^+]$ is large $[OH^-]$ is small, and when $[OH^-]$ is large $[H_3O^+]$ is small.

This small value of K_w means that in pure water the concentrations of $H_3O^+(aq)$ and $OH^-(aq)$ are low; that is, the equilibrium represented by Equation (17-1) lies far to the left. From the stoichiometry of Reaction (17-1) we note that if we start with pure water, then $H_3O^+(aq)$ and $OH^-(aq)$ are produced on a one-for-one basis. Therefore, *in pure water* we have the equality

$$[H_3O^+] = [OH^-]$$

Using this equation to eliminate $[OH^-]$ in Equation (17-3) yields

$$[H_3O^+]^2 = 1.00 \times 10^{-14}\ M^2$$

Taking the square root of both sides yields

$$[H_3O^+] = 1.00 \times 10^{-7}\ M$$

Because $[H_3O^+] = [OH^-]$, we also have

$$[OH^-] = 1.00 \times 10^{-7}\ M$$

Thus both $[H_3O^+]$ and $[OH^-]$ are equal to 1.00×10^{-7} M in pure water at 25°C. Although $[H_3O^+] = [OH^-]$ for pure water, this is not necessarily true when substances are dissolved in water.

A *neutral* aqueous solution is one for which

$$[H_3O^+] = [OH^-] \qquad \text{neutral solution}$$

An *acidic* aqueous solution is one for which

$$[H_3O^+] > [OH^-] \qquad \text{acidic solution}$$

A *basic* aqueous solution is one for which

$$[OH^-] > [H_3O^+] \qquad \text{basic solution}$$

17-3. STRONG ACIDS AND BASES ARE COMPLETELY DISSOCIATED IN AQUEOUS SOLUTIONS

Certain electrical measurements on dilute HCl(*aq*) solutions show that HCl in water is *completely dissociated,* that is, in water HCl exists as $H_3O^+(aq)$ and $Cl^-(aq)$. There are no undissociated HCl molecules in aqueous solution. Acids that are completely dissociated are referred to as *strong acids*. The term strong refers to the ability of such acids to donate protons to water molecules. Strong acids transfer all their dissociable protons to water molecules.

Example 17-1: (a) Compute $[H_3O^+]$, $[Cl^-]$, and $[OH^-]$ in an aqueous solution containing 0.15 $mol \cdot L^{-1}$ of $HCl(aq)$. (b) Is the solution acidic or basic?

Solution: (a) Because HCl is a strong acid in water, it is completely dissociated, and thus

$$[H_3O^+] = 0.15\text{ M} \qquad [Cl^-] = 0.15\text{ M}$$

The acid concentration in your stomach can be as high as 0.15 M $HCl(aq)$. The corresponding value of $[OH^-]$ can be computed from the K_w expression:

$$K_w = [H_3O^+][OH^-] = 1.00 \times 10^{-14}\text{ M}^2$$

$$[OH^-] = \frac{1.00 \times 10^{-14}\text{ M}^2}{[H_3O^+]} = \frac{1.00 \times 10^{-14}\text{ M}^2}{0.15\text{ M}} = 6.7 \times 10^{-14}\text{ M}$$

(b) Because $[H_3O^+] \gg [OH^-]$, the solution is strongly acidic.

Note that in part (a) we ignored the small contribution to $[H_3O^+]$ arising from the dissociation of water. From part (b) we see that the contribution to $[H_3O^+]$ from the dissociation of water, which is roughly equal to $[OH^-]$, about 6.7×10^{-14} M, is indeed very small compared with 0.15 M.

Conductivity measurements show that sodium hydroxide in water is completely dissociated, that is, it exists as $Na^+(aq)$ and $OH^-(aq)$:

$$NaOH(s) \xrightarrow[H_2O(l)]{} Na^+(aq) + OH^-(aq)$$

There is no undissociated NaOH present in aqueous solution. Sodium hydroxide is a base because $OH^-(aq)$ is a proton acceptor:

$$H^+(aq) + OH^-(aq) \rightarrow H_2O(l)$$

Strong acids and strong bases are completely dissociated in solution.

Completely dissociated bases are referred to as *strong bases*.

Example 17-2: (a) Compute $[OH^-]$, $[Na^+]$, and $[H_3O^+]$ in an aqueous solution containing 0.15 M $NaOH(aq)$. (b) Is the solution acidic or basic?

Solution: (a) Because NaOH is a strong base in water, it is completely dissociated, and thus

$$[OH^-] = 0.15\text{ M} \qquad [Na^+] = 0.15\text{ M}$$

MEMORIZE!!

Table 17-1 Inorganic water-soluble strong acids and bases

Strong acids in water		*Strong bases in water*	
$HClO_4$	perchloric	LiOH	lithium hydroxide
HNO_3	nitric	NaOH	sodium hydroxide
H_2SO_4	sulfuric	KOH	potassium hydroxide
HCl	hydrochloric	RbOH	rubidium hydroxide
HBr	hydrobromic	CsOH	cesium hydroxide
HI	hydroiodic	$Ca(OH)_2$	calcium hydroxide
		$Sr(OH)_2$	strontium hydroxide
		$Ba(OH)_2$	barium hydroxide
		TlOH	thallium(I) hydroxide

An acid not listed in Table 17-1 is a weak acid. A base not listed in Table 17-1 is a weak base. Most acids and bases are weak.

The value of $[H_3O^+]$ can be computed from the K_w expression:

$$K_w = [H_3O^+][OH^-] = 1.00 \times 10^{-14}\ M^2$$

$$[H_3O^+] = \frac{1.00 \times 10^{-14}\ M^2}{[OH^-]} = \frac{1.00 \times 10^{-14}\ M^2}{0.15\ M} = 6.7 \times 10^{-14}\ M$$

(b) Because $[OH^-] \gg [H_3O^+]$, the solution is strongly basic.

Problems 17-1 through 17-6 are similar to Examples 17-1 and 17-2.

There are only a few inorganic water-soluble strong acids and bases (Table 17-1). Most inorganic water-soluble acids and bases, when dissolved in water, are *weak,* that is, *incompletely dissociated.* The formulas of the strong acids and bases should be memorized because this information is essential in working problems in acid-base chemistry. Note that three of the six strong acids are halogen acids (HCl, HBr, and HI) and that five of the nine strong bases are alkali metal hydroxides (LiOH, NaOH, KOH, RbOH, and CsOH) and three are alkaline earth metal hydroxides [$Ca(OH)_2$, $Sr(OH)_2$, and $Ba(OH)_2$]. These groups can be remembered by referring to the halogen and alkali metal groups in the periodic table. Most organic acids and bases are weak. In contrast to the other halogen acids, HF(*aq*) is a weak acid. This is a result of the fact that HF has a much stronger bond than the other halogen acids.

Bond energies of the halogen acids/$kJ \cdot mol^{-1}$

HF	HCl	HBr	HI
565	431	368	297

17-4. pH IS A MEASURE OF THE ACIDITY OF AN AQUEOUS SOLUTION

You will find throughout your study of chemistry that many chemical reactions depend upon the concentration of $H_3O^+(aq)$ in the reaction mixture even though $H_3O^+(aq)$ may not be one of the reactants or products. The speed, or the rate, with which many reactions proceed can be altered dramatically by the addition of a small amount of $H_3O^+(aq)$. For example, the action of many proteins depends quite sensitively upon the concentration of $H_3O^+(aq)$. As we shall see, concentrations of $H_3O^+(aq)$ often lie in the range from 1 M to 10^{-14} M. Such a wide range of concentrations makes it difficult to plot these values on graphs. Because of this, it is convenient to use a logarithmic scale and to define a quantity called pH as

pH is simply a numerically convenient measure of the acidity of a solution.

$$\text{pH} \equiv -\log[H_3O^+] \tag{17-4}$$

The properties of *logarithms* are reviewed in Appendix A-2. The next example illustrates the use of Equation (17-4) to calculate the pH corresponding to various values of $[H_3O^+]$.

Example 17-3: Calculate the pH of a solution that has a $H_3O^+(aq)$ concentration of (a) 10^{-1} M and (b) 5.0×10^{-10} M.

Solution:
(a) By definition, the logarithm of 10^{-1} is -1, and so

$$\text{pH} = -\log(10^{-1}) = 1$$

(b) If you have a hand calculator, you enter 5.0×10^{-10}, press the log key, and change the sign to get

$$\text{pH} = -\log(5.0 \times 10^{-10}) = 9.30$$

If you use a table of logarithms instead of a calculator, you must use the fact that

$$\log ab = \log a + \log b$$

Using this property and the fact that $\log 5.0 = 0.70$, we have

$$\begin{aligned}\log(5.0 \times 10^{-10}) &= \log 5.0 + \log 10^{-10} \\ &= 0.70 - 10 \\ &= -9.30\end{aligned}$$

or

$$\text{pH} = 9.30$$

Problems 17-7 through 17-14 ask you to calculate the pH of some aqueous solutions.

Notice that the minus sign in the definition of pH causes pH values to be positive. Although it is easy to calculate pH using a calculator, you should understand how to use a table of logarithms and remember the principal properties of logarithms.

A table of logarithms is given in Appendix C.

Example 17-3 illustrates how to calculate pH from $[H_3O^+]$. It is often necessary to do the inverse calculation, that is, to calculate $[H_3O^+]$ from the pH, as shown in the next example.

Example 17-4: The pH of lemon juice is about 2.2. Compute the value of $[H_3O^+]$ for lemon juice.

Lemon juice is acidic, as shown by its sour taste. All citrus fruit juices are acidic.

Solution: From the definition of pH we have

$$pH = -\log[H_3O^+]$$

A property of logarithms is that if $y = \log x$, then $x = 10^y$. Thus

$$[H_3O^+] = 10^{-pH}$$

The pH of lemon juice is 2.2, and so

$$[H_3O^+] = 10^{-2.2}$$

The quantity $10^{-2.2}$ can be evaluated easily on your hand calculator by using the inverse logarithm operation (the 10^x key on some calculators):

$$[H_3O^+] = 10^{-2.2} = 6.3 \times 10^{-3}\,M$$

If you are using a table of logarithms instead of a hand calculator, then first you must write $10^{-2.2}$ as

$$10^{-2.2} = 10^{0.8} \times 10^{-3}$$

The antilogarithm of 0.8 is 6.3, and so once again we find that

$$[H_3O^+] = 6.3 \times 10^{-3}\,M$$

Problems 17-15 through 17-22 involve calculating $[H_3O^+]$ and $[OH^-]$ from pH.

Example 17-5: Compute the pH of an aqueous solution prepared by dissolving 0.26 g of calcium hydroxide in water and diluting to a final volume of 0.500 L.

Solution: We first compute the number of moles of $Ca(OH)_2$. The molecular mass of $Ca(OH)_2$ is 74.1, and so the number of moles is

$$(0.26\text{ g }Ca(OH)_2)\left(\frac{1\text{ mol }Ca(OH)_2}{74.1\text{ g }Ca(OH)_2}\right) = 3.51 \times 10^{-3}\text{ mol}$$

The molarity of the solution is the number of moles per liter of solution:

$$\text{molarity} = \frac{\text{moles of solute}}{\text{liters of solution}} = \frac{3.51 \times 10^{-3}\,\text{mol}}{0.500\,\text{L}} = 7.02 \times 10^{-3}\,\text{M}$$

Calcium hydroxide is a strong base and yields two $OH^-(aq)$ per mole of $Ca(OH)_2(aq)$. Therefore, the molarity of the $OH^-(aq)$ is

$$[OH^-] = (2)(7.02 \times 10^{-3}\,\text{M}) = 1.40 \times 10^{-2}\,\text{M}$$

The value of $[H_3O^+]$ is calculated using the ion product constant of water:

$$[H_3O^+] = \frac{1.00 \times 10^{-14}\,\text{M}^2}{[OH^-]} = \frac{1.00 \times 10^{-14}\,\text{M}^2}{1.40 \times 10^{-2}\,\text{M}} = 7.14 \times 10^{-13}\,\text{M}$$

The pH of the solution is

$$\text{pH} = -\log[H_3O^+] = -\log(7.14 \times 10^{-13}) = 12.15$$

Problems 17-11 and 17-12 are similar to Example 17-5.

The pH values of most aqueous solutions lie in the range 0 to 14, although pH values outside this range are occasionally encountered. The pH values of some common aqueous solutions are given in Figure 17-2.

At 25°C, pure water has a hydrogen ion concentration of $[H_3O^+] = 1.00 \times 10^{-7}$ M; and thus the pH of a neutral aqueous solution at 25°C is

$$\text{pH} = -\log[H_3O^+] = -\log[1.00 \times 10^{-7}] = 7.00$$

Figure 17-2 The range of pH values for some common aqueous solutions. The bars are spaced out to avoid overlapping.

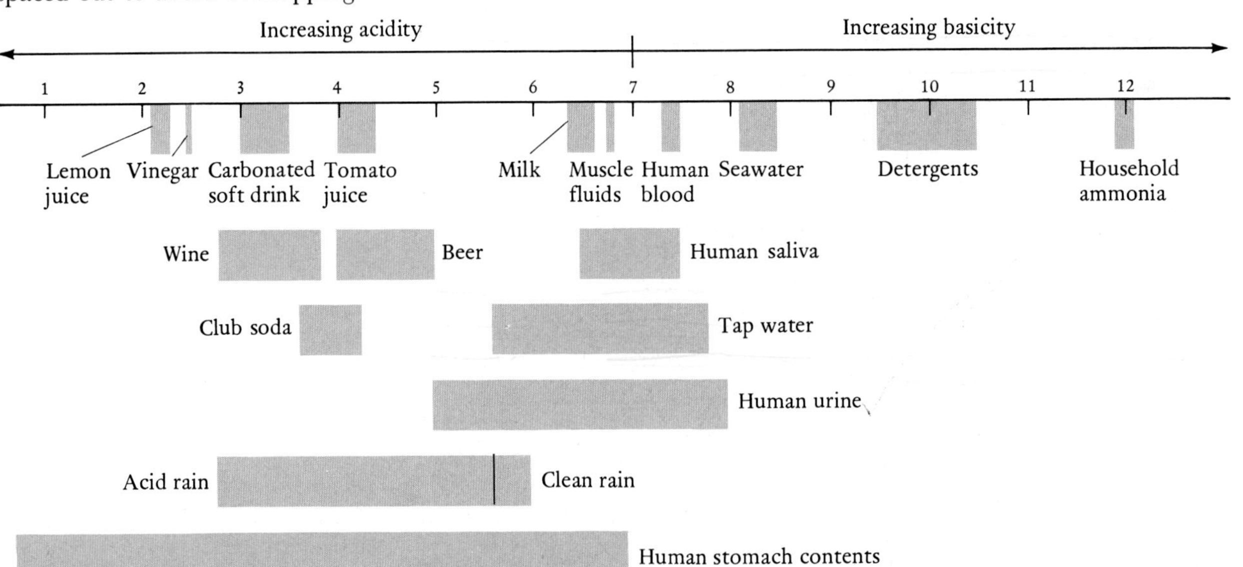

At 25 °C, acidic solutions have a $[H_3O^+]$ greater than 1.00×10^{-7} M, and thus acidic solutions have pH values less than 7.00. Basic solutions have a $[H_3O^+]$ less than 1.00×10^{-7} M and thus have pH values greater than 7.00. The pH scale is shown schematically as follows:

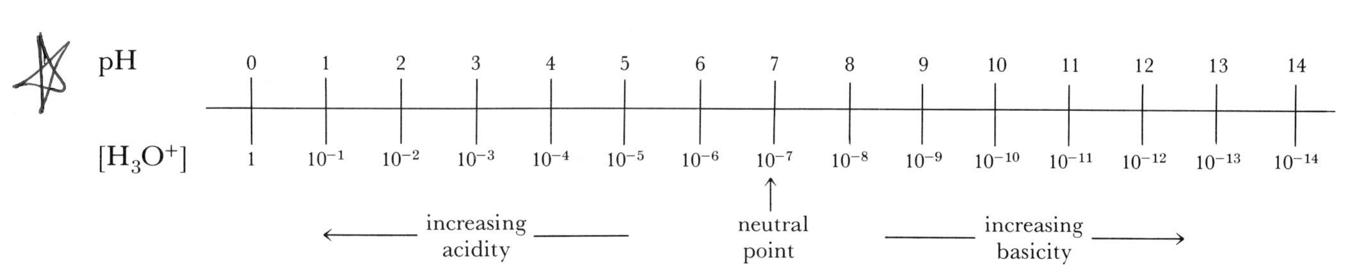

Note that a change in pH of one unit corresponds to a 10-fold change in $[H_3O^+]$.

The pH of a solution is conveniently measured in the laboratory with a *pH meter,* an electronic device that responds to the $[H_3O^+]$ of a solution (Figure 17-3). The meter scale or digital readout of the device is set up to display pH directly. How a pH meter works is discussed in Chapter 21.

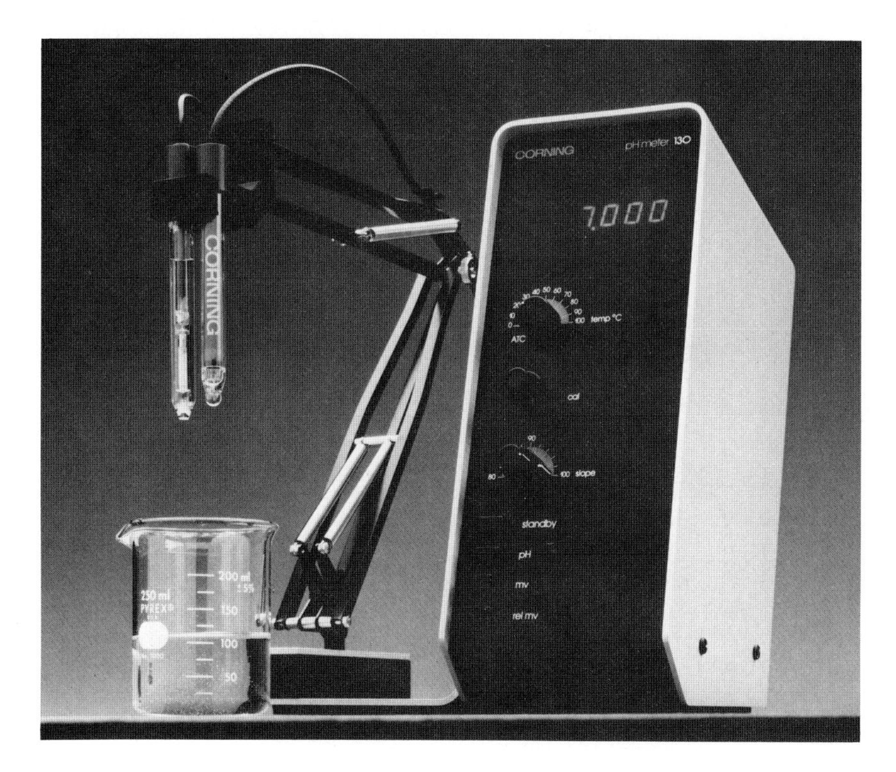

Figure 17-3 A pH meter and a pair of electrodes for measuring the pH of a solution.

17-5. WEAK ACIDS AND WEAK BASES REACT ONLY PARTIALLY WITH WATER

If 0.10 mol of hydrogen chloride gas is dissolved in enough water to make 1.00 L of aqueous solution, then the observed pH of the resulting solution is 1.00. If 0.10 mol of hydrogen fluoride gas is dissolved in enough water to make 1.00 L of aqueous solution, then the observed pH of the resulting solution is 2.10. Note that when either $HCl(g)$ or $HF(g)$ is dissolved in water, the pH of the water decreases, which means that the value of $[H_3O^+]$ increases. In the $HCl(g)$ case the pH decreases by 6.00 pH units (7.00 − 1.00), and in the $HF(g)$ case the pH decreases by 4.90 pH units (7.00 − 2.10).

We can calculate the value of $[H_3O^+]$ from the pH by using the relation

$$[H_3O^+] = 10^{-pH} \tag{17-5}$$

The values that we obtain using Equation (17-5) are

0.10 M $HCl(aq)$	0.10 M $HF(aq)$
pH = 1.00	pH = 2.10
$[H_3O^+] = 10^{-1.00} = 0.10\text{ M}$	$[H_3O^+] = 10^{-2.10} = 10^{0.90} \times 10^{-3}$
	$= 0.0079\text{ M}$

Comparison of these two values of $[H_3O^+]$ shows that, in contrast to hydrochloric acid, hydrofluoric acid is only partially dissociated in water. The *percent dissociation* of the hydrofluoric acid in the 0.10 M solution is only

$$\frac{[H_3O^+]}{0.10\text{ M}} \times 100 = \frac{0.0079\text{ M}}{0.10\text{ M}} \times 100 = 7.9\%$$

whereas hydrochloric acid is completely dissociated:

$$\frac{[H_3O^+]}{0.10\text{ M}} \times 100 = \frac{0.10\text{ M}}{0.10\text{ M}} \times 100 = 100\%$$

Ammonia is a base in aqueous solution because of the reaction

$$NH_3(aq) + H_2O(l) \rightleftharpoons NH_4^+(aq) + OH^-(aq)$$

Note that the base $NH_3(aq)$ accepts a proton from the acid $H_2O(l)$, resulting in the production of $NH_4^+(aq)$ and $OH^-(aq)$. Ammonia is said to be a weak base because not all the ammonia molecules are protonated in aqueous solution.

Example 17-6: The pH of a 0.10 M $NH_3(aq)$ solution is 11.10. Calculate the percentage of ammonia molecules that are protonated in this solution.

Solution: The reaction is

$$NH_3(aq) + H_2O(l) \rightleftharpoons NH_4^+(aq) + OH^-(aq)$$

The percentage of ammonia molecules that are protonated is

$$\% \text{ protonated} = \frac{[NH_4^+]}{0.10 \text{ M}} \times 100$$

Since $[NH_4^+] = [OH^-]$, we have

$$\% \text{ protonated} = \frac{[OH^-]}{0.10 \text{ M}} \times 100$$

We need to determine $[OH^-]$. The pH is 11.10, and so

$$[H_3O^+] = 10^{-11.10} = 10^{0.90} \times 10^{-12} = 7.94 \times 10^{-12} \text{ M}$$

The $[OH^-]$ can be calculated from the K_w expression:

$$[OH^-] = \frac{1.00 \times 10^{-14} \text{ M}^2}{7.94 \times 10^{-12} \text{ M}} = 1.26 \times 10^{-3} \text{ M}$$

The percentage of ammonia molecules that are protonated in a 0.10 M $NH_3(aq)$ solution is

$$\% \text{ protonated} = \frac{1.26 \times 10^{-3} \text{ M}}{0.10 \text{ M}} \times 100 = 1.3\%$$

17-6. THE EQUILIBRIUM CONSTANT FOR AN ACID DISSOCIATION REACTION IS DENOTED BY K_a

The equilibrium constant expression for an *acid dissociation reaction* is formulated according to the law of concentration action. Let's consider the weak acid acetic acid, $HC_2H_3O_2$, which has the Lewis formula

$$\begin{array}{c} \text{H} \quad \ddot{\text{O}}\text{:} \\ | \quad\; /\!/ \\ \text{H—C—C} \\ | \quad\; \backslash \\ \text{H} \quad \ddot{\text{O}}\text{—H} \end{array}$$

Only the proton attached to oxygen dissociates in water. The dissociation of acetic acid in water produces the acetate anion, $C_2H_3O_2^-$, and the hydronium ion:

$$CH_3{-}C(=\ddot{O}{:}){-}\ddot{O}{-}H\,(aq) + H_2O(l) \rightleftharpoons CH_3C(=\ddot{O}{:}){-}\ddot{O}{:}^{\ominus}\,(aq) + H_3O^+(aq)$$

The acetate ion is stabilized by delocalization of the negative charge over the $-\overset{O}{\overset{\|}{C}}-O^{\ominus}$ group:

$$CH_3{-}C(=\ddot{O}{:}){-}\ddot{O}{:}^{\ominus} \leftrightarrow CH_3{-}C({-}\ddot{O}{:}^{\ominus}){=}\ddot{O}{:}$$

The acetic acid dissociation reaction in water is

$$HC_2H_3O_2(aq) + H_2O(l) \rightleftharpoons H_3O^+(aq) + C_2H_3O_2^-(aq)$$

The law of concentration action is used to write K_a expressions.

The equilibrium constant expression for this reaction is

$$K_a = \frac{[H_3O^+][C_2H_3O_2^-]}{[HC_2H_3O_2]} \tag{17-6}$$

where the subscript a on K reminds us that K_a is an *acid dissociation constant.* Note that the $H_2O(l)$ concentration does not appear in the K_a expression. The experimental value of K_a at 25°C for acetic acid is

$$K_a = 1.74 \times 10^{-5}\ M = \frac{[H_3O^+][C_2H_3O_2^-]}{[HC_2H_3O_2]} \tag{17-7}$$

This small value of K_a reflects the fact that acetic acid is only slightly dissociated in aqueous solution.

The concentration of $H_3O^+(aq)$ in a 0.10 M acetic acid solution is less than 0.10 M.

Let us now consider the problem of calculating the pH of a 0.10 M acetic acid solution. The concentration of acetic acid at equilibrium is slightly less than 0.10 M because some of the $HC_2H_3O_2$ is converted to $C_2H_3O_2^-$ by the dissociation reaction. Thus we have the conservation condition

$$0.10\ M = [HC_2H_3O_2] + [C_2H_3O_2^-]$$

Because we started with pure acetic acid, we know from the reaction stoichiometry that

$$[H_3O^+] = [C_2H_3O_2^-]$$

where we have neglected the contribution to $[H_3O^+]$ from the dissociation of water because K_w ($= 1.0 \times 10^{-14}\ M^2$) is much smaller than K_a ($= 1.74 \times 10^{-5}$ M) for acetic acid. That is, $H_2O(l)$ is a much weaker acid than $HC_2H_3O_2(aq)$. We can set up a table for the initial and equilibrium concentrations of all the species in solution for the dissociation reaction:

	$HC_2H_3O_2(aq)$ +	$H_2O(l) \rightleftharpoons$	$H_3O^+(aq)$ +	$C_2H_3O_2^-(aq)$
initial concentration	0.10 M	—	≈0	0
equilibrium concentration	0.10 M − $[C_2H_3O_2^-]$	—	$[H_3O^+]$	$[C_2H_3O_2^-]$
or, using $[H_3O^+] = [C_2H_3O_2^-]$ from the reaction stoichiometry	0.10 M − $[H_3O^+]$	—	$[H_3O^+]$	$[H_3O^+]$

Note that we have set up this table so that the concentrations fall right under the species as they appear in the reaction. Combination of the results in the table with the K_a expression yields

$$1.74 \times 10^{-5}\ \text{M} = \frac{[H_3O^+][C_2H_3O_2^-]}{[HC_2H_3O_2]} = \frac{[H_3O^+]^2}{0.10\ \text{M} - [H_3O^+]} \qquad (17\text{-}8)$$

Equation (17-8) can be written in the standard form of a quadratic equation:

$$[H_3O^+]^2 + (1.74 \times 10^{-5}\ \text{M})[H_3O^+] - 1.74 \times 10^{-6}\ \text{M}^2 = 0$$

The two solutions to this equation are

$$[H_3O^+] = \frac{(-1.74 \times 10^{-5}\ \text{M}) \pm \sqrt{(1.74 \times 10^{-5}\ \text{M})^2 - 4(-1.74 \times 10^{-6}\ \text{M}^2)}}{2}$$

$$= 1.31 \times 10^{-3}\ \text{M} \quad \text{and} \quad -1.33 \times 10^{-3}\ \text{M}$$

Recall that the solutions to the quadratic equation $ax^2 + bx + c = 0$ are

$$x = \frac{-b \pm \sqrt{b^2 - 4ac}}{2a}.$$

We reject the physically unacceptable negative root and thus

$$[H_3O^+] = 1.31 \times 10^{-3}\ \text{M}$$

The pH of a 0.10 M acetic acid solution is

$$\begin{aligned} \text{pH} &= -\log[H_3O^+] \\ &= -\log[1.31 \times 10^{-3}] = 2.88 \end{aligned}$$

Although it is straightforward to solve Equation (17-8) using the quadratic formula, it is much easier to solve it in the following approximate manner. Acetic acid is a weak acid, and so we guess that the pH of a 0.10 M acetic acid solution is around 3, which is in the middle of the acid range. The value of $[H_3O^+]$ then would be around 10^{-3}. Note that 10^{-3} is small relative to 0.10, and so we guess that $[H_3O^+]$ in the denominator of Equation (17-8) is negligible *relative to* 0.10. This approximation converts Equation (17-8) to a simple quadratic equation for $[H_3O^+]$ that does not require the use of the quadratic formula. If we assume that $[H_3O^+]$ is small relative to 0.10 M, then Equation (17-8) becomes simply

$$\frac{[H_3O^+]^2}{0.10\ \text{M}} = 1.74 \times 10^{-5}\ \text{M}$$

from which we compute

$$[H_3O^+] = (0.10 \times 1.74 \times 10^{-5}\ \text{M}^2)^{1/2} = 1.32 \times 10^{-3}\ \text{M}$$

If is a simple matter to check if the approximation

$$K_a = \frac{x^2}{c - x} \approx \frac{x^2}{c}$$

(where c is the total concentration) is acceptable. Just solve for x as

$$x \approx (K_a c)^{1/2}$$

and compare the approximate value of x to the value of c.

We now must check the validity of our assumption that $[H_3O^+]$ is negligible relative to 0.10 M. Comparison of our value for $[H_3O^+]$ with 0.10 M shows that our result is good to two significant figures:

$$0.10 - 0.00132 = 0.099 \approx 0.10$$

In other words, the approximation $0.10\ \text{M} - [H_3O^+] \approx 0.10\ \text{M}$ yields a satisfactory solution to the quadratic equation. Using our approximate value, $[H_3O^+] = 1.32 \times 10^{-3}$ M, we find that the pH of the 0.10 M acetic acid solution is

$$\text{pH} = -\log[H_3O^+] = -\log(1.32 \times 10^{-3}) = 2.88$$

in excellent agreement with the result using the full quadratic formula.

The percentage dissociation of acetic acid in a 0.10 M solution of $HC_2H_3O_2(aq)$ is

$$\%\ \text{dissociation} = \frac{[H_3O^+]}{0.10\ \text{M}} \times 100 = \frac{1.3 \times 10^{-3}\ \text{M}}{0.10\ \text{M}} \times 100 = 1.3\%$$

Table 17-2 lists the K_a values for a number of weak acids. The following example illustrates a typical calculation involving K_a.

Example 17-7: The value of K_a for an aqueous solution of nitrous acid, HNO_2, is 4.47×10^{-4} M. Calculate the pH of a 0.50 M $HNO_2(aq)$ solution.

Table 17-2 The values of K_a and pK_a for some weak acids in water at 25°C

Acid	*Formula*	K_a/M	pK_a*
acetic	$HC_2H_3O_2$	1.74×10^{-5}	4.76
benzoic	$HC_7H_5O_2$	6.46×10^{-5}	4.19
chloroacetic	$HC_2H_2ClO_2$	1.35×10^{-3}	2.87
cyanic	HCNO	2.19×10^{-4}	3.66
formic	$HCHO_2$	1.78×10^{-4}	3.75
hydrazoic	HN_3	1.91×10^{-5}	4.72
hydrocyanic	HCN	4.79×10^{-10}	9.32
hydrofluoric	HF	6.76×10^{-4}	3.17
lactic	$HC_3H_5O_3$	1.40×10^{-4}	3.85
nitrous	HNO_2	4.47×10^{-4}	3.35
phenol	HC_6H_5O	1.0×10^{-10}	10.00

*pK_a is defined as p$K_a = -\log K_a$.

Solution: The reaction is

$$HNO_2(aq) + H_2O(l) \rightleftharpoons H_3O^+(aq) + NO_2^-(aq)$$

We can set up the following table for initial and equilibrium concentrations of all the species in solution:

	$HNO_2(aq)$	+ $H_2O(l)$	$\rightleftharpoons H_3O^+(aq)$	+ $NO_2^-(aq)$
initial concentration	0.50 M	—	≈ 0	0
equilibrium concentration	0.50 M − $[NO_2^-]$	—	$[H_3O^+]$	$[NO_2^-]$
or, using $[H_3O^+] = [NO_2^-]$ from the reaction stoichiometry	0.50 M − $[H_3O^+]$	—	$[H_3O^+]$	$[H_3O^+]$

If we combine the entries in this table with the expression for K_a, then we have

$$K_a = \frac{[H_3O^+][NO_2^-]}{[HNO_2]} = \frac{[H_3O^+]^2}{0.50\text{ M} - [H_3O^+]} = 4.47 \times 10^{-4}\text{ M}$$

The quadratic formula gives

$$[H_3O^+] = \frac{(-4.47 \times 10^{-4}\,M) \pm \sqrt{(4.47 \times 10^{-4}\,M)^2 + (4)(0.50\,M)(4.47 \times 10^{-4}\,M)}}{2}$$

$$= 1.47 \times 10^{-2}\,M \quad \text{and} \quad -1.52 \times 10^{-2}\,M$$

We reject the physically unreasonable negative concentration, so the pH is

$$pH = -\log[H_3O^+] = -\log(1.47 \times 10^{-2}) = 1.83$$

We can obtain almost the same answer by neglecting $[H_3O^+]$ relative to 0.50 in the denominator of the K_a expression. In this case,

$$[H_3O^+] \approx (0.50\,M \times 4.47 \times 10^{-4}\,M)^{1/2} = 1.50 \times 10^{-2}\,M$$

Note that 1.50×10^{-2} is only a few percent of 0.50. The pH is

$$pH = -\log(1.50 \times 10^{-2}) = 1.83$$

in excellent agreement with the result obtained using the quadratic formula.

Problems 17-33 through 17-40 involve calculating the pH of aqueous solutions of various weak acids.

17-7. THE LARGER THE VALUE OF K_a, THE STRONGER IS THE ACID

For a given concentration, the percent dissociation of an acid, or the strength of the acid, depends upon the value of the acid dissociation constant. The larger the value of K_a, the stronger is the acid. For example, consider the reactions

$$\underset{\text{formic acid}}{HCHO_2(aq)} + H_2O(l) \rightleftharpoons H_3O^+(aq) + CHO_2^-(aq)$$

$$K_a = 1.78 \times 10^{-4}\,M$$

$$\underset{\text{hydrocyanic acid}}{HCN(aq)} + H_2O(l) \rightleftharpoons H_3O^+(aq) + CN^-(aq)$$

$$K_a = 4.79 \times 10^{-10}\,M$$

The percent dissociation of each acid in a 0.10 M solution is

$$\%\text{ dissociation in } 0.10\,M\ HCHO_2(aq) = 4.13\%$$

$$\%\text{ dissociation in } 0.10\,M\ HCN(aq) = 0.0069\%$$

Another comparison of the relative strengths of two acids results from a consideration of an aqueous solution containing a mixture of the two. For example, chloroacetic acid, $HC_2H_2ClO_2$, is a stronger acid than acetic acid, $HC_2H_3O_2$:

$$:\ddot{Cl}-\underset{H}{\overset{H}{C}}-C\begin{matrix}\diagup\!\!\diagup \ddot{O}: \\ \diagdown :\ddot{O}-H\end{matrix}$$

chloroacetic acid
$K_a = 1.35 \times 10^{-3}$ M

$$H-\underset{H}{\overset{H}{C}}-C\begin{matrix}\diagup\!\!\diagup \ddot{O}: \\ \diagdown :\ddot{O}-H\end{matrix}$$

acetic acid
$K_a = 1.74 \times 10^{-5}$ M

The fact that chloroacetic acid is a stronger acid than acetic acid means that, in a solution that is 0.10 M in both $HC_2H_3O_2(aq)$ and $HC_2H_2ClO_2(aq)$, the reaction equilibrium

$$HC_2H_2ClO_2(aq) + C_2H_3O_2^-(aq) \rightleftharpoons C_2H_2ClO_2^-(aq) + HC_2H_3O_2(aq)$$

Notice that this reaction is simply a proton-transfer reaction.

lies to the right because $HC_2H_2ClO_2(aq)$ is a more potent proton donor than $HC_2H_3O_2(aq)$ and thus $HC_2H_2ClO_2(aq)$ can protonate $C_2H_3O_2^-(aq)$ more effectively than $HC_2H_3O_2(aq)$ can protonate $C_2H_2ClO_2^-(aq)$.

We can make this argument quantitative. The above equation can be obtained by adding the following two equations:

(1) $HC_2H_2ClO_2(aq) + H_2O(l) \rightleftharpoons H_3O^+(aq) + C_2H_2ClO_2^-(aq)$

(2) $H_3O^+(aq) + C_2H_3O_2^-(aq) \rightleftharpoons HC_2H_3O_2(aq) + H_2O(l)$

The sum of these two equations is

(3) $HC_2H_2ClO_2(aq) + C_2H_3O_2^-(aq) \rightleftharpoons HC_2H_3O_2(aq) + C_2H_2ClO_2^-(aq)$

The equilibrium constant expression for Reaction (3) is

$$K_3 = \frac{[HC_2H_3O_2][C_2H_2ClO_2^-]}{[HC_2H_2ClO_2][C_2H_3O_2^-]}$$

Because Reaction (3) is the sum of Reactions (1) and (2), the value of K_3 is equal to the product of K_1 and K_2:

Recall from Chapter 15 that $K_3 = K_2K_1$ where Reaction (3) is the sum of Reactions (1) and (2).

$$K_3 = K_1K_2 = \frac{K_a \text{ for } HC_2H_2ClO_2}{K_a \text{ for } HC_2H_3O_2} = \frac{1.35 \times 10^{-3}\text{ M}}{1.74 \times 10^{-5}\text{ M}} = 77.6$$

where we have used the fact that

$$K_2 = \frac{1}{K_a \text{ for } HC_2H_3O_2}$$

because Reaction (2) is the reverse of the K_a reaction.

The fact that $K_3 > 1$ means that Reaction (3) lies to the right at equilibrium, confirming that chloroacetic acid is a stronger acid than acetic acid.

17-8. THE SMALLER THE pK_a VALUE FOR AN ACID, THE STRONGER THE ACID

The larger the value of K_a for an acid, the stronger is the acid. Thus nitrous acid, for which $K_a = 4.47 \times 10^{-4}$ M, is a stronger acid than acetic acid, for which $K_a = 1.74 \times 10^{-5}$ M. Because K_a values for aqueous acids range over many powers of 10 (Table 17-2), it is convenient to define the quantity pK_a as

$$pK_a \equiv -\log K_a \tag{17-9}$$

Note the similarity in the definitions of pH (Equation 17-4) and pK_a. The pK_a values at 25°C for several weak acids in water are given in Table 17-2.

Example 17-8: Compute pK_a for acetic acid in water at 25°C.

Solution: In Table 17-2, we find that, for $HC_2H_3O_2(aq)$, $K_a = 1.74 \times 10^{-5}$ M. Using Equation 17-9 we have

$$\begin{aligned} pK_a &= -\log K_a \\ &= -\log(1.74 \times 10^{-5}) \\ &= 4.76 \end{aligned}$$

From the pK_a values given in Table 17-2, we see that the stronger the acid, the smaller the value of pK_a.

17-9. BASES REACT WITH WATER TO PRODUCE HYDROXIDE IONS

Ammonia is a base in aqueous solution because it reacts with water to produce hydroxide ions:

$$NH_3(aq) + H_2O(l) \rightleftharpoons NH_4^+(aq) + OH^-(aq)$$

In terms of Lewis formulas, the protonation of ammonia can be described by

$$H{-}\ddot{N}H_2 + H_2\ddot{O}: \rightleftharpoons H{-}N^{\oplus}H_3 + {}^{\ominus}:\ddot{O}{-}H$$

The lone pair of electrons on the nitrogen atom in the ammonia molecule is able to bond the proton donated by the water molecule. Ammonia is a weak base because the equilibrium constant for its reaction with water is 1.75×10^{-5} M, indicating that the reaction lies far to the left. The most common weak bases are ammonia and some organic compounds that are similar to ammonia. For example, if we substitute the hydrogen atoms in ammonia with methyl groups, then we obtain the weak bases

The methyl amines smell much like ammonia while aniline and pyridine have characteristic disagreeable odors.

$H{-}\ddot{N}(H){-}CH_3$ methyl amine

$H_3C{-}\ddot{N}(H){-}CH_3$ dimethyl amine

$H_3C{-}\ddot{N}(CH_3){-}CH_3$ trimethyl amine

Some other organic bases are

aniline

pyridine

Notice that the nitrogen atom in all these bases has a lone pair of electrons. Pyridine, for example, is basic in aqueous solution because of the reaction

$$\text{pyridine}(aq) + H_2O(l) \rightleftharpoons \text{pyridinium ion}(aq) + {}^{\ominus}OH(aq)$$

If we abbreviate pyridine by Py and the pyridinium ion by PyH^+, then the equilibrium-constant expression for this reaction is

$$K_b = \frac{[PyH^+][OH^-]}{[Py]}$$

The subscript b on K indicates that the reaction is a reaction of a weak base with water, that is, a base protonation reaction. Thus K_b is called the *base protonation constant.* The values of K_b for some weak

Table 17-3 The values of K_b and pK_b for some weak bases in water at 25°C

Base	*Formula*	K_b/M	pK_b*
ammonia	$H-\ddot{N}(H)-H$	1.75×10^{-5}	4.76
methyl amine	$H-\ddot{N}(H)-CH_3$	4.59×10^{-4}	3.34
dimethyl amine	$H_3C-\ddot{N}(H)-CH_3$	5.81×10^{-4}	3.24
trimethyl amine	$H_3C-\ddot{N}(CH_3)-CH_3$	6.11×10^{-5}	4.21
hydroxylamine	$H-\ddot{N}(H)-OH$	1.07×10^{-8}	7.97
aniline	$C_6H_5-\ddot{N}(H)-H$	4.17×10^{-10}	9.38
pyridine	$C_5H_5N:$	1.48×10^{-9}	8.83

*pK_b is defined as p$K_b = -\log K_b$.

bases are given in Table 17-3, which also gives pK_b values for the bases, where pK_b is defined, by analogy to pK_a, as

$$pK_b \equiv -\log K_b \tag{17-10}$$

The smaller the value of pK_b is, the stronger the base.

Aniline is a colorless, poisonous liquid.

Example 17-9: Aniline is used in the manufacture of dyes and various pharmaceuticals. The solubility of aniline in water at 25°C is 1.00 g per 28.6 mL of solution. Calculate the pH of a saturated aqueous solution of aniline at 25°C.

Solution: The concentration of a saturated solution of aniline in water at 25°C is

$$\left(\frac{1.0 \text{ g aniline}}{28.6 \text{ mL } H_2O}\right)\left(\frac{1000 \text{ mL } H_2O}{1 \text{ L } H_2O}\right)\left(\frac{1 \text{ mol aniline}}{93.13 \text{ g aniline}}\right) = 0.375 \text{ M}$$

Thus we must calculate the pH of a 0.375 M aqueous aniline solution. From Table 17-3, K_b is 4.17×10^{-10} M for the reaction

$$C_6H_5NH_2(aq) + H_2O(l) \rightleftharpoons C_6H_5\overset{+}{N}H_3(aq) + OH^-(aq)$$

If we let aniline be An and its protonated form be AnH^+, then

$$K_b = \frac{[AnH^+][OH^-]}{[An]} = 4.17 \times 10^{-10}\ M$$

A table of initial and equilibrium concentrations has the form

	$An(aq)$	+ $H_2O(l)$ ⇌	$AnH^+(aq)$ +	$OH^-(aq)$
initial concentration	0.375 M	—	0	0
equilibrium concentration	$0.375\ M - [AnH^+]$	—	$[AnH^+]$	$[OH^-]$
or, using $[AnH^+] = [OH^-]$ from the reaction stoichiometry	$0.375\ M - [OH^-]$	—	$[OH^-]$	$[OH^-]$

The expression for K_b is

$$K_b = \frac{[AnH^+][OH^-]}{[An]} = \frac{[OH^-]^2}{0.375\ M - [OH^-]} = 4.17 \times 10^{-10}\ M$$

Because K_b is so small, we expect $[OH^-]$ to be small also and thus negligible relative to 0.375. The K_b expression becomes

$$K_b = \frac{[OH^-]^2}{0.375\ M} = 4.17 \times 10^{-10}\ M$$

and

$$[OH^-] = 1.25 \times 10^{-5}\ M$$

We check to be sure that the value of $[OH^-]$ is negligible with respect to 0.375, and indeed it is. The $[H_3O^+]$ is

$$[H_3O^+] = \frac{1.00 \times 10^{-14}\ M^2}{1.25 \times 10^{-5}\ M} = 8.00 \times 10^{-10}\ M$$

and the pH is

$$pH = -\log[H_3O^+] = 9.10$$

Problems 17-43 through 17-46 ask you to calculate the pH of aqueous solutions of various weak bases.

17-10. THE ACID-BASE PAIR HB,B$^-$ IS CALLED A CONJUGATE ACID-BASE PAIR

When an acid such as $HC_2H_3O_2$ dissociates, water acts as a base by accepting a proton from the acid:

$$HC_2H_3O_2(aq) + H_2O(l) \rightleftharpoons H_3O^+(aq) + C_2H_3O_2^-(aq)$$

If we look at the reverse reaction, then we see that $H_3O^+(aq)$ donates a proton to $C_2H_3O_2^-(aq)$. In other words, $H_3O^+(aq)$ acts as an acid by donating a proton and $C_2H_3O_2^-$ acts as a base by accepting a proton. The base $C_2H_3O_2^-(aq)$ is called the *conjugate base* of the acid $HC_2H_3O_2(aq)$; the pair $HC_2H_3O_2(aq)$,$C_2H_3O_2^-(aq)$ is called a *conjugate acid-base pair*. Similarly, the hydronium ion, $H_3O^+(aq)$, is the *conjugate acid* of the base $H_2O(l)$. The conjugate acid-base pairs for the dissociation of acetic acid in water are

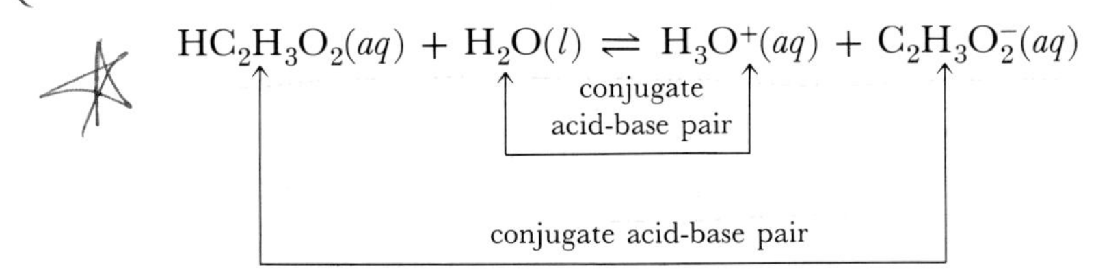

Problems 17-53 through 17-58 involve identifying conjugate acid-base pairs.

Note that a conjugate base has one fewer proton than its corresponding conjugate acid.

An acetate ion is a weak base because it is the anion of a weak acid, acetic acid. To see more explicitly that an acetate ion is a weak base, consider the reverse of the above reaction:

$$(1)\quad H_3O^+(aq) + C_2H_3O_2^-(aq) \rightleftharpoons HC_2H_3O_2(aq) + H_2O(l)$$

Because this reaction is the reverse of the reaction associated with the K_a of the acetic acid, we have for the equilibrium constant K_1 of Reaction (1)

$$K_1 = \frac{1}{K_a}$$

The dissociation reaction for water is

$$(2)\quad H_2O(l) + H_2O(l) \rightleftharpoons H_3O^+(aq) + OH^-(aq)$$

for which

$$K_2 = K_w$$

If we add Equations (1) and (2), then we obtain

$$(3)\quad C_2H_3O_2^-(aq) + H_2O(l) \rightleftharpoons HC_2H_3O_2(aq) + OH^-(aq)$$

The equilibrium constant for this reaction is the product of the equilibrium constants for Equations (1) and (2):

$$K_3 = \frac{K_w}{K_a}$$

Note that Equation (3) shows explicitly that an acetate ion reacts with water to yield OH^- ions. Equation (3) is the same form as the equation for the reaction between ammonia and water to form hydroxide ions. We denote the equilibrium constant of Reaction (3) by K_b instead of K_3:

$$C_2H_3O_2^-(aq) + H_2O(l) \rightleftharpoons HC_2H_3O_2(aq) + OH^-(aq)$$

$$K_b = \frac{K_w}{K_a} \tag{17-11}$$

The value of K_a for acetic acid is 1.74×10^{-5} M, and so

$$K_b = \frac{1.00 \times 10^{-14}\ \text{M}^2}{1.74 \times 10^{-5}\ \text{M}} = 5.75 \times 10^{-10}\ \text{M}$$

Generally, the anion of any weak acid is itself a weak base. If we denote a weak acid by HB and its conjugate base by B^-, then we have the reactions

$$HB(aq) + H_2O(l) \rightleftharpoons H_3O^+(aq) + B^-(aq)$$

with

$$K_a = \frac{[H_3O^+][B^-]}{[HB]} \tag{17-12}$$

and

$$B^-(aq) + H_2O(l) \rightleftharpoons HB(aq) + OH^-(aq)$$

with

$$K_b = \frac{[HB][OH^-]}{[B^-]} \tag{17-13}$$

The first reaction explicitly shows HB as an acid, and the second reaction explicitly shows its conjugate base B^- as a base. The relation between K_a and K_b for a conjugate acid-base pair is

$$K_a K_b = K_w \tag{17-14}$$

Equation (17-14) can be obtained by multiplying Equations (17-12) and (17-13).

Example 17-10: A fluoride ion is basic because it is the conjugate base of a weak acid. Use Table 17-2 to calculate K_b for $F^-(aq)$.

Solution: The value of K_a for HF, the acid that is conjugate to F^-, is 6.76×10^{-4} M. Using Equation (17-14), then, we find that

$$K_b = \frac{K_w}{K_a} = \frac{1.00 \times 10^{-14}\ \mathrm{M}^2}{6.76 \times 10^{-4}\ \mathrm{M}} = 1.48 \times 10^{-11}\ \mathrm{M}$$

The reaction associated with K_b is

Problems 17-59 and 17-60 are similar to Example 17-10.

$$F^-(aq) + H_2O(l) \rightleftharpoons HF(aq) + OH^-(aq)$$

Table 17-4 lists values of K_a and K_b for a number of conjugate acid-base pairs. Note that the stronger the acid is, the weaker its conjugate base. This reciprocal relation between the strength of an acid and the strength of its conjugate base is given in Equation (17-14).

It is interesting to note that water can act as either an acid or a base:

$$H_2O(l) + H_2O(l) \rightleftharpoons H_3O^+(aq) + OH^-(aq)$$

conjugate acid-base pair (second $H_2O(l)$ and $H_3O^+(aq)$)

conjugate acid-base pair (first $H_2O(l)$ and $OH^-(aq)$)

It is this property of water that gives it such a central role in acid-base chemistry.

17-11. AQUEOUS SOLUTIONS OF MANY SALTS ARE EITHER ACIDIC OR BASIC

Suppose we dissolve sodium acetate, $NaC_2H_3O_2(s)$, in water. The resulting solution contains sodium ions, $Na^+(aq)$, and acetate ions, $C_2H_3O_2^-(aq)$. However, the acetate ion is the conjugate base of the weak acid $HC_2H_3O_2$. Consequently, a fraction of the $C_2H_3O_2^-$ ions are protonated via the reaction

$$C_2H_3O_2^-(aq) + H_2O(l) \rightleftharpoons HC_2H_3O_2(aq) + OH^-(aq) \qquad (17\text{-}15)$$

The sodium ions do not react with water to produce hydronium ions or hydroxide ions, and so the net result is that the sodium acetate solution is basic because the protonation of acetate by water produces hydroxide ions.

Table 17-4 Values of K_a and K_b at 25°C for conjugate acid-base pairs

	Acid	K_a/M	Base	K_b/M
sulfurous acid	H_2SO_3	1.5×10^{-2}	HSO_3^-	6.7×10^{-13}
hydrogen sulfate ion	HSO_4^-	1.2×10^{-2}	SO_4^{2-}	8.3×10^{-13}
phosphoric acid	H_3PO_4	5.9×10^{-3}	$H_2PO_4^-$	1.7×10^{-12}
chloroacetic acid	$HC_2H_2ClO_2$	1.4×10^{-3}	$C_2H_2ClO_2^-$	7.1×10^{-12}
hydrofluoric acid	HF	6.8×10^{-4}	F^-	1.5×10^{-11}
nitrous acid	HNO_2	4.5×10^{-4}	NO_2^-	2.2×10^{-11}
cyanic acid	$HCNO$	2.2×10^{-4}	CNO^-	4.6×10^{-11}
formic acid	$HCHO_2$	1.8×10^{-4}	CHO_2^-	5.6×10^{-11}
benzoic acid	$HC_7H_5O_2$	6.5×10^{-5}	$C_7H_5O_2^-$	1.5×10^{-10}
acetic acid	$HC_2H_3O_2$	1.7×10^{-5}	$C_2H_3O_2^-$	5.9×10^{-10}
pyridinium ion	$C_5H_5NH^+$	6.8×10^{-6}	C_5H_5N	1.5×10^{-9}
hydrogen sulfide	H_2S	9.1×10^{-8}	HS^-	1.1×10^{-7}
dihydrogen phosphate ion	$H_2PO_4^-$	6.3×10^{-8}	HPO_4^{2-}	1.6×10^{-7}
hydrogen sulfite ion	HSO_3^-	6.2×10^{-8}	SO_3^{2-}	1.6×10^{-7}
hypochlorous acid	$HClO$	3.0×10^{-8}	ClO^-	3.3×10^{-7}
ammonium ion	NH_4^+	5.6×10^{-10}	NH_3	1.7×10^{-5}
boric acid	H_3BO_3	5.9×10^{-10}	$H_2BO_3^-$	1.7×10^{-5}
hydrocyanic acid	HCN	4.8×10^{-10}	CN^-	2.1×10^{-5}
phenol	HC_6H_5O	1.0×10^{-10}	$C_6H_5O^-$	1.0×10^{-4}
hydrogen carbonate ion	HCO_3^-	4.7×10^{-11}	CO_3^{2-}	2.1×10^{-4}
hydrogen phosphate ion	HPO_4^{2-}	4.8×10^{-13}	PO_4^{3-}	2.1×10^{-2}
hydrogen sulfide ion	HS^-	1.2×10^{-13}	S^{2-}	8.3×10^{-2}

decreasing acid strength ↓ (left margin); increasing basic strength ↓ (right margin)

For the acids: $HB(aq) + H_2O(l) \rightleftharpoons H_3O^+(aq) + B^-(aq)$ $\qquad K_a = \dfrac{[H_3O^+][B^-]}{[HB]}$

For the bases: $B^-(aq) + H_2O(l) \rightleftharpoons HB(aq) + OH^-(aq)$ $\qquad K_b = \dfrac{[HB][OH^-]}{[B^-]}$

$K_aK_b = K_w = 1.00 \times 10^{-14}\ M^2$ for a conjugate acid-base pair at 25°C.

Table 17-5 Acid-base properties of some common cations and anions in water

Cations

Acidic	*Neutral*		*Basic*
NH_4^+	Li^+	Mg^{2+}	none
Al^{3+}	Na^+	Ca^{2+}	
transition metal ions	K^+	Sr^{2+}	
	Rb^+	Ba^{2+}	

Anions

Acidic	*Neutral*		*Basic*	
HSO_4^-	Cl^-	ClO_4^-	F^-	CN^-
$H_2PO_4^-$	Br^-	NO_3^-	$C_2H_3O_2^-$	S^{2-}
	I^-		NO_2^-	SO_4^{2-}
			HCO_3^-	HPO_4^{2-}
			CO_3^{2-}	PO_4^{3-}
			plus many others	

Various ions react with water to produce hydronium ions or hydroxide ions. The acidic, neutral, or basic properties of a number of ions are given in Table 17-5, from which you should note that:

1. The conjugate bases of strong acids—for example, $Cl^-(aq)$—are *neutral anions;* they do not react with H_2O to produce $OH^-(aq)$. The conjugate bases of weak acids—for example, $C_2H_3O_2^-(aq)$—are *basic anions;* they react with H_2O to produce $OH^-(aq)$.
2. *Cations* are either *acidic* or *neutral.* The alkali and alkaline earth cations (except Be^{2+}) are all neutral. The conjugate acids of weak bases—$NH_4^+(aq)$, for example—are acidic. There are no basic cations.
3. Many transition-metal ions are acidic in aqueous solution. These ions exist in aqueous solution with a certain number of water molecules bonded to them and are said to be solvated. Two examples are iron(III), $Fe(H_2O)_6^{3+}(aq)$, and tin(II), $Sn(H_2O)_6^{2+}(aq)$. The acid dissociation reactions of these ions are

$$Fe(H_2O)_6^{3+}(aq) + H_2O(l) \rightleftharpoons Fe(OH)(H_2O)_5^{2+}(aq) + H_3O^+(aq)$$
$$K_a = 1 \times 10^{-3}\ \text{M}$$

$$Sn(H_2O)_6^{2+}(aq) + H_2O(l) \rightleftharpoons Sn(OH)(H_2O)_5^{+}(aq) + H_3O^+(aq)$$
$$K_a = 1 \times 10^{-2}\ \text{M}$$

Note that the acidity of these two metal ions arises from the loss of a proton from an attached water molecule. Note also that the hydrated iron(III) and tin(II) ions are significantly stronger acids than acetic acid, the K_a of which is only 1.74×10^{-5} M.

The K_a values of some other common acidic metal cations are given in Table 17-6.

Table 17-6 The K_a and pK_a values of some common acidic metal cations at 25°C in water

Cation	K_a/M	pK_a
$Pb(H_2O)_4^{2+}$	1×10^{-8}	8.0
$Al(H_2O)_6^{3+}$	1×10^{-5}	5.0
$In(H_2O)_6^{3+}$	2×10^{-4}	3.7
$Tl(H_2O)_6^{3+}$	6×10^{-2}	1.2

Salts of neutral cations and basic anions dissolved in water produce basic solutions. For example, sodium nitrite dissolved in water:

$$NaNO_2(s) \xrightarrow[H_2O(l)]{} Na^+(aq) + NO_2^-(aq)$$

The protonation reaction of the weak base $NO_2^-(aq)$

$$NO_2^-(aq) + H_2O(l) \rightleftharpoons HNO_2(aq) + OH^-(aq)$$

produces $OH^-(aq)$ and thus yields a basic solution.

Salts of acidic cations and neutral anions produce acidic solutions when dissolved in water. For example, for thallium(III) nitrate

$$Tl(NO_3)_3(s) \xrightarrow[H_2O(l)]{} Tl(H_2O)_6^{3+}(aq) + 3NO_3^-(aq)$$

the key reaction is

$$Tl(H_2O)_6^{3+}(aq) + H_2O(l) \rightleftharpoons Tl(OH)(H_2O)_5^{2+}(aq) + H_3O^+(aq)$$

which shows explicitly the production of $H_3O^+(aq)$.

Example 17-11: Predict whether the following salts produce acidic, neutral, or basic solutions when dissolved in water: KCl, NaCN, $AlBr_3$, NH_4NO_3, and NH_4CN.

Solution:

KCl	neutral cation, neutral anion	neutral solution
NaCN	neutral cation, basic anion	basic solution
$AlBr_3$	acidic cation, neutral anion	acidic solution
NH_4NO_3	acidic cation, neutral anion	acidic solution
NH_4CN	acidic cation, basic anion	cannot predict without calculations

Problems 17-67 through 17-72 are similar to Example 17-11.

In addition to predicting whether a salt solution is acidic or basic, we also can calculate its pH using the ideas of Section 17-9. Let's calculate the pH of a 0.050 M NH_4Cl solution. A chloride ion is neutral (Table 17-5). An ammonium ion is acidic, and its reaction with water is

$$NH_4^+(aq) + H_2O(l) \rightleftharpoons H_3O^+(aq) + NH_3(aq)$$

From Table 17-4, we find that $K_a = 5.6 \times 10^{-10}$ M for this reaction. A table of the initial and equilibrium ionic concentrations is

	$NH_4^+(aq)$	+ $H_2O(l)$ $\rightleftharpoons$	$H_3O^+(aq)$	+ $NH_3(aq)$
initial concentration	0.050 M	—	≈0	0
equilibrium concentration	0.050 M − $[NH_3]$	—	$[H_3O^+]$	$[NH_3]$
or, because $[H_3O^+] \approx [NH_3]$ from the reaction stoichiometry	0.050 M − $[H_3O^+]$	—	$[H_3O^+]$	$[H_3O^+]$

The expression for K_a becomes

$$\frac{[H_3O^+]^2}{0.050\text{ M} - [H_3O^+]} = 5.6 \times 10^{-10}\text{ M}$$

We neglect $[H_3O^+]$ relative to 0.050 M in the denominator and obtain

$$[H_3O^+] = (0.050\text{ M} \times 5.6 \times 10^{-10}\text{ M})^{1/2} = 5.3 \times 10^{-6}\text{ M}$$

After checking that $[H_3O^+]$ is indeed negligible compared with 0.050 M, we calculate the pH to be

$$pH = -\log[H_3O^+] = -\log(5.3 \times 10^{-6}) = 5.28$$

Notice that the solution is acidic; NH_4Cl consists of an acidic cation and a neutral anion.

Example 17-12: Compute the pH of a 0.050 M $NaC_2H_3O_2(aq)$ solution.

Solution: Sodium acetate consists of a neutral cation and a basic anion. The reaction of the anion with water is

$$C_2H_3O_2^-(aq) + H_2O(l) \rightleftharpoons HC_2H_3O_2(aq) + OH^-(aq)$$

From Table 17-4, we see that K_b for this reaction is 5.9×10^{-10} M. A table of the initial and equilibrium concentrations is

	$C_2H_3O_2^-(aq)$ +	$H_2O(l)$ ⇌	$HC_2H_3O_2(aq)$ +	$OH^-(aq)$
initial concentration	0.050 M	—	0	≈0
equilibrium concentration	0.050 M − $[HC_2H_3O_2]$	—	$[HC_2H_3O_2]$	$[OH^-]$
or, because $[OH^-] = [HC_2H_3O_2]$ from the reaction stoichiometry	0.050 M − $[OH^-]$	—	$[OH^-]$	$[OH^-]$

The expression for K_b is

$$5.9 \times 10^{-10}\,\text{M} = \frac{[OH^-]^2}{0.050\,\text{M} - [OH^-]}$$

Neglecting $[OH^-]$ relative to 0.050 M, we obtain

$$[OH] = (0.050\,\text{M} \times 5.9 \times 10^{-10}\,\text{M}^2)^{1/2} = 5.4 \times 10^{-6}\,\text{M}$$

The value of $[H_3O^+]$ can be obtained from the value of $[OH^-]$ using the K_w expression:

$$[H_3O^+] = \frac{K_w}{[OH^-]} = \frac{1.00 \times 10^{-14}\,\text{M}^2}{5.4 \times 10^{-6}\,\text{M}} = 1.9 \times 10^{-9}\,\text{M}$$

The pH of the solution is

$$pH = -\log[H_3O^+] = -\log(1.9 \times 10^{-9}) = 8.72$$

which shows that the solution is basic (pH > 7.0).

Problems 17-73 through 17-82 involve calculating the pH of various salt solutions.

17-12. A LEWIS ACID IS AN ELECTRON-PAIR ACCEPTOR

The most general acid-base classification system is that introduced by G. N. Lewis, who defined an *acid as an electron-pair acceptor* and a *base as an electron-pair donor.* For example, consider the reaction between ammonia and boron trifluoride:

$$\underset{\text{Lewis base}}{H_3N\!:} + \underset{\text{Lewis acid}}{BF_3} \rightleftharpoons \underset{\text{donor-acceptor complex}}{H_3N{-}BF_3}$$

The BF_3 reacts with the NH_3 by accepting the lone pair of electrons on nitrogen to form an N—B bond. Note that BF_3 is not an acid in the Brønsted-Lowry sense because it has no protons to donate, but it is a Lewis acid because it acts as an electron-pair acceptor. In general, an electron-deficient species (Chapter 10) can act as a Lewis acid and a species with a lone pair of electrons can act as a Lewis base.

Example 17-13: Classify the following species as either Lewis acids or Lewis bases: (a) H^+ (b) OH^- (c) $AlCl_3$ (d) Cl^-

Solution:

(a) A free proton is electron-deficient and combines with electron-pair donors:

$$H^+ + :\ddot{O}H^- \rightarrow H_2O$$

Thus H^+ is a Lewis acid.

(b) As shown in (a), OH^- is a Lewis base (electron-pair donor).

(c) $AlCl_3$ is an electron-deficient molecule:

$$AlCl_3 \text{ (Al bonded to three } :\ddot{Cl}: \text{ atoms)}$$

and thus can act as a Lewis acid; for example,

$$\underset{\text{Lewis base}}{:\ddot{Cl}:^{\ominus}} + \underset{\text{Lewis acid}}{AlCl_3} \rightarrow AlCl_4^{\ominus}$$

Problems 17-83 through 17-86 are similar to Example 17-13.

(d) As shown in (c), Cl^- is a Lewis base.

The Lewis acid-base classification scheme is useful in helping us to understand and predict reactions between electron-pair donors and electron-pair acceptors. It is especially useful in understanding the mechanisms of many organic reactions.

SUMMARY

The Brønsted-Lowry definition of acids and bases is that an acid is a species that donates protons and a base is a species that accepts protons. Acid-base reactions are proton-transfer reactions. Water plays a central role in acid-base reactions because it can act either as an acid or as a base. Water undergoes a self-protonation reaction to produce $H_3O^+(aq)$ and $OH^-(aq)$. The product of the equilibrium concentrations of $H_3O^+(aq)$ and $OH^-(aq)$ ions is fixed by the value of the ion product constant of water, $K_w = [H_3O^+][OH^-]$. The value of K_w is $1.00 \times 10^{-14}\ M^2$ at 25°C.

If $[H_3O^+] > [OH^-]$, then the solution is acidic; if $[H_3O^+] < [OH^-]$, then the solution is basic. The value of $[H_3O^+]$ is presented conveniently on a pH scale, and pH, a measure of the acidity of a solution, is defined through the equation $pH = -\log[H_3O^+]$. At 25°C, if $pH < 7$, then the solution is acidic; if $pH > 7$, then the solution is basic; if $pH = 7$, then the solution is neutral.

Strong acids and strong bases are completely dissociated in solution; weak acids and weak bases are only partially dissociated. The strength of a weak acid is governed by the value of K_a, its acid dissociation constant; the strength of a weak base is governed by the value of K_b, its base protonation constant. The larger the value of K_a or K_b, the stronger the acid or the base. The treatment of weak acids and weak bases can be unified by the introduction of the idea of a conjugate acid-base pair. The values of K_a and K_b for a conjugate acid-base pair obey the relation $K_w = K_a K_b$. The stronger an acid, the weaker is its conjugate base. The pK_a value of an acid is given by $pK_a = -\log K_a$; the smaller the value of pK_a, the stronger the acid. The pK_b value of a base is given by $pK_b = -\log K_b$; the smaller the value of pK_b, the stronger the base.

The aqueous solutions of many salts are acidic or basic. Many cations and anions have acidic or basic properties. Salts that consist of these cations or anions may be acidic or basic.

TERMS YOU SHOULD KNOW

Arrhenius acid and base 648
Brønsted-Lowry acid and base 648
proton donor 648
proton acceptor 648
hydronium ion, H_3O^+ 648
proton-transfer reaction 648
protonation reaction 648
ion product constant of water, K_w 649

neutral solution, $[H_3O^+] = [OH^-]$ 649
acidic solution, $[H_3O^+] > [OH^-]$ 649
basic solution, $[OH^-] > [H_3O^+]$ 649
strong acid 650
strong base 650
weak acid 651
weak base 651, 656
acidity 652
pH 652
logarithm 652
pH meter 655
percent dissociation 656
acid dissociation reaction 657
acid dissociation constant, K_a 658
pK_a 664
base protonation constant, K_b 665
pK_b 666
conjugate base 668
conjugate acid-base pair 668
conjugate acid 668
neutral anion 672
basic anion 672
acidic cation 672
neutral cation 672
acidic anion 672
Lewis acid 676
electron-pair acceptor 676
Lewis base 676
electron-pair donor 676

EQUATIONS YOU SHOULD KNOW HOW TO USE

$$K_w = [H_3O^+][OH^-] = 1.00 \times 10^{-14}\ M^2 \quad (17\text{-}3)$$

$$pH = -\log[H_3O^+] \quad (17\text{-}4)$$

$$[H_3O^+] = 10^{-pH} \quad (17\text{-}5)$$

$$pK_a = -\log K_a \quad (17\text{-}9)$$

$$pK_b = -\log K_b \quad (17\text{-}10)$$

$$K_a = \frac{[H_3O^+][B^-]}{[HB]} \quad (17\text{-}12)$$

$$K_b = \frac{[HB][OH^-]}{[B^-]} \quad (17\text{-}13)$$

$$K_w = K_a K_b \quad (17\text{-}14)$$

PROBLEMS

STRONG ACIDS AND BASES

17-1. Calculate $[H_3O^+]$ when $[OH^-]$ is 3.0×10^{-3} M. Is the solution acidic or basic?

17-2. Calculate $[OH^-]$ when $[H_3O^+]$ is 2.5×10^{-4} M. Is the solution acidic or basic?

17-3. Calculate $[H_3O^+]$, $[ClO_4^-]$, and $[OH^-]$ in an aqueous solution containing 0.050 mol·L^{-1} of $HClO_4(aq)$. Is the solution acidic or basic?

17-4. Calculate $[OH^-]$, $[K^+]$, and $[H_3O^+]$ in an aqueous solution containing 0.25 M KOH(*aq*). Is the solution acidic or basic?

17-5. Thallium(I) hydroxide is a strong base. Compute $[Tl^+]$, $[OH^-]$, and $[H_3O^+]$ for a solution that is prepared by dissolving 1.00 g of TlOH in enough water to make 500.0 mL of solution.

17-6. Calcium hydroxide is a strong base. Compute $[Ca^{2+}]$, $[OH^-]$, and $[H_3O^+]$ for a solution that is prepared by dissolving 0.60 g of $Ca(OH)_2$ in enough water to make 1500.0 mL of solution.

pH CALCULATIONS

17-7. Calculate the pH of an aqueous solution containing 0.050 M $HNO_3(aq)$. Is the solution acidic or basic?

17-8. Calculate the pH of an aqueous solution that is 0.10 M in CsOH(*aq*). Is the solution acidic or basic?

17-9. Calculate the pH of an aqueous solution that is 0.025 M in HCl(*aq*) and 0.025 M in HBr(*aq*). Is the solution acidic or basic?

17-10. Calculate the pH of an aqueous solution containing 0.0010 M $Ba(OH)_2(aq)$. Is the solution acidic or basic?

17-11. Calculate the pH of an aqueous solution prepared by dissolving 1.00 g of KOH pellets in water and diluting to a final volume of 0.500 L.

17-12. A solution of NaOH(*aq*) contains 6.25 g of NaOH per 100 mL of solution. Estimate the pH of the solution at 25°C.

17-13. The value of the ion product constant for water, K_w, at 0°C is $0.12 \times 10^{-14}\ M^2$. Compute the pH of a neutral aqueous solution at 0°C. Is an aqueous solution with a pH = 7.25 acidic or basic at 0°C?

$K_w = [H_3O^+][OH^-]$ $\quad \sqrt{0.12 \times 10^{-14}} = 3.4 \times 10^{-8}$ M
$[H_3O^+] = [OH^-]$ $\quad$ pH = 7.46 $\quad$ 7.25 - acidic

17-14. Given that K_w for water is $2.40 \times 10^{-14}\ M^2$ at 37°C, compute the pH of a neutral aqueous solution at 37°C, which is the normal human body temperature. Is a pH = 7.00 solution acidic or basic at 37°C?

CALCULATION OF $[H_3O^+]$ AND $[OH^-]$ FROM pH

17-15. The pH of human muscle fluids is 6.8. Compute the value of $[H_3O^+]$ in muscle fluid at 25°C.

pH = $-\log[H_3O^+]$
$[H_3O^+] = 1.6 \times 10^{-7}$ M

17-16. The pH of household ammonia is about 12. Calculate the value of $[OH^-]$ for the ammonia solution. Write the reaction for ammonia dissolved in water.

$NH_3(aq) + H_2O(l) \rightleftharpoons NH_4^+(aq) + OH^-(aq)$
pH = $-\log[H_3O^+]$
$[H_3O^+] = 1.0 \times 10^{-12}$ M $\quad \frac{1.0 \times 10^{-14}}{1.0 \times 10^{-12}} = 0.01$ M $[OH^-]$

17-17. The pH of the contents of the human stomach can be as low as 1.0. Calculate the value of $[H_3O^+]$ in the stomach at 25°C. The acid secreted into the stomach is HCl; calculate the concentration of an aqueous HCl solution that corresponds to a pH of 1.0.

$[H_3O^+] = 0.1$ M $\quad$ HCl strong acid $\quad [HCl] = [H_3O^+]$ $\quad [HCl] = 0.1$ M

17-18. Normal rainwater has a pH of about 5.6, whereas what is called acid rain has been observed to have pH values as low as 3.0. Compute the ratio of $[H_3O^+]$ in pH 3.0 acid rain to that in normal rain. What is the cause of the acidity of normal rain?

17-19. The pH of human blood is fairly constant at 7.4. Compute the hydronium ion concentration and the hydroxide ion concentration in human blood at 25°C.

pH = $-\log[H_3O^+]$ $\quad \frac{1.0 \times 10^{-14}}{3.9 \times 10^{-8}} = 2.6 \times 10^{-7}$
$[H_3O^+] = 3.9 \times 10^{-8}$ M $\quad [OH^-] = 2.6 \times 10^{-7}$ M

17-20. The pH of the world's oceans is remarkably constant at 8.15. Compute the hydronium ion and hydroxide ion concentrations in the ocean. Assume a temperature of 25°C.

17-21. A saturated solution of $Sr(OH)_2(aq)$ at 25°C has a measured pH of 13.5. Estimate the solubility of $Sr(OH)_2(s)$ in water at 25°C.

$Sr(OH)_2 \rightarrow Sr^{2+} + 2OH^-$
pH = $-\log[H_3O^+]$ $\quad \frac{1.0 \times 10^{-14}\ M^2}{3.16 \times 10^{-14}\ M} = 0.316$ M $[OH^-]$
$[H_3O^+] = 3.16 \times 10^{-14}$ M $\quad \frac{0.316\ M}{2} = 0.16$ M $Sr(OH)_2$
(0.16 mol/L)(0.1 L) = 1.6×10^{-2} mol $\quad 1.6 \times 10^{-2}$ mol × 121.64 g = 1.9 g

17-22. A saturated solution of $Mg(OH)_2(aq)$ at 25°C has a pH of 10.52. Estimate the solubility of $Mg(OH)_2(s)$ in water at 25°C.

$Mg(OH)_2 \rightarrow Mg^{2+} + 2(OH^-)$
$[H_3O^+] = 3.0 \times 10^{-11}$ M $\quad \frac{3.3 \times 10^{-4}\ M\ [OH^-]}{2} = 1.7 \times 10^{-4}$ M $Mg(OH)_2$
$\frac{1.0 \times 10^{-14}\ M^2}{3.0 \times 10^{-11}\ M} = 3.3 \times 10^{-4}$ M $[OH^-]$ $\quad (1.7 \times 10^{-4}$ M)(0.10 L) = 1.7×10^{-5} mol
1.7×10^{-5} mol × 58.33 g/mol = 9.9×10^{-4} g per 100 ml solution

CALCULATION OF K_a FROM pH

17-23. The pH of a 0.050 M aqueous solution of propionic acid, a component of milk, is found to be 3.09. Calculate K_a, the acid dissociation equilibrium constant, for propionic acid, $HC_3H_5O_2$.

$HC_3H_5O_2(aq) + H_2O(l) \rightleftharpoons H_3O^+(aq) + C_3H_5O_2^-(aq)$
$K_a = \frac{[H_3O^+][C_3H_5O_2^-]}{[HC_3H_5O_2]}$ $\quad [H_3O^+] = 8.13 \times 10^{-4}$ M $\quad [C_3H_5O_2^-] = 8.13 \times 10^{-4}$ M
$[HC_3H_5O_2] = 0.050\ M - 8.13 \times 10^{-4}$ M $\quad [HC_3H_5O_2] = 0.050$ M $\quad K_a = 1.3 \times 10^{-5}$ M

17-24. The pH of a 0.20 M aqueous solution of cyanic acid, HCNO, is found to be 2.19. Calculate K_a, the acid dissociation constant, for cyanic acid.

17-25. We are given that the measured pH of a 1.00×10^{-2} M solution of $HC_2H_3O_2(aq)$ is 3.39 at 25°C. Compute the value of K_a for $HC_2H_3O_2(aq)$ at 25°C.

$[H_3O^+] = 4.07 \times 10^{-4}$ M $\quad [HC_2H_3O_2] = 1.00 \times 10^{-2}\ M - 4.07 \times 10^{-4}$ M
$[C_2H_3O_2^-] = 4.07 \times 10^{-4}$ M $\quad [HC_2H_3O_2] = 0.96 \times 10^{-2}$ M
$K_a = \frac{(4.07 \times 10^{-4})^2}{0.96 \times 10^{-2}}$ $\quad K_a = 1.7 \times 10^{-5}$ M

17-26. The pH of a 0.10 M aqueous solution of formic acid, $HCHO_2$, is 2.38. Calculate the value of K_a for formic acid.

$HCHO_2 + H_2O \rightleftharpoons H_3O^+ + CHO_2^-$ $\quad K_a = \frac{[H_3O^+][CHO_2^-]}{[HCHO_2]}$
$[H_3O^+] = 4.17 \times 10^{-3}$ M $\quad [CHO_2^-] = 4.17 \times 10^{-3}$ M $\quad [HCHO_2] = 9.58 \times 10^{-2}$ M
$K_a = \frac{(4.17 \times 10^{-3})^2}{9.58 \times 10^{-2}}$ $\quad K_a = 1.8 \times 10^{-4}$ M

17-27. The highly toxic compound 2,4-dinitrophenol is used in biological research to inhibit energy production in cells. In an experiment, a solution of 2,4-dinitrophenol was prepared with the pH adjusted to 7.4. Estimate the ratio of the concentrations of the dissociated ion to the undissociated acid:

2,4 dinitrophenol

(OH, NO_2, NO_2 ring) $(aq) + H_2O(l) \rightleftharpoons$ (O^-, NO_2, NO_2 ring) $(aq) + H_3O^+(aq)$ $\quad K_a = 1.1 \times 10^{-4}$ M

undissociated acid — dissociated ion

pH = $-\log[H_3O^+]$
$[H_3O^+] = 4.0 \times 10^{-8}$ M
$K_a = \frac{(\text{dissociated ion})[H_3O^+]}{(2,4\ \text{dinitrophenol})}$
$\frac{(\text{dissociated ion})}{(2,4\ \text{dinitrophenol})} = \frac{1.1 \times 10^{-4}\ M}{4.0 \times 10^{-8}\ M}$ $\quad$ ratio = 2.8×10^3

17-28. The acid dissociation constant of chloroacetic acid, $HC_2H_2ClO_2$, is $K_a = 1.35 \times 10^{-3}$ M at 25°C. If the pH of an aqueous solution of chloroacetic acid is 4.61, then calculate the ratio $[C_2H_2ClO_2^-]/[HC_2H_2ClO_2]$ in the solution.

$HC_2H_2ClO_2 + H_2O \rightleftharpoons H_3O^+ + C_2H_2ClO_2^-$
$K_a = \frac{[H_3O^+][C_2H_2ClO_2^-]}{[HC_2H_2ClO_2]}$
pH = $-\log[H_3O^+]$
$[H_3O^+] = 2.45 \times 10^{-5}$ M
ratio = $\frac{K_a}{[H_3O^+]}$
ratio = 55.1

17-29. In its undissociated acid form, acetylsalicylic acid, aspirin, has a deleterious effect on the stomach. Only the undissociated acid can cross the stomach lining. Estimate the percentage of acid dissociated when the pH of the stomach contents is 2.0. Take $K_a = 2.75 \times 10^{-5}$ M for acetylsalicylic acid at 37°C.

$[H_3O^+] = 1.0 \times 10^{-2}$ $\quad 2.75 \times 10^{-5} = \frac{1.0 \times 10^{-2} \times \text{diss.}}{\text{undiss.}}$ ratio $= 2.75 \times 10^{-3}$ percent = 0.28%

17-30. Uric acid is an end product of the metabolism of certain biological substances and is excreted from the body in urine. The acid dissociation constant of uric acid is $K_a = 4.0 \times 10^{-6}$ M. The pH of a urine sample is 6.0. Estimate the ratio of urate ion to uric acid in the urine.

17-31. Nitrites, such as $NaNO_2$, are added to processed meats and hamburger both as a preservative and to give the meat a redder color by binding to hemoglobin in the red blood cells. When nitrite ion is ingested, it reacts with stomach acid to form nitrous acid, $HNO_2(aq)$, in the stomach. Given that $K_a = 4.47 \times 10^{-4}$ M for nitrous acid, compute the value of the ratio $[HNO_2]/[NO_2^-]$ in the stomach following ingestion of NO_2^- when $[H_3O^+]$ is 0.10 M.

$K_a = \frac{[H_3O^+][NO_2^-]}{[HNO_2]}$ $\quad \frac{[HNO_2]}{[NO_2^-]} = \frac{[H_3O^+]}{K_a}$ ratio = 220

17-32. Sodium benzoate, $NaC_7H_5O_2$, is used as a food preservative because of its antimicrobial action. The K_a of benzoic acid is 6.46×10^{-5} M. Estimate the ratio of the concentration of benzoic acid to the concentration of benzoate in a food with a pH of 3.0.

CALCULATION OF pH FROM K_a

17-33. The value of K_a in water at 25°C for benzoic acid, $HC_7H_5O_2$, is 6.46×10^{-5} M. Calculate the pH of an aqueous solution with a total concentration of $HC_7H_5O_2$ of 0.025 M.

$HC_7H_5O_2 + H_2O \rightleftharpoons C_7H_5O_2^- + H_3O^+$; 0.025 M; $0.025 - [H_3O^+]$, $[H_3O^+]$, $[H_3O^+]$

$6.46 \times 10^{-5} = \frac{[H_3O^+]^2}{0.025 - [H_3O^+]}$ $\quad [H_3O^+] = 1.24 \times 10^{-3}$ M pH = 2.91

17-34. The value of K_a in water at 25°C for hypochlorous acid, HClO, is 2.95×10^{-8} M. Calculate the pH of an aqueous solution with a total concentration of HClO of 0.15 M.

17-35. The value of K_a in water at 25°C for trichloroacetic acid, $HC_2Cl_3O_2$, is 2.3×10^{-1} M. Calculate the pH of an aqueous solution with a total concentration of $HC_2Cl_3O_2$ of 0.010 M.

$HC_2Cl_3O_2 + H_2O \rightleftharpoons C_2Cl_3O_2^- + H_3O^+$; 0.010 M; $0.010\text{ M} - [H_3O^+]$, $[H_3O^+]$, $[H_3O^+]$

$2.3 \times 10^{-1} = \frac{[H_3O^+]}{0.010\text{ M} - [H_3O^+]}$ $\quad [H_3O^+] = 1.0 \times 10^{-2}$ M pH = 2.0

17-36. Calculate the pH of a 0.10 M aqueous solution of chloroacetic acid given that $K_a = 1.35 \times 10^{-3}$ M at 25°C.

17-37. A 6.15-g sample of benzoic acid, $HC_7H_5O_2$, is dissolved in enough water to make 600 mL of solution. Calculate the pH of the solution. The value of K_a for benzoic acid at 25°C is 6.46×10^{-5} M.

$\frac{6.15\text{ g}}{122.12\text{ g/mole}} = 0.05036$ mol

$\frac{0.05036\text{ mol}}{0.6\text{ L}} = 0.0839$ M

$HC_7H_5O_2 + H_2O \rightleftharpoons C_7H_5O_2^- + H_3O^+$

0.0839; 0; 0

$0.0839 - [H_3O^+]$; $[H_3O^+]$; $[H_3O^+]$

$6.46 \times 10^{-5} = \frac{[H_3O^+]^2}{0.0839 - [H_3O^+]}$

– quadratic formula

$[H_3O^+] = 2.30 \times 10^{-3}$ M

pH = 2.64

17-38. Suppose two 5-grain (5 grains = 324 mg) aspirin tablets are dissolved in enough water to make 500 mL of solution at 25°C. Compute the pH of the resulting solution.

H ← acidic proton

aspirin $\quad K_a = 2.75 \times 10^{-5}$ M at 25°C

17-39. Sulfamic acid, HO_3SNH_2, is used as a stabilizer for chlorine in swimming pools. Calculate the pH of a 0.050 M sulfamic acid solution given that $K_a = 0.10$ M at 25°C.

$HO_3SNH_2 + H_2O \rightleftharpoons O_3SNH_2^- + H_3O^+$

0.050 M; 0; 0

$0.050\text{ M} - [H_3O^+]$; $[H_3O^+]$; $[H_3O^+]$

$0.10\text{ M} = \frac{[H_3O^+]^2}{0.050\text{ M} - [H_3O^+]}$

$[H_3O^+] = 0.04$ M

pH = 1.4

17-40. The compound sodium bisulfate, $NaHSO_4$, is used in cleaning metals and in leather treatment. Given that the K_a for $HSO_4^-(aq)$ is 0.012 M at 25°C, compute the pH of a $NaHSO_4(aq)$ solution prepared by dissolving 10.0 g of $NaHSO_4$ in sufficient water to make 100 mL of solution.

CALCULATIONS INVOLVING K_b

17-41. The measured pH of a 0.100 M solution of $NH_3(aq)$ at 25°C is 11.12. Compute K_b for $NH_3(aq)$ at 25°C.

17-42. The pH of a 0.50 M solution of the weak base ethyl amine, $C_2H_5NH_2$, is 12.20. Determine K_b for ethyl amine.

17-43. The organic solvent pyridine, C_5H_5N, has a strong, nauseating odor. Calculate the pH of a 0.600 M aqueous solution of pyridine. The value of K_b is 1.48×10^{-9} M at 25°C for the reaction

$$C_5H_5N(aq) + H_2O(l) \rightarrow C_5H_5NH^+(aq) + OH^-(aq)$$

17-44. The base hydroxylamine, NH_2OH, is used to synthesize a variety of organic compounds. Calculate the pH of a 0.125 M aqueous solution of NH_2OH. The value of K_b is 1.07×10^{-8} M at 25°C for the reaction

$$HONH_2(aq) + H_2O(l) \rightarrow HONH_3^+(aq) + OH^-(aq)$$

17-45. Calculate the pH of a 0.050 M aqueous solution of dimethyl amine, $(CH_3)_2NH$. The value of K_b is 5.81×10^{-4} M for the reaction

$$(CH_3)_2NH(aq) + H_2O(l) \rightarrow (CH_3)_2NH_2^+(aq) + OH^-(aq)$$

17-46. Compute the pH of a household ammonia cleaning solution prepared by dissolving $NH_3(g)$ in water to yield a 0.20 M $NH_3(aq)$ solution. Take $K_b = 1.8 \times 10^{-5}$ M for $NH_3(aq)$.

LE CHÂTELIER'S PRINCIPLE

17-47. Use Le Châtelier's principle to predict what happens to the equilibrium of the reaction

$$CO_2(g) + 2H_2O(l) \rightleftharpoons H_3O^+(aq) + HCO_3^-(aq)$$

if

(a) $[CO_2]$ is increased
(b) $[HCO_3^-]$ is increased
(c) $[H_3O^+]$ is increased
(d) the solution is diluted

17-48. Use Le Châtelier's principle to predict in which direction the equilibrium shifts if we

(a) add HCl to a 0.10 M NH_3 solution
(b) add NaOH to a 0.10 M NH_3 solution
(c) add HCl to a 0.10 M $HCHO_2$ solution
(d) add NaOH to a 0.10 M $HCHO_2$ solution

17-49. Use Le Châtelier's principle to predict the direction in which the following acid dissociation reaction shifts in response to the indicated change in conditions:

$$HCHO_2(aq) + H_2O(l) \rightleftharpoons H_3O^+(aq) + CHO_2^-(aq)$$

(a) addition of NaOH
(b) addition of $NaCHO_2$

17-50. Use Le Châtelier's principle to predict the direction in which the following acid dissociation reaction shifts in response to the indicated change in conditions:

$$HC_3H_5O_2(aq) + H_2O(l) \rightleftharpoons H_3O^+(aq) + C_3H_5O_2^-(aq)$$

(a) dilution of a 0.1 M solution to 0.01 M
(b) addition of HCl

17-51. Use Le Châtelier's principle to predict the direction in which the following equilibrium shifts in response to the indicated change in conditions:

$$HC_7H_5O_2(aq) + H_2O(l) \rightleftharpoons H_3O^+(aq) + C_7H_5O_2^-(aq) \qquad \Delta H^\circ_{rxn} \approx 0$$

(a) evaporation of water from the solution at a fixed temperature
(b) decrease in the temperature of the solution
(c) addition of $KC_7H_5O_2(s)$
(d) addition of $NH_3(g)$
(e) addition of $HCl(g)$

17-52. Use Le Châtelier's principle to predict the direction in which the following equilibrium shifts in response to the indicated change in conditions:

$$HNO_2(aq) + H_2O(l) \rightleftharpoons H_3O^+(aq) + NO_2^-(aq) \qquad \Delta H^\circ_{rxn} < 0$$

(a) increase in the temperature of the solution
(b) dissolution of $NaNO_2(s)$
(c) dissolution of $NaOH(s)$
(d) removal of $NO_2^-(aq)$ as $AgNO_2(s)$ by addition of $AgNO_3(s)$

CONJUGATE ACIDS AND BASES

17-53. Indicate the conjugate acid-base pairs in the following reactions:

(a) $HC_7H_5O_2(aq) + H_2O(l) \rightleftharpoons H_3O^+(aq) + C_7H_5O_2^-(aq)$
benzoic acid — benzoate ion

(b) $CH_3NH_2(aq) + H_2O(l) \rightleftharpoons CH_3NH_3^+(aq) + OH^-(aq)$
methyl amine — methyl ammonium ion

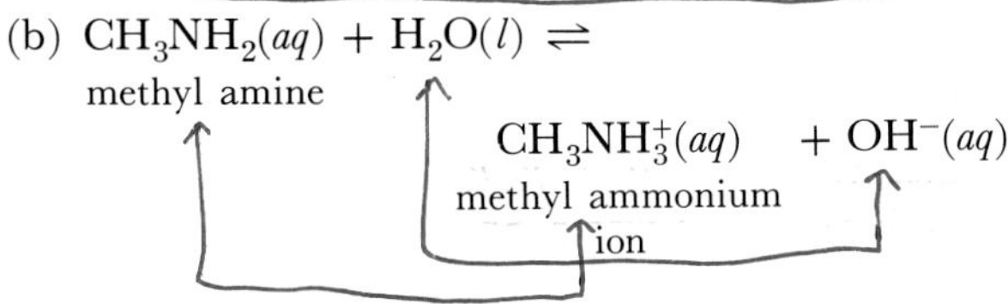

(c) $HCHO_2(aq) + H_2O(l) \rightleftharpoons H_3O^+(aq) + CHO_2^-(aq)$
formic acid — formate ion

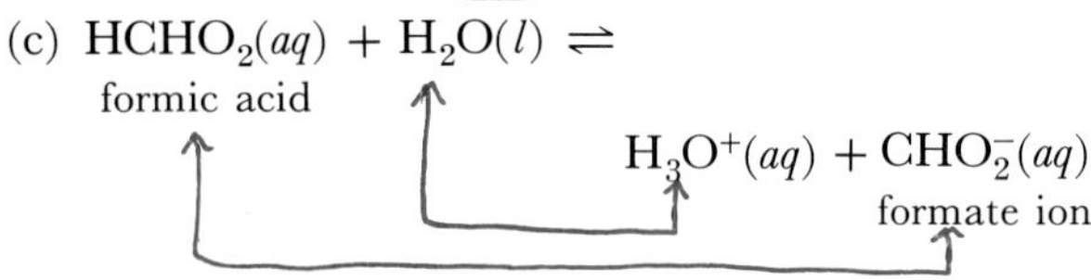

17-54. Indicate the conjugate acid-base pairs in the following reactions [(*am*) denotes the solvent NH_3]:

(a) $NH_3(l) + NH_3(l) \rightleftharpoons NH_4^+(am) + NH_2^-(am)$

(b) $HNO_2(aq) + H_2O(l) \rightleftharpoons H_3O^+(aq) + NO_2^-(aq)$

(c) $C_5H_5N(aq) + H_2O(l) \rightleftharpoons C_5H_5NH^+(aq) + OH^-(aq)$
pyridine — pyridinium ion

17-55. Give the conjugate base for each of the following acids:

(a) $HNO_2(aq)$ NO_2^-
(b) $HC_2H_2ClO_2(aq)$ $C_2H_2ClO_2^-$
(c) $H_3PO_4(aq)$ $H_2PO_4^-$
(d) $H_2PO_4^-(aq)$ HPO_4^{2-}

17-56. Give the conjugate base for each of the following acids:

(a) $HNO_3(aq)$
(b) $HCHO_2(aq)$
(c) $HC_6H_5O(aq)$
(d) $CH_3NH_3^+(aq)$

17-57. Identify which of the following species are Brønsted-Lowry acids and which are Brønsted-Lowry bases in water. In each case give the chemical formula for the conjugate member of the conjugate acid-base pair:

(a) $HCHO_2(aq)$ ← acid CHO_2^-
(b) $IO^-(aq)$ base HIO
(c) $F^-(aq)$ base HF
(d) $CH_3NH_3^+(aq)$ ← acid CH_3NH_2
(e) $ClNH_2(aq)$ base $ClNH_3^+$
(f) $CN^-(aq)$ base HCN

17-58. Identify which of the following species are Brønsted-Lowry acids and which are Brønsted-Lowry bases in water. In each case give the chemical formula for the conjugate member of the conjugate acid-base pair:

(a) $HC_2H_2ClO_2(aq)$
(b) $NH_3(aq)$
(c) $ClO^-(aq)$
(d) $CHO_2^-(aq)$
(e) $HN_3(aq)$
(f) $NO_2^-(aq)$

17-59. Given the following acids and their dissociation constants, calculate K_b for the conjugate bases:

Acid	K_a/M
(a) $HC_3H_5O_2$, propionic acid	1.34×10^{-5}
(b) HF	6.76×10^{-4}
(c) NH_4^+	5.6×10^{-10}
(d) H_3BO_3, boric acid	5.9×10^{-10}

17-60. Given the following bases and their values of K_b, calculate K_a for the conjugate acids:

Base	K_b/M
(a) C_5H_5N	1.5×10^{-9}
(b) CN^-	2.1×10^{-5}
(c) CNO^-	4.6×10^{-11}
(d) HS^-	1.1×10^{-7}

$K_b = \frac{K_w}{K_a}$

a. $K_b = \frac{1.0 \times 10^{-14}}{1.34 \times 10^{-5}} = 7.46 \times 10^{-10}$ M for $C_3H_5O_2^-$
b. 1.48×10^{-11} M for F^-
c. 1.8×10^{-5} M for NH_3
d. 1.7×10^{-5} M for $H_2BO_3^-$

a, $HCHO_2 + H_2O \rightleftharpoons H_3O^+ + CHO_2^-$ $K_a = 1.8 \times 10^{-4} = K_1$

$NO_2^- + H_2O \rightleftharpoons HNO_2 + OH^-$ $K_a = \frac{1}{4.5 \times 10^{-4}} = K_2$

$K = K_1 K_2 = \frac{1.8 \times 10^{-4}}{4.5 \times 10^{-4}} = \boxed{0.40}$

17-61. Use the data in Table 17-4 to compute the equilibrium constants for the following reactions at 25°C:

(a) $HCHO_2(aq) + NO_2^-(aq) \rightleftharpoons HNO_2(aq) + CHO_2^-(aq)$

(b) $NH_4^+(aq) + CN^-(aq) \rightleftharpoons NH_3(aq) + HCN(aq)$

$CN^- + H_2O \rightleftharpoons HCN + OH^-$ $K_b = 2.1 \times 10^{-5} = K_1$

$NH_4^+ + OH^- \rightleftharpoons NH_3 + H_2O$ $K_b = \frac{1}{1.8 \times 10^{-5}} = K_2$

$K = K_1 K_2 = \boxed{1.2}$

17-62. Use the data in Table 17-4 to compute the equilibrium constants for the following reactions at 25°C:

(a) $NH_4^+(aq) + C_2H_3O_2^-(aq) \rightleftharpoons NH_3(aq) + HC_2H_3O_2(aq)$

(b) $C_6H_5O^-(aq) + C_5H_5NH^+(aq) \rightleftharpoons HC_6H_5O(aq) + C_5H_5N(aq)$

17-63. Using the K_b data in Table 17-4, compute the pK_b values for

$pK_b = -\log K_b$

(a) $SO_3^{2-}(aq)$ $-\log(1.6 \times 10^{-7}) = \boxed{6.80}$

(b) $HSO_3^-(aq)$ $-\log(6.7 \times 10^{-13}) = \boxed{12.17}$

17-64. Using the K_a data in Table 17-4, compute the pK_a values for

(a) $NH_4^+(aq)$

(b) $HSO_4^-(aq)$

17-65. Given that pK_a for an acid is 5.00, compute K_b for the conjugate base.

$K_a = 1.0 \times 10^{-5}\,M$ $K_b = \frac{K_w}{K_a}$

$K_b = \frac{1.0 \times 10^{-14}}{1.0 \times 10^{-5}} = \boxed{1.0 \times 10^{-9}\,M}$

17-66. We know that $K_a K_b = K_w$ for a conjugate acid-base pair in water. Show that at 25°C

$$pK_a + pK_b = 14.00$$

for a conjugate acid-base pair.

ACID-BASE PROPERTIES OF SALTS

17-67. Predict whether the following salts, when dissolved in water, produce acidic, basic, or neutral solutions:

(a) $Fe(NO_3)_3$ acidic

(b) NH_4NO_2 cannot predict

(c) $NaHCO_3$ basic

(d) $CsClO_4$ neutral

(e) KNO_2 basic

17-68. Predict whether the following salts, when dissolved in water, produce acidic, basic, or neutral solutions:

(a) $CoBr_3$

(b) $NaNO_3$

(c) $KHSO_4$

(d) $NaNO_2$

(e) NH_4F

17-69. Predict whether the following salts, when dissolved in water, produce acidic, basic, or neutral solutions:

(a) Na_2CO_3 basic

(b) K_2S basic

(c) $SnCl_4$ acidic

(d) $KClO_4$ neutral

(e) $AlCl_3$ acidic

17-70. Predict whether the following salts, when dissolved in water, produce acidic, basic, or neutral solutions:

(a) $(NH_4)_2S$

(b) KCN

(c) $Pb(NO_3)_3$

(d) $NaHSO_4$

(e) $CaCl_2$

17-71. Soap was formerly produced by reacting caustic soda, NaOH, with animal fats, which contain a type of organic acid called a fatty acid. An example is stearic acid, $HC_{18}H_{35}O_2$. Fatty acids, like most organic acids, are weak acids. Is a soap solution acidic, basic, or neutral?

$NaOH + HC_{18}H_{35}O_2 \rightleftharpoons NaC_{18}H_{35}O_2 + H_2O$

soap is basic because of the basic anion.

17-72. Various aluminum salts, such as $Al_2(SO_4)_3$ and $KAl(SO_4)_2$, are used as additives to increase the acidity of soils for "acid-loving" plants such as azaleas and tomatoes. Explain how these salts increase soil acidity.

pH CALCULATIONS OF SALT SOLUTIONS

17-73. Sodium hypochlorite, NaClO, is a bleaching agent. Calculate the pH and the concentration of HClO in a solution that is 0.050 M in NaClO.

$ClO^- + H_2O \rightleftharpoons HClO + OH^-$

0.050M   0   0

0.050M − $[OH^-]$ $[OH^-]$ $[OH^-]$

$K_b = \frac{[OH^-]^2}{0.050 - [OH^-]} = 3.3 \times 10^{-7}\,M$ (from table)

− quadratic formula −

$[OH^-] = 1.3 \times 10^{-4}\,M$

$\boxed{[HClO] = 1.3 \times 10^{-4}\,M}$

$[H_3O^+] = \frac{1.0 \times 10^{-14}}{1.3 \times 10^{-4}} = 7.7 \times 10^{-11}\,M$

$pH = -\log[H_3O^+] = \boxed{10.11}$

17-74. Sodium propionate, $NaC_3H_7O_2$, is used as a food preservative. Calculate the pH of a 0.20 M solution of $NaC_3H_7O_2$, taking $K_a = 1.34 \times 10^{-5}\,M$ for propionic acid.

$CNO^- + H_2O \rightleftharpoons HCNO + OH^-$
0.10M
0.10M − [OH⁻]   [OH⁻]   [OH⁻] from tables
$K_b = \frac{[OH^-]^2}{0.10M - [OH^-]} = 4.57 \times 10^{-11}$
[OH⁻] = 2.14 × 10⁻⁶ M
[HCNO] = 2.14 × 10⁻⁶ M
[CNO⁻] = 0.10M − 2.14 × 10⁻⁶ M = 0.10M

17-75. A solution of sodium cyanate, NaCNO, is prepared at a concentration of 0.10 M. Calculate the equilibrium concentrations of $OH^-(aq)$, $HCNO(aq)$, $CNO^-(aq)$, and $H_3O^+(aq)$ and the pH of solution.

$[H_3O^+] = \frac{1.00 \times 10^{-14}}{2.14 \times 10^{-6}} = 4.67 \times 10^{-9}$ M   pH = −log[H₃O⁺] = 8.33

17-76. Calculate the equilibrium concentrations of $OH^-(aq)$, $HNO_2(aq)$, $NO_2^-(aq)$, and $H_3O^+(aq)$ and the pH of a solution that is 0.25 M in $NaNO_2$.

17-77. Calculate the pH of a 0.30 M aqueous solution of pyridinium chloride, C_5H_6NCl. $C_5H_5NH^+ + H_2O \rightleftharpoons C_5H_5N + H_3O^+$

$6.8 \times 10^{-6} M = \frac{[H_3O^+]^2}{0.30M - [H_3O^+]}$ [H₃O⁺] = 1.4 × 10⁻³ M   pH = 2.85

17-78. Calculate the pH of a 0.15 M aqueous solution of potassium cyanide, KCN.

17-79. A saturated aqueous solution of ammonium perchlorate contains 23.7 g of NH_4ClO_4 per 100 mL of solution at 25°C. Estimate the pH of this solution at 25°C.

$NH_4^+ + H_2O \rightleftharpoons H_3O^+ + NH_3$
2.02M
2.02M − [H₃O⁺]   [H₃O⁺]   [H₃O⁺]
n = 23.7g / 117.49 g/mole = 0.202 mol
0.202 mol / 0.100 L = 2.02M
$5.6 \times 10^{-10} = \frac{[H_3O^+]^2}{2.02M - [H_3O^+]}$ [H₃O⁺] = 3.4 × 10⁻⁵ M   pH = 4.47

17-80. Calculate the pH of a solution that is prepared by dissolving 25.0 g of barium acetate in enough water to make exactly one L of solution.

17-81. The acid dissociation constant at 25°C for the reaction

$$Fe(H_2O)_6^{3+}(aq) + H_2O(l) \rightleftharpoons H_3O^+(aq) + Fe(OH)(H_2O)_5^{2+}(aq)$$

0.10M   0.10M − [H₃O⁺]   [H₃O⁺]   [H₃O⁺]

is $K_a = 1.0 \times 10^{-3}$ M. Calculate the pH of a 0.10 M solution of $Fe(NO_3)_3(aq)$ at 25°C.

$1.0 \times 10^{-3} M = \frac{[H_3O^+]^2}{0.10M - [H_3O^+]}$ [H₃O⁺] = 9.5 × 10⁻⁵ M   pH = 2.02

17-82. Calculate the pH at 25°C of a solution that is 0.10 M in $TlBr_3$. The acid dissociation constant at 25°C for the reaction

$$Tl(H_2O)_6^{3+}(aq) + H_2O(l) \rightleftharpoons H_3O^+(aq) + Tl(OH)(H_2O)_5^{2+}(aq)$$

is 6×10^{-2} M.

LEWIS ACIDS AND BASES

17-83. Determine whether each of the following substances is an Arrhenius acid, a Brønsted-Lowry acid, or a Lewis acid (it is possible for each to be of more than one type):

(a) HCl Arrhenius acid, Brönsted-Lowry (b) $AlCl_3$ Lewis, Arrhenius acid, Brönsted-Lowry (c) BF_3 Lewis acid

17-84. Determine whether each of the following substances is an Arrhenius base, a Brønsted-Lowry base, or a Lewis base (it is possible for each to be of more than one type):

(a) NH_3 (b) Br^- (c) NaOH

17-85. Classify each of the following species as either a Lewis acid or a Lewis base:

(a) CH_3OCH_3 H–C(H)(H)–Ö–C(H)(H)–H Lewis base (b) $GaCl_3$ Lewis acid (c) H_2O Lewis base

17-86. Classify each of the following species as either a Lewis acid or a Lewis base:

(a) $CH_3-\ddot{N}(CH_3)-CH_3$ (b) BCl_3 (c) BeF_2

INTERCHAPTER J

Phosphorus

Phosphorus, in the form of P_4S_3, is one of the principal components of matches. The reaction that occurs when matches burn is discussed in Section J-5.

Although phosphorus (atomic number 15, atomic mass 30.97376) constitutes less than 0.1 percent by mass of the earth's crust, all living organisms contain this element. It is the sixth most abundant element in the human body. Plants require phosphorus as a nutrient, and most of the phosphorus and phosphorus compounds that are produced are used as fertilizers.

J-1. THERE ARE TWO PRINCIPAL ALLOTROPES OF SOLID PHOSPHORUS

Recall that allotropes are different forms of an element with different arrangements of the atoms.

There are several forms of elemental solid phosphorus, the most important of which are *white phosphorus* and *red phosphorus*. White phosphorus is a white, transparent, waxy crystalline solid (Figure J-1) that often appears pale yellow because of impurities. It is insoluble in water and alcohol but soluble in carbon disulfide. A characteristic property of white phosphorus is its high chemical reactivity. It ignites spontaneously in air at about 25°C. White phosphorus should

Figure J-1 White phosphorus, one of the principal allotropes of solid phosphorus, is very reactive and must be handled with care because it produces severe burns when it comes in contact with skin. Red phosphorus, on the other hand, is much less reactive than white phosphorus and does not require special handling.

never be allowed to come into contact with the skin because body temperature (37°C) is sufficient to ignite it spontaneously. Phosphorus burns are extremely painful and slow to heal. In addition, white phosphorus is very poisonous. The ingestion of even small amounts may produce severe gastrointestinal irritation, liver damage, convulsions, and death. The lethal dose is 50 to 100 mg. White phosphorus should always be kept under water and handled with forceps.

When white phosphorus is heated above 250°C in the absence of air, a form called red phosphorus is produced. Red phosphorus is a red to violet powder that is less reactive than white phosphorus. The chemical reactions that the red form undergoes are the same as those of the white form, but they generally occur only at higher temperatures. For example, red phosphorus must be heated to 260°C before it burns in air. The toxicity of red phosphorus is much lower than that of white phosphorus.

White phosphorus consists of tetrahedral P_4 molecules (Figure J-2), whereas red phosphorus consists of large, random aggregates of phosphorus atoms. The structure of red phosphorus is called *amorphous,* which means that it has no definite shape. Butter is an example of an amorphous substance.

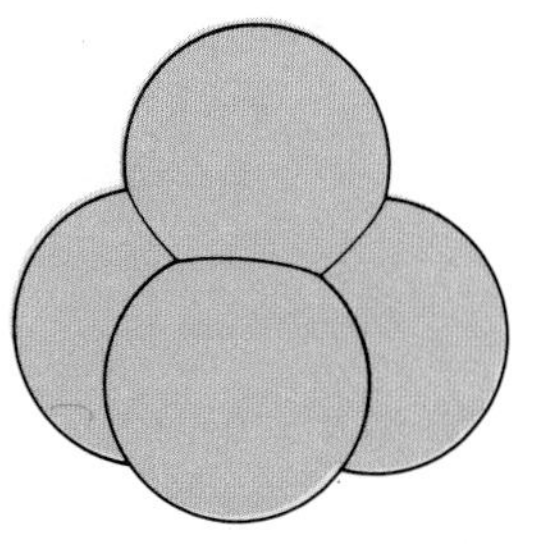

Figure J-2 White phosphorus consists of tetrahedral P_4 molecules.

Most of the phosphorus that is produced is used to make phosphoric acid or other phosphorus compounds. Elemental phosphorus, however, is used in the manufacture of pyrotechnics, matches, rat poisons, incendiary shells, smoke bombs, and tracer bullets.

Figure J-3 Hydroxyapatite is a common mineral that is found in many areas. It is used in the manufacture of phosphoric acid and phosphate fertilizers.

Phosphorus is not found as the free element in nature. The principal sources are the minerals calcium phosphate, hydroxyapatite, (Figure J-3), fluorapatite, and chlorapatite:

$Ca_3(PO_4)_2$	calcium phosphate
$Ca_{10}(OH)_2(PO_4)_6$	hydroxyapatite
$Ca_{10}F_2(PO_4)_6$	fluorapatite
$Ca_{10}Cl_2(PO_4)_6$	chlorapatite

These ores collectively are called phosphate rock. Large phosphate rock deposits occur in the USSR, in Morocco, and in Florida, Tennessee, and Idaho. An electric furnace is used to obtain phosphorus from phosphate rock. The furnace is charged with powdered phosphate rock, sand (SiO_2), and carbon in the form of coke. The source of heat is an electric current that produces temperatures of over 1000°C. A simplified version of the overall reaction that takes place is

$$\underset{\text{(phosphate rock)}}{2Ca_3(PO_4)_2(s)} + \underset{\text{(sand)}}{6SiO_2(s)} + \underset{\text{(coke)}}{10C(s)} \rightarrow 6CaSiO_3(l) + 10CO(g) + P_4(g)$$

The liquid calcium silicate, $CaSiO_3(l)$, called slag, is tapped off from the bottom of the furnace, and the phosphorus vapor produced solidifies to the white solid when the mixture of $CO(g)$ and $P_4(g)$ is passed through water (carbon monoxide does not dissolve in water).

J-2. PHOSPHORUS IS A KEY INGREDIENT OF FERTILIZERS

Although some phosphate rock is used to make elemental phosphorus, most phosphate rock is used in the production of fertilizers. Phosphorus is a required nutrient of all plants, and phosphorus compounds have long been used as fertilizer. In spite of its great abundance, phosphate rock cannot be used as a fertilizer because, as the name implies, it is insoluble in water. Consequently, plants are not able to assimilate the phosphorus from phosphate rock. To produce a water-soluble source of phosphorus, phosphate rock is reacted with sulfuric acid to produce a water-soluble product called *superphosphate,* $Ca(H_2PO_4)_2$, one of the world's most important fertilizers.

J-3. PHOSPHORUS FORMS TWO OXIDES

White phosphorus reacts directly with oxygen to produce the oxides P_4O_6 and P_4O_{10}. With excess phosphorus present, P_4O_6 is formed:

$$\underset{\text{excess}}{P_4(s)} + 3O_2(g) \rightarrow P_4O_6(s)$$

with excess oxygen present, P_4O_{10} is formed:

$$P_4(s) + \underset{\text{excess}}{5O_2(g)} \rightarrow P_4O_{10}(s)$$

The formulas for P_4O_6 and P_4O_{10} are often written P_2O_3 and P_2O_5, respectively. These obsolete (that is, now known to be incorrect) molecular formulas are the basis for the common names phosphorus *tri*oxide and phosphorus *pent*oxide.

It is interesting to compare the structures of P_4O_6 and P_4O_{10} (Figure J-4). The structure of P_4O_6 is obtained from that of P_4 by inserting an oxygen atom between each pair of adjacent phosphorus atoms; there are six edges on a tetrahedron, and thus a total of six oxygen atoms are required. The structure of P_4O_{10} is obtained from that of P_4O_6 by attaching an additional oxygen atom to each of the four phosphorus atoms.

J-4. PHOSPHORUS FORMS SEVERAL OXYACIDS

P_4O_6 produces a violent explosion when added to hot water.

The phosphorus oxides P_4O_6 and P_4O_{10} react with water to form the phosphorus oxyacids: phosphorous acid, H_3PO_3, and phosphoric acid, H_3PO_4:

$$P_4O_6(s) + 6H_2O(l) \rightarrow 4H_3PO_3(aq)$$
$$P_4O_{10}(s) + 6H_2O(l) \rightarrow 4H_3PO_4(aq)$$

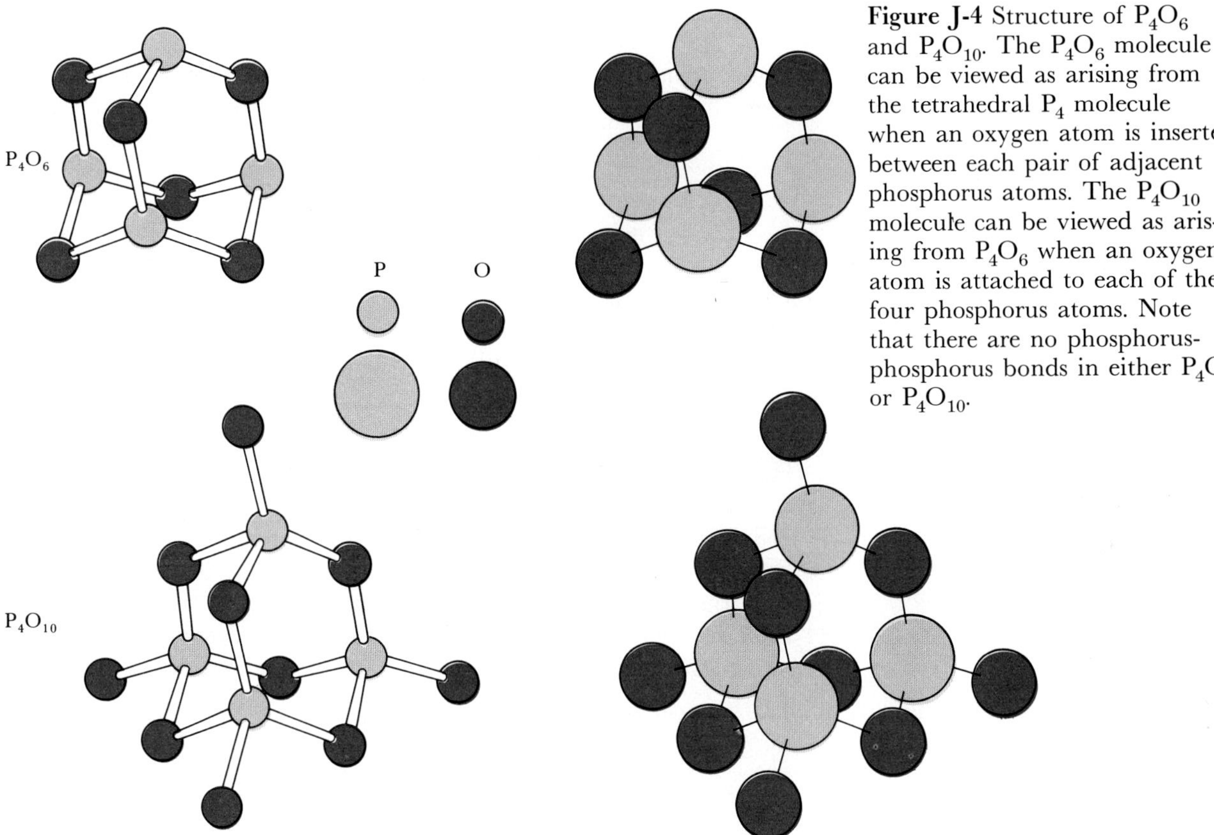

Figure J-4 Structure of P_4O_6 and P_4O_{10}. The P_4O_6 molecule can be viewed as arising from the tetrahedral P_4 molecule when an oxygen atom is inserted between each pair of adjacent phosphorus atoms. The P_4O_{10} molecule can be viewed as arising from P_4O_6 when an oxygen atom is attached to each of the four phosphorus atoms. Note that there are no phosphorus-phosphorus bonds in either P_4O_6 or P_4O_{10}.

Phosphorus pentoxide is a powerful dehydrating agent capable of removing water from concentrated sulfuric acid, which is itself a strong dehydrating agent. Thus phosphorus pentoxide is used as a drying agent in desiccators and dry boxes to remove water vapor.

Hypophosphorous acid, H_3PO_2, is prepared by reacting $P_4(g)$ with a warm aqueous solution of NaOH:

$$P_4(g) + 3OH^-(aq) + 3H_2O(l) \rightarrow 3H_2PO_2^-(aq) + PH_3(g)$$

$$H_2PO_2^-(aq) + H^+(aq) \rightarrow H_3PO_2(aq)$$

The structures of the phosphate anion, PO_4^{3-}, the phosphite ion, HPO_3^{2-}, and the hypophosphite ion, $H_2PO_2^-$, are

$$\left[\begin{matrix} O & & O \\ & P & \\ O & & O \end{matrix}\right]^{3-} \quad \left[\begin{matrix} H & & O \\ & P & \\ O & & O \end{matrix}\right]^{2-} \quad \left[\begin{matrix} H & & H \\ & P & \\ O & & O \end{matrix}\right]^{-}$$

Using VSEPR theory, we predict that these ions are tetrahedral.

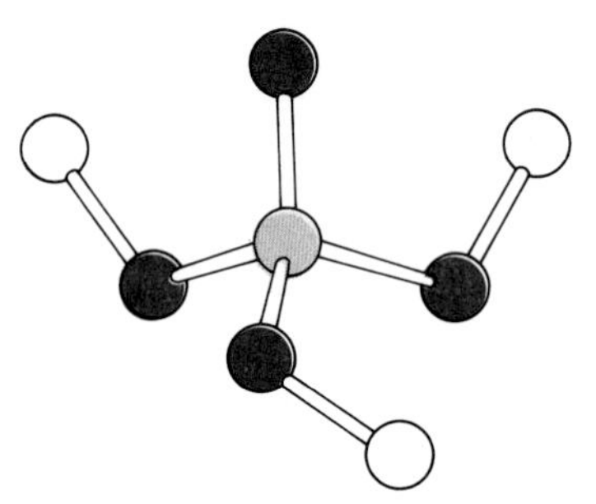

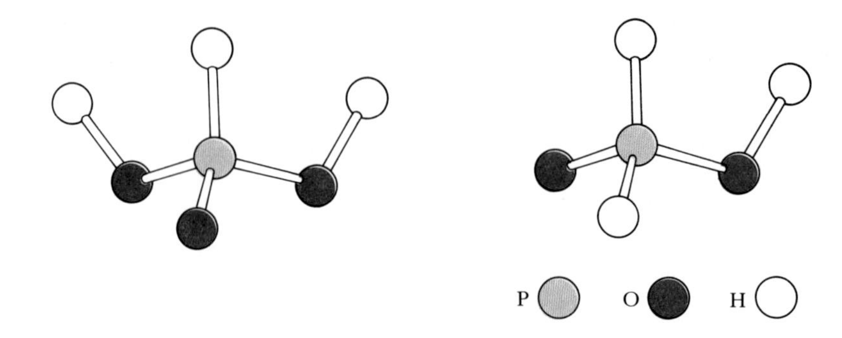

Figure J-5 The structure of phosphoric acid, phosphorous acid, and hypophosphorous acid. Note that all three hydrogen atoms of phosphoric acid are attached to oxygen atoms. One of the hydrogen atoms in phosphorous acid is attached directly to the phosphorus atom, and two of the hydrogen atoms in hypophosphorous acid are attached to the phosphorus atom. Only those hydrogen atoms attached to oxygen atoms are dissociable, and so phosphoric acid is triprotic, phosphorous acid is diprotic, and hypophosphorous acid is monoprotic.

The hydrogen atoms attached to phosphorus are not ionizable in aqueous solutions. Thus, phosphoric acid, H_3PO_4, is triprotic, phosphorous acid, $H_2(HPO_3)$, is diprotic, and hypophosphorous acid, $H(H_2PO_2)$, is monoprotic. The structures of these three acids are shown in Figure J-5.

Phosphoric acid (Figure J-6) is the ninth ranked industrial chemical, almost 22 billion pounds being produced annually in the United States. It is sold in various concentrations. The 85% by mass (85 g of H_3PO_4 to 15 g of H_2O) solution is a colorless, syrupy liquid. Laboratory phosphoric acid is sold as an 85% solution, which is equivalent to 15 M.

Phosphoric acid is used extensively in the production of soft drinks, and various of its salts are used in the food industry. For example, the monosodium salt, NaH_2PO_4, is used in a variety of foods to control acidity, and calcium dihydrogen phosphate, $Ca(H_2PO_4)_2$, is the acidic ingredient in baking powder. The evolu-

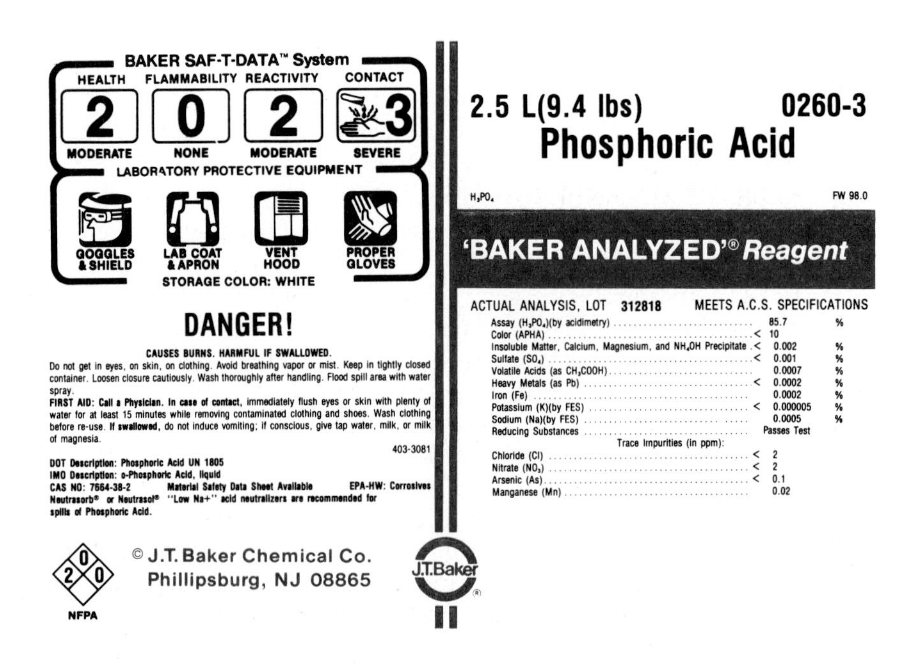

Figure J-6 Phosphoric acid label. Phosphoric acid is sold as a 15 M aqueous solution.

tion of carbon dioxide that takes place when baking powder is heated can be represented as

$$\underbrace{Ca(H_2PO_4)_2(s) + 2NaHCO_3(s)}_{\text{in baking powder}} \xrightarrow{300°C} 2CO_2(g) + 2H_2O(g) + CaHPO_4(s) + Na_2HPO_4(s)$$

The slowly evolving $CO_2(g)$ gets trapped in small gas pockets and thereby causes the cake or bread to rise.

Unleavened bread does not contain baking powder and thus does not rise when baked.

When phosphoric acid is heated gently, pyrophosphoric acid (pyro means heat) is obtained as a result of the elimination of a water molecule from a pair of phosphoric acid molecules:

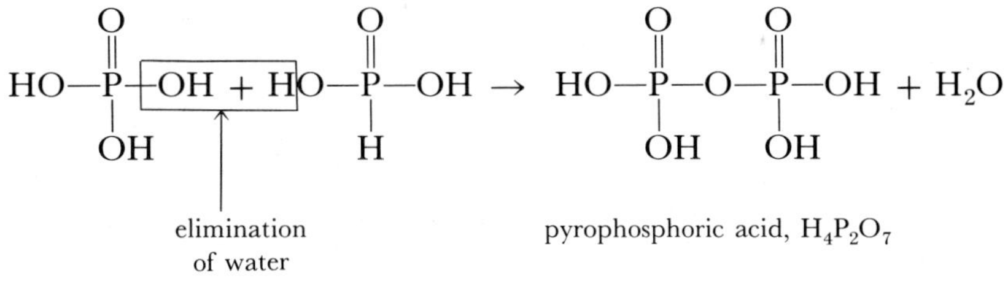

$$HO{-}\underset{OH}{\overset{O}{\overset{\|}{P}}}{-}\boxed{OH + H}O{-}\underset{H}{\overset{O}{\overset{\|}{P}}}{-}OH \rightarrow HO{-}\underset{OH}{\overset{O}{\overset{\|}{P}}}{-}O{-}\underset{OH}{\overset{O}{\overset{\|}{P}}}{-}OH + H_2O$$

elimination of water

pyrophosphoric acid, $H_4P_2O_7$

Pyrophosphoric acid is a viscous, syrupy liquid that tends to solidify on long standing. In aqueous solution, it slowly reverts to phosphoric acid.

Longer chains of phosphate groups can be formed. The compound sodium tripolyphosphate

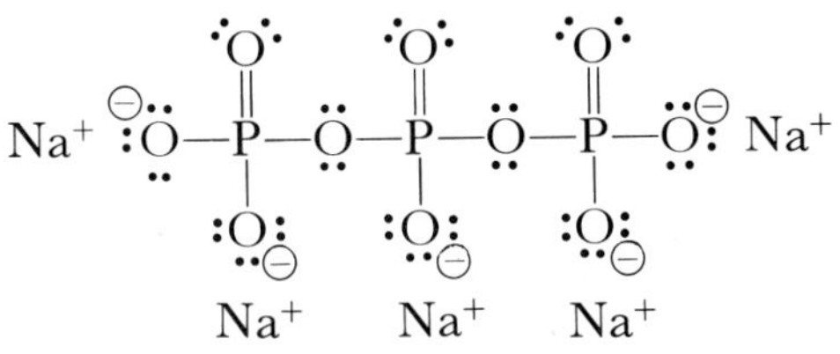

$$Na^+ \; {}^{\ominus}\!:\ddot{O}{-}\underset{:\ddot{O}:^{\ominus}}{\overset{\ddot{O}}{\overset{\|}{P}}}{-}\ddot{O}{-}\underset{:\ddot{O}:^{\ominus}}{\overset{\ddot{O}}{\overset{\|}{P}}}{-}\ddot{O}{-}\underset{:\ddot{O}:^{\ominus}}{\overset{\ddot{O}}{\overset{\|}{P}}}{-}\ddot{O}:^{\ominus} \; Na^+$$

$$Na^+ \quad Na^+ \quad Na^+$$

used to be the phosphate ingredient of detergents. Its role was to break up and suspend dirt and stains by forming water-soluble complexes with metal ions. (The formation of complexes is discussed in Chapter 23.) In the 1960's almost all detergents contained phosphates, sometimes as much as 50 percent by mass. It was discovered, however, that the phosphates led to a serious water pollution problem. The enormous quantity of phosphates discharged into rivers and lakes served as a nutrient for the rampant growth of algae and other organisms. When these organisms died, much of the oxygen dissolved in the water was consumed in the decay process, thus depleting the water's oxygen supply and destroying the ecological balance. This process is called *eutrophication* (Figure J-7). As a result of legislation in the 1970's, phosphates have been eliminated from detergents or their levels have been reduced markedly.

Figure J-7 Dead and dying algae from an algal bloom in a lake contaminated with phosphates from detergents.

J-5. PHOSPHORUS FORMS A NUMBER OF BINARY COMPOUNDS

Note the similarity of phosphides to nitrides.

Phosphorus reacts directly with reactive metals, such as sodium and calcium, to form phosphides. A typical reaction is

$$12Na(s) + P_4(s) \rightarrow 4Na_3P(s)$$

Most metal phosphides react vigorously with water to produce phosphine, $PH_3(g)$:

$$Ca_3P_2(s) + 6H_2O(l) \rightarrow 2PH_3(g) + 3Ca(OH)_2(aq)$$

Phosphine reacts violently with oxygen and the halogens.

Phosphine has a trigonal pyramidal structure with an H—P—H bond angle of 93.7°. It is a colorless, extremely toxic gas with an offensive odor like that of rotten fish. Unlike ammonia, phosphine does not act as a base toward water, and few phosphonium (PH_4^+) salts are stable. Phosphine can also be prepared by the reaction of white phosphorus with a strong base.

Phosphorus reacts directly with the halogens to form halides of the form PX_3 and PX_5. If an excess of phosphorus is used, then the trihalide is formed. For example,

$$\underset{\text{excess}}{P_4(s)} + 6Cl_2(g) \rightarrow 4PCl_3(l)$$

Table J-1 Physical states of the phosphorus halides at room temperature

Halide	*Physical state*	*Molecular description*
PF_3	colorless gas	trigonal pyramidal PF_3 molecules
PCl_3	clear, colorless, fuming liquid	trigonal pyramidal PCl_3 molecules
PBr_3	colorless, fuming liquid	trigonal pyramidal PBr_3 molecules
PI_3	red, crystalline, unstable solid	trigonal pyramidal PI_3 molecules
PF_5	colorless gas	trigonal bipyramidal PF_5 molecules
PCl_5	pale yellow, fuming crystals	$[PCl_4^+][PCl_6^-]$ ion pairs; PCl_4^+ tetrahedral and PCl_6^- octahedral
PBr_5	yellow, fuming, hygroscopic crystals	$[PBr_4^+]Br^-$ ion pairs; PBr_4^+ tetrahedral
PI_5	not known	presumably iodine atoms are too large to arrange five around a phosphorus atom

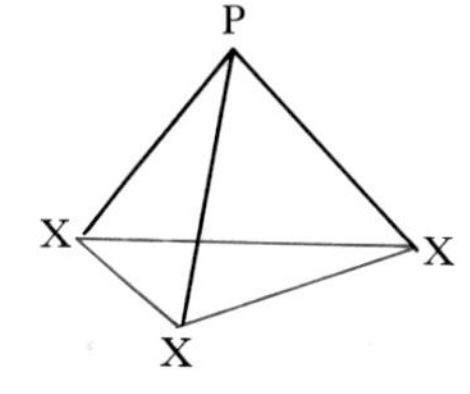

Figure J-8 The phosphorus trihalides, PX_3, have a trigonal pyramidal structure in the gas phase.

If an excess of halide is used, then the pentahalide is formed:

$$P_4(s) + \underset{\text{excess}}{10Cl_2(g)} \rightarrow 4PCl_5(s)$$

Recall from Chapter 11 that phosphorus trihalide molecules in the gas phase have a trigonal pyramidal structure (Figure J-8) and that phosphorus pentahalide molecules in the gas phase have a trigonal bipyramidal structure (Figure J-9). Table J-1 lists the physical states of the various phosphorus halides. Phosphorus halides react vigorously with water. For example,

$$PCl_3(l) + 3H_2O(l) \rightarrow H_3PO_3(aq) + 3HCl(aq)$$

$$PCl_5(s) + 4H_2O(l) \rightarrow H_3PO_4(aq) + 5HCl(aq)$$

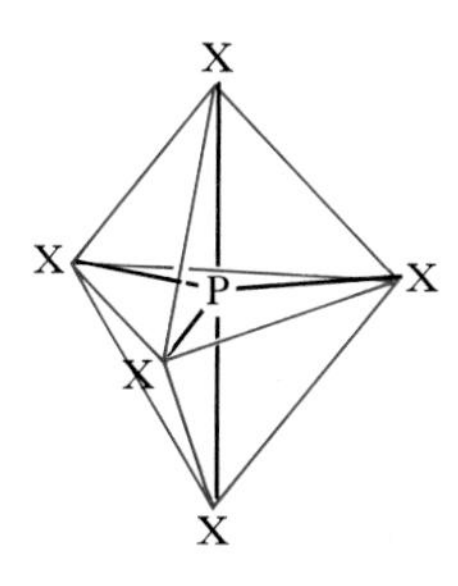

Figure J-9 The phosphorus pentahalides, PX_5, have a trigonal bipyramidal structure in the gas phase.

When phosphorus is heated with sulfur, the yellow crystalline compound tetraphosphorus trisulfide, P_4S_3, is formed (Frontispiece). Matches that can be ignited by striking on any rough surface contain a tip composed of the yellow P_4S_3 on top of a red portion that contains lead dioxide, PbO_2, together with antimony sulfide, Sb_2S_3.

Friction causes the P_4S_3 to ignite in air, and the heat produced then initiates a reaction between antimony sulfide and lead dioxide that produces a flame.

Safety matches consist of a mixture of potassium chlorate and antimony sulfide. The match is ignited by striking on a special rough surface composed of a mixture of red phosphorus, glue, and abrasive. The red phosphorus is ignited by friction and in turn ignites the reaction mixture in the matchhead.

J-6. MANY PHOSPHORUS COMPOUNDS ARE IMPORTANT BIOLOGICALLY

Ordinary tooth enamel is hydroxyapatite, $Ca_{10}(OH)_2(PO_4)_6$. If low concentrations of fluoride ion are added to the diets of children, then a substantial amount of the tooth enamel formed will consist of fluorapatite, $Ca_{10}F_2(PO_4)_6$, which is much harder and less affected by acidic substances than hydroxyapatite. Consequently, fluorapatite is more resistant to tooth decay than is hydroxyapatite. Small quantities of fluoride are added to most municipal water supplies, and the incidence of tooth decay among children has decreased markedly over the past 30 years.

The energy requirements for all biochemical reactions are supplied by a substance called adenosine triphosphate, or simply ATP. Under physiological conditions, the reaction of 1 mol of ATP with water releases 31 kJ of energy. The energy released is used by all living species to drive biochemical reactions, in which typically the products are at a higher energy than the reactants. ATP is a biological fuel. The formation and consumption of ATP occur on the average within about 1 min of each other. The amount of ATP used by the human body is truly remarkable: at rest over a 24-h period about 40 kg of ATP is consumed. For strenuous exercise the rate of utilization of ATP can reach 5 kg in 10 min.

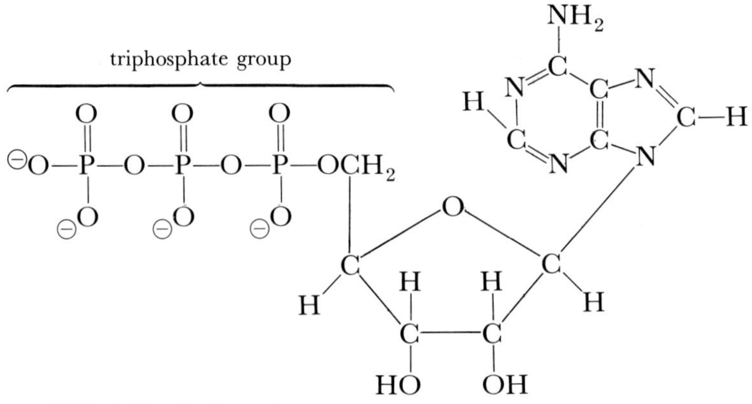

Figure J-10 A Lewis formula of ATP, the substance that is the energy source for all living species.

Many organic phosphates are potent insecticides that are also highly toxic to humans. These insecticides act by blocking the transmission of electrical signals in the respiratory system, thereby causing paralysis and death by suffocation. Fortunately, such poisons do not last for long in the environment because they are destroyed over a period of several days by reaction with water. An important example of an organophosphorus insecticide is malathion, which was used to combat the Mediterranean fruit fly infestation in California in the summer of 1981. Malathion is toxic to humans, but only at fairly large doses. There is an enzyme in human gastric juice that decomposes malathion (insects lack this enzyme). Thus malathion is most toxic to humans when it is absorbed directly into the bloodstream, as, for example, when it comes into contact with a cut in the skin.

$$(CH_3O)_2P(=S){-}S{-}CH(COOCH_2CH_3)CH_2COOCH_2CH_3$$

malathion

QUESTIONS

J-1. Discuss the difference in reactivity between white phosphorus and red phosphorus.

J-2. What is the structure of P_4?

J-3. What is phosphate rock? What is its most important use?

J-4. Why can't phosphate rock be used directly as fertilizer?

J-5. Compare the structures of P_4O_6 and P_4O_{10}.

J-6. What is a desiccant? Give an example of one.

J-7. Compare the structures of $H_2PO_2^-$, HPO_3^{2-}, and PO_4^{3-}. How many dissociable protons are there per mole of phosphorous acid? Of hypophosphorous acid?

J-8. Describe the action of baking powder.

J-9. Discuss the process of eutrophication.

J-10. Compare ammonia and phosphine as bases.

J-11. Describe two ways to prepare phosphine.

J-12. Discuss the difference between safety matches and strike-anywhere matches.

Two common acid-base indicators at various pH values. Across the top is methyl orange at pH = 2, 4, and 6. Across the bottom is bromthymol blue at pH = 5, 7, and 9.

18 / Titration, Buffers, and Polyprotic Acids

This chapter continues our treatment of acid-base chemistry. We begin with a discussion of colored acid-base indicators and of how they can provide a qualitative measure of the acidity of a solution. We then discuss acid-base titration curves, in which the pH of a solution is plotted against the volume of an added base or acid. Acid-base titrations are used to determine the total acidity of a solution by titration with a base or the total basicity of a solution by titration with an acid. Acid-base indicators are used to signal the equivalence point in an acid-base titration.

Buffers, which are mixtures of conjugate acid-base pairs, are used to control pH in chemical and biochemical systems. The chemical basis of buffer action is discussed in terms of Le Châtelier's principle, and the Henderson-Hasselbalch equation is used to calculate the pH of buffer solutions.

Finally, we describe polyprotic acids, which are acids having more than one dissociable proton. The acid-base properties of polyprotic acids are illustrated by means of distribution diagrams, which show how the composition of a solution of a polyprotic acid changes with pH.

18-1. AN INDICATOR IS A WEAK ORGANIC ACID WHOSE COLOR VARIES WITH pH

Numerous weak organic acids change color upon loss of a proton. These compounds are called *indicators* because they indicate the pH of a solution by their color. An example of an indicator is the compound methyl orange, which we denote by HIn to emphasize that it is a weak acid and an indicator. The acid form of methyl orange (HIn) is red, and the conjugate base form (In^-) is yellow (Frontispiece). We can represent the acid-base reaction of methyl orange by

$$\underset{\text{red}}{HIn(aq)} + H_2O(l) \rightleftharpoons H_3O^+(aq) + \underset{\text{yellow}}{In^-(aq)} \qquad (18\text{-}1)$$

According to Le Châtelier's principle, the reaction equilibrium lies to the left if $[H_3O^+]$ is large (low pH) and to the right if $[H_3O^+]$ is small (high pH). Consequently, a solution containing methyl orange is red at low pH and yellow at high pH.

We can make this discussion quantitative by considering the acid dissociation constant of methyl orange. The acid dissociation constant for Reaction (18-1) is

$$K_{ai} = \frac{[H_3O^+][In^-]}{[HIn]} \qquad (18\text{-}2)$$

where K_{ai} denotes the acid dissociation constant of an indicator. As mentioned above, an aqueous solution containing methyl orange in the acid form (HIn) is red and an aqueous solution containing methyl orange in the basic form (In^-) is yellow. If both the $HIn(aq)$ and $In^-(aq)$ forms are present simultaneously in comparable concentrations, that is, if

$$[HIn] \approx [In^-]$$

then the solution is orange (red + yellow) (Frontispiece). Substitution of the condition $[HIn] \approx [In^-]$ into Equation (18-2) yields

$$[H_3O^+] \approx K_{ai}$$

By taking the negative of the logarithms of this expression, we have

$$pH \approx pK_{ai}$$

The pH range over which an indicator changes color is approximately equal to $pK_{ai} \pm 1$.

In other words, if methyl orange is orange in a solution, then the pH of the solution is approximately equal to the pK_{ai} value for methyl orange. The value of pK_{ai} for methyl orange is 3.89, and so the pH of the orange solution is about 3.89.

If we place a few drops of a methyl orange solution into each of the following solutions:

0.010 M HCl (pH = 2) + methyl orange	0.00010 M HCl (pH = 4) + methyl orange	pure water (pH ≈ 7) + methyl orange
red solution	orange solution	yellow solution

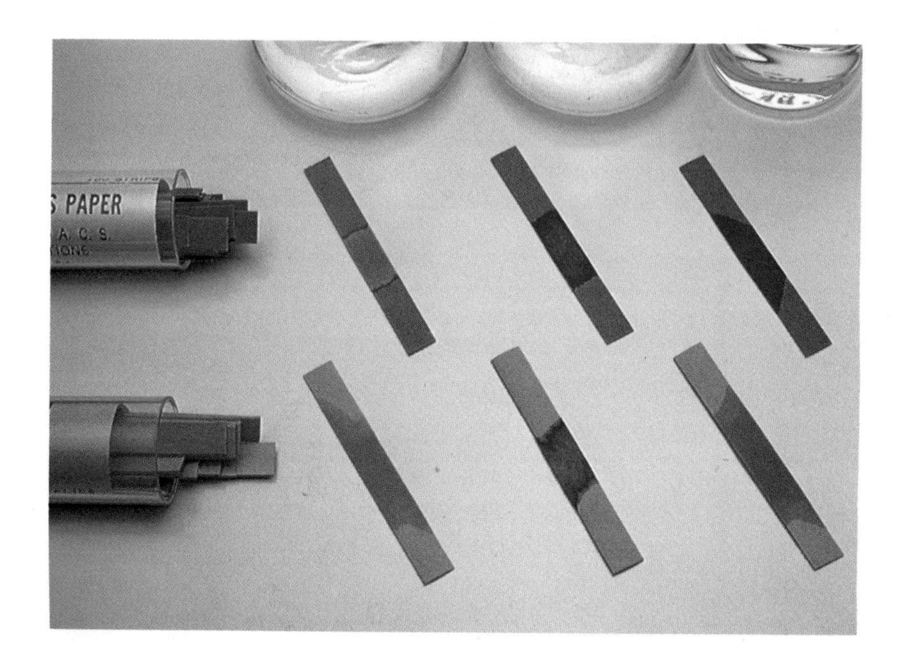

Figure 18-1 Litmus paper is used as a simple test for acidity or basicity. Blue litmus paper turns red when placed in an acidic solution. Red litmus paper turns blue when placed in a basic solution.

the resulting solutions will have the colors indicated. This simple experiment shows that acid-base indicators can be used to determine qualitatively the acidity of solutions.

Because of the intense color of indicators, only a very small concentration is necessary to produce a visible color. The contribution of the indicator acid, HIn, to the total acidity of the solution is negligible.

A common acid-base indicator is *litmus paper* (see Section 4-11). Litmus is red in acidic solutions and blue in basic solutions. An acidic aqueous solution turns blue litmus red, and a basic aqueous solution turns red litmus blue. The transition pH of litmus is at pH = 7; that is, a neutral solution does not cause a color change. The color change of litmus can be remembered as BRB: *B*ase turns *R*ed litmus *B*lue. The color changes of blue and red litmus are shown in Figure 18-1. Other indicators, together with their colors at various pH values, are given in Figure 18-2. Using the indicators in Figure 18-2 we can estimate the pH of an aqueous solution to within about 0.5 pH unit. Of course, the solution must be colorless initially, for otherwise the color change of the indicator may be obscured.

Litmus, which is a plant extract, is also used for coloring beverages.

Example 18-1: Estimate the pH of a colorless aqueous solution that turns blue when bromcresol green is added and red when litmus is added.

Solution: From Figure 18-2 we note that the pH at which bromcresol green is blue is >5.6, and that litmus is red at pH <7. Therefore the pH of the solution is about 6.

Figure 18-2 The color changes of various indicators with pH.

18-2. THE EQUIVALENCE POINT IS THE POINT AT WHICH AN ACID-BASE TITRATION IS COMPLETE

Suppose that an aqueous solution of a strong base, such as NaOH(*aq*), is added slowly to an aqueous solution of a strong acid, such as HCl(*aq*). This procedure is called *titration* (Figures 18-3 and 18-4), and a plot of the pH of the resulting solution as a function of the volume of added solution (the *titrant*) is called a *titration curve*. The titration curve of a 50.0-mL sample of 0.100 M HCl(*aq*) titrated with 0.100 M NaOH(*aq*) is shown in Figure 18-5. The *equivalence point* of a titration is the point at which stoichiometrically equivalent amounts of acid and base have been brought together. At the equivalence point, the number of moles of base added is equal to the number of moles of acid initially present.

The equation for the HCl–NaOH titration is

$$H_3O^+(aq) + Cl^-(aq) + Na^+(aq) + OH^-(aq) \rightarrow Na^+(aq) + Cl^-(aq) + 2H_2O(l) \quad (18\text{-}3)$$

At the equivalence point, the solution is simply an aqueous solution of NaCl. Because NaCl is a neutral salt, the pH is 7 at the equivalence point. We have written Equation (18-3) as an ionic equation to emphasize that HCl(*aq*) is a strong acid and that NaOH(*aq*) is a strong base. Ionic Equation (18-3) can be written more compactly as a net ionic equation:

reaction that takes place in titration (for strong acid and strong base)

$$H_3O^+(aq) + OH^-(aq) \rightarrow 2H_2O(l) \quad (18\text{-}4)$$

The point at which the indicator changes color is called the *end point* of the titration. The end point is the experimental estimate of the true equivalence point.

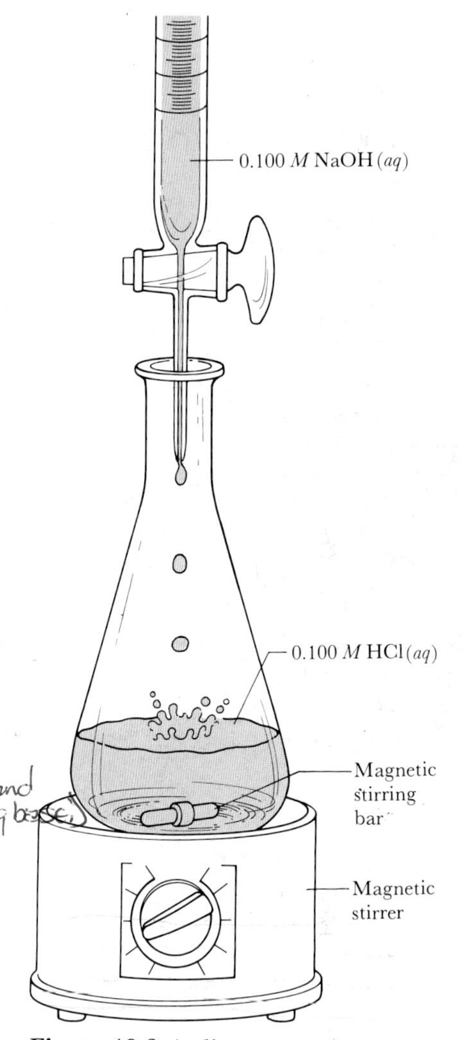

Figure 18-3 A diagrammatic representation of the set-up used for titration.

Example 18-2: By referring to Figures 18-2 and 18-5, choose an indicator to determine the equivalence point shown in Figure 18-5.

Solution: From Figure 18-5 we note that the equivalence point occurs at pH = 7.0. The titration curve is very steep in the vicinity of the equivalence point, however, and thus an indicator with a color transition range lying between pH = 5 and pH = 9 would be suitable. Referring to Figure 18-3, we see that there are several possible choices, for example, bromthymol blue or phenolphthalein. We must choose our indicator such that the end point and the equivalence point are the same within the required accuracy of the titration.

Problems 18-1 through 18-8 involve Figure 18-2 and are similar to Examples 18-1 and 18-2.

The calculation of points on the HCl + NaOH titration curve is straightforward. The key factor to recognize is that the equilibrium constant for the titration reaction (Equation 18-4) is very large. Since

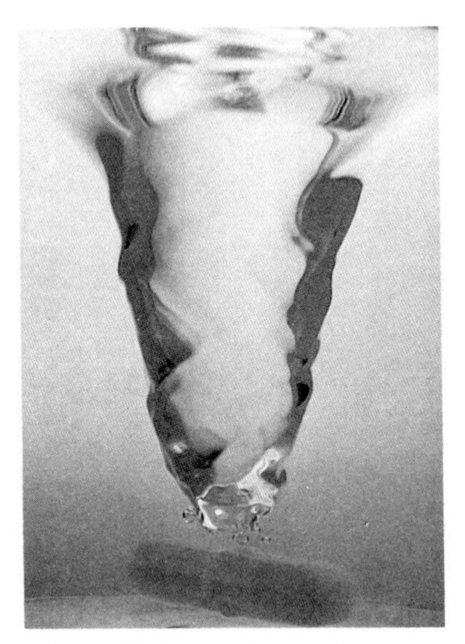
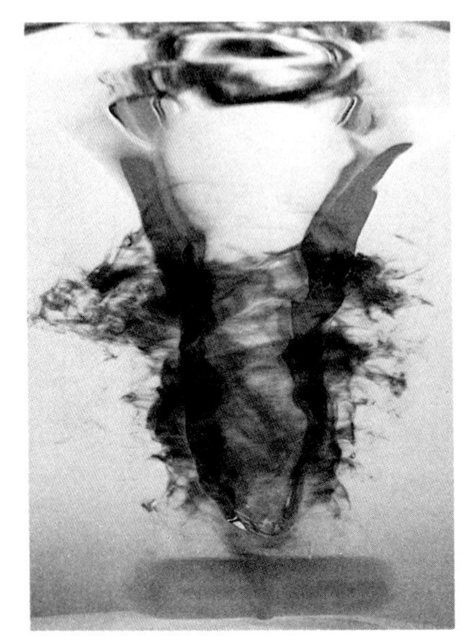

Figure 18-4 Experimental setup for the titration of $HCl(aq)$ with $NaOH(aq)$. A 50.0-mL sample of $HCl(aq)$ is placed in the reaction flask together with a magnetic bar for stirring the reaction mixture during the titration. Two or three drops of the acid-base indicator phenolphthalein is added to the $HCl(aq)$, and then $NaOH(aq)$ is added from a buret. As the equivalence point is approached, the base is added dropwise until a single drop turns the entire reaction mixture a pink color that does not revert to colorless with stirring.

it is the reverse of the water dissociation reaction, the equilibrium constant, K, for Equation (18-4) is equal to $1/K_w$, where $K_w = 1.0 \times 10^{-14}\ M^2$. Therefore, $K = 1.0 \times 10^{14}\ M^{-2}$. The large value of K for Equation (18-4) means that this reaction goes to completion.

The concentration of $H_3O^+(aq)$ in the 0.100 M $HCl(aq)$ solution before any base is added is $[H_3O^+] = 0.100$ M, and thus the pH of the solution initially is 1.00. After 10.0 mL of the 0.100 M $NaOH(aq)$

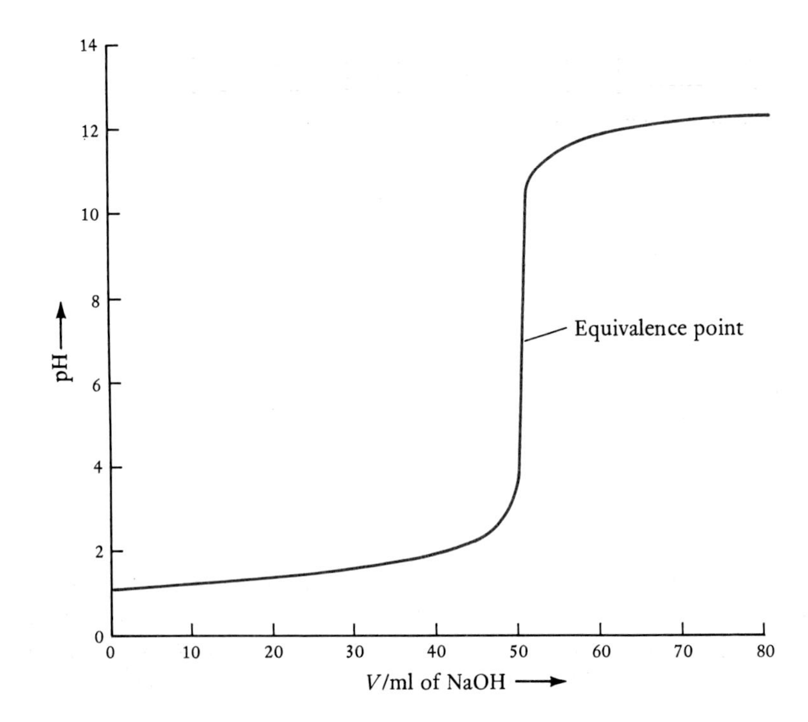

Figure 18-5 Curve for the titration of 50.0 mL of 0.100 M $HCl(aq)$ with 0.100 M $NaOH(aq)$. Note that the pH of the solution changes very slowly until the vicinity of the equivalence point is reached and then changes very rapidly in the vicinity of the equivalence point.

is added to the 50.0 mL of the initially 0.100 M $HCl(aq)$ solution, the total volume of the resulting solution is 50.0 mL + 10.0 mL = 60.0 mL. The $OH^-(aq)$ from the $NaOH(aq)$ reacts with the $H_3O^+(aq)$ from the $HCl(aq)$ to produce water:

$$H_3O^+(aq) + OH^-(aq) \rightarrow 2H_2O(l)$$

The number of moles of $H_3O^+(aq)$ that reacts is equal to the number of moles of $OH^-(aq)$ added:

$$\begin{pmatrix}\text{moles of}\\ OH^-(aq)\\ \text{added}\end{pmatrix} = MV = (0.100\ \text{mol}\cdot\text{L}^{-1})(0.0100\ \text{L})$$

$$= 1.00 \times 10^{-3}\ \text{mol} = \begin{pmatrix}\text{moles of}\\ H_3O^+(aq)\\ \text{reacted}\end{pmatrix}$$

The total number of moles of $H_3O^+(aq)$ present in 50.0 mL of 0.100 M $HCl(aq)$ is

$$\begin{pmatrix}\text{moles of}\\ H_3O^+(aq)\end{pmatrix} = MV = (0.100\ \text{mol}\cdot\text{L}^{-1})(0.0500\ \text{L}) = 5.00 \times 10^{-3}\ \text{mol}$$

The concentration of $H_3O^+(aq)$ that remains after the addition of 10.0 mL of 0.100 M NaOH is equal to the number of moles of $H_3O^+(aq)$ that remains unreacted divided by the total volume of resulting solution. The number of moles of $H_3O^+(aq)$ unreacted is

$$\begin{pmatrix}\text{moles of}\\ \text{unreacted}\\ H_3O^+(aq)\end{pmatrix} = \begin{pmatrix}\text{initial}\\ \text{moles}\\ \text{of } H_3O^+(aq)\end{pmatrix} - \begin{pmatrix}\text{moles of}\\ H_3O^+(aq)\\ \text{reacted}\end{pmatrix}$$

$$= (5.00 \times 10^{-3})\ \text{mol} - (1.00 \times 10^{-3})\ \text{mol}$$
$$= 4.00 \times 10^{-3}\ \text{mol}$$

and the molarity of unreacted $H_3O^+(aq)$ is

$$\text{molarity} = \frac{\text{moles}}{\text{liters}} = \frac{4.00 \times 10^{-3}\ \text{mol}}{0.060\ \text{L}} = 6.67 \times 10^{-2}\ \text{M}$$

Therefore, the pH of the solution after the addition of 10.0 mL of $NaOH(aq)$ is

$$\text{pH} = -\log[H_3O^+] = -\log(6.67 \times 10^{-2}) = 1.18$$

Proceeding in an analogous fashion, we can compute the pH of the resulting solution after the addition of other volumes of $NaOH(aq)$ (Table 18-1). Beyond the equivalence point, all the $H_3O^+(aq)$ has reacted and we simply have a diluted solution of $NaOH(aq)$.

Table 18-1 Calculation of various points for the titration of 50.0 mL of 0.100 M HCl(*aq*) with 0.100 M NaOH(*aq*)

Volume of 0.100 M NaOH(aq) added/mL	*Moles of $OH^-(aq)$ added*	*Moles of unreacted $H_3O^+(aq)$*	*Total volume of solution/L*	*Concentration of unreacted $H_3O^+(aq)/M$*	*pH*
20.0	2.00×10^{-3}	3.00×10^{-3}	0.070	4.29×10^{-2}	1.37
30.0	3.00×10^{-3}	2.00×10^{-3}	0.080	2.50×10^{-2}	1.60
40.0	4.00×10^{-3}	1.00×10^{-3}	0.090	1.11×10^{-2}	1.95
45.0	4.50×10^{-3}	0.50×10^{-3}	0.095	5.26×10^{-3}	2.28
49.0	4.90×10^{-3}	0.10×10^{-3}	0.099	1.01×10^{-3}	3.00
49.5	4.95×10^{-3}	0.05×10^{-3}	0.0995	5.03×10^{-4}	3.30
50.0 (equivalence point)	5.00×10^{-3}	1.00×10^{-8} (from H_2O dissociation)	0.100	1.00×10^{-7}	7.00
		Moles of unreacted $OH^-(aq)$		*Concentration of unreacted $OH^-(aq)/M$*	
50.5	5.05×10^{-3}	0.05×10^{-3}	0.1005	4.98×10^{-4}	10.70
51.0	5.10×10^{-3}	0.10×10^{-3}	0.101	9.90×10^{-4}	11.00
55.0	5.50×10^{-3}	0.50×10^{-3}	0.105	4.76×10^{-3}	11.68
60.0	6.00×10^{-3}	1.00×10^{-3}	0.110	9.09×10^{-3}	11.96
70.0	7.00×10^{-3}	2.00×10^{-3}	0.120	1.67×10^{-2}	12.22
80.0	8.00×10^{-3}	3.00×10^{-3}	0.130	2.31×10^{-2}	12.36
90.0	9.00×10^{-3}	4.00×10^{-3}	0.140	2.86×10^{-2}	12.46

*The number of moles is the key to solving titration problems.

Example 18-3: Compute the pH of a solution obtained by adding 60.0 mL of 0.100 M NaOH(*aq*) to 50.0 mL of 0.100 M HCl(*aq*).

Note in Table 18-1 the rapid change in pH in the range 49.5 mL to 50.5 mL of added base.

Solution: The total number of moles of $H_3O^+(aq)$ in 50.0 mL of 0.100 M HCl(*aq*) is

$$\begin{pmatrix}\text{moles of}\\ H_3O^+(aq)\end{pmatrix} = MV = (0.100\ \text{mol}\cdot\text{L}^{-1})(0.0500\ \text{L}) = 5.00 \times 10^{-3}\ \text{mol}$$

The total number of moles of $OH^-(aq)$ in 60.0 mL of 0.100 M NaOH(*aq*) is

$$\begin{pmatrix}\text{moles of added}\\ OH^-(aq)\end{pmatrix} = MV = (0.100\ \text{mol}\cdot\text{L}^{-1})(0.0600\ \text{L}) = 6.00 \times 10^{-3}\ \text{mol}$$

Note that the number of moles of $NaOH(aq)$ added (6.00×10^{-3} mol) exceeds the number of moles of $H_3O^+(aq)$ present (5.00×10^{-3} mol). The number of moles of $OH^-(aq)$ that remain unreacted after the NaOH addition is

$$\begin{pmatrix}\text{moles of}\\ \text{unreacted}\\ OH^-(aq)\end{pmatrix} = (6.00 \times 10^{-3}\ \text{mol}) - (5.00 \times 10^{-3}\ \text{mol}) = 1.00 \times 10^{-3}\ \text{mol}$$

The total volume of the solution is $50.0 + 60.0 = 110.0$ mL, and thus the concentration of unreacted $OH^-(aq)$ is

$$[OH^-] = \frac{1.00 \times 10^{-3}\ \text{mol}}{0.110\ \text{L}} = 9.09 \times 10^{-3}\ \text{M}$$

The $[H_3O^+]$ in the solution is calculated from the relation

$$K_w = [H_3O^+][OH^-] = 1.00 \times 10^{-14}\ \text{M}^2$$

Thus

$$[H_3O^+] = \frac{1.00 \times 10^{-14}\ \text{M}^2}{9.09 \times 10^{-3}\ \text{M}} = 1.10 \times 10^{-12}\ \text{M}$$

The pH of the solution is

$$\text{pH} = -\log(1.10 \times 10^{-12}) = 11.96$$

which is the result shown in Figure 18-5 and Table 18-1 for 60.0 mL of added base. You should try to reproduce some other values given in Table 18-1.

Problems 18-63 and 18-64 ask you to calculate points on the titration curve of a strong acid with a strong base.

18-3. WEAK ACIDS CAN BE TITRATED WITH STRONG BASES

Figure 18-5 shows the titration curve of a strong acid with a strong base. The titration curve of a weak acid with a strong base looks somewhat different from Figure 18-5. The titration curve for 50.0 mL of 0.100 M acetic acid, $HC_2H_3O_2(aq)$, titrated with 0.100 M $NaOH(aq)$ is shown in Figure 18-6. Note that the pH of the acetic acid solution is 2.9 initially and increases slowly until the equivalence point is reached. Around the equivalence point the pH changes from 6 to 11. Note also that the equivalence point occurs not at pH = 7 but around pH = 9.

The initial value of the pH is 2.9 because acetic acid is a weak acid. (Recall that we calculated the pH of a 0.100 M $HC_2H_3O_2(aq)$ solution in Section 17-6.) To see why the pH is 8.7 at the equivalence point, we must consider the principal species in the solution at the

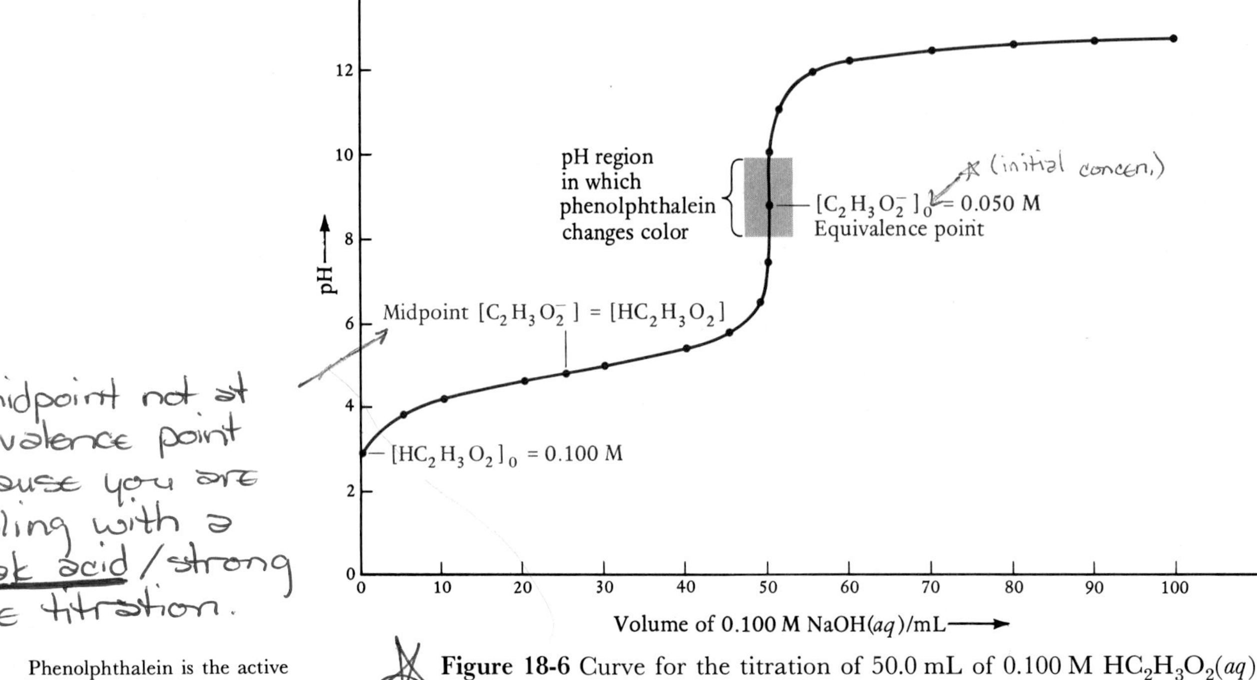

Phenolphthalein is the active ingredient in ExLax.

Figure 18-6 Curve for the titration of 50.0 mL of 0.100 M $HC_2H_3O_2(aq)$ with 0.100 M $NaOH(aq)$. The indicator used in this case is phenolphthalein, which changes from colorless to pink around pH = 9.

equivalence point. At the equivalence point, all the $HC_2H_3O_2(aq)$ has just been neutralized by the $NaOH(aq)$. The reaction is

$$HC_2H_3O_2(aq) + Na^+(aq) + OH^-(aq) \rightarrow Na^+(aq) + C_2H_3O_2^-(aq) + H_2O(l) \quad (18\text{-}5)$$

conjugate base of a weak acid

It is important to realize that Reaction (18-5) goes to completion even though $HC_2H_3O_2(aq)$ is a weak acid. As the $H_3O^+(aq)$ that results from the dissociation of $HC_2H_3O_2(aq)$ reacts with the added $OH^-(aq)$, more $HC_2H_3O_2(aq)$ dissociates to maintain its acid dissociation equilibrium. Eventually, essentially all the $HC_2H_3O_2(aq)$ dissociates and the $HC_2H_3O_2(aq) + NaOH(aq)$ reaction goes to completion. We can see this quantitatively by calculating the equilibrium constant of the reaction. The net ionic equation for the titration of $HC_2H_3O_2(aq)$ with $NaOH(aq)$ is

$$HC_2H_3O_2(aq) + OH^-(aq) \rightleftharpoons C_2H_3O_2^-(aq) + H_2O(l) \quad (18\text{-}6)$$

Equation (18-6) is the sum of the following two equations:

$$HC_2H_3O_2(aq) + H_2O(l) \rightleftharpoons H_3O^+(aq) + C_2H_3O_2^-(aq) \quad (18\text{-}7)$$

$$H_3O^+(aq) + OH^-(aq) \rightleftharpoons 2H_2O(l) \quad (18\text{-}8)$$

The equilibrium constant for Equation (18-7) is the K_a of $HC_2H_3O_2(aq)$, and the equilibrium constant for Equation (18-8) is $1/K_w$. Therefore, the equilibrium constant for Equation (18-6) is $K = K_a/K_w$. Using the fact that $K_a = 1.74 \times 10^{-5}$ M for $HC_2H_3O_2(aq)$ (Table 17-2) and that $K_w = 1.00 \times 10^{-14}$ M^2, we see that $K = 1.74 \times 10^9$ M^{-1}. This large value of K indicates that Reaction (18-6) lies far to the right, and thus $HC_2H_3O_2(aq)$ reacts essentially completely with a strong base added to the solution.

The titration of a weak acid with a strong base goes to completion, meaning that the value of its equilibrium constant is very large.

Because Reaction (18-5) goes essentially to completion, the equivalence point consists of a $NaC_2H_3O_2(aq)$ solution. To calculate the concentration of this $NaC_2H_3O_2(aq)$ solution, we use the fact that we started with 50.0 mL of 0.100 M $HC_2H_3O_2(aq)$. The initial number of moles of $HC_2H_3O_2(aq)$ is

$$\text{initial moles of } HC_2H_3O_2(aq) = MV = (0.100 \text{ mol}\cdot\text{L}^{-1})(0.0500 \text{ L}) = 5.00 \times 10^{-3} \text{ mol}$$

According to Equation (18-5)

$$\begin{pmatrix}\text{moles of } NaC_2H_3O_2(aq) \text{ at} \\ \text{the equivalence point}\end{pmatrix} = \begin{pmatrix}\text{initial moles of} \\ HC_2H_3O_2(aq)\end{pmatrix} = 5.00 \times 10^{-3} \text{ mol}$$

The total volume of the solution at the equivalence point is 100.0 mL, that is, 50.0 mL of $HC_2H_3O_2(aq)$ plus 50.0 mL of added $NaOH(aq)$, and so the concentration of the $NaC_2H_3O_2(aq)$ solution at the equivalence point is

$$M = \frac{\text{mol}}{V} = \frac{5.00 \times 10^{-3} \text{ mol}}{0.100 \text{ L}} = 0.0500 \text{ M}$$

We learned how to calculate the pH of a solution such as $NaC_2H_3O_2(aq)$ in Section 17-11. In fact, Example 17-12 involved the calculation of the pH of a 0.050 M $NaC_2H_3O_2(aq)$ solution, giving a value of pH = 8.72. Thus we see that the pH at the equivalence point in Figure 18-6 is 8.72.

We have calculated the pH at only two points (the initial point and the equivalence point) on the titration curve of a weak acid with a strong base. There is one other point that is relatively easy to calculate. The *midpoint* of a titration of a weak acid with a strong base is the point at which [HB] = [B$^-$]. The acid dissociation-constant expression

$$K_a = \frac{[H_3O^+][B^-]}{[HB]}$$

can be rearranged to the form

$$[H_3O^+] = K_a \frac{[HB]}{[B^-]}$$

At the midpoint, $[HB] = [B^-]$ and therefore

$$[H_3O^+] = K_a \qquad \text{at the midpoint}$$

Problems 18-19 to 18-26 deal with the titration of a weak acid with a strong base.

Upon taking logarithms, we obtain

$$pH = pK_a \qquad \text{at the midpoint}$$

The value of pK_a for acetic acid is 4.76. The midpoint for the titration of 50.0 mL of 0.100 M $HC_2H_3O_2(aq)$ with 0.100 M $NaOH(aq)$ occurs when 25.0 mL of $NaOH(aq)$ has been added, and so we find that the pH is 4.76 at this point (Figure 18-6).

The indicator is an acid and is titrated along with the acetic acid. The total moles of indicator, however, is usually negligible.

Example 18-4: Which indicator would you use to signal the equivalence point of the titration of 50.0 mL of 0.100 M $HC_2H_3O_2(aq)$ with 0.100 M $NaOH(aq)$?

Solution: The pH at the equivalence point in the titration is 8.7 (Example 17-12). Referring to Figure 18-2, we see that phenolphthalein is a suitable indicator for the titration.

We have considered the titration of a strong acid with a strong base and a weak acid with a strong base. Let's finish this section with a discussion of the titration of a weak base with a strong acid. The curve for the titration of 50.0 mL of 0.100 M $NH_3(aq)$ with 0.100 M $HCl(aq)$ is shown in Figure 18-7. Note that initially the pH is 11.1, which is the pH of a 0.100 M $NH_3(aq)$ solution, and that the pH is 5.3 at the equivalence point. The reaction is

$$NH_3(aq) + H_3O^+(aq) + Cl^-(aq) \rightleftharpoons NH_4^+(aq) + Cl^-(aq) + H_2O(l) \qquad (18\text{-}9)$$

As in the case of the reaction of a weak acid with a strong base, this reaction goes to completion.

Example 18-5: Calculate K for Reaction (18-9).

Solution: The net equation corresponding to Reaction (18-9) can be written as the sum of two equations.

$$NH_3(aq) + H_2O(l) \rightleftharpoons NH_4^+(aq) + OH^-(aq)$$
$$H_3O^+(aq) + OH^-(aq) \rightleftharpoons 2H_2O(l)$$

For the first of these equations, $K = K_b$; for the second, $K = 1/K_w$. Therefore, $K = K_b/K_w$ for Reaction (18-9). Using the fact that $K_b = 1.75 \times 10^{-5}$ M for $NH_3(aq)$ (Table 17-3) and that $K_w = 1.00 \times 10^{-14}$ M^2, we see that

$$K = \frac{K_b}{K_w} = \frac{1.75 \times 10^{-5}\ \text{M}}{1.00 \times 10^{-14}\ \text{M}^2} = 1.75 \times 10^{9}\ \text{M}^{-1}$$

This large value of K indicates that Reaction (18-9) goes to completion.

By the very same reasoning we used for the titration of 50.0 mL of 0.100 M $HC_2H_3O_2(aq)$ with 0.100 M NaOH(*aq*), we find that the solution at the equivalence point in the titration of 50.0 mL of $NH_3(aq)$ with 0.100 M HCl(*aq*) is a 0.050 M $NH_4Cl(aq)$ solution. In Section 17-11 we calculated the pH of a 0.050 M $NH_4Cl(aq)$ solution to be 5.3, in agreement with Figure 18-7. The pH at the midpoint of the titration is equal to pK_a for $NH_4^+(aq)$, which is 9.24, again in agreement with Figure 18-7.

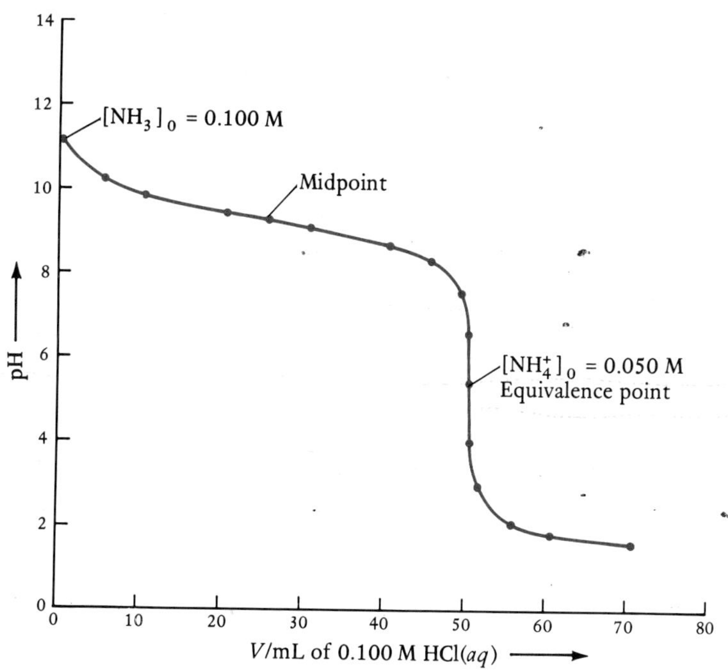

Figure 18-7 Curve for the titration of 50.0 mL of 0.100 M $NH_3(aq)$ with 0.100 M HCl(*aq*). The pH at the equivalence point is acidic because at that point we have a 0.050 M solution of $NH_4Cl(aq)$, which is acidic because $NH_4^+(aq)$ is a weak acid.

Table 18-2 Types of acid-base titrations

Type	*Typical reaction*	*K*
strong acid with strong base	$H_3O^+(aq) + OH^-(aq) \rightleftharpoons 2H_2O(l)$	$1.00 \times 10^{14}\ M^{-2}$
weak acid with strong base	$HC_2H_3O_2(aq) + OH^-(aq) \rightleftharpoons C_2H_3O_2^-(aq) + H_2O(l)$	$1.74 \times 10^{9}\ M^{-1}$
weak base with strong acid	$NH_3(aq) + H_3O^+(aq) \rightleftharpoons NH_4^+(aq) + H_2O(l)$	$1.75 \times 10^{9}\ M^{-1}$

Example 18-6: Which indicator would you use to signal the equivalence point in Figure 18-7?

Solution: The pH at the equivalence point in Figure 18-7 is 5.3. Reference to Figure 18-2 shows that a suitable indicator for the titration of $NH_3(aq)$ with $HCl(aq)$ is methyl red, which has a color transition range centered around pH = 5.

We have considered three types of acid-base titrations, which are summarized in Table 18-2. The common feature of the three types of titrations is that they all involve an acid-base reaction with a large equilibrium constant. In order for an acid-base titration to work, it is necessary that the reaction that occurs on addition of titrant go essentially to completion; that is, the titration reaction must have a large value of *K*.

18-4. THE pH OF A BUFFER SOLUTION CAN BE COMPUTED USING THE HENDERSON-HASSELBALCH EQUATION

A solution that contains both a weak acid and its conjugate base can resist a change in pH when either an acid or a base is added. A solution that is resistant to changes in pH upon the addition of an acid or a base is called a *buffer*. As an example of a buffer solution, consider a solution that contains a mixture of $HC_2H_3O_2(aq)$ and its conjugate base, $C_2H_3O_2^-(aq)$. If an acid is added to this buffer solution, then the added H_3O^+ ions are removed from the solution by the reaction

added acid is removed by $C_2H_3O_2^-$; added base is removed by $HC_2H_3O_2$. K in both reactions is large.

$$H_3O^+(aq) + C_2H_3O_2^-(aq) \rightleftharpoons HC_2H_3O_2(aq) + H_2O(l)$$

The value of *K* for this reaction is the reciprocal of K_a for $HC_2H_3O_2(aq)$, or $K = 5.75 \times 10^4\ M^{-1}$. Thus the reaction goes essentially to completion and removes essentially all the added H_3O^+.

Suppose now that a base instead of an acid is added to the buffer solution. The added OH^- ions are removed from solution by the reaction

$$HC_2H_3O_2(aq) + OH^-(aq) \rightleftharpoons C_2H_3O_2^-(aq) + H_2O(l)$$

We showed in Section 18-3 that $K = 1.74 \times 10^9\ M^{-1}$ for this reaction, and so we see that essentially all the OH^- ions are removed from solution by the above reaction. Thus a solution containing both $HC_2H_3O_2(aq)$ and $C_2H_3O_2^-(aq)$ is resistant to changes in pH on the addition of either an acid or a base.

The principal equilibrium for the $HC_2H_3O_2(aq)$–$C_2H_3O_2^-(aq)$ buffer is

$$HC_2H_3O_2(aq) + H_2O(l) \rightleftharpoons H_3O^+(aq) + C_2H_3O_2^-(aq) \qquad (18\text{-}10)$$

The basis of buffer action can be understood by applying Le Châtelier's principle to Equation (18-10). If a base is added to the $HC_2H_3O_2(aq)$–$C_2H_3O_2^-(aq)$ buffer, then the added $OH^-(aq)$ reacts with the $H_3O^+(aq)$ according to

$$H_3O^+(aq) + OH^-(aq) \rightarrow 2H_2O(l)$$

and the acid-base equilibrium (Equation 18-10) shifts from left to right, which tends to restore $[H_3O^+]$ to the initial value. If an acid is added to the $HC_2H_3O_2(aq)$–$C_2H_3O_2^-(aq)$ buffer, then the added $H_3O^+(aq)$ reacts with the $C_2H_3O_2^-(aq)$ according to

$$C_2H_3O_2^-(aq) + H_3O^+(aq) \rightarrow HC_2H_3O_2(aq) + H_2O(l)$$

and the acid-base equilibrium (Equation 18-10) shifts from right to left, which decreases $[H_3O^+]$ toward the initial value.

The general form of Equation (18-10) is

$$\underset{\text{conjugate acid}}{HB(aq)} + H_2O(l) \rightleftharpoons H_3O^+(aq) + \underset{\text{conjugate base}}{B^-(aq)} \qquad (18\text{-}11)$$

Buffer action can be described quantitatively in terms of the acid dissociation constant. For Equation (18-11), K_a is

$$K_a = \frac{[H_3O^+][B^-]}{[HB]} = \frac{[H_3O^+][\text{base}]}{[\text{acid}]} \qquad (18\text{-}12)$$

Taking the logarithm of both sides of Equation (18-12) yields

$$\log K_a = \log \frac{[H_3O^+][\text{base}]}{[\text{acid}]}$$

Using the fact that $\log ab = \log a + \log b$, this equation can be written in the form

$$\log K_a = \log[H_3O^+] + \log\frac{[\text{base}]}{[\text{acid}]}$$

Remember that $\log\frac{a}{b} = -\log\frac{b}{a}$.

Multiplying by -1 and using $pH = -\log[H_3O^+]$ and $pK_a = -\log K_a$, we write

$$pH = pK_a + \log\frac{[\text{base}]}{[\text{acid}]} \tag{18-13}$$

It is usually a good approximation to say that, in a buffer solution, the stoichiometric concentration of the acid is equal to its equilibrium concentration and the stoichiometric concentration of the conjugate base is equal to its equilibrium concentration (Problem 18-37). This assumption allows us to substitute

$$[\text{acid}] \approx [\text{acid}]_0 \qquad [\text{base}] \approx [\text{base}]_0$$

in Equation (18-13), where the subscript zeros denote stoichiometric concentrations, and obtain

$$pH \approx pK_a + \log\frac{[\text{base}]_0}{[\text{acid}]_0} \tag{18-14}$$

The Henderson-Hasselbalch equation is used extensively in biochemistry.

Equation (18-14) is known as the *Henderson-Hasselbalch equation* and is particularly useful in the analysis of buffer solutions. The advantage of using the Henderson-Hasselbalch equation is that we use the stoichiometric concentration *directly* without the need for a stepwise analysis of the equilibrium.

Example 18-7: Estimate the pH of a solution that is 0.20 M in $HC_2H_3O_2(aq)$ and 0.10 M in $NaC_2H_3O_2(aq)$. Take $K_a = 1.74 \times 10^{-5}$ M.

Solution: The stoichiometric concentrations of the acid and base forms are

$$[HC_2H_3O_2]_0 = [\text{acid}]_0 = 0.20\text{ M} \qquad [C_2H_3O_2^-]_0 = [\text{base}]_0 = 0.10\text{ M}$$

The pK_a of acetic acid is

$$pK_a = -\log K_a = -\log(1.74 \times 10^{-5}\text{ M}) = 4.76$$

From the Henderson-Hasselbalch equation, we have

$$pH = pK_a + \log\frac{[\text{base}]_0}{[\text{acid}]_0} = 4.76 + \log\left(\frac{0.10}{0.20}\right) = 4.76 - 0.30 = 4.46$$

Example 18-8: Estimate the pH of a buffer solution that is 0.20 M in $NH_4Cl(aq)$ and 0.050 M in $NH_3(aq)$. Take $pK_a = 9.24$ for $NH_4^+(aq)$ (Table 17-3).

Solution: The pH of the $NH_4^+(aq)$–$NH_3(aq)$ buffer is calculated using the Henderson-Hasselbalch equation. The stoichiometric concentrations of the conjugate acid and base are

$$[NH_4^+]_0 = [\text{acid}]_0 = 0.20\text{ M}$$
$$[NH_3]_0 = [\text{base}]_0 = 0.050\text{ M}$$

Thus, using Equation (18-14), we have

$$\text{pH} = pK_a + \log\frac{[\text{base}]_0}{[\text{acid}]_0} = 9.24 + \log\left(\frac{0.050\text{ M}}{0.20\text{ M}}\right) = 8.64$$

Note that the $NH_4^+(aq)$–$NH_3(aq)$ buffer is basic (pH > 7.0).

Problems 18-27 to 18-38 deal with the Henderson-Hasselbalch equation.

18-5. A BUFFER SOLUTION SUPPRESSES pH CHANGES WHEN ACID OR BASE IS ADDED

We now demonstrate the insensitivity of a buffer solution to the addition of small amounts of acid or base. Let the buffer be composed of a solution that is 0.10 M in $HC_2H_3O_2(aq)$ and 0.10 M in $C_2H_3O_2^-(aq)$. Suppose that 10.0 mL of 0.10 M $HCl(aq)$ is added to 100 mL of the buffer solution. The number of moles of acid in 10.0 mL of 0.10 M $HCl(aq)$ is

$$\text{moles of } H_3O^+ = MV = (0.10\,\text{mol}\cdot\text{L}^{-1})(0.010\text{ L}) = 0.0010\text{ mol}$$

The $H_3O^+(aq)$ reacts with $C_2H_3O_2^-(aq)$ in the buffer via the reaction

$$H_3O^+(aq) + C_2H_3O_2^-(aq) \rightarrow HC_2H_3O_2(aq) + H_2O(l)$$

Before the HCl is added, the number of moles of acetate ion in the buffer solution is

$$\begin{pmatrix}\text{moles of } C_2H_3O_2^- \\ \text{before HCl added}\end{pmatrix} = MV = (0.10\,\text{mol}\cdot\text{L}^{-1})(0.100\text{ L}) = 0.010\text{ mol}$$

The number of moles of $C_2H_3O_2^-(aq)$ after the addition of 10.0 mL of 0.10 M $HCl(aq)$ is

$$\begin{pmatrix}\text{moles of } C_2H_3O_2^- \\ \text{after HCl added}\end{pmatrix} = \begin{pmatrix}\text{moles of } C_2H_3O_2^- \\ \text{before HCl added}\end{pmatrix} - \begin{pmatrix}\text{moles of} \\ H_3O^+ \text{ added}\end{pmatrix}$$

$$= 0.010 \text{ mol} - 0.0010 \text{ mol} = 0.009 \text{ mol}$$

The number of moles of $HC_2H_3O_2(aq)$ after the addition of 0.0010 mol of $H_3O^+(aq)$ is

$$\begin{pmatrix}\text{moles of } HC_2H_3O_2 \\ \text{after HCl added}\end{pmatrix} = \begin{pmatrix}\text{moles of } HC_2H_3O_2 \\ \text{before HCl added}\end{pmatrix} + \begin{pmatrix}\text{moles of} \\ H_3O^+ \text{ added}\end{pmatrix}$$

$$= 0.010 \text{ mol} + 0.0010 \text{ mol} = 0.011 \text{ mol}$$

The pH of the buffer solution before the addition of the $HCl(aq)$ is

$$\text{pH} = \text{p}K_a + \log\frac{[\text{base}]_0}{[\text{acid}]_0} = 4.76 + \log\left(\frac{0.10 \text{ M}}{0.10 \text{ M}}\right) = 4.76$$

and the pH of the buffer solution after the addition of the $HCl(aq)$ is

$$\text{pH} = 4.76 + \log\frac{(0.009 \text{ mol}/0.110 \text{ L})}{(0.011 \text{ mol}/0.110 \text{ L})} = 4.67$$

Thus the pH of the solution changes by less than 0.10 pH unit. Notice that the volumes (0.110 L) cancel out in the log factor and thus we can work directly with the number of moles of the conjugate acid and base without the need to convert to molarities before taking the logarithm of the ratio.

In contrast to the above result, if we add 10.0 mL of 0.10 M $HCl(aq)$ to 100 mL of pure water, then the pH of the resulting solution is

$$[H_3O^+] = \frac{\text{moles of } H_3O^+}{\text{volume of solution}} = \frac{(0.10 \text{ mol}\cdot\text{L}^{-1})(0.010 \text{ L})}{(0.110 \text{ L})}$$

$$= 9.1 \times 10^{-3} \text{ M}$$

or $$\text{pH} = -\log[H_3O^+] = -\log(9.1 \times 10^{-3}) = 2.04$$

The pH of pure water is 7.0, and so we see that the pH changes by 5.0 units ($7.0 - 2.0 = 5.0$). The value of $[H_3O^+]$ changes by a factor of 10^5, or 100,000, whereas in the buffer solution the pH does not change significantly. Similar results are obtained when 10.0 mL of 0.10 M $NaOH(aq)$ is added to 100 mL of the 0.10 M acetic acid–acetate buffer.

Example 18-9: Compute the change in pH when 10.0 mL of 0.10 M $NaOH(aq)$ is added to 100 mL of a buffer that is 0.10 M in $HC_2H_3O_2(aq)$ and 0.10 M in $C_2H_3O_2^-(aq)$. Take $pK_a = 4.76$ for $HC_2H_3O_2(aq)$.

Solution: The number of moles of base in 10.0 mL of 0.10 M $NaOH(aq)$ is

$$\text{moles of OH}^- = MV = (0.10\ \text{mol}\cdot\text{L}^{-1})(0.010\ \text{L})$$
$$= 0.0010\ \text{mol}$$

The $OH^-(aq)$ reacts with $HC_2H_3O_2(aq)$ in the buffer via the reaction

$$\text{OH}^-(aq) + \text{HC}_2\text{H}_3\text{O}_2(aq) \rightarrow \text{H}_2\text{O}(l) + \text{C}_2\text{H}_3\text{O}_2^-(aq)$$

The number of moles of acetic acid in the buffer solution before the addition of $NaOH(aq)$ is

$$\begin{pmatrix}\text{moles of HC}_2\text{H}_3\text{O}_2\\ \text{before NaOH added}\end{pmatrix} = MV = (0.10\ \text{mol}\cdot\text{L}^{-1})(0.100\ \text{L})$$
$$= 0.010\ \text{mol}$$

The buffering action of the $HC_2H_3O_2(aq)$–$C_2H_3O_2^-(aq)$ buffer can be seen in Figure 18-6. It is the region between 5 and 45 mL of added base where the titration curve is relatively flat.

The number of moles of acetic acid after the addition of 10.0 mL of 0.10 M $NaOH(aq)$ is

$$\begin{pmatrix}\text{moles of HC}_2\text{H}_3\text{O}_2\\ \text{after NaOH added}\end{pmatrix} = \begin{pmatrix}\text{moles of HC}_2\text{H}_3\text{O}_2\\ \text{before NaOH added}\end{pmatrix} - \begin{pmatrix}\text{moles of}\\ \text{OH}^- \text{ added}\end{pmatrix}$$
$$= 0.010\ \text{mol} - 0.0010\ \text{mol} = 0.009\ \text{mol}$$

The number of moles of $C_2H_3O_2^-(aq)$ following the addition of 0.0010 mol of $NaOH(aq)$ is

$$\begin{pmatrix}\text{moles of C}_2\text{H}_3\text{O}_2^-\\ \text{after NaOH added}\end{pmatrix} = \begin{pmatrix}\text{moles of C}_2\text{H}_3\text{O}_2^-\\ \text{before NaOH added}\end{pmatrix} + \begin{pmatrix}\text{moles of}\\ \text{OH}^- \text{ added}\end{pmatrix}$$
$$= 0.010\ \text{mol} + 0.001\ \text{mol} = 0.011\ \text{mol}$$

The pH is

$$\text{pH} = \text{p}K_a + \log\frac{[\text{base}]_0}{[\text{acid}]_0} = 4.76 + \log\frac{0.011}{0.009} = 4.85$$

Once again, notice that the pH changes by less than 0.1 pH unit.

The capacity of a buffer to resist changes in pH is not unlimited. If sufficient acid (or base) is added to neutralize all the conjugate base (or all the conjugate acid), then the pH of the solution will change significantly. In such a case the buffer is simply overwhelmed.

Buffer capacity is an indication of the amount of acid or base that can be added to a buffer before it is no longer able to resist pH change.

Table 18-3 Resistance of a buffer to change in pH upon addition of acid or base and upon dilution

Buffer	*Initial pH*	*pH after addition of 10.0 mL of 0.10 M HCl to 100 mL of buffer*	*pH after addition of 10.0 mL of 0.10 M NaOH to 100 mL of buffer*	*pH after twofold dilution with water*
0.10 M $HC_2H_3O_2(aq)$ plus 0.10 M $C_2H_3O_2^-(aq)$	4.76	4.67	4.85	4.76
0.10 M $NH_4^+(aq)$ plus 0.10 M $NH_3(aq)$	9.24	9.15	9.33	9.24

Problems 18-39 through 18-48 deal with buffers.

Another property of buffers is their ability to resist changes in pH upon dilution with solvent. This property is readily understood from the Henderson-Hasselbalch equation. If we dilute a buffer solution by, say, a factor of 2, then the stoichiometric concentration of the base, $[\text{base}]_0$, and the stoichiometric concentration of the acid, $[\text{acid}]_0$, both decrease by a factor of 2, but the ratio of the stoichiometric concentrations does not change:

$$\frac{[\text{base}]_0/2}{[\text{acid}]_0/2} = \frac{[\text{base}]_0}{[\text{acid}]_0}$$

and thus the pH of the buffer solution does not change on dilution. The effects of the addition of $HCl(aq)$ and $NaOH(aq)$ and of dilution on the pH of $HC_2H_3O_2(aq)$–$C_2H_3O_2^-(aq)$ and $NH_4^+(aq)$–$NH_3(aq)$ buffers are summarized in Table 18-3.

18-6. BUFFERS CONTROL THE pH OF BLOOD

The principal utility of buffers is in the control of pH. Innumerable chemical and biochemical processes depend upon the value of $[H_3O^+]$, and the control of this key concentration is achieved by a buffer. The buffer must be chemically inert to the species in the system whose pH is to be controlled.

The activity of many enzymes depends upon pH, and numerous biological systems have natural buffers to control enzyme activities. Essentially constant pH values are required for maintaining the delicate balances in the complex sequences of biochemical reactions essential to the existence of life. For example, blood is buffered by a mixture of carbonates, phosphates, and proteins and exhibits a remarkably constant pH value of 7.4. At pH values lower than 7.3, the blood cannot efficiently remove CO_2 from the cells, and at pH values greater than 7.7 the blood cannot efficiently release CO_2 to the lungs. Blood pH values outside the range 7.0 to 7.8 cannot sustain life.

18-7. POLYPROTIC ACIDS CAN DONATE MORE THAN ONE PROTON

Acids that have more than one dissociable proton are called *polyprotic acids*. Examples are phosphoric acid, H_3PO_4, a triprotic acid, and carbonic acid, H_2CO_3, a diprotic acid. One mole of H_3PO_4 is capable of neutralizing 3 mol of strong base:

$$H_3PO_4(aq) + 3OH^-(aq) \rightarrow 3H_2O(l) + PO_4^{3-}(aq)$$

Phosphoric acid has three distinct acid dissociation constants:

1. Dissociation of phosphoric acid

$$H_3PO_4(aq) + H_2O(l) \rightarrow H_3O^+(aq) + H_2PO_4^-(aq)$$

$$K_{a1} = \frac{[H_3O^+][H_2PO_4^-]}{[H_3PO_4]} = 7.1 \times 10^{-3}\ \text{M}$$

2. Dissociation of dihydrogen phosphate

$$H_2PO_4^-(aq) + H_2O(l) \rightleftharpoons H_3O^+(aq) + HPO_4^{2-}(aq)$$

$$K_{a2} = \frac{[H_3O^+][HPO_4^{2-}]}{[H_2PO_4^-]} = 6.2 \times 10^{-8}\ \text{M}$$

3. Dissociation of monohydrogen phosphate

$$HPO_4^{2-}(aq) + H_2O(l) \rightleftharpoons H_3O^+(aq) + PO_4^{3-}(aq)$$

$$K_{a3} = \frac{[H_3O^+][PO_4^{3-}]}{[HPO_4^{2-}]} = 4.4 \times 10^{-13}\ \text{M}$$

As a rough rule of thumb, each successive acid dissociation constant of a polyprotic acid is 10^{-5} times the value of the preceding one. In other words, each succeeding proton of a polyprotic acid is more difficult to remove, primarily because of the extra energy required to separate a (positively charged) proton from a negatively charged ion. The dissociation constants for some polyprotic acids are given in Table 18-4.

Let's calculate the concentration of all the species in a 0.100 M $H_3PO_4(aq)$ solution. The relevant equilibria are the three acid dissociation equilibria given above, plus one additional equilibrium expression,

$$H_2O(l) + H_2O(l) \rightleftharpoons H_3O^+(aq) + OH^-(aq) \qquad K_w = 1.00 \times 10^{-14}\ \text{M}^2$$

Table 18-4 Acid dissociation constants for some polyprotic acids in water at 25°C

Acid	pK_{a1}	pK_{a2}	pK_{a3}
	Diprotic		
sulfuric H_2SO_4	strong	2.00	—
oxalic $H_2C_2O_4$	1.27	4.27	—
carbonic $H_2CO_3(aq)$	6.35	10.33	—
hydrogen sulfide H_2S	7.04	12.92	—
	Triprotic		
phosphoric H_3PO_4	2.15	7.21	12.36

We also have a *mass balance* of phosphorus,

$$0.100\text{ M} = [H_3PO_4] + [H_2PO_4^-] + [HPO_4^{2-}] + [PO_4^{3-}]$$

and an *electroneutrality condition*,

Electroneutrality means that the total amount of positive charge is equal to the total amount of negative charge; that is, the system in electrically neutral.

$$[H_3O^+] = [H_2PO_4^-] + 2[HPO_4^{2-}] + 3[PO_4^{3-}] + [OH^-]$$

The mass balance condition says that the total of all forms of phosphorus must equal the stoichiometric amount of phosphorus, and the electroneutrality condition says that the sum of all cation charges must equal the sum of all anion charges. This gives us six equations and six unknowns—$[H_3PO_4]$, $[H_2PO_4^-]$, $[HPO_4^{2-}]$, $[PO_4^{3-}]$, $[H_3O^+]$, and $[OH^-]$—and so the problem can be solved by brute force. However, this is not necessary because it can be solved much more easily by exploiting the fact that K_{a1} is much greater than K_{a2}, which in turn is much greater than K_{a3} or K_w. This allows us to neglect the second and third dissociations of H_3PO_4 and the dissociation of H_2O. The problem now becomes similar to the case of the dissociation of a monoprotic acid. We set up a table of concentrations for the first dissociation of H_3PO_4:

	$H_3PO_4(aq)$ +	$H_2O(l)$ $\rightleftharpoons$	$H_3O^+(aq)$ +	$H_2PO_4^-(aq)$
initial concentration	0.100 M	—	≈ 0	0
equilibrium concentration	$0.100\text{ M} - [H_2PO_4^-]$	—	$[H_3O^+]$	$[H_2PO_4^-]$
or, using $[H_3O^+] \approx [H_2PO_4^-]$ from the reaction stoichiometry	$0.100\text{ M} - [H_3O^+]$	—	$[H_3O^+]$	$[H_3O^+]$

The equilibrium constant expression gives

$$K_{a1} = \frac{[H_3O^+][H_2PO_4^-]}{[H_3PO_4]} \approx \frac{[H_3O^+]^2}{0.100\text{ M} - [H_3O^+]} = 7.1 \times 10^{-3}\text{ M}$$

The solution of the quadratic equation for $[H_3O]^+$ yields $[H_3O^+] =$ 0.023 M, for a pH of 1.64. The concentrations of $H_3PO_4(aq)$ and $H_2PO_4^-(aq)$ are

$$[H_2PO_4^-] \approx 0.023\text{ M}$$
$$[H_3PO_4] \approx 0.100\text{ M} - 0.023\text{ M} = 0.077\text{ M}$$

We have used approximate equalities here because we have neglected the second and third dissociations of H_3PO_4.

Let's calculate the concentrations of the neglected species using these results. For example, using the K_{a2} expression, we find that

$$[HPO_4^{2-}] \approx \frac{K_{a2}[H_2PO_4^-]}{[H_3O^+]} = \frac{(6.2 \times 10^{-8}\text{ M})(0.023\text{ M})}{(0.023\text{ M})} = 6.2 \times 10^{-8}\text{ M}$$

and using the K_{a3} expression,

$$[PO_4^{3-}] \approx \frac{K_{a3}[HPO_4^-]}{[H_3O^+]} = \frac{(4.4 \times 10^{-13}\text{ M})(6.2 \times 10^{-8}\text{ M})}{(0.023\text{ M})} = 1.2 \times 10^{-18}\text{ M}$$

These results show that $[HPO_4^{2-}]$ and $[PO_4^{3-}]$ are indeed negligible relative to $[H_2PO_4^-]$. We can check our results by appealing to the mass balance and electroneutrality conditions:

$$0.100 = [H_3PO_4] + [H_2PO_4^-] + [HPO_4^{2-}] + [PO_4^{3-}]$$
$$\approx 0.077\text{ M} + 0.023\text{ M} + (6.2 \times 10^{-8}\text{ M}) + (1.2 \times 10^{-18}\text{ M})$$

and

$$[H_3O^+] = [H_2PO_4^-] + 2[HPO_4^{2-}] + 3[PO_4^{3-}] + [OH^-]$$

or

$$0.023\text{ M} \approx 0.023\text{ M} + 2(6.2 \times 10^{-8}\text{ M}) + 3(1.2 \times 10^{-18}\text{ M}) + (4.3 \times 10^{-13}\text{ M})$$

Thus, we see that the assumptions that we have used are valid. Calculations involving other polyprotic acids are carried out in a similar manner (Problems 18-59 and 18-60).

Figure 18-8 Distribution diagrams for $HC_2H_3O_2(aq)$, $H_2CO_3(aq)$, and $H_3PO_4(aq)$. A distribution diagram gives the fraction of the total that each species constitutes at each pH. For example, in the acetic acid case, $f_{HC_2H_3O_2} + f_{C_2H_3O_2^-} = 1$ at any pH. At low pH values $f_{HC_2H_3O_2} = 1$, whereas at high pH values $f_{C_2H_3O_2^-} = 1$. At $pH = pK_a$, $f_{HC_2H_3O_2} = f_{C_2H_3O_2^-} = 0.50$. Notice that in the H_3PO_4 system there are at most two species present at appreciable concentrations at any particular pH. However, at any concentration $f_{H_3PO_4} + f_{H_2PO_4^-} + f_{HPO_4^{2-}} + f_{PO_4^{3-}} = 1$.

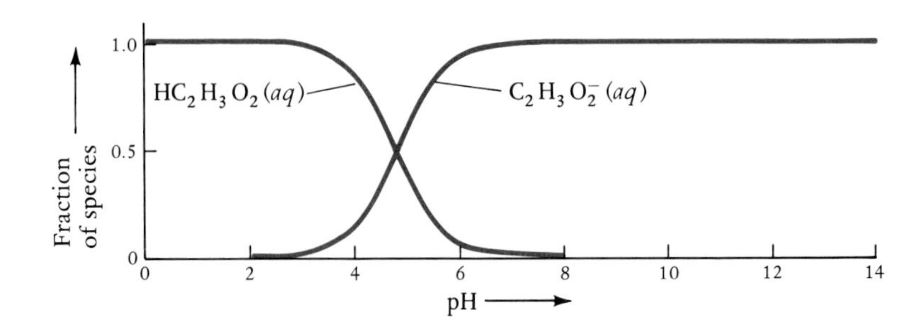

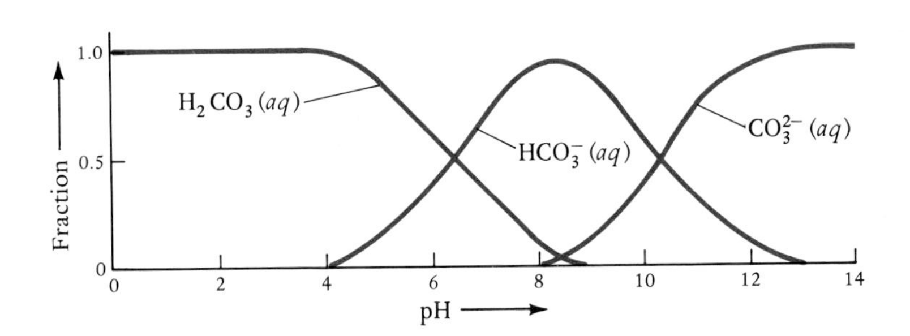

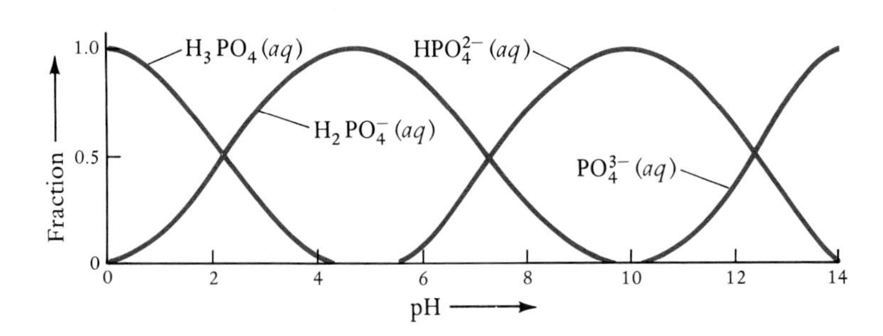

The simplest way to depict the variation with pH in the species arising from a polyprotic acid is by means of a *distribution diagram*, which shows how the relative amounts of an acid and its conjugate base change with pH (Figure 18-8). The distribution diagram for $H_3PO_4(aq)$ shows that, although three separate acid dissociation equilibria are involved, only one of the three acid dissociation reactions predominates at most pH values. This is a consequence of the fact that the three K_a values are widely separated in magnitude. For example, at pH values lower than 5, the concentrations of the major species are determined by the first acid dissociation equilibrium,

$$H_3PO_4(aq) + H_2O(l) \rightleftharpoons H_3O^+(aq) + H_2PO_4^-(aq)$$

and the species $HPO_4^{2-}(aq)$ and $PO_4^{3-}(aq)$ are present only at very low concentrations. The following example involves the calculation of parts of the distribution diagram of $H_3PO_4(aq)$.

Example 18-10: Calculate the concentrations of $H_3PO_4(aq)$, $H_2PO_4^-(aq)$, $HPO_4^{2-}(aq)$, and $PO_4^{3-}(aq)$ at pH = 2.00 and pH = 12.00 in a solution that is 0.100 M in total phosphorus species.

Solution: Using the three acid dissociation expressions for $H_3PO_4(aq)$, we have

$$\frac{[H_2PO_4^-]}{[H_3PO_4]} = \frac{K_{a1}}{[H_3O^+]} = \frac{7.1 \times 10^{-3}\text{ M}}{[H_3O^+]}$$

$$\frac{[HPO_4^{2-}]}{[H_2PO_4^-]} = \frac{K_{a2}}{[H_3O^+]} = \frac{6.2 \times 10^{-8}\text{ M}}{[H_3O^+]}$$

$$\frac{[PO_4^{3-}]}{[HPO_4^{2-}]} = \frac{K_{a3}}{[H_3O^+]} = \frac{4.4 \times 10^{-13}\text{ M}}{[H_3O^+]}$$

Substituting $[H_3O^+] = 1.00 \times 10^{-2}$ M (pH = 2.00) into these expressions gives

$$\frac{[H_2PO_4^-]}{[H_3PO_4]} = \frac{7.1 \times 10^{-3}\text{ M}}{1.00 \times 10^{-2}\text{ M}} = 0.71$$

$$\frac{[HPO_4^{2-}]}{[H_2PO_4^-]} = \frac{6.2 \times 10^{-8}\text{ M}}{1.00 \times 10^{-2}\text{ M}} = 6.2 \times 10^{-6}$$

$$\frac{[PO_4^{3-}]}{[HPO_4^{2-}]} = \frac{4.4 \times 10^{-13}\text{ M}}{1.00 \times 10^{-2}\text{ M}} = 4.4 \times 10^{-11}$$

Thus, we see that $[H_2PO_4^-]$ and $[H_3PO_4]$ are similar in magnitude but $[HPO_4^{2-}]$ and $[PO_4^{3-}]$ are extremely small in comparison. To a very good approximation, then, we can write

$$[H_3PO_4] + [H_2PO_4^-] \approx 0.100\text{ M}$$

and, using the first ratio, we have

$$[H_3PO_4] + 0.71[H_3PO_4] \approx 0.100\text{ M}$$

$$[H_3PO_4] \approx \frac{0.100\text{ M}}{1.71} = 0.058\text{ M}$$

Therefore,

$$[H_2PO_4^-] \approx 0.100\text{ M} - 0.058\text{ M} = 0.042\text{ M}$$

The ratio of each of these concentrations to total phosphorus concentration represents the fraction f of each species in the solution:

$$f_{H_3PO_4} = \frac{[H_3PO_4]}{0.100\text{ M}} = \frac{0.058\text{ M}}{0.100\text{ M}} = 0.58$$

$$f_{H_2PO_4^-} = \frac{[H_2PO_4^-]}{0.100\text{ M}} = \frac{0.042\text{ M}}{0.100\text{ M}} = 0.42$$

The concentrations of the other species are negligible at this pH. This result is in excellent agreement with Figure 18-8.

Now let's calculate the concentrations at pH = 12.00. If we substitute $[H_3O^+] = 1.00 \times 10^{-12}$ M (pH = 12.00) into the first three equations, then we obtain

$$\frac{[H_2PO_4^-]}{[H_3PO_4]} = \frac{7.1 \times 10^{-3}\text{ M}}{1.00 \times 10^{-12}\text{ M}} = 7.1 \times 10^9$$

$$\frac{[HPO_4^{2-}]}{[H_2PO_4^-]} = \frac{6.2 \times 10^{-8}\text{ M}}{1.00 \times 10^{-12}\text{ M}} = 6.2 \times 10^4$$

$$\frac{[PO_4^{3-}]}{[HPO_4^{2-}]} = \frac{4.4 \times 10^{-13}\text{ M}}{1.00 \times 10^{-12}\text{ M}} = 0.44$$

We interpret these ratios by saying that there is very little $H_3PO_4(aq)$ and $H_2PO_4^-(aq)$ present at pH = 12.00 and that $[HPO_4^{2-}]$ and $[PO_4^{3-}]$ are comparable in magnitude. To a very good approximation, then, we can write

$$[HPO_4^{2-}] + [PO_4^{3-}] \approx 0.100\text{ M}$$

and, using the third ratio, we have

$$[HPO_4^{2-}] + 0.44\,[HPO_4^{2-}] \approx 0.100\text{ M}$$

$$[HPO_4^{2-}] \approx \frac{0.100\text{ M}}{1.44} = 0.069\text{ M}$$

Therefore

$$[PO_4^{3-}] \approx 0.100\text{ M} - 0.069\text{ M} = 0.031\text{ M}$$

The ratio of each of these concentrations to the total phosphorus concentration is the fraction of each species in the solution:

$$f_{HPO_4^{2-}} = \frac{0.069\text{ M}}{0.100\text{ M}} = 0.69$$

$$f_{PO_4^{3-}} = \frac{0.031\text{ M}}{0.100\text{ M}} = 0.31$$

once again in good accord with Figure 18-8. The calculation of the fractions of various species at other values of pH is carried out in a similar manner.

Problems 18-49 through 18-62 deal with polyprotic acids.

The titration curve for $H_3PO_4(aq)$ titrated with $KOH(aq)$ is shown in Figure 18-9. Note that there are three equivalence points because $H_3PO_4(aq)$ has three dissociable protons.

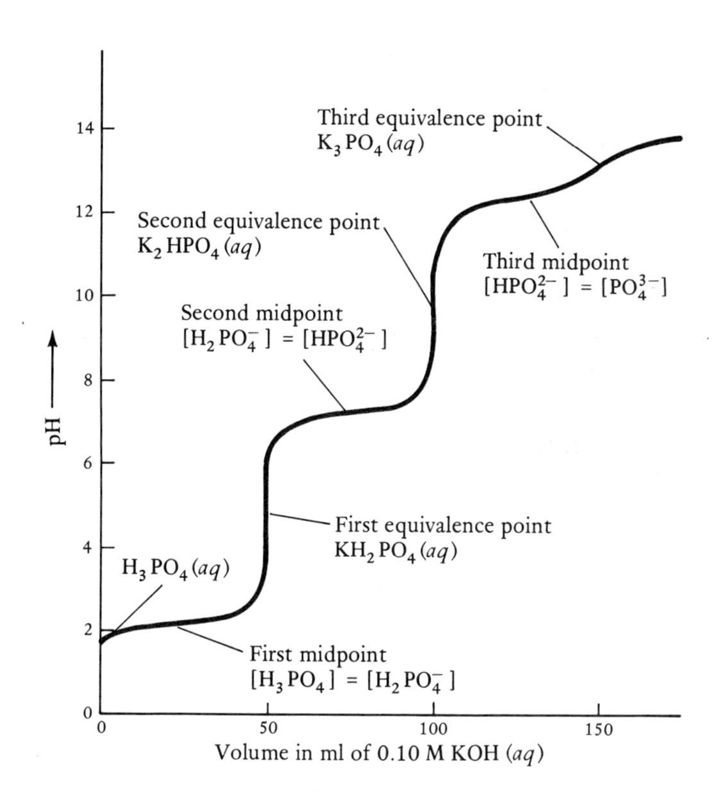

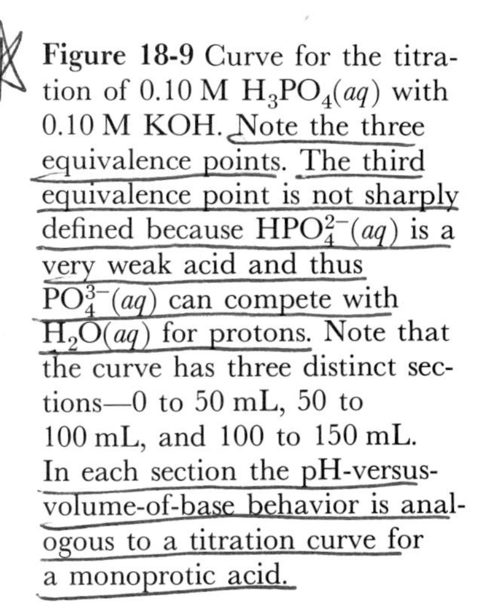
Figure 18-9 Curve for the titration of 0.10 M $H_3PO_4(aq)$ with 0.10 M KOH. Note the three equivalence points. The third equivalence point is not sharply defined because $HPO_4^{2-}(aq)$ is a very weak acid and thus $PO_4^{3-}(aq)$ can compete with $H_2O(aq)$ for protons. Note that the curve has three distinct sections—0 to 50 mL, 50 to 100 mL, and 100 to 150 mL. In each section the pH-versus-volume-of-base behavior is analogous to a titration curve for a monoprotic acid.

SUMMARY

A colored weak acid whose conjugate base has a different color can be used as a pH indicator. Indicators can be used to signal the equivalence point in an acid-base titration, that is, the point on the titration curve at which the number of moles of acid (or base) originally present is equal to the number of moles of base (or acid) added as titrant. The pH of a titration curve undergoes an especially rapid change in the vicinity of the equivalence point, and this rapid change in pH can be used to detect the equivalence point.

A buffer is a mixture of a weak acid and its conjugate base. A buffer suppresses the change in pH that would otherwise result from the addition of an acid or a base to a solution. The pH of a buffer solution can be calculated from the Henderson-Hasselbalch equation.

A polyprotic acid is an acid that can donate more than one proton per acid molecule. The variation in composition of a polyprotic acid solution with change in pH is displayed graphically on a distribution diagram, which shows the relative proportions of the various acids and conjugate bases as a function of pH.

TERMS YOU SHOULD KNOW

indicator 697	midpoint 707
litmus paper 699	buffer 710
titration 701	Henderson-Hasselbalch equation 712
titrant 701	polyprotic acid 717
titration curve 701	mass balance 718
equivalence point 701	electroneutrality 718
end point 701	distribution diagram 720

EQUATIONS YOU SHOULD KNOW HOW TO USE

$$pH = pK_a + \log\frac{[\text{base}]_0}{[\text{acid}]_0} \quad \text{(Henderson-Hasselbalch equation)} \qquad (18\text{-}14)$$

(for buffers)

PROBLEMS

INDICATORS

18-1. Estimate the pH of an aqueous ammonia solution that turns red when phenolphthalein is added and blue-violet when Poirrier's blue is added.

$10 < pH < 11$

18-2. Estimate the pH of a colorless aqueous solution that turns yellow when methyl orange is added and yellow when bromcresol purple is added.

18-3. A colorless aqueous solution was obtained from a soil sample. Litmus paper indicated that the solution was basic. The following indicators were added to separate samples of the solution and the colors observed were

nile blue—blue
neutral red—yellow
thymol blue—green

Estimate the pH of the solution. What indicator would you use to obtain a better estimate?

$8 < pH < 9.5$ tropeolin OOO (changes around 9)

18-4. We wish to estimate the pH of a colorless aqueous solution. Litmus paper indicates that the solution is acidic. The following indicators are added to separate samples of the solution, and the colors observed are

o-cresol red—yellow
methyl orange—orange
methyl red—red

What is the pH of the solution? What indicator would you use to obtain a better estimate?

18-5. A certain bacterium grows best in an acidic medium, and in an experiment with this bacterium the pH must be maintained at around 5. What indicator should be added to indicate a pH of 5 and to monitor changes in the pH of the medium?

bromcresol green (middle of range at 5)

18-6. The pH of the nutrient broth used to maintain cultures of tissue samples must be greater than 7 but should not rise above 8. What indicator can be added to the nutrient broth to monitor the pH?

18-7. The pH indicator congo red changes from blue to red over the pH range 3.0 to 5.0. Estimate the K_{ai} for congo red.

$[HIn] = [In^-]$ $K_{ai} \approx [H_3O^+]$ $K_{ai} = 1 \times 10^{-4}\ M$

18-8. The pH indicator nile blue changes from blue to pink over the pH range 10 to 11. Estimate the K_a of nile blue.

STRONG ACID–STRONG BASE TITRATIONS

18-9. Calculate the pH of a solution obtained by adding 51.0 mL of 0.100 M NaOH(*aq*) to 50.0 mL of 0.100 M HCl(*aq*).

strong acid / strong base
find moles of each.
$0.100\ M\ NaOH \times 0.051\ L = 5.10 \times 10^{-3}$ mol
$0.100\ M\ HCl \times 0.050\ L = 5.00 \times 10^{-3}$ mol
moles unreacted $OH^- = 5.10 \times 10^{-3} - 5.00 \times 10^{-3} = 1.0 \times 10^{-4}$ mol
$[OH^-] = \frac{1.0 \times 10^{-4}\ mol}{0.1010\ L} = 9.9 \times 10^{-4}\ M$ $[H_3O^+] = \frac{1.0 \times 10^{-14}}{9.9 \times 10^{-4}} = 1.00 \times 10^{-11}$ pH = 11

18-10. Calculate the pH of a solution obtained by adding 20.0 mL of 0.100 M HCl(*aq*) to 50.0 mL of 0.100 M NaOH(*aq*).

18-11. Sketch, but do not calculate, the titration curve for the titration of 50.0 mL of 0.100 M NaOH(*aq*) with 0.100 M HCl(*aq*).

18-12. Sketch, but do not calculate, the titration curve for the titration of 50.0 mL of 0.25 M HNO_3(*aq*) with 0.50 M KOH(*aq*).

18-13. Calculate the molarity of an aqueous nitric acid solution if it requires 32.1 mL of 0.150 M NaOH(*aq*) to neutralize 25.0 mL of the nitric acid.

18-14. Calculate the molarity of an aqueous sodium hydroxide solution if it requires 34.7 mL of 0.125 M HCl(*aq*) to neutralize 15.0 mL of the NaOH(*aq*) solution.

18-15. Various antacid tablets contain water-insoluble metal hydroxides, such as $Mg(OH)_2(s)$. Given that stomach acid is about 0.10 M in HCl(*aq*), compute the number of milliliters of stomach acid that can be neutralized by 1.00 g of $Mg(OH)_2(s)$.

18-16. Several types of commercial antacid tablets contain $Al(OH)_3(s)$ as the active ingredient. Given that stomach acid is about 0.10 M in HCl(*aq*), compute the number of milliliters of stomach acid that can be neutralized by 500 mg of $Al(OH)_3(s)$.

18-17. Given the following results for the titration of 30.0 mL of a 0.100 M HCl(*aq*) solution with a 0.100 M NaOH(*aq*) solution, plot the titration curve:

Volume of NaOH(aq) solution added/mL	*pH*
0.0	1.00
10.0	1.30
20.0	1.70
25.0	2.04
29.0	2.77
30.0	7.00
31.0	11.22
35.0	11.89
40.0	12.16
50.0	12.40

18-18. Suppose that you are titrating 25.0 mL of a 0.250 M HBr(*aq*) solution with a 0.250 M NaOH(*aq*) solution. The equivalence point occurs when 25.0 mL of the NaOH(*aq*) solution has been added. The pH at this point is 7.0. Calculate the pH if one more drop (0.05 mL) of the NaOH(*aq*) solution is added.

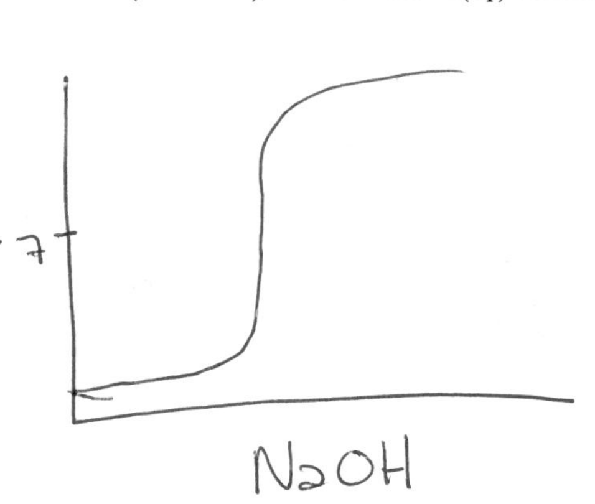

WEAK ACID NEUTRALIZATIONS

18-19. Calculate the pH of the solution obtained by titrating 50.0 mL of a 0.100 M solution of the herbicide cacodylic acid, $HC_2H_6AsO_2$, with 0.100 M NaOH(*aq*) to the equivalence point. Take $K_a = 5.4 \times 10^{-7}$ M. What indicator should be used to signal the equivalence point?

18-20. Calculate the pH of the solution obtained by titrating 50.0 mL of 0.100 M HNO_2(*aq*) with 0.100 M NaOH(*aq*) to the equivalence point. Take $K_a = 4.5 \times 10^{-4}$ M for HNO_2(*aq*). What indicator should be used to signal the equivalence point?

18-21. A 1.50-g sample of ascorbic acid, vitamin C, is dissolved in 100 mL of water and titrated with 0.250 M NaOH(*aq*) to the methyl orange equivalence point. The volume of base consumed is 34.1 mL. Calculate the molecular mass of vitamin C assuming one dissociable proton per molecule.

18-22. A 0.772-g sample of benzoic acid, which is found in most berries, is dissolved in 50.0 mL of water and titrated to the equivalence point with 0.250 M NaOH(*aq*). The volume of base consumed is 25.3 mL. Calculate the molecular mass of benzoic acid, assuming one dissociable proton.

18-23. An unknown sample is thought to be either benzoic acid, $HC_7H_5O_2$, or chlorobenzoic acid, $HC_7H_4ClO_2$. When 1.89 g is dissolved in water, 15.49 mL of 1.00 M NaOH(*aq*) is required to reach the equivalence point. Which acid is the unknown sample?

18-24. A 0.550-g sample of butyric acid is dissolved in 100 mL of water and titrated with 0.100 M NaOH(*aq*) to the equivalence point. The volume of base consumed is 62.4 mL. Calculate the molecular mass of butyric acid, which has one dissociable proton.

18-25. Vinegar is a dilute aqueous solution of acetic acid. A 21.0-mL sample of vinegar requires 38.5 mL of 0.400 M NaOH(*aq*) to neutralize the $HC_2H_3O_2(aq)$. Given that the density of the vinegar is 1.060 $g \cdot mL^{-1}$, calculate the mass percentage of acetic acid in the vinegar.

18-26. A 2.00-g sample of acetylsalicylic acid, better known as aspirin, is dissolved in 100 mL of water and titrated with 0.200 M NaOH(*aq*) to the equivalence point. The volume of base required is 55.5 mL. Calculate the molecular mass of the acetylsalicylic acid, which has one dissociable proton per molecule.

HENDERSON-HASSELBALCH EQUATION

18-27. Estimate the pH of a solution that is 0.050 M in $HC_2H_3O_2(aq)$ and 0.050 M in $NaC_2H_3O_2(aq)$. The value of K_a is 1.74×10^{-5} M.

18-28. Estimate the pH of a solution that is 0.10 M in $HC_2H_3O_2(aq)$ and 0.20 M in $NaC_2H_3O_2(aq)$. The value of K_a is 1.74×10^{-5} M.

18-29. Estimate the pH of a solution that is 0.25 M in $HCHO_2(aq)$ and 0.20 M in $NaCHO_2(aq)$. The value of K_a is 1.8×10^{-4} M.

18-30. Estimate the pH of a solution that is 0.15 M in $HNO_2(aq)$ and 0.25 M in $NaNO_2(aq)$. The value of K_a is 4.5×10^{-4} M.

18-31. The pK_a for the ammonium ion, $NH_4^+(aq)$, is 9.24 at 25°C. Compute the pH of an aqueous solution that is 0.20 M in $NH_3(aq)$ and 0.10 M in $NH_4Cl(aq)$.

18-32. The pK_a for the ammonium ion, $NH_4^+(aq)$, is 9.24 at 25°C. Compute the pH of a $NH_4^+(aq)$–$NH_3(aq)$ buffer solution that is 0.40 M in $NH_4Cl(aq)$ and 0.20 M in $NH_3(aq)$.

18-33. A commonly used buffer in biological experiments is a phosphate buffer containing NaH_2PO_4 and Na_2HPO_4. Estimate the pH of a solution that is

(a) 0.050 M NaH_2PO_4 and 0.050 M Na_2HPO_4
(b) 0.050 M NaH_2PO_4 and 0.10 M Na_2HPO_4
(c) 0.10 M NaH_2PO_4 and 0.050 M Na_2HPO_4

The relevant equation is

$$H_2PO_4^-(aq) + H_2O(l) \rightleftharpoons H_3O^+(aq) + HPO_4^{2-}(aq)$$

$$K_a = 6.2 \times 10^{-8}\ M$$

18-34. Calculate the pH of an aqueous solution that is 0.200 M in pyridine, C_5H_5N, and 0.250 M in pyridinium chloride, $C_5H_5NH^+Cl^-$, if pK_a = 5.17.

18-35. The electronic meters used to measure pH are calibrated using standard buffer solutions of known pH. For example, the directions for preparing a certain buffer at 25°C are as follows. Dissolve 3.40 ± 0.01 g of $KH_2PO_4(s)$ and 3.55 ± 0.01 g of $Na_2HPO_4(s)$ in sufficient water to make 1.00 L of solution. What pH do you calculate for this solution, using pK_a = 7.21?

18-36. Calculate the pH of a buffer solution obtained by dissolving 10.0 g of $KH_2PO_4(s)$ and 20.0 g of $Na_2HPO_4(s)$ in water and then diluting to 1.00 L. The relevant equation is $H_2PO_4^-(aq) + H_2O(l) \rightleftharpoons H_3O^+(aq) + HPO_4^{2-}(aq)$ with pK_a = 7.21.

18-37. It is important always to remember that the Henderson-Hasselbalch equation is based upon the assumption that [acid] ≈ $[\text{acid}]_0$ and [base] ≈ $[\text{base}]_0$. Calculate $[HC_2H_3O_2]$, $[HC_2H_3O_2]_0$, $[C_2H_3O_2^-]$, and $[C_2H_3O_2^-]_0$ for a solution that is 0.100 M in both $HC_2H_3O_2(aq)$ and $NaC_2H_3O_2(aq)$. Is $[HC_2H_3O_2] \approx [HC_2H_3O_2]_0$ and $[C_2H_3O_2^-] \approx [C_2H_3O_2^-]_0$? Can you see from your calculation why this is so?

18-38. Indicate for which of the following solutions the Henderson-Hasselbalch equation cannot be used to calculate the pH:

(a) 0.15 M $HNO_2(aq)$ plus 0.20 M $NaNO_2(aq)$
(b) 0.15 M $HNO_2(aq)$
(c) 0.20 M $NaNO_2(aq)$
(d) 0.10 M $Na_2HPO_4(aq)$ plus 0.20 M $KH_2PO_4(aq)$

18-39. Calculate the change in pH when 10.0 mL of 0.10 M $HCl(aq)$ is added to 100 mL of a solution of $HCl(aq)$ whose pH is 4.76.

0.10M × 0.010L = 0.001 mol

$[H_3O^+] = 1.74\times10^{-5}$ M

1.74×10^{-5} M × 0.100L = 1.74×10^{-6} mol

$\frac{1.002\times10^{-3}\ mol}{0.11\ L} = 9.11\times10^{-3}$ pH = 2.04   4.76 − 2.04 = 2.72

18-40. Calculate the change in pH when 10.0 mL of 0.10 M $NaOH(aq)$ is added to 100 mL of a solution of $HCl(aq)$ whose pH is 4.76.

18-41. Suppose you are performing an experiment during which the pH must be maintained at 3.29. What would be an appropriate buffer to use?

pH = pK_a pK_a = 3.29 $K_a = 5.1\times10^{-4}$ M HNO_2 and $NaNO_2$

18-42. Suppose you are performing an experiment during which the pH must be maintained at 5.17. What would be an appropriate buffer to use?

18-43. The higher the concentration of acid and conjugate base in a buffer, the smaller the pH change when acid or base is added. Compute the pH change in the following two buffers when 1.00 g of $KOH(s)$ is added to each (the pK_a of $NH_4^+(aq)$ is 9.24 at 25°C):

(a) 500 mL of a 0.10 M $NH_4Cl(aq)$–0.10 M $NH_3(aq)$ buffer

0.10M × 0.50L = 0.05 mol NH_4Cl NH_3

$\frac{1.00\ g}{56.11\ g/mol}$ = 0.0178 mol

pH = 9.24 + log $\frac{(0.0678\ mol)}{(0.0322\ mol)}$ pH = 9.56   change = 0.33

(b) 500 mL of a 1.00 M $NH_4Cl(aq)$–1.00 M $NH_3(aq)$ buffer

1.00M × 0.50L = 0.5 mol NH_4Cl NH_3

pH = 9.24 + log $\frac{(0.5178\ mol)}{(0.4822\ mol)}$ pH = 9.27   change = 0.03

18-44. Calculate the change in pH when 5.00 mL of 0.100 M $HCl(aq)$ is added to 100 mL of a buffer solution that is 0.100 M in $NH_3(aq)$ and 0.100 M in $NH_4Cl(aq)$. Calculate the change in pH when 5.00 mL of 0.100 M $NaOH(aq)$ is added to the original solution. The pK_a of $NH_4^+(aq)$ is 9.24 at 25°C.

18-45. Calculate the change in pH when 20.0 mL of 0.200 M $NaOH(aq)$ is added to 50.0 mL of a buffer solution that is 0.150 M in $HC_2H_3O_2(aq)$ and 0.150 M in $NaC_2H_3O_2(aq)$. Can you use the Henderson-Hasselbalch equation to do this problem? The pK_a of $HC_2H_3O_2$ is 4.76.

0.200M NaOH × 0.02L = 4.0×10^{-3} mol

0.150M × 0.05L = 7.50×10^{-3} mol $HC_2H_3O_2$, mol $C_2H_3O_2^-$

pH = 4.76 + log $\frac{(11.50\times10^{-3}/0.07L)}{(3.50\times10^{-3}/0.07L)}$ pH = 5.28

18-46. Suppose that you wish to determine whether a solution of unknown composition is buffered. Explain how you could do this with only two pH measurements.

18-47. The principal reaction that occurs when a salt containing an anion that can act as either an acid or a base, for example, $NaHCO_3$, is dissolved in water is of the type

$H_2CO_3 + H_2O \rightleftharpoons H_3O^+ + HCO_3^-$ K_{a1}

$+\ HCO_3^- + H_2O \rightleftharpoons H_3O^+ + CO_3^{2-}$ K_{a2}

$$2HCO_3^-(aq) \rightleftharpoons CO_3^{2-}(aq) + H_2CO_3(aq)$$

$H_2CO_3 + 2H_2O \rightleftharpoons 2H_3O^+ + CO_3^{2-}$ $K = K_{a1}K_{a2}$

(a) Show that the $[H_3O^+]$ of the solution is given by

$$[H_3O^+] = (K_{a1}K_{a2})^{1/2}$$

$K = \frac{[CO_3^{2-}][H_3O^+]^2}{[H_2CO_3]}$ can cancel because of stoichiometry of $2HCO_3^- \rightleftharpoons CO_3^{2-} + H_2CO_3$

$K = [H_3O^+]^2$ $[H_3O^+] = \sqrt{K_{a1}K_{a2}}$

where K_{a1} and K_{a2} are the first and second acid dissociation constants of $H_2CO_3(aq)$. Note that the pH of the solution is independent of the salt concentration.

(b) Explain the buffer property of $NaHCO_3(aq)$ solutions.

$NaHCO_3$ can act as a buffer because HCO_3^- can act as either an acid or a base.

$HCO_3^- + H_3O^+ \rightleftharpoons H_2CO_3 + H_2O$

$HCO_3^- + OH^- \rightleftharpoons CO_3^{2-} + H_2O$

18-48. The principal reaction when a salt composed of an acidic cation and a basic anion, for example, $NH_4C_2H_3O_2$, is dissolved in water is of the type

$$NH_4^+(aq) + C_2H_3O_2^-(aq) \rightleftharpoons NH_3(aq) + HC_2H_3O_2(aq)$$

(a) Show that the value of the equilibrium constant for this reaction is given by

$$K = \frac{K_{a,NH_4^+}}{K_{a,HC_2H_3O_2}}$$

(b) Given the above stoichiometry, show that the $[H_3O^+]$ of the solution is equal to

$$[H_3O^+] = (K_{a,NH_4^+}K_{a,HC_2H_3O_2})^{1/2}$$

Note that $[H_3O^+]$, and thus the pH of the solution, are independent of the concentration of the salt.

(c) Is an $NH_4C_2H_3O_2(aq)$ solution a buffer? Explain.

18-49. Refer to Figure 18-8 to determine, for phosphoric acid, which species are present at appreciable concentrations at

(a) pH = 3.0 (c) pH = 8.0
(b) pH = 5.0 (d) pH = 10.0

18-50. Refer to Figure 18-8 to determine, for $H_2CO_3(aq)$, which species are present at appreciable concentrations at

(a) pH = 3.0 (c) pH = 8.0
(b) pH = 6.0 (d) pH = 10.0

18-51. Sketch the titration curve for 0.10 M $H_2CO_3(aq)$ titrated with 0.10 M $KOH(aq)$. Refer to Figures 18-8 and 18-9.

18-52. Which indicator should be used to signal the complete neutralization by $NaOH(aq)$ of an aqueous solution of carbonic acid?

18-53. Calculate the volume of 0.10 M $NaOH(aq)$ required to neutralize completely 25.0 mL of 0.10 M oxalic acid, $H_2C_2O_4$.

18-54. Calculate the volume of 0.10 M $NaOH(aq)$ required to neutralize completely 100 mL of 0.10 M H_2S.

18-55. A 0.500-g sample of oxalic acid, a poisonous component of rhubarb leaves, is dissolved in 100 mL of water and titrated with 0.250 M NaOH to the second equivalence point. The volume of base is 44.4 mL. Calculate the molecular mass of oxalic acid, which has two dissociable protons per molecule.

18-56. A 1.20-g sample of fumaric acid, an essential component in the production of energy in living cells, is dissolved in water and titrated with 0.300 M $NaOH(aq)$ to the second equivalence point. The volume of base is 69.0 mL. Calculate the molecular mass of fumaric acid, which contains two dissociable protons.

18-57. Calculate the pH of a 0.100 M $H_3AsO_4(aq)$ solution given that $pK_{a1} = 2.22$, $pK_{a2} = 6.96$, and $pK_{a3} = 11.40$. Neglect the second and third dissociation reactions.

18-58. Ascorbic acid (vitamin C) has the structure

H, O; $HOCH_2CH{-}C$ (with OH), $C{=}O$; $C{=}C$; HO, OH

$pK_{a2} = 11.57$ (HO) $pK_{a1} = 4.17$ (OH)

Compute the pH of an aqueous vitamin C solution obtained by dissolving 500 mg of vitamin C in 1.00 L of water. Neglect the second dissociation reaction.

18-59. Sketch the distribution diagram for the NH_3–NH_4^+ conjugate acid-base pair. The value of pK_a is 9.24.

18-60. Hydrogen sulfide is used in numerous separation schemes for metal ions. The schemes are based in part on the variation in $[S^{2-}]$ with pH. Make a sketch of the distribution diagram for $H_2S(aq)$ solutions over the pH range 0 to 14. Refer to Table 18-4.

18-61. Alka-Seltzer contains sodium bicarbonate, $NaHCO_3(s)$, and the triprotic acid citric acid, $H_3C_6H_5O_7$, in addition to 324 mg of aspirin. Write the acid-base reaction that gives rise to the fizz (CO_2 evolution) when an Alka-Seltzer tablet is dissolved in water. $3HCO_3^- + H_3C_6H_5O_7 \rightleftharpoons 3H_2CO_3 + C_6H_5O_7^{3-}$

$H_2CO_3 \rightleftharpoons CO_2 + H_2O$

$\rightleftharpoons CO_2$

18-62. Compute the number of grams of $NaHCO_3(s)$ required to neutralize 2.00 g of citric acid in water. Citric acid is a triprotic acid, $H_3C_6H_5O_7$.

CALCULATION OF A TITRATION CURVE

18-63. Suppose that 80.0 mL of a 0.200 M $KOH(aq)$ solution is titrated with a 0.400 M $HCl(aq)$ solution. Fill in the following table and compute the pH of the resulting solution in each case. Plot the results in the form of a titration curve.

$0.200M \times 0.08 L = 0.0160 mol$ KOH

Volume of HCl soln added/mL	*Total volume of resulting solution/mL*	*Moles of $OH^-(aq)$ in solution*	*$[OH^-]$ /M*	*pH*
10.0	90.0	0.0120	0.133	13.12
20.0	100.0	0.0080	0.080	12.90
30.0	110.0	0.0040	0.036	12.56
35.0	115.0	0.0020	0.017	12.23
39.0	119.0	0.0004	0.003	11.48
40.0	120.0		1.0×10^{-7}	7.00
41.0	121.0		3×10^{-12}	2.48
50.0	130.0		3.3×10^{-13}	1.52
60.0	140.0		1.8×10^{-13}	1.26
70.0	150.0		1.25×10^{-13}	1.10

18-64. Suppose that 50.0 mL of a 0.100 M $HNO_3(aq)$ solution is titrated with a 0.100 M $NaOH(aq)$ solution. Fill in the following table and compute the pH of the resulting solution in each case. Plot the results in the form of a titration curve.

Volume of NaOH soln added/mL	*Total volume of resulting solution/mL*	*Moles of $H_3O^+(aq)$ in solution*	*$[H_3O^+]$ /M*	*pH*
10.0				
20.0				
30.0				
40.0				
45.0				
49.0				
50.0				
51.0				
60.0				
70.0				

$0.400M \times 0.01L = 0.004 mol$

$$\frac{0.0120 \text{ mol}}{0.090 L} = 0.133 M$$

$$\frac{1.0\times10^{-14} M^2}{0.133 M} = 13.12$$

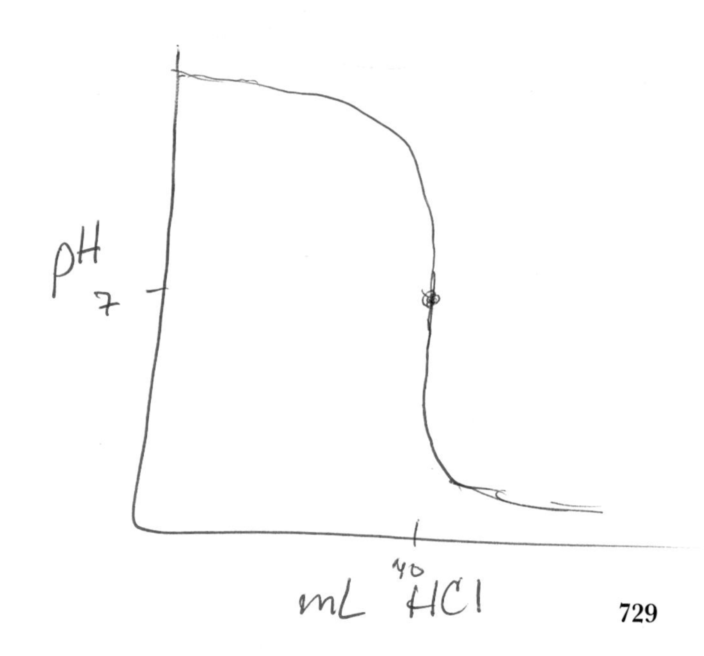

INTERCHAPTER K

Natural Waters

Seawater constitutes an enormous source of useful chemicals.

Approximately three fourths of the earth's surface is covered with water. The total amount of water on earth is estimated at 1.4×10^{12} kg, of which 97.3 percent is seawater, 2.1 percent is ice, and only 0.6 percent is fresh water. The fresh water is located in rivers and lakes and in the ground. Water is the solvent of life; the human body is about three fourths water by mass. More chemical and biochemical reactions take place in water than in all other solvents combined. In the absence of water, no known form of life is possible.

K-1. SEAWATER CONTAINS MANY DISSOLVED SALTS

We can classify water according to the amount of dissolved minerals it contains (Table K-1). Seawater averages 3.5 percent by mass of dissolved minerals, which puts it in the salty category. About 75 elements have been detected in seawater, but only 10 species (Table K-2) constitute over 99.9 percent of the mass of the various substances dissolved in seawater. Sodium ions plus chloride ions constitute about 86 percent by mass of the dissolved species in seawater.

Most of the ionic constituents of seawater enter the ocean in superheated (320°C), mineral-rich water that originates deep within the earth and flows through vent holes in the ocean floor (Figure K-1). Manganese nodules (Figure K-2), which are porous, roughly spherical chunks of metallic oxides ranging from 2 to 10 cm in diameter, form spontaneously in the vicinity of the vents. In addition to manganese, the nodules contain iron, cobalt, nickel, copper, zinc, chromium, vanadium, tungsten, and lead. Manganese nodules are a potentially rich source of scarce metals, such as cobalt and chromium, but they cannot be recovered economically at the present time.

Table K-1 Classification scheme for water

Type of water	*Quantity of dissolved minerals/% by mass*
fresh	0– 0.1
brackish	0.1– 1
salty	1–10
brine	>10

Table K-2 Concentration of the ten principal ionic constituents in seawater

Ion	*Name*	*Concentration/M*	*Grams of ion per kilogram of of seawater*
Cl^-	chloride	0.55	19.4
Na^+	sodium	0.46	10.6
SO_4^{2-}	sulfate	0.028	2.7
Mg^{2+}	magnesium	0.054	1.3
Ca^{2+}	calcium	0.010	0.41
K^+	potassium	0.010	0.39
HCO_3^-	hydrogen carbonate	0.0023	0.14
CO_3^{2-}	carbonate	0.0003	0.018
Br^-	bromide	0.00083	0.067
Sr^{2+}	strontium	9×10^{-5}	0.008

Figure K-1 Vents in the ocean floor through which hot, mineral-rich water enters the ocean. These minerals are the major source of dissolved solids in the oceans. The structure showing at the bottom of the photo is the bathyscope from which the picture was taken.

Figure K-2 Manganese nodules, which form in the vicinity of vents in the ocean floor, contain manganese, iron, cobalt, nickel, copper, zinc, chromium, vanadium, tungsten, and lead.

K-2. THE MAJOR NUTRIENTS IN THE OCEAN FOOD CHAIN ARE PHOSPHATE, NITRATE, CARBON DIOXIDE, AND OXYGEN

Temperature, oxygen concentration, carbon dioxide concentration, phosphate concentration as HPO_4^{2-}, and nitrate concentration play a key role in the chemistry and biochemistry of the oceans.

The temperature of ocean water ranges from a high of about 32°C near the surface in some regions to a low of about −2°C near an ice shelf. The average surface temperature is around 22°C and decreases to about 2°C at a depth of 2 km; below 2 km the ocean temperature is fairly constant at 2°C.

Variations in the concentration of O_2, NO_3^-, and HPO_4^{2-} with depth are shown in Figure K-3. All these constituents are essential to the development of plants and marine organisms such as phytoplankton, which serve as food sources for higher forms of marine life.

The concentration of oxygen near the ocean surface is high relative to that of nitrate and hydrogen phosphate (Figure K-3) because of the dissolution of oxygen from the air and its production in photosynthesis by phytoplankton. The value of $[O_2]$ initially decreases with depth because of the consumption of oxygen during the decomposition of animal and plant matter.

The phosphates and nitrates rise to the surface from the mineral-rich deep water and are depleted near the surface as they are utilized in the production of marine life. The growth of phytoplankton in the ocean is limited by the amount of available phosphate, nitrate, and trace mineral nutrients. There are certain places in the ocean where the deep water rises to the surface in large quantities, as a result of

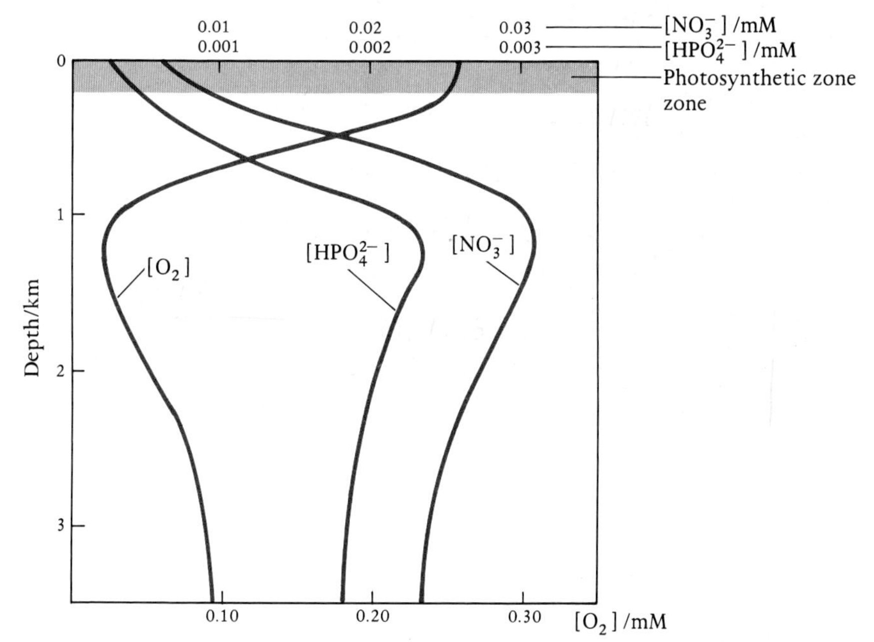

Figure K-3 Variation of $[O_2]$, $[HPO_4^{2-}]$, and $[NO_3^-]$ with ocean depth. Note that in the first kilometer the oxygen concentration decreases while the concentration of nitrate and hydrogen phosphate increases. The value of the ratio $[NO_3^-]/[HPO_4^{2-}]$ is about 15, which is equal to the nitrogen-to-phosphorus ratio in photoplankton.

prevailing winds that blow away the warmer and thus lighter surface water. These regions are especially rich in animal life and constitute the great fishing areas. Examples are regions off the coasts of Newfoundland, Chile, and Peru. Areas where the nutrient-rich deep water rises to the surface constitute only 0.1 percent of the ocean surface, but these areas supply over 50 percent of the total fish catch. Ninety percent of the ocean is a biological desert that yields only one percent of the total fish catch. Nutrient-poor tropical ocean waters, like those near Hawaii, for example, contain very little plant life and as a result are blue. Nutrient-rich ocean waters are teeming with microscopic plant life and as a result are greenish brown.

K-3. SOME CHEMICAL SUBSTANCES ARE OBTAINED COMMERCIALLY FROM SEAWATER

Seawater constitutes an enormous source of useful chemicals, but at the present time only four substances are obtained from seawater on a commercial scale: pure water, sodium chloride, bromine, and magnesium hydroxide.

Pure water can be obtained from seawater by reverse osmosis.

Pure water is obtained from seawater on a commercial scale by various distillation techniques. The process of removing dissolved solids from seawater is called *desalination*. Economical methods of desalination are of paramount importance in the arid regions of the world.

About 40 million metric tons (1 metric ton = 1000 kg = 2200 lb) of sodium chloride are obtained from seawater each year. The process involves filtration of the seawater to remove particulate matter followed by natural evaporation of the water from storage ponds until the NaCl crystallizes from solution.

Bromine is present in seawater as bromide ion at a concentration of 8.3×10^{-4} M. The economical recovery of bromine from seawater depends upon the fact that elemental bromine, Br_2, is a volatile liquid. The pH of seawater is decreased from 8.2 to 3.5 by the addition of sulfuric acid and the bromide ion is then converted to bromine by a replacement reaction with chlorine:

$$2Br^-(aq) + Cl_2(g) \rightarrow Br_2(l) + 2Cl^-(aq)$$

The Br_2 is removed from the seawater when a stream of air is passed through the solution to sweep out the bromine.

Magnesium ion is present in seawater at a concentration of 0.054 M. It is separated from seawater by the addition of lime, $CaO(s)$, which precipitates the $Mg^{2+}(aq)$ as the hydroxide, $Mg(OH)_2(s)$. The $Mg(OH)_2$ is collected by filtration and converted to magnesium chloride with hydrochloric acid. The magnesium chloride is recovered by evaporation of water from the solution until $MgCl_2(s)$ crystallizes.

K-4. FRESH WATER CONTAINS A VARIETY OF DISSOLVED SUBSTANCES

The major source of fresh water is rainfall. All rainwater contains dissolved oxygen, nitrogen, and carbon dioxide. Rain that has fallen through polluted air may contain a variety of contaminants, including sulfuric and sulfurous acids, carbon monoxide, oxides of nitrogen, dust, pollen, and salts of numerous trace metals, such as iron and lead. A beneficial feature of rain and snow is that they remove pollutants from the lower atmosphere. Rain that has fallen through clean air has a pH of about 5.6 because the dissolved carbon dioxide reacts with the water to produce $H^+(aq)$ and $HCO_3^-(aq)$ ions.

Acid rain (Interchapter E) sometimes has a pH less than 3.

Fresh water that has been in contact with the earth for some time contains a variety of anions and cations. Average values of the concentrations of the major ionic species in fresh water are given in Table K-3. *Hard water* contains appreciable amounts of divalent cations, primarily Ca^{2+}, Mg^{2+}, and Fe^{2+}. The major anions in hard water are the hydrogen carbonate ion (HCO_3^-) and the sulfate ion (SO_4^{2-}). These ions can arise in groundwater by reactions such as the interaction of water containing dissolved CO_2 with limestone ($CaCO_3$):

$$CaCO_3(s) + H_2O(l) + CO_2(aq) \rightleftharpoons Ca^{2+}(aq) + 2HCO_3^-(aq)$$

The reverse of this reaction results in the formation of stalagmites and stalactites.

The result of this reaction is the dissolution of $CaCO_3(s)$. Magnesium ions in groundwater are the result of a similar reaction between $CO_2(aq)$ and *dolomite,* a mineral containing $CaCO_3$ and $MgCO_3$.

The divalent cations in hard water form precipitates with soaps that appear as a scum in the wash water. Natural soaps are sodium

Table K-3 Comparison of concentrations of the major ionic constituents in seawater, fresh water, and blood serum[a]

	$[Na^+]$	$[K^+]$	$[Ca^{2+}]$	$[Mg^{2+}]$	$[Cl^-]$	$[SO_4^{2-}]$	$[HCO_3^-]$	*Total*
Seawater	460	10	10	54	550	28	2.3	1114
fresh water (average values)	0.27	0.06	0.38	0.34	0.22	0.12	0.96	2.35
human serum	145	5.1	2.5	1.2	103	2.5	12	271

[a]All concentrations are expressed in millimolar (mM) units.

salts of fatty acids, which are organic acids containing long hydrocarbon chains. A typical example is sodium stearate, $C_{17}H_{35}COO^-Na^+$:

$$\underbrace{CH_3CH_2CH_2CH_2CH_2CH_2CH_2CH_2CH_2CH_2CH_2CH_2CH_2CH_2CH_2CH_2CH_2}_{\text{hydrocarbon portion}}\overbrace{COO^-}^{\text{carboxyl portion}}Na^+$$

An example of scum formation is the reaction

$$\underset{\text{}}{Ca^{2+}(aq)} + \underset{\substack{\text{stearate ion}\\\text{(soap)}}}{2C_{17}H_{35}COO^-(aq)} \rightarrow \underset{\substack{\text{calcium stearate}\\\text{(scum)}}}{Ca(C_{17}H_{35}COO)_2(s)}$$

The amount of scum can be decreased by adding substances such as sodium tripolyphosphate to the soap. The tripolyphosphate combines with the divalent cations and keeps them in solution:

$$2Ca^{2+}(aq) + \underset{\text{tripolyphosphate}}{P_3O_{10}^{5-}(aq)} \rightleftharpoons \underset{\text{water soluble}}{Ca_2P_3O_{10}^-(aq)}$$

Tripolyphosphate additives in soaps have the disadvantage that they break down in water to produce HPO_4^{2-}, which in turn promotes the profuse growth of plants in lakes, a process called eutrophication (Interchapter J).

Hard water is classified as either *temporary* or *permanent* depending on which anions it contains. Temporary hard water contains $HCO_3^-(aq)$ anions, along with $Ca^{2+}(aq)$ and/or $Mg^{2+}(aq)$. When temporary hard water is heated, calcium carbonate or magnesium carbonate precipitates because of the reaction

$$M^{2+}(aq) + 2HCO_3^-(aq) \rightarrow MCO_3(s) + H_2O(l) + CO_2(g)$$

where M stands for either metal. The metal carbonate can be seen as *boiler scale* in boilers, hot water pipes, and tea kettles. Such deposits can clog pipes and, by acting as a heat insulator, increase the cost of heating water.

In permanent hard water the primary anion is sulfate. Both calcium sulfate and magnesium sulfate are soluble in hot water and are not precipitated by heating. Thus, the water is always hard.

Commercial water softeners contain ion-exchange resins.

Hard water can be *softened* (removal of divalent cations) or *deionized* (removal of all cations and anions) by chemical means. Water-insoluble *ion-exchange resins* are used for this purpose, as explained in Figure K-4. Ion exchangers are used to produce high-quality, essentially ion-free water.

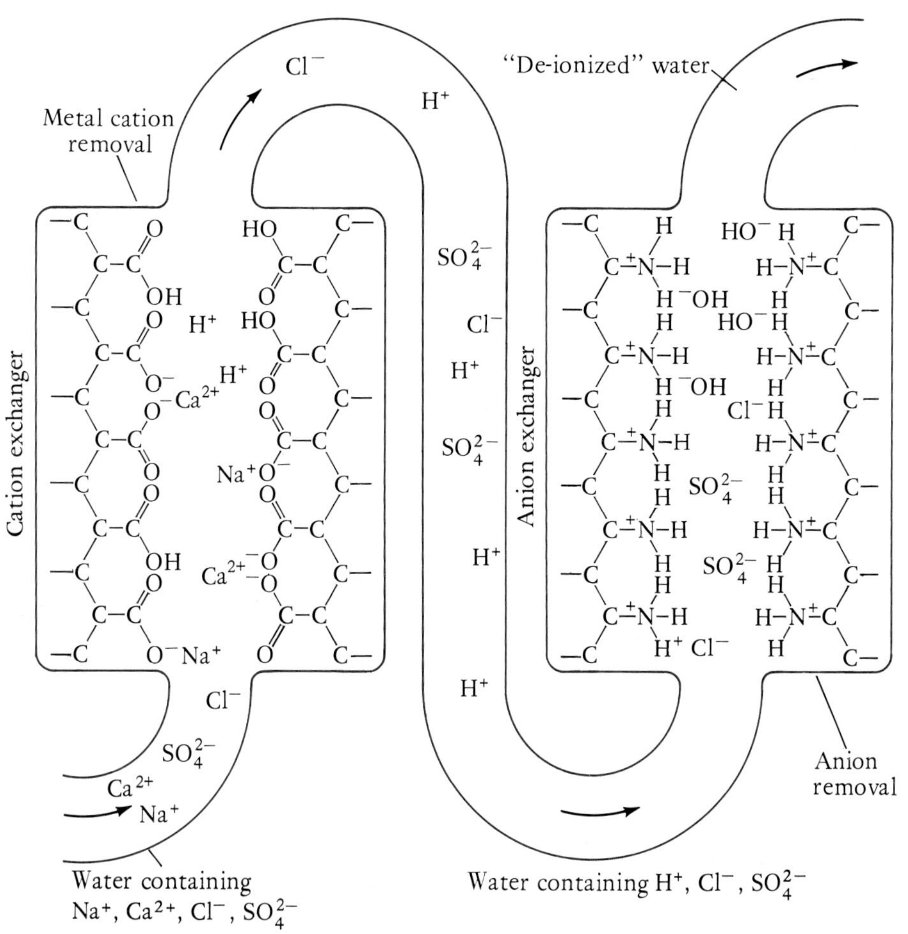

Figure K-4 Ions are removed from water by macromolecules called resins that contain acidic and basic groups. In the left-hand chamber, metal cations, such as Mg^{2+}, Ca^{2+}, and Na^+, displace $H^+(aq)$ ions from the cation-exchange resin and are bound to the resin. In the right-hand chamber, anions such as Cl^- and SO_4^{2-} displace $OH^-(aq)$ ions from the anion-exchange resin and are bound to the resin. The $H^+(aq)$ ions formed in the cation exchanger react with $OH^-(aq)$ ions formed in the anion exchanger to yield water. The two-stage ion-exchange process removes both cations and anions from the water. The resulting liquid is called deionized water.

K-5. SOAP CLEANS BY FORMING MICELLES

The cleaning action of soap is a consequence of the dual affinity of soap molecules for grease and water. Realize that water and hydrocarbon compounds do not dissolve in each other because they have very different electrical properties. Water is a polar molecule, whereas hydrocarbon chains are nonpolar (the electronegativities of carbon and hydrogen are about the same). Water molecules interact much more strongly with each other than with nonpolar molecules. Nonpolar molecules have no regions of net charge that can interact with the charged regions in a water molecule. Consequently, nonpolar molecules are excluded from water and so are not soluble in it. Polar substances, on the other hand, can interact electrostatically with water molecules and so are soluble in water. As a rule of thumb, polar substances dissolve polar substances and nonpolar substances dissolve nonpolar substances. More succinctly, *like dissolves like.*

Like dissolves like.

Soap is effective as a cleaning agent because the hydrocarbon portion of a soap molecule has a strong affinity for grease, whereas the charged portion of the molecule has a strong affinity for water. The anion of a soap molecule can be represented schematically as

When soap molecules that are dissolved in water come into contact with grease, the hydrocarbon portions stick into the grease, leaving the anion portions at the grease-water interface. The penetration of the grease by the soap molecules is followed by a remarkable phenomenon—the formation of *micelles* (Figure K-5), which are small, spherical grease-soap droplets that are soluble in water as a result of the polar groups on their surface. Micelles do not combine into larger drops because their surfaces are all negatively charged. Thus the cleaning action of soap is a consequence of the specialized molecular properties of the soap molecules and their interaction with grease that leads to the formation of water-soluble micelles. The micelles encapsulate small grease particles and are subsequently rinsed away, leaving a clean region behind.

K-6. MUNICIPAL SEWAGE IS TREATED IN STAGES

The effluents that are discharged from our homes and factories into the municipal sewer system are known as municipal sewage. About 90 percent of our sewage is cleaned up to varying degrees before it is released into streams, rivers, lakes, or the ocean. The treatment of municipal sewage is categorized as *primary, secondary,* or *tertiary.* Primary treatment of sewage (Figure K-6a) involves physical processes, such as the removal of large pieces of solid material by screening and settling in large sedimentation tanks, followed by the removal of fine solid particles by the addition of aluminum salts that produce a flocculent $Al(OH)_3$ precipitate. If only primary treatment is used,

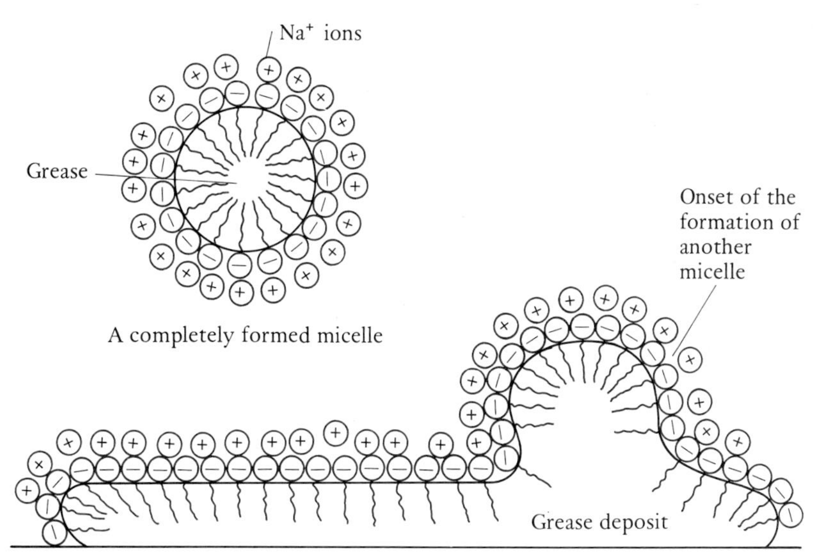

Figure K-5 The cleaning action of soap is due to the ability of soap molecules to form micelles that encapsulate grease and carry it away. The fatty acid portions of the soap molecules dissolve in each other and in the grease particles, thus forming a water-soluble spherical particle with charged group on its surface.

chlorine is added to the water to destroy bacteria. About 30 percent of the municipal sewage in the United States and Canada receives only primary treatment.

Secondary sewage treatment (Figure K-6b) involves biological processes. The water resulting from a primary treatment is run into aeration tanks where air in the presence of aerobic bacteria is bubbled through the water in order to remove organic wastes. The aeration is carried out over a bed of small stones to increase the surface area of the water and thus increase the rate of oxygen dissolution. The aeration step is followed by storage in a sedimentation tank and then chlorination to destroy microorganisms. About 60 percent of the municipal sewage in the United States and Canada receives secondary treatment. Secondary sewage treatment removes about 90 percent of the organic waste products.

Tertiary sewage treatment involves chemical processes. Tertiary sewage treatment is designed to remove specific chemical pollutants, such as nitrogen and phosphorus compounds. The cost of tertiary treatment is high, and so it is employed only rarely at the present time.

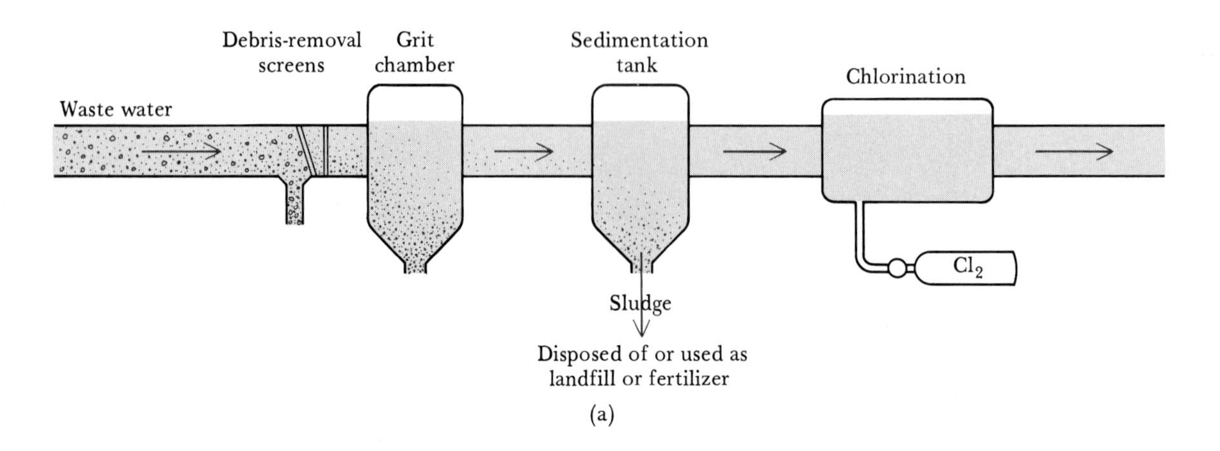

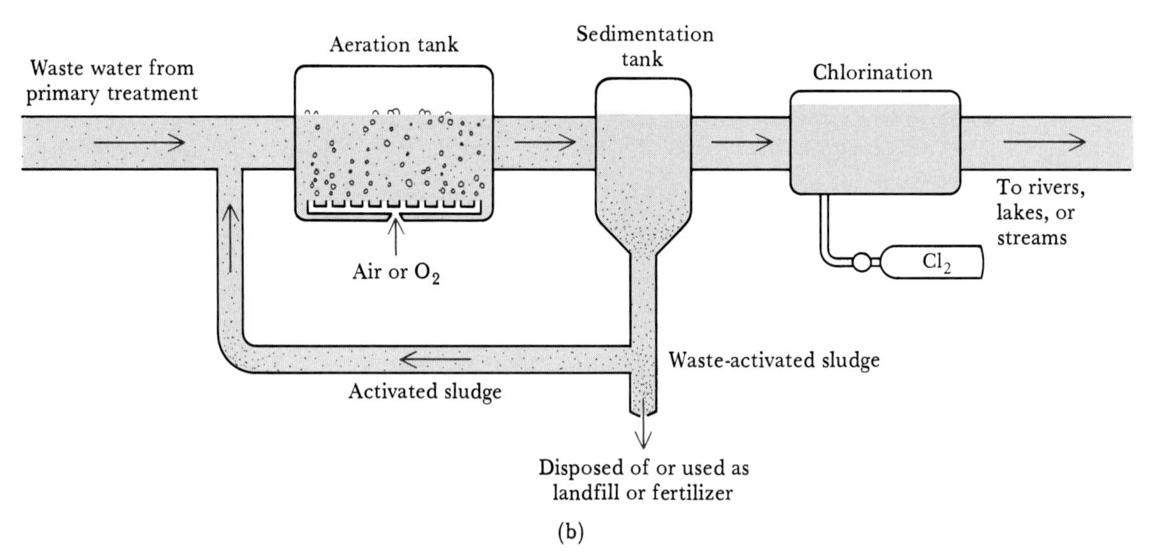

Figure K-6 A schematic description of the (a) primary and (b) secondary treatment of sewage.

QUESTIONS

K-1. What are the principal ionic constituents in seawater?

K-2. What is the major source of dissolved salts in the oceans?

K-3. Name four substances that are obtained from seawater on a commercial scale.

K-4. What is desalination?

QUESTIONS (continued)

K-5. Describe how bromine is obtained from seawater.

K-6. Distinguish between fresh water, brackish water, salty water, and brine.

K-7. What are the principal constituents of hard water?

K-8. Describe how soap acts as a cleaning agent.

K-9. What is a micelle? What are the special properties of molecules that form micelles?

K-10. Describe the difference between temporary and permanent hard water.

K-11. What is the cause of the formation of soap scum?

K-12. What does it mean to soften water?

K-13. Describe how an ion-exchange resin works.

K-14. Describe the primary, secondary, and tertiary treatments of sewage.

Many chemical reactions result in the formation of an insoluble product. Here we see a precipitate of silver bromate being formed from the addition of silver nitrate to sodium bromate.

19 / Solubility and Precipitation Reactions

Many substances are insoluble or only slightly soluble in water. In this chapter we discuss chemical equilibria involving slightly soluble substances. A saturated solution of a slightly soluble ionic compound involves an equilibrium between the solid and its constituent ions in solution. An equilibrium constant for a solubility equilibrium is defined according to the law of concentration action. The treatment of solubility as an equilibrium process allows us to calculate the solubility of a solid not only in pure water but in solutions of acids, bases, and salts as well. The solubility of many substances is shown to depend strongly upon pH and the presence of other salts.

19-1. THE DRIVING FORCE OF A DOUBLE REPLACEMENT REACTION IS THE FORMATION OF AN INSOLUBLE PRODUCT

In Chapter 4 we learned that one driving force of a double replacement reaction is the formation of a precipitate. For example, the reaction

$$BaCl_2(aq) + Na_2SO_4(aq) \rightarrow 2NaCl(aq) + BaSO_4(s)$$

The designations soluble, slightly soluble, and insoluble given here are arbitrary.

occurs readily, with the formation of a white $BaSO_4$ precipitate. To determine whether a precipitate will form, it is necessary to know the *solubility* of a compound. Although all compounds have a characteristic solubility in water, it is necessary to set up a few general rules for characterizing solubility. We call any substance whose solubility is less than $0.01\ mol \cdot L^{-1}$ insoluble. If its solubility is greater than $0.1\ mol \cdot L^{-1}$, we call it soluble. If its solubility is between 0.01 and $0.1\ mol \cdot L^{-1}$, we say that it is slightly soluble. The following *solubility rules* can be used to determine solubilities in water:

These solubility rules are neither infallible nor all inclusive, but nevertheless they are very useful.

1. All sodium, potassium, and ammonium salts are soluble.
2. All nitrates, acetates, and perchlorates are soluble.
3. All silver, lead, and mercury(I) salts are insoluble.
4. All chlorides, bromides, and iodides are soluble.
5. All carbonates, sulfides, oxides, and hydroxides are insoluble.
6. All sulfates are soluble except calcium sulfate and barium sulfate.

The solubility rules must be applied in the order given; the rule with the lower number takes precedence in case of a conflict. For example, Na_2S is soluble because rule 1 states that all sodium salts are soluble and the rule concerning sulfide solubilities is rule 5. Similarly, $PbSO_4$ is insoluble because rule 3 states that all Pb^{2+} salts are insoluble.

Example 19-1: Predict the solubility of the following compounds in water:

(a) $(NH_4)_2SO_4$ (b) $CaCO_3$ (c) Al_2O_3 (d) $Pb(NO_3)_2$ (e) $PbCl_2$

Solution: Apply the solubility rules in the order given:

Problems 19-1 through 19-10 deal with these solubility rules.

(a) soluble by rule 1: all ammonium salts are soluble
(b) insoluble by rule 5: all carbonates are insoluble
(c) insoluble by rule 5: all metal oxides are insoluble
(d) soluble by rule 2: all nitrates are soluble
(e) insoluble by rule 3: all lead salts are insoluble

These six solubility rules can be used to predict whether or not a double replacement reaction occurs. Consider the reaction between barium chloride, $BaCl_2(aq)$, and potassium carbonate, $K_2CO_3(aq)$. The possible double replacement products are KCl and $BaCO_3$. According to rule 5, $BaCO_3$ is insoluble, and so we have

$$BaCl_2(aq) + K_2CO_3(aq) \rightarrow 2KCl(aq) + BaCO_3(s)$$

The net ionic equation is

$$Ba^{2+}(aq) + CO_3^{2-}(aq) \rightarrow BaCO_3(s)$$

Example 19-2: Predict if there is a reaction when solutions of $Cd(NO_3)_2$ and Na_2S are mixed. If there is a reaction, then write the complete equation and the net ionic equation that describe the reaction.

Solution: The reaction we are considering is

$$Na_2S(aq) + Cd(NO_3)_2(aq) \rightarrow ?$$

The possible double replacement products are $NaNO_3$ and CdS. Although $NaNO_3$ is soluble, CdS is not (rule 5), and so we have

$$Na_2S(aq) + Cd(NO_3)_2(aq) \rightarrow 2NaNO_3(aq) + CdS(s)$$

The net ionic equation is

$$Cd^{2+}(aq) + S^{2-}(aq) \rightarrow CdS(s)$$

Cadmium sulfide is a yellow-orange solid. The addition of Na_2S to an unknown solution that may contain Cd^{2+} ions is a test for the presence of $Cd^{2+}(aq)$.

Calcium sulfide is used to color soaps, glass, textiles, paper, and rubber yellow.

19-2. THE LAW OF CONCENTRATION ACTION GOVERNS THE EQUILIBRIUM BETWEEN AN IONIC SOLID AND ITS CONSTITUENT IONS IN SOLUTION

Consider the equilibrium between solid silver bromate and its constituent ions in water:

$$AgBrO_3(s) \rightleftharpoons Ag^+(aq) + BrO_3^-(aq) \quad (19\text{-}1)$$

Application of the law of concentration action to Equation (19-1) yields the equilibrium constant expression for this reaction:

$$K_{sp} = [Ag^+][BrO_3^-] \quad (19\text{-}2)$$

The subscript *sp* stands for solubility product, and K_{sp} is called the *solubility-product constant*. Note that $AgBrO_3(s)$ does not appear in the K_{sp} expression because its concentration cannot vary. Recall from Chapter 15 that a pure solid does not appear in an equilibrium constant expression.

The experimental value of K_{sp} for Equation (19-1) at 25°C is

$$K_{sp} = [Ag^+][BrO_3^-] = 5.8 \times 10^{-5}\ M^2 \quad (19\text{-}3)$$

Equation (19-3) states that if $AgBrO_3(s)$ is in equilibrium with an aqueous solution of $AgBrO_3(aq)$ at 25°C, then the product of the concentrations of $Ag^+(aq)$ and $BrO_3^-(aq)$ at equilibrium must equal $5.8 \times 10^{-5}\ M^2$.

A K_{sp} expression can be used to estimate the solubility of a solid. For example, suppose that excess $AgBrO_3(s)$ is in contact with water at 25°C. Then at equilibrium

$$[Ag^+][BrO_3^-] = 5.8 \times 10^{-5}\,M^2$$

From the reaction stoichiometry, Equation (19-1), we have

$$[Ag^+] = [BrO_3^-]$$

because each formula unit of $AgBrO_3$ that dissolves produces one $Ag^+(aq)$ ion and one $BrO_3^-(aq)$ ion. Further, if we denote the solubility of $AgBrO_3(s)$ by s, then

$$s = \left(\begin{matrix}\text{solubility of } AgBrO_3(s)\\ \text{in water}\end{matrix}\right) = [Ag^+] = [BrO_3^-]$$

because the concentration of either $Ag^+(aq)$ or $BrO_3^-(aq)$ is equal to the number of moles of dissolved salt per liter of solution. Thus from the K_{sp} expression we have

$$K_{sp} = 5.8 \times 10^{-5}\,M^2 = [Ag^+][BrO_3^-] = s^2$$

and thus $$s = (5.8 \times 10^{-5}\,M^2)^{1/2} = 7.6 \times 10^{-3}\,M$$

The formula mass of $AgBrO_3(s)$ is 235.8, and thus the number of grams of $AgBrO_3(s)$ that dissolves in 1.00 L of solution at 25°C is

$$(7.6 \times 10^{-3}\,mol \cdot L^{-1})\left(\frac{235.8\ g\ AgBrO_3}{1\ mol\ AgBrO_3}\right) = 1.8\ g \cdot L^{-1}$$

Example 19-3: The K_{sp} for $BaSO_4(s)$ in equilibrium with an aqueous solution of its constituent ions at 25°C is $1.1 \times 10^{-10}\,M^2$. Write the chemical equation that represents the solubility equilibrium for $BaSO_4(s)$ and compute its solubility in water at 25°C.

Solution: We know that a barium ion has a charge of +2 (Group 2 metal) and that a sulfate ion has a charge of −2. Thus the solubility equilibrium is

$$BaSO_4(s) \rightleftharpoons Ba^{2+}(aq) + SO_4^{2-}(aq)$$

The K_{sp} expression is

$$K_{sp} = [Ba^{2+}][SO_4^{2-}] = 1.1 \times 10^{-10}\,M^2$$

If $BaSO_4(s)$ is equilibrated with pure water, then, from the reaction stoichiometry, we have at equilibrium

$$[Ba^{2+}] = [SO_4^{2-}] = s$$

where s is the solubility of $BaSO_4(s)$ in pure water. Thus

$$K_{sp} = s^2 = 1.1 \times 10^{-10}\ M^2$$

and thus $$s = (1.1 \times 10^{-10}\ M^2)^{1/2} = 1.0 \times 10^{-5}\ M$$

Problems 19-11 and 19-12 are similar to Example 19-3.

When excess $AgBrO_3(s)$ is equilibrated with pure water, we have at equilibrium $[Ag^+] = [BrO_3^-]$ because each $AgBrO_3$ unit that dissolves yields one $Ag^+(aq)$ and one $BrO_3^-(aq)$. Consider the problem of calculating the solubility in water of copper(II) iodate, $Cu(IO_3)_2(s)$, which yields *two* $IO_3^-(aq)$ and one $Cu^{2+}(aq)$ for each $Cu(IO_3)_2$ unit that dissolves. The chemical equation that represents the solubility equilibrium of $Cu(IO_3)_2(s)$ in water is

$$Cu(IO_3)_2(s) \rightleftharpoons Cu^{2+}(aq) + 2IO_3^-(aq) \qquad (19\text{-}4)$$

According to the law of concentration action, the K_{sp} expression for Equation (19-4) is

$$K_{sp} = [Cu^{2+}][IO_3^-]^2 \qquad (19\text{-}5)$$

The experimental value of K_{sp} at 25°C is $7.4 \times 10^{-8}\ M^3$, and so we have at 25°C

$$K_{sp} = [Cu^{2+}][IO_3^-]^2 = 7.4 \times 10^{-8}\ M^3 \qquad (19\text{-}6)$$

Note that it is the *square* of the concentration of $IO_3^-(aq)$ that appears in the K_{sp} expression for $Cu(IO_3)_2(s)$ because each formula unit of $Cu(IO_3)_2$ that dissolves produces *two* iodate ions. Thus, when $Cu(IO_3)_2(s)$ is in equilibrium with pure water, the concentration of iodate ion is twice as great as the concentration of copper(II) ion:

$$[IO_3^-] = 2[Cu^{2+}]$$

The solubility of $Cu(IO_3)_2(s)$ in pure water is equal to $[Cu^{2+}]$ because each mole of $Cu(IO_3)_2$ that dissolves yields one mol of $Cu^{2+}(aq)$. If we denote the solubility of $Cu(IO_3)_2(s)$ in pure water by s, then

For a salt with the formula A_xB_y, the solubility is given by

$$s = \frac{[A]}{x} = \frac{[B]}{y}$$

$$s = \begin{pmatrix}\text{solubility of } Cu(IO_3)_2(s) \\ \text{in water}\end{pmatrix} = [Cu^{2+}] = \frac{[IO_3^-]}{2}$$

It follows then that $[IO_3^-] = 2s$. Combining these results with the K_{sp} expression, Equation (19-6), yields

$$7.4 \times 10^{-8}\,M^3 = [Cu^{2+}][IO_3^-]^2 = (s)(2s)^2 = 4s^3$$

and therefore

$$s = \left(\frac{7.4 \times 10^{-8}\,M^3}{4}\right)^{1/3} = 2.6 \times 10^{-3}\,M$$

Note that $[Cu^{2+}] = s = 2.6 \times 10^{-3}\,M$ and that $[IO_3^-] = 2s = 5.2 \times 10^{-3}\,M$.

Example 19-4: The solubility product for silver sulfate in equilibrium with water at 25°C is $1.4 \times 10^{-5}\,M^3$. Compute the value of $[Ag^+]$ that results when pure water is saturated with $Ag_2SO_4(s)$.

Solution: The $Ag_2SO_4(s)$ solubility equilibrium is

$$Ag_2SO_4(s) \rightleftharpoons 2Ag^+(aq) + SO_4^{2-}(aq)$$

and the solubility-product expression is

$$K_{sp} = [Ag^+]^2[SO_4^{2-}] = 1.4 \times 10^{-5}\,M^3$$

Each Ag_2SO_4 unit that dissolves yields two $Ag^+(aq)$ and one $SO_4^{2-}(aq)$; thus

$$\frac{[Ag^+]}{2} = [SO_4^{2-}]$$

Substitution of this result into the K_{sp} expression for $Ag_2SO_4(s)$ yields

$$K_{sp} = [Ag^+]^2\frac{[Ag^+]}{2} = 1.4 \times 10^{-5}\,M^3$$

and so

$$[Ag^+]^3 = 2.8 \times 10^{-5}\,M^3$$

Solving for $[Ag^+]$ yields

$$[Ag^+] = (2.8 \times 10^{-5}\,M^3)^{1/3} = 3.0 \times 10^{-2}\,M$$

The solubility of $Ag_2SO_4(s)$ is not equal to $[Ag^+]$ because two $Ag^+(aq)$ ions result from each formula unit of Ag_2SO_4 that dissolves. The solubility of Ag_2SO_4 is $[Ag^+]/2$, or 0.015 M at 25°C.

Problems 19-15 and 19-16 are similar to Example 19-4.

Some solubility product constants are given in Table 19-1.

19-3. THE SOLUBILITY OF AN IONIC SOLID IS DECREASED WHEN A COMMON ION IS PRESENT IN THE SOLUTION

Consider the problem of calculating the solubility of silver bromate in an aqueous solution at 25°C that is 0.10 M in sodium bromate.

Table 19-1 Solubility product constants for salts in water at 25°C

Halides	K_{sp}	*Halides*	K_{sp}
AgCl	1.8×10^{-10} M^2	Hg_2I_2	4.5×10^{-29} M^3
AgBr	5.0×10^{-13} M^2	MgF_2	6.5×10^{-9} M^3
AgI	8.3×10^{-17} M^2	PbF_2	7.7×10^{-8} M^3
BaF_2	1.0×10^{-6} M^3	$PbCl_2$	1.6×10^{-5} M^3
CaF_2	5.3×10^{-9} M^3	$PbBr_2$	4.0×10^{-5} M^3
CuCl	1.2×10^{-6} M^2	PbI_2	7.1×10^{-9} M^3
CuBr	5.3×10^{-9} M^2	SrF_2	2.5×10^{-9} M^3
CuI	1.1×10^{-12} M^2	TlCl	1.7×10^{-4} M^2
Hg_2Cl_2	1.3×10^{-18} M^3	TlBr	3.4×10^{-6} M^2
Hg_2Br_2	5.6×10^{-23} M^3	TlI	6.5×10^{-8} M^2
Carbonates	K_{sp}	*Carbonates*	K_{sp}
Ag_2CO_3	8.1×10^{-12} M^3	$MgCO_3$	3.5×10^{-8} M^2
$BaCO_3$	5.1×10^{-9} M^2	$MnCO_3$	1.8×10^{-11} M^2
$CaCO_3$	2.8×10^{-9} M^2	$PbCO_3$	7.4×10^{-14} M^2
$CuCO_3$	1.4×10^{-10} M^2	$SrCO_3$	1.1×10^{-10} M^2
$FeCO_3$	3.2×10^{-11} M^2	$ZnCO_3$	1.4×10^{-11} M^2
Chromates	K_{sp}	*Oxalates*	K_{sp}
Ag_2CrO_4	1.1×10^{-12} M^3	$Ag_2C_2O_4$	3.4×10^{-11} M^3
$BaCrO_4$	1.2×10^{-10} M^2	CaC_2O_4	4×10^{-9} M^2
$PbCrO_4$	2.8×10^{-13} M^2	MgC_2O_4	7×10^{-7} M^2
Hydroxides	K_{sp}	*Hydroxides*	K_{sp}
$Al(OH)_3$	1.3×10^{-33} M^4	$Fe(OH)_3$	1.0×10^{-38} M^4
$Ca(OH)_2$	5.5×10^{-6} M^3	$Mg(OH)_2$	1.8×10^{-11} M^3
$Cd(OH)_2$	2.5×10^{-14} M^3	$Ni(OH)_2$	2.0×10^{-15} M^3
$Cr(OH)_3$	6.3×10^{-31} M^4	$Pb(OH)_2$	1.2×10^{-15} M^3
$Cu(OH)_2$	2.2×10^{-20} M^3	$Sn(OH)_2$	1.4×10^{-28} M^3
$Fe(OH)_2$	8.0×10^{-16} M^3	$Zn(OH)_2$	1.0×10^{-15} M^3
Sulfates	K_{sp}	*Sulfates*	K_{sp}
Ag_2SO_4	1.4×10^{-5} M^3	Hg_2SO_4	7.4×10^{-7} M^2
$BaSO_4$	1.1×10^{-10} M^2	$PbSO_4$	1.6×10^{-8} M^2
$CaSO_4$	9.1×10^{-6} M^2	$SrSO_4$	3.2×10^{-7} M^2
Sulfides	K_{sp}	*Sulfides*	K_{sp}
CdS	8.0×10^{-27} M^2	MnS	2.5×10^{-13} M^2
CuS	6.3×10^{-36} M^2	PbS	8.0×10^{-28} M^2
FeS	6.3×10^{-18} M^2	SnS	1.0×10^{-25} M^2
HgS	4×10^{-53} M^2	ZnS	1.6×10^{-24} M^2

As we learned in Section 19-2, the solubility equilibrium of $AgBrO_3(s)$ is

$$AgBrO_3(s) \rightleftharpoons Ag^+(aq) + BrO_3^-(aq)$$

and the solubility product expression is

$$K_{sp} = [Ag^+][BrO_3^-] = 5.8 \times 10^{-5}\ M^2 \qquad (19\text{-}7)$$

Recall that a spectator ion is an ion that is present but does not take part in the reaction.

The $Na^+(aq)$ from the $NaBrO_3(aq)$ is simply a spectator ion and does not enter into any of our calculations.

It is important to realize that Equation (19-7) fixes only the value of the product of the ionic concentrations, $[Ag^+][BrO_3^-]$. Any pair of values of $[Ag^+]$ and $[BrO_3^-]$ that when multiplied together equal $5.8 \times 10^{-5}\ M^2$ constitutes a pair of equilibrium ionic concentrations. It is important also to realize that the ionic concentrations in Equation (19-7) are the *total* ionic concentrations, *regardless of the source of each ionic species.* In the case of $AgBrO_3$ dissolved in a 0.10 M $NaBrO_3(aq)$ solution, the $Ag^+(aq)$ ions come only from the $AgBrO_3(s)$ that dissolves. The $BrO_3^-(aq)$ ions, on the other hand, come from two sources: from the 0.10 M $NaBrO_3(aq)$ initially present, which is completely dissociated into $Na^+(aq)$ and $BrO_3^-(aq)$, and from the $AgBrO_3(s)$ that dissolves. If we let s be the solubility of $AgBrO_3(s)$ in 0.10 M $NaBrO_3(aq)$, then (19-8)

$$[Ag^+] = s$$
$$[BrO_3^-] = s + 0.10\ M$$

If we substitute these two expressions into Equation (19-7), then we obtain

$$s(s + 0.10\ M) = 5.8 \times 10^{-5}\ M^2 \qquad (19\text{-}8)$$

Because $AgBrO_3(s)$ is only sparingly soluble in pure water, we expect the value of s in Equation (19-8) to be small. Therefore, we can neglect s relative to 0.10 M and write

$$s(0.10\ M) \approx 5.8 \times 10^{-5}\ M^2$$
$$s \approx 5.8 \times 10^{-4}\ M$$

The small value of $[Ag^+]$ justifies the assumption that $s = [Ag^+] \ll 0.10$ M. As a check, if we substitute $s = 5.8 \times 10^{-4}$ M into Equation (19-8), then we find that Equation (19-8) is satisfied. Note that the common ion BrO_3^- suppresses the solubility of $AgBrO_3(s)$. The solubility of $AgBrO_3(s)$ in pure water is 7.6×10^{-3} M, about 13 times greater than its solubility in a 0.10 M $NaBrO_3(aq)$ solution.

Example 19-5: Estimate the solubility of copper(II) iodate ($K_{sp} = 7.4 \times 10^{-8}\ M^3$) in an aqueous solution that is 0.20 M in copper(II) perchlorate.

Solution: The equilibrium expression that describes the solubility of $Cu(IO_3)_2$ in water is

$$Cu(IO_3)_2(s) \rightleftharpoons Cu^{2+}(aq) + 2IO_3^-(aq)$$

and the corresponding solubility product expression is

$$K_{sp} = [Cu^{2+}][IO_3^-]^2 = 7.4 \times 10^{-8}\ M^3 \tag{19-9}$$

The $ClO_4^-(aq)$ from the $Cu(ClO_4)_2(aq)$ is simply a spectator ion and does not enter into any of our calculations. The only source of $IO_3^-(aq)$ is from the $Cu(IO_3)_2(s)$ that dissolves. If we let s be the solubility of $Cu(IO_3)_2(s)$ in 0.20 M $Cu(ClO_4)_2(aq)$, then

$$[IO_3^-] = 2s$$

because two $IO_3^-(aq)$ ions are produced for each formula unit of $Cu(IO_3)_2(s)$ that dissolves. The $Cu^{2+}(aq)$ comes from the 0.20 M $Cu(ClO_4)_2(aq)$ *and* from the $Cu(IO_3)_2(s)$ that dissolves; thus

$$[Cu^{2+}] = 0.20\ M + s$$

If we substitute these values for $[Cu^{2+}]$ and $[IO_3^-]$ into Equation (19-9), then we obtain

$$(0.20\ M + s)(2s)^2 = 7.4 \times 10^{-8}\ M^3 \tag{19-10}$$

Because $Cu(IO_3)_2(s)$ is a slightly soluble salt, we expect the value of s to be small. Therefore, we can neglect s relative to 0.20 M in Equation (19-10) and write

$$(0.20\ M)(2s)^2 = 7.4 \times 10^{-8}\ M^3$$

Solving for s yields

$$s = \left(\frac{7.4 \times 10^{-8}\ M^3}{(0.20\ M)(4)}\right)^{1/2} = 3.0 \times 10^{-4}\ M$$

As a check, we note that the value of s that we obtain is indeed negligible relative to 0.20 M. The *common ion* $Cu^{2+}(aq)$ suppresses the solubility of $Cu(IO_3)_2(s)$. The solubility of $Cu(IO_3)_2(s)$ in pure water is 2.6×10^{-3} M, about 8 times greater than its solubility in a solution that is 0.20 M in $Cu^{2+}(aq)$.

Problems 19-21 through 19-30 deal with the common-ion effect.

The decrease in the solubility of an ionic solid in the presence of one of its constituent ions is called the *common-ion effect.* This phenomenon is understood readily in terms of Le Châtelier's principle. Consider the solubility equilibrium for silver bromate

$$AgBrO_3(s) \rightleftharpoons Ag^+(aq) + BrO_3^-(aq)$$

for which

$$K_{sp} = [Ag^+][BrO_3^-] = 5.8 \times 10^{-5}\ M^2$$

An increase in the concentration of $BrO_3^-(aq)$—for example, by adding $NaBrO_3$ to the solution—shifts the solubility equilibrium from right to left and thereby decreases the solubility of $AgBrO_3(s)$. The larger the value of $[BrO_3^-]$ at equilibrium, the smaller the value of $[Ag^+]$ because the product $[Ag^+][BrO_3^-]$ *must* equal $5.8 \times 10^{-5}\ M^2$ at equilibrium. Similarly, an increase in the concentration of $Ag^+(aq)$—for example, by adding $AgNO_3$, which is water-soluble—also shifts the solubility equilibrium from right to left and thereby decreases the solubility of $AgBrO_3(s)$. The common-ion effect for $AgBrO_3(s)$ is illustrated in Figure 19-1.

The solubility of $AgBrO_3(s)$ is *increased* by addition of any chemical species that decreases the concentration of either $Ag^+(aq)$ or $BrO_3^-(aq)$. For example, $Ag^+(aq)$ forms a diammine complex with $NH_3(aq)$:

The $Ag(NH_3)_2^+$ ion is called a complex ion. We shall study complex ions in Chapter 23, Transition Metal Complexes.

$$Ag^+(aq) + 2NH_3(aq) \rightleftharpoons Ag(NH_3)_2^+(aq)$$

Therefore, if $NH_3(aq)$ is added to an aqueous solution in equilibrium with $AgBrO_3(s)$, then the concentration of $Ag^+(aq)$ is reduced, owing

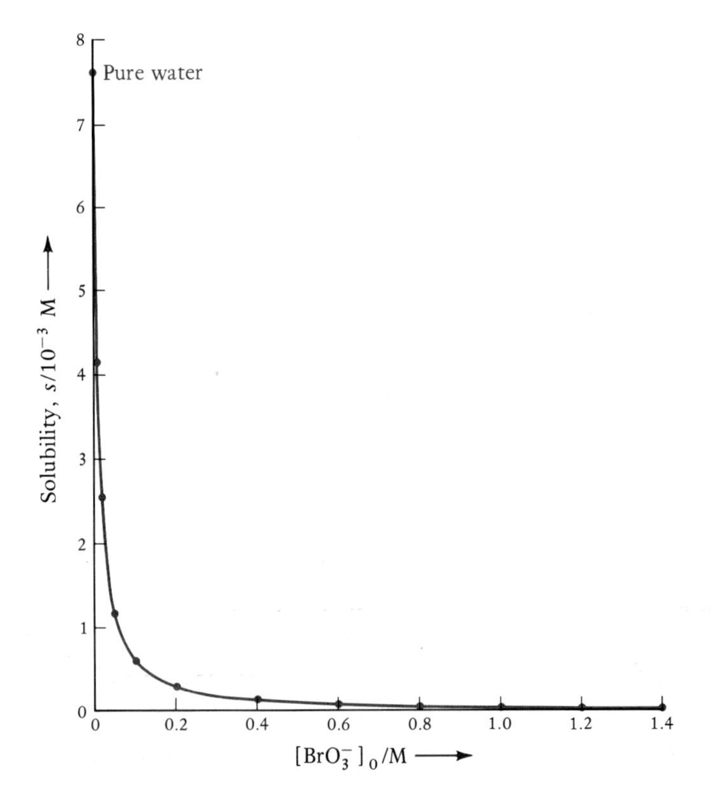

Figure 19-1 The solubility of $AgBrO_3(s)$ in water as a function of the bromate ion concentration. The bromate ion concentration can be controlled by adding a soluble salt such as $NaBrO_3$ to an aqueous solution of $AgBrO_3$. The plot illustrates the common ion effect, whereby the solubility of $AgBrO_3(s)$ is decreased by the addition of $BrO_3^-(aq)$.

Figure 19-2 Silver bromate is only slightly soluble in water at 25°C. If $NH_3(aq)$ is added to a saturated solution of $AgBrO_3(aq)$, then the solubility of $AgBrO_3(s)$ is increased as a result of the formation of $Ag(NH_3)_2^+(aq)$. Both test tubes initially contained 1.0 g of $AgBrO_3(s)$. Then 20 mL of water was added to the left-hand tube and 20 mL of 6 M $NH_3(aq)$ was added to the right-hand tube.

to the formation of the $Ag(NH_3)_2^+$ ion. The left-to-right solubility equilibrium shift leads to an increased amount of dissolved $AgBrO_3(s)$. Figure 19-2 illustrates the increase in solubility of $AgBrO_3$ by the addition of $NH_3(aq)$.

19-4. THE MAGNITUDE OF THE RATIO Q_{sp}/K_{sp} IS USED TO PREDICT WHETHER AN IONIC SOLID WILL PRECIPITATE

Consider the silver bromate solubility equilibrium,

$$AgBrO_3(s) \rightleftharpoons Ag^+(aq) + BrO_3^-(aq) \qquad (19\text{-}11)$$

for which

$$K_{sp} = [Ag^+][BrO_3^-]$$

The concentration quotient, Q_{sp}, for Reaction (19-11) is given by

$$Q_{sp} = [Ag^+]_0[BrO_3^-]_0 \qquad (19\text{-}12)$$

Recall from Section 15-9 that the distinction between K_{sp} and Q_{sp} is that only equilibrium values of $[Ag^+]$ and $[BrO_3^-]$ can be used in the K_{sp} expression, whereas arbitrary values of $[Ag^+]$ and $[BrO_3^-]$ can be used in the Q_{sp} expression. If we prepare a solution with arbitrary values of $[Ag^+]$ and $[BrO_3^-]$, then the criterion for whether or not

Table 19-2 Criterion for the formation of a precipitate from a solution prepared with the constituent ions

For any arbitrary ion concentrations:

$\frac{Q_{sp}}{K_{sp}} > 1$ precipitate forms

$\frac{Q_{sp}}{K_{sp}} < 1$ no precipitate forms

When equilibrium is disturbed:

$\frac{Q_{sp}}{K_{sp}} > 1$ more precipitate forms until $Q_{sp} = K_{sp}$

$\frac{Q_{sp}}{K_{sp}} < 1$ precipitate dissolves either until $Q_{sp} = K_{sp}$ or until solid phase disappears completely

precipitation occurs, that is, whether or not $AgBrO_3(s)$ separates out, is $Q_{sp}/K_{sp} > 1$. Recall that if $Q_{sp}/K_{sp} > 1$, then Reaction (19-11) is spontaneous from right to left, which leads to the formation of $AgBrO_3(s)$ precipitate. Conversely, if $Q_{sp}/K_{sp} < 1$, then no precipitate forms. If $Q_{sp}/K_{sp} > 1$, then the precipitation continues until $Q_{sp} = K_{sp}$, that is, until equilibrium is established. If Reaction (19-11) is at equilibrium and the equilibrium is disturbed in such a way that Q_{sp}/K_{sp} becomes less than unity, then additional $AgBrO_3(s)$ dissolves. These conditions are summarized in Table 19-2.

For example, suppose we mix 50.0 mL of 1.0 M $AgNO_3(aq)$ with 50.0 mL of 0.010 M $NaBrO_3(aq)$ at 25°C. Does $AgBrO_3(s)$ precipitate? The value of K_{sp} for $AgBrO_3(s)$ at 25°C is $5.8 \times 10^{-5}\ M^2$. The initial concentrations of $[Ag^+]$ and $[BrO_3^-]$ in the 100-mL mixture of the two solutions are

$$[Ag^+]_0 = \frac{(1.0\ M)(0.050\ L)}{(0.100\ L)} = 0.50\ M$$

$$[BrO_3^-]_0 = \frac{(0.010\ M)(0.050\ L)}{(0.100\ L)} = 0.0050\ M$$

The initial value of Q_{sp} for the mixture is

$$Q_{sp} = [Ag^+]_0[BrO_3^-]_0 = (0.500\ M)(0.0050\ M) = 2.5 \times 10^{-3}\ M^2$$

and

$$\frac{Q_{sp}}{K_{sp}} = \frac{2.5 \times 10^{-3}\ M^2}{5.8 \times 10^{-5}\ M^2} = 43 > 1$$

The fact that $Q_{sp}/K_{sp} > 1$ means that the precipitation of $AgBrO_3(s)$ from the mixture is a spontaneous process. Once started, the precipitation continues until $Q_{sp}/K_{sp} = 1$, that is, until equilibrium is attained.

Example 19-6: A 1.0×10^{-3} M $NaIO_3(aq)$ solution is made 0.010 M in $Cu^{2+}(aq)$ by dissolving the soluble salt $Cu(ClO_4)_2(s)$. Does $Cu(IO_3)_2(s)$ ($K_{sp} = 7.4 \times 10^{-8}\ M^3$) precipitate from the solution?

Solution: The value of Q_{sp} is

$$Q_{sp} = [Cu^{2+}]_0[IO_3^-]_0^2 = (1.0 \times 10^{-2}\ M)(1.0 \times 10^{-3}\ M)^2 = 1.0 \times 10^{-8}\ M^3$$

and

$$\frac{Q_{sp}}{K_{sp}} = \frac{1.0 \times 10^{-8}\ M^3}{7.4 \times 10^{-8}\ M^3} = 0.14 < 1$$

Hence no precipitate of $Cu(IO_3)_2(s)$ forms.

Problems 19-31 through 19-36 are similar to Example 19-6.

19-5. SALTS OF WEAK ACIDS ARE MORE SOLUBLE IN ACIDIC SOLUTIONS THAN IN NEUTRAL OR BASIC SOLUTIONS

Sodium benzoate, $NaC_7H_5O_2$, the salt of NaOH and the weak acid benzoic acid, has the formula

$$C_6H_5{-}C(=O){-}O^{\ominus}Na^{\oplus}$$

Sodium benzoate is a water-soluble food additive that functions as an antimicrobial agent in foods with pH values lower than about 4. Numerous acidic beverages, syrups, jams, jellies, and processed fruits contain about 0.1 percent sodium benzoate to prevent the growth of yeasts and harmful bacteria. Numerous benzoate salts are insoluble above pH = 4, and the formation of insoluble benzoate salts at these higher pH values removes benzoate ions from solution, rendering the added sodium benzoate solution ineffective as an antimicrobial agent. At pH $\leq$ 4, an appreciable fraction of the benzoate exists as benzoic acid, which is the biologically active form.

The amount of sodium benzoate in food products may not exceed 0.1 percent.

Let us use silver benzoate, $AgC_7H_5O_2(s)$, to illustrate the effect of pH on the solubility of a salt of a weak acid. The solubility equilibrium of silver benzoate is

$$AgC_7H_5O_2(s) \rightleftharpoons Ag^+(aq) + C_7H_5O_2^-(aq) \qquad (19\text{-}13)$$

The K_{sp} for Reaction (19-13) at 25°C is

$$K_{sp} = 2.5 \times 10^{-5}\ M^2 = [Ag^+][C_7H_5O_2^-]$$

The solubility equilibrium fixes the value of the product $[Ag^+][C_7H_5O_2^-]$. However, the solubility equilibrium in Reaction (19-13) has a feature that we have not yet discussed in this chapter, namely, that benzoate ion is the conjugate base of a weak acid. The solubility of silver benzoate increases as $[H_3O^+]$ increases because $H_3O^+(aq)$ reacts with benzoate ion to form undissociated benzoic acid via the reaction

$$C_7H_5O_2^-(aq) + H_3O^+(aq) \rightarrow HC_7H_5O_2(aq) + H_2O(l)$$

The protonation of benzoate ion shifts the solubility equilibrium, Reaction (19-13), from left to right, thereby increasing the solubility of the silver benzoate. This effect is understood readily from Le Châtelier's principle. Because at equilibrium

$$[Ag^+][C_7H_5O_2^-] = 2.5 \times 10^{-5}\ M^2$$

if we decrease $[C_7H_5O_2^-]$ by reacting $C_7H_5O_2^-(aq)$ with $H_3O^+(aq)$, then the $AgC_7H_5O_2(s)$ solubility equilibrium shifts from left to right ($Q_{sp}/K_{sp} < 1$) until $Q_{sp} = K_{sp}$. This left-to-right shift makes $[Ag^+] > [C_7H_5O_2^-]$ at the new equilibrium. Because $s = [Ag^+]$, the solubility of $AgC_7H_5O_2(s)$ is increased by the addition of $H_3O^+(aq)$.

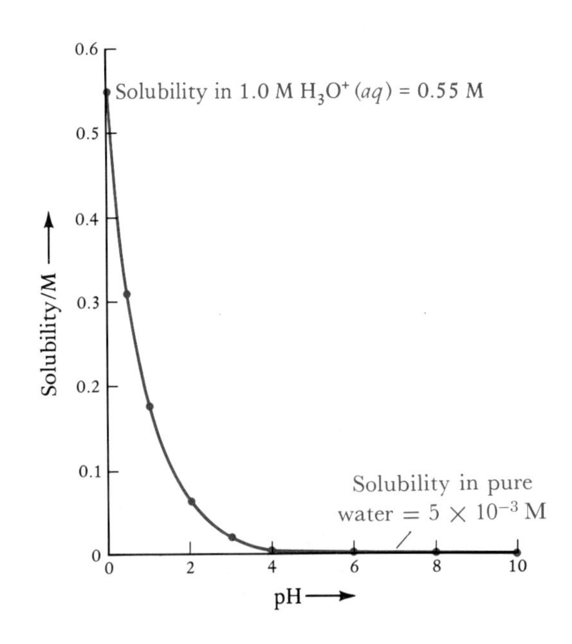

Figure 19-3 Solubility of silver benzoate in water as a function of pH. The addition of $H_3O^+(aq)$ shifts the solubility equilibrium to the right (increase in solubility) as $H_3O^+(aq)$ combines with $C_7H_5O_2^-(aq)$ to produce $HC_7H_5O_2(aq)$ and thereby decrease $[C_7H_5O_2^-]$.

Figure 19-3 shows the solubility of silver benzoate in water as a function of pH.

As Figure 19-3 shows, the solubility of a salt of a weak acid increases dramatically with decreasing pH. The variation in the solubility of salts of weak acids with pH is utilized in chemical analysis to separate metal ions, as we see in Section 19-7.

Above pH = 4, the benzoate precipitates out as metal benzoates.

19-6. SOME METAL CATIONS CAN BE SEPARATED FROM A MIXTURE BY THE FORMATION OF AN INSOLUBLE HYDROXIDE OF ONE OF THEM

It is possible to separate certain metal ions in aqueous solution by adjusting the pH of the solution such that the formation of an insoluble hydroxide of one of the ions is favored. For example, consider the hydroxide of $Zn^{2+}(aq)$, for which

$$Zn(OH)_2(s) \rightleftharpoons Zn^{2+}(aq) + 2OH^-(aq)$$

and

$$K_{sp} = 1.0 \times 10^{-15}\ M^3 = [Zn^{2+}][OH^-]^2$$

The solubility of $Zn(OH)_2(s)$ in water can be calculated from the K_{sp} expression:

$$s = [Zn^{2+}] = \frac{1.0 \times 10^{-15}\ M^3}{[OH^-]^2} \qquad (19\text{-}14)$$

The concentration of $OH^-(aq)$ can be related to $[H_3O^+]$ by using the ion product constant expression for water:

$$[OH^-] = \frac{K_w}{[H_3O^+]} = \frac{1.00 \times 10^{-14}\ M^2}{[H_3O^+]} \qquad (19\text{-}15)$$

Substitution of Equation (19-15) into Equation (19-14) yields

$$s = [Zn^{2+}] = \frac{1.0 \times 10^{-15}\ M^3}{(1.00 \times 10^{-14}\ M^2)^2}[H_3O^+]^2 \qquad (19\text{-}16)$$

$$= (1.0 \times 10^{13}\ M^{-1})[H_3O^+]^2$$

From this expression we can compute the solubility of $Zn(OH)_2(s)$, that is, $[Zn^{2+}]$, at various pH values, as shown in Table 19-3. These results are plotted in Figure 19-4, together with the analogous results for the solubility of $Fe(OH)_3(s)$. Note that $Fe^{3+}(aq)$ can be separated from $Zn^{2+}(aq)$ by adjusting the pH of the solution to about 5 with an acetic acid–acetate buffer. At pH $\approx$ 5, the $Fe(OH)_3(s)$ precipitates and the $Zn^{2+}(aq)$ remains in solution.

Table 19-3 Solubility of $Zn(OH)_2(s)$ at various pH values

pH	$[H_3O^+]/M$	$[H_3O^+]^2/M^2$	$s = [Zn^{2+}]/M$
6.5	3.2×10^{-7}	1.0×10^{-13}	1.0
6.8	1.6×10^{-7}	2.5×10^{-14}	0.25
7.0	1.0×10^{-7}	1.0×10^{-14}	0.10
7.5	3.2×10^{-8}	1.0×10^{-15}	0.010
8.0	1.0×10^{-8}	1.0×10^{-16}	0.0010
8.5	3.2×10^{-9}	1.0×10^{-17}	0.00010

Example 19-7: Compute the solubilities of $Zn(OH)_2(s)$ and $Fe(OH)_3(s)$ in an aqueous solution buffered at pH = 7.0.

Solution: The solubility of $Zn(OH)_2(s)$ in a solution buffered at pH = 7.0 can be computed from Equation (19-16):

$$s = [Zn^{2+}] = (1.0 \times 10^{13}\ M^{-1})[H_3O^+]^2$$

At pH = 7.0, we have $[H_3O^+] = 10^{-pH} = 10^{-7.0} = 1.0 \times 10^{-7}$ M, and thus

$$s = (1.0 \times 10^{13}\ M^{-1})(1.0 \times 10^{-7}\ M)^2 = 0.10\ M$$

The solubility of $Fe(OH)_3(s)$, the K_{sp} of which is $1.0 \times 10^{-38}\ M^4$ (Table 19-1), is computed in a manner analogous to that just described for $Zn(OH)_2(s)$. Thus, we write

$$Fe(OH)_3(s) \rightleftharpoons Fe^{3+}(aq) + 3OH^-(aq)$$

and

$$K_{sp} = [Fe^{3+}][OH^-]^3 = 1.0 \times 10^{-38}\ M^4$$

The solubility of $Fe(OH)_3(s)$ is equal to $Fe^{3+}(aq)$, and so

$$s = [Fe^{3+}] = \frac{1.0 \times 10^{-38}\ M^4}{[OH^-]^3} = \frac{1.0 \times 10^{-38}\ M^4[H_3O^+]^3}{K_w^3}$$

The final form here is obtained by using the relation $[OH^-] = K_w/[H_3O^+]$. Using the fact that $K_w = 1.0 \times 10^{-14}\ M^2$, we have

$$s = [Fe^{3+}] = \frac{1.0 \times 10^{-38}\ M^4[H_3O^+]^3}{(1.0 \times 10^{-14}\ M^2)^3} = (1.0 \times 10^4\ M^{-2})[H_3O^+]^3$$

At $[H_3O^+] = 1.0 \times 10^{-7}$ M, we calculate

$$s = [Fe^{3+}] = (1.0 \times 10^4\ M^{-2})(1.0 \times 10^{-7}\ M)^3 = 1.0 \times 10^{-17}\ M$$

At pH = 7.0, then, $Fe(OH)_3(s)$ is insoluble (see Figure 19-4).

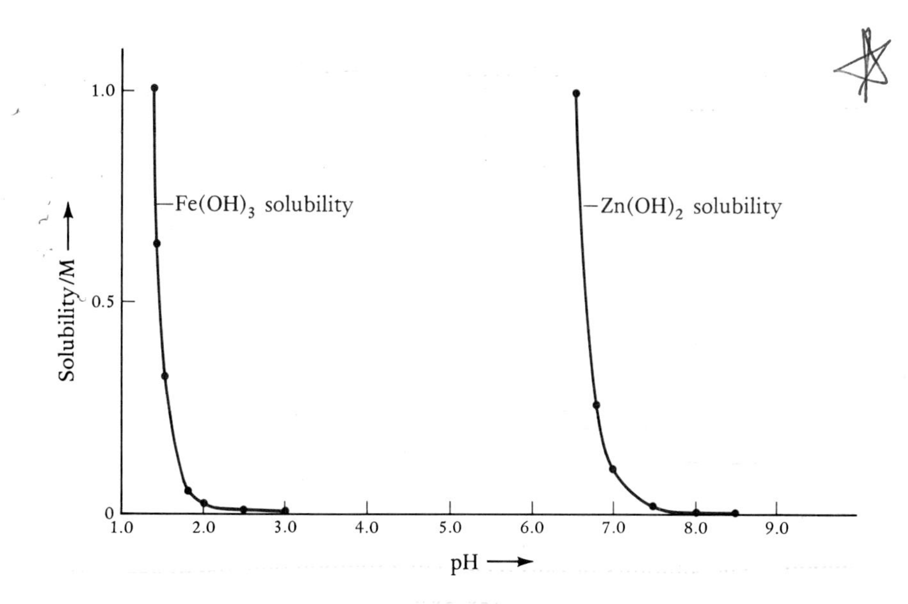

Figure 19-4 Solubilities of $Fe(OH)_3(s)$ and $Zn(OH)_2(s)$ as a function of pH. Note that a much lower solution pH is required to dissolve $Fe(OH)_3(s)$ than to dissolve $Zn(OH)_2(s)$. Therefore, at pH = 4.8, for example, $Fe(OH)_3(s)$ precipitates and $Zn^{2+}(aq)$ remains in solution. The $Fe(OH)_3(s)$ can be filtered off, thereby achieving a separation of $Fe^{3+}(aq)$ from $Zn^{2+}(aq)$.

19-7. INSOLUBLE SULFIDES ARE SEPARATED BY ADJUSTMENT OF SOLUTION pH

The solubility product values given in Table 19-1 show that the solubilities of metal sulfides vary over an enormous range. We can selectively precipitate certain metals by controlling the concentration of $S^{2-}(aq)$. This can be done by controlling the pH because $S^{2-}(aq)$ is the conjugate base of the weak acid HS^-. The relevant equilibrium expressions are

$$H_2S(aq) + H_2O(l) \rightleftharpoons H_3O^+(aq) + HS^-(aq) \qquad K_{a1} = 9.1 \times 10^{-8}\ \text{M}$$

$$HS^-(aq) + H_2O(l) \rightleftharpoons H_3O^+(aq) + S^{2-}(aq) \qquad K_{a2} = 1.2 \times 10^{-13}\ \text{M}$$

The sum of these two equations is

$$H_2S(aq) + 2H_2O(l) \rightleftharpoons 2H_3O^+(aq) + S^{2-}(aq)$$

Therefore the value of K for this reaction is

$$K = K_{a1}K_{a2} = (9.1 \times 10^{-8}\ \text{M})(1.2 \times 10^{-13}\ \text{M}) = 1.1 \times 10^{-20}\ \text{M}^2$$

$$= \frac{[H_3O^+]^2[S^{2-}]}{[H_2S]} \qquad (19\text{-}17)$$

The concentration of $H_2S(aq)$ in a saturated aqueous solution at 25°C is 0.10 M. If we substitute $[H_2S] = 0.10$ M into Equation (19-17) and solve for $[S^{2-}]$, then we obtain

Equation (19-18) is valid only when $[H_2S] = 0.10$ M.

$$[S^{2-}] = (1.1 \times 10^{-20}\ M^2)\left(\frac{0.10\ M}{[H_3O^+]^2}\right)$$

$$= \frac{1.1 \times 10^{-21}\ M^3}{[H_3O^+]^2} \qquad (19\text{-}18)$$

Equation (19-18) shows that the sulfide ion concentration in a saturated $H_2S(aq)$ solution can be controlled by the hydrogen ion concentration or, equivalently, by the pH. The distribution diagram (Section 18-7) for $H_2S(aq)$ is shown in Figure 19-5. Note that $S^{2-}(aq)$ becomes a major species at pH > 12.

To illustrate the use of Equation (19-18), we calculate the solubility of ZnS(*s*) and FeS(*s*) at pH = 2.0. According to Equation (19-18), at pH = 2.0, that is, at $[H_3O^+] = 1.0 \times 10^{-2}$ M,

$$[S^{2-}] = \frac{1.1 \times 10^{-21}\ M^3}{(1.0 \times 10^{-2}\ M)^2} = 1.1 \times 10^{-17}\ M$$

Using the K_{sp} values from Table 19-1, we have that

$$[Zn^{2+}][S^{2-}] = 1.6 \times 10^{-24}\ M^2$$
$$[Fe^{2+}][S^{2-}] = 6.3 \times 10^{-18}\ M^2$$

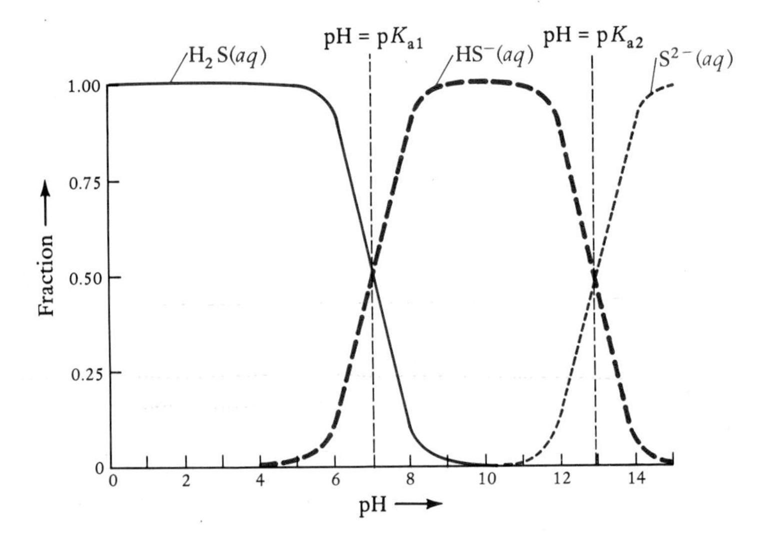

Figure 19-5 Distribution diagram for $H_2S(aq)$, giving the fraction of the total sulfide represented by each of the species $H_2S(aq)$, $HS^-(aq)$, and $S^{2-}(aq)$ as a function of pH. Note that $S^{2-}(aq)$ becomes a principal species at pH > 12.

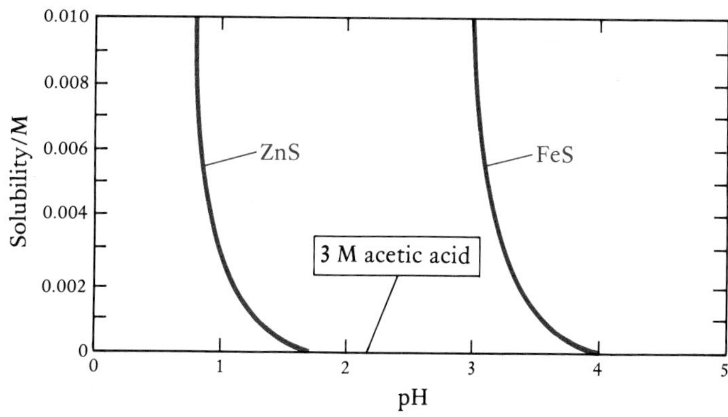

Figure 19-6 Solubilities of $ZnS(s)$ and $FeS(s)$ as a function of pH. Note that a much lower pH is required to dissolve $ZnS(s)$ than to dissolve $FeS(s)$.

Therefore, the concentrations of $Zn^{2+}(aq)$ and $Fe^{2+}(aq)$ in a saturated $H_2S(aq)$ solution at pH = 2.0 are

$$[Zn^{2+}] = \frac{1.6 \times 10^{-24}\ M^2}{[S^{2-}]} = \frac{1.6 \times 10^{-24}\ M^2}{1.1 \times 10^{-17}\ M} = 1.5 \times 10^{-7}\ M$$

$$[Fe^{2+}] = \frac{6.3 \times 10^{-18}\ M^2}{[S^{2-}]} = \frac{6.3 \times 10^{-18}\ M^2}{1.1 \times 10^{-17}\ M} = 0.57\ M$$

If the pH is adjusted to 2.0, then, the $Zn^{2+}(aq)$ precipitates out as $ZnS(s)$ and the $Fe^{2+}(aq)$ remains in solution. Thus we achieve a separation of Zn^{2+} from Fe^{2+}. The solubilities of $ZnS(s)$ and $FeS(s)$ as a function of pH are plotted in Figure 19-6.

19-8. AMPHOTERIC HYDROXIDES DISSOLVE IN BOTH HIGHLY ACIDIC AND HIGHLY BASIC SOLUTIONS

At high concentrations of $OH^-(aq)$, some hydroxides dissolve as the result of the formation of soluble hydroxy ions. For example,

$$Al(OH)_3(s) + OH^-(aq) \rightleftharpoons \underset{\text{hydroxy ion}}{Al(OH)_4^-(aq)} \qquad (19\text{-}19)$$

Such hydroxides are said to be *amphoteric*. An *amphoteric metal hydroxide* is one that is soluble in both strong acids and strong bases but insoluble in neutral solutions. The equilibrium constant expression at 25°C for Equation (19-19) is

$$K = \frac{[Al(OH)_4^-]}{[OH^-]} = 40 \qquad (19\text{-}20)$$

The solubility of $Al(OH)_3(s)$ in a basic solution is equal to $[Al(OH)_4^-]$, and so we can write Equation (19-20) as

$$K = \frac{s}{[OH^-]} = 40$$

or

$$s = 40[OH^-]$$

If $[OH^-] = 0.10$ M, then

$$s = 40(0.10\ \text{M}) = 4.0\ \text{M}$$

The K_{sp} for $Al(OH)_3(s)$ in pure water at 25°C is $1.3 \times 10^{-33}\ \text{M}^4$. Thus for the reaction

$$Al(OH)_3(s) \rightleftharpoons Al^{3+}(aq) + 3OH^-(aq) \qquad (19\text{-}21)$$

we have

$$K_{sp} = 1.3 \times 10^{-33}\ \text{M}^4 = [Al^{3+}][OH^-]^3$$

If Equation (19-19) did not occur, then the solubility of $Al(OH)_3(s)$ in a solution with $[OH^-] = 0.10$ M would be

$$s = [Al^{3+}] = \frac{1.3 \times 10^{-33}\ \text{M}^4}{[OH^-]^3} = \frac{1.3 \times 10^{-33}\ \text{M}^4}{(0.10\ \text{M})^3} = 1.3 \times 10^{-30}\ \text{M}$$

which is very small indeed. If we computed the solubility of $Al(OH)_3(s)$ in 0.10 M $OH^-(aq)$ without considering Equation (19-19), then our result would be in error by a factor of 3×10^{30}.

At high pH values the solubility of $Al(OH)_3(s)$ is determined by Equation (19-19). In neutral solution, its solubility is determined by Equation (19-21). At low pH, the $Al(OH)_3(s)$ dissolves readily because of the acid-base neutralization reaction:

$$Al(OH)_3(s) + 3H_3O^+(aq) \rightleftharpoons Al^{3+}(aq) + 6H_2O(l) \qquad (19\text{-}22)$$

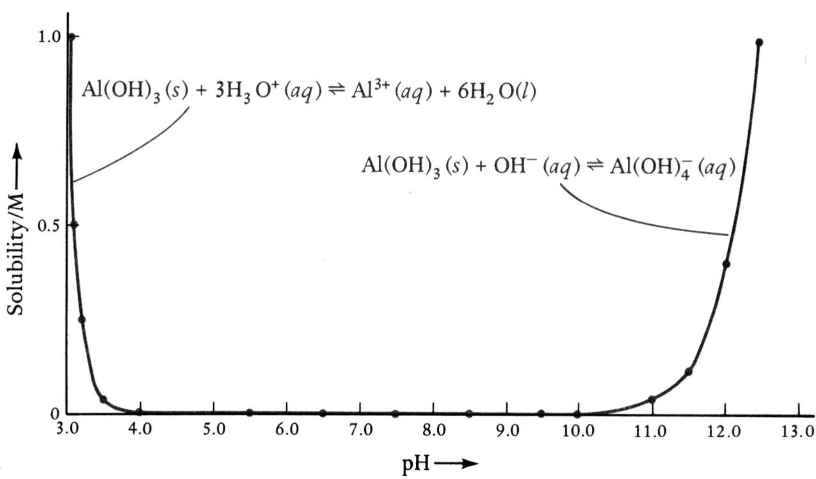

Figure 19-7 The solubility of $Al(OH)_3(s)$ as a function of pH. The amphoteric nature of $Al(OH)_3$ is clearly shown by its solubility in both highly acidic and highly basic solutions. Note that $Al(OH)_3(s)$ is essentially insoluble over the pH range 4 to 10.

Thus we see that $Al(OH)_3(s)$ dissolves in basic solutions (Equation 19-19) and in acidic solutions (Equation 19-22) but not in neutral solutions (Equation 19-21). The amphoteric behavior of $Al(OH)_3(s)$ is illustrated in Figure 19-7. Some other examples of amphoteric hydroxides are given in Table 19-4.

The hydroxide of gallium, which is under aluminum in the periodic table, is also amphoteric.

Example 19-8: Use the equilibrium constant data for zinc hydroxide in Table 19-4 to estimate its solubility in a solution buffered at pH = 14.0.

Solution: For the reaction

$$Zn(OH)_2(s) + 2OH^-(aq) \rightleftharpoons Zn(OH)_4^{2-}(aq)$$

we have

$$K = 0.05\ M^{-1} = \frac{[Zn(OH)_4^{2-}]}{[OH^-]^2} = \frac{s}{[OH^-]^2}$$

At pH = 14.0, $[H_3O^+] = 1.0 \times 10^{-14}$ M, and so

$$[OH^-] = \frac{K_w}{[H_3O^+]} = \frac{1.0 \times 10^{-14}\ M^2}{1.0 \times 10^{-14}\ M} = 1.0\ M$$

Substitution of this value for $[OH^-]$ into the expression for K yields for the solubility of $Zn(OH)_2(s)$

$$s = (0.05\ M^{-1})[OH^-]^2 = (0.05\ M^{-1})(1.0\ M)^2 = 0.05\ M$$

Problems 19-57 through 19-64 involve amphoteric hydroxides.

Table 19-4 Some amphoteric hydroxides

Reaction	*K*
$Al(OH)_3(s) + OH^-(aq) \rightleftharpoons Al(OH)_4^-(aq)$	40
$Pb(OH)_2(s) + OH^-(aq) \rightleftharpoons Pb(OH)_3^-(aq)$	0.08
$Zn(OH)_2(s) + 2OH^-(aq) \rightleftharpoons Zn(OH)_4^{2-}(aq)$	0.05 M^{-1}
$Cr(OH)_3(s) + OH^-(aq) \rightleftharpoons Cr(OH)_4^-(aq)$	0.04
$Sn(OH)_2(s) + OH^-(aq) \rightleftharpoons Sn(OH)_3^-(aq)$	0.01

19-9. QUALITATIVE ANALYSIS IS THE IDENTIFICATION OF THE SPECIES PRESENT IN A SAMPLE

The laboratory work in most introductory chemistry courses involves *qualitative analysis.* The objective of a qualitative analysis scheme is the identification of various cations and anions in a solution or in a mixture of solids. In qualitative analysis we seek only to identify the ions, not to determine their concentration. The determination of the amounts present or the percentage compositions is called *quantitative analysis.*

The sample to be analyzed is called the unknown, but you begin your analysis knowing the list of *possible* cations and anions in the unknown. You also know, on the basis of the results of a set of preliminary experiments, what tests to perform for the presence or absence of the various possible species. Qualitative analysis schemes are included in laboratory programs because they illustrate the application of numerous chemical principles, such as solubility product equilibria and buffer action. Furthermore, most students enjoy the challenge of identifying the ions in their unknowns.

The methods of qualitative analysis developed in general chemistry laboratories are not, in most cases, the same methods used by professional analytical chemists to identify the constituents of an unknown sample. An analytical chemist working in, say, a criminal investigations laboratory faces much more difficult challenges to his or her chemical ingenuity than those encountered in a general chemistry laboratory. The major differences are that, in the "real world," the possible constituents of an unknown are essentially unlimited and truly unknown. Furthermore, in many cases the available sample size is very small and irreplaceable. Consequently, the practicing analytical chemist depends on techniques that are much more sophisticated than those you will use in your qualitative analysis experiments. Nonetheless, all of the chemical principles you will use in your experiments are used by analytical chemists.

Figure 19-8 Although chemists now often use a variety of sophisticated electronic instruments in performing chemical analyses, they still employ "wet lab" chemical techniques in certain cases, as shown here at the Smithsonian Institution.

19-10. PRECIPITATION REACTIONS ARE USED TO SEPARATE MIXTURES OF IONS

The basic approach used in the qualitative analysis of a mixture of cations is the separation of some cations from the mixture by addition of a reagent that precipitates certain cations but not others. In this way the analytical problem is simplified because the number of cations involved in each of the two resulting samples (precipitate and remaining solution) is smaller than in the original sample. It is essential in qualitative analysis to know the solubilities of the various possible salts that may be formed as a result of the addition of a precipitating agent to the solution.

As a simple example, consider a solution that may contain $KNO_3(aq)$ and/or $AgNO_3(aq)$. Because AgCl is insoluble and KCl is soluble, the formation of a white precipitate upon addition of 6 M $HCl(aq)$ indicates that $Ag^+(aq)$ is present (Figure 19-9):

$$Ag^+(aq) + Cl^-(aq) \rightarrow AgCl(s)$$

$$K^+(aq) + Cl^-(aq) \rightarrow \text{no reaction}$$

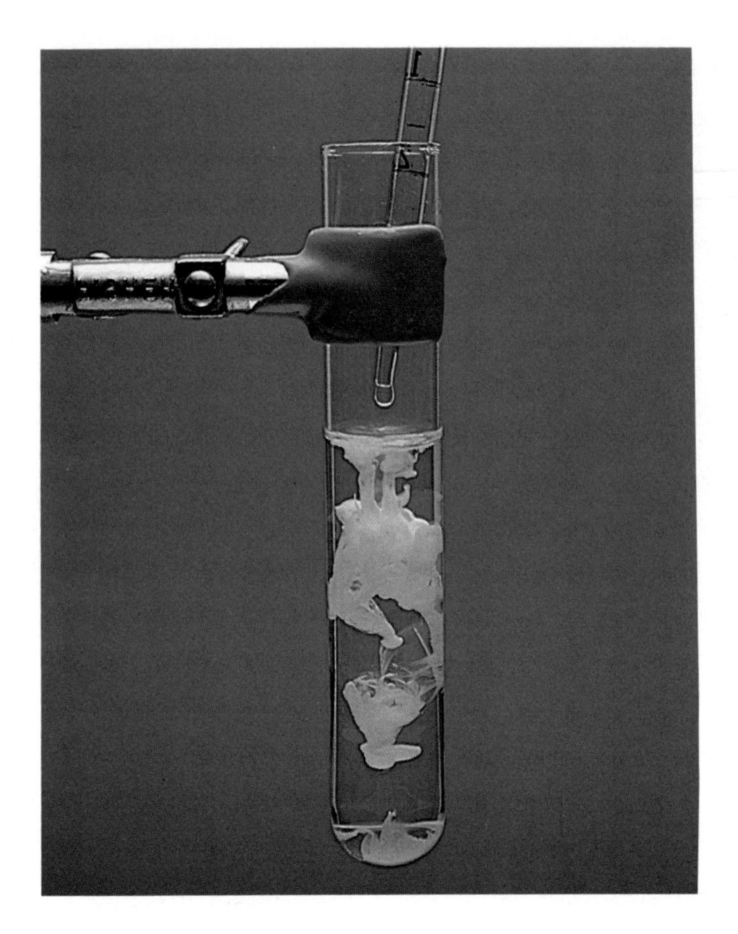

Figure 19-9 Precipitation of $Ag^+(aq)$ as $AgCl(s)$ by addition of $HCl(aq)$ to a solution containing $Ag^+(aq)$.

Addition of excess $HCl(aq)$ precipitates essentially all the $Ag^+(aq)$ as $AgCl(s)$. The settling of the $AgCl(s)$ is hastened by using a *centrifuge* (Figure 19-10). The resulting *supernatant* solution is then *decanted* (poured off) (Figure 19-11) and tested separately for the presence of $K^+(aq)$. Water-insoluble silver chloride is soluble in 6 M $NH_3(aq)$ (Figure 19-12) because of the formation of the $Ag(NH_3)_2^+(aq)$ ion:

$$AgCl(s) + 2NH_3(aq) \rightarrow Ag(NH_3)_2^+(aq) + Cl^-(aq)$$

The solubilization of the white $AgCl(s)$ in the $NH_3(aq)$ is a confirmatory test for $Ag^+(aq)$. Thus $Ag^+(aq)$ forms an insoluble white chloride that dissolves in 6 M $NH_3(aq)$.

You will learn how to name compounds such as potassium hexanitritocobaltate(III) in Chapter 23.

Most potassium salts are water-soluble. An exception is $K_3Co(NO_2)_6(s)$, potassium hexanitritocobaltate(III). Addition of $Na_3Co(NO_2)_6(aq)$, which is water-soluble, to a solution containing $K^+(aq)$ produces a pale yellow precipitate:

$$3K^+(aq) + Co(NO_2)_6^{3-}(aq) \rightarrow \underset{\text{pale yellow}}{K_3Co(NO_2)_6(s)}$$

Figure 19-10 A centrifuge hastens the settling of a precipitate by increasing the downward force on the solid particles.

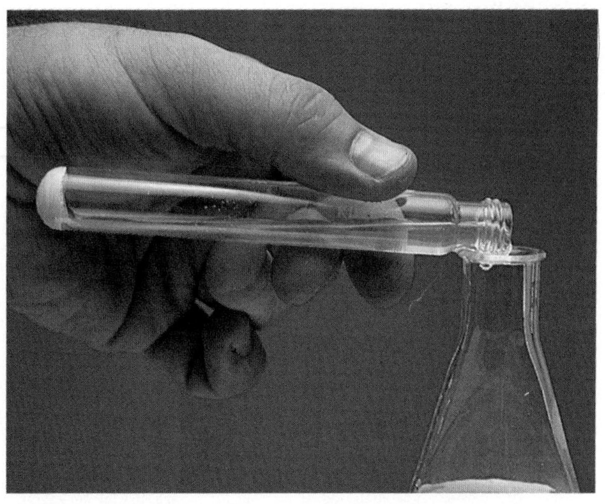

Figure 19-11 The supernatant solution from Figure 19-10 being decanted.

Silver ion, as well as most other cations except sodium, also forms an insoluble salt with $Co(NO_2)_6^{3-}(aq)$:

$$3Ag^+(aq) + Co(NO_2)_6^{3-}(aq) \rightarrow Ag_3Co(NO_2)_6(s)$$

but recall that $Ag^+(aq)$ was removed from the unknown by precipitation as $AgCl(s)$.

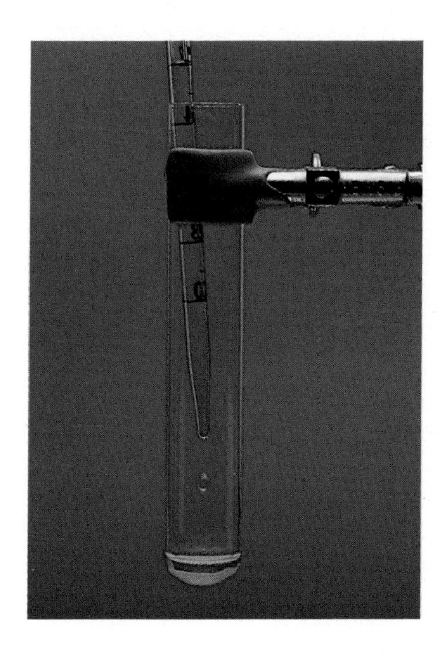

Figure 19-12 Dissolution of $AgCl(s)$ in 6 M $NH_3(aq)$ by formation of $Ag(NH_3)_2^+$.

19-11. MANY QUALITATIVE ANALYSIS SCHEMES INVOLVE THE SEPARATION OF CATIONS AS INSOLUBLE SULFIDES

Many of the common metal sulfides are colored, insoluble salts (Figure 19-13), and the different colors are useful in identifying cations. Metal sulfides are precipitated by adding H_2S to the solution. Many laboratories do not use H_2S directly because hydrogen sulfide is an extremely hazardous substance. In order to minimize H_2S exposure in the laboratory, H_2S is prepared in solution by the decomposition of thioacetamide on gentle heating:

A

B

C

D

E

Figure 19-13 Many metal sulfides are insoluble salts. Shown here in their powder and natural crystalline forms are chalcocite (CuS) in A, sphalerite (ZnS) in B, stannite (SnS_2) in C, cinnabar (HgS) in D, and galena (PbS) in E.

$$\underset{\text{thioacetamide}}{CH_3\underset{\underset{\textstyle \cdot\ddot{S}\cdot}{\|}}{C}NH_2(aq)} + 2H_2O(l) \xrightarrow{60°C} C_2H_3O_2^-(aq) + NH_4^+(aq) + H_2S(aq)$$

The use of thioacetamide as an H_2S source together with the use of laboratory ventilation hoods makes it possible to keep the H_2S concentration in the laboratory air at a safe level.

Selective precipitation of metal sulfides is achieved by adjusting the value of $[H_3O^+]$ in the solution because the value of $[H_3O^+]$ controls the value of $[S^{2-}]$ at a fixed value of $[H_2S]$ (Equation 19-18). The value of $[S^{2-}]$, in turn, controls the solubility of metal sulfides through the solubility equilibria. For example, the solubility equilibrium of $FeS(s)$ is

$$FeS(s) \rightleftharpoons Fe^{2+}(aq) + S^{2-}(aq)$$

for which $\quad K_{sp} = [Fe^{2+}][S^{2-}] = 6.3 \times 10^{-18}\ M^2$

Because $Q_{sp} = [Fe^{2+}]_0[S^{2-}]_0$, a precipitate will form if

$$\frac{Q_{sp}}{K_{sp}} = \frac{[Fe^{2+}]_0\,[S^{2-}]_0}{6.3 \times 10^{-18}\ M^2} > 1$$

That is, if

$$[S^{2-}]_0 > \frac{6.3 \times 10^{-8}\ M^2}{[Fe^{2+}]_0}$$

Thus with $[Fe^{2+}]_0 = 0.010$ M, a precipitate of $FeS(s)$ will form if

$$[S^{2-}]_0 > \frac{6.3 \times 10^{-18}\ M^2}{[Fe^{2+}]_0} = \frac{6.3 \times 10^{-18}\ M^2}{0.010\ M}$$

$$= 6.3 \times 10^{-16}\ M$$

At $[H_3O^+] = 0.3$ M, the sulfides of $Hg^{2+}(aq)$ and $Cu^{2+}(aq)$ are insoluble, whereas the sulfides of $Zn^{2+}(aq)$, $Al^{3+}(aq)$, $Ca^{2+}(aq)$, and $Fe^{2+}(aq)$ are soluble. Thus, by adjusting the value of $[H_3O^+]$ to 0.3 M, we can achieve a separation of $Hg^{2+}(aq)$ and $Cu^{2+}(aq)$ from $Zn^{2+}(aq)$, $Al^{3+}(aq)$, $Ca^{2+}(aq)$, and $Fe^{2+}(aq)$.

Copper(II) sulfide is separated from mercury(II) sulfide by reaction with hot 2 M $HNO_3(aq)$, which dissolves $CuS(s)$ but not $HgS(s)$ via the reaction (Figure 19-14):

$$3CuS(s) + 2HNO_3(aq) + 6H^+(aq) \rightarrow 3Cu^{2+}(aq) + 3S(s) + 2NO(g) + 4H_2O(l)$$

Figure 19-14 Separation of $CuS(s)$ from a mixture of $CuS(s)$ and $HgS(s)$ by reaction with hot 2 M $HNO_3(aq)$.

19-12. AMPHOTERIC METAL HYDROXIDES ARE SEPARATED BY DISSOLUTION IN STRONG BASE

Separations of certain metal ions are achieved by means of the formation of insoluble hydroxides at controlled pH values. A case in point is the $Al^{3+}(aq)$ ion, which precipitates as $Al(OH)_3(s)$ at pH = 12 along with $ZnS(s)$ and $FeS(s)$ from an aqueous solution containing $S^{2-}(aq)$. A mixture of $Al(OH)_3(s)$, $ZnS(s)$, and $FeS(s)$ is separated by treating it with an aqueous solution at pH = 2. The $FeS(s)$ and $Al(OH)_3(s)$ dissolve as $Fe^{2+}(aq)$ and $Al^{3+}(aq)$, leaving behind $ZnS(s)$, which is much less soluble at pH = 2 than $FeS(s)$ or $Al(OH)_3(s)$. The $Al^{3+}(aq)$ ion is separated from $Fe^{2+}(aq)$ by addition of 6 M $NaOH(aq)$. The $Fe^{2+}(aq)$ precipitates as $Fe(OH)_2(s)$, and the $Al^{3+}(aq)$ remains in solution as the aluminate ion:

$$Al^{3+}(aq) + 4OH^-(aq) \rightleftharpoons \underset{\text{aluminate ion}}{Al(OH)_4^-(aq)}$$

Recall that $Al(OH)_3(s)$ is amphoteric (Section 19-8).

An essential feature of a qualitative analysis scheme for a large group of ions is the *successive* removal of subgroups of the ions by precipitation reactions. It is essential to carry out the separation steps in a *systematic* fashion; otherwise ions that are presumed to have been removed may interfere with subsequent steps in the analytical scheme. If you are in doubt about how the test results look with certain mixtures of ions, then you should prepare a known mixture containing the ions in question and run the appropriate tests on the known for comparison to the unknown.

The keys to an accurate analysis of your unknown are careful observation and a thorough knowledge of the chemistry of the qualitative analysis scheme, such as solubilities at various pH values, possible complexation reactions, and so on. With a reasonable effort on your part to master the relevant chemistry, you will find your experience as a chemical sleuth both fun and informative.

19-13. QUALITATIVE ANALYSIS SCHEMES INVOLVE SEQUENTIAL SEPARATION OF SMALL GROUPS OF CATIONS

An outline of a simplified qualitative analysis scheme for ten cations is given in Figure 19-15 (page 772, 773). Study this scheme carefully so that you understand in chemical terms what happens at each step of the procedure.

The ten cations are divided into four subgroups (I, II, III, and IV). Each subgroup of cations is separated in successive precipitation

Table 19-5 Qualitative analysis separation scheme

Subgroup	*Constituents*	*Separation procedure*
I	Ag^+, Hg_2^{2+}	precipitation as chlorides AgCl, Hg_2Cl_2 with 6 M HCl(*aq*)
II	Hg^{2+}, Cu^{2+}	precipitation as sulfides HgS, CuS at $[H_3O^+] = 0.3$ M
III	Zn^{2+}, Fe^{2+}, Al^{3+}	precipitation as ZnS, FeS, $Al(OH)_3$
IV	Ca^{2+}, K^+, Na^+	remain in solution following separations of subgroups I, II, III; Ca^{2+} precipitated as $CaCO_3(s)$ with $(NH_4)_2CO_3(aq)$; K^+ precipitated as $K_3Co(NO_2)_6$ with $Na_3Co(NO_2)_6$

steps, as shown in Table 19-5 and Figure 19-15. The principal steps in the separation procedure are as follows.

Separation of subgroup I: The addition of 6 M HCl(*aq*) produces a mixed precipitate of AgCl(*s*) and $Hg_2Cl_2(s)$. The other eight cations have soluble chlorides. The AgCl(*s*) and $Hg_2Cl_2(s)$ are separated by reaction with 6 M $NH_3(aq)$ (Figure 19-15).

Separation of subgroup II: The addition of $H_2S(aq)$ in 0.3 M $H_3O^+(aq)$ forms a precipitate of the sulfides of Hg^{2+} and Cu^{2+}. (The sulfides of subgroups III and IV are soluble in 0.3 M $H_3O^+(aq)$.) The remaining mixture of CuS(*s*) and HgS(*s*) is separated by a reaction with hot $HNO_3(aq)$, which dissolves CuS(*s*) but not HgS(*s*).

Separation of subgroup III: The pH of the solution is adjusted to about pH = 12, where ZnS, FeS, and $Al(OH)_3$ precipitate (the subgroup IV cations remain in solution). This precipitate is then separated by adjusting the pH to 2, at which pH ZnS(*s*) is insoluble, whereas FeS(*s*) and $Al(OH)_3(s)$ dissolve. The mixture of Fe^{2+} and Al^{3+} is then separated by utilizing the amphoteric nature of $Al(OH)_3(s)$; that is, at high pH, $Fe(OH)_2$ precipitates, whereas Al^{3+} remains in solution as the aluminate ion, $Al(OH)_4^-(aq)$.

Separation of subgroup IV: Ammonium carbonate is added to precipitate calcium as $CaCO_3(s)$. Sodium can be detected by its characteristic yellow flame in a flame test on the solution. After $NH_3(aq)$ is removed by evaporating the solution to near dryness, potassium is detected by adding $Na_3Co(NO_2)_6$ to the solution obtained by redissolving the residue from the evaporation step.

There are many variations of qualitative analysis schemes used in general chemistry laboratories; you should consult your laboratory manual for the details of the procedure used in your course.

Figure 19-15 Qualitative analysis scheme for ten cations.

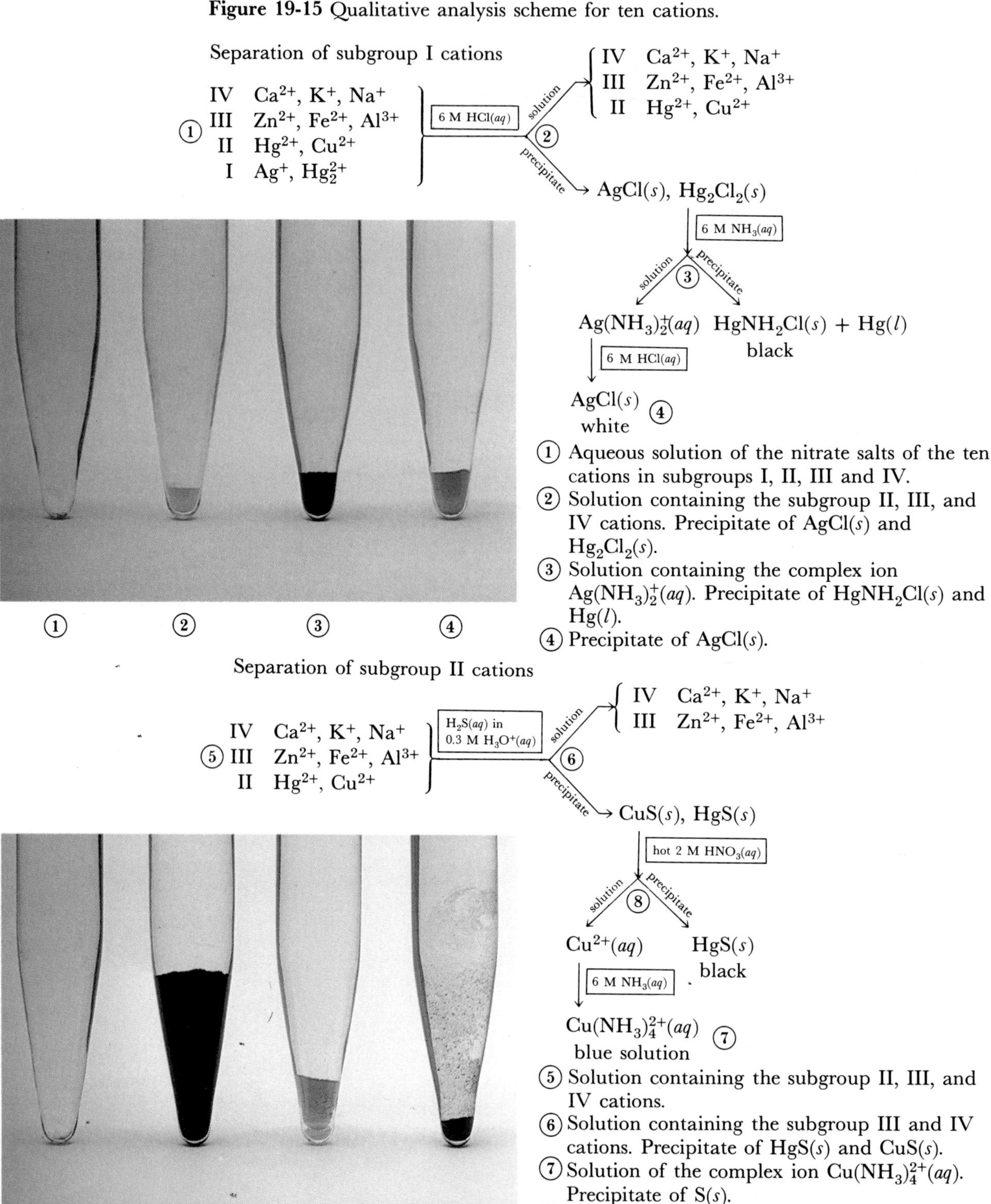

(1) Aqueous solution of the nitrate salts of the ten cations in subgroups I, II, III and IV.
(2) Solution containing the subgroup II, III, and IV cations. Precipitate of $AgCl(s)$ and $Hg_2Cl_2(s)$.
(3) Solution containing the complex ion $Ag(NH_3)_2^+(aq)$. Precipitate of $HgNH_2Cl(s)$ and $Hg(l)$.
(4) Precipitate of $AgCl(s)$.

(5) Solution containing the subgroup II, III, and IV cations.
(6) Solution containing the subgroup III and IV cations. Precipitate of $HgS(s)$ and $CuS(s)$.
(7) Solution of the complex ion $Cu(NH_3)_4^{2+}(aq)$. Precipitate of $S(s)$.
(8) Solution containing $Cu^{2+}(aq)$. Precipitate of $HgS(s)$.

Separation of subgroup III cations

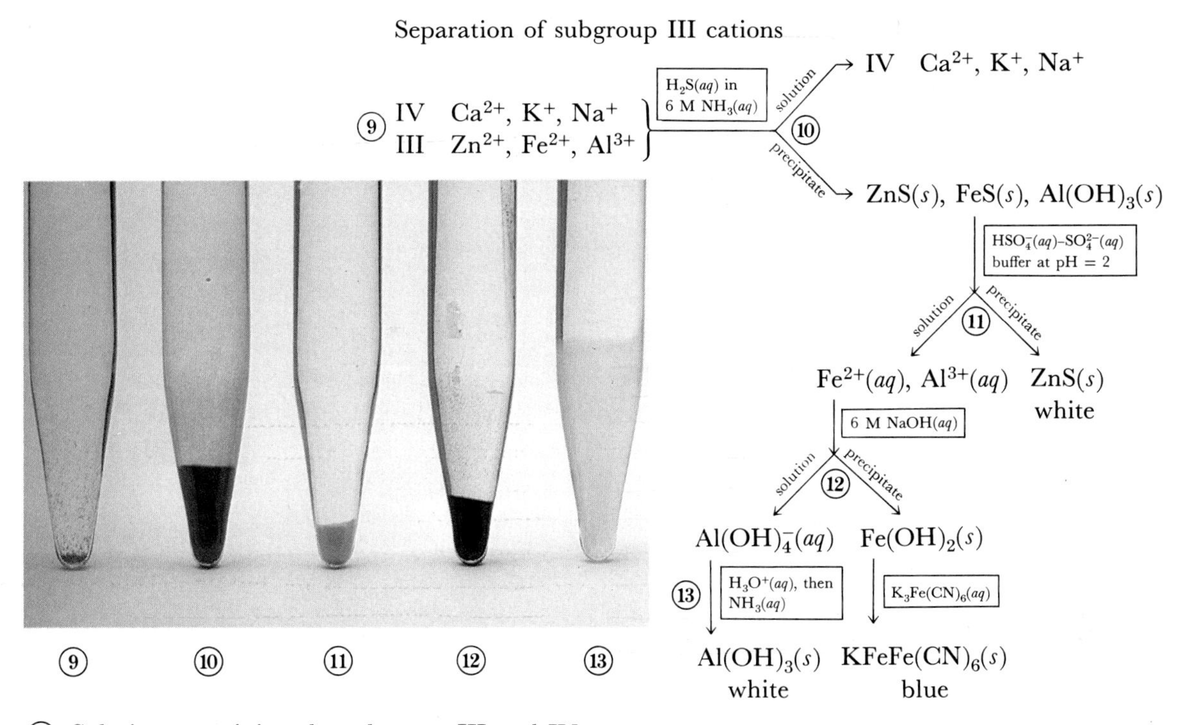

(9) Solution containing the subgroup III and IV cations.
(10) Solution containing the subgroup IV cations. Precipitate of $ZnS(s)$, $FeS(s)$ and $Al(OH)_3(s)$.
(11) Solution containing $Fe^{2+}(aq)$ and $Al^{3+}(aq)$. Precipitate of $ZnS(s)$.
(12) Solution containing $Al(OH)_4^-(aq)$. Precipitate of $Fe(OH)_2(s)$.
(13) Precipitate of $Al(OH)_3(s)$.

Separation of subgroup IV cations

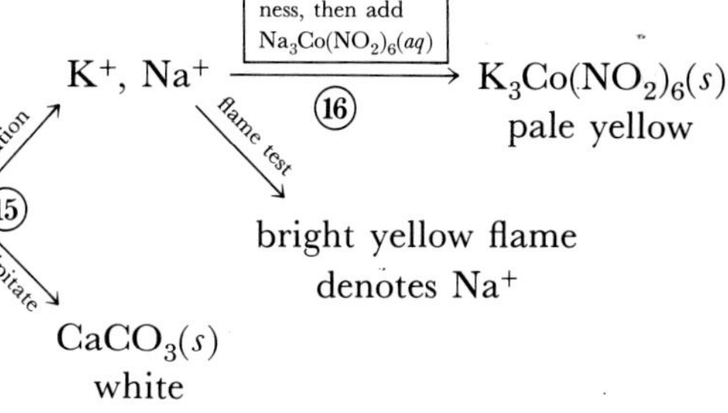

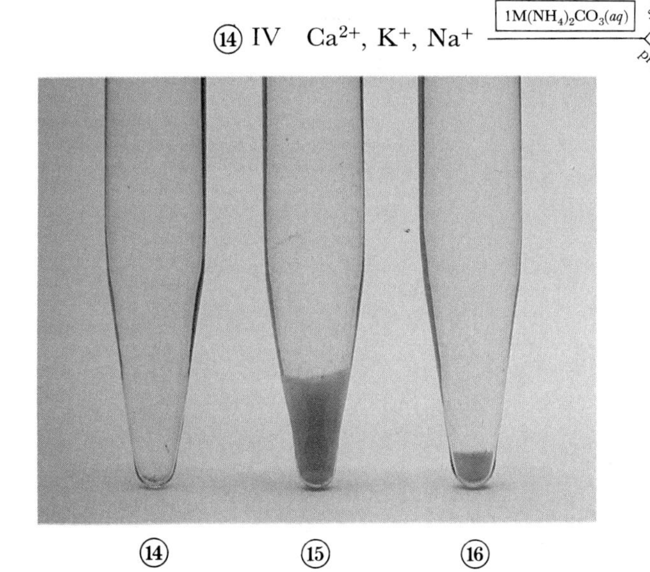

(14) Solution containing the subgroup IV cations.

(15) Solution containing $K^+(aq)$ and $Na^+(aq)$. Precipitate of $CaCO_3(s)$.

(16) Solution containing $Na^+(aq)$. Precipitate of $K_3Co(NO_2)_6(s)$.

SUMMARY

The solubility equilibrium between a salt and a solution saturated with its ions is characterized by a solubility product constant, K_{sp}. The algebraic form of K_{sp} is given by the law of concentration action. The common-ion effect is the suppression of the solubility of a salt by the addition of one of its ions to the solution in contact with the salt. The ratio Q_{sp}/K_{sp} can be used to predict whether a salt dissolves in or precipitates out of a solution.

Slightly soluble salts of weak acids become more soluble as the pH of the solution is lowered. The increase in solubility is a consequence of the protonation of the conjugate base to form the weak acid; this protonation reaction shifts the solubility equilibrium.

The dependence of the solubility of salts of weak acids on pH, especially hydroxides and sulfides, can be used to achieve separations of various metal ions by successive adjustments of pH and addition of precipitating agents. Amphoteric hydroxides are soluble in both acidic and basic solutions; the solubility of amphoteric hydroxides in strong bases is a consequence of the formation of a soluble hydroxy ion of the metal ion.

Mixtures of cations can be separated by selective precipitation reactions involving control of the solution pH and addition of precipitating agents. Ions can be identified by their characteristic chemical reactions and the color of their salts.

TERMS YOU SHOULD KNOW

solubility 744
solubility rules 744
solubility product constant, K_{sp} 745
common ion 751
common-ion effect 751
amphoteric metal hydroxide 761
qualitative analysis 764
quantitative analysis 764
centrifuge 766
supernatant 766
decant 766

EQUATIONS YOU SHOULD KNOW HOW TO USE

$K_{sp} = [A^+]^x[B^-]^y$ for the salt A_xB_y

$s = \dfrac{[A]}{x} = \dfrac{[B]}{y}$ for the salt A_xB_y

For any arbitrary ion concentrations:

$\frac{Q_{sp}}{K_{sp}} > 1$ precipitate forms

$\frac{Q_{sp}}{K_{sp}} < 1$ no precipitate forms

When equilibrium is disturbed:

$\frac{Q_{sp}}{K_{sp}} > 1$ more precipitate forms until $Q_{sp} = K_{sp}$

$\frac{Q_{sp}}{K_{sp}} < 1$ precipitate dissolves either until $Q_{sp} = K_{sp}$ or until solid phase disappears completely

$$[S^{2-}] = \frac{1.1 \times 10^{-21}\ M^3}{[H_3O^+]^2} \qquad (19\text{-}18)$$

PROBLEMS

SOLUBILITY RULES

19-1. Use the solubility rules to predict whether the following compounds are soluble or insoluble in water:

$AgBr$ $\quad PbSO_4$ $\quad NH_4Cl$ $\quad KNO_3$ $\quad BaSO_4$

19-2. Use the solubility rules to predict whether the following compounds are soluble or insoluble in water:

Fe_2O_3 $\quad Cu(NO_3)_2$ $\quad Na_2CO_3$ $\quad KOH$ $\quad Hg_2Cl_2$

19-3. Use the solubility rules to predict which of the following compounds are insoluble in water:

$CaCO_3$ $\quad Ca(NO_3)_2$ $\quad CaCl_2$ $\quad CaS$ $\quad CaSO_4$

19-4. Use the solubility rules to predict which of the following compounds are insoluble in water:

$AgCl$ $\quad AgNO_3$ $\quad Ag_2S$ $\quad AgClO_4$ $\quad Ag_2CO_3$

19-5. Use the solubility rules to predict the products of the following reactions. In each case, complete and balance the equation and also write the corresponding net ionic equation:

(a) $AgClO_4(aq) + NaI(aq) \rightarrow$

(b) $Hg_2(NO_3)_2(aq) + K_2SO_4(aq) \rightarrow$

(c) $Pb(NO_3)_2(aq) + Li_2CrO_4(aq) \rightarrow$

(d) $ZnCl_2(aq) + Na_2CO_3(aq) \rightarrow$

(e) $BaCl_2(aq) + ZnSO_4(aq) \rightarrow$

19-6. Use the solubility rules to predict the products of the following reactions. In each case complete and balance the equation and also write the net ionic equation. If no precipitate forms, then write no reaction:

(a) $LiNO_2(aq) + Pb(NO_3)_2(aq) \rightarrow$

(b) $H_2SO_4(aq) + Ca(ClO_4)_2(aq) \rightarrow$

(c) $AgNO_3(aq) + NaClO_4(aq) \rightarrow$

(d) $Hg_2(NO_3)_2(aq) + NaC_7H_5O_2(aq) \rightarrow$

(e) $Na_2SO_4(aq) + AgF(aq) \rightarrow$

19-7. In each of the following cases, the two solutions indicated are mixed. In each case for which a precipitate forms on mixing, write the complete equation. If no precipitate forms, then write "no reaction." Use the solubility rules and assume that all solutions before mixing are 0.20 M and that equal volumes of the two solutions are mixed:

(a) $Hg_2(ClO_4)_2(aq) + NaBr(aq) \rightarrow$
(b) $Fe(ClO_4)_3(aq) + NaOH(aq) \rightarrow$
(c) $CaCl_2(aq) + H_2SO_4(aq) \rightarrow$
(d) $Pb(NO_3)_2(aq) + LiIO_3(aq) \rightarrow$
(e) $H_2SO_4(aq) + Pb(NO_3)_2(aq) \rightarrow$

19-8. In each of the following cases, the two solutions indicated are mixed. In each case for which a precipitate forms on mixing, write the complete equation. If no precipitate forms, then write "no reaction." Use the solubility rules and assume that all solutions before mixing are 0.20 M and that equal volumes of the two solutions are mixed:

(a) $AgNO_3(aq) + NH_4ClO_4(aq) \rightarrow$
(b) $Hg_2(NO_3)_2(aq) + KCl(aq) \rightarrow$
(c) $Zn(ClO_4)_2(aq) + Na_2S(aq) \rightarrow$
(d) $CaCl_2(aq) + Na_2CO_3(aq) \rightarrow$
(e) $Cu(ClO_4)_2(aq) + LiOH(aq) \rightarrow$

19-9. One treatment for poisoning by soluble lead compounds is to give $MgSO_4(aq)$ or $Na_2SO_4(aq)$ as soon as possible. Explain in chemical terms why this procedure is effective.

19-10. Oxalic acid and soluble oxalates can cause death if swallowed. The recommended treatment for oxalic acid or oxalate poisoning is to give, as soon as possible, a glassful of limewater (saturated solution of calcium hydroxide) or a 1% calcium chloride solution, followed by inducing vomiting several times. Then give 15 to 30 g of Epsom salt ($MgSO_4$) in water and do not induce vomiting. Explain in chemical terms why this procedure is effective.

K_{sp} CALCULATIONS

19-11. The K_{sp} for $PbCrO_4(s)$ in equilibrium with water at 25°C is $2.8 \times 10^{-13}\ M^2$. Write the chemical equation that represents the solubility equilibrium for $PbCrO_4(s)$ and calculate its solubility in water at 25°C.

19-12. The K_{sp} for $TlCl(s)$ in equilibrium with water at 25°C is $1.7 \times 10^{-4}\ M^2$. Write the chemical equation that represents the solubility equilibrium for $TlCl(s)$ and calculate its solubility in water at 25°C.

19-13. The solubility of silver bromide in pure water at 18°C is $1.33 \times 10^{-4}\ g \cdot L^{-1}$. Calculate the value of K_{sp} for silver bromide at 18°C.

19-14. The solubility of lead(II) iodate in pure water is $2.24 \times 10^{-2}\ g \cdot L^{-1}$ at 25°C. Calculate the value of K_{sp} for lead(II) iodate at 25°C.

19-15. The solubility product for $Mg(OH)_2(s)$ in equilibrium with water at 25°C is $1.8 \times 10^{-11}\ M^3$. Calculate the solubility of $Mg(OH)_2(s)$ in water at 25°C.

19-16. The solubility product for $PbBr_2(s)$ in equilibrium with water at 25°C is $4.0 \times 10^{-5}\ M^3$. Calculate the solubility of $PbBr_2(s)$ in water at 25°C.

19-17. Potassium perchlorate is soluble in water to the extent of 0.75 g per 100 mL at 0°C. Calculate the K_{sp} of $KClO_4(aq)$ at 0°C.

19-18. Lithium fluoride dissolves in water to the extent of 0.27 g per 100 mL at 18°C. Estimate its K_{sp} at 18°C.

19-19. The solubility product of zinc hydroxide is $1.0 \times 10^{-15}\ M^3$ at 25°C. Calculate the pH of a saturated $Zn(OH)_2(aq)$ solution at 25°C.

19-20. Given that the pH of a saturated $Ca(OH)_2(aq)$ solution is 12.45, compute the solubility of $Ca(OH)_2(s)$ in water at 25°C.

COMMON-ION EFFECT

19-21. Calculate the solubility (in $g \cdot L^{-1}$) of silver sulfate in a 1.00 M silver nitrate solution at 25°C.

19-22. Calculate the solubility (in $g \cdot L^{-1}$) of barium chromate in a 0.0553 M ammonium chromate solution at 25°C.

19-23. The solubility product for BaF_2 at 25°C is $K_{sp} = 1.0 \times 10^{-6}\ M^3$. Estimate its solubility in an aqueous solution that is 0.25 M in NaF.

19-24. The solubility product for $PbI_2(s)$ at 25°C is $K_{sp} = 7.1 \times 10^{-9}\ M^3$. Compute its solubility in a 0.010 M $Pb(ClO_4)_2(aq)$ solution at 25°C.

19-25. Calculate the solubility of $AgCl(s)$ in 0.50 M $CaCl_2(aq)$ at 25°C.

19-26. Calculate the solubility of $CaSO_4(s)$ in 0.25 M $Na_2SO_4(aq)$ at 25°C.

19-27. Copper(I) ions in aqueous solution react with NH_3 according to

$$Cu^+(aq) + 2NH_3(aq) \rightleftharpoons Cu(NH_3)_2^+(aq) \qquad K_2 = 6.3 \times 10^{10}\ M^{-2}$$

Calculate the solubility of $CuBr(s)$ in a solution in which the *equilibrium* concentration of $NH_3(aq)$ is 0.185 M.

19-28. The equilibrium constant for the equilibrium

$$AgCl(s) + 2S_2O_3^{2-}(aq) \rightleftharpoons Ag(S_2O_3)_2^{3-}(aq) + Cl^-(aq)$$

is 5.20×10^3 at 25°C. Calculate the solubility of $AgCl(s)$ in a solution whose *equilibrium* concentration of $S_2O_3^{2-}(aq)$ is 0.015 M.

19-29. Consider the chemical equilibrium

$$AgBr(s) + 2S_2O_3^{2-}(aq) \rightleftharpoons Ag(S_2O_3)_2^{3-}(aq) + Br^-(aq)$$

Is the solubility of $AgBr(s)$ increased, decreased, or unchanged by

(a) an increase in the concentration of $Na_2S_2O_3(aq)$
(b) a decrease in the amount of $AgBr(s)$
(c) dissolution of $NaBr(s)$
(d) dissolution of $NaNO_3(s)$

19-30. Consider the chemical equilibrium

$$PbI_2(s) + 3OH^-(aq) \rightleftharpoons Pb(OH)_3^-(aq) + 2I^-(aq)$$

Is the solubility of $PbI_2(s)$ increased, decreased, or unaffected by

(a) an increase in the concentration of $OH^-(aq)$
(b) a decrease in the amount of $PbI_2(s)$
(c) a decrease in the concentration of $I^-(aq)$

Q_{sp} CALCULATIONS

19-31. A 100-mL sample of water from a salt lake has a chloride ion concentration of 0.25 M. To the sample is added 5.0 mL of a 0.10 M aqueous solution of $AgNO_3$. Does $AgCl(s)$ precipitate from solution?

19-32. Suppose we mix 50.0 mL of 0.20 M $AgNO_3(aq)$ with 150 mL of 0.10 M $H_2SO_4(aq)$. Does $Ag_2SO_4(s)$ precipitate from the solution?

19-33. If we mix 50.0 mL of 3.00 M $Pb(NO_3)_2(aq)$ with 25.0 mL of 2.00×10^{-3} M $NaI(aq)$, does $PbI_2(s)$ precipitate from the solution? If yes, then compute how many moles of $PbI_2(s)$ precipitate and the values of $[Pb^{2+}]$, $[I^-]$, $[NO_3^-]$, and $[Na^+]$ at equilibrium.

19-34. Suppose we mix 50.0 mL of 0.50 M $AgNO_3(aq)$ with 50.0 mL of 1.00×10^{-4} M $NaBr(aq)$. Does $AgBr(s)$ precipitate from the solution? If yes, then compute how many moles of $AgBr(s)$ precipitate and the values of $[Ag^+]$, $[Br^-]$, $[Na^+]$, and $[NO_3^-]$ at equilibrium.

19-35. Suppose we mix 100 mL of a 2.00 M $NaCl(aq)$ solution with 100 mL of a 0.020 M $AgNO_3(aq)$ solution. Determine

(a) the number of grams of $AgCl(s)$ that precipitate from the solution
(b) the concentration of $Ag^+(aq)$ at equilibrium following the precipitation of $AgCl(s)$

19-36. Suppose that 10.0 mL of a 0.30 M $Zn(NO_3)_2(aq)$ solution is added to 10.0 mL of a 2.00×10^{-4} M $Na_2S(aq)$ solution. Compute

(a) the number of milligrams of $ZnS(s)$ that precipitate
(b) the concentration of $Zn^{2+}(aq)$ and $S^{2-}(aq)$ at equilibrium

SOLUBILITY AND pH

19-37. Calcium fluoride is slightly soluble in water. Predict the effect on its solubility when

(a) the pH of the solution is decreased to 3
(b) the pH of the solution is increased to 11
(c) $CaCl_2(s)$ is added to the solution

19-39. Indicate for which of the following compounds the solubility increases as the pH of the solution is lowered:

$CaCO_3(s)$	$AgBr(s)$
$PbI_2(s)$	$Fe(OH)_3(s)$
$PbSO_3(s)$	$ZnS(s)$

19-41. Calculate the solubility of $Mg(OH)_2(s)$ in an aqueous solution buffered at pH = 9.0.

19-43. Calculate the solubility of $Cu(OH)_2(s)$ in an aqueous solution buffered at pH = 8.0.

19-45. Use Le Châtelier's principle to predict the effect on the solubility of

(a) $ZnS(s)$ when $HNO_3(aq)$ is added to a saturated $ZnS(aq)$ solution
(b) $MgCO_3(s)$ when $CO_2(g)$ is added to a saturated $MgCO_3(aq)$ solution
(c) $AgI(s)$ when $NH_3(g)$ is added to a saturated $AgI(aq)$ solution
(d) $CdS(s)$ when $NaOH(s)$ is added to a saturated $CdS(aq)$ solution

19-38. Magnesium oxalate is sparingly soluble in water. Predict the effect on its solubility when

(a) the solution is made more acidic
(b) the solution is made more basic
(c) $Mg(NO_3)_2(s)$ is added to the solution

19-40. Indicate for which of the following compounds the solubility increases as the pH of the solution is lowered:

$FeS(s)$	$Hg_2I_2(s)$
$ZnCO_3(s)$	$Ag_2C_2O_4(s)$
$PbCrO_4(s)$	$Ag_2O(s)$

19-42. Calculate the solubility of $AgC_7H_5O_2(s)$ in an aqueous solution buffered at pH = 4.0. Take $K_a = 6.5 \times 10^{-5}$ M for $HC_7H_5O_2(aq)$ and $K_{sp} = 2.5 \times 10^{-5}$ M^2 for $AgC_7H_5O_2(s)$.

19-44. Calculate the solubility of $Cd(OH)_2(s)$ in an aqueous solution buffered at pH = 9.0.

19-46. For the reaction

$$ZnS(s) + 2H_3O^+(aq) \rightleftharpoons Zn^{2+}(aq) + H_2S(aq) + 2H_2O(l)$$

in which direction does the equilibrium shift in response to each of the following changes in conditions (if the equilibrium is unaffected by the change, then write no change):

(a) bubbling in $HCl(g)$
(b) dissolving $NaOH(s)$
(c) diluting the solution
(d) increasing the pH of the solution

SEPARATION OF CATIONS AS HYDROXIDES

19-47. Calculate the solubility of $Cr(OH)_3(s)$ and $Ni(OH)_2(s)$ in an aqueous solution buffered at pH = 5.0. Can $Cr(OH)_3$ be separated from $Ni(OH)_2$ at this pH?

19-49. Insoluble $Pb(OH)_2$ and $Sn(OH)_2$ are formed when sodium hydroxide is added to a solution containing $Pb^{2+}(aq)$ and $Sn^{2+}(aq)$. At what pH can $Pb(OH)_2$ be separated from $Sn(OH)_2$? Assume that an effective separation requires a maximum concentration of the less soluble hydroxide of 1×10^{-6} M.

19-48. Calculate the solubility of $Cu(OH)_2(s)$ and $Zn(OH)_2(s)$ in an aqueous solution buffered at pH = 4.0. Can $Cu(OH)_2$ be separated from $Zn(OH)_2$ at this pH?

19-50. A deposit of limestone is analyzed for its calcium and magnesium content. A sample is dissolved, and then the calcium and magnesium are precipitated as $Ca(OH)_2$ and $Mg(OH)_2$. At what pH can $Ca(OH)_2$ be separated from $Mg(OH)_2$? Assume that an effective separation requires a maximum concentration of the less soluble hydroxide of 1×10^{-6} M.

19-51. Calculate the solubility of $CuS(s)$ in a solution buffered at pH = 2.0 and saturated with hydrogen sulfide so that $[H_2S] = 0.10$ M.

$[S^{2-}] = \frac{1.1 \times 10^{-21}\ M^3}{[H_3O^+]^2} = \frac{1.1 \times 10^{-21}\ M^3}{(1.0 \times 10^{-2}\ M)^2} = 1.1 \times 10^{-17}$ M

19-52. Calculate the solubility of $SnS(s)$ in a solution buffered at pH = 2.0 and saturated with hydrogen sulfide so that $[H_2S] = 0.10$ M.

$[Cu^{2+}] = s = 5.7 \times 10^{-19}$ M

$[Cu^{2+}][S^{2-}] = 6.3 \times 10^{-36}\ M^2$

19-53. What must the pH of a buffered solution saturated with H_2S ($[H_2S] = 0.10$ M) be in order to precipitate PbS leaving $[Pb^{2+}] = 1 \times 10^{-6}$ M, without precipitating any MnS? The original solution is 0.025 M in both $Pb^{2+}(aq)$ and $Mn^{2+}(aq)$.

$[Pb^{2+}][S^{2-}] = 8.0 \times 10^{-28}\ M^2$

$[Pb^{2+}] = s = \frac{8.0 \times 10^{-28}\ M}{[S^{2-}]} = \frac{(8.0 \times 10^{-28}\ M)[H_3O^+]^2}{1.1 \times 10^{-21}\ M^3} = 7.3 \times 10^{-7}[H_3O^+]^2$

19-54. Iron(II) sulfide is used as the pigment in black paint. A sample of $FeS(s)$ is suspected of containing lead(II) sulfide, which can cause lead poisoning if ingested. Suggest a scheme based on pH for separating FeS from PbS.

$1 \times 10^{-6}\ M = 7.3 \times 10^{-7}\ M^{-1}[H_3O^+]^2$

pH = −0.08

19-55. Calculate the solubility of HgS and CdS at pH = 3.0 and 6.0 for aqueous solutions that are saturated with H_2S ($[H_2S] = 0.10$ M).

$[Hg^{2+}][S^{2-}] = 4 \times 10^{-53}\ M^2$

$[S^{2-}] = \frac{1.1 \times 10^{-21}\ M^3}{(1.0 \times 10^{-6}\ M)^2} = 1.1 \times 10^{-9}$ M

$[Hg^{2+}] = s = \frac{4 \times 10^{-53}\ M^2}{1.1 \times 10^{-9}\ M} = 4 \times 10^{-38}$ M (pH = 3)

4×10^{-44} M (pH = 6)

19-56. A solution 0.30 M in $H_3O^+(aq)$ containing $Mn^{2+}(aq)$, $Cd^{2+}(aq)$, and $Fe^{2+}(aq)$ all at 0.01 M was saturated with $H_2S(g)$ at 25°C. Compute the equilibrium concentrations of $Mn^{2+}(aq)$, $Cd^{2+}(aq)$, and $Fe^{2+}(aq)$. Assume that the solution is continuously saturated with H_2S and that the pH remains constant.

$[Cd^{2+}][S^{2-}] = 8.0 \times 10^{-27}\ M^2$

$s = \frac{8.0 \times 10^{-27}\ M^2}{(1.0 \times 10^{-6}\ M)^2} = 7.3 \times 10^{-12}$ M (pH = 3)

7.3×10^{-18} M (pH = 6)

AMPHOTERIC HYDROXIDES

19-57. It is observed that a precipitate forms when 2.0 M $NaOH(aq)$ solution is added dropwise to a 0.10 M $Pb(NO_3)_2(aq)$ solution and that, on further addition of $NaOH(aq)$, the precipitate dissolves. Explain these observations using balanced chemical equations.

$Pb(NO_3)_2(aq) + 2NaOH(aq) \rightleftharpoons Pb(OH)_2(s) + 2NaNO_3(aq)$

$Pb(OH)_2(s) + OH^-(aq) \rightleftharpoons Pb(OH)_3^-(aq)$

19-58. It is observed that a precipitate forms when 2.0 M $KOH(aq)$ solution is added dropwise to a 0.20 M $Zn(ClO_4)_2(aq)$ solution and that, on further addition of $KOH(aq)$, the precipitate dissolves. Explain these observations using balanced chemical equations.

19-59. Use the equilibrium constant data in Table 19-4 to estimate the solubility of tin(II) hydroxide in a solution buffered at pH = 13.0.

$Sn(OH)_2(s) + OH^-(aq) \rightleftharpoons Sn(OH)_3^-(aq)$

$K = \frac{[Sn(OH)_3^-]}{[OH^-]} = 0.01$ $s = 1.0 \times 10^{-3}$ M

19-60. Use the equilibrium constant data in Table 19-4 to estimate the solubility of lead(II) hydroxide in a solution buffered at pH = 13.0.

19-61. The equilibrium constant for the reaction

$$Al(OH)_3(s) + OH^-(aq) \rightleftharpoons Al(OH)_4^-(aq)$$

is $K = 40$ at 25°C. Compute the solubility of $Al(OH)_3(s)$ in a solution buffered at pH = 12.0 at 25°C.

$K = \frac{[Al(OH)_4^-]}{[OH^-]} = 40$ $s = 0.40$ M

19-62. The equilibrium constant for the reaction

$$Zn(OH)_2(s) + 2OH^-(aq) \rightleftharpoons Zn(OH)_4^{2-}(aq)$$

is $K = 0.050\ M^{-1}$. Compute the solubility of $Zn(OH)_2(s)$ in a 0.10 M $NaOH(aq)$ solution.

19-63. Chromium(III) hydroxide is amphoteric (Tables 19-1 and 19-4). Compute its solubility in solutions buffered at pH = 3.0, 6.0, 9.0, and 12.0.

19-64. Lead(II) hydroxide is amphoteric (Tables 19-1 and 19-4). Compute its solubility in solutions buffered at pH = 3.0, 6.0, and 12.0.

$Cr(OH)_3(s) \rightleftharpoons Cr^{3+}(aq) + 3OH^-(aq)$

$K_{sp} = [Cr^{3+}][OH^-]^3$

$[Cr^{3+}] = s = \frac{K_{sp}}{[OH^-]^3} = \frac{6.3 \times 10^{-31}\ M^4}{[OH^-]^3}$

$s = 6.3 \times 10^{2}$ M pH = 3

$s = 6.3 \times 10^{-7}$ M pH = 6

$Cr(OH)_3(s) + OH^-(aq) \rightleftharpoons Cr(OH)_4^-(aq)$

$K = \frac{[Cr(OH)_4^-]}{[OH^-]} = 0.04$

$s = 4 \times 10^{-7}$ M pH = 9.0

$s = 4.0 \times 10^{-4}$ M pH = 12.0

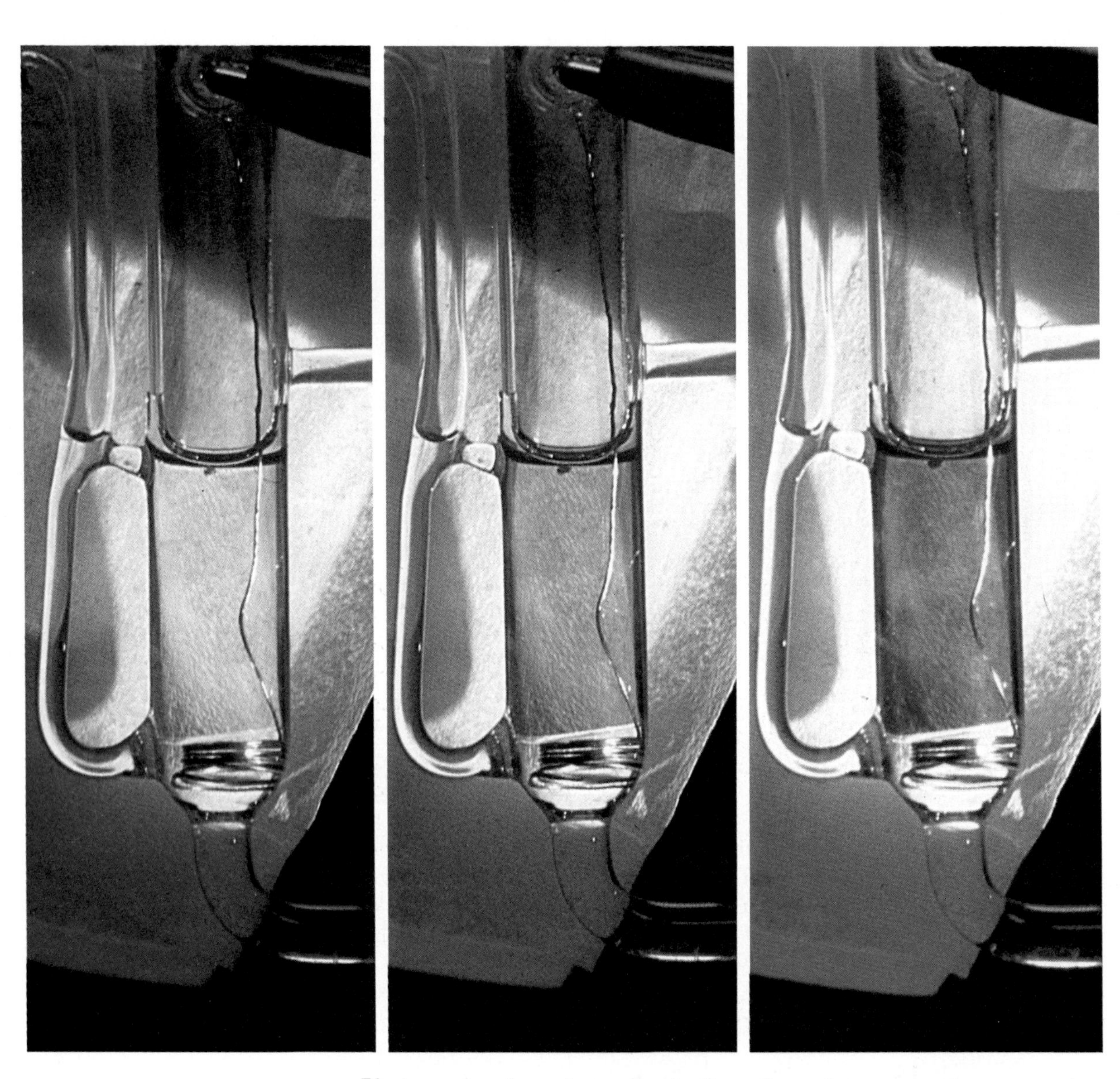

Photographs of an electrochemically active polymer coated on an optically transparent electrode (SnO_2). The color in the polymer is due to the ion RuL_3^{2+}, where L is a complex organic group. As the electrode potential is made progressively more negative, the oxidation state of the metal complex in the film changes and so does its color. The formal oxidation states of the ruthenium in the complex are as follows: pale orange, +2; blue, 0; and cherry red, −4.

20 / Oxidation-Reduction Reactions

All chemical reactions can be assigned to one of two classes: reactions in which electrons are transferred from one reactant to another and reactions in which electrons are not transferred. Reactions in which electrons are transferred from one reactant to another are called oxidation-reduction reactions or electron-transfer reactions. Most of the reactions that we have considered prior to this chapter (for example, acid-base reactions and precipitation reactions) do not involve electron transfer and thus are not oxidation-reduction reactions.

In this chapter we begin with a discussion of the concept of the oxidation states of elements in chemical species. Oxidation states are determined according to a set of rules and are used to balance equations for oxidation-reduction reactions in a systematic way. Much of this chapter is devoted to balancing equations for oxidation-reduction reactions. The final section of the chapter discusses the electron-transfer reactions involved in bleaches and disinfectants.

20-1. AN OXIDATION STATE CAN BE ASSIGNED TO EACH ATOM IN A CHEMICAL SPECIES

An *oxidation-reduction reaction,* commonly called a *redox reaction,* is one that involves the transfer of electrons from one chemical species to another. A simple example of an oxidation-reduction reaction is the reaction between zinc metal and a copper(II) salt in aqueous solution:

$$Zn(s) + Cu^{2+}(aq) \rightarrow Zn^{2+}(aq) + Cu(s) \qquad (20\text{-}1)$$

In Equation (20-1), zinc metal converts $Cu^{2+}(aq)$ to copper metal by transferring two electrons to each $Cu^{2+}(aq)$ ion. The loss of two electrons from a Zn atom yields a $Zn^{2+}(aq)$ ion.

The study of oxidation-reduction reactions can be systematized by assigning an *oxidation state* to each atom in a chemical species according to a set of rules. The rules originate from a consideration of the number of electrons and of the electronegativities of the various elements in a species. In some cases, an assigned oxidation state is the actual charge on an atom, but in general this is not the case. Oxidation states are used to balance equations for oxidation-reduction reactions.

The general procedure for assigning oxidation states to elements in chemical species containing two or more atoms is given by the following set of rules, *which take priority in the order given:*

Rules for Assigning Oxidation States

1. Free elements are assigned an oxidation state of zero.
2. The sum of the oxidation states of all the atoms in a species must be equal to the net charge on the species.
3. The alkali metals (Li, Na, K, Rb, and Cs) in compounds are always assigned an oxidation state of +1.
4. The alkaline earth metals (Be, Mg, Ca, Sr, Ba, and Ra) and also Zn and Cd in compounds are always assigned an oxidation state of +2.
5. Hydrogen in compounds is assigned an oxidation state of +1.
6. Oxygen in compounds is assigned an oxidation state of −2.

These rules cover many but not all cases. For cases not covered by the rules, the assignment of oxidation states is made by analogy, using the periodic table as a guide.

The +1 oxidation state of alkali metals (Group 1 metals) in compounds corresponds to the ionic charge of the alkali metal ions. The +1 state corresponds to the loss of an electron from the outermost *s* orbital in the neutral atoms. The +2 oxidation state of the alkaline earth metals (Group 2 metals) corresponds to the ionic charge of the alkaline earth metal ions. The +2 state corresponds to a loss of two electrons from the outermost *s* orbital in the neutral atoms.

The ionic charges of the metal ions discussed in Chapter 2 (Figure 2-18) correspond to the oxidation states of those elements (Table 20-1). Aluminum and gallium are always assigned an oxidation state of +3 in their compounds (Table 20-1).

Table 20-1 Comparison of ionic charge and oxidation state for some metal ions

Group	*Ionic charge*	*Oxidation state*
alkali metal ions (Li^+, Na^+, K^+, Cs^+, Rb^+)	+1	+1
alkaline earth ions (Be^{2+}, Mg^{2+}, Ca^{2+}, Sr^{2+}, Ba^{2+})	+2	+2
Group 3 ions (Al^{3+}, Ga^{3+})	+3	+3

Example 20-1: Determine the oxidation state of each atom in the following compounds:

(a) CsF
(b) NO_2
(c) $HClO_3$
(d) H_2O_2
(e) NaH

Solution:

(a) We assign cesium an oxidation state of +1 (rule 3), and thus fluorine is assigned an oxidation state of −1 (rule 2) because CsF is a neutral species and $+1 - 1 = 0$.

(b) We assign oxygen an oxidation state of −2 (rule 6). The oxidation state of nitrogen in NO_2, represented by x, is thus (rule 2)

$$x + 2(-2) = 0$$
$$x = +4$$

The oxidation state of nitrogen in NO_2 is +4.

(c) We assign hydrogen an oxidation state of +1 (rule 5) and oxygen an oxidation state of −2 (rule 6). Then the oxidation state x of chlorine is (rule 2)

$$+1 + x + 3(-2) = 0$$
$$x = +5$$

The oxidation state of chlorine in $HClO_3$ is +5.

(d) We assign hydrogen an oxidation state of +1 (rule 5), and the oxidation state x of oxygen is (rule 2)

$$2(+1) + 2x = 0$$
$$x = -1$$

Thus the oxidation state of oxygen in H_2O_2 is −1, which is characteristic of peroxides. This result does not contradict rule 6 because rules 2 and 5 take precedence.

Peroxides involve an oxygen-oxygen single bond.

(e) We assign sodium an oxidation state of +1 (rule 3), and, according to rule 2, the oxidation state of hydrogen is −1, which is characteristic of hydrides. This result does not violate rule 5 because rules 2 and 3 take precedence.

Example 20-1 involves only neutral molecules, whose net charge must be zero. For ionic species, the sum of the oxidation states for each atom must equal the net charge on the ion.

The O_2^- ion is called the super oxide ion.

Example 20-2: Determine the oxidation state of each atom in the following ions:

(a) CrO_4^{2-} (c) NH_4^+
(b) HF_2^- (d) O_2^-

Solution:

(a) We assign oxygen an oxidation state of −2 (rule 6), and thus the oxidation state x of chromium is (rule 2)

$$x + 4(-2) = -2$$
$$x = +6$$

The oxidation state of Cr in CrO_4^{2-} is +6.

(b) We assign hydrogen an oxidation state of +1 (rule 5), and thus the oxidation state x of fluorine is (rule 2)

$$+1 + 2x = -1$$
$$x = -1$$

The oxidation state of F in HF_2^- is −1.

(c) We assign hydrogen an oxidation state of +1 (rule 5), and thus the oxidation state x of nitrogen is (rule 2)

$$x + 4(+1) = +1$$
$$x = -3$$

The oxidation state of N in NH_4^+ is −3.

(d) Using rule 2, we assign oxygen an oxidation state of $-\frac{1}{2}$ (since $2(-\frac{1}{2}) = -1$), which is characteristic of superoxides (Interchapter F).

Problems 20-1 through 20-10 are similar to Examples 20-1 and 20-2.

20-2. OXIDATION STATES CAN BE ASSIGNED USING LEWIS FORMULAS

All the atoms in Examples 20-1 and 20-2 can be assigned oxidation states using the above rules. As we stated earlier, however, rules 1 through 6 do not cover all cases, and so we need a more general method for assigning oxidation states:

1. Write the Lewis formula for the molecule or ion.
2. Assign all the electrons in each bond to the more electronegative atom in the bond. If the two atoms in a bond are the same, then divide the bonding electrons equally between them.
3. Add up the total number of valence electrons around each atom and assign the oxidation state according to the formula

$$\begin{pmatrix}\text{oxidation}\\ \text{state}\end{pmatrix} = \begin{pmatrix}\text{group number of}\\ \text{element in}\\ \text{periodic table}\end{pmatrix} - \begin{pmatrix}\text{total number}\\ \text{of}\\ \text{valence electrons}\end{pmatrix}$$

The application of these rules is best illustrated by examples.

Example 20-3: Assign oxidation states to each atom in the following molecules:

(a) PCl_3 (b) ClF_3 (c) OF_2 (d) H_2O_2

Solution:

(a) The Lewis formula for PCl_3 is

```
 ..   ..  ..
:Cl—P—Cl:
 ..   |   ..
     :Cl:
      ..
```

Because chlorine is more electronegative than phosphorus (Table 20-2), we assign the electrons in each covalent bond to the chlorine atoms:

```
 ..      ..      ..
:Cl:     P      :Cl:
 ..              ..
         ..
        :Cl:
         ..
```

Therefore, the oxidation states are

$$\text{oxidation state of P in } PCl_3 = 5 - 2 = +3$$
$$\text{oxidation state of Cl in } PCl_3 = 7 - 8 = -1$$

Table 20-2 Electronegativities of selected elements

C 2.5	N 3.0	O 3.5	F 4.0
Si 1.8	P 2.1	S 2.5	Cl 3.0
	As 2.0	Se 2.4	Br 2.8
			I 2.5

(b) The Lewis formula for ClF_3 is

```
 ..  .  .  ..
:F—.Cl.—F:
 ..   |    ..
     :F:
      ..
```

Because fluorine is more electronegative than chlorine, we assign the electrons in each covalent bond to the fluorine atoms:

```
 ..     .  .      ..
:F:    .Cl.      :F:
 ..               ..
         ..
        :F:
         ..
```

Chlorine trifluoride is an extremely reactive gas. Its reactivity is comparable to that of fluorine.

Therefore, the oxidation states are

$$\text{oxidation state of Cl in ClF}_3 = 7 - 4 = +3$$
$$\text{oxidation state of F in ClF}_3 = 7 - 8 = -1$$

(c) The Lewis formula for OF_2 is

$$:\ddot{\underset{..}{F}}—\ddot{\underset{..}{O}}—\ddot{\underset{..}{F}}:$$

Because fluorine is more electronegative than oxygen, we assign the electrons in each covalent bond to the fluorine atoms:

$$:\ddot{\underset{..}{F}}: \quad \ddot{\underset{..}{O}} \quad :\ddot{\underset{..}{F}}:$$

Therefore, the oxidation states are

$$\text{oxidation state of O in OF}_2 = 6 - 4 = +2$$
$$\text{oxidation state of F in OF}_2 = 7 - 8 = -1$$

(d) The Lewis formula for H_2O_2 is

$$H—\ddot{\underset{..}{O}}—\ddot{\underset{..}{O}}—H$$

Because oxygen is more electronegative than hydrogen, we write

$$H \quad :\ddot{\underset{..}{O}}\cdot \quad \cdot\ddot{\underset{..}{O}}: \quad H$$

Notice that we assign one electron from the O—O bond to each oxygen atom. The oxidation states are

$$\text{oxidation state of H in H}_2\text{O}_2 = 1 - 0 = +1$$
$$\text{oxidation state of O in H}_2\text{O}_2 = 6 - 7 = -1$$

This case illustrates that oxygen is assigned an oxidation state of -1 in peroxides.

There are no known positive oxidation states of fluorine.

Cases (b) and (c) in Example 20-3 show that fluorine is assigned an oxidation state of -1 because it is more electronegative than chlorine and oxygen. In fact, because fluorine is the most electronegative element, it is always assigned an oxidation state of -1 when not in the elemental state, F_2. Fluorine has no positive oxidation states.

Our original six rules for the assignment of oxidation states result from the three basic rules presented here. The six rules, which are simpler to apply than the Lewis-formula method, cover most cases we shall encounter in this book. With a little experience (which you can gain by doing Problems 20-1 to 20-10), you will seldom have to resort to the Lewis-formula method.

Example 20-4: Working by analogy with other elements in the periodic table and also using the clues provided by the names, assign oxidation states to each of the elements in the following compounds:

(a) As_2S_5 arsenic pentasulfide
(b) SF_6 sulfur hexafluoride
(c) In_2Se_3 indium selenide

Solution:

(a) The name arsenic pentasulfide tells us that sulfur is more electronegative than arsenic, and so the sulfur is assigned a negative oxidation state. Arsenic is below nitrogen, and sulfur is below oxygen in the periodic table; thus the analogous compound is N_2O_5. Therefore, we assign sulfur an oxidation state of -2 (analogous to rule 6 for oxygen), and the oxidation state x of arsenic is

$$2x + 5(-2) = 0$$
$$x = +5$$

(b) Recall that fluorine, the most electronegative element, is always assigned an oxidation state of -1 in compounds; thus the oxidation state of sulfur is $+6$ (rule 2).

(c) Indium is below aluminum in the periodic table, and thus by analogy with Al^{3+} we assign indium an oxidation state of $+3$. The oxidation state x of selenium is

$$2(+3) + 3x = 0$$
$$x = -2$$

Note that Se is below S in the periodic table and recall that the oxidation state of sulfur in sulfides is -2. By analogy, the oxidation state of selenium in selenides is -2.

20-3. OXIDATION-REDUCTION REACTIONS INVOLVE THE TRANSFER OF ELECTRONS FROM ONE REACTANT TO ANOTHER

Consider the oxidation-reduction reaction between zinc metal and $Cu^{2+}(aq)$ ions in aqueous solution:

$$Zn(s) + Cu^{2+}(aq) \rightarrow Cu(s) + Zn^{2+}(aq)$$

The $Cu^{2+}(aq)$ is said to be *reduced* to Cu(s) because the process involves a decrease (reduction) in the oxidation state of copper (from $+2$ to 0):

Reduction is a gain of electrons.

$$Cu^{2+}(aq) + 2e^- \rightarrow Cu(s) \qquad \textit{reduction}$$

Oxidation is a loss of electrons.

The $Zn(s)$ is said to be *oxidized* because the process involves an increase in the oxidation state of zinc (from 0 to +2):

$$Zn(s) \rightarrow Zn^{2+}(aq) + 2e^- \qquad \textit{oxidation}$$

An essential feature of oxidation-reduction reactions is that in one reactant the oxidation state of an element increases, and in another reactant the oxidation state of an element decreases. Thus oxidation-reduction reactions involve a simultaneous oxidation and reduction. The simultaneous changes in oxidation states in oxidation-reduction reactions are the result of the transfer of electrons from one reactant to another, and thus oxidation-reduction reactions are also called *electron-transfer reactions*.

The reactant that contains the atom that is reduced is called the *oxidizing agent*. The reactant that contains the atom that is oxidized is called the *reducing agent*. The oxidizing agent gains electrons from the atom that is oxidized and thus is an *electron acceptor*. The reducing agent loses electrons to the atom that is reduced and therefore is an *electron donor*. In the reaction

$$\underset{\substack{\text{reducing agent}\\ \text{(oxidized)}}}{Zn(s)} + \underset{\substack{\text{oxidizing agent}\\ \text{(reduced)}}}{Cu^{2+}(aq)} \rightarrow Zn^{2+}(aq) + Cu(s)$$

the oxidizing agent (electron acceptor) is $Cu^{2+}(aq)$ because it accepts two electrons, and the reducing agent (electron donor) is $Zn(s)$ because it donates two electrons. In this reaction, as in all electron-transfer reactions, all the electrons donated by the reducing agent are accepted by the oxidizing agent.

Example 20-5: In the following reaction, identify the atom that is oxidized, the atom that is reduced, the oxidizing agent, and the reducing agent:

$$MnO_2(s) + 4HCl(aq) \rightarrow MnCl_2(aq) + Cl_2(g) + 2H_2O(l)$$

Solution: The oxidation state of Mn is +4 in MnO_2 and +2 in $MnCl_2$. The oxidation state of Cl is −1 in HCl and 0 in Cl_2. Therefore, Cl is oxidized and Mn is reduced in this reaction. The reactant that contains the atom that is reduced is MnO_2, and so MnO_2 is the oxidizing agent. The reactant that contains the atom that is oxidized is HCl, and so HCl is the reducing agent. Notice that two electrons are transferred in this reaction: one manganese atom accepts two electrons and each of two chlorine atoms donates one electron.

Oxidation involves an increase in oxidation state. Reduction involves a decrease in oxidation state.

Table 20-3 summarizes what we have learned so far about oxidation-reduction reactions.

Table 20-3 Summary of oxidation-reduction reactions

The reducing agent:
gives up electrons
contains the atom that is oxidized
contains the atom whose oxidation state increases
is the electron donor
The oxidizing agent:
accepts electrons
contains the atom that is reduced
contains the atom whose oxidation state decreases
is the electron acceptor

20-4. ELECTRON-TRANSFER REACTIONS CAN BE SEPARATED INTO TWO HALF-REACTIONS

The electron-transfer reaction

$$(1) \quad Zn(s) + Cu^{2+}(aq) \rightarrow Zn^{2+}(aq) + Cu(s)$$

can be written as the sum of two *half-reactions:*

$$(2) \quad Zn(s) \rightarrow Zn^{2+}(aq) + 2e^-$$

$$(3) \quad Cu^{2+}(aq) + 2e^- \rightarrow Cu(s)$$

We obtain Reaction (1) if we add Equations (2) and (3). The half-reaction in which electrons appear on the right-hand side (Reaction 2) is called the *oxidation half-reaction* (recall that oxidation is a *loss* of electrons). The half-reaction in which electrons appear on the left-hand side (Reaction 3) is called the *reduction half-reaction* (recall that reduction is a *gain* of electrons). The oxidation half-reaction supplies electrons to the reduction half-reaction.

Example 20-6: Identify the oxidizing and reducing agents and the oxidation and reduction half-reactions in the reaction

$$Tl^+(aq) + 2Ce^{4+}(aq) \rightarrow 2Ce^{3+}(aq) + Tl^{3+}(aq)$$

Solution: The oxidation state of thallium increases from +1 in Tl^+ to +3 in Tl^{3+}. Thus Tl^+ is oxidized and acts as the reducing agent (electron donor). The oxidation state of cerium decreases from +4 in Ce^{4+} to +3 in Ce^{3+}, thus Ce^{4+} is reduced and acts as the oxidizing agent (electron acceptor). The two half-reactions are obtained by writing the oxidation and the reduction reactions separately:

$$Tl^+(aq) \rightarrow Tl^{3+}(aq) + 2e^- \qquad \text{oxidation half-reaction}$$

$$Ce^{4+}(aq) + e^- \rightarrow Ce^{3+}(aq) \qquad \text{reduction half-reaction}$$

Note that Tl^+ is a two-electron reducing agent whereas Ce^{4+} is a one-electron oxidizing agent. Thus it requires 2 mol of Ce^{4+} to oxidize 1 mol of Tl^+. The number of electrons involved in the two half-reactions can be balanced by multiplying the cerium half-reaction by 2.

Problems 20-15 and 20-16 are similar to Example 20-6.

20-5. EQUATIONS FOR OXIDATION-REDUCTION REACTIONS CAN BE BALANCED BY BALANCING EACH HALF-REACTION SEPARATELY

Consider the reaction between iron metal and aqueous chlorine:

$$Fe(s) + Cl_2(aq) \rightarrow Fe^{3+}(aq) + Cl^-(aq)$$

This equation as it stands is not balanced. If we write

$$Fe(s) + Cl_2(aq) \rightarrow Fe^{3+}(aq) + 2Cl^-(aq)$$

then the equation is balanced with respect to the elements but not with respect to charge. The net charge on the left-hand side is zero, whereas the net charge on the right-hand side is $+3 + 2(-1) = +1$. The balanced equation is

$$2Fe(s) + 3Cl_2(aq) \rightarrow 2Fe^{3+}(aq) + 6Cl^-(aq)$$

The equation is now balanced both with respect to the elements and with respect to the charge; that is, the net charge is equal on both sides.

The balancing of equations for electron-transfer reactions must be done systematically because attempting to balance such equations by guessing the balancing coefficients can be a time-consuming and frustrating experience. The systematic procedure that we use is called the *method of half-reactions.* This method can be used to balance even the most complicated equation in a straightforward and systematic way. We illustrate the method of half-reactions by balancing the following equation:

$$Fe^{2+}(aq) + Cr_2O_7^{2-}(aq) \xrightarrow{H^+(aq)} Fe^{3+}(aq) + Cr^{3+}(aq) \quad (20\text{-}2)$$

which occurs in acid solution.

I. *Separate the equation into an oxidation half-reaction and a reduction half-reaction.*

The oxidation state of iron changes from $+2$ to $+3$, and the oxidation state of chromium changes from $+6$ (in $Cr_2O_7^{2-}$) to $+3$. Thus the two half-reactions are

$$Fe^{2+} \rightarrow Fe^{3+} \qquad \text{oxidation}$$

$$Cr_2O_7^{2+} \rightarrow Cr^{3+} \qquad \text{reduction}$$

II. *Balance each half-reaction with respect to all elements other than oxygen and hydrogen.*

The iron half-reaction is already balanced with respect to iron (one Fe on each side); the chromium half-reaction is balanced with respect to chromium when a 2 is placed in front of Cr^{3+}:

$$Fe^{2+} \rightarrow Fe^{3+}$$

$$Cr_2O_7^{2-} \rightarrow 2Cr^{3+}$$

III. *Balance each half-reaction with respect to oxygen by adding the appropriate number of H_2O to the side deficient in oxygen.*

Only the chromium half-reaction involves oxygen. There are seven oxygen atoms on the left and none on the right. Therefore we balance the oxygen by adding seven H_2O to the right-hand side of the chromium half-reaction:

$$Fe^{2+} \rightarrow Fe^{3+}$$

$$Cr_2O_7^{2-} \rightarrow 2Cr^{3+} + 7H_2O$$

IV. *Balance each half-reaction with respect to hydrogen by adding the appropriate number of H^+ to the side deficient in hydrogen.*

Only the chromium half-reaction involves hydrogen. There are 14 hydrogens on the right and none on the left. Therefore we balance the hydrogen by adding 14 H^+ to the left-hand side of the chromium half-reaction:

$$Fe^{2+} \rightarrow Fe^{3+}$$

$$14H^+ + Cr_2O_7^{2-} \rightarrow 2Cr^{3+} + 7H_2O$$

The two half-reactions are now balanced with respect to atoms, but they are not balanced with respect to charge.

V. *Balance each half-reaction with respect to charge by adding the appropriate number of electrons to the side with the excess positive charge.*

The iron half-reaction has a charge of $+2$ on the left and $+3$ on the right. Thus we balance the charge by adding one electron to the right-hand side:

$$Fe^{2+} \rightarrow Fe^{3+} + e^- \qquad \text{oxidation}$$

The chromium half-reaction has a net charge of $+12$ $[= 14(+1) + (-2)]$ on the left and $+6[= 2(+3)]$ on the right. Thus we balance the charge by adding six electrons to the left-hand side:

$$14H^+ + Cr_2O_7^{2-} + 6e^- \rightarrow 2Cr^{3+} + 7H_2O \qquad \text{reduction}$$

Recall that all the electrons consumed by the oxidizing agent must be supplied by the reducing agent (conservation of electrons).

The two half-reactions are now balanced. Note that the iron half-reaction donates electrons (electrons on the right-hand side) and the chromium half-reaction accepts electrons (electrons on the left-hand side).

VI. *Multiply each half-reaction by an integer that makes the number of electrons supplied by the oxidation half-reaction equal to the number of electrons consumed by the reduction half-reaction.*

The iron half-reaction supplies one electron for each Fe^{2+} that is oxidized to Fe^{3+}, and the chromium half-reaction consumes six electrons for each $Cr_2O_7^{2-}$ that is reduced to Cr^{3+}. Therefore, we multiply the iron half-reaction through by 6:

$$6Fe^{2+} \rightarrow 6Fe^{3+} + 6e^-$$

$$14H^+ + Cr_2O_7^{2-} + 6e^- \rightarrow 2Cr^{3+} + 7H_2O$$

VII. *Obtain the complete balanced equation by adding the two half-reactions and canceling any like terms.*

Adding the two half-reactions and canceling the $6e^-$ that appear on both sides yields

$$\begin{array}{r} 6Fe^{2+} \rightarrow 6Fe^{3+} + 6e^- \\ 14H^+ + Cr_2O_7^{2-} + 6e^- \rightarrow 2Cr^{3+} + 7H_2O \\ \hline 6Fe^{2+} + 14H^+ + Cr_2O_7^{2-} \rightarrow 6Fe^{3+} + 2Cr^{3+} + 7H_2O \end{array}$$

Note that the electrons cancel. No electrons ever appear in the complete balanced equation. This fact serves as a nice intermediate check. You should also check your result by making sure that the final equation is balanced with respect to each element and with respect to charge. As a final step, we rewrite the balanced equation with phases indicated:

$$6Fe^{2+}(aq) + 14H^+(aq) + Cr_2O_7^{2-}(aq) \rightarrow 6Fe^{3+}(aq) + 2Cr^{3+}(aq) + 7H_2O(l)$$

Although the method of half-reactions involves numerous steps, it is actually simple to use and with a little practice becomes straightforward.

Example 20-7: Use the method of half-reactions to balance the equation

$$Fe(s) + Cl_2(aq) \rightarrow Fe^{3+}(aq) + Cl^-(aq)$$

Solution: Iron changes oxidation state from 0 to +3, and chlorine changes oxidation state from 0 to −1. Thus the oxidation and reduction half-reactions are

$$Fe \rightarrow Fe^{3+} \qquad \text{oxidation}$$

$$Cl_2 \rightarrow Cl^- \qquad \text{reduction}$$

The iron half-reaction is balanced with respect to Fe; the chlorine half-reaction can be balanced with respect to chlorine by multiplying Cl^- by 2:

$$Fe \rightarrow Fe^{3+}$$

$$Cl_2 \rightarrow 2Cl^-$$

Oxygen and hydrogen are not involved, and so we can proceed to balance the charge in the two half-reactions. The iron half-reaction has an excess charge of +3 on the right, and therefore we add three electrons:

$$Fe \rightarrow Fe^{3+} + 3e^- \qquad \text{oxidation of Fe}$$

The chlorine half-reaction has a deficiency of two negative charges on the left, and therefore we add two electrons:

$$2e^- + Cl_2 \rightarrow 2Cl^- \qquad \text{reduction of Cl}$$

The iron half-reaction supplies three electrons for each Fe oxidized, and the chlorine half-reaction consumes two electrons for each Cl_2 reduced. If we multiply the iron half-reaction by 2 and the chlorine half-reaction by 3, then both half-reactions will involve six electrons:

$$2Fe \rightarrow 2Fe^{3+} + 6e^-$$

$$6e^- + 3Cl_2 \rightarrow 6Cl^-$$

Addition of the half-reactions and cancellation of the six electrons on each side yields

$$2Fe + 3Cl_2 \rightarrow 2Fe^{3+} + 6Cl^-$$

Finally, we indicate the various phases and write

$$2Fe(s) + 3Cl_2(aq) \rightarrow 2Fe^{3+}(aq) + 6Cl^-(aq)$$

Note that this equation is balanced with respect to each element and with respect to charge.

The next example illustrates the balancing of a single half-reaction.

Example 20-8: Balance the following half-reaction, which takes place in acidic solution:

$$Cl^-(aq) \rightarrow ClO_3^-(aq)$$

Problems 20-33 and 20-34 ask you to balance half-reactions in acidic solution.

Solution: Note that chlorine changes oxidation state from -1 to $+5$. The chlorine is already balanced, and so we proceed directly to balance the oxygen, using water. We add three H_2O to the left-hand side to obtain

$$3H_2O + Cl^- \rightarrow ClO_3^-$$

We next balance the hydrogens by adding six H^+ to the right-hand side because there is an excess of six hydrogens on the left-hand side:

$$3H_2O + Cl^- \rightarrow ClO_3^- + 6H^+$$

The half-reaction is now balanced with respect to each element, and so now we balance it with respect to charge. The net charge on the left is -1, and that on the right is $+5$ [$= 6(+1) - 1$]. Addition of six electrons to the right makes the net charge on the right equal to -1 [$= +5 + 6(-1)$] and therefore balances the half-reaction with respect to charge:

$$3H_2O(l) + Cl^-(aq) \rightarrow ClO_3^-(aq) + 6H^+(aq) + 6e^-$$

The half-reaction is now balanced.

20-6. CHEMICAL EQUATIONS FOR REACTIONS OCCURRING IN BASIC SOLUTION SHOULD NOT CONTAIN $H^+(aq)$

The half-reactions considered up to this stage have all taken place in acidic aqueous solution, where $H^+(aq)$ and H_2O are readily available and thus can be used in balancing the equation for the reaction. However, in basic solution $H^+(aq)$ is not available at significant concentrations and thus should not appear in the final balanced equation.

To balance the equation for a redox reaction that takes place in basic solution, we proceed as if the reaction were taking place in acidic solution and then eliminate any $H^+(aq)$ appearing in the balanced equation by combining the $H^+(aq)$ with $OH^-(aq)$ to form H_2O. For example, consider the reaction

$$Fe(OH)_2(s) + CrO_4^{2-}(aq) \xrightarrow{OH^-(aq)} Fe(OH)_3(s) + Cr(OH)_4^-(aq)$$

which occurs in basic solution. The oxidation and reduction half-reactions are

$$Fe(OH)_2 \rightarrow Fe(OH)_3$$
$$CrO_4^{2-} \rightarrow Cr(OH)_4^-$$

The iron half-reaction has one more oxygen on the right than on the left, and thus the half-reaction is balanced with respect to oxygen when one H_2O is added to the left:

$$H_2O + Fe(OH)_2 \rightarrow Fe(OH)_3$$

There are now four hydrogens on the left and only three on the right, and so we balance hydrogen by adding one H^+ to the right:

$$H_2O + Fe(OH)_2 \rightarrow Fe(OH)_3 + H^+$$

The excess charge is $+1$ on the right, and therefore the addition of one electron to the right balances the iron half-reaction with respect to charge:

$$H_2O + Fe(OH)_2 \rightarrow Fe(OH)_3 + H^+ + e^-$$

The chromium half-reaction has four oxygens on each side, and so we do not need to add H_2O. There are four hydrogens on the right and none on the left, and so we add four H^+ to the left:

$$4H^+ + CrO_4^{2-} \rightarrow Cr(OH)_4^-$$

The net charge on the left is $+2$ and that on the right is -1, and so we add three electrons to the left to balance the charge:

$$4H^+ + CrO_4^{2-} + 3e^- \rightarrow Cr(OH)_4^-$$

The iron half-reaction as written supplies one electron, and the chromium half-reaction consumes three electrons. Therefore, we must multiply the iron half-reaction by 3 to balance the number of electrons in the two half-reactions:

$$3H_2O + 3Fe(OH)_2 \rightarrow 3Fe(OH)_3 + 3H^+ + 3e^-$$
$$4H^+ + CrO_4^{2-} + 3e^- \rightarrow Cr(OH)_4^-$$

Adding these two half-reactions yields

$$3H_2O + 3Fe(OH)_2 + 4H^+ + CrO_4^{2-} \rightarrow 3Fe(OH)_3 + 3H^+ + Cr(OH)_4^-$$

Cancellation of like terms yields

$$3H_2O + 3Fe(OH)_2 + H^+ + CrO_4^{2-} \rightarrow 3Fe(OH)_3 + Cr(OH)_4^-$$

We now convert the balanced equation to one representing a reaction in basic solution by adding OH^- to both sides to convert the H^+ on the left side to H_2O:

$$3H_2O + 3Fe(OH)_2 + \underbrace{H^+ + OH^-}_{H_2O} + CrO_4^{2-} \rightarrow 3Fe(OH)_3 + Cr(OH)_4^- + OH^-$$

Indicating the phases in the balanced equation yields the final result

$$4H_2O(l) + 3Fe(OH)_2(s) + CrO_4^{2-}(aq) \rightarrow 3Fe(OH)_3(s) + Cr(OH)_4^-(aq) + OH^-(aq)$$

Example 20-9: Balance the following half-reaction, which occurs in basic solution:

$$ClO^-(aq) \rightarrow Cl^-(aq)$$

Solution: The half-reaction is already balanced with respect to chlorine. Therefore we proceed to balance it with respect to oxygen, using H_2O:

$$ClO^- \rightarrow Cl^- + H_2O$$

We now use H^+ to balance the half-reaction with respect to hydrogen:

$$2H^+ + ClO^- \rightarrow Cl^- + H_2O$$

The half-reaction is now balanced with respect to the elements. It has a net charge of $+1$ on the left and -1 on the right; the addition of two electrons to the left balances the charge:

$$2H^+ + ClO^- + 2e^- \rightarrow Cl^- + H_2O$$

We next convert the equation to one representing a half-reaction in basic solution by adding OH^-. Because there are two H^+ on the left, the addition of two OH^- to each side yields

$$\underbrace{2H^+ + 2OH^-}_{2H_2O} + ClO^- + 2e^- \rightarrow Cl^- + H_2O + 2OH^-$$

Cancellation of H_2O terms yields

$$H_2O(l) + ClO^-(aq) + 2e^- \rightarrow Cl^-(aq) + 2OH^-(aq)$$

which is the balanced half-reaction for the reduction of hypochlorite, ClO^-, to chloride in basic solution.

Problems 20-19 through 20-30 deal with balancing oxidation-reduction equations.

20-7. BLEACHES AND DISINFECTANTS ARE STRONG OXIDIZING AGENTS

Common household *bleach* is an approximately 5% by mass solution of sodium hypochlorite, NaClO, in water. *Bleaching powder* contains calcium hypochlorite, $Ca(ClO)_2$, which, in contrast to NaClO, is not explosive when dry. The bleaching and *disinfectant* action of hypochlorite ion is a consequence of its oxidative attack on pigments and microorganisms, respectively. The half-reaction for ClO^- acting as an oxidizing agent in basic solution is (Example 20-9)

$$ClO^-(aq) + H_2O(l) + 2e^- \rightarrow Cl^-(aq) + 2OH^-(aq)$$

Hydrogen peroxide is also used as a bleach and disinfectant and is sold as aqueous solutions of various percentages. A 3% solution is used as an antiseptic, and a 6% solution is used to bleach hair. More concentrated solutions, which are dangerously explosive in contact with many solids, are used in the chemical industry as potent oxidizing agents. The explosive decomposition of H_2O_2 proceeds via the reaction

$$2H_2O_2(l) \rightarrow 2H_2O(l) + O_2(g)$$

Example 20-10: Write the half-reaction for H_2O_2 acting as an oxidizing agent in acidic aqueous solution.

Solution: The oxidation state of oxygen in H_2O_2 is -1 (Example 20-3). Because H_2O_2 is an oxidizing agent, oxygen must be reduced to an oxidation state of -2, which is the lowest possible oxidation state of oxygen. Thus

$$H_2O_2 \rightarrow H_2O$$

Balancing the oxygen atoms by adding H_2O to the right-hand side yields

$$H_2O_2 \rightarrow 2H_2O$$

and balancing the hydrogen atoms gives

$$2H^+(aq) + H_2O_2(aq) + 2e^- \rightarrow 2H_2O(l)$$

Another example of an oxidizing agent used to destroy bacteria is benzoyl peroxide (in Oxy-5, for instance):

$$C_6H_5-\overset{O}{\overset{\|}{C}}-\underbrace{O-O}_{\text{peroxide bond}}-\overset{O}{\overset{\|}{C}}-C_6H_5$$

benzoyl peroxide

which is used to control acne by destroying bacteria on the skin.

SUMMARY

In order to determine whether a reaction involves electron transfer, we first assign oxidation states to each element on both sides of the equation. In an electron-transfer reaction, the oxidation state of one element increases and the oxidation state of another element decreases.

Electron-transfer reactions can be separated into two half-reactions, the oxidation half-reaction (electrons on the right) and the reduction half-reaction (electrons on the left). The oxidation half-reaction supplies electrons to the reduction half-reaction. The equations for electron-transfer reactions can be balanced by a systematic procedure once the oxidation half-reaction and the reduction half-reaction have been identified by the assignment of oxidation states to the atoms involved in the reaction. The procedure for balancing oxidation-reduction reactions in acidic solutions involves seven steps:

I. Separate the equation into an oxidation half-reaction and a reduction half-reaction.

II. Balance each half-reaction with respect to all elements other than oxygen and hydrogen.

III. Balance each half-reaction with respect to oxygen by adding the appropriate number of H_2O to the side deficient in oxygen.

IV. Balance each half-reaction with respect to hydrogen by adding the appropriate number of H^+ to the side deficient in hydrogen.

V. Balance each half-reaction with respect to charge by adding the appropriate number of electrons to the side with the excess positive charge.

VI. Multiply each half-reaction by an integer that makes the number of electrons supplied by the oxidation half-reaction equal to the number of electrons accepted by the reduction half-reaction.

VII. Obtain the complete balanced equation by adding the two half-reactions and canceling any like terms.

To balance the equation for a redox reaction that takes place in basic solution, we proceed as if the reaction were taking place in acidic solution and then eliminate any $H^+(aq)$ appearing in the balanced equation by combining the $H^+(aq)$ with $OH^-(aq)$ to form $H_2O(l)$.

TERMS YOU SHOULD KNOW

oxidation-reduction reaction 781
redox reaction 781
oxidation state 782
reduction 787
oxidation 788
electron-transfer reaction 788
oxidizing agent 788
reducing agent 788
electron acceptor 788
electron donor 788
half-reaction 789
oxidation half-reaction 789
reduction half-reaction 789
method of half-reactions 790
bleach 797
bleaching powder 797
disinfectant 797

EQUATIONS YOU SHOULD KNOW HOW TO USE

$$\begin{pmatrix}\text{oxidation}\\ \text{state}\end{pmatrix} = \begin{pmatrix}\text{group number of the}\\ \text{element in the periodic}\\ \text{table}\end{pmatrix} - \begin{pmatrix}\text{total number of}\\ \text{valence electrons}\end{pmatrix}$$

PROBLEMS

ASSIGNMENT OF OXIDATION STATES

20-1. Assign an oxidation state to the metal in each of the following species:

(a) $Cr_2O_7^{2-}$
(b) MoO_4^{2-}
(c) $Cr(OH)_4^-$
(d) VO_2^+
(e) MnO_4^-

20-2. Assign an oxidation state to sulfur in each of the following species:

(a) S^{2-}
(b) SO_3
(c) S
(d) SO_2
(e) SO_4^{2-}

20-3. Assign an oxidation state to the underlined element in each of the following compounds:

(a) $Na_2\underline{Se}O_3$
(b) $H_3\underline{P}O_4$
(c) $Na_2\underline{Si}F_6$
(d) $Na\underline{Br}O_3$
(e) $H\underline{As}O_2$

20-4. Assign an oxidation state to the underlined element in each of the following compounds:

(a) $Na\underline{Cl}O$
(b) $K\underline{Mn}O_4$
(c) $Ca\underline{H}_2$
(d) $K\underline{O}_2$
(e) $\underline{Fe}_2O_3$

20-5. Assign an oxidation state to nitrogen in each of the following species:

(a) NO_2
(b) N_2
(c) N_2O
(d) NO_3^-
(e) NO_2^-

20-6. Determine the oxidation state of chlorine in each of the following chlorine oxides:

(a) Cl_2O
(b) Cl_2O_3
(c) ClO_2
(d) Cl_2O_4
(e) Cl_2O_6

20-7. Determine the oxidation state of carbon in each of the following compounds:

(a) H_2CO
(b) CH_4
(c) CH_3OH
(d) $HCHO_2$
(e) CO_2

20-8. Phosphorus forms a number of oxides. Assign an oxidation state to phosphorus in

(a) P_4O_6
(b) P_4O_7
(c) P_4O_8
(d) P_4O_9
(e) P_4O_{10}

20-9. Assign an oxidation state to antimony in each of the following species:

(a) $SbCl_3$
(b) Sb_4O_6
(c) SbF_5^{2-}
(d) $SbCl_6^{3-}$
(e) Sb_2O_4

20-10. Assign an oxidation state to xenon in each of the following compounds:

(a) XeF_2
(b) XeF_6
(c) Xe
(d) $XeOF_4$
(e) XeO_2F_2

OXIDIZING AGENTS AND REDUCING AGENTS

20-11. Identify the oxidizing and reducing agents in the reaction

$$I_2O_5(s) + 5CO(g) \rightarrow 5CO_2(g) + I_2(s)$$

20-12. Sodium sulfide is manufactured by reacting sodium sulfate with carbon in the form of coke:

$$Na_2SO_4(s) + 4C(s) \rightarrow Na_2S(s) + 4CO(g)$$

Identify the oxidizing and reducing agents in this reaction.

20-13. Sodium nitrite, an important chemical in the dye industry, is manufactured by the reaction

$$NaNO_3(aq) + Pb(s) \rightarrow NaNO_2(aq) + PbO(s)$$

Identify the oxidizing and reducing agents in this reaction.

20-14. Sodium chlorite, an industrial bleaching agent, is prepared by the reaction

$$4NaOH(aq) + Ca(OH)_2(aq) + C(s) + 4ClO_2(g) \rightarrow 4NaClO_2(aq) + CaCO_3(s) + 3H_2O(l)$$

Identify the oxidizing and reducing agents in this reaction.

20-15. Identify the oxidizing and reducing agents and the oxidation and reduction half-reactions in the following reactions:

(a) $V(s) + Ni^{2+}(aq) \rightarrow V^{2+}(aq) + Ni(s)$
(b) $2Ti^{2+}(aq) + Co^{2+}(aq) \rightarrow 2Ti^{3+}(aq) + Co(s)$

20-16. Identify the oxidizing and reducing agents and the oxidation and reduction half-reactions in the following reactions:

(a) $Cu^{2+}(aq) + 2Cr^{2+}(aq) \rightarrow Cu(s) + 2Cr^{3+}(aq)$
(b) $In^{+}(aq) + 2Fe^{3+}(aq) \rightarrow 2Fe^{2+}(aq) + In^{3+}(aq)$

20-17. Potassium superoxide, KO_2, is a strong oxidizing agent. Explain why.

20-18. Lithium aluminum hydride, $LiAlH_4$, is a strong reducing agent. Explain why.

BALANCING OXIDATION-REDUCTION REACTIONS

20-19. Use the method of half-reactions to balance the following equations:

(a) $CrO_4^{2-}(aq) + Cl^-(aq) \rightarrow Cr^{3+}(aq) + ClO_2^-(aq)$ (acidic)
(b) $IO_3^-(aq) + Cr(OH)_4^-(aq) \rightarrow I^-(aq) + CrO_4^{2-}(aq)$ (basic)
(c) $HAsO_2(aq) + I_3^-(aq) \rightarrow H_3AsO_4(aq) + I^-(aq)$ (acidic)

20-20. Use the method of half-reactions to balance the following equations:

(a) $Co(OH)_2(s) + SO_3^{2-}(aq) \rightarrow SO_4^{2-}(aq) + Co(s)$ (basic)
(b) $GeO(s) + IO_3^-(aq) \rightarrow GeO_2(s) + I^-(aq)$ (acidic)
(c) $NO_3^-(aq) + PbO(s) \rightarrow NO_2^-(aq) + PbO_2(s)$ (basic)

20-21. Use the method of half-reactions to balance the following equations:

(a) $BrO_3^-(aq) + NO_2^-(aq) \rightarrow Br^-(aq) + NO_3^-(aq)$ (acidic)

(b) $MnO_4^-(aq) + Sb(s) \rightarrow Sb_2O_3(s) + MnO_2(s)$ (acidic)

(c) $H_2MoO_4(aq) + Cr^{2+}(aq) \rightarrow Mo(s) + Cr^{3+}(aq)$ (acidic)

20-23. Complete and balance the following equations:

(a) $Fe(OH)_2(s) + O_2(g) \rightarrow Fe(OH)_3(s)$ (basic)

(b) $Cu(s) + NO_3^-(aq) \rightarrow Cu^{2+}(aq) + NO(g)$ (acidic)

(c) $Bi_2O_3(s) + ClO^-(aq) \rightarrow BiO_3^-(aq) + Cl^-(aq)$ (basic)

20-25. Complete and balance the following equations:

(a) $NH_4^+(aq) + NO_3^-(aq) \rightarrow N_2O(g) + 2H_2O(l)$ (acidic)

(b) $Hg(l) + H_2SO_4(aq) + Cr_2O_7^{2-}(aq) \rightarrow Hg_2SO_4(s) + Cr^{3+}(aq)$ (acidic)

(c) $Fe(s) + O_2(g) \rightarrow Fe_2O_3 \cdot 3H_2O(s)$ (basic)

20-27. Balance the following equations for reactions that occur in acidic solution:

(a) $MnO(s) + PbO_2(s) \rightarrow MnO_4^-(aq) + Pb^{2+}(aq)$

(b) $As_2S_5(s) + NO_3^-(aq) \rightarrow H_3AsO_4(aq) + HSO_4^-(aq) + NO_2(g)$

For each of these reactions, identify the

electron donor	reducing agent
electron acceptor	species oxidized
oxidizing agent	species reduced

20-29. Balance the following equation for the reaction occurring in basic solution:

$$Zn(s) + MnO_2(s) \rightarrow Mn_2O_3 \cdot H_2O(s) + ZnO_2^{2-}(aq)$$

20-31. For the strong of heart, balance

$CrI_3(s) + Cl_2(g) \rightarrow CrO_4^{2-}(aq) + IO_4^-(aq) + Cl^-(aq)$ (basic)

20-22. Use the method of half-reactions to balance the following equations:

(a) $BrO_3^-(aq) + F_2(g) \rightarrow BrO_4^-(aq) + F^-(aq)$ (basic)

(b) $H_3AsO_3(aq) + I_2(aq) \rightarrow H_3AsO_4(aq) + I^-(aq)$ (acidic)

(c) $MnO_4^-(aq) + C_2O_4^{2-}(aq) \rightarrow Mn^{2+}(aq) + CO_2(aq)$ (acidic)

20-24. Complete and balance the following equations:

(a) $MnO_2(s) + Cl^-(aq) \rightarrow Mn^{2+}(aq) + Cl_2(g)$ (acidic)

(b) $Cr_2O_7^{2-}(aq) + I^-(aq) \rightarrow Cr^{3+}(aq) + I_2(s)$ (acidic)

(c) $CuS(s) + NO_3^-(aq) \rightarrow Cu^{2+}(aq) + S(s) + NO(g)$ (acidic)

20-26. Complete and balance the following equations:

(a) $CoCl_2(s) + Na_2O_2(aq) \rightarrow Co(OH)_3(s) + Cl^-(aq) + Na^+(aq)$ (basic)

(b) $N_2H_4(aq) + Cu(OH)_2(s) \rightarrow N_2(g) + Cu(s)$ (basic)

(c) $C_2O_4^{2-}(aq) + MnO_2(s) \rightarrow Mn^{2+}(aq) + CO_2(g)$ (acidic)

20-28. Balance the following equations for reactions that occur in acidic solution:

(a) $ZnS(s) + NO_3^-(aq) \rightarrow Zn^{2+}(aq) + S(s) + NO(g)$

(b) $MnO_4^-(aq) + HNO_2(aq) \rightarrow NO_3^-(aq) + Mn^{2+}(aq)$

For each of these reactions, identify the

electron donor	reducing agent
electron acceptor	species oxidized
oxidizing agent	species reduced

20-30. Balance the following equation for the reaction occurring in basic solution:

$$NiO_2H(s) + Cd(s) \rightarrow Ni(OH)_2(s) + Cd(OH)_2(s)$$

20-32. For the strong of heart, balance

$C_2H_5OH(aq) + I_3^-(aq) \rightarrow CO_2(g) + CHO_2^-(aq) + CHI_3(aq) + I^-(aq)$ (acidic)

BALANCING HALF-REACTIONS

20-33. Balance the following half-reactions, which occur in acidic solution:

(a) $PbO_2(s) \rightarrow Pb^{2+}(aq)$
(b) $MnO_2(s) \rightarrow Mn^{2+}(aq)$
(c) $NO(g) \rightarrow NO_3^-(aq)$

20-34. Balance the following half-reactions, which occur in acidic solution:

(a) $H_2BO_3^-(aq) \rightarrow BH_4^-(aq)$
(b) $ClO_3^-(aq) \rightarrow Cl_2(g)$
(c) $Cl_2(g) \rightarrow HClO(aq)$

20-35. Balance the following half-reactions:

(a) $WO_3(s) \rightarrow W_2O_5(s)$ (acidic)
(b) $U^{4+}(aq) \rightarrow UO_2^+(aq)$ (acidic)
(c) $Zn(s) \rightarrow Zn(OH)_4^{2-}(aq)$ (basic)

20-36. Balance the following half-reactions:

(a) $OsO_4(s) \rightarrow Os(s)$ (acidic)
(b) $S(s) \rightarrow SO_3^{2-}(aq)$ (basic)
(c) $Sn(s) \rightarrow HSnO_2^-(aq)$ (basic)

20-37. Balance the following half-reactions, which occur in basic solution:

(a) $SO_3^{2-}(aq) + H_2O(l) + e^- \rightarrow S_2O_4^{2-}(aq)$
(b) $Cu(OH)_2(s) + e^- \rightarrow Cu_2O(s) + H_2O(l)$
(c) $AgO(s) + H_2O(l) + e^- \rightarrow Ag_2O(s)$
(d) $HgO(s) + H_2O(l) + e^- \rightarrow Hg(l)$

20-38. Balance the following half-reactions:

(a) $Au(CN)_2^-(aq) + e^- \rightarrow Au(s) + CN^-(aq)$
(b) $PbO_2(s) + SO_4^{2-}(aq) + H^+(aq) + e^- \rightarrow PbSO_4(s) + H_2O(l)$
(c) $MnO_4^-(aq) + H^+(aq) + e^- \rightarrow MnO_2(s) + H_2O(l)$
(d) $Cr(OH)_3(s) + OH^-(aq) \rightarrow CrO_4^{2-}(aq) + H_2O(l) + e^-$

PRACTICAL APPLICATIONS

20-39. *Aqua regia* (royal water) is a mixture of concentrated $HNO_3(aq)$ and $HCl(aq)$ that dissolves the noble metals gold and platinum. Balance the following equation:

$$Au(s) + NO_3^-(aq) + Cl^-(aq) \rightarrow AuCl_4^-(aq) + NO_2(g)$$

20-40. Gunpowder is a mixture of potassium nitrate, charcoal (carbon), and sulfur that explodes when ignited or detonated. The reaction is an oxidation-reduction with the final products $N_2(g)$, $CO_2(g)$, and $K_2S(s)$. Write a balanced equation for the gunpowder reaction. What are the reducing agents and the oxidizing agents?

20-41. Solid phosphorus reacts with $BaSO_4$ under oxygen-free, anhydrous conditions to produce $P_4O_{10}(s)$ and $BaS(s)$; write a balanced equation for this process.

20-42. Potassium superoxide, $KO_2(s)$, is used as a source of oxygen in the cannisters of the breathing equipment used by firefighters. The oxygen is produced when moisture comes into contact with the $KO_2(s)$. Solid KOH is also produced in the reaction and absorbs the $CO_2(g)$ in the exhaled air by forming $KHCO_3(s)$. Thus such an apparatus enables a person to breathe the same air for an extended period. Write balanced equations for these two reactions.

20-43. Polluted air frequently contains $SO_2(g)$, which is oxidized in the presence of moisture to sulfuric acid. The SO_2 oxidation is catalyzed by some metals, such as iron, and the sulfuric acid that forms reacts with the metal to form metal sulfates. Write a sequence of two balanced equations leading to the corrosion of iron by sulfur dioxide.

20-44. One simple method for passive solar energy storage is to use sunlight to heat water in steel drums. Sodium chromate crystals, $Na_2CrO_4(s)$, are often dissolved in the water to protect the inside of the drums from corrosion by assisting in the formation of a protective oxide film on the metal surface. The equation for the reaction is

$$Fe(s) + CrO_4^{2-}(aq) \rightarrow Fe_2O_3(s) + Cr_2O_3(s)$$

Balance this equation as it occurs in basic solution.

20-45. Sodium perborate, $NaBO_3(s)$, is a mild oxidizing agent that is used as a household bleach for colorfast fabrics and also finds use in medicine and dentistry as a mouthwash. It is produced by the action of peroxides on borax:

$$\underset{\text{borax}}{Na_2B_4O_7(aq)} + Na_2O_2(aq) + H_2O_2(aq) \rightarrow NaBO_3(aq)$$

Balance this equation.

20-46. The drain cleaner Drano consists of a mixture of aluminum metal and sodium hydroxide pellets. When water is added to the mixture, hydrogen gas is evolved. The cleaner works by a combination of evolved heat, which melts grease, concentrated alkali, which solubilizes grease, and evolved hydrogen bubbles, which break up the plug in the drain. Write a balanced equation for the hydrogen evolution.

CALCULATIONS INVOLVING OXIDATION-REDUCTION REACTIONS

20-47. The quantity of antimony in a sample can be determined by an oxidation-reduction titration with an oxidizing agent. A 7.70-g sample of stibnite, an ore of antimony, is dissolved in hot, concentrated $HCl(aq)$ and passed over a reducing agent so that all the antimony is in the form Sb^{3+}. The $Sb^{3+}(aq)$ is completely oxidized by 43.7 mL of a 0.125 M solution of $KBrO_3$. The equation for the reaction is

$$BrO_3^-(aq) + 6H^+(aq) + 3Sb^{3+}(aq) \rightarrow Br^-(aq) + 3Sb^{5+}(aq) + 3H_2O(l)$$

Calculate the amount of antimony in the sample and its percentage in this ore.

20-48. An ore is to be assayed for its iron content by an oxidation-reduction titration with permanganate ion. A 5.00-g sample of the ore is dissolved in hydrochloric acid and passed over a reducing agent so that all the iron is in the form Fe^{2+}. The $Fe^{2+}(aq)$ is completely oxidized by 39.5 mL of a 0.0512 M solution of $KMnO_4$. The equation for the reaction is

$$KMnO_4(aq) + 8HCl(aq) + 5FeCl_2(aq) \rightarrow MnCl_2(aq) + 5FeCl_3(aq) + 4H_2O(l) + KCl(aq)$$

Calculate the amount of iron in the sample and its percentage in the ore.

20-49. A rock sample is to be assayed for its tin content by an oxidation-reduction titration with $I_3^-(aq)$. A 10.0-g sample of the rock is crushed, dissolved in sulfuric acid, and passed over a reducing agent so that all the tin is in the form Sn^{2+}. The $Sn^{2+}(aq)$ is completely oxidized by 49.5 mL of a 0.400 M solution of NaI_3. The equation for the reaction is

$$I_3^-(aq) + Sn^{2+}(aq) \rightarrow Sn^{4+}(aq) + 3I^-(aq)$$

Calculate the amount of tin in the sample and its percentage in the rock.

20-50. Sodium chlorite, $NaClO_2(s)$, is a powerful but stable oxidizing agent used in the paper industry, especially for the final whitening of paper. Sodium chlorite is capable of bleaching materials containing cellulose without oxidizing the cellulose. Sodium chlorite is made by the reaction

$$NaOH(aq) + Ca(OH)_2(s) + C(s) + ClO_2(g) \rightarrow NaClO_2(aq) + CaCO_3(s)$$

Balance the equation for the reaction and compute the number of kilograms of $ClO_2(g)$ required to make 1.00 metric ton of $NaClO_2$.

20-51. The amount of $I_3^-(aq)$ in a solution can be determined by titration with a solution containing a known concentration of $S_2O_3^{2-}(aq)$ (thiosulfate ion). The determination is based on the reaction

$$I_3^-(aq) + 2S_2O_3^{2-}(aq) \rightarrow 3I^-(aq) + S_4O_6^{2-}(aq)$$

Given that it requires 48.7 mL of 0.250 M $Na_2S_2O_3(aq)$ to titrate the $I_3^-(aq)$ in a 15.0-mL sample, compute $[I_3^-]$ in the solution.

20-52. The amount of $Fe^{2+}(aq)$ in an $FeSO_4(aq)$ solution can be determined by titration with a solution containing a known concentration of $Ce^{4+}(aq)$. The determination is based on the reaction

$$Fe^{2+}(aq) + Ce^{4+}(aq) \rightarrow Fe^{3+}(aq) + Ce^{3+}(aq)$$

Given that it requires 29.5 mL of 0.105 M $Ce^{4+}(aq)$ to oxidize the Fe^{2+} in a 25.0-mL sample to $Fe^{3+}(aq)$, compute the value of $[Fe^{2+}]$ and the number of milligrams of iron in the sample.

Michael Faraday, an English scientist, was one of the great pioneers in electrochemistry. In the 1820's Faraday discovered the basic laws of electrolysis. One of his electrolysis cells is shown above.

21 / Electrochemical Cells

An electric current can be drawn from an oxidation-reduction, or electron-transfer, reaction by means of a device called an electrochemical cell. An electrochemical cell can be characterized by its voltage, which is a measure of how strongly an electric current is driven through a wire. The voltage of a cell depends upon the concentration of the various species in the cell reaction. In this chapter we shall learn how to calculate equilibrium constants from cell voltages. The operation of batteries and fuel cells is based upon electrochemical cells, and we shall learn how lead storage batteries and nickel-cadmium rechargeable batteries work.

The passage of an electric current through an electrolyte is called electrolysis. Chemical changes occur at the electrodes during electrolysis. In particular, many electrolytes can be decomposed by electrolysis. In industry, many substances are produced electrolytically, and we shall discuss how to calculate how much electric current is required to produce a given quantity of product by electrolysis.

21-1. AN ELECTROCHEMICAL CELL PRODUCES ELECTRICITY DIRECTLY FROM A CHEMICAL REACTION

Electrons are transferred from one substance to another in oxidation-reduction, or electron-transfer, reactions. It is possible to carry out electron-transfer reactions in such a way that an electric current is produced. A setup in which an electric current can be obtained

Figure 21-1 A simple electrochemical cell. The pieces of zinc and copper are separated by a felt cloth soaked with NaCl(*aq*). The current flow is indicated by the lit bulb.

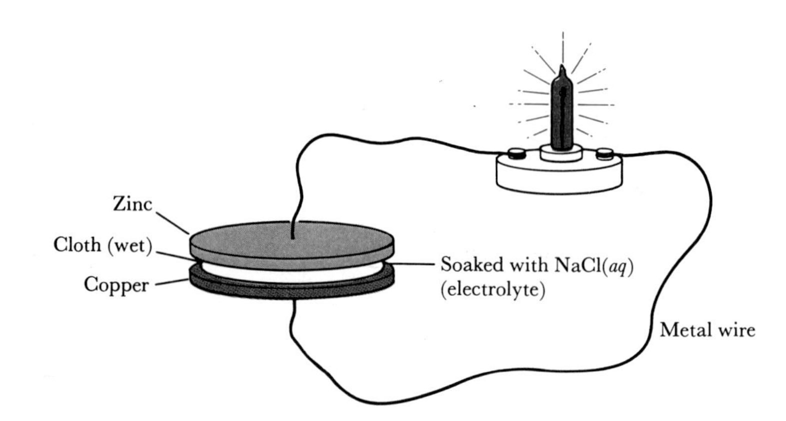

from a chemical reaction is called an *electrochemical cell*. The simplest type of an electrochemical cell consists of pieces of two different metals, such as Zn(*s*) and Cu(*s*), separated by a piece of cloth that has been soaked in an electrolyte. Recall from Chapter 14 that an electrolyte is a substance that conducts an electric current when it is dissolved in water or when it is molten. The simple arrangement of chemicals shown in Figure 21-1 has the remarkable property of being able to supply an electric current to power a light bulb or an electric motor. The two pieces of metal in contact with the electrolyte solution are called *electrodes*. Metal wires attached to the electrodes enable the device to deliver electric current to an external circuit—*electricity from chemical reactions*.

A current can be drawn from an electrochemical cell of the type shown in Figure 21-1 only when the external circuit is complete, that is, only when there is a continuous path for the current to flow along from one electrode to the other. When current is drawn from the cell, chemical changes take place: the zinc electrode is oxidized and hydrogen gas is evolved at the copper electrode.

A circuit must be closed, that is, continuous, in order for a current to flow.

The operation of the cell shown in Figure 21-1 can be greatly improved upon. A particularly effective improvement is to immerse the zinc electrode in an aqueous zinc sulfate solution and the copper electrode in an aqueous copper sulfate solution (Figure 21-2). Electrical contact between the two solutions is then made through a *salt bridge* consisting of a saturated KCl(*aq*) solution mixed with agar, a substance that forms a gel similar to Jell-O. The purpose of the gel is to hold the salt solution in the tube and thus prevent mixing. The salt bridge provides an ionic current path between the $ZnSO_4$(*aq*) and the $CuSO_4$(*aq*) solutions while simultaneously preventing their mixing.

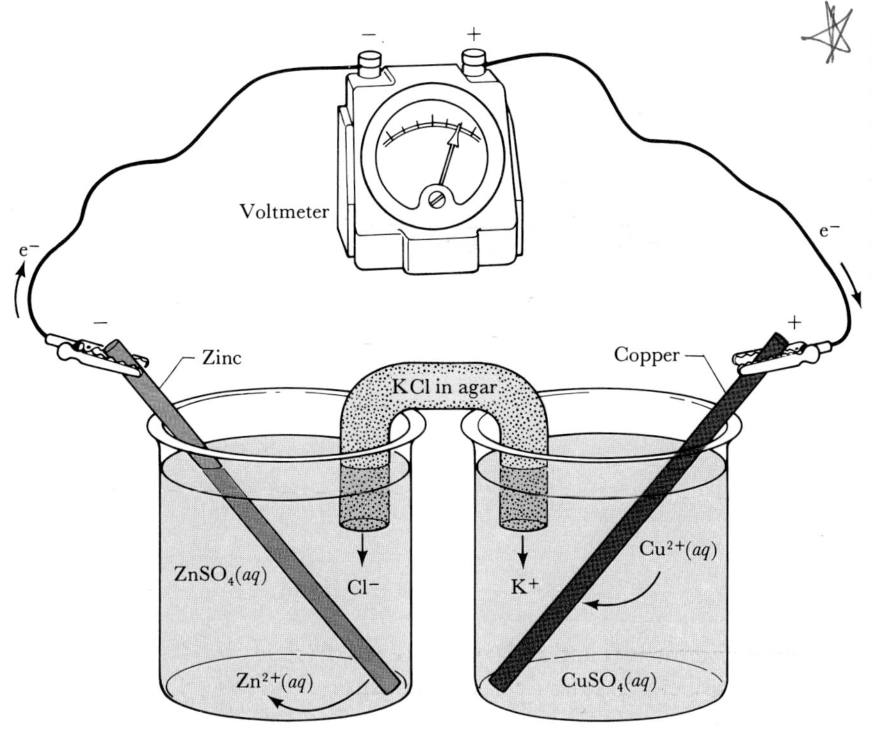

Figure 21-2 A zinc-copper electrochemical cell. The polarity of the cell (zinc electrode negative, copper electrode positive) is determined by experiment with a voltmeter. When the cell operates, electrons flow through the external circuit from the zinc electrode to the copper electrode.

When the cell shown in Figure 21-2 is used as a current source, the following changes are found by chemical analysis:

1. The concentration of $Zn^{2+}(aq)$ in the $ZnSO_4(aq)$ solution increases.
2. The mass of the zinc metal electrode decreases.
3. The concentration of $Cu^{2+}(aq)$ in the $CuSO_4(aq)$ solution decreases.
4. The mass of the copper metal electrode increases.

Observations 1 and 2 can be interpreted as a consequence of the oxidation of Zn(*s*) to $Zn^{2+}(aq)$. Observations 3 and 4 can be interpreted as a consequence of the reduction of $Cu^{2+}(aq)$ to Cu(*s*). When the cell is used as a source of electricity, the cell reaction is

$$Zn(s) + Cu^{2+}(aq) \xrightarrow{\text{discharge}} Zn^{2+}(aq) + Cu(s)$$

where the word *discharge* indicates that current is drawn from the cell.

The two electrode reactions on discharge are

$$Zn(s) \rightarrow Zn^{2+}(aq) + 2e^- \qquad \text{oxidation of } Zn(s)$$

$$Cu^{2+}(aq) + 2e^- \rightarrow Cu(s) \qquad \text{reduction of } Cu^{2+}(aq)$$

Note that the zinc-copper cell is based on the coupling of two half-reactions. When the external circuit is completed by an electric connection between the Zn(*s*) and Cu(*s*) electrodes, electrons can flow in the *external circuit,* that is, through the wire leads and the voltmeter (Figure 21-2). The electrons produced at the zinc electrode by the oxidation of Zn(*s*) travel through the external circuit to the Cu(*s*) electrode, where they are consumed in the reduction of $Cu^{2+}(aq)$ to Cu(*s*).

The current in the electrolyte solutions and in the salt bridge is carried by ions and not by electrons. Because $Zn^{2+}(aq)$ ions are produced in the solution containing the Zn(*s*) electrode, negative ions—in this case, $Cl^{-}(aq)$—must enter the $ZnSO_4(aq)$ from the salt bridge in order to maintain electrical neutrality in the solution. Because $Cu^{2+}(aq)$ ions are removed from the solution containing the Cu(*s*) electrode, positive ions—in this case, $K^{+}(aq)$—must enter the $CuSO_4(aq)$ from the salt bridge in order to maintain electrical neutrality in the solution. Thus the current through the cell solutions is carried by moving ions, with negative ions moving toward the negative (zinc) electrode and positive ions moving toward the positive (copper) electrode.

The components of an electrochemical cell are set up in a manner that forces the electron-transfer reaction to proceed via the external circuit. Electrons are transferred from the reducing agent, Zn(*s*), to the oxidizing agent, $Cu^{2+}(aq)$, through the metal part (wire leads and voltmeter) of the circuit. If the two electrolyte solutions were mixed together, then the cell would not work because $Cu^{2+}(aq)$ would be spontaneously deposited as copper metal on the zinc electrode. In other words, the cell would be internally short-circuited. The reducing agent, Zn(*s*), must be physically separated from the oxidizing agent, $Cu^{2+}(aq)$, in order for the cell to operate.

The spontaneous reaction between Zn(*s*) and $Cu^{2+}(aq)$ is shown on page 136.

A current of electrons flows spontaneously through a metal conductor from a region of negative electric potential to a region of positive electric potential. Thus during the spontaneous discharge of the cell shown in Figure 21-2, the zinc electrode is negative and the copper electrode is positive. The total current through the cell electrolyte, which must equal the current in the external circuit, is carried by anions moving toward the zinc electrode and cations moving toward the copper electrode.

Example 21-1: Consider the electrochemical cell shown at the top of p. 809. Indicate the current flow in the external circuit and in the cell electrolytes.

Solution: The lead electrode is negative, and the silver electrode is positive. Therefore electrons flow in the external circuit from the lead electrode

$$Pb(s) \rightarrow Pb^{2+}(aq) + 2e^{-}$$

to the silver electrode

$$2Ag^{+}(aq) + 2e^{-} \rightarrow 2Ag(s)$$

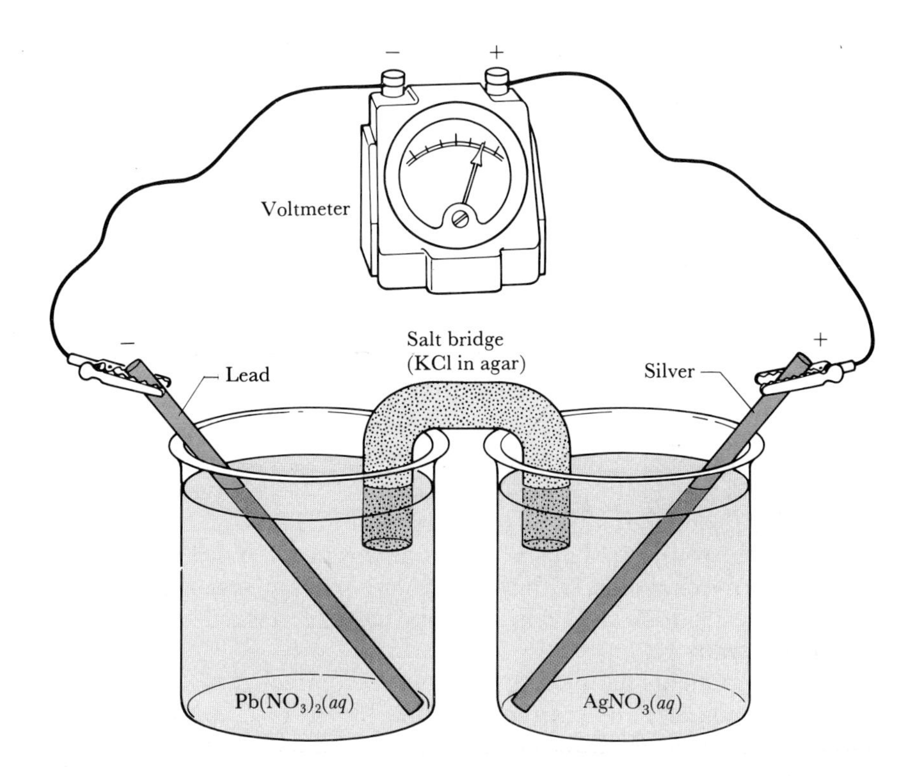

Positive Pb^{2+} ions are produced at the lead electrode, and thus $Cl^-(aq)$ ions flow from the salt bridge into the $Pb(NO_3)_2(aq)$ solution to maintain electrical neutrality. Positive Ag^+ ions are consumed at the silver electrode, and thus $K^+(aq)$ ions flow from the salt bridge into the $AgNO_3(aq)$ solution to maintain electrical neutrality.

21-2. A CELL DIAGRAM REPRESENTS AN ELECTROCHEMICAL CELL

The pictorial representation of an electrochemical cell can be made more compact by the use of a *cell diagram*. For example, the cell diagram for the cell shown in Figure 21-2 is

$$Zn(s)|ZnSO_4(aq)||CuSO_4(aq)|Cu(s)$$

where the single vertical bars indicate boundaries of phases that are in contact and the double vertical bars indicate a salt bridge. Thus in the cell represented by this diagram, $Zn(s)$ and $ZnSO_4(aq)$ are separate phases in physical contact, as are $CuSO_4(aq)$ and $Cu(s)$, and a salt bridge separates the $ZnSO_4(aq)$ and $CuSO_4(aq)$ solutions.

The basic convention for writing cell reactions is to *write the half-reaction of the left-hand electrode as an oxidation half-reaction and the half-reaction of the right-hand electrode as a reduction half-reaction*. This convention enables us to write the cell reaction unambiguously. For the above cell, then, we have, with oxidation occurring at the left-hand electrode,

$$Zn(s) \rightleftharpoons Zn^{2+}(aq) + 2e^- \qquad \text{oxidation at left-hand electrode}$$

If oxidation takes place at the left-hand electrode, then reduction must take place at the right-hand electrode:

$$Cu^{2+}(aq) + 2e^- \rightleftharpoons Cu(s) \qquad \text{reduction at right-hand electrode}$$

note r correspondence

Problems 21-5 through 21-10 ask you to write half-reactions from a cell diagram.

The net cell reaction is given by the sum of the two electrode half-reactions:

$$Zn(s) + Cu^{2+}(aq) \rightleftharpoons Cu(s) + Zn^{2+}(aq)$$

Consider the redox reaction

$$Zn(s) + 2H^+(aq) \rightleftharpoons H_2(g) + Zn^{2+}(aq)$$

Let's construct an electrochemical cell that utilizes this reaction. We first note that $Zn(s)$ is oxidized to $Zn^{2+}(aq)$, and thus the oxidation half-reaction of the cell is

$$Zn(s) \rightarrow Zn^{2+}(aq) + 2e^- \qquad \text{oxidation, left-hand electrode}$$

One of the electrode wires is connected to the zinc metal.

We next note that $H^+(aq)$ is reduced to hydrogen gas, and thus the reduction half-reaction of the cell is

$$2H^+(aq) + 2e^- \rightarrow H_2(g) \qquad \text{reduction, right-hand electrode}$$

This half-reaction requires a metal electrode that can provide a pathway for the electrons from the zinc electrode. The reduction of the $H^+(aq)$ to $H_2(g)$ takes place on the surface of this metal electrode, which acts as a source of electrons. The necessary electrode is provided by inserting a platinum coil into the cell compartment containing $H^+(aq)$ and $H_2(g)$. Platinum is a relatively unreactive metal and simply provides a metallic surface on which the reduction half-reaction occurs. An electrode involving a gaseous species is called a *gas electrode.*

An electrochemical cell that incorporates the zinc oxidation and $H^+(aq)$ reduction half-reactions is shown in Figure 21-3. Hydrogen gas is bubbled continuously through the right-hand compartment in order to provide $H_2(g)$ at a known pressure.

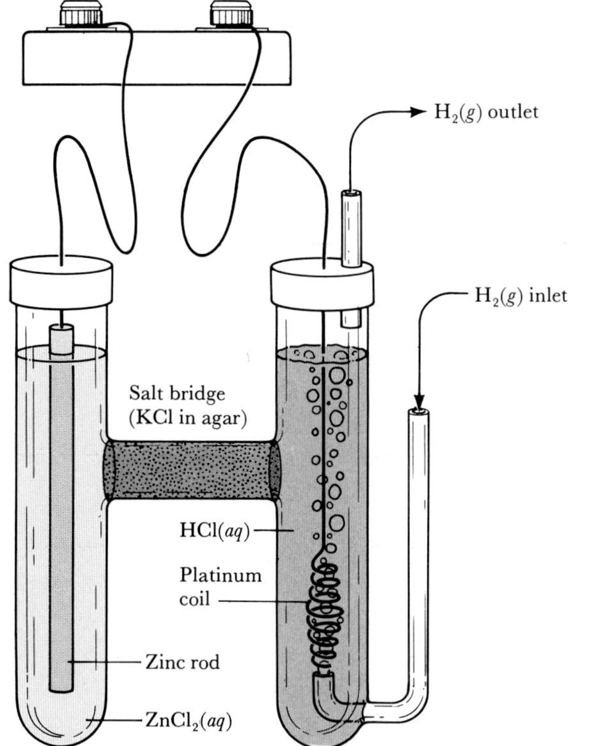

Figure 21-3 A zinc-hydrogen electrochemical cell. The cell diagram is

$$Zn(s)\,|\,ZnCl_2(aq)\,||\,HCl(aq)\,|\,H_2(g)\,|\,Pt(s)$$

and the cell reaction is

$$Zn(s) + 2H^+(aq) \rightleftharpoons H_2(g) + Zn^{2+}(aq)$$

Note the H-type geometry of the cell, which is used together with the salt bridge to separate the two electrolyte solutions.

Example 21-2: Give the cell diagram for the following cell:

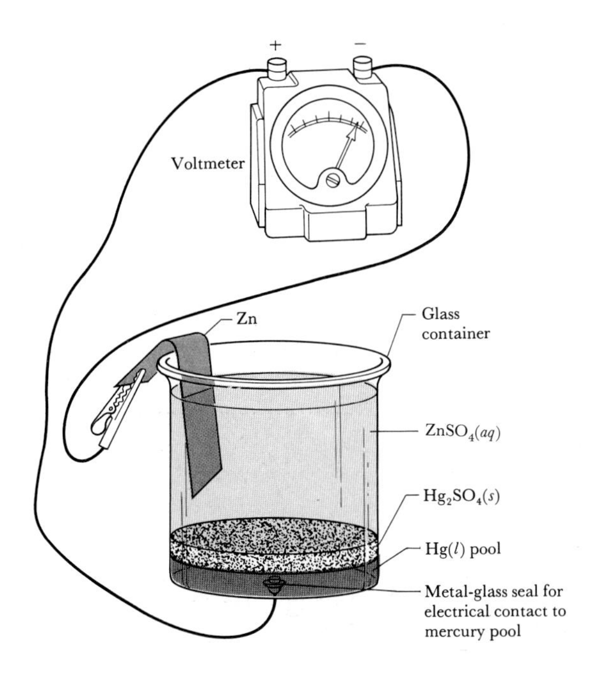

Solution: One connecting wire of the cell is attached to a zinc metal electrode that dips into a solution of zinc sulfate. Thus, this part of the cell diagram is represented as

$$Zn(s)\,|\,ZnSO_4(aq)$$

The other wire is connected to a liquid mercury metal electrode that is covered with solid mercury(I) sulfate; thus, this electrode can be represented as

$$Hg_2SO_4(s)\,|\,Hg(l)$$

Note that the construction of the cell in Example 21-2 physically separates the $Zn(s)$ from $Hg_2SO_4(s)$ and thereby prevents the direct reaction of these two substances.

The $ZnSO_4(aq)$ solution also is in contact with the $Hg_2SO_4(s)$, and thus the two electrodes are in contact with the same electrolyte solution. The complete cell diagram is therefore

$$Zn(s)\,|\,ZnSO_4(aq)\,|\,Hg_2SO_4(s)\,|\,Hg(l)$$

Note that this cell contains only a single electrolyte solution, $ZnSO_4(aq)$, and therefore a salt bridge is not necessary.

Example 21-3: Give the electrode half-reactions and the net cell reaction for the electrochemical cell shown in Example 21-2.

Solution: Oxidation is assumed to take place at the left-hand electrode. The oxidation of $Zn(s)$ yields $Zn^{2+}(aq)$, and thus

$$Zn(s) \rightarrow Zn^{2+}(aq) + 2e^- \qquad \text{oxidation}$$

Oxidation occurs at the left-hand electrode, and therefore reduction must occur at the right-hand electrode. The only element besides zinc that appears in two different oxidation states in the cell is mercury—zero in $Hg(l)$ and $+1$ in $Hg_2SO_4(s)$. The reduction at the mercury electrode is therefore

$$Hg_2SO_4(s) \rightarrow 2Hg(l) \qquad \text{not balanced}$$

The balanced electrode half-reaction is

$$Hg_2SO_4(s) + 2e^- \rightarrow 2Hg(l) + SO_4^{2-}(aq) \qquad \text{reduction}$$

The sum of the oxidation and reduction half-reactions is the net cell reaction

$$Zn(s) + Hg_2SO_4(s) \rightarrow 2Hg(l) + Zn^{2+}(aq) + SO_4^{2-}(aq)$$

21-3. CELL VOLTAGE DEPENDS ON THE CONCENTRATIONS OF THE REACTANTS AND PRODUCTS OF THE CELL REACTION

Walther Nernst, who received a Nobel Prize in chemistry for his pioneering work in electrochemistry, investigated the dependence of cell voltage on the concentration of the electrolytes, size of the electrodes, and other factors. He found, for example, that the voltage of the cell

$$Zn(s)|ZnSO_4(aq)||CuSO_4(aq)|Cu(s)$$

was *independent* of the size of the cell, the size of the electrodes, and the volume of the $ZnSO_4(aq)$ and $CuSO_4(aq)$ solutions. He also found that the voltage of the cell increased when the concentration of $Cu^{2+}(aq)$ increased and decreased when the concentration of $Zn^{2+}(aq)$ increased.

The effect of a change in reactant or product concentration on cell voltage is easily understood in qualitative terms by applying Le Châtelier's principle to the cell reaction, given that the *cell voltage is a measure of the driving force of the reaction.* We have seen that the reaction of the zinc-copper cell (Figure 21-2) is

Cell voltage is often referred to as the electromotive force, or simply emf.

$$Zn(s) + Cu^{2+}(aq) \rightleftharpoons Cu(s) + Zn^{2+}(aq) \qquad (21\text{-}1)$$

An increase in the value of $[Cu^{2+}]$ drives the reaction from left to right, thereby increasing the cell voltage, whereas an increase in the value of $[Zn^{2+}]$ drives the reaction from right to left, thereby decreasing the cell voltage.

Example 21-4: Consider an electrochemical cell in which the cell reaction

$$H_2(g) + 2AgCl(s) \rightleftharpoons 2Ag(s) + 2H^+(aq) + 2Cl^-(aq)$$

takes place. In this reaction $H_2(g)$ reduces $AgCl(s)$ to silver metal. Predict the effect of the following changes on the observed cell voltage:
(a) increase in the $H_2(g)$ pressure
(b) increase in the value of $[H^+]$
(c) increase in the amount of $AgCl(s)$

Solution: By using Le Châtelier's principle, we predict that
(a) an increase in P_{H_2} corresponds to an increase in the concentration of $H_2(g)$, by Henry's law, and this increases the reaction driving force from left to right. The cell voltage thus increases.
(b) an increase in $[H^+]$ increases the reaction driving force from right to left and thus decreases the cell voltage.
(c) an increase in the amount of $AgCl(s)$ has no effect on the cell voltage.

A change in the amount of $AgCl(s)$ has no effect on the concentration of $AgCl(s)$ and thus has no effect on the voltage.

Recall from Chapter 15 that the value of Q/K, where Q is the reaction concentration quotient and K is the reaction equilibrium constant, can be used to predict the direction in which a reaction proceeds toward equilibrium. The *Nernst equation* expresses the quantitative relationship between the cell voltage, E, and the value of Q/K:

$$E = -\frac{2.303\,RT}{nF}\log\left(\frac{Q}{K}\right) \qquad (21\text{-}2)$$

The value of n is readily obtained as the number of electrons that cancel out from the two half-reactions when they are combined to give the balanced oxidation-reduction reaction. See steps VI and VII on page 792.

where R is the gas constant, T is the Kelvin temperature, n is the number of electrons transferred from the reducing agent to the oxidizing agent for the cell reaction as written, and F is *Faraday's constant*, named for the English scientist Michael Faraday.

Faraday's constant is the magnitude of the charge of one mol of electrons. Recall from Chapter 7 that the charge on an electron is 1.602×10^{-19} C (coulombs). There are 6.022×10^{23} electrons in 1 mol of electrons, and so

$$\begin{aligned} F &= (1.602 \times 10^{-19}\,\text{C}\cdot\text{electron}^{-1})(6.022 \times 10^{23}\,\text{electron}\cdot\text{mol}^{-1}) \\ &= 96{,}500\,\text{C}\cdot\text{mol}^{-1} \end{aligned}$$

Before we can use the Nernst equation to calculate the effect of concentration on cell voltage, we must discuss the units involved. You will learn in physics that a charged particle gains or loses energy when it is driven by a voltage difference. The energy change, the charge, and the voltage difference are related by the formula

$$\text{energy change} = \text{electric charge} \times \text{voltage difference} \qquad (21\text{-}3)$$

The SI units corresponding to Equation (21-3) are

$$\text{joule} = \text{coulomb} \times \text{volt} \qquad (21\text{-}4)$$

or

$$\text{J} = \text{C}\cdot\text{V}$$

In Equation (21-2), RT has units of $\text{J}\cdot\text{mol}^{-1}$ and nF has units of $\text{C}\cdot\text{mol}^{-1}$, and so E has units of $\text{J}\cdot\text{C}^{-1}$, or V (volts).

We often apply the Nernst equation at 25.0°C, and so let's evaluate $2.303\,RT/F$ at 25.0°C:

$$\frac{2.303\,RT}{F} = \frac{(2.303)(8.314\,\text{J}\cdot\text{K}^{-1}\cdot\text{mol}^{-1})(298.2\,\text{K})}{96{,}500\,\text{C}\cdot\text{mol}^{-1}}$$
$$= 0.0592\,\text{J}\cdot\text{C}^{-1} = 0.0592\,\text{V}$$

Thus at 25.0°C the Nernst equation becomes

$$E = -\left(\frac{0.0592\,\text{V}}{n}\right)\log\left(\frac{Q}{K}\right) \qquad (21\text{-}5)$$

where E is in volts.

Recall from Chapter 15 that, if $(Q/K) < 1$, then the reaction is spontaneous from left to right. If $(Q/K) < 1$, then $\log(Q/K)$ is negative because the logarithm of a number less than 1 is negative. Thus, if $(Q/K) < 1$, then Equation (21-5) shows that the cell voltage is positive:

Log X is positive if $X > 1$.
Log $X = 0$ if $X = 1$.
Log X is negative if $X < 1$.

$$E = -\left(\frac{0.0592\,\text{V}}{n}\right) \times (\text{negative quantity}) > 0$$

The various possibilities for the values of Q/K and the corresponding voltages are

$\frac{Q}{K} < 1$	$\log\left(\frac{Q}{K}\right) < 0$	$E > 0$	cell reaction is spontaneous from left to right
$\frac{Q}{K} = 1$	$\log\left(\frac{Q}{K}\right) = 0$	$E = 0$	cell reaction is at equilibrium
$\frac{Q}{K} > 1$	$\log\left(\frac{Q}{K}\right) > 0$	$E < 0$	cell reaction is spontaneous from right to left

A reaction at equilibrium has no net driving force.

Thus we see that the sign of the cell voltage tells us the direction in which the cell reaction is spontaneous. Furthermore, the value of the cell voltage is a quantitative measure of the driving force of the cell reaction toward equilibrium. The larger the voltage, the greater is the reaction driving force. Note that, if the reaction is at equilibrium, then the corresponding cell voltage is zero.

21-4. THE NERNST EQUATION CAN BE WRITTEN IN TERMS OF A STANDARD CELL VOLTAGE

Equation (21-5) can be written in a form that is often more convenient by using the fact that

$$\log\left(\frac{Q}{K}\right) = \log Q - \log K \qquad (21\text{-}6)$$

Combining this equation with Equation (21-5) yields

$$E = -\left(\frac{0.0592\ \text{V}}{n}\right)\log\left(\frac{Q}{K}\right)$$

$$= \left(\frac{0.0592\ \text{V}}{n}\right)\log K - \left(\frac{0.0592\ \text{V}}{n}\right)\log Q \qquad (21\text{-}7)$$

If $Q = 1$ for the cell reaction, that is, if all solution species involved are at a concentration of 1.00 M and all gaseous species involved are at 1.00 atm, then the resulting cell voltage is called the *standard cell voltage* and is denoted by E^0. Substituting $Q = 1$ into Equation (21-7) and noting that $\log 1 = 0$, we obtain

$$E(\text{at } Q = 1) = \underset{\substack{\uparrow \\ \text{standard} \\ \text{cell voltage}}}{E^0} = \left(\frac{0.0592\ \text{V}}{n}\right)\log K \qquad (21\text{-}8)$$

Equation (21-8) enables us to compute the value of the equilibrium constant, K, for the cell reaction if the value of E^0 is known and to compute the value of E^0 if K is known.

Example 21-5: The value of E^0 at 25.0°C for the reaction

$$Zn(s) + Cu^{2+}(aq) \rightleftharpoons Cu(s) + Zn^{2+}(aq)$$

is 1.10 V. Compute the equilibrium constant for this reaction at 25.0°C.

Solution: From Equation (21-8), we have

$$E^0 = \left(\frac{0.0592\ \text{V}}{n}\right)\log K$$

Solving this equation for $\log K$, we obtain

$$\log K = \left(\frac{nE^0}{0.0592\ \text{V}}\right)$$

The value of n is 2 for the above reaction, and thus

$$\log K = \frac{2(1.10\ \text{V})}{0.0592\ \text{V}} = 37.2$$

and

$$K = 10^{37.2} = 1.6 \times 10^{37} = \frac{[Zn^{2+}]_{eq}}{[Cu^{2+}]_{eq}}$$

This very large value of K means that at equilibrium the ratio of $[Zn^{2+}]$ to $[Cu^{2+}]$ is very large, or, in other words, that the value of $[Cu^{2+}]$ at equilibrium is very small indeed. (The subscript eq on a concentration term denotes an *eq*uilibrium value of that concentration.)

Substitution of Equation (21-8) into Equation (21-7) yields an especially useful form of the Nernst equation:

$$E = E^0 - \left(\frac{0.0592\text{ V}}{n}\right)\log Q \tag{21-9}$$

where E and E^0 are in volts and all values are at 25.0°C. Equation (21-9) tells us that the cell voltage differs from the standard cell voltage when the reaction quotient is not equal to 1.00. If $Q < 1$, then $E > E^0$; if $Q > 1$, then $E < E^0$.

Example 21-6: The measured voltage at 25.0°C of a cell in which the reaction

$$Zn(s) + Cu^{2+}(aq,\ 1.00\text{ M}) \rightleftharpoons Cu(s) + Zn^{2+}(aq,\ 0.100\text{ M})$$

occurs at the concentrations shown is 1.13 V. Compute E^0 for the cell reaction.

Solution: From Equation (21-9) with $n = 2$, we have

$$E = E^0 - \left(\frac{0.0592\text{ V}}{n}\right)\log Q = E^0 - \left(\frac{0.0592\text{ V}}{2}\right)\log\frac{[Zn^{2+}]}{[Cu^{2+}]}$$

or

$$1.13\text{ V} = E^0 - \left(\frac{0.0592\text{ V}}{2}\right)\log\left(\frac{0.100\text{ M}}{1.00\text{ M}}\right)$$

from which we compute

$$E^0 = 1.13\text{ V} + (0.0296\text{ V})\log 0.100 = 1.10\text{ V}$$

Note that the measurement of the cell voltage when the reactant and product concentrations are known yields the value of E^0 for the cell reaction and hence, from Equation (21-8), the value of the equilibrium constant for the cell reaction. Note also that $Zn(s)$ and $Cu(s)$ do not appear in the Q expression because they are both solids.

Problems 21-19 through 21-26 are similar to Example 21-6.

Figure 21-4 Plot of E (in volts) versus log Q for the cell reaction

$$Zn(s) + Cu^{2+}(aq) \rightleftharpoons Cu(s) + Zn^{2+}(aq)$$

The plot is based on the Nernst equation applied to the above cell reaction, that is,

$$E = 1.10\ \text{V} - (0.030\ \text{V}) \log \frac{[Zn^{2+}]}{[Cu^{2+}]}$$

Note that a 10-fold increase in the value of $Q = [Zn^{2+}]/[Cu^{2+}]$ decreases the cell voltage by -0.030 V. The plot illustrates the linear dependence of cell voltage on log Q.

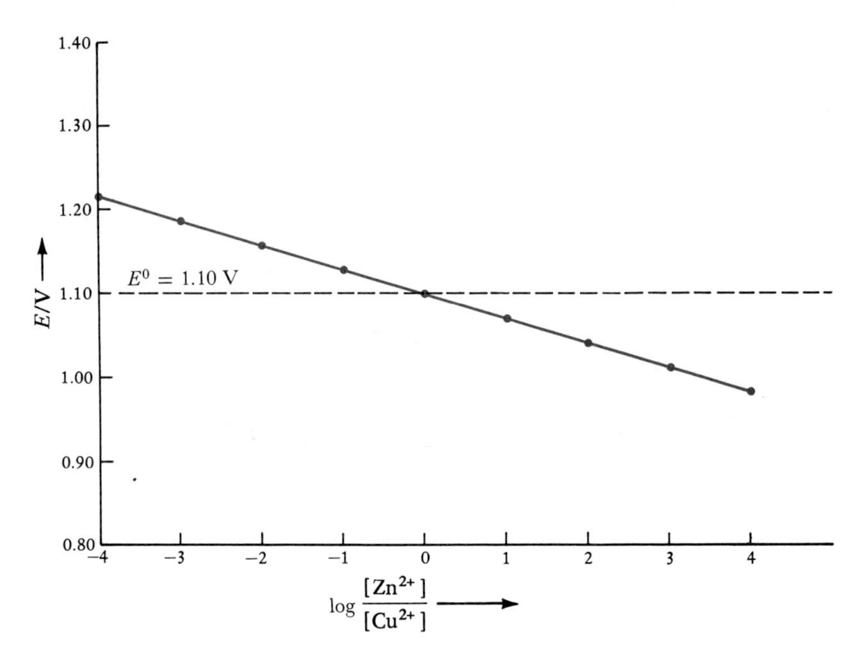

Application of the Nernst equation in the form of Equation (21-9) to the cell reaction

$$Zn(s) + Cu^{2+}(aq) \rightleftharpoons Cu(s) + Zn^{2+}(aq)$$

at 25.0°C with $n = 2$ and $Q = [Zn^{2+}]/[Cu^{2+}]$ yields

$$E = E^0 - \left(\frac{0.0592\ \text{V}}{2}\right) \log \frac{[Zn^{2+}]}{[Cu^{2+}]}$$

The value of E^0 at 25.0°C is 1.10 V. Figure 21-4 shows cell voltage plotted versus the logarithm of Q, that is, versus $\log\{[Zn^{2+}]/[Cu^{2+}]\}$ for this reaction.

21-5. E^0 VALUES CAN BE ASSIGNED TO HALF-REACTIONS

It is not possible to measure the voltage of a single electrode; only the difference in voltage between two electrodes can be measured. By convention, we shall use *standard reduction voltages* for half-reactions. Thus, when discussing voltages of half-reactions, we shall always write the half-reactions with electrons on the left-hand side of the equation. This convention allows us to write the E^0 value of any cell as

$$E^0_{\text{cell}} = E^0_{\text{right}} - E^0_{\text{left}} \tag{21-10}$$

where E^0_{right} denotes the standard reduction voltage of the right-hand electrode and E^0_{left} denotes the standard reduction voltage of the left-hand electrode. Equation (21-10) shows that a cell voltage represents a difference in voltage between two electrodes. Recall that, with our convention for the direction of current, *r*eduction occurs at the *r*ight-hand electrode and oxidation occurs at the left-hand electrode. This is why the right-hand side of Equation 21-10 is written E^0_{right} (reduction) $- E^0_{left}$ (oxidation).

In order to set up a table of standard reduction voltages for half-reactions in aqueous solution, it is necessary to choose arbitrarily a numerical value for the standard voltage of a particular electrode. The convention is to set the standard reduction voltage of the hydrogen electrode equal to zero; that is, we choose $E^0 = 0$ for the electrode half-reaction

$$2H^+(aq, 1\text{ M}) + 2e^- \rightarrow H_2(g, 1\text{ atm}) \qquad E^0 = 0 \text{ by convention} \tag{21-11}$$

The experimental value for the standard voltage of the cell (Figure 21-3)

$$Zn(s)|ZnCl_2(aq)||HCl(aq)|H_2(g)|Pt(s)$$

is

$$E^0_{cell} = E^0_{right} - E^0_{left} = 0.76\text{ V}$$

The right-hand electrode involves the H^+/H_2 pair, and the left-hand electrode involves the Zn^{2+}/Zn pair. Thus we write

$$E^0_{cell} = E^0_{H^+/H_2} - E^0_{Zn^{2+}/Zn} = 0.76\text{ V} \tag{21-12}$$

Notice that the subscripts on E^0 in Equation 21-12 indicate oxidized and reduced forms.

where the subscripts indicate the oxidized and reduced species for each electrode. Substitution of $E^0_{H^+/H_2} = 0$ into Equation (21-12) yields

$$E^0_{cell} = 0 - E^0_{Zn^{2+}/Zn} = 0.76\text{ V}$$

or

$$E^0_{Zn^{2+}/Zn} = -0.76\text{ V}$$

Therefore, the standard reduction voltage for the electrode half-reaction

$$Zn^{2+}(aq, 1\text{ M}) + 2e^- \rightarrow Zn(s)$$

is $E^0 = -0.76$ V at 25°C.

Once we have determined the standard reduction voltage of an electrode relative to the hydrogen electrode, we can use that electrode to determine the standard reduction voltage of other electrodes. This procedure is illustrated in Example 21-7.

Example 21-7: The standard voltage of the cell

$$Zn(s)|ZnSO_4(aq)||CuSO_4(aq)|Cu(s)$$

is $E^0_{cell} = 1.10$ V. Given that E^0 for the electrode half-reaction

$$Zn^{2+}(aq, 1\ M) + 2e^- \rightarrow Zn(s)$$

is $E^0_{Zn^{2+}/Zn} = -0.76$ V, compute E^0 for the electrode half-reaction

$$Cu^{2+}(aq, 1\ M) + 2e^- \rightarrow Cu(s)$$

Solution: Assuming, as is our convention, that oxidation takes place at the left-hand electrode, we have for the half-reactions

$$Zn(s) \rightarrow Zn^{2+}(aq) + 2e^- \qquad \text{oxidation, left-hand electrode}$$

$$Cu^{2+}(aq) + 2e^- \rightarrow Cu(s) \qquad \text{reduction, right-hand electrode}$$

Therefore, we have

$$E^0_{cell} = E^0_{right} - E^0_{left} = E^0_{Cu^{2+}/Cu} - E^0_{Zn^{2+}/Zn} = 1.10\ V$$

Because $E^0_{Zn^{2+}/Zn} = -0.76$ V, we have

$$E^0_{cell} = E^0_{Cu^{2+}/Cu} - (-0.76\ V) = 1.10\ V$$

and thus

$$E^0_{Cu^{2+}/Cu} = 1.10\ V - 0.76\ V = +0.34\ V$$

The standard reduction voltage of the $Cu^{2+}(aq)/Cu(s)$ electrode is $E^0_{Cu^{2+}/Cu} = +0.34$ V.

Thus we see that the standard reduction voltage of an electrode can be obtained from the standard cell voltage of a cell for which the standard reduction voltage of the other electrode is known.

Proceeding in the manner just outlined, we can construct a table of standard reduction voltages. In the arrangement used in Table 21-1, more positive E^0 values indicate more powerful oxidizing agents. Thus we can represent Table 21-1 in schematic form as follows:

increasing strength of oxidizing agents ↑	Reaction	E^0/V	increasing strength of reducing agents ↓
	$F_2(g) + 2e^- \rightarrow 2F^-(aq)$	+2.87	
	⋮	⋮	
	$Li^+(aq) + e^- \rightarrow Li(s)$	−3.05	

Table 21-1 Standard reduction voltages at 25.0°C for aqueous solutions

Electrode half-reaction	E^0/V
Acidic solution	
$F_2(g) + 2e^- \rightarrow 2F^-(aq)$	+2.87
$O_3(g) + 2H^+(aq) + 2e^- \rightarrow O_2(g) + H_2O(l)$	+2.07
$S_2O_8^{2-}(aq) + 2e^- \rightarrow 2SO_4^{2-}(aq)$	+2.01
$Co^{3+}(aq) + e^- \rightarrow Co^{2+}(aq)$	+1.81
$H_2O_2(aq) + 2H^+(aq) + 2e^- \rightarrow 2H_2O(l)$	+1.78
$PbO_2(s) + SO_4^{2-}(aq) + 4H^+(aq) + 2e^- \rightarrow PbSO_4(s) + 2H_2O(l)$	+1.68
$O_2(g) + 4H^+(aq) + 4e^- \rightarrow 2H_2O(l)$	+1.23
$Pd^{2+}(aq) + 2e^- \rightarrow Pd(s)$	+0.99
$Ag^+(aq) + e^- \rightarrow Ag(s)$	+0.80
$Hg_2^{2+}(aq) + 2e^- \rightarrow 2Hg(l)$	+0.79
$Fe^{3+}(aq) + e^- \rightarrow Fe^{2+}(aq)$	+0.77
$O_2(g) + 2H^+(aq) + 2e^- \rightarrow H_2O_2(aq)$	+0.68
$Hg_2SO_4(s) + 2e^- \rightarrow 2Hg(l) + SO_4^{2-}(aq)$	+0.62
$Cu^+(aq) + e^- \rightarrow Cu(s)$	+0.52
$Fe(CN)_6^{3-}(aq) + e^- \rightarrow Fe(CN)_6^{4-}(aq)$	+0.36
$Cu^{2+}(aq) + 2e^- \rightarrow Cu(s)$	+0.34
$Hg_2Cl_2(s) + 2e^- \rightarrow 2Hg(l) + 2Cl^-(aq)$	+0.27
$AgCl(s) + e^- \rightarrow Ag(s) + Cl^-(aq)$	+0.22
$Cu^{2+}(aq) + e^- \rightarrow Cu^+(aq)$	+0.15
$2H^+(aq) + 2e^- \rightarrow H_2(g)$	0
$Pb^{2+}(aq) + 2e^- \rightarrow Pb(s)$	−0.13
$Cd^{2+}(aq) + 2e^- \rightarrow Cd(s)$	−0.40
$Cr^{3+}(aq) + e^- \rightarrow Cr^{2+}(aq)$	−0.41
$Fe^{2+}(aq) + 2e^- \rightarrow Fe(s)$	−0.44
$Zn^{2+}(aq) + 2e^- \rightarrow Zn(s)$	−0.76
$Al^{3+}(aq) + 3e^- \rightarrow Al(s)$	−1.66
$H_2(g) + 2e^- \rightarrow 2H^-(aq)$	−2.25
$Mg^{2+}(aq) + 2e^- \rightarrow Mg(s)$	−2.36
$Na^+(aq) + e^- \rightarrow Na(s)$	−2.71
$Ca^{2+}(aq) + 2e^- \rightarrow Ca(s)$	−2.87
$K^+(aq) + e^- \rightarrow K(s)$	−2.93
$Li^+(aq) + e^- \rightarrow Li(s)$	−3.05
Basic solution	
$O_3(g) + H_2O(l) + 2e^- \rightarrow O_2(g) + 2OH^-(aq)$	+1.24
$O_2(g) + 2H_2O(l) + 4e^- \rightarrow 4OH^-(aq)$	+0.40
$HgO(s) + H_2O(l) + 2e^- \rightarrow Hg(l) + 2OH^-(aq)$	+0.10
$O_2(g) + e^- \rightarrow O_2^-(aq)$	−0.56
$Fe(OH)_3(s) + e^- \rightarrow Fe(OH)_2(s) + OH^-(aq)$	−0.56
$2H_2O(l) + 2e^- \rightarrow H_2(g) + 2OH^-(aq)$	−0.83
$2SO_3^{2-}(aq) + 2H_2O(l) + 2e^- \rightarrow S_2O_4^{2-}(aq) + 4OH^-(aq)$	−1.12
$ZnS(s) + 2e^- \rightarrow Zn(s) + S^{2-}(aq)$	−1.41

The more positive the E^0 value for a half-reaction, the stronger the oxidizing agent (electron acceptor) in the half-reaction. The more negative the E^0 value for a half-reaction, the stronger the reducing agent (electron donor) in the half-reaction. Electrons appear on the left-hand side of the half-reactions in Table 21-1 because the values given refer to reduction half-reactions. The arrangement of half-reactions in order of the standard reduction voltages is sometimes called the *electromotive force* (*emf*) series.

Example 21-8: Use the data in Table 21-1 to compute E^0 for the reaction

$$4Fe^{2+}(aq) + O_2(g) + 4H^+(aq) \rightarrow 4Fe^{3+}(aq) + 2H_2O(l)$$

Solution: We can see from the oxidation states of the elements involved in the reaction that Fe^{2+} is oxidized to Fe^{3+},

$$Fe^{2+}(aq) \rightarrow Fe^{3+}(aq) + e^- \text{ (oxidation)}$$

and O_2 is reduced to water,

$$O_2(g) + 4H^+(aq) + 4e^- \rightarrow 2H_2O(l) \text{ (reduction)}$$

The E^0_{cell} value for the complete cell reaction is

$$E^0_{cell} = E^0_{right} \text{ (reduction)} - E^0_{left} \text{ (oxidation)}$$
$$= E^0_{O_2/H_2O} - E^0_{Fe^{3+}/Fe^{2+}}$$

From Table 21-1 we obtain

$$E^0_{O_2/H_2O} = +1.23 \text{ V} \quad \text{and} \quad E^0_{Fe^{3+}/Fe^{2+}} = +0.77 \text{ V}$$

Thus

$$E^0 = 1.23 \text{ V} - 0.77 \text{ V} = 0.46 \text{ V}$$

Notice that we do *not* multiply $E^0_{Fe^{3+}/Fe^{2+}}$ by 4, even though the overall reaction has 4 $Fe^{2+}(aq)$ and 4 $Fe^{3+}(aq)$.

The positive value of E^0 tells us that the oxidation of Fe^{2+} in an acidic solution in contact with air is a spontaneous process under standard conditions.

An amalgam is a solution of a metal in mercury.

Dental amalgam for silver tooth fillings is prepared by adding liquid mercury to a silver alloy that is 70 percent silver, 26 percent tin, 2 percent copper, and 2 percent zinc. When freshly prepared, the amalgam is soft and easily worked, but it hardens with time owing to the formation of metallic crystals. You may be familiar with the acute pain that is felt when a piece of aluminum foil from a gum or candy wrapper comes into contact with an amalgam dental filling.

The pain is a result of the excitation of the tooth nerves by the transitory electric current that flows from the electrochemical cell that is established in the mouth. The cell consists of an aluminum electrode and a silver amalgam electrode, with saliva acting as the electrolyte. The voltage developed is about 3 V.

21-6. ELECTROCHEMICAL CELLS ARE USED TO DETERMINE CONCENTRATIONS OF IONS

We have seen how the Nernst equation is used to calculate the voltage of an electrochemical cell when the concentrations of the species involved in the cell reaction are known. Conversely, a measured cell voltage can be used to determine the concentration of a species in solution. Consider the cell reaction

$$Zn(s) + Cu^{2+}(aq) \rightarrow Cu(s) + Zn^{2+}(aq) \qquad (21\text{-}13)$$

Application of the Nernst equation in the form of Equation (21-9) to the cell reaction at 25.0°C yields

$$E = E^0 - \left(\frac{0.0592\ \mathrm{V}}{2}\right)\log\frac{[Zn^{2+}]}{[Cu^{2+}]} \qquad (21\text{-}14)$$

The value of E^0 for Reaction (21-13) is 1.10 V; thus we have from Equation (21-14)

$$E = 1.10\ \mathrm{V} - (0.0296\ \mathrm{V})\log\frac{[Zn^{2+}]}{[Cu^{2+}]} \qquad (21\text{-}15)$$

Electrochemical cells are extensively used in analytical chemistry to measure the concentrations of ions in solution.

We see from Equation (21-15) that if we measure E at a known value of, say, $[Cu^{2+}]$, then we can use the equation to compute the value of $[Zn^{2+}]$. For example, suppose that when $[Cu^{2+}] = 0.10$ M, we find that $E = 1.20$ V. Substitution of these values for E and $[Cu^{2+}]$ into Equation (21-15) yields

$$1.20\ \mathrm{V} = 1.10\ \mathrm{V} - (0.0296\ \mathrm{V})\log\frac{[Zn^{2+}]}{0.10\ \mathrm{M}}$$

from which we compute

$$\log\frac{[Zn^{2+}]}{0.10\text{ M}} = \frac{1.10\text{ V} - 1.20\text{ V}}{0.0296\text{ V}} = -3.38$$

or $$[Zn^{2+}] = 0.10\text{ M} \times 10^{-3.38} = 4.17 \times 10^{-5}\text{ M}$$

The foregoing calculations show that electrochemical cells can be used to determine the concentrations of ions in solution.

Example 21-9: Given that the measured voltage of the cell reaction

$$H_2(g) + 2AgCl(s) \rightarrow 2H^+(aq) + 2Cl^-(aq, 1.0\text{ M}) + 2Ag(s)$$

is $+0.34$ V at 25°C when the pressure of $H_2(g)$ is 1.00 atm, compute the value of $[H^+]$.

Solution: Application of the Nernst equation to the cell reaction, for which $n = 2$, yields

$$E = E^0 - \left(\frac{0.0592\text{ V}}{2}\right)\log\frac{[H^+]^2[Cl^-]^2}{P_{H_2}}$$

The value of E^0 for the cell reaction is computed from the E^0 values for the half-reactions (Table 21-1) using Equation (21-10):

$$E^0_{cell} = E^0_{right} - E^0_{left} = E^0_{AgCl/Ag} - E^0_{H^+/H_2}$$
$$= +0.22\text{ V} - 0 = +0.22\text{ V}$$

Substitution of the known values of E, E^0, $[Cl^-]$, and P_{H_2} into the Nernst equation yields

$$+0.34\text{ V} = +0.22\text{ V} - (0.0296\text{ V})\log\frac{[H^+]^2(1.0)^2}{1.00}$$

Problems 21-37 and 21-38 are similar to Example 21-9.

Solving for $\log[H^+]$ gives

$$\log[H^+]^2 = \frac{+0.22\text{ V} - 0.34\text{ V}}{0.0296\text{ V}} = -4.05$$

and $$[H^+]^2 = 10^{-4.05} = 8.9 \times 10^{-5}$$

$$[H^+] = 9.4 \times 10^{-3}\text{ M}$$

21-7. THE pH OF A SOLUTION CAN BE DETERMINED BY ELECTROCHEMICAL CELL MEASUREMENTS

The voltage of the hydrogen electrode

$$\tfrac{1}{2}H_2(g) \rightleftharpoons H^+(aq) + e^-$$

depends on the magnitude of $[H^+]$ in the solution in which the electrode is placed. Application of Equation (21-9) to the hydrogen electrode with $E^0 = 0$ yields

$$E = -(0.0592\ \text{V}) \log \frac{[H^+]}{P_{H_2}^{1/2}}$$

Taking $P_{H_2} = 1.00$ atm and converting to millivolts (1000 mV = 1 V), we obtain

$$E = -(59.2\ \text{mV}) \log[H^+]$$

Since the pH of a solution is given by

$$\text{pH} = -\log[H^+]$$

we can write

$$E = (59.2\ \text{mV})\ \text{pH at } 25\,^\circ\text{C}$$

This equation shows that the voltage of the hydrogen electrode is directly proportional to pH. If the pH of a solution increases by 1.00 pH unit, then the voltage of a hydrogen electrode immersed in the solution increases by 59.2 mV. Thus, a hydrogen electrode can be used to measure pH.

In practice, electrochemical pH measurements are made with a *hydrogen glass electrode* rather than a hydrogen gas electrode because of the convenience of the glass electrode. A hydrogen glass electrode consists of a silver/silver chloride electrode that dips into a NaCl(*aq*) plus HCl(*aq*) solution inside of a special thin-walled glass bulb (Figure 21-5). The voltage drop across the glass bulb depends on the value of $[H^+(aq)]$ in the solution into which the electrode is

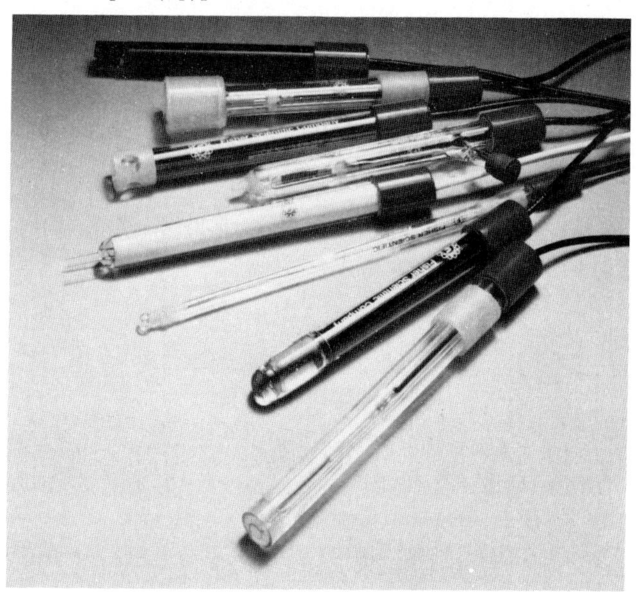

Figure 21-5 Examples of hydrogen glass electrodes. Different types of electrodes are used for various experimental conditions.

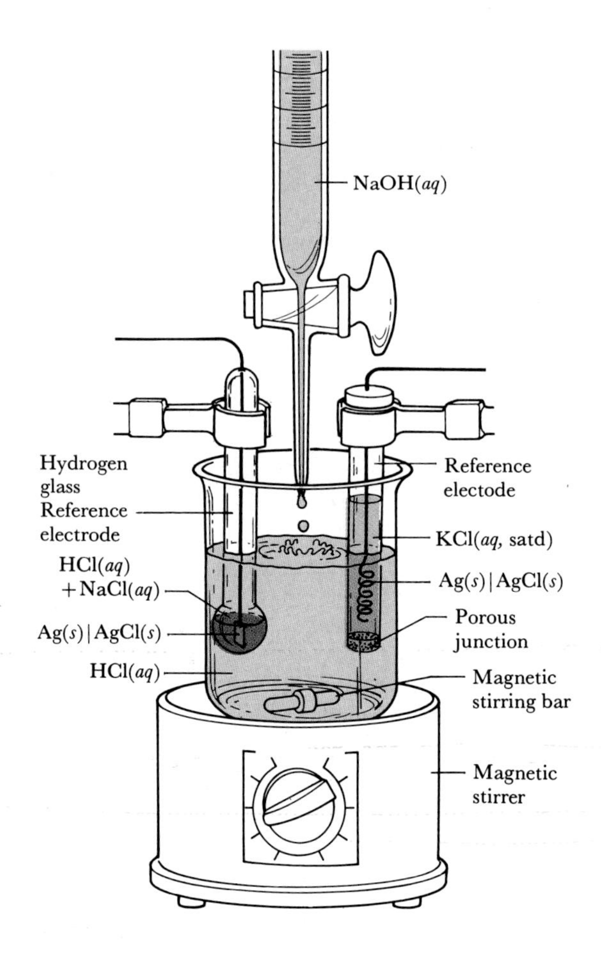

Figure 21-6 A hydrogen glass electrode can be used to monitor the titration of HCl(*aq*) with NaOH(*aq*) by measuring pH. A titration that is monitored electrochemically is called a *potentiometric titration.*

placed. The details of the operation of the hydrogen glass electrode are complex, but its utilization is simple. A hydrogen glass electrode responds to changes in solution pH in the same manner as a hydrogen gas electrode.

A cell that uses a hydrogen glass electrode to monitor an acid-base titration is shown in Figure 21-6. This electrode has made electrochemical pH measurements a routine laboratory procedure, which in turn has made possible innumerable scientific discoveries in chemistry, biology, and medicine.

21-8. A BATTERY IS AN ELECTROCHEMICAL CELL OR GROUP OF CELLS DESIGNED FOR USE AS A POWER SOURCE

There are many types of batteries. Some of the better-known examples are the lead storage battery, the dry cell, the alkaline cell, the mercury battery, the nickel-cadmium battery, and the lithium battery. Batteries are especially useful as power sources when mobility is a prime consideration. They are adaptable to a wide range of power requirements, including the production of very high currents for short periods or stable voltages under low current drain for long periods. Batteries are classed as *primary* if they are not rechargeable and *secondary* if they are rechargeable. Examples of primary batteries are the dry cell and the mercury battery, and examples of secondary batteries are the lead storage battery and the nickel-cadmium battery.

The 12-V *lead storage battery* consists of six of the following cells arranged in series:

$$^{\ominus}Pb(s)|PbSO_4(s)|H_2SO_4(aq)|PbO_2(s),PbSO_4(s)|Pb(s)^{\oplus}$$

where the charges show the electrode polarities. The battery electrolyte is a concentrated $H_2SO_4(aq)$ solution (about 10 M). The electrode reactions on discharge are

$$Pb(s) + SO_4^{2-}(aq) \rightarrow PbSO_4(s) + 2e^- \quad \text{oxidation}$$

$$2e^- + PbO_2(s) + 4H^+(aq) + SO_4^{2-}(aq) \rightarrow PbSO_4(s) + 2H_2O(l) \quad \text{reduction}$$

The electrons used in the reduction half-reaction are obtained from the inert $^{\oplus}Pb(s)$ electrode, which is in contact with the $PbO_2(s)$, $PbSO_4(s)$ mixture. The overall cell reaction is obtained by adding together the two half-reactions:

$$Pb(s) + PbO_2(s) + 2H_2SO_4(aq) \underset{\text{recharge}}{\overset{\text{discharge}}{\rightleftharpoons}} 2PbSO_4(s) + 2H_2O(l)$$

A 12-V lead storage automobile battery can provide over 300 amperes for short periods.

A single cell of this type develops about 2 V when fully charged, and the six cells in series in an ordinary 12-V automobile battery develop about 12 V. The cell voltage decreases on discharge because sulfuric acid is consumed and water is produced, and both of these reactions dilute the $H_2SO_4(aq)$ battery electrolyte.

On rapid recharge of a lead storage battery, appreciable amounts of $H_2(g)$ and $O_2(g)$ may be formed by the reactions

$$2H^+(aq) + 2e^- \rightarrow H_2(g) \quad \text{at the } Pb(s)/PbSO_4(s) \text{ electrode}$$

$$H_2O(l) \rightarrow \tfrac{1}{2}O_2(g) + 2H^+(aq) + 2e^- \quad \text{at the } PbO_2(s), PbSO_4(s)/Pb(s) \text{ electrode}$$

There is danger of explosion because of the possible reaction

$$2H_2(g) + O_2(air) \xrightarrow{\text{spark}} 2H_2O(g)$$

Figure 21-7 Various commercially available batteries.

Thus sparks and flames should not be brought near a lead storage battery.

The decomposition of water on battery recharge leads to a loss of water from the battery electrolyte, which in turn requires that water be added to the battery. For maximum battery life only distilled or deionized water should be added. Ordinary tap water containing traces of dissolved iron or copper salts can shorten battery life by promoting corrosion of the lead electrode. Completely sealed lead storage batteries have lead-calcium alloy electrodes, on which the evolution of H_2 and O_2 on recharge is much slower than on pure lead. Because not much water is electrolyzed, there is usually no need to add water to the battery.

Another rechargeable battery is the *nickel-cadmium (nicad) battery* (Figure 21-7), which is used, for example, in hand calculators, rechargeable flashlight batteries, and cordless electric shavers, knives, and toothbrushes. The cell diagram for the nickel-cadmium battery is

$$^{\ominus}\text{steel}\,|\,\text{Cd}(s)\,|\,\text{Cd(OH)}_2(s)\,|\,\text{LiOH}(aq)\,|\,\text{NiOOH}(s),\text{Ni(OH)}_2(s)\,|\,\text{steel}^{\oplus}$$

The cell reaction for the nickel-cadmium battery is

$$2\text{NiOOH}(s) + \text{Cd}(s) + 2\text{H}_2\text{O}(l) \underset{\text{recharge}}{\overset{\text{discharge}}{\rightleftharpoons}} 2\text{Ni(OH)}_2(s) + \text{Cd(OH)}_2(s)$$

Note that the reaction consumes only water from the cell electrolyte, and thus the voltage on discharge is fairly constant. A completely sealed nicad battery is much more stable than a lead storage battery and can be left inactive for long periods without adverse effects. The steel electrodes in a nicad battery are inert; their function is to provide a current path for electrons that leave and re-enter the battery.

Mercury batteries (Figure 21-7) are used in heart pacemakers, hearing aids, some types of electric watches, and other electronic instruments that require a constant voltage during discharge. The cell diagram for the mercury battery is

$$^{\ominus}\text{steel}\,|\,\text{Zn}(s)\,|\,\text{ZnO}(s)\,|\,\text{KOH}(aq,\ 40\%)\,|\,\text{HgO}(s)\,|\,\text{Hg}(l)\,|\,\text{steel}^{\oplus}$$

The main advantage of the mercury battery is a constant voltage (1.35 V) during discharge, which is a consequence of the fact that the cell reaction does not change the cell electrolyte composition. Mercury batteries are not safe to recharge because the container might rupture as a result of gas evolution in the sealed battery.

Example 21-10: The cell reaction for the mercury battery is

$$\text{Zn}(s) + \text{HgO}(s) \rightarrow \text{ZnO}(s) + \text{Hg}(l)$$

Use the Nernst equation to explain why the voltage of the mercury battery remains constant during discharge.

Solution: The Nernst equation at 25°C is

$$E = E^0 - \left(\frac{0.0592\ \text{V}}{n}\right)\log Q$$

All the reactants and products in the cell reaction have fixed concentrations and therefore do not appear in Q. Thus $Q = 1$ for the cell reaction and hence

$$E = E^0 - \left(\frac{0.0592\ \text{V}}{2}\right)\log 1 = E^0$$

Because E^0 is a constant, E is a constant. The voltage of a mercury battery remains at 1.35 V until all the $HgO(s)$ is used up (Zn is in excess), at which point the battery voltage drops to zero because the cell reaction can no longer occur.

The *dry cell* and *alkaline manganese cell* (Figure 21-7), both of which are widely used in flashlights and battery-powered toys, are closely related. Both contain a zinc metal negative electrode and an inert positive electrode on which $MnO_2(s)$ is reduced to $Mn_2O_3(s)$. The positive electrode is a carbon rod in the dry cell and a steel rod in the alkaline manganese cell. The electrolyte in the dry cell is a paste of NH_4Cl; the electrolyte in the alkaline manganese cell is a concentrated solution of $NaOH(aq)$.

The cell diagram of the alkaline manganese cell is

$$^{\ominus}\text{Zn}(s)\,|\,\text{ZnO}(s)\,|\,\text{NaOH}(aq)\,|\,\text{MnO}_2(s),\text{Mn}_2\text{O}_3(s)\,|\,\text{steel}^{\oplus}$$

and the cell reaction is

$$\text{Zn}(s) + 2\text{MnO}_2(s) \rightarrow \text{ZnO}(s) + \text{Mn}_2\text{O}_3(s)$$

The alkaline manganese battery has twice the capacity, a steadier voltage under heavy current drain, and a higher available current than its major marketplace competitor, the dry cell. Its main disadvantage is that it costs roughly three times as much as the dry cell because of the more elaborate internal construction necessary to prevent leakage of the concentrated NaOH solution. No attempt should be made to recharge dry cells or alkaline manganese batteries because of the high explosion hazard arising from gas formation inside the sealed battery during recharge.

21-9. ELECTROLYSIS IS A CHEMICAL REACTION THAT OCCURS AS A RESULT OF THE PASSAGE OF AN ELECTRIC CURRENT

Faraday was the first to study the effect of passing an electric current through ionic solutions. His primary observation was that the passage of an electric current through a solution causes chemical reactions that could not occur in the absence of the current. For example, the passage of an electric current through molten NaCl results in the decomposition of NaCl(l) into Na(l) and $Cl_2(g)$ according to the equation

$$2NaCl(l) \xrightarrow{\text{electrolysis}} 2Na(l) + Cl_2(g)$$

The half-reactions for this process are

$$Na^+(l) + e^- \rightarrow Na(l) \quad \text{reduction}$$

$$Cl^-(l) \rightarrow \tfrac{1}{2}Cl_2(g) + e^- \quad \text{oxidation}$$

The electrodes for this decomposition are platinum strips, which do not react with either Na(l) or $Cl_2(g)$. The current through the wires and Pt electrodes of the electrolysis apparatus is carried by electrons, but electrons do not pass through the NaCl(l) phase. The current through the NaCl(l) is carried by Na^+ and Cl^- moving in opposite directions toward the two platinum electrodes. The electrode at which the reduction half-reaction takes place is called the *cathode*, and the electrode at which the oxidation half-reaction takes place is called the *anode*. In electrolysis, the cations in the solution move toward the cathode and the anions move toward the anode. This is the origin of the names cathode and anode.

The definitions of anode and cathode are conveniently remembered with the aid of the mnemonic

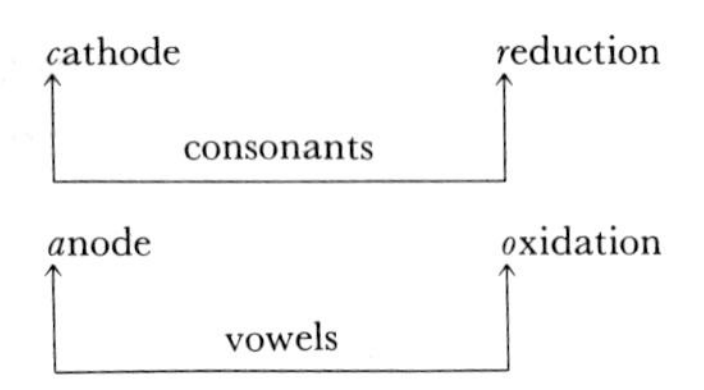

As another example of electrolysis, consider the electrolysis of $H_2O(l)$. Water containing the salt Na_2SO_4 is decomposed into hydrogen and oxygen as a result of the passage of an electric current through the solution (Figure 21-8):

$$2H_2O(l) \xrightarrow{\text{electrolysis}} 2H_2(g) + O_2(g)$$

The Na_2SO_4 does not appear in the net electrolysis reaction because Na^+ and SO_4^{2-} are not changed chemically by the current. However, the Na^+ and SO_4^{2-} ions are the current carriers in the water phase.

The cathode half-reaction in the electrolysis of water is

$$4H^+(aq) + 4e^- \xrightarrow{\text{cathode}} 2H_2(g) \qquad \text{reduction, at cathode}$$

The anode half-reaction in the electrolysis of water is

$$2H_2O(l) \xrightarrow{\text{anode}} O_2(g) + 4H^+(aq) + 4e^- \qquad \text{oxidation, at anode}$$

The net reaction is the sum of the anode and cathode half-reactions:

$$2H_2O(l) \xrightarrow{\text{electrolysis}} 2H_2(g) + O_2(g)$$

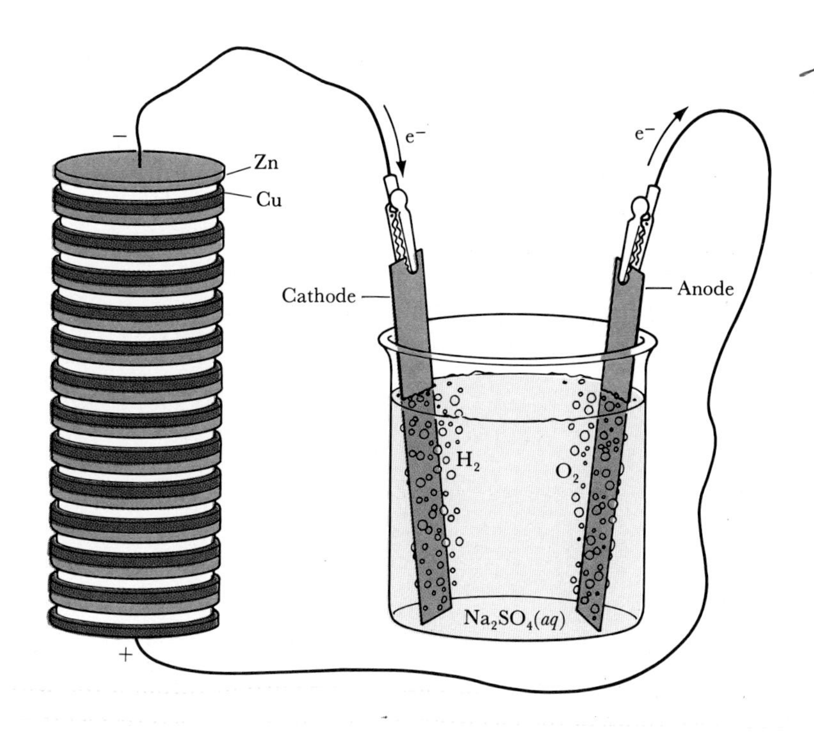

Figure 21-8 Electrolysis of water containing $Na_2SO_4(aq)$. The gas evolved from the platinum strip connected to the top zinc plate of the set of cells (a volta pile) is hydrogen. The gas evolved from the platinum strip connected to the bottom copper plate of the volta pile is oxygen. The current through the solution is carried by $Na^+(aq)$ and $SO_4^{2-}(aq)$ ions.

The electrons that arrive at the platinum cathode are consumed at the metal-solution interface by reaction with $H^+(aq)$ to produce H_2. An equal number of electrons is introduced back into the external circuit at the anode; these electrons are produced in the oxidation of H_2O to O_2, which occurs at the anode metal-solution interface.

Water electrolysis requires energy to break the O—H bonds:

$$2H_2O(l) \rightarrow 2H_2(g) + O_2(g) \qquad \Delta H^\circ_{rxn} = +571.6 \text{ kJ}$$

The water-splitting reaction is driven uphill by the voltage applied across the electrodes. The electrochemical decomposition of water cannot occur unless the voltage is large enough to overcome the strength of the O—H bonds. The minimum voltage necessary to electrochemically decompose a sample is called the *decomposition voltage.* The minimum decomposition voltage for water at 25°C is about 1.2 V. In practice, more than 1.2 V is usually required in order to make the reaction proceed at an acceptable rate.

Example 21-11: (a) Calculate the value of E^0 for the reaction

$$2H_2(g, 1 \text{ atm}) + O_2(g, 1 \text{ atm}) \rightarrow 2H_2O(l)$$

(b) Given the results in part (a), estimate the decomposition voltage for the reaction

$$2H_2O(l) \rightarrow 2H_2(g) + O_2(g)$$

Solution: (a) The half-reactions are (E^0 values from Table 21-1)

$$4H^+(aq) + 4e^- \rightarrow 2H_2(g) \qquad E^0 = 0$$
$$O_2(g) + 4H^+(aq) + 4e^- \rightarrow 2H_2O(l) \qquad E^0 = +1.23\ V$$

Thus the value of E^0 for the reaction is

$$E^0_{cell} = E^0_{right} - E^0_{left} = 1.23\ V - 0 = 1.23\ V$$

(b) The reaction between $H_2(g)$ and $O_2(g)$ to produce water develops about 1.23 V. Thus the decomposition voltage is 1.23 V, which is the minimum voltage that must be applied to cause $H_2O(l)$ to decompose into $H_2(g)$ and $O_2(g)$.

21-10. METALS ARE OBTAINED FROM SALTS BY ELECTROCHEMICAL PROCESSES

Faraday discovered that the metal ions of many salts are deposited as the metal when an electric current is passed through aqueous solutions of the salts. For example, silver metal is deposited from a solution of $AgNO_3(aq)$ and copper metal is deposited from $Cu(NO_3)_2(aq)$. The relevant half-reactions are

$$Ag^+(aq) + e^- \rightarrow Ag(s)$$
$$Cu^{2+}(aq) + 2e^- \rightarrow Cu(s)$$

Note that 1 mol of electrons deposits 1 mol of silver from $Ag^+(aq)$ but 2 mol of electrons is needed to deposit 1 mol of copper from $Cu^{2+}(aq)$. The number of electrons can be controlled by controlling the electric current through the solution. Electric currents are measured in *amperes* (A), and one ampere is the flow of one coulomb of charge per second. In an equation, we have

An ampere is the SI unit of electric current.

$$\text{current} = \text{charge/time}$$

A = C/t

The total charge that flows is

$$\text{charge} = \text{current} \times \text{time} \qquad (21\text{-}16)$$

Using SI units, we have

$$\text{coulombs} = \text{amperes} \times \text{seconds}$$

Representing charge by Z, current by I, and time by t, we have

$$Z = It$$

Example 21-12: If a current of 1.50 A flows for 5.00 min, then what quantity of charge has flowed?

Solution: The charge is equal to the current multiplied by time, and so

$$\text{charge} = (1.5\ \text{A})(5.00\ \text{min})$$
$$= (1.50\ \text{C}\cdot\text{s}^{-1})(5.00\ \text{min})(60\ \text{s}\cdot\text{min}^{-1}) = 450\ \text{C}$$

Consider the passage of an electric current through a solution of $AgNO_3(aq)$. The charge that flows through the solution is directly related to the number of electrons that participate in the electrochemical reaction

$$Ag^+(aq) + e^- \rightarrow Ag(s) \qquad (21\text{-}17)$$

Suppose that a current of 0.850 A flows through the solution for 20.0 min. The total charge is

$$\text{charge} = (0.850\ \text{A})(20.0\ \text{min})$$
$$= (0.850\ \text{C}\cdot\text{s}^{-1})(20.0\ \text{min} \times 60\ \text{s}\cdot\text{min}^{-1}) = 1020\ \text{C}$$

The number of moles of electrons that corresponds to 1020 C can be determined by using the fact that the charge on 1 mol of electrons is Faraday's constant, 96,500 C (Section 21.3). The number of moles of electrons that corresponds to 1020 C is

$$\text{mol electrons} = \frac{1020\ \text{C}}{96{,}500\ \text{C}\cdot\text{mol}^{-1}} = 1.06 \times 10^{-2}\ \text{mol}$$

Because one electron deposits one atom of silver (Equation 21-17), we have

$$\text{mol Ag deposited} = \text{mol electrons}$$
$$= 1.06 \times 10^{-2}\ \text{mol}$$

The number of grams of silver deposited by the passage of 0.850 C through an $AgNO_3(aq)$ solution for 20.0 min is

$$\text{mass of Ag deposited} = (1.06 \times 10^{-2}\ \text{mol})\left(\frac{107.9\ \text{g}}{\text{mol}}\right) = 1.14\ \text{g}$$

Example 21-13: Calculate the number of grams of copper deposited from a $Cu(NO_3)_2(aq)$ solution by a current of 0.500 A that flows for 10.2 min.

Solution: The total charge is given by Equation (21-16):

$$\text{charge} = (0.500\ \text{A})(10.2\ \text{min})$$
$$= (0.500\ \text{C}\cdot\text{s}^{-1})(10.2\ \text{min} \times 60\ \text{s}\cdot\text{min}^{-1}) = 306\ \text{C}$$

The number of moles of electrons that corresponds to 306 C is

$$\text{mol electrons} = \frac{306\text{ C}}{96{,}500\text{ C}\cdot\text{mol}^{-1}} = 3.17 \times 10^{-3}\text{ mol}$$

The relevant equation,

$$Cu^{2+}(aq) + 2e^- \rightarrow Cu(s)$$

shows that it requires 2 mol of electrons to deposit 1 mol of copper. Therefore, the number of moles of copper deposited is

$$\left(\begin{matrix}\text{mol Cu}\\ \text{deposited}\end{matrix}\right) = \left(\frac{1.00\text{ mol Cu}}{2.00\text{ mol electrons}}\right)(3.17 \times 10^{-3}\text{ mol electrons})$$
$$= 1.59 \times 10^{-3}\text{ mol Cu}$$

The number of grams of copper deposited is

$$\left(\begin{matrix}\text{g Cu}\\ \text{deposited}\end{matrix}\right) = (1.59 \times 10^{-3}\text{ mol Cu})\left(\frac{63.55\text{ g Cu}}{\text{mol Cu}}\right)$$
$$= 0.101\text{ g Cu}$$

Example 21-14: Magnesium is produced commercially by the electrolysis of molten $MgCl_2$ (melting point 650°C) in cells that have a capacity of about 8 metric tons of molten $MgCl_2$. The currents used are between 80,000 and 100,000 A. Calculate how many metric tons of magnesium metal can be produced per day by the electrolysis of molten $MgCl_2$ using a 100,000-A current.

Solution: The equation for the production of magnesium by electrolysis of molten $MgCl_2$ is

$$MgCl_2(l) \xrightarrow{710°C} Mg(l) + Cl_2(g)$$

or

$$Mg^{2+}(l) + 2e^- \rightarrow Mg(l)$$

The charge passed through the molten $MgCl_2$ in 24 h is

$$\text{charge} = (100{,}000\text{ A})(24\text{ h})$$
$$= (1.0 \times 10^5\text{ C}\cdot\text{s}^{-1})(24\text{ h} \times 60\text{ min}\cdot\text{h}^{-1} \times 60\text{ s}\cdot\text{min}^{-1})$$
$$= 8.6 \times 10^9\text{ C}$$

The number of moles of electrons in 8.6×10^9 C is

$$\text{mol electrons} = \frac{8.6 \times 10^9\text{ C}}{96{,}500\text{ C}\cdot\text{mol}^{-1}} = 9.0 \times 10^4\text{ mol}$$

The number of moles of magnesium produced is

$$\text{mol Mg produced} = \left(\frac{1.0 \text{ mol Mg}}{2.0 \text{ mol electrons}}\right)(9.0 \times 10^4 \text{ mol electrons})$$
$$= 4.5 \times 10^4 \text{ mol}$$

The number of grams of magnesium produced is

$$\text{g Mg produced} = (4.5 \times 10^4 \text{ mol Mg})\left(\frac{24.3 \text{ g Mg}}{\text{mol Mg}}\right)$$
$$= 1.1 \times 10^6 \text{ g}$$

The number of metric tons of magnesium produced daily is

$$\left(\begin{matrix}\text{metric tons of Mg}\\ \text{produced}\end{matrix}\right) = (1.1 \times 10^6 \text{ g})\left(\frac{1 \text{ kg}}{10^3 \text{ g}}\right)\left(\frac{1 \text{ metric ton}}{10^3 \text{ kg}}\right)$$
$$= 1.1 \text{ metric tons}$$

Problems 21-49 through 21-56 deal with Faraday's laws.

The principles on which the calculations in this section are based are known as *Faraday's laws*. Faraday's laws are summarized as

$$\text{mass deposited} = \frac{\text{current} \times \text{time} \times \text{atomic mass}}{\text{Faraday constant} \times \text{ionic charge}} \tag{21-18}$$

The calculations involved in Examples 21-13 and 21-14 are simple applications of this equation.

21-11. MANY CHEMICALS ARE PREPARED ON AN INDUSTRIAL SCALE BY ELECTROLYSIS

Chlorine is prepared by the *chlor-alkali* process, which involves the electrolysis of an NaCl(aq) brine solution. The chlor-alkali process cell (Figure 21-9) consists of a horizontal liquid mercury cathode

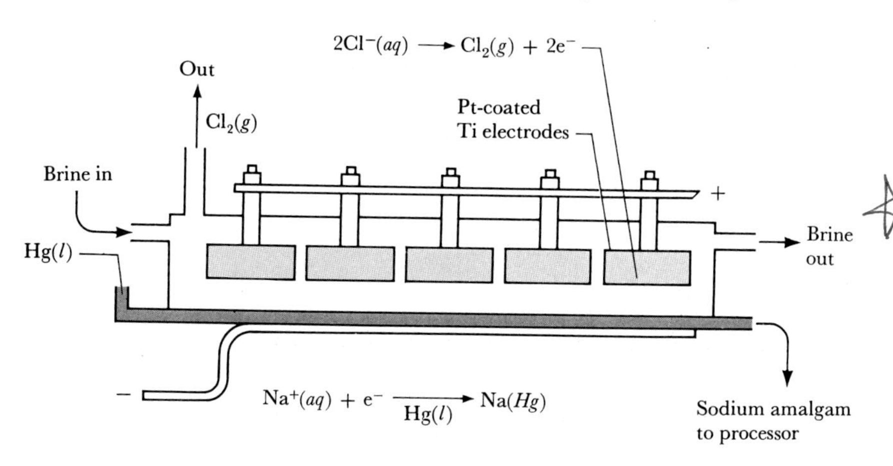

Figure 21-9 The chlor-alkali process produces chlorine gas and sodium-mercury amalgam from the electrolysis of a brine solution. The cathode is a liquid mercury pool at the bottom of the cell; the anode is titanium coated with either ruthenium oxide or platinum.

positioned below anodes made of titanium metal coated with either ruthenium oxide or platinum. The mercury pool sits in a steel-bottomed container that serves as the electrical contact to the mercury cathode. The NaCl(*aq*) solution, which is prepared from rock salt, is continuously passed between the electrodes. The electrode reactions are

$$\text{cathode:} \qquad Na^{+}(aq) + e^{-} \xrightarrow[\text{Hg}(l)]{} \text{Na [in Hg}(l)]$$

where Na [in Hg(*l*)] denotes a sodium *amalgam,* that is, sodium metal dissolved in mercury, and

$$\text{anode:} \qquad Cl^{-}(aq) \rightarrow \tfrac{1}{2}Cl_2(g) + e^{-}$$

The overall reaction is

$$NaCl(aq) \xrightarrow[\text{Hg}(l)]{} \text{Na [in Hg}(l)] + \tfrac{1}{2}Cl_2(g)$$

The high-purity chlorine gas is collected, and the sodium amalgam is cycled out of the cell and reacted with pure water to produce high-purity sodium hydroxide and hydrogen gas:

$$\text{Na [in Hg}(l)] + H_2O(l) \rightarrow Na^{+}(aq) + OH^{-}(aq) + \tfrac{1}{2}H_2(g)$$

The mercury is then returned to the cell. The decomposition voltage of NaCl(*aq*) is about 2.3 V, but the cells are run at about 3.6 V to increase the rate of the reaction. The products of the chlor-alkali process are high-purity chlorine gas, high-purity sodium hydroxide, and hydrogen gas. Over 21 billion pounds of chlorine are produced annually in the United States, primarily by electrochemical processes.

All aluminum metal is produced by the *Hall process,* which was patented by Charles Hall in 1889, when he was 26 years old. Hall conceived his process for the electrochemical production of aluminum while a student at Oberlin College. A schematic of the electrolysis apparatus used in the Hall process, which is essentially the same as that used in the original process, is shown in Figure 21-10. The electrolysis is carried out at about 980°C, where aluminum is a liquid and can be siphoned off from the cathode compartment. Electrical contact to the molten aluminum cathode is made through a steel shell that constitutes the bottom of the electrode compartment. The consumable anodes are composed of a petroleum coke that is obtained by heating to dryness the heavy petroleum fraction remaining from petroleum refining.

The key to the Hall process is the molten salt electrolyte consisting of powdered aluminum oxide dissolved in the mineral cryolite,

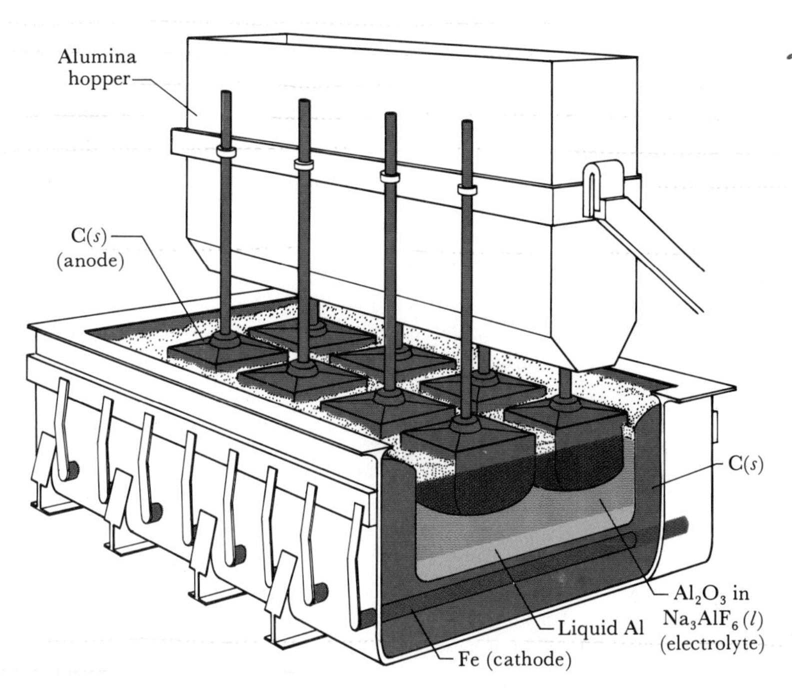

Figure 21-10 Hall process for the electrochemical production of aluminum. The cathodes are iron rods, and the anodes are carbon attached to metal rods. The electrolyte is alumina, Al_2O_3, dissolved in cryolite, Na_3AlF_6. The electrolysis is carried out at 980°C, where the aluminum is a liquid that can be drained from the cell.

Na_3AlF_6. The overall electrochemical reaction is

$$2Al_2O_3(soln) + 3C(s) \rightarrow 4Al(l) + 3CO_2(g)$$

Aluminum oxide cannot be electrolyzed directly because its melting point is too high (2050°C). Over 5 million tons of 99.7 percent pure aluminum metal are produced each year in the United States by the Hall process.

The electrolytic production of aluminum consumes from 4 to 7 percent of the total U.S. production of electricity.

SUMMARY

An electrochemical cell provides the means for obtaining electricity from an electron-transfer reaction. The cell consists of a pair of metal electrodes in contact with an electrolyte solution. The dependence of the cell voltage on the concentrations of the reactants and products of the cell reaction is described quantitatively by the Nernst equation.

The standard cell voltage, E^0, is the voltage of the cell when the reaction quotient $Q = 1$. The equilibrium constant of the cell reaction can be computed from the value of E^0. Electrode reactions can be arranged in a series of decreasing electrode reduction voltages. The assignment of standard reduction voltages to electrode reactions is achieved by setting $E^0 = 0$ for the hydrogen gas electrode reaction. The E^0 values for electrode reactions given in Table 21-1 can be used to compute E^0 values for cells. Electrochemical cells can be used to determine the concentration of an ion in solution.

Batteries are electrochemical cells designed to function as power sources. Their principal advantages are portability and the ability to supply large currents. Electrolysis is the process by which a chemical reaction is driven uphill on the energy scale by the application of a voltage across electrodes placed in a solution. The extent of the electrochemical reaction that occurs is proportional to the current that flows through the solution and is given quantitatively by Faraday's laws of electrolysis. The minimum applied voltage necessary to cause electrochemical decomposition of a solution is called the decomposition voltage and can be estimated using the Nernst equation.

TERMS YOU SHOULD KNOW

electrochemical cell 806
electrode 806
salt bridge 806
discharge 807
external circuit 808
cell diagram 809
gas electrode 810
cell voltage 813
Nernst equation 814
Faraday's constant, F 814
standard cell voltage 816
standard reduction voltage 818
emf series 822
hydrogen glass electrode 825
battery 826
primary battery 827
secondary battery 827
lead storage battery 827
nickel-cadmium battery 828
mercury battery 828
dry cell 829
alkaline manganese cell 829
electrolysis 830
cathode 830
anode 830
decomposition voltage 831
ampere 832
Faraday's laws 835
chlor-alkali process 835
amalgam 836
Hall process 836

EQUATIONS YOU SHOULD KNOW HOW TO USE

1 joule = 1 volt × 1 coulomb (J = V · C) (21-4)

$$E^0 = \left(\frac{0.0592\text{ V}}{n}\right)\log K \text{ at } 25^\circ\text{C} \qquad (21\text{-}8)$$

$$E = E^0 - \left(\frac{0.0592\text{ V}}{n}\right)\log Q \quad \text{Nernst equation at } 25^\circ\text{C} \qquad (21\text{-}9)$$

$$E^0_{\text{cell}} = E^0_{\text{right}} - E^0_{\text{left}} \qquad (21\text{-}10)$$

1 ampere = 1 coulomb per second

$$\text{mass deposited} = \frac{\text{current} \times \text{time} \times \text{atomic mass}}{\text{Faraday constant} \times \text{ionic charge}} \qquad (21\text{-}18)$$

PROBLEMS

CELL SETUPS

21-1. Consider an iron-nickel electrochemical cell. The negative electrode is an iron rod immersed in a $FeCl_2(aq)$ solution, and the positive electrode a nickel rod in a $NiCl_2(aq)$ solution. The two solutions are connected by a salt bridge. Sketch a diagram of the cell, indicating the flow of electrons. Write the reaction that occurs at each electrode, and draw the cell diagram.

21-3. Consider a vanadium-zinc electrochemical cell. The negative electrode is a vanadium rod immersed in a $VI_2(aq)$ solution, and the positive electrode is a zinc rod in a $ZnSO_4(aq)$ solution. The two solutions are connected by a salt bridge. Sketch a diagram of the cell, indicating the flow of electrons. Write the reaction that occurs at each electrode and draw the cell diagram.

21-2. Consider a manganese-chromium electrochemical cell. The negative electrode is a manganese rod immersed in a $MnSO_4(aq)$ solution, and the positive electrode is a chromium rod in a $CrSO_4(aq)$ solution. The two solutions are connected by a salt bridge. Sketch a diagram of the cell, indicating the flow of electrons. Write the reaction that occurs at each electrode, and draw the cell diagram.

21-4. Consider a cobalt-lead electrochemical cell. The negative electrode is a cobalt rod immersed in a $Co(NO_3)_2(aq)$ solution, and the positive electrode is a lead rod in a $Pb(NO_3)_2(aq)$ solution. The two solutions are connected by a salt bridge. Sketch a diagram of the cell, indicating the flow of electrons. Write the reaction that occurs at each electrode and draw the cell diagram.

CELL DIAGRAMS

21-5. Write the electrode reactions and the net cell reaction for the electrochemical cell

$$Ag(s)|AgI(s)|HI(aq)|H_2(g)|Pt(s)$$

21-7. The cell diagram for an electrochemical cell is given as

$$Cd(s)|CdSO_4(aq)||ReCl_3(aq)|Re(s)$$

Write the half-reactions that occur at the electrodes. Make a sketch of the cell.

21-9. Write the cell reaction for the cell

$$Pt(s)|H_2(g)|HCl(aq)|PbSO_4(s)|Pb(s)$$

21-11. Draw a cell diagram that corresponds to the reaction

$$2La(s) + 3CuCl_2(aq) \rightarrow 3Cu(s) + 2LaCl_3(aq)$$

21-6. Write the electrode reactions and the net cell reaction for the electrochemical cell

$$Cu(s)|Cu(ClO_4)_2(aq)||AgClO_4(aq)|Ag(s)$$

21-8. The cell diagram for an electrochemical cell is given as

$$Sn(s)|SnCl_2(aq)||AgNO_3(aq)|Ag(s)$$

Write the half-reactions that occur at the electrodes. Make a sketch of the cell.

21-10. Write the cell reaction for the cell

$$Pb(s)|PbSO_4(s)|K_2SO_4(aq)|Hg_2SO_4(s)|Hg(l)$$

21-12. Draw the cell diagram that corresponds to the following reaction without using any salt bridges:

$$H_2(g) + 2AgCl(s) \rightleftharpoons 2HCl(aq) + 2Ag(s)$$

LE CHÂTELIER'S PRINCIPLE

21-13. Consider an electrochemical cell in which the following reaction occurs:

$$Cl_2(g) + Ni(s) \rightleftharpoons 2Cl^-(aq) + Ni^{2+}(aq)$$

Predict the effect of the following changes on the observed cell voltage:

(a) increase in $[Cl^-]$
(b) increase in $Cl_2(g)$ pressure
(c) increase in amount of $Ni(s)$

21-14. Consider an electrochemical cell in which the following reaction occurs:

$$Pb(s) + 2Ag^+(aq) + SO_4^{2-}(aq) \rightleftharpoons PbSO_4(s) + 2Ag(s)$$

Predict the effect of the following changes on the observed cell voltage:

(a) increase in $[Ag^+]$
(b) increase in amount of $PbSO_4(s)$
(c) increase in $[SO_4^{2-}]$

21-15. Consider an electrochemical cell in which the following reaction occurs:

$$2HCl(aq) + Ca(s) \rightarrow CaCl_2(aq) + H_2(g)$$

Predict the effect of the following changes on the cell voltage:

(a) increase in amount of $Ca(s)$ no effect
(b) increase in pressure of $H_2(g)$ rxn ←; voltage ↓
(c) decrease in $[HCl(aq)]$ rxn ←; voltage ↓
(d) dissolution of $Ca(NO_3)_2$ in the $CaCl_2(aq)$ solution rxn ←; voltage ↓

21-16. Consider the following reaction taking place in an electrochemical cell:

$$2Cr^{2+}(aq) + HClO(aq) + H^+(aq) \rightleftharpoons 2Cr^{3+}(aq) + Cl^-(aq) + H_2O(l)$$

Predict the effect of the following changes on the cell voltage:

(a) increase in $[HClO]$
(b) increase in size of inert electrodes
(c) increase in pH of cell solution
(d) dissolution of KCl in cell solution containing $Cl^-(aq)$

21-17. Given that the reaction in an electrochemical cell is

$$H_2(g) + PbSO_4(s) \rightleftharpoons 2H^+(aq) + SO_4^{2-}(aq) + Pb(s)$$

predict the effect of the following changes (increase, decrease, or no change) on the cell voltage:

(a) increase in pressure of $H_2(g)$ rxn →; voltage ↑
(b) increase in size of $Pb(s)$ electrode no effect
(c) decrease in pH of cell electrolyte rxn ←; voltage ↓
(d) dilution of cell electrolyte with water rxn →; volt. ↑
(e) dissolution of $Na_2SO_4(s)$ in cell electrolyte rxn ←; volt ↓
(f) decrease in amount of $PbSO_4(s)$ no effect as long as some remains
(g) dissolution of small amount of $NaOH(s)$ in cell electrolyte rxn →; voltage ↑

21-18. Given that the reaction in an electrochemical cell is

$$Fe^{2+}(aq) + Ag^+(aq) \rightleftharpoons Ag(s) + Fe^{3+}(aq)$$

predict the effect of the following changes (increase, decrease, or no change) on the cell voltage:

(a) increase in $[Ag^+]$
(b) increase in $[Fe^{3+}]$
(c) twofold decrease in $[Fe^{3+}]$ and $[Fe^{2+}]$
(d) decrease in amount of $Ag(s)$
(e) decrease in $[Fe^{2+}]$
(f) addition of $NaCl(aq)$ to $Ag^+(aq)$ solution

NERNST EQUATION

21-19. Determine the value of n in the Nernst equation for the following reactions:

$2Fe^{3+} + 6e^- \rightarrow 2Fe$; $3Cu \rightarrow 6e^- + 3Cu^{2+}$

(a) $2Fe^{3+}(aq) + 3Cu(s) \rightleftharpoons 3Cu^{2+}(aq) + 2Fe(s)$ $n = 6$
(b) $Cr^{2+}(aq) + Cu^{2+}(aq) \rightleftharpoons Cr^{3+}(aq) + Cu^+(aq)$ $n = 1$

$Cr^{2+} \rightarrow Cr^{3+} + e^-$; $Cu^{2+} + e^- \rightarrow Cu^+$

21-20. Determine the value of n in the Nernst equation for the following reactions:

(a) $Cu(s) + Mg^{2+}(aq) \rightarrow Cu^{2+}(aq) + Mg(s)$
(b) $2H_2O(l) + 2Na(s) \rightarrow 2Na^+(aq) + 2OH^-(aq) + H_2(g)$

21-21. The value of E^0 at 25°C for the reaction

$$Pb(s) + Cu^{2+}(aq) \rightleftharpoons Pb^{2+}(aq) + Cu(s)$$

$Pb \rightarrow Pb^{2+} + 2e^-$; $Cu^{2+} + 2e^- \rightarrow Cu$; $n = 2$

is 0.463 V. Calculate the equilibrium constant for this reaction at 25°C.

$E^\circ = \left(\frac{0.0592\text{ V}}{n}\right)\log K$; $0.463 = \frac{0.0592\text{ V}}{2}\log K$; $K = 4.38 \times 10^{15}$

21-22. The value of E^0 at 25°C for the reaction

$$H_2(g) + 2AgCl(s) \rightleftharpoons 2Ag(s) + 2HCl(aq)$$

is +0.22 V. Calculate the equilibrium constant for this reaction at 25°C.

21-23. The measured voltage at 25°C of a cell in which the reaction

$$Cd(s) + Pb^{2+}(aq, 0.150\text{ M}) \rightleftharpoons Pb(s) + Cd^{2+}(aq, 0.0250\text{ M})$$

takes place at the concentrations shown is 0.293 V. Calculate E^0 and K, the equilibrium constant, for the cell reaction.

$E = E^\circ - \left(\frac{0.0592\text{ V}}{n}\right)\log Q$

$0.293 = E^\circ - \left(\frac{0.0592\text{ V}}{2}\right)\log\frac{[Cd^{2+}]}{[Pb^{2+}]}$

$0.293 = E^\circ - \left(\frac{0.0592\text{ V}}{2}\right)\log\frac{[0.0250]}{[0.150]}$

$E^\circ = 0.270\text{ V}$

21-24. The measured voltage at 25°C of a cell in which the reaction

$$Co(s) + Sn^{2+}(aq, 0.18\text{ M}) \rightleftharpoons Sn(s) + Co^{2+}(aq, 0.020\text{ M})$$

takes place at the concentrations shown is 0.168 V. Calculate E^0 and K, the equilibrium constant, for the cell reaction.

$E^\circ = \left(\frac{0.0592\text{ V}}{n}\right)\log K$

$0.270 = \left(\frac{0.0592\text{ V}}{2}\right)\log K$

$K = 1.32 \times 10^9$

21-25. The measured voltage at 25°C of a cell in which the reaction

$$Al(s) + Fe^{3+}(aq, 0.0050\ M) \rightleftharpoons Al^{3+}(aq, 0.250\ M) + Fe(s)$$

takes place at the concentrations shown is 1.59 V. Calculate E^0 and K, the equilibrium constant, for the cell reaction.

21-27. Consider the electrochemical cell

$$Zn(s)|ZnCl_2(aq)|Hg_2Cl_2(s)|Hg(l)$$

(a) Determine the balanced equation for the cell reaction.

(b) The standard cell voltage for the cell at 25°C is $E^0 = +1.03$ V. Use the Nernst equation to compute the cell voltage when the concentration of $ZnCl_2(aq)$ is 0.010 M.

21-29. Suppose the leads of an electrochemical cell are connected together external to the cell and the cell is allowed to come to equilibrium. What will be the value of the cell voltage at equilibrium?

21-26. The measured voltage at 25°C of a cell in which the reaction

$$Zn(s) + Hg_2^{2+}(aq, 0.30\ M) \rightleftharpoons 2Hg(l) + Zn^{2+}(aq, 0.50\ M)$$

takes place at the concentrations shown is 1.54 V. Calculate E^0 and K, the equilibrium constant, for the cell reaction.

21-28. The voltage of the cell

$$Pt(s)|H_2(g)|H^+(aq)||Cd^{2+}(aq)|Cd(s)$$

is observed to be negative. Write the chemical reaction that occurs spontaneously.

21-30. Suppose a zinc rod is dipped into 1.0 M $CuSO_4(aq)$ solution containing a copper rod and the system is allowed to stand for several hours. What do you predict for the voltage measured between the Zn(s) and Cu(s) rods?

USE OF TABULATED E^0 VALUES

21-31. The standard voltage for the reaction

$$HClO(aq) + H^+(aq) + 2Cr^{2+}(aq) \rightleftharpoons 2Cr^{3+}(aq) + Cl^-(aq) + H_2O(l)$$

is $E^0 = 1.80$ V. Use data from Table 21-1 to calculate E^0 for the half-reaction

$$HClO(aq) + H^+(aq) + 2e^- \rightleftharpoons Cl^-(aq) + H_2O(l)$$

21-33. Use data from Table 21-1 to calculate E^0 for the reaction

$$Cd^{2+}(aq) + Pb(s) \rightleftharpoons Pb^{2+}(aq) + Cd(s)$$

Will lead displace cadmium from the compound $Cd(C_2H_3O_2)_2(aq)$?

21-35. Compute the E^0 value and the equilibrium constant at 25°C for the cell reaction

$$V^{2+}(aq) + H^+(aq) \rightleftharpoons V^{3+}(aq) + \tfrac{1}{2}H_2(g)$$

The value of E^0 for the half-reaction

$$V^{3+}(aq) + e^- \rightarrow V^{2+}(aq)$$

is $E^0 = -0.24$ V.

21-32. The standard voltage for the reaction

$$NO_3^-(aq) + 2H^+(aq) + Cu^+(aq) \rightleftharpoons NO_2(g) + H_2O(l) + Cu^{2+}(aq)$$

is $E^0 = 0.65$ V. Use data from Table 21-1 to calculate E^0 for the half-reaction

$$NO_3^-(aq) + 2H^+(aq) + e^- \rightleftharpoons NO_2(g) + H_2O(l)$$

21-34. Use data from Table 21-1 to calculate E^0 for the reaction

$$S_2O_8^{2-}(aq) + 2H_2O(l) \rightleftharpoons H_2O_2(aq) + 2SO_4^{2-}(aq) + 2H^+(aq)$$

Is an aqueous solution of potassium peroxosulfate $(K_2S_2O_8)$ stable over a long period of time?

21-36. Compute the E^0 value and the equilibrium constant at 25°C for the cell reaction

$$Cu(s) + 2Ag^+(aq) \rightleftharpoons 2Ag(s) + Cu^{2+}(aq)$$

21-37. Calculate the voltage at 25°C of an electrochemical cell in which the following reaction occurs:

$$2Zn(s) + O_2(g, 0.20\ \text{atm}) + 4H^+(aq, 0.20\ \text{M}) \rightleftharpoons 2Zn^{2+}(aq, 0.0010\ \text{M}) + 2H_2O(l)$$

See Table 21-1 for the necessary E^0 data.

21-39. The standard voltage for the reaction

$$S_2O_3^{2-}(aq) + 2OH^-(aq) + O_2(g) \rightleftharpoons 2SO_3^{2-}(aq) + H_2O(l)$$

is $E^0 = 0.98$ V. Write the two half-reactions for this reaction and use data from Table 21-1 to determine the value of E^0 for the $SO_3^{2-}(aq)/S_2O_3^{2-}(aq)$ half-reaction in basic solution.

BATTERIES

21-41. In the Edison battery, an early version of the nickel-cadmium cell, iron is used in place of cadmium. Write the cell diagram for an Edison battery. In contrast to the nickel-cadmium cell, the Edison battery has heavy-duty industrial and railway uses.

21-43. The cell reaction for the Clark cell, which is sometimes used as a reference voltage in the laboratory, is

$$Zn(s) + Hg_2SO_4(s) \rightleftharpoons ZnSO_4(aq) + 2Hg(l)$$

Draw the cell diagram for the cell.

21-38. Calculate the voltage at 25°C of an electrochemical cell in which the following reaction occurs:

$$Cd(s) + Pb(NO_3)_2(aq, 0.10\ \text{M}) \rightleftharpoons Cd(NO_3)_2(aq, 0.010\ \text{M}) + Pb(s)$$

See Table 21-1 for the necessary E^0 data.

21-40. The standard voltage for the reaction

$$BH_4^-(aq) + 8OH^-(aq) + 8O_2(g) \rightleftharpoons H_2BO_3^-(aq) + 5H_2O(l) + 8O_2^-(aq)$$

is $E^0 = 0.68$ V. Write the two half-reactions for the reaction and use data from Table 21-1 to determine the value of E^0 for the $H_2BO_3^-(aq)/BH_4^-(aq)$ half-reaction in basic solution.

21-42. The Weston standard cell is given by

$$\ominus Cd(Hg)|CdSO_4(aq, satd)|Hg_2SO_4(s)|Hg(l)\oplus$$
(12.5% Cd)

Write the reaction that occurs in the cell. Ten Weston standard cells that use a saturated $CdSO_4(aq)$ solution are maintained at the U.S. Bureau of Standards as the official unit of voltage. The voltage of each cell is virtually constant at 1.01875 V. Explain why the voltage remains constant.

21-44. The silver-zinc battery has the cell diagram

$$\ominus Zn(s)|K_2ZnO_2(s)|KOH(aq, 40\%)|Ag_2O_2(s)|Ag(s)\oplus$$

This battery is used in space satellites because of its exceptional compactness, high current capacity, and constant voltage during discharge. The high cost of silver precludes the use of this battery in routine applications, however.

(a) Give the balanced half-reaction that occurs at the left-hand electrode.
(b) Give the balanced half-reaction that occurs at the right-hand electrode.
(c) Give the net cell reaction.

21-45. A battery that operates at 500°C was developed for the exploration of Venus. The electrodes are a magnesium metal anode and a mixture of copper(I) and copper(II) oxides in contact with an inert steel cathode. The electrolyte is a mixture of LiCl and KCl, which is melted to activate the cell. The MgO that is produced is sparingly soluble in the molten salt mixture and precipitates. Draw the cell diagram and write the equation for the reaction for the cell.

21-46. A battery that operates at −50°C was developed for the exploration of the moon and Mars. The electrodes are magnesium metal/magnesium chloride and silver chloride/silver. The electrolyte is potassium thiocyanate, KSCN, in liquid ammonia. Draw the cell diagram and write the equation for the reaction for the cell.

ELECTROLYSIS

21-47. How long will it take an electric current of 3.64 A to deposit all the copper from 740 mL of 0.250 M $CuSO_4(aq)$?

21-48. How much silver is deposited if an electric current of 0.150 A flows through a silver nitrate solution for 20.0 min?

21-49. Cesium metal is produced by the electrolysis of molten cesium cyanide. Calculate how much Cs(*s*) is deposited from CsCN(*l*) in 1.0 h by a current of 1.0 A.

21-50. Beryllium occurs naturally in the form of beryl. The metal is produced from its ore by electrolysis after the ore has been converted to the oxide and then to the chloride. How much Be(*s*) is deposited from a $BeCl_2$ solution by a current of 5.0 A that flows for 1.0 h?

21-51. Fluorine is manufactured by the electrolysis of HF dissolved in molten KF. The reaction is

$$2HF(KF) \rightarrow H_2(g) + F_2(g)$$

The KF acts as a solvent for HF and as the conductor of electricity. A commercial cell for producing fluorine operates at a current of 1500 A. How much F_2 can be produced per 24 h? Why isn't the electrolysis of liquid HF alone used?

21-52. From 1882 to 1895 home electricity was provided as direct current rather than as alternating current, as is now the case. Thomas Edison invented a meter to measure the amount of electricity used by a consumer. A small amount of current was diverted to an electrolysis cell that consisted of zinc electrodes in a zinc sulfate solution. Once a month the cathode was removed, washed, dried, and weighed. The bill was figured in ampere-hours. In 1888 Boston Edison Company had 800 chemical meters in service. In one case, in one 30-day period, 65 g of zinc was deposited on the cathode. The meter used 11 percent of the current into the house. How many coulombs were used in the month? Calculate the current used in ampere-hours.

21-53. Suppose that it is planned to electrodeposit 200 mg of gold onto the surface of a steel object via the reaction

$$Au(CN)_2^-(aq) + e^- \rightarrow Au(s) + 2CN^-(aq)$$

If the electric current in the circuit is set at 30 mA, for how long should the current be passed?

21-54. Electrolysis can be used to determine atomic masses. A current of 0.600 A deposits 2.38 g of a certain metal in exactly 1 h. Calculate the atomic mass of the metal if $n = 1$. What is the metal?

21-55. Many metals can be refined electrolytically. The impure metal is used as the anode, and the cathode is made of the pure metal. When an electric current is passed between these electrodes, the metal leaves the impure anode and is deposited in a pure form on the cathode. How many ampere-hours of electricity are required to refine electrolytically 1 metric ton of copper? (Use $n = 2$.) An ampere-hour is an ampere times an hour.

21-56. Bauxite, the principle source of aluminum oxide, contains about 55 percent Al_2O_3 by mass. How much bauxite is required to produce the 5 million tons of aluminum metal produced each year by electrolysis?

21-57. Given that $E^0 = -0.728$ V for the cell

$$Ag(s)|AgBr(s)|Br^-(aq)||Ag^+(aq)|Ag(s)$$

$Ag \rightarrow Ag^+ + e^-$ ox. rxn $\quad AgBr + e^- \rightarrow Ag + Br^-$ red. rxn

write the cell reaction and determine the solubility product of AgBr. cell rxn = $AgBr \rightleftharpoons Ag^+ + Br^-$

$K = K_{sp} = [Ag^+][Br^-]$

$E^\circ = \left(\frac{0.0592V}{n}\right)\log K$

$-0.728V = (0.0592V)\log K_{sp}$

$K_{sp} = 5.0 \times 10^{-13} M^2$

21-58. Devise a cell that you could use to determine the solubility product of AgCl.

21-59. The half-cell

$$Hg(l)|Hg_2Cl_2(s)|KCl(aq)$$

is called a calomel electrode. If the KCl solution is saturated, then $E^0(satd) = 0.2415$ V at 25°C. Explain how the cell

$$Pt(s)|H_2(g)|H^+(aq)||KCl(aq, satd)|Hg_2Cl_2(s)|Hg(l)$$

can be used to measure pH. Derive an equation for the voltage of this cell as a function of pH.

$H_2 \rightarrow 2H^+ + 2e^-$ ox. rxn ; 0.00V

$+ \quad 2e^- + Hg_2Cl_2 \rightarrow 2Hg + 2Cl^-$ red. rxn

$H_2(g) + Hg_2Cl_2(s) \rightarrow 2H^+(aq) + 2Hg(l) + 2Cl^-(s)$

$E^\circ_{cell} = E^\circ_{red} - E^\circ_{ox}$

$E^\circ_{cell} = 0.2415V - 0 = 0.2415V$

$E = E^\circ - \left(\frac{0.0592V}{2}\right)\log\frac{[H^+]^2}{P_{H_2}}$

when pressure H_2 held at 1.00 atm:

$E = 0.2415V - \left(\frac{0.0592V}{2}\right)\log[H^+]^2$

$E = 0.2415V - (0.0592V)\log[H^+]$

$0.0592V\,pH = E - 0.2415V$

$$pH = \frac{E - 0.2415V}{0.0592V}$$

21-60. Show that the cell

$$Pt(s)|H_2(g)|HC_2H_3O_2(aq),NaC_2H_3O_2(aq)||KCl(aq, satd)|Hg_2Cl_2(s)|Hg(l)$$

can be used to determine K_a for $HC_2H_3O_2(aq)$.

INTERCHAPTER L

The Halogens

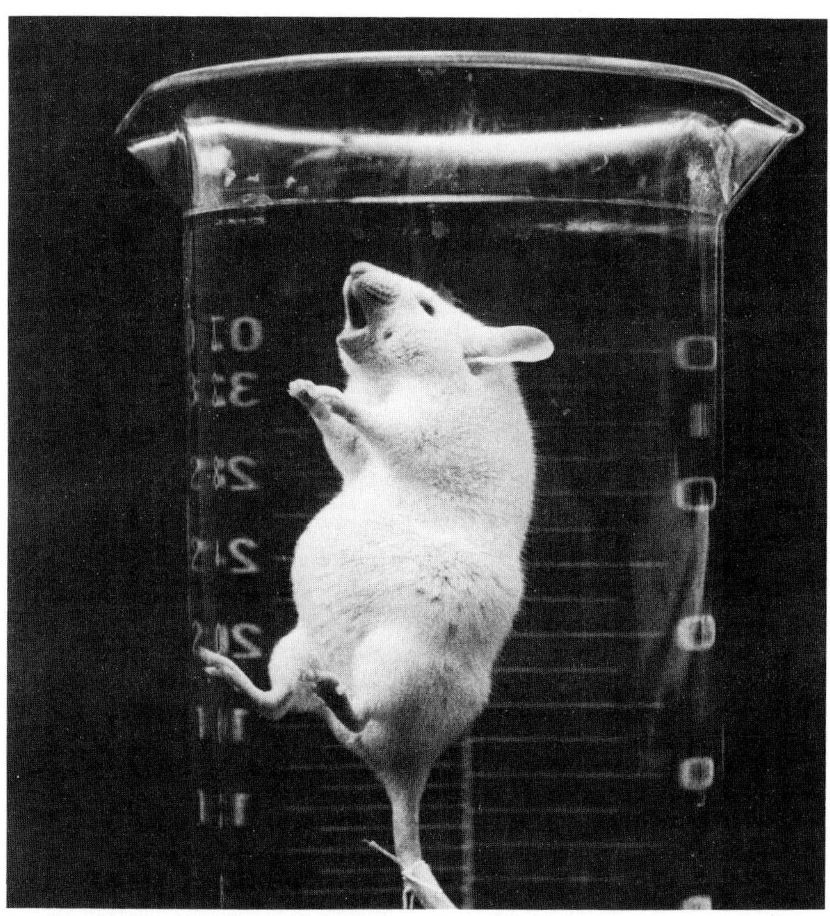

This submerged mouse is breathing oxygen dissolved in a liquid fluorocarbon. The solubility of oxygen in this liquid is so great that the mouse is able to breath by absorbing oxygen from the oxygen-containing fluorocarbon that fills its lungs. When the mouse is removed from the liquid, the fluorocarbon vaporizes from its lungs and normal breathing resumes.

The elements fluorine, chlorine, bromine, iodine, and astatine are collectively called the halogens. These nonmetals, which occur in Group 7 of the periodic table, are all highly reactive. They have a pungent, irritating odor, and all are poisonous to breathe. The elements exist as reactive diatomic molecules and are not found free in nature. The halogens (salt formers) occur in nature primarily as halide (F^-, Cl^-, Br^-, I^-) salts.

Table L-1 Physical properties of the halogens

Halogen	*Molecular mass of diatomic species*	*Melting point/°C*	*Boiling point/°C*	*Bond length/pm*	*Bond energy/kJ·mol^{-1}*	*Atomic electronegativity*
fluorine, F_2	38.0	−220	−188	142	155	4.0
chlorine, Cl_2	70.9	−101	−35	198	243	3.2
bromine, Br_2	159.8	−7	59	228	192	3.0
iodine, I_2	253.8	114	184	266	150	2.7

Table L-1 shows that the properties of the halogens vary smoothly from the lightest, F_2, to the heaviest, I_2. Astatine is not included in Table L-1 because it has no stable isotopes. Halogen compounds have many uses and are produced commercially on a large scale.

L-1. THE HALOGENS ARE REACTIVE NONMETALS

At 25°C fluorine and chlorine are gases, bromine is a liquid, and iodine is a solid (see Figure 2-8, page 51). As already mentioned, there are no stable isotopes of astatine. The relative sizes of the diatomic halogen molecules and the halide ions are shown in Figure L-1. Note that the size increases with increasing atomic number.

The diatomic halogens, X_2, are strong oxidizing agents. They combine directly with almost all metals. Some representative reactions of halogens with metals are

$$2Al(s) + 3Cl_2(g) \rightarrow 2AlCl_3(s)$$
$$2Ag(s) + Cl_2(g) \rightarrow 2AgCl(s)$$
$$U(s) + 3F_2(g) \rightarrow UF_6(g)$$

Note that in each case the metal is oxidized by the halogen and the oxidation state of the halogen is −1 in the resulting compounds.

The halogens also react directly with many nonmetals. Some representative examples of the reactions of halogens with nonmetals are

$$\underset{\text{(excess)}}{2P(s)} + 3Cl_2(g) \rightarrow 2PCl_3(l)$$
$$2P(s) + \underset{\text{(excess)}}{5Cl_2(g)} \rightarrow 2PCl_5(s)$$
$$\underset{\text{(excess)}}{S(s)} + 2F_2(g) \rightarrow SF_4(g) \qquad \text{(also with } Cl_2\text{)}$$
$$S(s) + \underset{\text{(excess)}}{3F_2(g)} \rightarrow SF_6(g) \qquad \text{(only with } F_2\text{)}$$

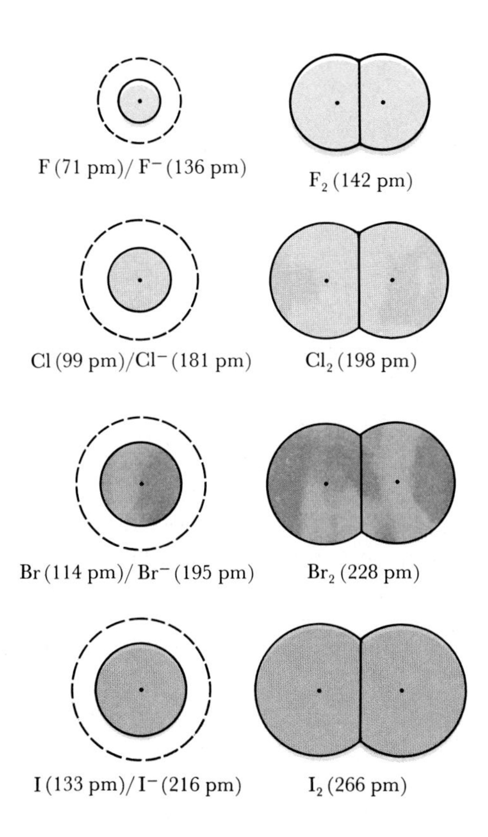

Figure L-1 Relative sizes of neutral halogen atoms (single spheres), halide ions (dashed circles), and diatomic halogen molecules (attached pairs). Distances are in picometers.

Note that the product obtained varies with the mole ratio of reactants. An excess of halogen may yield a higher oxidation state of the nonmetal.

L-2. FLUORINE IS THE MOST REACTIVE ELEMENT

Fluorine is a pale yellow, corrosive gas that is the strongest oxidizing agent known and the most reactive of all the elements. It reacts directly, and in most cases vigorously, with all the elements except helium and neon. The extremely corrosive nature of fluorine is shown by its reactions with glass, ceramics, and carbon; even water burns vigorously in fluorine:

$$2F_2(g) + H_2O(g) \rightarrow OF_2(g) + 2HF(g)$$

Many of the known noble-gas compounds are fluorides, such as XeF_2, XeF_4, and XeF_6.

Because of its high electronegativity, fluorine is capable of stabilizing unusually high oxidation states of other elements. Some examples are

$$OF_2 \quad O(II) \qquad AgF_2 \quad Ag(II) \qquad IF_7 \quad I(VII)$$

Xenon tetrafluoride crystals. Xenon tetrafluoride was first prepared in 1962 by the direct combination of Xe(*g*) and $F_2(g)$ at 6 atm and 400°C.

Fluorine is the most powerful molecular oxidizing agent known.

In effect, fluorine stabilizes high oxidation states in other elements because its electronegativity is so high that even strong oxidizing agents, such as Ag(II), cannot remove electrons from it. Because its electronegativity is higher than that of any other element, fluorine does not occur with a positive oxidation state in any compound. All fluorine compounds are fluorides, where the oxidation state of fluorine is -1.

Fluorides are widely distributed in nature. The main natural sources of fluorine are the minerals fluorspar, CaF_2; cryolite, Na_3AlF_6; and fluorapatite, $Ca_{10}F_2(PO_4)_6$. Because of its extreme reactivity, elemental fluorine wasn't isolated until 1886. The French chemist Henri Moisson, who first isolated fluorine, received the 1906 Nobel Prize in chemistry for his work. Elemental fluorine is obtained by the electrolysis of hydrogen fluoride dissolved in molten potassium fluoride.

$$2HF\ (\textit{in KF melt}) \xrightarrow{\text{electrolysis}} H_2(g) + F_2(g)$$

The modern method of producing F_2 is essentially a variation of the method first used by Moisson. Prior to World War II, there was no commercial production of fluorine. The atomic bomb project required huge quantities of fluorine for the production of uranium hexafluoride, UF_6, a gaseous compound that is used in the separation of uranium-235 from uranium-238. It is uranium-235 that is used in nuclear devices. The production of uranium hexafluoride for the preparation of fuel for nuclear power plants is today a major commercial use of fluorine.

$^{235}UF_6$ is separated from $^{238}UF_6$ by gaseous effusion. The lighter $^{235}UF_6$ effuses more rapidly than does the heavier $^{238}UF_6$.

Hydrogen fluoride is used in petroleum refining and in the production of fluorocarbon polymers, such as Teflon (Interchapter N) and Freons. It is also used to etch, or "frost," glass for light bulbs and decorative glassware via the reaction

$$SiO_2(s) + 6HF(aq) \rightarrow H_2SiF_6(s) + 2H_2O(l)$$

Because hydrofluoric acid, $HF(aq)$, dissolves glass via this reaction, it must be stored in plastic bottles. Various fluorides, such as tin(II) fluoride, SnF_2, and sodium monofluorophosphate, Na_2PO_3F, are used as toothpaste additives, and sodium fluoride is added to some municipal water supplies to aid in the prevention of tooth decay. The fluoride converts tooth enamel from hydroxyapatite, $Ca_{10}(OH)_2(PO_4)_6$, to fluorapatite, $Ca_{10}F_2(PO_4)_6$, which is harder and more resistant to acids than hydroxyapatite. Many organofluoride compounds are used as refrigerants. Two common ones are dichlorodifluoromethane (Freon 12), CCl_2F_2, which is used in automobile air conditioners, and chlorodifluoromethane (Freon 21), $CHClF_2$, which is used in home air conditioners. Fluorocarbons have displaced refrigerants such as ammonia and sulfur dioxide in refrigerators because of their much lower toxicity and greater chemical stability.

Tin(II) fluoride is also known as stannous fluoride.

Oxygen is very soluble in numerous liquid fluorocarbons, and this unusual property of fluorocarbons has led to their study for potential use as artificial blood fluids. The Frontispiece (page 845) shows a mouse totally submerged in dichlorofluoromethane saturated with oxygen. The mouse is able to breathe by absorbing oxygen from the oxygen-containing fluorocarbon that fills its lungs.

L-3. CHLORINE IS OBTAINED FROM CHLORIDES BY ELECTROLYSIS

Chlorine is the most abundant of the halogens. The major source of chlorine in nature is the chloride ion in the oceans and salt lakes. The major mineral sources of chloride are rock salt, NaCl, sylvite, KCl, and carnallite, $KMgCl_3 \cdot 6H_2O$.

Chlorine is a green-yellow, poisonous, corrosive gas that is prepared commercially by the electrolysis of either brines or molten rock salt:

$$\underset{\text{molten rock salt}}{2NaCl(l)} \xrightarrow{\text{electrolysis}} 2Na(l) + Cl_2(g)$$

About 10 million metric tons of chlorine are produced annually in the United States, making it the eighth ranked chemical in terms of production. It is prepared on a laboratory scale by heating a mixture of hydrochloric acid and manganese dioxide:

$$MnO_2(s) + 4H^+(aq) + 2Cl^-(aq) \rightarrow Mn^{2+}(aq) + Cl_2(g) + 2H_2O(l)$$

This reaction is the one used by the Swedish chemist Karl Scheele in 1774 in the first laboratory preparation of Cl_2.

Most chlorine produced in the United States is used as a bleaching agent in the pulp and paper industry. It is also used extensively as a germicide in water purification (Interchapter K) and in the production of insecticides (DDT and chlordane) and herbicides (2,4-D).

Chlorinated hydrocarbons present a serious health hazard to humans and other mammals, fishes, and birds. Such compounds are not biodegradable and, because of their high solubility in nonpolar solvents, accumulate in fatty tissues, where their presence may lead to irreversible liver damage and, in some cases, cancer. They also tend to work their way up the food chain to humans in increasing concentrations. Chlorinated hydrocarbons can be absorbed directly through the skin.

An especially insidious group of chlorinated hydrocarbons are the carcinogenic PCBs, *p*oly*c*hlorinated *b*iphenyls. PCBs are inexpensive, nonflammable, very stable compounds with excellent insulation properties and, as a consequence, were once widely used in transformers and capacitors on electric power lines. The discovery of the health hazards of PCBs has led to an extensive effort to remove them from power grids.

This photo shows the extent which is now taken with PCB cleanup operations. This spill was the result of a fire that damaged a transformer in the basement of a San Francisco highrise.

The use of chlorinated hydrocarbons as insecticides, herbicides, dry-cleaning agents, and anesthetics is in rapid decline because of the associated health hazards. For example, carbon tetrachloride is no longer sold as a home-use spot remover and chloroform, which is a powerful, rapidly acting general anesthetic, is no longer used in anesthesia. Ethyl chloride, C_2H_5Cl, is still used as an externally applied local anesthetic, however. Its rapid evaporation from the skin causes a rapid temperature drop in the tissue because of the heat absorbed in the vaporization. Ethyl chloride is a fairly potent general anesthetic.

L-4. BROMINE AND IODINE ARE OBTAINED BY OXIDATION OF BROMIDES AND IODIDES WITH CHLORINE

The major source of bromine in the United States is from brines that contain bromide ions. The pH of the brine is adjusted to 3.5, and chlorine is added; the chlorine oxidizes bromide ion to bromine, which is swept out of the brine with a current of air:

$$2Br^-(aq) + Cl_2(aq) \xrightarrow{pH = 3.5} 2Cl^-(aq) + Br_2(aq)$$

About 1 kg of bromine can be obtained from 15,000 L of seawater. About 200,000 metric tons of bromine were produced in the United States during 1980. It is a dense, red-brown, corrosive liquid with a very pungent odor. It attacks skin and tissue and produces painful, slow-healing sores. Bromine vapor and solutions of bromine in nonpolar solvents are red.

Bromine is used to prepare a wide variety of metal bromide and organobromide compounds. Its major uses are in the production of dibromoethane, $BrCH_2CH_2Br$, which is added to leaded gasolines as a lead scavenger, and in the production of silver bromide emulsions for black-and-white photographic films. Bromine is also used as a fumigant and in the synthesis of fire retardants, dyes, and pharmaceuticals, especially sedatives.

Iodide ion is present in seawater and is assimilated and concentrated by many marine animals and by seaweed. Certain seaweeds are an especially rich source of iodine. The iodide ion in seaweed is converted to iodine by oxidation with chlorine. Iodine is also obtained from Chilean mineral deposits of sodium iodate, $NaIO_3$, and sodium periodate, $NaIO_4$. Iodine is the only halogen to occur naturally in a positive oxidation state. The free element is obtained by reduction of IO_3^- and IO_4^- with sodium hydrogen sulfite:

$$2IO_3^-(aq) + 5HSO_3^-(aq) \rightarrow I_2(aq) + 5SO_4^{2-}(aq) + 3H^+(aq) + H_2O(l)$$

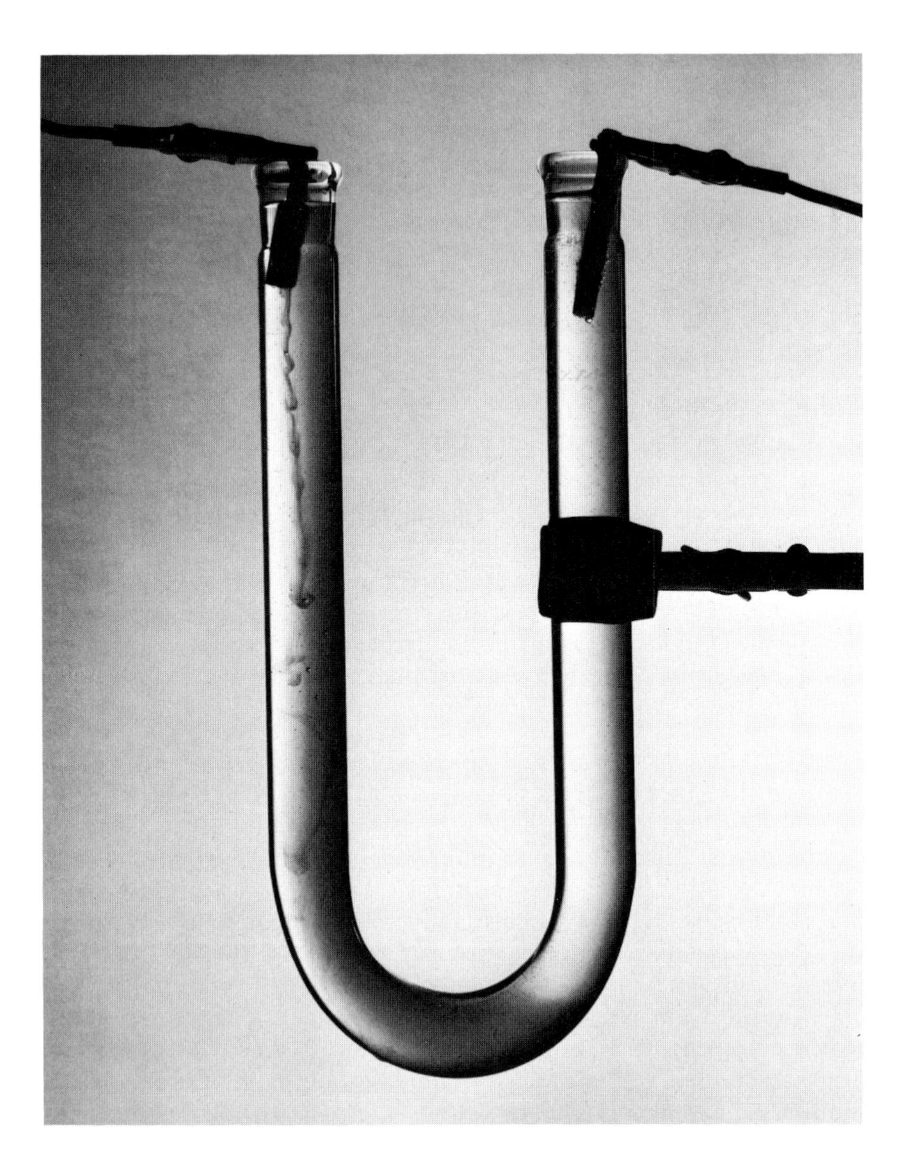

This U-tube is filled with a potassium iodide solution. When a current is passed through the solution, it breaks down into potassium hydroxide, iodine (which can be seen dropping from the positive anode on the left), and hydrogen (which forms from the negative cathode on the right).

Solid iodine is dark gray with a slight metallic luster. It sublimes at atmospheric pressure. Iodine vapor and solutions of iodine in nonpolar solvents are purple. Solutions of iodine in water and alcohols are brown as a result of the specific polar interactions between I_2 and the O—H bond. Iodine is only slightly soluble in pure water, but when I^- is present, the iodine forms a linear, colorless triiodide complex,

$$I_2(s) + I^-(aq) \rightleftharpoons I_3^-(aq) \qquad K_c = 720\ M^{-1}\ (25\,°C)$$

and the solubility of $I_2(s)$ is greatly enhanced. The presence of very low concentrations of aqueous triiodide can be detected by adding starch to the solution. The triiodide ion combines with starch to form a brilliant deep-blue species.

Iodide ion is essential for the proper functioning of the human thyroid gland, which is located in the base of the throat. Iodide deficiency manifests itself as the disease *goiter,* which causes an enlargement of the thyroid gland. Small quantities of potassium iodide are added to ordinary table salt, which is then marketed as iodized salt. Alcohol solutions of iodine, known as *tincture of iodine,* were once used as an antiseptic.

L-5. HALOGENS FORM NUMEROUS COMPOUNDS WITH OXYGEN-HALOGEN BONDS

The best known and most important oxygen-halogen compounds are the halogen oxyacids. The halogens form a series of oxyacids in which the oxidation state of the halogen atom can be +1, +3, +5, or +7. For example, the oxyacids of chlorine are

$HClO$	hypochlorous acid	+1
$HClO_2$	chlorous acid	+3
$HClO_3$	chloric acid	+5
$HClO_4$	perchloric acid	+7

The numbers after the names give the oxidation state of the chlorine in the acid. The Lewis formulas for these acids are

$$:\ddot{\underset{..}{Cl}}-\ddot{\underset{..}{O}}-H \qquad O{=}\ddot{Cl}-\ddot{\underset{..}{O}}-H \qquad (O{=})_2Cl-\ddot{\underset{..}{O}}-H \qquad (O{=})_3Cl-\ddot{\underset{..}{O}}-H$$

Notice that in each case the hydrogen is attached to an oxygen atom. The anions of the chlorine oxyacids are

ClO^-	hypochlorite
ClO_2^-	chlorite
ClO_3^-	chlorate
ClO_4^-	perchlorate

The shapes of these ions are predicted correctly by VSEPR theory (Chapter 11) and are shown in Figure L-2. Table L-2 gives the known halogen oxyacids and their anions. Note that there are no oxyacids of fluorine; the only oxidation state of fluorine is −1.

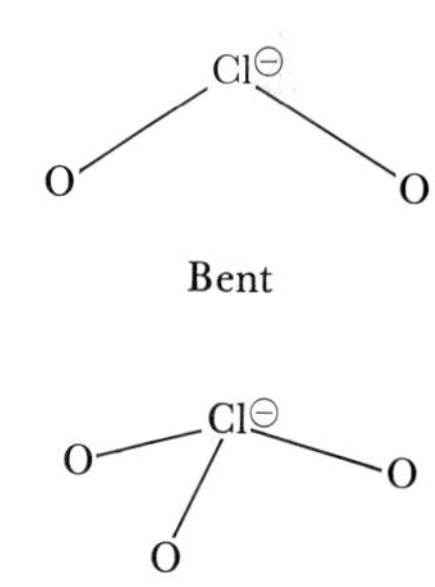

Figure L-2 The shapes of the oxyacid anions of chlorine.

Table L-2 The halogen oxyacids

Halogen oxidation state	*Chlorine*	*Bromine*	*Iodine*	*Acid*	*Salt*
+1	$HClO$	$HBrO$	HIO	hypohalous	hypohalite
+3	$HClO_2$	—	—	halous	halite
+5	$HClO_3$	$HBrO_3$	HIO_3	halic	halate
+7	$HClO_4$	$HBrO_4$	HIO_4	perhalic	perhalate

When Cl_2, Br_2, or I_2 is dissolved in aqueous alkaline solution, the following type of reaction occurs:

$$Cl_2(g) + 2OH^-(aq) \rightleftharpoons Cl^-(aq) + ClO^-(aq) + H_2O(l)$$

A solution of $NaClO^-(aq)$ is a bleaching agent, and many household bleaches are a 5.25% aqueous solution of sodium hypochlorite, NaClO. Commercially, solutions of NaClO are manufactured by the electrolysis of cold aqueous solutions of sodium chloride:

$$2Cl^-(aq) \rightarrow Cl_2(g) + 2e^- \qquad \text{anode}$$

$$2H_2O(l) + 2e^- \rightarrow H_2(g) + 2OH^-(aq) \qquad \text{cathode}$$

The products of the two electrode reactions are allowed to mix, producing $ClO^-(aq)$ by the above reaction. Sodium hypochlorite is also employed as a disinfectant and deodorant in dairies, creameries, water supplies, and sewage disposals.

Hypohalite ions decompose in basic solution via reactions of the type

$$3IO^-(aq) \xrightarrow{OH^-} 2I^-(aq) + IO_3^-(aq)$$

The analogous reaction with $ClO^-(aq)$ is slow, which makes it possible to use hypochlorite as a bleach in basic solutions. The rate of decomposition of $ClO^-(aq)$, $IO^-(aq)$, and $BrO^-(aq)$ in hot alkaline aqueous solution is sufficiently fast that when Cl_2, Br_2, or I_2 is dissolved in basic solution and the resulting solution is heated to 60°C, the following type of reaction goes essentially to completion:

$$3Br_2(aq) + 6OH^-(aq) \xrightarrow{60^\circ C} 5Br^-(aq) + BrO_3^-(aq) + 3H_2O(l)$$

Chlorates, bromates, and iodates also can be prepared by the reaction of the appropriate halogen with concentrated nitric acid or hydrogen peroxide or (commercially) by electrolysis of the halide. For example, the reaction for the oxidation of I_2 by H_2O_2 is

$$I_2(s) + 5H_2O_2(aq) \rightarrow 2IO_3^-(aq) + 4H_2O(l) + 2H^+(aq)$$

Perchlorate and periodate are prepared by the electrochemical oxidation of chlorate and iodate, respectively. For example,

$$ClO_3^-(aq) + H_2O(l) \xrightarrow{\text{electrolysis}} ClO_4^-(aq) + 2H^+(aq) + 2e^-$$

The perchlorate is obtained from the electrolyzed cell solution by adding potassium chloride in order to precipitate potassium perchlorate, which is only moderately soluble in water. Perchloric acid, $HClO_4$, also can be obtained from the electrolyzed solution containing perchlorate by adding sulfuric acid and then distilling. The distillation is very dangerous and often results in a violent explosion. Concentrated perchloric acid should not be allowed to come into contact with reducing agents, such as organic matter, because of the extreme danger of a violent explosion. Solutions containing perchlorates should not be evaporated because of their treacherously explosive nature. Perchlorates are used in explosives, fireworks, and matches.

QUESTIONS

L-1. Explain briefly why fluorine is capable of stabilizing unusually high oxidation states in many elements.

L-2. Describe how each of the halogens is prepared on a commercial scale.

L-3. Describe how the glass used in frosted light bulbs is etched.

L-4. Give the chemical formula for each of the following minerals:

(a) rock salt
(b) sylvite
(c) carnallite
(d) fluorspar
(e) cryolite
(f) fluorapatite

L-5. Give the chemical formula and a major use for each of the following compounds:

(a) ethyl chloride
(b) sodium hypochlorite

L-6. Give the chemical formulas and names of the oxyacids of chlorine.

L-7. What is household bleach?

L-8. Describe by balanced chemical equations how you would prepare $KIO_3(s)$ starting with $I_2(s)$.

L-9. Name the following oxyacids:

(a) $HBrO_2$
(b) HIO
(c) $HBrO_4$
(d) HIO_3

L-10. Given the following key oxyacids

H_2SO_4 sulfuric acid
HNO_3 nitric acid
H_3PO_4 phosphoric acid

name the following oxyacids

(a) HNO_2
(b) H_2SO_3
(c) H_3PO_2
(d) H_3PO_3
(e) $H_2N_2O_2$

and the following salts

(a) K_2SO_3
(b) $Ca(NO_2)_2$
(c) KIO_2
(d) $Mg(BrO)_2$

Ink spontaneously dispersing throughout the entire volume of water is an example of an entropy-driven process.

22 / Entropy and Gibbs Free Energy

It is natural to ask why some substances react with each other and others do not. When a reaction occurs on its own, we say that it is spontaneous, and the condition that must be met in order for a reaction to be spontaneous is called the criterion of spontaneity. The observation that all highly exothermic reactions are spontaneous led the French chemist Pierre Berthelot to put forth, in the 1860's, the hypothesis that all spontaneous reactions are exothermic. Berthelot's criterion of spontaneity for a chemical reaction was based on the sign of ΔH_{rxn}. According to Berthelot, if ΔH_{rxn} is negative, then a reaction is spontaneous.

Berthelot's criterion of reaction spontaneity was shown to be incorrect, and it was superseded by the criterion of reaction spontaneity developed by the American thermodynamicist J. Willard Gibbs, who showed that reaction spontaneity is not just a matter of energetics. There is another property, called entropy, that also must be considered when determining whether a reaction is spontaneous or not. We see in this chapter that entropy is a measure of the randomness or disorder of a system, and we see how the spontaneity of a chemical reaction is governed by both the energy change and the entropy change. The central idea of the chapter is a quantity called the Gibbs free energy, which combines the energy change and the entropy change of a reaction.

22-1. AN EVOLUTION OF ENERGY DOES NOT GUARANTEE REACTION SPONTANEITY

In Chapter 6 we discussed enthalpy changes for chemical reactions. Recall that an exothermic chemical reaction is one in which energy

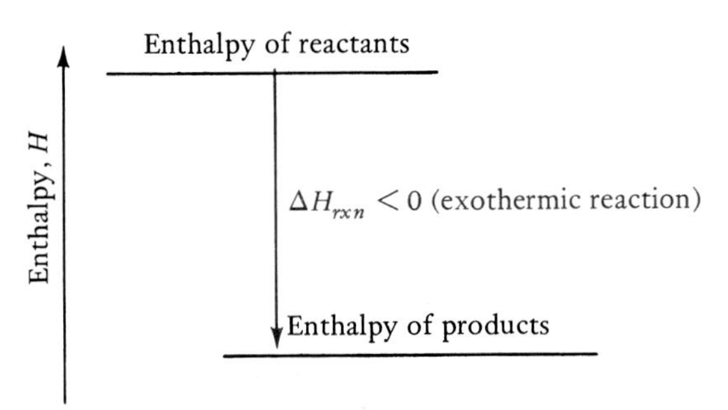

Figure 22-1 In an exothermic reaction, the total enthalpy of the products is less than the total enthalpy of the reactants.

is evolved as heat. Exothermic reactions are energetically downhill in the sense that the total enthalpy of the products is less than the total enthalpy of the reactants (Figure 22-1). For an exothermic reaction, the *enthalpy change*, ΔH_{rxn}, is always negative:

$$\Delta H_{rxn} = \Delta H(\text{products}) - \Delta H(\text{reactants}) < 0$$

In Chapter 6 we learned to calculate ΔH_{rxn} for a reaction by using a table of heats of formation (Table 6-1). Recall from Chapter 6 that the value of ΔH_{rxn} is approximately equal to the value of the energy change, ΔU_{rxn}, for a reaction.

The natural tendency of a simple mechanical system is to undergo processes that lead to a decrease in energy of the system. For example, water flows spontaneously downhill without any help from us. Water at the bottom of a waterfall has a lower potential energy than water at the top. To get water back to the top of the waterfall, we have to use a pump, which requires energy in order to function. If we release the ball shown at the top of the hill in Figure 22-2, then it spontaneously rolls down the hill and comes to rest at the bottom of the valley. The ball at the lowest point of the valley has the lowest possible potential energy for this system.

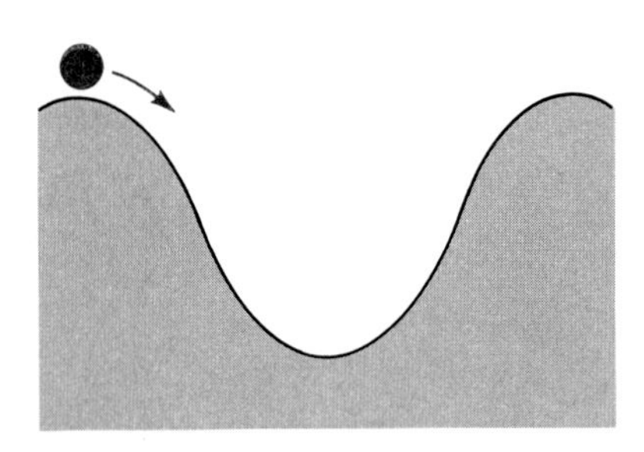

(a) Initial state

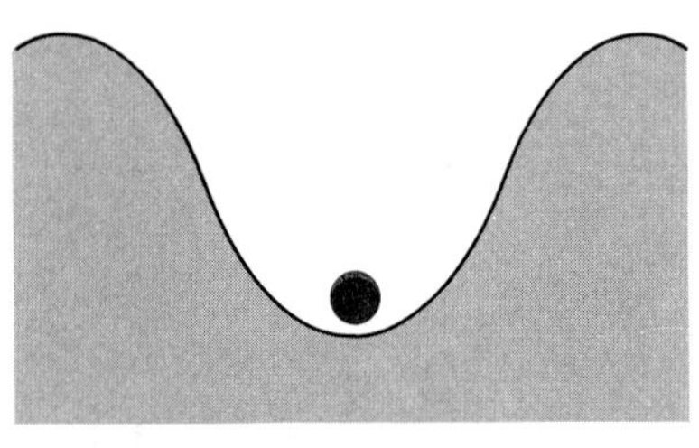

(b) Final state

Figure 22-2 A ball initially at the top of a hill will roll down the hill and come to rest. The potential energy of the ball at the bottom of the hill is less than at the top. The ball spontaneously goes from a state of high potential energy to a state of low potential energy.

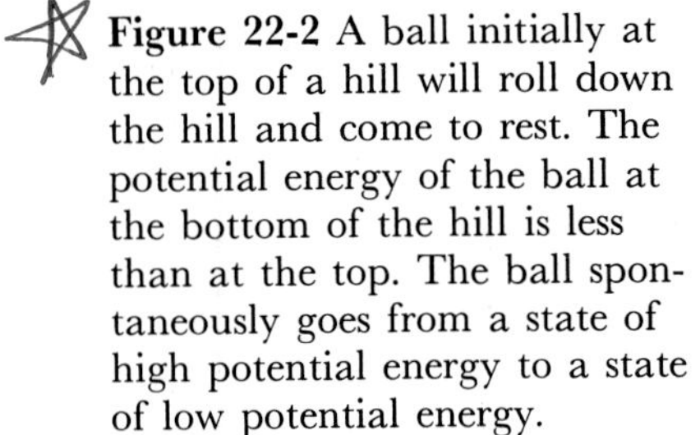

We have all observed a wide variety of spontaneous chemical processes. For example, natural gas, CH_4, once ignited, burns spontaneously in air to yield carbon dioxide and water:

$$CH_4(g) + 2O_2(g) \rightarrow CO_2(g) + 2H_2O(g) \qquad \Delta H_{rxn} = -802\ \text{kJ}$$

Iron, on exposure to air and moisture, spontaneously rusts:

$$2Fe(s) + \tfrac{3}{2}O_2(g) \xrightarrow[H_2O(l)]{} Fe_2O_3(s) \qquad \Delta H_{rxn} = -824\ \text{kJ}$$

Zinc reacts spontaneously with 1.0 M $HCl(aq)$ to yield $H_2(g)$ and $ZnCl_2(aq)$:

$$Zn(s) + 2HCl(aq) \rightarrow H_2(g) + ZnCl_2(aq) \qquad \Delta H_{rxn} = -150\ \text{kJ}$$

Furthermore, these reactions do not occur spontaneously in the reverse direction. For example, if $CO_2(g)$ and $H_2O(g)$ reacted spontaneously to produce $CH_4(g)$, then there would be no energy crisis because $CO_2(g)$ and $H_2O(g)$ are plentiful and inexpensive and $CH_4(g)$ is a fuel.

The three chemical reactions we have considered so far have been highly exothermic, and it may not be surprising that they occur spontaneously. There are many spontaneous processes, however, that are not highly exothermic. Let's consider two gases, such as $N_2(g)$ and $I_2(g)$, occupying two separate containers, as shown in Figure 22-3. We all know that, if allowed to, the gases will mix spontaneously. Furthermore, simple gaseous mixtures do not separate spontaneously. It would be a disastrous occurrence if the air in a room all of a sudden separated so that part of the room contained pure oxygen and the rest contained nitrogen. Although the mixing of two gases is what we could call a spontaneous process, it turns out that $\Delta H_{rxn} \approx 0$ for such a process. Surely it is not the energy change that drives the process.

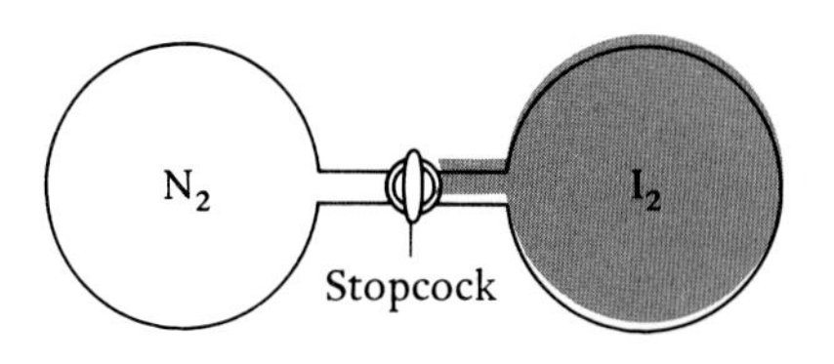

(a) Stopcock closed

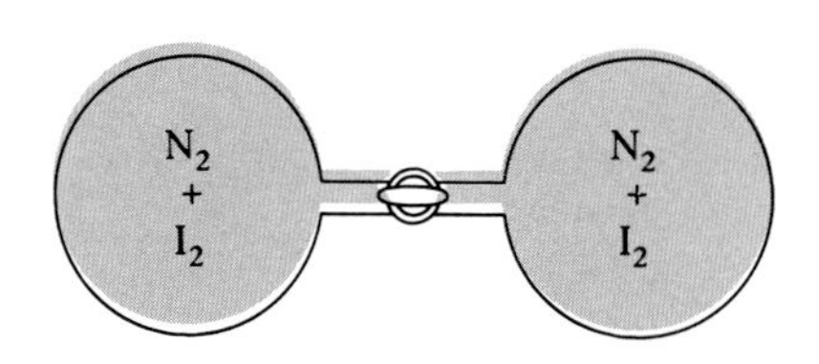

(b) After stopcock is opened

Figure 22-3 A simple example of a spontaneous process is the mixing of two gases.

Not only do some spontaneous processes have $\Delta H_{rxn} \approx 0$, but others are even endothermic. The energy of the products is *greater* than that of the reactants. For example, ice at any temperature greater than 0°C spontaneously melts:

$$H_2O(s) \rightarrow H_2O(l) \qquad \Delta H_{rxn} = +6.0\text{ kJ}$$

and table salt spontaneously dissolves in water at 20°C:

$$NaCl(s) \xrightarrow[H_2O(l)]{} Na^+(aq) + Cl^-(aq) \qquad \Delta H_{rxn} = +6.4\text{ kJ}$$

The fact that spontaneous processes can occur with negative, zero, or positive values of ΔH_{rxn} clearly indicates that heat evolution, that is, $\Delta H_{rxn} < 0$, is *not* a generally suitable criterion of reaction spontaneity. Reaction spontaneity is not determined solely by the energy change for a reaction. There is an additional factor involved. This factor is the entropy change for the reaction.

22-2. ENTROPY IS A MEASURE OF THE AMOUNT OF DISORDER OR RANDOMNESS IN A SYSTEM

In order to develop a suitable criterion of spontaneity, we need to introduce the concept of entropy. The *entropy* of a substance is a quantitative measure of the amount of disorder in the substance. The disorder is of two types: *positional disorder,* the distribution of the particles in space, and *thermal disorder,* the distribution of the available energy among the particles. Any process that produces a more random distribution of the particles in space or any constant-pres-

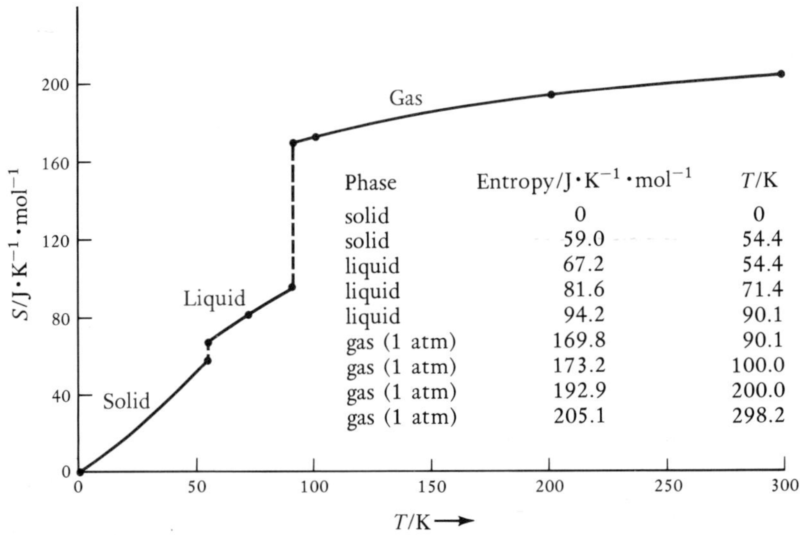

Phase	Entropy/$J \cdot K^{-1} \cdot mol^{-1}$	T/K
solid	0	0
solid	59.0	54.4
liquid	67.2	54.4
liquid	81.6	71.4
liquid	94.2	90.1
gas (1 atm)	169.8	90.1
gas (1 atm)	173.2	100.0
gas (1 atm)	192.9	200.0
gas (1 atm)	205.1	298.2

Figure 22-4 The molar entropy of oxygen as a function of temperature at one atm. Note that the entropy of each phase increases smoothly with increasing temperature. This gradual increase in entropy with increase in temperature is a result of the increase in thermal disorder. The jumps in entropy that occur when the solid melts and the liquid vaporizes are the result of an increase in positional disorder of the O_2 molecules.

sure process that increases the temperature of the particles gives rise to an *increase* in the total entropy of the substance.

When energy as heat is added to a system, the entropy of the system always increases (second law of thermodynamics).

Entropy, which is customarily denoted by S, probably is a new concept to you, and so to help you understand entropy, let's look at Figure 22-4, where the molar entropy of oxygen is plotted against temperature from 0 K to 300 K. The entropy is zero at 0 K. At this temperature (called absolute zero), there is neither positional disorder—the O_2 molecules in the crystal are perfectly arrayed in the lattice—nor thermal disorder—all the molecules are in the lowest possible energy state. Entropy is associated with disorder—no disorder, no entropy. The entropy of a (perfect) crystalline substance is zero at absolute zero.

The statement that the entropy of a perfect crystalline substance is zero at absolute zero is known as the third law of thermodynamics.

As the temperature is increased from 0 K in Figure 22-4, the oxygen molecules begin to vibrate more freely about their lattice sites. The increase in temperature causes an increase in thermal disorder. When the temperature of a substance held at constant pressure increases, the entropy of the substance increases. The increase in entropy with an increase in temperature is associated with the larger amount of energy that must be distributed among the molecules of the substance. Molecules, like atoms, are restricted to discrete, quantized energy levels. The greater the amount of energy stored in a substance, the greater the number of ways in which the energy can be distributed and thus the greater the thermal disorder and the greater the entropy.

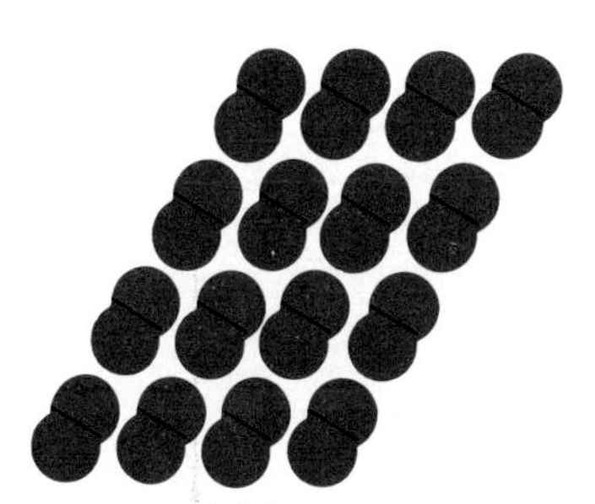

Solid oxygen

Liquid oxygen

Figure 22-5 When a solid melts, the molecules become free to move throughout the volume of the liquid. This increased positional disorder in the liquid implies that, under the same conditions, the entropy of a liquid is always greater than the entropy of the solid.

22-3. THERE IS A JUMP IN ENTROPY AT A PHASE TRANSITION

Figure 22-4 shows that the entropy in a particular phase increases smoothly with increasing temperature. At a phase transition, however, there is a jump in entropy. Let's investigate the origin of the two jumps in Figure 22-4. When solid oxygen melts to liquid oxygen (at 54.4 K), the ordered lattice of the solid breaks down into a liquid, where the molecules no longer are confined to lattice sites. Each molecule moves throughout the liquid volume, and so there is an increase in the positional disorder of the oxygen (Figure 22-5). This increase in positional disorder leads to an increase in entropy. Because all the entropy increase occurs at the melting point, ΔS of melting (fusion) appears as a vertical jump in Figure 22-4.

We can calculate the change in the molar entropy of a substance upon fusion. If we let $\Delta\bar{S}_{fus}$ be the *molar entropy change upon fusion*, then

$$\Delta\bar{S}_{fus} = \frac{\Delta\bar{H}_{fus}}{T_m} \qquad (22\text{-}1)$$

where $\Delta\bar{H}_{fus}$ is the molar enthalpy of fusion (Chapter 13) and T_m is the melting (fusion) point in kelvins. Using Equation (22-1), we find that the entropy increase of 1.00 mol of oxygen upon melting at 54.4 K is

$$\Delta\bar{S}_{fus} = \frac{\Delta\bar{H}_{fus}}{T_m} = \frac{440\ \text{J}\cdot\text{mol}^{-1}}{54.4\ \text{K}} = 8.09\ \text{J}\cdot\text{K}^{-1}\cdot\text{mol}^{-1}$$

The units of entropy are joule per kelvin per mole. Note that $\Delta\bar{S}_{fus}$ is positive, that is, the entropy of liquid oxygen is greater than the entropy of solid oxygen by $8.09\ \text{J}\cdot\text{K}^{-1}\cdot\text{mol}^{-1}$.

When liquid oxygen is heated from its melting point to its boiling point, the entropy increases smoothly, all due to an increase in thermal disorder. At the boiling point, there is another jump in entropy. This jump is due to the fact that the molecules in the gas phase can move throughout a much larger volume than the molecules in the liquid phase. Therefore, there is a sudden increase in positional disorder in going from the liquid phase to the gas phase (Figure 22-6).

We can calculate the change in the entropy of a substance upon

Liquid oxygen

Gaseous oxygen

Figure 22-6 When a liquid vaporizes, there is a large increase in positional disorder and hence a large increase in entropy.

vaporization by using a formula that is analogous to Equation (22-1). If we let $\Delta\bar{S}_{vap}$ be the *molar entropy change upon vaporization*, then

$$\Delta\bar{S}_{vap} = \frac{\Delta\bar{H}_{vap}}{T_b} \tag{22-2}$$

where $\Delta\bar{H}_{vap}$ is the molar enthalpy (heat) of vaporization (Chapter 6) and T_b is the boiling point in kelvins. Using Equation (22-2), we find that the change in the entropy of 1.0 mol of liquid oxygen upon being vaporized at 90.1 K is

$$\Delta\bar{S}_{vap} = \frac{\Delta\bar{H}_{vap}}{T_b} = \frac{6810\,\text{J}\cdot\text{mol}^{-1}}{90.1\,\text{K}} = 75.6\,\text{J}\cdot\text{K}^{-1}\cdot\text{mol}^{-1}$$

Note that $\Delta\bar{S}_{vap}$ is positive, that is, the molar entropy of gaseous oxygen is greater than the molar entropy of liquid oxygen.

Note also that $\Delta\bar{S}_{vap} > \Delta\bar{S}_{fus}$; this is a consequence of the fact that the molecules in the gas phase are much more disordered than the molecules in the liquid or the solid phase.

Example 22-1: The molar enthalpy of fusion of water is $\Delta\bar{H}_{fus} = 6.02\,\text{kJ}\cdot\text{mol}^{-1}$ at 0°C, and the molar enthalpy of vaporization of water is $\Delta\bar{H}_{vap} = 40.7\,\text{kJ}\cdot\text{mol}^{-1}$ at 100°C. Compute $\Delta\bar{S}_{fus}$ and $\Delta\bar{S}_{vap}$ for water.

Solution: The value of $\Delta\bar{S}_{fus}$ is computed using Equation (22-1). The melting point of water is 0°C, or 273 K, and so we have

$$\Delta\bar{S}_{fus} = \frac{\Delta\bar{H}_{fus}}{T_m} = \frac{6.02\times10^3\,\text{J}\cdot\text{mol}^{-1}}{273\,\text{K}} = 22.1\,\text{J}\cdot\text{K}^{-1}\cdot\text{mol}^{-1}$$

The value of $\Delta\bar{S}_{vap}$ is computed using Equation (22-2) with $T_b = 373$ K:

$$\Delta\bar{S}_{vap} = \frac{\Delta\bar{H}_{vap}}{T_b} = \frac{40.7\times10^3\,\text{J}\cdot\text{mol}^{-1}}{373\,\text{K}} = 109\,\text{J}\cdot\text{K}^{-1}\cdot\text{mol}^{-1}$$

Here, as in all cases, both $\Delta\bar{S}_{fus}$ and $\Delta\bar{S}_{vap}$ are positive and $\Delta\bar{S}_{vap} > \Delta\bar{S}_{fus}$.

Problems 22-1 through 22-8 deal with $\Delta\bar{S}_{vap}$ and $\Delta\bar{S}_{fus}$.

22-4. THE MOLAR ENTROPY OF A SUBSTANCE DEPENDS ON THE NUMBER, MASSES, AND ARRANGEMENT OF THE ATOMS IN THE MOLECULE

The standard molar entropies of a variety of substances are given in Table 22-1. Remember that the superscript circle on a thermodynamic quantity tells us that the value given is the standard value of the quantity, which, for our purposes, is the value at 1 atm pressure. Thus, $S°$ is the *standard entropy* of a substance, that is, the entropy at 1 atm pressure. Notice that there is a column headed $\Delta \bar{G}_f^\circ$ in Table 22-1. We shall learn about this column in Section 22.10.

Let's look at the molar entropy values in Table 22-1 and try to determine some trends. First notice that the molar entropies of the gaseous substances are the largest and the molar entropies of the solid substances are the smallest. We have already discussed the reason for this.

Now consider the molar entropies of the noble gases, whose values in $J \cdot K^{-1} \cdot mol^{-1}$ at 25°C and 1 atm are: He(*g*), 126.0; Ne(*g*), 146.2; Ar(*g*), 154.7; Kr(*g*), 164.0; and Xe(*g*), 169.6. The increase in molar entropy of the noble gases is a consequence of their increasing mass as we move down the periodic table. The increased mass leads to an increased capacity to take up energy and thus to a greater entropy. The same trend can be seen by comparing the molar entropies of the gaseous halogens, which are, in $J \cdot mol^{-1} \cdot K^{-1}$: $F_2(g)$, 204; $Cl_2(g)$, 223; $Br_2(g)$, 245; and $I_2(g)$, 261.

Table 22-1 Standard molar entropies ($\bar{S}°$), heats of formation ($\Delta \bar{H}_f^\circ$), and Gibbs free energies of formation ($\Delta \bar{G}_f^\circ$) of various substances at 25°C and 1 atm

Substance	$\bar{S}°/J \cdot K^{-1} \cdot mol^{-1}$	$\Delta \bar{H}_f^\circ/kJ \cdot mol^{-1}$	$\Delta \bar{G}_f^\circ/kJ \cdot mol^{-1}$
Ag(*s*)	42.6	0	0
AgBr(*s*)	107.1	−100.4	−96.90
AgCl(*s*)	96.2	−127.1	−109.8
Al(*s*)	28.3	0	0
$Al_2O_3(s)$	50.9	−1676	−1582
Ar(*g*)	154.7	0	0
Ba(*s*)	62.8	0	0
$BaCO_3(s)$	112.2	−1216	−1138
BaO(*s*)	70.2	−553.5	−525.1
Br(*g*)	174.9	111.9	82.43
$Br_2(g)$	245.4	30.91	3.14
$Br_2(l)$	152.2	0	0
C(*s, diamond*)	2.38	1.90	2.900
C(*s, graphite*)	5.74	0	0

Table 22-1 (continued) Standard molar entropies ($\overline{S}^\circ$), heats of formation ($\Delta\overline{H}_f^\circ$), and Gibbs free energies of formation ($\Delta\overline{G}_f^\circ$) of various substances at 25°C and 1 atm

Substance	$\overline{S}^\circ/J\cdot K^{-1}\cdot mol^{-1}$	$\Delta\overline{H}_f^\circ/kJ\cdot mol^{-1}$	$\Delta\overline{G}_f^\circ/kJ\cdot mol^{-1}$
$CH_4(g)$	186.2	−74.86	−50.75
$C_2H_2(g)$	200.8	226.7	209.2
$C_2H_4(g)$	219.6	52.28	68.12
$C_2H_6(g)$	229.5	−84.68	−32.89
$C_3H_8(g)$	269.9	−103.8	−23.49
$C_6H_6(l)$	172.8	49.03	124.5
$CH_3OH(l)$	126.9	−238.7	−166.3
$C_2H_5OH(l)$	160.8	−277.7	−174.8
$CH_3Cl(g)$	234.8	−80.83	−57.40
$CH_3Cl(l)$	145.3	−102	−51.5
$CH_2Cl_2(g)$	270.4	−92.47	−65.90
$CH_2Cl_2(l)$	178.1	−121	−67.4
$CHCl_3(g)$	294.9	−103.1	−70.37
$CHCl_3(l)$	202.6	−134.5	−73.72
$CCl_4(g)$	308.7	−103.0	−60.63
$CCl_4(l)$	215.4	−135.4	−65.27
$CO(g)$	197.8	−110.5	−137.2
$CO_2(g)$	213.6	−393.5	−394.4
$Ca(s)$	41.5	0	0
$CaC_2(s)$	69.9	−59.8	−64.8
$CaCO_3(s)$	92.9	−1207	−1129
$CaO(s)$	39.7	−635.1	−604.0
$Cl(g)$	165.1	121.7	105.7
$Cl_2(g)$	222.9	0	0
$Cu(s)$	33.2	0	0
$CuO(s)$	43.6	−157	−130
$Cu_2O(s)$	93.1	−169	−146
$F(g)$	158.6	78.99	61.92
$F_2(g)$	203.7	0	0
$Fe(s)$	27.3	0	0

Table 22-1 (continued) Standard molar entropies ($\bar{S}°$), heats of formation ($\Delta\bar{H}_f°$), and Gibbs free energies of formation ($\Delta\bar{G}_f°$) of various substances at 25°C and 1 atm

Substance	$\bar{S}°/J\cdot K^{-1}\cdot mol^{-1}$	$\Delta\bar{H}_f°/kJ\cdot mol^{-1}$	$\Delta\bar{G}_f°/kJ\cdot mol^{-1}$
$Fe_2O_3(s)$	87.9	−824.2	−742.2
$Fe_3O_4(s)$	146.3	−1118	−1015
$H(g)$	114.6	218.0	203.3
$H_2(g)$	130.6	0	0
$H_2O(g)$	188.7	−241.8	−228.6
$H_2O(l)$	69.9	−285.8	−237.2
$H_2O_2(l)$	110.0	−187.8	−120.4
$HF(g)$	173.6	−271.1	−273
$HCl(g)$	186.8	−92.31	−95.30
$HBr(g)$	198.6	−36.4	−53.43
$HI(g)$	206.4	26.1	1.7
$H_2S(g)$	205.7	−20.1	−33.0
$He(g)$	126.0	0	0
$I(g)$	180.7	106.8	70.28
$I_2(g)$	260.6	62.4	19.36
$I_2(s)$	116.5	0	0
$Kr(g)$	164.0	0	0
$N(g)$	153.2	472.6	455.5
$N_2(g)$	191.5	0	0
$NH_3(g)$	192.5	−46.19	−16.64
$N_2H_4(l)$	121	50.6	149
$NO(g)$	210.6	90.37	86.69
$NO_2(g)$	240.4	33.85	51.84
$N_2O(g)$	220.2	81.55	103.6
$N_2O_4(g)$	304.3	9.66	98.29
$N_2O_5(s)$	113.1	−41.8	134.2
$NOCl(g)$	262	51.9	66.1
$Na(g)$	153.6	107.1	77.30
$Na(s)$	51.3	0	0
$NaHCO_3(s)$	102	−947.7	−851.9

Table 22-1 (continued) Standard molar entropies ($\bar{S}°$), heats of formation ($\Delta\bar{H}_f°$), and Gibbs free energies of formation ($\Delta\bar{G}_f°$) of various substances at 25°C and 1 atm

Substance	$\bar{S}°/J\cdot K^{-1}\cdot mol^{-1}$	$\Delta\bar{H}_f°/kJ\cdot mol^{-1}$	$\Delta\bar{G}_f°/kJ\cdot mol^{-1}$
$Na_2CO_3(s)$	138.8	−1131.1	−1048.2
$Na_2O(s)$	75.06	−418.0	−379.1
$NaCl(s)$	72.13	−411.2	−384.0
$NaBr(s)$	86.82	−361.4	−349.3
$NaI(s)$	98.49	−287.8	−284.6
$Ne(g)$	146.2	0	0
$O(g)$	161.0	247.5	230.1
$O_2(g)$	205.0	0	0
$O_3(g)$	238.8	142	163.4
P(*s, white*)	41.1	0	0
P(*s, red*)	22.8	−18.4	−12.6
$P_4O_{10}(s)$	228.9	−2984	−2698
$POCl_3(g)$	325.3	−558.5	−513.0
$POCl_3(l)$	222	−597.0	−520.9
$PCl_3(g)$	311.7	−306.4	−286.3
$PCl_5(g)$	354.5	−375.0	−305.0
$PH_3(g)$	210.1	5.4	13.1
S(*s, rhombic*)	31.8	0	0
S(*s, monoclinic*)	32.6	0.30	0.10
$SO_2(g)$	248.4	−296.8	−300.2
$SO_3(g)$	256.3	−395.7	−371.1
$SF_6(g)$	291.7	−1209.3	−1105.2
$Xe(g)$	169.6	0	0
$Zn(s)$	41.63	0	0
$ZnO(s)$	43.64	−348.3	−318.3
$ZnS(s)$	57.7	−206.0	−201.3

Generally speaking, the more atoms of a given type there are in a molecule, the greater the capacity of the molecule to take up energy and thus the greater the entropy. This trend is illustrated by the series $C_2H_2(g)$, $C_2H_4(g)$, and $C_2H_6(g)$, whose Lewis formulas are

H—C≡C—H acetylene (201)

$H_2C{=}CH_2$ ethylene (220)

$H_3C{-}CH_3$ ethane (230)

The molar entropies in $J \cdot K^{-1} \cdot mol^{-1}$ are shown in parentheses. For molecules with the same number of atoms, the standard entropy is greater, the greater the mass of the molecules.

Example 22-2: Arrange the following molecules in order of increasing standard molar entropy:

$$CH_4(g) \qquad CCl_4(g) \qquad CH_3Cl(g)$$

Solution: The number of atoms is the same in each case, but chlorine has a greater mass than hydrogen. Thus we predict that

$$\bar{S}°(CH_4) < \bar{S}°(CH_3Cl) < \bar{S}°(CCl_4)$$

The values of the standard molar entropies in $J \cdot K^{-1} \cdot mol^{-1}$ at 25°C are 186.2, 234.8, and 308.7, respectively.

Problems 22-9 through 22-14 are similar to Example 22-2.

Another interesting comparison involves the isomeric molecules acetone and trimethylene oxide. At 25°C and 1 atm their molar entropies are

acetone: $(H_3C)_2C{=}O$ — $\bar{S}° = 295\ J \cdot K^{-1} \cdot mol^{-1}$

trimethylene oxide: ring of CH_2, H_2C, CH_2, O — $\bar{S}° = 274\ J \cdot K^{-1} \cdot mol^{-1}$

The entropy of acetone is higher than the entropy of trimethylene oxide because of the greater freedom of movement of the methyl groups in acetone. The relatively rigid ring structure of trimethylene oxide restricts the movement of the ring atoms. This restriction gives rise to a lower entropy because the capacity of the rigid isomer to take up energy is less than that of the more flexible acetone molecule.

For molecules with approximately the same molecular masses, the more compact the molecule, the smaller is its entropy.

22-5. ΔS°_{rxn} EQUALS THE ENTROPY OF THE PRODUCTS MINUS THE ENTROPY OF THE REACTANTS

We can use the values of the standard molar entropies in Table 22-1 to calculate standard entropy changes of reactions, ΔS°_{rxn}. For example, let's consider the reaction

$$S(s) + 3F_2(g) \rightarrow SF_6(g)$$

The total standard entropy of the products is (Table 22-1)

$$S^\circ(\text{products}) = \bar{S}^\circ[SF_6(g)] = (1\ \text{mol})(291.7\ \text{J}\cdot\text{K}^{-1}\cdot\text{mol}^{-1}) = 291.7\ \text{J}\cdot\text{K}^{-1}$$

The total standard entropy of the reactants is (Table 22-1)

$$\begin{aligned} S^\circ(\text{reactants}) &= \bar{S}^\circ[S(s)] + 3\bar{S}^\circ[F_2(g)] \\ &= (1\ \text{mol})(31.8\ \text{J}\cdot\text{K}^{-1}\cdot\text{mol}^{-1}) \\ &\quad + (3\ \text{mol})(203.7\ \text{J}\cdot\text{K}^{-1}\cdot\text{mol}^{-1}) \\ &= 642.9\ \text{J}\cdot\text{K}^{-1} \end{aligned}$$

Just as we were able to define ΔH°_{rxn} in terms of *standard molar enthalpies of formation*, $\Delta \bar{H}^\circ_f$, we can define the standard entropy change of a reaction by

$$\Delta S^\circ_{rxn} = S^\circ(\text{products}) - S^\circ(\text{reactants})$$

Thus for the sulfur-fluorine reaction, we have

The greater the difference between the total number of moles of gaseous products and the total number of moles of gaseous reactants, the greater is the value of ΔS°_{rxn}.

$$\begin{aligned} \Delta S^\circ_{rxn} &= 291.7\ \text{J}\cdot\text{K}^{-1} - 642.9\ \text{J}\cdot\text{K}^{-1} \\ &= -351.2\ \text{J}\cdot\text{K}^{-1} \end{aligned}$$

There is a large negative change in entropy for this reaction because there is only 1 mol of gaseous product but 3 mol of gaseous reactants. Under the same conditions, the volume of the products will be only about one third the volume of the reactants (we are neglecting the volume of the sulfur because it is a solid). The product state is more ordered than the reactant state, and so the entropy of the products is less than that of the reactants.

By the same argument, the entropy change for the reaction

$$2H_2O(l) \rightarrow 2H_2(g) + O_2(g)$$

should be positive. The volume, and hence the disorder (entropy), of the products is much greater than the volume, and the entropy, of the reactants. Using Table 22-1 we obtain

$$\begin{aligned}\Delta S^\circ_{rxn} &= 2\bar{S}^\circ[H_2(g)] + \bar{S}^\circ[O_2(g)] - 2\bar{S}^\circ[H_2O(l)]\\ &= (2\text{ mol})(130.6\text{ J}\cdot\text{K}^{-1}\cdot\text{mol}^{-1})\\ &\quad + (1\text{ mol})(205.0\text{ J}\cdot\text{K}^{-1}\cdot\text{mol}^{-1})\\ &\quad - (2\text{ mol})(69.9\text{ J}\cdot\text{K}^{-1}\cdot\text{mol}^{-1})\\ &= +326.4\text{ J}\cdot\text{K}^{-1}\end{aligned}$$

We see that ΔS°_{rxn} is positive, just as we predicted.

The standard entropy change, ΔS°_{rxn}, for the general chemical reaction

$$aA + bB \rightarrow yY + zZ$$

is given by

$$\begin{aligned}\Delta S^\circ_{rxn} &= S^\circ(\text{products}) - S^\circ(\text{reactants})\\ &= (y\bar{S}^\circ_Y + z\bar{S}^\circ_Z) - (a\bar{S}^\circ_A + b\bar{S}^\circ_B)\end{aligned} \qquad (22\text{-}3)$$

where $\bar{S}^\circ_Y$ is the standard molar entropy at 1 atm of Y and y is the number of moles of Y consumed in the reaction, and similarly for the other substances involved. Equation (22-3) can be used together with the data in Table 22-1 to compute ΔS°_{rxn} values.

Example 22-3: Use data in Table 22-1 to calculate ΔS°_{rxn} for the reaction

$$CaCO_3(s) \rightarrow CaO(s) + CO_2(g)$$

What sign do you predict for ΔS°_{rxn}?

Solution: There is 1 mol of gaseous product and zero moles of gaseous reactants. Therefore, we predict that the entropy of the products will be greater than the entropy of the reactants, and so $\Delta S^\circ_{rxn} > 0$. Using data in Table 22-1, we obtain

$$\begin{aligned}\Delta S^\circ_{rxn} &= \bar{S}^\circ[CaO(s)] + \bar{S}^\circ[CO_2(g)] - \bar{S}^\circ[CaCO_3(s)]\\ &= (1\text{ mol})(39.7\text{ J}\cdot\text{K}^{-1}\cdot\text{mol}^{-1}) + (1\text{ mol})(213.6\text{ J}\cdot\text{K}^{-1}\cdot\text{mol}^{-1})\\ &\quad - (1\text{ mol})(92.9\text{ J}\cdot\text{K}^{-1}\cdot\text{mol}^{-1})\\ &= +160.4\text{ J}\cdot\text{K}^{-1}\end{aligned}$$

As we predicted, ΔS°_{rxn} is positive.

Problems 22-21 through 22-24 ask you to calculate ΔS°_{rxn} for various reactions.

Example 22-4: Predict the sign of ΔS°_{rxn} for (a) the melting of ice and (b) the dissolution of sodium chloride in water.

Solution:
(a) Liquid water is more disordered than ice, and so

$$\Delta S^\circ_{rxn} = S^\circ(\text{products}) - S^\circ(\text{reactants}) > 0$$

Problems 22-19 and 22-20 are similar to Example 22-4.

(b) Upon dissolving, the sodium and chloride ions are dispersed throughout the solvent, and so

$$\Delta S^\circ_{rxn} = S^\circ(\text{products}) - S^\circ(\text{reactants}) > 0$$

22-6. NATURE ACTS TO MINIMIZE THE ENERGY AND TO MAXIMIZE THE ENTROPY OF ALL PROCESSES

The ill-fated principle of Berthelot that a reaction has to be exothermic to be spontaneous was based on the notion that reactions that evolve energy are spontaneous because the products are energetically downhill from the reactants. However, we have seen that there are spontaneous reactions in which the products are either energetically uphill from the reactions or have essentially the same energy as the reactants. Such reactions are *entropy-driven.* A reaction for which the total entropy of the products is greater than the total entropy of the reactants ($\Delta S_{rxn} > 0$) is *entropy-favored.*

A reaction for which the total energy of the products is less than the total energy of the reactants ($\Delta H_{rxn} < 0$) is *energy-favored.* If there is little or no change in entropy ($\Delta S_{rxn} \approx 0$), then a system will change in such a way that its energy is decreased. When the energy of the system is minimized, the system no longer changes and is at equilibrium. The simple mechanical examples of energy minimization for spontaneous processes that we discussed in Section 22-1 (for example, water flowing downhill) are processes for which there is little or no change in entropy.

If there is no change in energy ($\Delta H_{rxn} \approx 0$), then a system will change in such a way that its entropy is increased. When the entropy of the system is maximized, the system no longer changes and is at equilibrium. The mixing of two gases is a good example of this case. The gases mix spontaneously because the entropy of the final state (the mixture) is greater than the entropy of the initial state (separate gases). You may have had some idea that systems try to minimize energy, but the idea of entropy maximization may be new to you. Both ideas are equally important, however.

All processes that are *both* energy-favored and entropy-favored are spontaneous. That is, if $\Delta H_{rxn} < 0$ and $\Delta S_{rxn} > 0$, then the process is spontaneous. Conversely, if $\Delta H_{rxn} > 0$ *and* $\Delta S_{rxn} < 0$, then the process does not occur spontaneously. We must now consider processes in which the energy (enthalpy) and entropy factors oppose each other.

22-7. THE SIGN OF THE GIBBS FREE ENERGY CHANGE DETERMINES WHETHER OR NOT A REACTION IS SPONTANEOUS

Many reactions have values of ΔH_{rxn} and ΔS_{rxn} that oppose each other. For such a case we have

$$\Delta H_{rxn} > 0 \qquad \text{and} \qquad \Delta S_{rxn} > 0$$

or

$$\Delta H_{rxn} < 0 \qquad \text{and} \qquad \Delta S_{rxn} < 0$$

The first case is entropy-favored but energy-disfavored, and the second case is energy-favored but entropy-disfavored. Such reactions may or may not be spontaneous, depending on the temperature and on the relative magnitudes of ΔH_{rxn} and ΔS_{rxn}. The spontaneity of such reactions was explained about 100 years ago by J. Willard Gibbs, one of America's greatest scientists. Gibbs introduced a quantity now called the *Gibbs free energy, G,* which, as we shall see, serves as a compromise function between the enthalpy change and the entropy change of a reaction. For a reaction run at a constant temperature, the *Gibbs free energy change* is given by

The officially approved IUPAC (International Union of Pure and Applied Chemistry) name of G is Gibbs energy, but the common usage is Gibbs free energy.

$$\Delta G_{rxn} = \Delta H_{rxn} - T\Delta S_{rxn} \qquad (22\text{-}4)$$

where T is the kelvin temperature.

Chemical reactions are forced by nature to seek a compromise between energy minimization and entropy maximization. For a reaction that occurs at a constant temperature and pressure, the nature of the compromise is given by the sign and magnitude of ΔG_{rxn}. The *Gibbs criteria* are

$\Delta G_{rxn} < 0$	reaction is spontaneous and additional products can form
$\Delta G_{rxn} > 0$	reaction is not spontaneous and no additional products can form without energy input
$\Delta G_{rxn} = 0$	reaction is at equilibrium and no further net change occurs

Table 22-2 Reaction spontaneity for various possible values of ΔH_{rxn} and ΔS_{rxn}

ΔH_{rxn}	ΔS_{rxn}	$\Delta G_{rxn} = \Delta H_{rxn} - T\Delta S_{rxn}$	*Comment*
$-$	$+$	$-$	spontaneous
0	$+$	$-$	spontaneous
$-$	0	$-$	spontaneous
$+$	$-$	$+$	not spontaneous
0	$-$	$+$	not spontaneous
$+$	0	$+$	not spontaneous
$+$	$+$	$+$ or $-$	spontaneity depends on relative magnitudes of ΔH_{rxn} and $-T\Delta S_{rxn}$
$-$	$-$	$+$ or $-$	
$\Delta H_{rxn} = T\Delta S_{rxn}$		0	equilibrium

The various possibilities for ΔG_{rxn} given different values of ΔH_{rxn} and ΔS_{rxn} are listed in Table 22-2.

The spontaneous mixing of two gases and the spreading of a drop of ink throughout a volume of water (Frontispiece, page 856) are examples of processes for which $\Delta H_{rxn} \approx 0$ and $\Delta S_{rxn} > 0$. For these processes,

$$\Delta G_{rxn} = \Delta H_{rxn} - T\Delta S_{rxn} = 0 - T\Delta S_{rxn} < 0$$

and so the Gibbs criteria predict that two gases mix spontaneously and that a drop of ink becomes uniformly dispersed throughout a volume of water. Furthermore, note that $\Delta G_{rxn} > 0$ for the reverse reactions. Therefore, the Gibbs criteria predict that we shall never see the reverse processes occur spontaneously.

The Gibbs criteria apply to chemical reactions as well as to the mixing of gases, and, in fact, their great utility is the prediction of whether or not a given chemical reaction occurs spontaneously. Let's consider the reaction

$$2H_2O(l) \rightarrow 2H_2(g) + O_2(g)$$

at 25°C. Using data in Table 22-1, we calculate the entropy change for this reaction to be

$$\begin{aligned}\Delta S^\circ_{rxn} &= 2\bar{S}^\circ[H_2(g)] + \bar{S}^\circ[O_2(g)] - 2\bar{S}^\circ[H_2O(l)] \\ &= (2\text{ mol})(130.6\text{ J}\cdot\text{K}^{-1}\cdot\text{mol}^{-1}) \\ &\qquad + (1\text{ mol})(205.0\text{ J}\cdot\text{K}^{-1}\cdot\text{mol}^{-1}) \\ &\qquad - (2\text{ mol})(69.9\text{ J}\cdot\text{K}^{-1}\cdot\text{mol}^{-1}) \\ &= 326.4\text{ J}\cdot\text{K}^{-1}\end{aligned}$$

The entropy change is positive, and so the reaction is entropy-favored. Let's calculate ΔH°_{rxn}, using the data in Table 22-1. We find that

$$\begin{aligned}\Delta H^\circ_{rxn} &= 2\Delta \bar{H}^\circ_f[H_2(g)] + \Delta \bar{H}^\circ_f[O_2(g)] - 2\Delta \bar{H}^\circ_f[H_2O(l)] \\ &= (2\ \text{mol})(0) + (1\ \text{mol})(0) - (2\ \text{mol})(-285.8\ \text{kJ}\cdot\text{mol}^{-1}) \\ &= +571.6\ \text{kJ}\end{aligned}$$

Thus, the reaction is energy-disfavored. To tell whether or not the reaction is spontaneous, we must use the Gibbs criteria and determine whether ΔG°_{rxn}, the *standard Gibbs free energy change,* for the reaction, is positive or negative. According to Equation (22-4), at 1 atm,

$$\Delta G^\circ_{rxn} = \Delta H^\circ_{rxn} - T\Delta S^\circ_{rxn}$$

At 25°C, we have

$$\begin{aligned}\Delta G^\circ_{rxn} &= 571.6 \times 10^3\ \text{J} - (298\ \text{K})(326.4\ \text{J}\cdot\text{K}^{-1}) \\ &= +4.74 \times 10^5\ \text{J}\end{aligned}$$

We see that ΔG°_{rxn} is greater than zero, and so the reaction is *not* spontaneous at 25°C and 1 atm. Certainly this result is in accord with our everyday observations; we know that water is a very stable substance and does not decompose spontaneously into hydrogen and oxygen.

Example 22-5: For the reaction at 25°C

$$S(s) + O_2(g) \rightleftharpoons SO_2(g)$$

$\Delta H^\circ_{rxn} = -297$ kJ and $\Delta S^\circ_{rxn} = 11.6\ \text{J}\cdot\text{K}^{-1}$. Compute ΔG°_{rxn} at 25°C and indicate the direction in which the reaction is spontaneous with the gases O_2 and SO_2 both at 1.0 atm pressure.

Solution: We compute ΔG°_{rxn} by using Equation (22-4):

$$\begin{aligned}\Delta G^\circ_{rxn} = \Delta H^\circ_{rxn} - T\Delta S^\circ_{rxn} &= -297 \times 10^3\ \text{J} - (298\ \text{K})(11.6\ \text{J}\cdot\text{K}^{-1}) \\ &= -300\ \text{kJ}\end{aligned}$$

The value of ΔG°_{rxn} is large and negative, and thus the reaction is spontaneous left to right. Sulfur (when ignited) burns rapidly in air to form sulfur dioxide. Note, however, that in the absence of some form of ignition, $S(s)$ is stable in air. A negative value of ΔG°_{rxn} is no guarantee that the reaction *actually* occurs. Thermodynamics has nothing to say about *the rate* of a chemical reaction.

Problems 22-25 through 22-30 are similar to Example 22-5.

One nice application of the Gibbs criteria is the determination of the temperature range in which a reaction is spontaneous. For example, consider the reaction

$$N_2(g) + 3H_2(g) \rightleftharpoons 2NH_3(g)$$

Using the data in Table 22-1, we see that

$$\Delta H^\circ_{rxn} = (2\text{ mol})(-46.19\text{ kJ}\cdot\text{mol}^{-1}) - (1\text{ mol})(0) - (3\text{ mol})(0) = -92.38\text{ kJ}$$

and

$$\begin{aligned}\Delta S^\circ_{rxn} &= (2\text{ mol})(192.5\text{ J}\cdot\text{K}^{-1}\cdot\text{mol}^{-1}) \\ &\quad - (1\text{ mol})(191.5\text{ J}\cdot\text{K}^{-1}\cdot\text{mol}^{-1}) \\ &\quad - (3\text{ mol})(130.6\text{ J}\cdot\text{K}^{-1}\cdot\text{mol}^{-1}) \\ &= -198.3\text{ J}\cdot\text{K}^{-1}\end{aligned}$$

Note that this reaction is energy-favored ($\Delta H^\circ_{rxn} < 0$) but entropy-disfavored ($\Delta S^\circ_{rxn} < 0$) at 1 atm. Substitution of the values of ΔH°_{rxn} and ΔS°_{rxn} into Equation (22-4) yields, at 25°C and 1 atm,

$$\begin{aligned}\Delta G^\circ_{rxn} &= \Delta H^\circ_{rxn} - T\Delta S^\circ_{rxn} \\ &= -92.38 \times 10^3\text{ J} - (298\text{ K})(-198.3\text{ J}\cdot\text{K}^{-1}) = -33.3\text{ kJ}\end{aligned}$$

Thus, we find that, at 25°C and 1 atm, the ammonia synthesis reaction is spontaneous left to right as written.

We can calculate ΔG°_{rxn} at temperatures other than 25°C if we assume that ΔH°_{rxn} and ΔS°_{rxn} do not change appreciably with temperature. Usually this is a good assumption, and we shall always make it. Assuming, then, that ΔH°_{rxn} and ΔS°_{rxn} do not change with temperature, ΔG°_{rxn} for the ammonia synthesis reaction at 500°C is

$$\Delta G^\circ_{rxn} = -92.38 \times 10^3\text{ J} - (773\text{ K})(-198.3\text{ J}\cdot\text{K}^{-1}) = +60.9\text{ kJ}$$

Thus at 500°C and 1 atm, the ammonia synthesis reaction is spontaneous right to left. Note that the $-T\Delta S^\circ_{rxn}$ term becomes more important as the temperature increases. In the ammonia synthesis reaction, the increasing importance of the $-T\Delta S^\circ_{rxn}$ term with increasing temperature is sufficiently great to reverse the direction of spontaneity at 500°C.

Reactions with positive values of ΔS°_{rxn} always become spontaneous at high temperature.

It is generally true, as shown for the ammonia synthesis reaction, that the energy change for a chemical reaction is the dominant factor in determining the equilibrium distribution of species at low temperatures and the entropy change is the dominant factor at high temperatures. At intermediate temperatures, energy and entropy factors often oppose each other and a compromise between minimization of energy and maximization of entropy is reached.

22-8. THE VALUE OF ΔG_{rxn} IS EQUAL TO THE MAXIMUM AMOUNT OF WORK THAT CAN BE OBTAINED FROM THE REACTION

We saw in Chapter 15 that the magnitude of the ratio of the reaction quotient to the equilibrium constant, Q/K, can be used to determine whether a reaction is spontaneous. If $Q/K < 1$, then the reaction is spontaneous left to right as written. We saw in Chapter 21 that a reaction in an electrochemical cell is spontaneous if the cell voltage, E, is positive. Further, E and Q/K for a reaction are related through the Nernst equation:

$$E = -\left(\frac{2.30RT}{nF}\right)\log\frac{Q}{K}$$

The Nernst equation shows that if $Q/K < 1$, then $\log(Q/K) < 0$ and $E > 0$ (Section 21-3).

The Gibbs free energy change for an oxidation-reduction reaction is related to the corresponding cell voltage by the equation (see Equation 21-3)

$$\text{Gibbs free energy change} = \text{charge transferred} \times \text{voltage}$$

or

$$\Delta G_{rxn} = -nFE \qquad (22\text{-}5)$$

where n is the number of moles of electrons transferred from the reducing agent to the oxidizing agent for the reaction as written and F is Faraday's constant (96,500 C per mole of electrons). Recall from Section 21-3 that charge × voltage has units of energy. If we use $F = 96{,}500\ \text{C}\cdot\text{mol}^{-1}$ and E in volts in Equation (22-5), then the units of ΔG_{rxn} will be joules.

Example 22-6: An electrochemical cell is set up in which the reaction

$$2H_2(g) + O_2(g) \rightleftharpoons 2H_2O(l)$$

occurs. At 25°C the measured cell voltage is 1.23 V. Compute the value of ΔG_{rxn} for the reaction.

Solution: The value of ΔG_{rxn} can be computed using Equation (22-5), but we first must determine the value of n for the reaction. In the reaction 2 mol of oxygen *atoms* (1 mol of O_2) are reduced from an oxidation state of zero to an oxidation state of −2. Thus the reaction requires 4 mol of electrons per mole of O_2 and so $n = 4$ mol. Therefore

$$\Delta G_{rxn} = -(4\ \text{mol})(96{,}500\ \text{C}\cdot\text{mol}^{-1})(1.23\ \text{V}) = -4.75 \times 10^5\ \text{C}\cdot\text{V}$$
$$= -4.75 \times 10^5\ \text{J}$$

The negative value of ΔG_{rxn} indicates that the reaction is spontaneous, which we already know from the fact that mixtures of H_2 and O_2 can react explosively.

The word "free" in Gibbs free energy is used in the sense of "available to perform work."

The value of ΔG_{rxn} is related to the energy that can be obtained from a process. If ΔG_{rxn} is negative (spontaneous reaction), then its magnitude is equal to the maximum energy that can be obtained from the reaction. In Example 22-6, the reaction of 2 mol of $H_2(g)$ with 1 mol of $O_2(g)$ can provide a maximum of 4.75×10^5 J of energy to an external device, such as an electric motor. The maximum amount of energy that we calculate for a reaction is an ideal value. In practice we would obtain less than the maximum amount.

If ΔG_{rxn} is positive, then its value is the minimum energy that must be supplied to the reaction in order to make it occur. For example, Example 22-6 shows that $\Delta G_{rxn} = +4.75 \times 10^5$ J for the decomposition of water into $H_2(g)$ and $O_2(g)$ at 1 atm. Thus we must supply at least 4.75×10^5 J to decompose 2 mol of water. This energy could be supplied by an electric current.

22-9. THE VALUE OF ΔG_{rxn} DEPENDS UPON THE RATIO Q/K

In Chapter 21 we related the voltage produced by a cell reaction to the ratio Q/K. We know that if

$$\frac{Q}{K} > 1 \qquad \text{reaction is not spontaneous}$$

$$\frac{Q}{K} < 1 \qquad \text{reaction is spontaneous}$$

$$\frac{Q}{K} = 1 \qquad \text{reaction is at equilibrium}$$

Because the sign of ΔG_{rxn} also determines whether a reaction is spontaneous or not, we should expect a relation between ΔG_{rxn} and Q/K. This relation is

$$\Delta G_{rxn} = 2.30RT \log \frac{Q}{K} \qquad (22\text{-}6)$$

Equation (22-6) can be obtained by combining Equation (21-2) with Equation (22-5). The following example illustrates the use of Equation (22-6).

Example 22-7: The reaction system

$$2SO_2(g) + O_2(g) \rightleftharpoons 2SO_3(g)$$

is prepared at 100°C such that $Q/K = 1.0 \times 10^{-3}$. Compute ΔG_{rxn} under these conditions.

Solution: We know from the fact that $Q/K < 1$ that the reaction is spontaneous. From Equation (22-6) we have

$$\Delta G_{rxn} = 2.30RT \log \frac{Q}{K}$$

Thus

$$\Delta G_{rxn} = (2.30)(8.31\ \text{J}\cdot\text{K}^{-1})(373\ \text{K}) \log(1.0 \times 10^{-3}) = -21.4\ \text{kJ}$$

The negative value of ΔG_{rxn} shows that the reaction is spontaneous left to right as written. A maximum of 21.4 kJ of utilizable work can be obtained from the reaction per 2 mol of $SO_2(g)$ consumed at 100°C with $Q/K = 1.0 \times 10^{-3}$.

We can express an equilibrium constant in terms of a standard Gibbs free energy change. We start with Equation (22-6) and utilize the fact that the logarithm of a ratio is equal to the logarithm of the numerator minus the logarithm of the denominator. Equation (22-6) thus can be rewritten as

$$\Delta G_{rxn} = 2.30RT \log \frac{Q}{K}$$

$$= 2.30RT \log Q - 2.30RT \log K \qquad (22\text{-}7)$$

$\text{Log} \frac{Q}{K} = \log Q - \log K.$

When $Q = 1$, all reactants and products are at 1.00 M concentration for solution species or 1.00 atm pressure for gaseous species. Under these conditions, ΔG_{rxn} is equal to ΔG°_{rxn}. Setting $Q = 1$ in Equation (22-7) thus yields

$$\Delta G^\circ_{rxn} = -2.30RT \log K \qquad (22\text{-}8)$$

The value of ΔG°_{rxn} for a cell reaction is given by $\Delta G^\circ_{rxn} = -nFE^\circ_{cell}$

Equation (22-8) shows that the magnitude of the standard Gibbs free energy change for a reaction at a particular temperature is determined by the magnitude of the equilibrium constant for the reaction at that temperature. If K is greater than unity, then $\log K$ is positive and ΔG°_{rxn} is negative. In other words, if $K > 1$, then the reaction is spontaneous *under standard conditions,* that is, for 1.00 M concentration for solution species and 1.00 atm pressure for gas species. If K is less than unity, then $\log K$ is negative, ΔG°_{rxn} is positive, and the reaction is not spontaneous under standard conditions.

Example 22-8: The equilibrium constant at 25°C for the reaction

$$HC_2H_3O_2(aq) + H_2O(l) \rightleftharpoons H_3O^+(aq) + C_2H_3O_2^-(aq)$$

is $K = 1.74 \times 10^{-5}$. Compute the value of ΔG°_{rxn} at 25°C.

Solution: The value of ΔG°_{rxn} can be computed from the value of K by using Equation (22-8):

$$\begin{aligned}\Delta G^{\circ}_{rxn} &= -2.30RT \log K \\ &= -(2.30)(8.31\ \mathrm{J\cdot K^{-1}})(298\ \mathrm{K}) \log(1.74 \times 10^{-5}) \\ &= +2.71 \times 10^{4}\ \mathrm{J}\end{aligned}$$

Problems 22-41 through 22-54 involve the relation between equilibrium constants and ΔG°_{rxn}

Note that the acetic acid dissociation reaction is not spontaneous under standard conditions ($[H_3O^+] = [C_2H_3O_2^-] = [HC_2H_3O_2] = 1.00$ M) because ΔG°_{rxn} is positive.

The criteria for reaction spontaneity are summarized in Table 22-3. In applying any of these criteria to a reaction, it is worth remembering that, even though a reaction may be spontaneous under the prevailing conditions, it may not occur at a detectable rate. *Spontaneous is not synonymous with immediate*. On the other hand, if $\Delta G_{rxn} > 0$, then the reaction will not occur under the prevailing conditions. The *no* of thermodynamics is emphatic; the *yes* of thermodynamics is actually a *maybe*. For a reaction to occur, it is *absolutely necessary* that $\Delta G_{rxn} < 0$, but a negative value of ΔG_{rxn} is not sufficient to guarantee that the reaction will occur at a detectable rate.

Table 22-3 Criteria for reaction spontaneity

	Value of Q/K	*Cell voltage*	*Gibbs free energy change*
spontaneous reaction	< 1	$E > 0$	$\Delta G_{rxn} < 0$
nonspontaneous reaction	> 1	$E < 0$	$\Delta G_{rxn} > 0$
reaction at equilibrium	$= 1$	$E = 0$	$\Delta G_{rxn} = 0$

22-10. ΔG°_{rxn} CAN BE CALCULATED FROM TABULATED $\Delta \bar{G}^\circ_f$ VALUES

In Chapter 6 we described the procedure for setting up a table of standard molar enthalpies of formation of compounds from their constituent elements. The procedure for setting up a table of *standard molar Gibbs free energies of formation* of compounds from their constituent elements is exactly the same as for $\Delta \bar{H}^\circ_f$ values. Table 22-1 lists $\Delta \bar{G}^\circ_f$ values for a variety of compounds. These values are used just like the $\Delta \bar{H}^\circ_f$ values in Table 22-1. For example, for the reaction

$$2H_2(g) + O_2(g) \rightarrow 2H_2O(l)$$

we have

$$\begin{aligned}\Delta G^\circ_{rxn} &= 2\Delta \bar{G}^\circ_f[H_2O(l)] - 2\Delta \bar{G}^\circ_f[H_2(g)] - \Delta \bar{G}^\circ_f[O_2(g)] \\ &= (2\text{ mol})(-237.2\text{ kJ}\cdot\text{mol}^{-1}) - (2\text{ mol})(0) - (1\text{ mol})(0) \\ &= -474.4\text{ kJ}\cdot\text{mol}^{-1}\end{aligned}$$

The role of $\Delta \bar{G}^\circ_f$ values in the calculation of ΔG°_{rxn} values can be seen from the following diagram for the general reaction

$$aA + bB \rightarrow yY + zZ$$

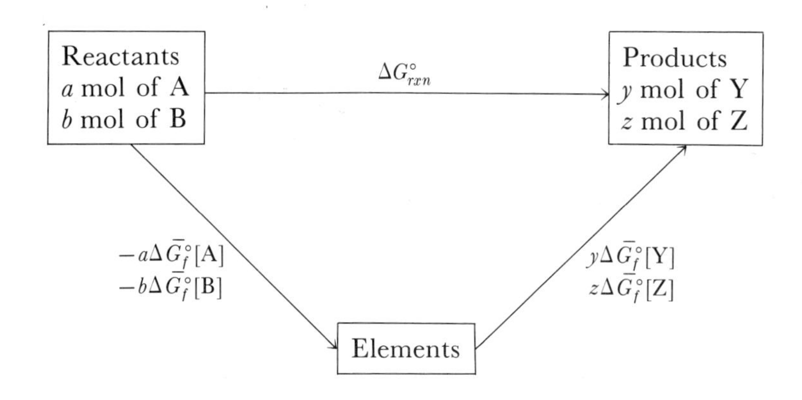

Note that one obtains the same value of ΔG°_{rxn} regardless of how we go from reactants to products

from which we note that

$$\Delta G^\circ_{rxn} = \{y\Delta \bar{G}^\circ_f[Y] + z\Delta \bar{G}^\circ_f[Z]\} - \{a\Delta \bar{G}^\circ_f[A] + b\Delta \bar{G}^\circ_f[B]\} \quad (22\text{-}9)$$

Equation 22-9 is similar to Equation 6-6 for ΔH°_{rxn}

or, more simply,

$$\Delta G^\circ_{rxn} = \Delta \bar{G}^\circ_f\text{ (products)} - \Delta \bar{G}^\circ_f\text{ (reactants)}$$

Schematically the reactants are converted to the elements and then the elements are converted to the products. By convention, we take $\Delta \bar{G}_f^\circ$ of an element in its normal physical state at 25°C and 1 atm to be zero. This is similar to the convention that we introduced for $\Delta \bar{H}_f^\circ$ in Chapter 6. Let's consider the reaction

$$N_2(g) + 3H_2(g) \rightleftharpoons 2NH_3(g)$$

for which at 298 K

$$\Delta G_{rxn}^\circ = -33.2 \text{ kJ}$$
$$\Delta H_{rxn}^\circ = -92.4 \text{ kJ}$$

Application of Equation (22-9) yields

$$\begin{aligned}\Delta G_{rxn}^\circ &= -33.2 \text{ kJ} \\ &= 2\Delta \bar{G}_f^\circ[NH_3(g)] - \Delta \bar{G}_f^\circ[N_2(g)] - 3\Delta \bar{G}_f^\circ[H_2(g)]\end{aligned}$$

By convention, however, $\Delta \bar{G}_f^\circ[N_2(g)] = 0$ and $\Delta \bar{G}_f^\circ[H_2(g)] = 0$ (Table 22-1), and therefore

$$\Delta \bar{G}_f^\circ[NH_3(g)] = -\frac{33.2 \text{ kJ}}{2 \text{ mol}} = -16.6 \text{ kJ}\cdot\text{mol}^{-1}$$

We can calculate $\Delta \bar{H}_f^\circ[NH_3(g)]$ by the same procedure. Application of Equation (6-6) yields

$$\begin{aligned}\Delta H_{rxn}^\circ &= -92.4 \text{ kJ} \\ &= 2\Delta \bar{H}_f^\circ[NH_3(g)] - \Delta \bar{H}_f^\circ[N_2(g)] - 3\Delta \bar{H}_f^0[H_2(g)]\end{aligned}$$

but $\Delta \bar{H}_f^\circ[N_2(g)] = 0$ and $\Delta \bar{H}_f^\circ[H_2(g)] = 0$ (Table 22-1), and so

$$\Delta \bar{H}_f^\circ[NH_3(g)] = \frac{-92.4 \text{ kJ}}{2 \text{ mol}} = -46.2 \text{ kJ}\cdot\text{mol}^{-1}$$

Tables of $\Delta \bar{G}_f^\circ$ and $\Delta \bar{H}_f^\circ$ values can be built up and then used to compute ΔG_{rxn}° and ΔH_{rxn}° values for reactions involving species for which the $\Delta \bar{G}_f^\circ$ and $\Delta \bar{H}_f^\circ$ values are known.

We can use Equation (22-8) to calculate the equilibrium constant for the ammonia synthesis reaction. Using the value of ΔG_{rxn}° that we have calculated ($-33.2 \text{ kJ}\cdot\text{mol}^{-1}$), we have at 25°C

$$\Delta G^\circ_{rxn} = -2.30RT \log K_p$$

$$\log K_p = -\frac{\Delta G^\circ_{rxn}}{2.30RT} = -\frac{-33.2 \times 10^3\ \text{J} \cdot \text{mol}^{-1}}{(2.30)(8.31\ \text{J} \cdot \text{mol}^{-1} \cdot \text{K}^{-1})(298\ \text{K})}$$

$$= 5.83$$

$$K_p = 6.76 \times 10^5$$

Example 22-9: Calculate the equilibrium constant for the reaction

$$2NH_3(g) + \tfrac{7}{2}O_2(g) \rightleftharpoons 2NO_2(g) + 3H_2O(g)$$

Solution: Application of Equation (22-9) to this reaction, which is used in the production of nitric acid by the Ostwald process, yields

$$\Delta G^\circ_{rxn} = 2\Delta \bar{G}^\circ_f[NO_2(g)] + 3\Delta \bar{G}^\circ_f[H_2O(g)] - 2\Delta \bar{G}^\circ_f[NH_3(g)] - \tfrac{7}{2}\Delta \bar{G}^\circ_f[O_2(g)]$$

Referring to Table 22-1 for values of $\Delta \bar{G}^\circ_f$, we obtain at 25°C

$$\Delta G^\circ_{rxn} = (2\ \text{mol})(51.84\ \text{kJ} \cdot \text{mol}^{-1}) + (3\ \text{mol})(-228.6\ \text{kJ} \cdot \text{mol}^{-1}) - (2\ \text{mol})(-16.64\ \text{kJ} \cdot \text{mol}^{-1}) - (\tfrac{7}{2}\ \text{mol})(0\ \text{kJ} \cdot \text{mol}^{-1})$$

$$= -548.8\ \text{kJ}$$

The value of K_p at 25°C is

$$\log K_p = -\frac{\Delta G^\circ_{rxn}}{2.30RT} = -\frac{-548.8 \times 10^3\ \text{J} \cdot \text{mol}^{-1}}{(2.30)(8.31\ \text{J} \cdot \text{K}^{-1} \cdot \text{mol}^{-1})(298\ \text{K})}$$

$$= 96.35$$

$$K_p = 2.24 \times 10^{96}$$

which implies that the reaction essentially goes to completion.

Problems 22-55 through 22-62 ask you to calculate ΔG°_{rxn} and K using data in Table 22-1

Tables of $\Delta \bar{G}^\circ_f$, $\Delta \bar{H}^\circ_f$, and $\bar{S}^\circ$ values are especially useful for the thermodynamic analysis of chemical reactions and, in particular, for the prediction of reaction spontaneity under standard conditions.

22-11. IT IS THE SIGN OF ΔG_{rxn} AND NOT ΔG°_{rxn} THAT DETERMINES REACTION SPONTANEITY

It is important to remember the difference between ΔG_{rxn} and ΔG°_{rxn}. We can write Equation (22-6) in the form

$$\Delta G_{rxn} = -2.30RT \log K + 2.30RT \log Q$$

Using Equation (22-8), we have

$$\Delta G_{rxn} = \Delta G^\circ_{rxn} + 2.30RT \log Q \qquad (22\text{-}10)$$

If all the reactants and products are at standard conditions, that is, if all concentrations are 1.00 M and all pressures are 1.00 atm, then $Q = 1$ in Equation (22-10) and

$$\Delta G_{rxn} = \Delta G^\circ_{rxn} \qquad \text{standard conditions}$$

We emphasize standard conditions here to point out that ΔG°_{rxn} is equal to ΔG_{rxn} only if all reactants and products are at standard conditions. Recall that ΔG°_{rxn} is called the *standard* Gibbs free energy change. If reactants and products are not at standard conditions, then we must calculate ΔG_{rxn} from Equation (22-10).

To see clearly the distinction between ΔG_{rxn} and ΔG°_{rxn}, consider the familiar reaction of the dissociation of the weak acid acetic acid:

$$HC_2H_3O_2(aq) + H_2O(l) \rightleftharpoons H_3O^+(aq) + C_2H_3O_2^-(aq)$$
$$K_a = 1.74 \times 10^{-5}\ \text{M at } 25\,^\circ\text{C}$$

The standard Gibbs free energy change for this reaction at 25 °C is (Example 22-8) $\Delta G^\circ_{rxn} = +27.1$ kJ. The fact that ΔG°_{rxn} is positive does *not* mean that no acetic acid dissociates when we dissolve acetic acid in water at 25 °C. Some acetic acid does dissociate because it is ΔG_{rxn}, not ΔG°_{rxn}, that determines whether or not a reaction occurs. The value of ΔG_{rxn} at 25 °C for the dissociation of acetic acid in water is

$$\begin{aligned}\Delta G_{rxn} &= \Delta G^\circ_{rxn} + 2.30RT \log Q \\ &= 27.1\ \text{kJ} + (5.70\ \text{kJ}) \log \frac{[H_3^+O]_0[C_2H_3O_2^-]_0}{[HC_2H_3O_2]_0}\end{aligned} \qquad (22\text{-}11)$$

The logarithm of a very small number is a large negative number

Let's consider a 0.10 M $HC_2H_3O_2(aq)$ solution. With $[HC_2H_3O_2]_0 = 0.10$ M and $[H_3^+O]_0 = [C_2H_3O_2^-]_0 \approx 0$, the value of ΔG_{rxn} is very large and *negative*. Therefore, the dissociation of $HC_2H_3O_2(aq)$ takes place spontaneously. The concentration of $HC_2H_3O_2(aq)$ decreases and the concentrations of $H_3^+O(aq)$ and $C_2H_3O_2^-(aq)$ increase until equilibrium is reached. The equilibrium state is determined by the condition $\Delta G_{rxn} = 0$. If we set $\Delta G_{rxn} = 0$ in Equation (22-10), then we obtain

$$Q = K \qquad \text{equilibrium}$$

At equilibrium, $\Delta G_{rxn} = 0$, but, in general, $\Delta G^\circ_{rxn} \neq 0$ at equilibrium.

Thus, initially ΔG_{rxn} has a large negative value, but this value decreases to zero as the reaction goes to equilibrium.

The value of ΔG_{rxn} depends upon the concentrations of the reactants and products through the quantity Q in Equation (22-10). The sign of ΔG_{rxn} depends upon the ratio Q/K. The value of ΔG°_{rxn}, on the other hand, is fixed at any given temperature and requires that all reactants and products be at standard conditions.

SUMMARY

Not all chemical reactions that release energy are spontaneous. Reaction spontaneity is determined by both energy and entropy changes. Entropy is a measure of the disorder, or randomness, of a system. In general, the position of equilibrium in a chemical reaction involves a compromise between a minimization of the energy and a maximization of the entropy.

The entropy of a compound increases with increasing temperature. Both fusion and vaporization processes lead to an increase in the entropy of a compound because molecules in a liquid are more disordered than the same molecules in the solid phase and molecules in a gas are more disordered than the same molecules in the liquid phase.

The entropy of a perfect crystalline substance is zero at 0 K. This property of crystals is the basis of the absolute entropy scale, on which the entropy of a compound at $T > 0$ K is always positive and increases with increasing temperature.

The Gibbs criterion for reaction spontaneity is that ΔG_{rxn} must be less than zero. The value of ΔG_{rxn} depends on the value of ΔH_{rxn} and ΔS_{rxn}. For a reaction with $\Delta G_{rxn} < 0$, the value of ΔG_{rxn} equals the maximum amount of work that can be obtained from the reaction under the stated conditions.

Spontaneous is not synonymous with immediate. The fact that $\Delta G_{rxn} < 0$ is not sufficient to guarantee that a reaction proceeds toward equilibrium at a detectable rate.

The standard Gibbs free energy change ΔG°_{rxn} is equal to the value of ΔG_{rxn} when all products and reactants are at standard conditions (1 atm, 1 M).

Summary of criteria for reaction spontaneity

	Value of Q/K	*Cell voltage*	*Gibbs free energy change*
spontaneous reaction	< 1	$E > 0$	$\Delta G_{rxn} < 0$
nonspontaneous reaction	> 1	$E < 0$	$\Delta G_{rxn} > 0$
reaction at equilibrium	$= 1$	$E = 0$	$\Delta G_{rxn} = 0$

If $\Delta H_{rxn} \leqslant 0$ and $\Delta S_{rxn} > 0$, then $\Delta G_{rxn} < 0$ and the reaction is spontaneous.

If $\Delta H_{rxn} \geqslant 0$ and $\Delta S_{rxn} < 0$, then $\Delta G_{rxn} > 0$ and the reaction is not spontaneous.

If ΔH_{rxn} and ΔS_{rxn} have the same sign, then the spontaneity of the reaction depends on the relative magnitudes of ΔH_{rxn} and $-T\Delta S_{rxn}$, because $\Delta G_{rxn} = \Delta H_{rxn} - T\Delta S_{rxn}$.

TERMS YOU SHOULD KNOW

enthalpy change, ΔH_{rxn} 858
entropy, S 859
positional disorder 859
thermal disorder 859
molar entropy of fusion, $\Delta \bar{S}_{fus}$ 861
molar entropy of vaporization, $\Delta \bar{S}_{vap}$ 862
standard entropy, $S°$ 863
entropy change, ΔS_{rxn} 863
standard enthalpy change, ΔH°_{rxn} 868
standard molar enthalpy of formation, $\Delta \bar{H}^\circ_f$ 868
standard entropy change, ΔS°_{rxn} 868
entropy-driven reaction 870
entropy-favored reaction 870
energy-favored reaction 870
Gibbs free energy, G 871
Gibbs free energy change, ΔG_{rxn} 871
Gibbs criteria 871
standard Gibbs free energy change, ΔG°_{rxn} 873
standard molar Gibbs free energy of formation, $\Delta \bar{G}^\circ_f$ 879

EQUATIONS YOU SHOULD KNOW HOW TO USE

$$\Delta \bar{S}_{fus} = \frac{\Delta \bar{H}_{fus}}{T_m} \quad (22\text{-}1)$$

$$\Delta \bar{S}_{vap} = \frac{\Delta \bar{H}_{vap}}{T_b} \quad (22\text{-}2)$$

$$\Delta S^\circ_{rxn} = (y\bar{S}^\circ_Y + z\bar{S}^\circ_Z) - (a\bar{S}^\circ_A + b\bar{S}^\circ_B) \quad (22\text{-}3)$$

$$\Delta G_{rxn} = \Delta H_{rxn} - T\Delta S_{rxn} \quad (22\text{-}4)$$

$$\Delta G^\circ_{rxn} = \Delta H^\circ_{rxn} - T\Delta S^\circ_{rxn}$$

$$\Delta G_{rxn} = -nFE \quad (22\text{-}5)$$

$$\Delta G_{rxn} = 2.30RT\log(Q/K) \quad (22\text{-}6)$$

$$\Delta G^\circ_{rxn} = -2.30RT\log K \quad (22\text{-}8)$$

$$\Delta G^\circ_{rxn} = (y\,\Delta G^\circ_f[Y] + z\,\Delta G^\circ_f[Z]) - (a\,\Delta G^\circ_f[A] + b\,\Delta G^\circ_f[B]) \quad (22\text{-}9)$$

PROBLEMS

ENTROPIES OF FUSION AND VAPORIZATION

22-1. From the following data, calculate $\Delta \bar{S}_{fus}$ and $\Delta \bar{S}_{vap}$ for lithium chloride, sodium chloride, and potassium chloride:

	$t_m/°C$	$\Delta \bar{H}_{fus}/kJ\cdot mol^{-1}$	$t_b/°C$	$\Delta \bar{H}_{vap}/kJ\cdot mol^{-1}$
LiCl	614	19.8	1325	161.4
NaCl	800	28.2	1413	181.7
KCl	770	26.8	1500	172.8

22-2. From the following data, calculate $\Delta \bar{S}_{fus}$ and $\Delta \bar{S}_{vap}$ for hydrogen fluoride, hydrogen chloride, hydrogen bromide, and hydrogen iodide:

	$t_m/°C$	$\Delta \bar{H}_{fus}/kJ\cdot mol^{-1}$	$t_b/°C$	$\Delta \bar{H}_{vap}/kJ\cdot mol^{-1}$
HF	−83.11	4.577	19.54	25.18
HCl	−114.3	1.991	−84.9	17.53
HBr	−86.96	2.406	−67.0	19.27
HI	−50.91	2.871	−35.38	21.16

22-3. From the following data, calculate $\Delta\bar{S}_{fus}$ and $\Delta\bar{S}_{vap}$ for dichloromethane, chloroform, and carbon tetrachloride:

	$t_m/°C$	$\Delta\bar{H}_{fus}/kJ \cdot mol^{-1}$	$t_b/°C$	$\Delta\bar{H}_{vap}/kJ \cdot mol^{-1}$
CH_2Cl_2	−95.14	1.434	40.0	31.68
$CHCl_3$	−63.6	8.800	61.7	31.38
CCl_4	−23.0	3.275	76.54	31.92

CH_2Cl_2 8.056 J/K·mol 101.2 J/K·mol
$CHCl_3$ 41.99 J/K·mol 93.70 J/K·mol
CCl_4 13.09 J/K·mol 91.28 J/K·mol

22-4. From the following data, calculate $\Delta\bar{S}_{fus}$ and $\Delta\bar{S}_{vap}$ for methyl alcohol, CH_3OH; ethyl alcohol, C_2H_5OH; and *n*-propyl alcohol, C_3H_7OH:

	$t_m/°C$	$\Delta\bar{H}_{fus}/kJ \cdot mol^{-1}$	$t_b/°C$	$\Delta\bar{H}_{vap}/kJ \cdot mol^{-1}$
CH_3OH	−97.8	3.177	64.96	37.57
C_2H_5OH	−114.5	5.021	78.5	40.48
C_3H_7OH	−126.1	5.195	97.4	43.60

22-5. Calculate $\Delta\bar{S}_{fus}$ for the alkali metals:

Metal	t_m/K	$\Delta\bar{H}_{fus}/kJ \cdot mol^{-1}$	
Li	454	2.99	6.59 J/K·mol
Na	371	2.60	7.01 J/K·mol
K	336	2.33	6.93 J/K·mol
Rb	312	2.34	7.50 J/K·mol
Cs	302	2.10	6.95 J/K·mol

22-6. Calculate $\Delta\bar{S}_{vap}$ for the alkali metals:

Metal	t_b/K	$\Delta\bar{H}_{vap}/kJ \cdot mol^{-1}$
Li	1615	134.7
Na	1156	89.6
K	1033	77.1
Rb	956	69
Cs	942	66

22-7. From the following data, calculate $\Delta\bar{S}_{fus}$ and $\Delta\bar{S}_{vap}$ for hydrogen sulfide.

	$t_m/°C$	$\Delta\bar{H}_{fus}/kJ \cdot mol^{-1}$	$t_b/°C$	$\Delta\bar{H}_{vap}/kJ \cdot mol^{-1}$
H_2S	−85.6	2.38	−60.7	18.7

Note the differences between these values and those for water (Example 22-1). Give a simple molecular interpretation for the differences.

$\Delta\bar{S}_{fus} = \frac{2.38 \times 10^3 \text{ J/mol}}{187.6 \text{ K}} = 12.7$ J/K·mol

$\Delta\bar{S}_{vap} = \frac{18.7 \times 10^3 \text{ J/mol}}{212.5 \text{ K}} = 88.0$ J/K·mol

H_2O $\Delta\bar{S}_{fus} = 22.1$ J/K·mol

$\Delta\bar{S}_{vap} = 109$ J/K·mol

Values are larger for H_2O because of the breaking of the H bonds. The H bonds make for a higher degree of order; their breaking produces a greater increase in disorder

22-8. Arrange the compounds NH_3, CH_4, and H_2O in order of increasing $\Delta\bar{S}_{vap}$ values. Describe the reasoning that you used to reach your conclusions.

ENTROPY AND STRUCTURE

22-9. In each case, predict which molecule of the pair has the greater molar entropy under the same conditions:

(a) PCl_3 PF_3

(b) $CH_3CH_2CH_3$ (propane) $H_2C—CH_2$ / CH_2 (cyclopropane)

(c)

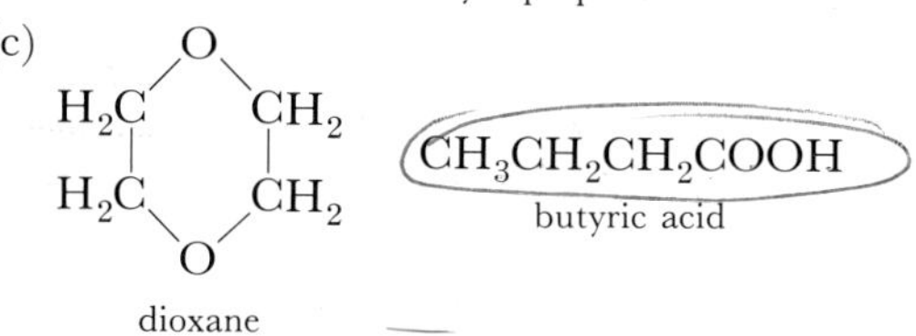

dioxane $CH_3CH_2CH_2COOH$ (butyric acid)

(d) $CH_3CH_2CH_2CH_2CH_3$ (*n*-pentane) $H_3C—C(CH_3)_2—CH_3$ (neopentane)

22-10. In each case, predict which molecule of the pair has the greater molar entropy under the same conditions:

(a) PCl_3 PCl_5

(b) CH_3CH_2OH (ethanol) $H_2C—CH_2$ / O (ethylene oxide)

(c) $CH_3CH_2CH_2CH_2NH_2$ (butyl amine) pyrrolidine

22-11. Arrange the following molecules in order of increasing values of total molar entropy:

$C_2Cl_4(g)$ $\quad$ $C_2Cl_6(g)$ $\quad$ $C_2Cl_2(g)$

22-12. Arrange the following molecules in order of increasing values of total molar entropy:

$CH_3Cl(g)$ $\quad$ $CH_4(g)$ $\quad$ $CH_3OH(g)$

22-13. Explain why the total molar entropy of $SO_2(g)$ is less than that of $SO_3(g)$ at 25°C and 1 atm pressure.

22-14. Explain why the total molar entropy of $Fe_3O_4(s)$ is greater than that of $Fe_2O_3(s)$ at 25°C and 1 atm pressure.

ENTROPY AND PHYSICAL STATE

22-15. Explain why the total molar entropy of $I_2(g)$ is greater than that of $I_2(s)$ when both are at 25°C and the same pressure.

22-16. Explain why the total molar entropy of $H_2O(g)$ is greater than that of $H_2O(l)$ when both are at 100°C and 1 atm pressure.

22-17. Predict whether the entropy increases, decreases, or remains the same in the following processes:

(a) $Ar(l) \rightarrow Ar(g)$
(b) $O_2(g)$ (1.00 atm, 300 K) $\rightarrow$ $O_2(g)$ (10.0 atm, 300 K)
(c) $Cu(s)$ (100 K) $\rightarrow$ $Cu(s)$ (800 K)
(d) $H_2O(g) \rightarrow H_2O(s)$

22-18. Predict whether the entropy increases, decreases, or remains the same in the following processes:

(a) $H_2O(g)$ (75 torr, 300 K) $\rightarrow$ $H_2O(g)$ (150 torr, 300 K)
(b) $Br_2(l)$ (1 atm, 25°C) $\rightarrow$ $Br_2(g)$ (1 atm, 25°C)
(c) $I_2(g)$ (1 atm, 125°C) $\rightarrow$ $I_2(g)$ (1 atm, 200°C)
(d) $Fe(s)$ (250°C, 1 atm) $\rightarrow$ $Fe(s)$ (25°C, 1 atm)

VALUES OF ΔS°_{rxn}

22-19. Arrange the following reactions according to increasing ΔS°_{rxn} values (do not consult any references):

(a) $C(s) + O_2(g) \rightarrow CO_2(g)$
(b) $2H_2(g) + O_2(g) \rightarrow 2H_2O(g)$
(c) $CO(g) + 3H_2(g) \rightarrow CH_4(g) + H_2O(g)$
(d) $C(s) + H_2O(g) \rightarrow CO(g) + H_2(g)$

22-20. Arrange the following reactions according to increasing ΔS°_{rxn} values (do not consult any references):

(a) $2H_2O_2(l) \rightarrow O_2(g) + 2H_2O(l)$
(b) $NH_3(g) + HCl(g) \rightarrow NH_4Cl(s)$
(c) $K(s) + O_2(g) \rightarrow KO_2(s)$
(d) $N_2(g) + 3H_2(g) \rightarrow 2NH_3(g)$

22-21. Use the data in Table 22-1 to compute ΔS°_{rxn} values for the following reactions at 25°C:

(a) $2NH_3(g) + \frac{7}{2}O_2(g) \rightarrow 2NO_2(g) + 3H_2O(g)$
(b) $CO(g) + 2H_2(g) \rightarrow CH_3OH(l)$
(c) $C(s, \textit{graphite}) + H_2O(g) \rightarrow CO(g) + H_2(g)$
(d) $CO(g) + 3H_2(g) \rightarrow CH_4(g) + H_2O(g)$

22-22. Use the data in Table 22-1 to compute ΔS°_{rxn} values for the following reactions at 25°C:

(a) $2H_2O_2(l) + N_2H_4(l) \rightarrow N_2(g) + 4H_2O(g)$
(b) $N_2(g) + O_2(g) \rightarrow 2NO(g)$
(c) $2CH_4(g) + O_2(g) \rightarrow 2CH_3OH(l)$
(d) $C_2H_4(g) + H_2(g) \rightarrow C_2H_6(g)$

22-23. Use the data in Table 22-1 to calculate ΔS°_{rxn} for the following reactions at 25°C:

(a) $C(s, \textit{graphite}) + O_2(g) \rightarrow CO_2(g)$
(b) $2SO_2(g) + O_2(g) \rightarrow 2SO_3(g)$
(c) $CH_4(g) + 2O_2(g) \rightarrow CO_2(g) + 2H_2O(l)$
(d) $C_2H_2(g) + H_2(g) \rightarrow C_2H_4(g)$

22-24. Use the data in Table 22-1 to calculate ΔS°_{rxn} for the following reactions at 25°C:

(a) $I_2(s) \rightarrow I_2(g)$
(b) $BaCO_3(s) \rightarrow BaO(s) + CO_2(g)$
(c) $CH_4(g) + Cl_2(g) \rightarrow CH_3Cl(g) + HCl(g)$
(d) $2NaBr(s) + Cl_2(g) \rightarrow 2NaCl(s) + Br_2(l)$

SPONTANEITY AND ΔG_{rxn}

22-25. Water slowly evaporates at 25°C. Is the reaction

$$H_2O(l) \rightarrow H_2O(g)$$

spontaneous? What are the signs of ΔG_{rxn}, ΔH_{rxn}, and $T\Delta S_{rxn}$ at 25°C? What drives the reaction?

22-27. For the reaction

$$3C_2H_2(g) \rightleftharpoons C_6H_6(l) \qquad \Delta H^\circ_{rxn} = -631 \text{ kJ}$$

use the data in Table 22-1 to calculate ΔS°_{rxn} at 25°C. Combine your calculated value of ΔS°_{rxn} with the value of ΔH°_{rxn} and compute ΔG°_{rxn}. Indicate the direction in which the reaction is spontaneous at 1 atm pressure.

22-29. For the reaction

$$C_2H_4(g) + 3O_2(g) \rightleftharpoons 2CO_2(g) + 2H_2O(g) \qquad \Delta H^\circ_{rxn} = -1323 \text{ kJ}$$

use the data in Table 22-1 to calculate ΔS°_{rxn} at 25°C. Combine your calculated value of ΔS°_{rxn} with the value of ΔH°_{rxn} and compute ΔG°_{rxn}. Indicate the direction in which the reaction is spontaneous at 1 atm pressure.

22-31. Suppose that you see an advertisement for a catalyst that decomposes water into hydrogen and oxygen at room temperature. Would you be skeptical of this claim?

22-26. Naphthalene, the active component of one variety of mothballs, sublimes at room temperature. Is the reaction

$$\text{naphthalene}(s) \rightarrow \text{naphthalene}(g)$$

spontaneous? What are the signs of ΔG_{rxn}, ΔH_{rxn}, and $T\Delta S_{rxn}$ at 25°C? What drives the reaction?

22-28. For the reaction

$$C_2H_4(g) + H_2O(l) \rightleftharpoons C_2H_5OH(l) \qquad \Delta H^\circ_{rxn} = -35.4 \text{ kJ}$$

use the data in Table 22-1 to calculate ΔS°_{rxn} at 25°C. Combine your calculated value of ΔS°_{rxn} with the value of ΔH°_{rxn} and compute ΔG°_{rxn}. Indicate the direction in which the reaction is spontaneous at 1 atm pressure.

22-30. For the reaction

$$C(s, \text{graphite}) + CO_2(g) \rightarrow 2CO(g) \qquad \Delta H^\circ_{rxn} = +172 \text{ kJ}$$

Use the data in Table 22-1 to calculate ΔS°_{rxn} at 25°C. Combine your calculated value of ΔS°_{rxn} with the value of ΔH°_{rxn} and compute ΔG°_{rxn}. Indicate the direction in which the reaction is spontaneous at 1 atm pressure.

22-32. Discuss the possible effects of a catalyst on the value of ΔG°_{rxn}.

CELLS AND ΔG_{rxn}

22-33. An electrochemical cell is set up so that the reaction

$$Zn(s) + Cu^{2+}(aq) \rightarrow Zn^{2+}(aq) + Cu(s)$$

occurs. At 25°C the measured cell voltage is 1.05 V. Calculate the value of ΔG_{rxn} for the reaction.

22-35. An electrochemical cell is set up so that the reaction

$$2NO_3^-(aq) + 4H^+(aq) + Cu(s) \rightarrow 2NO_2(g) + 2H_2O(l) + Cu^{2+}(aq)$$

occurs. The standard voltage is $E^0 = 0.65$ V. Calculate the value of ΔG°_{rxn} for the reaction.

22-34. An electrochemical cell is set up so that the reaction

$$H_2O_2(aq) + Fe(s) + 2H^+(aq) \rightarrow Fe^{2+}(aq) + 2H_2O(l)$$

occurs. At 25°C the measured cell voltage is 2.03 V. Calculate the value of ΔG_{rxn}.

22-36. An electrochemical cell is set up so that the reaction

$$Cr_2O_7^{2-}(aq) + 14H^+(aq) + 6Fe^{2+}(aq) \rightarrow 2Cr^{3+}(aq) + 6Fe^{3+}(aq) + 7H_2O(l)$$

occurs. At 25°C, the standard cell voltage is 0.56 V. Calculate the value of ΔG°_{rxn} for the reaction.

22-37. Use the relationship

$$\Delta G^\circ_{rxn} = -nFE^0_{\text{cell}}$$

together with the E^0 data in Table 21-1 to compute the ΔG°_{rxn} values for the following reactions:

(a) $2Ag(s) + F_2(g) \rightarrow 2Ag^+(aq) + 2F^-(aq)$
(b) $\frac{1}{2}H_2(g) + Fe^{3+}(aq) \rightarrow Fe^{2+}(aq) + H^+(aq)$

22-39. For the following electrochemical cell

$$Zn(s)|Zn^{2+}(aq, 0.010\ M)||Cd^{2+}(aq, 0.050\ M)|Cd(s)$$

use the data in Table 21-1 to calculate E^0, ΔG°_{rxn}, ΔG_{rxn} and E for the cell reaction at 25°C. What is the cell reaction?

EQUILIBRIUM CONSTANTS AND ΔG°_{rxn}

22-41. For each of the following reactions at 25°C, determine whether ΔG°_{rxn} is positive, negative, or zero:

(a) $2NH_3(g) \rightleftharpoons 3H_2(g) + N_2(g)$ $\quad K = 1.47 \times 10^{-6}\ atm^2$
(b) $CH_4(g) + 4Cl_2(g) \rightleftharpoons CCl_4(l) + 4HCl(g)$ $\quad K = 1 \times 10^{70}\ atm^{-1}$

22-43. The equilibrium constant at 250°C for the reaction

$$PCl_5(g) \rightleftharpoons PCl_3(g) + Cl_2(g)$$

is $K_c = 1.8$ M. Calculate the value of ΔG°_{rxn} at 250°C. In which direction is the reaction spontaneous when PCl_3, Cl_2, and PCl_5 are at standard conditions? Calculate ΔG_{rxn} when $[PCl_3] = 0.25$ M, $[Cl_2] = 0.25$ M, and $[PCl_5] = 0.010$ M. Is the reaction to the right spontaneous?

22-45. The equilibrium constant for the reaction

$$HNO_2(aq) \rightleftharpoons H^+(aq) + NO_2^-(aq)$$

is $K_a = 4.5 \times 10^{-4}$ M at 25°C. Calculate the value of ΔG°_{rxn} at 25°C. Will nitrous acid spontaneously dissociate when $[NO_2^-] = [H^+] = [HNO_2] = 1.00$ M? When $[NO_2^-] = [H^+] = 1.0 \times 10^{-5}$ M and $[HNO_2] = 1.0$ M?

22-38. Use the relationship

$$\Delta G^\circ_{rxn} = -nFE^0_{\text{cell}}$$

together with the E^0 data in Table 21-1 to compute the ΔG°_{rxn} values for the following reactions:

(a) $Zn(s) + Cu^{2+}(aq) \rightarrow Zn^{2+}(aq) + Cu(s)$
(b) $Ag(s) + Fe^{3+}(aq) \rightarrow Fe^{2+}(aq) + Ag^+(aq)$

22-40. The standard voltage of the following cell at 25°C is $E^0 = 1.08$ V:

$$Co(s)|Co^{2+}(aq, 0.0155\ M)||Ag^+(aq, 1.50\ M)|Ag(s)$$

Calculate ΔG°_{rxn}, ΔG_{rxn} and E. What is the cell reaction?

22-42. For each of the following reactions at 25°C, determine whether ΔG°_{rxn} is positive, negative, or zero:

(a) $H_2(g) + Br_2(g) \rightleftharpoons 2HBr(g)$ $\quad K = 4.8 \times 10^{18}$
(b) $3O_2(g) \rightleftharpoons 2O_3(g)$ $\quad K = 5.2 \times 10^{-58}\ atm^{-1}$

22-44. The equilibrium constant at 527°C for the reaction

$$COCl_2(g) \rightleftharpoons CO(g) + Cl_2(g)$$

is $K_c = 4.63 \times 10^{-3}$ M. Calculate the value of ΔG°_{rxn} at 527°C. In which direction is the reaction spontaneous when CO_2, Cl_2, and $COCl_2$ are at standard conditions? Calculate ΔG_{rxn} when $[CO] = 1.0$ M, $[Cl_2] = 1.0$ M, and $[COCl_2] = 0.025$ M. In which direction is the reaction spontaneous?

22-46. The equilibrium constant for the reaction

$$HClO(aq) \rightleftharpoons H^+(aq) + ClO^-(aq)$$

is $K_a = 3.0 \times 10^{-8}$ M at 25°C. Calculate the value of ΔG°_{rxn} at 25°C. Will hypochlorous acid spontaneously dissociate when $[ClO^-] = [H^+] = [HClO] = 1.0$ M? When $[ClO^-] = [H^+] = 1.0 \times 10^{-6}$ M and $[HClO] = 0.10$ M?

22-47. The equilibrium constant for the reaction

$$HC_2H_2ClO_2(aq) \rightleftharpoons H^+(aq) + C_2H_2ClO_2^-(aq)$$

is $K_a = 1.35 \times 10^{-3}$ M at 25°C. Calculate ΔG°_{rxn} at 25°C. Will chloroacetic acid spontaneously dissociate when $[C_2H_2ClO_2^-] = [H^+] = [HC_2H_2ClO_2] = 1.0$ M? When $[C_2H_2ClO_2^-] = 0.0010$ M, $[H^+] = 1.0 \times 10^{-5}$ M, and $[HC_2H_2ClO_2] = 0.10$ M?

$\Delta G_{rxn} = 1.63 \times 10^{4}$ J

no.

22-49. The equilibrium constant for the reaction

$$AgCl(s) \xrightarrow[H_2O(l)]{} Ag^+(aq) + Cl^-(aq)$$

is the solubility product constant, $K_{sp} = 1.78 \times 10^{-10}$ M^2 at 25°C. Calculate the value of ΔG°_{rxn} at 25°C. Is it possible to prepare a solution that is 1.0 M in $Ag^+(aq)$ and $Cl^-(aq)$? spont. ←

$\Delta G^\circ_{rxn} = 55.5$ kJ no, AgCl will form.

22-51. The equilibrium constant at 25°C for the reaction

$$Ag^+(aq) + 2NH_3(aq) \rightleftharpoons Ag(NH_3)_2^+(aq)$$

is $K_c = 2.5 \times 10^3$ M^{-2}. Calculate the value of ΔG°_{rxn} at 25°C. In which direction is the reaction spontaneous when $Ag^+(aq)$, $NH_3(aq)$, and $Ag(NH_3)_2^+(aq)$ are at standard conditions? Calculate the value of ΔG_{rxn} when $[Ag^+] = 1.0 \times 10^{-3}$ M, $[NH_3] = 0.10$ M, and $[Ag(NH_3)_2^+] = 1.0 \times 10^{-3}$ M. In which direction is the reaction spontaneous under these conditions?

$\Delta G^\circ_{rxn} = -1.94 \times 10^{4}$ J spon. →

22-53. Is it possible to have a reaction for which K is infinite? Explain in terms of ΔG°_{rxn}.

No, because if K is infinite then ΔG°_{rxn} is infinite, and thus an infinite amount of work could be obtained from the reaction, which is impossible

22-48. The equilibrium constant for the reaction

$$NH_3(aq) + H_2O(l) \rightleftharpoons NH_4^+(aq) + OH^-(aq)$$

is $K_b = 1.75 \times 10^{-5}$ M at 25°C. Calculate the value of ΔG°_{rxn} at 25°C. In which direction is the reaction spontaneous when $NH_3(aq)$, $NH_4^+(aq)$, and $OH^-(aq)$ are at standard conditions? Will ammonia react with water when both $[NH_4^+]$ and $[OH^-] = 1.0 \times 10^{-6}$ M and $[NH_3] = 0.050$ M?

22-50. The equilibrium constant for the reaction

$$CaCO_3(s) \xrightarrow[H_2O(l)]{} Ca^{2+}(aq) + CO_3^{2-}(aq)$$

is the solubility product constant, $K_{sp} = 2.8 \times 10^{-9}$ M^2 at 25°C. Calculate the value of ΔG°_{rxn} at 25°C. What happens when a solution is prepared in which $[Ca^{2+}] = [CO_3^{2-}] = 1.0$ M?

22-52. The equilibrium constant at 25°C for the reaction

$$Co^{3+}(aq) + 6NH_3(aq) \rightleftharpoons Co(NH_3)_6^{3+}(aq)$$

is $K_c = 2.0 \times 10^7$ M^{-6}. Calculate the value of ΔG°_{rxn} at 25°C. In which direction is the reaction spontaneous when $Co^{3+}(aq)$, $NH_3(aq)$, and $Co(NH_3)_6^{3+}(aq)$ are at standard conditions? Calculate the value of ΔG_{rxn} when $[Co^{3+}] = 0.025$ M, $[NH_3] = 0.50$ M, and $[Co(NH_3)_6^{3+}] = 0.0010$ M. In which direction is the reaction spontaneous under these conditions?

22-54. Is it possible to have a reaction for which $K = 0$? Explain in terms of ΔG°_{rxn}.

CALCULATION OF ΔG°_{rxn} FROM TABULATED DATA

22-55. Use the data in Table 22-1 to compute ΔG°_{rxn} and K at 25°C for the following reactions:

(a) $CO(g) + 2H_2(g) \rightarrow CH_3OH(l)$
(b) $C(s) + H_2O(g) \rightarrow CO(g) + H_2(g)$
(c) $CO(g) + 3H_2(g) \rightarrow CH_4(g) + H_2O(g)$

$\Delta G^\circ_{rxn} = G^\circ_f$ (products) $- G^\circ_f$ (reactants) then $G_{rxn} = -RT \ln K$

22-57. Use the data in Table 22-1 to calculate ΔG°_{rxn} and ΔH°_{rxn} at 25°C for the reaction

$$2HCl(g) + F_2(g) \rightleftharpoons 2HF(g) + Cl_2(g)$$

Calculate the equilibrium constant for the reaction.

22-56. Use the data in Table 22-1 to compute ΔG°_{rxn} and K at 25°C for the following reactions:

(a) $2H_2O_2(l) + N_2H_4(l) \rightarrow N_2(g) + 4H_2O(g)$
(b) $N_2(g) + O_2(g) \rightarrow 2NO(g)$
(c) $2CH_4(g) + O_2(g) \rightarrow 2CH_3OH(l)$

22-58. Use the data in Table 22-1 to calculate ΔG°_{rxn} and ΔH°_{rxn} at 25°C for the reaction

$$Fe_3O_4(s) + 4C(s, \textit{graphite}) \rightarrow 3Fe(s) + 2CO_2(g)$$

Calculate the equilibrium constant for the reaction.

22-59. The reaction

$$2SO_2(g) + O_2(g) \rightleftharpoons 2SO_3(g)$$

is an important reaction in the manufacture of sulfuric acid. Use the data in Table 22-1 to calculate the values of ΔG°_{rxn} and ΔH°_{rxn} at 25°C for the reaction. Calculate the equilibrium constant for the reaction. Estimate the equilibrium constant for the reaction at 400°C. (Assume that ΔH°_{rxn} and ΔS°_{rxn} are constant over the temperature range of interest.)

22-60. Use the data in Table 22-1 to calculate ΔG°_{rxn} and ΔH°_{rxn} at 25°C for the reaction

$$H_2(g) + I_2(g) \rightleftharpoons 2HI(g)$$

Calculate the equilibrium constant for the reaction at 25°C. Estimate the equilibrium constant for the reaction at 100°C (assume that ΔH°_{rxn} and ΔS°_{rxn} are constant over the temperature range).

22-61. Use the data in Table 22-1 to calculate the value of ΔG°_{rxn}, ΔH°_{rxn}, and ΔS°_{rxn} at 25°C for the reaction

$$H_2(g) + CO_2(g) \rightleftharpoons H_2O(g) + CO(g)$$

What drives the reaction and in what direction?

22-62. Use the data in Table 22-1 for $CH_4(g)$, $Cl_2(g)$, and $HCl(g)$ to calculate ΔG_f° and ΔH_f° at 25°C for $CCl_4(l)$, given that $\Delta G^\circ_{rxn} = -395.7$ kJ and $\Delta H^\circ_{rxn} = -429.8$ kJ for the reaction

$$CH_4(g) + 4Cl_2(g) \rightleftharpoons CCl_4(l) + 4HCl(g)$$

Compare your results for $CCl_4(l)$ with those given in Table 22-1.

22-63. Given the following possibilities for ΔG°_{rxn}, what can you say in each case about the magnitude of the equilibrium constant for the reaction:

(a) $\Delta G^\circ_{rxn} > 0$ (b) $\Delta G^\circ_{rxn} = 0$ (c) $\Delta G^\circ_{rxn} < 0$

22-64. Compute the value (in kilojoules) of ΔG°_{rxn} that corresponds to a 10-fold change in K at 25°C.

22-65. Hydrogen peroxide can be prepared in several ways. One method is the reaction between hydrogen and oxygen:

$$H_2(g) + O_2(g) \rightleftharpoons H_2O_2(l)$$

Another method is the reaction between water and oxygen:

$$2H_2O(l) + O_2(g) \rightleftharpoons 2H_2O_2(l)$$

Calculate the value of ΔG°_{rxn} for both reactions. Predict which method is the more efficient with respect to energy.

22-66. Glucose is a primary fuel in the production of energy in biological systems. Given that $\Delta G_f^\circ = -916\ \mathrm{kJ \cdot mol^{-1}}$ for glucose, calculate the maximum amount of work that can be obtained from the complete combustion of 1.00 mol of glucose under standard conditions:

$$C_6H_{12}O_6(s) + 6O_2(g) \rightleftharpoons 6CO_2(g) + 6H_2O(l)$$

22-67. Calculate the maximum amount of work that can be obtained from the combustion of 1.00 mol of ethane, $C_2H_6(g)$, at 25°C and 1 atm.

22-68. Calculate the maximum amount of work that can be obtained from the combustion of 1.00 mol of methane, $CH_4(g)$, at 25°C and 1 atm.

22-69. Calculate the standard voltage that can be obtained from an ethane-oxygen ($C_2H_6(g)$–$O_2(g)$) fuel cell at 25°C.

22-70. Calculate the standard voltage that can be obtained from a methane-oxygen fuel cell at 25°C.

22-71. Given the following Gibbs free energies at 25°C:

Substance	$\Delta \bar{G}_f^\circ/kJ\cdot mol^{-1}$
$Ag^+(aq)$	77.1
$Cl^-(aq)$	−131.2
$AgCl(s)$	−109.7

calculate the solubility product of AgCl.

22-72. Given the following standard Gibbs free energies at 25°C:

Substance	$\Delta \bar{G}_f^\circ/kJ\cdot mol^{-1}$
$Ag^+(aq)$	77.1
$Br^-(aq)$	−102.8
$AgBr(s)$	−96.8

calculate the solubility product of AgBr.

TEMPERATURE DEPENDENCE OF EQUILIBRIUM CONSTANTS

22-73. Given that ΔH°_{rxn} and ΔS°_{rxn} for a particular reaction are independent of T, start from the relations

$$\Delta G^\circ_{rxn,2} = \Delta H^\circ_{rxn} - T_2\,\Delta S^\circ_{rxn}$$

$$\Delta G^\circ_{rxn,1} = \Delta H^\circ_{rxn} - T_1\,\Delta S^\circ_{rxn}$$

and use the relation $\Delta G^\circ_{rxn} = -2.30RT\log K$ to show that

$$\log\frac{K_2}{K_1} = \frac{\Delta H^\circ_{rxn}}{2.30R}\left(\frac{T_2 - T_1}{T_1 T_2}\right)$$

This equation is known as the van't Hoff equation and is used to compute, say, K_2 at T_2 given K_1 at T_1 and ΔH°_{rxn} for the reaction.

22-74. Aluminum chloride acts as a catalyst for the isomerization reaction

cyclohexane (C_6H_{12}: ring of six CH_2 groups) $\rightleftharpoons$ methylcyclopentane (CH_3–CH in a five-membered ring with four CH_2 groups)

At 25°C, $K = 0.140$, and at 100°C, $K = 0.365$. Calculate ΔH°_{rxn} and ΔS°_{rxn} for this reaction (see Problem 22-73).

22-75. For the reaction $N_2(g) + O_2(g) \rightleftharpoons 2\,NO(g)$, use the following data to calculate ΔH°_{rxn}

T/K	$K_p/10^{-4}$
2000	4.08
2100	6.86
2200	11.0
2300	16.9
2400	25.1

22-76. For the dissociation of $Br_2(g)$ into $2\,Br(g)$, use the following data to calculate ΔH°_{rxn}

$t/°C$	$K_p/10^{-3}$ atm
850	0.600
900	1.45
950	3.26
1000	6.88

22-77. For the reaction $H_2(g) + CO_2(g) \rightleftharpoons CO(g) + H_2O(g)$, use the following data to calculate ΔH°_{rxn}

$t/°C$	K
600	0.39
700	0.64
800	0.96
900	1.34
1000	1.77

22-78. For the reaction $2SO_2(g) + O_2(g) \rightleftharpoons 2SO_3(g)$, use the following data to calculate ΔH°_{rxn}

T/K	K_p/atm^{-1}
900	43.1
1000	3.46
1100	0.44
1170	0.13

22-79. Use Table 22-1 to calculate the value of ΔH°_{rxn} for the reaction

$$PCl_3(g) + Cl_2(g) \rightleftharpoons PCl_5(g)$$

Given that $K_p = 0.562\ \text{atm}^{-1}$ at 250°C, calculate the value of K_p at 400°C.

22-80. Use Table 22-1 to calculate the value of ΔH°_{rxn} for the reaction

$$H_2(g) + I_2(g) \rightleftharpoons 2HI(g)$$

Given that $K = 58.0$ at 400°C, calculate K at 500°C.

22-81. Use the van't Hoff equation (Problem 22-73) to show that the value of an equilibrium constant decreases with increasing temperature for an exothermic reaction.

22-82. Use the van't Hoff equation (Problem 22-73) to show that the value of an equilibrium constant increases with increasing temperature for an endothermic reaction.

22-83. Use the van't Hoff equation (Problem 22-73) to show that a plot of $\log K$ versus $1/T$ is linear.

22-84. Using the result of Problem 21-83, plot the data in Problem 21-75 and show that $\log K$ versus $1/T$ is a straight line. Evaluate ΔH°_{rxn} from this plot.

INTERCHAPTER M

The Transition Metals

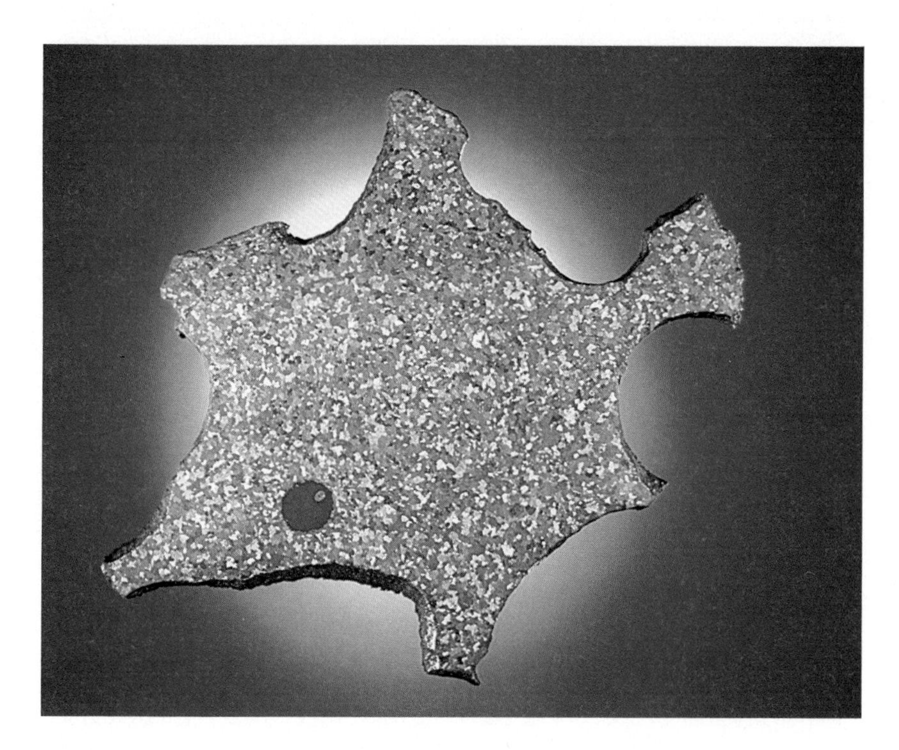

A nickel-iron meteorite with an eye of troilite. This is a slice of the Willamette meteorite, the largest object of its kind found in the Western Hemisphere; it was found in Willamette, Oregon in 1902. The granular pattern indicates reheating and secondary crystallization after its original formation in its parent body.

In this interchapter we describe the sources, properties, and uses of the transition metals. Transition metal alloys are the structural backbone of modern civilization. Human development progressed from the Stone Age, to the Bronze Age and then to the Iron Age. The Industrial Revolution was powered by steam engines made from steels. The Space and Computer Age utilizes a truly remarkable variety of exotic alloys developed to meet a wide range of specialized requirements.

M-1. MANY OF THE TRANSITION METALS ARE USED TO MAKE ALLOYS

Many of the transition metals are probably familiar to you. Iron, nickel, chromium, tungsten, and titanium are widely used in alloys for structural materials and play a key role in the world's technology. The precious metals—gold, platinum, and silver—are used as hard currency, and to make jewelry and electrical components. Copper is the most widely used metal for electric wiring.

The transition metals vary greatly in abundance. Iron and titanium are the fourth and tenth most abundant elements in the earth's crust, whereas rhenium (Re) and hafnium (Hf) are unfamiliar even to many chemists because they are so rare. The characteristics of the

Palladium crystals as seen through an electron microscope. Palladium is a silver-white, ductile, malleable metallic element that resembles platinum. It is used for ornamentation, in dentistry, in electrical devices, and as a catalyst.

transition metals vary from family to family, and yet they are all characterized by high densities and high melting points. The two metals with the greatest densities (iridium, Ir, $22.65\ g \cdot cm^{-3}$, and osmium, Os, $22.61\ g \cdot cm^{-3}$) and the metal with the highest melting point (tungsten, W, 3410°C) are transition metals.

The physical properties of the transition metals vary greatly. Iron, the most common and most important transition metal, is discussed in the next section. The second most common transition metal is titanium, which constitutes 0.6 percent of the earth's crust by mass. Pure titanium is a lustrous, white metal (Figure M-1). It is used to

Figure M-1 Titanium has a relatively low density, high strength, excellent corrosion resistance and is easily machined.

make lightweight alloys that are stable at high temperatures for use in missiles and high-performance aircraft. Titanium is as strong as most steels but 50 percent lighter. It is 60 percent heavier than aluminum but twice as strong. In addition, it has excellent resistance to corrosion. The most important ore of titanium is rutile (Figure M-2), which is primarily TiO_2. Pure titanium metal is produced by first converting TiO_2 to $TiCl_4$ and then reducing the $TiCl_4$ by reacting it with magnesium. Most titanium is used in the production of titanium steels, but TiO_2, which is white when pure, is used as the white pigment in many paints. Titanium tetrachloride is also used to make smoke screens; when it is sprayed into the air it reacts with moisture to produce a dense and persistent white cloud of TiO_2.

Gold is a very dense, soft, yellow metal with a high luster (Figure M-3). It is found in nature as the free element and in tellurides. It occurs in veins and alluvial deposits and is often separated from rocks and other minerals by sluicing or panning. Over two thirds of the gold produced by the noncommunist world comes from South Africa. In many mining operations, about 5 g of gold is

Figure M-2 The most important ore of titanium is rutile, which is primarily TiO_2. Deposits of rutile are found in Georgia, Virginia, Australia, Brazil, Italy, and Mexico.

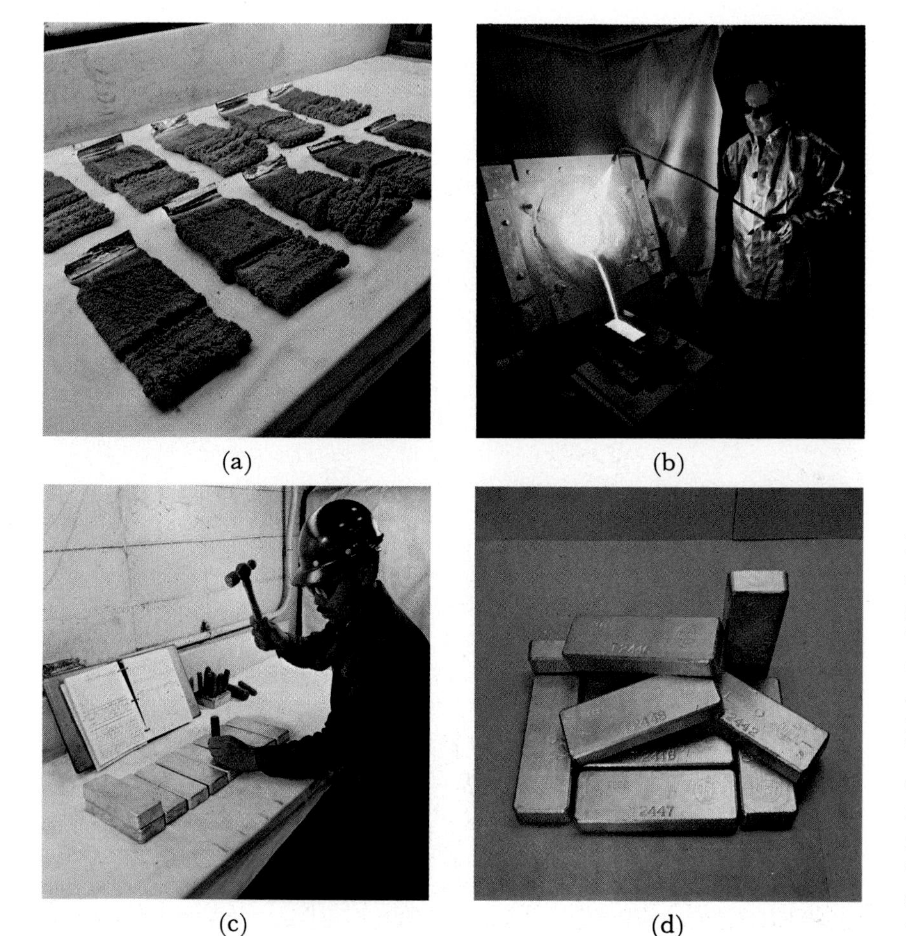

(a) (b) (c) (d)

Figure M-3 (a) In the process of electrolytic gold refining, sponge gold adheres to cathodes, the thin sheets of gold metal. (b) The cathodes are melted, and the liquid gold is poured into 400-ounce bars. (c) Each bar is then weighed, double-checked for 99.999% purity, registered and stamped. (d) Apart from its intrinsic psychological value as a world monetary standard, gold has many industrial uses.

recovered from 1 ton of rock. Gold is very unreactive and has a remarkable resistance to corrosion. Pure gold is soft and often alloyed to make it harder. The amount of gold in an alloy is expressed in *karats*. Pure gold is 24 karat. Coinage gold is 22 karat, or $(22/24) \times 100 = 92$ percent. White gold, which is used in jewelry, is usually an alloy of gold and nickel. In addition to its use in jewelry and as a world monetary standard, gold is an excellent conductor of electricity and is used in microelectronic devices. It is also used extensively in dentistry and medicine.

Gold compounds are used to treat arthritis.

Copper is important because of its use as an electric conductor. Silver is the only metal that is a better conductor than copper, but the price of silver precludes its widespread use. Copper is reddish and takes on a bright metallic luster. It occurs occasionally as the free metal but usually in the minerals chalcopyrite, cuprite, and malachite (Figure M-4). Brass is an alloy of copper with zinc, and bronze is an alloy of copper with tin. Brass and bronze are among the earliest known alloys. Bronze usually contains from 5 to 10 percent tin and is very resistant to corrosion. It is used for casting, marine equipment, fine arts work, and spark-resistant tools. Yellow brasses contain about 35 percent zinc and have good ductility and high strength. Brass is used for piping, hose nozzles, marine equipment, and jewelry and in the fine arts. The physical properties, principal sources, and major uses of the *d* transition series metals are given in Tables M-1, M-2, and M-3.

Figure M-4 Malachite, $Cu_2CO_3(OH)_2$, is a widely distributed and important ore of copper. Large deposits of malachite are found in the Ural mountains, Namibia, South Australia, and the southwestern United States.

Table M-1 Properties of the 3*d* transition series metals

Element	*Density/g·cm^{-3}*	*Melting point/°C*	*Principal sources*	*Main uses*
Sc	3.0	1541	thortveitite, $(Sc,Y)_2Si_2O_7$	no major industrial uses
Ti	4.5	1660	rutile, TiO_2	high-temperature, lightweight steel alloys, TiO_2 in white paints
V	6.0	1890	vanadinite, $(PbO)_9(V_2O_5)_3PbCl_2$	vanadium steels (rust-resistant)
Cr	7.2	1860	chromite, $FeCr_2O_4$	stainless steels, chrome plating
Mn	7.4	1244	pyrolusite, MnO_2 manganosite, MnO nodules on ocean floor	alloys
Fe	7.9	1535	hematite, Fe_2O_3 magnetite, Fe_3O_4	steels
Co	8.9	1490	cobaltite, $CoS_2 \cdot CoAs_2$ linnaetite, Co_3S_4	alloys, cobalt-60 medicine
Ni	8.9	1455	pentlandite, $(Fe, Ni)_9S_8$ pyrrhotite, $F_{0.8}S$	nickel plating, coins, magnets, catalysts
Cu	9.0	1083	chalcopyrite, $CuFeS_2$ cuprite, Cu_2O malachite, $Cu_2(CO_3)(OH)_2$	bronzes, brass, coins, electric conductors
Zn	7.1	420	zinc blende, ZnS smithsonite, $ZnCO_3$	galvanizing, bronze, brass, dry cells

The chemistry of the transition metals varies greatly. Zinc and iron are fairly reactive metals:

$$Zn^{2+}(aq) + 2e^- \rightarrow Zn(s) \qquad E^0_{Zn^{2+}/Zn} = -0.76\ V$$

$$Fe^{2+}(aq) + 2e^- \rightarrow Fe(s) \qquad E^0_{Fe^{2+}/Fe} = -0.43\ V$$

Recall that the more negative the value of E^0 for a half-reaction, the stronger is the reducing agent.

whereas silver and gold are unreactive:

$$Ag^+(aq) + e^- \rightarrow Ag(s) \qquad E^0_{Ag^+/Ag} = +0.80\ V$$

$$Au^+(aq) + e^- \rightarrow Au(s) \qquad E^0_{Au^+/Au} = +1.68\ V$$

Although there are similarities within groups, the chemistry of each *d* transition metal is distinctive. The variable oxidation states of the metals (Table M-4) add a richness to their chemistry.

The reactions of the transition metals are usually oxidation-reduction reactions, which are governed by standard reduction potentials. Using chromium as an example, we have the standard reduction voltages

Table M-2 Properties of the 4*d* transition series metals

Element	*Density/g·cm*$^{-3}$	*Melting point/°C*	*Principal sources*	*Principal uses*
Y	4.5	1522	rare-earth ores	coating on high-temperature alloys, microwave ferrites, special semiconductors
Zr	6.5	1852	monazite (with rare earths) zircons, $ZrSO_4$	fuel-rod cladding for nuclear reactors, explosive primers, formerly in antiperspirants
Nb	8.6	2468	columbite, $(Fe,Mn)/(Nb,Ta)_2O_6$	stainless steels, welding rods, nuclear reactor alloys, superconductors
Mo	10.3	2622	molybdenite, MoS_2 wulfenite, $PbMoO_4$	tool steels, boiler plate, rifle barrels, spark plugs, X-ray tube filaments
Tc	11.5	2172	does not occur in nature	brain and thyroid scans (Tc-99)
Ru	12.4	2310	osmiridium platinum ores	platinum substitute in jewelry, pen nibs, electrical contact alloys, catalyst, superconductors
Rh	12.4	1966	platinum ores rhodite	platinum alloys, plating for kitchen utensils, reflectors, catalyst
Pd	12.0	1555	Pt and Au ores PdSe ore NiS ores	dental and watch alloys, astronomical instruments (mirrors), catalyst
Ag	10.5	961	as free metal or with Cu, Au, and Pb ores sulfide ores	coins, mirrors, jewelry, silverware, electroplating, dental alloys, photography (AgBr)
Cd	8.7	321	impurity in zinc ores CdS $CdCO_3$	low-melting alloys, photoelectric cells, Ni-Cd batteries, dentistry

$$Cr^{2+}(aq) + 2e^- \rightarrow Cr(s) \qquad E^0_{Cr^{2+}/Cr} = -0.86\ V$$

$$Cr^{3+}(aq) + e^- \rightarrow Cr^{2+}(aq) \qquad E^0_{Cr^{3+}/Cr^{2+}} = -0.41\ V$$

$$Cr_2O_7^{2-}(aq) + 14H^+(aq) + 6e^- \rightarrow 2Cr^{3+}(aq) + 7H_2O(l) \qquad E^0_{Cr_2O_7^{2-}/Cr^{3+}} = +1.36\ V$$

The first two reactions show that chromium is easily oxidized to the $+3$ state; the second reaction shows that $Cr^{2+}(aq)$ is a strong reducing agent; the third reaction shows that $Cr_2O_7^{2-}(aq)$ in acidic solution is a strong oxidizing agent. Standard reduction voltages are especially useful in understanding and predicting the chemistry of the transition metals.

Table M-3 Properties of the 5*d* transition series metals

Element	*Density/g·cm^{-3}*	*Melting point/°C*	*Principal sources*	*Principal uses*
Lu	9.8	1663	monazite	
Hf	13.3	2227	zirconium minerals	"getter" for oxygen and nitrogen, incandescent lamps
Ta	16.7	2996	niobium ores, especially columbite	pen nibs, balance weights, surgical, dental, and chemical instruments, optical glass
W	19.3	3410 (highest melting point of any metal)	wolframite, (Fe,Mn)WO scheelite, $CaWO_4$	steel toughening, incandescent lamp filaments, electrical contact points, metal cutting tools, catalysis
Re	21.0	3180	rare-earth minerals some sulfide ores	filaments, alloys for electrical contacts, jewelry plating, superconductors
Os	22.61	3045	osmiridium platinum ores	alloy with indium for pen points and machine bearings, catalysis
Ir	22.65 (densest element)	2450	free metal with osmium as alloy platinum ores	pen points, crucibles, platinum hardening
Pt	21.5	1774	natural alloys	jewelry, dentistry, thermometers, electroplating, catalysis
Au	18.9	1065	free metal	jewelry, coins, currency, electrical contacts, dentistry
Hg	13.5	−39 (lowest melting point of any metal)	cinnabar, HgS	barometers, thermometers, Hg arc lamps, silent switches, blasting caps (as fulminate), pharmaceuticals

Table M-4 The common oxidation states of the 3*d* transition metals

Element	*Atomic number*	*Oxidation state*	*Element*	*Atomic number*	*Oxidation state*
Sc	21	+3	Fe	26	+2, +3
Ti	22	+4	Co	27	+2, +3
V	23	+2, +3, +4, +5	Ni	28	+2
Cr	24	+2, +3, +6	Cu	29	+1, +2
Mn	25	+2, +4, +6, +7	Zn	30	+2

M-2. IRON IS THE FOURTH MOST ABUNDANT ELEMENT IN THE EARTH'S CRUST

Iron constitutes 4.7 percent by mass of the earth's crust. It is the cheapest metal and, in the form of steel, the most useful. Pure iron is a silvery-white, soft metal that rusts rapidly in moist air. It has little use as the pure element but is strengthened greatly by the addition of small amounts of carbon and of various other transition metals. It occurs in nature as hematite, Fe_2O_3, magnetite, Fe_3O_4, siderite, $FeCO_3$, and iron pyrite, FeS_2 (fool's gold). Millions of tons of iron are produced annually in the United States by the reaction of Fe_2O_3 with coke, which is carried out in a blast furnace (Figure M-5). A modern blast furnace is about 100 ft high and 25 ft wide and produces about 5000 tons of iron daily (Figure M-6). A mixture of iron ore, coke, and limestone ($CaCO_3$) is loaded into the top, and preheated compressed air and oxygen are blown in near the bottom. The reaction of the coke and the oxygen to produce carbon dioxide gives off a great deal of heat, and the temperature in the lower region of a blast furnace is around 1900°C. As the CO_2 rises, it reacts with more coke to produce hot carbon monoxide, which reduces the iron ore to iron. The molten iron metal is denser than the other substances and drops to the bottom, where it can be drained off to form ingots of what is called *pig iron.*

The function of the limestone is to remove the sand and gravel that normally occur with iron ore. The intense heat decomposes the limestone to CaO and CO_2. The CaO(s) combines with the sand and gravel (both of which are primarily silicon dioxide) to form calcium silicate:

$$\underset{\text{calcium oxide}}{CaO(s)} + \underset{\text{sand, gravel}}{SiO_2(s)} \rightarrow \underset{\text{molten calcium silicate}}{CaSiO_3(l)}$$

The molten calcium silicate, called *slag,* floats on top of the molten iron and is drained off periodically. It is used in building materials, such as cement and concrete aggregate, rock-wool insulation, and cinder block, and as railroad ballast.

Pig iron contains about 4 or 5 percent carbon together with lesser amounts of silicon, manganese, phosphorus, and sulfur. It is brittle, difficult to weld, and not strong enough for structural applications. To be useful, pig iron must be converted to steel, which is an alloy of iron with small but definite amounts of other metals and between 0.1 and 1.5 percent carbon. Steel is made from pig iron in several different processes, all of which use oxygen to oxidize most of the impurities. One such process is the *basic oxygen process,* in which hot, pure O_2 gas is blown through molten pig iron (Figure M-7). The oxidation of carbon and phosphorus is complete in less than 1 h. The desired carbon content of the steel is then achieved by adding high-carbon steel alloy.

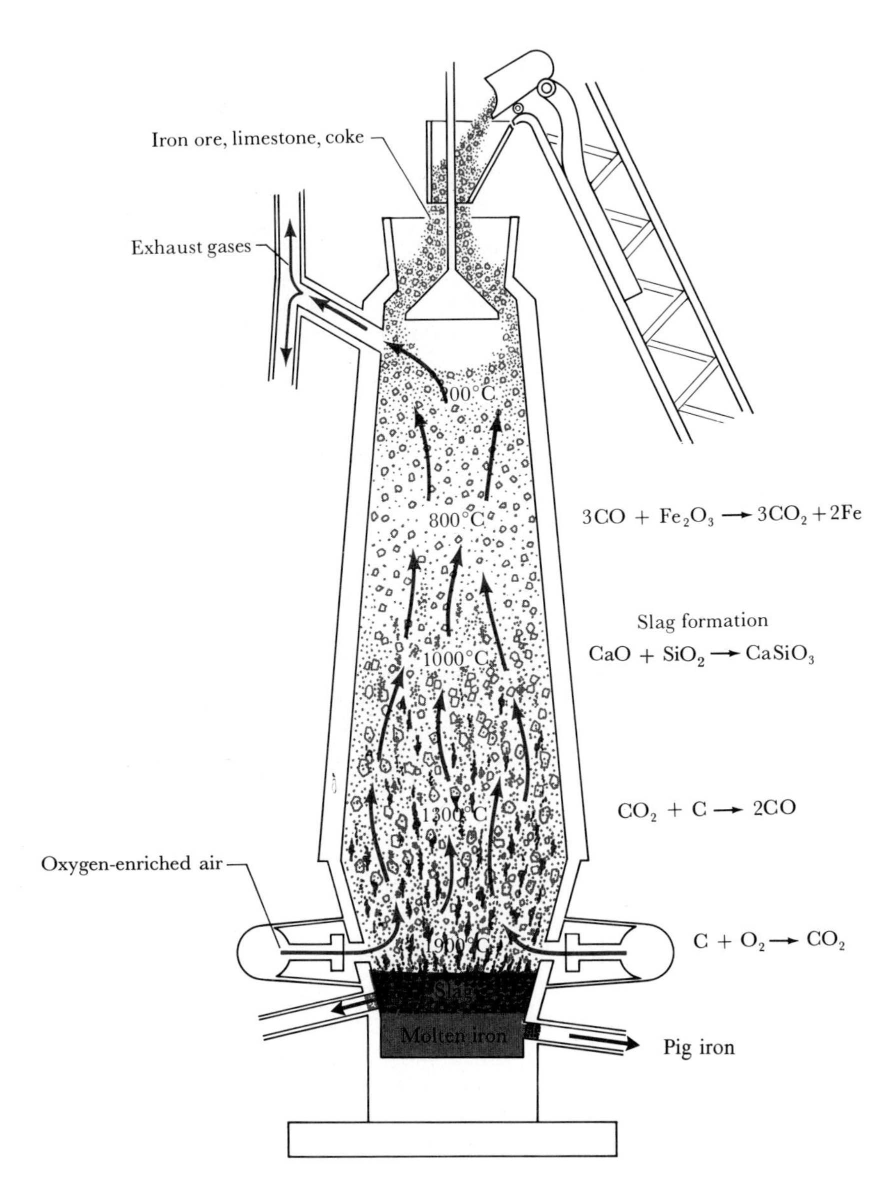

Figure M-5 A diagram of a blast furnace.

Figure M-6 These two blast furnaces at Bethlehem Steel's Burns Harbor plant have a combined total production of over 10,000 tons of iron daily. Ore, limestone, and coke are fed directly into the top of the furnaces by an automatic conveyor system.

Figure M-7 Molten iron being charged into a basic oxygen furnace. Most steel is produced by a process called the basic oxygen process. A typical basic oxygen furnace is charged with about 200 tons of molten pig iron, 100 tons of scrap iron, and 20 tons of limestone (to form a slag). A stream of hot oxygen is blown through the molten mixture, where the impurities are oxidized and blown out of the iron. High quality steel is produced in an hour or less.

There are two types of steels, carbon steels and alloy steels. Both types contain carbon, but carbon steels contain essentially no other metals besides iron. About 90 percent of all steel produced is carbon steel. Carbon steel that contains less than 0.2 percent carbon is called *mild steel.* Mild steels are malleable and ductile and are used where load-bearing ability is not a consideration. *Medium steels,* which contain 0.2 to 0.6 percent carbon, are used for such structural materials as beams and girders and for railroad equipment. *High-carbon steels* contain 0.8 to 1.5 percent carbon and are used to make drill bits, knives, and other tools in which hardness is important.

Alloy steels contain other metals in small amounts. Different metals give different properties to steels. For example,

Chromium improves hardness and resistance to corrosion.
Tungsten and molybdenum increase heat resistance.
Nickel adds toughness, as in armor plating.
Vanadium adds springiness.
Manganese improves resistance to wear.

The alloy steels called stainless steels contain high percentages of chromium and nickel. Stainless steels resist corrosion and are used for cutlery and hospital equipment. The most common stainless steel contains 18 percent chromium and 8 percent nickel.

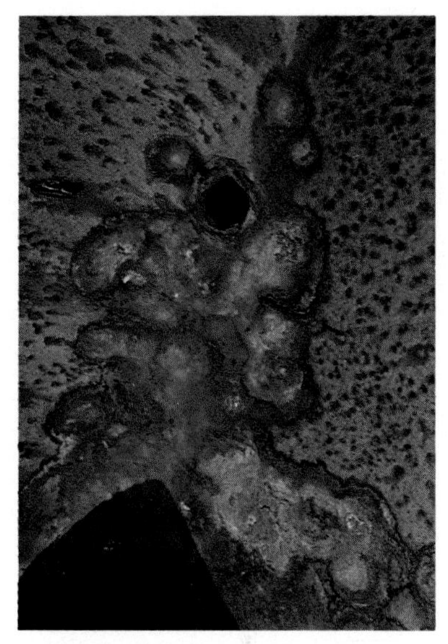

Iron(III) oxide results when iron is oxidized in the air.

M-3. BILLIONS OF DOLLARS ARE SPENT EACH YEAR TO PROTECT METALS FROM CORROSION

We are all familiar with corrosion, the best-known example of which is the rusting of iron and steel. Rust is iron(III) oxide, and the corrosion of iron proceeds by air oxidation of the iron:

$$4Fe(s) + 3O_2(g) \xrightarrow{H_2O(l)} \underset{\text{rust}}{2Fe_2O_3(s)}$$

Most metals, when exposed to air, develop an oxide film. In some cases this film is very thin and protects the metal, and the metal maintains its luster. However, depending on the humidity, the presence of certain anions—such as Cl^-, which promotes corrosion through the formation of chloro complexes and is present in sea spray and on roads treated with rock salt—or the presence of certain gaseous species—such as the oxides of sulfur and nitrogen, which combine with water to form acids that attack metal—can lead to corrosion that can completely destroy a metal.

Corrosion is a major problem costing billions of dollars annually for replacements for corroded parts. Research is directed toward corrosion mechanisms because a detailed understanding of these mechanisms can provide important clues as to how to prevent the process.

Corrosion involves electron-transfer reactions between different sections of the same piece of metal or between two dissimilar metals in electrical contact with each other. One metal piece acts as the anode, and the other acts as the cathode. For example, iron in contact with air and moisture corrodes according to the mechanism sketched in Figure M-8. The anodic process is

$$2Fe(s) + 4OH^-(aq) \rightarrow 2Fe(OH)_2(s) + 4e^-$$

and the cathodic process is

$$2H_2O(l) + O_2(g) + 4e^- \rightarrow 4OH^-(aq)$$

where the $O_2(g)$ comes from the air. The iron(II) hydroxide formed is rapidly air-oxidized in the presence of water to iron(III) hydroxide:

$$4Fe(OH)_2(s) + O_2(g) + 2H_2O(l) \rightarrow 4Fe(OH)_3(s)$$

The Statue of Liberty has extensive copper covering. The statue now appears green because of the oxidation of the copper metal to form various copper compounds.

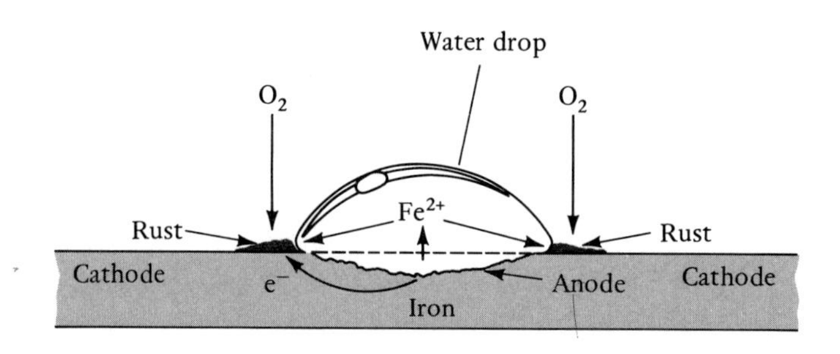

Figure M-8 Corrosion of iron. A drop of water on an iron surface can act as a corrosion center. The iron is oxidized by oxygen from the air. Moisture is necessary for corrosion because the mechanism involves the formation of dissolved $Fe^{2+}(aq)$ ions. Salt promotes the corrosion by enabling a larger current flow between anode and cathode.

which in turn converts spontaneously to iron(III) oxide:

$$2Fe(OH)_3(s) \rightarrow Fe_2O_3 \cdot 3H_2O(s)$$

The corrosion of aluminum in air is not so pronounced as that of iron because the Al_2O_3 film that forms is tough and adherent and impervious to oxygen. The same is true for chromium and nickel.

The simplest method of corrosion prevention is to provide a protective layer of paint or of a corrosion-resistant metal, such as chromium or nickel. The weakness of such methods is that any scratch or crack in the protective layer exposes the metal surface. The exposed surface, even though small in area, can act as an anode in conjunction with other exposed metal parts, which act as cathodes. This combination then leads to corrosion of the metal under the no-longer-protective layer.

Another anticorrosion technique uses a replaceable *sacrificial anode,* which is a piece of metal electrically connected to a less active metal (Figure M-9). The more active metal is the stronger reducing agent and is thus preferentially oxidized; oxygen is reduced on the surface of the less active metal. Sacrificial anodes are used to protect water pipes and ship propellers. This method is also the electrochemical

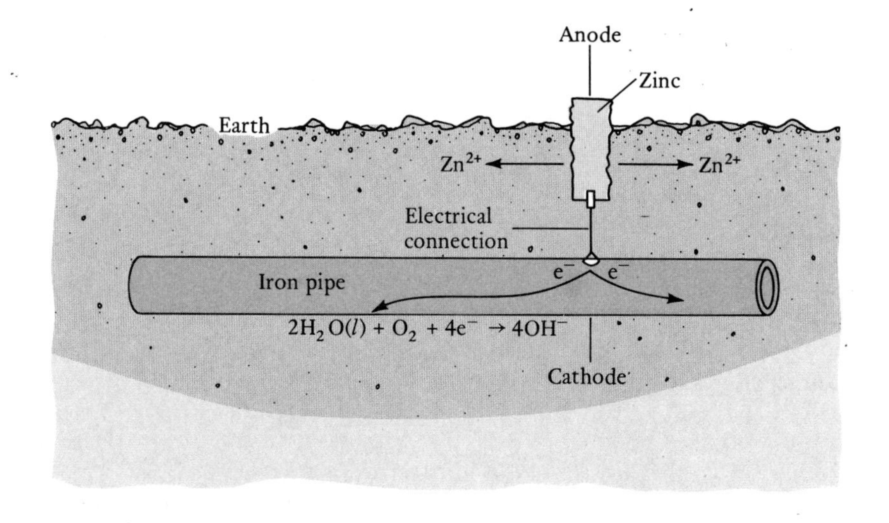

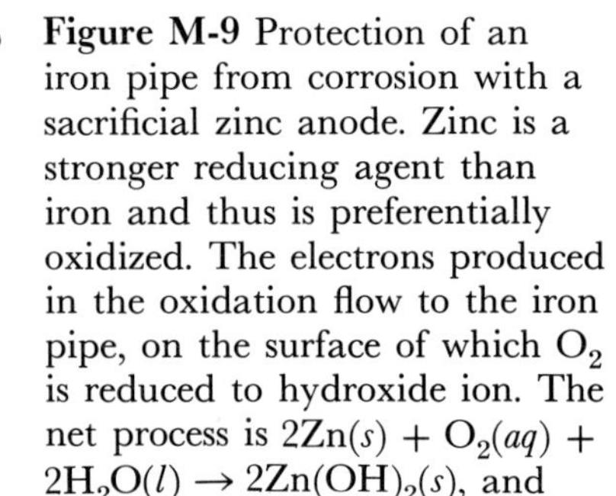

Figure M-9 Protection of an iron pipe from corrosion with a sacrificial zinc anode. Zinc is a stronger reducing agent than iron and thus is preferentially oxidized. The electrons produced in the oxidation flow to the iron pipe, on the surface of which O_2 is reduced to hydroxide ion. The net process is $2Zn(s) + O_2(aq) + 2H_2O(l) \rightarrow 2Zn(OH)_2(s)$, and the iron remains intact.

basis of galvanization, in which iron is protected from corrosion by a zinc coating, a process used in the manufacture of automobile bodies. A crack in the zinc coating does not affect the corrosion protection provided. Note that the less active metal (iron) promotes the corrosion of the more active metal (zinc).

QUESTIONS

M-1. Identify the *d* transition metals in the periodic table.

M-2. Which metal has the highest melting point?

M-3. Which are the two densest metals?

M-4. Which is the most abundant transition metal?

M-5. Describe how titanium metal is produced.

M-6. What is the percentage of gold in 14-karat gold?

M-7. Describe how iron is produced in a blast furnace.

M-8. What is pig iron?

M-9. What is slag? What is it used for?

M-10. Describe the basic oxygen process.

M-11. Discuss the differences between mild steel, medium steel, and high-carbon steel.

M-12. Describe the reactions that occur in the corrosion of iron.

M-13. Why isn't the corrosion of aluminum as serious a problem as the corrosion of iron?

M-14. Describe how a sacrificial anode works.

The colors of many gemstones are due to small quantities of transition metal ions. Shown here are (1) kunzite, (2) garnet, (3) zircon, (4) aquamarine, (5) amethyst, (6) peridot, (7) morganite, (8) topaz, (9) ruby, (10) tourmaline (indicolite), (11) chrome tourmaline, (12) rose quartz, (13) rubellite tourmaline, (14) kyanite, (15) citrine, and (16) green tourmaline.

23 / Transition Metal Complexes

Transition metal complexes are species in which several anions or neutral molecules, called ligands, bond to a transition metal atom or ion. The chemistry of the transition metals is especially rich and interesting because of these complexes, which occur in a variety of geometries and oxidation states. Much of the chemistry of transition metal complexes can be understood in terms of the electron occupancy of the *d* orbitals of the metal ion. In this chapter, we use *d*-orbital electron configurations to explain many of the spectral, magnetic, and structural properties of transition metal complexes.

23-1. THERE ARE 10 ELEMENTS IN EACH *d* TRANSITION METAL SERIES

The 3*d*, 4*d*, and 5*d* *transition metal series* are shown as colored rows in Figure 23-1. There are 10 metals in each *d* series because the five *d orbitals,* each of which can hold a total of two electrons, are being filled as we move from left to right through a series. Recall from Chapter 7 that the angular momentum quantum number, ℓ, for a *d* orbital is equal to 2. For $\ell = 2$, the magnetic quantum number, m_ℓ, can be 2, 1, 0, -1, or -2. Thus there are five *d* orbitals for each value of the principal quantum number for $n \geqslant 3$. Each of the five *d* orbitals can hold a maximum of two electrons with opposite spins, giving a total of 10 electrons in a *d* subshell. For $n = 3$, we have the 3*d* series (Sc $\rightarrow$ Zn); for $n = 4$, we have the 4*d* series (Y $\rightarrow$ Cd); and for $n = 5$, we have the 5*d* series (Lu $\rightarrow$ Hg).

Figure 23-1 The periodic table with the $3d$ transition metal series shown in blue, the $4d$ series shown in yellow and the $5d$ series shown in green.

	1	2											3	4	5	6	7	8
	1 H																	2 He
	3 Li	4 Be											5 B	6 C	7 N	8 O	9 F	10 Ne
	11 Na	12 Mg											13 Al	14 Si	15 P	16 S	17 Cl	18 Ar
3d	19 K	20 Ca	21 Sc	22 Ti	23 V	24 Cr	25 Mn	26 Fe	27 Co	28 Ni	29 Cu	30 Zn	31 Ga	32 Ge	33 As	34 Se	35 Br	36 Kr
4d	37 Rb	38 Sr	39 Y	40 Zr	41 Nb	42 Mo	43 Tc	44 Ru	45 Rh	46 Pd	47 Ag	48 Cd	49 In	50 Sn	51 Sb	52 Te	53 I	54 Xe
5d	55 Cs	56 Ba	71 Lu	72 Hf	73 Ta	74 W	75 Re	76 Os	77 Ir	78 Pt	79 Au	80 Hg	81 Tl	82 Pb	83 Bi	84 Po	85 At	86 Rn
	87 Fr	88 Ra	103 Lr	104 Unq	105 Unp	106 Unh	107 Uns	108	109 Une									

Lanthanum series

57 La	58 Ce	59 Pr	60 Nd	61 Pm	62 Sm	63 Eu	64 Gd	65 Tb	66 Dy	67 Ho	68 Er	69 Tm	70 Yb

Actinium series

89 Ac	90 Th	91 Pa	92 U	93 Np	94 Pu	95 Am	96 Cm	97 Bk	98 Cf	99 Es	100 Fm	101 Md	102 No

Example 23-1: Explain why there is no $2d$ transition metal series.

Solution: For a given value of the principal quantum number, n, the possible values of the angular momentum quantum number, ℓ, are

$$\ell = 0, 1, 2, \ldots, n - 1$$

where $n - 1$ is the maximum possible value of ℓ. For $n = 2$, the maximum value of ℓ is 1. Therefore, ℓ cannot have the value of 2 for $n = 2$. Since d orbitals are orbitals with $\ell = 2$, there are no $2d$ orbitals.

The shape and relative spatial orientation of the five d orbitals are shown in Figure 23-2. These orbitals are distinguished by x, y, and z subscripts that define the orientation of the orbitals with respect to the x, y, and z coordinate axes. In the absence of any external electric or magnetic field, the energies of the five d orbitals for a given value of the principal quantum number are equal.

Consider the electron configuration of a neutral gaseous iron atom. From Figure 23-1, we note that iron has an atomic number $Z = 26$, and thus there are 26 electrons in a neutral iron atom. From Chapter 8 we have for the electron configuration of iron

$$\text{Fe} \quad 1s^2 2s^2 2p^6 3s^2 3p^6 4s^2 3d^6$$

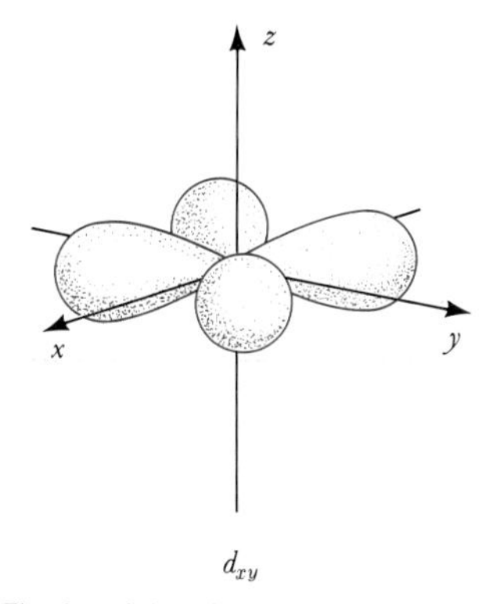

d_{xy}

The four lobes lie between the x and y axes in the four quadrants on the xy plane

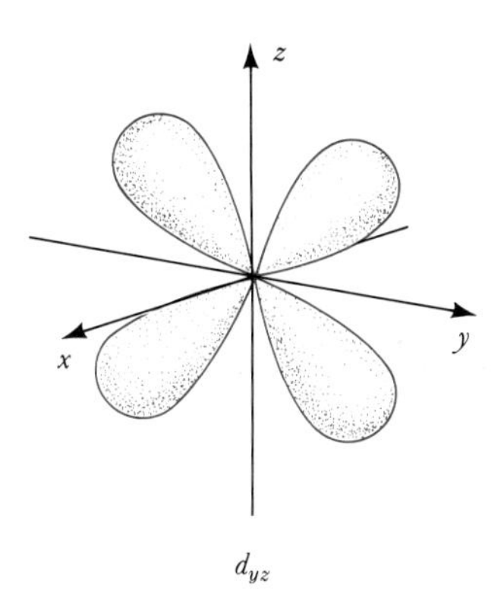

d_{yz}

The four lobes lie between the y and z axes in the four quadrants on the yz plane

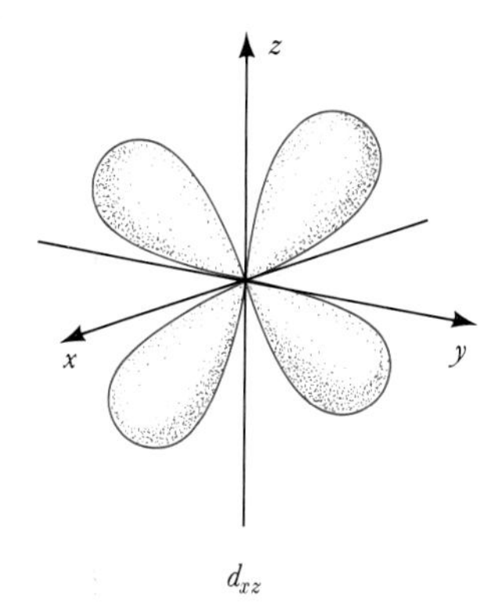

d_{xz}

The four lobes lie between the x and z axes in the four quadrants on the xz plane

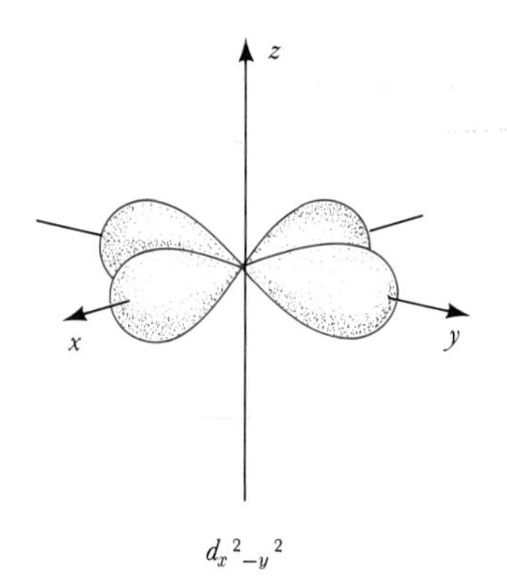

$d_{x^2-y^2}$

The four lobes lie along the x and y axes

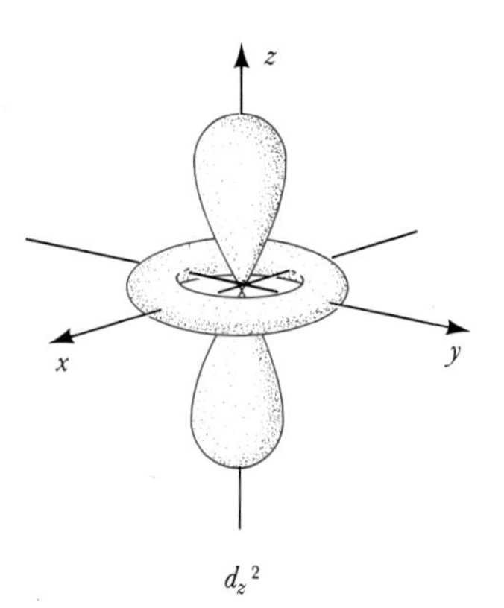

d_{z^2}

Two lobes are on the z axis and a donut-shaped lobe is symmetrically placed on the xy plane

Figure 23-2 The shape and relative orientation of the five d orbitals.

In writing the ground-state electron configuration of an iron atom, we fill the 4s orbital before the 3d orbitals. However, in all 3d transition metal ions, the energy of the 3d orbitals lies below the energy of the 4s orbital. Consequently, the ground-state electron configuration of Fe^{2+}, which is written Fe(II) in transition metal notation, is

$$\text{Fe(II)} \quad 1s^2 2s^2 2p^6 3s^2 3p^6 3d^6$$

Note that, for the 3d transition metal ions, the order of filling of the orbitals follows the arithmetic sequence: $n = 1$ is filled first, then $n = 2$, then $n = 3$, and so on, until all the electrons are used up.

Example 23-2: Give the ground-state electron configuration for the 4d transition metal ion Pd(II).

Solution: The roman numeral II denotes a palladium atom from which two electrons have been removed. For palladium, $Z = 46$, and thus there are $46 - 2 = 44$ electrons in Pd(II). The electron configuration is

$$1s^2 2s^2 2p^6 3s^2 3p^6 3d^{10} 4s^2 4p^6 4d^8$$

Note that the order of filling is strictly regular; that is, $n = 1$ is filled first, then $n = 2$, then $n = 3$, then $n = 4$, until the available electrons are used up.

23-2. TRANSITION METAL IONS WITH x ELECTRONS IN THE OUTER d ORBITALS ARE CALLED d^x IONS

As we move from left to right across a particular d transition metal series, we add d electrons one at a time for each successive element. For example, for the $3d$ transition metal series, the $1s^2 2s^2 2p^6 3s^2 3p^6$ core is filled in all cases, and thus we can concentrate our attention on the electron occupancy of the $3d$ orbitals. Similarly, we can concentrate on the electron occupancy of the $4d$ orbitals for the $4d$ series and on the $5d$ orbitals for the $5d$ series. For the M(II) ions of the $3d$ series we have

d^1	d^2	d^3	d^4	d^5
Sc(II)	Ti(II)	V(II)	Cr(II)	Mn(II)
21	22	23	24	25
d^6	d^7	d^8	d^9	d^{10}
Fe(II)	Co(II)	Ni(II)	Cu(II)	Zn(II)
26	27	28	29	30

The M(III) ions of the $3d$ series have one fewer electron than the M(II) ions, and so we have

d^0	d^1	d^2	d^3	d^4
Sc(III)	Ti(III)	V(III)	Cr(III)	Mn(III)
d^5	d^6	d^7	d^8	d^9
Fe(III)	Co(III)	Ni(III)	Cu(III)	Zn(III)

Note that Fe(II) and Co(III) both have six d electrons; for this reason they are called d^6 ions. Removal of an electron from Fe(II) yields Fe(III), which is a d^5 ion. This variety of possible oxidation states in transition metals is one reason for their unusually rich and interesting chemistry.

We can use the atomic numbers of the elements in the $3d$ series to determine the number of d electrons in the ions. For example, scandium is element 21 and Sc(II) is a d^1 ion; manganese is element 25 and Mn(II) is a d^5 ion; iron is element 26 and Fe(III), which has one fewer electron than Fe(II), is a d^5 ($6 - 1 = 5$) ion. Thus, the second digit of the atomic number of an element in the 3*d* series is equal to the number of *d* electrons in the M(II) ion of that element. (Note, however, that Zn(II), element 30 (10th of the series), is a d^{10} ion.

For ions in the 4*d* and 5*d* series, we can quickly determine the number of *d* electrons by noting the position of the element relative to the 3*d* series. For example, rhodium is directly below cobalt (element 27) and thus Rh(II) is a d^7 ion.

element	*atomic number*	d^x
Sc	21	d^1
Ti	22	d^2
V	23	d^3
Cr	24	d^4
Mn	25	d^5
Fe	26	d^6
Co	27	d^7
Ni	28	d^8
Cu	29	d^9
Zn	30	d^{10}

Example 23-3: Determine the number of outer-shell *d* electrons in Ir(II), Pt(IV), and Mo(III).

Solution: Iridium is directly below cobalt ($Z = 27$) in the periodic table, and so Ir(II) is a d^7 ion. Platinum is below nickel ($Z = 28$), and so Pt(II) is a d^8 ion. Platinum(IV) has two electrons fewer than Pt(II), and so Pt(IV) is a d^6 ion. Molybdenum is below chromium ($Z = 24$), and so Mo(II) is a d^4 ion. Consequently, Mo(III), which has one electron fewer than Mo(II), is a d^3 ion.

Example 23-4: Give three examples of d^6 ions with an oxidation state of +3.

Solution: For an M(III) ion to be a d^6 ion, the corresponding M(II) ion must be a d^7 ion. The M(II) d^7 ions are Co(II), Rh(II), and Ir(II), and so the M(III) d^6 ions are Co(III), Rh(III), and Ir(III).

23-3. LIGANDS ARE ANIONS OR NEUTRAL MOLECULES THAT BIND TO METAL IONS TO FORM COMPLEXES

Simple cyanide salts, such as sodium cyanide, NaCN, are deadly poisons. A solution of NaCN contains the ions $Na^+(aq)$ and $CN^-(aq)$, and the solution is colorless. The toxicity is due to the $CN^-(aq)$ ions, which block the oxygen-carrying capacity of hemoglobin in red blood cells. If excess iron(II) nitrate, $Fe(NO_3)_2$, is added to an aqueous solution of NaCN, the solution turns yellow. Chemical tests show that $CN^-(aq)$ is no longer present and that the solution is no longer poisonous. What happens is that cyanide ions react with iron(II) to form the *complex ion* $[Fe(CN)_6]^{4-}(aq)$, in which six cyanide ions are bound directly to the iron in an octahedral structure (Figure 23-3). The $[Fe(CN)_6]^{4-}$ ion occurs in solution as a single ion. The charge on the $[Fe(CN)_6]^{4-}$ ion is -4, and its sodium salt has the formula

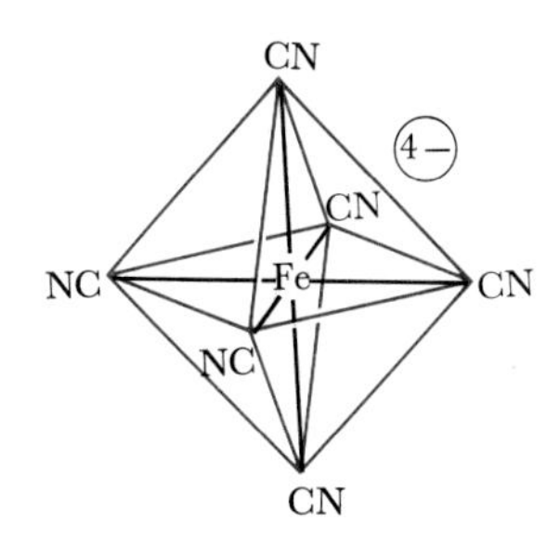

Figure 23-3 The yellow complex ion $[Fe(CN)_6]^{4-}$ is octahedral, with the six cyanide ions bonded to the central iron atom.

The complex ion $[Fe(CN)_6]^{4-}$ was the first complex ion synthesized in the laboratory. It was made in 1691 by the artist Dreisbach, who used it to make blue compounds for paints.

$Na_4[Fe(CN)_6]$. If $Na_4[Fe(CN)_6]$ is dissolved in water, then the resulting solution contains $Na^+(aq)$ and $[Fe(CN)_6]^{4-}(aq)$ ions:

$$Na_4[Fe(CN)_6](s) \xrightarrow[H_2O(l)]{} 4Na^+(aq) + [Fe(CN)_6]^{4-}(aq)$$

Generally, a complex ion contains a central metal ion to which are attached anions or neutral molecules. A complex ion is a distinct chemical species with properties different from those of its constituent species. Transition metal ions are capable of bonding to a wide variety of anions and neutral molecules to form complex ions. An anion or neutral molecule attached directly to a metal ion is called a *ligand*.

Transition metal ions in aqueous solution form complex ions by bonding with the water molecules. For example, a nickel(II) ion forms the octahedral complex ion $[Ni(H_2O)_6]^{2+}$ in aqueous solution (Figure 23-4). A solution of nickel(II) perchlorate, $Ni(ClO_4)_2$, in water is brilliant green because of the presence of $[Ni(H_2O)_6]^{2+}$ ions. If $NH_3(aq)$ is added to the solution, then the color changes from green to blue-violet. The chemical reaction responsible for this color change involves a change in the ligands attached to the nickel(II) ion:

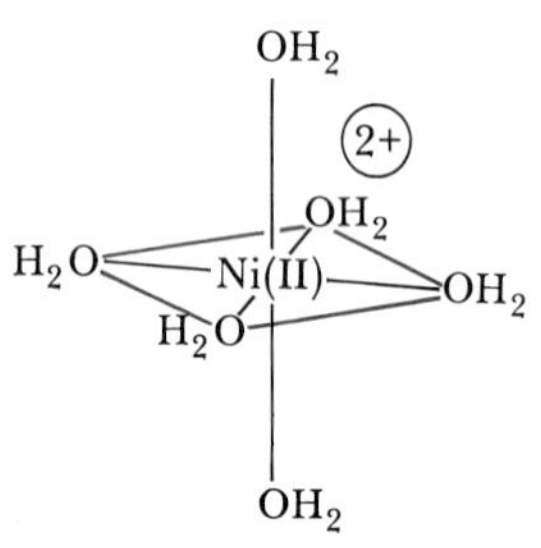

Figure 23-4 Octahedral structure of the green complex ion $[Ni(OH_2)_6]^{2+}$. All six Ni—O bonds are equivalent.

$$\underset{\text{green}}{[Ni(H_2O)_6]^{2+}(aq)} + \underset{\text{colorless}}{6NH_3(aq)} \rightleftharpoons \underset{\text{blue-violet}}{[Ni(NH_3)_6]^{2+}(aq)} + \underset{\text{colorless}}{6H_2O(l)}$$

The H_2O ligands—that is, the water molecules attached to the Ni(II) ion—are displaced by the NH_3 molecules, which become the new ligands attached to the nickel(II) ion. The $[Ni(NH_3)_6]^{2+}$ ion is octahedral like $[Ni(H_2O)_6]^{2+}$, with NH_3 molecules instead of H_2O molecules surrounding the central nickel atom. A reaction involving a change in the ligands attached to the central metal ion in a complex ion is called a *ligand-substitution reaction.*

The change in ligands from H_2O to NH_3 around nickel(II) produces a modified electrical environment around the nickel(II) ion, and this changes the energies of the eight *d* electrons on the nickel(II) ion. This change in energy of the *d* electrons causes a change in the wavelength of the light absorbed by the ion and thus results in a color change. In general, changes in the energy of the *d* electrons give rise to the color of transition metal complex ions.

The colors of transition metal complexes are due to electronic transitions of their *d* electrons.

Not all complex ions are octahedral. For example, if we add $NaCN(aq)$ to a solution containing the blue-violet $[Ni(NH_3)_6]^{2+}(aq)$ ion, then cyanide ions displace the NH_3 ligands from the nickel(II) ion to form the bright yellow $[Ni(CN)_4]^{2-}$ ion:

$$\underset{\text{blue-violet}}{[Ni(NH_3)_6]^{2+}(aq)} + \underset{\text{colorless}}{4CN^-(aq)} \rightleftharpoons \underset{\text{yellow}}{[Ni(CN)_4]^{2-}(aq)} + \underset{\text{colorless}}{6NH_3(aq)}$$

The structure of the $[Ni(CN)_4]^{2-}(aq)$ ion is square-planar; the four CN^- ligands are arranged in a plane at the four corners of an imaginary square around the nickel(II) ion.

It is the carbon end of the cyanide ion that is bonded to the nickel(II) ion. Note also that the overall charge on the $[Ni(CN)_4]^{2-}$ complex is -2 because the nickel(II) ion contributes a charge of $+2$ and the four CN^- ligands each contribute a charge of -1.

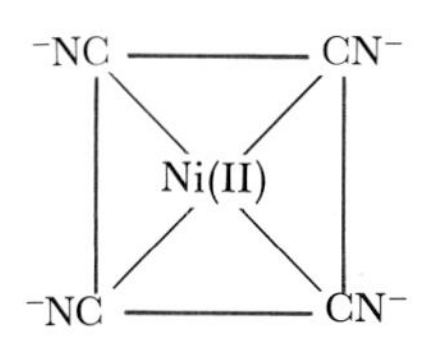

The $[Ni(CN)_4]^{2-}(aq)$ ion is square planar.

Example 23-5: Determine the oxidation state of platinum in $[Pt(NH_3)_4Cl_2]^{2+}$.

Solution: The $[Pt(NH_3)_4Cl_2]^{2+}$ complex ion has two kinds of ligands around the central metal atom: four NH_3 ligands and two Cl^- ligands. The charge on the NH_3 ligands is zero, the charge on each Cl^- is -1, and the overall charge is $+2$. Denoting the charge on the Pt ion as x, we have

$$\underbrace{x}_{\text{charge on Pt}} + \underbrace{4(0)}_{4NH_3} + \underbrace{2(-1)}_{2Cl^-} = \underbrace{+2}_{\text{overall charge on ion}}$$

or

$$x + 2(-1) = +2$$
$$x = +4$$

Thus the oxidation state of platinum in the complex is $+4$.

Devices that indicate humidity levels by changing color are based on the ligand-substitution reaction

$$\underset{\text{pink}}{2[Co(H_2O)_6]Cl_2(s)} \rightleftharpoons \underset{\text{blue}}{Co[CoCl_4](s)} + 12H_2O(g)$$

The $[Co(H_2O)_6]^{2+}$ ion is pink, and the $[CoCl_4]^{2-}$ ion is blue. When the humidity is high, $Co[CoCl_4]$ reacts with the water vapor in the air to form the pink $[Co(H_2O)_6]^{2+}$ ion. When the humidity is low, the reaction shifts from left to right, forming the blue $[CoCl_4]^{2-}$ ion. The $[CoCl_4]^{2-}$ ion is tetrahedral, with the cobalt(II) ion at the center of a tetrahedron formed by the four chloride ligands.

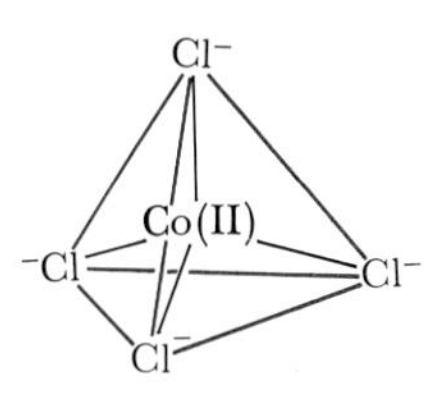

The $[CoCl_4]^{2-}(aq)$ ion is tetrahedral.

In many qualitative analysis schemes, $AgCl(s)$ is separated from other insoluble chlorides by the addition of $NH_3(aq)$ to form the soluble salt $[Ag(NH_3)_2]Cl$ which involves the complex $Ag(NH_3)_2^+$:

$$AgCl(s) + 2NH_3(aq) \rightarrow [Ag(NH_3)_2]^+(aq) + Cl^-(aq)$$

The $[Ag(NH_3)_2]^+$ ion is linear, with the N—Ag—N atoms arranged in a straight line.

$$[H_3N\text{—}Ag\text{—}NH_3]^{\oplus}$$

Table 23-1 Some examples of transition metal complexes of various geometries

Octahedral	*Tetrahedral*	*Square-planar*	*Linear*
$[Fe(H_2O)_6]^{2+}$	$[Zn(NH_3)_4]^{2+}$	$[Pt(CN)_4]^{2-}$	$[AgCl_2]^-$
$[Fe(CN)_6]^{4-}$	$[CoCl_4]^{2-}$	$[AuCl_4]^-$	$[CuI_2]^-$
$[Co(NO_2)_6]^{3-}$	$[Pb(OH)_4]^{2-}$	$[Rh(CN)_4]^{3-}$	
$[Ru(NH_3)_6]^{3+}$	$[CrO_4]^{2-}$	$[Pt(NH_3)_4]^{2+}$	
$[Pt(NH_3)_6]^{4+}$	$[FeCl_4]^-$	$[PdCl_4]^{2-}$	
$[Cr(H_2O)_6]^{3+}$	$[Ni(CO)_4]$	$[Ni(CN)_4]^{2-}$	
$[Cr(CO)_6]$	$[MnO_4]^-$	$[Cu(NH_3)_4]^{2+}$	
$[Fe(C_2O_4)_3]^{3-}$	$[HgI_4]^{2-}$		

The most common structures for transition metal complexes are octahedral, tetrahedral, square-planar, and linear. Of these four geometries, octahedral is by far the most common. Examples of transition metal complexes with these geometries are given in Table 23-1.

23-4. TRANSITION METAL COMPLEXES HAVE A SYSTEMATIC NOMENCLATURE

The wide variety and large number of possible transition metal complexes make a systematic procedure for naming them essential. An example of a systematic name for a transition metal complex is

$[Ni(NH_3)_6](NO_3)_2$ hexaamminenickel(II) nitrate

Let's analyze the name of this compound. As with any salt, the cation is named first. Thus this name tells us that the compound consists of a hexaamminenickel(II) cation and nitrate anions. The Greek prefix hexa denotes six, and ammine denotes the ligand NH_3. Thus the hexaammine part of the name tells us that there are six NH_3 ligands in the cation. The roman numeral II tells us that the nickel is in the $+2$ oxidation state. Because ammonia is a neutral molecule and nickel is in a $+2$ oxidation state, the charge on the complex cation is $+2$. Its formula is $[Ni(NH_3)_6]^{2+}$.

A simplified set of nomenclature rules for complexes is as follows:

1. *Name the cation first and then the anion:* for example potassium tetracyanonickelate(II), $K_2[Ni(CN)_4]$.

2. *In any complex ion or neutral molecule, name the ligands first and then the metal:* for example, hexaamminenickel(II) or tetracyanonickelate(II). If there are more than one type of ligand in the complex, then name them in alphabetical order: for example, diamminedichloroplatinum(II), $[Pt(NH_3)_2Cl_2]$.
3. *End the names of negative ligands in the letter o, but give neutral ligands the name of the ligand molecule.* Some common neutral ligands have special names, such as aqua for H_2O, ammine for NH_3, and carbonyl for CO. Table 23-2 lists the names of a number of ligands.
4. *Denote the number of ligands of a particular type by a Greek prefix, such as di-, tri-, tetra-, penta-, or hexa-.*
5. *If the complex ion is a cation, then use the ordinary name for the metal; if the complex ion is an anion, then end the name of the metal in -ate:* for example, tetrachlorocobaltate(II), $[CoCl_4]^{2-}$, where the suffix -ate on the metal name tells us that the complex ion is an anion. Table 23-3 lists some exceptions to this rule.
6. *Denote the oxidation state of the metal by a roman numeral in parentheses following the name of the metal.*

Note that there are two m's in the word ammine.

The application of these rules is best illustrated through examples.

Table 23-2 Names for some common ligands

Ligand[a]	*Name*	*Name as ligand*
F^-	fluoride	fluoro
Cl^-	chloride	chloro
Br^-	bromide	bromo
I^-	iodide	iodo
$\underline{C}N^-$	cyanide	cyano
$\underline{N}CS^-$	isothiocyanate	isothiocyanato
$\underline{S}CN^-$	thiocyanate	thiocyanato
$\underline{O}H^-$	hydroxide	hydroxo
O^{2-}	oxide	oxo
$\underline{O}NO^-$	nitrite	nitrito
$\underline{C}O$	carbon monoxide	carbonyl
$H_2\underline{O}$	water	aqua
$\underline{N}H_3$	ammonia	ammine

[a]For ligands with two or more different atoms, the underlined atom is the one bonded to the metal.

The names in Table 23-3 derive from the original Latin names of the metals.

Table 23-3 Some exceptions to rule 5 for naming complexes

Metal	*Name in complex anion*	*Complex anion*	*Name of complex anion*
silver	argentate	$[AgCl_2]^-$	dichloroargentate(I)
gold	aurate	$[Au(CN)_4]^-$	tetracyanoaurate(III)
copper	cuprate	$[CuCl_4]^{2-}$	tetrachlorocuprate(II)
iron	ferrate	$[Fe(CN)_6]^{3-}$	hexacyanoferrate(III)

Example 23-6: The water-soluble yellow-orange compound $Na_3[Co(NO_2)_6]$ is used in some qualitative analysis schemes to test for $K^+(aq)$. Almost all potassium salts are water-soluble, but $K_3[Co(NO_2)_6]$ is only slightly soluble in water. The precipitation reaction is

$$3K^+(aq) + [Co(NO_2)_6]^{3-}(aq) \rightarrow K_3[Co(NO_2)_6](s)$$

Name the $[Co(NO_2)_6]^{3-}$ ion.

Solution: The oxidation state of cobalt in the complex ion is determined as follows. The overall charge on the ion is -3, and there are six nitrite ions in the complex, each with a charge of -1. Denoting the oxidation state of cobalt as x, we have

$$\underbrace{x}_{\text{Co}} + \underbrace{6(-1)}_{6NO_2^-} = \underbrace{-3}_{\substack{\text{net charge} \\ \text{on complex ion}}}$$

or $x = +3$. The ion is called hexanitritocobaltate(III), where the -ate ending tells us that the complex is an anion.

Some other examples of transition metal compound names are

$K_2[Ni(CN)_4]$ — potassium (cation) tetracyanonickelate(II) (complex anion)
- tetracyano: $4CN^-$ ligands
- nickelate(II): Ni in $+2$ oxidation state

$[Cr(CO)_6]$ — hexacarbonylchromium(0) (neutral molecule)
- hexacarbonyl: 6CO ligands
- chromium(0): Cr in zero oxidation state

$[Co(H_2O)_4Cl_2]Cl$ — tetraaquadichlorocobalt(III) (complex cation) chloride (anion)
- tetraaqua: $4H_2O$ ligands
- dichloro: $2Cl^-$ ligands
- cobalt(III): Co in $+3$ oxidation state

The rules for writing a chemical formula from the name of a complex follow from the nomenclature rules. For example, the formula for the compound named potassium hexacyanoferrate(II) is determined as follows:

1. The cation is potassium, K^+.
2. The complex anion contains six CN^- (hexacyano) ions and an iron atom. The oxidation state of the iron is +2, as indicated by the roman numeral. The ending -ate tells us that the complex is an anion.
3. The charge on the anion is computed by adding up the charges on the metal ion and the ligands:

$$\underbrace{(+2)}_{\text{Fe(II)}} + \underbrace{6(-1)}_{\text{six CN}^-} = \underbrace{-4}_{\substack{\text{net charge}\\\text{on complex}}}$$

4. The formula for a complex ion is enclosed in brackets, and so we write the formula for the anion as $[Fe(CN)_6]^{4-}$. The formula for the salt is $K_4[Fe(CN)_6]$ because four K^+ ions are required to balance the −4 charge on the anion.

Example 23-7: Give the formula for the compound

hexaamminecobalt(III) hexachlorocobaltate(III)

Solution: In this case both the cation and the anion are complex ions. The cation has six NH_3 ligands with zero charge and one cobalt in a +3 oxidation state. Therefore, the formula for the cation is

$$[Co(NH_3)_6]^{3+}$$

Note that we write the formulas for complexes in brackets.

The anion has six Cl^- with a total ligand charge of −6 plus one cobalt in a +3 oxidation state. Therefore, the net charge on the complex anion is

$$6(-1) + (+3) = -3$$

and the formula is

$$[CoCl_6]^{3-}$$

The magnitudes of the charges on the cation and the anion are equal and thus appear in the formula for the salt on a one-to-one basis:

$$[Co(NH_3)_6][CoCl_6]$$

When this compound is dissolved in water, two ions are formed per formula unit:

$$[Co(NH_3)_6][CoCl_6](s) \xrightarrow[H_2O(l)]{} [Co(NH_3)_6]^{3+}(aq) + [CoCl_6]^{3-}(aq)$$

The nomenclature of transition metal complexes may appear cumbersome at first because of the length of the names, but with a little practice you will find it straightforward and ultimately much

Table 23-4 Some examples of nomenclature for transition metal compounds

Compound	*Name*
$[Co(NH_3)_6]Cl_3$	hexaamminecobalt(III) chloride
$K[AuCl_4]$	potassium tetrachloroaurate(III)
$Cu_2[Fe(CN)_6]$	copper(II) hexacyanoferrate(II)
$[Pt(NH_3)_6]Cl_4$	hexaammineplatinum(IV) chloride
$[Cu(NH_3)_4(H_2O)_2]Cl_2$	tetraamminediaquacopper(II) chloride
$[Cr(CO)_6]$	hexacarbonylchromium(0)
$K_3[CoF_6]$	potassium hexafluorocobaltate(III)

simpler than memorizing a host of nonsystematic common names. Table 23-4 gives several additional examples of names of transition metal complexes. You should try to name each one from the formula and write the formula from each name.

23-5. CERTAIN OCTAHEDRAL AND SQUARE-PLANAR TRANSITION METAL COMPLEXES CAN EXIST IN ISOMERIC FORMS

Consider the compound

$$[Pt(NH_3)_2Cl_2] \qquad \text{diamminedichloroplatinum(II)}$$

Platinum(II) complexes are invariably square-planar. There are two possible arrangements of the four ligands around the central platinum(II) ion:

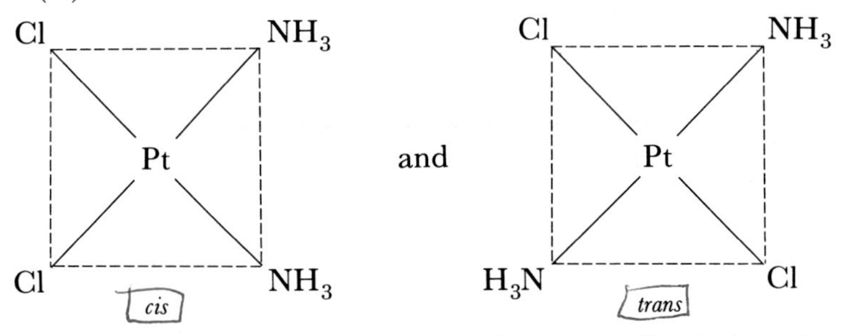

The *cis* and *trans* compounds are *geometric isomers*. The designation *cis* (the same) tells us that the identical ligands are placed adjacent to each other (on the same side) in the structure. The designation *trans* (opposite) tells us that the identical ligands are placed directly opposite each other in the structure. This nomenclature is similar to the nomenclature of the 1,2-dichloroethene isomers we discussed in Section 12-6. The *cis* and *trans* isomers of $[Pt(NH_3)_2Cl_2]$ are different compounds with different physical and chemical properties. For example, the *cis*-diamminedichloroplatinum(II) isomer is a potent anticancer drug that is manufactured under the name *Cisplatin,* whereas the *trans* isomer does not exhibit anticancer activity. How the *cis* isomer destroys cancer cells is not understood, and this is an important area of chemical research.

Cis and *trans* isomers are also found in certain octahedral complexes. Consider the octahedral ion tetraamminedichlorocobalt(III), $[Co(NH_3)_4Cl_2]^+$. The two Cl^- ligands can be placed in adjacent (*cis*) or opposite (*trans*) positions around the central cobalt(III) ion:

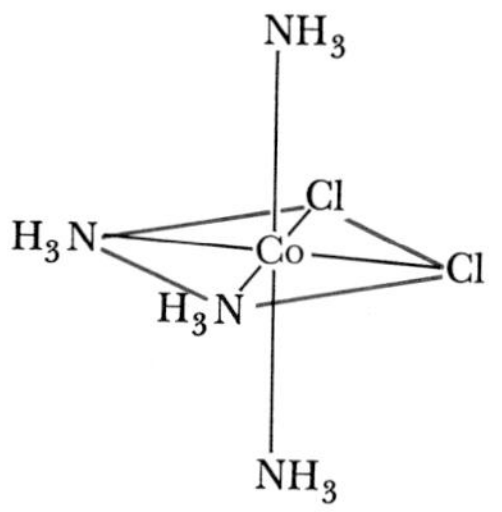

cis-$[Co(NH_3)_4Cl_2]$ (violet)

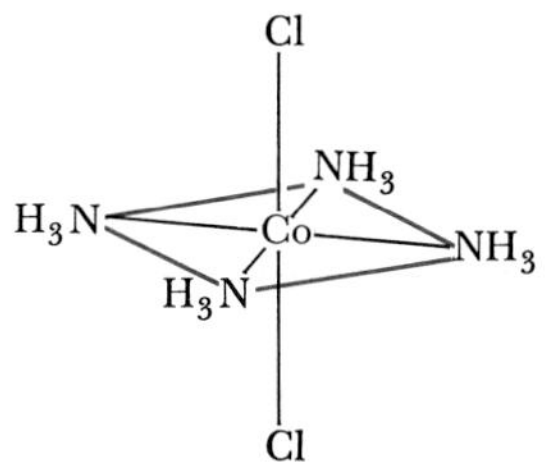

trans-$[Co(NH_3)_4Cl_2]$ (green)

Note that because the six *coordination positions* (that is, points of attachment) in an octahedral complex are equivalent, any other *cis* placement of the two Cl^- ligands around the cobalt(III) ion yields a structure identical to the *cis* structure shown here; this is also true for the *trans* placement of the two Cl^- ligands.

Example 23-8: The compound $[Co(NH_3)_3Cl_3]$ exists in two isomeric forms. Draw the structures of the two isomers.

Solution: The structures of the two isomers are shown in the margin. One compound is called *cis,cis* because each Cl^- is adjacent to the two others. The other compound is called *cis,trans* because one Cl^- is adjacent to and one Cl^- is opposite the other chloride.

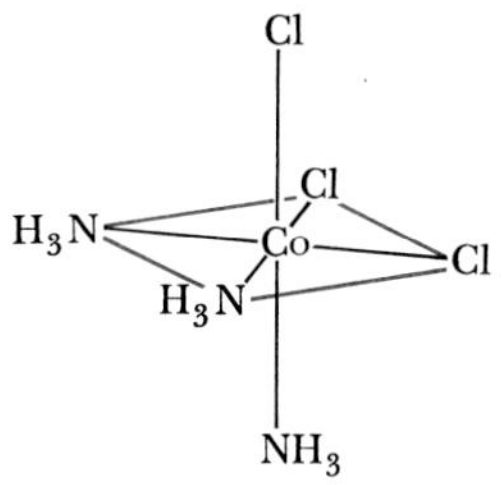

cis, cis-$[Co(NH_3)_3Cl_3]$

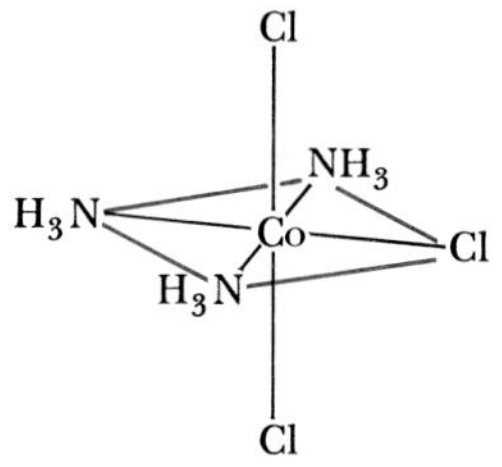

cis, trans-$[Co(NH_3)_3Cl_3]$

The structure of transition metal complexes was worked out by the Swiss chemist Alfred Werner in the late nineteenth and early twentieth centuries, without the aid of modern X-ray structure determination methods. In 1893, at the age of 26, he proposed a correct structural theory based on the number of different types of complexes, including isomers, that could be prepared for platinum(II), platinum(IV), and cobalt(III) amminechloro complexes. In 1913 Werner was awarded the Nobel Prize for his research in transition metal chemistry.

23-6. SOME LIGANDS BIND TO MORE THAN ONE COORDINATION POSITION AROUND THE METAL ION

Certain ligands can attach to a central metal cation at more than one coordination position. Two examples of such ligands are the oxalate ion (abbreviated ox) and ethylenediamine (abbreviated en):

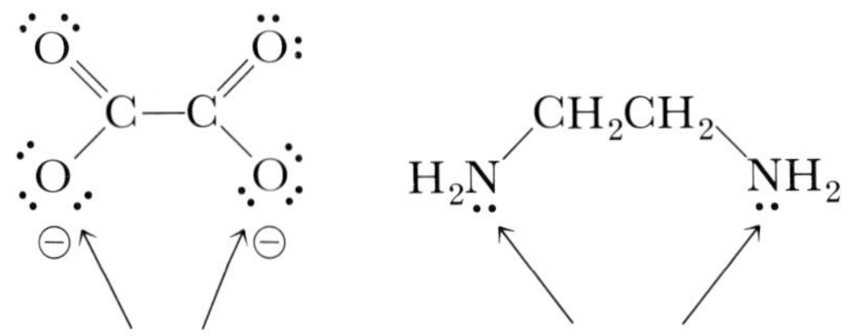

ligating atoms
oxalate ion (ox)

ligating atoms
ethylenediamine (en)

Note that the atoms of the ligand that attach to the metal ion are called *ligating atoms*. Two complexes involving these two ligands are

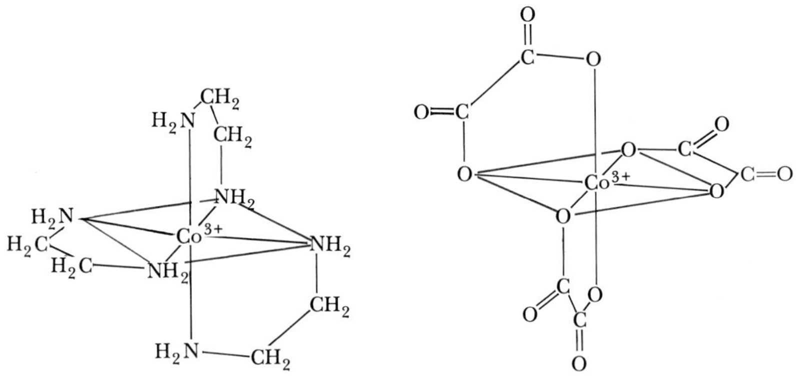

$[Co(en)_3]^{3+}$

$[Co(ox)_3]^{3-}$

Ligands that attach to a metal ion at more than one coordination position are called *polydentate ligands* or *chelating ligands*. The resulting complex is called a *chelate*. The word chelate comes from the Greek word meaning claw—the attachment of a chelating ligand can be thought of as a grasping of the metal ion with molecular claws. A chelating ligand that attaches to two metal coordination positions is called *bidentate* (two teeth), one that attaches to three positions is *tridentate* (three teeth), and so on. To simplify the presentation of structures involving chelating ligands, it is convenient to use a shorthand notation. For example, oxalate and ethylenediamine ligands are often written

Thus the structure of the complex ion $[Co(en)_3]^{3+}$, shown in the preceding diagram with the complete ligand composition, can also be represented as

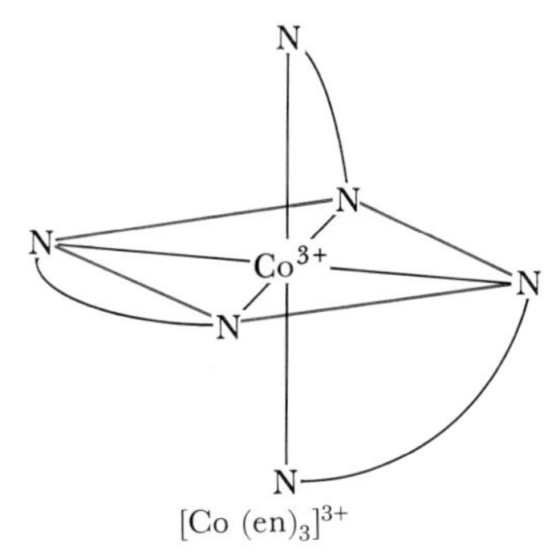

$[Co\ (en)_3]^{3+}$

The nomenclature for complex ions and molecules that have polydentate ligands follows the rules listed in Section 23-4, with one additional rule:

7. *If the ligand attached to the metal ion is a polydentate ligand, then enclose the ligand name in parentheses and use the prefix bis- for two ligands and tris- for three ligands:* for example, tris(ethylenediamine) cobalt(III), $[Co(H_2NCH_2CH_2NH_2)_3]^{3+}$.

Example 23-9: Give the chemical formula for ammonium tris(oxalato)ferrate(III).

Solution: The cation is the ammonium ion, NH_4^+. The anion is a complex ion with three (tris) oxalate ions, $C_2O_4^{2-}$, and an iron atom in the +3 oxidation state. The net charge on the complex anion is

$$\underbrace{(+3)}_{\text{Fe(III)}} + \underbrace{3(-2)}_{3C_2O_4^{2-}} = \underbrace{-3}_{\text{net charge}}$$

Therefore, the formula for the complex anion is $[Fe(C_2O_4)_3]^{3-}$ and the formula for the ammonium tris(oxalato)ferrate(III) is

$$(NH_4)_3[Fe(C_2O_4)_3]$$

Note that we do not say triammonium because the number of NH_4^+ ions is unambiguously fixed by the net charge of −3 on the complex anion. There are four ions per formula unit in $(NH_4)_3[Fe(C_2O_4)_3]$.

23-7. THE FIVE *d* ORBITALS OF A TRANSITION METAL ION IN AN OCTAHEDRAL COMPLEX ARE SPLIT INTO TWO GROUPS BY THE LIGANDS

The five *d* orbitals (Figure 23-2) in a gas-phase transition metal or metal ion without any attached ligands all have the same energy. However, when six identical ligands are attached to the transition metal ion to form an octahedral complex, the d orbitals on the metal ion are split into two sets. The lower set, called t_{2g}, consists of three orbitals (d_{xy}, d_{xz}, and d_{yz}), and the upper set, called e_g, consists of two orbitals ($d_{x^2-y^2}$, d_{z^2}). The magnitude of the splitting of the d orbitals depends upon both the central metal ion and the ligands. We shall let Δ_o (where the o stands for octahedral) be the energy separation between the t_{2g} and e_g orbitals. The t_{2g} orbitals can accommodate up to six electrons, and the e_g orbitals can accommodate up to four electrons:

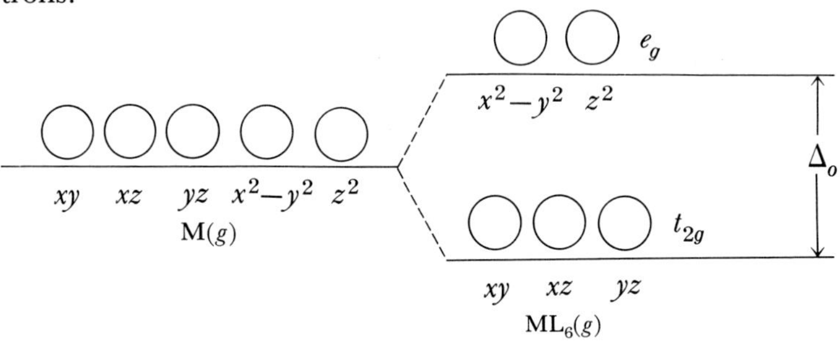

The reason for this particular d-orbital splitting pattern can be understood from the placement of the six ligands relative to the d orbitals in the complex. Consider the octahedral transition metal complex shown in Figure 23-5. The key point is the placement of the ligands relative to the lobes of the d orbitals.

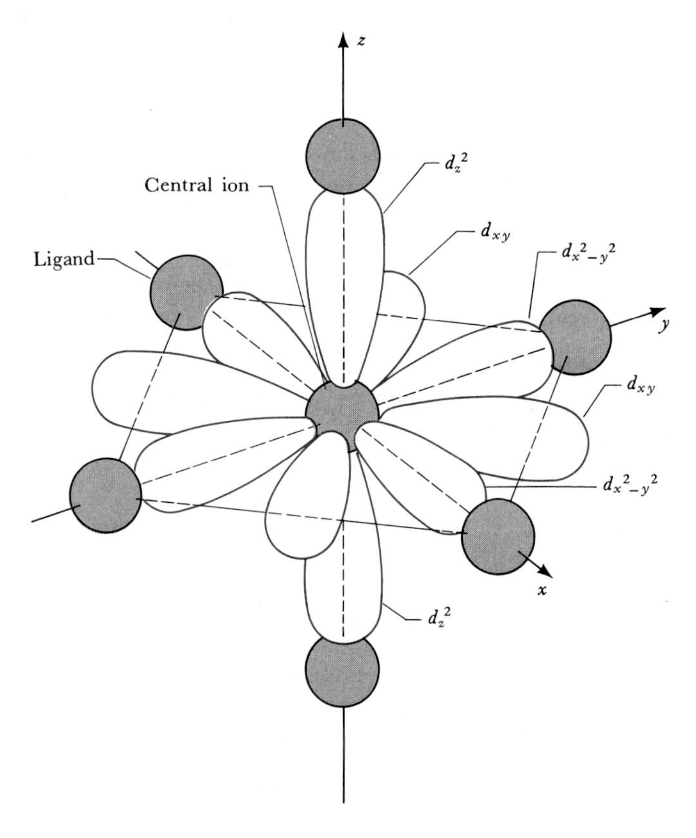

Figure 23-5 A regular octahedral complex, showing the orientation of the d_{z^2}, $d_{x^2-y^2}$, and d_{xy} orbitals relative to the ligands, which are brought in along the x, y, and z axes toward the central metal ion. For simplicity, the d_{xz} and d_{yz} orbitals are not shown. The d_{xy}, d_{xz}, and d_{yz} lobes point toward positions between the ligands, as shown here for d_{xy}.

The ligand electrons, especially those in lone pairs directed at the central metal ion, provide an electric field. This electric field is not spherically symmetric around the central metal ion because the ligands are positioned in an octahedral array. Therefore, the energies of the metal *d* orbitals are changed by different amounts, depending on their positions with respect to the ligands. In the simplified picture described here, the ligand lone pairs remain with the ligand. Figure 23-5 shows that the lobes of the $d_{x^2-y^2}$ and d_{z^2} orbitals point directly at the ligands and that the lobes of the d_{xy}, d_{xz}, and d_{yz} orbitals all point between the ligands. Thus electrons placed in $d_{x^2-y^2}$ and d_{z^2} orbitals experience a greater electrostatic repulsion (like charges repel) and thus have a higher energy in the complex than electrons placed in the d_{xy}, d_{xz}, and d_{yz} orbitals. The *d*-orbital splitting pattern in an octahedral complex is thus seen to be a consequence of the positions of the ligands relative to the *d* orbitals.

23-8. THE COLORS OF MOST TRANSITION METAL COMPLEXES ARISE FROM TRANSITIONS OF ELECTRONS IN THE METAL *d* ORBITALS

Although Δ_o, the magnitude of the separation between the t_{2g} and e_g orbitals, depends upon the central metal ion and the ligands, for most cases Δ_o is such that the energy difference corresponds to the visible region of the electromagnetic spectrum. In other words, the frequency of the radiation that is absorbed, which obeys the relation

$$h\nu = \Delta_o$$

The equation $E = h\nu$ was introduced in Section 7-5.

is in the visible region of the spectrum. Thus, many complexes absorb light in the visible region and are colored.

It is possible to understand the variety of colors of complex ions in terms of Δ_o and the electron occupancy of the t_{2g} and e_g orbitals. For example, consider the red-purple ion $[Ti(H_2O)_6]^{3+}(aq)$. Reference to Figure 23-1 shows that titanium is the second member of the 3*d* transition series, and thus Ti(II) has two *d* electrons. The $[Ti(H_2O)_6]^{3+}$ complex, which contains Ti(III), has one fewer *d* electron than Ti(II), and therefore $[Ti(H_2O)_6]^{3+}$ is a d^1 ion. The ground-state and excited-state *d* electron configurations of the d^1 $[Ti(H_2O)_6]^{3+}$ ion are

x^2-y^2 z^2 $\qquad$ x^2-y^2 z^2

Δ_o $\quad$ $\xrightarrow{h\nu}$

xy xz yz $\qquad$ xy xz yz

t_{2g}^1 ground state $\qquad$ e_g^1 excited state

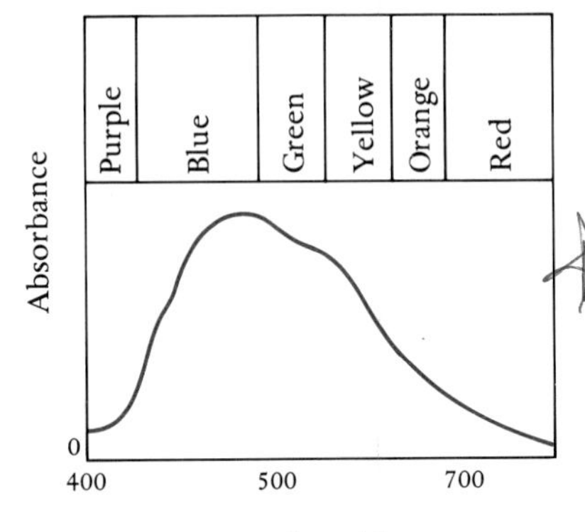

Figure 23-6 Absorption spectrum of the red-purple ion $[Ti(H_2O)_6]^{3+}$ in aqueous solution. The complex absorbs light in the blue, green, yellow, and orange regions, but most of the light in the red and purple regions passes through the sample and so is detected by the eye. Thus, the $[Ti(H_2O)_6]^{3+}$ complex is red-purple.

The absorption spectrum of $[Ti(H_2O)_6]^{3+}$ in the visible region is shown in Figure 23-6. The absorption of a photon in the blue-to-orange region excites the *d* electron from the lower (t_{2g}) set of *d* orbitals to the upper (e_g) set.

Example 23-10: Predict the color of the octahedral complex ion $[Zn(H_2O)_6]^{2+}$.

Solution: By referring to Figure 23-1, we see that Zn(II) is a d^{10} ion. Thus, the *d*-orbital electron configuration of $[Zn(H_2O)_6]^{2+}$ is

$$\underset{x^2-y^2\quad z^2}{\underline{(\uparrow\downarrow)\ (\uparrow\downarrow)}}$$

$$\underset{xy\quad xz\quad yz}{\underline{(\uparrow\downarrow)\ (\uparrow\downarrow)\ (\uparrow\downarrow)}}$$

or $t_{2g}^6e_g^4$. Because the t_{2g} and e_g orbitals of $[Zn(H_2O)_6]^{2+}$ are completely filled, the ion is unable to absorb light. Therefore, we predict (correctly) that a Zn(II) is colorless in aqueous solution.

23-9. THE COLORS OF MANY GEMSTONES ARE DUE TO TRANSITION METAL IONS

The colors of many familiar gemstones (Frontispiece) are due to transition metal ions. For example, the mineral corundum, Al_2O_3, is colorless when pure. However, when various M^{3+} transition metals are present in trace amounts, various gemstones result (Table 23-5).

Emeralds are a rare variety of the mineral beryl, $Be_3Al_2Si_6O_{18}$, with Cr^{3+} substituted for Al^{3+} in trace amounts, and aquamarines are beryl with Fe^{3+} in place of Al^{3+} (Table 23-5).

Garnets (red), jade (green), and peridot (light green) are silicates with trace impurities of Fe^{3+}, Cr^{3+}, and Fe^{2+}, respectively. Turquoise is a blue-to-aqua copper aluminum hydroxyphosphate, the blue-green color being due to the Cu^{2+} ions.

All the corundum and beryl gemstones in Table 23-5 can be made in the laboratory. Very large ruby rods are used in ruby lasers. The quality of synthetic rubies is at least as good as that of any natural ruby. Synthetic emeralds differ slightly in water content from natural emeralds, but the difference cannot be detected by eye. Synthetic emeralds can withstand much higher temperatures than natural

Table 23-5 Transition metal ion substitutions for various gemstones

Trace transition metal	*Color*	*Name*
	Corundum	
Cr^{3+}	brilliant red	ruby
Mn^{3+}	deep purple	amethyst
Fe^{3+}	yellow to reddish yellow	topaz (yellow sapphire)
V^{3+} or Co^{3+}	rich blue	blue sapphire
	Beryl	
Cr^{3+}	green	emerald
Fe^{3+}	pale blue to light greenish blue	aquamarine

emeralds without shattering. The values of many types of gemstones are artificially maintained by limiting production of the synthetic ones.

23-10. *d*-ORBITAL ELECTRON CONFIGURATION IS THE KEY TO UNDERSTANDING THE CHEMISTRY OF THE *d* TRANSITION METALS

We saw in Section 23-8 that the *d*-orbital electron configuration of an octahedral Ti(III) complex is t_{2g}^1 and that of an octahedral Zn(II) complex is $t_{2g}^6e_g^4$. To determine *d*-orbital electron configurations, we place *d* electrons in the t_{2g} and e_g orbitals, using the diagram

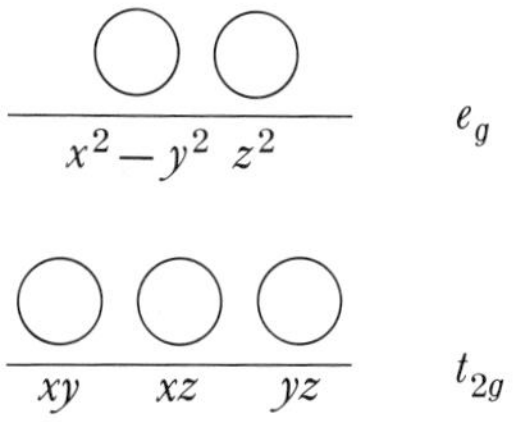

$$\underset{x^2-y^2\quad z^2}{\underline{\bigcirc\ \bigcirc}}\qquad e_g$$

$$\underset{xy\quad xz\quad yz}{\underline{\bigcirc\ \bigcirc\ \bigcirc}}\qquad t_{2g}$$

For example, an octahedral complex ion of chromium(III) has three *d* electrons, and we place them in the t_{2g} orbitals in accord with Hund's rule (Section 8-4), which says that we should place one electron in each t_{2g} orbital such that the spins are aligned. Thus, the ground-state d-electron configuration of chromium(III) is

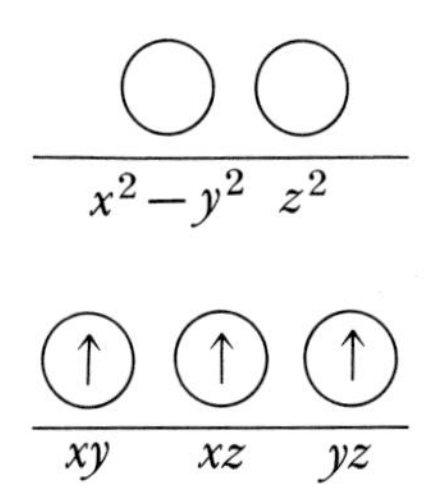

which we write as t_{2g}^3.

A difficulty arises when we consider octahedral complex ions with more than three d electrons. Let's consider the ion $[Fe(H_2O)_6]^{2+}$. Iron(II) has six d electrons, and so there are two possibilities for the d-electron configuration of an octahedral iron(II) complex ion:

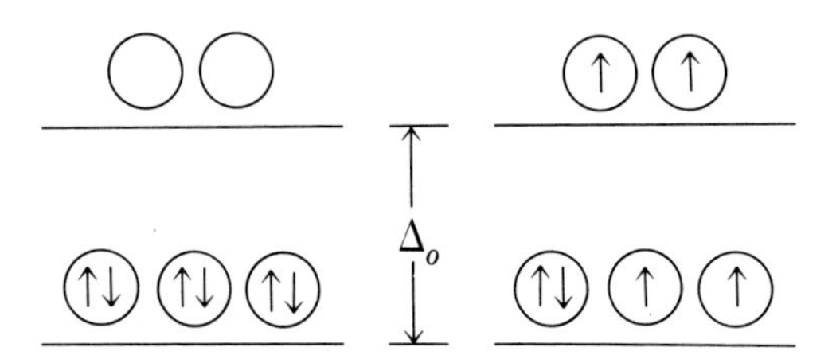

In one case the d-electron configuration is t_{2g}^6, and in the other case it is $t_{2g}^4e_g^2$. In the t_{2g}^6 configuration, the spins of all the electrons are paired; in the $t_{2g}^4e_g^2$ configuration, the electrons are in different orbitals with four unpaired electrons, in accord with Hund's rule. The t_{2g}^6 configuration is said to be a *low-spin configuration,* and the $t_{2g}^4e_g^2$ configuration is said to be a *high-spin configuration.*

The value of Δ_o determines whether a d-electron configuration is low-spin or high-spin. If Δ_o is small, then the d electrons occupy the e_g orbitals before they pair up in the t_{2g} orbitals. If Δ_o is large, then the d electrons fill the t_{2g} orbitals completely before occupying the higher-energy e_g orbitals. For example, the d^6 ion $[Fe(H_2O)_6]^{2+}$ is high-spin,

e_g^2

$\leftarrow \Delta_o(H_2O)$ $\qquad [Fe(H_2O)_6]^{2+}$ $\qquad t_{2g}^4e_g^2$

t_{2g}^4

because Δ_o is small relative to the energy required to pair up the electrons in the t_{2g} orbitals. The d^6 ion $[Fe(CN)_6]^{4-}$, on the other hand, is low-spin:

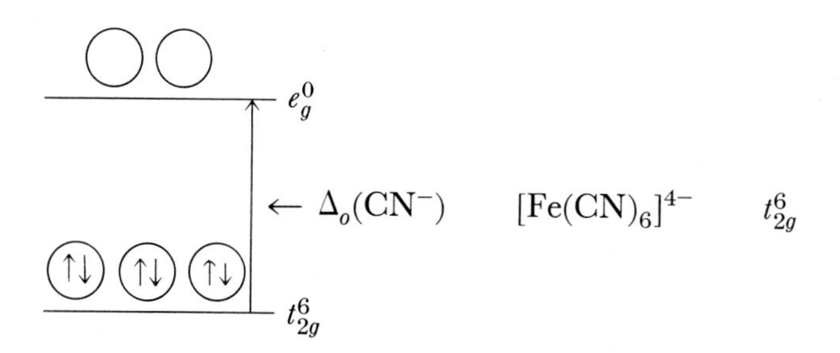

because Δ_o is large relative to the energy required to pair up the electrons in the t_{2g} orbitals. Both $[Fe(H_2O)_6]^{2+}$ and $[Fe(CN)_6]^{4-}$ contain iron(II), and both are therefore d^6 ions. The Δ_o values obtained from the spectra of the complexes show that the CN^- ligands interact much more strongly with iron(II) than do the H_2O ligands; that is, $\Delta_o(CN^-) \gg \Delta_o(H_2O)$. The increased *d*-orbital splitting energy for CN^- is sufficiently great to overcome the additional electron-electron repulsions that result from pairing up the electrons.

low-spin complex
$\Delta_o >$ pairing energy
high-spin complex
$\Delta_o <$ pairing energy

In summary, then, a low-spin complex results whenever the energy difference between the t_{2g} and e_g orbitals is greater than the *electron-pairing energy.*

Example 23-11: Give the *d*-electron configuration of the low-spin complex $[Pt(NH_3)_6]^{4+}$.

Solution: Referring to Figure 23-1, we note that platinum is the eighth member of the 5*d* transition series. Therefore, platinum(II) is a d^8 ion. The platinum in $[Pt(NH_3)_6]^{4+}$ is platinum(IV), and thus platinum(IV) is a d^6 ion [two fewer *d* electrons than platinum(II)]. The *d*-electron configuration of a low-spin d^6 ion is t_{2g}^6:

e_g^0: ○ ○ (x^2-y^2, z^2)

t_{2g}^6: (↑↓) (↑↓) (↑↓) (xy, xz, yz)

The various possible *d*-electron configurations for octahedral transition metal ions are shown in Figure 23-7. Note that for d^4, d^5, d^6, and d^7 octahedral complexes there are two possible *d*-electron configurations, namely, the high-spin (maximum possible number of unpaired *d* electrons) and the low-spin (minimum possible number of unpaired *d* electrons).

A particularly interesting case of a conversion of a transition metal ion from a high-spin to a low-spin configuration occurs in hemoglo-

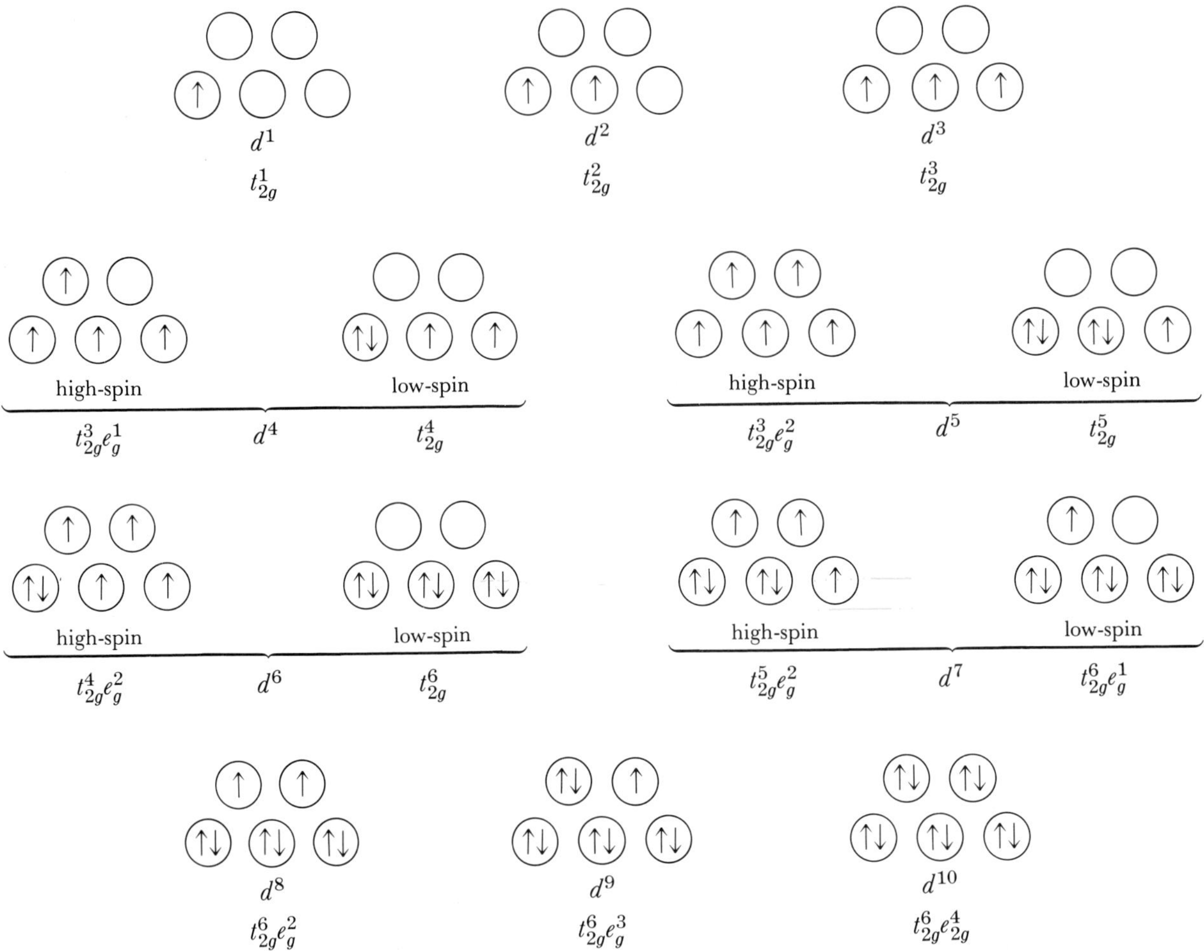

Figure 23-7 Possible ground-state *d*-electron configurations for octahedral d^x ions, where x is the total number of *d* electrons in the transition metal. Note that for d^4, d^5, d^6 and d^7 ions there are two possibilities: the *high-spin* and the *low-spin* configurations.

bin. The heme group (Figure 23-8) in hemoglobin contains the d^6 ion iron(II). When oxygen molecules bind to the iron(II) ions, the molecule is red and has a low-spin *d*-electron configuration. When oxygen is removed, the blue high-spin deoxyhemoglobin is formed:

$$\underset{\text{red (low-spin)}}{\text{Hmb}\cdot 4\text{O}_2} \rightleftharpoons \underset{\text{blue (high-spin)}}{\text{Hmb}} + 4\text{O}_2$$

Δ_{red} Δ_{blue}

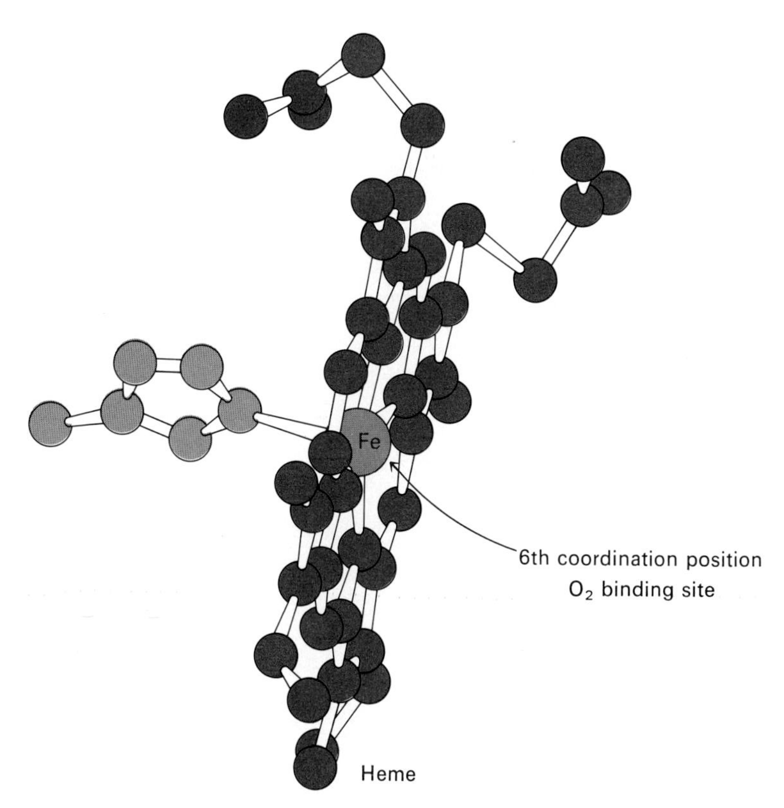

Figure 23-8 Model of the oxygen-binding site in a heme portion of hemoglobin. The iron atom is already attached to five atoms. The O_2 molecule attaches to the vacant sixth coordination position of the iron atom.

Arterial blood is red because it contains O_2 coordinated to iron(II); venous blood, which has passed through the capillaries and given up its oxygen to the tissues, is blue (look at the bluish veins near the surface of your skin). Deprivation of oxygen, as occurs when the air passages to the lungs are blocked, produces a characteristic blue coloration in the affected person. The toxicity of carbon monoxide gas and cyanide salts arises from the ability of these two ligands to bind more tightly than O_2 to the iron(II) in hemoglobin and thereby block the uptake of O_2 from the lungs.

23-11. LIGANDS CAN BE ORDERED ACCORDING TO THEIR ABILITY TO SPLIT THE TRANSITION METAL *d* ORBITALS

The Japanese chemist Ryutaro Tsuchida first noted that for octahedral complexes with the same metal oxidation state in the same transition series, the value of Δ_o depends primarily on the type of ligands and is independent of the metal ion. Tsuchida's observation enabled him to arrange common ligands in a *spectrochemical series* in order of increasing ability of ligands to split the metal *d* orbitals. An abbreviated version of Tsuchida's spectrochemical series is shown in Figure 23-9.

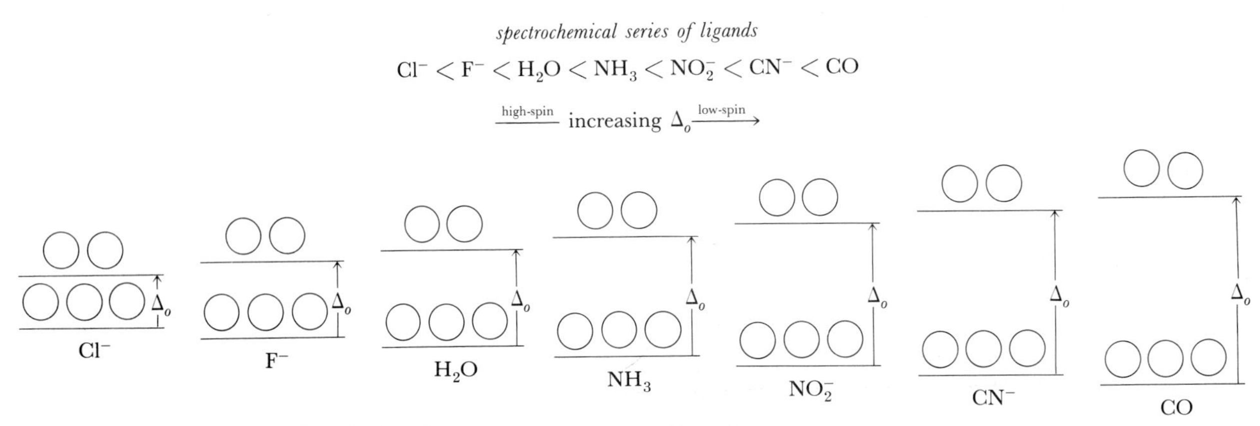

Figure 23-9 The spectrochemical series for some common ligands.

The ligands NO_2^-, CN^-, and CO cause large *d*-orbital splitting (large Δ_o), which results in low-spin complexes. The halide ions cause relatively small *d*-orbital splitting (small Δ_o), which results in high-spin complexes. Thus the chloride complex of the d^6 ion cobalt(III), $[CoCl_6]^{3-}$, is high-spin ($t_{2g}^4e_g^2$) and the cyanide complex, $[Co(CN)_6]^{3-}$, is low-spin ($t_{2g}^6e_g^0$).

Example 23-12: Predict whether the complex ions $[FeF_6]^{3-}$ and $[Fe(CN)_6]^{3-}$ are high-spin or low-spin and give the *d*-electron configuration of each.

Solution: Both complexes involve iron(III), which is a d^5 ion. Referring to the spectrochemical series (Figure 23-9), we note that F^- produces a relatively small Δ_o value, and thus we predict that $[FeF_6]^{3-}$ is a high-spin complex with the *d*-electron configuration $t_{2g}^3e_g^2$. The CN^- ligand produces a large Δ_o value, and thus we predict that $[Fe(CN)_6]^{3-}$ is a low-spin complex with the *d*-electron configuration t_{2g}^5.

The ligands H_2O and NH_3 give rise to Δ_o values for 3*d* metal ions that are roughly equal to the pairing energy, and thus some H_2O and NH_3 complexes are high-spin and some are low-spin.

23-12. TRANSITION METAL COMPLEXES ARE CLASSIFIED AS EITHER INERT OR LABILE

Many of the chemical properties of complex ions can be understood or predicted from the *d*-electron configuration. Here we consider the stability of complex ions. An *inert complex* is one that only slowly exchanges its ligands with other available ligands. A *labile complex* is one that rapidly exchanges its ligands with other available ligands. As a dramatic example of the contrasting ligand exchange rates of labile and inert complexes, consider the following ligand-substitution reactions:

The word labile means readily open to change.

$$(1)\quad [Co(NH_3)_6]^{2+}(aq) + 6H_3O^+(aq) \rightleftharpoons [Co(H_2O)_6]^{2+}(aq) + 6NH_4^+(aq)$$

$$(2)\quad [Co(NH_3)_6]^{3+}(aq) + 6H_3O^+(aq) \rightleftharpoons [Co(H_2O)_6]^{3+}(aq) + 6NH_4^+(aq)$$

The equilibrium for both reactions lies far to the right. Nonetheless, Reaction (1), which involves the d^7 $[Co(NH_3)_6]^{2+}(aq)$ complex, attains equilibrium in 1 M $H^+(aq)$ in about 10 s, whereas Reaction (2), which involves the d^6 $[Co(NH_3)_6]^{3+}(aq)$ complex, requires over a month to reach equilibrium at 25°C under the same conditions. In other words, a difference of only one *d* electron causes the reaction rates to differ by a factor of over 350,000.

Henry Taube, then at the University of Chicago, was the first to note that t_{2g}^3, t_{2g}^4, t_{2g}^5, and t_{2g}^6 octahedral complexes are inert and all other octahedral complexes are labile. Thus $[Co(NH_3)_6]^{2+}(aq)$, an octahedral high-spin d^7 ion ($t_{2g}^5\,e_g^2$), is labile and $[Co(NH_3)_6]^{3+}(aq)$, an octahedral, low-spin d^6 ion (t_{2g}^6), is inert.

Henry Taube received the 1983 Nobel Prize in Chemistry for his work in transition metal chemistry.

Example 23-13: Predict whether the following complex ions are labile or inert:

$$[Fe(CN)_6]^{4-} \qquad [Cr(H_2O)_6]^{3+} \qquad [V(H_2O)_6]^{3+}$$

Solution: The *d*-electron configurations of the three complexes are

$[Fe(CN)_6]^{4-}$, d^6 low-spin	$[Cr(H_2O)_6]^{3+}$, d^3	$[V(H_2O)_6]^{3+}$, d^2
t_{2g}^6	t_{2g}^3	t_{2g}^2
e_g: ○ ○	e_g: ○ ○	e_g: ○ ○
t_{2g}: (↑↓) (↑↓) (↑↓)	t_{2g}: (↑) (↑) (↑)	t_{2g}: (↑) (↑) ○
inert	inert	labile

Problems 23-45 through 23-48 ask you to predict whether certain complexes are labile or inert.

The d^6 low-spin $[Fe(CN)_6]^{4-}$ ion is t_{2g}^6 and thus inert; the d^3 $[Cr(H_2O)_6]^{3+}$ ion is t_{2g}^3 and thus inert; the d^2 $[V(H_2O)_6]^{3+}$ ion is t_{2g}^2 and thus labile. These three predictions are confirmed by experiment.

Although Taube's rules can be given a theoretical foundation, we treat them here only as useful empirical rules because the theoretical interpretation is not simple. Some additional examples of labile and inert octahedral complexes are given in Table 23-6.

Table 23-6 Some examples of inert and labile octahedral complexes

d^x	*d-electron configuration*	*Kinetic classification*	*Example*
d^0	t_{2g}^0	labile	$[Sc(H_2O)_6]^{3+}$
d^1	t_{2g}^1	labile	$[Ti(H_2O)_6]^{3+}$
d^2	t_{2g}^2	labile	$[V(en)_3]^{3+}$
d^3	t_{2g}^3	inert	$[Cr(C_2O_4)_3]^{3-}$
d^4	$t_{2g}^3e_g^1$	high-spin, labile	$[Cr(H_2O)_6]^{2+}$
d^4	t_{2g}^4	low-spin, inert	$[Cr(CN)_6]^{4-}$
d^5	$t_{2g}^3e_g^2$	high-spin, labile	$[Mn(H_2O)_6]^{2+}$
d^5	t_{2g}^5	low-spin, inert	$[Fe(CN)_6]^{3-}$
d^6	$t_{2g}^4e_g^2$	high-spin, labile	$[Fe(H_2O)_6]^{2+}$
d^6	t_{2g}^6	low-spin, inert	$[Pt(NH_3)_6]^{4+}$
d^7	$t_{2g}^5e_g^2$	high-spin, labile	$[Co(C_2O_4)_3]^{4-}$
d^7	$t_{2g}^6e_g^1$	low-spin, labile	$[Co(CN)_6]^{4-}$
d^8	$t_{2g}^6e_g^2$	labile	$[Ni(NH_3)_6]^{2+}$
d^9	$t_{2g}^6e_g^3$	labile	$[Cu(NH_3)_6]^{2+}$
d^{10}	$t_{2g}^6e_g^4$	labile	$[Zn(H_2O)_6]^{2+}$

23-13. THE *d*-ORBITAL SPLITTING PATTERNS IN SQUARE-PLANAR AND TETRAHEDRAL COMPLEXES ARE DIFFERENT FROM THOSE IN OCTAHEDRAL COMPLEXES

The splitting pattern of the d orbitals in an octahedral complex is explained in Figure 23-5. The splitting patterns in square-planar and tetrahedral complexes differ from the splitting pattern in octahedral complexes because the ligands are brought in from different directions toward the central metal ion.

The splitting pattern of a square-planar complex is shown in Figure 23-10. The closer the lobes of a particular d orbital to the ligands, the higher the energy of that orbital. In a square-planar complex, the four ligands are placed on the x and y axes, and thus the lobes of the $d_{x^2-y^2}$ orbital point directly at the ligands and the $d_{x^2-y^2}$ orbital has the highest energy of the five d orbitals. The relative energies of the other d orbitals can be deduced by an analysis of their interaction with the ligands, as we did in Figure 23-6 for the octahedral case.

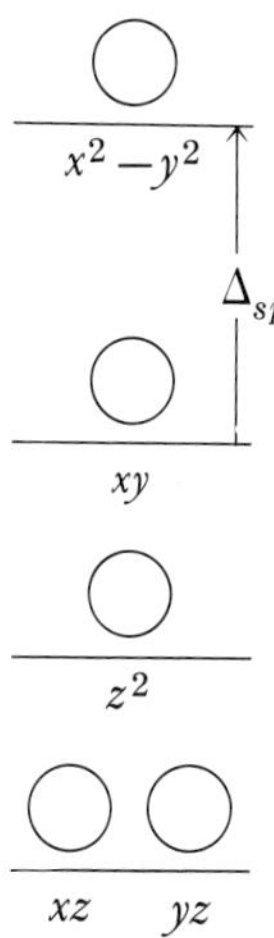

Figure 23-10 The d-orbital splitting pattern for a square-planar transition metal complex. The value of Δ_{sp} (sp stands for square-planar) is the energy separation between the two highest-energy d orbitals.

Square planar complexes are formed most often with d^8 ions.

Example 23-14: Give the d-electron configuration for the square-planar complex $[Pd(CN)_4]^{2-}$.

Solution: Referring to Figure 23-1, we note that palladium is the eighth member of the $4d$ transition series. The oxidation state of palladium in $[Pd(CN)_4]^{2-}$ is Pd(II), and thus palladium(II) is a d^8 ion. Referring to Figure 23-10, we have two possible configurations:

orbital	low-spin	high-spin
x^2-y^2	○	(↑)
xy	(↑↓)	(↑)
z^2	(↑↓)	(↑↓)
xz yz	(↑↓) (↑↓)	(↑↓) (↑↓)

The actual configuration is the one on the left (no unpaired electrons). The $d_{x^2-y^2}$ orbital is so high in energy relative to the d_{xy} orbital that all d^8 square-planar complexes are low-spin.

The d-orbital splitting pattern in a tetrahedral complex is shown in Figure 23-11. The lower set of two d orbitals is called the e set, and the upper set of three d orbitals is called the t_2 set. The value of Δ_t (the d-orbital splitting energy for a tetrahedral complex, where t denotes tetrahedral) is much smaller than the value of Δ_o (the d-orbital splitting energy for an octahedral complex) because none of the d-orbital lobes in tetrahedral complexes point directly at the ligands. A consequence of the relatively small value of Δ_t is that there are no low-spin tetrahedral complexes.

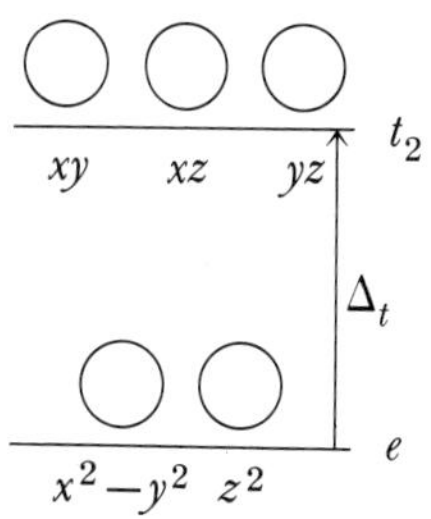

Figure 23-11 The d-orbital splitting pattern for a tetrahedral complex. Note that the pattern is the reverse of that for an octahedral complex. The difference in splitting patterns for tetrahedral and octahedral complexes is a direct consequence of the different placement of the ligands relative to the five d orbitals.

Example 23-15: Give the *d*-electron configuration in the tetrahedral complex $[Co(NCS)_4]^{2-}$.

Solution: The oxidation state of cobalt in the complex is Co(II) because there are four NCS^- ligands and the net charge on the complex is -2. Cobalt is the seventh member of the 3*d* transition series, and thus cobalt(II) is a d^7 ion. Referring to Figure 23-11, we have, for the *d*-electron configuration of a tetrahedral d^7 ion, $e^4t_2^3$:

(↑) (↑) (↑) t_2^3
xy xz yz

(↑↓) (↑↓) e^4
x^2-y^2 z^2

The *d*-orbital splitting patterns for octahedral, tetrahedral, and square-planar complexes are summarized in Figure 23-12.

23-14. TRANSITION METAL COMPLEXES WITH UNPAIRED *d* ELECTRONS ARE PARAMAGNETIC

Moving charges give rise to a magnetic field. Unpaired electrons act like tiny magnets as a result of their spinning about their axes. The magnetic fields from paired electrons, which have opposite spins, cancel out. Molecules with no unpaired electrons cannot be magnetized by an external magnetic field and are called *diamagnetic*. Molecules with unpaired electrons can be magnetized by an external field and are called *paramagnetic*. In paramagnetic molecules, an external magnetic field lines up the electron spins parallel to the ap-

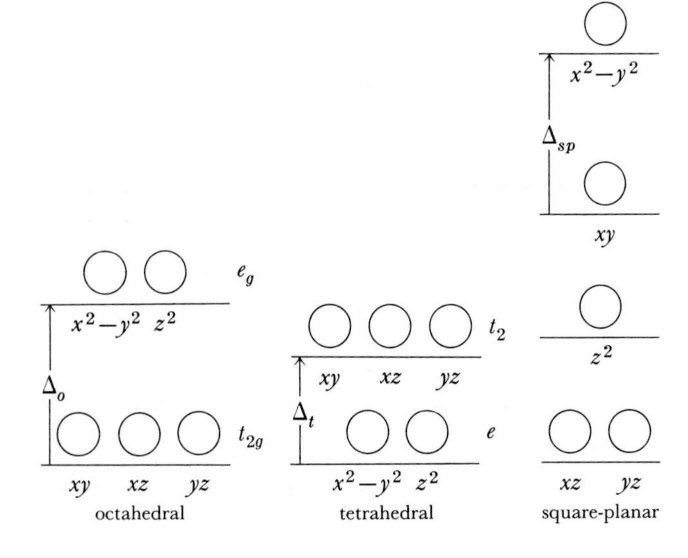

Figure 23-12 Comparison of the *d*-orbital splitting patterns in octahedral, tetrahedral, and square-planar complexes. The relative magnitudes of the *d*-orbital splitting energy for a particular metal ion with a particular ligand are $\Delta_{sp} \approx \Delta_o \approx 2\Delta_t$.

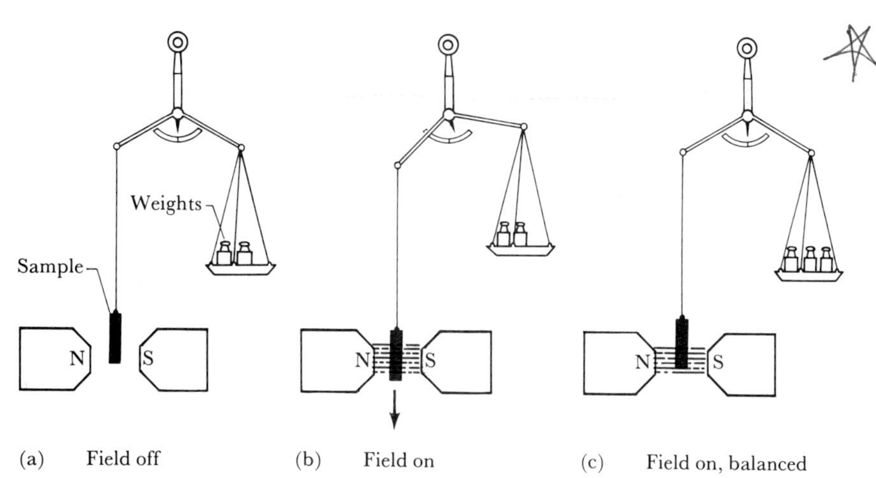

Figure 23-13 Attraction of a paramagnetic substance into a magnetic field. The substance is pulled into the field by the attraction between it and the applied magnetic field. The magnetic pull on the sample makes it appear heavier. The number of unpaired electrons in the sample can be calculated from the apparent mass gain. Masses are added to the balance pan until balance is restored with the field on.

plied field, and thus a paramagnetic substance behaves like a magnet in an externally applied magnetic field (Figure 23-13). A diamagnetic substance is not drawn into an applied magnetic field and thus can be distinguished from a paramagnetic substance.

The magnetic behavior of transition metal complexes can, in some cases, be used for structure determination. For example, experiments show that the compound $K_2[NiBr_4]$ is paramagnetic and that the compound $K_2[Ni(CN)_4]$ is diamagnetic. Using this information, we can predict the structures of the $[NiBr_4]^{2-}$ and $[Ni(CN)_4]^{2-}$ ions. Both of these complexes contain nickel(II), a d^8 ion. Because there are four ligands, the two structure possibilities are tetrahedral and square-planar. The d^8 electron distributions for tetrahedral and square-planar complexes are

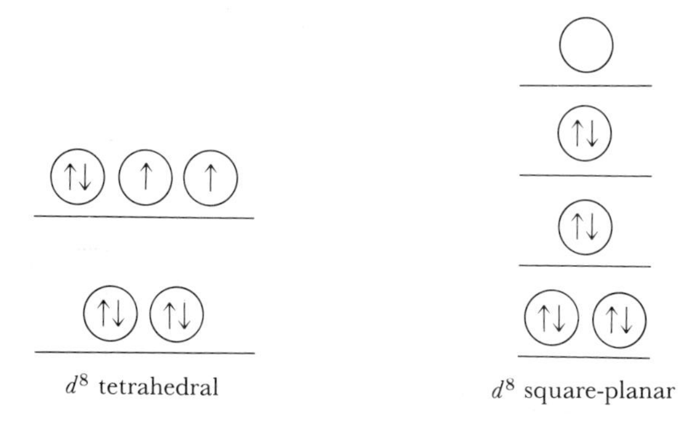

A tetrahedral d^8 ion has two unpaired d electrons and is therefore paramagnetic. A square-planar d^8 ion has no unpaired d electrons and is therefore diamagnetic. Thus $[NiBr_4]^{2-}$, which is paramagnetic, must have a tetrahedral structure and $[Ni(CN)_4]^{2-}$, which is diamagnetic, must have a square-planar structure.

Example 23-16: The salt $[Pt(NH_3)_6]Cl_4$ is diamagnetic. Determine its ground-state electron configuration.

Solution: Platinum is the eighth member of the 5*d* transition series (Figure 23-1), and thus platinum(II) is a d^8 ion. The $[Pt(NH_3)_6]^{4+}$ ion contains platinum(IV), which has two fewer electrons than platinum(II). Thus, platinum(IV) is a d^6 ion. The high-spin and low-spin *d*-electron configurations for octahedral d^6 ions are

d^6 octahedral high-spin, $t_{2g}^4 e_g^2$ d^6 octahedral low-spin, t_{2g}^6

Problems 23-49 through 23-54 deal with paramagnetism in complex ions.

The high-spin configuration has four unpaired electrons and so yields paramagnetic complexes. The low-spin configuration has no unpaired electrons and thus yields diamagnetic complexes. The $[Pt(NH_3)_6]^{4+}$ complex is diamagnetic and therefore is a low-spin complex. Its ground-state electron configuration is t_{2g}^6.

SUMMARY

There are three *d* transition metal series (3*d*, 4*d*, and 5*d*), each having 10 members. The keys to understanding the chemistry of the *d* transition series metals are the electron occupancy of the five *d* orbitals of the metal ion and the influence of the ligands on the relative energies and splitting patterns of the orbitals. Ligands are anions or neutral molecules that bind to metal ions to form complexes. Chelating ligands are ligands that attach to two or more coordination positions on the metal ion.

The structure of a complex may be of several different types: octahedral (the most common), tetrahedral, square-planar, or linear. Certain octahedral and square-planar complexes exist as *cis* and *trans* isomers. The *d*-orbital splitting pattern is different for each geometry. The determining factor of the splitting of the *d* orbitals is the placement of the ligands in the complex ion relative to the positions of the *d* orbitals. The splitting of the *d* orbitals gives rise to the possibility of low-spin and high-spin complexes for octahedral d^4, d^5, d^6, and d^7 ions. A paramagnetic complex ion has some electrons that are unpaired. Magnetic measurements can be used to detect the presence of unpaired electrons in a complex ion.

The color of most transition metal complexes arises from electronic transitions between the *d* orbitals of the metal ion. Transition metal complexes are classified as either labile or inert depending on whether they undergo ligand substitution reactions rapidly or slowly. Inert octahedral complexes are t_{2g}^3, t_{2g}^4, t_{2g}^5, and t_{2g}^6 ions.

TERMS YOU SHOULD KNOW

transition metal 907
d transition metal series 907
d orbital 907
d_{xy}, d_{xz}, d_{yz}, $d_{x^2-y^2}$, d_{z^2} orbitals 909
d^x ion 910
complex ion 911
ligand 912
ligand-substitution reaction 912
cis and *trans* isomers 918
geometric isomers 918
ligating atoms 920
polydentate ligand 920
chelating ligand 920
chelate 920
bidentate 920
tridentate 920
t_{2g} and e_g orbitals 922
Δ_o 922
low-spin configuration 926
high-spin configuration 926
electron pairing energy 927
spectrochemical series 929
inert complex 931
labile complex 931
Δ_{sp} 933
e and t_2 orbitals 933
Δ_t 933
diamagnetic 934
paramagnetic 934

PROBLEMS

ELECTRON CONFIGURATIONS

23-1. Write the ground-state electron configuration for each of the following transition metal ions:

(a) $Cr^{2+}(g)$
(b) $Cu^{+}(g)$
(c) $Zr^{2+}(g)$
(d) $Mo^{3+}(g)$
(e) $Pt^{4+}(g)$

23-2. Write the ground-state electron configuration for each of the following transition metal ions:

(a) $Ni^{2+}(g)$
(b) $Ru^{3+}(g)$
(c) $Ta^{4+}(g)$
(d) $Au^{+}(g)$
(e) $Cd^{2+}(g)$

23-3. How many outer-shell d electrons are there in each of the following transition metal ions:

(a) Fe^{3+}
(b) Pd^{2+}
(c) Mo^{3+}
(d) Ir^{3+}
(e) Cu^{2+}

23-4. How many outer-shell d electrons are there in each of the following transition metal ions:

(a) Ag^{+}
(b) V^{3+}
(c) Zr^{3+}
(d) Ru^{4+}
(e) Hg^{2+}

23-5. Give three examples of each of the following:

(a) d^7 ions with an oxidation state of $+3$
(b) d^3 ions with an oxidation state of $+4$
(c) d^{10} ions with an oxidation state of $+2$

23-6. Give three examples of each of the following:

(a) d^4 ions with an oxidation state of $+2$
(b) d^8 ions with an oxidation state of $+3$
(c) d^0 ions with an oxidation state of $+4$

OXIDATION STATES

23-7. Give the oxidation state of the central metal in each of the following complex ions:

(a) $[Co(H_2O)_6]^{2+}$
(b) $[Cr(NH_3)_4Cl_2]^{+}$
(c) $[Cr(CN)_6]^{3-}$
(d) $[CoF_6]^{3-}$
(e) $[Ag(NH_3)_2]^{+}$

23-8. Give the oxidation state of the central metal in each of the following complex ions:

(a) $[Hg(SO_4)_2]^{2-}$
(b) $[CuCl_4]^{2-}$
(c) $[AlOH]^{2+}$
(d) $[FeCl_4]^{-}$
(e) $[Ni(CN)_4]^{2-}$

23-9. Give the oxidation state of the central metal in each of the following complex ions:

(a) $[Zn(CN)_4]^{2-}$
(b) $[Ni(NH_3)_6]^{2+}$
(c) $[CdI_4]^{2-}$
(d) $[Fe(CN)_6]^{3-}$
(e) $[Fe(CN)_6]^{4-}$

23-10. Give the oxidation state of the central metal in each of the following complex ions:

(a) $[Ir(NH_3)_4Br_2]^{+}$
(b) $[RhCl_6]^{3-}$
(c) $[IrCl_6]^{3-}$
(d) $[V(NO_2)_6]^{3-}$
(e) $[Co(NH_3)_3(CO)_3]^{3+}$

IONS FROM COMPLEX SALTS

23-11. How many moles of ions result when 1 mol of $[Pt(NH_3)_5Cl]Cl_3$ is dissolved in water?

23-12. How many moles of ions result when 1 mol of $K_2[Ni(CN)_4]$ is dissolved in water?

23-13. What major species are present when the following compounds are dissolved in water and how many moles of each species are present if 1 mol is dissolved:

(a) $K_3[Fe(CN)_6]$
(b) $[Ir(NH_3)_2Cl_2]NO_3$
(c) $[Pt(NH_3)_4Cl_2]Cl_2$
(d) $[Ru(NH_3)_6]Br_3$

23-14. What major species are present when the following compounds are dissolved in water and how many moles of each species are present if 1 mol is dissolved:

(a) $[Co(NH_3)_6]Br_3$
(b) $[Pt(NH_3)_3Cl_3]Cl$
(c) $[Cr(H_2O)_6]Br_3$
(d) $K_4[Cr(CN)_6]$

23-15. Some of the first complexes discovered by Werner in the 1890's had the empirical formulas given below. Also given are the number of chloride ions per formula unit precipitated by the addition of $Ag^+(aq)$. Explain these observations.

Empirical formula	*Number of Cl^- per formula unit precipitated by $Ag^+(aq)$*
$PtCl_4 \cdot 6NH_3$	4
$PtCl_4 \cdot 5NH_3$	3
$PtCl_4 \cdot 4NH_3$	2
$PtCl_4 \cdot 3NH_3$	1
$PtCl_4 \cdot 2NH_3$	0

23-16. Some of the first complexes discovered by Werner in the 1890's had the empirical formulas given below. Also given are the number of chloride ions per formula unit precipitated by the addition of $Ag^+(aq)$. Explain these observations.

Empirical formula	*Number of Cl^- per formula unit precipitated by $Ag^+(aq)$*
$PtCl_2 \cdot 4NH_3$	2
$PtCl_2 \cdot 3NH_3$	1
$PtCl_2 \cdot 2NH_3$	0

WRITING NAMES FROM FORMULAS

23-17. Give the systematic name for each of the following complex ions:

(a) $[Cr(NH_3)_5Cl]^{2+}$
(b) $[W(NH_3)_3(H_2O)_3]^{2+}$
(c) $[Fe(H_2O)_6]^{3+}$
(d) $[FeCl_4]^-$

23-18. Give the systematic name for each of the following complex ions:

(a) $[Fe(CN)_6]^{3-}$
(b) $[Ni(CO)_4]$
(c) $[Fe(H_2O)_5SCN]^{2+}$
(d) $[Co(NH_3)_2(H_2O)_4]^{2+}$

23-19. Give the systematic name for each of the following compounds:

(a) $(NH_4)_3[Co(NO_2)_6]$
(b) $[Ir(NH_3)_4Br_2]Br$
(c) $K_2[CuCl_4]$
(d) $[Ru(CO)_5]$

23-20. Give the systematic name for each of the following compounds:

(a) $Na[Au(CN)_4]$
(b) $[Cr(H_2O)_6]Cl_3$
(c) $[Mn(CO)_5]Cl$
(d) $[Cu(NH_3)_6]Cl_2$

23-21. Give the systematic name for each of the following compounds:

(a) $Na_3[Fe(CN)_5CO]$
(b) *trans*-$[Pt(NH_3)_4Cl_2]I_2$
(c) $K_2[Mo(NCS)_6]$
(d) $(NH_4)_2[MnCl_4]$

23-22. Give the systematic name for each of the following compounds:

(a) $[Ru(NH_3)_5Cl]Cl_2$
(b) $Na_3[Co(CN)_6]$
(c) $Na_2[ReF_6]$
(d) $[Fe(H_2O)_5SCN](NO_3)_2$

CHEMICAL FORMULAS FROM NAMES

23-23. Give the chemical formula for each of the following complex ions:

(a) pentacyanocarbonylferrate(II)
(b) *trans*-dichlorodiiodoaurate(III)
(c) hexacyanocobaltate(III)
(d) hexanitrocobaltate(III)

23-24. Give the chemical formula for each of the following complex ions:

(a) bromochlorodicyanonickelate(II)
(b) tetrathiocyanatocobaltate(II)
(c) hexachlorovanadate(III)
(d) pentamminechlorochromium(III)

23-25. Give the chemical formula for each of the following complexes:

(a) hexaamminecobalt(III)
(b) tetracarbonylnickel(0)
(c) hexathiocyanatomanganate(II)
(d) diamminecopper(I)

23-26. Give the chemical formula for each of the following complexes:

(a) hexaaquacobalt(II)
(b) tetraamminecopper(II)
(c) hexacarbonylchromium(0)
(d) dichlorodioxotungsten(VI)

23-27. Give the chemical formula for each of the following compounds:

(a) potassium tetraoxochromate(VI)
(b) tetraamminenickel(II) nitrate
(c) diamminesilver(I) chloride
(d) calcium hexacyanomanganate(III)
(e) sodium tetrathiocyanatocadmate(II)

23-28. Give the chemical formula for each of the following compounds:

(a) ammonium tetrachlorocobaltate(II)
(b) potassium hexacyanoferrate(II)
(c) sodium tetrahydroxodioxoosmate(VI)
(d) pentaaquathiocyanatoiron(III) chloride
(e) potassium hexachloroiridate(III)

POLYDENTATE LIGANDS/NOMENCLATURE

23-29. Name each of the following complex ions:

(a) $[Ru(en)_3]^{3+}$
(b) $[Co(en)_2]^{2+}$
(c) $[Mo(C_2O_4)_2I_2]^{3-}$
(d) $[Co(NH_3)_2(en)Cl_2]^+$

23-30. Name each of the following complex ions:

(a) *trans*-$[CoCl_2(en)_2]^+$
(b) $[Ru(en)(NH_3)_4]^{3+}$
(c) $[Mo(C_2O_4)_3]^{3-}$
(d) $[Fe(en)_2(H_2O)_2]^{3+}$

23-31. Give the chemical formula for each of the following complex ions:

(a) tris(oxalato)ferrate(III)
(b) tris(ethylenediamine)chromium(III)
(c) tris(ethylenediamine)ruthenium(II)
(d) dichlorobis(ethylenediamine)platinum(IV)

23-32. Give the chemical formula for each of the following complex ions:

(a) chlorohydroxobis(ethylenediamine)cobalt(III)
(b) bis(ethylenediamine)oxalatocadmium(II)
(c) dinitritobis(oxalato)platinate(IV)
(d) bis(ethylenediamine)oxalatovanadium(III)

GEOMETRIC ISOMERS

23-33. Draw all the geometric isomers with the composition $[PtCl_4Br_2]^{2-}$.

23-34. The SCN^- ion is capable of bonding through sulfur (thiocyanato) or nitrogen (isothiocyanato). Draw the structures of all the possible isomers of $[Pt(NH_3)_2(SCN)_2]^{2+}$.

23-35. Draw all the geometric isomers for the following complexes:

(a) $[Co(en)_2Br_2]$
(b) $[RuCl_2Br_2(NO_2)_2]^{3-}$

23-36. Draw all the geometric isomers for the following complexes:

(a) $[Pd(C_2O_4)_2I_2]^{2-}$
(b) $[PtCl_3Br_3]^{2-}$

23-37. Draw the structure for each of the following complexes:

(a) *trans*-dichloroethylenediamineplatinum(IV) chloride (square-planar)
(b) potassium *trans*-dichlorodiodoaurate(III) (square-planar)
(c) *cis,cis*-triamminetrichlorocobalt(III)
(d) *cis,trans*-triamminetrichloroplatinum(IV) chloride

23-38. Which of the following complexes have geometric isomers:

(a) $[Cr(NH_3)_4Cl_2]^+$
(b) $[Cr(NH_3)_5Cl]^{2+}$
(c) $[Co(NCS)_2Cl_2]^{2-}$ (tetrahedral)
(d) $[Pt(NH_3)_2Cl_2]$ (square-planar)

HIGH AND LOW SPIN

23-39. Write the *d*-orbital electron configurations for the following octahedral complex ions:

(a) a V^{3+} complex
(b) a Cr^{2+} complex if Δ_o is greater than the electron pairing energy
(c) a Mn^{2+} complex if Δ_o is less than the electron pairing energy
(d) a Cu^{+} complex
(e) a Co^{3+} complex if Δ_o is greater than the electron pairing energy

23-40. Write the *d*-orbital electron configurations for the following octahedral complex ions:

(a) a high-spin Co^{2+} complex
(b) a high-spin Mn^{2+} complex
(c) a low-spin Fe^{3+} complex
(d) a Ti^{4+} complex
(e) a Ni^{2+} complex

23-41. Classify the following complex ions as high-spin or low-spin:

(a) $[Fe(CN)_6]^{4-}$ (no unpaired electrons)
(b) $[Fe(CN)_6]^{3-}$ (one unpaired electron)
(c) $[Co(NH_3)_6]^{2+}$ (three unpaired electrons)
(d) $[CoF_6]^{3-}$ (four unpaired electrons)
(e) $[Mn(H_2O)_6]^{2+}$ (five unpaired electrons)

23-42. Classify the following complex ions as high-spin or low-spin:

(a) $[Mn(NH_3)_6]^{3+}$ (two unpaired electrons)
(b) $[Rh(CN)_6]^{3-}$ (no unpaired electrons)
(c) $[Co(C_2O_4)_3]^{4-}$ (three unpaired electrons)
(d) $[IrBr_6]^{4-}$ (three unpaired electrons)
(e) $[Ru(NH_3)_6]^{3+}$ (one unpaired electron)

23-43. Classify the following complex ions as high-spin or low-spin and write the *d*-orbital electron configurations:

(a) $[Fe(CN)_6]^{4-}$
(b) $[MnF_6]^{4-}$
(c) $[Co(NO_2)_6]^{3-}$
(d) $[FeCl_4]^{-}$ (tetrahedral)

23-44. Classify the following complex ions as high-spin or low-spin and write the *d*-orbital electron configurations:

(a) $[Cr(NO_2)_6]^{4-}$
(b) $[CoF_6]^{3-}$
(c) $[Rh(CN)_6]^{3-}$
(d) $[MnCl_4]^{2-}$ (tetrahedral)

INERT AND LABILE COMPLEXES

23-45. Indicate which of the following complexes would be expected to be inert to ligand substitution:

(a) $[FeF_6]^{3-}$
(b) $[Ni(H_2O)_6]^{2+}$
(c) $[Cr(H_2O)_6]^{3+}$
(d) $[V(H_2O)_6]^{3+}$

23-46. Indicate which of the following complexes would be expected to be inert to ligand substitution:

(a) $[Fe(CN)_6]^{3-}$
(b) $[CrF_6]^{3-}$
(c) $[V(NH_3)_6]^{3+}$
(d) $[Cu(en)_3]^{2+}$

23-47. Indicate which of the following complexes would be expected to be inert to ligand substitution:

(a) $[Cr(NH_3)_6]^{3+}$
(b) $[Co(en)_3]^{3+}$ (low-spin)
(c) $[Mo(NCS)_6]^{2-}$
(d) $[Sc(H_2O)_6]^{3+}$

23-48. Indicate which of the following complexes would be expected to be inert to ligand substitution:

(a) $[MnCl_6]^{2-}$
(b) $[Ti(H_2O)_6]^{3+}$
(c) $[Co(H_2O)_6]^{2+}$ (high-spin)
(d) $[Zr(H_2O)_6]^{2+}$

PARAMAGNETISM IN COMPLEX IONS

23-49. Predict the number of unpaired electrons in each of the following complex ions:

(a) $[Pd(CN)_4]^{2-}$ (square-planar)
(b) $[Ru(NH_3)_6]^{3+}$ (low-spin)
(c) $[Mn(CN)_6]^{4-}$
(d) $[FeF_6]^{3-}$

23-50. Predict the number of unpaired electrons in each of the following complex ions:

(a) $[V(H_2O)_6]^{3+}$
(b) $[CoF_4]^{2-}$ (tetrahedral)
(c) $[Cr(CO)_6]^{2+}$
(d) $[PdCl_4]^{2-}$ (square-planar)

23-51. Which of the following complexes are paramagnetic:

(a) $[Co(en)_3]^{3+}$ (low-spin)
(b) $[Fe(CN)_6]^{4-}$
(c) $[NiF_4]^{2-}$ (tetrahedral)
(d) $[CoBr_4]^{2-}$ (tetrahedral)

23-52. Which of the following complexes is paramagnetic:

(a) $[Cu(NH_3)_6]^{2+}$
(b) $[CrF_6]^{3-}$
(c) $[CoCl_4]^{2-}$ (tetrahedral)
(d) $[Zn(H_2O)_6]^{2+}$

23-53. The complex $[NiF_4]^{2-}$ is paramagnetic but $[Ni(CN)_4]^{2-}$ is diamagnetic. Explain the difference.

23-54. The complex $[Fe(H_2O)_6]^{2+}$ is paramagnetic whereas $[Fe(CN)_6]^{4-}$ is diamagnetic. Explain the difference.

In autumn of 1938, the German nuclear physicist Otto Hahn and his student Fritz Strassman discovered that when uranium-235 nuclei are bombarded with neutrons, they split into two roughly equal-sized fragments. The apparatus used by Hahn and Strassman is shown above.

24 / Nuclear and Radiochemistry

Many nuclei are unstable and spontaneously emit subatomic particles. The process of nuclear decomposition is called radioactivity, which is the principal topic of this chapter. All elements with more than 83 protons are radioactive. In addition, many radioactive isotopes that do not occur in nature have been made in the laboratory. Radioactive isotopes are used in chemistry, physics, medicine, biology, agriculture, geology, and criminal investigations. Radioactivity is used to determine the age of rocks that are billions of years old as well as the age of archaeological findings a few thousand years old. As little as 10^{-12} g of an impurity in a sample can be measured by radiochemical methods. Although the emphasis of this chapter is on the chemical aspects of nuclear processes, we also study nuclear reactions, the source of enormous amounts of energy.

24-1. MOST ELEMENTS CONSIST OF MIXTURES OF ISOTOPES

Recall from Chapter 1 that nuclei consist of protons and neutrons, which are collectively called *nucleons*. The number of protons in a nucleus is the atomic number, Z, which specifies the element. The total number of nucleons (protons plus neutrons) in a nucleus is the mass number, A. The number of neutrons in two atoms of a given element need not be the same. Atoms that contain the same number of protons but different numbers of neutrons are called isotopes,

Isotopes are denoted by

$$^{A}_{Z}X$$

where X is the symbol of the element, Z is its atomic number, and A is its mass number.

which are denoted $^{A}_{Z}X$. For example, chlorine has two naturally occurring isotopes, $^{35}_{17}Cl$ and $^{37}_{17}Cl$, called chlorine-35 and chlorine-37. Both undergo the same chemical reactions.

Most elements found in nature consist of mixtures of isotopes, as is shown in Table 1-9 for several elements. Example 1-8 illustrates the calculation of the atomic mass of neon from the masses of the various neon isotopes and their observed mass percentages.

24-2. MANY NUCLEI SPONTANEOUSLY EMIT SMALL PARTICLES

Uranium-238 nuclei emit helium nuclei, $^{4}_{2}He$. When a $^{238}_{92}U$ nucleus emits a helium nucleus, the mass number of the uranium decreases by four and the atomic number decreases by two. The change in atomic number tells us that a different element has been produced. We can describe the process by a *nuclear equation:*

$$^{238}_{92}U \rightarrow \, ^{234}_{90}Th + \, ^{4}_{2}He$$

The product here is thorium because the atomic number of the resulting nucleus is 90. This nuclear equation is balanced: the total charge (number of protons) and the number of nucleons are the same on both sides. The spontaneous disintegration of a nucleus is called *radioactivity,* and such a nucleus is said to be *radioactive.* An isotope that is radioactive is called a *radioisotope.* Uranium-238 is an example of a radioisotope.

Radioactivity was discovered by the French scientist Henri Becquerel in 1896. He found that the compound potassium uranyl sulfate, $K_2UO_2(SO_4)_2$, emitted radiation and was able to show that the uranium was the source of the emissions. Becquerel referred to the radiation as α-rays (alpha rays) or *α-particles.* It was subsequently shown that α-particles are helium-4 nuclei. Uranium-238 is said to be an *α-emitter.* All elements beyond lead ($Z = 82$) in the periodic table have been observed to emit α-particles.

Example 24-1: Radon-222 is an α-emitter. Write a balanced nuclear equation for the reaction.

Nuclear equations must be balanced with respect to charge and number of nucleons.

Solution: The symbol for radon-222 is $^{222}_{86}Rn$. The mass number decreases by four and the atomic number decreases by two when an α-particle is emitted. The product has $Z = 84$ and $A = 218$. Polonium is the element that has $Z = 84$, and so the balanced nuclear equation is

$$^{222}_{86}Rn \rightarrow \, ^{218}_{84}Po + \, ^{4}_{2}He$$

Note that $86 = 84 + 2$ (charge balance) and $222 = 218 + 4$ (nucleon balance).

24-3. THERE ARE SEVERAL TYPES OF RADIOACTIVE EMISSIONS

Not all radioactive nuclei emit α-particles. Two other common types of emissions were found in early experiments on radioactivity, and they were called *β-particle* (beta-particle) emission and *γ-ray* (gamma-ray) emission. It was soon discovered that β-particles are electrons. Electrons in nuclear equations are denoted by the symbol ${}_{-1}^{0}\mathrm{e}$. The superscript zero refers to the small mass of an electron relative to that of a nucleon, and the subscript -1 refers to the negative charge on an electron. Two examples of β-emission reactions are

Radioactive decay is a spontaneous process.

$$ {}_{49}^{116}\mathrm{In} \rightarrow {}_{50}^{116}\mathrm{Sn} + {}_{-1}^{0}\mathrm{e} $$

$$ {}_{90}^{234}\mathrm{Th} \rightarrow {}_{91}^{234}\mathrm{Pa} + {}_{-1}^{0}\mathrm{e} $$

Note that the mass numbers do not change but the atomic number increases by one as a result of the emission of a β-particle.

There are no electrons in nuclei. The emission of a β-particle results from the conversion of a neutron to a proton within the nucleus, which can be represented as

$$ {}_{0}^{1}\mathrm{n} \rightarrow {}_{1}^{1}\mathrm{H} + {}_{-1}^{0}\mathrm{e} $$

Gamma rays are high-energy electromagnetic waves, similar to X-rays. The emission of γ-rays causes no change in either Z or A. The emission of an α-particle or a β-particle frequently leaves the product nucleus in an excited state. The excited nucleus can relax to its ground state by emitting a photon, whose frequency is given by the equation (Section 7-6)

$$ \Delta E = h\nu $$

where ΔE is the energy difference between the excited state and the ground state. Gamma-ray emission is analogous to an atom's emitting electromagnetic radiation when an electron falls from an excited state to a ground state. We often do not indicate γ-radiation in writing nuclear equations.

Several other types of radioactive emissions have been observed. For example, some nuclei emit a *positron,* which is a particle that has the same mass as an electron but a positive charge. The symbol for a positron in nuclear equations is ${}_{+1}^{0}\mathrm{e}$. Two examples of positron emission are

A positron is said to be the antiparticle of an electron.

$$ {}_{19}^{38}\mathrm{K} \rightarrow {}_{18}^{38}\mathrm{Ar} + {}_{+1}^{0}\mathrm{e} $$

$$ {}_{51}^{120}\mathrm{Sb} \rightarrow {}_{50}^{120}\mathrm{Sn} + {}_{+1}^{0}\mathrm{e} $$

The emission of a positron can be viewed as the conversion of a proton to a neutron:

$$^{1}_{1}H \rightarrow ^{1}_{0}n + ^{0}_{+1}e$$

Positrons exist for only a very short time. They combine with electrons in about 10^{-9} s and are converted to γ-radiation.

Free electrons do not exist in nuclei.

Another type of nuclear transformation is called *electron capture*. In an electron-capture process, one of the innermost electrons of an atom is absorbed by the nucleus and a proton is converted to a neutron. An example of an electron-capture process is

$$^{195}_{79}Au + ^{0}_{-1}e \rightarrow ^{195}_{78}Pt$$

Positron emission and electron capture lead to the same result: they both decrease Z by one and leave A unchanged. Electron capture usually occurs in nuclei with Z greater than 80, whereas positron emission usually occurs in nuclei with values of Z less than about 30. Both processes are observed for Z values in the range $30 < Z < 80$. Table 24-1 summarizes the various types of radioactive processes.

When a radioactive nucleus emits a particle and transforms to another nucleus, we say that it decays to that nucleus. Thus, the expression *radioactive decay* refers to a radioactive process.

Table 24-1 The various particles emitted in radioactive processes

		Change in nucleus		
Emission	*Symbol*	*Mass number*	*Atomic number*	*Example*
α	$^{4}_{2}He$	decreases by 4	decreases by 2	$^{238}_{92}U \rightarrow ^{234}_{90}Th + ^{4}_{2}He$
β	$^{0}_{-1}e$	no change	increases by 1	$^{14}_{6}C \rightarrow ^{14}_{7}N + ^{0}_{-1}e$
γ	$^{0}_{0}\gamma$	no change	no change	$^{16}_{7}N \rightarrow ^{16}_{8}O + ^{0}_{-1}e + \gamma$
positron	$^{0}_{+1}e$	no change	decreases by 1	$^{38}_{19}K \rightarrow ^{38}_{18}Ar + ^{0}_{+1}e$
electron capture	EC	no change	decreases by 1	$^{195}_{79}Au + ^{0}_{-1}e \rightarrow ^{195}_{78}Pt$
neutron	$^{1}_{0}n$	decreases by 1	no change	$^{137}_{53}I \rightarrow ^{136}_{53}I + ^{1}_{0}n$

Example 24-2: Fill in the missing symbols in the following nuclear equations:

(a) ${}^{214}_{82}\text{Pb} \rightarrow ? + {}^{214}_{83}\text{Bi}$

(b) ${}^{137}_{53}\text{I} \rightarrow {}^{1}_{0}\text{n} + ?$

(c) ${}^{11}_{6}\text{C} \rightarrow {}^{0}_{+1}\text{e} + ?$

(d) $? \rightarrow {}^{0}_{-1}\text{e} + {}^{97}_{41}\text{Nb}$

Solution:

(a) The missing particle has a charge of -1 $(82 = 83 - 1)$, and A does not change; thus the particle is a β-particle, ${}^{0}_{-1}\text{e}$.
(b) The missing particle has $Z = 53$ and $A = 136$; because $Z = 53$ corresponds to iodine, the missing particle is ${}^{136}_{53}\text{I}$.
(c) The missing particle has $Z = 5$ and $A = 11$ and so is ${}^{11}_{5}\text{B}$.
(d) The missing particle has $Z = 40$ and $A = 97$. The element that has $Z = 40$ is zirconium, and so the nucleus that decays is ${}^{97}_{40}\text{Zr}$.

Problems 24-3 through 24-10 are similar to Example 24-2.

24-4. THE PROTON-TO-NEUTRON RATIO DETERMINES THE STABILITY OF A NUCLEUS

Although many nuclei are radioactive, many others are stable. There is no general theory that we can use to predict the stability of nuclei, but there are a number of empirical observations that help. Figure 24-1 is a plot of all the known stable nuclei as a function of their number of protons, Z, and their number of neutrons, $N = A - Z$. It

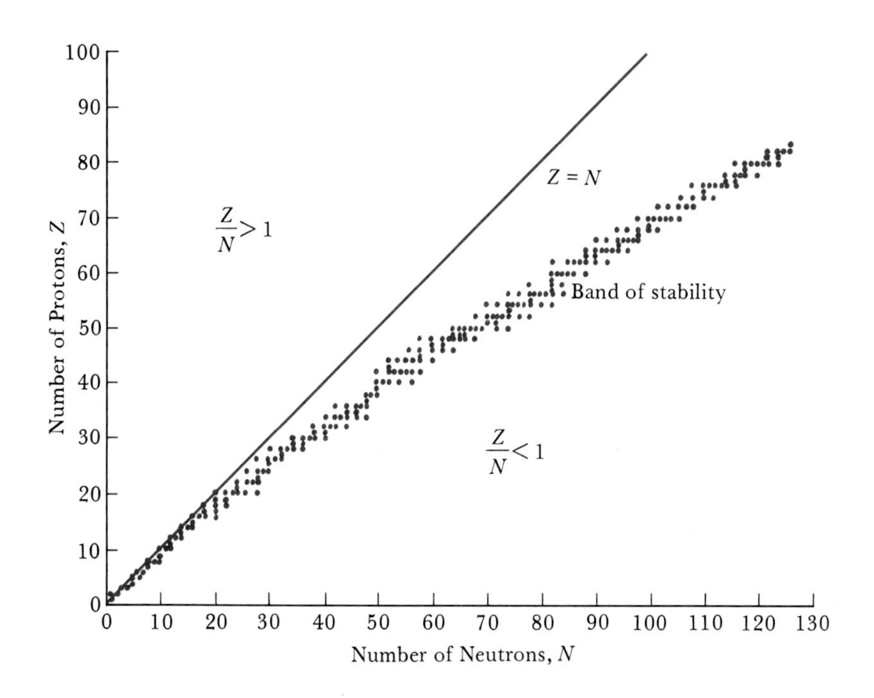

Figure 24-1 Known stable (nonradioactive) nuclei plotted on a graph of number of protons versus number of neutrons. The stable nuclei fall within the band of stability. The solid straight line is the line for which $Z = N$. Nuclei that lie above the band of stability emit positrons or undergo electron capture because by doing so the product nucleus lies closer to the band of stability. Similarly, nuclei that lie below the band of stability emit β-particles. All nuclei with $Z \geqslant 83$ are radioactive.

shows that the stable nuclei form what is called a *band of stability*. This band is determined by the ratio of protons to neutrons. For elements with atomic number lower than 21, the proton-to-neutron ratio is unity or slightly less (the only exceptions are ^1_1H and ^3_2He). Helium-4, carbon-12, oxygen-16, neon-20, and calcium-40 are examples of stable nuclei with equal numbers of neutrons and protons. As Z increases, the proton-to-neutron ratio decreases below unity and reaches a value of 0.7 at $^{209}_{83}Bi$, which is the stable nucleus with the largest value of Z. All nuclei with 83 or more protons are unstable.

Example 24-3: We shall see later in this chapter that carbon-14 can be used to determine the age of objects made from plant and animal materials. Predict the mode of decay of carbon-14.

Problems 24-11 through 24-16 ask you to predict the modes of decay of various nuclei.

Solution: The proton-to-neutron ratio in carbon-14 is 6/8, or 0.75. At Z values lower than 21, the band of stability lies close to the $Z/N = 1$ line. Thus a value of $Z/N = 0.75$ places carbon-14 well below the band of stability in Figure 24-1. Referring to the legend of Figure 24-1, we predict that carbon-14 decays by β-emission, according to

$$^{14}_{6}C \rightarrow ^{14}_{7}N + ^{0}_{-1}e$$

which is in agreement with experiment.

24-5. ONLY A FEW STABLE NUCLEI HAVE ODD NUMBERS OF PROTONS AND NEUTRONS

Elements with an odd number of protons usually have only one or two stable isotopes, whereas elements with an even number of protons may have several stable isotopes (Table 24-2). Most stable nuclei contain an even number of protons or neutrons. In fact, there are only four stable nuclei that contain an odd number of both protons and neutrons (^2_1H, ^6_3Li, $^{10}_5B$, $^{14}_7N$). Table 24-3 lists the numbers of stable nuclei with even and odd numbers of protons and neutrons. It can be seen from this table that nature seems to prefer nuclei with even numbers of protons and neutrons.

Nuclei that contain 2, 8, 20, 28, 50, 82, or 126 protons or neutrons are particularly stable and abundant in nature. The numbers 2, 8, 20, 28, 50, 82, and 126 are called *magic numbers,* and they have the same significance in nuclear structure theory that the numbers 2, 10, 18, 32, 54, and 86, the numbers of electrons in the noble gases, have in the theory of the electronic structure of atoms.

Table 24-2 Illustration of the fact that elements with odd values of Z usually have only one or two stable isotopes and elements with even values of Z have several stable isotopes

Z	*Element*	*Number of stable isotopes*
40	zirconium (Zr)	5
41	niobium (Nb)	1
42	molybdenum (Mo)	7
43	technetium (Tc)	0
44	ruthenium (Ru)	7
45	rhodium (Rh)	1
46	palladium (Pd)	6
47	silver (Ag)	2
48	cadmium (Cd)	7
49	indium (In)	1
50	tin (Sn)	10
51	antimony (Sb)	2
52	tellurium (Te)	7
53	iodine (I)	1
54	xenon (Xe)	9
55	cesium (Cs)	1
56	barium (Ba)	7

Table 24-3 The number of stable nuclei with even and odd numbers of protons and neutrons

Number of protons	*Number of neutrons*	*Number of stable nuclei*	*Examples*
odd	odd	4	${}^{2}_{1}H$ ${}^{6}_{3}Li$ ${}^{10}_{5}B$ ${}^{14}_{7}N$
even	odd	57	${}^{13}_{6}C$ ${}^{25}_{12}Mg$ ${}^{47}_{22}Ti$
odd	even	50	${}^{19}_{9}F$ ${}^{23}_{11}Na$ ${}^{27}_{13}Al$
even	even	168	${}^{12}_{6}C$ ${}^{16}_{8}O$ ${}^{20}_{10}Ne$

Example 24-4: Predict whether or not the following nuclei are radioactive (unstable):

Problems 24-17 through 24-20 are similar to Example 24-4.

(a) $^{120}_{50}Sn$ (b) $^{213}_{87}Fr$ (c) $^{16}_{8}O$ (d) $^{15}_{8}O$

Solution: (a) Tin-120 has an even number of protons and neutrons and a magic number of protons; thus it is a stable isotope. (b) Although francium-213 has a magic number of neutrons (126), its atomic number is greater than 83. Since there are no stable nuclei with more than 83 protons, francium-213 is an unstable nucleus. (c) Oxygen-16 has a magic number (8) of both protons and neutrons and so is a stable isotope. (d) Oxygen-15 has a magic number of protons, but its proton-to-neutron ratio (8/7) is greater than unity. We predict that oxygen-15 is a radioactive isotope. Because it lies above the band of stability in Figure 24-1, we predict (correctly) that it decays by positron emission.

24-6. THE RATE OF DECAY OF A RADIOACTIVE ISOTOPE IS SPECIFIED BY ITS HALF-LIFE

Not all radioactive nuclei decay at the same rate. Some radioactive samples decay in a few millionths of a second; the same amount of another may take billions of years. An important aspect of radioactive decay is that its rate is independent of external factors, such as temperature and pressure, at least under normal conditions. At the present time, we are unable to alter the rate of radioactive decay processes. This is a serious problem in the disposal of radioactive waste.

Radioactive decay is a first-order rate process (Chapter 16). This means that the number of radioactive nuclei, N, remaining in a sample at time t—given that N_0 is the number of nuclei present initially—is given by (Section 16-4)

Recall that

$$t_{1/2} = \frac{0.693}{k}$$

for a first-order reaction.

$$\log \frac{N_0}{N} = \frac{kt}{2.30} = \frac{0.301t}{t_{1/2}} \tag{24-1}$$

where $t_{1/2}$ is the *half-life* for the decay of the radioactive nucleus. Recall from Chapter 16 that the half-life is the time required for one half of the reacting particles to undergo reaction. Thus the half-life for a radioactive isotope is the time it takes for one half of a sample of the isotope to undergo radioactive decay. Figure 24-2 shows a plot of N/N_0, which is the fraction of the nuclei remaining, versus time for the decay of sodium-25, which has a half-life of 1 min.

Different radioisotopes have different half-lives. The half-life is characteristic of a given radioisotope and is a direct measure of how

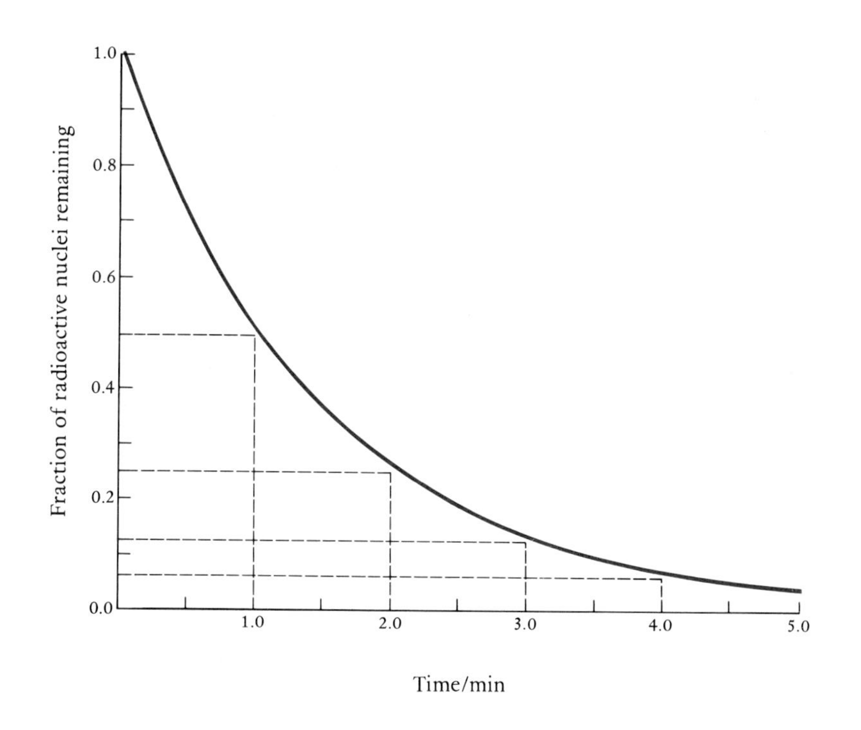

Figure 24-2 A graph of the decay of sodium-25 over time. The solid line is a plot of the fraction of sodium-25 nuclei present in a sample versus time. This curve is representative of a first-order kinetic process; the rate of decay is proportional to the number of sodium-25 nuclei present in the sample. The half-life of sodium-25 is 1.0 min. One half of the sample decays in the first minute; one half of what remains decays in the next minute, and so on.

rapidly it decays. Table 24-4 lists the half-lives of some common radioisotopes, and the next two examples illustrate the use of Equation (24-1).

Recall for Chapter 16 that the fraction remaining, N/N_0 after n half-lives is given by

$$\frac{N}{N_0} = \left(\frac{1}{2}\right)^n$$

Table 24-4 Half-lives of some radioisotopes

Isotope	*Half-life*	*Mode of decay*
$^{3}_{1}H$	12.33 years	β
$^{14}_{6}C$	5730 years	β
$^{25}_{11}Na$	60 s	β
$^{32}_{15}P$	14.28 days	β
$^{60}_{27}Co$	5.27 years	β
$^{87}_{37}Rb$	4.9×10^{10} years	β
$^{90}_{38}Sr$	29 years	β
$^{131}_{53}I$	8.04 days	β
$^{214}_{84}Po$	164 μs	α
$^{226}_{88}Ra$	1600 years	α
$^{238}_{92}U$	4.47×10^{9} years	α
$^{239}_{93}Pu$	2.41×10^{4} years	α

Example 24-5: The radioisotope cobalt-60 is used to destroy cancerous cells by directing γ-rays into the cancerous cell tissue. Calculate the fraction of a cobalt-60 sample left after 20.0 years.

Solution: From Table 24-4, we see that the half-life of cobalt-60 is 5.27 years. If we use Equation (24-1), then we have

$$\log\frac{N_0}{N} = \frac{0.301t}{t_{1/2}} = \frac{(0.301)(20.0\text{ years})}{(5.27\text{ years})} = 1.14$$

$$\frac{N_0}{N} = 10^{1.14} = 13.8$$

$$\frac{N}{N_0} = 0.0725$$

Thus, only 7.25% of the original sample remains after 20 years.

Example 24-6: A 2.00-mg sample of the radioisotope phosphorus-32 is found to contain 0.40 mg of phosphorus-32 after 33.3 days. Calculate the half-life of this isotope.

Solution: Solving Equation (24-1) for $t_{1/2}$, we have

$$t_{1/2} = \frac{0.301t}{\log(N_0/N)}$$

The value of N_0/N after 33.3 days is

$$\frac{N_0}{N} = \frac{(2.00 \times 10^{-3}\text{ g})\left(\dfrac{1\text{ mol}}{32.0\text{ g}}\right)(6.022 \times 10^{23}\text{ atom}\cdot\text{mol}^{-1})}{(0.40 \times 10^{-3}\text{ g})\left(\dfrac{1\text{ mol}}{32.0\text{ g}}\right)(6.022 \times 10^{23}\text{ atom}\cdot\text{mol}^{-1})}$$

$$= \frac{2.00}{0.400} = 5.00$$

Notice that we have just shown that

$$\frac{N_0}{N} = \frac{\text{initial mass of phosphorus-32}}{\text{present mass of phosphorus-32}}$$

Problems 24-21 through 24-32 deal with rates of nuclear decay.

Using the value of 5.00 for N_0/N, we compute for the half-life of phosphorus-32

$$t_{1/2} = \frac{(0.301)(33.3\text{ days})}{\log 5.00} = 14.3\text{ days}$$

which agrees with the value given in Table 24-4.

24-7. RADIOACTIVITY CAN BE USED TO DETERMINE THE AGE OF ROCKS

Naturally occurring radioactive uranium-238 decays through a series of processes to lead-206. The half-life for this decay process is 4.51×10^9 years.

Now suppose that a sample of uranium ore is found to contain equal molar quantities of uranium-238 and lead-206 and that all the lead-206 arose from the decay of the uranium-238. Because one half of the uranium-238 initially present has decayed since the rock was formed, we conclude that the rock is 4.51×10^9 years old. The following example illustrates this *uranium-lead dating* method.

Example 24-7: A sample of uranium ore is found to contain 5.20 mg of uranium-238 and 1.85 mg of lead-206. Calculate the age of the rock.

Solution: If we solve Equation (24-1) for t, then we have

$$t = \frac{t_{1/2}}{0.301} \log \frac{N_0}{N} \qquad (24\text{-}2)$$

The half-life of the conversion of uranium-238 to lead-206 is 4.51×10^9 years. In order to use Equation (24-2), we must now determine N_0/N. The initial amount of uranium-238 present in the ore is 5.20 mg plus whatever amount has been transformed into lead. The amount of uranium that has been transformed into lead is given by

$$\begin{pmatrix} \text{mg } ^{238}_{92}\text{U transformed} \\ \text{into } ^{206}_{82}\text{Pb} \end{pmatrix} = \left(\frac{238}{206}\right)(1.85 \text{ mg}) = 2.14 \text{ mg } ^{238}_{92}\text{U}$$

Therefore, the initial amount of uranium-238 is

$$\text{initial mg } ^{238}_{92}\text{U} = 2.14 \text{ mg} + 5.20 \text{ mg} = 7.34 \text{ mg}$$

As in Example 24-6, we can use the fact that

$$\frac{N_0}{N} = \frac{\text{initial mass of radioactive isotope}}{\text{present mass of radioactive isotope}} = \frac{7.34 \text{ mg}}{5.20 \text{ mg}} = 1.41$$

Using Equation (24-2), we get

$$t = \left(\frac{4.51 \times 10^9 \text{ years}}{0.301}\right)(\log 1.41) = 2.24 \times 10^9 \text{ years}$$

The rock is over two billion years old. The oldest rocks analyzed by this method are 3.6 billion years old, which is about 1 billion years less than the present estimate of the age of the earth. Rocks that have been obtained from the moon in the Apollo program indicate that the moon is about the same age as the earth.

Other nuclear reactions can be used to date rocks. The reaction

$$^{87}_{37}\text{Rb} \rightarrow \, ^{87}_{38}\text{Sr} + \, ^{0}_{-1}\text{e} \qquad t_{1/2} = 4.9 \times 10^{10} \text{ years}$$

can be used for ores containing rubidium. For rocks that contain potassium, the reaction

$$^{40}_{19}\text{K} \xrightarrow{\text{electron capture}} \, ^{40}_{18}\text{Ar} \qquad t_{1/2} = 1.3 \times 10^{9} \text{ years}$$

is used. The agreement obtained by the various radiodating methods usually is good. Problems 24-41 through 24-44 illustrate age determinations involving the rubidium-strontium method and the potassium-argon method. The potassium-argon method has been used to date skulls found by Mary and Louis Leakey in Africa. The ages of the rocks in which the skulls are embedded range from about 1 to 4 million years.

24-8. CARBON-14 CAN BE USED TO DATE CERTAIN ARCHAEOLOGICAL OBJECTS

Willard Libby was a professor of chemistry at UCLA.

The radiodating methods that we have described for determining the age of rocks and ores are not useful for ages less than about a million years. The half-lives of the reactions are more than 1 billion years, and thus not enough decay occurs in less than 1 million years to be measured accurately. In the 1940's, Willard Libby developed a method of using carbon-14 to date carbon-containing objects derived from formerly living materials. The method is useful for dating objects that are less than about 30,000 years old. *Carbon-14 dating* has found wide use in archaeology, and Libby received the 1960 Nobel Prize in chemistry for his work.

The idea behind carbon-14 dating, or *radiocarbon dating,* is as follows. The earth's upper atmosphere is being bombarded constantly by radiation from the sun and other parts of the universe. As a result, a small but fairly constant amount of carbon-14 is produced in the reaction between cosmic-ray neutrons and atmospheric nitrogen:

$$^{14}_{7}\text{N} + \, ^{1}_{0}\text{n} \rightarrow \, ^{14}_{6}\text{C} + \, ^{1}_{1}\text{H}$$

Carbon-14 is radioactive and decays by the reaction

$$^{14}_{6}\text{C} \rightarrow \, ^{14}_{7}\text{N} + \, ^{0}_{-1}\text{e}$$

The half-life of carbon-14 is 5730 years. The carbon-14 occurs in $^{14}CO_2$. The $^{14}CO_2$ diffuses throughout the earth's atmosphere, and so a small but fairly constant fraction of atmospheric CO_2 contains carbon-14.

Living plants absorb CO_2 to build carbohydrates through photosynthesis, and so the carbon-14 is incorporated into plants and into the food chain of animals. As a result, all living plants and animals contain the same fraction of carbon-14 atoms as in atmospheric CO_2. The radiation due to the carbon-14 in all living organisms is 15.3 disintegrations of carbon-14 nuclei per minute per gram of total carbon. When the organism dies, it no longer incorporates carbon-14, and so the quantity of carbon-14 decreases with a half-life of 5730 years.

The equation that we use to determine the age of an object is the one we used in Example 24-7:

$$t = \frac{t_{1/2}}{0.301} \log \frac{N_0}{N} \tag{24-2}$$

The half-life of carbon-14 is 5730 years, and if we substitute this value into Equation (24-2), then we have

$$t = (1.90 \times 10^4 \text{ years}) \log \frac{N_0}{N}$$

The ratio N_0/N can be converted to a more convenient form by realizing that the 15.3 disintegrations per minute per gram of carbon is proportional to N_0, the initial number of carbon-14 nuclei per gram of carbon. We can write this as

$$N_0 \propto 15.3$$

Similarly, if R is the present disintegration rate per minute per gram of carbon, then

$$N \propto R$$

The ratio N_0/N then is

$$\frac{N_0}{N} = \frac{15.3}{R} \tag{24-3}$$

If we substitute Equation (24-3) into Equation (24-2), then we get

$$t = (1.90 \times 10^4 \text{ years}) \log \frac{15.3}{R} \tag{24-4}$$

If we assume that the atmospheric level of carbon-14 is the same now as when the artifact was living matter, then the age of the object can be determined by measuring R. The following example illustrates the use of Equation (24-4).

The assumptions of carbon-14 dating have been tested extensively against other archeological dating techniques.

Example 24-8: Stonehenge is an ancient megalithic site in southern England. Some archaeologists believe that it was designed to make astronomical observations, but its actual purpose is still controversial. Charcoal samples taken from a series of holes at Stonehenge have a disintegration rate of 9.65 disintegrations per minute per gram of carbon. Calculate the age of the charcoal sample.

Solution: Using Equation (24-4), we have

$$\begin{aligned} t &= (1.90 \times 10^4 \text{ years}) \log \frac{15.3}{9.65} \\ &= (1.90 \times 10^4 \text{ years})(0.200) \\ &= 3.80 \times 10^3 \text{ years} \end{aligned}$$

Thus we estimate that the charcoal pits at Stonehenge are about 3800 years old (about 1800 B.C.). This result is in agreement with evidence based upon other archaeological data.

Radiocarbon dating has been used to date many archaeological objects. Figure 24-3 shows Equation (24-4) plotted as log R versus t, indicating a few of the many dates that have been determined by the carbon-14 method.

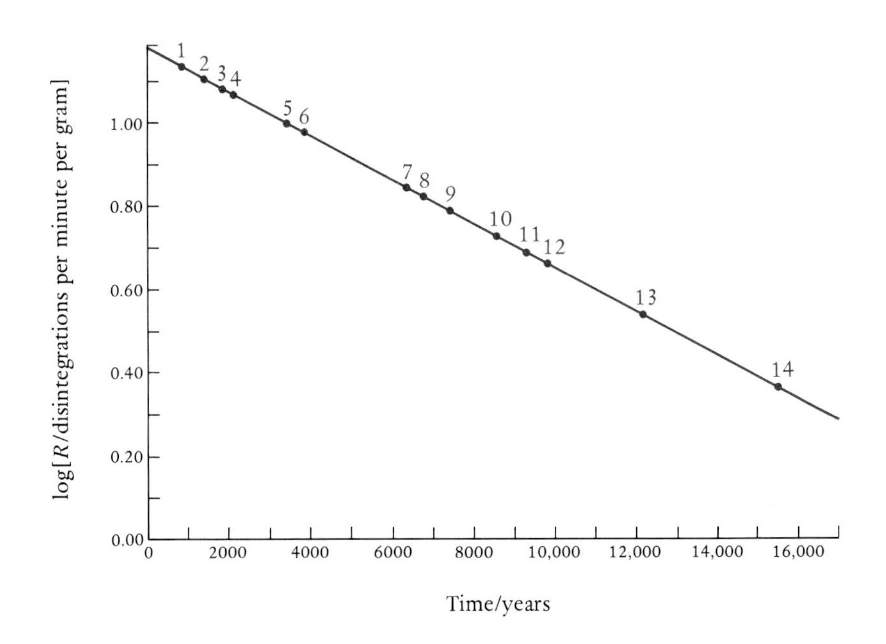

Figure 24-3 Plot of the logarithm of R, the number of disintegrations per minute per gram of carbon, versus time for the carbon-14 dating method. As Equation (24-4) indicates, this plot is a straight line. The numbers on the curve designate archaeological objects that have been dated by the carbon-14 method: (1) Charcoal from earliest Polynesian culture in Hawaii (946 ± 180 years). (2) Wooden lintels from a Mayan site in Tikal, Guatemala (1503 ± 110 years). (3) Linen wrappings from the Dead Sea Scrolls (1917 ± 200 years). (4) Wood from coffin from the Egyptian Ptolemaic period (2190 ± 450 years). (5) Samples of oak from an ancient cooking place at Killeens, County Cork (3506 ± 230 years). (6) Charcoal sample from Stonehenge (3798 ± 275 years). (7) Charcoal from a tree destroyed by the explosion of Mount Mazama, the explosion that formed Crater Lake in Oregon (6453 ± 250 years). (8) Land-snail shells found at Jarmo, Iraq (6707 ± 320 years). (9) Charcoal from an archaeological site near Beer-Sheba, Israel (7420 ± 520 years). (10) Burned animal bones found near a site inhabited by humans in Palli Aike Cave in southern Chile (8639 ± 450 years). (11) Woven rope sandals found in Fork Rock Cave, Oregon (9053 ± 350 years). (12) Buried bison bone from Folsom Man site near Lubbock, Texas (9883 ± 350 years). (13) Glacial wood found near Skunk River, Iowa (12,200 ± 500 years). (14) Charcoal from the Lascaux cave in France, which contains many cave paintings (15,516 ± 900 years).

24-9. RADIOISOTOPES CAN BE PRODUCED IN THE LABORATORY

In 1919, the New Zealand scientist Ernest Rutherford was the first to carry out a nuclear reaction in the laboratory. He bombarded nitrogen with a beam of α-particles and was able to detect the reaction

$$^{14}_{7}N + ^{4}_{2}He \rightarrow ^{17}_{8}O + ^{1}_{1}H$$

This was the first laboratory synthesis of a nucleus, $^{17}_{8}O$. Since then, hundreds of different nuclear reactions have been achieved. The products of many of these reactions are radioactive isotopes that are not found in nature and so are called *artificial radioisotopes*. If these isotopes ever did exist in nature, they have long since disappeared because their half-lives are so much shorter than the age of the earth. For example, the element francium, which is the Group 1 metal with the largest atomic number, is not found in nature. Twenty-seven isotopes of francium have been made; the longest half-life among these isotopes is less than 20 min.

Many of the radioisotopes produced have applications in medicine, agriculture, insect control, oil exploration, and many other fields. For example, iodine-131, which has a half-life of 8 days, is used to measure the activity of the thyroid gland. A few of the many other radioisotopes that are used in medicine are sodium-24 (to follow blood circulation), technetium-99 (for brain, liver, and spleen scans) and phosphorus-32 (for treatment of chronic leukemia). Table 24-5 lists some other radioisotopes that are used in medicine.

Figure 24-4 Marie Curie with her daughter, Irene. Both were pioneers in nuclear chemical research. Marie and her husband, Pierre, discovered radium and several other radioactive elements in the early 1900's. Irene and her husband, Frédéric Joliot, made the first artificial radioisotope, which has led to numerous applications in chemistry, biology, and medicine. Marie shared the 1903 Nobel Prize in physics with her husband and Henri Becquerel, and she received the Nobel Prize in chemistry singly in 1911. Irene Joliot-Curie shared the 1935 Nobel Prize in chemistry with her husband.

24-10. NUCLEAR CHEMISTRY CAN BE USED TO DETECT EXTREMELY SMALL QUANTITIES OF THE ELEMENTS

One of the most important uses of radioisotopes in chemistry is in *neutron activation analysis,* which is one of the most sensitive analytical methods. It is capable of detecting as little as 10^{-12} g of some elements and is particularly useful for measuring trace quantities. In a neutron activation analysis, a sample is irradiated by a beam of neutrons. Various nuclei in the sample undergo nuclear reactions by absorbing neutrons. The product nuclei are usually radioactive and emit γ-rays. The energies of the γ-rays emitted by a radioisotope are characteristic of that radioisotope. Because each radioisotope emits γ-rays of only certain, well-defined energies, the frequencies of the γ-rays emitted by a sample after neutron irradiation can be used to identify the elements present. The advantages of neutron activation analysis over more conventional analytical techniques are that the sample does not have to be pretreated, the method is nondestructive,

Table 24-5 Some radioactive isotopes used in medicine

	Half-life		
Isotope	*Biological*[a]	*Physical*	*Use*
fluorine-18	808 days	110 min	bone imaging
potassium-43	58 days	22.4 h	myocardial imaging
chromium-51	616 days	27.8 days	red blood cell studies
cobalt-58	9.5 days	71.4 days	Schilling test for pernicious anemia
gallium-67	4.8 days	78 h	imaging for Hodgkin's disease, lymphomas, malignant tumors
rubidium-81	45 days	4.6 h	myocardial imaging
indium-111	10 h	2.8 days	spinal and cranial fluid imaging
indium-113	10 h	1.73 h	liver, lung, brain, blood pool scan
iodine-123	138 days	13.3 h	thyroid function, thyroid imaging
iodine-131	138 days	8.04 days	radioimmunoassay thyroid uptake, thyroid cancer, metastases scan, blood plasma volume, lung blood flow, kidney function, liver function
xenon-133	—	5.3 days	lung ventilation, lung inhalation
ytterbium-169	1000 days	32 days	spinal and cranial imaging
gold-198	280 days	2.7 days	liver structure scans
mercury-203	14.5 days	46.9 days	kidney scans

[a]Biological half-life is the time it requires for one half of an administered sample to be eliminated by the organism.

many elements can be analyzed simultaneously, and the sensitivity is very great.

Neutron activation analysis has numerous applications. The authenticity of paintings can be established by determining the mineral content of the paint used. In early times, each school of artists prepared its own paints from distinctive and individual recipes. The paint used in a painting in question is compared with the paint used in a painting known to be done by the artist. Another application of neutron activation analysis showed that Napoleon was poisoned by arsenic. When arsenic is ingested, it is concentrated in the hair. Analysis of several of Napoleon's hairs showed an abnormally large concentration of arsenic.

The results of neutron activation analysis in criminal investigations are at times dramatic. The basic idea is to match the distribution of elements in soil, paint, cosmetics, and so on found at the scene of a crime with those found with a criminal suspect. Neutron activation analysis of a wiping taken from a suspect's hand can reveal not only whether the person has fired a gun recently but also the type of ammunition used.

Another analytical method that involves nuclear techniques is *PIXE* (pronounced pixie), for *particle-induced X-ray emission.* In the PIXE method, a sample is irradiated with a beam of protons, which strike the atoms in the sample and eject inner-shell electrons, such as K- and L-shell electrons. As electrons from the other shells fill these empty innermost shells, the atoms emit characteristic X-rays. The energies of the X-rays can be measured accurately and used to identify the atomic composition of the sample. Figure 24-5 shows the results of a PIXE analysis of a sample of ore. This analysis was done with less than 1 μg of the ore, an amount too small to be seen by the naked eye. One of the advantages of PIXE is that it requires only very small samples. It does not have the great sensitivity that neutron activation analysis has, however. PIXE and neutron activation analysis are just two of a number of analytical methods that use nuclear chemistry.

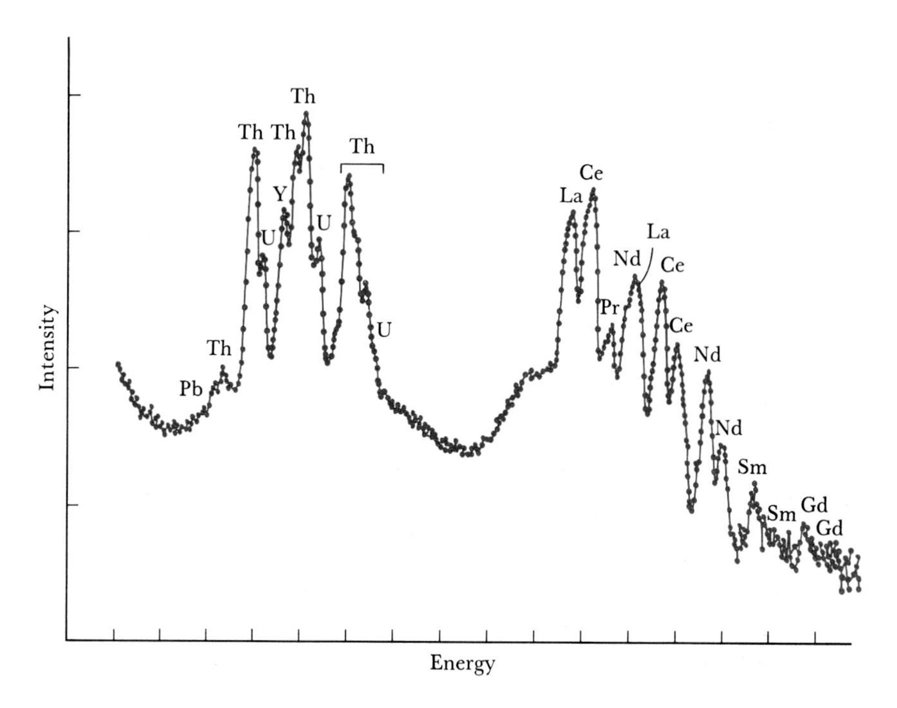

Figure 24-5 The results of PIXE analysis of a sample of the uranium-containing ore monazite. The intensity of the emitted X-rays is plotted versus their energies. Most elements emit X-rays of several different energies, and so several peaks belong to each element, as indicated.

24-11. ELEMENTS BEYOND URANIUM IN THE PERIODIC TABLE ARE PRODUCED ARTIFICIALLY

The production of radioisotopes involves directing a beam of particles onto a sample that acts as a target. Sample reactions are

$$(1)\quad {}^{27}_{13}\mathrm{Al} + {}^{4}_{2}\mathrm{He} \rightarrow {}^{30}_{15}\mathrm{P} + {}^{1}_{0}\mathrm{n}$$

$$(2)\quad {}^{12}_{6}\mathrm{C} + {}^{1}_{1}\mathrm{H} \rightarrow {}^{13}_{7}\mathrm{N} + {}^{0}_{0}\gamma$$

In the first reaction, a beam of α-particles is directed onto an ${}^{27}_{13}\mathrm{Al}$ target, and in the second reaction, a beam of protons is directed onto a ${}^{12}_{6}\mathrm{C}$ target. Both reactants in these equations are nuclei and so are positively charged. For the two nuclei to come together and react, the α-particles or protons must be traveling fast enough to overcome the electric repulsion between the target nucleus and the projectile nucleus. This can be accomplished by a machine called a *particle accelerator,* such as a cyclotron. These machines produce beams of high-velocity particles. With the development of more and more powerful accelerators, it has become possible to accelerate relatively large particles, such as ${}^{12}_{6}\mathrm{C}$ and ${}^{18}_{8}\mathrm{O}$ nuclei. Beams of these nuclei are being used to produce elements with atomic numbers greater than 100.

Before the development of nuclear science, uranium lay at the end of the periodic table. Since the 1940's, elements with $Z = 93$ to 107 and 109 have been produced (Table 24-6). Many of these have been produced at the Lawrence Radiation Laboratory in Berkeley, California, as the names of several indicate. Many of these *transuranium elements* can be made in commercial quantities. For example, americium-241 is used in many home smoke detectors (Figure 24-6). The α-particles emitted by the americium-241, as most α-particles, have a very low penetrating power and so do not escape from the ionization chamber. Thus, the isolated radioactive source offers no health hazard unless the ionization chamber is opened or broken apart.

The transuranium elements up to lawrencium are inner-transition metals.

24-12. ENORMOUS AMOUNTS OF ENERGY ACCOMPANY NUCLEAR REACTIONS

Consider the nuclear reaction

$$\begin{array}{lcccccccc} & {}^{6}_{3}\mathrm{Li} & + & {}^{1}_{1}\mathrm{H} & \rightarrow & {}^{3}_{2}\mathrm{He} & + & {}^{4}_{2}\mathrm{He} \\ \text{atomic mass:} & 6.01512 & & 1.00782 & & 3.01603 & & 4.00260 \end{array}$$

The total mass of the reactants is 7.02294 amu, and the total mass of the products is 7.01863 amu. The mass is not the same on the two sides of the equation. The difference is

Table 24-6 The transuranium elements and the reactions used to produce them

Atomic number	*Name*	*Symbol*	*Reaction*
93	neptunium	Np	${}^{238}_{92}U + {}^{1}_{0}n \rightarrow {}^{239}_{93}Np + {}^{0}_{-1}e$
94	plutonium	Pu	${}^{238}_{92}U + {}^{2}_{1}H \rightarrow {}^{238}_{93}Np + 2{}^{1}_{0}n$
			${}^{238}_{93}Np \rightarrow {}^{238}_{94}Pu + {}^{0}_{-1}e$
95	americium	Am	${}^{239}_{94}Pu + {}^{1}_{0}n \rightarrow {}^{240}_{95}Am + {}^{0}_{-1}e$
96	curium	Cm	${}^{239}_{94}Pu + {}^{4}_{2}He \rightarrow {}^{242}_{96}Cm + {}^{1}_{0}n$
97	berkelium	Bk	${}^{241}_{95}Am + {}^{4}_{2}He \rightarrow {}^{243}_{97}Bk + 2{}^{1}_{0}n$
98	californium	Cf	${}^{242}_{96}Cm + {}^{4}_{2}He \rightarrow {}^{245}_{98}Cf + {}^{1}_{0}n$
99	einsteinium	Es	${}^{238}_{92}U + 15{}^{1}_{0}n \rightarrow {}^{253}_{99}Es + 7{}^{0}_{-1}e$
100	fermium	Fm	${}^{238}_{92}U + 17{}^{1}_{0}n \rightarrow {}^{255}_{100}Fm + 8{}^{0}_{-1}e$
101	mendelevium	Md	${}^{253}_{99}Es + {}^{4}_{2}He \rightarrow {}^{256}_{101}Md + {}^{1}_{0}n$
102	nobelium	No	${}^{246}_{96}Cm + {}^{12}_{6}C \rightarrow {}^{254}_{102}No + 4{}^{1}_{0}n$
103	lawrencium	Lr	${}^{252}_{98}Cf + {}^{10}_{5}B \rightarrow {}^{257}_{103}Lr + 5{}^{1}_{0}n$
104	—	Unq	${}^{249}_{98}Cf + {}^{12}_{6}C \rightarrow {}^{257}_{104}Unq + 4{}^{1}_{0}n$
105	—	Unp	${}^{249}_{98}Cf + {}^{15}_{7}N \rightarrow {}^{260}_{105}Unp + 4{}^{1}_{0}n$
106	—	Unh	${}^{249}_{98}Cf + {}^{18}_{8}O \rightarrow {}^{263}_{106}Unh + 4{}^{1}_{0}n$

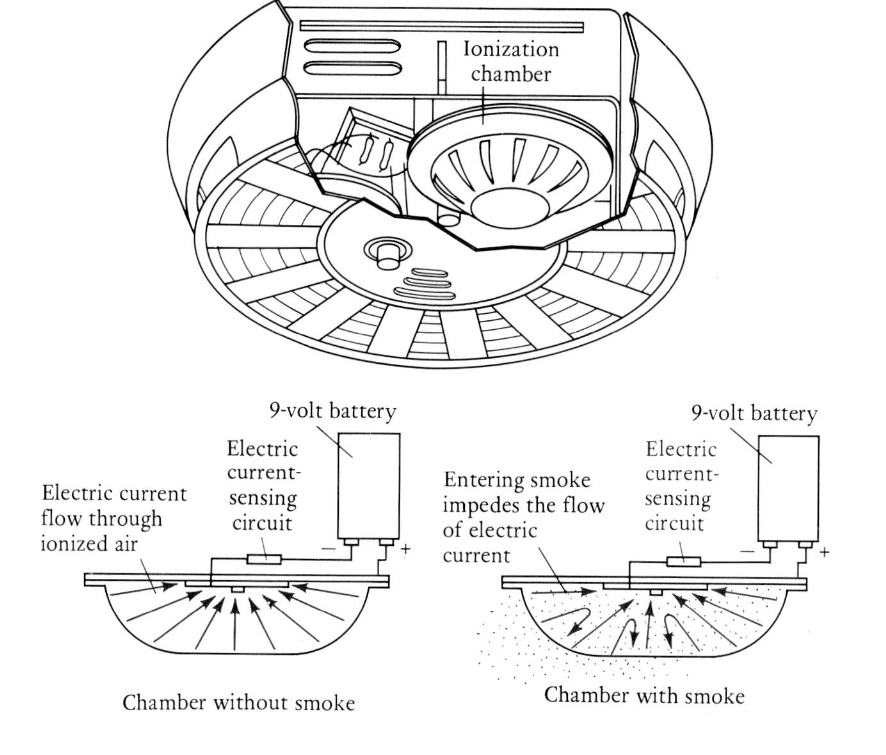

Figure 24-6 Diagram of a home smoke detector. A small quantity of americium-241 ($t_{1/2} = 432$ years) ionizes the air in the ionization chamber. An electric voltage is applied across the ionization chamber, and the ions in the ionized air conduct an electric current, which is constantly monitored electronically. When smoke particles pass through the ionization chamber, they impede the flow of electricity, which is detected by electronic circuitry and signals an alarm.

$$\Delta m = 3.01603 + 4.00260 - (6.01512 + 1.00782)$$
$$= -0.00431 \text{ amu}$$

This number is far larger than the uncertainties of the masses of the reactants and products. Since 1 amu $= 1.66 \times 10^{-27}$ kg, -0.00431 amu is equal to

$$-0.00431 \text{ amu} = (-0.00431 \text{ amu})(1.66 \times 10^{-27} \text{ kg}\cdot\text{amu}^{-1})$$
$$= -7.15 \times 10^{-30} \text{ kg}$$

This amount of mass is lost when just one $^{6}_{3}\text{Li}$ nucleus reacts with one $^{1}_{1}\text{H}$ nucleus. If we calculate the mass lost on a per mole basis, then we get

$$-0.00431 \text{ amu}\cdot\text{atom}^{-1}$$
$$= (-7.15 \times 10^{-30} \text{ kg}\cdot\text{atom}^{-1})(6.02 \times 10^{23} \text{ atom}\cdot\text{mol}^{-1})$$
$$= -4.30 \times 10^{-6} \text{ kg}\cdot\text{mol}^{-1}$$

Up to this point, we have always stated that mass is conserved in reactions; yet in this nuclear reaction there is a discrepancy that lies far outside experimental error. The explanation is that the missing mass has been converted to energy. The relation between mass and energy is given by Einstein's famous formula,

The equation $E = mc^2$ comes from Einstein's theory of relativity.

$$E = mc^2 \tag{24-5}$$

where m is the mass lost and c is the speed of light. Because we are referring to energy changes and mass changes, we write Equation (24-5) in the form

$$\Delta E = c^2 \Delta m \tag{24-6}$$

Using the fact that the speed of light is 3.00×10^{8} m$\cdot$s^{-1}, we can write Equation (24-6) in the form

$$\Delta E = (9.00 \times 10^{16} \text{ m}^2\cdot\text{s}^{-2})\Delta m \tag{24-7}$$

The mass lost in the

$$^{6}_{3}\text{Li} + ^{1}_{1}\text{H} \rightarrow ^{3}_{2}\text{He} + ^{4}_{2}\text{He}$$

reaction is -4.30×10^{-6} kg$\cdot$mol^{-1}. Therefore

$\text{J} = \text{kg}\cdot\text{m}^2\cdot\text{s}^{-2}$

$$\Delta E = c^2 \Delta m$$
$$= (9.00 \times 10^{16} \text{ m}^2\cdot\text{s}^{-2})(-4.30 \times 10^{-6} \text{ kg}\cdot\text{mol}^{-1})$$
$$= -3.87 \times 10^{11} \text{ J}\cdot\text{mol}^{-1}$$

where we have used the fact that 1 joule is 1 $kg \cdot m^2 \cdot s^{-2}$ (Section 5-11). This value of ΔE is typical for nuclear reactions. Values of ΔE for ordinary chemical reactions are about $10^5\ J \cdot mol^{-1}$, and so we see that nuclear reactions involve energies 1 million times greater than conventional chemical reactions. This is why enormous amounts of energy are produced by nuclear explosions and nuclear reactors; a small amount of mass is converted to a large amount of energy.

Example 24-9: The value of ΔE for the reaction

$$H_2(g) + \tfrac{1}{2}O_2(g) \rightarrow H_2O(l)$$

is $-282.1\ kJ \cdot mol^{-1}$ at 298 K and 1 atm. Calculate the loss in mass when 1 mol of liquid water is formed from hydrogen and oxygen at 298 K.

Solution: The negative sign of ΔE means that energy is evolved and therefore that mass is lost. The mass corresponding to $-282.1\ kJ \cdot mol^{-1}$ is

ΔE for chemical reactions is about 10^{-6} as large as ΔE for nuclear reactions.

$$\Delta m = \frac{\Delta E}{c^2} = \frac{-282.1 \times 10^3\ J \cdot mol^{-1}}{9.00 \times 10^{16}\ m^2 \cdot s^{-2}}$$
$$= -3.13 \times 10^{-12}\ kg \cdot mol^{-1}$$

This is the mass lost when 1 mol of liquid water is formed. It is a very small mass and not directly measurable. The total energy evolved upon the formation of each molecule of water is equivalent to about 1 millionth of the mass of an electron. Mass changes in ordinary chemical reactions are negligible.

24-13. ENERGY IS REQUIRED TO BREAK UP A NUCLEUS

The energy required to break up a nucleus into its constituent protons and neutrons is called the *binding energy*. Let's calculate the binding energy of ${}^4_2H$. The reaction that we are considering is

$$\begin{array}{lccccc} & {}^4_2He & \rightarrow & 2\,{}^1_1H & + & 2\,{}^1_0n \\ \text{mass/amu:} & 4.0026 & & 1.0078 & & 1.0087 \end{array}$$

The mass difference is

$$\Delta m = 2(1.0078\ amu) + 2(1.0087\ amu) - 4.0026\ amu$$
$$= 4.0330\ amu - 4.0026\ amu$$
$$= 0.0304\ amu$$

This value of Δm corresponds to an energy of

$$\begin{aligned}\Delta E &= c^2\Delta m \\ &= (9.00 \times 10^{16}\ \mathrm{m^2 \cdot s^{-2}})(0.0304\ \mathrm{amu})(1.66 \times 10^{-27}\ \mathrm{kg \cdot amu^{-1}}) \\ &= 4.54 \times 10^{-12}\ \mathrm{J}\end{aligned}$$

The binding energy of $^{4}_{2}\mathrm{He}$ is 4.54×10^{-12} J.

Example 24-10: Given that the mass of $^{55}_{25}\mathrm{Mn}$ is 54.9380 amu, calculate the binding energy and the binding energy per nucleon.

Solution: The list of physical constants on the inside back cover gives the following masses:

$^{1}_{1}\mathrm{H}$	1.0078 amu
$^{1}_{0}\mathrm{n}$	1.0087 amu

The mass difference between $^{55}_{25}\mathrm{Mn}$ and its constituent particles is

$$\begin{aligned}\Delta m &= (25 \times 1.0078\ \mathrm{amu}) + (30 \times 1.0087\ \mathrm{amu}) - 54.9380\ \mathrm{amu} \\ &= 55.4560\ \mathrm{amu} - 54.9380\ \mathrm{amu} \\ &= 0.518\ \mathrm{amu}\end{aligned}$$

which corresponds to an energy of

$$\begin{aligned}\Delta E = c^2\Delta m &= (9.00 \times 10^{16}\ \mathrm{m^2 \cdot s^{-2}})(0.518\ \mathrm{amu})(1.66 \times 10^{-27}\ \mathrm{kg \cdot amu^{-1}}) \\ &= 7.74 \times 10^{-11}\ \mathrm{J}\end{aligned}$$

Problems 24-45 through 24-50 involve binding energy.

This is the binding energy of $^{55}_{25}\mathrm{Mn}$. There are 55 nucleons in $^{55}_{25}\mathrm{Mn}$, and so the binding energy per nucleon is

$$\begin{aligned}\text{binding energy per nucleon} &= \frac{7.74 \times 10^{-11}\ \mathrm{J}}{55\ \text{nucleons}} \\ &= 1.41 \times 10^{-12}\ \mathrm{J \cdot nucleon^{-1}}\end{aligned}$$

Figure 24-7 shows the binding energy per nucleon versus the number of nucleons (mass number), a curve we use in the next section.

24-14. SOME NUCLEI FRAGMENT IN NUCLEAR REACTIONS

Because a neutron is uncharged, it is relatively difficult to detect and so was not discovered until 1932. As soon as the neutron was discovered, it was apparent that it would be ideal to use to bombard nuclei.

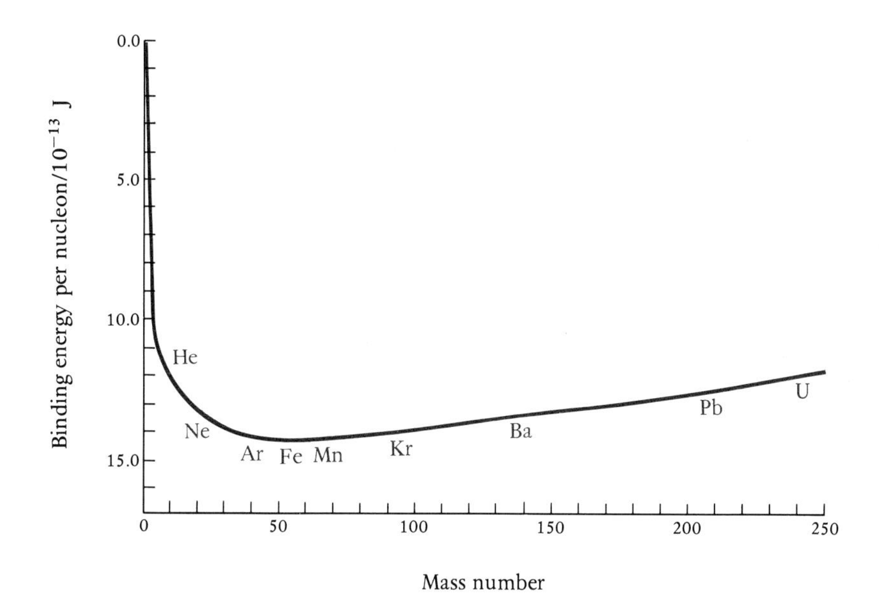

Figure 24-7 Binding energy per nucleon plotted versus mass number of a nucleus. This plot is called a *binding energy curve.*

The most stable nucleus is $^{56}_{26}Fe$.

Being uncharged, neutrons can burrow into a nucleus more readily than can protons or α-particles. In the mid-1930's, neutrons were used to bombard uranium in an effort to produce elements beyond uranium in the periodic table. For example, the first element beyond uranium, neptunium (the planet Neptune lies beyond Uranus in the solar system), is produced by the reaction

$$^{238}_{92}U + ^{1}_{0}n \rightarrow ^{239}_{93}Np + ^{0}_{-1}e$$

Neptunium-239 has a half-life of about 2 days and decays to plutonium-239 (the planet Pluto lies beyond Neptune) by β-decay:

$$^{239}_{93}Np \rightarrow ^{239}_{94}Pu + ^{0}_{-1}e$$

Plutonium-239 has a half-life of about 24,000 years.

During the search for transuranium elements in the 1930's, it was discovered that when uranium-235 absorbs a neutron, it breaks into two fragments of roughly the same size (Figure 24-8). A typical reaction is

$$^{235}_{92}U + ^{1}_{0}n \rightarrow ^{92}_{36}Kr + ^{141}_{56}Ba + 3^{1}_{0}n$$

A nuclear reaction in which a nucleus splits into two smaller fragments is called *fission.* By referring to the binding energy curve in Figure 24-7, we see that barium and krypton are more stable than uranium. Consequently, energy is released in the fission of uranium-235. The amount of energy released per mole of uranium is 2×10^{13} J. This is an enormous amount of energy, about 1 million times greater than that for an ordinary chemical reaction.

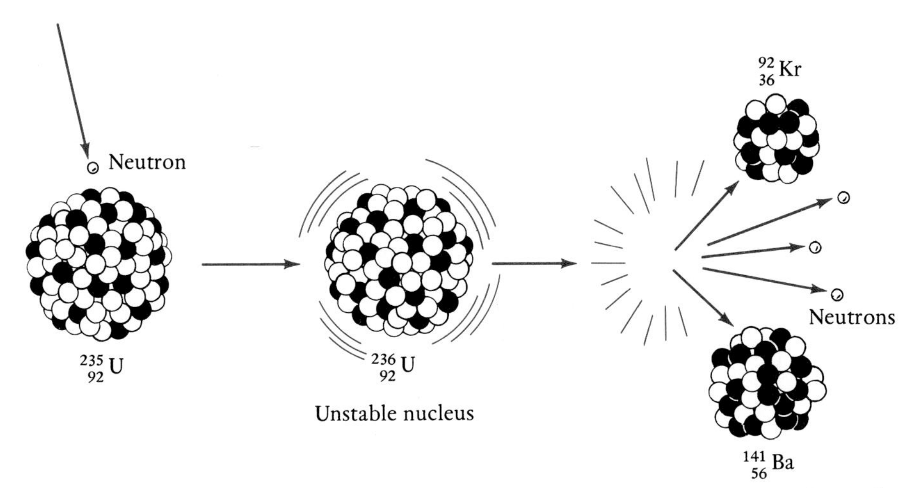

Figure 24-8 A fission reaction occurs when a uranium-235 nucleus absorbs a neutron. The fission products shown here are krypton-92 and barium-141, but in practice a distribution of fission products is obtained. The most important feature of the uranium-235 fission reaction is the simultaneous production of several neutrons.

24-15. THE FISSION OF URANIUM-235 CAN INITIATE A CHAIN REACTION

Nuclear reactions often release great amounts of energy per mole, but usually so few nuclei are involved that the total energy output is small. The fission reaction of uranium-235 is special, however. Note that three neutrons are produced in the fission reaction given above. Each of these neutrons can cause another uranium-235 nucleus to fission, producing nine neutrons. This process can continue, producing what is called a *chain reaction* (Figure 24-9).

There are several conditions that must be met in order for a chain reaction to occur. The level of impurities that absorb neutrons and do not produce any other neutrons must be kept very low. In addition, the quantity of fissionable material must be large enough to make the rate at which neutrons are lost through the surface of the material less than the rate at which neutrons are produced. The smallest mass for which more neutrons are produced than lost through the surface of the material is called the *critical mass.* A quantity of uranium-235 that is less than the critical mass will not support a chain reaction; a quantity that is greater than the critical mass will support a chain reaction. The critical mass of uranium-235 is just a few kilograms, about the size of a softball. Under the right conditions, a neutron chain reaction can occur extremely rapidly; it can be over in less than a microsecond. The energy is released as an enormous explosion, known to all of us.

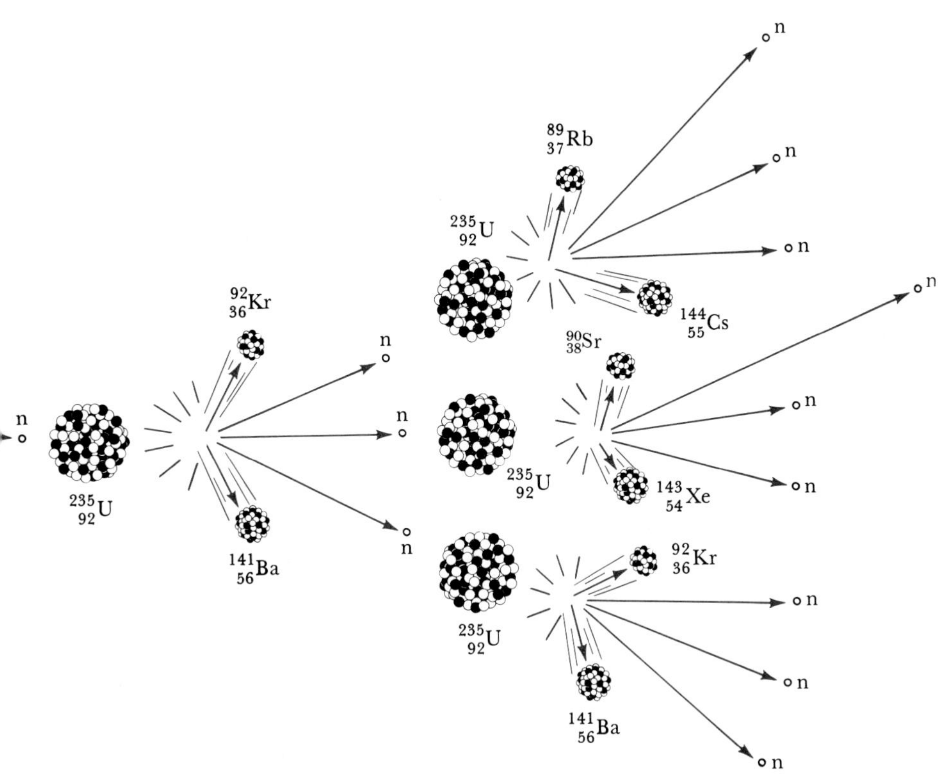

Figure 24-9 The chain reaction produced when uranium-235 undergoes fission after absorbing a neutron. The key point is that several (three here) neutrons are produced in each reaction. Consequently, after the first step, three other nuclei absorb a neutron and undergo fission. These three nuclei produce 9 neutrons, which lead to the production of 27 neutrons in the third step, and so on. The number of fission reactions increases very rapidly with the number of steps, and the result can be an explosive release of energy.

The nuclear research that led to these discoveries was being carried out when the world was in turmoil and on the brink of World War II. A multinational effort went into using a neutron chain reaction to develop the atomic bomb. The theory behind the atomic bomb is very simple. Two *subcritical masses* of uranium-235 are brought together very rapidly, producing a *supercritical mass,* which then explodes. A simplified diagram of an atomic bomb is shown in Figure 24-10. The actual production of the first atomic bomb was a monumental technological task.

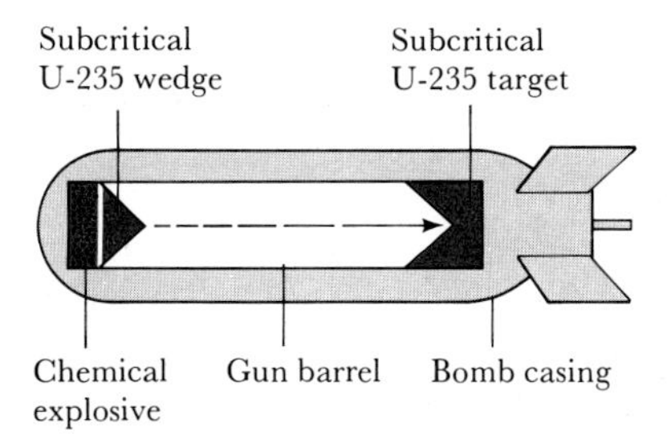

Figure 24-10 Simplified schematic of an atomic bomb. Two subcritical masses of a fissionable isotope, such as uranium-235, are located at the ends of the cylinder. A conventional chemical explosive charge is located behind one of the masses. When the charge is detonated, the left-hand subcritical mass is rapidly propelled toward the right-hand one. When the two collide, the total mass is supercritical and can sustain a neutron chain reaction and release the energy as an explosion.

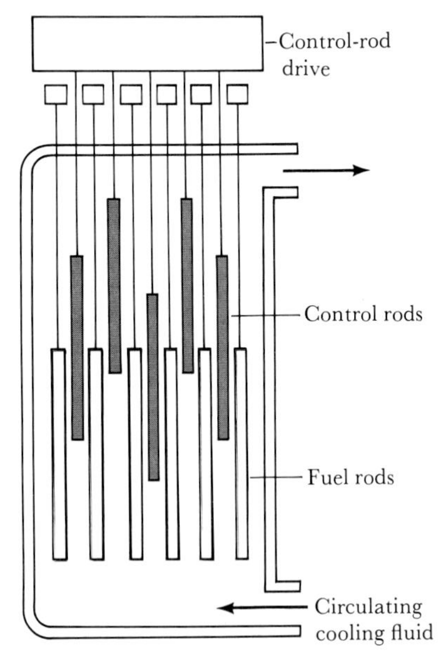

Figure 24-11 Diagram of the core of a nuclear reactor. The *fuel rods* contain the fissionable material. The control rods consist of a material that is a good absorber of neutrons. By raising and lowering the control rods, the density of neutrons in the core, and thus the rate of production of energy, can be controlled.

24-16. A NUCLEAR REACTOR UTILIZES A CONTROLLED CHAIN REACTION

In a *nuclear reactor,* it is possible to tame a neutron chain reaction and produce the energy in a controlled manner. One way to do this is to insert into the uranium core a material that strongly absorbs neutrons (Figure 24-11). Cadmium and boron both are strong absorbers of neutrons. The density of neutrons can be controlled by changing the height of the cadmium or boron rods, called *control rods.* In this manner, the uranium-235 chain reaction can be maintained at a steady, controlled rate, producing energy in the form of heat. The heat produced is used to generate steam, which can run a turbine and produce electricity. A diagram of a nuclear reactor is shown in Figure 24-12.

One of several serious problems associated with the generation of electricity by nuclear reactors is the disposal of fission products and

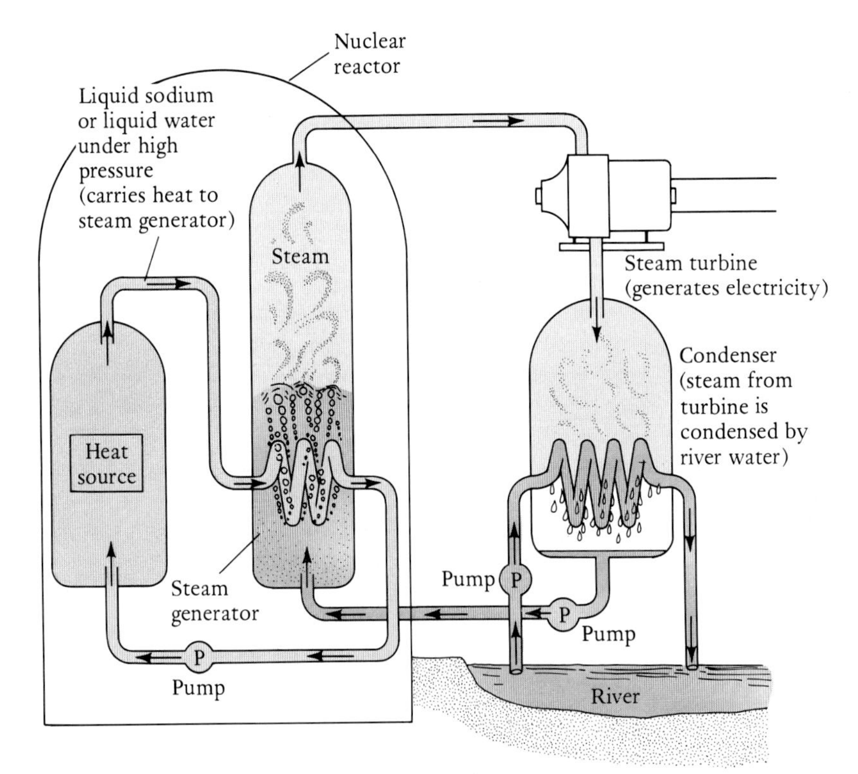

Figure 24-12 Diagram of a nuclear reactor. The heat source is the core of the reactor (Figure 24-11). The heat produced by the core is transferred by a closed loop of liquid sodium or liquid water to a steam generator. The steam produced runs a steam turbine, which produces electricity. The steam from the turbine is cooled by water from a nearby source, such as a river, or through a forced-flow cooling tower, and is pumped back into the steam generator.

spent fuel. The high neutron densities generated in reactor cores produce large quantities of radioisotopes, which must be dealt with when the uranium fuel rods need replenishing. Just what to do with the tons of highly radioactive waste produced has become an intense political issue. At the present time, the spent fuel rods are being stored under water in vaults, usually at the reactor site.

24-17. A BREEDER REACTOR IS DESIGNED TO PRODUCE MORE FISSIONABLE MATERIAL THAN IT CONSUMES

It has been projected that the world's supply of uranium-235 will be exhausted sometime in the twenty-first century. Uranium-235 constitutes less than 1 percent of naturally occurring uranium, most of which (99.3 percent) is nonfissionable uranium-238. Some uranium-238 is converted to fissionable plutonium-239 by the series of reactions

$$^{238}_{92}\text{U} + ^{1}_{0}\text{n} \rightarrow ^{239}_{92}\text{U}$$

$$^{239}_{92}\text{U} \rightarrow ^{239}_{93}\text{Np} + ^{0}_{-1}\text{e} \qquad t_{1/2} = 24 \text{ min}$$

$$^{239}_{93}\text{Np} \rightarrow ^{239}_{94}\text{Pu} + ^{0}_{-1}\text{e} \qquad t_{1/2} = 2.35 \text{ days}$$

The result of this process is plutonium-239, which has a half-life of 24,000 years. The high neutron flux in nuclear reactors provides the neutrons needed to initiate the $^{238}_{92}\text{U} \rightarrow ^{239}_{94}\text{Pu}$ process. If this process can be made to occur to such an extent that more than one plutonium-239 nucleus is produced for every uranium-235 that fissions, then more fissionable material will be produced than consumed. A reactor that is designed to accomplish this task is called a *breeder reactor*. It would take about 20 breeders running for 1 year to make enough fuel to fuel one additional breeder for 1 year. Breeder reactor technology is inherently a very large-scale operation.

The design of breeder reactors has been fraught with technical and political difficulties. The units must be run at higher temperatures than conventional reactors, and controlling them is much more difficult. Furthermore, not only do they have the possibility of a core meltdown accident, but also, in contrast to ordinary fission reactors, an out-of-control breeder reactor can lead to a nuclear explosion.

24-18. FUSION REACTIONS RELEASE MORE ENERGY THAN FISSION REACTIONS

Figure 24-6 shows that the binding energy per nucleon for uranium is less than that of its fission products. The energy released in the fission of uranium-235 is the difference between its binding energy and that of its fission products. Figure 24-7 suggests that the *fusion* of four protons into a helium nucleus would release much more energy than the fission of uranium-235.

The sun is powered by nuclear fusion.

The major source of the sun's energy is the fusion of four protons into a helium-4 nucleus and two positrons. The net reaction is

$$4\,{}^{1}_{1}\mathrm{H} \rightarrow {}^{4}_{2}\mathrm{He} + 2\,{}^{\;0}_{+1}\mathrm{e}$$

The energy released in this process is enormous. The sun produces about $10^{26}\,\mathrm{J \cdot s^{-1}}$.

Fusion is the basis of the hydrogen bomb, in which lithium-6 deuteride, ${}^{6}\mathrm{Li}{}^{2}\mathrm{H}$, is surrounded by an ordinary fission bomb. Upon detonation, the fission bomb produces the temperatures of millions of degrees that are necessary to initiate fusion reactions. The principal fusion reactions that occur are

$${}^{2}_{1}\mathrm{H} + {}^{2}_{1}\mathrm{H} \rightarrow {}^{3}_{2}\mathrm{He} + {}^{1}_{0}\mathrm{n}$$
$${}^{6}_{3}\mathrm{Li} + {}^{1}_{0}\mathrm{n} \rightarrow {}^{4}_{2}\mathrm{He} + {}^{3}_{1}\mathrm{H}$$
$${}^{2}_{1}\mathrm{H} + {}^{3}_{1}\mathrm{H} \rightarrow {}^{4}_{2}\mathrm{He} + {}^{1}_{0}\mathrm{n}$$
$${}^{2}_{1}\mathrm{H} + {}^{2}_{1}\mathrm{H} \rightarrow {}^{3}_{1}\mathrm{H} + {}^{1}_{1}\mathrm{H}$$

The quantity of energy released by 1 ton of ${}^{6}\mathrm{Li}{}^{2}\mathrm{H}$ is equivalent to that released by 60 million tons of TNT.

The control of fusion reactions for the generation of electricity is one of the most exciting and important technological problems of our time. Excluding the use of the radioactive isotope tritium, the best possibilities seem to be the reactions

$${}^{2}_{1}\mathrm{H} + {}^{2}_{1}\mathrm{H} \rightarrow {}^{4}_{2}\mathrm{He} \qquad \text{and} \qquad {}^{1}_{1}\mathrm{H} + {}^{2}_{1}\mathrm{H} \rightarrow {}^{3}_{2}\mathrm{He}$$

Since hydrogen and deuterium are plentiful, the successful development of a fusion reactor would totally change the world we live in by allowing all countries to have an essentially inexhaustible supply of energy. The technological problems that bar the way are staggering, however. In addition to being heated to millions of degrees, the hydrogen and deuterium must be contained long enough for the reaction to occur. Of course, there is no structural material that can withstand such temperatures, but present research is aimed at confining the nuclei by using very strong magnetic and electric fields. The enormous temperatures required also can be achieved with high-power laser beams. Controlled fusion is the subject of intensive research. Some progress appears to have been made, but an operational fusion reactor is not yet close to realization.

24-19. EXPOSURE TO RADIATION DAMAGES CELLS, TISSUES, AND GENES

Example 24-5 notes that γ-radiation from cobalt-60 can be used to destroy cancerous cells. Actually, γ-radiation destroys healthy cells as well, but it is more destructive of cancerous cells because they grow and divide more rapidly than normal cells and so are more vulnerable. The effect of various kinds of radiation on living systems has received much study and has resulted in governments' placing limits on exposure to radiation.

One measure of the activity of a radioactive substance is its *specific activity,* which is the number of nuclei that disintegrate per second per gram of radioactive isotope. The specific activity of a radioisotope is related to its half-life and atomic mass M by the equation.

$$\text{specific activity} = \left(\frac{4.2 \times 10^{23}\ \text{disintegrations}\cdot\text{g}^{-1}}{Mt_{1/2}}\right) \qquad (24\text{-}8)$$

Equation (24-8) is obtained as follows: The number of disintegrations per second is the rate of the decay process

$$\text{rate} = kN$$

The specific activity is the rate divided by the mass m.

$$\text{specific activity} = \frac{\text{rate}}{m} = \frac{kN}{m} = \frac{0.693}{t_{1/2}}\frac{N}{m} = \frac{0.693}{t_{1/2}}\frac{N_0}{M}$$

where N_0 is Avogadro's number.

Because specific activity is defined as the number of disintegrations per second per gram, the value of $t_{1/2}$ in Equation (24-8) should be in seconds. As an example, let's compute the specific activity of radium-226, which has a half-life of 1600 years. We first convert the half-life to seconds:

$$(1600\ \text{yr})(365\ \text{day}\cdot\text{yr}^{-1})(24\ \text{h}\cdot\text{day}^{-1})(60\ \text{min}\cdot\text{h}^{-1})(60\ \text{s}\cdot\text{min}^{-1}) = 5.05 \times 10^{10}\ \text{s}$$

The specific activity of radium-226 is thus

$$\text{specific activity} = \frac{4.2 \times 10^{23}\ \text{disintegrations}\cdot\text{g}^{-1}}{(226)(5.05 \times 10^{10}\ \text{s})} = 3.7 \times 10^{10}\ \text{disintegrations}\cdot\text{s}^{-1}\cdot\text{g}^{-1}$$

The quantity 3.7×10^{10} disintegrations$\cdot$s^{-1} is called a *curie,* Ci, after Marie Curie, one of the pioneers in research in radioactivity (Figure 24-4). She received a Nobel Prize in physics in 1903 and one in chemistry in 1911. She is one of only four scientists to earn two Nobel prizes, and she is the only scientist to earn prizes in two different sciences.

Table 24-7 Specific activities of some important radioactive isotopes

Isotope	*Half-life*	*Specific activity/ disintegrations·s⁻¹·g⁻¹*	*Specific activity/ Ci·g⁻¹*
radium-226	1600 years	3.7×10^{10}	1.00
uranium-238	4.47×10^{9} years	1.3×10^{4}	3.5×10^{-7}
plutonium-239	2.41×10^{4} years	2.3×10^{9}	6.2×10^{-2}
iodine-131	8.04 days	4.6×10^{15}	1.2×10^{5}
cobalt-60	5.27 years	4.2×10^{13}	1.1×10^{3}
strontium-90	29 years	5.1×10^{12}	1.4×10^{2}
cesium-137	30.2 years	3.2×10^{12}	87
carbon-14	5730 years	1.7×10^{11}	4.6

Table 24-7 lists the specific activities of a number of important radioactive isotopes.

Example 24-11: In reading about the use of uranium-238 in radiodating rocks, you may have wondered how it is possible to determine a half-life as long as 4.47×10^{9} years. The answer lies in our ability to count individual radioactive decay events, coupled with the fact that Avogadro's number is enormous. There are 6.02×10^{23} uranium-238 nuclei in a sample that contains 238 g of uranium-238. Compute the number of uranium-238 nuclei that disintegrate in 10 s in a sample that contains 2.0 g of uranium-238.

Solution: Using Equation (24-8), we find that the specific activity of uranium-238, which has a half-life of 4.47×10^{9} years, is

$$\text{specific activity} = \frac{4.2 \times 10^{23}\ \text{disintegrations}\cdot\text{g}^{-1}}{(238)(4.47 \times 10^{9}\ \text{years})(3.15 \times 10^{7}\ \text{s}\cdot\text{year}^{-1})}$$
$$= 1.3 \times 10^{4}\ \text{disintegrations}\cdot\text{s}^{-1}\cdot\text{g}^{-1}$$

Thus the number of disintegrations in 2.0 g in 10 s is

$$\left(1.3 \times 10^{4}\frac{\text{disintegrations}}{\text{g}\cdot\text{s}}\right)(2.0\ \text{g})(10\ \text{s}) = 2.6 \times 10^{5}\ \text{disintegrations}$$

Thus, about 260,000 uranium-238 nuclei disintegrate in a 2.0-g sample in only 10 s. It is easy with a modern radiocounting apparatus to measure such a large number of events in a 10-s interval. The half-life of uranium-238 can be measured by counting the number of disintegrations in a known mass of the radioisotope and then carrying out the reverse of the above calculation with $t_{1/2}$ as the unknown, which is calculated from the measured activity level.

The damage that is produced by radiation depends on more than just the specific activity. As the radiation passes through tissue, it ionizes molecules and breaks chemical bonds, leaving behind a trail of molecular damage. The extent of the damage produced depends upon the energy and type of radiation. The different types of radiation vary in their ability to penetrate matter. Alpha particles can be stopped by a sheet of paper or by the skin. Beta particles penetrate deeper than α-particles, but β-particles of moderate energy are stopped by about 1 cm of water. Gamma rays are the most penetrating form of radiation, requiring walls of lead bricks to protect against them.

SUMMARY

Many nuclei are radioactive, meaning that they spontaneously emit subatomic particles such as α-particles, β-particles, and positrons. The rate of decay of a radioactive substance is a first-order kinetic process. An important property of a first-order kinetic process is that its rate is characterized by a half-life, which is the time required for one half of the amount of a sample to react. The half-lives of radioisotopes vary from less than 1 picosecond to billions of years. Rates of radioactive decay are used to determine the age of rocks and archaeological objects.

Many radioisotopes that are not found in nature can be produced by nuclear reactions in the laboratory. Such artificial radioisotopes have found wide application in chemistry, biology, forensic science, and medicine. All the elements beyond uranium in the periodic table have been artificially produced, allowing an extension of the table to atomic number 109.

The energies associated with nuclear reactions are about 1 million times greater than the energies associated with chemical reactions. Nuclear energy can be released in a controlled manner, as in nuclear reactors, or in an uncontrolled manner, as in atomic weapons. The future of nuclear energy as a commercial source of energy is clouded by uncertainty about the effect of radiation on living matter. It is well known that exposure to radiation causes damage to biological systems, and safe and acceptable levels of exposure (if any) have yet to be unequivocally determined.

TERMS YOU SHOULD KNOW

nucleon 943
nuclear equation 944
radioactivity 944
radioactive 944
radioisotope 944
α-particle 944
α-emitter 944
β-particle 945
positron 945
electron capture 946
radioactive decay 946
band of stability 948
magic numbers (2, 8, 20, 28, 50, 82, 126) 948
half-life, $t_{1/2}$ 950
uranium-lead dating 953

EQUATIONS YOU SHOULD KNOW HOW TO USE

$$\log\frac{N_0}{N} = \frac{0.301t}{t_{1/2}} \qquad (24\text{-}1)$$

$$t = (1.90 \times 10^4 \text{ years}) \log\frac{15.3}{R} \qquad (24\text{-}4)$$

$$\Delta E = c^2 \Delta m \qquad (24\text{-}6)$$

$$\text{specific activity} = \left(\frac{4.2 \times 10^{23} \text{ disintegrations} \cdot \text{g}^{-1}}{Mt_{1/2}}\right) \qquad (24\text{-}8)$$

PROBLEMS

PROTONS AND NEUTRONS

24-1. How many neutrons are there in each of the following nuclei:

(a) ^{177}Lu
(b) ^{233}Th
(c) ^{92}Zr
(d) ^{68}Ge
(e) ^{252}Cf

24-2. How many protons and neutrons are there in each of the following nuclei:

(a) ^{82}Se
(b) ^{70}Zn
(c) ^{180}W
(d) ^{223}Ra
(e) ^{245}Bk

NUCLEAR EQUATIONS

24-3. Fill in the missing symbols in the following nuclear equations:

(a) $^{32}_{15}\text{P} \rightarrow {}^{0}_{-1}\text{e} + ?$
(b) $^{210}_{84}\text{Po} \rightarrow ? + {}^{206}_{82}\text{Pb}$
(c) $^{52}_{26}\text{Fe} \rightarrow ? + {}^{52}_{25}\text{Mn}$
(d) $^{67}_{31}\text{Ga} + {}^{0}_{-1}\text{e} \rightarrow ?$

24-4. Fill in the missing symbols in the following nuclear equations:

(a) $^{43}_{19}\text{K} \rightarrow ? + {}^{43}_{20}\text{Ca}$
(b) $^{37}_{18}\text{Ar} + {}^{0}_{-1}\text{e} \rightarrow ?$
(c) $^{208}_{87}\text{Fr} \rightarrow {}^{4}_{2}\text{He} + ?$
(d) $^{83}_{38}\text{Sr} \rightarrow {}^{0}_{+1}\text{e} + ?$

24-5. Fill in the missing symbols in the following nuclear equations:

(a) $^{12}_{6}\text{C} + {}^{12}_{6}\text{C} \rightarrow ? + {}^{1}_{1}\text{H}$
(b) $^{12}_{6}\text{C} + {}^{12}_{6}\text{C} \rightarrow ? + {}^{4}_{2}\text{He}$
(c) $? + {}^{1}_{1}\text{H} \rightarrow {}^{15}_{7}\text{N} + {}^{4}_{2}\text{He}$
(d) $^{17}_{9}\text{F} \rightarrow ? + {}^{0}_{+1}\text{e}$
(e) $^{249}_{98}\text{Cf} + {}^{18}_{8}\text{O} \rightarrow ? + 4{}^{1}_{0}\text{n}$

24-6. Fill in the missing symbols in the following nuclear equations:

(a) $^{15}_{7}\text{N} + ? \rightarrow {}^{12}_{6}\text{C} + {}^{4}_{2}\text{He}$
(b) $? + {}^{4}_{2}\text{He} \rightarrow {}^{7}_{4}\text{Be} + \gamma$
(c) $^{9}_{4}\text{Be} + ? \rightarrow {}^{10}_{5}\text{B} + {}^{1}_{0}\text{n}$
(d) $^{81}_{37}\text{Rb} \rightarrow ? + {}^{0}_{+1}\text{e}$
(e) $^{238}_{92}\text{U} + {}^{16}_{8}\text{O} \rightarrow ? + 5{}^{1}_{0}\text{n}$

24-7. Equations for nuclear reactions are often written in a condensed form, for example, $^{14}_{7}N(\alpha, p)$. This notation means that when $^{14}_{7}N$ is bombarded by an α-particle, a proton is produced. The complete equation is $^{14}_{7}N + ^{4}_{2}He \rightarrow ^{1}_{1}H + ^{17}_{8}O$, where $^{17}_{8}O$ is deduced from the rules for balancing nuclear equations. Write out complete equations for the following processes (d stands for a deuteron, $^{2}_{1}H$):

(a) $^{9}_{4}Be(\alpha, n)$
(b) $^{7}_{3}Li(p, \alpha)$
(c) $^{63}_{29}Cu(p, n)$
(d) $^{14}_{7}N(n, p)$
(e) $^{54}_{26}Fe(d, n)$

24-8. As explained in Problem 24-7, equations for nuclear reactions are often written in a condensed notation. Write out complete nuclear equations for the following processes:

(a) $^{16}_{8}O(p, \gamma)$
(b) $^{20}_{10}Ne(p, \gamma)$
(c) $^{17}_{8}O(p, \alpha)$
(d) $^{22}_{11}Na(p, \alpha)$
(e) $^{18}_{8}O(\alpha, n)$

24-9. Fill in the missing symbols in the following sequence of nuclear equations:

(a) $^{12}_{6}C + ? \rightarrow ^{13}_{7}N + \gamma$
(b) $? \rightarrow ^{13}_{6}C + ^{0}_{+1}e$
(c) $^{13}_{6}C + ^{1}_{1}H \rightarrow ? + \gamma$
(d) $^{14}_{7}N + ? \rightarrow ^{15}_{8}O + \gamma$
(e) $^{15}_{8}O \rightarrow ^{15}_{7}N + ?$
(f) $^{15}_{7}N + ^{1}_{1}H \rightarrow ^{12}_{6}C + ?$

This sequence of reactions is believed to power stars. What is the net reaction?

24-10. The first artificial radioisotope was produced by Irene Curie and Frédéric Joliot in 1934. They bombarded $^{27}_{13}Al$ with α-particles to produce $^{30}_{15}P$ and $^{1}_{0}n$. Write the equation for this nuclear reaction. Predict the mode of decay of phosphorus-30.

PREDICTION OF THE MODE OF NUCLEAR DECAY

24-11. Sodium-20 is an artificial radioisotope that has been produced in a cyclotron. Predict its mode of decay.

24-12. Potassium-37 is an artificial radioisotope that has been produced in a reactor. Predict its mode of decay.

24-13. The element technetium has no stable isotopes. Predict the mode of decay of the radioisotope technetium-93.

24-14. The element promethium has no known stable isotopes. One radioisotope, $^{140}_{61}Pm$, has a half-life of 9.2 s. Predict its mode of decay.

24-15. Predict the mode of decay of the following nuclei, all of which are produced in nuclear reactors:

(a) ^{130}Sn
(b) ^{52}Fe
(c) ^{239}Pu
(d) ^{34}Ar
(e) ^{50}Ca

24-16. Predict the mode of decay of the following nuclei:

(a) ^{236}Pu
(b) ^{11}C
(c) ^{18}N
(d) ^{12}N
(e) ^{3}H

PREDICTION OF WHETHER A NUCLEUS IS STABLE OR RADIOACTIVE

24-17. Which of the following nuclei would you predict to be radioactive:

(a) ^{215}Th
(b) ^{24}Mg
(c) ^{12}N
(d) ^{22}Na
(e) ^{72}Ge

24-18. Which of the following nuclei would you predict to be stable:

(a) ^{36}Ar
(b) ^{40}K
(c) ^{58}Ni
(d) ^{8}B
(e) ^{20}Ne

24-19. Predict whether or not the following nuclei are radioactive:

(a) $^{24}_{11}Na$
(b) $^{118}_{51}Sb$
(c) $^{118}_{50}Sn$
(d) $^{26}_{13}Al$

24-20. Predict whether or not the following nuclei are radioactive:

(a) $^{38}_{17}Cl$
(b) $^{95}_{42}Mo$
(c) $^{40}_{20}Ca$
(d) $^{225}_{87}Fr$

RATES OF NUCLEAR DECAY

24-21. Astatine-211 is used for the treatment of certain types of thyroid cancer. If 0.10 mg is administered to a patient at 10 a.m., how much remains at 10 p.m. the same day? The half-life of astatine-211 is 7.2 h.

24-22. A sample of sodium-24 chloride containing 0.055 mg of sodium-24 is injected into an animal to study sodium balance. How much sodium-24 remains 6.0 h later? The half-life of sodium-24 is 15.0 h.

24-23. The radioisotope argon-41 is used to measure the rate of the flow of gases from smokestacks. It is a γ-emitter with a half-life of 110 min. Calculate the fraction of an argon-41 sample that remains after 1 day.

24-24. The radioisotope bromine-82 is used as a tracer for organic materials in environmental studies. Its half-life is 36 h. Calculate the fraction of a sample of bromine-82 that remains after one day.

24-25. The radioisotope bromine-82 is used as a tracer for organic materials in environmental studies. Its half-life is 36 h. Calculate how long it takes a sample of bromine-82 to decay so that only 10 percent remains.

24-26. Cesium-137 is produced in nuclear reactors. If this isotope has a half-life of 30.2 years, how many years will it take for it to decay to one percent of its initial amount?

24-27. The radioisotope sulfur-38 can be incorporated into proteins to follow certain aspects of protein metabolism. If a protein sample initially has an activity of 10,000 disintegrations$\cdot$min^{-1}, then what is the activity 10 h later? The half-life of sulfur-38 is 2.84 h.

24-28. The radioisotope hydrogen-3 is used in fusion reactors. It is a β-emitter with a half-life of 12.3 years. Calculate the fraction of a hydrogen-3 sample that will remain after 100 years.

24-29. A sample of KI containing some of the radioisotope potassium-42 has an activity of 8500 disintegrations$\cdot$min^{-1} initially and 2780 disintegrations$\cdot$min^{-1} 20.0 h later. Calculate the half-life of potassium-42.

24-30. A sample of H_2S containing some of the radioisotope sulfur-38 has an activity of 12,000 disintegrations$\cdot$min^{-1} initially and 1710 disintegrations$\cdot$min^{-1} 8.0 h later. Calculate the half-life of sulfur-38.

24-31. The radioisotope phosphorus-32 can be incorporated into nucleic acids to follow certain aspects of their metabolism. If a nucleic acid sample initially has an activity of 40,000 disintegrations$\cdot$min^{-1}, then what is the activity 220 h later? The half-life of phosphorus-32 is 14.28 days.

24-32. A sample of radioactive NaI (iodine-128 is the radioisotope here) is injected into a patient as part of radioiodine treatment of a thyroid condition. If the sample has an activity of 10,000 disintegrations$\cdot$min^{-1} at 8 a.m., the time of injection, then what is the activity at 2 p.m. the same day? The half-life of iodine-128 is 25.0 min.

CARBON-14 DATING

24-33. Wooden lintels from the Mayan site in Tikal, Guatemala, have a disintegration rate of 12.8 disintegrations$\cdot$min^{-1} per gram of carbon. Estimate the age of the lintels.

24-34. Charcoal from an archaeological site near Beer-Sheba, Israel, has a disintegration rate of 6.24 disintegrations$\cdot$min^{-1} per gram of carbon. Estimate the age of the charcoal.

24-35. The French explorer Fernand Navarra claims to have discovered Noah's Ark. In 1955 he discovered a log on Mt. Ararat in eastern Turkey, the legendary resting spot of the ark. Navarra claims that the log is a beam from the ark. Samples of the wood have a disintegration rate of 13.19 disintegrations$\cdot$min^{-1} per gram of carbon. Calculate the age of the log.

24-36. An 80-kg human contains about 15 kg of carbon in various chemical compounds. Compute the number of ^{14}C disintegrations per second in the human body.

24-37. A buried bison bone found at the Folsom Man site near Lubbock, Texas, has a carbon-14 content equal to 30 percent that in living matter. Estimate the age of the bone.

24-38. Samples of oak from an ancient Irish cooking site at Killeens, County Cork, have a carbon-14 content equal to 65 percent that in living matter. Estimate the age of the wood.

GEOLOGICAL DATING

24-39. A sample of uranite is found to have a $^{206}Pb/^{238}U$ mass ratio of 0.420. Estimate the age of the uranite. The half-life of the conversion of uranium-238 to lead-206 is 4.51×10^{9} years.

24-40. A sample of ocean sediment is found to contain 1.50 mg of uranium-238 and 0.460 mg of lead-206. Estimate the age of the sediment. The half-life for the conversion of uranium-238 to lead-206 is 4.51×10^{9} years.

24-41. A recently fallen meteorite is found to contain a $^{87}Sr/^{87}Rb$ mass ratio of 0.0038. Estimate the age of the meteorite. The half-life of rubidium-87 is 4.9×10^{10} years.

24-42. One of the oldest terrestrial rocks known, a gneiss from Greenland, has a $^{87}Sr/^{87}Rb$ mass ratio of 0.056. Estimate the age of the rock. The half-life of rubidium-87 is 4.9×10^{10} years.

24-43. Potassium-40 decays by two different modes:

$$^{40}_{19}K \xrightarrow{\text{electron capture}} {}^{40}_{18}Ar \quad (10.7 \text{ percent})$$

$$^{40}_{19}K \rightarrow {}^{40}_{20}Ca + {}^{\ 0}_{-1}e \qquad (89.3 \text{ percent})$$

$$t_{1/2} = 1.3 \times 10^{9} \text{ years}$$

The mass ratio $^{40}Ar/^{40}K$ in a lunar rock is 1.05. Estimate its age.

24-44. The mass ratio $^{40}Ar/^{40}K$ observed for sedimentary rocks surrounding skulls found near Olduvai Gorge in Africa is 6.0×10^{-5}. Using the information given in Problem 24-43, estimate the age of the skulls.

BINDING ENERGY

24-45. Given that the atomic mass of cobalt-57 is 56.9363 amu, calculate the binding energy and the binding energy per nucleon in this isotope.

24-46. Iron-56 (atomic mass = 55.9346 amu) is the most stable nucleus. Calculate its binding energy and the binding energy per nucleon.

24-47. Given that the atomic mass of $^{35}_{17}Cl$ is 34.9689 amu, calculate the binding energy and the binding energy per nucleon.

24-48. Given that the atomic mass of $^{20}_{10}Ne$ is 19.9924 amu, calculate the binding energy and the binding energy per nucleon.

24-49. Calculate the energy released per mole of helium by the reaction

$$2\,^{2}_{1}H \rightarrow {}^{4}_{2}He$$

The atomic mass of $^{2}_{1}H$ is 2.0141 amu, and the atomic mass of $^{4}_{2}He$ is 4.0026 amu. How many kilograms of octane would have to be burned to produce as much energy as is produced by the formation of 1.0 g of helium in this reaction? For the combustion of octane, C_8H_{18}, $\Delta H^{\circ}_{rxn} = -5450\ kJ \cdot mol^{-1}$.

24-50. Calculate the energy released per mole of lithium by the reaction

$$^{7}_{3}Li + {}^{1}_{1}H \rightarrow 2\,^{4}_{2}He$$

The required atomic masses are $^{1}_{1}H = 1.0078$ amu, $^{7}_{3}Li = 7.0160$ amu, and $^{4}_{2}He = 4.0026$ amu. How many kilograms of octane would have to be burned to produce as much energy as is produced when 1.0 g of lithium is consumed in this reaction? For the combustion of octane, C_8H_{18}, $\Delta H^{\circ}_{rxn} = -5450\ kJ \cdot mol^{-1}$.

NUCLEAR REACTIONS AND ENERGY

24-51. A major source of the sun's energy is the net reaction

$$4\,^{1}_{1}H \rightarrow {}^{4}_{2}He + 2\,^{\ 0}_{+1}e$$

If the sun produces energy at the rate of $8.0 \times 10^{22}\ kJ \cdot s^{-1}$, then how much helium is produced per second by this reaction? The necessary atomic masses are $^{1}_{1}H = 1.0078$ amu, $^{4}_{2}He = 4.0026$ amu,

(*Continued on p. 978*)

24-52. Suppose that uranium-235 undergoes the fission reaction

$$^{235}_{92}U + {}^{1}_{0}n \rightarrow {}^{141}_{56}Ba + {}^{88}_{36}Kr + 7\,^{1}_{0}n$$

Given the atomic masses $^{235}_{92}U = 235.0439$ amu, $^{141}_{56}Ba = 140.9137$ amu, $^{88}_{36}Kr = 87.9142$ amu, and $^{1}_{0}n = 1.0087$ amu, calculate the energy released when one uranium-235 nucleus undergoes this reaction.

(*Continued on p. 978*)

and $_{+1}^{0}e = 0.00055$ amu. Realize that the masses given are *atomic* masses, and so the number of electrons on the two sides of the reaction must be considered in this case.

Calculate the mass of uranium-235 that reacts in a 50-kiloton bomb. The designation 50-kiloton means that the bomb has the explosive equivalent of 50 kilotons of TNT. Assume that 2500 kJ is liberated per kilogram of TNT. For simplicity, assume that the tons here are metric tons, that is, 1000 kg.

24-53. Marie Curie observed that 1.0 g of pure radium-226 generates $148\ J\cdot h^{-1}$. Given that 1.0 g of radium-226 has a disintegration rate of 3.7×10^{10} disintegrations$\cdot s^{-1}$, calculate the energy released in each disintegration. Compare this with what you calculate given the reaction

$$^{226}_{88}Ra \rightarrow\ ^{222}_{86}Rn + ^{4}_{2}He$$

and the atomic masses $^{226}_{88}Ra = 226.0254$ amu, $^{222}_{86}Rn = 222.0154$ amu, and $^{4}_{2}He = 4.0026$ amu.

24-54. The half-life of a positron is very short. It reacts with an electron, and the masses of both are converted to two γ-rays:

$$_{+1}^{0}e + _{-1}^{0}e \rightarrow 2\gamma$$

This reaction is called an annihilation reaction. Calculate the energy produced by the reaction between one electron and one positron. Assuming that the two γ-rays have the same frequency, calculate this frequency.

24-55. Calculate the difference in mass between the products and reactants for the reaction

$$C(diamond) + O_2(g) \rightarrow CO_2(g) \qquad \Delta H^\circ_{rxn} = -396\ kJ\cdot mol^{-1}$$

24-56. Calculate the difference in mass between the products and reactants for the reaction

$$S(s) + O_2(g) \rightarrow SO_2(g) \qquad \Delta H^\circ_{rxn} = -297\ kJ\cdot mol^{-1}$$

NUCLEAR REACTORS AND ENERGY

24-57. The fusion of 1.0 g of atomic hydrogen (^1H) to ^4He releases 6.4×10^{11} J of energy. Compute the number of metric tons of water that could supply the total annual U.S. energy requirement of 85×10^{15} kJ if all the hydrogen were converted to ^4He.

24-58. The world's energy consumption is estimated to be $3 \times 10^{17}\ kJ\cdot year^{-1}$. How much uranium-235 would be required to produce all this energy by fission? The amount of energy released by the fission of uranium-235 is $2 \times 10^{13}\ J\cdot mol^{-1}$. The world supply of uranium is estimated to be about 10^6 metric tons. About 0.7 percent of naturally occurring uranium is uranium-235. At a world energy consumption rate of $3 \times 10^{17}\ kJ\cdot year^{-1}$, how long will the supply of uranium last?

24-59. A nuclear power plant producing 1000 megawatts (a watt is a $J\cdot s^{-1}$) of electricity produces 3000 megawatts of heat. Given that the available thermal energy per ^{235}U disintegration is 2.9×10^{-14} kJ, compute the number of moles of ^{235}U burned per month when the nuclear reactor operates continuously at maximum power.

24-60. In one type of nuclear reactor, the uranium fuel is about 3 percent uranium-235. Only about one third of this uranium-235 can be utilized before released fission products eventually poison the fuel. If each fission reaction releases 2.9×10^{-14} kJ, calculate the energy released by the "burning" of 1 kg of naturally occurring uranium. Assuming that 30 percent of this energy is available for use, calculate the quantity of fuel required to operate a 1000-megawatt electric power plant at full power for 1 year. One watt is equal to one joule per second.

SPECIFIC ACTIVITY

24-61. Thorium-232 has a half-life of 1.40×10^{10} years. Calculate the specific activity of thorium-232 in units of curies per gram.

24-62. Cerium-144 has a half-life of 284 days. Calculate the specific activity of cerium-144 in units of curies per gram.

24-63. The radioisotope sulfur-35 is used extensively in biological and environmental studies. It is prepared in a cyclotron and is free of all other isotopes of sulfur. Its half-life is 87 days. Calculate the specific activity (in curies per gram) of a freshly prepared sample of sulfur-35.

24-64. The only radioisotope of aluminum that is useful for tracer studies is aluminum-26, which has a half-life of 7.4×10^5 years. Calculate the specific activity (expressed in curies per gram) of a sample that contains only aluminum-26.

24-65. The radioisotope gallium-67, which has a half-life of 78 h, is used as a diagnostic tool in tumor location. How many milligrams of $^{67}GaCl_3$ is equivalent to 200 mCi?

24-66. The radioisotope iodine-128 (half-life 25 min) is used as a diagnostic tool for thyroid imaging. A typical dose is 100 μCi. Compute the number of milligrams of $Na^{128}I$ equivalent to 100 μCi.

APPLICATION OF RADIOACTIVE DECAY RATES

24-67. A 10.0-g sample of SO_2 containing the radioisotope sulfur-35 has an activity of 3.23×10^{11} disintegrations$\cdot s^{-1}$. Given that sulfur-35 has a half-life of 87.2 days, calculate the fraction of radioactive sulfur atoms in the SO_2 sample.

24-68. A 1.00-mL sample of blood is withdrawn from an animal, and the red blood cells are labeled with phosphorus-32 ($t_{1/2} = 14.3$ days). The activity of this sample is 50,000 disintegrations$\cdot min^{-1}$. The sample is then reinjected into the animal. A few hours later, another 1.00-mL sample is withdrawn, and its activity is 10.0 disintegrations$\cdot min^{-1}$. Determine the volume of blood in the animal. Assume that the phosphorus-32 is uniformly distributed throughout the blood and that the activity due to phosphorus-32 remains constant during the experiment. For dogs, an approximately linear relationship has been established between blood volume and body weight. By using similar methods, it has been found that the human body contains 73 to 75 mL of blood per kilogram of body weight.

24-69. The amount of oxygen dissolved in a sample of water can be measured by using the radioisotope thallium-204. Solid thallium reacts with oxygen according to the equation

$$4Tl(s) + O_2(aq) + 2H_2O(l) \rightarrow 4Tl^+(aq) + 4OH^-(aq)$$

The amount of oxygen can be determined by measuring the radioactivity due to thallium-204 in the water sample. In one experiment 10.0 mL of water is reacted with some thallium metal whose activity is 1.13×10^8 disintegrations$\cdot min^{-1}$ per mole of thallium. The radioactivity of water is 563 disintegrations$\cdot min^{-1}$. Calculate the concentration of oxygen in the sample.

24-70. Low concentrations of sulfate ion in aqueous samples can be measured by precipitating $SO_4^{2-}(aq)$ with the radioisotope barium-131 in the form of $Ba^{2+}(aq)$. The radioactivity of the precipitate, $BaSO_4(s)$, is then measured. In one experiment, sulfate ion was precipitated with barium-131 whose specific activity was 7.6×10^7 disintegrations$\cdot min^{-1} \cdot g^{-1}$. The radioactivity of $BaSO_4(s)$ precipitated from a 10.0-mL sample of the aqueous solution was 3270 disintegrations$\cdot min^{-1}$. Calculate the concentration of SO_4^{2-} in the aqueous solution.

24-71. A 50.0-mL sample of a 0.075 M $Pb(NO_3)_2(aq)$ solution is mixed with 50.0 mL of a 0.15 M $NaI(aq)$ solution that has been labeled with the radioisotope iodine-131. The activity of the radioactive NaI solution is 20,000 disintegrations$\cdot min^{-1} \cdot mL^{-1}$. After precipitation, the mixture is filtered, and the supernatant liquid has an activity of 320 disintegrations$\cdot min^{-1} \cdot mL^{-1}$. Calculate the solubility product of $PbI_2(s)$.

24-72. A 75-mL sample of a 0.010 M $Pb(NO_3)_2(aq)$ solution is mixed with 75 mL of a 0.010 M $Na_2SO_4(aq)$ solution that has been labeled with the radioisotope sulfur-35. The activity of the Na_2SO_4 solution is 14,000 disintegrations$\cdot min^{-1} \cdot mL^{-1}$. After precipitation, the mixture is filtered, and the supernatant liquid has an activity of 183 disintegrations$\cdot min^{-1} \cdot mL^{-1}$. Calculate the solubility product of $PbSO_4(s)$.

Much of our oil is obtained from vast deposits under the ocean. This photo shows an offshore oil-drilling complex in the North Sea.

25 / Organic Chemistry

The carbon atom is unusual in that it can bond covalently to other carbon atoms to form long chains of carbon atoms. The vast majority of all compounds, and particularly all biologically important compounds, consist of molecules that contain many carbon atoms. Because it was once believed that all compounds containing carbon had their origins in living sources, compounds that contain carbon are called organic compounds. In this chapter we discuss a variety of organic compounds: saturated, unsaturated, and aromatic hydrocarbons, alcohols, aldehydes, ketones, organic acids, and esters.

25-1. ALKANES ARE HYDROCARBONS THAT CONTAIN ONLY SINGLE BONDS

Methane, CH_4, and ethane, C_2H_6 (Figure 25-1), belong to the class of *organic compounds* called *hydrocarbons* because they consist of only hydrogen and carbon. Methane and ethane are the first two members of the series of hydrocarbons called the *alkanes*. Alkanes are hydrocarbons that contain only single carbon-carbon bonds, and their bonding is described in terms of sp^3 hybrid orbitals on the carbon atoms (Section 12-3).

The next member of the alkane series contains three carbon atoms in a row and is called propane, C_3H_8. Propane is a gas that is com-

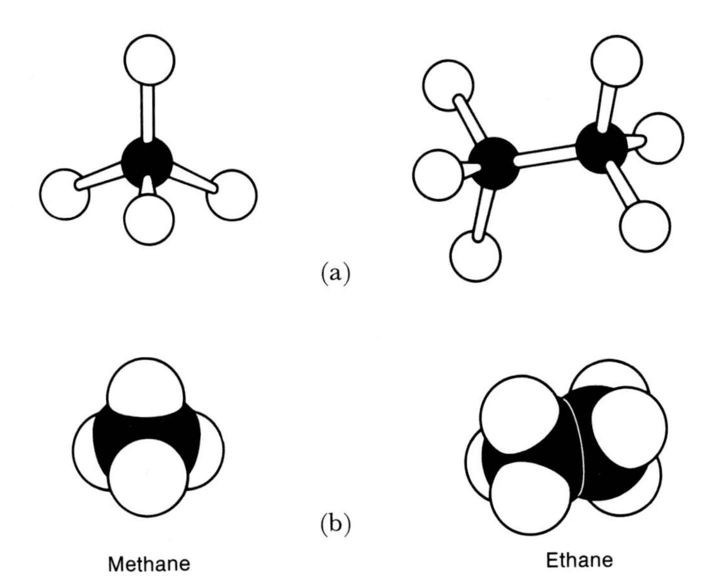

Figure 25-1 (a) Ball-and-stick and (b) space-filling models of methane and ethane. The hydrogen atoms in methane point to the vertices of a regular tetrahedron. Each of the carbon atoms in ethane is surrounded by a tetrahedral array of atoms. The C—H bond length (110 pm) in ethane is the same as in methane, and the H—C—H and H—C—C bond angles are equal to the tetrahedral bond angle (109.5°).

monly used as fuel in areas not serviced by gas mains. The Lewis formula for propane is

```
    H   H   H
    |   |   |
H — C — C — C — H
    |   |   |
    H   H   H
```

which can be written as the *structural formula* $CH_3CH_2CH_3$. Each of the carbon atoms in propane is surrounded by a tetrahedral array of atoms (Figure 25-2).

A structural formula is less detailed than a Lewis formula, but still conveys enough information so that the structure of the molecule can be deduced.

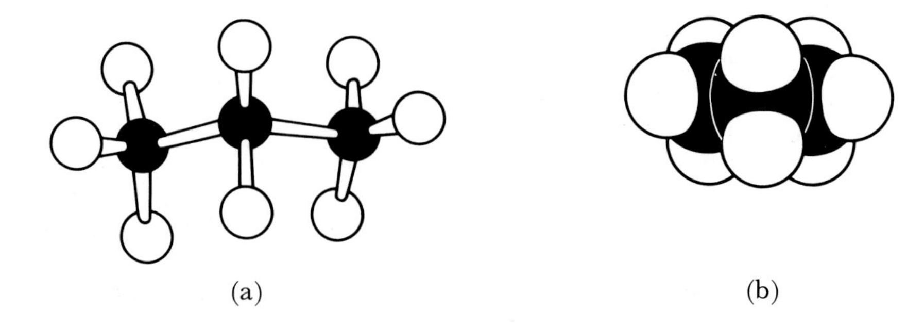

Figure 25-2 (a) Ball-and-stick and (b) space-filling models of propane. Note that the bonds about each carbon atom are tetrahedrally oriented. The bond lengths and angles are the same as in ethane.

25-2. MOLECULES THAT HAVE THE SAME CHEMICAL FORMULA BUT DIFFERENT STRUCTURES ARE CALLED STRUCTURAL ISOMERS

The fourth member of the alkane series, butane, C_4H_{10}, is interesting because there are two different types of butane molecules (Figure 25-3). The Lewis formulas for the two forms of butane are

$$\begin{array}{ccccccccccc} & & \text{H} & & \text{H} & & \text{H} & & \text{H} & & \\ & & | & & | & & | & & | & & \\ \text{H} & - & \text{C} & - & \text{C} & - & \text{C} & - & \text{C} & - & \text{H} \\ & & | & & | & & | & & | & & \\ & & \text{H} & & \text{H} & & \text{H} & & \text{H} & & \end{array} \qquad \text{and} \qquad \begin{array}{ccccccccc} & & \text{H} & & \text{H} & & \text{H} & & \\ & & | & & | & & | & & \\ \text{H} & - & \text{C} & - & \text{C} & - & \text{C} & - & \text{H} \\ & & | & & | & & | & & \\ & & \text{H} & & | & & \text{H} & & \\ & & & & | & & & & \\ & & & \text{H} & - \text{C} - & \text{H} & & & \\ & & & & | & & & & \\ & & & & \text{H} & & & & \end{array}$$

n-butane isobutane

and their structural formulas are

$$CH_3CH_2CH_2CH_3 \qquad \text{and} \qquad \begin{array}{c} CH_3\underset{|}{C}HCH_3 \\ \;\; CH_3 \end{array}$$

n-butane isobutane

The straight-chain molecule is called normal butane (written *n*-butane), and the branched molecule is called isobutane. It is im-

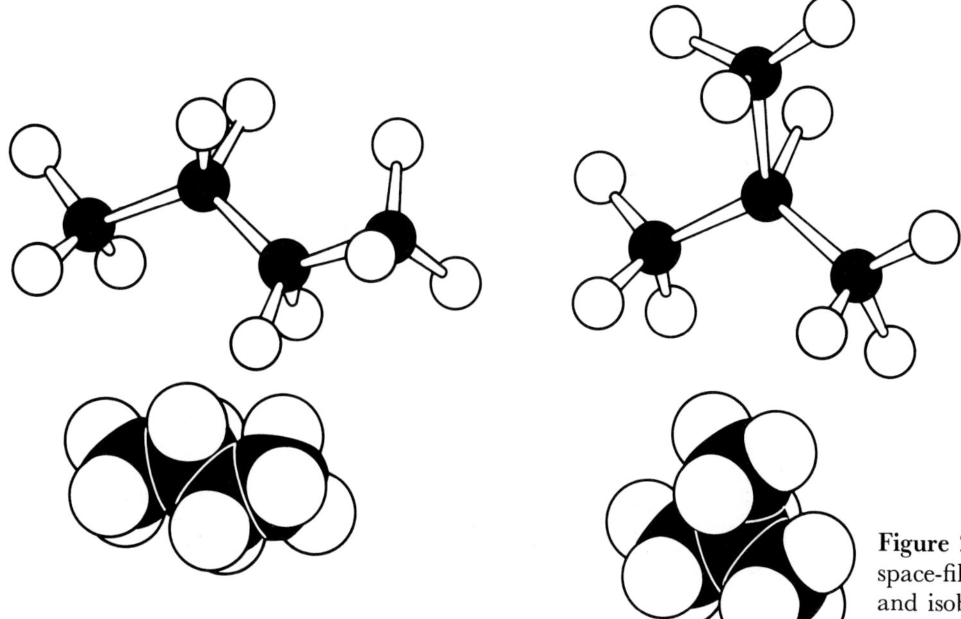

Figure 25-3 Ball-and-stick and space-filling models of *n*-butane and isobutane. As in ethane and propane, the bonds from each carbon atom are tetrahedrally oriented.

portant to realize that *n*-butane and isobutane are different compounds, with different physical and chemical properties. For example, the boiling point of *n*-butane is $-0.5\,^\circ C$ and that of isobutane is $-10.2\,^\circ C$. Compounds that have the same molecular formula but different structures are called *structural isomers.*

In alkanes, the arrangement of the four bonds about each carbon atom is tetrahedral. All the H—C—H, H—C—C, and C—C—C bond angles are 109.5°C. In addition, the H—C bond lengths are all equal (110 pm) and the C—C bond lengths are all equal (154 pm). Each of the four bonds from a carbon atom is connected to a different atom. It is not possible to bond any additional atoms directly to carbon atoms in alkanes. The bonding about each carbon atom is saturated, and so alkanes are called *saturated hydrocarbons.*

Because the carbon-carbon bond in ethane is a σ bond (Figure 12-7), its energy is unaffected by rotation around the carbon-carbon axis. Thus one $-CH_3$ group in ethane can rotate relative to the other. All carbon-carbon single bonds are formed the same way as in ethane, and so all carbon-carbon single bonds are σ bonds. Consequently, we find that *rotation can occur about carbon-carbon single bonds.*

It is important to realize that, because there is rotation about carbon-carbon single bonds, the formula

$$\begin{array}{c} \qquad\qquad\quad H \\ \qquad\qquad\quad | \\ CH_3{-}CH_2{-}C{-}H \\ \qquad\qquad\quad | \\ \qquad\qquad\quad CH_3 \end{array}$$

does *not* represent a third isomer of butane. It simply represents four carbon atoms joined in a chain and is just a somewhat misleading way of writing the formula for *n*-butane.

25-3. THE NUMBER OF STRUCTURAL ISOMERS INCREASES WITH THE NUMBER OF CARBON ATOMS IN AN ALKANE

The first four members of the saturated hydrocarbon series are methane, ethane, propane, and butane. After butane, the alkanes are assigned systematic names that indicate the number of carbon atoms in the molecule. For example, C_5H_{12} is called pentane. The prefix pent- indicates that there are five carbon atoms, and the ending -ane denotes an alk*ane*. The names of the first 10 straight-chain alkanes are given in Table 25-1, which also gives their melting points and boiling points. The first four *n*-alkanes are gases at room temperature (20°C), whereas *n*-pentane through *n*-decane are liquids at room temperature. Higher alkanes are waxy solids at room temperature. Note that the boiling points of the *n*-alkanes increase with molecular

The melting point of *n*-hexadecane, $C_{16}H_{34}$, is 18.1°C, and the melting point of *n*-heptadecane, $C_{17}H_{36}$, is 22.0°C.

Table 25-1 The names, melting points, and boiling points of the first 10 straight-chain alkanes

Name	*Number of carbon atoms*	*Molecular formula*[a]	*mp/°C*	*bp/°C (at atmospheric pressure)*
methane	1	CH_4	−183	−162
ethane	2	C_2H_6	−172	−89
propane	3	C_3H_8	−188	−42
n-butane	4	C_4H_{10}	−135	−0.5
n-pentane	5	C_5H_{12}	−130	36
n-hexane	6	C_6H_{14}	−95	69
n-heptane	7	C_7H_{16}	−91	98
n-octane	8	C_8H_{18}	−57	126
n-nonane	9	C_9H_{20}	−54	151
n-decane	10	$C_{10}H_{22}$	−30	174

[a]The molecular formulas for the alkanes fit the general formula C_nH_{2n+2}, where n represents the number of carbon atoms in the molecule.

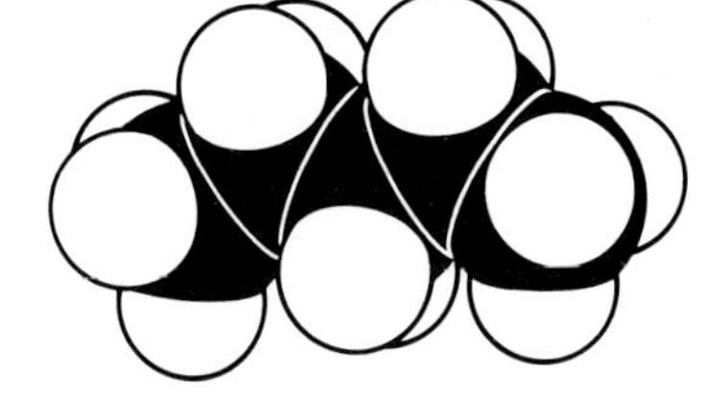

n-Pentane

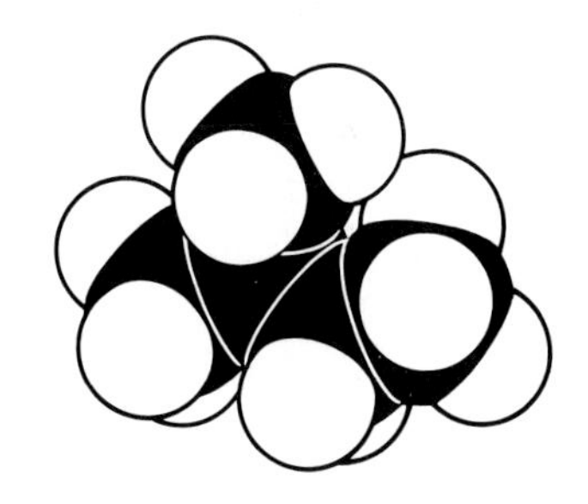

Isopentane

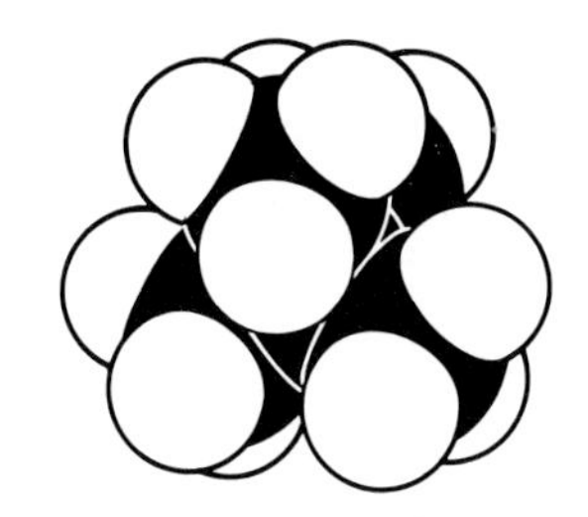

Neopentane

Figure 25-4 Space-filling models of the three structural isomers of pentane.

mass. This is in accord with the fact that the van der Waals forces between molecules increase with the size of the molecules (Section 13-4).

The number of structural isomers increases with the number of carbon atoms in an alkane. There are three isomers of pentane (Figure 25-4):

$$CH_3CH_2CH_2CH_2CH_3 \qquad CH_3\underset{}{\overset{\displaystyle CH_3}{\overset{|}{C}}}HCH_2CH_3$$

n-pentane isopentane

$$CH_3-\overset{\displaystyle CH_3}{\underset{\displaystyle CH_3}{\overset{|}{\underset{|}{C}}}}-CH_3$$

neopentane

Note that all three isomers have the molecular formula C_5H_{12}. Table 25-2 shows the number of structural isomers versus the number of carbon atoms in an alkane. Note that the number of structural isomers increases very rapidly with the number of carbon atoms.

Table 25-2 The number of structural isomers vs. the number of carbon atoms in an alkane

Number of carbon atoms	*Alkane*	*Number of structural isomers*
1	methane	1
2	ethane	1
3	propane	1
4	butane	2
5	pentane	3
6	hexane	5
7	heptane	9
8	octane	18
9	nonane	35
10	decane	75

25-4. THE PRINCIPAL SOURCES OF SATURATED HYDROCARBONS ARE PETROLEUM AND NATURAL GAS DEPOSITS

Petroleum is a complex mixture of hydrocarbons, organosulfur, and organonitrogen compounds.

Table 25-3 lists some common petroleum products. All the normal alkanes through $C_{33}H_{68}$, as well as many branched-chain hydrocarbons, have been isolated from petroleum. A few alkanes occur elsewhere in nature. For example, the skin of an apple contains the C_{27} and C_{29} *n*-alkanes. These are the waxes that produce the waxy feel of an apple when it is polished. Long-chain saturated hydrocarbons often form part of the protective coating on leaves and fruits. Similar hydrocarbons are found in beeswax. Apparently the major function of these waxes is to retard water loss.

An exciting, but still speculative, new source of hydrocarbons has been proposed by Melvin Calvin at the University of California. He has found that the leaves of the gopher plant (*Euphorbia lathyrus*), a shrub from Africa that grows in arid regions, produce a milky substance that is rich in hydrocarbons. He has proposed that vast arid areas of the southwestern United States be devoted to growing gopher plants and similar plants, which can be harvested as a renewable source of hydrocarbons. It has been estimated that *petroleum plantations* could produce about 10 barrels of hydrocarbons per acre per year at a price per barrel comparable to that of crude oil. Field research along these lines has been initiated by Calvin and others.

Table 25-3 Some common petroleum products

Product	*Alkanes present*	*Boiling range/°C*
natural gas	C_1	−162
liquefied petroleum gas (LPG), propane, butane	C_3–C_4	−42 to 0
petroleum ether (solvent)	C_5–C_7	30 to 98
gasoline	C_5–C_{10}	36 to 175
kerosene, jet fuel	C_{10}–C_{18}	175 to 275
diesel fuel	C_{12}–C_{20}	190 to 330
fuel oil	C_{14}–C_{22}	230 to 360
lubricating oil	C_{20}–C_{30}	above 350
mineral oil (refined)	C_{20}–C_{30}	above 350
		Melting range/°C
petroleum jelly	C_{22}–C_{40}	40 to 60
paraffin	C_{25}–C_{50}	50 to 65

25-5. ALKANES ARE RELATIVELY UNREACTIVE

Alkanes react with F_2, Cl_2, and Br_2 and burn in O_2 to form CO_2 and H_2O. These reactions are all highly exothermic. For example, ΔH°_{rxn} for the reaction

$$C_3H_8(g) + 5O_2(g) \rightarrow 3CO_2(g) + 4H_2O(g)$$

is −2040 kJ. Reactions of this type constitute the basis for the use of hydrocarbons as heating fuels.

A mixture of an alkane and chlorine can be kept indefinitely in the dark. If such a mixture is heated or exposed to sunlight (ultraviolet radiation), however, a reaction occurs in which one or more of the hydrogen atoms in the alkane are replaced by chlorine atoms. Such a reaction is called a *substitution reaction* because the alkane hydrogen atoms are substituted by chlorine atoms. For example,

$$CH_4(g) + Cl_2(g) \xrightarrow{\text{UV light}} CH_3Cl(g) + HCl(g)$$

The product in this case can be considered to be a derivative of methane and so is named chloromethane. The function of the ultraviolet radiation is to break the bond in the Cl_2 molecule and produce free chlorine atoms. The free chlorine atoms are free radicals—hence highly reactive—and react with methane:

$$:\ddot{\underset{..}{Cl}}-\ddot{\underset{..}{Cl}}: \xrightarrow{UV} 2:\ddot{\underset{..}{Cl}}\cdot$$

$$:\ddot{\underset{..}{Cl}}\cdot + H-\overset{\displaystyle H}{\underset{\displaystyle H}{\overset{|}{\underset{|}{C}}}}-H \rightarrow H\cdot + H-\overset{\displaystyle H}{\underset{\displaystyle H}{\overset{|}{\underset{|}{C}}}}-\ddot{\underset{..}{Cl}}:$$

$$H\cdot + :\ddot{\underset{..}{Cl}}\cdot \rightarrow H-\ddot{\underset{..}{Cl}}:$$

As the concentration of CH_3Cl builds up during the reaction, it reacts with chlorine atoms to produce dichloromethane, CH_2Cl_2.

For simplicity, we shall not always include lone pairs in Lewis formulas.

$$:\ddot{\underset{..}{Cl}}\cdot + H-\overset{\displaystyle H}{\underset{\displaystyle H}{\overset{|}{\underset{|}{C}}}}-\ddot{\underset{..}{Cl}}: \rightarrow H\cdot + H-\overset{\displaystyle :\ddot{Cl}:}{\underset{\displaystyle H}{\overset{|}{\underset{|}{C}}}}-\ddot{\underset{..}{Cl}}:$$

The dichloromethane can react with $:\ddot{\underset{..}{Cl}}\cdot$ to produce trichloromethane, $CHCl_3$, and so on. Thus, this *free radical reaction* leads to more than a single product. By varying the relative concentrations of CH_4 and Cl_2, however, it is possible to favor one product over another.

An alkane in which hydrogen atoms are replaced by halogen atoms is called an *alkyl halide* or a *haloalkane*. Alkyl halides are named by prefixing the name of the parent alkane to designate the attached halogen atoms. For example, the molecule

$$Br-\overset{\displaystyle H}{\underset{\displaystyle H}{\overset{|}{\underset{|}{C}}}}-H$$

is called bromomethane and the molecule

$$H-\overset{\displaystyle H}{\underset{\displaystyle H}{\overset{|}{\underset{|}{C}}}}-\overset{\displaystyle H}{\underset{\displaystyle H}{\overset{|}{\underset{|}{C}}}}-Cl$$

Chloroethane is also called ethyl chloride and is used as a local anesthetic.

is called chloroethane.

If two hydrogen atoms in ethane are replaced by chlorine atoms, then there are two distinct products:

1,2-Dichloroethane is extensively used as a dry-cleaning fluid.

$$H-\overset{\displaystyle H}{\underset{\displaystyle Cl}{\overset{1|}{\underset{|}{C}}}}-\overset{\displaystyle H}{\underset{\displaystyle Cl}{\overset{2|}{\underset{|}{C}}}}-H \qquad \text{and} \qquad H-\overset{\displaystyle Cl}{\underset{\displaystyle Cl}{\overset{1|}{\underset{|}{C}}}}-\overset{\displaystyle H}{\underset{\displaystyle H}{\overset{2|}{\underset{|}{C}}}}-H$$

1,2-dichloroethane and 1,1-dichloroethane

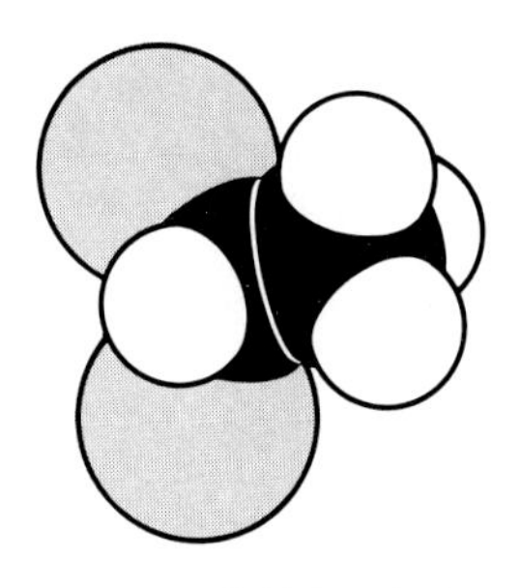

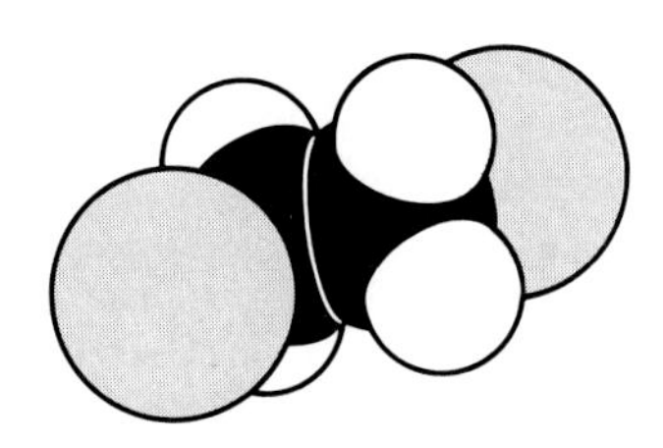

Figure 25-5 Space-filling models of 1,1-dichloroethane and 1,2-dichloroethane, the two structural isomers of dichloroethane.

This is another example of structural isomerism. We must distinguish between these two dichloroethanes. We can do this by numbering the carbon atoms along the alkane chain and designating which carbon atoms have attached chlorine atoms. Here, the dichloroethane shown on the left is called 1,2-dichloroethane and the right-hand one is called 1,1-dichloroethane. Models of these two molecules are shown in Figure 25-5.

Although alkanes react with O_2 and Cl_2, they are relatively inert and are known for their lack of chemical reactivity. For example, some reactions that do *not* occur are

$$\left.\begin{array}{l}\text{alkane} + \text{strong acid } (H_2SO_4) \\ \text{alkane} + \text{strong base } (NaOH) \\ \text{alkane} + \text{oxidizing agent } (KMnO_4) \\ \text{alkane} + \text{reducing agent } (Na)\end{array}\right\}\ \text{no reaction}$$

Example 25-1: Complete and balance the following equations. If there is no reaction, then write N.R.

(a) $C_3H_8(g) + O_2(g) \rightarrow$

(b) $C_3H_8(g) + Cl_2(g) \xrightarrow{\text{dark}}$

(c) $C_3H_8(g) + Cl_2(g) \xrightarrow{\text{UV}}$

Solution:

(a) The equation for the combustion of propane is:

$$C_3H_8(g) + 5O_2(g) \rightarrow 3CO_2(g) + 4H_2O(g)$$

(b) Chlorine and propane do not react in the dark:

$$C_3H_8(g) + Cl_2(g) \xrightarrow{\text{dark}} \text{N.R.}$$

(c) In ultraviolet light, we have

$$C_3H_8(g) + Cl_2(g) \xrightarrow{\text{UV}} C_3H_7Cl(g) + HCl(g)$$

Because this is a free radical reaction, dichloropropane and polychloropropanes are formed also.

25-6. ALKANES AND SUBSTITUTED ALKANES CAN BE NAMED SYSTEMATICALLY ACCORDING TO IUPAC RULES

Structural isomerism leads to an enormous number of alkanes and substituted alkanes. Consequently, it is necessary to have a systematic method of naming alkanes and their derivatives simply and unambiguously. A system of nomenclature for organic molecules has been devised and is used by chemists throughout the world. The system has been recommended by the International Union of Pure and Applied Chemistry (IUPAC).

The *IUPAC nomenclature* rules for alkanes and their derivatives are as follows:

1. For straight-chain alkanes, use the names in Table 25-1 *without* the *n* prefix. Thus, the straight-chain alkane containing eight carbon atoms is called octane.
2. To name a branched or a substituted alkane, first identify the longest chain (the main chain) of consecutive carbon atoms in the molecule. Name this main chain according to rule 1. For example, the main chain in the following molecule has five carbon atoms (shown in color):

$$\begin{array}{ccccccccc} & & \text{Cl} & & \text{CH}_3 & & & & \\ & & | & & | & & & & \\ \text{CH}_3 & — & \text{CH} & — & \text{CH} & — & \text{CH}_2 & — & \text{CH}_3 \end{array}$$

   Because there are five carbon atoms in the main chain, this substituted alkane is named as a substituted pentane, even though the molecule has six carbon atoms in all.
3. Number the carbon atoms in the main chain consecutively, starting at the end that gives the *lowest* numbers to the carbon atoms that have attached groups. For our substituted pentane, we have

$$\begin{array}{ccccccccc} & & \text{Cl} & & \text{CH}_3 & & & & \\ & & | & & | & & & & \\ {}^{1}\text{CH}_3 & — & {}^{2}\text{CH} & — & {}^{3}\text{CH} & — & {}^{4}\text{CH}_2 & — & {}^{5}\text{CH}_3 \end{array}$$

We number the carbon atoms from left to right so that the attached groups are on the lowest-numbered carbon atoms, 2 and 3. If we number the chain from right to left, then the carbon atoms with attached groups would be 3 and 4.

4. Name the groups attached to the main chain according to Table 25-4 and indicate their position along the chain by showing the number of the carbon atom to which they are attached. For example, the substituted alkane we are using as our example is 2-chloro-3-methylpentane. Punctuation is important in assigning IUPAC names. Numbers are separated from letters by hyphens, and the name is written as one word.
5. When two or more different groups are attached to the main chain, list them in alphabetical order. For example, as we saw above,

$$\begin{array}{c} \quad\quad\ \ Cl\quad\ CH_3 \\ CH_3-\overset{|}{C}H-\overset{|}{C}H-CH_2-CH_3 \end{array}$$

is called 2-chloro-3-methylpentane, whereas

$$\begin{array}{c} \quad\quad\ \ CH_3\ \ Cl \\ CH_3-\overset{|}{C}H-\overset{|}{C}H-CH_2-CH_3 \end{array}$$

is called 3-chloro-2-methylpentane.

6. When two or more identical groups are attached to the main chain, use prefixes such as *di-*, *tri-*, or *tetra-*. For example,

$$\begin{array}{c} \quad\quad\ \ CH_3\ \ CH_3 \\ {}^{1}CH_3-{}^{2}\overset{|}{C}H-{}^{3}\overset{|}{C}H-{}^{4}CH_2-{}^{5}CH_3 \end{array}$$

is 2,3-dimethylpentane. Note that the numbers are separated by commas. Every attached group must be named and numbered, even if two identical groups are attached to the same carbon atom. For example, the IUPAC name for

$$\begin{array}{c} \quad\quad CH_3\,CH_3 \\ {}^{1}CH_3-{}^{2}\overset{|}{\underset{|}{C}}-{}^{3}\overset{|}{C}H-{}^{4}CH_2-{}^{5}CH_3 \\ CH_3\quad\quad\quad\quad \end{array}$$

is 2,2,3-trimethylpentane.

The assignment of IUPAC names is best learned by example. With experience and practice, you will become proficient in the IUPAC nomenclature.

Table 25-4 The names of some common groups

Group	*Name*[a]
$-CH_3$	methyl
$-CH_2CH_3$	ethyl
$-CH_2CH_2CH_3$	propyl
$CH_3\underset{\vert}{C}HCH_3$	isopropyl
$-F$	fluoro
$-Cl$	chloro
$-Br$	bromo
$-I$	iodo
$-NH_2$	amino
$-NO_2$	nitro

[a]Groups that are derived from alkanes are called *alkyl groups*. The first four groups here are alkyl groups, and they are named by dropping the -ane ending from the name of the alkane and adding -yl.

Example 25-2: Assign IUPAC names to isobutane, isopentane, and neopentane.

Solution: The structural formula for isobutane is

$$\begin{array}{c} \quad CH_3 \\ \quad | \\ CH_3{-}CH{-}CH_3 \end{array}$$

Its main chain is

$$\begin{array}{c} | \\ {}^{1}CH_3{-}{}^{2}CH{-}{}^{3}CH_3 \end{array}$$

and so we name it as a derivative of propane. The IUPAC name is 2-methylpropane.

The structural formula for isopentane is

$$\begin{array}{l} \qquad\;\; CH_3 \\ \qquad\;\; | \\ CH_3{-}CH{-}CH_2{-}CH_3 \end{array}$$

Its main chain is

$$\begin{array}{l} \qquad\quad | \\ {}^{1}CH_3{-}{}^{2}CH{-}{}^{3}CH_2{-}{}^{4}CH_3 \end{array}$$

and so we name it as a derivative of butane. The IUPAC name is 2-methylbutane.

The structural formula for neopentane is

$$\begin{array}{c} CH_3 \\ | \\ CH_3{-}C{-}CH_3 \\ | \\ CH_3 \end{array}$$

Its main chain is

$$\begin{array}{c} | \\ {}^{1}CH_3{-}{}^{2}C{-}{}^{3}CH_3 \\ | \end{array}$$

and so we name it as a derivative of propane. The IUPAC name is 2,2-dimethylpropane.

Example 25-3: Assign a IUPAC name to

$$\begin{array}{l} \qquad\qquad\quad Cl \quad\;\; Br \quad\; CH_3 \\ \qquad\qquad\quad | \qquad\; | \qquad\; | \\ CH_3{-}CH{-}CH{-}CH{-}CH{-}CH_2{-}CH_3 \\ \qquad\quad | \\ \qquad\; CH_3 \end{array}$$

Solution: The longest chain of consecutive carbon atoms has seven carbon atoms

$$^{1}CH_3—^{2}CH—^{3}CH—^{4}CH—^{5}CH—^{6}CH_2—^{7}CH_3$$

and so we name this molecule as a derivative of heptane. We list the groups in alphabetical order and write

4-bromo-3-chloro-2,5-dimethylheptane

Example 25-4: Write the IUPAC name for

$$\begin{array}{c} H \\ | \\ CH_3—C—CH_3 \\ | \\ CH_2 \\ | \\ CH_2 \\ | \\ CH_3 \end{array}$$

Solution: If you think that the main chain consists of three carbon atoms, then you are making an error made by many beginners. The main chain consists of *five* carbon atoms:

$$\begin{array}{c} H \\ | \\ ^{1}CH_3—^{2}C— \\ | \\ ^{3}CH_2 \\ | \\ ^{4}CH_2 \\ | \\ ^{5}CH_3 \end{array}$$

The IUPAC name is 2-methylpentane.

Problems 25-7 through 25-16 involve the nomenclature of alkanes and related compounds.

Example 25-5: Using Table 25-4, draw structural formulas for (a) 1-chloro-2-methylbutane and (b) 2,4-dimethyl-3-nitropentane.

Solution: (a) The parent alkane is butane. The name indicates that there is a chlorine atom bonded to the first carbon atom and a methyl group bonded to the second carbon atom in the butane chain. Thus, the structural formula is

$$\begin{array}{ll} H_2C—CH—CH_2CH_3 \\ \;\;|\quad\;\;| \\ \;Cl\;\; CH_3 \end{array}$$

(b) The parent alkane is pentane. The name indicates that there are two methyl groups, one attached to the second carbon atom and one to the fourth, and an $—NO_2$ group attached to the third carbon atom in the pentane chain. The formula is

$$\begin{array}{c} \quad CH_3 \quad CH_3 \\ \quad | \qquad | \\ CH_3CHCHCHCH_3 \\ | \\ NO_2 \end{array}$$

25-7. HYDROCARBONS THAT CONTAIN DOUBLE BONDS ARE CALLED ALKENES

All the hydrocarbons that we have discussed so far have been saturated. Each carbon atom has been bonded to four atoms. There is another class of hydrocarbons called *unsaturated hydrocarbons,* in which not all the carbon atoms are bonded to four other atoms. These molecules necessarily contain double or triple bonds. Unsaturated hydrocarbons that contain one or more double bonds are called *alkenes.* The simplest alkene is called ethene, or more commonly, ethylene:

Ethylene is the starting material for polyethylene (Interchapter N).

$$\begin{array}{ccc} H & & H \\ & C{=}C & \\ H & & H \end{array}$$

We learned in Section 12-5 that the bonding in ethene is described by sp^2 orbitals on each carbon atom (Figure 12-14). The double bond consists of a σ bond and a π bond. The σ bond results from the combination of two sp^2 orbitals, one from each carbon atom, and the π bond results from the combination of two p orbitals, also one from each carbon atom. The π orbital maintains the σ-bond framework in a planar shape and prevents rotation about the double bond. Consequently, all six atoms in an ethylene molecule lie in one plane (Figure 12-10) and there are *cis* and *trans* isomers of 1,2-dichloroethene:

$$\begin{array}{ccc} Cl & & Cl \\ & C{=}C & \\ H & & H \end{array} \qquad \begin{array}{ccc} Cl & & H \\ & C{=}C & \\ H & & Cl \end{array}$$

cis *trans*

Alkenes are named by identifying the longest chain of carbon atoms, dropping the final -ane from the name of the corresponding alkane, and then adding the ending -ene. Thus, we have

$$\underset{\text{ethene}}{\mathrm{H_2C{=}CH_2}} \qquad \underset{\text{propene}}{\mathrm{H_2C{=}CH(CH_3)}}$$

If there is more than one possible position of a double bond in an alkene, then that position must be designated. Consider butene, an alkene that contains four carbon atoms. There are two positions for the double bond in the chain, and so we have

$$\underset{\text{1-butene}}{\mathrm{H_2C{=}CH(CH_2CH_3)}} \qquad \underset{\text{2-butene}}{\mathrm{H_3C(H)C{=}C(H)CH_3}}$$

The digits in the names of these two structural isomers of butene designate the carbon atom after which the double bond occurs.

The planar >C=C< portion of each of these molecules is shaded.

This emphasizes that the name 2-butene is ambiguous because of *cis-trans* isomerism. The *cis-trans* isomers of 2-butene are

$$\underset{\substack{cis\text{-2-butene} \\ (\text{mp } -139^\circ\text{C})}}{\mathrm{H_3C(H)C{=}C(H)CH_3}} \qquad \underset{\substack{trans\text{-2-butene} \\ (\text{mp } -106^\circ\text{C})}}{\mathrm{H(H_3C)C{=}C(H)CH_3}}$$

Molecular models of the three isomers of butene are shown in Figure 25-6. An important feature of a carbon-carbon double bond is the planar geometry it imposes in the region around it.

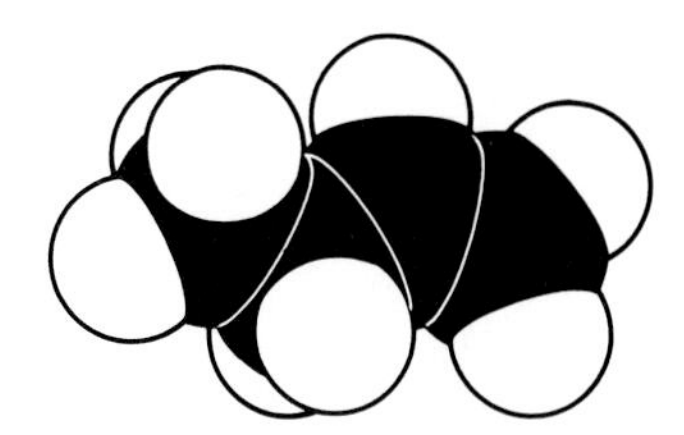
1-butene

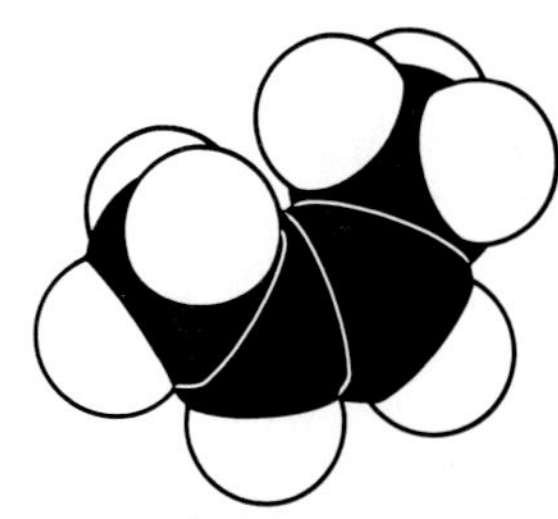
cis-2-butene

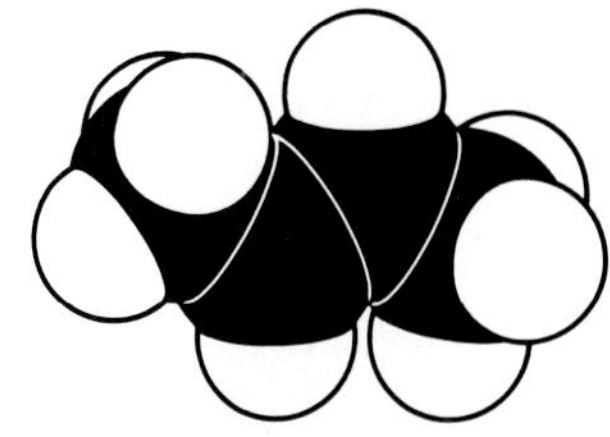
trans-2-butene

Figure 25-6 Space-filling models of 1-butene, *cis*-2-butene, and *trans*-2-butene. It is the lack of rotation about the carbon-carbon double bond that leads to the occurrence of *cis* and *trans* isomers for 2-butene. Note that the four carbon atoms in *cis*-2-butene and *trans*-2-butene lie in the same plane.

Example 25-6: Discuss the shape of a propene molecule, whose structural formula is

$$\mathrm{H_2C{=}CH(CH_3)}$$

Solution: The >C=C< portion of propene is planar, and so all three carbon atoms and the three hydrogen atoms attached to the double-bonded carbon atoms lie in one plane, as the shaded region shows:

$$\begin{matrix} H & & CH_3 \\ & C{=}C & \\ H & & H \end{matrix}$$

The methyl group can rotate about the carbon-carbon single bond. A molecular model of propene is shown in Figure 25-7.

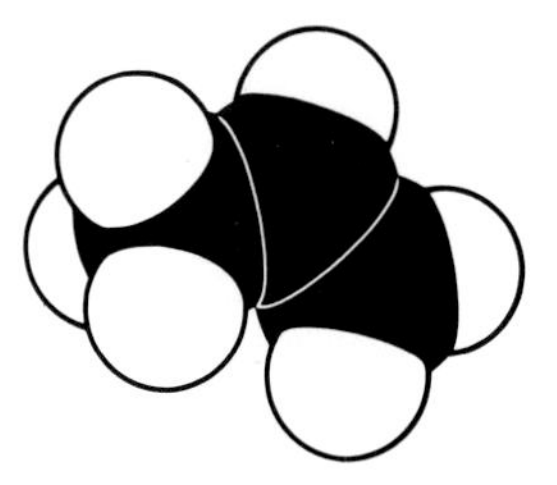

Figure 25-7 Space-filling model of propene. Note that all three carbon atoms and the three hydrogen atoms bonded to the double-bonded carbon atoms lie in the same plane. The segment $\rangle C{=}C\langle$ gives rise to planar regions in molecules.

The IUPAC nomenclature for alkenes uses the longest chain of consecutive carbon atoms *containing the double bond* as the parent compound. The parent compound is named by dropping the -ane and adding -ene to the name of the corresponding alkane and using a number to designate the carbon atom preceding the double bond. For example, the IUPAC name for

$$H_2C{=}CH{-}\underset{}{\overset{CH_3}{\overset{|}{C}}}H{-}CH_2CH_3$$

is 3-methyl-1-pentene. Two other examples are

$$^5CH_3{-}^4CH_2{-}^3\underset{CH_3}{\underset{|}{\overset{CH_3}{\overset{|}{C}}}}{-}^2CH{=}^1CH_2 \qquad ^4CH_3{}^3CH_2{}^2\overset{CH_3}{\overset{|}{C}}{=}^1CH_2$$

3,3-dimethyl-1-pentene 2-methyl-1-butene

25-8. ALKENES UNDERGO ADDITION REACTIONS

Alkenes are more reactive than alkanes because the carbon-carbon double bond provides a reactive center in the molecule. In a sense, the double bond has "extra" electrons available for reaction. In addition to the combustion and substitution reactions that alkanes undergo, alkenes undergo *addition reactions.* Some examples of addition reactions are

1. Addition of hydrogen (*hydrogenation*):

$$\begin{matrix} H_3C & & H \\ & C{=}C & \\ H & & H \end{matrix}(g) + H_2(g) \xrightarrow[\text{high pressure}]{\text{catalyst}} CH_3{-}\underset{H}{\underset{|}{\overset{H}{\overset{|}{C}}}}{-}\underset{H}{\underset{|}{\overset{H}{\overset{|}{C}}}}{-}H(g)$$

This reaction requires a catalyst and high pressure and temperature. Usually powdered nickel or platinum is used as the catalyst. "Hydrogenated vegetable oils" are made by hydrogenating the double bonds in vegetable oils. This hydrogenation makes vegetable oils solid at room temperature.

2. Addition of chlorine or bromine:

$$\begin{matrix} H_3C & & H \\ & C{=}C & \\ H & & H \end{matrix}(g) + Br_2(l) \rightarrow CH_3{-}\underset{Br}{\overset{H}{C}}{-}\underset{Br}{\overset{H}{C}}{-}H(l)$$

This reaction can be carried out either with pure chlorine or bromine or by dissolving the halogen in some solvent, such as carbon tetrachloride. The addition reaction with bromine is a useful qualitative test for the presence of double bonds. A solution of bromine in carbon tetrachloride is red, whereas alkenes and bromoalkanes are usually colorless. As the bromine adds to the double bond, the red color disappears, giving a simple test for the presence of double bonds.

3. Addition of hydrogen chloride:

$$\begin{matrix} H_3C & & H \\ & C{=}C & \\ H & & H \end{matrix}(g) + HCl(g) \rightarrow CH_3{-}\underset{Cl}{CH}{-}\underset{H}{CH_2}(g) \quad \text{sole product}$$

$$\rightarrow CH_3{-}\underset{H}{CH}{-}\underset{Cl}{CH_2}(g) \quad \text{none produced}$$

Although two different products might seem possible in this reaction, only one is found. There is a simple rule for determining which product is produced. *Markovnikov's rule* states that, when HX adds to an alkene, the hydrogen atom attaches to the carbon atom in the double bond already bearing the larger number of hydrogen atoms. More succinctly, the rich get richer.

Example 25-7: Use Markovnikov's rule to predict the product of the reaction

$$\begin{matrix} H_3C & & CH_3 \\ & C{=}C & \\ H_3C & & H \end{matrix} + HCl \rightarrow$$

Solution: According to Markovnikov's rule, the H of HCl will end up on the carbon atom of the double bond that already has the greater number of hydrogen atoms. This gives the sole product

$$CH_3{-}\underset{Cl}{\overset{CH_3}{C}}{-}\underset{H}{\overset{H}{C}}{-}CH_3$$

4. Addition of water. In the presence of acid, water adds to the more reactive alkenes:

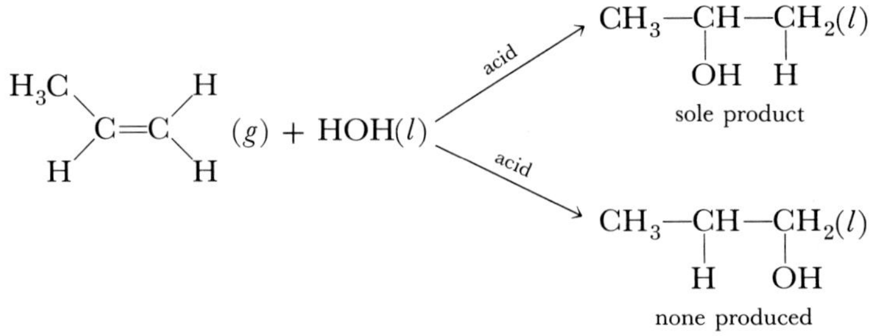

$$\begin{matrix} H_3C & & H \\ & C{=}C & \\ H & & H \end{matrix}(g) + HOH(l) \xrightarrow{acid} \underset{\text{sole product}}{CH_3{-}\underset{\large OH}{\underset{|}{C}}H{-}\underset{\large H}{\underset{|}{C}}H_2(l)}$$

$$\begin{matrix} H_3C & & H \\ & C{=}C & \\ H & & H \end{matrix}(g) + HOH(l) \xrightarrow{acid} \underset{\text{none produced}}{CH_3{-}\underset{\large H}{\underset{|}{C}}H{-}\underset{\large OH}{\underset{|}{C}}H_2(l)}$$

Note that the addition of water to an alkene obeys Markovnikov's rule. Simply picture the water as H—OH.

Example 25-8: Write structural formulas for and assign IUPAC names to the products of the following reactions:
(a) addition of Br_2 to *cis*-2-butene
(b) addition of HBr to 1-butene

Solution:
(a)

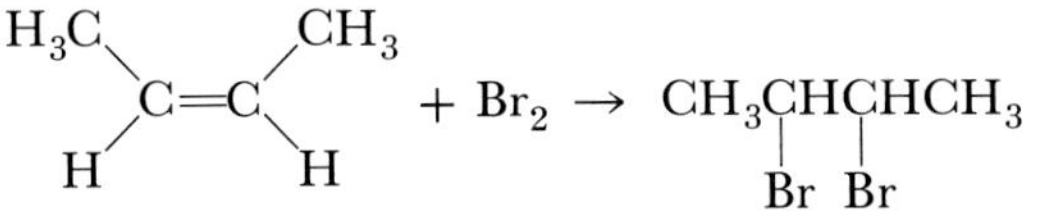

$$\begin{matrix} H_3C & & CH_3 \\ & C{=}C & \\ H & & H \end{matrix} + Br_2 \rightarrow \underset{\text{2,3-dibromobutane}}{CH_3\underset{\large Br}{\underset{|}{C}}H\underset{\large Br}{\underset{|}{C}}HCH_3}$$

Problems 25-19 through 25-28 involve the reactions of alkenes.

(b) We must use Markovnikov's rule in this case:

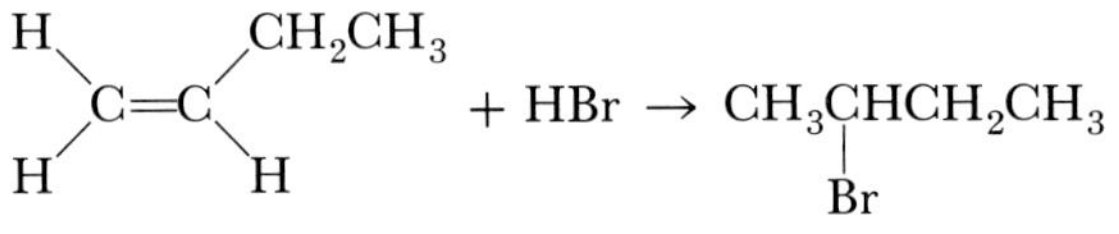

$$\begin{matrix} H & & CH_2CH_3 \\ & C{=}C & \\ H & & H \end{matrix} + HBr \rightarrow \underset{\text{2-bromobutane}}{CH_3\underset{\large Br}{\underset{|}{C}}HCH_2CH_3}$$

25-9. ALCOHOLS ARE ORGANIC COMPOUNDS THAT CONTAIN AN —OH GROUP

The class of organic compounds called *alcohols* are characterized by an —OH group attached to a hydrocarbon chain. Some important common alcohols are

CH_3OH	CH_3CH_2OH	$CH_3\underset{\large OH}{\underset{\mid}{C}}HCH_3$
methanol (methyl alcohol)	ethanol (ethyl alcohol)	2-propanol (isopropyl alcohol)

These alcohols have both common names (in parentheses) and IUPAC names. The common names are formed by naming the alkyl group to which the —OH is attached and adding the word alcohol. The IUPAC name is formed by dropping the -e from the end of the alkane name of the longest chain of carbon atoms containing the —OH group and adding the suffix -ol. Thus, we have methanol, ethanol, and propanol. If the —OH group can be attached at more than one position, then its position is denoted by a number, as in

$$\underset{\text{1-propanol}}{{}^{3}CH_3{}^{2}CH_2{}^{1}CH_2OH} \qquad \underset{\text{2-propanol}}{{}^{3}CH_3{}^{2}\underset{|}{\overset{}{C}}\!\!\underset{OH}{}H^{1}CH_3}$$

Problems 25-31 and 25-32 develop the nomenclature of alcohols.

The bonding in methanol is described in Figure 25-8. This alcohol sometimes is called wood alcohol because it can be produced from wood. Over 8 billion pounds of methanol are produced annually in the United States by the reaction

$$CO(g) + 2H_2(g) \xrightarrow[\text{high pressure}]{\text{catalyst}} CH_3OH(l)$$

Methanol is highly toxic and can cause blindness and death if taken internally. During Prohibition in the United States, many people died or became seriously ill from drinking methyl alcohol, either because they were not aware of the difference between methanol and

All alcohols are toxic. The least toxic is ethyl alcohol.

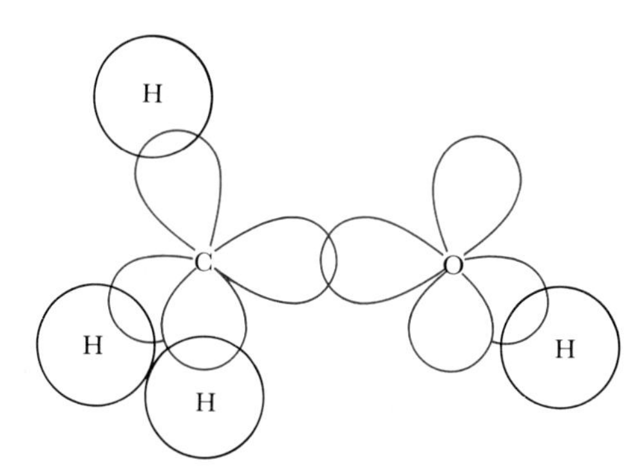

Figure 25-8 A schematic representation of the bonding orbitals in methanol. The bonding can be described in terms of sp^3 orbitals on both the carbon atom and the oxygen atom. The carbon-oxygen bond orbital results from the combination of an sp^3 orbital from each atom. The carbon-hydrogen and oxygen-hydrogen bonds result from the combination of sp^3 and hydrogen $1s$ orbitals. Note that there are five bond orbitals in CH_3OH. Ten of the fourteen valence electrons occupy the five bond orbitals, and the other four occupy the two remaining sp^3 orbitals as lone pairs on the oxygen atom.

ethanol or because the alcohol they purchased contained methanol as a major impurity.

Ethanol is an ingredient in all fermented beverages. Various sugars and the starch in potatoes, grains, and similar substances can be used to produce ethanol by fermentation, a process in which yeast is used to convert sugar to ethanol. Although ethanol may be best known as an ingredient in alcoholic beverages, it is an important industrial chemical as well. An aqueous solution of 70% by volume 2-propanol (isopropyl alcohol) is sold as rubbing alcohol.

25-10. HYDROCARBONS THAT CONTAIN A TRIPLE BOND ARE CALLED ALKYNES

The simplest *alkyne* is acetylene, C_2H_2,

$$\mathrm{H{-}C{\equiv}C{-}H}$$

Cave explorer with a calcium carbide lamp on his helmet.

Acetylene is a colorless gas with a penetrating odor. One of its most important uses is in oxyacetylene torches. We learned in Section 12-8 that acetylene is a linear molecule whose bonding can be described in terms of *sp* hybrid orbitals on the carbon atoms (Figure 12-19).

Acetylene can be produced by the reaction of calcium carbide and water:

$$\mathrm{CaC_2}(s) + 2\mathrm{H_2O}(l) \rightarrow \mathrm{Ca(OH)_2}(s) + \mathrm{C_2H_2}(g)$$

This reaction is used by spelunkers as a light source in caves. Acetylene is produced by allowing H_2O to drop slowly onto CaC_2 in a canister. The C_2H_2 pressure builds up, leaks out of the canister through a nozzle, and is burned according to the reaction

$$2\mathrm{C_2H_2}(g) + 5\mathrm{O_2}(air) \rightarrow 4\mathrm{CO_2}(g) + 2\mathrm{H_2O}(g)$$

In addition to the combustion reaction that all hydrocarbons undergo, acetylene and related hydrocarbons undergo the same type of addition reactions as alkenes.

Example 25-9: Predict the product in the reaction

$$\mathrm{H{-}C{\equiv}C{-}H} + 2\mathrm{HCl} \rightarrow$$

Solution: This reaction can be broken down into two steps. The first step is

$$\mathrm{H{-}C{\equiv}C{-}H} + \mathrm{HCl} \rightarrow \begin{matrix} \mathrm{Cl} & & \mathrm{H} \\ & \mathrm{C{=}C} & \\ \mathrm{H} & & \mathrm{H} \end{matrix}$$

The second step is

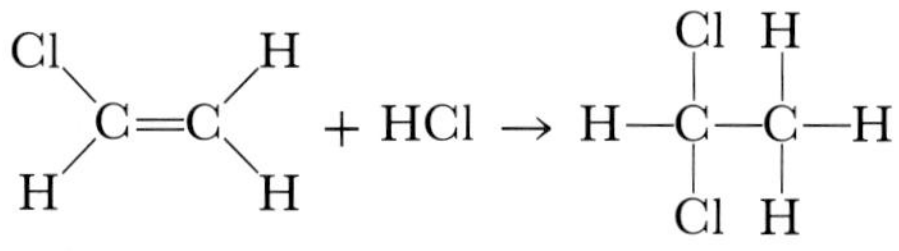

$$\underset{H}{\overset{Cl}{}}\!\!\!\!\!\!\;\;C{=}C\;\underset{H}{\overset{H}{}} + HCl \rightarrow H{-}\underset{Cl}{\overset{Cl}{C}}{-}\underset{H}{\overset{H}{C}}{-}H$$

Note that we have used Markovnikov's rule to predict the product in the second step. The sole product is 1,1-dichloroethane.

The IUPAC nomenclature for alkynes uses the longest chain of consecutive carbon atoms *containing the triple bond* as the parent compound. The parent compound is named by dropping the -ane and adding -yne to the name of the corresponding alkane, and using a number to designate the carbon atom preceding the triple bond. (Compare with the nomenclature of alkenes given in Section 25-7.) For example, the name of

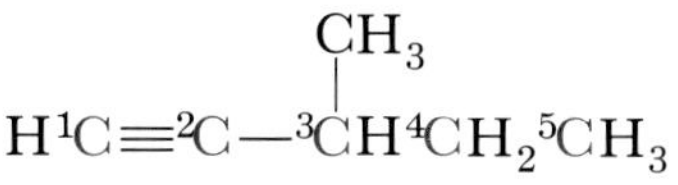

$$H^1C{\equiv}^2C{-}^3\overset{CH_3}{\overset{|}{C}}H^4CH_2{}^5CH_3$$

is 3-methyl-1-pentyne. Some other examples are

$^4CH_2Cl{-}^3CH_2{-}^2C{\equiv}^1CH$	4-chloro-1-butyne
$^5CH_3{-}^4CH_2{-}^3C{\equiv}^2C{-}^1CH_2Cl$	1-chloro-2-pentyne

Problems 25-39 and 25-40 involve the nomenclature of alkynes.

25-11. BENZENE BELONGS TO A CLASS OF HYDROCARBONS CALLED AROMATIC HYDROCARBONS

Benzene is a colorless, poisonous, highly flammable liquid (bp 80.1°C) with a characteristic odor. We learned in Section 12-9 that the π electrons in benzene are delocalized (Figure 12-22). Recall that the two principal resonance forms of benzene are

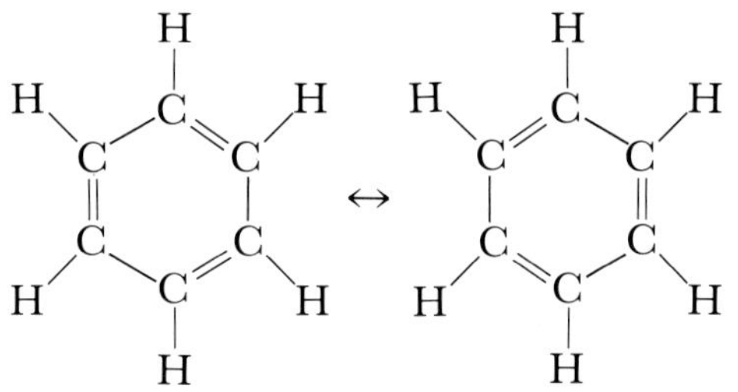

[Two Kekulé structures of benzene, C_6H_6, each carbon bearing one H, connected by ↔]

whose superposition gives

H
H C H
C C
C C
H C H
H

or, more compactly,

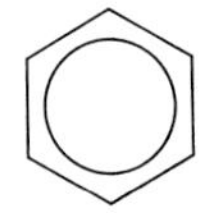

Although each resonance form of benzene shows double bonds, their superposition does not. Benzene does not have double bonds and does not react as an unsaturated hydrocarbon. In fact, the π-electron delocalization causes the ring to be so stable that most of the reactions that benzene undergoes are substitution reactions in which the hydrogen atoms on the ring are replaced by other atoms or groups. For example, the usual reaction of benzene with bromine is a substitution reaction:

$$C_6H_6(l) + Br_2(l) \xrightarrow{FeBr_3} C_6H_5Br(l) + HBr(g)$$

The iron(III) bromide is a catalyst for this reaction. The fact that only one monobromobenzene has ever been isolated indicates that all the hydrogen atoms in benzene are equivalent.

Example 25-10: Draw structural formulas for all the isomers of dibromobenzene.

Solution: Because the benzene ring is a regular hexagon, there are three isomers of dibromobenzene:

Br Br Br Br Br Br

I II III

We can name these compounds by numbering the carbon atoms:

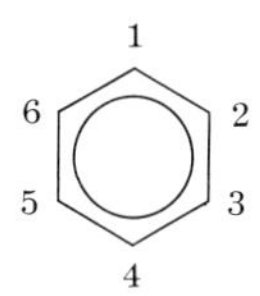

Therefore, we have

compound I	1,2-dibromobenzene
compound II	1,3-dibromobenzene
compound III	1,4-dibromobenzene

There is a less systematic but more common way to designate the positions of the bromine atoms in the three disubstituted benzenes in Example 25-10. Substituents at the 1,2 positions are designated *ortho-* (*o*-), those at the 1,3 positions are designated *meta-* (*m*-), and those at the 1,4 positions are designated *para-* (*p*-). Thus we write

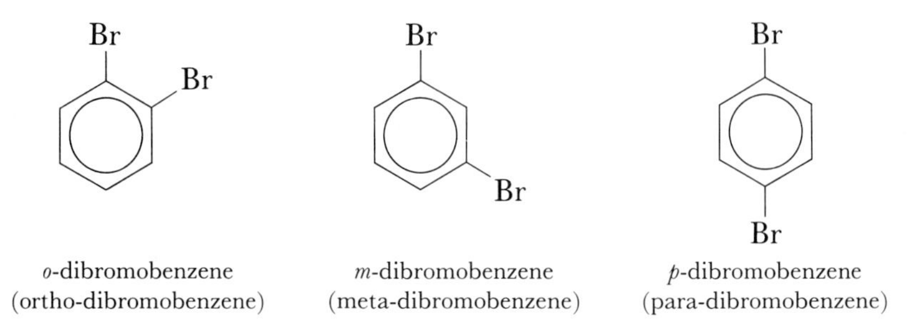

o-dibromobenzene (ortho-dibromobenzene) | *m*-dibromobenzene (meta-dibromobenzene) | *p*-dibromobenzene (para-dibromobenzene)

As another example of this nomenclature, consider xylene, which is a benzene ring with two substituent methyl groups. Xylene is used as a solvent and as a starting material for the production of many other compounds. The three isomers of xylene are

CH_3 CH_3 | CH_3 CH_3 | CH_3 CH_3

o-xylene | *m*-xylene | *p*-xylene

Benzene belongs to the class of hydrocarbons called *aromatic hydrocarbons*. Aromatic hydrocarbons have relatively stable rings as a result of π-electron delocalization. Benzene and many other aromatic hydrocarbons are obtained from petroleum and coal tar. Almost 10 billion pounds of benzene are produced annually in the United

States, making it the fifteenth ranked industrial chemical. Benzene is used in the manufacture of medicinal chemicals, dyes, plastics, varnishes, lacquers, linoleum, and many other products. Some important derivatives of benzene are

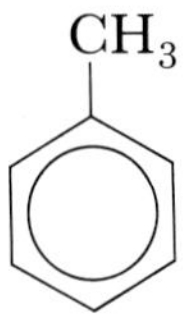

CH_3 toluene

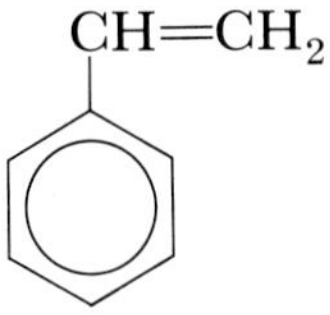

$CH{=}CH_2$ styrene

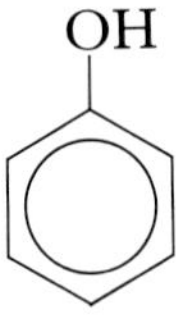

OH phenol (carbolic acid)

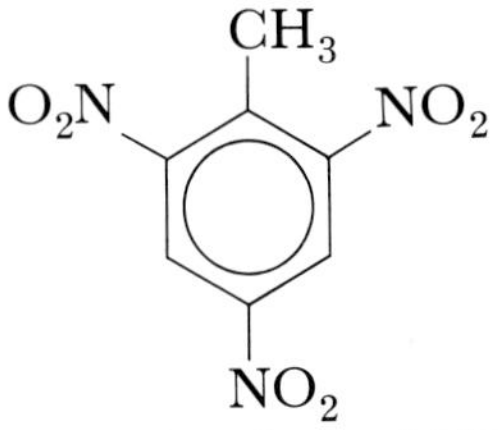

CH_3, O_2N, NO_2, NO_2 2,4,6-trinitrotoluene (TNT)

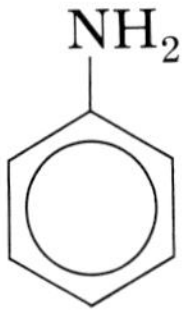

NH_2 aniline

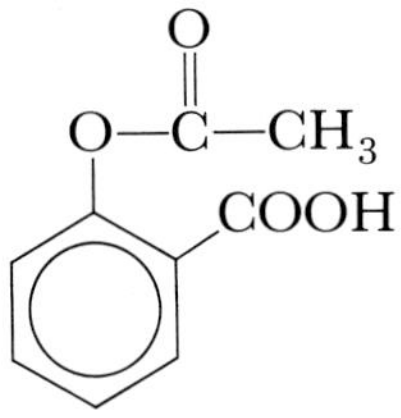

$O{-}\overset{O}{\overset{\|}{C}}{-}CH_3$, COOH acetylsalicylic acid (aspirin)

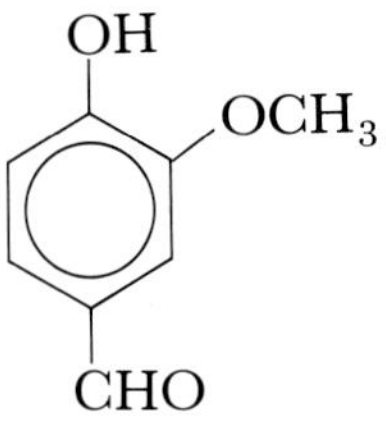

OH, OCH_3, CHO vanillin

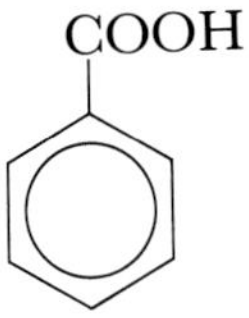

COOH benzoic acid

25-12. ALDEHYDES AND KETONES CONTAIN A CARBON-OXYGEN DOUBLE BOND

An important class of compounds that have a carbon-oxygen double bond has the general formula RCHO,

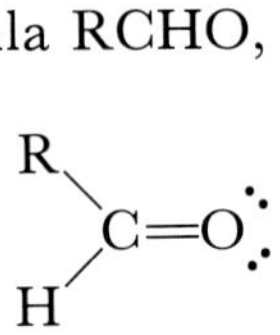

$$\begin{matrix} R \\ & C{=}\ddot{O}: \\ H \end{matrix}$$

where R is a hydrogen atom or an alkyl group. The compounds that have this general formula are called *aldehydes,* and the —CHO group is called the *aldehyde group.* The two simplest aldehydes are

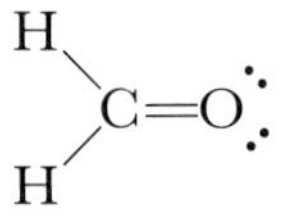

$HCHO$
formaldehyde

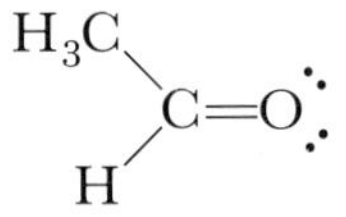

CH_3CHO
acetaldehyde

Problems 25-47 and 25-48 develop the IUPAC nomenclature of aldehydes.

Formaldehyde is a gas with an offensive and characteristic odor. It is one of the irritants in photochemical smog. Approximately 1 million tons of formaldehyde are manufactured annually in the United States. Most of this is used in the production of plastic materials such as Melmac and Bakelite. Melmac is used in plastic dinnerware, and Bakelite is used in the plastic casing of radios and telephones. A solution of formaldehyde in water is called formalin and is used to preserve biological specimens.

Acetaldehyde is widely used in the plastics and rubber industries. An interesting simple derivative of acetaldehyde is chloral (trichloroacetaldehyde),

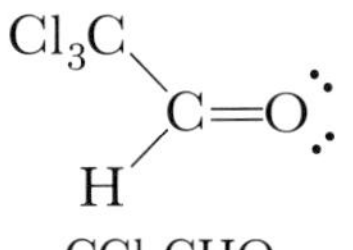

CCl_3CHO
chloral

Chloral forms a crystalline hydrate that is used in veterinary medicine as an anesthetic for horses, cattle, and poultry.

A Mickey Finn, containing what are often referred to as knock-out drops in old movies, is a solution of ethyl alcohol in chloral hydrate.

Example 25-11: Which atoms in an acetaldehyde molecule, CH_3CHO, lie in the same plane?

Solution: The σ-bond framework of acetaldehyde is similar to that of formaldehyde. The σ bond orbitals in acetaldehyde are

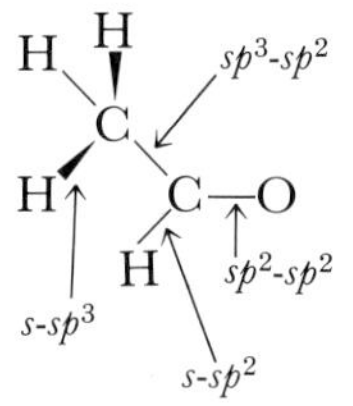

In addition to the σ carbon-oxygen bond, the $2p$ orbitals on these two atoms overlap to give a π bond orbital. There are seven bond orbitals in acetaldehyde: six σ bond orbitals and one π bond orbital. Fourteen of the eighteen valence electrons in acetaldehyde occupy the seven bond orbitals. The other four occupy the two oxygen sp^2 orbitals, constituting two lone electron pairs on the oxygen atom. The planar region induced by the π bond is shown by the shaded region:

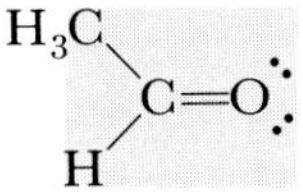

$$\begin{array}{l} H_3C \\ \quad\;\; \diagdown \\ \quad\;\;\; C{=}\ddot{O}: \\ \quad\;\; \diagup \\ H \end{array}$$

The oxidation state of the carbon atom in CH_3OH is -2, and the oxidation state of the carbon atom in HCHO is 0.

Aldehydes are obtained from the oxidation of *primary alcohols,* which are alcohols in which the —OH-bearing carbon atom is bonded to only one other carbon atom. Examples of primary alcohols are

CH_3CH_2OH	ethanol
$CH_3CH_2CH_2OH$	1-propanol
$CH_3CH_2CH_2CH_2OH$	1-butanol

A commonly used oxidizing agent for the oxidation of primary alcohols is $K_2Cr_2O_7$ dissolved in an aqueous acidic solution:

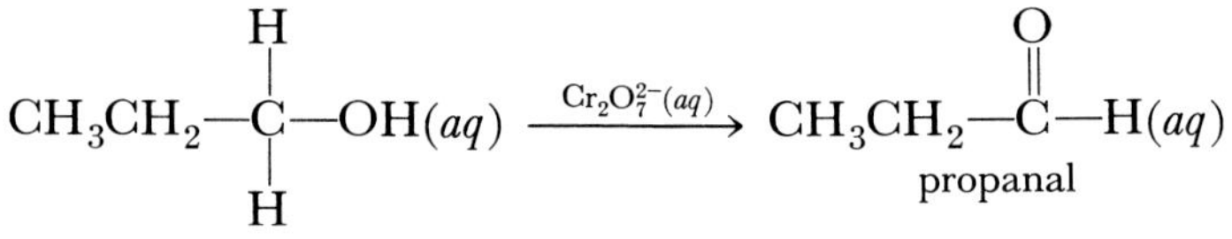

$$\underset{\text{1-propanol}}{CH_3CH_2{-}\overset{\displaystyle H}{\underset{\displaystyle H}{\overset{|}{\underset{|}{C}}}}{-}OH(aq)} \xrightarrow{Cr_2O_7^{2-}(aq)} \underset{\text{propanal}}{CH_3CH_2{-}\overset{\displaystyle O}{\overset{\|}{C}}{-}H(aq)}$$

The systematic nomenclature for aldehydes is developed in Problem 25-47.

Secondary alcohols are alcohols in which the —OH-bearing carbon atom is bonded to two other carbon atoms. Examples of secondary alcohols are

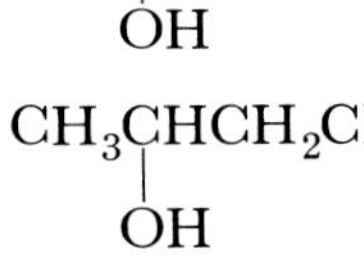

$CH_3\underset{\displaystyle OH}{\underset{|}{C}}HCH_3$	2-propanol
$CH_3\underset{\displaystyle OH}{\underset{|}{C}}HCH_2CH_3$	2-butanol
$CH_3CH_2\underset{\displaystyle OH}{\underset{|}{C}}HCH_2CH_3$	3-pentanol

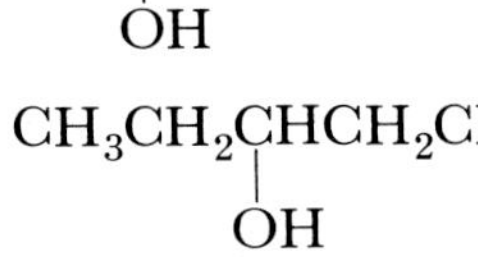

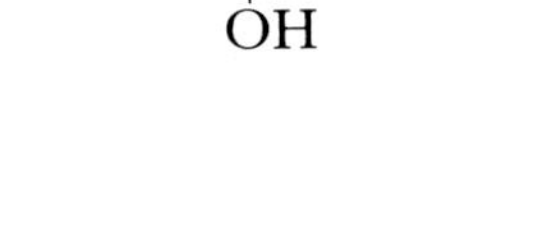

Secondary alcohols can be oxidized to compounds called *ketones,* whose general formula is

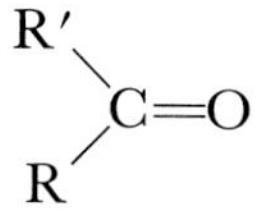

$$\begin{matrix} R' \\ \quad \diagdown \\ \quad\quad C{=}O \\ \quad \diagup \\ R \end{matrix}$$

where R and R′ are alkyl groups. The simplest ketone, acetone, $(CH_3)_2CO$, is prepared from the oxidation of 2-propanol:

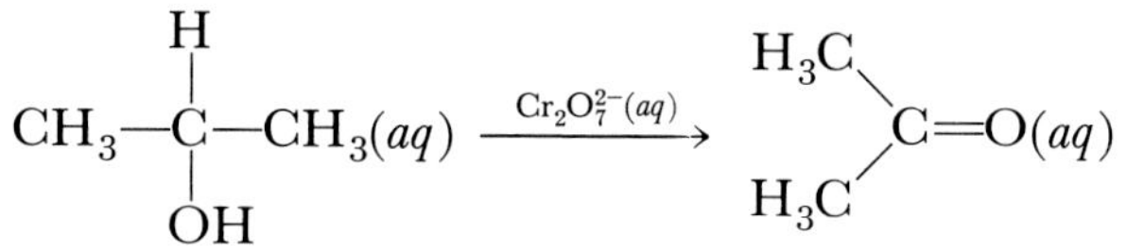

$$\underset{\text{2-propanol}}{CH_3{-}\underset{\underset{OH}{|}}{\overset{\overset{H}{|}}{C}}{-}CH_3(aq)} \xrightarrow{Cr_2O_7^{2-}(aq)} \underset{\text{acetone}}{(H_3C)_2C{=}O(aq)}$$

Tertiary alcohols, which are alcohols in which the —OH-bearing carbon atom is attached to three other carbon atoms, are resistant to oxidation. The —OH-bearing carbon atom is already bonded to three other carbon atoms and so cannot form a double bond with an oxygen atom:

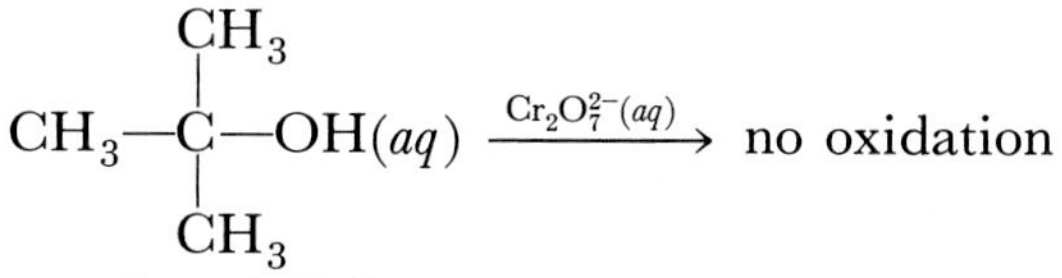

$$\underset{\substack{\text{tertiary alcohol} \\ \text{2-methyl-2-propanol}}}{CH_3{-}\underset{\underset{CH_3}{|}}{\overset{\overset{CH_3}{|}}{C}}{-}OH(aq)} \xrightarrow{Cr_2O_7^{2-}(aq)} \text{no oxidation}$$

Example 25-12: Which alcohol would you use to produce $CH_3CH_2\underset{\underset{O}{\|}}{C}CH_3$?

Solution: The product here is a ketone, and so we must oxidize a secondary alcohol. The secondary alcohol should have a methyl group and an ethyl group attached to the —OH-bearing carbon atom. Thus, we have

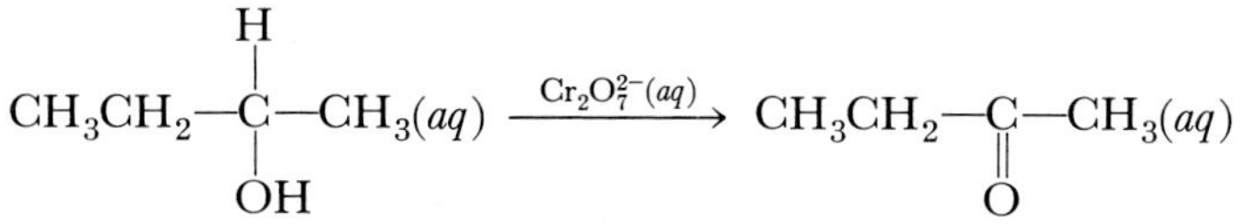

$$\underset{\text{2-butanol}}{CH_3CH_2{-}\underset{\underset{OH}{|}}{\overset{\overset{H}{|}}{C}}{-}CH_3(aq)} \xrightarrow{Cr_2O_7^{2-}(aq)} \underset{\text{methyl ethyl ketone}}{CH_3CH_2{-}\underset{\underset{O}{\|}}{C}{-}CH_3(aq)}$$

Problems 25-49 and 25-50 are similar to Example 25-12.

25-13. ORGANIC CARBOXYLIC ACIDS CONTAIN A —COOH GROUP

alcohol ROH
aldehyde RCHO
ketone $R(R')C{=}O$
carboxylic acid RCOOH

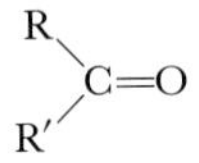

One important use of aldehydes is in the production of *organic acids:*

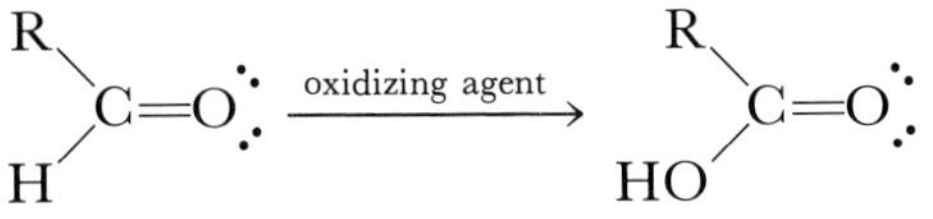

$$\underset{\text{aldehyde}}{R(H)C{=}\ddot{O}\text{:}} \xrightarrow{\text{oxidizing agent}} \underset{\text{organic acid}}{R(HO)C{=}\ddot{O}\text{:}}$$

For example,

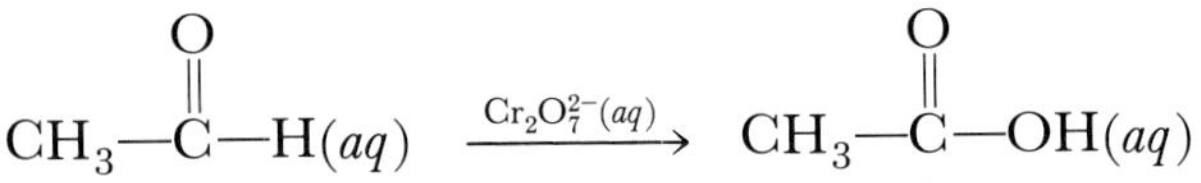

$$\underset{\text{acetaldehyde}}{CH_3{-}\overset{O}{\overset{\|}{C}}{-}H(aq)} \xrightarrow{Cr_2O_7^{2-}(aq)} \underset{\text{acetic acid}}{CH_3{-}\overset{O}{\overset{\|}{C}}{-}OH(aq)}$$

We can also obtain organic acids by the direct oxidation of primary alcohols. In this oxidation, the aldehyde occurs as an intermediate species:

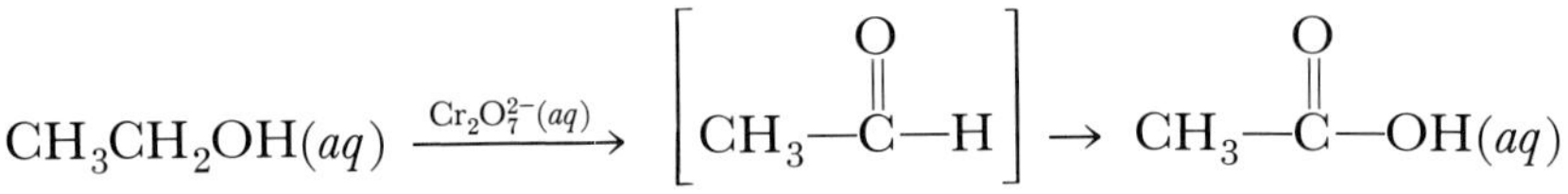

$$CH_3CH_2OH(aq) \xrightarrow{Cr_2O_7^{2-}(aq)} \left[CH_3{-}\overset{O}{\overset{\|}{C}}{-}H\right] \rightarrow CH_3{-}\overset{O}{\overset{\|}{C}}{-}OH(aq)$$

It requires carefully controlled conditions to isolate the aldehyde in good yield from this reaction because it is readily oxidized to the acid.

The general formula for organic carboxylic acids is RCOOH. The —COOH group, called the *carboxyl group,* is characteristic of organic acids, or *carboxylic acids.* The two simplest carboxylic organic acids are formic acid and acetic acid:

Problems 25-51 and 25-52 develop the IUPAC nomenclature of carboxylic acids.

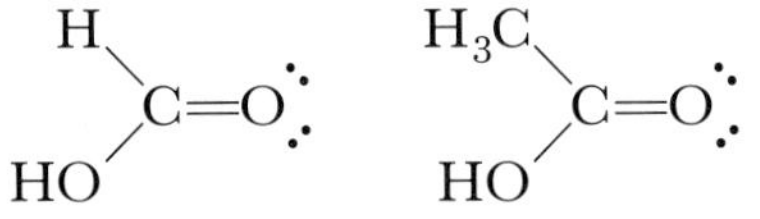

$$\underset{\substack{HCOOH\\ \text{formic acid}}}{H(HO)C{=}\ddot{O}\text{:}} \qquad \underset{\substack{CH_3COOH\\ \text{acetic acid}}}{H_3C(HO)C{=}\ddot{O}\text{:}}$$

Formic acid is one of the irritants in the fluid secreted by ants when they bite. Vinegar is a dilute aqueous solution of acetic acid.

The carboxyl group produces hydronium ions in water:

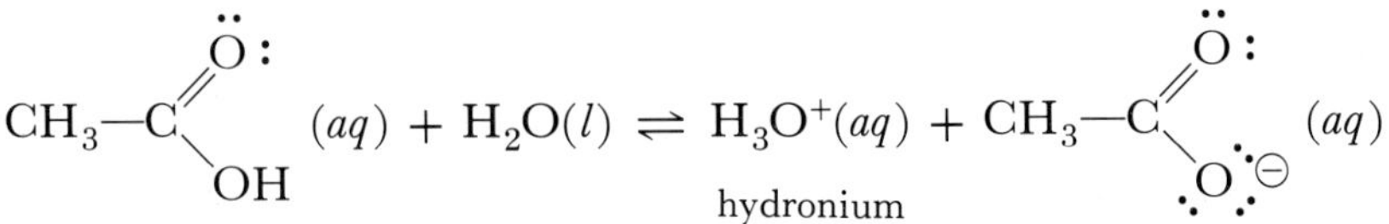

$$\underset{\text{acetic acid}}{CH_3{-}C(\ddot{O}\text{:})(OH)(aq)} + H_2O(l) \rightleftharpoons \underset{\text{hydronium ion}}{H_3O^+(aq)} + \underset{\text{acetate ion}}{CH_3{-}C(\ddot{O}\text{:})(\ddot{O}\text{:}^{\ominus})(aq)}$$

Organic acids also react with bases to produce salts and water:

$$\mathrm{H{-}C({=}\ddot{O}{:})OH} + \mathrm{Na^{+}OH^{-}} \rightleftharpoons \mathrm{Na^{+}H{-}C({=}\ddot{O}{:})\ddot{\underset{\cdot\cdot}{O}}{:}^{\ominus}} + \mathrm{H_2O}$$

formic acid sodium formate

Example 25-13: Complete and balance the following equation and name the product:

$$\mathrm{CH_3COOH}(aq) + \mathrm{Ca(OH)_2}(aq) \rightarrow$$

Solution: Each acetic acid contributes one hydrogen ion, and so 2 mol of acetic acid is required to neutralize 1 mol of $Ca(OH)_2$. The balanced equation is

$$2\mathrm{CH_3COOH}(aq) + \mathrm{Ca(OH)_2}(aq) \rightarrow \mathrm{Ca(CH_3COO)_2}(aq) + 2\mathrm{H_2O}(l)$$

The product is a salt of calcium hydroxide and acetic acid. To name the salt, we change the -ic ending of the acid to -ate to give calcium acetate.

The reaction between organic acids and water or bases occurs because the resulting anion, called a *carboxylate ion*, is stabilized by charge delocalization (Section 12-9). The carboxylate ion is described by the two resonance formulas

$$\mathrm{R{-}C({=}\ddot{O}{:})\ddot{\underset{\cdot\cdot}{O}}{:}^{\ominus}} \leftrightarrow \mathrm{R{-}C(\ddot{\underset{\cdot\cdot}{O}}{:}^{\ominus}){=}\ddot{O}{:}}$$

The superposition of these two formulas gives

$$\left[\mathrm{R{-}C}\begin{matrix}\overset{\cdot\cdot}{O}{:}\\ \underset{\cdot\cdot}{O}{:}\end{matrix}\right]^{-}$$

showing that the negative charge is distributed equally between the two oxygen atoms. The delocalization of the negative charge over the two oxygen atoms confers a degree of stability to a carboxylate ion. The two carbon-oxygen bonds in formic acid have different lengths, but in sodium formate the two carbon-oxygen bond lengths are identical and are intermediate between those of single and double carbon-oxygen bonds:

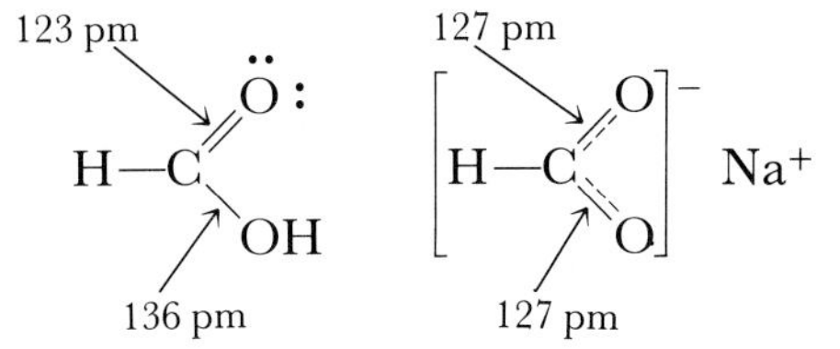

$$\text{H—C}(\text{=}\ddot{\text{O}}\text{:})(\text{—OH}) \quad \text{(C=O: 123 pm; C—OH: 136 pm)}$$

formic acid

$$\left[\text{H—C}(\text{O})(\text{O})\right]^{-}\ \text{Na}^{+} \quad \text{(C—O: 127 pm; C—O: 127 pm)}$$

sodium formate

25-14. ESTERS ARE PRODUCED BY THE REACTION OF ORGANIC ACIDS WITH ALCOHOLS

When an organic acid is heated with an alcohol, the following reaction takes place:

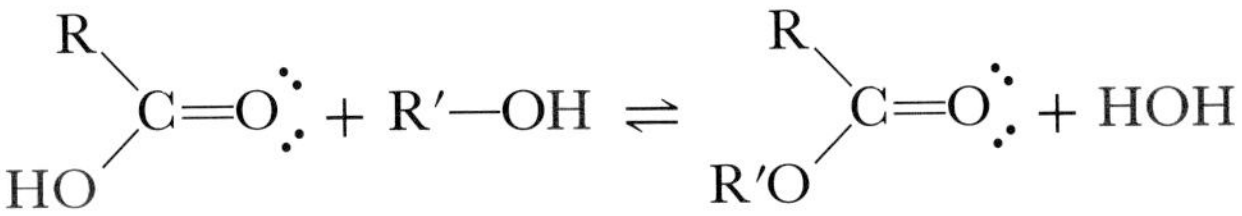

$$\text{R(HO)C=O} + \text{R'—OH} \rightleftharpoons \text{R(R'O)C=O} + \text{HOH}$$

where R and R′ represent possibly different alkyl groups. The coloring used in this reaction emphasizes that the water is formed from the —OH group of the acid and the hydrogen atom of the alcohol. This is quite unlike an acid-base neutralization reaction, in which the water is formed by the reaction between a hydroxide ion from the base and a hydronium ion from the acid. Alcohols do not react like bases. The similarity in their chemical formulas, for example, CH_3OH versus NaOH, is completely superficial.

The product of the reaction between an organic acid and an alcohol is called an *ester*. As the above reaction indicates, the general formula for an ester is

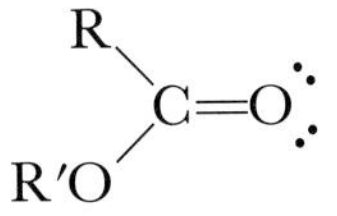

$$\text{R(R'O)C=O}$$

where the R′ group comes from the alcohol. The pleasant odors of flowers and fruits are due to esters. Esters are named by first naming the alkyl group from the alcohol and then designating the acid, with its -ic ending changed to -ate. For example,

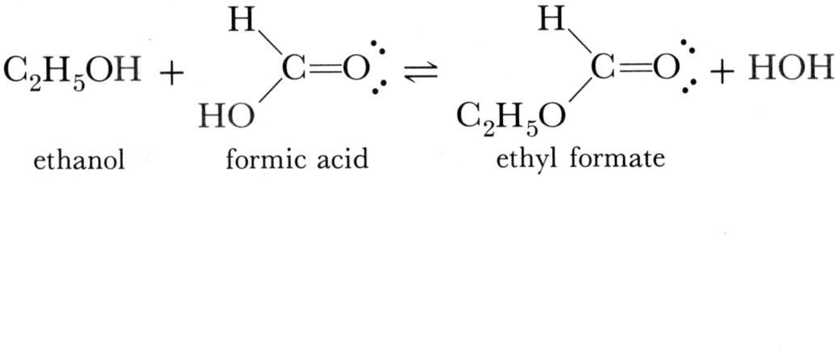

$$\underset{\text{ethanol}}{C_2H_5OH} + \underset{\text{formic acid}}{\text{H(HO)C=O}} \rightleftharpoons \underset{\text{ethyl formate}}{\text{H}(C_2H_5O)\text{C=O}} + \text{HOH}$$

$$C_5H_{11}OH + \begin{array}{l} CH_3 \diagdown \\ \quad\quad C{=}\ddot{\underset{..}{O}} \\ HO \diagup \end{array} \rightleftharpoons \begin{array}{l} \quad\quad CH_3 \diagdown \\ \quad\quad\quad\quad C{=}\ddot{\underset{..}{O}} \\ C_5H_{11}O \diagup \end{array} + HOH$$

1-pentanol acetic acid pentyl acetate

Table 25-5 lists some naturally occurring esters and their odors.

Table 25-5 Some esters and their odors

Name	*Odor*	*Name*	*Odor*
ethyl formate	rum	methyl butyrate	apples
pentyl acetate	bananas	ethyl butyrate	pineapples
octyl acetate	oranges	pentyl butyrate	apricots

Example 25-14: Complete and balance the equation for the reaction between ethanol and acetic acid. Name the product.

Solution: The reaction between an organic acid and an alcohol yields an ester. In this case, the reaction is

$$CH_3CH_2OH + \begin{array}{l} H_3C \diagdown \\ \quad\quad C{=}\ddot{\underset{..}{O}} \\ HO \diagup \end{array} \rightarrow \begin{array}{l} \quad\quad\quad H_3C \diagdown \\ \quad\quad\quad\quad\quad C{=}\ddot{\underset{..}{O}} \\ CH_3CH_2O \diagup \end{array} + H_2O$$

ethanol acetic acid ethyl acetate

SUMMARY

Organic chemistry is the chemistry of compounds that contain carbon atoms. The simplest organic compounds are hydrocarbons, which are compounds that consist of only hydrogen and carbon. Hydrocarbons are divided into alkanes, alkenes, alkynes, and aromatic hydrocarbons. Alkanes have only single bonds, alkenes have one or more double bonds, alkynes have one or more triple bonds, and aromatic hydrocarbons have rings with delocalized electrons. Alkanes, which are also called saturated hydrocarbons, are relatively unreactive and undergo primarily combustion and substitution reactions. Alkenes and alkynes, which are called unsaturated hydrocarbons, undergo the same reactions as saturated hydrocarbons, as well as addition reactions, in which small molecules, such as H_2, Cl_2, Br_2, HCl, and H_2O, add across the double or triple bonds.

Other classes of organic compounds are alcohols (ROH), aldehydes (RCHO), ketones (RCOR′), carboxylic acids (RCOOH), and esters (RCOOR′). Compounds in each class undergo reactions that are characteristic of the class. For example, carboxylic acids and alcohols react to form esters and aldehydes can be oxidized to carboxylic acids.

Because of the great diversity of organic compounds, a systematic nomenclature is necessary. An internationally recognized system (IUPAC) has been developed that allows each organic compound to be named unambiguously.

TERMS YOU SHOULD KNOW

organic compound 981
hydrocarbon 981
alkane 981
structural formula 982
structural isomer 984
saturated hydrocarbon 984
petroleum plantation 986
substitution reaction 987
free radical reaction 988
alkyl halide 988
haloalkane 988
IUPAC nomenclature 990
alkyl group 991
unsaturated hydrocarbon 994
alkene 994
addition reaction 996
hydrogenation 996
Markovnikov's rule 997
alcohol 998
alkyne 1000
ortho 1003
meta 1003
para 1003
aromatic hydrocarbon 1003
aldehyde 1004
aldehyde group 1004
primary alcohol 1006
secondary alcohol 1006
ketone 1007
tertiary alcohol 1007
organic acid 1008
carboxyl group 1008
carboxylic acid 1008
carboxylate ion 1009
ester 1010

PROBLEMS

REACTIONS OF ALKANES

25-1. Complete and balance the following equations (if no reaction, then write N.R.):

(a) $C_4H_{10}(g) + O_2(g) \rightarrow$
(b) $C_6H_{14}(l) + NaOH(aq) \rightarrow$
(c) $C_3H_8(g) + Cl_2(g) \xrightarrow{\text{dark}}$
(d) $C_2H_6(g) + Cl_2(g) \xrightarrow{\text{UV}}$

25-2. Complete and balance the following equations (if no reaction, then write N.R.):

(a) $C_3H_8(g) + H_2SO_4(aq) \rightarrow$
(b) $C_5H_{12}(g) + O_2(g) \rightarrow$
(c) $C_2H_6(g) + HCl(aq) \rightarrow$
(d) $C_5H_{12}(l) + Cl_2(g) \xrightarrow{\text{UV}}$

ISOMERS OF CHLOROALKANES

25-3. Which of the following pairs of molecules are identical and which are different:

(a) $ClCH_2CH_2CH_3$ and $CH_3CH_2CH_2Cl$

(b) $CH_3CH_2\underset{CH_3}{\underset{|}{C}}HCl$ and $CH_3\underset{Cl}{\underset{|}{C}}HCH_2CH_3$

(c) $Cl-CH_2-\overset{CH_3}{\overset{|}{C}}H-CH_3$ and $CH_3-\overset{CH_3}{\overset{|}{\underset{Cl}{\underset{|}{C}}}}-CH_3$

(d) $CH_3-\overset{CH_2Cl}{\overset{|}{C}}H-CH_3$ and $ClCH_2-\overset{CH_3}{\overset{|}{C}}H-CH_3$

25-4. Which of the following pairs of molecules are identical and which are different:

(a) $CH_3CH_2\underset{Cl}{\underset{|}{C}}HCH_3$ and $CH_3\underset{Cl}{\underset{|}{C}}HCH_2CH_3$

(b) $H_2\overset{Cl}{\overset{|}{C}}CH_2CH_3$ and $CH_3CH_2CH_2Cl$

(c) $CH_3\overset{CH_3}{\overset{|}{C}}HCH_2CH_2CH_3$ and $CH_3CH_2\overset{CH_3}{\overset{|}{C}}HCH_2CH_3$

(d) $CH_3-\overset{H}{\overset{|}{\underset{CH_2CH_3}{\underset{|}{C}}}}-Cl$ and $CH_3-\overset{Cl}{\overset{|}{\underset{H}{\underset{|}{C}}}}-CH_2CH_3$

25-5. If you substitute a chlorine atom for a hydrogen atom in *n*-pentane, how many isomers do you get? Give IUPAC names for all these isomers.

25-6. Write structural formulas for all the isomers of dichloropropane. Assign IUPAC names to these isomers.

IUPAC NOMENCLATURE OF ALKANES AND RELATED COMPOUNDS

25-7. Give IUPAC names for the following compounds:

(a) $CH_3-\overset{Cl}{\overset{|}{C}}H-CH_2-CH_3$

(b) $Cl-CH_2-\underset{Cl}{\underset{|}{C}}H-\underset{Cl}{\underset{|}{C}}H-CH_3$

(c) $CH_3CH_2\overset{CH_3}{\overset{|}{\underset{I}{\underset{|}{C}}}}CH_2CH_2CH_3$

(d) $\overset{CH_3}{\overset{|}{C}}H_2-CH_2-CH_2-\overset{CH_3}{\overset{|}{C}}H_2$

25-8. Give IUPAC names for the following compounds:

(a) $CH_3-CH_2-\overset{Cl}{\overset{|}{C}}H-CH_2-CH_3$

(b) $CH_3-\overset{NO_2}{\overset{|}{C}}H-\overset{CH_3}{\overset{|}{\underset{CH_3}{\underset{|}{C}}}}-CH_3$

(c) $CH_3CH_2\underset{Cl}{\underset{|}{C}}HCH_3$

(d) $CH_3-\underset{CH_2CH_3}{\underset{|}{C}}H-CH_3$

25-9. Explain why the following names are incorrect and give a correct IUPAC name in each case:

(a) 4-methylpentane
(b) 2-ethylbutane
(c) 2-propylhexane
(d) 2-dimethylpropane

25-10. Explain why the following names are incorrect and give a correct IUPAC name in each case:

(a) 2,3-dichloropropane
(b) 3-bromo-2-ethylpropane
(c) 1,1,3-trimethylpropane
(d) 2-dimethylbutane

25-11. Draw Lewis formulas for all the isomers of hexane. Assign IUPAC names to these isomers.

25-12. Do 2-chlorobutane and 3-chlorobutane differ? Do 2-chloropentane and 3-chloropentane differ? What about 2-chloropentane and 4-chloropentane?

25-13. Write the Lewis formula for and assign an IUPAC name to each of the following compounds:

(a) $CH_3CH(CH_3)CH_2CH_3$
(b) CH_2ClCH_2Br
(c) $CH_3CCl_2CCl_3$
(d) $CH_3C(CH_3)_2CH_3$

25-14. Write the Lewis formula for and assign an IUPAC name to each of the following compounds:

(a) $CH_3(CH_2)_4CH_3$
(b) $(CH_3)_2CHCH_2CH_2CH(CH_3)_2$
(c) $(CH_3)_4C$
(d) $CH_3CHClCCl_2CCl_3$

25-15. Write the structural formula for each of the following alkanes:

(a) 2,3-dimethylbutane
(b) 2,2,3-trimethylbutane
(c) 3,3-dimethyl-4-ethylhexane
(d) 4-isopropyloctane

25-16. Write the structural formula for each of the following trichloroalkanes:

(a) 1,1,2-trichlorobutane
(b) 1,1,1-trichloroethane
(c) 1,2,3-trichloropentane
(d) 2,2,4-trichlorohexane

IUPAC NOMENCLATURE OF ALKENES

25-17. Assign an IUPAC name to each of the following alkenes:

(a) $CH_3-\overset{\overset{\large CH_3}{|}}{C}H-CH=CH_2$

(b) $CH_3-\overset{\overset{\large CH_3}{|}}{C}=CH-CH_3$

(c) $CH_2=\overset{\overset{\large CH_3}{|}}{C}-CH_2CH_3$

(d) $CH_3CH_2\overset{\overset{\large CH_3}{|}}{C}HCH=CHCH_3$

25-18. Write a structural formula for each of the following alkenes:

(a) 2-methyl-3-hexene
(b) 2-methyl-1-pentene
(c) 2,3-dimethyl-2-butene
(d) 3-methyl-2-pentene

REACTIONS OF ALKENES

25-19. Write the structural formula for and assign an IUPAC name to the product when each of the following compounds reacts with 1 mol of bromine:

(a) 1-pentene
(b) 2-pentene
(c) 1,3-pentadiene (try this one)

25-20. Write the structural formula for and assign an IUPAC name to the product when each of the following compounds reacts with 1 mol of bromine:

(a) 1-butene
(b) 2-butene
(c) 1,3-hexadiene (try this one)

25-21. Write the structural formula for and assign an IUPAC name to the product when each of the following compounds reacts with 1-butene:

(a) Cl_2 (addition)
(b) HCl
(c) H_2O (acid catalyst) (see Section 25-9 for nomenclature)
(d) H_2 (platinum catalyst)

25-22. Write the structural formula for and assign an IUPAC name to the product when each of the following compounds reacts with 2-pentene:

(a) Cl_2 (addition)
(b) HCl
(c) H_2O (acid catalyst) (see Section 25-9 for nomenclature)
(d) H_2 (platinum catalyst)

25-23. Complete and balance the following equations (assume addition to all double bonds):

(a) $$\mathrm{H_2C{=}CH{-}CH{=}CH_2}(g) + \mathrm{H_2}(g) \xrightarrow{\text{Pt}}$$

(b) $$\mathrm{H_2C{=}CH{-}CH{=}CH_2}(g) + \mathrm{Cl_2}(g) \rightarrow$$

(c) $$\mathrm{H_2C{=}CH{-}CH{=}CH_2}(g) + \mathrm{HCl}(g) \rightarrow$$

25-24. Complete and balance the following equations:

(a) $$\mathrm{(H_3C)_2C{=}C(H)CH_3}(l) + \mathrm{Br_2}(l) \rightarrow$$

(b) $$\mathrm{(H_3C)_2C{=}C(H)CH_3}(l) + \mathrm{HCl}(g) \rightarrow$$

(c) $$\mathrm{(H_3C)_2C{=}C(H)CH_3}(l) + \mathrm{H_2O}(l) \xrightarrow{\text{acid}}$$

25-25. In each case, determine which unsaturated hydrocarbon (alkene or alkyne) reacts with what reagent to form the given product:

(a) $\mathrm{CH_3{-}C(CH_3)(OH){-}CH_2CH_3}$

(b) $\mathrm{CH_3{-}CH(OH){-}CH_3}$

(c) $\mathrm{CH_3CH{=}CHCH_3}$

(d) $\mathrm{CH_3{-}C(Br){=}CH_2}$

25-26. In each case, determine which unsaturated hydrocarbon (alkene or alkyne) reacts with what reagent to form the given product:

(a) $\mathrm{CH_3CHBrCHBrCH_3}$
(b) $\mathrm{CH_3CCl_2CCl_2CH_3}$
(c) $\mathrm{(CH_3)_3COH}$
(d) $\mathrm{CH_3CH(CH_3)CHBrCH_2Br}$

25-27. Write a structural formula for the alcohol formed when H_2O is added to each of the following alkenes:

(a) 3-methyl-1-butene
(b) 2-methyl-2-butene
(c) 2-pentene
(d) 2-butene

25-28. Write a structural formula for the product when HCl is added to each of the following alkenes:

(a) 3-chloro-1-butene
(b) 1-bromo-2-butene
(c) 2-methyl-1-propene
(d) 1-chloro-1-propene

CIS-TRANS ISOMERISM IN ALKENES

25-29. Which of the following alkenes show *cis-trans* isomerism:

(a) $\mathrm{CH_2{=}CHCH_2CH_3}$
(b) $\mathrm{CH_3CH{=}CHCH_2CH_3}$
(c) $\mathrm{CH_2{=}C(CH_3)CH_2CH_3}$

25-30. Which of the following alkenes show *cis-trans* isomerism:

(a) $\mathrm{CH_2{=}C(CH_3){-}CH_2CH_2CH_3}$
(b) $\mathrm{CH_3{-}CH(CH_3){-}CH{=}CH{-}CH_3}$
(c) $\mathrm{CH_3{-}C(CH_3){=}CH{-}CH_2CH_3}$

IUPAC NOMENCLATURE OF ALCOHOLS

25-31. The IUPAC nomenclature for alcohols uses the longest chain of consecutive carbon atoms *containing the —OH group* as the parent compound. The parent compound is named by dropping the terminal *-e* and adding *-ol* to the name of the corresponding alkane, and using a number to indicate the location of the —OH along the chain. For example, the IUPAC name for

$$\begin{array}{c} CH_3CHCH_2OH \\ | \\ CH_3 \end{array}$$

is 2-methyl-1-propanol. Assign IUPAC names to the following compounds:

(a) $\begin{array}{c} CH_3 \\ | \\ CH_3—C—CH_3 \\ | \\ OH \end{array}$ (b) $\begin{array}{c} Cl \\ | \\ CH_3—C—CH_2OH \\ | \\ H \end{array}$

(c) $\begin{array}{c} CH_3—CH—CH_2Cl \\ | \\ OH \end{array}$ (d) $\begin{array}{c} CH_3 \\ | \\ CH_3—CH_2—C—OH \\ | \\ CH_3 \end{array}$

25-32. Using the IUPAC nomenclature for alcohols presented in Problem 25-31, write a structural formula for each of the following alcohols:

(a) 2,2-dimethyl-1-butanol
(b) 5-methyl-2-hexanol
(c) 1-chloro-2-hexanol
(d) 1,2-dichloro-3-pentanol

CLASSIFICATION OF ALCOHOLS

25-33. Classify the following alcohols as primary, secondary, or tertiary:

(a) CH_3CH_2OH (b) $\begin{array}{c} H \\ | \\ CH_3—C—CH_3 \\ | \\ OH \end{array}$

(c) $\begin{array}{c} CH_3 \\ | \\ CH_3CH_2—C—H \\ | \\ OH \end{array}$ (d) $\begin{array}{c} CH_3 \\ | \\ CH_3—C—OH \\ | \\ CH_3 \end{array}$

25-34. Classify the following alcohols as primary, secondary, or tertiary:

(a) $\begin{array}{c} CH_3 \\ | \\ CH_3CH_2—C—OH \\ | \\ CH_3 \end{array}$ (b) $\begin{array}{c} OH \\ | \\ CH_3CH_2CHCH_3 \end{array}$

(c) $CH_3CH_2CH_2CH_2OH$ (d) $\begin{array}{c} CH_3 \\ | \\ CH_3—C—CH_2OH \\ | \\ CH_3 \end{array}$

REACTIONS OF ALCOHOLS

25-35. Alcohols react with alkali metals to produce hydrogen and a salt called an *alkoxide*. For example,

$$2CH_3OH(l) + 2Na(s) \rightarrow \underset{\text{sodium methoxide}}{2Na^+CH_3O^-(s)} + H_2(g)$$

Compare this reaction with that between water and an alkali metal. Complete and balance the following equations:

$$CH_3CH_2OH(l) + Na(s) \rightarrow$$

$$CH_3CH_2CH_2OH(l) + Na(s) \rightarrow$$

25-36. Metal alkoxides (see Problem 25-35) are bases that are comparable in strength to sodium hydroxide. Write an equation that illustrates the basic character of alkoxides.

AMINES

25-37. A class of organic compounds called *amines* can be viewed as derived from NH_3 with one or more hydrogen atoms replaced by alkyl groups. Examples of amines are

CH_3NH_2	$(CH_3)_2NH$	$(CH_3)_3N$
methylamine	dimethylamine	trimethylamine

Aqueous solutions of ammonia and amines are basic because $OH^-(aq)$ is produced when NH_3 and amines are dissolved in H_2O. For example,

$$(CH_3)_2NH(g) + H_2O(l) \rightarrow (CH_3)_2NH_2^+(aq) + OH^-(aq)$$

Like ammonia, amines react with acids to form salts. For example,

$$CH_3NH_2(aq) + HCl(aq) \rightarrow CH_3NH_3^+Cl^-(aq)$$

methylamine (bp −6.3°C) → methylammonium chloride (mp 227°C)

Complete and balance the following equations:

(a) $C_2H_5NH_2(aq) + HBr(aq) \rightarrow$
(b) $(CH_3)_2NH(aq) + H_2SO_4(aq) \rightarrow$
(c) 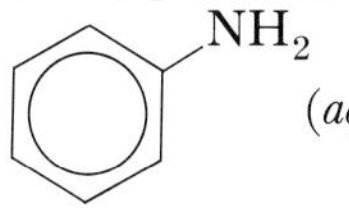 $(aq) + HCl(aq) \rightarrow$
(d) $(C_2H_5)_3N(aq) + HCl(aq) \rightarrow$

25-38. It is possible for a molecule to contain two amino groups, $—NH_2$. Such molecules are called *diamines.* The simplest diamine is ethylenediamine, whose formula is

$$H_2NCH_2CH_2NH_2$$

Predict the product of the reaction of 1 mol of ethylenediamine with 2 mol of HBr.

IUPAC NOMENCLATURE OF ALKYNES

25-39. Assign IUPAC names to the following alkynes:

(a) $CH_3—CH_2—C{\equiv}CH$
(b) $CH_3—C{\equiv}C—CH_3$
(c) $CH_3—C{\equiv}C—C(CH_3)_2—CH_2CH_2CH_3$ (the quaternary carbon bears two CH_3 groups, one above and one below)
(d) $HC{\equiv}C—CH(CH_3)CH(CH_3)—CH_2CH_3$ (one CH_3 below the first CH, one CH_3 above the second CH)

25-40. Write a structural formula for each of the following alkynes:

(a) propyne
(b) 3-hexyne
(c) 5-ethyl-3-octyne
(d) 2,2-dimethyl-3-hexyne

REACTIONS OF ALKYNES

25-41. Predict the product in the addition of 2 mol of HBr to 1 mol of propyne.

25-42. Predict the product in the addition of 2 mol of HBr to 1 mol of 1-butyne.

25-43. Complete and balance the following equations (assume complete saturation of the triple bonds):

(a) $CH_3C{\equiv}CH(g) + O_2(g) \xrightarrow{\text{combustion}}$
(b) $CH_3C{\equiv}CH(g) + HCl(g) \rightarrow$
(c) $CH_3C{\equiv}CH(g) + Br_2(l) \rightarrow$

25-44. Complete and balance the following equations (assume complete saturation of the triple bonds):

(a) $CH_3C{\equiv}CCH_3(g) + HCl(g) \rightarrow$
(b) $CH_3C{\equiv}CCH_3(g) + H_2(g) \xrightarrow{Ni(s)}$
(c) $CH_3C{\equiv}CCH_3(g) + Cl_2(g) \rightarrow$

BENZENE DERIVATIVES

25-45. Write the structural formula for each of the following benzene derivatives:

(a) ethylbenzene
(b) 1,3,5-trichlorobenzene
(c) 2-chlorotoluene
(d) 1-chloro-3-bromobenzene

25-46. Name the following benzene derivatives:

(a) NO_2, Cl
(b) CH_3, H_3C, CH_3, H_3C, CH_3, CH_3
(c) Cl, Cl
(d) Cl, Cl

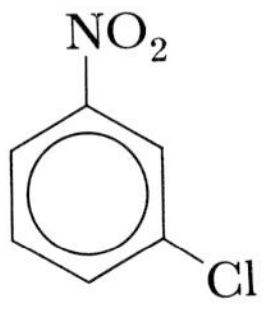

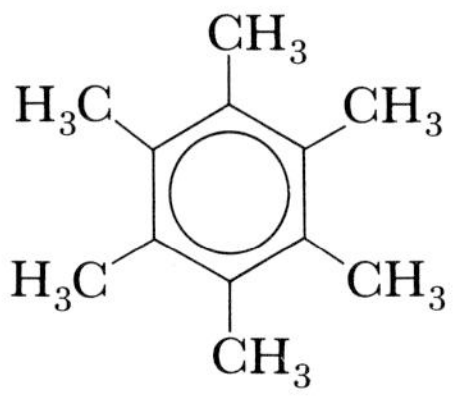

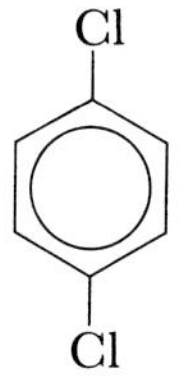

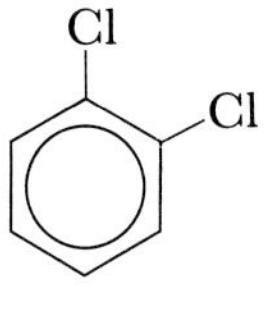

ALDEHYDES

25-47. The IUPAC nomenclature for aldehydes uses the longest chain of consecutive carbon atoms *containing the —CHO group* as the parent compound. The parent compound is named by dropping the terminal *-e* and adding *-al* to the name of the corresponding alkane (compare with the nomenclature for alcohols given in Problem 25-31). For example, the IUPAC name for acetaldehyde, CH_3CHO, is ethanal. Assign IUPAC names to the following aldehydes:

(a) $CH_3CH_2CH_2CHO$
(b) $CH_3\underset{}{\overset{CH_3}{\overset{|}{C}}}HCH_2CHO$
(c) HCHO
(d) $CH_3\overset{CH_3}{\overset{|}{C}}H\underset{CH_3}{\underset{|}{C}}HCH_2CHO$

25-48. Using the IUPAC nomenclature for aldehydes presented in Problem 25-47, write the structural formula for each of the following aldehydes:

(a) propanal
(b) 2-methylpentanal
(c) 4-methylpentanal
(d) 3,3-dimethylhexanal

25-49. Primary alcohols can be oxidized to aldehydes by acidic solutions of $K_2Cr_2O_7$. Determine which alcohol you would use to produce each of the following aldehydes:

(a) ethanal
(b) 2-methylpropanal
(c) 2,2-dimethylpropanal

25-50. Aldehydes can be reduced to primary alcohols by the strong reducing agent sodium borohydride, $NaBH_4$. Determine which aldehyde you would use to produce each of the following primary alcohols:

(a) 1-propanol
(b) 2-methyl-1-propanol
(c) 2,2-dimethyl-1-butanol

IUPAC NOMENCLATURE OF CARBOXYLIC ACIDS

25-51. The IUPAC nomenclature for carboxylic acids uses the longest chain of consecutive carbon atoms *containing the —COOH group* as the parent compound. The parent compound is named by changing the ending *-e* in the name of the corresponding alkane to *-oic acid.* For example, the IUPAC name for formic acid is methanoic acid and that for acetic acid is ethanoic acid. Write the structural formula for each of the following carboxylic acids:

(a) propanoic acid
(b) 2-methylpropanoic acid
(c) 3,3-dimethylbutanoic acid
(d) 3-methylpentanoic acid

25-52. Using the IUPAC nomenclature for carboxylic acids presented in Problem 25-51, name the following compounds:

(a) $CH_3{-}CH(Cl){-}CH_2{-}COOH$

(b) $CH_3{-}C(CH_3)_2{-}COOH$

(c) $CH_3{-}CH(Cl){-}CH(CH_3){-}CH_2{-}COOH$

(d) $Cl{-}CCl_2{-}CCl_2{-}COOH$

REACTIONS OF CARBOXYLIC ACIDS

25-53. Write a balanced chemical equation for the formation of octyl acetate from acetic acid and 1-octanol.

25-54. Write a balanced chemical equation for the formation of ethyl butyrate from butyric acid, $CH_3CH_2CH_2COOH$, and ethanol.

25-55. Complete and balance the following equations:

(a) $HCOOH(aq) + NaOH(aq) \rightarrow$
(b) $HCOOH(aq) + CH_3OH(aq) \rightarrow$
(c) $HCOOH(aq) + Ca(OH)_2(aq) \rightarrow$

25-56. Complete and balance the following equations:

(a) $CH_3CH_2COOH(aq) + NH_3(aq) \rightarrow$
(b) $CH_3CH_2COOH(aq) + CH_3OH(aq) \rightarrow$
(c) $CH_3CH_2COOH(aq) + CH_3CH_2OH(aq) \rightarrow$

25-57. To form the IUPAC name of anions of carboxylic acids, the ending *-ic acid* is replaced by *-ate.* For example, the IUPAC name for sodium acetate is sodium ethanoate. Complete and balance the following equations and name the salt in each case:

(a) $CH_3CH_2COOH(aq) + KOH(aq) \rightarrow$
(b) $CH_3CH(CH_3)COOH(aq) + KOH(aq) \rightarrow$
(c) $Cl_2CHCOOH(aq) + Ca(OH)_2(aq) \rightarrow$

25-58. Write the structural formula for each of the following salts (see Problem 25-57):

(a) sodium 2-chloropropanoate
(b) rubidium methanoate
(c) strontium 2,2-dimethylpropanoate
(d) lanthanum ethanoate

INTERCHAPTER N

Synthetic Polymers

The formation of nylon by a condensation polymerization reaction at the interface of two immiscible solvents, one of which contains the diamine monomer $H_2N(CH_2)_6NH_2$ and the other of which contains the compound $\mathrm{Cl}\underset{\underset{\mathrm{O}}{\|}}{\mathrm{C}}(\mathrm{CH_2})_4\underset{\underset{\mathrm{O}}{\|}}{\mathrm{C}}\mathrm{Cl}$.

The small molecule that splits out in the reaction is HCl.

Polymers are long, chainlike molecules that are formed by the bonding together of relatively small molecules called monomers. Monomers are the subunits, or links, of the molecular chains. *Poly* means many, and *mer* means members. Molecules composed of long chains of atoms are the basis of all synthetic and natural polymers, which are also called *macromolecules.* Macromolecules, polymers which may contain thousands of atoms, typically have lengths in the range 100 to 100,000 nm, whereas nonpolymeric molecules typically have lengths of about 1 nm. Plastics, rubber, fibers, proteins, starch, cellulose, and DNA are polymers.

Synthetic polymers of incredibly diverse properties have been made by chemists over the last 50 years. Because of their superior physical and chemical properties, synthetic fibers and synthetic rubber have displaced natural materials for many purposes. Polymers with twice the strength of steel on a volume basis and 15 times the strength of steel on a mass basis have been made. Plastics are especially useful where low mass and durability are essential. Synthetic rubber, such as that found in modern automobile tires, resists wear far better than natural rubber.

Polymer chemistry has led to the incorporation of numerous new words into our vocabulary. There are very few of us who have not heard the terms nylon, rayon, polyester, polyethylene, polyurethane, polystyrene, Teflon, Formica, Tygon, and Saran, all of which are synthetic polymers. The technological impact of polymer chemistry on our daily lives is immense and continues to increase. There is no question that synthetic polymers have enriched the quality of life for most of us.

N-1. POLYMERS ARE COMPOSED OF MANY MOLECULAR SUBUNITS JOINED END TO END

Recall that the simplest unsaturated hydrocarbon is ethylene, $H_2C{=}CH_2$. Polyethylene is formed by the repeated joining of ethylene molecules. This repeated addition of small molecules to form a long, continuous chain is called *polymerization*.

The polymerization of ethylene can be initiated by a free radical, such as $HO\cdot$, the hydroxyl radical. (A free radical is a species having an unpaired electron.) The first step in the polymerization is

$$HO\cdot + H_2C{=}CH_2 \rightarrow HO{-}CH_2CH_2\cdot$$

The product is a free radical that can react with another ethylene molecule to give

$$HOCH_2CH_2\cdot + H_2C{=}CH_2 \rightarrow HOCH_2CH_2CH_2CH_2\cdot$$

The product of this step is also a free radical that can react with another ethylene molecule:

$$HOCH_2CH_2CH_2CH_2\cdot + H_2C{=}CH_2 \rightarrow HOCH_2CH_2CH_2CH_2CH_2CH_2\cdot$$

The product here is a reactive chain that can grow longer by the sequential addition of more ethylene molecules. The chain continues to grow until some *termination reaction,* such as the combination of two free radicals, occurs. The polyethylene molecules formed in this manner typically contain thousands of carbon atoms. The monomers of polyethylene are ethylene.

The polymerization of ethylene can be written schematically as

$$\underset{\text{monomer}}{nH_2C{=}CH_2} \rightarrow \underset{\text{polymer}}{-\!\!\left(CH_2CH_2\right)\!\!-_n}$$

The notation $-\!\!(CH_2CH_2)\!\!-_n$ means that the group enclosed in the parentheses is repeated n times. The free radical initiator is not indicated because n is large and thus the end group constitutes only a trivial fraction of the molecule. The precise number of monomer molecules incorporated into a polymer molecule is not important for typically large values of n. It makes little difference whether a polyethylene molecule consists of 5000 or 5100 monomer units, for example.

Polyethylene is a tough, flexible plastic that is used in the manufacture of packaging films and sheets, wire and cable insulation, ice cube trays, refrigerator dishes, bags for foods and clothes, and many other articles. Other well-known polymeric materials are made from other monomers. For example, Teflon is produced from the monomer tetrafluoroethylene:

$$\underset{\text{tetrafluoroethylene}}{n\,F_2C{=}CF_2} \rightarrow \underset{\text{Teflon}}{\left(\begin{array}{cc} F & F \\ | & | \\ -C & -C- \\ | & | \\ F & F \end{array}\right)_n}$$

Teflon is a tough, nonflammable, and exceptionally inert polymer that is used for nonstick surfaces in pots and pans, electrical insulation, plastic pipes, and cryogenic bearings. Some other polymers that are produced from alkenes are given in Table N-1.

N-2. NYLON AND DACRON ARE MADE BY CONDENSATION REACTIONS

The polymerization reactions of ethylene and the substituted ethylenes listed in Table N-1 are called *addition polymerization reactions* because they involve the direct addition of monomer molecules. Another type of polymerization reaction is a *condensation reaction.* In a condensation reaction, a small molecule, such as water, is split out as each monomer is added to the polymer chain. The formation of nylon is an example of a condensation polymerization reaction.

Table N-1 Some common polymers prepared from substituted alkenes

Monomer	*Polymer*	*Typical uses*
ethylene, $CH_2{=}CH_2$	polyethylene, $\left(CH_2—CH_2\right)_n$	squeeze bottles, bags for food and clothes, packaging, refrigerator dishes, trash bags
propylene, $CH_2{=}CHCH_3$	polypropylene, $\left(CH(CH_3)—CH_2\right)_n$	indoor-outdoor carpeting, pipes, valves
vinyl chloride, $CH_2{=}CHCl$	polyvinylchloride (PVC), $\left(CH(Cl)—CH_2\right)_n$	pipes, floor tiles, insulation, records
vinylidene chloride, $CH_2{=}CCl_2$	Saran, $\left(CH_2—CCl_2\right)_n$	food packaging, fibers
tetrafluoroethylene, $CF_2{=}CF_2$	Teflon, $\left(CF_2—CF_2\right)_n$	nonstick coating for cooking utensils, electrical insulation
acrylonitrile, $CH_2{=}CHCN$	Orlon, Acrilan, $\left(CH—CH(CN)\right)_n$	fabrics
styrene, 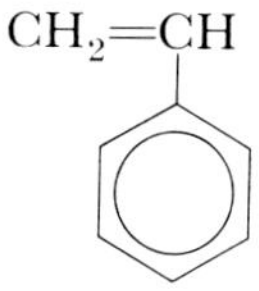 $CH_2{=}CH(C_6H_5)$	polystyrene, $\left(CH_2—CH(C_6H_5)\right)_n$ 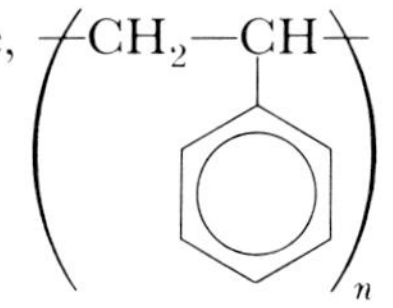	molded plastics, household articles, Styrofoam, picnic jugs, hot-drink cups

Nylon is formed by the reaction of a diamino compound, such as

$$H_2NCH_2CH_2CH_2CH_2CH_2CH_2NH_2$$

and a dicarboxylic acid, such as

$$HO—\underset{\underset{\ddot{O}}{\|}}{C}—CH_2CH_2CH_2CH_2—\underset{\underset{\ddot{O}}{\|}}{C}—OH$$

These two molecules can be linked by the reaction

$$H_2NCH_2CH_2CH_2CH_2CH_2CH_2\underset{\underset{H}{|}}{N}\boxed{H + HO}\underset{\underset{\ddot{O}}{\|}}{C}—CH_2CH_2CH_2CH_2—\underset{\underset{\ddot{O}}{\|}}{C}—OH \rightarrow$$

$$H_2NCH_2CH_2CH_2CH_2CH_2CH_2\underset{\underset{H}{|}}{N}—\underset{\underset{\ddot{O}}{\|}}{C}—CH_2CH_2CH_2CH_2—\underset{\underset{\ddot{O}}{\|}}{C}—OH + HOH$$

The product in this reaction is called a *dimer*. Note that one end of the dimer is an amino group and the other end is a carboxyl group. The dimer can grow by the reaction of its amino end with a dicarboxylic acid monomer or by the reaction of its carboxyl end with a diamine monomer. This process can continue to produce nylon, whose general formula is

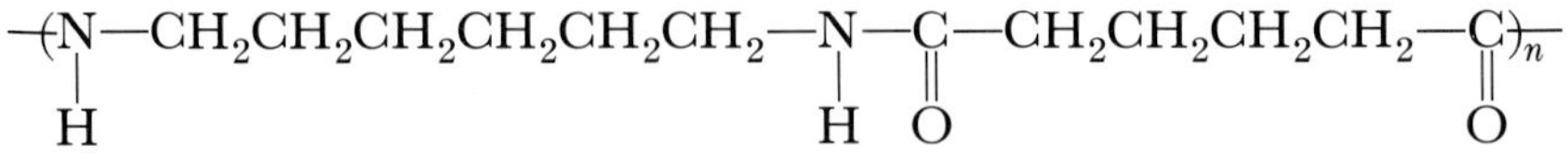

$$\left(\underset{\displaystyle H}{\underset{|}{N}}-CH_2CH_2CH_2CH_2CH_2CH_2-\underset{\displaystyle H}{\underset{|}{N}}-\underset{\displaystyle O}{\underset{\|}{C}}-CH_2CH_2CH_2CH_2-\underset{\displaystyle O}{\underset{\|}{C}}\right)_n$$

Nylon is called a *polyamide* because the monomers are linked together through amide bonds:

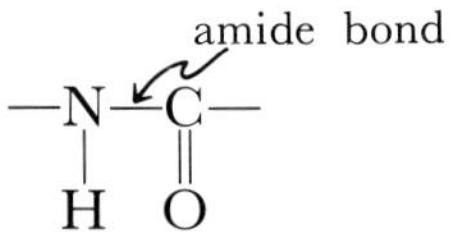

$$-\underset{\displaystyle H}{\underset{|}{N}}-\underset{\displaystyle O}{\underset{\|}{C}}-$$

(amide bond: the N—C bond)

Over 2 billion pounds of nylon are produced annually in the United States. It is used to make strong, long-wearing fibers that find extensive use in rugs and in hosiery, sweaters, and other clothing. Nylon resembles silk in many of its properties but is cheaper to produce. The resemblance to silk is not at all coincidental: silk is a protein, and proteins are polymers in which the monomers are linked together by amide bonds in the same manner as in nylon (Chapter 26).

Dacron is a condensation polymer formed from the monomers ethylene glycol (a dialcohol), $HOCH_2CH_2OH$, and *para*-terephthalic acid (a dicarboxylic acid),

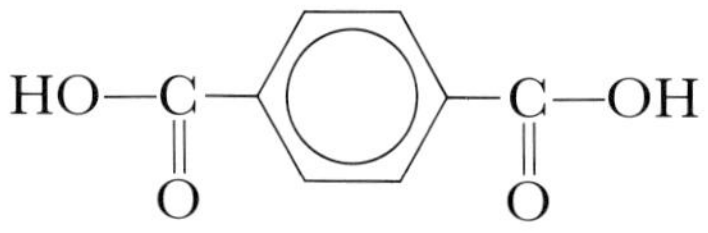

$$HO-\underset{\displaystyle O}{\underset{\|}{C}}-C_6H_4-\underset{\displaystyle O}{\underset{\|}{C}}-OH$$

The basic polymer unit in Dacron is

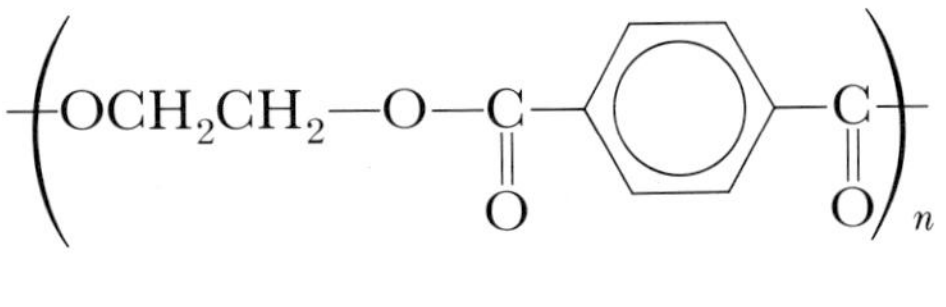

$$\left(OCH_2CH_2-O-\underset{\displaystyle O}{\underset{\|}{C}}-C_6H_4-\underset{\displaystyle O}{\underset{\|}{C}}\right)_n$$

Dacron is called a polyester because the monomers are linked together through ester bonds:

$$-O-\overset{\text{ester bond}}{\underset{\|_{O}}{C}}-$$

Dacron, which is light and tough, is used to make clear films, skis, boat and aircraft components, surgical components (Figure N-1), and permanent-press clothing. When used for clothing, it is usually blended in a roughly two-to-one ratio with cotton because the resulting blended fibers are softer and pass moisture more readily than pure Dacron.

In addition to nylon and Dacron, there are many other condensation polymers, including polycarbonates (safety helmets, lenses, electrical components, photographic film), polyurethanes (insulation, furniture), and phenolics (brake linings, structural components). The essential feature of the monomers used to synthesize condensation polymers is the presence of reactive groups at both ends.

Plastics of incredibly diverse physical and chemical properties have been created (Figure N-2), and the commercial market for plastics is enormous. The major plastics ranked by annual sales in 1981 are shown in the margin. The market for plastics continues to expand, especially in the automotive industry, where weight reduction resulting from the use of polymers translates directly into fuel savings.

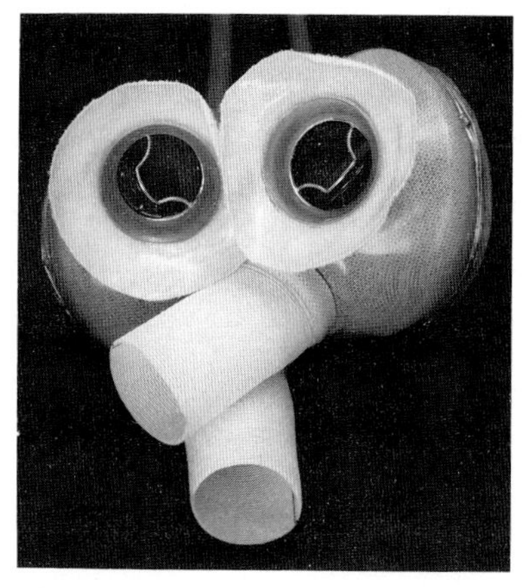

Figure N-1 This is the Jarvik-7 artificial heart, which was implanted in a human being in the first artificial heart implant at the University of Utah Medical Center.

Plastic	*Annual sales/ millions of pounds*
polyethylene	12388
polyvinylchloride	5707
polyester	4176
polypropylene	3955
polystyrene	3621
phenolics	2333
nylon	2332
polyesters	997

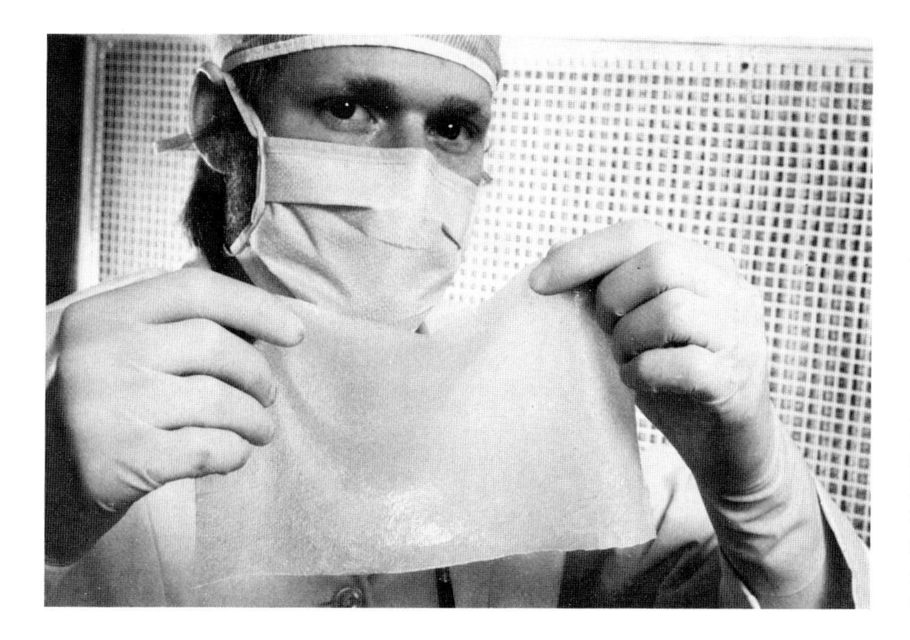

Figure N-2 Membranes made of silicone or polyurethane are being used as a skin substitute for burn victims. Membrane skin substitutes help patients recover more rapidly by passing only oxygen and water.

N-3. POLYMERS WITH CROSS-LINKED CHAINS ARE ELASTIC

Depending on the length of the polymer chains and on the temperature, a particular type of polymer may occur as a viscous liquid, a rubbery solid, a glass, or a partially crystalline solid. Generally, the longer the average length of the polymer chains, the less liquidlike the polymer at a given temperature. For chain lengths involving more than 10,000 bonds, liquid flow is negligible at normal temperatures but elastic-like deformations are possible.

The elasticity of polymers can be explained in terms of structure. Polymer chains are coiled (Figure N-3) and a collection of polymer chains resembles somewhat a pile of cooked spaghetti. If the polymer is stretched, then the chains slowly untangle and the sample appears to flow. The relative movement of polymer chains that occurs when the sample is stretched can be decreased by connecting the polymer chains through chemical bonds called cross-links (Figure N-4a). When the cross-linked network is stretched, the coils become elongated (Figure N-4b). When the stress is released, the polymer network returns to its original coiled state. A cross-linked polymer that exhibits elastic behavior is called an *elastomer*. If the cross-links occur at average intervals of about 100 to 1000 bonds along the chain, then

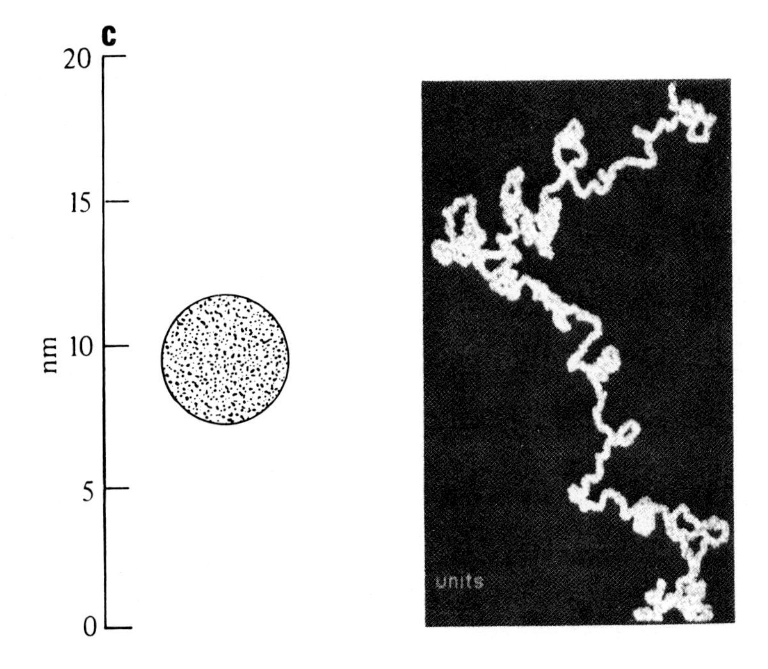

Figure N-3 A computer-generated random configuration of a polyethylene chain containing 3000 carbon-carbon bonds. Note the kinks and coils in subsections of the chain. The sphere is the minimum volume that would be occupied by the fully compacted polymer chain.

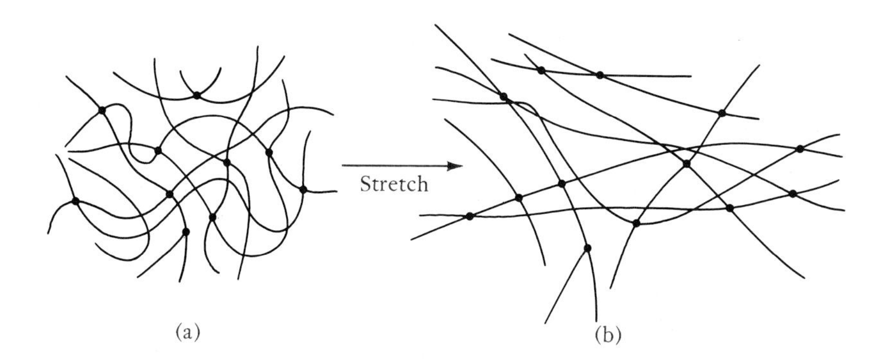

Figure N-4 Schematic representation of a cross-linked polymer network. The cross-links between chains are represented by dots. (a) In the natural (unstretched) state, the polymer chains are coiled. (b) When the polymer is stretched, the chains becomes elongated.

the polymer may be stretched to several times its unstretched length without breaking. The stretched polymer returns to its initial length when the force is released. The resistance of an elastomer to stretching can be increased by increasing the number of cross-links between chains. High elasticity is found in substances composed of long polymer chains joined by sparsely distributed cross-links.

N-4. RUBBER IS A CROSS-LINKED POLYMER

Rubber is of considerable economic and strategic importance. Rubber is used extensively in transportation and in numerous other industries, such as chemicals, electronics, sporting equipment, and footwear.

Rubber was discovered by the native peoples of the West Indies, who used rubber balls in their games. Columbus brought rubber back to Europe, where the British chemist Joseph Priestley, the discoverer of oxygen, coined the word rubber on finding that the substance could be used to rub out pencil marks.

The West Indians obtained their rubber from rubber trees. These trees are native to the Americas, but the world's major rubber tree plantations are located in Malaysia, Indonesia, and Liberia, where trees brought from the Americas grow better than in their natural habitat. Rubber yields of 1500 to 3000 pounds per acre are obtained from rubber tree plantations.

Natural rubber is composed of chains of *cis*-1,4-isoprene units (see margin) with an average chain length of 5000 isoprene units. The major problem associated with natural rubber is tackiness. This problem was solved in 1839 by Charles Goodyear, who discovered how to vulcanize, or cure, natural rubber with sulfur. The vulcanization of natural rubber involves the formation of —S—S— cross-links between the polyisoprene chains (see margin).

$-CH_2-C(CH_3)=CH-CH_2-$ $|$ $-CH_2-C(CH_3)=CH-CH_2-$

cis-isoprene unit *cis*-isoprene unit

$—CH_2—C(CH_3)(S—)—CH(S—)—CH_2—CH_2—C(CH_3)=CH—CH_2—$
$\quad\quad\quad\quad |\quad\quad\quad |$
$\quad\quad\quad\quad S\quad\quad\quad S$
$—CH_2—C(CH_3)(S—)—CH(S—)—CH_2—CH_2—C(CH_3)=CH—CH_2—$

sulfur cross-links

The U.S. sources of natural rubber were cut off by Japanese territorial expansions in 1941. As a result, research in the United States directed toward the production of synthetic rubber underwent a major expansion that culminated in the development of several varieties of synthetic rubber, including a product essentially identical to natural rubber. An example of a synthetic rubber is polybutadiene, which is formed from butadiene, $H_2C{=}CHCH{=}CH_2$.

$$\left(CH_2{-}HC{=}CH{-}CH_2CH_2{-}HC{=}CH{-}CH_2\right)_n$$

polybutadiene

Cross-linking in polybutadiene is achieved by vulcanization. Numerous other rubbers are produced by copolymerization, which involves two different monomers. For example, butadiene and styrene form a copolymer rubber with chains of the type

$$-CH_2{-}HC{=}CH{-}CH_2CH_2\underset{\displaystyle C_6H_5}{\underset{|}{C}H}CH_2{-}HC{=}CH{-}CH_2-$$

The usual monomer ratio is three parts butadiene to one part styrene. An increase in styrene content leads to an increase in plasticity, and the resulting high-styrene polymer is used in latex paints. Butadiene-styrene copolymers are used extensively in the production of automobile tires. As with all rubbers, vulcanization is used to produce the cross-links that give rise to the desired degree of elasticity. The annual production of synthetic rubber in the United States exceeds 2,000,000 metric tons.

QUESTIONS

N-1. Give an example of an addition polymerization reaction.

N-2. Give an example of a condensation polymerization reaction.

N-3. Give the chemical formula for Teflon and Dacron and state whether each is formed in an addition or condensation polymerization reaction.

N-4. Give the chemical formula for nylon and polystyrene and state whether each is formed in an addition or condensation polymerization reaction.

N-5. Write an equation for the condensation reaction that occurs in the reaction shown in the Frontispiece.

N-6. Give the chemical formula of the polymer unit in natural rubber.

N-7. Most synthetic rubbers are copolymers. Write the basic polymer unit in a one-to-one copolymer formed from ethylene and propylene:

$$\underset{\text{ethylene}}{H_2C{=}CH_2} \qquad \underset{\text{propylene}}{\underset{\displaystyle CH_3}{HC\!\!\mid}{=}CH_2}$$

N-8. Describe in chemical terms the vulcanization of natural rubber.

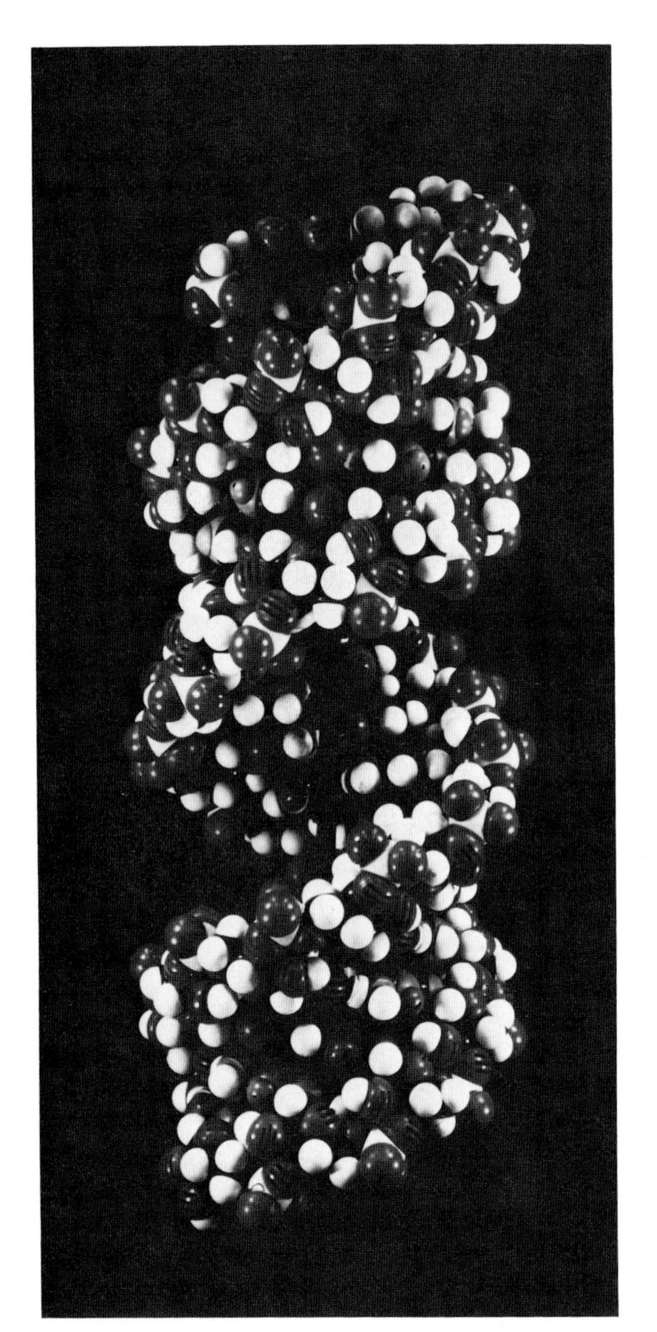

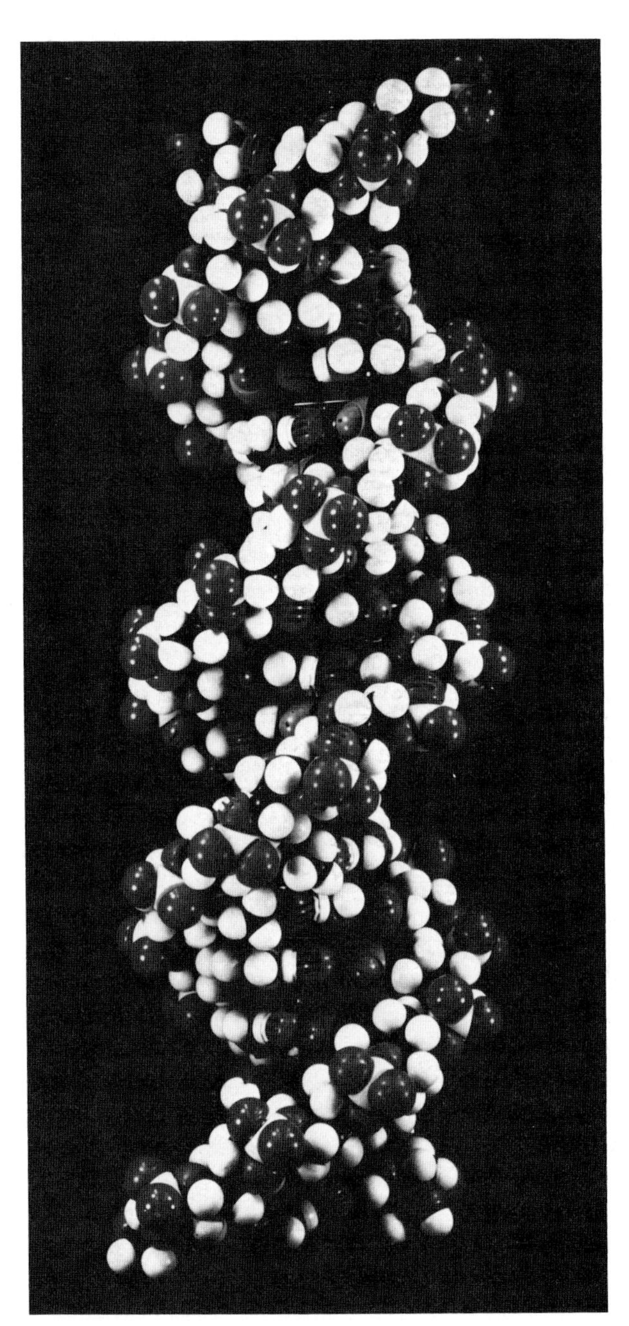

RNA (ribonucleic acid, *left*) and DNA (deoxyribonucleic acid, *right*) play central roles in the storage and transmission of hereditary information from one generation to another.

26 / Biochemistry

As you know from Interchapter N, polymers are large molecules that are built by the joining together of small molecules called monomers. Many biologically important substances, such as proteins, starch, cellulose and nucleic acids, are polymers. In this chapter we describe the structure and function of proteins, carbohydrates, and nucleic acids. About 50 percent of the body's dry mass is protein. Proteins are found in all cells and in virtually all parts of cells. The variety of functions of proteins is staggering. Some proteins are major structural components of skin, tendons, muscles, hair, and connective tissue. Others control the regulation and transmission of neural impulses. Still others, such as enzymes, hormones, and gene regulators, direct and control the diverse chemical reactions that occur in the body. All biological reactions involve one or more protein catalysts called enzymes. The understanding of the structure and function of proteins is one of the great achievements of modern science and still an active area of research.

The nucleic acids DNA and RNA are the molecules in which organisms store genetic information and through which they pass this information from generation to generation. DNA is the principal component of chromosomes, and RNA plays a variety of roles in the biosynthesis of proteins. We see at the end of the chapter how molecules are able to store and transmit genetic information.

26-1. AMINO ACIDS HAVE AN AMINO GROUP AND A CARBOXYL GROUP ATTACHED TO A CENTRAL CARBON ATOM

The word protein was coined by the Swedish chemist Jöns Berzelius in 1838. The word is derived from the Greek word *proteios,* which means "of the first rank."

The monomers from which *proteins* are built are called *amino acids.* The general formula of an amino acid is

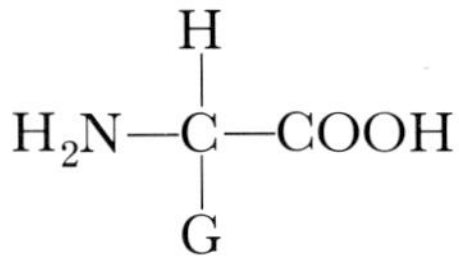

$$\begin{array}{c} \mathrm{H} \\ | \\ \mathrm{H_2N{-}C{-}COOH} \\ | \\ \mathrm{G} \end{array}$$

Amino acids differ from each other only in the *side group,* G, attached to the central carbon atom. These compounds are called amino acids because they contain both an amino group, $-NH_2$, and an acidic group, —COOH. There are 20 amino acids commonly found in proteins (Table 26-1). This set of 20 amino acids occurs in proteins at all levels of life, from the simplest bacteria to humans.

Except for glycine, which is the simplest amino acid,

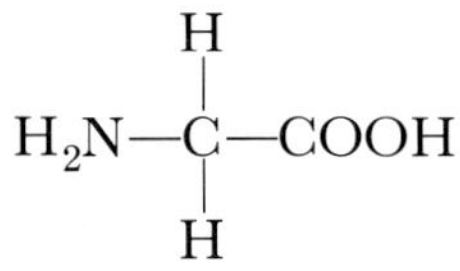

$$\begin{array}{c} \mathrm{H} \\ | \\ \mathrm{H_2N{-}C{-}COOH} \\ | \\ \mathrm{H} \end{array}$$

glycine

all the amino acids have four different groups attached to a central carbon atom. For example, the structural formula for the amino acid alanine is

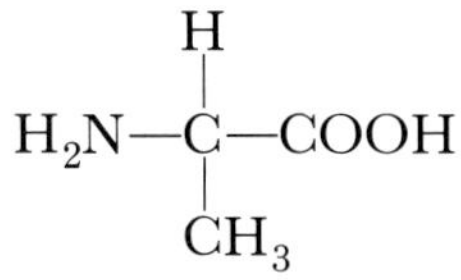

$$\begin{array}{c} \mathrm{H} \\ | \\ \mathrm{H_2N{-}C{-}COOH} \\ | \\ \mathrm{CH_3} \end{array}$$

alanine

The directions for constructing a tetrahedron are given in Appendix E.

The four bonds about the central carbon atom in an amino acid are tetrahedrally oriented, which can be represented as

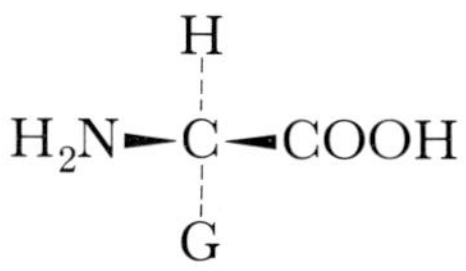

$$\begin{array}{c} \mathrm{H} \\ \vdots \\ \mathrm{H_2N} \blacktriangleright \mathrm{C} \blacktriangleleft \mathrm{COOH} \\ \vdots \\ \mathrm{G} \end{array}$$

The dashed bonds indicate that the —H and —G groups lie below the page, and the dark, wedge-shaped bonds indicate that the $-NH_2$ and —COOH groups lie above the page.

Table 26-1 The side groups and names of the 20 amino acids of proteins

Side group	*Amino acid*
Nonpolar side groups	
$-H$	glycine (gly)
$-CH_3$	alanine (ala)
$-\underset{\displaystyle CH_3}{\underset{\vert}{C}}HCH_3$	valine (val)
$-CH_2\underset{\displaystyle CH_3}{\underset{\vert}{C}}HCH_3$	leucine (leu)
$-\underset{\displaystyle CH_3}{\underset{\vert}{C}}HCH_2CH_3$	isoleucine (ile)
CH_2 — CH_2 — CH_2 — N(H) — C(H)$-$ (ring)	proline (pro)
$-CH_2-C_6H_5$ (benzene ring)	phenylalanine (phe)
$-CH_2-C$ (indole ring: HC, N–H, C, C, C–H, C–H, C–H, C–H)	tryptophan (trp)
$-CH_2CH_2SCH_3$	methionine (met)
Uncharged polar side groups	
$-CH_2OH$	serine (ser)
$-\underset{\displaystyle OH}{\underset{\vert}{C}}HCH_3$	threonine (thr)
$-CH_2SH$	cysteine (cys)
$-CH_2-C_6H_4-OH$ (benzene ring)	tyrosine (tyr)
$-CH_2\underset{\displaystyle O}{\underset{\Vert}{C}}-NH_2$	asparagine (asn)
$-CH_2CH_2\underset{\displaystyle O}{\underset{\Vert}{C}}-NH_2$	glutamine (gln)
Acidic side groups	
$-CH_2C\begin{matrix}\diagup\!\!\!\!\diagup O \\ \diagdown OH\end{matrix}$	aspartic acid (asp)
$-CH_2CH_2C\begin{matrix}\diagup\!\!\!\!\diagup O \\ \diagdown OH\end{matrix}$	glutamic acid (glu)
Basic side groups	
$-CH_2CH_2CH_2CH_2NH_2$	lysine (lys)
$-CH_2CH_2CH_2\overset{\displaystyle H}{\overset{\vert}{N}}\underset{\displaystyle NH}{\underset{\Vert}{C}}NH_2$	arginine (arg)
$-CH_2-C=CH$, HN, N, $\underset{\displaystyle H}{\underset{\vert}{C}}$ (imidazole ring)	histidine (his)

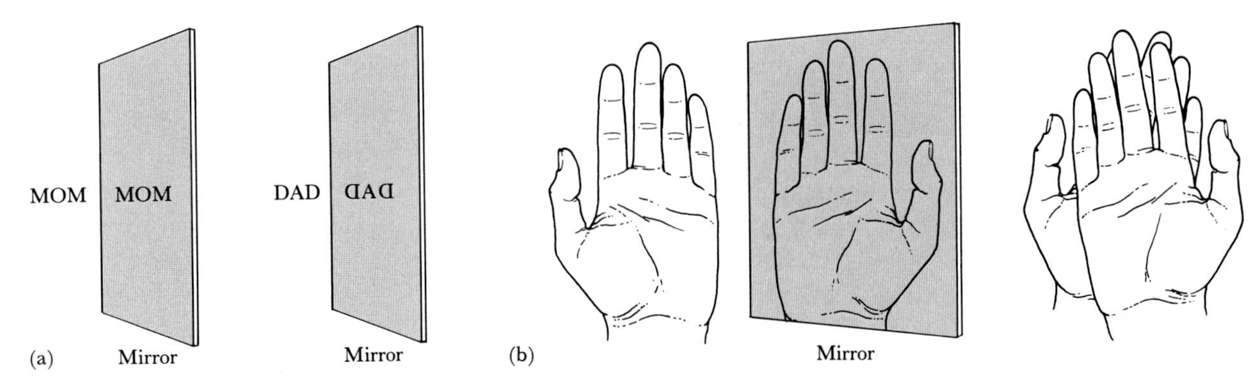

Figure 26-1 (a) Mirror images of the words MOM and DAD. The mirror image of MOM is superimposable on the original, but the mirror image of DAD is not superimposable on the original. (b) Your two hands are an excellent example of nonsuperimposable mirror images. Your right hand is not superimposable on your left hand.

26-2. THE AMINO ACIDS HAVE OPTICAL ISOMERS

The amino acids display a type of isomerism that we have not studied so far in this text. They exist as *optical isomers,* which are nonsuperimposable isomers that are mirror images of each other. To get an idea of what a nonsuperimposable mirror image means, consider the mirror images of the words MOM and DAD (Figure 26-1a). The mirror image of the word MOM is superimposable on the word MOM, whereas the mirror image of the word DAD, that is, ꓷAꓷ, is not superimposable on the word DAD. Similarly, your right hand is a mirror image of your left hand, and the two cannot be superimposed (Figure 26-1b). A molecule can exist in optically isomeric forms if the mirror image of the molecule cannot be superimposed onto itself. If we let the four groups attached to the central carbon atom in alanine be H, X, Y, and Z for simplicity, then we see

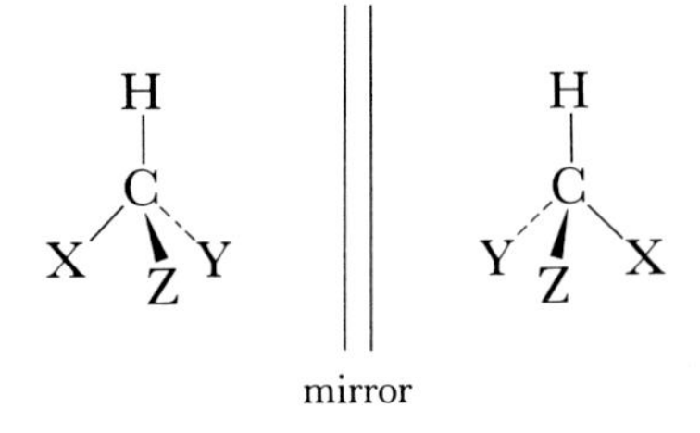

These two mirror images cannot be superimposed, just as a right hand cannot be superimposed onto a left hand. Two optical isomers are distinguished from each other by a D or L placed in front of the name of the amino acid:

The D and the L are derived from dextro- (right) and levo- (left).

$$H_2N\text{—}\overset{H}{\underset{CH_3}{C}}\text{—}COOH \qquad HOOC\text{—}\overset{H}{\underset{CH_3}{C}}\text{—}NH_2$$

D-alanine L-alanine

Optical isomers ordinarily display the same chemical properties, but, with very few exceptions, only the L isomers of the amino acids occur in biological systems. Biochemical reactions are exceptionally *stereospecific;* that is, they are exceptionally dependent upon the shape of the reactants. Apparently life on earth originated from L amino acids, and once the process started, it continued to use only L isomers.

26-3. AMINO ACIDS ARE THE MONOMERS OF PROTEINS

Proteins are formed by condensation reactions between the carboxyl group on one amino acid and the amino group on another. For example,

A condensation reaction is a reaction in which two molecules join together accompanied by the elimination of a small molecule such as HOH.

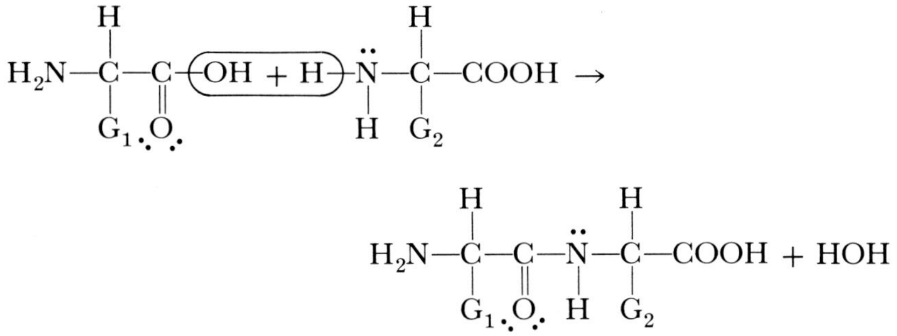

$$H_2N-\underset{G_1}{\overset{H}{C}}-\underset{\|}{\overset{}{C}}(O)-OH + H-\underset{H}{\ddot{N}}-\underset{G_2}{\overset{H}{C}}-COOH \rightarrow H_2N-\underset{G_1}{\overset{H}{C}}-\underset{\ddot{O}}{\overset{\|}{C}}-\underset{H}{\ddot{N}}-\underset{G_2}{\overset{H}{C}}-COOH + HOH$$

The two amino acid monomers are linked by a *peptide bond:*

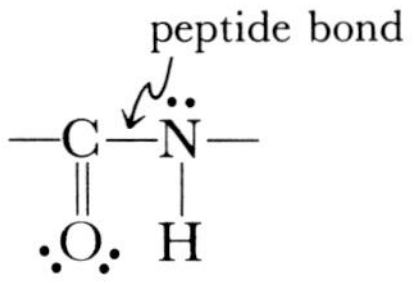

$$-\underset{\ddot{O}}{\overset{}{C}}-\underset{H}{\ddot{N}}- \quad \text{(peptide bond: C–N)}$$

The product of the reaction is called a *dipeptide* because it contains two amino acids.

Example 26-1: Write an equation for the reaction between alanine and serine.

Solution: The side groups in alanine and serine are given in Table 26-1. The reaction between them can be written as

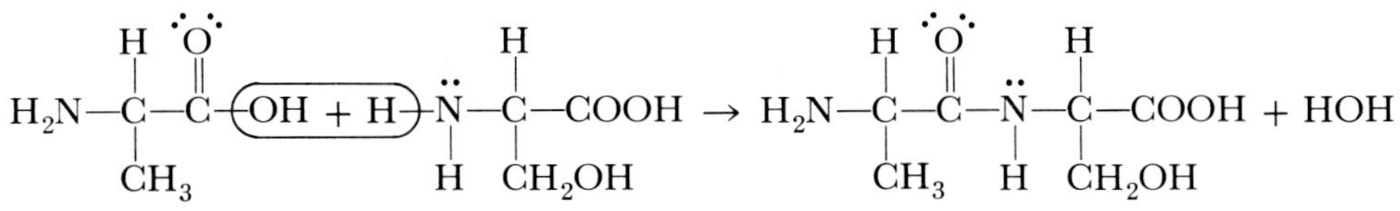

$$H_2N-\underset{CH_3}{\overset{H}{C}}-\overset{\ddot{O}}{\overset{\|}{C}}-OH + H-\underset{H}{\ddot{N}}-\underset{CH_2OH}{\overset{H}{C}}-COOH \rightarrow H_2N-\underset{CH_3}{\overset{H}{C}}-\overset{\ddot{O}}{\overset{\|}{C}}-\underset{H}{\ddot{N}}-\underset{CH_2OH}{\overset{H}{C}}-COOH + HOH$$

This is not the only possible result, however. A different dipeptide is formed when the carboxyl group on serine reacts with the amino group on alanine:

$$\mathrm{H_2N{-}\underset{HOCH_2}{\overset{H}{C}}{-}\overset{\ddot{O}}{\overset{\|}{C}}{-}\boxed{OH + H}{-}\underset{H}{\ddot{N}}{-}\underset{CH_3}{\overset{H}{C}}{-}COOH \rightarrow H_2N{-}\underset{HOCH_2}{\overset{H}{C}}{-}\overset{\ddot{O}}{\overset{\|}{C}}{-}\underset{H}{\ddot{N}}{-}\underset{CH_3}{\overset{H}{C}}{-}COOH + HOH}$$

Thus we see that it is necessary to specify the order of the amino acids in a dipeptide.

Further condensation reactions of a dipeptide with additional amino acid molecules produce a *polypeptide,* which is a polymer having amino acids as monomers. For example,

For simplicity, we do not always indicate lone pairs of electrons in structural formulas.

$$\mathrm{{-}\underset{H}{N}{-}\underset{G_1}{\overset{H}{C}}{-}\underset{O}{\underset{\|}{C}}{-}\underset{H}{N}{-}\underset{G_2}{\overset{H}{C}}{-}\underset{O}{\underset{\|}{C}}{-}\underset{H}{N}{-}\underset{G_3}{\overset{H}{C}}{-}\underset{O}{\underset{\|}{C}}{-}\underset{H}{N}{-}\underset{G_4}{\overset{H}{C}}{-}\underset{O}{\underset{\|}{C}}{-}}$$

a portion of a polypeptide

where the carbon atoms that are bonded to the amino acid side groups are shown in color and the peptide bonds are shown in black. Polypeptides are long chains of amino acids joined together by peptide bonds. The chain to which the amino acid side groups are attached is called the *polypeptide backbone.*

The great diversity of protein structure and function is achieved by variations in the side groups, which fall into one of four classes (Table 26-1): nonpolar, uncharged polar, acidic, and basic. The nonpolar groups are hydrocarbon-like and are said to be *hydrophobic* (water fearing). Like hydrocarbons, nonpolar side groups tend to cluster together and orient themselves away from an aqueous phase. The polar side groups are said to be *hydrophilic* (water loving) and tend to surround themselves with water molecules by orienting themselves toward an aqueous phase.

Recall that the word phobia means a fear or dislike of something.

26-4. A PROTEIN IS UNIQUELY SPECIFIED BY ITS PRIMARY STRUCTURE

Amino acids may be thought of as the letters of an alphabet and proteins as words, sentences, and paragraphs. Just as the two letters n and o can be arranged to give two different words, *no* and *on,* two different amino acids can be joined in two different ways to give two *different* dipeptides.

Table 26-1 gives the three-letter symbols for the 20 amino acids that occur in natural proteins. These three-letter designations are used to write abbreviated structures of polypeptides. By convention, we start with the free amino end and write the three-letter symbols of each amino acid, with each symbol separated by a dash. Thus the tripeptide gly-val-cys represents the three amino acids glycine, valine, and cysteine linked by peptide bonds. Thus, referring to Table 26-1, we write for the tripeptide gly-val-cys

```
           H              H      O  H   H
           |              |      ||  |   |
     H2N—C—C—N——C——C—N—C—COOH
           |   ||  |      |              |
           H   O   H   HC(CH3)2        CH2SH
           ↑              ↑              ↑
        glycine         valine        cysteine
         side            side           side
         group           group          group
```

Example 26-2: Draw the structure of ala-ser-leu.

Solution: Referring to Table 26-1, we find that ala, ser, and leu are the amino acids alanine, serine, and leucine. Therefore, the tripeptide ala-ser-leu represents the three amino acids alanine, serine, and leucine linked by peptide bonds. By convention, alanine contains the free amino end and leucine contains the free carboxyl end. Referring to Table 26-1, we write

```
           H   O   N      H      O  H   H
           |   ||  |      |      ||  |   |
     H2N—C—C—N——C——C—N—C—COOH
           |              |              |
          CH3           CH2OH        CH2CHCH3
           ↑              ↑              ↑   |
           |              |              |  CH3
        alanine         serine        leucine
         side            side           side
         group           group          group
```

Problems 26-11 through 26-16 are similar to Example 26-2.

Proteins are naturally occurring polypeptides. Each protein has a specific number of amino acid units and a specific order of these units along the polypeptide backbone. The number of amino acid units in a protein varies from a few to hundreds. Table 26-2 lists some important proteins and the number of amino acid units in each.

The order of the amino acid units in a polypeptide is called the *primary structure* of the polypeptide. The primary structure of a protein characterizes the protein uniquely. The primary structures of hundreds of proteins have been determined since the 1950's. Figure 26-2 shows the primary structure of the protein beef insulin.

Insulin is a polypeptide hormone that regulates carbohydrate metabolism. A deficiency of insulin leads to diabetes mellitus.

NH_2-terminal ends

A chain	B chain
Gly	Phe
He	Val
Val	Asn
Glu	Gln
Gln	His
Cys	Leu
Cys—S—S—	Cys
Ala	Gly
Ser	Ser
Val	His
Cys	Leu
Ser	Val
Leu	Glu
Tyr	Ala
Gln	Leu
Leu	Tyr
Glu	Leu
Asn	Val
Tyr	Cys
Cys—S—S—	Gly
Asn	Glu
	Arg
	Gly
	Phe
	Phe
	Tyr
	Thr
	Pro
	Lys
	Ala

S—S (linking A-chain Cys 6 and Cys 11)

A chain

B chain

Table 26-2. Number of amino acids in and formula mass of some common proteins

Protein	*Number of amino acids*	*Formula mass*	*Number of polypeptide chains*
insulin (hormone)	51	5700	2
cobratoxin (snake toxin)	62	7000	1
lysozyme (destroys cell walls)	129	13,900	1
myoglobin (carries oxygen in muscles)	153	16,900	1
keratin (wool protein)	204	21,000	1
chymotrypsin (hydrolyzes proteins in small intestine)	241	22,600	3
actin (muscle protein)	410	46,000	1
hemoglobin (transports oxygen in bloodstream)	574	64,500	4
alcohol dehydrogenase (metabolism of alcohol)	748	80,000	2
γ-globulin (antibody)	1250	150,000	4
collagen (skin, tendons, cartilage)	3000	300,000	3

Figure 26-2 The primary structure of the protein beef insulin. The determination of the primary structure of a protein is like a complicated chemical jigsaw puzzle. The protein is hydrolyzed into shorter chains, which are separated and analyzed separately. The first primary structure determination was completed by the British chemist Frederick Sanger in 1953. Sanger received the 1958 Nobel Prize for chemistry for this work.

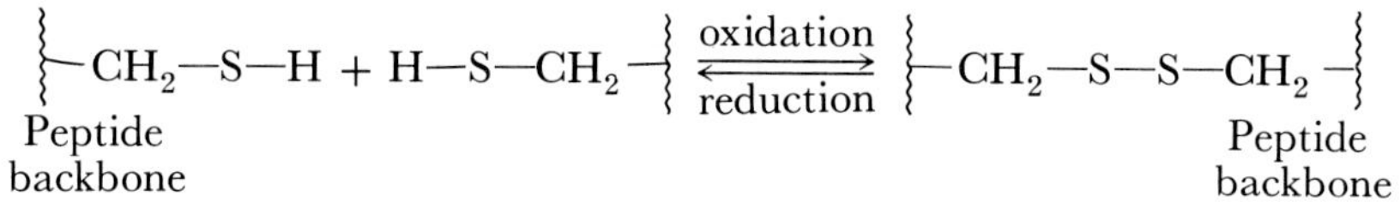

$$\text{—}CH_2\text{—}S\text{—}H + H\text{—}S\text{—}CH_2\text{—} \underset{\text{reduction}}{\overset{\text{oxidation}}{\rightleftharpoons}} \text{—}CH_2\text{—}S\text{—}S\text{—}CH_2\text{—}$$

Peptide backbone   Peptide backbone

Figure 26-3 If two polypeptide chains contain the amino acid cysteine, then the sulfur atoms in the two cysteine side groups can be linked to form a disulfide bond between the two chains. Many important proteins consist of two or more polypeptide chains held together by disulfide bonds.

26-5. DISULFIDE BONDS STRONGLY INFLUENCE THE SHAPE OF PROTEINS

The amino acid cysteine,

$$H_2N\text{—}\underset{\underset{SH}{|}}{\underset{CH_2}{\underset{|}{\overset{\overset{H}{|}}{C}}}}\text{—}COOH$$

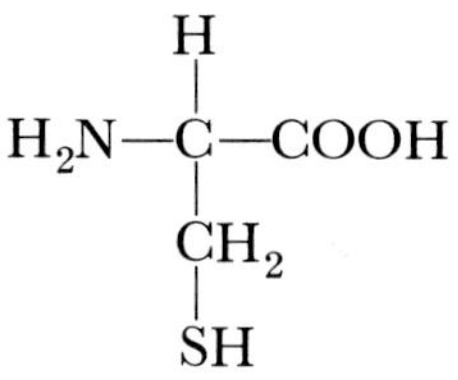

plays a special role in protein structure because two cysteine side groups on neighboring polypeptide chains can be linked to form a *disulfide bond* (Figure 26-3). If the two cysteine side groups are on the same polypeptide chain, the result is a loop (Figure 26-4).

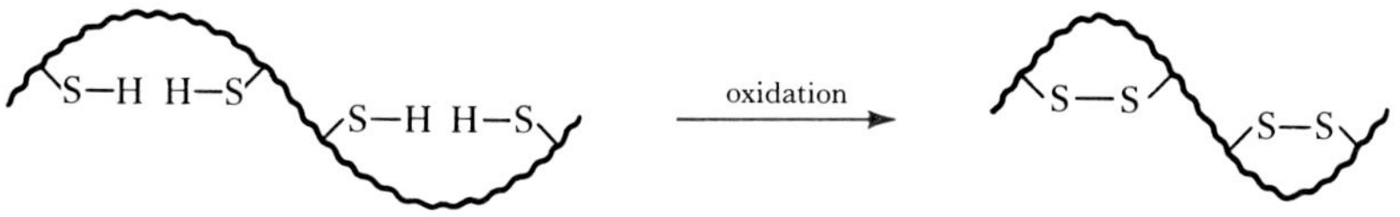

S—H H—S S—H H—S oxidation S—S S—S

Figure 26-4 Disulfide bonds can be formed between two cysteine side groups on the same polypeptide chain. When this occurs, the chain is bent into loops. Disulfide bonds either between chains or within chains greatly affect the shape of protein molecules.

Hair is about 10 percent cysteine. A permanent curls hair by first breaking the disulfide bonds. When the hair is then curled as desired and treated with a "neutralizer," new disulfide bonds form that hold the curls. Many important polypeptides besides hair contain disulfide bonds. For example, each of the two relatively simple hormone polypeptides, oxytocin and vasopressin, consists of a chain that is forced into a loop by a disulfide bond:

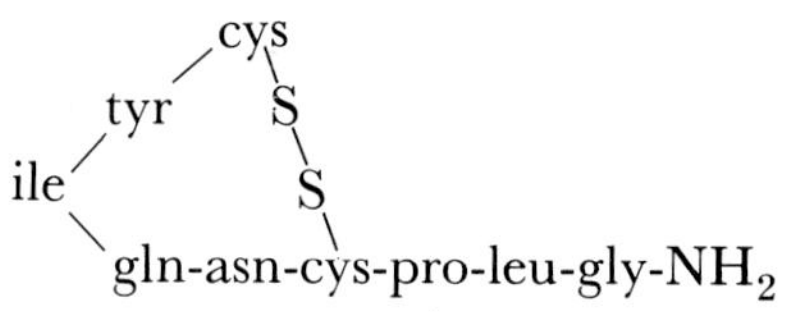

```
      cys                        cys
    tyr  S                     tyr  S
ile       S                phe       S
   gln-asn-cys-pro-leu-gly-NH2   gln-asn-cys-pro-arg-gly-NH2
          oxytocin                      vasopressin
```

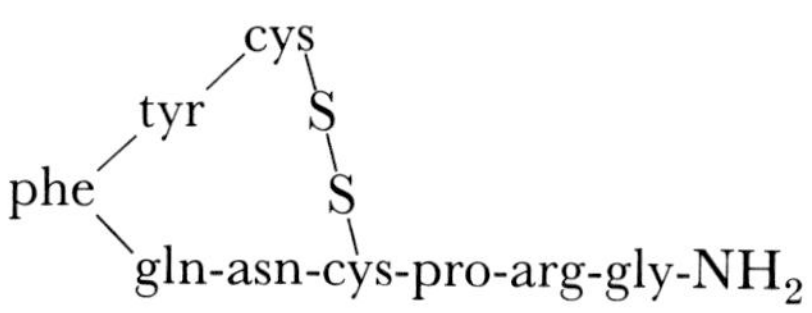

Both polypeptides have nine amino acid units, with seven of them being the same and in the same position in the chain. The similarity is only superficial, however. Vasopressin is a hormone that is produced by the pituitary gland. It increases blood pressure and also regulates kidney action, preventing the loss of too much water in urine. The disease diabetes insipidus results when the body produces an insufficient quantity of vasopressin. A person suffering from this disease excretes a large volume of urine and suffers from constant thirst. Oxytocin, on the other hand, does not increase blood pressure, as vasopressin does. Instead, oxytocin induces contraction of the smooth muscles, particularly those of the uterus, and is used to induce labor. Why vasopressin and oxytocin should have such different functions but such similar chemical composition is still unknown. This is one of many examples that show that the function of a protein depends crucially upon its structure.

26-6. THE SHAPE OF A PROTEIN MOLECULE IS CALLED TERTIARY STRUCTURE

A first step in understanding how a particular protein functions is a determination of its shape. Because many proteins are such large molecules, this is a very complex task. The ultimate step in the determination of a protein's structure is X-ray crystallography. We saw in Chapter 13 that X-ray patterns can be used to determine the arrangement of atoms in crystalline solids. The X-ray patterns obtained from proteins are more difficult to analyze and interpret because there are so many atoms involved.

In the 1950's two American chemists, Linus Pauling and R. B. Corey, were able to interpret X-ray patterns of proteins to show that many proteins have regions in which the chain twists into a helix (which is the shape of a spiral staircase). Pauling and Corey called the helix an *α-helix* (Figure 26-5). The helical shape results from the formation of hydrogen bonds between amino acids in the peptide chain. Individually, these hydrogen bonds are relatively weak, but collectively they combine to bend the protein chain into an α-helix. This coiled, helical shape in different regions of a protein chain is called *secondary structure*.

The overall shape of a protein is determined by a number of factors. A protein molecule may be pictured as at least one long polypeptide backbone with amino acid side groups attached at uniform intervals. Realize that many proteins in living systems occur in aqueous solution, and so we must consider the interactions between the amino acid side groups and water. Some of these side groups are nonpolar. The molecules of water form a hydrogen-bonded network that excludes nonpolar substances. Consequently, nonpolar side groups minimize their contact with water by clustering together on

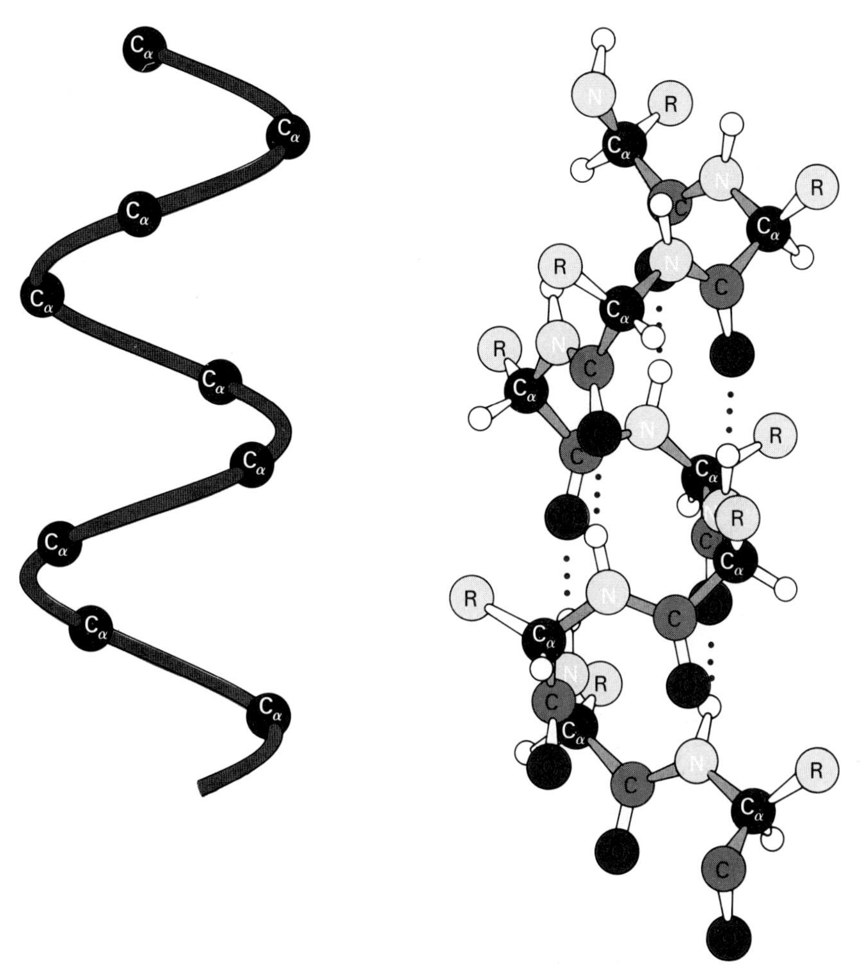

Figure 26-5 A segment of an α-helical region along a polypeptide chain. The chain is held in a helical shape by hydrogen bonds (dotted lines), which are formed between hydrogen and oxygen atoms in peptide bonds that are separated by three other peptide bonds along the polypeptide chain. C_α refers to the carbon atoms to which the side groups are attached, and R represents the side groups.

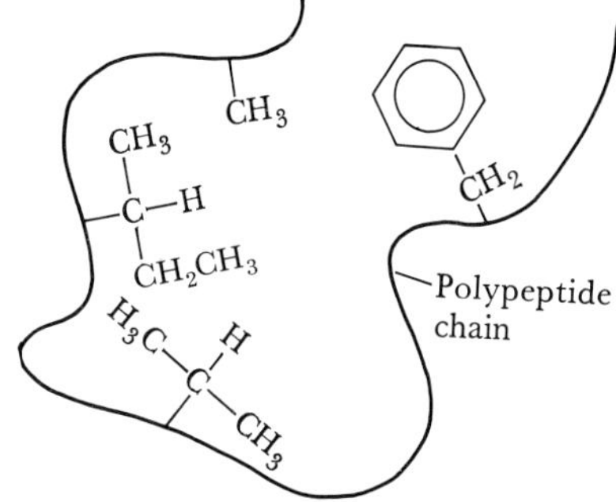

Figure 26-6 Nonpolar amino-acid side groups are excluded by the hydrogen-bonded network of water and so cluster together in the interior of proteins.

the inside of the polypeptide chain, away from the water phase (Figure 26-6). Other amino acid side groups are polar and so can form hydrogen bonds with water. These side groups point outward toward the aqueous phase (Figure 26-7). Consequently, a region along a protein chain that contains amino acids with polar side groups might be relatively stretched out.

Thus, the overall shape of a protein molecule in water results from a complicated interplay between the amino acid side groups along the protein chain and the solvent, water. This interplay causes the protein to coil, fold, and bend into a three-dimensional overall shape called the *tertiary structure* (Figure 26-8). The tertiary structure of a protein is obtained by X-ray analysis.

Myoglobin (Figure 26-8) is an oxygen-binding protein that stores and transports oxygen in muscles. It appears to be tailor-made by nature to bind, transport, and release oxygen molecules. The tertiary structure of a protein depends upon the nature and order of the

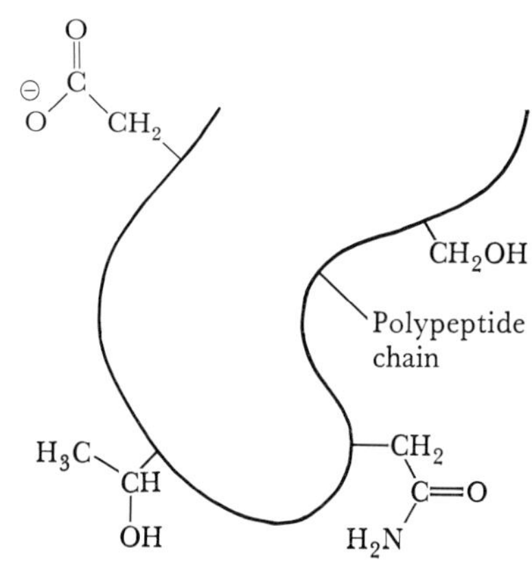

Figure 26-7 Polar amino acid side groups are able to form hydrogen bonds with water and so lie on the surfaces of proteins.

amino acid units making up the protein chain. Nature can build proteins to perform specific tasks by incorporating the appropriate amino acids. These cause the protein to take on a specific shape that is suited to the function of the protein.

Example 26-3: Referring to Table 26-1, determine which amino acids have side groups that cluster in the interior of a protein and which have side groups that occur at the surface of the protein and hydrogen-bond with water molecules.

Solution: Table 26-1 is arranged according to the nature of the side groups of the amino acids. The first nine amino acids (glycine through methionine) have nonpolar side groups and cluster in the interior. The other 12 (serine through histidine) have either polar or charged side groups and occur at the surface of the protein.

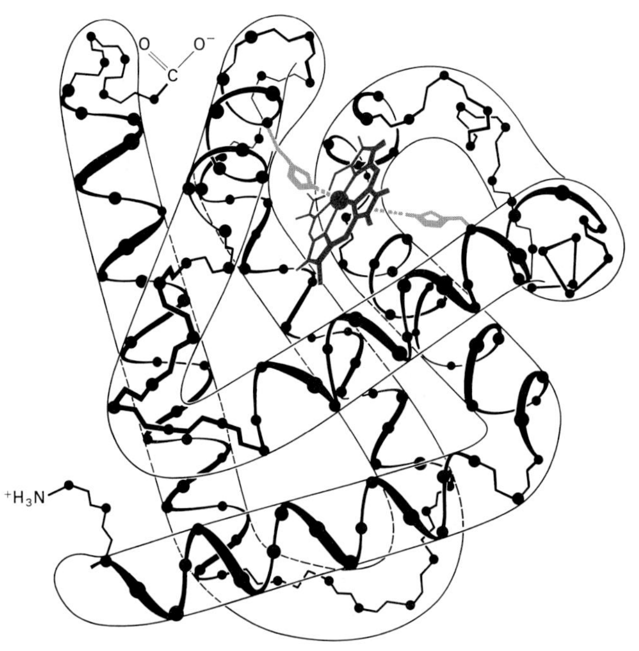

Primary structure–order of the amino acids along polypeptide chains.

Secondary structure–steric relationship of amino acid groups that are near each other along a polypeptide chain (α-helical structure.

Tertiary structure–overall three-dimensional shape of a protein.

Figure 26-8 The tertiary structure—that is, the overall three-dimensional shape—of the protein myoglobin, which stores and transports oxygen in muscles. Myoglobin has 153 amino acid units and consists of one polypeptide chain. The black dots represent the carbon atoms that are linked by peptide bonds. The oxygen-binding site, which is buried inside the molecule, is shown in red. The straight, cylindrical regions are α-helical regions. It is these regions that are packed into the three-dimensional arrangement that forms the tertiary structure.

26-7. ENZYMES ARE PROTEIN CATALYSTS

The importance of tertiary structure is well illustrated by *enzymes,* which are proteins that catalyze chemical reactions in living systems. Without enzymes, most biochemical reactions are too slow to be of any consequence. The absence or an insufficient quantity of an enzyme can lead to serious physiological disorders. Albinism, cystinosis, and phenylketonuria are three of many examples. Cells contain thousands of different enzymes. More than 1000 enzymes have been identified, and the primary structures (that is, the amino acid sequences) of hundreds have been determined. The tertiary structures of a few enzymes have been determined by X-ray analysis.

A remarkable property of enzymes is their extraordinary specificity. An enzyme usually catalyzes a single chemical reaction or a set of closely related reactions. The exact manner in which enzymes function is still a topic of research, but one simple picture of enzyme activity is the *lock-and-key theory* (Figure 26-9). In this picture, the enzyme binds the *substrate,* which is the substance that is reacting, to its surface. The binding site on the enzyme is such that it can bind only one substrate or a closely related one. The particular shape of and the nature and location of atoms at the binding site account for

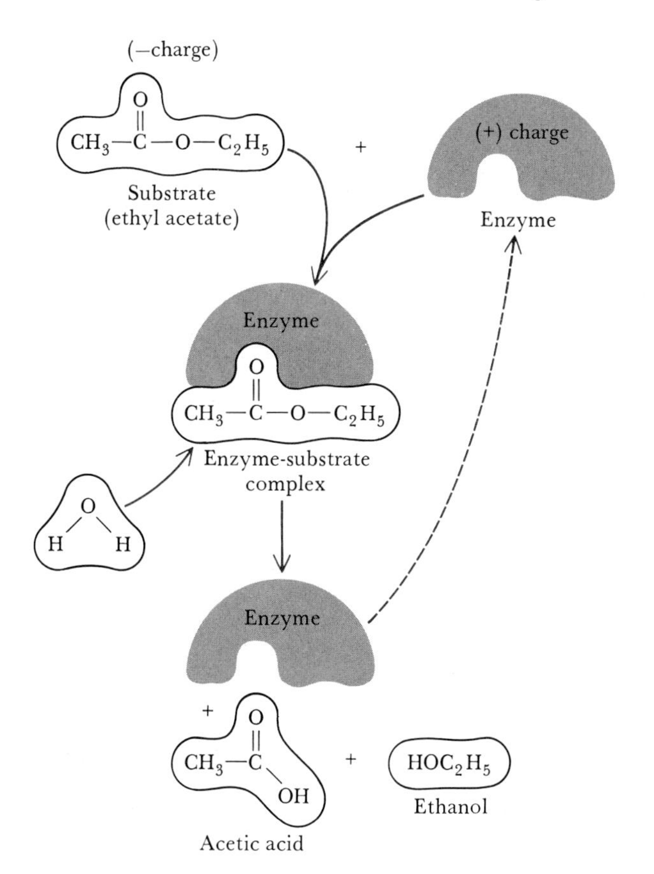

Figure 26-9 The lock-and-key theory of enzyme catalyzed reactions. The reaction in this case is the hydrolysis of ethyl acetate: $CH_3COOC_2H_5 + H_2O \rightarrow C_2H_5OH + HC_2H_3O_2$. The shape of and the charges at the binding site of the enzyme are perfectly suited to accommodate the substrate, ethyl acetate. Because of the specific way the enzyme binds the ethyl acetate, the ester bond is exposed and readily attacked by a water molecule. By holding the substrate this way, the enzyme speeds up the reaction rate.

Figure 26-10 A space-filling model of ribonuclease, an enzyme that breaks down RNA in cells. The groove at the top of the molecular model is the substrate binding site. The model was built in accord with the known tertiary structure determined by X-ray diffraction.

the extraordinary specificity of enzymes. The substrate molecule is bound to the enzyme in such a way that the substrate is susceptible to chemical attack. Figure 26-10 shows models of the enzyme lysozyme, which attacks the cell walls of certain bacteria and causes them to burst.

A small variation in the chemical composition of a hemoglobin molecule is sufficient to change the shape of red blood cells from normal (circular) to sickled (irregular). The difference in shape is caused by a difference of only 2 amino acids out of the 574 in a hemoglobin molecule.

26-8. NORMAL HEMOGLOBIN AND SICKLE-CELL HEMOGLOBIN DIFFER BY 2 OUT OF 574 AMINO ACIDS

Hemoglobin, the protein that transports oxygen through the bloodstream from the lungs to the various tissues in the body, serves as a striking example of how the shape, and hence the function, of a protein depend in exacting detail upon its amino acid sequence. The photograph at the left shows an electron micrograph of normal and sickled red blood cells. The sickled red blood cells become trapped in small blood vessels. This impairs circulation and results in damage to a number of organs, particularly the bones and kidneys. In the 1950's it was determined that the hemoglobin in sickled cells differs from that found in normal cells.

Hemoglobin consists of two sets of two identical chains, called α and β chains. The two α chains each contain 141 amino acid units, and the two β chains each contain 146 amino acid units. The α chains of normal and sickled cells are the same. The β chains of sickled cells differ from the β chains of normal cells by just one amino acid. The difference is

	Position in β chain
	1 2 3 4 5 6 7 8
hemoglobin from normal cell	H_2N-val-his-leu-thr-pro-glu-glu-lys-
hemoglobin from sickled cell	H_2N-val-his-leu-thr-pro-val-glu-lys-

Only 2 amino acids out of a total of 574 in the hemoglobins differ. The difference lies in the sixth position from the amino end in the β chains. Normal hemoglobin has a glutamic acid unit there, and sickle-cell hemoglobin has a valine unit. Table 26-1 shows that a change from glutamic acid to valine is a change from a hydrophilic group to one that is hydrophobic. This is apparently enough to alter the shape of the protein molecule in a profound way and is responsible for the sickle shape of the red blood cells that contain the abnormal hemoglobin. Sickle-cell anemia is a molecular disease. The shape versus function of proteins is an illustration of the importance of molecular structure. An understanding of the underlying causes of disease is a first step in their eradication.

26-9. STARCH AND CELLULOSE ARE POLYMERIC CARBOHYDRATES

Carbohydrates serve as a source of energy (as starch) and as structural material (as cellulose). They are synthesized from CO_2 and H_2O by green plants in a process called photosynthesis:

$$nCO_2 + nH_2O \xrightarrow{\text{photosynthesis}} \underset{\text{carbohydrate}}{\left(CH_2O\right)_n} + nO_2$$

Some typical small carbohydrate molecules are ribose, $C_5H_{10}O_5$; fructose, $C_6H_{12}O_6$; glucose, $C_6H_{12}O_6$; and sucrose, $C_{12}H_{22}O_{11}$.

Carbohydrates have the general formula $C_x(H_2O)_y$ and were once thought to be hydrates of carbon (hence the name). We now know, however, that carbohydrates are predominantly polyhydroxy ring compounds, as is shown by the structure of, for example, glucose, which exists as two forms, called α-glucose and β-glucose:

It has been estimated that the per capita consumption of sucrose in the United States exceeds 100 pounds annually.

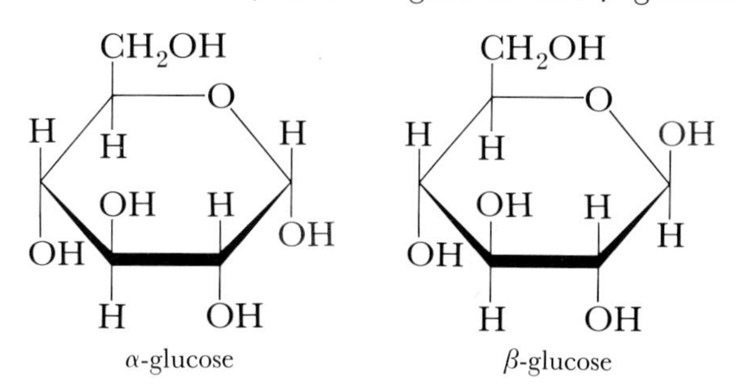

α-glucose β-glucose

The ring carbon atoms are not indicated in these formulas, but there is one at each unmarked vertex. The α and β designations indicate the position of the —OH group shown in color. The geometry about each atom in the glucose ring is tetrahedral, and so the ring is not planar as these formulas might imply. A more realistic, but more difficult to draw, representation is

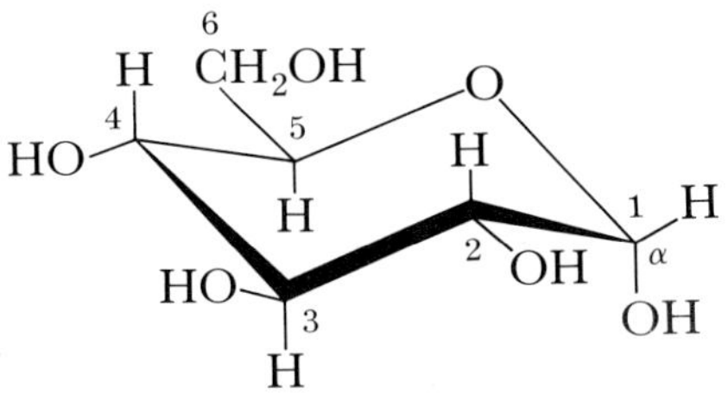

α-glucose 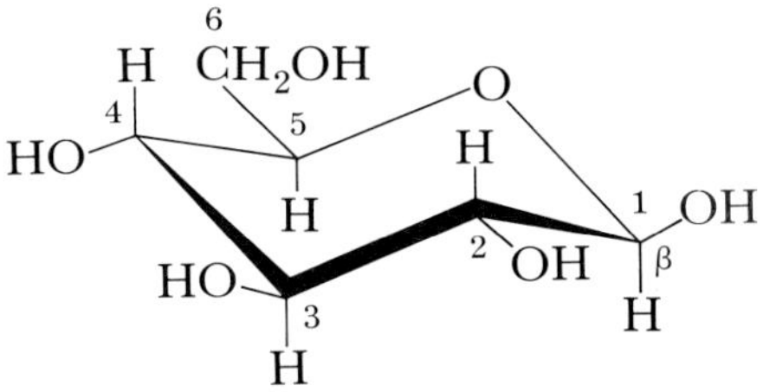 β-glucose

Fructose, which has the same molecular formula ($C_6H_{12}O_6$) as glucose, has the structural formula

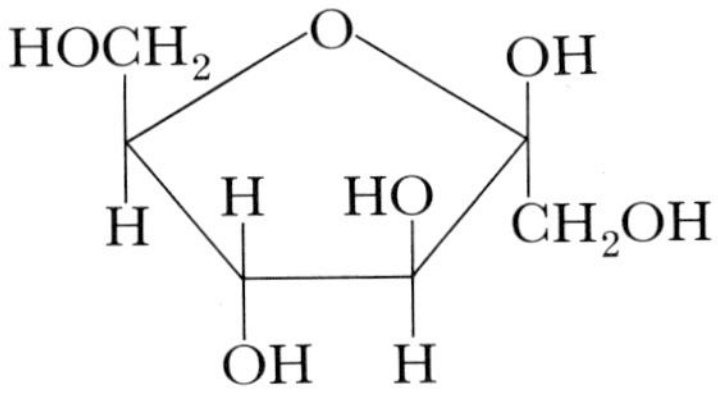

β-fructose

Glucose and fructose, which consist of just one ring, are called *monosaccharides*. Most carbohydrates are made up of monosaccharides joined together to form *polysaccharides*. Although there are many monosaccharides, only a few of them account for most biologically important substances.

Two monosaccharides can be joined together to form a *disaccharide*. A common disaccharide is sucrose (table sugar), which is formed from a condensation reaction between α-glucose and β-fructose:

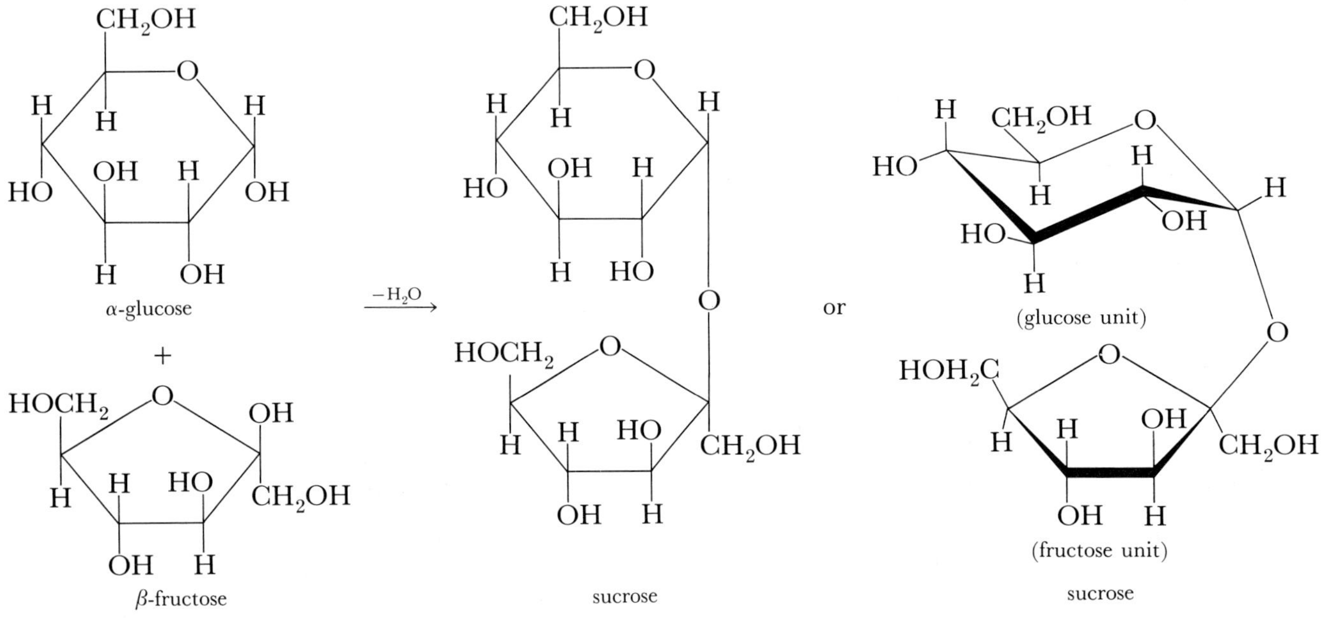

α-glucose + β-fructose $\xrightarrow{-H_2O}$ sucrose or sucrose

(glucose unit)

(fructose unit)

Monosaccharides and disaccharides are usually sweet-tasting and are called *sugars*. Two other common disaccharide sugars are maltose, which occurs in baby food and malted milk, and lactose, which occurs in milk and milk products. The structural formulas for maltose and lactose are

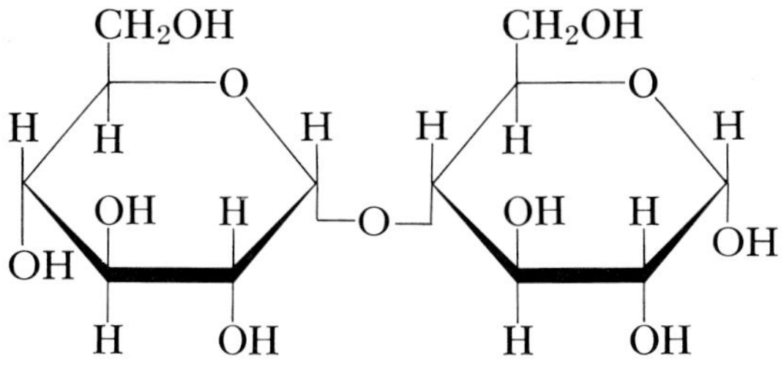

maltose

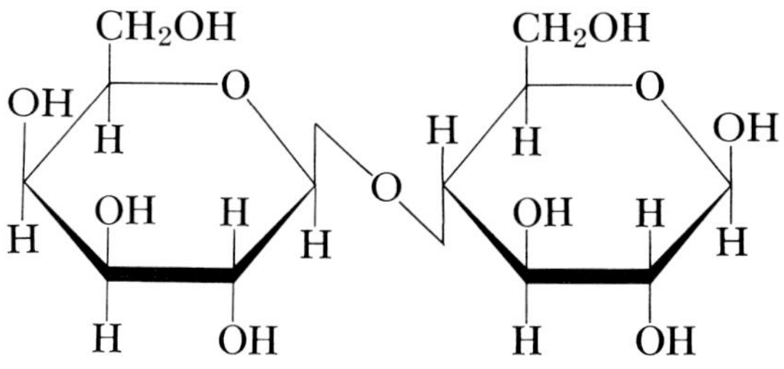

lactose

or, more realistically,

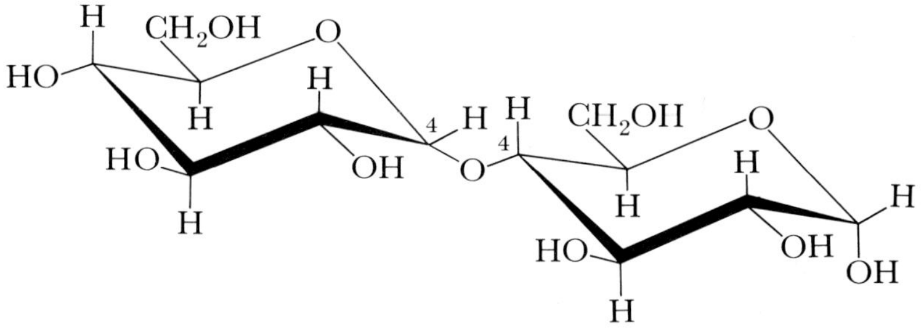

maltose

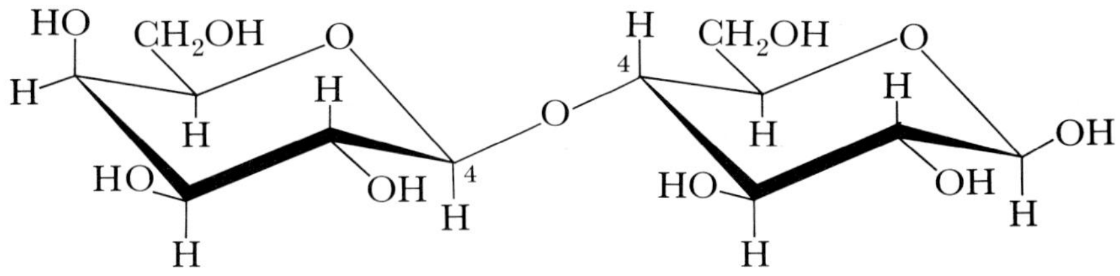

lactose

A majority of adults in certain population groups are intolerant to milk because they are deficient in the enzyme (lactase) that is needed to hydrolyze lactose.

Example 26-4: Draw the condensation reaction that produces the disaccharide maltose, which is a disaccharide of two α-glucose molecules.

Solution: The reaction between two α-glucose molecules to produce maltose is

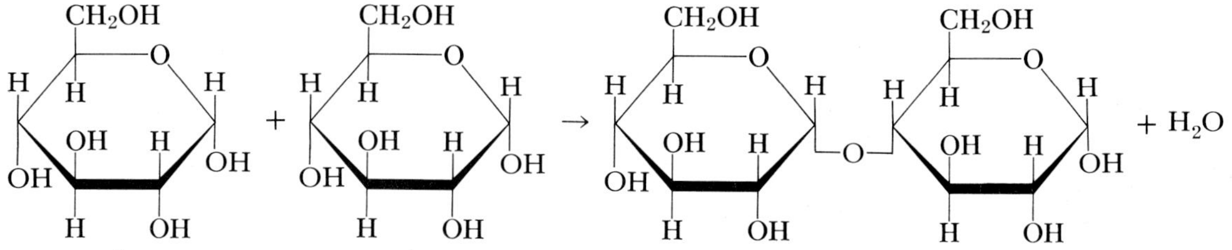

α-glucose + α-glucose → maltose + H_2O

The polysaccharides starch and cellulose are both polymers of glucose; starch is a polymer of α-glucose, and cellulose is a polymer of β-glucose. A major component of starch is amylose, which is a straight-chain polymer of α-glucose:

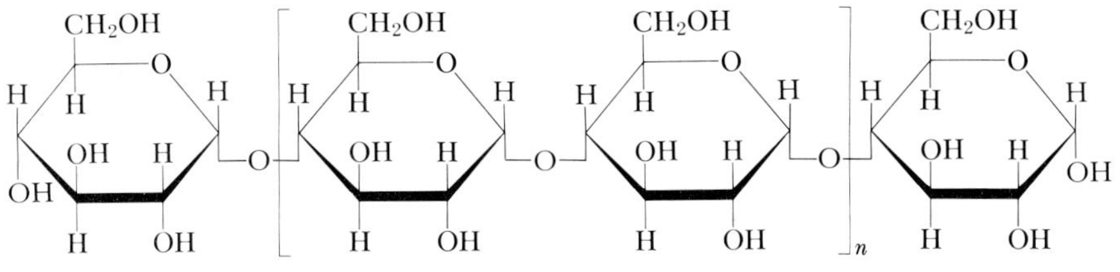

or

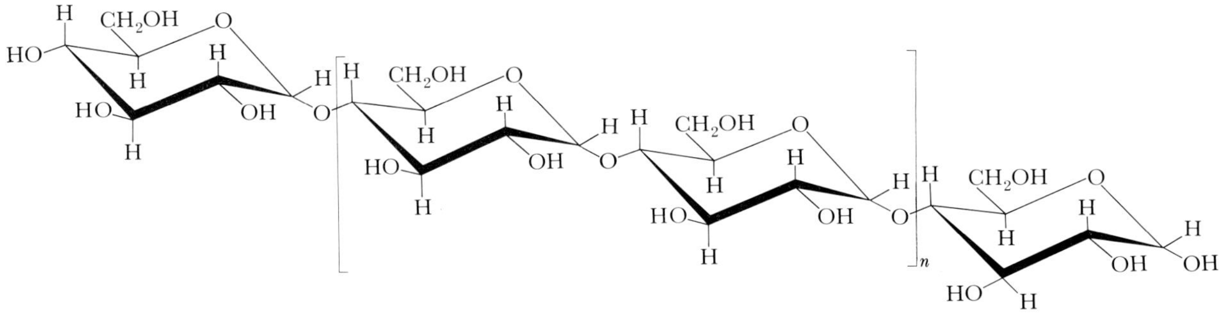

amylose

The digestion of starch begins in the mouth, where the enzyme amylase, which is found in saliva, breaks down the starch molecules into maltose units. The maltose is then broken down into two glucose molecules when it reaches the intestine. The glucose is absorbed into the bloodstream and carried to cells throughout the body, where it is the source of energy for various cellular processes. The glucose is ultimately converted in cells to CO_2 and H_2O through a series of enzymatic reactions and yields 2820 $kJ \cdot mol^{-1}$. The net overall reaction is

$$C_6H_{12}O_6(aq) + 6O_2(aq) \rightarrow 6CO_2(aq) + 6H_2O(l) + 2820\ kJ$$

The structural formula for cellulose, which is a straight-chain polymer of β-glucose, is

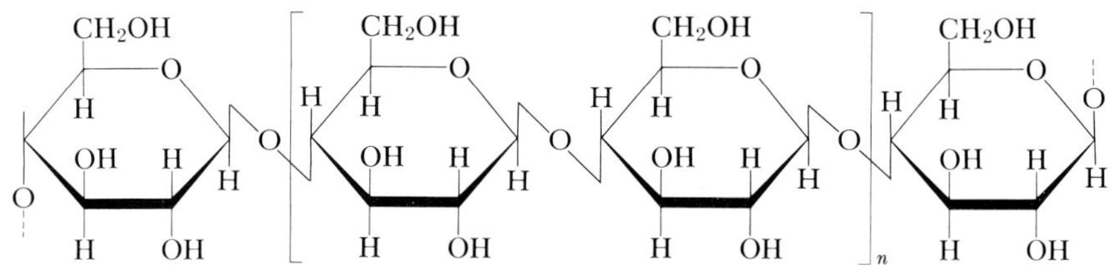

or

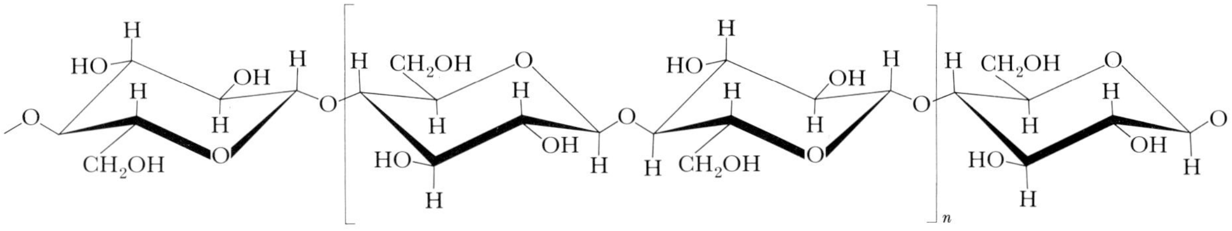

cellulose

The linkage between the glucose units in cellulose is completely different than in amylose. Cellulose is the major component of wood, cotton, and other plant materials. Cellulose accounts for more than one half of all the matter in living things on earth. Humans do not possess the enzymes that are necessary for *hydrolyzing* the β-linkages between the glucose units in cellulose and thus breaking it down into usable glucose. Consequently, we are unable to use cellulose as food. Animals such as cows and deer have intestinal bacteria that possess the proper enzymes for breaking down cellulose into glucose. Cellulose often consists of chains that contain over 10,000 β-glucose units and have molecular masses of over 1 million.

Example 26-5: Calculate the formula mass of a cellulose polymer consisting of 10,000 β-glucose units.

Solution: The formula mass of β-glucose, $C_6H_{12}O_6$, is 180. Realize that one water molecule is eliminated when two glucose molecules react, and so the formula mass of a cellulose molecule containing 10,000 β-glucose units is

$$\begin{aligned}\text{formula mass} &= (10{,}000)(180) - (9999)(18)\\ &= 1{,}620{,}018\\ &= 1.62 \times 10^6\end{aligned}$$

26-10. DNA AND RNA ARE POLYNUCLEOTIDES

The final class of biopolymers that we study in this chapter are the *polynucleotides*. The two most important polynucleotides are *DNA* (deoxyribonucleic acid) and *RNA* (ribonucleic acid). DNA occurs in nuclei of cells and is the principal component of chromosomes. The genetic information that is passed from one generation to another is stored in DNA molecules. The discovery of just how this is done was made in 1953 and has led to a revolution in biology that is as profound and far-reaching as the harnessing of nuclear energy in the 1940's and 1950's. An understanding of heredity, reproduction, and aging at the molecular level has produced an entirely new field of science, molecular biology, which has given birth to genetic engineering, with its awesome possibilities. In order to see how DNA can store and pass on information, we must learn about its molecular structure.

DNA is a polynucleotide, which implies that it is a polymer made up of nucleotides. *Nucleotides,* the monomers of DNA and RNA, consist of three parts: a sugar (monosaccharide), a phosphate group, and a nitrogen-containing ring compound called a *base.* The sugar in DNA is 2-deoxyribose and that in RNA is ribose:

Notice that the difference between 2-deoxyribose and ribose is that 2-deoxyribose is lacking an oxygen atom to the carbon atom labeled 2.

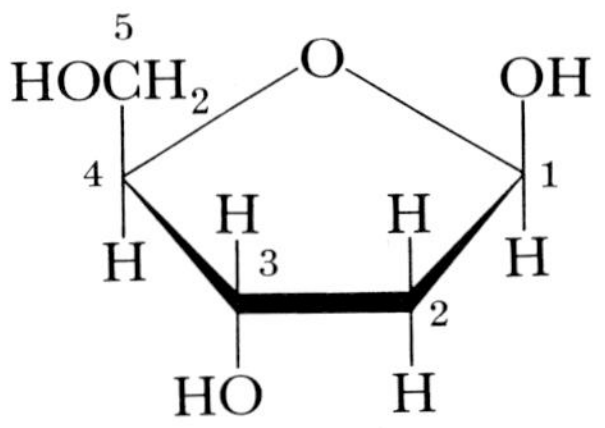

2-deoxyribose

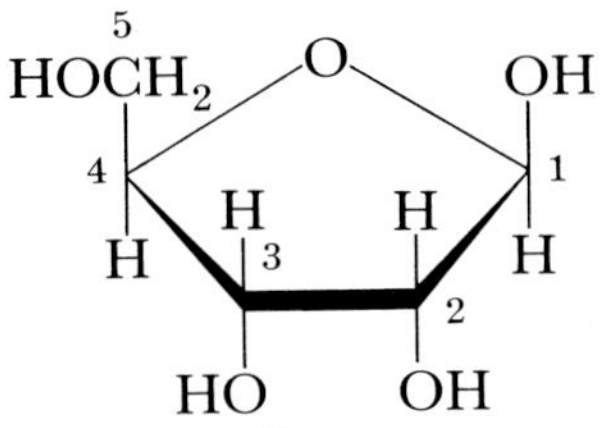

ribose

The carbon atoms are numbered 1 to 5 in the formulas because we are going to refer to them later. In both DNA and RNA the phosphate group is attached to the number 5 carbon atom in the sugar:

The phosphate group is bonded to the monosaccharide group by the condensation reaction

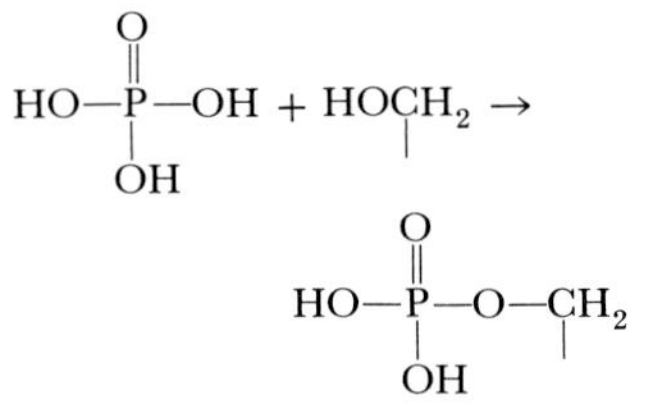

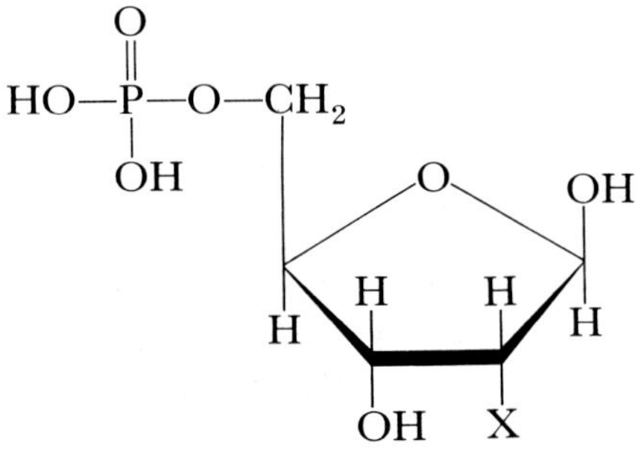

$$\text{HO—P(=O)(OH)—OH} + \text{HOCH}_2\text{—} \rightarrow \text{HO—P(=O)(OH)—O—CH}_2\text{—}$$

The group labeled X is —OH in ribose and —H in deoxyribose.
The five bases that occur in DNA and RNA are

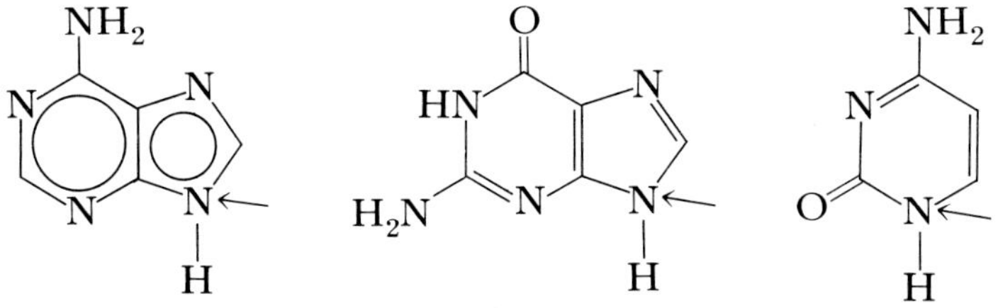

adenine guanine cytosine

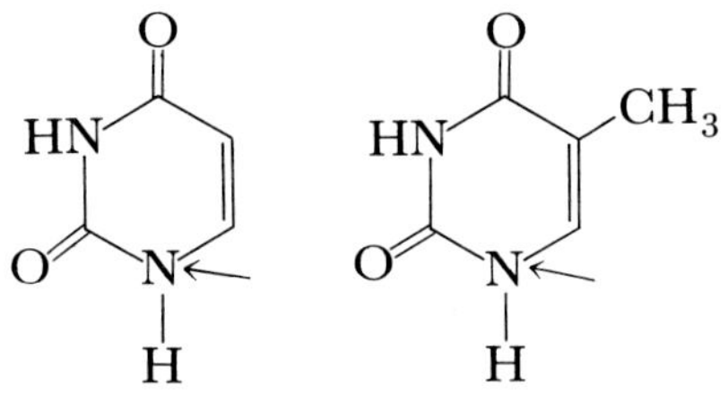

uracil thymine

The bases are bonded to the ribose or deoxyribose at the positions indicated by the arrows. DNA contains only the four bases adenine (A), guanine (G), cytosine (C), and thymine (T), and RNA contains adenine (A), guanine (G), cytosine (C), and uracil (U). The bases are bonded to the sugar-phosphates at the number 1 carbon atom. For example, we have

The bases that occur in DNA and RNA are

DNA	RNA
A, G, C, T	A, G, C, U

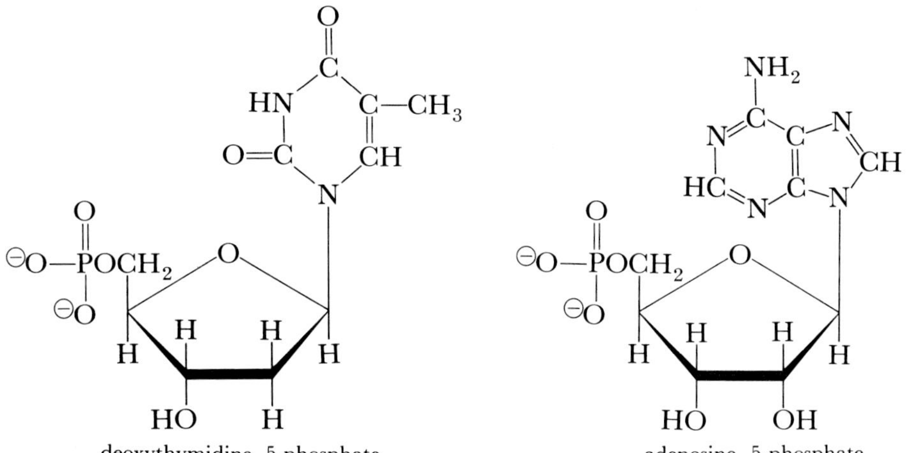

deoxythymidine 5-phosphate adenosine 5-phosphate

Both of these molecules are *nucleotides*. Deoxythymidine 5-phosphate is one of four monomers of DNA, and adenosine 5-phosphate is one of four monomers of RNA.

Nucleotides can be joined together by a condensation reaction between the phosphate group of one nucleotide and the 3-hydroxyl group of another. The result is a polynucleotide, part of which might look like this:

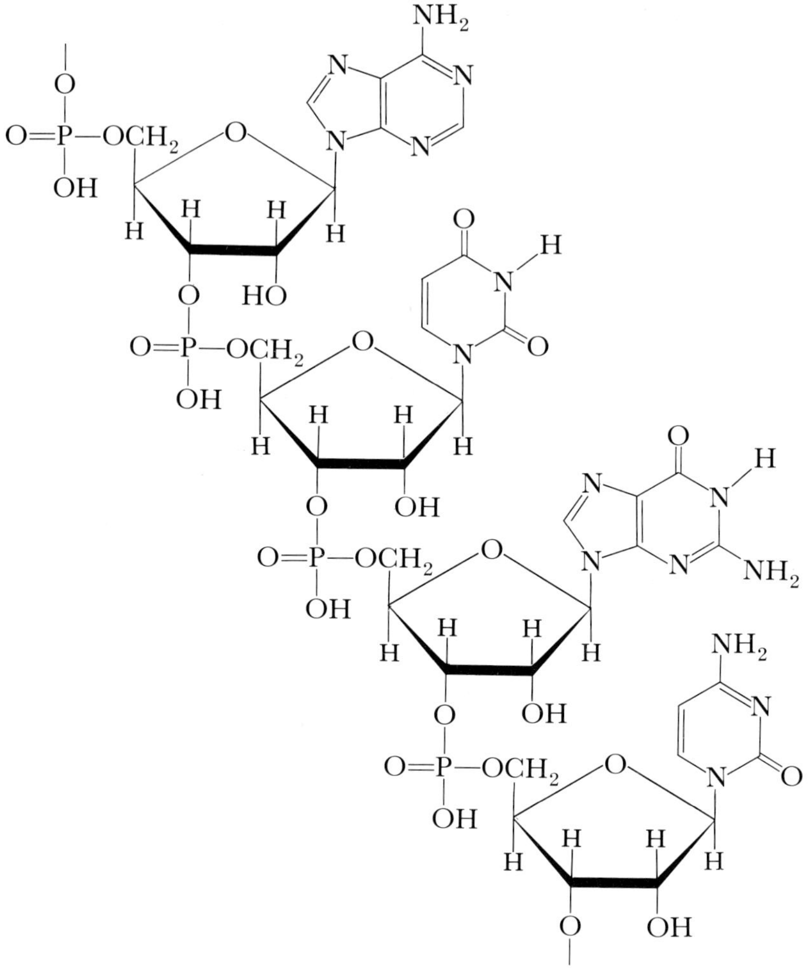

Thus we see that DNA and RNA consist of a *sugar-phosphate backbone* with bases attached at intervals. Let's see now how a molecule like DNA can store and pass on genetic information.

26-11. DNA IS A DOUBLE HELIX

$G = C$, $A = T$ } regardless of DNA source

The key to understanding how DNA works lies in its three-dimensional structure. In 1953, James Watson and Francis Crick (Figure 26-11) proposed that DNA consists of two polynucleotide chains intertwined in a *double helix* (Figure 26-12). Their proposal of such a structure was based on two principal observations: X-ray data indicated that DNA is helical, and chemical analysis revealed that, regardless of the source of DNA, be it a simple bacterium or the higher vertebrates, the amount of guanine is always equal to the amount of

Figure 26-11 In the early 1950's, James Watson (left), who had recently received his Ph.D. in zoology from Indiana University, went to Cambridge University on a postdoctoral research fellowship. He and the British physicist Francis Crick worked together on the molecular structure of DNA. In 1953 they proposed the double helical model of DNA, which explains beautifully how DNA can store and transmit genetic information. Their proposal is touted as one of the most important scientific breakthroughs of modern times. Watson and Crick were awarded the Nobel Prize in physiology and medicine in 1962. The details of their discovery are given by Watson in his book *The Double Helix.*

cytosine and the amount of adenine is always equal to the amount of thymine.

Watson and Crick realized that the bases in DNA must somehow be paired. Working with molecular models, they discovered that A and T were of the right shape and size to form two hydrogen bonds:

H
CH_3 O···H—N
N—H···N
N
N
N
N—
Sugar
Sugar
O
1.1 nm
T A

Similarly, G and C fit together at the same distance, forming three hydrogen bonds:

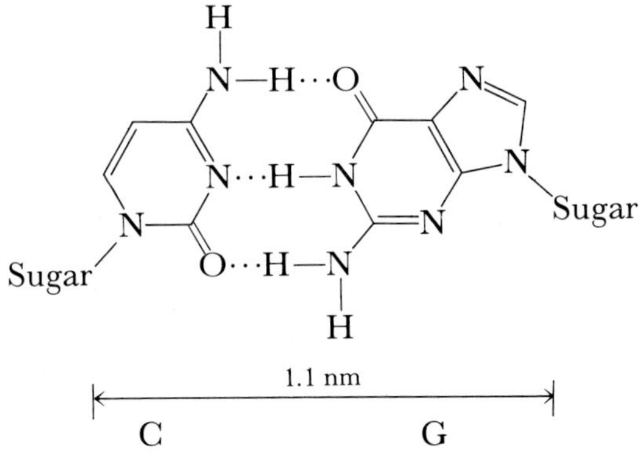

1.1 nm

C G

Notice that both base-pairing schemes allow the two strands of the double helix to have a separation of 1.1 nm. Other possible base pairs, such as C—C, T—T, and C—T, are too small, and A—A, G—G, and A—G are too large. Others that are the right size (A—C and G—T) cannot pair because their atoms are not in suitable positions to form hydrogen bonds:

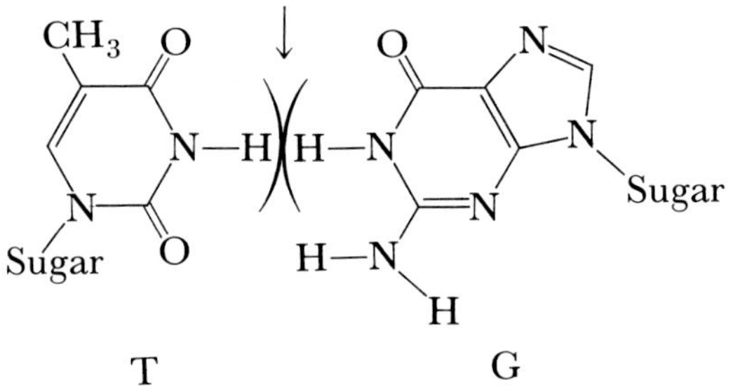

Unfavorable for hydrogen bonding

T G

Thus only A—T and G—C base pairs can form, and this fact accounts for the structure of DNA. The two strands of the double helix are *complementary* to each other.

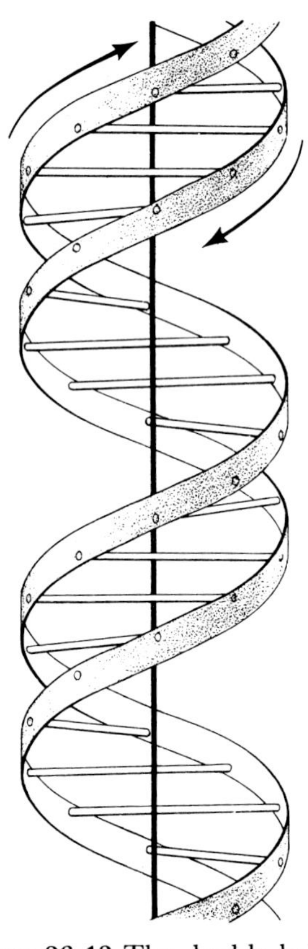

Figure 26-12 The double helical structure of DNA consists of two polynucleotide strands twisted about each other.

Example 26-6: If the base sequence along a portion of one strand of a double helix is . . . AGCCTCG . . . , what must be the corresponding sequence on the other strand?

Solution: The two sequences must be complementary to each other, meaning that a T and an A must be opposite each other and a G and a C must be opposite each other. Thus, the other sequence must have the base sequence . . . TCGGAGC . . . .

The two strands of the DNA double helix are held together by hydrogen bonds between complementary bases (Figure 26-13). Figure 26-14 shows a molecular model of a segment of a DNA double helix, and Figure 26-15 gives a cross-sectional view.

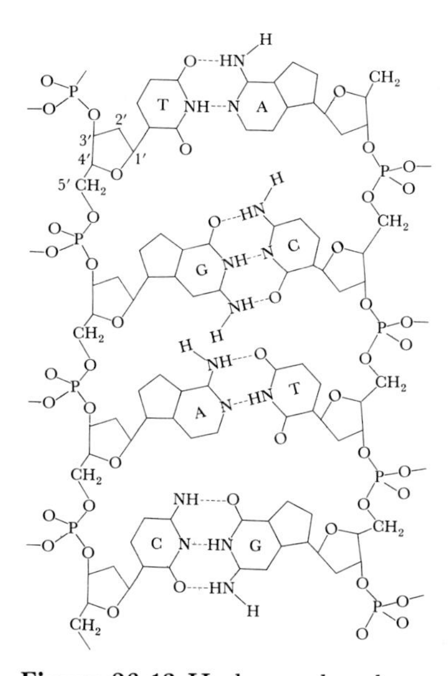

Figure 26-13 Hydrogen bonds between complementary base pairs hold the two strands of DNA together in a double helical configuration.

26-12. DNA CAN REPRODUCE ITSELF

The Watson-Crick model for DNA explains DNA replication. The two strands of the double helix are held together by hydrogen bonds, not covalent bonds. Hydrogen bonds are weak enough to allow the double helix to uncoil into two separate strands. Each strand can then act as a template for building a complementary strand, and the result will be two double helices that are identical to the first. Example 26-7 illustrates this process.

Example 26-7: Suppose a segment along a double helix is

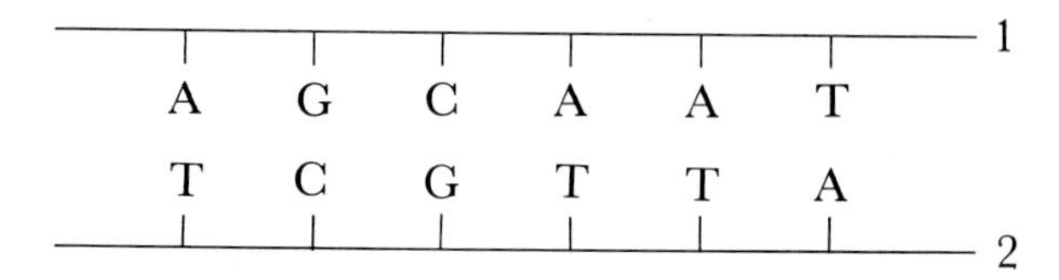

Show that if the double helix uncoils, then each separate strand can act as a template to build two identical double helices.

Solution: The two strands come apart to give

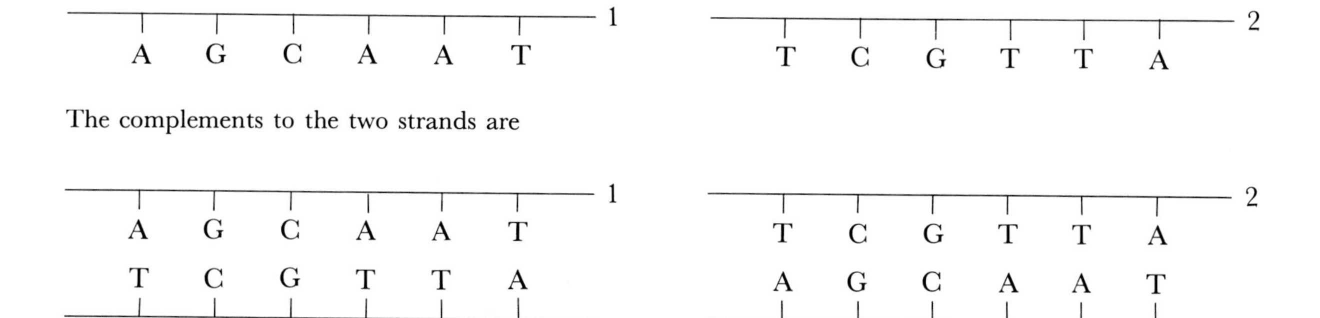

The complements to the two strands are

These two double helices are identical to each other and to the first double helix. Thus, the original double helix has repeated itself.

Figure 26-14 A molecular model of a segment of a DNA double helix. The double helical structure gives DNA the appearance of a long cylinder. The bases are stacked parallel to each other and perpendicular to the axis of the cylinder. The bases lie inside the cylinder, and the two sugar-phosphate backbone chains are wrapped around the outside.

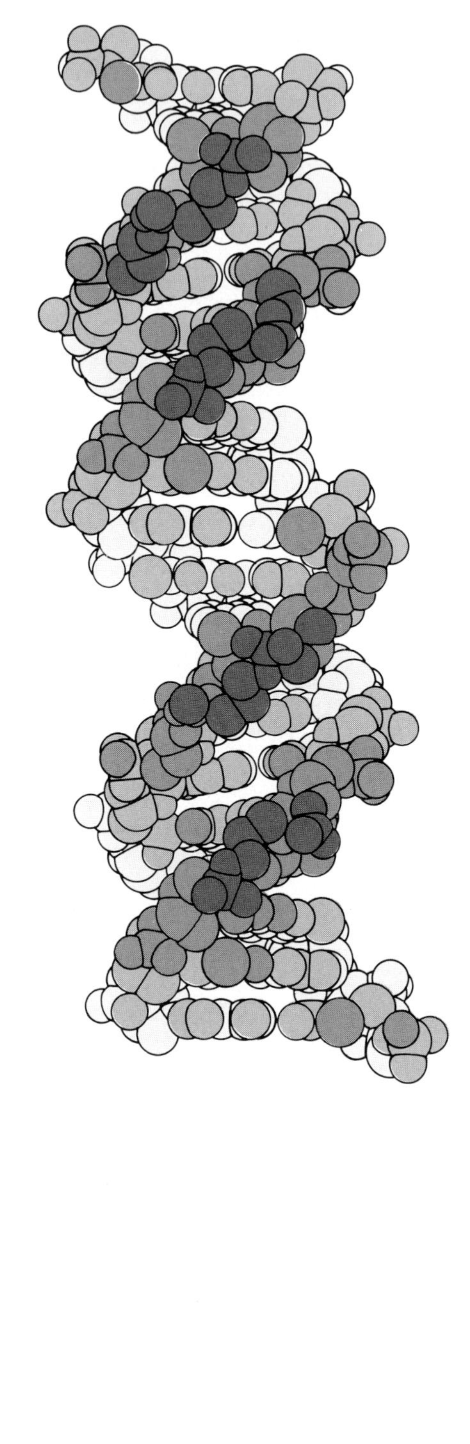

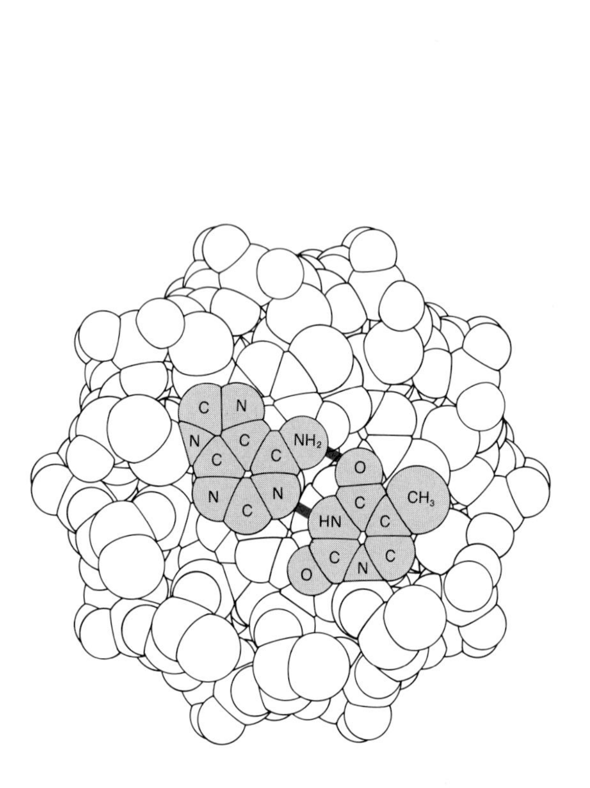

Figure 26-15 A cross-sectional view of a DNA double helix, showing how the bases lie inside.

26-13. GENETIC INFORMATION IS STORED IN A TRIPLET CODE

Living systems differ from one another by the myriad biochemical processes characteristic of each system. All these biochemical reactions are controlled by enzymes, and many of them involve other proteins as well. In a sense, each living system is a reflection of its various proteins. What we mean by genetic information is the information that calls for the production of all the proteins characteristic of a given organism. This is the information that is stored in DNA.

It was discovered in the 1950's that each series of three bases along a DNA segment is a code that leads to a particular amino acid. For example, the triplet AAA is a DNA code for phenylalanine and TTC is a DNA code for lysine. Thus the segment AAATTC in DNA would give rise to a segment phe-lys in a protein. Because a sequence of three bases is the code for one particular amino acid, the code is called the *triplet code*.

The chemical reactions that are involved in transcribing the base sequence along a DNA strand into a protein molecule are complicated but fairly well understood. They involve several types of RNA and numerous enzymes and other proteins. If you go on to take a course in biochemistry or biology, you will study the DNA-protein pathway in detail.

In summary, then, DNA is a double helix that is held together by hydrogen bonds between complementary bases. During replication, the two strands come apart and act as templates for the production of two identical double helices. It is in this way that genetic information is transmitted. The genetic information is stored in the form of the sequence of bases along each strand of the double helix. Each sequence of three bases codes a particular amino acid during protein synthesis. Because the nature and the order of the bases are equivalent to genetic information, the bases are in the interior of the double helix for protection. A *gene* is a segment along a DNA molecule that codes the synthesis of one polypeptide. DNA can have a molecular mass of over 100,000,000.

SUMMARY

Proteins are naturally occurring polypeptides, which are polymers whose monomers are the 20 amino acids that are common to all living species. The order of the amino acids in a polypeptide chain is called the primary structure of the chain and specifies the polypeptide uniquely. A polypeptide chain can be pictured as a polypeptide backbone with amino acid side groups attached at intervals. The side groups interact with each other and with solvent to effect a three-dimensional shape that is characteristic of the polypeptide and called its tertiary structure. The function and efficacy of a protein are extraordinarily sensitive to its tertiary structure.

Another class of biopolymers are the polysaccharides. Both starch and cellulose have the same monomer (monosaccharide), glucose. The glucose units are joined in different ways in starch and cellulose. Humans possess the enzymes that are required to break down starch into glucose but not those required to break down cellulose. Consequently, starch can be used as a food for humans but cellulose cannot.

A third class of biopolymers are polynucleotides. The polynucleotide DNA stores and transmits genetic information. The Watson and Crick double helical model pictures DNA as two complementary strands joined together by hydrogen bonds. This model explains how DNA can replicate itself and transmit genetic information from generation to generation. The genetic information is stored in DNA as a triplet code.

TERMS YOU SHOULD KNOW

protein 1032
amino acid 1032
side group 1032
optical isomer 1034
stereospecific 1035
peptide bond 1035
dipeptide 1035
polypeptide 1036
polypeptide backbone 1036
hydrophobic 1036
hydrophilic 1036
primary structure 1037
disulfide bond 1039
α-helix 1040
secondary structure 1040
tertiary structure 1041
enzyme 1043
lock-and-key theory 1043
substrate 1043
carbohydrate 1045
monosaccharide 1046
polysaccharide 1046
disaccharide 1046
sugar 1047
hydrolyze 1049
polynucleotide 1049
DNA 1050
RNA 1050
nucleotide 1050
base 1050
sugar-phosphate backbone 1052
double helix 1052
complementary 1054
triplet code 1057
gene 1057

PROBLEMS

OPTICAL ISOMERS

26-1. Which of the following compounds can exist as optical isomers:

(a) CH_3Cl

(b) $H_2C{=}CCl_2$

(c) H_2NCHCH_2OH (with COOH on the CH carbon)

(d) $Cl{-}C(H)(Br){-}COOH$

(e) 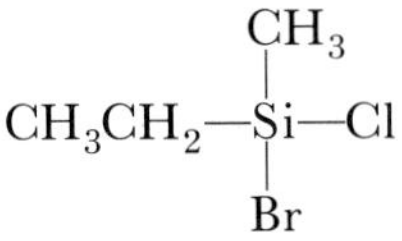 $CH_3CH_2{-}Si(CH_3)(Br){-}Cl$

26-2. Which of the following compounds can exist as optical isomers:

(a) $Br{-}C(NH_2)(H){-}COOH$

(b) $F_2C{=}CCl_2$

(c) $CH_3{-}C(CH_2OH)(OH){-}CH{=}CH_2$

(d) $(CH_3)_2SiCl_2$

(e) 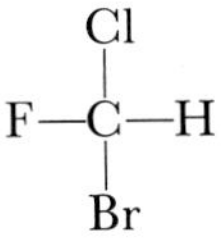 $F{-}C(Cl)(Br){-}H$

PROPERTIES OF AMINO ACID SIDE GROUPS

26-3. According to Table 26-1, aspartic acid has an acidic side group. Show the reaction that gives the acidic character to aspartic acid.

26-4. According to Table 26-1, lysine has a basic side group. Show the reaction that gives the basic character to lysine.

FORMATION OF PEPTIDES FROM AMINO ACIDS

26-5. Write the equation for the reaction between isoleucine and methionine.

26-6. Write the equation for the reaction between threonine and lysine.

26-7. Draw the structural formulas for the two possible dipeptides that can be formed from the reaction between glycine and alanine.

26-8. Draw the structural formulas for the two possible dipeptides that can be formed from the reaction between valine and asparagine.

26-9. How many different tripeptides can be formed from two different amino acids? Draw the structural formula for each of your tripeptides.

26-10. How many different tripeptides can be formed from three different amino acids? Draw the structural formula for each of your tripeptides.

STRUCTURAL FORMULAS FOR PEPTIDES

26-11. Draw the structural formula for the tripeptide glu-arg-trp.

26-12. Draw the structural formula for the tripeptide asn-val-cys.

26-13. The pentapeptide leucine enkephalin occurs in the brain and is involved in the perception of pain. Draw the structural formula for this compound, which has the sequence tyr-gly-gly-phe-leu.

26-14. The pentapeptide methionine enkephalin occurs in the brain and regulates pain perception. Draw the structural formula for this compound, which has the sequence tyr-gly-gly-phe-met.

26-15. Draw the structural formulas for the hormones oxytocin and vasopressin (see text for sequences).

26-16. The antibiotic enniatin is a cyclic compound composed of three monomer units of the type

```
H3C     CH2CH3    CH3
   \   /      \   /
    CH          CH
    |           |
—C—C—O—C—C—N—
 ‖  |     ‖  |  |
 O  H     O  H  CH3
```

Draw the structural formula for enniatin.

SECONDARY AND TERTIARY STRUCTURE

26-17. Which of the following amino acids have side groups that occur on the surface of a protein:

(a) cys (b) leu (c) phe (d) arg (e) gln

26-18. Which of the following amino acids have side groups that occur on the surface of a protein:

(a) lys (b) trp (c) asn (d) met (e) ala

26-19. Which of the following amino acids have side groups that cluster in the interior of a protein:

(a) thr (b) leu (c) arg (d) phe (e) tyr

26-20. Which of the following amino acids have side groups that cluster in the interior of a protein:

(a) glu (b) lys (c) ser (d) val (e) pro

26-21. Identify the atom(s), if any, in the following amino acid side groups that can form a hydrogen bond with water:

(a) $-\underset{\displaystyle OH}{\underset{|}{C}}HCH_3$

(b) $-\underset{\displaystyle CH_3}{\underset{|}{C}}HCH_3$

(c) $-CH_2CH_2CH_2CH_2NH_2$

(d) $-CH_2CH_2C(=O)O^{\ominus}$

(e) $-CH_2CH_2SCH_3$

26-22. Identify the atom(s), if any, in the following amino acid side groups that can form a hydrogen bond with water:

(a) $-CH_2OH$

(b) $-CH_2-C_6H_4-OH$

(c) $-CH_2C(=O)-NH_2$

(d) $-CH_2C(=O)O^{\ominus}$

(e) $-CH_3$

26-23. What is the difference between primary, secondary, and tertiary structure of proteins?

26-24. What are the factors that govern the secondary and tertiary structure of proteins?

CARBOHYDRATES

26-25. Write the equation for the condensation reaction that produces the disaccharide lactose.

26-26. Write the equation for the condensation reaction that produces the disaccharide cellobiose, which is the repeating disaccharide unit of cellulose.

26-27. Draw the structural formula for the trisaccharide melezitose (α-glucose, β-fructose, α-glucose), which is found in the sap of some coniferous trees.

26-28. Draw the structural formula for the trisaccharide raffinose (β-fructose, α-glucose, galactose), which is found in sugar beets and other plants.

26-29. An amylose polymer was found to have a molecular mass of 200,000. Estimate the number of α-glucose units in the molecule.

26-30. A cellulose polymer was found to have a molecular mass of 500,000. Estimate the number of β-glucose units in the polymer.

26-31. The first step in the digestion of sucrose is its hydrolysis to glucose and fructose. Write the equation for this reaction in terms of structural formulas.

26-32. The first step in the digestion of maltose is its hydrolysis to two glucose molecules. Write the equation for this reaction in terms of structural formulas.

STRUCTURAL FORMULAS FOR TRINUCLEOTIDES

26-33. Draw the structural formula for the DNA triplet GAT.

26-34. Draw the structural formula for the DNA triplet ATC.

26-35. Draw the structural formula for the RNA triplet UCU.

26-36. Draw the structural formula for the RNA triplet CUG.

COMPLEMENTARY BASE PAIRS

26-37. If the base sequence along a portion of one strand of a double helix is GTTCCCAAG, what must the corresponding sequence on the other strand be?

26-38. If the base sequence along a portion of one strand of a double helix is CATGGCTAA, what must the corresponding sequence on the other strand be?

26-39. Determine the complementary base sequence that corresponds to the following sequence of DNA bases:

G C T C G C A

26-40. Determine the complementary base sequence that corresponds to the following sequence of DNA bases:

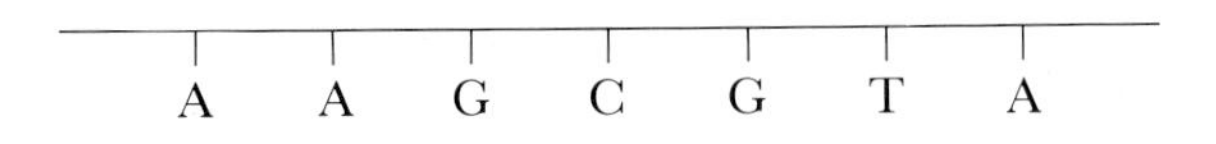

26-41. Suppose a segment along a double helix is

```
——┬———┬———┬———┬———┬———┬———┬——— 1
  G   C   T   T   A   C   G
  C   G   A   A   T   G   C
——┴———┴———┴———┴———┴———┴———┴——— 2
```

Draw the segments obtained when the DNA reproduces itself.

26-42. Suppose a segment along a double helix is

```
——┬———┬———┬———┬———┬———┬———┬——— 1
  T   C   G   T   A   C   G
  A   G   C   A   T   G   C
——┴———┴———┴———┴———┴———┴———┴——— 2
```

Draw the segments obtained when the DNA reproduces itself.

26-43. Determine the number of hydrogen bonds that must be broken to separate the strands of the DNA segment shown in Example 26-7.

26-44. Determine the number of hydrogen bonds that must be broken to separate the strands of the DNA segment shown in Problem 26-42.

CHARGED FORMS OF AMINO ACIDS

26-45. We have shown the structural formulas for amino acids as

$$H_2N-\underset{G}{\overset{H}{C}}-COOH$$

In solution, however, amino acids exist in the form

$$H_3\overset{\oplus}{N}-\underset{G}{\overset{H}{C}}-COO^{\ominus}$$

which is called a *zwitterion*. Draw the structural formula for the zwitterionic form of alanine.

26-46. Problem 26-45 describes the zwitterionic form of amino acids. Draw the structural formula for the zwitterionic form of valine.

ENZYME KINETICS

26-47. The Michaelis-Menten mechanism for enzyme-catalyzed reactions is

$$E(aq) + S(aq) \underset{k_{-1}}{\overset{k_1}{\rightleftharpoons}} ES(aq) \xrightarrow{k_2} E(aq) + P(aq)$$

where E denotes the enzyme, S the substrate, ES the enzyme-substrate complex, and P the product. The rate law for this mechanism is

$$\text{rate} = \frac{k_2[E_0][S]}{(k_{-1} + k_2)/k_1 + [S]}$$

where $[E_0]$ is the total enzyme concentration. Sketch a graph of the initial rate versus [S] for a fixed value of $[E_0]$. Show that this rate law is linear in [S] for small [S] and levels off for large [S]. Interpret this behavior.

26-48. Another form for the rate law for the Michaelis-Menten mechanism shown in Problem 26-47 is

$$\text{rate} = \frac{k[E_0][S]}{K + [S]}$$

where k and K are constants and $[E_0]$ is the total enzyme concentration. Show that a plot of 1/rate versus 1/[S] is linear. What is the slope? What is the intercept? Such a plot is called a Lineweaver-Burk plot.

26-49. Show that the Michaelis-Menten rate law (Problem 26-47) becomes

$$\text{rate} = k_2[\text{E}_0]$$

for large [S]. The constant k_2 is called the *turnover number* of the enzyme reaction. Argue that the turnover number is the number of substrate molecules converted to products per second by one enzyme molecule when all available active sites of all enzyme molecules are occupied by substrate molecules. The turnover numbers for some enzymes are given in Table 26-3.

26-50. Use the data in Table 26-3 to compute

(a) the limiting reaction rate (large values of [S]) for acetylcholine esterase hydrolysis of the substrate acetylcholine for an enzyme concentration of 1.0×10^{-5} M
(b) the time required under the limiting-rate conditions for a given enzyme-substrate complex to undergo reaction (see Problem 26-49)

Table 26-3 Rate constant data (turnover numbers) for some enzyme-catalyzed reactions

Enzyme	*Substrate*	*Turnover number/s^{-1}*
carbonic anhydrase	carbon dioxide	600,000
acetylcholine esterase	acetylcholine	25,000
penicillinase	benzylpenicillin	2000
chymotrypsin	acetyl-L-tryptophanamide	100
DNA polymerase I	DNA and nucleosides	15
lysozyme	hexa-*N*-acetylglucosamine	0.5

26-51. A 1.0-mL sample containing acetylcholine esterase hydrolyzed 2.5×10^{-5} mol of acetylcholine per second. Calculate the concentration of the enzyme, using the turnover number given in Table 26-3.

26-52. The hydrolysis of urea, $H_2NC(=O)NH_2$, in aqueous solution,

$$H_2N{-}C({=}O){-}NH_2(aq) + H_2O(l) \rightarrow CO_2(aq) + 2NH_3(aq)$$

is catalyzed by the enzyme urease. The uncatalyzed hydrolysis of urea has a reaction half-life at 25°C of 32 years (10^9 s). The half-life for the urease·urea (enzyme·substrate) complex at 25°C is 10^{-4} s, which means that the enzyme-catalyzed reaction goes 10^{13} times faster. Suppose that a biochemist prepares two solutions, one containing the enzyme urease from the liver and the other containing enzyme from the kidney. The rates of urea hydrolysis are measured using the same urea concentrations and the same volumes of protein solution. The rate of hydrolysis in the solution from the liver is $0.032\ \text{M}\cdot\text{min}^{-1}$ and that in the kidney solution is $0.67\ \text{M}\cdot\text{min}^{-1}$. Which organ contains more urease and how much more?

26-53. An important source of energy for cells is the combustion of glucose:

$$C_6H_{12}O_6(aq) + 6O_2(g) \rightarrow 6CO_2(g) + 6H_2O(l)$$
$$\Delta G^\circ_{rxn} = -2.87 \times 10^3\ kJ \cdot mol^{-1} \text{ at pH} = 7.0,\ 25^\circ C$$

Calculate the maximum amount of work that can be obtained from 1.0 g of glucose under standard conditions and pH = 7.0.

26-54. Plants synthesize carbohydrates from CO_2 and H_2O by the process of photosynthesis. For example,

$$6CO_2(g) + 6H_2O(l) \rightarrow \underset{\text{glucose}}{C_6H_{12}O_6(aq)} + 6O_2(g)$$

Calculate ΔG°_{rxn} for this reaction (see Problem 26-53). Is the reaction spontaneous?

26-55. Given that at 25°C and pH = 7.0

$$\text{glucose}(aq) + 6O_2(g) \rightarrow 6CO_2(g) + 6H_2O(l) \quad \Delta G^\circ_{rxn} = -2.87 \times 10^3\ kJ \cdot mol^{-1}$$

$$\text{sucrose}(aq) + H_2O(l) \rightarrow \text{glucose}(aq) + \text{fructose}(aq) \quad \Delta G^\circ_{rxn} = -29.3\ kJ \cdot mol^{-1}$$

$$\text{fructose}(aq) \rightarrow \text{glucose}(aq) \quad \Delta G^\circ_{rxn} = -1.6\ kJ \cdot mol^{-1}$$

calculate the value of ΔG°_{rxn} for the combustion of sucrose. Calculate the maximum amount of work that can be obtained from the combustion of 1.0 g of sucrose.

26-56. An important biochemical reaction is the hydrolysis of adenosine triphosphate, ATP:

$$ATP(aq) + H_2O(l) \rightarrow ADP(aq) + HPO_4^{2-}(aq)$$
$$\Delta G^\circ_{rxn} = -29\ kJ \cdot mol^{-1} \text{ at pH} = 7.0 \text{ and } 25^\circ C$$

Calculate the value of ΔG°_{rxn} for the reaction

$$\text{glucose}(aq) + 38HPO_4^{2-}(aq) + 38ADP(aq) + 6O_2(g) \rightarrow 6CO_2(g) + 44H_2O(l) + 38ATP(aq)$$

Is the reaction spontaneous?

26-57. Much of the energy from the combustion of glucose is used to synthesize adenosine triphosphate, ATP. The hydrolysis of ATP provides the energy to drive unfavorable reactions, perform work, and carry out other functions of the cells. The reaction is

$$ATP(aq) + H_2O(l) \rightarrow ADP(aq) + HPO_4^{2-}(aq)$$
$$\Delta G^\circ_{rxn} = -31\ kJ \cdot mol^{-1} \text{ (cellular conditions, } 37^\circ C)$$

Calculate the equilibrium constant for the hydrolysis of ATP.

26-58. The hydrolysis of ATP provides the energy needed for the contraction of muscles. In a resting muscle, the concentration of ATP is 50 mM, the concentration of ADP is 0.5 mM, and the concentration of HPO_4^{2-} is 1.0 mM. Calculate the value of ΔG°_{rxn} for the hydrolysis of ATP under these conditions. Take $\Delta G^\circ_{rxn} = -31\ kJ \cdot mol^{-1}$.

26-59. Another important source of energy in biological systems is glycolysis, the process by which glucose is broken down to lactic acid:

$$C_6H_{12}O_6(aq) \rightarrow 2CH_3CHOHCOOH(aq)$$
$$\Delta G^\circ_{rxn} = -200\ kJ \cdot mol^{-1} \text{ at pH} = 7.0 \text{ and } 25^\circ C$$

Calculate the equilibrium constant for this reaction.

26-60. The source of energy in human erythrocytes (red blood cells) is glycolysis. In erythrocytes the concentration of glucose is 5.0 mM and the concentration of lactic acid is 2.9 mM. Calculate the value of ΔG°_{rxn} under these conditions at 25°C. See Problem 26-59 for the necessary data.

APPENDIX A

A Mathematical Review

A1. SCIENTIFIC NOTATION AND EXPONENTS

The numbers encountered in chemistry are often extremely large (such as Avogadro's number) or extremely small (such as the mass of an electron in kilograms). When working with such numbers, it is convenient to express them in *scientific notation,* where we write the number as a number between 1 and 10 multiplied by 10 raised to the appropriate power. For example, the number 171.3 is $1.713 \times 100 = 1.713 \times 10^2$ in scientific notation. Some other examples are

$$7320 = 7.32 \times 10^3$$
$$1{,}624{,}000 = 1.624 \times 10^6$$

The zeros in these numbers are not regarded as significant figures and are dropped in scientific notation. Notice that in each case the power of 10 is the number of places that the decimal point has been moved to the left:

$$\underbrace{7320.}_{\text{3 places}} \qquad \underbrace{1624000.}_{\text{6 places}}$$

When numbers that are smaller than 1 are expressed in scientific notation, the 10 is raised to a negative power. For example, 0.614 becomes 6.14×10^{-1}. Recall that a negative exponent is governed by the relation

$$10^{-n} = \frac{1}{10^n} \tag{A1-1}$$

Some other examples are

$$0.0005 = 5 \times 10^{-4}$$
$$0.000000000446 = 4.46 \times 10^{-10}$$

Notice that the power of 10 in each case is the number of places that the decimal point has been moved to the right:

$$\underbrace{0.0005}_{\text{4 places}} \qquad \underbrace{0.000000000446}_{\text{10 places}}$$

It is necessary to be able to work with numbers in scientific notation. To add or subtract two or more numbers expressed in scientific notation, the power of 10 must be the same in both. For example, consider the sum

$$5.127 \times 10^4 + 1.073 \times 10^3$$

We rewrite the first number so that we have 10^3:

To change a number such as 51.27×10^3 to 5.127×10^4, we make the number in front one factor of 10 smaller, and so we must make 10^3 one factor of 10 larger.

$$5.127 \times 10^4 = 51.27 \times 10^3$$

Note that we have changed the 10^4 factor to 10^3, and so we must make the factor in front of 10^3 one power of 10 larger. Thus we have

$$\begin{aligned} 5.127 \times 10^4 + 1.073 \times 10^3 &= (51.27 + 1.073) \times 10^3 \\ &= 52.34 \times 10^3 \\ &= 5.234 \times 10^4 \end{aligned}$$

Similarly, we have

Note that in changing 2.156×10^{-7} to 0.2156×10^{-6}, we make 2.156 one factor of 10 smaller and 10^{-7} one factor of 10 larger.

$$\begin{aligned} (4.728 \times 10^{-6}) - (2.156 \times 10^{-7}) &= (4.728 - 0.2156) \times 10^{-6} \\ &= 4.512 \times 10^{-6} \end{aligned}$$

In multiplying two numbers, we add the powers of 10 because of the relation

$$(10^x)(10^y) = 10^{x+y} \tag{A1-2}$$

For example,

$$\begin{aligned} (5.00 \times 10^2)(4.00 \times 10^3) &= (5.00)(4.00) \times 10^5 \\ &= 20.0 \times 10^5 \\ &= 2.00 \times 10^6 \\ (3.014 \times 10^3)(8.217 \times 10^{-6}) &= (3.014)(8.217) \times 10^{-3} \\ &= 24.77 \times 10^{-3} \\ &= 2.477 \times 10^{-2} \end{aligned}$$

To divide, we subtract the power of 10 of the number in the denominator from the power of 10 of the number in the numerator because of the relation

$$\frac{10^x}{10^y} = 10^{x-y} \tag{A1-3}$$

For example,

$$\begin{aligned} \frac{4.0 \times 10^{12}}{8.0 \times 10^{23}} &= \left(\frac{4.0}{8.0}\right) \times 10^{12-23} \\ &= 0.50 \times 10^{-11} \\ &= 5.0 \times 10^{-12} \\ \frac{2.80 \times 10^{-4}}{4.73 \times 10^{-5}} &= \left(\frac{2.80}{4.73}\right) \times 10^{-4+5} \\ &= 0.592 \times 10^{1} \\ &= 5.92 \end{aligned}$$

To raise a number to a power, we use the fact that

$$(10^x)^n = 10^{nx} \tag{A1-4}$$

For example,

$$\begin{aligned}(2.187 \times 10^2)^3 &= (2.187)^3 \times 10^6\\ &= 10.46 \times 10^6\\ &= 1.046 \times 10^7\end{aligned}$$

To take a root of a number, we use the relation

$$\sqrt[n]{10^x} = (10^x)^{1/n} = 10^{x/n} \tag{A1-5}$$

Thus, the power of 10 must be written such that it is divisible by the root. For example,

$$\begin{aligned}\sqrt[3]{2.70 \times 10^{10}} &= (2.70 \times 10^{10})^{1/3} = (27.0 \times 10^9)^{1/3}\\ &= (27.0)^{1/3} \times 10^3 = 3.00 \times 10^3\end{aligned}$$

and

$$\begin{aligned}\sqrt{6.40 \times 10^5} &= (6.40 \times 10^5)^{1/2} = (64.0 \times 10^4)^{1/2}\\ &= (64.0)^{1/2} \times 10^2 = 8.00 \times 10^2\end{aligned}$$

A2. LOGARITHMS

You know that $100 = 10^2$, $1000 = 10^3$, and so on. You also know that

$$\sqrt{10} = 10^{1/2} = 10^{0.50} = 3.16$$

By taking the square root of both sides of

$$10^{0.50} = 3.16$$

we find that

$$\sqrt{10^{0.50}} = 10^{(1/2)(0.50)} = 10^{0.25} = \sqrt{3.16} = 1.78$$

Furthermore, because

$$(10^x)(10^y) = 10^{x+y}$$

we can write

$$10^{0.25} \times 10^{0.50} = 10^{0.75} = (3.16)(1.78) = 5.62$$

By continuing this process, we can express any number y as

$$y = 10^x \tag{A2-1}$$

The number x to which 10 must be raised to get y is called the *logarithm* of y and is written as

$$x = \log y \tag{A2-2}$$

Equations (A2-1) and (A2-2) are equivalent. For example, we have just shown that

$$\log 1.78 = 0.25$$
$$\log 3.16 = 0.50$$
$$\log 5.62 = 0.75$$
$$\log 10.00 = 1.00$$

Logarithms of other numbers may be obtained from tables (Appendix C) or, more conveniently, from a hand calculator. If you use tables, you must always write the number y in standard scientific notation. Thus, for example, you must write 782 as 7.82×10^2 and 0.000465 as 4.65×10^{-4}. To take the logarithm of such numbers, we use the fact that

$$\log ab = \log a + \log b \tag{A2-3}$$

Thus we write

$$\begin{aligned}\log 782 &= \log (7.82 \times 10^2) = \log 7.82 + \log 10^2 \\ &= \log 7.82 + 2.000\end{aligned}$$

Logarithm tables are set up such that the number a in $\log a$ is between 1 and 10 and the numbers in the tables are between 0 and 1. Thus from Appendix C we find, for example, that

$$\log 4.12 = 0.6149$$
$$\log 8.37 = 0.9227$$

and so on. If we look up log 7.82 in Appendix C, then we find that it is equal to 0.8932. Therefore,

$$\begin{aligned}\log 782 = \log (7.82 \times 10^2) &= \log 7.82 + \log 10^2 \\ &= 0.8932 + 2.000 = 2.8932\end{aligned}$$

If you use your calculator, you simply enter 782 and push a log key to get 2.893 directly.

To find log 0.000465, we write

$$\begin{aligned}\log 0.000465 &= \log (4.65 \times 10^{-4}) \\ &= \log 4.65 + \log 10^{-4} \\ &= \log 4.65 - 4.000\end{aligned}$$

We find $\log 4.65 = 0.6675$ from Appendix C, and so

$$\log 0.000465 = 0.6675 - 4.000$$
$$= -3.3325$$

If you use your calculator, simply enter 0.000465 and push the log key to get -3.3325 directly. Although a hand calculator is much more convenient than a table of logarithms, you should be able to handle logarithms by either method.

Because logarithms are exponents ($y = 10^x$), they have certain special properties, such as

$$\log ab = \log a + \log b \quad \text{(A2-3)}$$

$$\log \frac{a}{b} = \log a - \log b \quad \text{(A2-4)}$$

$$\log a^n = n \log a \quad \text{(A2-5)}$$

$$\log \sqrt[n]{a} = \log a^{1/n} = \frac{1}{n} \log a \quad \text{(A2-6)}$$

If we let $a = 1$ in Equation (A2-4), then we have

$$\log \frac{1}{b} = \log 1 - \log b$$

or, because $\log 1 = 0$,

$$\log \frac{1}{b} = -\log b \quad \text{(A2-7)}$$

Thus we change the sign of a logarithm by taking the reciprocal of its argument. Notice that because $\log 1 = 0$,

$$\log y > 0 \quad \text{if} \quad y > 1$$
$$\log y < 0 \quad \text{if} \quad y < 1$$

Up to this point we have found the value of x in

$$y = 10^x$$

when y is given. It is often necessary to find the value of y when x is given. Because x is called the logarithm of y, y is called the *antilogarithm* of x. For example, suppose that $x = 6.1303$ and we wish to find y. We write

$$y = 10^{6.1303} = 10^{0.1303} \times 10^6$$

From the logarithm table, we see that the number whose logarithm is 0.1303 is 1.35. Thus we find that

$$10^{6.1303} = 1.35 \times 10^6$$

You can obtain this result directly from your calculator. On a calculator having inv and log keys, for example, enter 6.1303, press the inv key (for inverse) and then the log key.

To obtain the antilogarithm of y using logarithm tables, you must express y as

$$y = 10^a \times 10^n \tag{A2-8}$$

where n is a positive or negative integer and a is between 0 and 1. The quantity a is found in the table and the antilog of a is then read.

As another example, let's find the antilog of 1.9509. We write

$$N = 10^{1.9509} = 10^{0.9509} \times 10^1$$

We find the value 0.9509 in the log table and see that its antilog is 8.93. Thus we have

$$N = 10^{1.9509} = 8.93 \times 10^1 = 89.3$$

You should be able to obtain this result directly from your calculator. If your calculator has a 10^x key, then you can obtain the antilog of 1.9509 by entering 1.9509 and pressing the 10^x key. This operation is equivalent to using the inv key followed by the log key.

In many problems, it is necessary to find the antilogarithm of negative numbers. For example, let's find the antilogarithm of -4.167, or the value of y in

$$y = 10^{-4.167}$$

Even though the exponent is negative, we still must express y in the form of Equation (A2-8). To do this, we write $-4.167 = 0.833 - 5.000$ so that

$$y = 10^{0.833} \times 10^{-5}$$

Now we find 0.833 in a logarithm table and see that its antilogarithm is 6.81. Thus

$$y = 6.81 \times 10^{-5}$$

You should be able to obtain this same result from your calculator by entering -4.167 and finding the inverse logarithm directly.

A3. THE QUADRATIC FORMULA

The standard form for a quadratic equation in x is

$$ax^2 + bx + c = 0 \tag{A3-1}$$

where a, b, and c are constants. The two solutions to the quadratic equation are

$$x = \frac{-b \pm \sqrt{b^2 - 4ac}}{2a} \tag{A3-2}$$

Equation (A3-2) is called the *quadratic formula* and is used to obtain the solutions to a quadratic equation expressed in the standard form.

For example, let's find the solutions to the quadratic equation

$$2x^2 - 3x - 1 = 0$$

In this case, $a = 2$, $b = -3$, and $c = -1$ and Equation (A3-2) gives

$$\begin{aligned} x &= \frac{3 \pm \sqrt{(3)^2 - 4(2)(-1)}}{(2)(2)} \\ &= \frac{3 \pm 4.123}{4} \\ &= 1.781 \quad \text{and} \quad -0.2808 \end{aligned}$$

To use the quadratic formula, it is first necessary to put the quadratic equation in the standard form so that we know the values of the constants a, b, and c. For example, consider the problem of solving for x in the quadratic equation

$$\frac{x^2}{0.50 - x} = 0.040$$

To identify the constants a, b, and c, we must write this equation in the standard quadratic form. Multiplying both sides by $0.50 - x$ yields

$$\begin{aligned} x^2 &= (0.50 - x)0.040 \\ &= 0.020 - 0.040x \end{aligned}$$

Rearrangement to the standard quadratic form yields

$$x^2 + 0.040x - 0.020 = 0$$

Thus $a = 1$, $b = 0.040$, and $c = -0.020$. Using Equation (A3-2), we have

$$x = \frac{-0.040 \pm \sqrt{(0.040)^2 - 4(1)(-0.020)}}{2(1)}$$

from which we compute

$$\begin{aligned} x &= \frac{-0.040 \pm \sqrt{0.0816}}{2} \\ &= \frac{-0.040 \pm 0.286}{2} \end{aligned}$$

Thus the solutions for x are

$$x = \frac{-0.040 + 0.286}{2} = 0.123$$

and

$$x = \frac{-0.040 - 0.286}{2} = -0.163$$

If x represents, say, a concentration or gas pressure, then the only physically possible value is $+0.123$ because concentrations and pressures cannot have negative values.

A4. PLOTTING DATA

The human eye and brain are quite sensitive to recognizing straight lines, and so it is always desirable to plot equations or experimental data such that a straight line is obtained. The mathematical equation for a straight line is of the form

$$y = mx + b \tag{A4-1}$$

In this equation, m and b are constants: m is the *slope* of the line and b is its *intercept* with the y axis. The slope of a straight line is a measure of its steepness; it is defined as the ratio of its vertical rise to the corresponding horizontal distance.

Let's plot the two straight lines

$$\text{I} \quad y = x + 1$$

$$\text{II} \quad y = 2x - 2$$

We first make a table of values of x and y:

I		II	
x	y	x	y
−3	−2	−3	−8
−2	−1	−2	−6
−1	0	−1	−4
0	1	0	−2
1	2	1	0
2	3	2	2
3	4	3	4
4	5	4	6
5	6	5	8

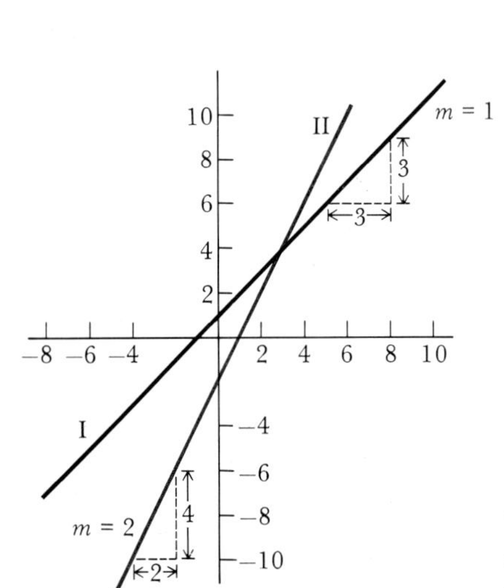

Figure A-1 Plots of the equations (I) $y = x + 1$ and (II) $y = 2x - 2$.

These results are plotted in Figure A-1. Note that curve I intersects the y axis at $y = 1$ ($b = 1$) and has a slope of 1 ($m = 1$). Curve II intersects the y axis at $y = -2$ ($b = -2$) and has a slope of 2 ($m = 2$).

Usually the equation to be plotted will not appear to be of the form of Equation (A4-1) at first. For example, consider the Boyle's law (Chapter 5) relation between the volume of a gas and its pressure:

$$V = \frac{c}{P} \qquad \text{constant temperature} \qquad \text{(A4-2)}$$

where c is a proportionality constant whose value depends upon the temperature for a given sample. For example, for a 0.29-g sample of air, $c = 0.244\ \text{L}\cdot\text{atm}$ at 25°C. Some results for such a sample are presented in Table A-1, and the data in Table A-1 are plotted as volume versus pressure in Figure A-2.

It may appear at first sight that Equation (A4-2) is not of the form $y = mx$. However, if we let $V = y$ and $1/P = x$, then Equation (A4-2) becomes

$$y = cx$$

Thus, if we plot V versus $1/P$ instead of versus P, then we get a straight line. The data in Table A-1 are plotted as V versus $1/P$ in Figure A-3. Note that a straight line is obtained.

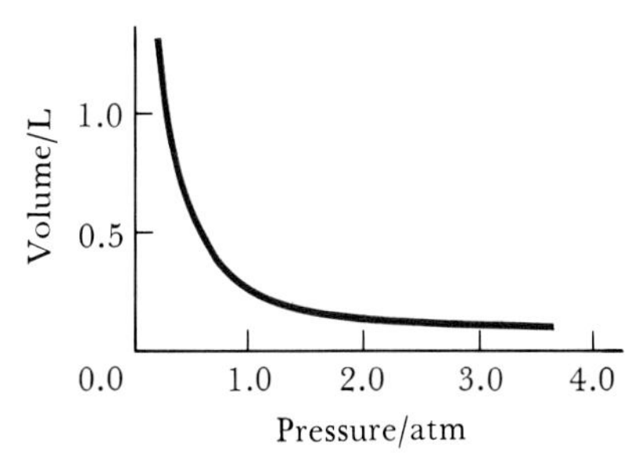

Figure A-2 The volume of 0.29 g of air plotted versus pressure at 25°C. The data are given in Table A-1. The curve in this figure is from the Boyle's law equation

$$V = \frac{0.244\ \text{L}\cdot\text{atm}}{P}$$

Table A-1 Pressure-volume data for 0.29 g of air at 25°C

P/atm	V/L	$\frac{1}{P}/atm^{-1}$
0.26	0.938	3.85
0.41	0.595	2.44
0.83	0.294	1.20
1.20	0.203	0.83
2.10	0.116	0.48
2.63	0.093	0.38
3.14	0.078	0.32

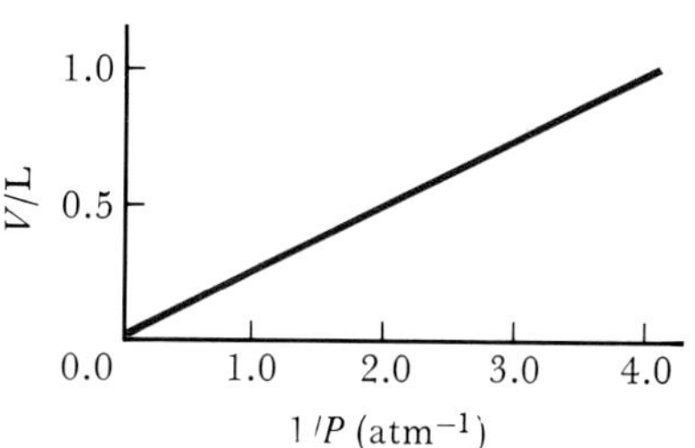

Figure A-3 The volume of 0.29 g of air plotted versus the reciprocal of the pressure ($1/P$) at 25°C. If we compare this curve with Figure A-2, we see that a straight line results when V is plotted versus $1/P$ instead of versus P. Straight lines are much easier to work from than other curves, and so it is always desirable to plot equations and data in the form of a straight line.

APPENDIX B

SI Units and Conversion Factors

Measurements and physical quantities in the sciences are expressed in the *metric system,* which is a system of units that was formalized by the French National Academy in 1790. There are several subsystems of units in the metric system, and in an international effort to achieve uniformity, the International System of Units (abbreviated SI from the French Système Internationale d'Unites) was adopted by the General Conference of Weights and Measures in 1960 as *the* recommended units for science and technology. The SI is constructed from the seven basic units given in Table B-1. The first five units in Table B-1 are used frequently in general chemistry. Each has a technical definition that serves to define the unit in an unambiguous, reproducible way, but here we simply relate the SI units to the English system.

In 1983, one meter was redefined as the distance that light travels through space in $\frac{1}{299,792,458}$ s.

1. Length: One meter is equivalent to 1.0936 yards, or to 39.370 in. Thus, a meter stick is about 3 in. longer than a yardstick.
2. Mass: One kilogram is equivalent to 2.2046 lb. The mass of a substance is determined by balancing it against a set of standard masses using a balance.
3. Temperature: The Kelvin temperature scale is related to the Celsius, or centigrade, temperature scale. On the Celsius scale, the freezing point of water is set at 0°C and its boiling point at sea level is set at 100°C. The Kelvin and Celsius scales are related by the equation (Chapter 5)

$$K = {}^\circ C + 273.15$$

   Recall that the freezing point of water is 32°F and its boiling point (at sea level) is 212°F on the Fahrenheit scale. The relation between the Celsius and Fahrenheit scales is given by

$$^\circ C = \frac{5}{9}(^\circ F - 32) \qquad \text{(B-1)}$$

   Thus, for example, 50°F corresponds to 10°C and 86°F corresponds to 30°C.

   Note that the symbol for kelvin is K and not °K.
4. Amount of substance: One mole is the amount of substance that contains as many elementary entities as there are atoms in exactly 0.012 kg of carbon-12 (Chapter 4).

Table B-1 The seven SI basic units

Physical quantity	*Name of unit*	*Symbol*
length	meter	m
mass	kilogram	kg
time	second	s
temperature	kelvin	K
amount	mole	mol
electric current	ampere	A
luminous intensity	candela	cd

An important feature of the metric system and the SI is the use of prefixes to designate multiples of the basic units (Table B-2).

The units of all quantities not listed in Table B-1 involve combinations of the basic SI units and are called *derived units.* The derived units frequently used in general chemistry are given in Table B-3. Many of these units may not be familiar to you unless you have had a course in physics. For example, the SI unit of force is a *newton,* which is defined as the force required to give a 1-kg body a speed of $1\ m \cdot s^{-1}$ when the force is applied for 1 s. The SI unit of pressure is the *pascal.* Pressure is force per area, and a pascal is defined as the pressure produced by a force of 1 N acting on an area of $1\ m^2$. The SI unit of energy is the *joule,* which is the energy that a 1-kg mass has when it

Table B-2 Prefixes used for multiples and fractions of SI units

Prefix	*Symbol*	*Multiple*	*Example*
tera-	T	10^{12}	terawatt, 1 TW = 10^{12} W
giga-	G	10^{9}	gigavolt, 1 GV = 10^{9} V
mega-	M	10^{6}	megawatt, 1 MW = 10^{6} W
kilo-	k	10^{3}	kilometer, 1 km = 10^{3} m
deci-	d	10^{-1}	decimeter, 1 dm = 10^{-1} m
centi-	c	10^{-2}	centimeter, 1 cm = 10^{-2} m
milli-	m	10^{-3}	millisecond, 1 ms = 10^{-3} s
micro-	μ[a]	10^{-6}	microsecond, μs = 10^{-6} s
nano-	n	10^{-9}	nanosecond, 1 ns = 10^{-9} s
pico-	p	10^{-12}	picometer, 1 pm = 10^{-12} m
femto-	f	10^{-15}	femtometer, 1 fm = 10^{-15} m
atto-	a	10^{-18}	attomole, 1 amol = 10^{-18} mol

[a]This is the Greek letter mu, pronounced mew.

Chemists can now measure processes that occur in one picosecond.

Table B-3 Names and symbols for some SI derived units

Quantity	*Unit*	*Symbol*	*Definition*
area	square meter	m^2	
volume	cubic meter	m^3	
density	kilogram per cubic meter	$kg \cdot m^{-3}$	
speed	meter per second	$m \cdot s^{-1}$	
frequency	hertz	Hz	s^{-1} (cycles per second)
force	newton	N	$kg \cdot m \cdot s^{-2}$
pressure	pascal	Pa	$N \cdot m^{-2} = kg \cdot m^{-1} \cdot s^{-2}$
energy	joule	J	$kg \cdot m^2 \cdot s^{-2} = N \cdot m$
electric charge	coulomb	C	$A \cdot s$
electric potential difference	volt	V	$J \cdot A^{-1} \cdot s^{-1} = kg \cdot m^2 \cdot s^{-3} \cdot A^{-1}$

is traveling at a speed of 1 $m \cdot s^{-1}$. A joule is also the energy that a mass gains when it is acted upon by a force of 1 N through a distance of 1 m. Thus, we have $J = N \cdot m$.

Although the SI is gradually becoming the universally accepted system of units, there are a number of older units that are used frequently (Table B-4). For example, volume is usually expressed in *liters,* L. A liter is defined as a cubic decimeter and is slightly larger than a quart, being equivalent to 1.0567 qt. The glassware in your laboratory is measured in milliliters (mL). One milliliter is equivalent to one cubic centimeter (cm^3).

The SI unit of pressure, the pascal, is rarely used in the United States. The most commonly used units of pressure are the *atmosphere* (atm) and the *torr* (Chapter 5).

Table B-4 Some commonly used non-SI units

Quantity	*Unit*	*Symbol*	*SI definition*
length	angstrom	Å	10^{-10} m
length	micron	μ	10^{-6} m = 1 μm
volume	liter	L	10^{-3} m^3
energy	calorie	cal	4.184 J
pressure	atmosphere	atm	101.325 kPa
pressure	torr	torr	133.322 Pa
pressure	bar	bar	10^5 Pa

We can use the relation between atmospheres and pascals to derive a relation between liter-atmospheres and joules. We use this relationship in Section 5-11. We start with

$$1 \text{ atm} = 101.32 \text{ kPa} = 1.0132 \times 10^5 \text{ Pa}$$

and multiply by L

$$1 \text{ L}\cdot\text{atm} = 1.0132 \times 10^5 \text{ L}\cdot\text{Pa}$$

Using the relations

$$\text{Pa} = \text{N}\cdot\text{m}^{-2} \qquad \text{J} = \text{N}\cdot\text{m} \qquad \text{L} = \text{dm}^3 = 10^{-3}\text{ m}^3$$

we obtain

$$\begin{aligned} 1 \text{ L}\cdot\text{atm} &= (1.0132 \times 10^5 \text{ N}\cdot\text{m}^{-2})(10^{-3}\text{ m}^3) \\ &= 101.32 \text{ N}\cdot\text{m} = 101.32 \text{ J} \end{aligned}$$

or writing this result as a unit conversion factor

$$\frac{101.32 \text{ J}}{\text{L}\cdot\text{atm}} = 1$$

In particular, in Section 5-11 we need the relation (also see Problem 5-70)

$$0.08206 \text{ L}\cdot\text{atm} = (0.08206 \text{ L}\cdot\text{atm})(101.32 \text{ J}\cdot\text{L}^{-1}\cdot\text{atm}^{-1}) = 8.314 \text{ J}$$

The SI units and their conversion factors are given on the inside back cover of this book.

APPENDIX C

Four-Place Logarithms

N	0	1	2	3	4	5	6	7	8	9
10	0000	0043	0086	0128	0170	0212	0253	0294	0334	0374
11	0414	0453	0492	0531	0569	0607	0645	0682	0719	0755
12	0792	0828	0864	0899	0934	0969	1004	1038	1072	1106
13	1139	1173	1206	1239	1271	1303	1335	1367	1399	1430
14	1461	1492	1523	1553	1584	1614	1644	1673	1703	1732
15	1761	1790	1818	1847	1875	1903	1931	1959	1987	2014
16	2041	2068	2095	2122	2148	2175	2201	2227	2253	2279
17	2304	2330	2355	2380	2405	2430	2455	2480	2504	2529
18	2533	2577	2601	2625	2648	2672	2695	2718	2742	2765
19	2788	2810	2833	2856	2878	2900	2923	2945	2967	2989
20	3010	3032	3054	3075	3096	3118	3139	3160	3181	3201
21	3222	3243	3263	3284	3304	3324	3345	3365	3385	3404
22	3424	3444	3464	3483	3502	3522	3541	3560	3579	3598
23	3617	3636	3655	3674	3692	3711	3729	3747	3766	3784
24	3802	3820	3838	3856	3874	3892	3909	3927	3945	3962
25	3979	3997	4014	4031	4048	4065	4082	4099	4116	4133
26	4150	4166	4183	4200	4216	4232	4249	4265	4281	4298
27	4314	4330	4346	4362	4378	4393	4409	4425	4440	4456
28	4472	4487	4502	4518	4533	4548	4564	4579	4594	4609
29	4624	4639	4654	4669	4683	4698	4713	4728	4742	4757
30	4771	4786	4800	4814	4829	4843	4857	4871	4886	4900
31	4914	4928	4942	4955	4969	4983	4997	5011	5024	5038
32	5051	5065	5079	5092	5105	5119	5132	5145	5159	5172
33	5185	5198	5211	5224	5237	5250	5263	5276	5289	5302
34	5315	5328	5340	5353	5366	5378	5391	5403	5416	5428
35	5441	5453	5465	5478	5490	5502	5514	5527	5539	5551
36	5563	5575	5587	5599	5611	5623	5635	5647	5658	5670
37	5682	5694	5705	5717	5729	5740	5752	5763	5775	5786
38	5798	5809	5821	5832	5843	5855	5866	5877	5888	5899
39	5911	5922	5933	5944	5955	5966	5977	5988	5999	6010
40	6021	6031	6042	6053	6064	6075	6085	6096	6107	6117
41	6128	6138	6149	6160	6170	6180	6191	6201	6212	6222
42	6232	6243	6253	6263	6274	6284	6294	6304	6314	6325
43	6335	6345	6355	6365	6375	6385	6395	6405	6415	6425
44	6435	6444	6454	6464	6474	6484	6493	6503	6513	6522
45	6532	6542	6551	6561	6571	6580	6590	6599	6609	6618
46	6628	6637	6646	6656	6665	6675	6684	6693	6702	6712
47	6721	6730	6739	6749	6758	6767	6776	6785	6794	6803
48	6812	6821	6830	6839	6848	6857	6866	6875	6884	6893
49	6902	6911	6920	6928	6937	6946	6955	6964	6972	6981
50	6990	6998	7007	7016	7024	7033	7042	7050	7059	7067
51	7076	7084	7093	7101	7110	7118	7126	7135	7143	7152
52	7160	7168	7177	7185	7193	7202	7210	7218	7226	7235
53	7243	7251	7259	7267	7275	7284	7292	7300	7308	7316
54	7324	7332	7340	7348	7356	7364	7372	7380	7388	7396

N	0	1	2	3	4	5	6	7	8	9
55	7404	7412	7419	7427	7435	7443	7451	7459	7466	7474
56	7482	7490	7497	7505	7513	7520	7528	7536	7543	7551
57	7559	7566	7574	7582	7589	7597	7604	7612	7619	7627
58	7634	7642	7649	7657	7664	7672	7679	7686	7694	7701
59	7709	7716	7723	7731	7738	7745	7752	7760	7767	7774
60	7782	7789	7796	7803	7810	7818	7825	7832	7839	7846
61	7853	7860	7868	7875	7882	7889	7896	7903	7910	7917
62	7924	7931	7938	7945	7952	7959	7966	7973	7980	7987
63	7993	8000	8007	8014	8021	8028	8035	8041	8048	8055
64	8062	8069	8075	8082	8089	8096	8102	8109	8116	8122
65	8129	8136	8142	8149	8156	8162	8169	8176	8182	8189
66	8195	8202	8209	8215	8222	8228	8235	8241	8248	8254
67	8261	8267	8274	8280	8287	8293	8299	8306	8312	8319
68	8325	8331	8338	8344	8351	8357	8363	8370	8376	8382
69	8388	8395	8401	8407	8414	8420	8426	8432	8439	8445
70	8451	8457	8463	8470	8476	8482	8488	8494	8500	8506
71	8513	8519	8525	8531	8537	8543	8549	8555	8561	8567
72	8573	8579	8585	8591	8597	8603	8609	8615	8621	8627
73	8633	8639	8645	8651	8657	8663	8669	8675	8681	8686
74	8692	8698	8704	8710	8716	8722	8727	8733	8739	8745
75	8751	8756	8762	8768	8774	8779	8785	8791	8797	8802
76	8808	8814	8820	8825	8831	8837	8842	8848	8854	8859
77	8865	8871	8876	8882	8887	8893	8899	8904	8910	8915
78	8921	8927	8932	8938	8943	8949	8954	8960	8965	8971
79	8976	8982	8987	8993	8998	9004	9009	9015	9020	9025
80	9031	9036	9042	9047	9053	9058	9063	9069	9074	9079
81	9085	9090	9096	9101	9106	9112	9117	9122	9128	9133
82	9138	9143	9149	9154	9159	9165	9170	9175	9180	9186
83	9191	9196	9201	9206	9212	9217	9222	9227	9232	9238
84	9243	9248	9253	9258	9263	9269	9274	9279	9284	9289
85	9294	9299	9304	9309	9315	9320	9325	9330	9335	9340
86	9345	9350	9355	9360	9365	9370	9375	9380	9385	9390
87	9395	9400	9405	9410	9415	9420	9425	9430	9435	9440
88	9445	9450	9455	9460	9465	9469	9474	9479	9484	9489
89	9494	9499	9504	9509	9513	9518	9523	9528	9533	9538
90	9542	9547	9552	9557	9562	9566	9571	9576	9581	9586
91	9590	9595	9600	9605	9609	9614	9619	9624	9628	9633
92	9638	9643	9647	9652	9657	9661	9666	9671	9675	9680
93	9685	9689	9694	9699	9703	9708	9713	9717	9722	9727
94	9731	9736	9741	9745	9750	9754	9759	9763	9768	9773
95	9777	9782	9786	9791	9795	9800	9805	9809	9814	9818
96	9823	9827	9832	9836	9841	9845	9850	9854	9859	9863
97	9868	9872	9877	9881	9886	9890	9894	9899	9903	9908
98	9912	9917	9921	9926	9930	9934	9939	9943	9948	9952
99	9956	9961	9965	9969	9974	9978	9983	9987	9991	9996

APPENDIX D

Instructions for Building a Tetrahedron and an Octahedron

Tetrahedron: To construct a tetrahedron, trace out the following figure on a piece of light cardboard:

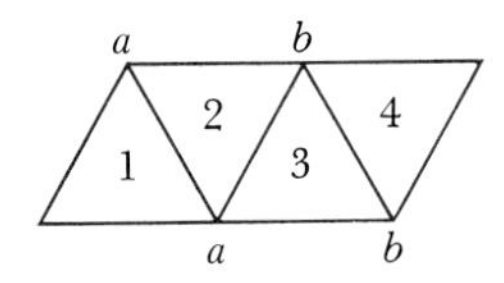

Bend face 1 upward about the line *aa*, and bend face 4 upward about the line *bb*:

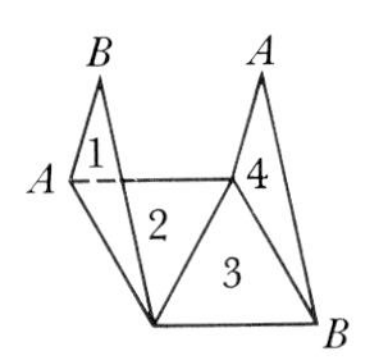

Now bend face 3 upward about the line between faces 2 and 3 and connect points *A* to *A* and *B* to *B* to get

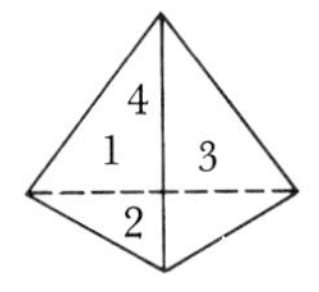

Octahedron: To construct an octahedron, trace out the following figure on a piece of light cardboard:

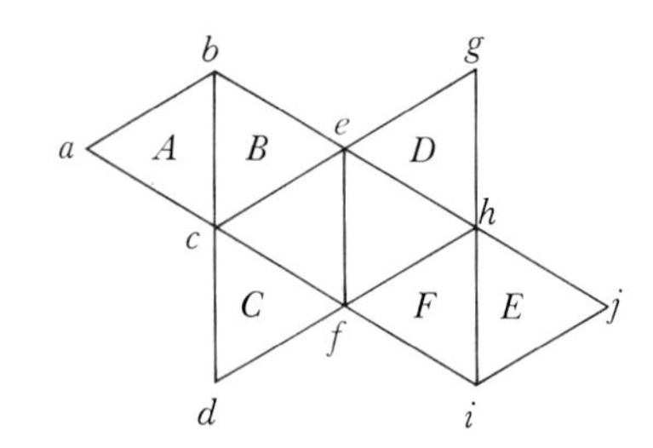

Bend face A upward about the line bc and face C upward about the line cf. Then bend face B upward about line ce and join points a and d. Lines ac and cd are now aligned and should be taped together. Now bend face D upward about the line eh and face E upward about the line hi. Then bend face F upward about the line fh and join points g and j. Lines gh and hj are now aligned and should be taped in place. Lastly, bend both sides upward about the line ef and tape the resulting octahedron together to get

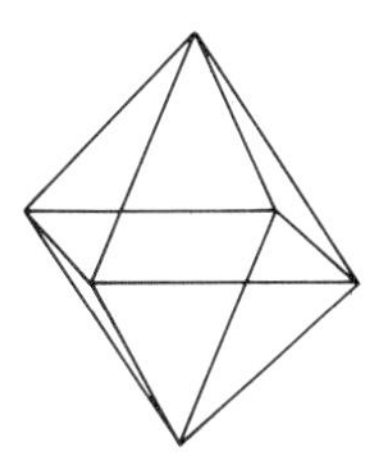

APPENDIX E

Symbols and Names of Elements with Atomic Number Greater Than 100

The elements with atomic numbers 93 and greater (called transuranium elements) are produced by particle accelerators such as cyclotrons, and traditionally, the research group that first produced an element assigned the name to that element. Conflict first arose when two groups claimed credit for the discovery of elements 104 and 105. For example, for element 105 a group in Berkeley, California, proposed the name rutherfordium, while another group, located in Dubna, USSR, proposed the name kurchatovium. Consequently, the International Union of Pure and Applied Chemistry (IUPAC) has recommended that newly discovered elements be assigned names that reflect their atomic numbers until the priority of their discovery has been determined. The symbols of these elements are formed according to their atomic numbers and consist of three letters, each being the first letter of the following roots:

0	nil	4	quad	8	oct
1	un	5	pent	9	enn
2	bi	6	hex		
3	tri	7	sept		

Thus, the symbols of the elements 104 through 110 are

104	Unq	106	Unh	108	Uno
105	Unp	107	Uns	109	Une
				110	Uun

To name an element, the three roots are joined in the order of the digits in the atomic number and the ending *ium* is added. Thus, for example, element 104 is called unnilquadium and element 109 is called unnilennium. For a further discussion, see the Journal of Chemical Education article by Mary V. Orna (J. Chem. Educ., 59:123, 1982).

APPENDIX F
Table of Atomic Masses

Element	*Symbol*	*Atomic Number*	*Atomic Mass*[a]
actinium	Ac	89	(227)
aluminum	Al	13	26.98154
americium	Am	95	(243)
antimony	Sb	51	121.75
argon	Ar	18	39.948
arsenic	As	33	74.9216
astatine	At	85	(210)
barium	Ba	56	137.33
berkelium	Bk	97	(247)
beryllium	Be	4	9.01218
bismuth	Bi	83	208.9804
boron	B	5	10.81
bromine	Br	35	79.904
cadmium	Cd	48	112.41
calcium	Ca	20	40.08
californium	Cf	98	(251)
carbon	C	6	12.011
cerium	Ce	58	140.12
cesium	Cs	55	132.9054
chlorine	Cl	17	35.453
chromium	Cr	24	51.996
cobalt	Co	27	58.9332
copper	Cu	29	63.546
curium	Cm	96	(247)
dysprosium	Dy	66	162.50
einsteinium	Es	99	(252)

[a]A value given in parentheses denotes the mass number of the longest-lived isotope.

Element	*Symbol*	*Atomic Number*	*Atomic Mass*[a]
erbium	Er	68	167.26
europium	Eu	63	151.96
fermium	Fm	100	(257)
fluorine	F	9	18.998403
francium	Fr	87	(223)
gadolinium	Gd	64	157.25
gallium	Ga	31	69.72
germanium	Ge	32	72.59
gold	Au	79	196.9665
hafnium	Hf	72	178.49
helium	He	2	4.00260
holmium	Ho	67	164.9304
hydrogen	H	1	1.0079
indium	In	49	114.82
iodine	I	53	126.9045
iridium	Ir	77	192.22
iron	Fe	26	55.847
krypton	Kr	36	83.80
lanthanum	La	57	138.9055
lawrencium	Lr	103	(260)
lead	Pb	82	207.2
lithium	Li	3	6.941
lutetium	Lu	71	174.967
magnesium	Mg	12	24.305
manganese	Mn	25	54.9380
mendelevium	Md	101	(258)
mercury	Hg	80	200.59
molybdenum	Mo	42	95.94
neodymium	Nd	60	144.24
neon	Ne	10	20.179
neptunium	Np	93	(237)
nickel	Ni	28	58.70

[a]A value given in parentheses denotes the mass number of the longest-lived isotope.

Element	*Symbol*	*Atomic Number*	*Atomic Mass*[a]
niobium	Nb	41	92.9064
nitrogen	N	7	14.0067
nobelium	No	102	(259)
osmium	Os	76	190.2
oxygen	O	8	15.9994
palladium	Pd	46	106.4
phosphorus	P	15	30.97376
platinum	Pt	78	195.09
plutonium	Pu	94	(244)
polonium	Po	84	(209)
potassium	K	19	39.0983
praseodymium	Pr	59	140.9077
promethium	Pm	61	(145)
protactinium	Pa	91	(231)
radium	Ra	88	226.0254
radon	Rn	86	(222)
rhenium	Re	75	186.207
rhodium	Rh	45	102.9055
rubidium	Rb	37	85.4678
ruthenium	Ru	44	101.07
samarium	Sm	62	150.4
scandium	Sc	21	44.9559
selenium	Se	34	78.96
silicon	Si	14	28.0855
silver	Ag	47	107.868
sodium	Na	11	22.98977
strontium	Sr	38	87.62
sulfur	S	16	32.06
tantalum	Ta	73	180.9479
technetium	Tc	43	(98)
tellurium	Te	52	127.60
terbium	Tb	65	158.9254

[a]A value given in parentheses denotes the mass number of the longest-lived isotope.

Element	*Symbol*	*Atomic Number*	*Atomic Mass*[a]
thallium	Tl	81	204.37
thorium	Th	90	232.0381
thulium	Tm	69	168.9342
tin	Sn	50	118.69
titanium	Ti	22	47.90
tungsten	W	74	183.85
uranium	U	92	238.029
vanadium	V	23	50.9415
xenon	Xe	54	131.30
ytterbium	Yb	70	173.04
yttrium	Y	39	88.9059
zinc	Zn	30	65.38
zirconium	Zr	40	91.22

[a]A value given in parentheses denotes the mass number of the longest-lived isotope.

APPENDIX G

Answers To Odd-Numbered Problems

CHAPTER 1

1-1. (a) Cd (b) In (c) Pb (d) Sn (e) Hg (f) Xe (g) Cu (h) K (i) U (j) P **1-3.** (a) selenium (b) gold (c) silver (d) palladium (e) lithium (f) hydrogen (g) silicon (h) vanadium (i) plutonium (j) helium **1-5.** 52.5% K, 47.5% Cl **1-7.** 71.5% Ca, 28.5% O **1-9.** 60.0% K, 18.4% C, 21.5% N **1-11.** (a) lithium chloride (b) magnesium oxide (c) aluminum fluoride (d) sodium phosphide (e) potassium iodide **1-13.** (a) calcium carbide (b) gallium arsenide (c) beryllium nitride (d) potassium oxide (e) strontium fluoride

1-15.

(a) iodine trifluoride
iodine pentafluoride

(b) iodine monochloride
iodine trichloride

(c) nitrogen dioxide
dinitrogen tetraoxide

(d) arsenic trifluoride
arsenic pentafluoride

(e) chlorine monoxide
chlorine dioxide

1-17. (a) 16.04 (b) 44.09 (c) 58.12 (d) 114.22 **1-19.** (a) 288.0 (b) 231.55 (c) 209.95 (d) 537.54 (e) 222.85 **1-21.** 67.10% Zn, 32.90% S **1-23.** 74.87% C, 25.13% H **1-25.** 42.10% C, 6.479% H, 51.42% O **1-27.** NH_3 **1-29.** The ratio of the masses of fluorine to xenon in the two compounds is 2 to 1. **1-31.** The ratio of the masses of hydrogen to oxygen in the two compounds is 2 to 1. **1-33.** The ratio of the masses of copper to oxygen in the two compounds is 2 to 1. **1-35.** $^{131}_{53}I$ (53 protons, 53 electrons, 78 neutrons); $^{125}_{53}I$ (53 protons, 53 electrons, 72 neutrons) **1-37.** (a) 94 protons, 94 electrons, 145 neutrons (b) 27 protons, 27 electrons, 33 neutrons (c) 19 protons, 19 electrons, 24 neutrons (d) 92 protons, 92 electrons, 143 neutrons **1-39.** $^{12}_{6}C$ (6, 6, 12) $^{32}_{16}S$ (16, 16, 32); $^{196}_{79}Au$ (79, 117, 196); $^{20}_{10}Ne$ (10, 10, 20) **1-41.** $^{23}_{11}Na$ (11, 12, 23); $^{202}_{80}Hg$ (80, 122, 202); $^{239}_{94}Pu$ (94, 145, 239); $^{249}_{98}Cf$ (98, 151, 249) **1-43.** 15.999 **1-45.** 12.011 **1-47.** 50.65% Br-79, 49.35% Br-81 **1-49.** 0.36% **1-51.** 4.67% Si-29, 3.10% Si-30 **1-53.** 33 g $^{2}_{1}H$, 360 g $^{17}_{8}O$, 2040 g $^{18}_{8}O$ **1-55.** (a) 18 (b) 18 (c) 18 (d) 18 (e) 10 **1-57.** (a) 38 (b) 89 (c) 47 (d) 36 (e) 18 **1-59.** (a) Mg^{2+}, F^-, O^{2-} (b) Cl^-, K^+, Ca^{2+} (c) I^-, Cs^+, Ba^{2+} (d) N^+, O^{2+}, B^- (e) P^{3-}, S^{2-}, K^+ **1-61.** C-12: 2.6×10^4, Ar-40: 2.1×10^4, Ag-108: 2.3×10^4, Ra-226: 2.8×10^4 **1-63.** (a) 4 (b) 3 (c) 3 (d) 3 (e) exact **1-65.** (a) 18.0152 (b) 239.2 (c) 133.341 (d) 258 **1-67.** (a) 56.77 (b) 219.499 (c) 19.12 (d) 0.36 (e) 7.6 **1-69.** (a) 0.946 L (b) 6.71×10^8 mi/h (c) $8.314\ J \cdot K^{-1} \cdot mol^{-1}$ **1-71.** 9.46×10^{15} m, 5.88×10^{12} mi **1-73.** The eight-pack is a better buy (79¢/L vs. 89¢/L).

CHAPTER 2

2-1. (a) $2AgClO_3(aq) + CaBr_2(aq) \rightarrow 2AgBr(s) + Ca(ClO_3)_2(aq)$
(b) $Ba(s) + 2HNO_3(aq) \rightarrow Ba(NO_3)_2(aq) + H_2(g)$
(c) $H_2SO_4(aq) + 2KOH(aq) \rightarrow K_2SO_4(aq) + 2H_2O(l)$
(d) $2C_3H_8(g) + 7O_2(g) \rightarrow 6CO(g) + 8H_2O(g)$
(e) $CaH_2(s) + 2H_2O(l) \rightarrow Ca(OH)_2(aq) + 2H_2(g)$
2-3. (a) $2NO(g) + Br_2(g) \rightarrow 2NOBr(g)$
(b) $2Na(s) + O_2(g) \rightarrow Na_2O_2(s)$ (c) $2P(s) + 3Br_2(l) \rightarrow 2PBr_3(l)$
(d) $3H_2(g) + N_2(g) \rightarrow 2NH_3(g)$ (e) $MgO(s) + SiO_2(s) \rightarrow MgSiO_3(s)$
2-5. (a) $CH_4(g) + 2O_2(g) \rightarrow CO_2(g) + 2H_2O(l)$
(b) $2CO(g) + O_2(g) \rightarrow 2CO_2(g)$
(c) $C_3H_8(g) + 5O_2(g) \rightarrow 3CO_2(g) + 4H_2O(l)$
(d) $C_6H_{12}O_6(s) + 6O_2(g) \rightarrow 6CO_2(g) + 6H_2O(l)$
(e) $2Sr(s) + O_2(g) \rightarrow 2SrO(s)$ **2-7.** (a) $6Li(s) + N_2(g) \rightarrow 2Li_3N(s)$
(b) $4Li(s) + O_2(g) \rightarrow 2Li_2O(s)$
(c) $2Li(s) + 2H_2O(l) \rightarrow 2LiOH(s) + H_2(g)$
(d) $2Li(s) + H_2(g) \rightarrow 2LiH(s)$
(e) $2Li(s) + 2CO_2(g) + 2H_2O(g) \rightarrow 2LiHCO_3(s) + H_2(g)$
2-9. (a) $2Al(s) + 3Cl_2(g) \rightarrow 2AlCl_3(s)$
aluminum chlorine aluminum chloride
(b) $4Al(s) + 3O_2(g) \rightarrow 2Al_2O_3(s)$
aluminum oxygen aluminum oxide
(c) $COCl_2(g) + 2Na(s) \rightarrow CO(g) + 2NaCl(s)$
phosgene sodium carbon monoxide sodium chloride
(d) $2Be(s) + O_2(g) \rightarrow 2BeO(s)$
beryllium oxygen beryllium oxide
(e) $2K(s) + S(l) \rightarrow K_2S(s)$
sodium sulfur sodium sulfide
2-11. (a) $2K(s) + 2H_2O(l) \rightarrow 2KOH(s) + H_2(g)$
(b) $3Mg(s) + N_2(g) \rightarrow Mg_3N_2(s)$
(c) $2Ca(s) + O_2(g) \rightarrow 2CaO(s)$
(d) $2Na(s) + I_2(s) \rightarrow 2NaI(s)$
(e) $Sr(s) + 2HCl(g) \rightarrow SrCl_2(s) + H_2(g)$
2-13. (a) $Mg(s) + Br_2(l) \rightarrow MgBr_2(s)$
(b) $2Ba(s) + O_2(g) \rightarrow 2BaO(s)$
(c) $Ba(s) + S(s) \rightarrow BaS(s)$
(d) $Mg(s) + 2HCl(g) \rightarrow MgCl_2(s) + H_2(g)$
(e) $Sr(s) + H_2(g) \rightarrow SrH_2(s)$
2-15. (a) $2Na(s) + H_2(g) \rightarrow 2NaH(s)$
(b) $Ba(s) + H_2(g) \rightarrow BaH_2(s)$
(c) $2K(s) + F_2(g) \rightarrow 2KF(s)$
(d) $Ba(s) + Br_2(l) \rightarrow BaBr_2(s)$
(e) $3Ca(s) + N_2(g) \rightarrow Ca_3N_2(s)$
2-17. Iodine belongs to the halogen family. **2-19.** (a) solid (b) NaAt
(c) white (d) At_2 (e) black
2-21. In, a main-group (3) metal; Er, an inner transition metal; Ar, a main-group (8) nonmetal; Y, a transition metal; Rh, a transition metal; Cf, an inner transition metal; Be, a main-group (2) metal.
2-25. $4Sb(s) + 3O_2(g) \rightarrow Sb_4O_6(s)$
2-27. (a) $2Ra(s) + O_2(g) \rightarrow 2RaO(s)$
(b) $Ra(s) + Cl_2(g) \rightarrow RaCl_2(s)$

(c) $Ra(s) + 2HCl(g) \rightarrow RaCl_2(s) + H_2(g)$
(d) $Ra(s) + H_2(g) \rightarrow RaH_2(s)$
(e) $Ra(s) + S(s) \rightarrow RaS(s)$
2-29. (a) 24 (b) 36 (c) 10 (d) 18 (e) 10 **2-31.** (a) yes (Ar) (b) no (c) no (d) yes (Ne) (e) yes (Ne) (f) yes (Xe) **2-33.** (a) K^+ S^{2-} (b) Al^{3+} N^{3-} (c) Al^{3+} F^- (d) Cs^+ I^- (e) Mg^{2+} Se^{2-}
2-35. (a) Ga_2O_3 (b) $AlCl_3$ (c) Li_2O (d) HBr (e) SrI_2 **2-37.** (a) Li_3N (b) RbI (c) Ga_2S_3 (d) BaO (e) MgI_2 **2-39.** (a) Ga_2O_3 (b) ZnI_2 (c) FeS (d) $RuCl_3$ (e) Ag_2S
2-41. (a) $2K(s) + 2H_2O(l) \rightarrow 2KOH(s) + H_2(g)$
(b) $KH(s) + H_2O(l) \rightarrow KOH(s) + H_2(g)$
(c) $SiO_2(s) + 3C(s) \rightarrow SiC(s) + 2CO(g)$
(d) $SiO_2(s) + 4HF(g) \rightarrow SiF_4(g) + 2H_2O(l)$
(e) $2P(s) + 3Cl_2(g) \rightarrow 2PCl_3(l)$ **2-43.** (a) $2Na(s) + H_2(g) \rightarrow 2NaH(s)$
(b) $2Al(s) + 3S(s) \rightarrow Al_2S_3(s)$ (c) $H_2O(g) + C(s) \rightarrow CO(g) + H_2(g)$
(d) $C(s) + 2H_2(g) \rightarrow CH_4(g)$ (e) $PCl_3(l) + Cl_2(g) \rightarrow PCl_5(s)$
2-45. (a) Calcium is oxidized and chlorine is reduced. (b) Aluminum is oxidized and oxygen is reduced. (c) Rubidium is oxidized and bromine is reduced. (d) Sodium is oxidized and sulfur is reduced. **2-47.** (a) 2 (b) 6 (c) 1 (d) 2

CHAPTER 3

3-1. (a) 18.02 (b) 151.91 (c) 208.2 (d) 180.16 **3.3** (a) 1.52 mol (b) 0.898 mol (c) 1.60 mol **3-5.** (a) 1.55 mol (b) 1.67×10^{-2} mol (c) 7.77 mol **3-7.** 8.55×10^{22} kg **3-9.** 6.02×10^{23}
3-11. (a) 2.829×10^{-23} g (b) 2.992×10^{-22} g (c) 9.274×10^{-23} g
3-13. (a) 2.111×10^{-20} g (b) 2.828×10^{-7} g (c) 3.155×10^{-17} g (d) 6.310×10^{-17} g **3-15.** 0.555 mol; 3.34×10^{23} molecules; 1.00×10^{24} atoms **3-17.** 2.54×10^6 molecule$\cdot$cm^{-3} **3-19.** CaC_2 **3-21.** $CuCl_2$
3-23. BrF_5 **3-25.** CF_2Cl_2 **3-27.** 144.3 (Nd) **3-29.** 137; barium
3-31. C_3H_6O **3-33.** $Na_6P_6O_{18}$ **3-35.** 0.168 metric tons **3-37.** 0.650 kg of NaOH; 0.576 kg of Cl_2; 0.0164 kg of H_2 **3-39.** 21.7 g **3-41.** 9.380 g of Pb; 17.40% Pb **3-43.** 1.34×10^5 kg of Zn **3-45.** 1.58×10^5 kg of HNO_3 **3-47.** 65.5 kg of KNO_3; 23.0 kg of Cl_2 **3-49.** 33.4 g
3-51. 2.07 g **3-53.** 0.0250 M **3-55.** 14.3 M **3-57.** Dissolve 85.6 g of sucrose in about 500 mL of H_2O and then dilute to exactly one liter of solution. **3-59.** 44.2 mL **3-61.** 1770 g **3-63.** 267 mL **3-65.** 0.14 g
3-67. 12 μg

CHAPTER 4

4-1. (a) rubidium cyanide (b) silver perchlorate (c) lanthanum chromate (d) cesium permanganate (e) aluminum hydroxide
4-3. (a) sodium nitrate (b) ammonium sulfate (c) ammonium phosphate (d) calcium phosphate (e) potassium phosphate
4-5. (a) $Na_2S_2O_3$ (b) $Al(HCO_3)_3$ (c) $KClO_4$ (d) $CaCO_3$ (e) $Ba(C_2H_3O_2)_2$ **4-7.** (a) Na_2CrO_4 (b) K_3PO_4 (c) RbOH (d) $LiClO_4$ (e) NH_4NO_3 **4-9.** (a) mercury(II) oxide (b) tin(IV) oxide (c) copper(I) iodide (d) cobalt(II) bromide (e) chromium(III) cyanide
4-11. (a) CuO (b) AuCl (c) $Sn_3(PO_4)_2$ (d) $Fe(C_2H_3O_2)_2$ (e) $Co_2(SO_4)_3$

4-13. (a) $3Mg(s) + N_2(g) \rightarrow Mg_3N_2(s)$ (b) $H_2(g) + S(s) \rightarrow H_2S(g)$ (c) $2K(s) + Br_2(l) \rightarrow 2KBr(s)$ (d) $4Al(s) + 3O_2(g) \rightarrow 2Al_2O_3(s)$ (e) $MgO(s) + SO_2(g) \rightarrow MgSO_3(s)$ **4-15.** (a) $Li_2O(s) + H_2O(l) \xrightarrow[H_2O(l)]{} 2LiOH(aq) \rightarrow 2Li^+(aq) + 2OH^-(aq)$ (b) $SO_3(g) + H_2O(l) \xrightarrow[H_2O(l)]{} H_2SO_4(aq) \rightarrow 2H^+(aq) + SO_4^{2-}(aq)$ (c) $H_2SO_4(l) \xrightarrow[H_2O(l)]{} 2H^+(aq) + SO_4^{2-}(aq)$ (d) $HBr(g) \xrightarrow[H_2O(l)]{} HBr(aq) \rightarrow H^+(aq) + Br^-(aq)$ **4-17.** (a) $CaCO_3(s) \xrightarrow{\text{high T}} CaO(s) + CO_2(g)$ (b) $2H_2O_2(aq) \xrightarrow{\text{light}} 2H_2O(l) + O_2(g)$ (c) $PbSO_3(s) \xrightarrow{\text{high T}} PbO(s) + SO_2(g)$ (d) $2Ag_2O(s) \xrightarrow{\text{high T}} 4Ag(s) + O_2(g)$ **4-19.** (a) $3Na(l) + AlCl_3(s) \rightarrow Al(s) + 3NaCl(s)$ (b) $Cu(s) + Fe_2O_3(s) \rightarrow$ N.R. (c) $Zn(s) + K_2SO_4(aq) \rightarrow$ N.R. (d) $Fe(s) + SnCl_2(aq) \rightarrow FeCl_2(aq) + Sn(s)$ (e) $Br_2(l) + 2NaI(aq) \rightarrow 2NaBr(aq) + I_2(s)$
4-21. (a) $Cu(NO_3)_2(aq) + Au(s) \rightarrow$ N.R. (b) $NaCl(aq) + I_2(aq) \rightarrow$ N.R. (c) $2AgNO_3(aq) + Mg(s) \rightarrow Mg(NO_3)_2(aq) + 2Ag(s)$ (d) $Ag(s) + HCl(aq) \rightarrow$ N.R. (e) $Pb(s) + H_2O(l) \rightarrow$ N.R.
4-23. Na Fe Sn Au **4-25.** $2PbS(s) + 3O_2(g) \rightarrow 2PbO(s) + 2SO_2(g)$; $2PbO(s) + C(s) \rightarrow 2Pb(s) + CO_2(g)$; $2Ag_2O(s) \xrightarrow{\text{high T}} 4Ag(s) + O_2(g)$ **4-27.** (a) $2H^+(aq) + S^{2-}(aq) \rightarrow H_2S(g)$ (b) $Pb^{2+}(aq) + S^{2-}(aq) \rightarrow PbS(s)$ (c) $H^+(aq) + OH^-(aq) \rightarrow H_2O(l)$
(d) $H^+(aq) + OH^-(aq) \rightarrow H_2O(l)$; Na_2O is a basic
(e) $NH_3(g) + H^+(aq) \rightarrow NH_4^+(aq)$ **4-29.** (a) $Fe(NO_3)_2(aq) + 2NaOH(aq) \rightarrow 2NaNO_3(aq) + Fe(OH)_2(s)$; $Fe^{2+}(aq) + 2OH^-(aq) \rightarrow Fe(OH)_2(s)$ (b) $Zn(ClO_4)_2(aq) + K_2S(aq) \rightarrow 2KClO_4(aq) + ZnS(s)$; $Zn^{2+}(aq) + S^{2-}(aq) \rightarrow ZnS(s)$ (c) $Pb(NO_3)_2(aq) + 2KOH(aq) \rightarrow 2KNO_3(aq) + Pb(OH)_2(s)$; $Pb^{2+}(aq) + 2OH^-(aq) \rightarrow Pb(OH)_2(s)$ (d) $Zn(NO_3)_2(aq) + Na_2CO_3(aq) \rightarrow 2NaNO_3(aq) + ZnCO_3(s)$; $Zn^{2+}(s) + CO_3^{2-}(aq) \rightarrow ZnCO_3(s)$ (e) $Cu(ClO_4)_2(aq) + Na_2CO_3(aq) \rightarrow CuCO_3(s) + 2NaClO_4(aq)$; $Cu^{2+}(aq) + CO_3^{2-}(aq) \rightarrow CuCO_3(s)$ **4-31.** (a) acidic (b) acidic (c) basic (d) acidic (e) basic
4-33. (a) $2HClO_3(aq) + Ba(OH)_2(aq) \rightarrow Ba(ClO_3)_2(aq) + 2H_2O(l)$ — $Ba(ClO_3)_2$: barium chlorate
(b) $HC_2H_3O_2(aq) + KOH(aq) \rightarrow KC_2H_3O_2(aq) + H_2O(l)$ — $KC_2H_3O_2$: potassium acetate
(c) $2HI(aq) + Mg(OH)_2(s) \rightarrow MgI_2(aq) + 2H_2O(l)$ — MgI_2: magnesium iodide
(d) $H_2SO_4(aq) + 2RbOH(aq) \rightarrow Rb_2SO_4(aq) + 2H_2O(l)$ — Rb_2SO_4: rubidium sulfate
(e) $H_3PO_4(aq) + 3LiOH(aq) \rightarrow Li_3PO_4(aq) + 3H_2O(l)$ — Li_3PO_4: lithium phosphate
4-35. (a) decomposition (b) combination
(c) single-replacement (d) double-replacement **4-37.** (a) $ZnBr_2(s) + Cl_2(g) \rightarrow ZnCl_2(s) + Br_2(l)$; single-replacement (b) $2HCl(aq) + Mg(OH)_2(aq) \rightarrow MgCl_2(aq) + 2H_2O(l)$; double-replacement
(c) $BaO(s) + CO_2(g) \rightarrow BaCO_3(s)$; combination
(d) $2Ag_2O(s) \xrightarrow{\text{high T}} 4Ag(s) + O_2(g)$; decomposition (e) $2Li(s) + 2H_2O(l) \rightarrow 2LiOH(s) + H_2(g)$; single-replacement
4-39. (a) $HCl(aq) + KCN(aq) \rightarrow KCl(aq) + HCN(g)$ (b) $2K(s) + 2H_2O(l) \rightarrow 2KOH(aq) + H_2(g)$ (c) $2H_2O_2(aq) \rightarrow 2H_2O(l) + O_2(g)$

(d) $H_2(g) + Br_2(l) \rightarrow 2HBr(g)$
4-41. $2Pb(l) + O_2(g) \rightarrow 2PbO(s)$; $Ag(l) + O_2(g) \rightarrow$ no reaction
4-43. 0.115 M **4-45.** (a) 1.2 μL (b) 4.17 mL **4-47.** 0.104 M
4-49. 52.1% NaOH **4-51.** 60.1 **4-53.** 90.1

CHAPTER 5

5-1. 7.60×10^4 torr; 101 bar **5-3.** 0.76 atm; 0.77 bar **5-5.** 38.3 torr
5-9. 2.25 mL **5-11.** 6.0 atm **5-13.** 20.1 mL **5-15.** 10 L **5-17.** 32.3 L
5-19. 1.7 atm; 18 mL **5-21.** 1.0×10^{20} molecules; 2.7×10^{22} molecules
5-23. 4.1×10^{15} molecules **5-25.** 780 K, 510°C **5-27.** 35.0 L; 50.3 L
5-29. 848 mL **5-31.** 52.4 L **5-33.** $0.588\ g \cdot L^{-1}$; $958\ g \cdot L^{-1}$
5-35. $74.2\ g \cdot mol^{-1}$ **5-37.** C_2H_4 **5-39.** $C_3H_6O_3$ **5-41.** $P_{N_2} = 518$ torr; $P_{O_2} = 222$ torr **5-43.** 5.1 L; 10 atm **5-45.** 1.02 atm **5-47.** $324\ m \cdot s^{-1}$
5-49. by a factor of $\sqrt{2}$ **5-51.** $v_{av,O_2}/v_{av,H_2} = 0.354$ **5-53.** 1.2×10^5 pm; 1.6×10^{10} collisions/s **5-55.** 2.2×10^{31} pm, or 2.2×10^{19} m
5-57. 7.2×10^9 collisions/s **5-59.** 200 mL/h **5-61.** 70.5 **5-63.** 244 atm; 227 atm **5-65.** 2.49×10^{19} molecules **5-67.** 5.67 μL; 56.2 μg
5-69. $17.7\ g \cdot m^{-3} = 1.77 \times 10^{-2}\ g \cdot L^{-1}$

CHAPTER 6

6-1. 804 kJ **6-3.** 89.3 kJ **6-5.** (a) −44.2 kJ (b) −1322.9 kJ
(c) −136.96 kJ All three reactions are exothermic.
6-7. (a) −1234.7 kJ (b) −1427.7 kJ The combustion of one gram of $C_2H_6(g)$ produces almost twice as much heat per gram as does $C_2H_5OH(l)$: −47.48 kJ vs. −26.80 kJ. **6-9.** $-1260.0\ kJ \cdot mol^{-1}$
6-11. +37 kJ **6-13.** (a) $472.6\ kJ \cdot mol^{-1}$ (b) $79.0\ kJ \cdot mol^{-1}$
(c) $218.0\ kJ \cdot mol^{-1}$ (d) $121.7\ kJ \cdot mol^{-1}$ N_2 has the greatest bond strength. **6-15.** 62.4 kJ **6-17.** +145.4 kJ **6-19.** −88.5 kJ
6-21. +229.27 kJ **6-23.** −4.2 kJ **6-25.** +90.0 kJ **6-27.** $171\ kJ \cdot mol^{-1}$
6-29. −420 kJ **6-31.** $50\underline{2}\ kJ \cdot mol^{-1}$ **6-33.** $414\ kJ \cdot mol^{-1}$
6-35. $C_p = 180\ J \cdot K^{-1}$; $\bar{C}_p = 114\ J \cdot K^{-1} \cdot mol^{-1}$; $c_{sp} = 2.47\ J \cdot K^{-1} \cdot g^{-1}$
6-37. 5.2×10^6 J **6-39.** 39.7°C **6-41.** 450°C **6-43.** −56.2 kJ
6-45. $-602\ kJ \cdot mol^{-1}$ **6-47.** $+17.2\ kJ \cdot mol^{-1}$ **6-49.** $-46.3\ kJ \cdot g^{-1}$; $-2040\ kJ \cdot mol^{-1}$ **6-51.** $-246\ kJ \cdot mol^{-1}$; $-827\ kJ \cdot mol^{-1}$
6-53. 49 mol **6-55.** 150 kJ; 4.5 g

CHAPTER 7

7-1. Ar, Ne, He, Li^+ **7-3.** The plot clearly shows two shells.
7-5. Li·, Na·, K·, Rb·, Cs·; :F̤̈·, :C̤̈l·, :B̤̈r·, :Ï̤· All the alkali metal atoms have one outer electron and all the halogen atoms have seven outer electrons. **7-7.** $\cdot B \cdot^{+}$, $:\ddot{N}:^{3-}$, $:\ddot{F}:^{-}$, $:\ddot{O}:^{2-}$, $:\ddot{Na}:^{+}$ or Na^+ **7-9.** $4.74 \times 10^{14}\ s^{-1}$ **7-11.** 2.41×10^{-7} m = 241 nm
7-13. $1200\ kJ \cdot mol^{-1}$ **7-15.** $5.70 \times 10^5\ kJ \cdot mol^{-1}$ **7-17.** 2.6×10^{17} photons per second **7-19.** 1.7×10^{16} photons per pulse **7-21.** 6000 K
7-23. blue **7-25.** 3.97 pm **7-27.** 2.85×10^{-11} m **7-29.** $7.3 \times 10^7\ m \cdot s^{-1}$
7-33. 3.03×10^{-19} J; 656 nm **7-35.** 91.2 nm **7-37.** He^+:

5.25 MJ·mol^{-1}; Li^{2+}: 11.8 MJ·mol^{-1}; Be^{3+}: 21.0 MJ·mol^{-1} These values are in excellent agreement with those in Table 7-1. **7-39.** The density of the dots in any region gives an indication of the probability of finding an electron in that region. **7-41.** 16 (one $4s$ orbital, three $4p$ orbitals, five $4d$ orbitals, and seven $4f$ orbitals) **7-43.** b and d are impossible.

7-45.

n	ℓ	m_ℓ	m_s
4	3	3	$+\frac{1}{2}$
4	3	3	$-\frac{1}{2}$
4	3	2	$+\frac{1}{2}$
4	3	2	$-\frac{1}{2}$
4	3	1	$+\frac{1}{2}$
4	3	1	$-\frac{1}{2}$
4	3	0	$+\frac{1}{2}$
4	3	0	$-\frac{1}{2}$
4	3	-1	$+\frac{1}{2}$
4	3	-1	$-\frac{1}{2}$
4	3	-2	$+\frac{1}{2}$
4	3	-2	$-\frac{1}{2}$
4	3	-3	$+\frac{1}{2}$
4	3	-3	$-\frac{1}{2}$

There are 14 possible sets.

7-47. (a) $2p$ (b) $4s$ (c) $5f$ (d) $3d$ **7-49.** If $\ell = 2$, then n must be at least 3. If $m_\ell = 3$, then ℓ must be at least 3. **7-51.** a and c are allowed. **7-53.** (a) +0.02 MJ (b) −0.71 MJ (c) −5.14 MJ

CHAPTER 8

8-1. (a) $4p$ (b) $4p$ (c) $6s$ (d) $5p$ (e) $4f$
8-3. $(3, 2, 2, +\frac{1}{2})$ $(3, 2, 2, -\frac{1}{2})$
$(3, 2, 1, +\frac{1}{2})$ $(3, 2, 1, -\frac{1}{2})$
$(3, 2, 0, +\frac{1}{2})$ $(3, 2, 0, -\frac{1}{2})$
$(3, 2, -1, +\frac{1}{2})$ $(3, 2, -1, -\frac{1}{2})$
$(3, 2, -2, +\frac{1}{2})$ $(3, 2, -2, -\frac{1}{2})$
Note that there is a total of 10.
8-5. 2 (s orbital), 6 (p orbitals), 10 (d orbitals), 14 (f orbitals)
8-7. Because the five d orbitals can be occupied by a maximum of 10 electrons (see Problem 8-3). **8-9.** b, c and d. **8-11.** (a) ground state (b) excited state (c) ground state (d) ground state
8-13. (a) $1s^22s^22p^63s^23p^2$ silicon (b) $1s^22s^22p^63s^23p^64s^13d^5$ chromium

(c) $1s^22s^22p^63s^23p^64s^23d^{10}4p^2$ germanium (d) $1s^22s^22p^63s^23p^64s^23d^{10}4p^5$ bromine (e) $1s^22s^22p^1$ boron **8-15.** (a) $1s^22s^22p^63s^23p^64s^23d^2$ (b) $1s^22s^22p^63s^23p^64s^23d^{10}4p^65s^14d^5$ (c) $1s^22s^22p^63s^23p^64s^13d^{10}$ (d) $1s^22s^22p^63s^23p^64s^23d^{10}$ (e) $1s^22s^22p^63s^23p^64s^23d^{10}4p^65s^14d^{10}$ **8-17.** $1s^22s^22p^63s^23p^3$ (b) $1s^22s^22p^63s^23p^64s^23d^8$ (c) $1s^22s^22p^63s^23p^64s^23d^{10}4p^65s^24d^{10}5p^5$ (d) $1s^22s^22p^63s^23p^64s^23d^{10}4p^65s^24d^{10}$ (e) $1s^22s^22p^63s^23p^64s^23d^{10}4p^65s^24d^{10}5p^66s^24f^4$ **8-19.** (a) nickel (b) silver (c) sulfur (d) lead **8-21.** (a) $[\text{Ne}]3s^2$ (b) $[\text{Ar}]4s^23d^{10}4p^5$ (c) $[\text{Xe}]6s^14f^{14}5d^{10}$ (d) $[\text{Xe}]6s^1$ (e) $[\text{Ar}]4s^23d^{10}4p^1$ **8-23.** (a) 8, 34, 52, 84 (b) 1, 3, 19, 37, 55, 87 **8-25.** (a) the alkali metals (Li–Fr) and the alkaline earths (Be–Ra) (b) Groups 3, 4, 5, 6, 7 and 8 (c) the transition metal series, Sc–Zn, Y–Cd, Lu–Hg (d) the lanthanide series and the actinide series

8-27. (a) $\cdot\dot{\underset{\cdot}{\text{Si}}}\cdot$ (b) $\cdot\ddot{\underset{\cdot}{\text{P}}}\cdot$ (c) $\cdot\ddot{\underset{\cdot\cdot}{\text{Se}}}\cdot$ (d) $:\ddot{\underset{\cdot\cdot}{\text{I}}}\cdot$ (e) $\cdot\text{Sr}\cdot$

8-29. (a) $1s^22s^22p_x^12p_y^12p_z^1$ (b) $1s^22s^22p_x^12p_y^12p_z^1$ (c) $1s^22s^22p_x^22p_y^12p_z^1$ (d) $1s^22s^22p_x^12p_y^1$ (e) $1s^22s^22p_x^22p_y^12p_z^1$ **8-31.** (a) 2 (b) 3 (c) 2 (d) 2 (e) 2 **8-33.** $\cdot\ddot{\underset{\cdot\cdot}{\text{O}}}\cdot$, $\cdot\ddot{\underset{\cdot\cdot}{\text{S}}}\cdot$, $\cdot\ddot{\underset{\cdot\cdot}{\text{Se}}}\cdot$, $\cdot\ddot{\underset{\cdot\cdot}{\text{Te}}}\cdot$, $\cdot\ddot{\underset{\cdot\cdot}{\text{Po}}}\cdot$ **8-35.** (a) $\cdot\text{H}\cdot^-$ (gain 1 electron to achieve a helium-like electron configuration)

(b) $:\ddot{\underset{\cdot\cdot}{\text{O}}}:^{2-}$ (gain 2 electrons for a neon-like electron configuration)

(c) $:\ddot{\underset{\cdot\cdot}{\text{C}}}:^{4-}$ (gain 4 electrons for a neon-like electron configuration)

(d) $:\ddot{\underset{\cdot\cdot}{\text{S}}}:^{2-}$ (gain 2 electrons for an argon-like electron configuration)

(e) $:\ddot{\underset{\cdot\cdot}{\text{Cl}}}:^-$ (gain 1 electron for an argon-like electron configuration)

8-37. (a) $1s^22s^22p^63s^23p^6$ (b) $1s^22s^22p^63s^23p^64s^23d^{10}4p^6$ (c) $1s^22s^22p^63s^23p^6$ (d) $1s^22s^22p^63s^23p^64s^23d^{10}4p^65s^24d^{10}5p^6$ (e) $1s^22s^22p^6$ They all have a noble-gas electron configuration. **8-39.** (a) 0 (b) 0 (c) 0 (d) 1 (e) 1 **8-41.** (F^-, O^{2-}, Na^+) (Sc^{3+}, Ti^{4+}, Ar) (Be^{2+}, B^{3+}, He) (Rb^+, Se^{2-}, Y^{3+}) **8-43.** (a) $\text{O}(1s^22s^22p^4) + 2e^- \rightarrow \text{O}^{2-}(1s^22s^22p^6)$ (b) $\text{Ca}([\text{Ar}]4s^2) + \text{Sr}^{2+}([\text{Kr}]) \rightarrow \text{Ca}^{2+}([\text{Ar}]) + \text{Sr}([\text{Kr}]5s^2)$ **8-45.** (a) $1s^22s^22p^53s^1$ (b) $2s^1$ (c) $1s^22s^22p^53s^1$ (d) $1s^22s^22p^23s^1$ **8-47.** a, b, c and d **8-49.** (a) P; (b) P; (c) S; (d) Kr **8-51.** (a) Li, Na, Rb, Cs (b) P, Al, Mg, Na (c) Ca, Sr, Ba **8-53.** The atoms increase in size in going from Be to Ba.

8-55. $\text{Cs} < \text{K} < \text{Br} < \text{Cl} < \text{F}$

CHAPTER 9

9-1. (a) $\text{K}([\text{Ar}]4s^1) + \text{Cl}([\text{Ne}]3s^23p^5) \rightarrow \text{K}^+([\text{Ar}]) + \text{Cl}^-([\text{Ar}])$ (b) $\text{Sr}([\text{Kr}]5s^2) + 2\text{Br}([\text{Ar}]4s^23d^{10}4p^5) \rightarrow \text{Sr}^{2+}([\text{Kr}]) + 2\text{Br}^-([\text{Kr}])$ (c) $2\text{Al}([\text{Ne}]3s^23p^1) + 3\text{O}([\text{He}]2s^22p^4) \rightarrow 2\text{Al}^{3+}([\text{Ne}]) + 3\text{O}^{2-}([\text{Ne}])$ **9-3.** (a) $\text{Ga}([\text{Ar}]4s^23d^{10}4p^1) + 3\text{F}([\text{He}]2s^22p^5) \rightarrow \text{Ga}^{3+}([\text{Ar}]3d^{10}) + 3\text{F}^-([\text{Ne}])$ (b) $\text{Ca}([\text{Ar}]4s^2) + \text{S}([\text{Ne}]3s^23p^4) \rightarrow \text{Ca}^{2+}([\text{Ar}]) + \text{S}^{2-}([\text{Ar}])$ (c) $3\text{Li}([\text{He}]2s^1) + \text{N}([\text{He}]2s^22p^3) \rightarrow 3\text{Li}^+([\text{He}]) + \text{N}^{3-}([\text{Ne}])$

9-5. (a) $3\text{Li}\cdot + \cdot\dot{\underset{\cdot}{\text{N}}}\cdot \rightarrow 3\text{Li}^+ + :\ddot{\underset{\cdot\cdot}{\text{N}}}:^{3-}$ (b) $\text{Na}\cdot + \text{H}\cdot \rightarrow \text{Na}^+ + \text{H}:^-$

(c) $\cdot\dot{\text{Al}}\cdot + 3:\ddot{\underset{\cdot}{\text{I}}}: \rightarrow \text{Al}^{3+} + 3:\ddot{\underset{\cdot\cdot}{\text{I}}}:^-$ **9-7.** (a) CaS (b) $AlCl_3$ (c) BeO

9-9. (a) $[\text{Ar}]3d^4$ (b) $[\text{Ar}]3d^9$ (c) $[\text{Ar}]3d^6$ (d) $[\text{Ar}]3d^5$ (e) $[\text{Ar}]3d^7$

9-11. (a) Fe, Ru, Os (b) Zn, Cd, Hg (c) Sc, Y, Lu (d) Mn, Tc, Re (e) Ni, Pd, Pt The number of *d* electrons in the +2 ion is equal to the position of the element in the transition metal series.
9-13. (a) $[Kr]4d^{10}$ (b) $[Kr]4d^{10}$ (c) $[Xe]4f^{14}5d^{10}$ (d) $[Ar]3d^{10}$ **9-15.** b, d, e, and f **9-17.** (a) Y_2S_3 (b) $LaBr_3$ (c) MgTe (d) Rb_3N (e) Al_2Se_3
9-19. (a) AgCl (b) CdS (c) Zn_3N_2 (d) CuBr (e) Ga_2O_3 **9-21.** (a) Cl^- (b) Ag^+ (c) Cu^+ (d) O^{2-} (e) P^{3-} **9-23.** (a) 307 pm (b) 291 pm (Hg–Cl); 582 pm (Cl–Cl) (c) 214 pm
9-25. % increase = $(V_{ion} - V_{atom})/V_{atom} = 492\%$
9-27. (a) $Na^+(aq)$ and $F^-(aq)$ (b) $Zn^{2+}(aq)$ and $Cl^-(aq)$ (c) $Ag^+(aq)$ and $F^-(aq)$ (d) $Sr^{2+}(aq)$ and $I^-(aq)$ (e) $Sc^{3+}(aq)$ and $I^-(aq)$ **9-29.** (a) strong (b) weak (c) weak (d) non **9-31.** H, I, Br, Cl **9-33.** approximately $270\ kJ\cdot mol^{-1}$ **9-35.** (a) $79\ kJ\cdot mol^{-1}$ (b) $172\ kJ\cdot mol^{-1}$ (c) $86\ kJ\cdot mol^{-1}$
9-37. (a) $397\ kJ\cdot mol^{-1}$ (b) $-53\ kJ\cdot mol^{-1}$ (c) $398\ kJ\cdot mol^{-1}$
9-39. -7.36×10^{-19} J **9-41.** $-329\ kJ\cdot mol^{-1}$ **9-43.** $-134\ kJ\cdot mol^{-1}$
9-47. $-585\ kJ\cdot mol^{-1}$ **9-49.** $-303\ kJ\cdot mol^{-1}$
9-51. $-799\ kJ\cdot mol^{-1}$

CHAPTER 10

10-1. :F̤̈—F̤̈: :Ï̤—Ï̤:

10-3. (a) :Cl—P—Cl: with :Cl: below P (b) :F—Si—F: with :F: above and below Si (c) :I—N—I: with :I: below N

10-5. H—P—H (H below P)   H—As—H (H below As)   H—Sb—H (H below Sb)

10-7. H—S—H   H—Se—H   H—Te—H   **10-9.** H—O—O—H

10-11. :Cl—N—Cl: with :Cl: below N

10-13. (a) H—C—H with H above and below C (b) H—C—H with :F: above and H below C (c) H—C—N—H with H above C, H below C and H below N

10-15. (a) H—C—C—C—H with H above and below each C (b) H—C—C—C—C—H with H above and below each C
(c) H—C—C—C—C—C—C—C—C—H with H above and below each C

```
           H  H                       H  H  H
           |  |  ..                   |  |  |  ..
10-17. (a) H—C—C—O—H (b) H—C—C—C—O—H
           |  |  ..                   |  |  |  ..
           H  H                       H  H  H

          H
          |
       H :O: H
       |  |  |
(c) H—C—C—C—H
       |  |  |
       H  H  H

              H  H                 H :Cl:
         ..   |  |   ..            |  |    ..
10-19. :Cl—C—C—Cl:          H—C—C—Cl:
         ..   |  |   ..            |  |    ..
              H  H                 H  H

                         H  H
                         |  |                   ..  ..
10-21. b  10-23. (a) H—C=C—C≡N:  (b) :O=N—O—H
                                                    ..
(c) :O=Si=O:
```

```
         H                    H     H               H   H
          \                    \   /                 \ /
10-25.     C=O:       10-27.     C          10-29.    C          H
          /                     / \                  / \        /
     H—O:                      H   \                H   C=C
                               H    C=O:                /    \
                                \  /                   H      H
                                 C
                                / \
                               H   H

              ..   ..
10-31. H—N=N—H
```

```
          :O:                  :O: ⊖
          ||                    |
10-33.     C            ↔       C
          / \                  / \\
         H   O:⊖              H   O:
```

The two carbon-oxygen bonds are equivalent (same bond length and bond energy).

```
           :O:               :O: ⊖            :O: ⊖
           ||                 |                 |
10-35.      C           ↔     C          ↔      C
          /   \              /  \\            //  \
     ⊖:O:      :O:⊖    ⊖:O:      O:        :O      :O:⊖
```

The three carbon-oxygen bonds are equivalent (same bond length and bond energy).

```
         ⊖    ⊕               ⊖   (2+)   ⊖
10-37.  :N=N=O:              :N=O=N:  .
```

The formal charges in NON are greater than that in NNO. Consequently, we predict (correctly) that the actual structure is NNO.

10-39. (a) 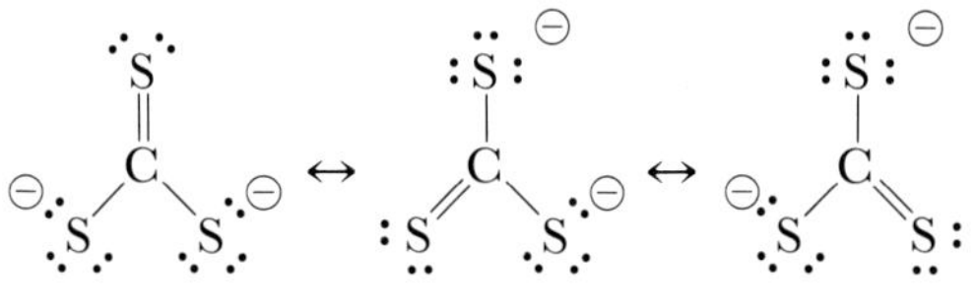 [Resonance structures of CS_3^{2-}]

(b) 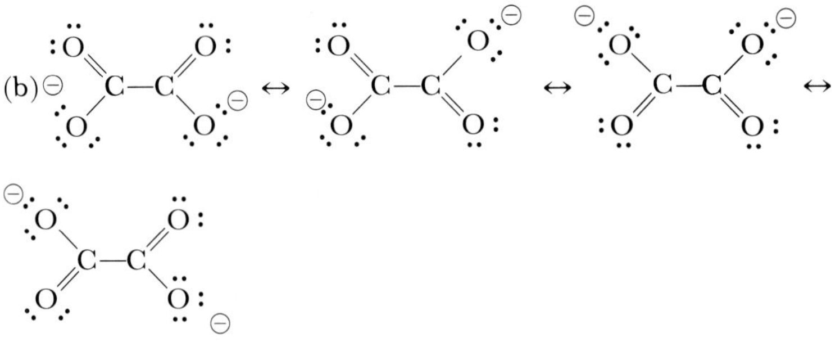 [Resonance structures of $C_2O_4^{2-}$]

10-41. a, c, and d

10-43. $H_3C{-}\dot{N}{-}N{=}O$ Methylnitrosamine is a free radical.

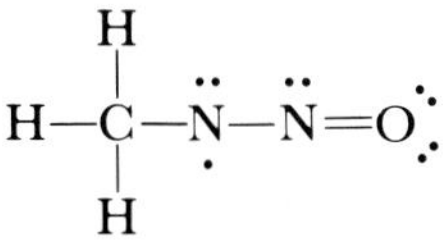

10-45. (a) 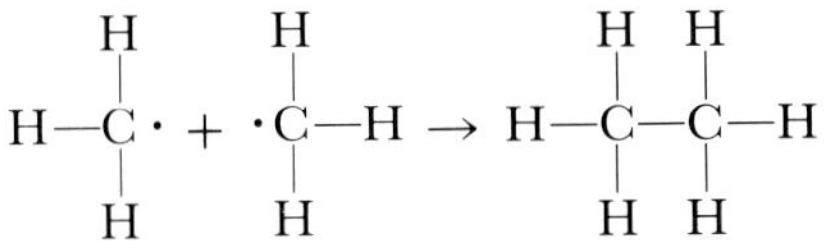 $H_3C\cdot + \cdot CH_3 \rightarrow H_3C{-}CH_3$

(b) 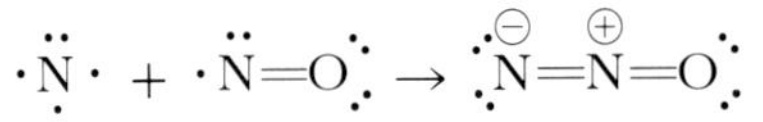 $\cdot\dot{N}\cdot + \cdot N{=}O \rightarrow N^{\ominus}{=}N^{\oplus}{=}O$

10-47. (a) 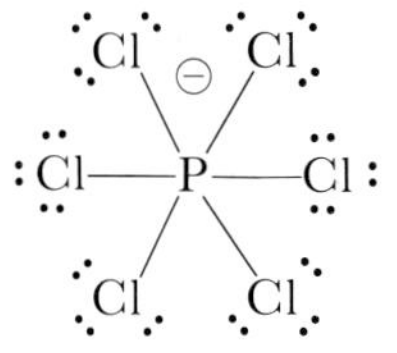 PCl_6^{-} (b) 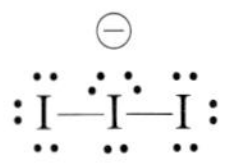 I_3^{-} (c) 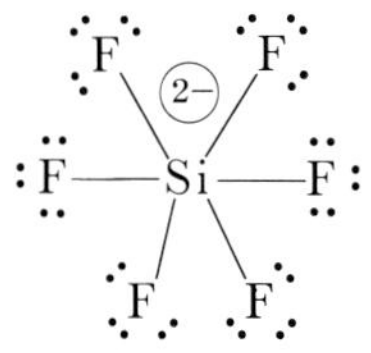 SiF_6^{2-}

10-49. XeF_2, XeF_4, XeF_6, $XeOF_4$, XeO_2F_2

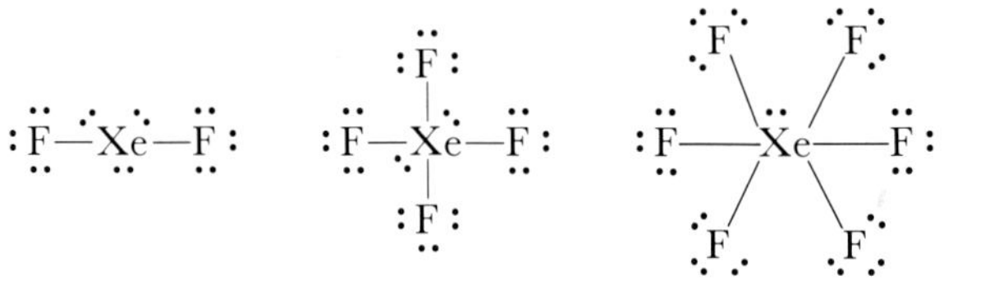

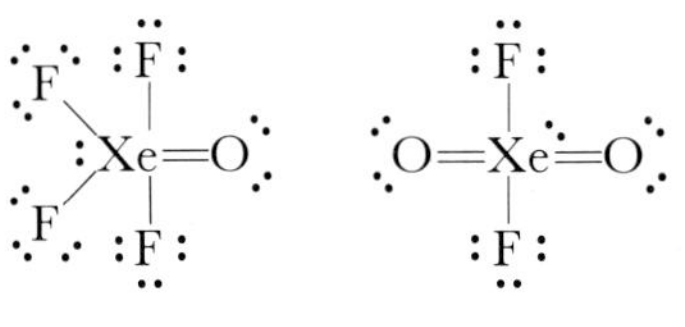

10-51. (a) :F̤—N(⊕)—F̤: with :F̤: above and below N (b) :F̤—Cl(⊕)—F̤: with :F̤: above and below Cl (c) H—P(⊕)—H with H above and below P

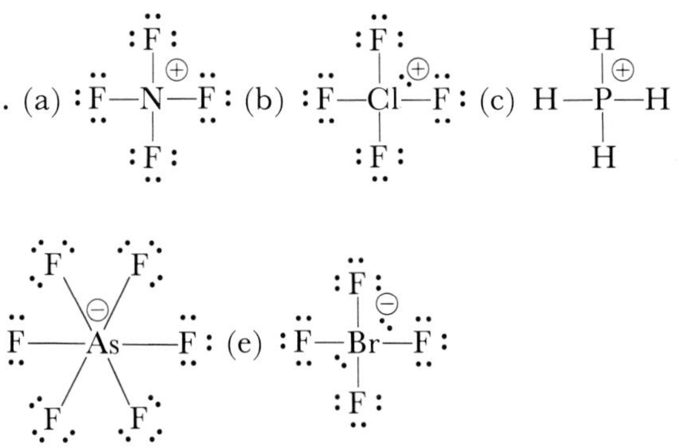

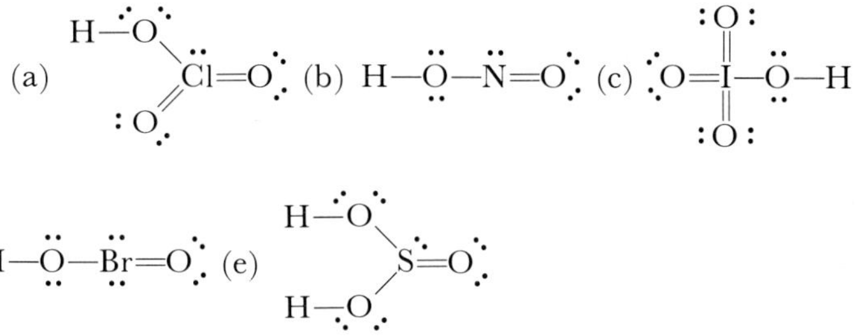

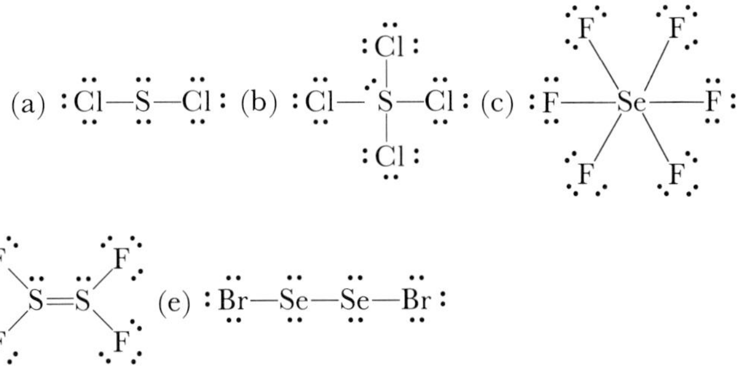

(d) As(⊖) with six F atoms, each with three lone pairs (e) :F̤—Br(⊖)—F̤: with :F̤: above and below Br

10-53. (a) H—O̤—Cl(=O)(=O) (b) H—O̤—N̤=O̤ (c) O=I(=O)(=O)—O̤—H (d) H—O̤—Br̤=O̤ (e) (H—O)₂S=O

10-55. (a) :Cl̤—S̤—Cl̤: (b) :Cl̤—S̤(—Cl̤:)(—Cl̤:)—Cl̤: (c) Se with six F atoms, each with three lone pairs

(d) F₂S=SF₂ (e) :Br̤—Se̤—Se̤—Br̤:

10-57. O=Cr(⊖)(=O)(=O)—O̤—Cr(⊖)(=O)(=O)=O plus other resonance forms

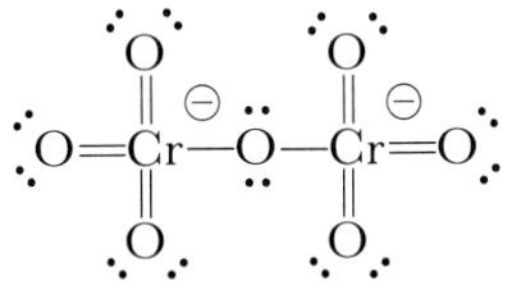

10-59. :Br̤—Cl̤:

⟼ (dipole arrow pointing toward Cl)

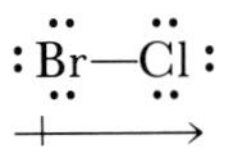

10-61. (a) 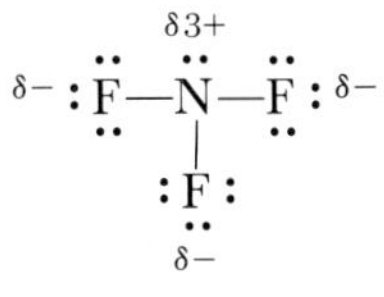 $\delta-$:F̤—N̤($\delta3+$)—F̤: $\delta-$, with :F̤: ($\delta-$) below N

The fluorine atoms are slightly negatively charged and the nitrogen atom is slightly positively charged.

(b) $\overset{\delta-}{:\ddot{\underset{..}{F}}}-\overset{\delta 2+}{\ddot{\underset{..}{O}}}-\overset{\delta-}{\ddot{\underset{..}{F}}:}$ The fluorine atoms are slightly negatively charged and the oxygen atom is slightly positively charged.

(c) $\overset{\delta+}{:\ddot{\underset{..}{Br}}}-\overset{\delta 2-}{\ddot{\underset{..}{O}}}-\overset{\delta+}{\ddot{\underset{..}{Br}}:}$ The oxygen atom is slightly negatively charged and the bromine atoms are slightly positively charged.

CHAPTER 11

11-1. a, c **11-3.** c, d **11-5.** a, c are linear; b, d, e are bent **11-7.** b, c, e **11-9.** b, c **11-11.** (a) seesaw (b) square planar (c) tetrahedral (d) tetrahedral (e) seesaw **11-13.** (1) 90°: a, b, c (2) 109.5°: none (3) 120°: b, d **11-15.** (1) 90°: b, c, (2) 109.5°: a, d (3) 120°: none **11-17.** (1): c (2): d, e (3): none (4): a (5): none (6): d (7): none (8): b, c (9): d, e **11-19.** AB_3 is T-shaped and AB_5 is square pyramidal. **11-21.** (a) H_2O (b) NH_2^- (c) NH_4^+ (d) SF_6 (e) H_3O^+

11-23. (a) :Cl—Hg—Cl: linear (b) 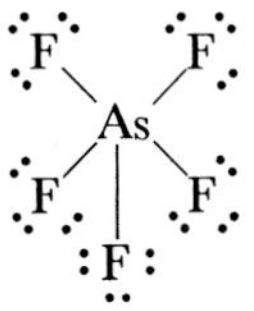 AsF_5 (As bonded to five F atoms, each with three lone pairs) trigonal bipyramidal

(c) :Cl—S—Cl: bent (d) :Cl—O—Cl: bent

(e) 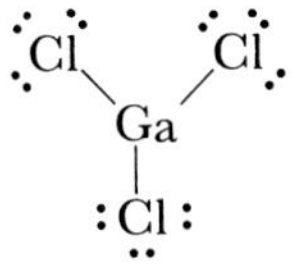 $GaCl_3$ (Ga bonded to three Cl atoms, each with three lone pairs) trigonal planar

c and d have dipole moments.

11-25. (a) 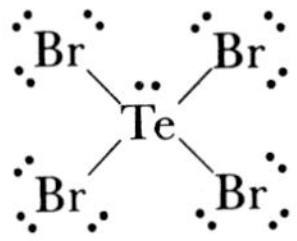 $TeBr_4$ (Te with one lone pair bonded to four Br atoms) seesaw (b) :I—Cd—I: linear

(c) 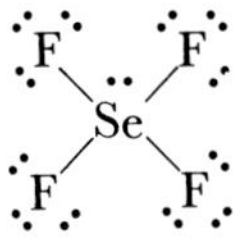 SeF_4 (Se with one lone pair bonded to four F atoms) seesaw (d) 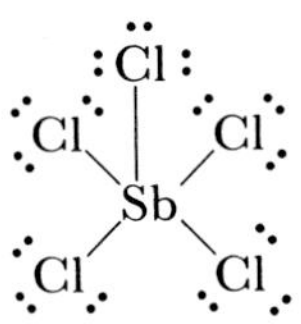 $SbCl_5$ (Sb bonded to five Cl atoms) trigonal bipyramidal

(e) 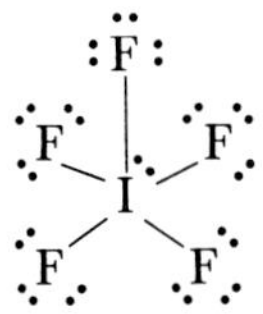 IF_5 (I with one lone pair bonded to five F atoms) square pyramidal

a, c and e have dipole moments.
11-27. Only SeF_4 is polar.

11-29. (a) 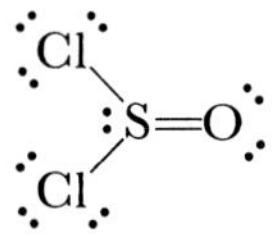 Cl—S(—Cl)=O trigonal pyramidal (b) 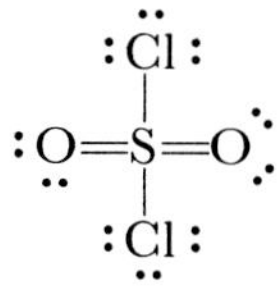 O=S(Cl)(Cl)=O tetrahedral

(c) 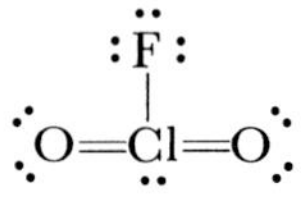 O=Cl(—F)=O trigonal pyramidal (d) 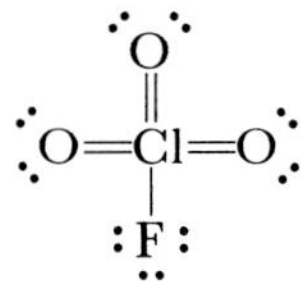 O=Cl(=O)(—F)=O tetrahedral

(e) 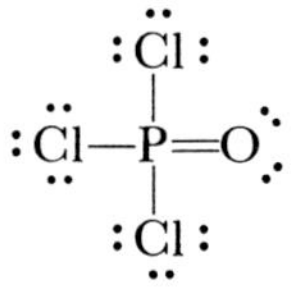 Cl—P(Cl)(Cl)=O tetrahedral

11-31. (a) 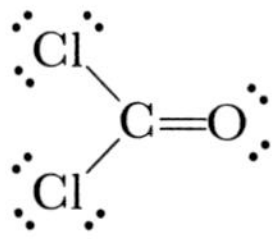 Cl₂C=O trigonal planar (b) 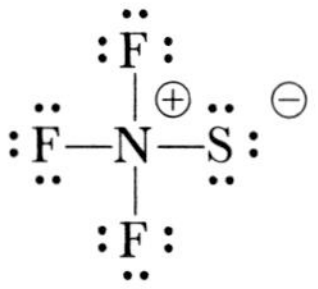 F—N⁺(F)(F)—S⁻ tetrahedral

(c) 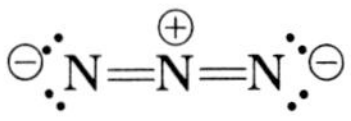 ⁻N=N⁺=N⁻ linear (d) 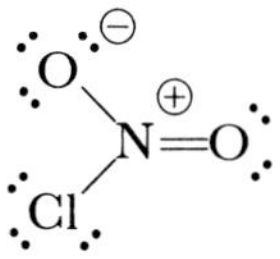 ⁻O—N⁺(Cl)=O trigonal planar

(e) 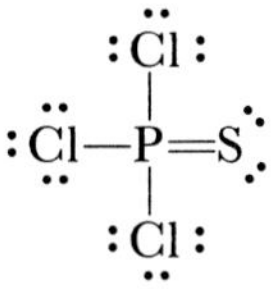 Cl—P(Cl)(Cl)=S tetrahedral

11-33. (a) bent (b) linear (c) tetrahedral (d) tetrahedral (e) tetrahedral **11-35.** (a) trigonal pyramidal (b) tetrahedral (c) linear (d) octahedral (e) square pyramidal **11-37.** (a) bent (b) trigonal pyramidal (c) tetrahedral

11-39. (a) 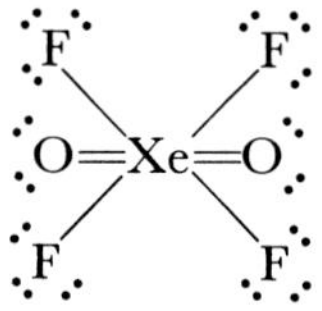 O=Xe(F)₄=O octahedral (b) 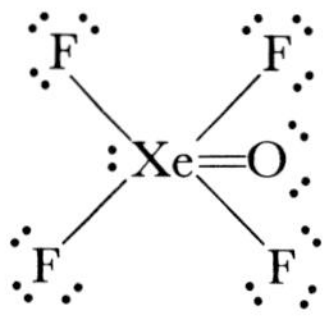 :Xe(F)₄=O square pyramidal

(c) 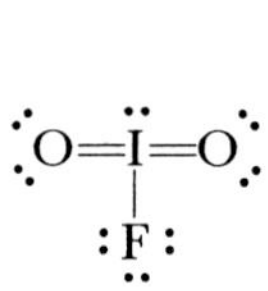 O=I(—F)=O trigonal pyramidal (d) 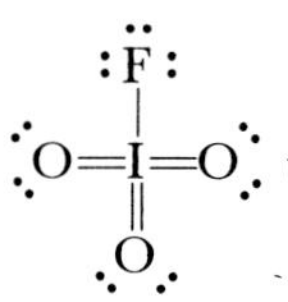 O=I(—F)(=O)=O tetrahedral

(e) 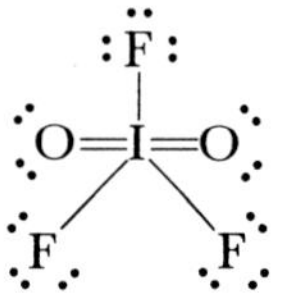 O=I(F)(F)(F)=O trigonal bipyramidal

11-41. $:\!\ddot{O}=\overset{\oplus}{N}=\ddot{O}\!:$ linear $[\ddot{O}{=\!=}\ddot{N}{=\!=}\ddot{O}]^-$ bent

11-43. (a) one isomer (b) only one isomer (c) only one isomer (d) two isomers

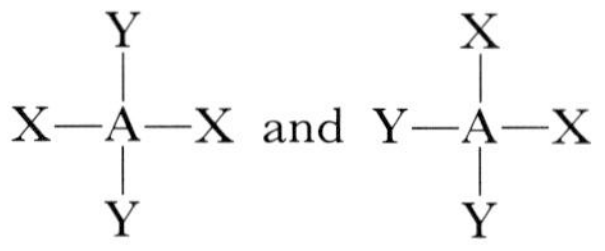

```
      Y              X
      |              |
  X—A—X   and   Y—A—X
      |              |
      Y              Y
```

11-45. (a) only one isomer (b) two isomers, Y's at opposite vertices or at adjacent vertices (c) two isomers **11-47.** (a) 1 (b) 1 (c) 2 (d) 2

CHAPTER 12

12-1. The interaction of two widely separated hydrogen atoms is zero because they are so far apart that they do not interact with each other. When the two hydrogen atoms are separated by 74 pm, they are joined by a covalent bond, and hence attract each other. Because they attract each other, their interaction energy is negative. **12-3.** (a) 3 localized bonds and 10 lone pairs (b) 4 localized bonds and 13 lone pairs (c) 4 localized bonds and no lone pairs (d) 2 localized bonds and 8 lone pairs (e) 2 localized bonds and 5 lone pairs **12-5.** There are 10 localized bonds in propane. There are $3 \times 4 = 12$ valence electrons from the three carbon atoms and $8 \times 1 = 8$ valence electrons from the eight hydrogen atoms, giving a total of 20 valence electrons. The 20 valence electrons occupy the 10 localized bond orbitals.
12-7. We form four equivalent localized bond orbitals by combining each sp^3 orbital on the carbon atom with a $2p$ orbital on each fluorine atom. There are $4 + (4 \times 7) = 32$ valence electrons in CF_4.
12-9. We form four equivalent localized bond orbitals by combining each sp^3 orbital on the carbon atom with a $3p$ orbital on each chlorine atom.
12-11. Each of the three σ bond orbitals in H_3O^+ is formed by combining an sp^3 orbital on the oxygen atom with a $1s$ hydrogen orbital. The lone electron pair occupies the remaining sp^3 orbital.
12-13. The three sigma bonds on each nitrogen atom are formed by combining two of the sp^3 orbitals on the nitrogen atom with two $1s$ hydrogen orbitals, and one of the sp^3 nitrogen orbitals with an sp^3 orbital on the other nitrogen atom. A lone electron pair occupies the remaining sp^3 orbital on each nitrogen atom.
12-15. There are three covalent bonds and one lone pair on the nitrogen atom. Two bonds are formed by combining an sp^3 on the nitrogen atom with a $1s$ orbital on a hydrogen atom. The remaining bond is formed by combining an sp^3 orbital on the nitrogen atom with an sp^3 orbital on the carbon atom. The lone electron pair occupies the remaining sp^3 orbital. There are four covalent bonds on the carbon atom. Three bonds are formed by combining an sp^3 orbital on the carbon atom with a $1s$ orbital on each of the three hydrogen atoms. The shape around the carbon atom is tetrahedral; the shape around the nitrogen atom is trigonal pyramidal.
12-17. There are two covalent bonds and two lone electron pairs on the

oxygen atom. One bond is formed by combining an sp^3 orbital on the oxygen atom with a $1s$ orbital on a hydrogen atom. The other bond is formed by combining an sp^3 orbital on the oxygen atom with an sp^3 orbital on the carbon atom. Each lone electron pair occupies an sp^3 orbital on the oxygen atom. There are four covalent bonds on the carbon atom. Three bonds are formed by combining an sp^3 orbital on the carbon atom with a $1s$ orbital on a hydrogen atom. The shape around the carbon atom is tetrahedral; the shape around the oxygen atom is bent.

12-19. There are two covalent bonds and two lone electron pairs on the oxygen atom. Each bond is formed by combining an sp^3 orbital on the oxygen atom with an sp^3 orbital on a carbon atom. Each lone electron pair occupies an sp^3 orbital on the oxygen atom. There are four covalent bonds on each carbon atom. Three bonds are formed by combining an sp^3 orbital on a carbon atom and a $1s$ orbital on a hydrogen atom.

12-21. The electron configuration of an arsenic atom is $[Ar](4s)^2(3d)^{10}(4p_x)^1(4p_y)^1(4p_z)^1$. To describe the bonding in AsH_3, use the $1s$ orbitals on the hydrogen atoms and the three $4p$ orbitals on the arsenic atom.

12-23. Each bond in BH_3 is formed by combining an sp^2 hybrid orbital on the boron atom with a $1s$ hydrogen orbital.

12-25. There are eight σ bonds and one π bond in propene. There are $(6 \times 1) + (3 \times 4) = 18$ valence electrons, which occupy the nine bond orbitals.

12-27. (a) 5 σ bonds and 1 π bond (b) 9 σ bonds and 2 π bonds (c) 6 σ bonds and 2 π bonds (d) 14 σ bonds and 2 π bonds

12-29. The two C—H bonds on one carbon atom are formed by combining an sp^2 orbital on the carbon atom with a $1s$ orbital on a hydrogen atom. The C—H bond on the middle carbon atom is formed by combining an sp^2 orbital on the carbon atom and a $1s$ orbital on the hydrogen atom. The C—C single bond is formed by combining an sp^2 orbital on the middle carbon atom and an sp^3 orbital on the $—CH_3$ carbon atom. The C=C double bond is formed by combining sp^2 orbitals on each carbon atom and the remaining $2p$ orbitals on each carbon atom. The three C—H bonds on the last carbon atom are formed by combining an sp^3 orbital on the carbon atom and a $1s$ orbital on a hydrogen atom.

12-31. The central carbon atom in acetaldehyde forms two σ bonds to the other carbon atom and the hydrogen atom and a σ bond and π bond to the oxygen atom. The carbon-carbon σ bond is formed by combining an sp^2 orbital on the central carbon atom with an sp^3 orbital on the other carbon atom. The carbon-oxygen σ bond is formed by combining the remaining sp^2 orbital of the central carbon atom with an sp^2 orbital on the oxygen atom. The π bond is formed by combining the remaining p orbital on the central carbon atom with the remaining p orbital on the oxygen atom. The two lone electron pairs on the oxygen atom occupy the other two sp^2 orbitals. The shape is trigonal planar around the central carbon atom.

12-33. The nitrogen atom uses sp^3 orbitals and the carbon atom uses sp^2 orbitals. The nitrogen-carbon bond results from a combination of an sp^3 orbital on the nitrogen atom and an sp^2 orbital on the carbon atom. The oxygen atom uses sp^2 orbitals. The carbon-oxygen double bond results from combining an sp^2 on the carbon atom with an sp^2 on the oxygen atom (σ bond) and from combining a p orbital on the carbon atom with a p orbital on the oxygen atom (π bond). The lone pair on the nitrogen atom occupies an sp^3 orbital and the two lone pairs on the oxygen atom occupy sp^2 orbitals.

12-35. There are nine σ bonds, two π bonds and $(6 \times 1) + (4 \times 4) = 22$ valence electrons.

12-37. The σ bond between the carbon and oxygen atoms is formed by combining an sp orbital on the carbon atom and an sp orbital on the oxygen atom. The two π bonds are formed by combining a $2p$ orbital on the carbon atom and a $2p$ orbital on the oxygen atom. One lone electron pair occupies an sp orbital on the carbon atom; the other pair occupies an sp orbital on the oxygen atom.

12-39. The σ bond framework of methylacetylene is

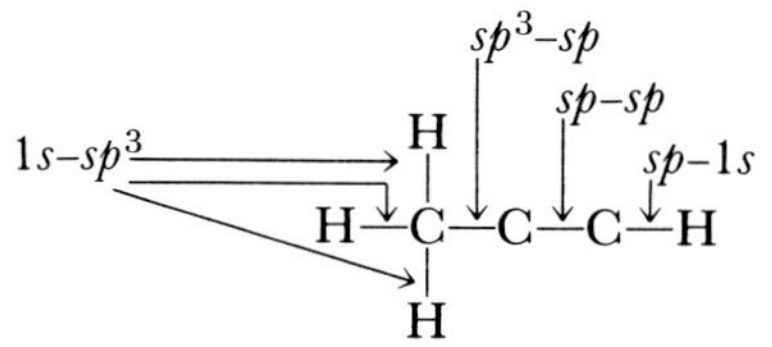

sp^3–sp
sp–sp
$1s$–sp^3 H sp–$1s$
H–C–C–C–H
H

The remaining bonds between the carbon atoms are two π bonds that are formed from the $2p$ orbitals on each atom. There are six σ bonds and two π bonds. There are $(4 \times 1) + (3 \times 4) = 16$ valence electrons which occupy the eight bond orbitals.

12-41. The Lewis formula of phenol is

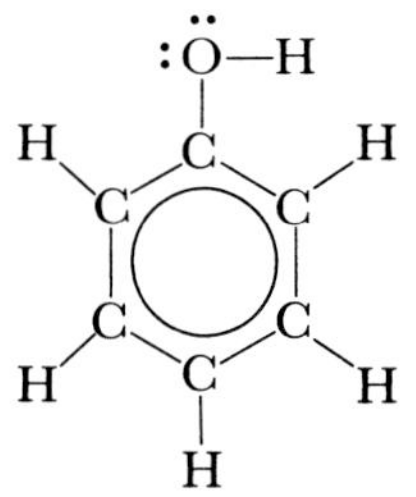

:Ö–H
H C H
C C
C C
H C H
H

The σ bond framework is

sp^3–$1s$
:Ö–H
sp^2–sp^3
H C H
C C sp^2–$1s$
sp^2–sp^2
C C
H C H
H

The three π bonds are delocalized over the entire ring.

12-43. The Lewis formula of naphthalene is

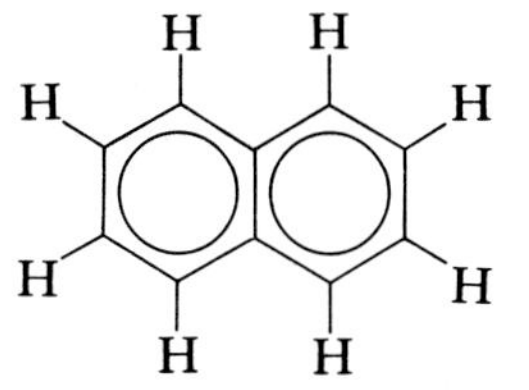

The five π bonds are delocalized over the entire two rings as indicated by the circles in the Lewis formula.
The σ bond framework is

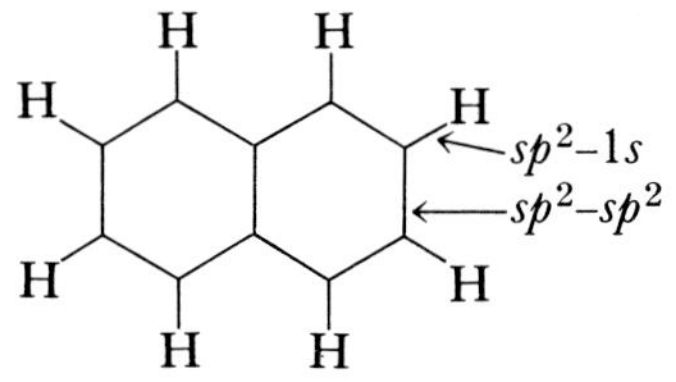

sp^2–$1s$
sp^2–sp^2

12-45. The Lewis formula for hexamethylbenzene is

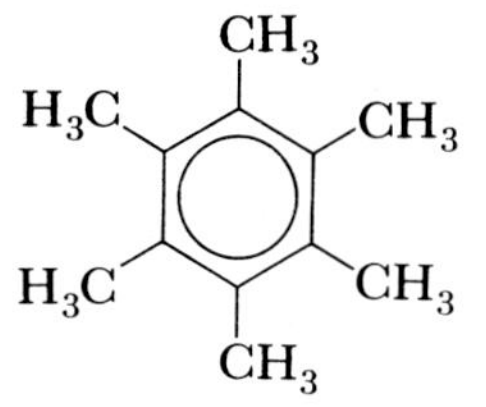

The σ bond framework is

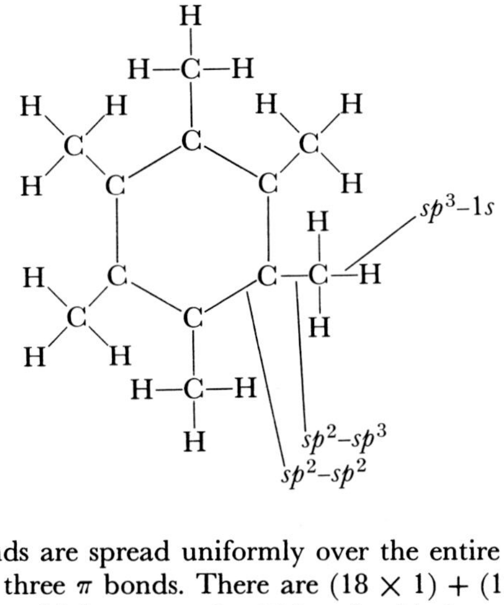

sp^3–$1s$
sp^2–sp^3
sp^2–sp^2

The three π bonds are spread uniformly over the entire ring. There are 30 σ bonds and three π bonds. There are $(18 \times 1) + (12 \times 4) = 66$ valence electrons, which occupy the 33 bond orbitals.
12-47. The ground-state electron configuration of diatomic beryllium is $(1\sigma)^2(1\sigma^*)^2(2\sigma)^2(2\sigma^*)^2$. There are four electrons in bonding orbitals and four electrons in antibonding orbitals and so Be_2 has no net bonding.

12-49. We find that the bond order of N_2 is 3 while the bond order of N_2^+ is $2\frac{1}{2}$ (Example 12-8). The bond energy increases as the bond order increases; therefore, the bond energy of N_2 is greater than that of N_2^+. However, we find that the bond order of O_2 is 2 while the bond order of O_2^+ is $2\frac{1}{2}$ (p. 456). Therefore, the bond energy of O_2 is less than that of O_2^+.
12-51. The ground-state electron configurations and bond orders of C_2 and C_2^- are

	ground-state electron configuration	bond order
C_2	$(1\sigma)^2(1\sigma^*)^2(2\sigma)^2(2\sigma^*)^2(1\pi)^4$	2
C_2^{2-}	$(1\sigma)^2(1\sigma^*)^2(2\sigma)^2(2\sigma^*)^2(1\pi)^4(3\sigma)^2$	3

Because the bond order of C_2^{2-} is greater than that of C_2, we predict that C_2^{2-} has a larger bond energy and a shorter bond length than C_2.
12-53. Carbon monoxide has $6 + 8 = 14$ electrons, and so the ground-state electron configuration of carbon monoxide is $(1\sigma)^2(1\sigma^*)^2(2\sigma)^2(2\sigma^*)^2(1\pi)^4(3\sigma)^2$. The bond order of CO is 3. The Lewis formula of CO is $:\overset{-}{C}\equiv\overset{+}{O}:$. Both molecular orbital theory and the Lewis formula predict that there is a triple bond in CO.

CHAPTER 13

13-1. 810 kJ **13-3.** $34.1\ kJ\cdot mol^{-1}$ **13-5.** 219 kJ **13-7.** 0.410 kJ
13-9. 70.1 kJ
13-13. $4.77\ kJ\cdot mol^{-1}$ **13-15.** $T_b[He] < T_b[C_2H_6] < T_b[C_2H_5OH] < T_b[KBr]$ **13-17.** ClF and NF_3 **13-19.** $\Delta\bar{H}_{vap}[CH_4] < \Delta\bar{H}_{vap}[C_2H_6] < \Delta\bar{H}_{vap}[CH_3OH] < \Delta\bar{H}_{vap}[C_2H_5OH]$ **13-21.** $7.4\ kJ\cdot mol^{-1}$ **13-23.** 57.3 kJ
13-25. (a) $29.1\ kJ\cdot mol^{-1}$ (b) $9.11\ kJ\cdot mol^{-1}$ **13-27.** between 45 torr and 50 torr **13-29.** liquid will be present **13-31.** between 80 and 85°C
13-33. Denver, 95°C; Mt. Kilimanjaro, 80°C; Mt. Washington, 95°C; the Matterhorn, 85°C **13-35.** 51 torr **13-37.** approximately 20°C
13-39. The water molecules attract each other relatively strongly due to hydrogen bonding. **13-41.** 0.2 μJ **13-43.** 392.4 pm **13-45.** $2.70\ g\cdot cm^{-3}$ or $2.70\ g\cdot mL^{-1}$ **13-47.** 6.02×10^{23} **13-49.** 537.8 pm; 268.9 pm
13-51. four formula units in a unit cell; face-centered cubic, like NaCl
13-53. $3.995\ g\cdot cm^{-3}$ **13-55.** (a) gas (b) gas (c) liquid (d) liquid
13-57. There are three triple points; sulfur will go from rhombic to monoclinic to liquid; 10^{-5} atm **13-59.** 1.4 atm, or 1070 torr
13-61. 337.2 K and 1140 torr

CHAPTER 14

14-1. $X_{H_2O} = 0.803$; $X_{C_2H_5OH} = 0.197$ **14-3.** Dissolve 760 g acetone in 1 kg of water. **14-5.** 46.6 torr; 0.5 torr **14-7.** 23.57 torr; 0.19 torr
14-9. The total pressure is the sum of the partial pressures, $P_{total} = P_1 + P_2$, and from Raoult's law, we have that $P_{total} = X_1P_1^0 + X_2P_2^0$.
14-11. In a two-component system, $X_1 + X_2 = 1$. Substituting $X_2 = 1 - X_1$ into the expression for P_{total} in the answer to Problem 14-9, we obtain

$$\begin{aligned} P_{\text{total}} &= X_1P_1^0 + (1 - X_1)P_2^0 \\ &= X_1P_1^0 + P_2^0 - X_1P_2^0 \\ &= P_2^0 + X_1(P_1^0 - P_2^0) \end{aligned}$$

The expression for P_{total} has the same form as the equation for a straight line, $y = mx + b$. The slope of the line is $P_1^0 - P_2^0$ and the y intercept at $X_1 = 0$ is P_2^0. The slope is 78 torr and the y intercept is 270 torr. (See Study Guide/Solutions Manual for graph.) **14-13.** 0.103 m **14-15.** 0.11 m **14-17.** (a) 2.0 m_c (b) 3.0 m_c (c) 1.0 m_c
14-19. (a) 0.16 torr (b) 0.23 torr (c) 0.08 torr **14-21.** 2.26 M
14-23. 1.2 K = 1.2°C; 101.2°C **14-25.** 0.61 K = 0.61°C; 81.3°C
14-27. Ethyl alcohol is a temporary antifreeze because its equilibrium vapor pressure is much greater than one atmosphere at 100°C. Thus ethyl alcohol is much more readily lost by evaporation from the coolant system than a relatively high boiling liquid like ethylene glycol.
14-29. −0.62°C **14-31.** 4.97°C **14-33.** 450 **14-35.** 10 m **14-37.** 52%; 55% **14-39.** $HgCl_2$ is a weak electrolyte. **14-41.** 13 atm **14-43.** 5730
14-45. 1.50×10^5 **14-47.** 4.16 M_c; 13.6 L **14-49.** 6.3×10^{-4} mol·L^{-1}
14-51. 0.069 M **14-53.** 130 ft

CHAPTER 15

15-1. (a) $\dfrac{[H_2O]^2}{[H_2]^2}$ (b) $[HCl]^2[POCl_3]$

15-3. (a) $\dfrac{[H_2]^2[S_2]}{[H_2S]^2}$ (b) $\dfrac{[C_2H_6]}{[C_2H_4][H_2]}$

15-5. (a) $\dfrac{[H_2O]^2}{[H_2]^2[O_2]}$ (b) $\dfrac{[NO_2]^4[O_2]}{[N_2O_5]^2}$ (c) $[H_2O]^2$

15-7. (a) $\dfrac{P_{H_2O}^2}{P_{H_2}^2 P_{O_2}}$ (b) $\dfrac{P_{NO_2}^4 P_{O_2}}{P_{N_2O_5}^2}$ (c) $P_{H_2O}^2$

15-9. Because concentrations and pressures are always positive.
15-11. 4.63×10^{-3} M **15-13.** 3800 torr^2, or 6.7×10^{-3} atm^2 **15-15.** 50
15-17. $K_c = (RT)^2K_p = 170\ M^{-2}$ **15-19.** $K_c = 7.99 \times 10^{-2}$ M; $K_p = 7.04$ atm **15-21.** 0.38 M **15-23.** $[PCl_3] = 1.6$ M, $[Cl_2] = 1.6$ M, $[PCl_5] = 1.4$ M **15-25.** 0.494 mol **15-27.** 0.658 atm
15-29. $[ICl] = 0.26$ M; $[Cl_2] = [I_2] = 0.085$ M
15-31. $[H_2] = [I_2] = 0.52$ M; $[HI] = 4.33$ M **15-33.** $P_{NO_2} = 1.76$ atm, $P_{N_2O_4} = 0.64$ atm **15-35.** The gaseous mixture is almost all CO_2 at equilibrium. **15-37.** (a) 0.090 atm (b) $P_{CH_3OH} = 8.98$ atm, $P_{CO} = 1.00$ atm **15-39.** more CO_2 and less CO will be formed
15-41. more NOBr and less NO will be formed **15-43.** (a) to the left (b) to the left (c) to the right (d) to the right (e) no change
15-45. (a) to the right (b) to the left (c) to the right (d) to the left
15-47. (a) no change (b) to the left (c) to the right (d) to the right

15-49. (a) $K_1 = \dfrac{[CO][H_2]}{[H_2O]}$, $K_2 = \dfrac{[CO_2][H_2]}{[CO][H_2O]}$, $K_3 = \dfrac{[H_2O][CH_4]}{[CO][H_2]^3}$
(b) (i) reaction 1 shifts to the right
reaction 2 shifts to the left
reaction 3 shifts to the left

(ii) reaction 1 shifts to the left
reaction 2: no change
reaction 3 shifts to the right

15-51. (a) to the right (b) to the left **15-53.** The reaction is not at equilibrium, and will proceed to the left. **15-55.** (a) 20 atm^{-1} (b) to the left **15-57.** 36.9 atm^2
15-59. 2.85 $atm^{1/2}$ **15-61.** 210 atm **15-63.** 5.9×10^{-5} atm^{-1}
15-65. 56.6 kJ

CHAPTER 16

16-1. (a) $M^{-1} \cdot s^{-1}$ (b) $M^{-1/2} \cdot s^{-1}$ **16-3.** (a) monitor the color of the reaction mixture (b) monitor the pressure of the reaction mixture **16-5.** 1.2×10^{-5} $M \cdot s^{-1}$ **16-7.** 9.5×10^{-3} $atm \cdot s^{-1}$ **16-9.** 9.1×10^{-4} $M \cdot s^{-1}$; 5.0×10^{-4} $M \cdot s^{-1}$ **16-11.** 3.0×10^{-2} $torr \cdot s^{-1}$; 2.3×10^{-2} $torr \cdot s^{-1}$; 1.8×10^{-2} $torr \cdot s^{-1}$. The average rate of appearance of CO_2 is one-half of the average rate of disappearance of CO. **16-13.** first order **16-15.** rate = $(7.3 \times 10^{-30}\ s^{-1})[C_2H_5Cl]$ **16-17.** rate = $(2.8 \times 10^{-5}\ M^{-1} \cdot s^{-1})[NOCl]^2$ **16-19.** rate = $(2.0 \times 10^{-6}\ M^{-1} \cdot s^{-1})[Cr(H_2O)_6^{3+}] \times [SCN^-]$ **16-21.** rate = $k[I^-][ClO_3^-][H^+]^2$. **16-23.** (a) initial rate = 3.6×10^{-4} $M \cdot s^{-1}$ (b) 1.3 $mol \cdot L^{-1} \cdot h^{-1}$ **16-25.** rate = $(5.0 \times 10^4\ M^{-1} \cdot s^{-1})[NO_2][O_3]$ **16-27.** 0.23*s* **16-29.** 0.67 **16-31.** first order; there will be 25,600 bacteria after two hours; the rate constant is 0.046 min^{-1}. **16-33.** 1.00 **16-35.** Use the equation $\log([A]/[A]_0) = -kt/2.30$ **16-37.** rate = $(0.041\ min^{-1})[S_2O_8^{2-}]$
16-39. 230 days **16-41.** 0.015 M **16-43.** 1.1×10^{-2} $M^{-1} \cdot s^{-1}$
16-45. (a) first order (b) zero order (c) $\frac{1}{2}$ order
16-47. (a) rate = $k[N_2O][O]$ (b) rate = $k[O][O_3]$ (c) rate = $k[ClCO][Cl_2]$ **16-49.** $2O_3(g) \rightleftarrows 3O_2(g)$ **16-51.** Yes. The rate law of the overall reaction is given by the slow step.
16-53. $2N_2O_5(g) \rightarrow 4NO_2(g) + O_2(g)$; rate = $k[N_2O_5]$
16-57. rate = $k[NO]^2[O_2]$ **16-59.** CH_3 is a free radical, and two CH_3 radicals can join to form a stable (C_2H_6) molecule. The activation energy for the process is essentially zero. **16-61.** 100 $kJ \cdot mol^{-1}$
16-63. 3.13×10^{-2} s^{-1} **16-65.** 0.017 h **16-67.** The plot is a straight line, and E_a = 120 $kJ \cdot mol^{-1}$, $k = 5 \times 10^{-3}$ $M^{-1} \cdot s^{-1}$ at 500 k **16-69.** No. A catalyst affects only the rate of attainment of equilibrium, not the position of the equilibrium. **16-71.** (a) $H^+(aq)$ and $Br^-(aq)$ (b) third order (c) $[H^+]$ and $[Br^-]$ do not change with time because $H^+(aq)$ and $Br^-(aq)$ are catalysts. For $[H^+]$ fixed at 1.00×10^{-3} M and $[Br^-]$ fixed at 1.00×10^{-3} M, $[H_2O_2]$ is governed by a first-order rate law with $k = 1.0 \times 10^{-3}$ s^{-1}. **16-73.** Change the surface area of the wall material to see if the rate of the reaction changes. **16-75.** The surface sites for adsorption of O_2 are occupied except at very low O_2 pressures. The rate-determining step is the dissociation of the adsorbed O_2 molecules to O atoms. **16-77.** 4.8×10^{-6} $M^{-1} \cdot s^{-1}$ **16-79.** rate = $k_r[H_2][I_2]$

16-81. reverse rate = $\dfrac{k_r[Cr^{3+}]^2[SO_4^{2-}]^3}{[HCrO_4^-][HSO_3^-][H^+]^4}$

CHAPTER 17

17-1. $[H_3O^+] = 3.3 \times 10^{-12}$ M; basic

17-3. $[H_3O^+] = [ClO_4^-] = 0.050$ M; $[OH^-] = 2.0 \times 10^{-13}$ M; acidic
17-5. $[Tl^+] = [OH^-] = 9.03 \times 10^{-3}$ M; $[H_3O^+] = 1.11 \times 10^{-12}$ M
17-7. 1.30; acidic **17-9.** 1.30; acidic **17-11.** 12.55 **17-13.** 7.46; acidic
17-15. 1.6×10^{-7} M **17-17.** 0.10 M; 0.10 M
17-19. $[H_3O^+] = 4.0 \times 10^{-8}$ M; $[OH^-] = 2.5 \times 10^{-7}$ M **17-21.** 0.16 M or 19 $g \cdot L^{-1}$ **17-23.** 1.3×10^{-5} M **17-25.** 1.7×10^{-5} M
17-27. 2.8×10^3 **17-29.** 0.28% **17-31.** 220 **17-33.** 2.91 **17-35.** 2.0
17-37. 2.64 **17-39.** 1.4 **17-41.** 1.8×10^{-5} M **17-43.** 9.47 **17-45.** 11.71
17-47. (a) shifts to the right (b) shifts to the left (c) shifts to the left (d) shifts to the left **17-49.** (a) shifts to the right (b) shifts to the left
17-51. (a) shifts to the left (b) no change (c) shifts to the left (d) shifts to the right (e) shifts to the left **17-53.** (a) $HC_7H_5O_2(aq)$–$C_7H_5O_2^-(aq)$ and $H_3O^+(aq)$–$H_2O(aq)$ (b) $CH_3NH_3^+(aq)$–$CH_3NH_2(aq)$ and $H_2O(aq)$–$OH^-(aq)$ (c) $HCHO_2(aq)$–$CHO_2^-(aq)$ and $H_3O^+(aq)$–$H_2O(aq)$
17-55. (a) $NO_2^-(aq)$ (b) $C_2H_2ClO_2^-(aq)$ (c) $H_2PO_4^-(aq)$ (d) $HPO_4^{2-}(aq)$
17-57. (a) acid, $CHO_2^-(aq)$ (b) base, $HIO(aq)$ (c) base, $HF(aq)$ (d) acid, $CH_3NH_2(aq)$ (e) base, $ClNH_3^+(aq)$; (f) base, $HCN(aq)$
17-59. (a) 7.46×10^{-10} M (b) 1.48×10^{-11} M (c) 1.8×10^{-5} M (d) 1.7×10^{-5} M **17-61.** (a) 0.40 (b) 1.2 **17-63.** (a) 6.80 (b) 12.17
17-65. 1.00×10^{-9} M **17-67.** (a) acidic (b) cannot tell without a detailed calculation (c) basic (d) neutral (e) basic **17-69.** (a) basic (b) basic (c) acidic (d) neutral (e) acidic **17-71.** basic **17-73.** 10.11; $[HClO] = 1.3 \times 10^{-4}$ M **17-75.** $[HCNO] = [OH^-] = 2.1 \times 10^{-6}$ M; $[CNO^-] = 0.10$ M; $[H_3O^+] = 4.7 \times 10^{-9}$ M; pH = 8.33 **17-77.** 2.85
17-79. 4.47 **17-81.** 2.02 **17-83.** (a) Arrhenius acid and a Brønsted-Lowry acid (b) Arrhenius acid, a Brønsted-Lowry acid, and a Lewis acid (c) Lewis acid **17-85.** (a) Lewis base (b) Lewis acid (c) Arrhenius acid and base, Brønsted-Lowry acid and base, Lewis base

CHAPTER 18

18-1. 10 to 11 **18-3.** 8 to 9.5, tropeolin 000 **18-5.** bromcresol green
18-7. $K_{ai} \approx 10^{-4}$ M **18-9.** 11.00
18-13. 0.193 M **18-15.** 340 mL
18-19. 9.48, thymolphthalein **18-21.** 176 **18-23.** benzoic acid
18-25. 4.15% **18-27.** 4.76 **18-29.** 3.64 **18-31.** 9.54 **18-33.** (a) 7.21 (b) 7.51 (c) 6.91 **18-35.** 7.21 **18-37.** $[HC_2H_3O_2] \approx [HC_2H_3O_2]_0$ and $[C_2H_3O_2^-] \approx [C_2H_3O_2^-]_0$ because K_a and K_w/K_a are both small.
18-39. 4.76 to 2.04 **18-41.** A buffer for which $pK_a = 3.29$, or for which $K_a = 5.13 \times 10^{-4}$ M. From Table 17-4, HNO_2–NO_2^- is well suited.
18-43. (a) 9.57 (b) 9.27 **18-45.** 5.28; 5.28 (Henderson-Hasselbalch equation) **18-47.** (b) A $NaHCO_3(aq)$ solution acts as a buffer through the equations $HCO_3^-(aq) + H_3O^+(aq) \rightleftarrows H_2CO_3(aq) + H_2O(l)$ and $HCO_3^-(aq) + OH^-(aq) \rightleftarrows CO_3^{2-}(aq) + H_2O(l)$ **18-49.** (a) $H_3PO_4(aq)$ and $H_2PO_4^-(aq)$ (b) $H_2PO_4^-(aq)$ (c) $H_2PO_4^-(aq)$ and $HPO_4^{2-}(aq)$ (d) $HPO_4^{2-}(aq)$ **18-51.** There are two equivalence points.
18-53. 50.0 mL **18-55.** 90.1 **18-57.** 1.66
18-61. $3HCO_3^-(aq) + H_3C_6H_5O_7(aq) \rightleftarrows 3H_2CO_3(aq) + C_6H_5O_7^{3-}(aq)$
$H_2CO_3(aq) \rightleftarrows CO_2(aq) + H_2O(l)$; $CO_2(aq) \rightleftarrows CO_2(g)$
18-63. See following table

Volume of HCl solution added/mL	*Total volume of resulting solution/mL*	*Moles of $OH^-(aq)$ in solution*	*$[OH^-]/M$*	*pH*
10.0	90.0	0.0120	0.133	13.12
20.0	100.0	0.0080	0.080	12.90
30.0	110.0	0.0040	0.036	12.56
35.0	115.0	0.0020	0.017	12.23
39.0	119.0	0.0004	0.003	11.48
40.0	120.0		1.00×10^{-7}	7.00
41.0	121.0		3.0×10^{-12}	2.48
50.0	130.0		3.3×10^{-13}	1.52
60.0	140.0		1.8×10^{-13}	1.26
70.0	150.0		1.3×10^{-13}	1.10

CHAPTER 19

19-1. AgBr, $PbSO_4$ and $BaSO_4$ are insoluble. **19-3.** $CaCO_3$, CaS and $CaSO_4$ are insoluble.
19-5. (a) $AgClO_4(aq) + NaI(aq) \rightarrow NaClO_4(aq) + AgI(s)$
$Ag^+(aq) + I^-(aq) \rightarrow AgI(s)$
(b) $Hg_2(NO_3)_2(aq) + K_2SO_4(aq) \rightarrow 2KNO_3(aq) + Hg_2SO_4(s)$
$Hg_2^{2+}(aq) + SO_4^{2-}(aq) \rightarrow Hg_2SO_4(s)$
(c) $Pb(NO_3)_2(aq) + Li_2CrO_4(aq) \rightarrow 2LiNO_3(aq) + PbCrO_4(s)$
$Pb^{2+}(aq) + CrO_4^{2-}(aq) \rightarrow PbCrO_4(s)$
(d) $ZnCl_2(aq) + Na_2CO_3(aq) \rightarrow 2NaCl(aq) + ZnCO_3(s)$
$Zn^{2+}(aq) + CO_3^{2-}(aq) \rightarrow ZnCO_3(s)$
(e) $BaCl_2(aq) + ZnSO_4(aq) \rightarrow ZnCl_2(aq) + BaSO_4(s)$
$Ba^{2+}(aq) + SO_4^{2-}(aq) \rightarrow BaSO_4(s)$
19-7. (a) $Hg_2(ClO_4)_2(aq) + 2NaBr(aq) \rightarrow 2NaClO_4(aq) + Hg_2Br_2(s)$
(b) $Fe(ClO_4)_3(aq) + 3NaOH(aq) \rightarrow 3NaClO_4(aq) + Fe(OH)_3(s)$
(c) $CaCl_2(aq) + H_2SO_4(aq) \rightarrow 2HCl(aq) + CaSO_4(s)$
(d) $Pb(NO_3)_2(aq) + 2LiIO_3(aq) \rightarrow 2LiNO_3(aq) + Pb(IO_3)_2(s)$
(e) $H_2SO_4(aq) + Pb(NO_3)_2(aq) \rightarrow 2HNO_3(aq) + PbSO_4(s)$ **19-9.** The lead precipitates out of solution, and so is no longer physiologically active. **19-11.** $PbCrO_4(s) \rightleftarrows Pb^{2+}(aq) + CrO_4^{2-}(aq)$; $s = 5.3 \times 10^{-7}$ M, or $1.7 \times 10^{-4}\ g \cdot L^{-1}$ **19-13.** $5.02 \times 10^{-13}\ M^2$ **19-15.** 1.7×10^{-4} M, or $9.9 \times 10^{-3}\ g \cdot L^{-1}$ **19-17.** $2.9 \times 10^{-3}\ M^2$ **19-19.** 9.11
19-21. $4.4 \times 10^{-3}\ g \cdot L^{-1}$ **19-23.** 1.6×10^{-5} M, or $2.8 \times 10^{-3}\ g \cdot L^{-1}$
19-25. 1.8×10^{-10} M, or $2.6 \times 10^{-8}\ g \cdot L^{-1}$ **19-27.** 3.4 M
19-29. (a) increased (b) unchanged (c) decreased (d) unchanged
19-31. The chloride ions precipitate as AgCl(*s*). **19-33.** $PbI_2(s)$ precipitates from solution. $[Pb^{2+}] = 2.00$ M, $[I^-] = 6.0 \times 10^{-5}$ M, $[NO_3^-] = 4.00$ M, $[Na^+] = 6.67 \times 10^{-4}$ M; 2.28×10^{-5} mol $PbI_2(s)$ precipitate. **19-35.** (a) 0.29 g (b) 1.8×10^{-10} M **19-37.** the solubility (a) increases (b) decreases (c) decreases **19-39.** $CaCO_3$, $PbSO_3$, $Fe(OH)_3$ and ZnS. **19-41.** 0.18 M, or $10\ g \cdot L^{-1}$ **19-43.** 2.2×10^{-8} M, or

2.1×10^{-6} g·L^{-1} **19-45.** the solubility (a) increases (b) decreases (c) increases (d) decreases **19-47.** $[Cr(OH)_3] = 6.3 \times 10^{-4}$ M; $[Ni(OH)_2] = 2.0 \times 10^3$ M (very soluble). Yes, $Cr(OH)_3$ can be separated from $Ni(OH)_2$ at this pH. **19-49.** 3.1 **19-51.** 5.7×10^{-19} M **19-53.** A low pH, slightly less than zero. **19-55.** $[HgS] = 4 \times 10^{-38}$ M and $[CdS] = 7.3 \times 10^{-12}$ M at pH = 3; $[HgS] = 4 \times 10^{-44}$ M and $[CdS] = 7.3 \times 10^{-18}$ M at pH = 6. **19-57.** $Pb(NO_3)_2(aq) + 2NaOH(aq) \rightarrow 2NaNO_3(aq) + Pb(OH)_2(s)$; $Pb(OH)_2(s) + OH^-(aq) \rightarrow Pb(OH)_3^-(aq)$ **19-59.** 1×10^{-3} M **19-61.** 0.40 M **19-63.** $s = 6.3 \times 10^2$ M (very soluble) at pH = 3.0; $s = 6.3 \times 10^{-7}$ M at pH = 6.0; $s = 4 \times 10^{-7}$ M at pH = 9.0; $s = 4 \times 10^{-4}$ M at pH = 12.0.

CHAPTER 20

20-1. (a) +6 (b) +6 (c) +3 (d) +5 (e) +7 **20-3.** (a) +4 (b) +5 (c) +4 (d) +5 (e) +3 **20-5.** (a) +4 (b) 0 (c) +1 (d) +5 (e) +3 **20-7.** (a) 0 (b) −4 (c) −2 (d) +2 (e) +4 **20-9.** (a) +3 (b) +3 (c) +3 (d) +3 (e) +4 **20-11.** $I_2O_5(s)$ is the oxidizing agent and $CO(g)$ is the reducing agent. **20-13.** $NaNO_3(aq)$ is the oxidizing agent and $Pb(s)$ is the reducing agent. **20-15.** (a) $V(s) \rightarrow V^{2+}(aq) + 2e^-$, oxidation; $Ni^{2+}(aq) + 2e^- \rightarrow Ni(s)$, oxidation; $Ni^{2+}(aq)$ is the oxidizing agent and $V(s)$ is the reducing agent. (b) $Ti^{2+}(aq) \rightarrow Ti^{3+}(aq) + e^-$, oxidation; $Co^{2+}(aq) + 2e^- \rightarrow Co(s)$, reduction. $Co^{2+}(aq)$ is the oxidizing agent and $Ti^{2+}(aq)$ is the reducing agent. **20-17.** Oxygen in superoxides is in an unusual oxidation state of $-\frac{1}{2}$, and readily converts to its more normal (oxide) oxidation state of −2.

20-19. (a) $4CrO_4^{2-}(aq) + 3Cl^-(aq) + 2OH^+(aq) \rightarrow 3ClO_2^-(aq) + 4Cr^{3+}(aq) + 10H_2O(l)$ (b) $IO_3^-(aq) + 2Cr(OH)_4^-(aq) + 2OH^-(aq) \rightarrow I^-(aq) + 2CrO_4^{2-}(aq) + 5H_2O(l)$ (c) $HAsO_2(aq) + I_3^-(aq) + 2H_2O(l) \rightarrow H_3AsO_4(aq) + 3I^-(aq) + 2H^+(aq)$ **20-21.** (a) $BrO_3^-(aq) + 3NO_2^-(aq) \rightarrow 3NO_3^-(aq) + Br^-(aq)$ (b) $2MnO_4^-(aq) + 2Sb(s) + 2H^+(aq) \rightarrow Sb_2O_3(s) + 2MnO_2(s) + H_2O(l)$ (c) $H_2MoO_4(aq) + 6Cr^{2+}(aq) + 6H^+(aq) \rightarrow Mo(s) + 6Cr^{3+}(aq) + 4H_2O(l)$

20-23. (a) $4Fe(OH)_2(s) + O_2(g) + 2H_2O(l) \rightarrow 4Fe(OH)_3(s)$ (b) $3Cu(s) + 2NO_3^-(aq) + 8H^+(aq) \rightarrow 2NO(g) + 3Cu^{2+}(aq) + 4H_2O(l)$ (c) $Bi_2O_3(s) + 2ClO^-(aq) + 2OH^-(aq) \rightarrow 2BiO_3^-(aq) + 2Cl^-(aq) + H_2O(l)$

20-25. (a) $NH_4^+(aq) + NO_3^-(aq) \rightarrow N_2O(g) + 2H_2O(l)$ (b) $6Hg(l) + 3H_2SO_4(aq) + Cr_2O_7^{2-}(aq) + 8H^+(aq) \rightarrow 3Hg_2SO_4(s) + 2Cr^{3+}(aq) + 7H_2O(l)$ (c) $4Fe(s) + 3O_2(g) + 6H_2O(l) \rightarrow 2Fe_2O_3 \cdot 3H_2O(s)$

20-27. (a) $2MnO(s) + 5PbO_2(s) + 8H^+(aq) \rightarrow 2MnO_4^-(aq) + 5Pb^{2+}(aq) + 4H_2O(l)$; electron donor and reducing agent, MnO; electron acceptor and oxidizing agent, PbO_2; species oxidized, Mn; species reduced, Pb (b) $As_2S_5(s) + 40NO_3^-(aq) + 35H^+(aq) \rightarrow 5HSO_4^-(aq) + 2H_3AsO_4(aq) + 40NO_2(g) + 12H_2O(l)$; electron donor and reducing agent, As_2S_5; electron acceptor and oxidizing agent, NO_3^-; species oxidized, S; species reduced, N

20-29. $Zn(s) + 2MnO_2(s) + 2OH^-(aq) \rightarrow ZnO_2^{2-}(aq) + Mn_2O_3 \cdot H_2O(s)$

20-31. $2CrI_3(s) + 27Cl_2(g) + 64OH^-(aq) \rightarrow 2CrO_4^{2-}(aq) + 6IO_4^-(aq) + 54Cl^-(aq) + 32H_2O(l)$ **20-33.** (a) $PbO_2(s) + 4H^+(aq) + 2e^- \rightarrow Pb^{2+}(aq) + 2H_2O(l)$ (b) $MnO_2(s) + 4H^+(aq) + 2e^- \rightarrow Mn^{2+}(aq) + 2H_2O(l)$ (c) $NO(g) + 2H_2O(l) \rightarrow NO_3^-(aq) + 4H^+(aq) + 3e^-$ **20-35.** (a) $2WO_3(s) + 2H^+(aq) + 2e^- \rightarrow$

$W_2O_5(s) + H_2O(l)$ (b) $U^{4+}(aq) + 2H_2O(l) \rightarrow UO_2^+(aq) + 4H^+(aq) + e^-$ (c) $Zn(s) + 4OH^-(aq) \rightarrow Zn(OH)_4^{2-}(aq) + 2e^-$
20-37. (a) $2SO_3^{2-}(aq) + 2H_2O(l) + 2e^- \rightarrow S_2O_4^{2-}(aq) + 4OH^-(aq)$
(b) $2Cu(OH)_2(s) + 2e^- \rightarrow Cu_2O(s) + H_2O(l) + 2OH^-(aq)$
(c) $2AgO(s) + H_2O(l) + 2e^- \rightarrow Ag_2O(s) + 2OH^-(aq)$
(d) $HgO(s) + H_2O(l) + 2e^- \rightarrow Hg(l) + 2OH^-(aq)$
20-39. $Au(s) + 3NO_3^-(aq) + 4Cl^-(aq) + 6H^+(aq) \rightarrow AuCl_4^-(aq) + 3NO_2(g) + 3H_2O(l)$ **20-41.** $8P(s) + 5BaSO_4(s) \rightarrow 2P_4O_{10}(s) + 5BaS(s)$; $8P(s)$ can be written as $2P_4(s)$. **20-43.** $2SO_2(g) + O_2(g) + 2H_2O(l) \rightarrow 2H_2SO_4(aq)$
$H_2SO_4(aq) + Fe(s) \rightarrow FeSO_4(aq) + H_2(g)$
20-45. $Na_2B_4O_7(aq) + Na_2O_2(aq) + 3H_2O_2(aq) \rightarrow 4NaBO_3(aq) + 3H_2O(l)$
20-47. 2.00 g; 25.9% **20-49.** 2.35 g; 23.5% **20-51.** 0.407 M

CHAPTER 21

21-1. $Fe(s) \rightarrow Fe^{2+}(aq) + 2e^-$ (negative electrode); $Ni^{2+}(aq) + 2e^- \rightarrow Ni(s)$ (positive electrode); $Fe(s)|FeCl_2(aq)||NiCl_2(aq)|Ni(s)$
21-3. $V(s) \rightarrow V^{2+}(aq) + 2e^-$ (negative electrode); $Zn^{2+}(aq) + 2e^- \rightarrow Zn(s)$ (positive electrode); $V(s)|VI_2(aq)||ZnSO_4(aq)|Zn(s)$ **21-5.** $Ag(s) + I^-(aq) \rightarrow AgI(s) + e^-$; $H^+(aq) + e^- \rightarrow \frac{1}{2}H_2(g)$; $2Ag(s) + 2HI(aq) \rightarrow 2AgI(s) + H_2(g)$ **21-7.** $Cd(s) \rightarrow Cd^{2+}(aq) + 2e^-$; $Re^{3+}(aq) + 3e^- \rightarrow Re(s)$ **21-9.** $H_2(g) + PbSO_4(s) \rightarrow 2H^+(aq) + Pb(s) + SO_4^{2-}(aq)$
21-11. $La(s)|LaCl_3(aq)||CuCl_2(aq)|Cu(s)$ **21-13.** the cell voltage (a) decreases, (b) increases, (c) is unaffected **21-15.** the cell voltage (a) is unaffected, (b) decreases, (c) decreases, (d) decreases **21-17.** the cell voltage (a) increases, (b) is unaffected, (c) decreases, (d) increases, (e) decreases, (f) is unaffected, (g) increases **21-19.** (a) 6 (b) 1
21-21. $\log K = 15.64$; $K = 4.38 \times 10^{15}$ **21-23.** $E^0 = 0.270$ V; $\log K = 9.12$; $K = 1.32 \times 10^9$ **21-25.** $E^0 = 1.62$ V; $\log K = 82.09$
21-27. (a) $Zn(s) + Hg_2Cl_2(s) \rightarrow ZnCl_2(aq) + 2Hg(l)$ (b) 1.19 V
21-29. 0 V **21-31.** 1.39 V **21-33.** −0.27 V; lead will not displace the cadmium from $Cd(C_2H_3O_2)_2(aq)$ under standard conditions.
21-35. $E^0 = +0.24$ V; $\log K = 4.05$; $K = 1.12 \times 10^4$ **21-37.** 2.03 V
21-39. The oxidation half-reaction in basic solution is $S_2O_3^{2-}(aq) + 6OH^-(aq) \rightarrow 2SO_3^{2-}(aq) + 3H_2O(l) + 4e^-$. The reduction half-reaction in basic solution is $O_2(g) + 2H_2O(l) + 4e^- \rightarrow 4OH^-(aq)$; $E^0_{SO_3^{2-}/S_2O_3^{2-}} = -0.58$ V **21-41.** $\ominus$steel$|Fe(s)|Fe(OH)_2(s)|LiOH(s)|NiOOH(s), Ni(OH)_2(s)|$steel$\oplus$ **21-43.** $\ominus Zn(s)|ZnSO_4(aq)|Hg_2SO_4(s)|Hg(l)\oplus$ **21-45.** $\ominus Mg(s)|MgO(s)|LiCl + KCl(soln)|Cu_2O(s), CuO(s)|$steel$\oplus$ **21-47.** 2.73 h **21-49.** 4.96 g **21-51.** 25.52 kg; hydrogen fluoride is a covalent compound and so is a poor conductor of an electric current. **21-53.** 54 min **21-55.** 8.44×10^5 A·h
21-57. $AgBr(s) \rightarrow Ag^+(aq) + Br^-(aq)$; $\log K_{sp} = -12.30$; $K_{sp} = 5.0 \times 10^{-13}$ M^2 **21-59.** $\text{pH} = \dfrac{E - 0.2415\text{ V}}{0.0592\text{ V}}$

CHAPTER 22

22-1. LiCl: $\Delta\bar{S}_{fus} = 22.3$ J·K^{-1}·mol^{-1}; $\Delta\bar{S}_{vap} = 101$ J·K^{-1}·mol^{-1}. NaCl: $\Delta\bar{S}_{fus} = 26.3$ J·K^{-1}·mol^{-1}; $\Delta\bar{S}_{vap} = 107.8$ J·K^{-1}·mol^{-1}. KCl: $\Delta\bar{S}_{fus} = 25.7$ J·K^{-1}·mol^{-1}; $\Delta\bar{S}_{vap} = 97.46$ J·K^{-1}·mol^{-1} **22-3.** CH_2Cl_2: $\Delta\bar{S}_{fus} = 8.056$ J·K^{-1}·mol^{-1}; $\Delta\bar{S}_{vap} = 101.2$ J·K^{-1}·mol^{-1}. $CHCl_3$: $\Delta\bar{S}_{fus} =$

41.99 $J \cdot K^{-1} \cdot mol^{-1}$; $\Delta\bar{S}_{vap}$ = 93.70 $J \cdot K^{-1} \cdot mol^{-1}$. CCl_4: $\Delta\bar{S}_{fus}$ = 13.09 $J \cdot K^{-1} \cdot mol^{-1}$; $\Delta\bar{S}_{vap}$ = 91.28 $J \cdot K^{-1} \cdot mol^{-1}$ **22-5.** Li: 6.59 $J \cdot K^{-1} \cdot mol^{-1}$; Na: 7.01 $J \cdot K^{-1} \cdot mol^{-1}$; K: 6.93 $J \cdot K^{-1} \cdot mol^{-1}$; Rb: 7.50 $J \cdot K^{-1} \cdot mol^{-1}$; Cs: 6.95 $J \cdot K^{-1} \cdot mol^{-1}$. **22-7.** $\Delta\bar{S}_{fus}$ = 12.7 $J \cdot K^{-1} \cdot mol^{-1}$; $\Delta\bar{S}_{vap}$ = 88.0 $J \cdot K^{-1} \cdot mol^{-1}$ **22-9.** (a) PCl_3 (b) $CH_3CH_2CH_3$ (c) $CH_3CH_2CH_2COOH$ (d) $CH_3CH_2CH_2CH_2CH_3$ **22-11.** $\bar{S}[C_2Cl_6(g)] > \bar{S}[C_2Cl_4(g)] > \bar{S}[C_2Cl_2(g)]$ **22-13.** SO_3 has more atoms per molecule than SO_2 and so is able to take up energy more easily. **22-15.** $I_2(g)$ has much more positional disorder than $I_2(s)$. **22-17.** (a) increases (b) decreases (c) increases (d) decreases **22-19.** (c) < (b) < (a) < (d) **22-21.** (a) −55.6 $J \cdot K^{-1}$ (b) −332.1 $J \cdot K^{-1}$ (c) 134.0 $J \cdot K^{-1}$ (d) −214.7 $J \cdot K^{-1}$ **21-23.** (a) 2.9 $J \cdot K^{-1}$ (b) −189.2 $J \cdot K^{-1}$ (c) −242.8 $J \cdot K^{-1}$ (d) −111.8 $J \cdot K^{-1}$ **22-25.** $\Delta G_{rxn} = -8.6\ kJ \cdot mol^{-1} < 0$; $\Delta H_{rxn} > 0$, $T\Delta S_{rxn} > 0$ The reaction is entropy-driven. **22-27.** ΔS°_{rxn} = −429.6 $J \cdot K^{-1}$; ΔG°_{rxn} = −503 kJ The reaction is spontaneous from left to right at 1 atm pressure. **22-29.** ΔS°_{rxn} = −30.0 $J \cdot K^{-1}$; $\Delta G^\circ_{rxn} = -1.314 \times 10^3$ kJ The reaction is spontaneous from left to right at 1 atm pressure. **22-31.** Yes. ΔG°_{rxn} is positive for this reaction. **22-33.** ΔG_{rxn} = −203 kJ **22-35.** ΔG_{rxn} = −130 kJ **22-37.** (a) −400 kJ (b) −74 kJ **22-39.** E^0 = +0.36 V; ΔG°_{rxn} = −69 kJ; ΔG_{rxn} = −73 kJ; E = +0.38 V; $Zn(s) + Cd^{2+}(aq) \rightarrow Zn^{2+}(aq) + Cd(s)$ **22-41.** (a) $\Delta G^\circ_{rxn} > 0$; (b) $\Delta G^\circ_{rxn} < 0$ **22-43.** −2.55 kJ. The reaction is spontaneous from left to right under standard conditions; 5400 kJ. The reaction is spontaneous from right to left. **22-45.** +19.1 kJ; nitrous acid will not dissociate spontaneously under standard conditions. It will dissociate when $[H^+] = [NO_2^-] = 1.0 \times 10^{-5}$ M and $[HNO_2]$ = 1.0 M. **22-47.** +16.3 kJ; chloroacetic acid will not dissociate spontaneously under standard conditions. It will dissociate spontaneously when $[H^+] = 1.0 \times 10^{-5}$ M, $[C_2H_2ClO_2^-]$ = 0.001 M and $[HC_2H_2ClO_2]$ = 0.10 M. **22-49.** +55.5 kJ; no **22-51.** −19.3 kJ; from left to right; −7.96 kJ; left to right **22-53.** No. If K is infinite, then ΔG°_{rxn} is infinite, and an infinite amount of work could be obtained from the reaction. **22-55.** (a) −29.1 kJ; 1.29×10^5 (b) 91.4 kJ; 8.97×10^{-17} (c) −142.2 kJ; 9.25×10^{24} **22-57.** ΔG°_{rxn} = −355 kJ; ΔH°_{rxn} = −357.6 kJ; $K = 2.31 \times 10^{62}$ **22-59.** ΔG°_{rxn} = −141.8 kJ; ΔH°_{rxn} = −197.8 kJ; $K = 7.87 \times 10^{24}$; $K = 3.01 \times 10^5$ at 400°C. **22-61.** ΔG°_{rxn} = 28.6 kJ; ΔH°_{rxn} = 41.2 kJ; ΔS°_{rxn} = 42.3 $J \cdot K^{-1}$ The reaction is enthalpy-driven to the left. **22-63.** (a) $K < 1$ (b) $K = 1$ (c) $K > 1$ **22-65.** ΔG°_{rxn} = −120.4 kJ for the first reaction and ΔG°_{rxn} = 233.6 kJ for the second reaction. The first reaction is energetically more favorable. **22-67.** 1468 kJ **22-69.** 1.09 V **22-71.** $1.73 \times 10^{-10}\ M^2$ **22-73.** See Study Guide **22-75.** 181 kJ **22-77.** 34.5 kJ **22-79.** ΔH°_{rxn} = −68.6 kJ; K_p = 0.0166 atm^{-1} at 400°C **22-81.** $\log(K_2/K_1) < 0$ when $T_2 < T_1$.

CHAPTER 23

23-1. (a) $1s^22s^22p^63s^23p^63d^4$ (b) $1s^22s^22p^63s^23p^63d^{10}$ (c) $1s^22s^22p^63s^23p^63d^{10}4s^24p^64d^2$ (d) $1s^22s^22p^63s^23p^63d^{10}4s^24p^64d^3$ (e) $1s^22s^22p^63s^23p^63d^{10}4s^24p^64d^{10}4f^{14}5s^25p^65d^6$ **23-3.** (a) 5 (b) 8 (c) 3

(d) 6 (e) 9 **23-5.** (a) Ni(III), Pd(III), Pt(III) (b) Mn(IV), Tc(IV), Re(IV) (c) Zn(II), Cd(II), Hg(II) **23-7.** (a) +2 (b) +3 (c) +3 (d) +3 (e) +1 **23-9.** (a) +2 (b) +2 (c) +2 (d) +3 (e) +2 **23-11.** 4 **23-13.** (a) 3 mol $K^+(aq)$ and 1 mol $[Fe(CN)_6]^{3-}(aq)$ (b) 1 mol $[Ir(NH_3)_2Cl_2]^+(aq)$ and 1 mol $NO_3^-(aq)$ (c) 1 mol $[Pt(NH_3)_4Cl_2]^{2+}(aq)$ and 2 mol $Cl^-(aq)$ (d) 1 mol $[Ru(NH_3)_6]^{3+}(aq)$ and 3 mol $Br^-(aq)$ **23-15.** The given empirical formulas correspond to $[Pt(NH_3)_6]Cl_4$, $[Pt(NH_3)_5Cl]Cl_3$, $[Pt(NH_3)_4Cl_2]Cl_2$, $[Pt(NH_3)_3Cl_3]Cl$ and $[Pt(NH_3)_2Cl_4]$.
23-17. (a) pentaamminechlorochromium(III) (b) triamminetriaquatungsten(II) (c) hexaaquairon(III) (d) tetrachloroferrate(III) **23-19.** (a) ammonium hexanitritocobaltate(III) (b) tetraamminedibromoiridium(III) bromide (c) potassium tetrachlorocuprate(II) (d) pentacarbonylruthenium(0) **23-21.** (a) sodium carbonylpentacyanoferrate(II) (b) *trans*-tetraamminedichloroplatinum(IV) iodide (c) potassium hexaisothiocyanatomolybdate(IV) (d) ammonium tetrachloromanganate(II) **23-23.** (a) $[Fe(CN)_5CO]^{3-}$ (b) $[AuCl_2I_2]^-$ (c) $[Co(CN)_6]^{3-}$ (d) $[Co(NO_2)_6]^{3-}$ **23-25.** (a) $[Co(NH_3)_6]^{3+}$ (b) $[Ni(CO)_4]$ (c) $[Mn(SCN)_6]^{4-}$ (d) $[Cu(NH_3)_2]^+$
23-27. (a) $K_2[CrO_4]$ (b) $[Ni(NH_3)_4](NO_3)_2$ (c) $[Ag(NH_3)_2]Cl$ (d) $Ca_3[Mn(CN)_6]_2$ (e) $Na_2[Cd(SCN)_4]$
23-29. (a) tris(ethylenediamine)ruthenium(III) (b) bis(ethylenediamine)cobalt(II) (c) diiodobis(oxalato)molybdate(III) (d) diamminedichloroethylenediaminecobalt(III)
23-31. (a) $[Fe(ox)_3]^{3-}$ (b) $[Cr(en)_3]^{3+}$ (c) $[Ru(en)_3]^{2+}$ (d) $[PtCl_2(en)_2]^{2+}$
23-33. There are two geometric isomers, *cis* and *trans*. **23-35.** (a) There are 3 geometric isomers. (b) There are 6 geometric isomers.
23-39. (a) t_{2g}^2 (b) t_{2g}^4 (c) $t_{2g}^3e_g^2$ (d) $t_{2g}^6e_g^4$ (e) t_{2g}^6 **23-41.** (a) low-spin (b) low-spin (c) high-spin (d) high-spin (e) high-spin **23-43.** (a) t_{2g}^6, low-spin (b) $t_{2g}^3e_g^2$, high-spin (c) t_{2g}^6, low-spin (d) $e^2t_2^3$, high-spin
23-45. c **23-47.** a, b **23-49.** (a) 0 (b) 1 (c) 1 (d) 5 **23-51.** c and d
23-53. $[NiF_4]^{2-}$ is tetrahedral with a $e^4t_2^4$ d-electron configuration and $[Ni(CN)_4]^{2-}$ is square planar with no unpaired electrons.

CHAPTER 24

24-1. (a) 106 (b) 143 (c) 52 (d) 36 (e) 154 **24-3.** (a) $^{32}_{16}S$ (b) ^4_2He (c) $^{0}_{+1}e$ (d) $^{67}_{30}Zn$ **24-5.** (a) $^{23}_{11}Na$ (b) $^{20}_{10}Ne$ (c) $^{18}_{8}O$ (d) $^{17}_{8}O$ (e) $^{263}_{106}Unh$ (see Appendix E) **24-7.** (a) $^9_4Be + ^4_2He \rightarrow ^1_0n + ^{12}_6C$ (b) $^7_3Li + ^1_1H \rightarrow ^4_2He + ^4_2He$ (c) $^{63}_{29}Cu + ^1_1H \rightarrow ^1_0n + ^{63}_{30}Zn$ (d) $^{14}_7N + ^1_0n \rightarrow ^1_1H + ^{14}_6C$ (e) $^{54}_{26}Fe + ^2_1H \rightarrow ^1_0n + ^{55}_{27}Co$
24-9. (a) ^1_1H (b) $^{13}_7N$ (c) $^{14}_7N$ (d) ^1_1H (e) $^{0}_{+1}e$ (f) ^4_2He The net reaction is $4^1_1H \rightarrow ^4_2He + 2\ ^{0}_{+1}e + 3\gamma$ **24-11.** positron emission **24-13.** positron emission **24-15.** (a) beta emission (b) positron emission (c) alpha emission (d) positron emission (e) beta emission **24-17.** (a) radioactive (b) stable (c) radioactive (d) radioactive (e) stable
24-19. (a) radioactive (b) radioactive (c) stable (d) radioactive
24-21. 0.032 mg **24-23.** 1.15×10^{-4} **24-25.** 120 h
24-27. 870 disintegrations$\cdot$min^{-1} **24-29.** 12.4 h
24-31. 25,600 disintegrations$\cdot$min^{-1} **24-33.** 1470 yr **24-35.** 1220 yr

24-37. 9900 yr **24-39.** 2.57×10^9 yr **24-41.** 2.7×10^8 yr **24-43.** 4.5×10^9 yr **24-45.** 8.00×10^{-11} J; 1.40×10^{-12} J$\cdot$nucleon^{-1} **24-47.** 4.79×10^{-11} J; 1.37×10^{-12} J$\cdot$nucleon^{-1} **24-49.** 2.30×10^{12} J$\cdot$mol^{-1}; 1.21×10^4 kg **24-51.** 1.3×10^{14} g $= 1.3 \times 10^8$ metric tons **24-53.** 1.1×10^{-12} J$\cdot$disintegration^{-1}; 1.1×10^{-12} J$\cdot$disintegration^{-1} **24-55.** 4.40×10^{-12} kg per mol carbon **24-57.** 1.2×10^9 g $= 1.2 \times 10^3$ metric tons **24-59.** 4.4×10^2 mol **24-61.** 1.1×10^{-7} Ci$\cdot$g^{-1} **24-63.** 4.3×10^4 Ci$\cdot$g^{-1} **24-65.** 8.6×10^{-4} mg **24-67.** 4.0×10^{-5}, or 0.0040% **24-69.** 1.25×10^{-4} M **24-71.** 6.9×10^{-9} M^3 **24-73.** 1.8×10^{14} g$\cdot$cm^{-3}; the density of the nucleus is 8×10^{12} times greater than the density of the element. **24-75.** 1.2×10^6 bonds

CHAPTER 25

25-1. (a) $2C_4H_{10}(g) + 13O_2(g) \rightarrow 8CO_2(g) + 10H_2O(l)$ (b) N.R. (c) N.R. (d) $C_2H_6(g) + Cl_2(g) \rightarrow C_2H_5Cl(l) + HCl(g)$ **25-3.** a, b, and d are identical pairs **25-5.** 3 isomers: 1-chloropentane, 2-chloropentane, 3-chloropentane **25-7.** (a) 2-chlorobutane (b) 1,2,3-trichlorobutane (c) 3-iodo-3-methylhexane (d) hexane **25-9.** (a) 2-methylpentane (b) 3-methylpentane (c) 4-methyloctane (d) 2,2-dimethylpropane

25-11.

```
        H  H  H  H  H  H
        |  |  |  |  |  |
25-11. H—C——C——C——C——C——C—H      hexane
        |  |  |  |  |  |
        H  H  H  H  H  H

        H  H  H  H  H
        |  |  |  |  |
     H—C——C——C——C——C—H        2-methylpentane
        |  |  |  |  |
        H  |  H  H  H
           |
        H—C—H
           |
           H

        H  H  H  H  H
        |  |  |  |  |
     H—C——C——C——C——C—H        3-methylpentane
        |  |  |  |  |
        H  H  |  H  H
              |
           H—C—H
              |
              H

              H
              |
           H—C—H
              |
        H  H  |  H
        |  |  |  |
     H—C——C——C——C—H           2,2-dimethylbutane
        |  |  |  |
        H  H  |  H
              |
           H—C—H
              |
              H
```

```
H   H             H   H
|   |             |   |
H—C—C—————————————C—C—H      2,3-dimethylbutane
|   |             |   |
H   |             |   H
    |             |
  H—C—H         H—C—H
    |             |
    H             H
```

```
              H
              |
            H—C—H
              |
          H   |   H   H
          |   |   |   |
25-13. (a) H—C—C—C—C—H      2-methylbutane
          |   |   |   |
          H   H   H   H
```

```
         H   H
   ..    |   |   ..
(b) :Cl—C—C—Br:      1-bromo-2-chloroethane
   ..    |   |   ..
         H   H
```

```
        ..    ..
    H  :Cl:  :Cl:
    |   |     |    ..
(c) H—C—C—————C—Cl:      1,1,1,2,2-pentachloropropane
    |   |     |    ..
    H  :Cl:  :Cl:
        ..    ..
```

```
          H
          |
        H—C—H
          |
      H   |   H
      |   |   |
(d) H—C—C—C—H      2,2-dimethylpropane
      |   |   |
      H   |   H
          |
        H—C—H
          |
          H
```

```
              H   H             H   H
              |   |             |   |
25-15. (a) H—C—C—————————————C—C—H
              |   |             |   |
              H   |             |   H
                  |             |
                H—C—H         H—C—H
                  |             |
                  H             H
```

```
        H
        |
      H-C-H
        |
      H |     H H
      | |     | |
(b) H-C-C-----C-C-H
      | |     | |
      H |     | H
        |     |
      H-C-H H-C-H
        |     |
        H     H
```

```
          H
          |
        H-C-H
          |
      H H |     H H H
      | | |     | | |
(c) H-C-C-C-----C-C-C-H
      | | |     | | |
      H H |     | H H
          |     |
          |   H-C-H
          |     |
          |   H-C-H
          |     |
        H-C-H   H
          |
          H
```

```
      H H H H H H H H
      | | | | | | | |
(d) H-C-C-C-C-C-C-C-C-H
      | | | | | | | |
      H H H | H H H H
            |
          H | H
          | | |
        H-C-C-C-H
          | | |
          H H H
```

25-17. (a) 3-methyl-1-butene (b) 2-methyl-2-butene (c) 2-methyl-1-butene (d) 4-methyl-2-hexene

```
             H   H   H H H
             |   |   | | |
25-19. (a) H-C---C---C-C-C-H     1,2-dibromopentane
             |   |   | | |
           :Br: :Br: H H H
            ..   ..

             H  H    H    H H
             |  |    |    | |
       (b) H-C--C----C----C-C-H     2,3-dibromopentane
             |  |    |    | |
             H :Br: :Br:  H H
                ..   ..
```

(c)

```
     H     H   H   H   H                 H   H   H     H     H
     |     |   |   |   |                 |   |   |     |     |
 H—C——C—C=C—C—H             H—C=C—C——C——C—H
     |     |           |                         |     |     |
   :Br: :Br:          H                       :Br: :Br:    H
    ••    ••                                   ••    ••
```

1,2-dibromo-3-pentene (4,5-dibromo-2-pentene)

3,4-dibromo-1-pentene (2,3-dibromo-4-pentene)

25-21. (a)

```
      H     H      H   H
      |     |      |   |
  H—C——C——C—C—H
      |     |      |   |
    :Cl: :Cl:    H   H
     ••    ••
```

1,2-dichlorobutane

(b)

```
      H     H      H   H
      |     |      |   |
  H—C——C——C—C—H
      |     |      |   |
      H   :Cl:     H   H
            ••
```

2-chlorobutane

(c)

```
      H   H   H   H
      |   |   |   |
  H—C—C—C—C—H
      |   |   |   |
      H :O: H   H
          |
          H
```

2-butanol

(d)

```
      H   H   H   H
      |   |   |   |
  H—C—C—C—C—H        butane
      |   |   |   |
      H   H   H   H
```

25-23. (a)

```
           H       H                                      H   H   H   H
            \     /                                       |   |   |   |
    H        C=C      (g) + 2H2(g) --Pt--> H—C—C—C—C—H
     \      /     \                                       |   |   |   |
      C=C          H                                      H   H   H   H
     /    \
    H      H
```

$(g) + 2H_2(g) \xrightarrow{Pt}$

(b)

```
           H       H                               H     H     H     H
            \     /                                |     |     |     |
    H        C=C      (g) + 2Cl2(g) → H—C——C——C——C—H
     \      /     \                                |     |     |     |
      C=C          H                             :Cl: :Cl: :Cl: :Cl:
     /    \                                       ••    ••    ••    ••
    H      H
```

$(g) + 2Cl_2(g) \rightarrow$

(c)

```
           H       H                               H     H     H     H
            \     /                                |     |     |     |
    H        C=C      (g) + 2HCl(g) → H—C——C——C——C—H
     \      /     \                                |     |     |     |
      C=C          H                               H   :Cl: :Cl:   H
     /    \                                              ••    ••
    H      H
```

$(g) + 2HCl(g) \rightarrow$

25-25. (a)

```
    H3C       H                      H       H
       \     /                        \     /
        C=C        + H2O   (b)         C=C        + H2O
       /     \                        /     \
    H3C       CH3                    H       CH3
```

(a) $(H_3C)_2C{=}CH(CH_3) + H_2O$ (b) $H_2C{=}CH(CH_3) + H_2O$

(c) $CH_3—C{\equiv}C—CH_3 + H_2$ (d) $CH_3—C{\equiv}CH + HBr$

25-27. (a)

```
     H  H  H  H
     |  |  |  |
  H—C—C—C—C—H
     |  |  |  |
     H :O:  |  H
        |   |
        H   |
            |
         H—C—H
            |
            H
```

(b)

```
       ..   ..
   H :O—H   H  H
   |  |     |  |
H—C—C—————C—C—H
   |  |     |  |
   H  |     H  H
      |
   H—C—H
      |
      H
```

(c)

```
   H  H  H  H  H               H  H  H  H  H
   |  |  |  |  |               |  |  |  |  |
H—C—C—C—C—C—H   and   H—C—C—C—C—C—H
   |  |  |  |  |               |  |  |  |  |
   H :O: H  H  H               H  H :O: H  H
      |                              |
      H                              H
```

(d)

```
   H  H  H  H
   |  |  |  |
H—C—C—C—C—H
   |  |  |  |
   H :O: H  H
      |
      H
```

25-29. b **25-31.** (a) 2-methyl-2-propanol (b) 2-chloro-1-propanol (c) 1-chloro-2-propanol (d) 2-methyl-2-butanol
25-33. (a) primary (b) secondary (c) secondary (d) tertiary
25-35. $2CH_3CH_2OH(l) + 2Na(s) \rightarrow 2Na^+CH_3CH_2O^-(s) + H_2(g)$; $2CH_3CH_2CH_2OH(l) + 2Na(s) \rightarrow 2Na^+CH_3CH_2CH_2O^-(s) + H_2(g)$
25-37. (a) $C_2H_5NH_2(aq) + HBr(aq) \rightarrow C_2H_5NH_3^+(aq) + Br^-(aq)$

(b) $2(CH_3)_2NH(aq) + H_2SO_4(aq) \rightarrow 2(CH_3)_2NH_2^+(aq) + SO_4^{2-}(aq)$

(c) $C_6H_5NH_2(aq) + HCl(aq) \rightarrow C_6H_5NH_3^+(aq) + Cl^-(aq)$ (drawn with benzene rings bearing NH_2 and NH_3^+)

(d) $(C_2H_5)_3N(aq) + HCl(aq) \rightarrow (C_2H_5)_3NH^+(aq) + Cl^-(aq)$

25-39. (a) 1-butyne (b) 2-butyne (c) 4,4-dimethyl-2-heptyne (d) 3,4-dimethyl-1-hexyne **25-41.** 2,2-dibromopropane
25-43. (a) $CH_3C{\equiv}CH(g) + 4O_2(g) \rightarrow 3CO_2(g) + 2H_2O(l)$

(b) $CH_3C{\equiv}CH(g) + 2HCl(g) \rightarrow CH_3CCl_2CH_3(l)$
(c) $CH_3C{\equiv}CH(g) + 2Br_2(l) \rightarrow CH_3CBr_2CHBr_2(l)$

25-45. (a) benzene ring with C_2H_5 (b) benzene ring with Cl, Cl, Cl (1,3,5)

(c) benzene ring with CH_3 and adjacent Cl (d) benzene ring with Cl and Br (1,3)

25-47. (a) butanal (b) 3-methylbutanal (c) methanal (d) 3,4-dimethylpentanal **25-49.** (a) ethanol (b) 2-methylpropanol (c) 2,2-dimethylpropanol

25-51. (a) $CH_3CH_2{-}C({=}O){-}OH$ (b) $CH_3CH(CH_3){-}C({=}O){-}OH$

(c) $CH_3C(CH_3)_2{-}CH_2{-}C({=}O){-}OH$ (d) $CH_3CH_2CH(CH_3){-}CH_2{-}C({=}O){-}OH$

25-53. $CH_3COOH(aq)$ + $CH_3CH_2CH_2CH_2CH_2CH_2CH_2CH_2OH(aq) \rightarrow$
acetic acid 1-octanol

$CH_3CH_2CH_2CH_2CH_2CH_2CH_2CH_2O(CH_3)C{=}\ddot{O}\!: (aq) + H_2O(l)$
octyl acetate

25-55. (a) $HCOOH(aq) + NaOH(aq) \rightarrow Na^+HCOO^-(aq) + H_2O(l)$
(b) $HCOOH(aq) + CH_3OH(aq) \rightarrow HCOOCH_3(aq) + H_2O(l)$
(c) $2HCOOH(aq) + Ca(OH)_2(aq) \rightarrow Ca(OOCH)_2(aq) + 2H_2O(l)$
25-57. (a) $CH_3CH_2COOH(aq) + KOH(aq) \rightarrow$

$K^+CH_3CH_2COO^-(aq) + H_2O(l)$
potassium propanoate

(b) $CH_3CH(CH_3)COOH(aq) + KOH(aq) \rightarrow K^+CH_3CH(CH_3)COO^-(aq) + H_2O(l)$
potassium 2-methylpropanoate

(c) $2Cl_2CHCOOH(aq) + Ca(OH)_2(aq) \rightarrow Ca(Cl_2CHCOO)_2(aq) + 2H_2O(l)$
calcium dichloroethanoate

CHAPTER 26

26-1. c, d, and e

26-3. $$-CH_2-C(=O)OH\,(aq) + H_2O(l) \rightleftharpoons -CH_2C(=O)O^-\,(aq) + H_3O^+(aq)$$

26-5. One possible reaction is

$$H_2N-\underset{\underset{\underset{CH_3}{|}}{CHCH_2CH_3}}{\overset{H}{C}}-\overset{\ddot{O}}{\overset{\|}{C}}-\boxed{OH + H}-\underset{H}{\ddot{N}}-\underset{CH_2CH_2SCH_3}{\overset{H}{C}}-COOH \rightarrow$$

$$H_2N-\underset{\underset{\underset{CH_3}{|}}{CH_2CH_2CH_3}}{\overset{H}{C}}-\overset{\ddot{O}}{\overset{\|}{C}}-\underset{H}{\ddot{N}}-\underset{CH_2CH_2SCH_3}{\overset{H}{C}}-COOH \quad + H_2O$$

The second possible reaction is

$$H_2N-\underset{CH_2CH_2SCH_3}{\overset{H}{C}}-\overset{\ddot{O}}{\overset{\|}{C}}-\boxed{OH + H}-\underset{H}{\ddot{N}}-\underset{\underset{\underset{CH_3}{|}}{CHCH_2CH_3}}{\overset{H}{C}}-COOH \rightarrow$$

$$H_2N-\underset{CH_2CH_2SCH_3}{\overset{H}{C}}-\overset{\ddot{O}}{\overset{\|}{C}}-\underset{H}{\ddot{N}}-\underset{\underset{\underset{CH_3}{|}}{CHCH_2CH_3}}{\overset{H}{C}}-COOH \quad + H_2O$$

26-7. $$\underset{\text{gly}}{H_2N-\underset{H}{\overset{H}{C}}-\overset{\ddot{O}}{\overset{\|}{C}}}-\underset{\text{ala}}{\underset{H}{\ddot{N}}-\underset{CH_3}{\overset{H}{C}}-COOH} \quad \text{or} \quad \underset{\text{ala}}{H_2N-\underset{CH_3}{\overset{H}{C}}-\overset{\ddot{O}}{\overset{\|}{C}}}-\underset{\text{gly}}{\underset{H}{\ddot{N}}-\underset{H}{\overset{H}{C}}-COOH}$$

26-9. 6;

$H_2N-CH(G_1)-C(=O)-NH-CH(G_1)-C(=O)-NH-CH(G_2)-COOH$

$H_2N-CH(G_1)-C(=O)-NH-CH(G_2)-C(=O)-NH-CH(G_1)-COOH$

$H_2N-CH(G_2)-C(=O)-NH-CH(G_1)-C(=O)-NH-CH(G_1)-COOH$

$H_2N-CH(G_2)-C(=O)-NH-CH(G_2)-C(=O)-NH-CH(G_1)-COOH$

$H_2N-CH(G_2)-C(=O)-NH-CH(G_1)-C(=O)-NH-CH(G_2)-COOH$

$H_2N-CH(G_1)-C(=O)-NH-CH(G_2)-C(=O)-NH-CH(G_2)-COOH$

26-11.

$H_2N-CH(CH_2CH_2COOH)-C(=O)-NH-CH(CH_2CH_2CH_2-NH-C(=NH)NH_2)-C(=O)-NH-CH(CH_2-C_8H_5NH)-COOH$

(side chains: $CH_2-CH_2-C(=O)OH$; $CH_2-CH_2-CH_2-N(H)-C(=NH)NH_2$; CH_2–indole ring)

26-13.

$H_2N-CH(CH_2-C_6H_4-OH)-C(=O)-NH-CH_2-C(=O)-NH-CH_2-C(=O)-NH-CH(CH_2-C_6H_5)-C(=O)-NH-CH(CH_2CHCH_3CH_3)-COOH$

26-15.

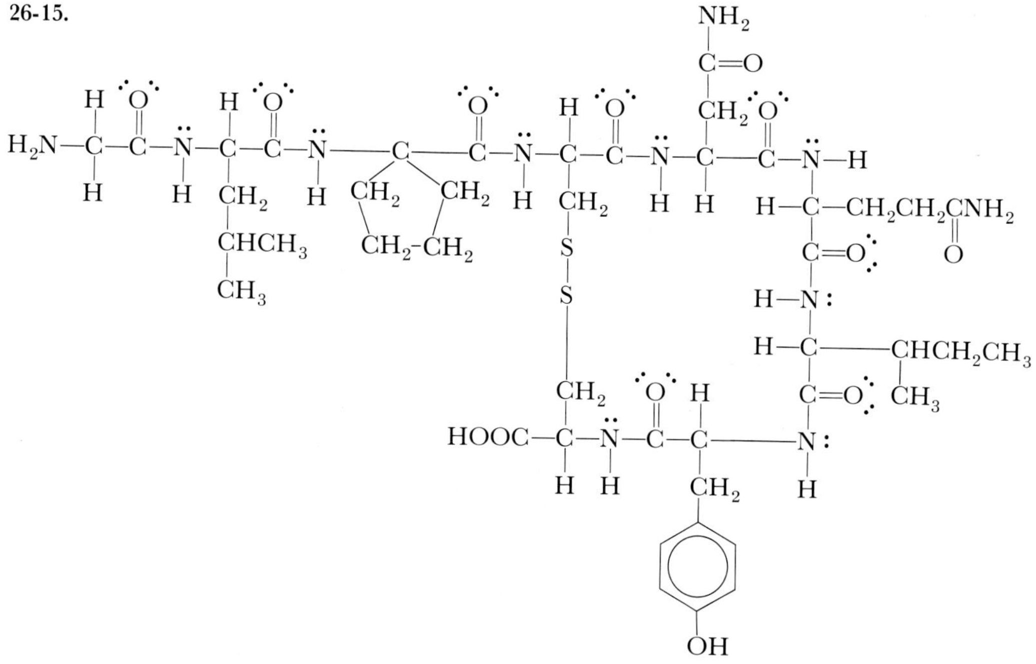

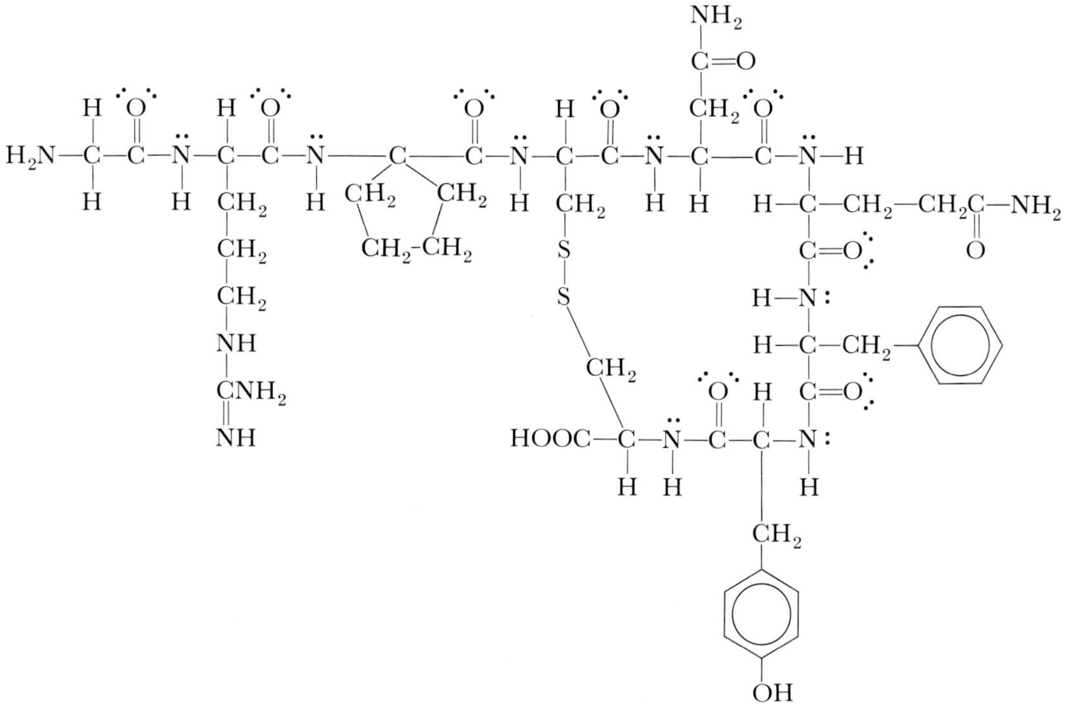

NH$_2$
C=O
H O H O O H O CH$_2$ O
H$_2$N—C—C—N—C—C—N—C—C—N—C—C—N—C—C—N—H
H H CH$_2$ H CH$_2$ CH$_2$ H CH$_2$ H H H—C—CH$_2$CH$_2$CNH$_2$
CHCH$_3$ CH$_2$-CH$_2$ S C=O O
CH$_3$ S H—N:
H—C—CHCH$_2$CH$_3$
CH$_2$ O H C=O CH$_3$
HOOC—C—N—C—C—N:
H H CH$_2$ H
OH

NH$_2$
C=O
H O H O O H O CH$_2$ O
H$_2$N—C—C—N—C—C—N—C—C—N—C—C—N—C—C—N—H
H H CH$_2$ H CH$_2$ CH$_2$ H CH$_2$ H H H—C—CH$_2$—CH$_2$C—NH$_2$
CH$_2$ CH$_2$-CH$_2$ S C=O O
CH$_2$ S H—N:
NH H—C—CH$_2$—
CH$_2$
CNH$_2$ O H C=O
NH HOOC—C—N—C—C—N:
H H H
CH$_2$
OH

26-17. a, d and e **26-19.** b and d **26-21.** (a) The oxygen atom can form a hydrogen bond to a hydrogen atom in H_2O. (b) There are none. (c) The nitrogen atom can form a hydrogen bond to a hydrogen atom in H_2O. (d) The two hydrogen atoms can form hydrogen bonds to an oxygen atom in H_2O. (e) There are none. **26-23.** The primary structure of a protein is the order of amino acid units in the protein. The secondary structure is the structure within subsections of the protein. The tertiary structure is the overall three-dimensional shape of the protein.

26-25.

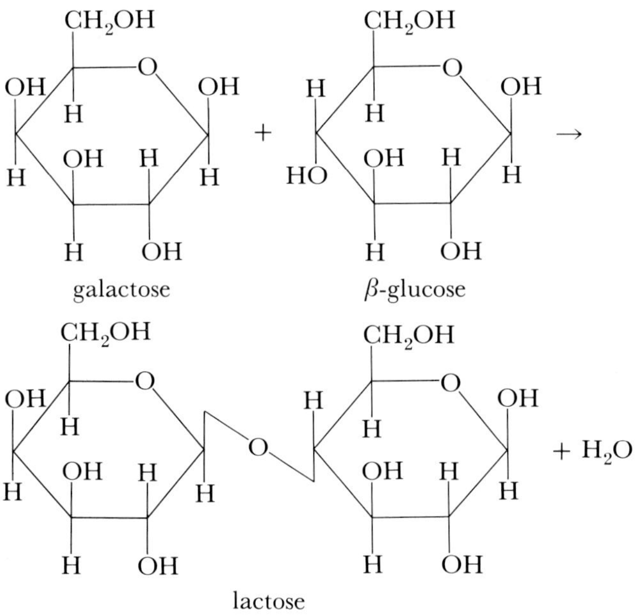

galactose + β-glucose →

lactose + H_2O

26-27.

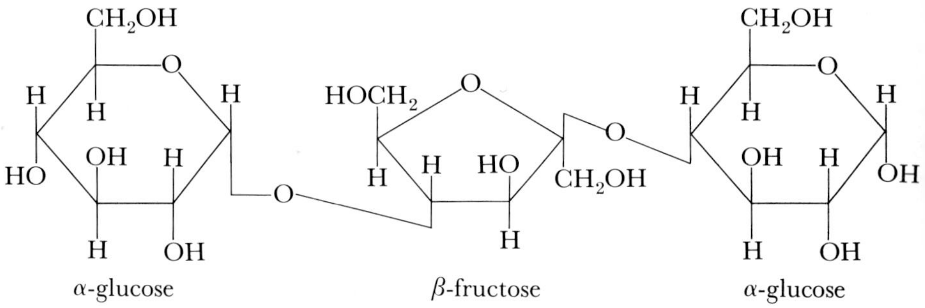

α-glucose — β-fructose — α-glucose

26-29. ≈ 1100

26-31.

CH_2OH
O
H H H
HO OH H
H OH
O
$HOCH_2$ O
H H HO CH_2OH
HO H

$+ H_2O \rightarrow$

CH_2OH
O
H H H
HO OH H OH
H OH

+

$HOCH_2$ O OH
H H HO CH_2OH
HO H

sucrose α-glucose β-fructose

26-33.

O
HN C C N
CH
H_2N—C N C N
O
^-O—P—O—CH_2 O
$_-O$
H H H H
H
NH_2
N C C N
CH
HC N C N
O
O=P—O—CH_2 O
O_-
H H H H
H
O
HN C C—CH_3
O=C N CH
O
O=P—O—CH_2 O
$_-O$
H H H H
OH H

26-35.

26-37. CAAGGGTTC 26-39. CGAGCGT

26-41.

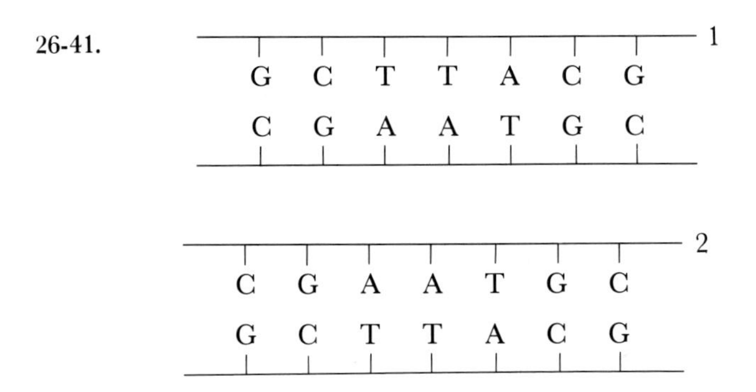

26-43. 14 **26-45.**

$$H_3\overset{+}{N}-\underset{\underset{CH_3}{|}}{\overset{\overset{H}{|}}{C}}-COO^-$$

26-47. When $[S]$ is small, the value of $\dfrac{k_{-1}+k_2}{k_1}$ will be much larger than $[S]$. Thus we can neglect $[S]$ in the denominator. For small $[S]$, the rate law is approximately

$$\text{rate} \simeq \frac{k_2[E_0][S]}{\dfrac{k_{-1}+k_2}{k_1}} = C[S]$$

The rate is proportional to $[S]$, thus the curve is a straight line. When $[S]$ is large, the value of $\dfrac{k_{-1}+k_2}{k_1}$ will be much smaller than $[S]$. Thus we can neglect $\dfrac{k_{-1}+k_2}{k_1}$ in the denominator. For large $[S]$, the rate law is approximately

$$\text{rate} \simeq \frac{k_2[E_0][S]}{[S]} = k_2[E_0] = \text{constant}$$

The rate is constant for a fixed value of $[E_0]$. We connect the two sections of the plot by a curved line. So the plot looks like

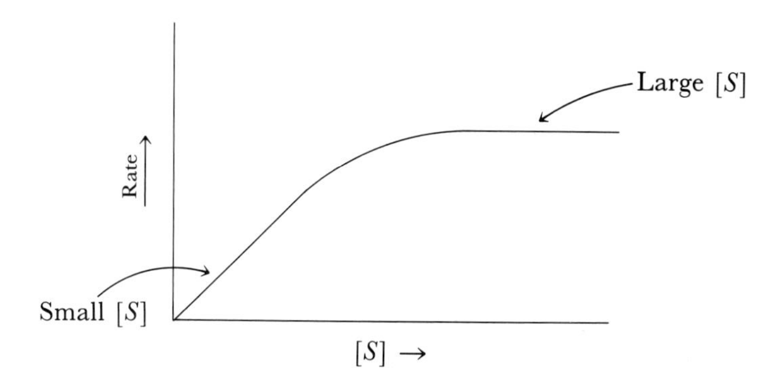

26-49. For large $[S]$

$$\frac{k_{-1}+k_2}{k_1} + [S] \simeq [S]$$

thus

$$\text{rate} = \frac{k_2[E_0][S]}{\dfrac{k_{-1} + k_2}{k_1} + [S]} \approx \frac{k_2[E_0][S]}{[S]} = k_2[E_0]$$

26-51. 1.0×10^{-6} M **26-53.** 15.9 kJ **26-55.** 16.9 kJ

Illustration Credits

All photos by Chip Clark except the following:

p. xx, Michael Isaacson/Cornell University; p. 5, Metropolitan Museum of Art; p. 6, Burndy Library of Medicine; p. 8, The Science Museum, London; p. 16, Cavendish Laboratory; p. 25, Mettler Instrument Corp.; p. 44 (Fig. 2-1), N.Y.P.L.; p. 44 (Fig. 2-2), Hans Pfletschinger/Peter Arnold; p. 49 (top), Bill Tronca/Tom Stack and Assoc.; pp. 49 (bottom) and 50, Alexander Boden; p. 56, Werner Muller/Peter Arnold; p. 57 (bottom), Lawrence Berkeley Laboratory; p. 60 (top), U.S. Borax; p. 60 (bottom), Philips; p. 79, Tom Stack/Tom Stack and Assoc.; p. 82, J.T. Baker Chemical Co.; p. 114, Texasgulf; p. 116, Historical Photography Collection, University of Washington Archives; p. 121, Chromatronix; p. 129, Manfred Kage/Peter Arnold; p. 132, Grant Heilman/Grant Heilman; p. 133, W.H. Fahrenbach; p. 138, Ethyl Corp.; p. 154, Kontes Glassware; p. 165, IVAC Corp.; p. 194, Union 76; p. 196, Grant Heilman/ Grant Heilman; p. 198, Allied Chemical Company; p. 199, USDA; p. 202, NASA; p. 238 (Frontispiece), Grant Heilman/Grant Heilman; p. 238, NASA; p. 240 (Fig. D-3), Texasgulf; p. 240 (Fig. D-4a), Division of Mineral Sciences, Smithsonian Institution; p. 240 (Fig. D-4b), Runk/Schoenberger/Grant Heilman; p. 240 (Fig. D-4c), Ron Tesla/Field Museum of Natural History; p. 241 (Fig. D-5), Division of Mineral Sciences, Smithsonian Institution; p. 241 (Fig. D-6a), Alan Pitcairn/Grant Heilman; p. 241 (Fig. D-6b), Ron Tesla/Field Museum of Natural History; p. 242 (Fig. D-7a), Division of Mineral Sciences, Smithsonian Institution; p. 242 (D-7b), National Park Service; p. 245, J.T. Baker Chemical Co.; p. 248, Bell Laboratories; p. 250, Bausch and Lomb; p. 267 (Fig. 7-11), Education Development Center, Inc., Newton, MA; p. 267 (Fig. 7-12), A.K. Kleinschmidt/ Elsevier; p. 268, Photo Pfaundler/A.I.P.; p. 314, NASA; p. 322, Harvey Lloyd/Peter Arnold; p. 330, Fundamental Photos; p. 364, University of California Archives, The Bancroft Library; p. 421, Institute of Geological Sciences; p. 432, Pauling Institute; p. 486 (Figs. 13-22 and 13-23), Nancy Rodgers/Exploratorium; p. 488, Håkon Hope; pp. 500 (top) and 502, General Electric; p. 510, Philips; p. 512, Lawrence Berkeley Laboratory; p. 514, Philips; p. 517, Runk/Schoenberger/Grant Heilman; p. 519, Steven Smale; p. 522, Kenrick Day; p. 537, Tom Bean/Tom Stack; p. 541, Envirogenics Systems; p. 544, Runk/Schoenberger/Grant Heilman; p. 546, Ken Lucas; p. 554, Division of Mineral Sciences, Smithsonian Institution; p. 559 (Fig. I-5), Tom Carroll/Martin Marietta; p. 559 (right), Manfred Kage/Peter Arnold; p. 561, Steven Smale; p. 562, ASARCO Inc.; p. 600, Fritz Goro; p. 646, Bill Wunsch/Denver Post; p. 655, Corning Glass Works; p. 685, Alexander Boden; p. 687, Bill Tronca/Tom Stack and Assoc.; p. 690, J.T. Baker Chemical Co.; p. 692, Tom Stack/Tom Stack and Assoc.; p. 730, Warren Bolster/Oceanic Society; p. 732 (Fig. K-1), Dudley Foster, Woods Hole Oceanographic Institution; p. 732 (Fig. K-2), Deepsea Ventures, Inc.; p. 765, Kjell Sandved; p. 780, Mike Elliot/Colorado State University; p. 804, The Royal Institution; p. 825, Fisher Scientific Instruments; p. 845, Leland C. Clark; p. 848, Argonne National Laboratory; p. 850, Mike Maloney/S.F. Chronicle; p. 851, Fundamental Photos; p. 894 (top), Manfred Kage/Peter Arnold; p. 894 (Fig. M-1), Martin Marietta; p. 895, Ray Manley/AMAX; p. 896, Ron Tesla/Field Museum of Natural History; pp. 901 and 902, Bethlehem Steel Corp.; p. 903 (top), Fundamental Photos; p. 903 (bottom), Peter Arnold; p. 929, From L. Stryer, *Biochemistry*, 2nd ed, p. 51, Freeman, 1981; p. 942, Deutsches Museum; p. 971, Archives of the Institute of Radium. From E. Segrè, *From X-rays to Quarks*, Freeman, 1980; p. 980, Atlantic Richfield; p. 1020, Richard Byrnes; p. 1025 (Fig. N-1), University of Utah Medical Center; p. 1025 (Fig. N-2), Dan McCoy, Rainbow; p. 1030, Sung-Hou Kim; p. 1044, Wallace Jensen and Panpit Klug, The University of Miami Medical School; p. 1053, A.C. Barrington Brown. From J.D. Watson, *The Double Helix*, Atheneum, N.Y., p. 215. Copyright© 1968 by J.D. Watson.

Index

Physical Constants

Constant	*Symbol*	*Value*
atomic mass unit	amu	1.66056×10^{-27} kg
Avogadro's number	N	6.02205×10^{23} mol^{-1}
Bohr radius	a_0	5.292×10^{-11} m
Boltzmann constant	k	1.38066×10^{-23} J·K^{-1}
charge of a proton	e	1.60219×10^{-19} C
Faraday constant	F	96,485 C·mol^{-1}
gas constant	R	8.31441 J·K^{-1}·mol^{-1}
		0.08206 L·atm·K^{-1}·mol^{-1}
mass of an electron	m_e	9.10953×10^{-31} kg
		5.48580×10^{-4} amu
mass of a neutron	m_n	1.67495×10^{-27} kg
		1.00866 amu
mass of a proton	m_p	1.67265×10^{-27} kg
		1.00728 amu
Planck's constant	h	6.62618×10^{-34} J·s
speed of light	c	2.997925×10^{8} m·s^{-1}

SI Prefixes

Prefix	*Multiple*	*Symbol*	*Prefix*	*Multiple*	*Symbol*
tera	10^{12}	T	deci	10^{-1}	d
giga	10^{9}	G	centi	10^{-2}	c
mega	10^{6}	M	milli	10^{-3}	m
kilo	10^{3}	k	micro	10^{-6}	μ
			nano	10^{-9}	n
			pico	10^{-12}	p
			femto	10^{-15}	f
			atto	10^{-18}	a